THE INSECTS

Fifth Edition

THE INSECTS

AN OUTLINE OF ENTOMOLOGY

P.J. Gullan and P.S. Cranston

Research School of Biology, The Australian National University, Canberra, Australia &
Department of Entomology and Nematology, University of California, Davis, USA

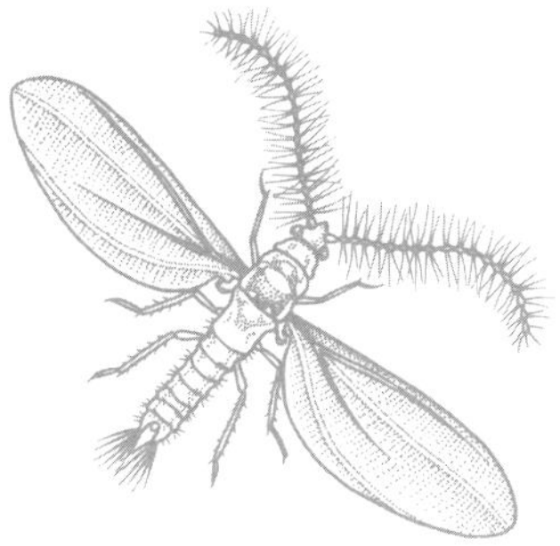

With illustrations by
Karina H. McInnes

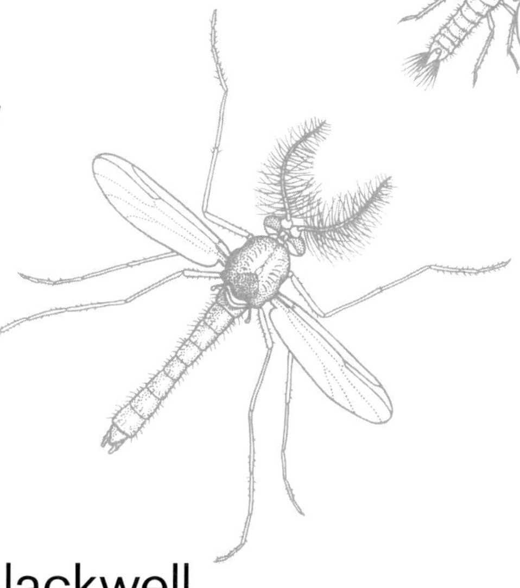

WILEY Blackwell

This edition first published 2014 © 2014 by John Wiley & Sons, Ltd
Fourth edition published 2010 © 2010 by P.J. Gullan and P.S. Cranston
Third edition published 2005 by Blackwell Publishing Ltd
Second edition published 2000 by Blackwell Science
First edition published 1994 by Chapman & Hall

Registered office: John Wiley & Sons, Ltd, The Atrium, Southern Gate, Chichester, West Sussex, PO19 8SQ, UK

Editorial offices: 9600 Garsington Road, Oxford, OX4 2DQ, UK
The Atrium, Southern Gate, Chichester, West Sussex, PO19 8SQ, UK
111 River Street, Hoboken, NJ 07030-5774, USA

For details of our global editorial offices, for customer services and for information about how to apply for permission to reuse the copyright material in this book please see our website at www.wiley.com/wiley-blackwell.

Library of Congress Cataloging-in-Publication Data

Gullan, P. J.
The insects : an outline of entomology / P.J. Gullan and P.S. Cranston ; with illustrations by Karina H. McInnes. – Fifth edition.
pages cm
Includes bibliographical references and index.
ISBN 978-1-118-84615-5 (cloth)
1. Insects. 2. Entomology. I. Cranston, P. S. II. Title.
QL463.G85 2014
595.7– dc23

2014013797

A catalogue record for this book is available from the British Library.

Wiley also publishes its books in a variety of electronic formats. Some content that appears in print may not be available in electronic books.

Cover image: A cartoon depicting some accidental passenger insects – a consequence of increased global commerce.
Cover design by Peter Cranston and Karina McInness.

Set in 9/11pt, PhotinaMTStd by Laserwords Private Limited, Chennai, India
Printed and bound in Singapore by Markono Print Media Pte Ltd

1 2014

CONTENTS

List of colour plates, ix

List of boxes, xiii

Preface to the fifth edition, xv

Preface to the fourth edition, xvii

Preface to the third edition, xix

Preface to the second edition, xxi

Preface and acknowledgments for first edition, xxiii

About the companion website, xxv

1 THE IMPORTANCE, DIVERSITY AND CONSERVATION OF INSECTS, 1

1.1 What is entomology?, 2
1.2 The importance of insects, 2
1.3 Insect biodiversity, 6
1.4 Naming and classification of insects, 10
1.5 Insects in popular culture and commerce, 11
1.6 Culturing insects, 13
1.7 Insect conservation, 14
1.8 Insects as food, 20
Further reading, 25

2 EXTERNAL ANATOMY, 26

2.1 The cuticle, 27
2.2 Segmentation and tagmosis, 33
2.3 The head, 35
2.4 The thorax, 45
2.5 The abdomen, 52
Further reading, 55

3 INTERNAL ANATOMY AND PHYSIOLOGY, 56

3.1 Muscles and locomotion, 57
3.2 The nervous system and co-ordination, 63
3.3 The endocrine system and the function of hormones, 66
3.4 The circulatory system, 69
3.5 The tracheal system and gas exchange, 73
3.6 The gut, digestion and nutrition, 77
3.7 The excretory system and waste disposal, 86
3.8 Reproductive organs, 90
Further reading, 93

4 SENSORY SYSTEMS AND BEHAVIOUR, 95

4.1 Mechanical stimuli, 96
4.2 Thermal stimuli, 105
4.3 Chemical stimuli, 107
4.4 Insect vision, 117
4.5 Insect behaviour, 122
Further reading, 124

5 REPRODUCTION, 125

5.1 Bringing the sexes together, 126
5.2 Courtship, 128
5.3 Sexual selection, 128
5.4 Copulation, 131
5.5 Diversity in genitalic morphology, 136
5.6 Sperm storage, fertilization and sex determination, 139
5.7 Sperm competition, 140
5.8 Oviparity (egg-laying), 144

5.9 Ovoviviparity and viviparity, 150
5.10 Other modes of reproduction, 150
5.11 Physiological control of reproduction, 153
Further reading, 154

6 INSECT DEVELOPMENT AND LIFE HISTORIES, 156
6.1 Growth, 157
6.2 Life-history patterns and phases, 158
6.3 Process and control of moulting, 169
6.4 Voltinism, 172
6.5 Diapause, 173
6.6 Dealing with environmental extremes, 174
6.7 Migration, 178
6.8 Polymorphism and polyphenism, 180
6.9 Age-grading, 181
6.10 Environmental effects on development, 183
Further reading, 188

7 INSECT SYSTEMATICS: PHYLOGENY AND CLASSIFICATION, 190
7.1 Systematics, 191
7.2 The extant Hexapoda, 201
7.3 Informal group Entognatha: Collembola (springtails), Diplura (diplurans) and Protura (proturans), 202
7.4 Class Insecta (true insects), 203
Further reading, 224

8 INSECT EVOLUTION AND BIOGEOGRAPHY, 227
8.1 Relationships of the Hexapoda to other Arthropoda, 228
8.2 The antiquity of insects, 229
8.3 Were the first insects aquatic or terrestrial?, 236
8.4 Evolution of wings, 238
8.5 Evolution of metamorphosis, 241
8.6 Insect diversification, 242
8.7 Insect biogeography, 244
8.8 Insect evolution in the Pacific, 245
Further reading, 247

9 GROUND-DWELLING INSECTS, 249
9.1 Insects of litter and soil, 250
9.2 Insects and dead trees or decaying wood, 260
9.3 Insects and dung, 261
9.4 Insect–carrion interactions, 264
9.5 Insect–fungal interactions, 265
9.6 Cavernicolous insects, 268
9.7 Environmental monitoring using ground-dwelling hexapods, 268
Further reading, 270

10 AQUATIC INSECTS, 271
10.1 Taxonomic distribution and terminology, 272
10.2 The evolution of aquatic lifestyles, 275
10.3 Aquatic insects and their oxygen supplies, 277
10.4 The aquatic environment, 282
10.5 Environmental monitoring using aquatic insects, 284
10.6 Functional feeding groups, 285
10.7 Insects of temporary waterbodies, 286
10.8 Insects of the marine, intertidal and littoral zones, 287
Further reading, 288

11 INSECTS AND PLANTS, 289
11.1 Coevolutionary interactions between insects and plants, 291
11.2 Phytophagy (or herbivory), 293
11.3 Insects and plant reproductive biology, 313
11.4 Insects that live mutualistically in specialized plant structures, 318
Further reading, 320

12 INSECT SOCIETIES, 322
12.1 Subsociality in insects, 323
12.2 Eusociality in insects, 327
12.3 Inquilines and parasites of social insects, 345
12.4 Evolution and maintenance of eusociality, 348
12.5 Success of social insects, 351
Further reading, 353

13 INSECT PREDATION AND PARASITISM, 354

13.1 Prey/host location, 355
13.2 Prey/host acceptance and manipulation, 361
13.3 Prey/host selection and specificity, 364
13.4 Population biology – predator/parasitoid and prey/host abundance, 372
13.5 The evolutionary success of insect predation and parasitism, 375
Further reading, 376

14 INSECT DEFENCE, 377

14.1 Defence by hiding, 379
14.2 Secondary lines of defence, 380
14.3 Mechanical defences, 382
14.4 Chemical defences, 384
14.5 Defence by mimicry, 388
14.6 Collective defences in gregarious and social insects, 392
Further reading, 396

15 MEDICAL AND VETERINARY ENTOMOLOGY, 397

15.1 Insects as causes and vectors of disease, 398
15.2 Generalized disease cycles, 399
15.3 Pathogens, 399
15.4 Forensic entomology, 413
15.5 Insect nuisance and phobia, 414
15.6 Venoms and allergens, 416
Further reading, 417

16 PEST MANAGEMENT, 418

16.1 Insects as pests, 419
16.2 The effects of insecticides, 425
16.3 Integrated pest management, 428
16.4 Chemical control, 429
16.5 Biological control, 435
16.6 Host-plant resistance to insects, 447
16.7 Physical control, 451
16.8 Cultural control, 451
16.9 Pheromones and other insect attractants, 452
16.10 Genetic manipulation of insect pests, 454
Further reading, 455

17 INSECTS IN A CHANGING WORLD, 457

17.1 Models of change, 458
17.2 Economically significant insects under climate change, 463
17.3 Implications of climate change for insect biodiversity and conservation, 467
17.4 Global trade and insects, 468
Further reading, 473

18 METHODS IN ENTOMOLOGY: COLLECTING, PRESERVATION, CURATION AND IDENTIFICATION, 474

18.1 Collection, 475
18.2 Preservation and curation, 478
18.3 Identification, 488
Further reading, 491

TAXOBOXES, 493

1 Entognatha: non-insect hexapods (Collembola, Diplura and Protura), 493
2 Archaeognatha (Microcoryphia: archaeognathans or bristletails), 495
3 Zygentoma (silverfish), 496
4 Ephemeroptera (mayflies), 497
5 Odonata (damselflies and dragonflies), 498
6 Plecoptera (stoneflies), 500
7 Dermaptera (earwigs), 500
8 Zoraptera (zorapterans or angel insects), 501
9 Orthoptera (grasshoppers, locusts, katydids and crickets), 502
10 Embioptera (Embiidina, Emboidea: embiopterans or webspinners), 503
11 Phasmatodea (phasmids, stick-insects or walking sticks), 503
12 Grylloblattodea (Grylloblattaria or Notoptera: grylloblattids, ice crawlers or rock crawlers), 504
13 Mantophasmatodea (heelwalkers), 505
14 Mantodea (mantids, mantises or praying mantids), 506
15 Blattodea: roach families (cockroaches or roaches), 507

16 Blattodea: epifamily Termitoidae (former order Isoptera; termites, “white ants”), 508
17 Psocodea: “Psocoptera” (bark lice and book lice), 509
18 Psocodea: “Phthiraptera” (chewing lice and sucking lice), 510
19 Thysanoptera (thrips), 511
20 Hemiptera (bugs, moss bugs, cicadas, leafhoppers, planthoppers, spittle bugs, treehoppers, aphids, jumping plant lice, scale insects and whiteflies), 512
21 Neuropterida: Neuroptera (lacewings, owlflies and antlions), Megaloptera (alderflies, dobsonflies and fishflies) and Raphidioptera (snakeflies), 514
22 Coleoptera (beetles), 516
23 Strepsiptera (strepsipterans), 517
24 Diptera (true flies), 519
25 Mecoptera (hangingflies, scorpionflies and snowfleas), 520
26 Siphonaptera (fleas), 521
27 Trichoptera (caddisflies), 522
28 Lepidoptera (butterflies and moths), 523
29 Hymenoptera (ants, bees, wasps, sawflies and wood wasps), 524

Glossary, 526

References, 555

Index, 563

Appendix: A reference guide to orders, 589

LIST OF COLOUR PLATES

PLATE 1

(a) An atlas moth, *Attacus atlas* (Lepidoptera: Saturniidae), one of the largest of all lepidopterans, with a wingspan of about 24 cm and a larger wing area than any other moth; southern India and Southeast Asia (P.J. Gullan).

(b) The moon moth, *Argema maenas* (Lepidoptera: Saturniidae), is found in Southeast Asia and India; this female, from rainforest in Borneo, has a wingspan of about 15 cm (P.J. Gullan).

(c) Lord Howe Island stick-insect, *Dryococelus australis* (Phasmatodea: Phasmatidae), Lord Howe Island, Pacific Ocean, Australia (N. Carlile).

(d) A female of the Stephens Island giant weta, *Deinacrida rugosa* (Orthoptera: Anostostomatidae), Mana Island, New Zealand (G.H. Sherley; courtesy of New Zealand Department of Conservation).

(e) A Richmond birdwing, *Ornithoptera richmondia* (Lepidoptera: Papilionidae), and its cast pupal exuviae on a native pipevine (*Pararistolochia* sp.), eastern Australia (D.P.A. Sands).

(f) Owl butterfly, *Caligo memnon*, with two common morpho butterflies, *Morpho peleides* (both Nymphalidae), Cali Zoo, Colombia (P.J. Gullan).

(g) A cage of butterfly pupae awaiting eclosion, Penang Butterfly Farm, Malaysia (P.J. Gullan).

PLATE 2

(a) Palm weevil grubs, *Rhynchophorus ferrugineus* (Coleoptera: Curculionidae), reared for human consumption from ground palm mash and pig pellets, Thailand (M.S. Hoddle).

(b) A "worm" or "phane"–the caterpillar of *Gonimbrasia belina* (Lepidoptera: Saturnidae)–feeding on the foliage of *Schotia brachypetala*, Limpopo Province, South Africa (R.G. Oberprieler).

(c) Witchety (witjuti) grub, a larva of *Endoxyla* (Lepidoptera: Cossidae) from a desert *Acacia* tree, Flinders Ranges, South Australia (P. Zborowski).

(d) Food insects at a market stall displaying silk-moth pupae (*Bombyx mori*), beetle pupae, and adult hydrophiloid beetles and water bugs (*Lethocerus indicus*), Lampang Province, northern Thailand (R.W. Sites).

(e) A dish of edible water bugs, *Lethocerus indicus* (Hemiptera: Belostomatidae), Lampang Province, northern Thailand (R.W. Sites).

(f) Edible stink bugs (Hemiptera: Tessaratomidae), at an insect market, Thailand (A.L. Yen).

(g) Repletes (see Fig. 2.4) of the honeypot ant, *Camponotus inflatus* (Hymenoptera: Formicidae), on an Aboriginal wooden dish, Northern Territory, Australia (A.L. Yen).

(h) Repletes of the honeypot ant, *Camponotus inflatus*, Northern Territory, Australia (A.L. Yen).

PLATE 3

(a) A tropical butterfly, the five-bar swordtail, *Graphium antiphates* (Lepidoptera: Papilionidae), obtaining salts by imbibing sweat from a training shoe, Borneo (P.J. Gullan).

(b) A female katydid of an undescribed species of *Austrosalomona* (Orthoptera: Tettigoniidae), with a large spermatophore attached to her genital opening, northern Australia (D.C.F. Rentz).

(c) Pupa of a Christmas beetle, *Anoplognathus* sp. (Coleoptera: Scarabaeidae), removed from its pupation site in the soil, Canberra, Australia (P.J. Gullan).

(d) A teneral (newly moulted) giant burrowing cockroach, *Macropanesthia rhinoceris* (Blattodea: Blaberidae), Queensland, Australia (M.D. Crisp).

(e) Egg mass of *Tenodera australasiae* (Mantodea: Mantidae) with young mantid nymphs emerging, Queensland, Australia (D.C.F. Rentz).

(f) Eclosing (moulting) adult katydid of an *Elephantodeta* species (Orthoptera: Tettigoniidae), Northern Territory, Australia (D.C.F. Rentz).

(g) Overwintering monarch butterflies, *Danaus plexippus* (Lepidoptera: Nymphalidae), Mill Valley, California, USA (D.C.F. Rentz).

PLATE 4

(a) A fossilized worker ant of *Pseudomyrmex oryctus* (Hymenoptera: Formicidae) in Dominican amber from the Miocene (P.S. Ward).

(b) Female (long snout) and male (short snout) of the cycad weevil, *Antliarhinus zamiae* (Coleoptera: Curculionidae), on seeds of *Encephalartos altensteinii* (Zamiaceae), South Africa (P.J. Gullan).

(c) The common milkweed grasshopper, *Phymateus morbillosus* (Orthoptera: Pyrgomorphidae), for which bright colours advertise toxicity acquired by feeding on milkweed foliage, Northern Cape, South Africa (P.J. Gullan).

(d) Mine of a scribbly gum moth, *Ogmograptis racemosa* (Lepidoptera: Bucculatricidae), on trunk of *Eucalyptus racemosa*, New South Wales, Australia (P.J. Gullan).

(e) Euglossine bees (Hymenoptera: Apidae: Euglossini) collecting fragrances from spadix of *Anthurium* sp. (Araceae), Ecuador (P.J. Gullan).

(f) A bush coconut or bloodwood apple gall of *Cystococcus pomiformis* (Hemiptera: Eriococcidae), cut open to show the cream-coloured adult female and her numerous, tiny nymphal male offspring covering the gall wall, northern Australia (P.J. Gullan).

(g) Close-up of the second-instar male nymphs of *C. pomiformis* feeding from the nutritive tissue lining the cavity of the maternal gall, northern Australia (P.J. Gullan).

PLATE 5

(a) Coccoid-induced gall of *Apiomorpha pharetrata* (Hemiptera: Eriococcidae): dark compound gall of males attached to green gall of female, with ants collecting honeydew at orifice of female's gall, eastern Australia (P.J. Gullan).

(b) Aphid-induced galls of *Baizongia pistaciae* (Hemiptera: Aphididae: Fordinae) on turpentine tree, *Pistacia teredinthus*, Bulgaria (P.J. Gullan).

(c) Rose bedeguar gall of *Diplolepis rosae* (Hymenoptera: Cynipidae) on *Rosa* sp. (wild rose), Bulgaria (P.J. Gullan).

(d) A female thynnine wasp of *Zaspilothynnus trilobatus* (Hymenoptera: Tiphiidae) (on right) compared with flower of the sexually deceptive orchid *Drakaea glyptodon*, which attracts pollinating male wasps by mimicking the female wasp, Western Australia (R. Peakall).

(e) A male thynnine wasp of *Neozeleboria cryptoides* (Hymenoptera: Tiphiidae) attempting to copulate with the sexually-deceptive orchid *Chiloglottis trapeziformis*, Australian Capital Territory (R. Peakall).

(f) Myophily – pollination of mango flowers by a flesh fly, *Australopierretia australis* (Diptera: Sarcophagidae), northern Australia (D.L. Anderson).

(g) Hummingbird hawk moth, *Macroglossum stellatarum* (Lepidoptera: Sphingidae), on a thistle, Bulgaria (P.J. Gullan).

(h) Honey bee, *Apis mellifera* (Hymenoptera: Apidae), pollinating a passion flower, *Passiflora edulis*, Colombia (T. Kondo).

PLATE 6

(a) Ovipositing parasitic wasps (Hymenoptera): a eurytomid (Eurytomidae, top) and cynipid (Cynipidae, right), on an oak apple gall on *Quercus*, Illinois, USA (A.L. Wild).

(b) Weaver ants, *Oecophylla smaragdina* (Hymenoptera: Formicidae), tending *Rastococcus* mealybugs (Hemiptera: Pseudococcidae), Thailand (T. Kondo).

(c) The huge queen termite (approx. 7.5 cm long) of *Odontotermes transvaalensis* (Blattodea: Termitoidae: Termitidae: Macrotermitinae) surrounded by her king (mid front), soldiers and workers, South Africa (the late J.A.L. Watson).

(d) A parasitic *Varroa* mite on a pupa of *Apis cerana* (Hymenoptera: Apidae) in a hive, Irian Jaya, New Guinea (D.L. Anderson).

(e) Ant (Hymenoptera: Formicidae) interactions: the smaller Argentine ant (*Linepithema humile*) attacks the much larger red imported fire ant (*Solenopsis invicta*), Austin, Texas, USA (A.L. Wild).

(f) An egg-parasitoid wasp, *Telenomus* sp. (Hymenoptera: Scelionidae), oviposits into an egg of an owl butterfly, *Caligo* sp. (Lepidoptera: Nymphalidae), Belize (A.L. Wild).

PLATE 7

(a) A cryptic grasshopper, *Calliptamus* sp. (Orthoptera: Acrididae), Bulgaria (T. Kondo).

(b) A camouflaged late-instar caterpillar of *Plesanemma fucata* (Lepidoptera: Geometridae) resting on a eucalypt leaf so that its red dorsal line resembles the leaf midrib, eastern Australia (P.J. Gullan).

(c) A female webspinner of *Antipaluria urichi* (Embioptera: Clothodidae) defending the entrance of her gallery from an approaching male, Trinidad (J.S. Edgerly-Rooks).

(d) A snake-mimicking caterpillar of the spicebush swallowtail, *Papilio troilus* (Lepidoptera: Papilionidae), New Jersey, USA (D.C.F. Rentz).

(e) An adult moth of *Utetheisa ornatrix* (Lepidoptera: Arctiidae) emitting defensive froth containing pyrrolizidine alkaloids sequestered by larval feeding on *Crotalaria* (Fabaceae) (the late T. Eisner).

(f) A blister beetle, *Lytta polita* (Coleoptera: Meloidae), reflex-bleeding from the knee joints; the haemolymph contains the toxin cantharidin (the late T. Eisner).

(g) The cryptic adult moths of four species of *Acronicta* (Lepidoptera: Noctuidae): *A. alni*, the alder moth (top left); *A. leporina*, the miller (top right); *A. aceris*, the sycamore (bottom left); and *A. psi*, the grey dagger (bottom right) (D. Carter and R.I. Vane-Wright).

(h) Aposematic or mechanically protected caterpillars of the same four species of *Acronicta: A. alni* (top left); *A. leporina* (top right); *A. aceris* (bottom left); and *A. psi* (bottom right); showing the divergent appearance of the larvae compared with their drab adults (D. Carter and R.I. Vane-Wright).

PLATE 8

(a) One of Bates' mimicry complexes from the Amazon Basin involving species from three different lepidopteran families – the butterflies *Methona confusa confusa* (Nymphalidae: Ithomiinae) (top), *Lycorea ilione ilione* (Nymphalidae: Danainae) (second top) and *Patia orise orise* (Pieridae) (second from bottom), and a day-flying moth of *Gazera heliconioides* (Castniidae) (R.I. Vane-Wright).

(b) A mature cottony-cushion scale, *Icerya purchasi* (Hemiptera: Monophlebidae), with a partly formed ovisac, on the stem of an *Acacia* host, attended by meat ants of *Iridomyrmex* sp. (Formicidae), New South Wales, Australia (P.J. Gullan).

(c) Adult male gypsy moth, *Lymantria dispar* (Lepidoptera: Lymantriidae), New Jersey, USA (D.C.F. Rentz).

(d) A biological control wasp *Aphidius ervi* (Hymenoptera: Braconidae) attacking pea aphids, *Acyrthosiphon pisum* (Hemiptera: Aphididae), Arizona, USA (A.L. Wild).

(e) A circular lerp of the red gum lerp psyllid, *Glycaspis brimblecombei*, and a white lace lerp of *Cardiaspina albitextura* (Hemiptera: Psyllidae), on *Eucalyptus blakelyi*, Canberra, Australia; note the small brown eggs of *C. albitextura* attached to the leaf (M.J. Cosgrove).

(f) An adult of the eucalypt-damaging weevil, *Gonipterus platensis* (Coleoptera: Curculionidae), Western Australia (M. Matsuki).

(g) Adult beetle of the goldspotted oak borer, *Agrilus auroguttatus* (Coleoptera: Buprestidae), which threatens native oaks, southern California, USA (M. Lewis).

LIST OF BOXES

Box 1.1 Citizen entomologists – community participation, 3
Box 1.2 Butterfly houses, 12
Box 1.3 Tramp ants and biodiversity, 15
Box 1.4 Conservation of the large blue butterfly, 18
Box 1.5 Palmageddon? Weevils in the palms, 21
Box 3.1 Molecular genetic techniques and their application to neuropeptide research, 67
Box 3.2 Tracheal hypertrophy in mealworms at low oxygen concentrations, 76
Box 3.3 The filter chamber of Hemiptera, 79
Box 3.4 Cryptonephric systems, 88
Box 4.1 Aural location of host by a parasitoid fly, 102
Box 4.2 Reception of communication molecules, 109
Box 4.3 The electroantennogram, 110
Box 4.4 Biological clocks, 118
Box 5.1 Courtship and mating in Mecoptera, 129
Box 5.2 Mating in katydids and crickets, 133
Box 5.3 Cannibalistic mating in mantids, 135
Box 5.4 Puddling and gifts in Lepidoptera, 135
Box 5.5 Sperm precedence, 141
Box 5.6 Control of mating and oviposition in a blow fly, 143
Box 5.7 Does mother know best?, 146
Box 5.8 Egg-tending fathers – the giant water bugs, 147
Box 6.1 Molecular insights into insect development, 163
Box 6.2 Calculation of day-degrees, 185
Box 7.1 *Gonipterus* weevils – recognition of a species complex, 198
Box 7.2 Integrative taxonomy of woodroaches, 199
Box 7.3 DNA barcoding and species discovery, 200
Box 8.1 The difficulties with dating, 230
Box 8.2 There were giants – evolution of insect gigantism, 232
Box 9.1 Soldier flies can recycle waste, 252
Box 9.2 Antimicrobial tactics to protect the brood of ground-nesting wasps, 255
Box 9.3 Ecosystem engineering by southern African termites, 257
Box 9.4 Ground pearls, 259
Box 10.1 Aquatic immature Diptera (true flies), 273
Box 10.2 Aquatic Hemiptera (true bugs), 274
Box 10.3 Aquatic Coleoptera (beetles), 275
Box 10.4 Aquatic Neuropterida, 276
Box 10.5 Aquatic–terrestrial insect fluxes, 282
Box 11.1 Figs and fig wasps, 291
Box 11.2 The grape phylloxera, 296
Box 11.3 Insects and the wood of live trees, 303
Box 11.4 Salvinia and phytophagous weevils, 310
Box 12.1 The dance language of bees, 331
Box 12.2 Africanized honey bees, 334
Box 12.3 Colony collapse disorder, 337
Box 12.4 Social insects as urban pests, 352
Box 13.1 Viruses, wasp parasitoids and host immunity, 368
Box 14.1 Avian predators as selective agents for insects, 381
Box 14.2 Backpack bugs – dressed to kill?, 383
Box 14.3 Chemically protected eggs, 386
Box 14.4 Insect binary chemical weapons, 388
Box 15.1 Life cycle of *Plasmodium*, 400
Box 15.2 *Anopheles gambiae* complex, 404
Box 15.3 Bed nets, 406
Box 15.4 Dengue – an emerging insect-borne disease, 408
Box 15.5 West Nile virus – an arbovirus disease emergent in North America, 409
Box 15.6 Bed bugs resurge, 415
Box 16.1 Exotic insect pests of crops in the United States, 422

Box 16.2 *Bemisia tabaci* – a pest species complex, 424
Box 16.3 The cottony-cushion scale, 426
Box 16.4 Neonicotinoid insecticides, 431
Box 16.5 Taxonomy and biological control of the cassava mealybug, 435
Box 16.6 Glassy-winged sharpshooter biological control – a Pacific success, 436
Box 16.7 The Colorado potato beetle, 448
Box 17.1 Modelling distributions of fruit flies, 461
Box 17.2 Trouble brewing? A beetle threat to coffee, 462
Box 17.3 Global eucalypts and their pests, 470
Box 17.4 Alien insects change landscapes, 471
Box 17.5 Insects and biosecurity – an Australian perspective, 472

PREFACE TO THE FIFTH EDITION

In the preface to the previous (fourth) edition of this textbook, we predicted several changes to the discipline of the study of insects – entomology. For many, we were correct, for good and bad. Most changes are associated with human activities, including warming of the planet but also including global trade, such that we have written a new chapter entitled "Insects in a changing world". Insects clearly respond to changes in climate, and this is of immediate concern for the spread of insect-borne diseases affecting our crops, our domestic animals and us. However, at least of equal significance are the insect range expansions associated with trade, as depicted on the cover of this book. Increased global commerce ("free trade") brings with it many accidental passenger insects that impact agriculture and the natural environment, but we did not foresee how many of these would damage our timber industries and landscape trees. Biosecurity, involving increased surveillance at our ports, airports and land borders, is a growing industry requiring personnel trained in entomology. Without such vigilance, we will have pests "without borders" in a homogeneous agricultural world.

We foresaw that increasingly sophisticated molecular genetic techniques would transform many areas of entomology. These studies have particularly informed our ideas of evolutionary relationships at all levels. Our understanding of the relationships among orders has been strengthened in the past five years, and there is much less uncertainty. For example, we are now confident that the hexapods, to which the insects belong, evolved from within the Crustacea, thus forming a group Pancrustacea. Closer to tips of insect phylogenetic trees, the true diversity at species level is being revealed with molecular techniques, including the use of "DNA barcoding". Modelling techniques of increasing sophistication allow exploration of the rate of molecular evolution, which in conjunction with critically examined fossil insects, provide increasingly reliable estimates of the tempo of insect evolution over the past 400 million years.

The relative ease and ever-lower costs of genomics have seen an "explosion" in "whole genomes" across a diversity of species, allowing comparative studies and also to understand the genetic cascades and controllers that shape the physiology and morphology of insects. We are selective in presenting these studies – they are so numerous and yet remain exploratory and not definitive across the whole of the hexapods.

Chapter texts are updated and supplemented with many new boxes and illustrations. In the opening chapter we report some of the dynamism given to the field by "non-mainstream" insect lovers, from recording schemes and citizen scientists to the managers of insect houses. Inevitably, we have to document the effects of insects spreading and damaging plants of interest to us, namely ornamental and environmental trees, including palms, and most worryingly, high-quality Arabica coffee.

This fifth edition was written in Australia, where we are grateful for academic status in the Division of Evolution, Ecology and Genetics of the Research School of Biology at the Australian National University (ANU), Canberra. We retain emeritus status at the Department of Entomology and Nematology at the University of California, Davis, which provides us with remote access to the wonderful online resources of the University of California library system. We thank both institutions for their support.

We are grateful to the following colleagues worldwide (listed alphabetically) for providing information, literature or ideas on many aspects of insect biology and phylogeny: Richard Cornette, Jeff Garnas, Penny

Greenslade, Mark Hoddle, Matt Krosch, Laurence Mound, Karen Meusemann, Mike Picker, Kathy Su, You-Ning Su, Gary Taylor, Alice Wells, Shaun Winterton and Andreas Zwick. The following people kindly gave permission to use one or more of their photographs in the reinstated and expanded colour plates: Denis Anderson, Nicholas Carlile, David Carter, Meredith Cosgrove, Mike Crisp, Janice Edgerly-Rooks, Mark Hoddle, Takumasa Kondo, Mike Lewis, Mamoru Matsuki, Rolf Oberprieler, Rod Peakall, David Rentz, Don Sands, Greg Sherley, Robert Sites, Richard Vane-Wright, Phil Ward, Alex Wild, Alan Yen and Paul Zborowski.

As with every edition of this textbook since 1996, were are so pleased that Karina McInnes (http://www.spilt-ink.com) revived her entomological pen-and-ink skills to provide us with a series of wonderful illustrations to capture the essence of so many insects, both friends and foes of the human world, going about their daily business. As always, we are grateful to the staff at Wiley-Blackwell, especially Ward Cooper and Kelvin Matthews for their continued support and excellent service, and Audrie Tan who was our production editor in Singapore. We were fortunate to have freelance editor Katrina Rainey to copy edit, as she did our 3rd edition.

PREFACE TO THE FOURTH EDITION

In the five years since the previous (third) edition of this textbook, the discipline of entomology has seen some major changes in emphasis. The opening up of global commerce (free trade) has brought with it many accidental passengers, including both potential and actual pestilential insects of our crops and ornamental plants, and our health. Efforts to prevent further incursions include increased surveillance, in what has become known as biosecurity, at our ports, airports and land borders. Entomologists increasingly are employed in quarantine and biosecurity, where they predict threats, and are expected to use diagnostics to recognize pests and distinguish those that are new arrivals. The inevitable newly arrived and established pests must be surveyed and control measures planned. In this edition we discuss several of these "emergent" threats from insects and the diseases that some can carry.

Molecular techniques of ever-increasing sophistication are now commonplace in many aspects of entomological study, ranging from genomic studies seeking to understand the basis of behaviors, to molecular diagnostics and the use of sequences to untangle the phylogeny of this most diverse group of organisms. Although this book is not the place to detail this fast evolving field, we present the results of many molecular studies, particularly in relation to our attempts to reconcile different ideas on evolutionary relationships, where much uncertainty remains despite a growing volume of nucleotide sequence data from a cadre of informative markers. In addition, ever more insects have their complete mitochondrial genomes sequenced, and the whole nuclear genome is available for an increasing diversity of insects, providing much scope for in-depth comparative studies. Important insights have already come from the ability to "silence" particular genes to observe the outcome, for example in aspects of development and communication. Inevitably, new molecular information will change some views on insect relationships, physiology and behavior, even in the short time between completion of this new revision and its publication. Thus we present only well established views.

In this edition of the textbook, we have updated and relocated the boxes concerning each major grouping (the traditional orders) from the chapter in which their generalized ecology placed them, to the end of the book, where they can be located more easily. We have used the best current estimates of relationships and implement a new ordinal classification for several groups. Strong evidence suggests that (a) termites ("Isoptera") are actually cockroaches (Blattodea), (b) the parasitic lice ("Phthiraptera") arose from within the free-living bark and book lice ("Psocoptera") forming order Psocodea, and (c) the fleas ("Siphonaptera") perhaps arose within Mecoptera. We discuss (and illustrate with trees) the evolutionary and classificatory significance and applications of these and other findings.

The updated chapter texts are supplemented with an additional 18 new boxes, including on the topical subjects of the *African honey bee* and *Colony Collapse Disorder* (of bees) in the sphere of apiary, *beewolf microbial defense*, and the use of *bed nets* and resurgence of *bed bugs*, *Dengue fever* and *West Nile virus* in relation to human health. New boxes are provided on *how entomologists recognize species*, on important *aquatic insects* and *energy fluxes*, and on *evolutionary relationships of flamingo lice*. Some case studies in emergent plant pests are presented, including the *Emerald ash borer* that is destroying North American landscape trees, and other insects (*light brown apple moth*, *citrus psyllid* and *fruit flies*) that threaten US crops. We relate the astonishing success story in classical biological control of the *glassy-winged sharpshooter in the Pacific* that provides hope for rejuvenation of this method of pest control.

Much of this fourth edition was written in Australia. We acknowledge the generosity and companionship of everyone in "Botany and Zoology" within the School of Biology at the Australian National University (ANU), Canberra, where we spent 10 weeks as ANU Visiting Fellows in early 2009. We also appreciate the hospitality of Frances FitzGibbon, with whom we stayed for the period that we were based at ANU. Thank you Frances for the use of your spare bedroom with the view of the Australian bush with its many birds and the occasional group of kangaroos. Our home Department of Entomology and the academic administration at UC Davis approved our three-month sabbatical leave to allow us quality time to revise this book. We are grateful to the staff in our home department for logistic support during the time we worked on the book and especially Kathy Garvey for assistance with an illustration. We are grateful to the following colleagues worldwide (listed alphabetically) for providing information and ideas on many aspects of insect biology and phylogeny: Eldon Ball, Stephen Cameron, Jason Cryan, Mark Hoddle, Kevin Johnson, Bob Kimsey, Karl Kjer, Klaus-Dieter Klass, Ed Lewis, Jim Marden, Jenny Mordue, Geoff Morse, Laurence Mound, Eric Mussen, Laurence Packer, Brad Sinclair, Vince Smith and Shaun Winterton. However, any errors of interpretation or fact are our responsibility alone. We thank our students Haley Bastien, Sarah Han, Nick Herold, and Scott McCluen for their assistance in compiling the index.

Most importantly, we were so pleased that Karina McInnes was able to recall her entomological pen-and-ink skills to provide us with a series of wonderful illustrations to capture the essence of so many insects, both friends and foes of the human world, going about their daily business.

We are grateful to the staff at Wiley-Blackwell, especially Ward Cooper, Rosie Hayden, and Delia Sandford for their continued support and excellent service. We were fortunate to have Nik Prowse edit the text of this edition.

PREFACE TO THE THIRD EDITION

Since writing the earlier editions of this textbook, we have relocated from Canberra, Australia, to Davis, California, where we teach many aspects of entomology to a new cohort of undergraduate and graduate students. We have come to appreciate some differences which may be evident in this edition. We have retained the regional balance of case studies for an international audience. With globalization has come unwanted, perhaps unforeseen, consequences, including the potential worldwide dissemination of pest insects and plants. A modern entomologist must be aware of the global status of pest control efforts. These range from insect pests of specific origin, such as many vectors of disease of humans, animals, and plants, to noxious plants, for which insect natural enemies need to be sought. The quarantine entomologist must know, or have access to, global databases of pests of commerce. Successful strategies in insect conservation, an issue we cover for the first time in this edition, are found worldwide, although often they are biased towards Lepidoptera. Furthermore, all conservationists need to recognize the threats to natural ecosystems posed by introduced insects such as crazy, big-headed, and fire ants. Likewise, systematists studying the evolutionary relationships of insects cannot restrict their studies to a regional subset, but also need a global view.

Perhaps the most publicized entomological event since the previous edition of our text was the "discovery" of a new order of insects – named as Mantophasmatodea – based on specimens from 45-million-year-old amber and from museums, and then found living in Namibia (south-west Africa), and now known to be quite widespread in southern Africa. This finding of the first new order of insects described for many decades exemplifies several aspects of modern entomological research. First, existing collections from which mantophasmatid specimens initially were discovered remain important research resources; second, fossil specimens have significance in evolutionary studies; third, detailed comparative anatomical studies retain a fundamental importance in establishing relationships, even at ordinal level; fourth, molecular phylogenetics usually can provide unambiguous resolution where there is doubt about relationships based on traditional evidence.

The use of molecular data in entomology, notably (but not only) in systematic studies, has grown apace since our last edition. The genome provides a wealth of characters to complement and extend those obtained from traditional sources such as anatomy. Although analysis is not as unproblematic as was initially suggested, clearly we have developed an ever-improving understanding of the internal relationships of the insects as well as their relationships to other invertebrates. For this reason we have introduced a new chapter (Chapter 7) describing methods and results of studies of insect phylogeny, and portraying our current understanding of relationships. Chapter 8, also new, concerns our ideas on insect evolution and biogeography. The use of robust phylogenies to infer past evolutionary events, such as origins of flight, sociality, parasitic and plant-feeding modes of life, and biogeographic history, is one of the most exciting areas in comparative biology.

Another growth area, providing ever more challenging ideas, is the field of molecular evolutionary development in which broad-scale resemblances (and unexpected differences) in genetic control of developmental processes are being uncovered. Notable studies provide evidence for identity of control for development of gills, wings, and other appendages across phyla. However, details of this field are beyond the scope of this textbook.

We retain the popular idea of presenting some tangential information in boxes, and have introduced seven new boxes: Box 1.1 Collected to extinction?; Box 1.2 Tramp ants and biodiversity; Box 1.3 Sustainable use of mopane worms; Box 4.3 Reception of communication molecules; Box 5.5 Egg-tending fathers – the giant water bugs; Box 7.1 Relationships of the Hexapoda to other Arthropoda; Box 14.2 Backpack bugs – dressed to kill?, plus a taxonomic box (Box 13.3) concerning the Mantophasmatodea (heel walkers).

We have incorporated some other boxes into the text, and lost some. The latter include what appeared to be a very neat example of natural selection in action, the peppered moth *Biston betularia*, whose melanic *carbonaria* form purportedly gained advantage in a sooty industrial landscape through its better crypsis from bird predation. This interpretation has been challenged lately, and we have reinterpreted it in Box 14.1 within an assessment of birds as predators of insects.

Our recent travels have taken us to countries in which insects form an important part of the human diet. In southern Africa we have seen and eaten mopane, and have introduced a box to this text concerning the sustainable utilization of this resource. Although we have tried several of the insect food items that we mention in the opening chapter, and encourage others to do so, we make no claims for tastefulness. We also have visited New Caledonia, where introduced ants are threatening the native fauna. Our concern for the consequences of such worldwide ant invasives, that are particularly serious on islands, is reflected in Box 1.2.

Once again we have benefited from the willingness of colleagues to provide us with up-to-date information and to review our attempts at synthesizing their research. We are grateful to Mike Picker for helping us with Mantophasmatodea and to Lynn Riddiford for assisting with the complex new ideas concerning the evolution of holometabolous development. Matthew Terry and Mike Whiting showed us their unpublished phylogeny of the Polyneoptera, from which we derived part of Fig. 7.2. Bryan Danforth, Doug Emlen, Conrad Labandeira, Walter Leal, Brett Melbourne, Vince Smith, and Phil Ward enlightened us or checked our interpretations of their research speciality, and Chris Reid, as always, helped us with matters coleopterological and linguistic. We were fortunate that our updating of this textbook coincided with the issue of a compendious resource for all entomologists: *Encyclopedia of Insects*, edited by Vince Resh and Ring Cardé for Academic Press. The wide range of contributors assisted our task immensely: we cite their work under one header in the "Further reading" following the appropriate chapters in this book.

We thank all those who have allowed their publications, photographs, and drawings to be used as sources for Karina McInnes' continuing artistic endeavors. Tom Zavortink kindly pointed out several errors in the second edition. Inevitably, some errors of fact and interpretation remain, and we would be grateful to have them pointed out to us.

This edition would not have been possible without the excellent work of Katrina Rainey, who was responsible for editing the text, and the staff at Blackwell Publishing, especially Sarah Shannon, Cee Pike, and Rosie Hayden.

PREFACE TO THE SECOND EDITION

Since writing the first edition of this textbook, we have been pleasantly surprised to find that what we consider interesting in entomology has found a resonance amongst both teachers and students from a variety of countries. When invited to write a second edition we consulted our colleagues for a wish list, and have tried to meet the variety of suggestions made. Foremost we have retained the chapter sequence and internal arrangement of the book to assist those that follow its structure in their lecturing. However, we have added a new final (16th) chapter covering methods in entomology, particularly preparing and conserving a collection. Chapter 1 has been radically reorganized to emphasize the significance of insects, their immense diversity and their patterns of distribution. By popular request, the summary table of diagnostic features of the insect orders has been moved from Chapter 1 to the end pages, for easier reference. We have expanded insect physiology sections with new sections on tolerance of environmental extremes, thermoregulation, control of development and changes to our ideas on vision. Discussion of insect behaviour has been enhanced with more information on insect–plant interactions, migration, diapause, hearing and predator avoidance, "puddling" and sodium gifts. In the ecological area, we have considered functional feeding groups in aquatic insects, and enlarged the section concerning insect–plant interactions. Throughout the text we have incorporated new interpretations and ideas, corrected some errors and added extra terms to the glossary.

The illustrations by Karina McInnes that proved so popular with reviewers of the first edition have been retained and supplemented, especially with some novel chapter vignettes and additional figures for the taxonomic and collection sections. In addition, 41 colour photographs of colourful and cryptic insects going about their lives have been chosen to enhance the text.

The well-received boxes that cover self-contained themes tangential to the flow of the text are retained. With the assistance of our new publishers, we have more clearly delimited the boxes from the text. New boxes in this edition cover two resurging pests (the phylloxera aphid and *Bemisia* whitefly), the origins of the aquatic lifestyle, parasitoid host-detection by hearing, the molecular basis of development, chemically protected eggs, and the genitalia-inflating phalloblaster.

We have resisted some invitations to elaborate on the many physiological and genetic studies using insects – we accept a reductionist view of the world appeals to some, but we believe that it is the integrated whole insect that interacts with its environment and is subject to natural selection. Breakthroughs in entomological understanding will come from comparisons made within an evolutionary framework, not from the technique-driven insertion of genes into insect and/or host.

We acknowledge all those who assisted us with many aspects of the first edition (see Preface for first edition following) and it is with some regret that we admit that such a breadth of expertise is no longer available for consultation in one of our erstwhile research institutions. This is compensated for by the following friends and colleagues who reviewed new sections, provided us with advice, and corrected some of our errors. Entomology is a science in which collaboration remains the norm – long may it continue. We are constantly surprised at the rapidity of freely given advice, even to electronic demands: we hope we haven't abused the rapidity of communication. Thanks to, in alphabetical order: Denis Anderson – varroa mites; Andy Austin – wasps and polydnaviruses; Jeff Bale – cold tolerance; Eldon Ball – segment development; Paul Cooper – physiological updates; Paul De Barro – *Bemisia*; Hugh Dingle – migration; Penny Greenslade – collembola facts; Conrad Labandeira – fossil insects; Lisa Nagy – molecular basis for limb development; Rolf Oberprieler – edible insects; Chris Reid – reviewing Chapter 1 and coleopteran factoids; Murray Upton – reviewing collecting methods; Lars-Ove Wikars – mycangia information and illustration; Jochen Zeil – vision. Dave Rentz supplied many excellent colour photographs, which we supplemented with some photos by Denis Anderson, Janice Edgerly-Rooks, Tom Eisner, Peter Menzel, Rod Peakall, Dick Vane-Wright, Peter Ward, Phil Ward and the late Tony Watson. Lyn Cook and Ben Gunn provided help with computer graphics. Many people assisted by supplying current names or identifications for particular insects, including from photographs. Special thanks to John Brackenbury, whose photograph of a soldier beetle in preparation for flight (from Brackenbury, 1990) provided the inspiration for the cover centerpiece.

When we needed a break from our respective offices in order to read and write, two Dons, Edward and Bradshaw, provided us with some laboratory space in the Department of Zoology, University of Western Australia, which proved to be rather too close to surf, wineries and wildflower sites – thank you anyway.

It is appropriate to thank Ward Cooper of the late Chapman & Hall for all that he did to make the first edition the success that it was. Finally, and surely not least, we must acknowledge that there would not have been a second edition without the helping hand put out by Blackwell Science, notably Ian Sherman and David Frost, following one of the periodic spasms in scientific publishing when authors (and editors) realize their minor significance in the "commercial" world.

PREFACE AND ACKNOWLEDGMENTS FOR FIRST EDITION

Insects are extremely successful animals and they affect many aspects of our lives, despite their small size. All kinds of natural and modified, terrestrial and aquatic, ecosystems support communities of insects that present a bewildering variety of life-styles, forms and functions. Entomology covers not only the classification, evolutionary relationships and natural history of insects, but also how they interact with each other and the environment. The effects of insects on us, our crops and domestic stock, and how insect activities (both deleterious and beneficial) might be modified or controlled, are amongst the concerns of entomologists.

The recent high profile of biodiversity as a scientific issue is leading to increasing interest in insects because of their astonishingly high diversity. Some calculations suggest that the species richness of insects is so great that, to a near approximation, all organisms can be considered to be insects. Students of biodiversity need to be versed in entomology.

We, the authors, are systematic entomologists teaching and researching insect identification, distribution, evolution and ecology. Our study insects belong to two groups – scale insects and midges – and we make no apologies for using these, our favourite organisms, to illustrate some points in this book.

This book is not an identification guide, but addresses entomological issues of a more general nature. We commence with the significance of insects, their internal and external structure, and how they sense their environment, followed by their modes of reproduction and development. Succeeding chapters are based on major themes in insect biology, namely the ecology of ground-dwelling, aquatic and plant-feeding insects, and the behaviours of sociality, predation and parasitism, and defence. Finally, aspects of medical and veterinary entomology and the management of insect pests are considered.

Those to whom this book is addressed, namely students contemplating entomology as a career, or studying insects as a subsidiary to specialized disciplines such as agricultural science, forestry, medicine or veterinary science, ought to know something about insect systematics – this is the framework for scientific observations. However, we depart from the traditional order-by-order systematic arrangement seen in many entomological textbooks. The systematics of each insect order are presented in a separate section following the ecological–behavioural chapter appropriate to the predominant biology of the order. We have attempted to keep a phylogenetic perspective throughout, and one complete chapter is devoted to insect phylogeny, including examination of the evolution of several key features.

We believe that a picture is worth a thousand words. All illustrations were drawn by Karina Hansen McInnes, who holds an Honours degree in Zoology from the Australian National University, Canberra. We are delighted with her artwork and are grateful for her hours of effort, attention to detail and skill in depicting the essence of the many subjects that are figured in the following pages. Thank you Karina.

This book would still be on the computer without the efforts of John Trueman, who job-shared with Penny in second semester 1992. John delivered invertebrate zoology lectures and ran lab classes while Penny revelled in valuable writing time, free from undergraduate teaching. Aimorn Stewart also assisted Penny by keeping her research activities alive during book preparation and by helping with labelling of figures. Eva Bugledich acted as a library courier and brewed hundreds of cups of coffee.

The following people generously reviewed one or more chapters for us: Andy Austin, Tom Bellas, Keith Binnington, Ian Clark, Geoff Clarke, Paul Cooper, Kendi Davies, Don Edward, Penny Greenslade, Terry Hillman, Dave McCorquodale, Rod Mahon, Dick Norris, Chris Reid, Steve Shattuck, John Trueman and Phil Weinstein. We also enjoyed many discussions on hymenopteran phylogeny and biology with Andy. Tom sorted out our chemistry and Keith gave expert advice on insect cuticle. Paul's broad knowledge of insect physiology was absolutely invaluable. Penny put us straight with springtail facts. Chris' entomological knowledge, especially on beetles, was a constant source of information. Steve patiently answered our endless questions on ants. Numerous other people read and commented on sections of chapters or provided advice or helpful discussion on particular entomological topics. These people included John Balderson, Mary Carver, Lyn Cook, Jane Elek, Adrian Gibbs, Ken Hill, John Lawrence, Chris Lyal, Patrice Morrow, Dave Rentz, Eric Rumbo, Vivienne Turner, John Vranjic and Tony Watson. Mike Crisp assisted with checking on current host-plant names. Sandra McDougall inspired part of Chapter 15. Thank you everyone for your many comments which we have endeavoured to incorporate as far as possible, for your criticisms which we hope we have answered, and for your encouragement.

We benefited from discussions concerning published and unpublished views on insect phylogeny (and fossils), particularly with Jim Carpenter, Mary Carver, Niels Kristensen, Jarmila Kukalová-Peck and John Trueman. Our views are summarized in the phylogenies shown in this book and do not necessarily reflect a consensus of our discussants' views (this was unattainable).

Our writing was assisted by Commonwealth Scientific and Industrial Research Organization (CSIRO) providing somewhere for both of us to work during the many weekdays, nights and weekends during which this book was prepared. In particular, Penny managed to escape from the distractions of her university position by working in CSIRO. Eventually, however, everyone discovered her whereabouts. The Division of Entomology of the CSIRO provided generous support: Carl Davies gave us driving lessons on the machine that produced reductions of the figures, and Sandy Smith advised us on labelling. The Division of Botany and Zoology of the Australian National University also provided assistance in aspects of the book production: Aimorn Stewart prepared the SEMs from which Fig. 4.7 was drawn, and Judy Robson typed the labels for some of the figures.

ABOUT THE COMPANION WEBSITE

This book is accompanied by a companion website:

www.wiley.com/go/gullan/insects

The website includes:

- Powerpoints of all figures from the book for downloading
- PDFs of tables from the book

Chapter 1

THE IMPORTANCE, DIVERSITY AND CONSERVATION OF INSECTS

Charles Darwin inspecting beetles collected during the voyage of the *Beagle*. (After various sources, especially Huxley & Kettlewell 1965 and Futuyma 1986.)

Curiosity alone as to the identities and lifestyles of the fellow inhabitants of our planet justifies the study of insects. Some of us have used insects as totems and symbols in spiritual life, and we portray them in art and music. If we consider economic factors, the effects of insects are enormous. Few human societies lack honey, which is provided by bees (or specialized ants). Insects pollinate our crops. Many insects share our houses, agriculture and food stores. Others live on us, on our domestic pets or our livestock, and more visit to feed on us, where they may transmit disease. Clearly, we should understand these pervasive animals.

Although there are millions of kinds of insects, we do not know exactly (or even approximately) how many. This ignorance as to how many organisms we share our planet with is remarkable considering that astronomers have listed, mapped and uniquely identified a comparable diversity of galactic objects. Some estimates, which we discuss in detail later, imply that the species richness of insects is so great that, to a near approximation, all organisms can be considered to be insects. Although dominant on land and in freshwater, few insects are found beyond the tidal limit of oceans.

In this opening chapter, we outline the significance of insects and discuss their diversity and classification, and their roles in our economic and wider lives. First, we outline the field of entomology and the role of entomologists, and then introduce the ecological functions of insects. Next, we explore insect diversity, and then discuss how we name and classify this immense diversity. Sections follow in which we consider some cultural and economic aspects of insects, their aesthetic and tourism appeal, their conservation, and how and why they may be reared. We conclude with a section on insects as food for humans and animals. In text boxes we discuss citizen involvement in entomology (Box 1.1), the phenomenal growth of butterfly houses (Box 1.2), the effects of tramp ants on biodiversity (Box. 1.3), the conservation of the large blue butterfly in England (Box 1.4) and insect threats to palm trees (Box 1.5).

1.1 WHAT IS ENTOMOLOGY?

Entomology is the study of insects. Entomologists are the people who study insects, and observe, collect, rear and experiment with insects. Research undertaken by entomologists covers the total range of biological disciplines, including evolution, ecology, behaviour, anatomy, physiology, biochemistry and genetics. The unifying feature is that the study organisms are insects. Biologists work with insects for many reasons: ease of culturing in a laboratory, rapid population turnover, and availability of many individuals are important factors. The minimal ethical concerns regarding responsible experimental use of insects, as compared with vertebrates, are a significant consideration.

Modern entomological study commenced in the early 18th century when a combination of rediscovery of the classical literature, the spread of rationalism, and the availability of ground-glass optics made the study of insects acceptable for the thoughtful privately wealthy. Although today many people working with insects hold professional positions, some aspects remain suitable for informed citizens (Box 1.1). Charles Darwin's initial enthusiasm in natural history was as a collector of beetles (as shown in the vignette for this chapter) and throughout his life he communicated with amateur entomologists throughout the world. Much of our present understanding of worldwide insect diversity is derived from studies of non-professionals. Many such contributions come from collectors of attractive insects such as butterflies and beetles, but others with patience and ingenuity continue the tradition of Jean-Henri Fabre in observing close-up the activities of insects. We can discover much of scientific interest at little expense regarding the natural history of even "well-known" insects. The variety of size, structure and colour in insects (see **Plate 1a–f**) is striking, whether depicted in drawings, photographs or movies.

A popular misperception is that professional entomologists emphasize killing or controlling insects, but numerous entomological studies document their beneficial roles.

1.2 THE IMPORTANCE OF INSECTS

There are many reasons why we should study insects. Their ecologies are incredibly variable. Insects may dominate food chains and food webs in terms of both

The Insects: An Outline of Entomology, Fifth Edition. P.J. Gullan and P.S. Cranston.

Companion Website: www.wiley.com/go/gullan/insects

Box 1.1 Citizen entomologists – community participation

The involvement of non-professional "citizen scientists" in biodiversity studies dates back at least to the 18th century, especially in the United Kingdom. Published guides to the fauna became best-sellers – Victorian ladies of leisure studied the flora and collected shells and fossils, wealthy gentlemen shot rare birds and collected their eggs, and the rich assembled "cabinets of curios" that became the world-renowned natural history collections of the great museums. Darwin, portrayed in the vignette at the beginning of this chapter, was a skilled collector and student of the Coleoptera, and many church curates, with little to do between Sunday sermons, were serious entomologists at a time when few were paid for such studies.

Despite the transformation of natural history into a professional science, areas such as floristics and ornithology continue to benefit from amateur involvement. The ever-increasing availability of internet-based guides to images, distribution maps, bird-songs, etc., encourages involvement of citizens in recording many facets of the biota of their local areas. The more popular insects are also subjects of interest to a wider public, and there is substantial participation in reporting occurrences, particularly for butterflies, dragonflies, wasps and bees, and beetles. Especially in Europe and North America, many can be identified (some more easily than others) without killing the insect, by eye or by using a hand lens. With digital photography, excellent macro-images can be passed to experts for confirmation of identification, used to "voucher" observations, and validated records then can be entered into databases. Citizen-collected records are valuable in establishing distributions and temporal presence (e.g. early and late dates for appearance) – and have assisted in documentation for conservation and for assessing effects of climate change.

The longest lasting, and surely largest, participatory survey, the Rothamsted Insect Survey, has used light-traps at over 430 sites in the United Kingdom, many operated by volunteers, since the 1960s. More than 730 species of macrolepidoptera have been recognized since the survey began, and more than 10 million data points (species identity × location × date × abundance) have been databased. Although this resource is used widely to infer effects of climate change, most studies infer that observed declines in lepidopteran populations relate more to the staggering loss of natural habitats to agriculture, with climate effects evident most in the previous years' summer conditions.

In the United States, citizen scientists have been recruited to collect long-term data on sightings of migrating butterflies, eggs and larval populations of the monarch butterfly (*Danaus plexippus*) and its milkweed habitat. Volunteers contribute to monarch conservation by regular monitoring of their local sites. One major goal is to understand how and why monarch populations vary in time and space, particularly during the breeding season in North America.

The United Kingdom's Ladybird Survey provides another example of public participation in insect recording. Ladybirds (called lady beetles or ladybugs in the United States; a pair of copulating adults is illustrated here) are common, colourful, and can be identified with appropriate guidance (using a "Ladybird Atlas") built on several pre-existing specialist recording schemes. A website (http://www.ladybird-survey.org/) provides much information to help find and identify species, and provides online forms to record observations. There is great value in this scheme, as evidenced by the recent invasion of the United Kingdom by the East Asian harlequin ladybird (*Harmonia axyridis*); documentation of its spread and the impact on native ladybirds has been possible due to existing community surveys. Such a project requires a good existing database: the impacts of this introduced species in the United States can only be surmised as pre-invasion data are inadequate.

Global recording of odonates (dragonflies and damselflies) by amateurs is very popular, especially in Asia. As with small birds, identifications can be made at some distance using binoculars (as illustrated here). Photographic vouchers can be taken while the insects are sedentary (e.g. at dawn). Care must be taken though not to assume that the presence of adult odonates indicates water conditions suitable for nymphal development since adults are strong flyers (see section 10.5 on biomonitoring).

As with all observational data, appropriate checks on identifications are important and observer biases often occur. Thus citizen-science data should be interpreted with caution, although there is little doubt as to the value of data collection by an interested and informed public.

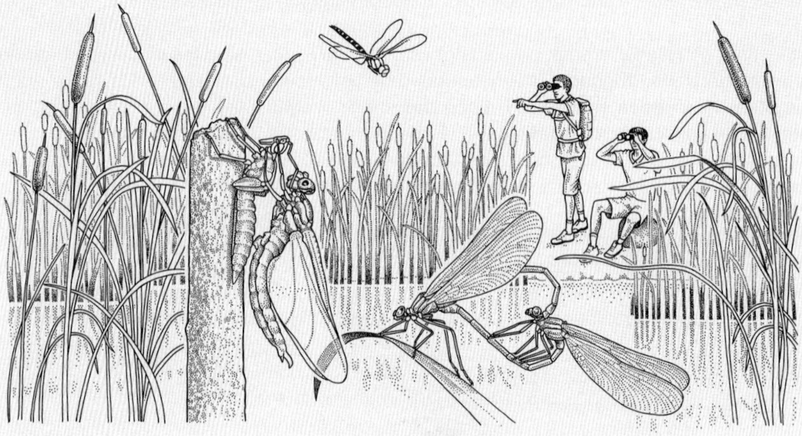

volume and numbers. Feeding specializations of different insect groups include ingestion of detritus, rotting materials, wood and fungus (Chapter 9), aquatic filter feeding and grazing (Chapter 10), herbivory (= phytophagy), including sap feeding (Chapter 11), and predation and parasitism (Chapter 13). Insects may live in water, on land, or in soil, during part or all of their lives. Their lifestyles may be solitary, gregarious, subsocial or highly social (Chapter 12). They may be conspicuous, mimics of other objects, or concealed (Chapter 14), and may be active by day or by night. Insect life cycles (Chapter 6) allow survival under a wide range of conditions, such as extremes of heat and cold, wet and dry, and unpredictable climates.

Insects are essential to the following ecosystem functions:

- nutrient recycling, via leaf-litter and wood degradation, dispersal of fungi, disposal of carrion and dung, and soil turnover;
- plant propagation, including pollination and seed dispersal;
- maintenance of plant community composition and structure, via phytophagy, including seed feeding;

• food for insectivorous vertebrates, such as many birds, mammals, reptiles and fish;
• maintenance of animal community structure, through transmission of diseases of large animals, and predation and paratization of smaller ones.

Each insect species is part of a greater assemblage and its loss affects the complexities and abundance of other organisms. Some insects are considered "**keystone species**" because loss of their critical ecological functions could lead to collapse of the wider ecosystem. For example, termites convert cellulose in tropical soils (section 9.1), suggesting that they are keystones in tropical soil structuring. In aquatic ecosystems, a comparable service is provided by the guild of mostly larval insects that breaks down and releases the nutrients from wood and leaves derived from the surrounding terrestrial environment.

Insects are associated intimately with our survival, in that certain insects damage our health and that of our domestic animals (Chapter 15) and others adversely affect our agriculture and horticulture (Chapter 16). Certain insects greatly benefit human society, either by providing us with food directly or by contributing to our food or materials that we use. For example, honey bees not only provide us with honey but also are valuable agricultural pollinators worth an additional estimated US$15 billion annually in increased crop yields in the United States alone. Also the quality of bee-pollinated fruits can exceed that of fruits pollinated by wind or selfing (see section 11.3.1). Furthermore, estimates of the value of pollination by wild, free-living bees is US$1.0–2.4 billion per year for California alone. The total economic value of pollination services estimated for the 100 crops used directly for human food globally exceeds US$200 billion annually. Furthermore, valuable services, such as those provided by predatory beetles and bugs or parasitic wasps that control pests, often go unrecognized, especially by city-dwellers, and yet such ecosystem services are worth billions of US$ annually.

Insects contain a vast array of chemical compounds, some of which can be collected, extracted or synthesized for our use. Chitin, a component of insect cuticle, and its derivatives act as anticoagulants, enhance wound and burn healing, reduce serum cholesterol, serve as non-allergenic drug carriers, provide strong biodegradable plastics, and enhance removal of pollutants from waste water, to mention just a few developing applications. Silks from the cocoons of silkworm moths, *Bombyx mori*, and related species have been used for fabric for centuries, and two endemic South African species may be increasing in local value. The red dye cochineal is obtained commercially from scale insects of *Dactylopius coccus* cultured on *Opuntia* cacti. Another scale insect, the lac insect *Kerria lacca*, is a source of a commercial varnish called shellac. Given this range of insect-produced chemicals, and accepting our ignorance of most insects, there is a high likelihood that novel chemicals await our discovery and use.

Insects provide more than economic or environmental benefits; characteristics of certain insects make them useful models for understanding general biological processes. For instance, the short generation time, high fecundity, and ease of laboratory rearing and manipulation of the vinegar or common fruit fly, *Drosophila melanogaster*, have made it a model research organism. Studies of *D. melanogaster* have provided the foundations for our understanding of genetics and cytology, and these flies continue to provide the experimental materials for advances in molecular biology, embryology and development. Outside the laboratories of geneticists, studies of social insects, notably hymenopterans such as ants and bees, have allowed us to understand the evolution and maintenance of social behaviours such as altruism (section 12.4.1). The field of sociobiology owes its existence to entomologists' studies of social insects. Several theoretical ideas in ecology have derived from the study of insects. For example, our ability to manipulate the food supply (cereal grains) and number of individuals of flour beetles (*Tribolium* spp.) in culture, combined with their short life history (compared to most vertebrates), has provided insights into how populations are regulated. Some concepts in ecology, for example the ecosystem and niche, came from scientists studying freshwater systems where insects dominate. Alfred Wallace (depicted in the vignette of Chapter 18), the independent and contemporaneous discoverer with Charles Darwin of the theory of evolution by natural selection, based his ideas on observations of tropical insects. Hypotheses concerning the many forms of mimicry and sexual selection derive from observations of insect behaviour, which continue to be investigated by entomologists.

Finally, the sheer numbers of insects means that their impact upon the environment, and hence our lives, is highly significant. Insects are the major component of macroscopic biodiversity and, for this reason alone, we should try to understand them better.

1.3 INSECT BIODIVERSITY

1.3.1 The described taxonomic richness of insects

Probably slightly over one million species of insects have been described, that is, have been recorded in a taxonomic publication as "new" (to science that is), accompanied by a description and often with illustrations or some other means of recognizing the particular insect species (section 1.4). Since some insect species have been described as new more than once, due to failure to recognize variation or through ignorance of previous studies, the actual number of described species is uncertain.

The described species of insects are distributed unevenly amongst the higher taxonomic groupings called orders (section 1.4). Five "major" orders stand out for their high species richness: the beetles (Coleoptera); flies (Diptera); wasps, ants and bees (Hymenoptera); butterflies and moths (Lepidoptera); and the true bugs (Hemiptera). J.B.S. Haldane's jest – that "God" (evolution) shows an inordinate "fondness" for beetles – appears to be confirmed since they comprise almost 40% of described insects (more than 350,000 species). The Hymenoptera have more than 150,000 described species, with the Diptera and Lepidoptera having at least 150,000 described species each, and Hemiptera over 100,000. Of the remaining orders of living insects, none exceed the approximately 24,000 described species of the Orthoptera (grasshoppers, locusts, crickets and katydids). Most of the "minor" orders comprise some hundreds to a few thousands of described species. Although an order may be described as "minor", this does not mean that it is insignificant – the familiar earwig belongs to an order (Dermaptera) with fewer than 2000 described species, and the ubiquitous cockroaches belong to an order (Blattodea, which includes termites) with only about 7500 species. Moreover, there are only twice as many species described in Aves (birds) as in the "small" order Blattodea.

1.3.2 The estimated taxonomic richness of insects

Surprisingly, the figures given above, which represent the cumulative effort by many insect taxonomists from all parts of the world over some 250 years, appear to represent something less than the true species richness of the insects. Just how far short is the subject of continuing speculation. Given the very high numbers and the patchy distributions of many insects in time and space, it is impossible in our time-scales to inventory (count and document) all species even for a small area. Extrapolations are required to estimate total species richness, which range from some three million to as many as 80 million species. These various calculations either extrapolate ratios for richness in one taxonomic group (or area) to another unrelated group (or area), or use a hierarchical scaling ratio, extrapolated from a subgroup (or subordinate area) to a more inclusive group (or wider area).

Generally, ratios derived from temperate/tropical species numbers for well-known groups such as vertebrates provide rather conservatively low estimates if used to extrapolate from temperate insect taxa to essentially unknown tropical insect faunas. The most controversial estimation, based on hierarchical scaling and providing the highest estimated total species numbers, was an extrapolation from samples from a single tree species to global rainforest insect species richness. Sampling used insecticidal fog to assess the little-known fauna of the upper layers (the canopy) of Neotropical rainforest. Much of this estimated increase in species richness was derived from arboreal beetles (Coleoptera), but several other canopy-dwelling groups were found to be much more numerous than believed previously. Key factors in calculating tropical diversity included identification of the number of beetle species found, estimation of the proportion of novel (previously unseen) groups, estimation of the degree of host-specificity to the surveyed tree species, and the ratio of beetles to other arthropods. Certain assumptions have been tested and found to be suspect, notably, host-plant specificity of herbivorous insects, at least in some tropical forests, seems very much less than estimated early on in this debate.

Estimates of global insect diversity calculated from experts' assessments of the proportion of undescribed versus described species amongst their study insects tend to be comparatively low. Belief in lower numbers of total species comes from our general inability to confirm the prediction, which is a logical consequence of the high species-richness estimates, that insect samples ought to contain very high proportions of previously unrecognized and/or undescribed ("novel") taxa. Obviously, any expectation of an even spread of novel species is unrealistic, since some groups and regions of the world are poorly known compared to others. However, amongst the minor (less species-rich) orders there is little or no scope for dramatically increased, unrecognized species richness. Very high levels of novelty, if they do exist, realistically could only be amongst the Coleoptera, Diptera, Lepidoptera and parasitic Hymenoptera. Molecular techniques, for example **DNA barcoding**, sometimes in conjunction with trained biodiversity technicians (**parataxonomists**), reveal high levels of cryptic (hidden) diversity of the latter two groups in Costa Rica (see Box 7.3).

Nevertheless, some (but not all) recent re-analyses tend towards the lower end of the range of estimates derived from taxonomists' calculations and extrapolations from regional sampling rather than those derived from ecological scaling. A figure of between two and six million species of insects appears realistic.

1.3.3 The location of insect species richness

The regions in which additional undescribed insect species might occur (i.e. up to an order of magnitude greater number of novel species than described) cannot be in the northern hemisphere, where such hidden diversity in the well-studied faunas is unlikely. For example, the British Isles inventory of about 22,500 species of insect is likely to be within 5% of being complete, and the 30,000 or so described from Canada must represent over half of the total species. Any hidden diversity is not in the Arctic, with some 3000 species present in the American Arctic, nor in Antarctica, the southern polar mass, which supports a bare handful of insects. Evidently, just as species-richness patterns are uneven across groups, so too is their geographic distribution.

Despite the lack of necessary local species inventories to prove it, tropical species richness appears to be much higher than that of temperate areas. For example, a single tree surveyed in Peru produced 26 genera and 43 species of ants: a tally that equals the total ant diversity from all habitats in Britain. Our inability to be certain about finer details of geographical patterns stems in part from entomologists interested in biodiversity issues being based mostly in the temperate northern hemisphere, whereas the centres of richness of the insects themselves are in the tropics and southern hemisphere.

Studies in tropical American rainforests suggest that much undescribed novelty in insects does come from the beetles, which provided the basis for the original high species-richness estimate. Although beetle dominance may be true in places such as the Neotropics, this might be an artefact of research biases of entomologists. In some well-studied temperate regions such as the United Kingdom and Canada, species of true flies (Diptera) appear to outnumber beetles. Studies of canopy insects on the tropical island of Borneo have shown that both Hymenoptera and Diptera can be more species rich at particular sites than the Coleoptera. Comprehensive regional inventories or credible estimates of insect faunal diversity may eventually tell us which order of insects is globally most diverse.

Whether we estimate 30–80 million species or an order of magnitude less, insects constitute at least half of global species diversity (Fig. 1.1). If we consider only life on land, insects comprise an even greater proportion of extant species, since the radiation of insects is a predominantly terrestrial phenomenon. The relative contribution of insects to global diversity will be somewhat lessened if marine diversity, to which insects make a negligible contribution, actually is higher than currently understood.

1.3.4 Some reasons for insect species richness

Whatever the global estimate is, insects surely are remarkably speciose. This high species richness has been attributed to several factors. The small size of insects, a limitation imposed by their method of gas

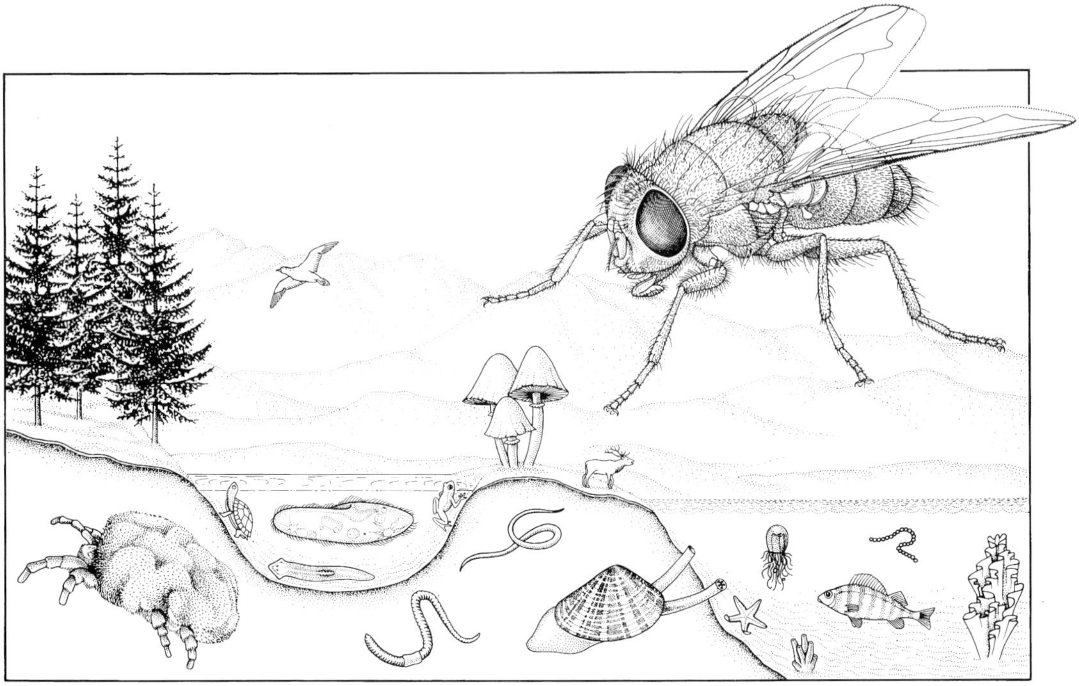

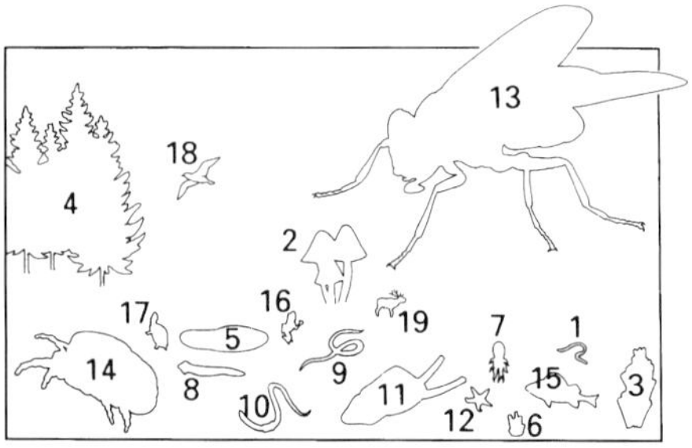

1 Prokaryotes
2 Fungi
3 Algae
4 Plantae (multicellular plants)
5 Protozoa
6 Porifera (sponges)
7 Cnidaria (jellyfish, corals, etc.)
8 Platyhelminthes (flatworms)
9 Nematoda (roundworms)
10 Annelida (earthworms, leeches, etc.)
11 Mollusca (snails, bivalves, octopus, etc.)
12 Echinodermata (starfish, sea urchins, etc.)
13 Insecta
14 Non-insect Arthropoda
15 Pisces (fish)
16 Amphibia (frogs, salamanders, etc.)
17 Reptilia (snakes, lizards, turtles)
18 Aves (birds)
19 Mammalia (mammals)

Fig. 1.1 Speciescape, in which the size of individual organisms is approximately proportional to the number of described species in the higher taxon that it represents. (After Wheeler 1990.)

exchange via tracheae, is an important determinant. Many more niches exist in any given environment for small organisms than for large organisms. Thus, a single acacia tree, that feeds one giraffe, supports the complete life cycle of dozens of insect species: a lycaenid butterfly larva chews the leaves; a bug sucks the stem sap; a longicorn beetle bores into the wood; a midge galls the flower buds; a bruchid beetle destroys the seeds; a mealybug sucks the root sap; and several wasp species parasitize each host-specific phytophage. An adjacent acacia of a different species feeds the same giraffe but may have a very different suite of phytophagous insects. The environment is more fine-grained from an insect's perspective compared to that of a mammal or bird.

Small size alone is insufficient to allow exploitation of this environmental heterogeneity, since organisms must be capable of recognizing and responding to environmental differences. Insects have highly organized sensory and neuromotor systems, which are more comparable to those of vertebrate animals than to those of other invertebrates. However, insects differ from vertebrates both in size and in how they respond to environmental change. Generally, vertebrate animals are longer lived than insects and individuals can adapt to change by some degree of learning. Insects, on the other hand, normally respond to, or cope with, altered conditions (e.g. the application of insecticides to their host plant) by genetic change between generations (e.g. leading to insecticide resistant insects). High genetic heterogeneity or elasticity within insect species allows persistence in the face of environmental change. Persistence exposes species to processes that promote speciation, predominantly involving phases of range expansion and/or subsequent fragmentation. Stochastic processes (genetic drift) and/or selection pressures provide the genetic alterations that may become fixed in spatially or temporally isolated populations.

Insects possess characteristics that expose them to other potentially diversifying influences that enhance their species richness. Interactions between certain groups of insects and other organisms, such as plants in the case of herbivorous insects, or hosts for parasitic insects, may promote the genetic diversification of eater and eaten (section 8.6). These interactions are often called coevolutionary and are discussed in more detail in Chapters 11 and 13. The reciprocal nature of such interactions may speed up evolutionary change in one or both partners or sets of partners, perhaps even leading to major radiations in certain groups. Such a scenario involves increasing specialization of insects, at least on plant hosts. Evidence from phylogenetic studies suggests that this has happened – but also that generalists may arise from within a specialist radiation, perhaps after some plant chemical barrier has been overcome. Waves of specialization followed by breakthrough and radiation must have been a major factor in promoting the high species richness of phytophagous insects.

Another explanation for the high species numbers of insects is the role of sexual selection in the diversification of many insects. The propensity of insects to become isolated in small populations (because of the fine scale of their activities) in combination with sexual selection (sections 5.3 and 8.6) may lead to rapid alteration in intra-specific communication. When (or if) the isolated population rejoins the larger parental population, altered sexual signalling deters hybridization and the identity of each population (incipient species) is maintained despite the sympatry. This mechanism is seen to be much more rapid than genetic drift or other forms of selection, and need involve little if any differentiation in terms of ecology or non-sexual morphology and behaviour.

Comparisons among insect groups and between insects and their close relatives may suggest some reasons for insect diversity. Which characteristics are shared by the most speciose insect orders: the Coleoptera, Hymenoptera, Diptera and Lepidoptera? Which features of insects do other arthropods, such as arachnids (spiders, mites, scorpions and their allies) lack? No simple explanation emerges from such comparisons; probably design features, flexible life-cycle patterns and feeding habits play a part (some factors are explored in Chapter 8). In contrast to the most speciose insect groups, arachnids lack winged flight, lack complete transformation of body form during development (metamorphosis) and lack dependence on specific food organisms, and generally do not feed on plants. Exceptionally, mites, the most diverse and abundant of arachnids, have many very specific associations with other living organisms, including with plants.

High persistence of species or lineages, or the numerical abundance of individual species are considered as indicators of insect success. However, insects differ from vertebrates by at least one popular measure of success: body size. Miniaturization is the insect success story: most insects have body lengths of 1–10 mm, with a body length of around 0.3 mm in mymarid wasps (which are parasitic on eggs of insects) being unexceptional. At the other extreme, the greatest wingspan in a living insect belongs to the tropical American owlet moth, *Thysania agrippina* (Noctuidae), with a span of up to 30 cm, although fossils show that some extinct insects were appreciably larger than their living relatives. For example, an Upper Carboniferous silverfish, *Ramsdelepidion schusteri* (Zygentoma), had a body length of 6 cm, as compared to a modern maximum of less than 2 cm. The wingspans of many Carboniferous insects exceeded 45 cm, and a Permian dragonfly, *Meganeuropsis americana* (Meganisoptera), had a wingspan of 71 cm. Notably, amongst these large insects, the great size is predominantly associated with a narrow, elongate body, although one of the heaviest extant insects, the Hercules beetle, *Dynastes hercules* (Scarabaeidae), with a body up 17 cm long, is an exception in having a bulky body. The heaviest recorded insect is a weta (Anostostomatidae; see **Plate 1d**), with a female of the bulky Little Barrier Island giant weta, *Deinacrida heteracantha*, weighing 71 g.

Barriers to large size include the inability of the tracheal system to diffuse gases across extended distances from active muscles to and from the external environment (see Box 3.2). Further elaborations of the tracheal system would jeopardize water balance in a large insect. Most large insects are narrow and have not greatly extended the maximum distance between the external oxygen source and the muscular site of gaseous exchange, compared with smaller insects. A possible explanation for the gigantism of some Palaeozoic insects is considered in Box 8.2.

In summary, many insect radiations probably depended upon: (i) the small size of individuals, combined with (ii) short generation time, (iii) sensory and neuromotor sophistication, (iv) evolutionary interactions with plants and other organisms, (v) metamorphosis, and (vi) mobile winged adults. The substantial time since the origin of each major insect group has allowed many opportunities for lineage diversification (Chapter 8). Present-day species diversity results from either higher rates of speciation (for which there is limited evidence) and/or lower rates of species extinction (higher persistence) than other organisms. The high species richness seen in some (but not all) groups in the tropics may result from the combination of higher rates of species formation with high species accumulation in equable climates.

1.4 NAMING AND CLASSIFICATION OF INSECTS

The formal naming of insects follows the rules of nomenclature developed for all animals (plants have a slightly different system). Formal scientific names are required for unambiguous communication between all scientists, no matter what their native language. Vernacular (common) names do not fulfil this need: the same insects may even have different vernacular names amongst people who speak the same language. For instance, the British refer to "ladybirds", whereas the same coccinellid beetles are "ladybugs" to many people in the United States. Many insects have no vernacular name, or a common name is given to multiple species as if only one species is involved. These difficulties are addressed by the Linnaean system, which provides every described species with two given names (the binomen). The first is the generic (genus) name, used for a broader grouping than the second name, which is the specific (species) name. These Latinized names always are used together and are italicized, as in this book. The combination of genus and species names provides each organism with a unique name. Thus, the name *Aedes aegypti* is recognized by any medical entomologist, anywhere, whatever the local name (and there are many) for this disease-transmitting mosquito. Ideally, all taxa should have such a Latinized binomen, but in practice some alternatives may be used prior to naming formally (section 18.3.2).

In scientific publications, the species name often is followed by the name of the original describer of the species and perhaps the year in which the name first was published legally. In this book, we do not follow this practice but, in discussion of particular insects, we give the order and family names to which the species belongs. In publications, after the first citation of the combination of genus and species names in the text, it is common practice in subsequent citations to abbreviate the genus to the initial letter only (e.g. *A. aegypti*). However, where this might be ambiguous, such as for the two mosquito genera *Aedes* and *Anopheles*, the initial two letters *Ae.* and *An.* are used, as in Chapter 15.

Various taxonomically defined groups, also called taxa (singular: **taxon**), are recognized amongst the insects. As for all other organisms, the basic biological taxon, lying above the individual and population, is the species, which is both the fundamental nomenclatural unit in taxonomy and, arguably, a unit of evolution. Multi-species studies allow recognition of genera, which are discrete higher groups. In a similar manner, genera can be grouped into tribes, tribes into subfamilies, and subfamilies into families. The families of insects are placed in relatively large but easily recognized groups called orders. This hierarchy of ranks (or categories) thus extends from the species

Table 1.1 Taxonomic categories (obligatory categories are shown in **bold**).

Taxon category	Standard suffix	Example
Order		Hymenoptera
Suborder		Apocrita
Superfamily	-oidea	Apoidea
Epifamily	-oidae	Apoidae
Family	-idae	Apidae
Subfamily	-inae	Apinae
Tribe	-ini	Apini
Genus		*Apis*
Subgenus		
Species		*A. mellifera*
Subspecies		*A. m. mellifera*

level through a series of "higher" levels of greater and greater inclusivity until all true insects are included in one class, the Insecta. There are standard suffixes for certain ranks in the taxonomic hierarchy, so that the rank of most group names can be recognized by inspection of the ending (Table 1.1).

Depending on the classification system used, some 25 to 30 orders of Insecta may be recognized. Differences arise principally because there are no fixed rules for deciding the taxonomic ranks referred to above – only general agreement that groups should be monophyletic, comprising all the descendants of a common ancestor (section 7.1.1). Over time, a relatively stable classification system has developed, but differences of opinion remain as to the boundaries around and among groups, with "splitters" recognizing a greater number of groups and "lumpers" favouring broader categories. For example, some North American taxonomists group ("lump") the alderflies, dobsonflies, snakeflies and lacewings into one order, the Neuroptera, whereas others, including ourselves, "split" the group and recognize three separate (but clearly closely related) orders, Megaloptera, Raphidioptera, and a more narrowly defined Neuroptera (see Fig. 7.2). The order Hemiptera has sometimes been divided into two orders, Homoptera and Heteroptera, but the homopteran grouping is invalid (non-monophyletic) and we advocate a different classification for these bugs as shown in Fig. 7.6 and discussed in section 7.4.2 and Taxobox 20. New data and methods of analysis are further causes of instability in the recognition of insect orders. As we show in Chapter 7, two groups (termites and parasitic lice) previously treated as orders, belong within each of two other orders and thus the ordinal count is reduced by two.

We recognize 28 orders of insects, with relationships considered in section 7.4 and the physical characteristics and biologies of their constituent taxa described in taxoboxes near the end of the book. A summary of the diagnostic features of all orders and a few subgroups, plus cross references to fuller identificatory and ecological information, appear in tabular form in the reference guide to orders in the Appendix at the end of the book.

1.5 INSECTS IN POPULAR CULTURE AND COMMERCE

People have been attracted to the beauty or mystique of certain insects throughout history. We know the importance of scarab beetles to the Egyptians as religious items, and earlier shamanistic cultures elsewhere in the Old World made ornaments that represent scarabs and other beetles including buprestids (jewel beetles). In Old Egypt the scarab, which shapes dung into balls, is identified as a potter; similar insect symbolism extends also further east. Egyptians, and subsequently the Greeks, made ornamental scarabs from many materials including lapis lazuli, basalt, limestone, turquoise, ivory, resins, and even valuable gold and silver. Such adulation may have been the pinnacle that an insect lacking economic importance ever gained in popular and religious culture, although many human societies recognized insects in their ceremonial lives. The ancient Chinese regarded cicadas as symbolizing rebirth or immortality. In Mesopotamian literature the *Poem of Gilgamesh* alludes to odonates (dragonflies/damselflies) as signifying the impossibility of immortality. In martial arts the swaying and sudden lunges of a praying mantis are evoked in Chinese praying mantis kung fu. The praying mantis carries much cultural symbolism, including creation and patience in zen-like waiting for the San ("bushmen") of the Kalahari. Honeypot ants *(yarumpa)* and witchety grubs (*udnirringitta*) figure amongst the personal or clan totems of Aboriginal Australians of the Arrernte language groups. Although important as food in the arid central Australian environment (see section 1.8.1), these insects were not to be eaten by clan members belonging to that particular totem.

Totemic and food insects are represented in many Aboriginal artworks in which they are associated with cultural ceremonies and the depiction of important locations. Insects have had a place in many societies for their symbolism – such as ants and bees representing hard workers throughout the Middle Ages of Europe, where they even entered heraldry. Crickets, grasshoppers, cicadas, and scarab and lucanid beetles have long been valued as caged pets in Japan. Ancient Mexicans observed butterflies in detail, and lepidopterans were

Box 1.2 Butterfly houses

It used to be that seeing tropical butterflies entailed an expensive visit to an exotic location in order to appreciate their living diversity of shape and colours. Now many children are not far from a tropical butterfly house displaying, for a quite modest entry fee, many examples of some of the showiest lepidopterans. These live insects fly free in spacious walk-in cages or greenhouses replete with tropical vegetation and the appropriate mood sounds of a remote rainforest.

In just 35 years, the number of butterfly houses worldwide has gone from none to several hundred, attracting an estimated 40 million annual visitors and with a global turnover valued at some US$100 million. Equally dramatic has been the conversion from networks of local suppliers to industrial-scale insect production facilities. Although an estimated 4000 species of butterflies have been reared in the tropics, for the past decade some 500 have been listed for sale, and worldwide a core of only 50 species constitute most of those traded, primarily as live pupae. Papilionidae, including the well-known swallowtails, graphiums and birdwings, and Nymphalidae, including *Caligo*, *Danaus*, *Heliconius* and the electric blue *Morpho* species (see **Plate 1f**), are most popular.

Early on, when butterfly houses emphasized education and conservation, the exhibited butterflies were reared by local people, who sometimes were organized into co-operatives. Production came from tropical countries, including Costa Rica, Kenya and Papua New Guinea. "Ranching" butterflies for export in the pupal stage provides economic benefits and revenue flows to local communities and assists in natural habitat conservation. In East Africa, the National Museums of Kenya, in collaboration with many biodiversity programmes, supported local people of the Arabuko-Sukoke forest-edge in the Kipepeo Project to export harvested butterflies for live overseas exhibit. Self-sustaining since 1999, the project has enhanced income for impoverished people, and supported further nature-based projects including honey production. In Papua New Guinea, village farmers enhance ("ranch") the appropriate vine hosts for butterflies, often on land cleared at the forest edge for their vegetable gardens. Wild adult butterflies emerge from the forest to feed and lay their eggs; hatched larvae feed on the managed vines until harvested at pupation. According to species and conservation legislation, butterflies can be exported live as pupae, or dead as high-quality collector specimens.

Ranching evidently is a "small volume with high unit cost" process, suited best to rare species, and providing dead specimens for collectors in high-value trade. However, the much expanded modern insect zoos and butterfly houses require mass production rearing techniques in order to satisfy the demand for a limited range of species of selected butterflies in high numbers and continuous availablity (see **Plate 1g**). In some tropical countries, such as Costa Rica and Malaysia, commercial facilities have been constructed to provide continuous breeding in confinement from a few founder individuals, providing high-quality pupae in volume for air-freighting to purchasers. Although this is now "large volume with small unit value" production, major farms are likely to be more sustainable ecologically, due to their ability to maintain continuous culture, than are small individual breeders. However, larger suppliers often maintain the link to local butterfly farmers or ranchers. It might be argued that ranching, by potentially depleting wild populations, carries a risk of damage to target species. However, in the Kenyan Kipepeo Project, although preferred lepidopteran species originated from the wild as eggs or early larvae, walk-through visual assessment of butterflies in flight suggested that the relative abundance rankings of species was unaffected despite many years of selective harvest for export. Furthermore, ranching there and in New Guinea builds local support for intact forest as a valuable resource, rather than as "wasted" land to clear for subsistence agriculture.

Of course, irrespective of production method, the translocation of non-native insects for the public to view carries inherent risks. Escape, breeding and establishment outside the native range is a potential problem – not every butterfly is "harmless" and several have larvae that feed on our crops. For example, the attractive Australian orchard swallowtail (*Papilio aegeus*) has larvae that defoliate most species of citrus (lemons, limes, oranges and grapefruit and significant natives), and must be constrained to Australian butterfly houses.

well represented in mythology, including in poem and song. Amber has a long history as jewellery, and the inclusion of insects can enhance the value of the piece.

Most urbanized humans have lost much of this contact with insects, excepting those that share our domicile, such as the cockroaches, tramp ants and hearth crickets that generally arouse antipathy. Nonetheless, exhibits of insects, notably in butterfly farms and insect zoos, are very popular, with millions of people per year visiting such attractions throughout the world (Box 1.2). Insects remain part of Japanese culture, and not only for children; there are insect video games, numerous suppliers of entomological equipment, thousands of personal insect collections, and beetle breeding and rearing is so popular that it can be called beetlemania. In other countries, natural occurrences of certain insects attract ecotourism, including aggregations of overwintering monarch butterflies in coastal central California and in Mexico, the famous glow-worm caves of Waitomo, New Zealand, and Costa Rican locations such as Selva Verde rich in tropical insect biodiversity.

Although insect ecotourism may be limited, other economic benefits are associated with an interest in insects. This is especially so amongst children in Japan, where native rhinoceros beetles (Scarabaeidae, *Allomyrina dichotoma*) sell for a few US$ each, and longer-lived common stag beetles for up to US$10, and may be purchased from automatic vending machines. Adults collect and rear insects with a passion: at the peak of a craze for the large Japanese stag beetle (Lucanidae, *Dorcus curvidens*, called *o-kuwagata*) one could sell for between 40,000 and 150,000 yen (US$300 and US$1250), depending on whether captive reared or taken from the wild. Largest specimens, even if captive reared, fetched several million yen >US $10,000 at the height of the fashion. Such enthusiasm by Japanese collectors can lead to a valuable market for insects from outside Japan. According to official statistics, in 2002 some 680,000 beetles, including over 300,000 each of rhinoceros and stag beetles, were imported, predominantly originating from southern and Southeast Asia. Enthusiasm for valuable specimens extends outside Coleoptera: Japanese and German tourists are reported in the past decade to have purchased rare butterflies in Vietnam for US$1000–2000, which is a huge income for the generally poor local people. Unfortunately, some collectors ignore other countries' legislation (including that of Australia, New Zealand and Himalayan countries) by collecting without permits/licences, taking large numbers of insects and damaging the environment in their desire to harvest them rapidly.

In Asia, particularly in Malaysia, there is interest in rearing, exhibiting and trading in mantises (Mantodea), including orchid mantises (*Hymenopus* species; see sections 13.1.1 and 14.1) and stick-insects (Phasmatodea). Hissing cockroaches from Madagascar and burrowing cockroaches from tropical Australia are reared readily in captivity and can be kept as domestic pets as well as being displayed in insect zoos in which handling the exhibits is encouraged.

Questions remain as to whether wild insect collection, either for personal interest or commercial trade and display, is sustainable. In Japan, although expertise in captive rearing has increased and thus undermined the very high prices paid for certain wild-caught beetles, wild harvesting continues over an ever-increasing region. The possibility of over-collection for trade is discussed in section 1.7, together with other conservation issues.

1.6 CULTURING INSECTS

Many species of insects are maintained routinely in culture for purposes ranging from commercial sale, to scientific research, and even to conservation and reintroduction into the wild. As mentioned in section 1.2, much of our understanding of genetics and developmental biology comes from *Drosophila melanogaster* – a species with a short generation time of about 10 days, high fecundity with hundreds of eggs in a lifetime, and ease of culture in simple yeast-based media. These characteristics allow for large-scale research studies across many generations in an appropriate timescale. Other species of *Drosophila* can be reared in a similar manner, although often requiring more particular dietary requirements, including micronutrients and sterols. *Tribolium* flour beetles (section 1.2) are reared solely on flour. However, many phytophagous insects can be reared only on a particular host plant, in a time- and space-consuming programme, and the search for artificial diets is an important component of applied entomological research. Thus, *Manduca sexta*, the tobacco hornworm, which has provided many physiological insights, including how metamorphosis is controlled, is reared en masse on artificial diets of wheatgerm, casein, agar, salts and vitamins, rather than on any of its diverse natural host plants.

The situation is more complex if host-specific insect parasitoids of pests are to be reared for biological control purposes. In addition to maintaining the pest in quarantine to avoid accidental release, the appropriate life stage must be available for the mass production of parasitoids. The rearing of egg parasitoid *Trichogramma* wasps for biological control of caterpillar pests, which commenced over a century ago, relies on the availability of large numbers of moth eggs. Typically, these come from one of two species, the Angoumois grain moth, *Sitotroga cerealella*, and the Mediterranean flour moth, *Ephestia kuehniella*, which are reared easily and inexpensively on wheat or other grains. The use of artificial media, including insect haemolymph and artificial moth eggs, have been patented as being more efficient egg production methods. However, if host location by parasitoids involves chemical odours produced by damaged tissues (section 4.3.3), such signals are unlikely to be produced by an artificial diet. Thus, mass production of parasitoids against troublesome wood-mining beetle larvae must involve rearing the beetles from egg to adult on appropriately conditioned wood of the correct plant species.

Insects such as crickets, mealworms (tenebrionid beetle larvae) and bloodworms (midge larvae) are mass-reared commercially for feeding to pets, or as bait for anglers. Immature soldier flies can recycle domestic "green" waste and provide feed for chickens and certain pets (see Box 9.1). Furthermore, hobbyists and insect pet owners form an increasing clientele for captive-reared insects such as scarabs and lucanid beetles, mantises, phasmids and tropical cockroaches, many of which can be bred with ease by children following on-line instructions.

Zoos, particularly those with butterfly houses (Box 1.2) or petting facilities, maintain some of the larger and more charismatic insects in captivity. Indeed, some zoos have captive-breeding programmes for certain insects that are endangered in the wild – such as the endangered Lord Howe Island phasmid (*Dryococelus australis*; see **Plate 1c**), a large, flightless stick-insect captive-reared on the island and in Melbourne Zoo (in Australia). In New Zealand, several species of charismatic wetas (outsized, flightless orthopterans; see **Plate 1d**) have been reared in captivity and successfully reintroduced to predator-free offshore islands. Among the greatest successes have been the rearing of several endangered species of butterfly in captivity in Europe and North America, for example by the Oregon Zoo, with eventual releases and reintroductions into restored habitat proving quite successful as interim conservation strategies.

1.7 INSECT CONSERVATION

The major threats to insect biodiversity are similar to those affecting other organisms, namely habitat loss and fragmentation, climate change, and invasive species. Introductions of alien social insects, especially ants (Box 1.3), invasive plants, generalist biological control agents (section 16.5), pathogens, and vertebrate grazers and predators often have led to threats to native insect species. Human-induced changes to climate affect the ranges and phenology of some insect species (section 17.3), but more research is needed on threats to non-pest insects other than butterflies. However, the prime cause of insect declines and extinctions, at least of local populations if not whole species, is the loss of their natural habitats.

Biological conservation typically involves either setting aside large tracts of land for "nature", or addressing and remediating specific processes that threaten large and charismatic vertebrates, such as endangered mammals and birds, or plant species or communities. The concept of conserving habitat for insects, or species thereof, seems of low priority on a threatened planet. Nevertheless, land is reserved and plans exist specifically to conserve certain insects. Such conservation efforts often are associated with human aesthetics, and many (but not all) involve the "charismatic megafauna" of entomology – the butterflies and large, showy beetles. Such charismatic insects can act as "**flagship species**" to enhance wider public awareness and engender financial support for conservation efforts. Flagship species are chosen for their vulnerability, distinctiveness or public appeal, and support for their conservation may help to protect all species that live in the flagship's habitat. Thus, single-species conservation, not necessarily of an insect, is argued to preserve many other species by default, in what is known as the "**umbrella effect**". Somewhat complementary to this is advocacy of a habitat-based approach, which argues for increases in the number and size of protected areas to conserve many insects, which are not (and arguably "do not need to be") understood on a species-by-species basis. No doubt efforts to conserve habitats of native fish globally will preserve, as a spin-off, the much more diverse aquatic insect fauna that depends also upon waters being maintained in

Box 1.3 Tramp ants and biodiversity

No ants are native to Hawai'i, yet there are more than 40 species on the island – all have been brought from elsewhere within the last hundred or so years. In fact, all social insects (honey bees, yellow jackets, paper wasps, termites and ants) on Hawai'i have arrived as the result of human commerce.

Almost 150 species of ants have hitchhiked with us on our global travels and managed to establish themselves outside their native ranges. The invaders of Hawai'i belong to the same suite of ants that have invaded the rest of the world, or seem likely to do so in the near future. From a conservation perspective, one particular behavioural subset of ants is very important, the so-called invasive "tramp" ants. They rank amongst the world's most serious pest species, and local, national and international agencies are concerned with their surveillance and control. The big-headed ant (*Pheidole megacephala*), the long-legged or yellow crazy ant (*Anoplolepis gracilipes*), the Argentine ant (*Linepithema humile*; see Plate 6e), the "electric" or little fire ant (*Wasmannia auropunctata*) and tropical fire ants (*Solenopsis* species, especially *S. geminata* and *S. invicta*; see Plate 6e) are considered the most serious of these ant pests.

Invasive ant behaviour threatens biodiversity, especially on islands such as Hawai'i, the Galapagos and other Pacific Islands (section 8.7). Interactions with other insects include the protection and tending of aphids and scale insects for their carbohydrate-rich honeydew secretions. This boosts densities of these insects, which include invasive agricultural pests. Interactions with other arthropods are predominantly negative, resulting in aggressive displacement and/or predation on other species, even other tramp ant species encountered. Initial founding is often associated with unstable environments, including those created by human activity. The tendency for tramp ants to be small and short-lived is compensated by year-round increase and rapid production of new queens. Nest-mate queens show no hostility to each other. Colonies reproduce by one or more mated queens plus some workers relocating only a short distance from the original nest – a process known as budding. When combined with the absence of intraspecific antagonism between newly founded and natal nests, colony budding ensures the spreading of a "supercolony" across the ground. Furthermore, some invasive ant species exhibit female parthenogenesis (thelytoky) (section 5.10.1), which is beneficial in founding populations in new areas.

Although initial nest foundation is associated with human or naturally disturbed environments, most invasive tramp species can move into more natural habitats and displace the native biota. Ground-dwelling insects, including many native ants, do not survive the encroachment, and arboreal species may follow into local extinction. Surviving insect communities tend to be skewed towards subterranean species and those with especially thick cuticle such as carabid beetles and cockroaches, which also are chemically defended. Such an impact can be seen from the effects of big-headed ants during the monitoring of rehabilitated sand-mining sites, using ants as indicators (section 9.7). Six years into rehabilitation, as seen in the graph (from Majer 1985), ant diversity neared that found in unimpacted control sites, but the arrival of *Pheidole megacephala* dramatically restructured the system, seriously reducing diversity relative to controls. Even large animals can be threatened by ants – land crabs on Christmas Island, horned lizards in southern California, hatchling turtles in the southeastern United States, and ground-nesting birds everywhere. Invasion by Argentine ants of fynbos, a mega-diverse South African plant assemblage, eliminates ants that specialize in carrying and burying large seeds, but not those that carry smaller seeds (section 11.3.2). Since the vegetation originates by germination after periodic fires, the shortage of buried large seeds is predicted to cause dramatic change to vegetation structure.

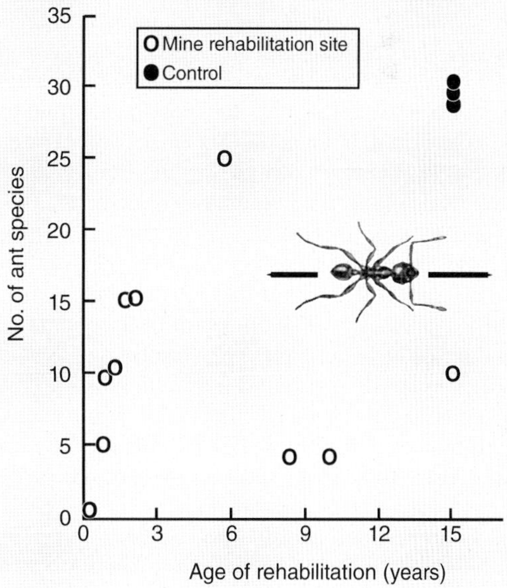

Introduced ants are very difficult to eradicate: all attempts to eliminate the red imported fire ant, *Solenopsis invicta*, in the United States have failed, and now a few billion US$ are spent annually on control. This ant also has invaded the West Indies, China and Taiwan, and has spread rapidly. In contrast, it is hoped that an ongoing campaign, costing nearly A$200 million (more than US$150 million) in the first eight years, may prevent *S. invicta* from establishing as an "invasive" species in Australia. The first fire ant sites were found around Brisbane in February 2001, although this ant is suspected to have been present for a number of years prior to its detection. At the height of surveillance, the area infested by fire ants extended to some 80,000 ha. Potential economic damage in excess of A$100 billion over 30 years was estimated if control failed, with inestimable damage to native biodiversity continent-wide. Although intensive searching, baiting, and destruction of nests appear to have been successful in eliminating major infestations, all nests must be eradicated to prevent resurgence, and thus continual monitoring and containment measures are essential. Recent surveying has included the innovative use of an infrared and thermal camera mounted to a helicopter to capture image data during flight that later are analysed by a ground-based computer system "trained" to recognize a fire ant mound's "signature" of reflected energy. Fortunately, incursions of *S. invicta* into New Zealand were detected rapidly and the populations eradicated. Undoubtedly, the best strategy for control of invasive ants is quarantine diligence to prevent their entry, and public awareness to detect accidental entry.

natural condition. Equally, preservation of old-growth forests to protect tree-hole nesting birds such as owls or parrots also will conserve habitat for wood-mining insects that use timber across a complete range of wood species and states of decomposition.

Land that once supported diverse insect communities has been transformed for human agriculture, urban development and to extract resources such as timber and minerals. Many remaining insect habitats have been degraded by the invasion of alien species, both plants and animals, including invasive insects (Box 1.3). The habitat approach to insect conservation aims to maintain healthy insect populations by supporting large patch (habitat) size, good patch quality, and reduced patch isolation. Six basic, interrelated principles serve as guidelines for the conservation management of insects: (i) maintain reserves; (ii) protect land outside of reserves; (iii) maintain quality heterogeneity of the landscape; (iv) reduce contrast between remnant patches of habitat and nearby disturbed patches; (v) simulate natural conditions, including disturbance; and (vi) connect patches of quality habitat. Habitat-based conservationists accept that single-species oriented conservation is important, but argue that it may be of limited value for insects because there are so many species. Furthermore, rarity of insect species may be due to populations being localized in just one or a few places, or in contrast, may be widely dispersed but with low density over a wide area. Clearly, different conservation strategies are required for each case.

Migratory species, such as the monarch butterfly (*Danaus plexippus*), require special conservation. Although the monarch butterfly is not a threatened species, its migration in North America is considered an endangered biological phenomenon by the IUCN (the International Union for the Conservation of Nature). Monarchs from east of the Rockies overwinter in Mexico and then migrate northwards as far as Canada throughout the summer (section 6.7). Critical to the conservation of these monarchs is the safeguarding of the overwintering habitat at Sierra Chincua and elsewhere in Mexico. A highly significant insect conservation measure implemented in recent years is the decision of the Mexican government to support the Monarch Butterfly Biosphere Reserve (Mariposa Monarca Biosphere Reserve), which was established to protect this phenomenon. Another major threat to monarch butterflies is loss of larval breeding sites in North America (discussed in section 16.6.1). Efforts to monitor monarch populations in North America involve citizen scientists (Box 1.1). Successful conservation of this flagship butterfly requires collaborations involving the United States, Canada and Mexico, to ensure protection of both overwintering sites and migration flyway habitats. However, preservation of western overwintering populations in coastal California conserves no other native species. The reason for this is that the major resting sites are in groves of large, introduced eucalypt trees, especially blue gums, which have a depauperate fauna in their non-native habitat.

A successful example of single-species conservation involves the endangered El Segundo blue butterfly, *Euphilotes battoides* ssp. *allyni*, whose principal colony in sand dunes near Los Angeles airport was threatened by urban sprawl and golf-course development. Protracted negotiations with many interested parties resulted in designation of 80 hectares as a reserve, sympathetic management of the golf course "rough" for the larval food plant *Erigonum parvifolium* (buckwheat), and control of alien plants plus limitation on human disturbance. Southern Californian coastal dune systems are seriously endangered habitats, and management of this reserve for the El Segundo blue conserves other threatened species.

Land conservation for butterflies is not an indulgence of affluent southern Californians: the world's largest butterfly, the Queen Alexandra's birdwing, *Ornithoptera alexandrae*, of Papua New Guinea, provides a success story from the developing world. This spectacular species, whose caterpillars feed only on *Aristolochia dielsiana* vines, is endangered and limited to a small area of lowland rainforest in northern Papua New Guinea. Under Papua New Guinean law, this birdwing species has been protected since 1966, and international commercial trade was banned by listing it on Appendix I of the Convention on International Trade in Endangered Species of Wild Fauna and Flora (CITES). The Queen Alexandra's birdwing has acted as a flagship species for conservation in Papua New Guinea, and its initial conservation success attracted external funding for surveys and reserve establishment. Conserving Papua New Guinean forests for this and related birdwings undoubtedly results in conservation of much diversity under the umbrella effect, but mining and corrupt large-scale logging in Papua New Guinea in the past two decades threatens much of rainforests there.

As discussed in Box 1.2, Kenyan and New Guinean insect conservation efforts have some commercial incentives and provide impoverished people with some recompense for their protection of natural environments. However, commerce need not be the sole motivation: the aesthetic appeal of having native birdwing butterflies flying wild in local neighbourhoods, combined with local education programmes in schools and communities, are helping to save the subtropical Australian Richmond birdwing butterfly, *Ornithoptera richmondia* (see **Plate 1e**). Larval Richmond birdwings develop on two species of native *Pararistolochia* vines and require large plants to satisfy their appetite. However, about two-thirds of the original coastal rainforest habitat supporting native vines has been lost, and the alien South American *Aristolochia elegans* ("Dutchman's pipe"), introduced as an ornamental plant and escaped from gardens, lures females to lay eggs on it as a prospective host. This oviposition mistake is deadly since toxins of this plant kill young caterpillars. This conservation problem has been addressed by an education programme to encourage the removal of Dutchman's pipe vines from native vegetation, from sale in nurseries, and from gardens and yards. Replacement in bush and gardens with native *Pararistolochia* was encouraged after a massive effort to propagate the vines. Birdwing populations isolated by habitat fragmentation also suffer inbreeding depression, which is being alleviated by planting corridors of suitable host plants and by captive breeding and reintroduction of genetically diverse individuals. Although recovery of the birdwing population was impacted by continued loss of habitat and years of drought, wetter conditions from 2010 improved food-plant quality and, combined with a renewed cultivation effort, led to the first population increases following a century of decline. However, habitat loss continues and ongoing community action throughout the native range of the Richmond birdwing is necessary to reverse its decline.

The idea that concerned citizens can conserve insects by management of their gardens (backyards) has gained acceptance, notably with regard to bee decline. Guidance is available, including from "show gardens", for developing a pollinator-friendly garden by growing selected plants, emphasizing nectar-producers such as borage (which is best for honey bees), lavender (for bumblebees) and marjoram (an all-round attractant to all bees and hoverflies). With increasing enthusiasm for local/urban honey-bee hives, sustainability can be attained only by increasing flowering plants in gardens and public spaces to provide food sources for a diversity of insects.

Butterflies and bees, being familiar insects with non-threatening lifestyles, are flagships for invertebrate conservation. However, certain orthopterans, including New Zealand wetas, have been afforded legal protection, and conservation plans exist for dragonflies and other freshwater insects in the context of conservation and management of aquatic environments, and there are plans for firefly (beetle) and glow worm (fungus gnat) habitats. Agencies in certain countries have recognized the importance of retention of fallen dead wood as insect habitat, particularly for long-lived wood-feeding beetles.

Designation of reserves for conservation, seen by some as the answer to threat, rarely is successful without understanding species requirements and responses to management. The butterfly family Lycaenidae (blues, coppers and hairstreaks), with some 6000 species, comprises over 30% of the butterfly diversity. Many have relationships with ants (e.g. as inquilines; section 12.3), some being obliged to pass some or all of their immature development inside ant nests, others are tended on their preferred host plant by ants, yet others are predators on ants and scale insects, while being tended by ants. These relationships can be very complex, and may be rather easily disrupted by environmental changes, leading to threats to the butterflies. Certainly in Western Europe, species of Lycaenidae figure prominently on lists of threatened insect taxa. Notoriously, the decline of the large blue butterfly *Phengaris* (formerly *Maculinea*) *arion* in England was blamed upon over-collection, but see Box 1.4 for a different interpretation. Action plans

Box 1.4 Conservation of the large blue butterfly

The large blue butterfly, *Phengaris* (formerly *Maculinea*) *arion* (Lepidoptera: Lycaenidae), was reported to be in serious decline in southern England in the late 19th century, a phenomenon ascribed then to poor weather. By the mid-20th century, this attractive species was restricted to some 30 colonies in southwest England. Few colonies remained by 1974 and the estimated adult population had declined from about 100,000 in 1950 to 250 in some 20 years. Final extinction of the species in England in 1979 followed two successive hot, dry breeding seasons. Since the butterfly is beautiful and highly sought by collectors, excessive collecting was presumed to have caused at least the long-term decline that made the species vulnerable to deteriorating climate. This decline occurred even though a reserve was established in the 1930s to exclude both collectors and domestic livestock, in an attempt to protect the butterfly and its habitat.

Evidently, habitat had changed through time, including a reduction of wild thyme (*Thymus praecox*), which provides the food for early instars of the large blue's caterpillar. Shrubbier vegetation replaced short-turf grassland because of loss of grazing rabbits (through disease) and the exclusion of grazing cattle and sheep from the reserved habitat. Thyme survived, however, but the butterflies continued to decline to extinction in Britain.

A more complex story has since been revealed by research associated with the reintroduction of the large blue to England from continental Europe. The larva of the large blue butterfly in England and on the European continent is an obligate predator in colonies of red ants belonging to species of *Myrmica*. Larval large blues must enter a *Myrmica* nest, in which they feed on larval ants. Similar predatory behaviour, and/or tricking ants into feeding them as if they were the ants' own brood, are features in the natural history of many Lycaenidae (blues and coppers) worldwide (section 1.7 and section 12.3). After hatching from an egg laid on the larval food plant, the large blue's caterpillar feeds on thyme flowers until the moult into the final (fourth) larval instar, around August. At dusk, the caterpillar drops to the ground from the natal plant, where it waits inert until a *Myrmica* ant finds it. The worker ant attends the larva for an extended period, perhaps more than an hour, during which it feeds from a sugar gift secreted from the caterpillar's dorsal nectary organ. At some stage, the caterpillar becomes turgid and adopts a posture that seems to convince the tending ant that it is dealing with an escaped ant brood, and it is carried into the nest. Until this stage, immature growth has been modest, but in the ant nest the caterpillar becomes predatory on ant eggs and larvae and grows rapidly. The caterpillar spends winter in the ant nest and, 9–10 months after it entered the nest, pupates in early summer of the following year. The caterpillar requires an average 230 immature ants for successful pupation. It apparently escapes predation by the ants by secreting surface chemicals that mimic those of the ant brood, and probably receives special treatment in the colony by producing sounds that mimic those of the queen ant (section 12.3). The adult butterfly emerges from the pupal cuticle in summer, and departs rapidly from the nest before the ants identify it as an intruder.

Adoption and incorporation into the ant colony turns out to be the critical stage in the life history. The complex system involves the "correct" ant, *Myrmica sabuleti*, being present, and this in turn depends on the appropriate microclimate associated with short-turf grassland. Longer grass causes cooler near-soil microclimate conditions, favouring other *Myrmica* species, including *M. scabrinodes*, which may displace *M. sabuleti*. Although caterpillars associate apparently indiscriminately with any *Myrmica* species, survivorship differs dramatically: with *M. sabuleti* approximately 15% survive, but an unsustainable reduction to <2% survivorship occurs with *M. scabrinodes*. Successful maintenance of large blue populations requires that >50% of the adoption by ants must be by *M. sabuleti*.

Other factors affecting survivorship include the requirements for the ant colony to have no alate (winged) queens and have at least 400 well-fed workers to provide enough larvae for the caterpillar's feeding needs, and to lie within 2 m of the host thyme plant. Such nests are associated with newly burnt grasslands, which are rapidly colonized by *M. sabuleti*. Nests should not be so old as to have developed more than the founding queen: the problem here being that with numerous alate queens in the nest, the caterpillar can be mistaken for a queen and attacked and eaten by nurse ants.

Now that we understand the intricacies of the relationship, we can see that the well-meaning creation of reserves that lacked rabbits and excluded other grazers created vegetation and microhabitat changes that altered the dominance of ant species, to the detriment of the butterfly's complex relationships. Over-collecting is not implicated, although on a broader scale climate change must be significant. Now, five populations originating from Sweden have been reintroduced to habitat and conditions appropriate for *M. sabuleti*, thus leading to thriving populations of the large blue butterfly. Interestingly, other rare species of insects in the same habitat have responded positively to this informed management, suggesting perhaps an umbrella role for the butterfly species.

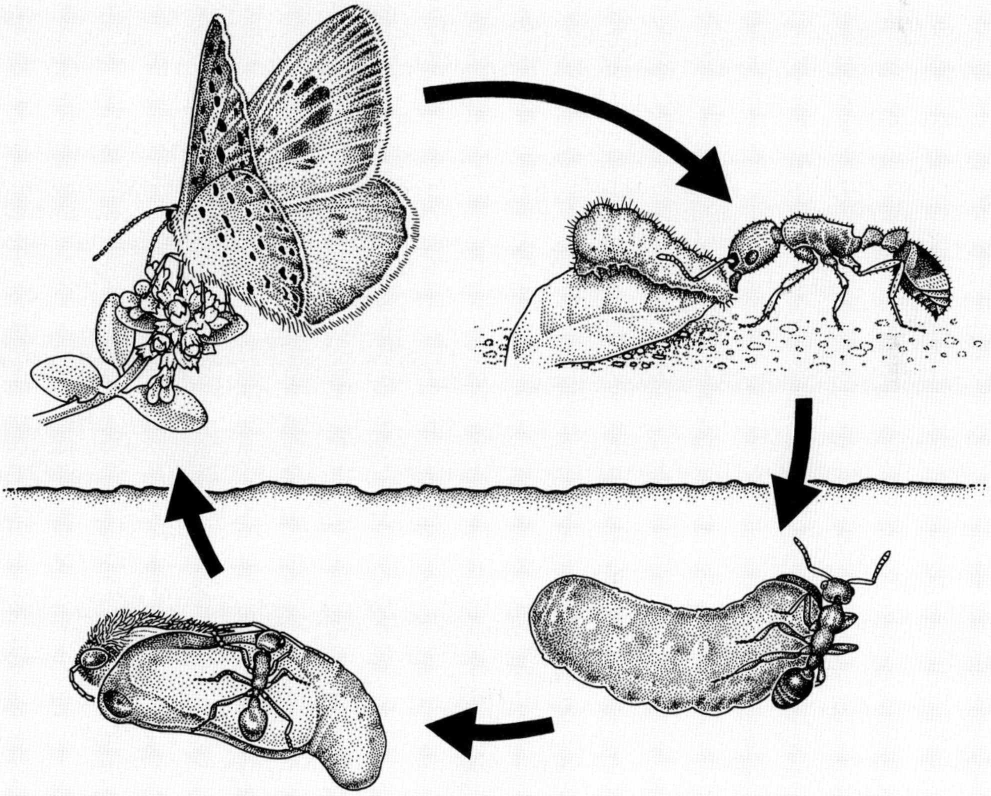

in Europe for the reintroduction of this and related species, and appropriate conservation management of *Phengaris* species, have been put in place: these depend vitally upon a species-based approach. Only with understanding of general and specific ecological requirements of conservation targets can appropriate management of habitat be implemented.

The impediments to insect conservation are multifaceted and include poor public perception of the ecological importance of invertebrates, limited knowledge of their species diversity, distributions and abundance in space and time, and paucity of information on their sensitivity to changes in habitat. As discussed earlier, policymakers and land managers often assume that the protection of habitat for vertebrates will conserve resources for invertebrates under an umbrella effect. Clearly, we need more information on the effectiveness for insects and other invertebrates of conservation measures designed primarily for vertebrates and plants. Funding for comprehensive entomological surveys or experimental studies is always limited, but the identification and detailed study of suitable indicator or surrogate taxa can provide helpful guidelines for conservation decisions involving insects. Furthermore, valuable data on insect abundance, distributions and phenology can derive from citizen science programmes (Box 1.1).

1.8 INSECTS AS FOOD

1.8.1 Insects as human food: entomophagy

In this section we review the increasingly popular topic of insects as human food. Nearly 2000 species of insects in more than 100 families are, or have been, used for food somewhere in the world, especially in central and southern Africa, Asia, Australia and Latin America. Food insects generally feed on either living or dead plant matter, and chemically protected species are avoided. Termites, crickets, grasshoppers, locusts, beetles, ants, bee brood and moth larvae are frequently consumed insects. It is estimated that insects form part of the traditional diets of at least two billion people, but the ever-increasing human population and demand for food is causing over-exploitation of some wild edible insects. Although insects are high in protein, energy and various vitamins and minerals, and can form 5–10% of the annual animal protein consumed by certain indigenous peoples, western society essentially overlooks entomological cuisine.

Typical "western" repugnance of entomophagy is cultural rather than scientific or rational. After all, other invertebrates such as certain crustaceans and molluscs are considered to be desirable culinary items. Objections to eating insects cannot be justified on the grounds of taste or food value. Many have a nutty flavour and studies report favourably on the nutritional content of insects, although their amino acid and fatty acid compositions vary considerably among different food-insect species.

Mature larvae of palm weevils, *Rhynchophorus* species (Coleoptera: Curculionidae), have been appreciated by people in tropical Africa, Asia and the Neotropics for centuries. These fat, legless "palmworms" (see **Plate 2a**) provide a rich source of animal fat, plus substantial amounts of riboflavin, thiamine, zinc and iron. Primitive cultivation systems involve wounding or felling palm trees to provide suitable food for the weevils. Such cultivation occurs in South America (Brazil, Colombia, Paraguay and Venezuela) and parts of Southeast Asia. Commercial facilities have been constructed in Thailand to raise the grubs on macerated ("chipped") palm trunk and leaf-base tissues. However, throughout Asia, palmworms are pests that damage and kill coconut and oil palm trees in plantations. From this part of the world a *Rhynchophorus* species entered California, USA, threatening ornamental and date palms (Box 1.5).

In Africa, people eat many species of moth caterpillars that provide a rich source of iron and have high calorific value, with a protein content ranging from 45 to 80%. In Zambia, the edible caterpillars of emperor moths (Saturniidae), locally called *mumpa*, provide a valuable dietary supplement of fresh, fried, boiled or sun-dried larvae that offsets malnutrition caused by protein deficiency. Caterpillars of *Gonimbrasia belina* (Saturniidae) (see **Plate 2b**), called mopane, mopanie, mophane or phane, are utilized widely. They develop on the widespread mopane, a leguminous tree (*Colophospermum mopane*) that grows in a wide belt of open forest ("mopane woodland" landscape) across southern Africa (Fig. 1.2). Caterpillars are hand picked by poor rural women for family subsistence. The de-gutted insects are boiled or sometimes salted and dried, after which they contain about 50% protein and 15% fat – approximately twice the values for cooked beef. However, organized teams, largely of men,

Box 1.5 Palmageddon? Weevils in the palms

Residents of upmarket Laguna Beach in Orange County, southern California, are losing the ornamental palm trees that provide landscape features in this and other Mediterranean-climate communities (as illustrated here). In August 2010, a large dying Canary Island date palm (*Phoenix canariensis*) was felled by arborists and found to be infested with weevils. The culprit was believed initially to be *Rhynchophorus ferrugineus*, the red palm weevil (see **Plate 2a**), that was causing problems already in Asia, the Middle East and Mediterranean Europe, but this was the first report of the pest in North America. Given that California's date industry is valued at US$30 million, and ornamental palm tree sales at $70 million per year in California and $127 million in Florida, a research programme was commenced without delay.

The programme exemplifies many issues in insect biology associated with control of introduced pests. First, it was necessary to establish the size of the affected area. It is very unlikely that the first insects to be noticed are those that invaded: the public is rarely vigilant to early infestations, especially when symptoms may occur several years after arrival, and skilled specialists are ever-more thinly spread. Usually, arrival times of first invaders are estimated to have been several (too many) years prior to official recognition. Recruits had to be mobilized and trained to recognize the damage symptoms, and some form of automated trapping had to be developed and distributed in Laguna Beach and surrounds.

Since these weevils already were killing palms elsewhere in the world, researchers travelled to see how others coped with the problem, and also to observe the weevils in their native areas. The bad news was that the invasions in the Middle East and Mediterranean Europe were serious, spreading and uncontrolled. For example, since 2004 the south of France had been dealing with the weevils, which had spread along the Riviera to the Pyrenees and even to Corsica. Even the iconic palms that provide the backdrop to images of the Cannes Film Festival are dying. The good news was that an airborne lure that combined essence of damaged palms and weevil communication chemicals

(pheromones; section 4.3.2) had been developed already. This cocktail could be used to monitor the flying adult weevils in California, and potentially trap enough to exert some level of population control (section 16.9). California had some advantages, notably that the problems seemed restricted to Laguna Beach, and despite a crippling recession, a research budget was found quickly and implemented. Although total destruction by chipping the complete palm was the only method of control, residents were "on-side" with the programme.

In an unexpected twist, it turned out that the weevil that had entered the Golden State was not the globally invasive red palm weevil *R. ferrugineus*, but the differently pigmented (a black form with a red stripe) but otherwise similar, *R. vulneratus*. This revised identification, supported by molecular evidence for its distinction from red palm weevil, allows better identification of the natural source area, where its biology, including control can be studied (for more on biological control, see section 16.5), and explains some subtle differences in behaviour relative to the red palm weevil. The red stripe form of the weevil is native to Bali, Indonesia, where the larvae are considered to be such a "live" food delicacy that coconuts, sago and nipa (nypa) palms are damaged deliberately to induce weevil infestation, from which the tasty larvae are harvested. However, the occurrence of *R. vulneratus* in California does raise the question of how the weevils got there – this might have been explained more easily had the invader been the widespread and clearly invasive red palm weevil. It seems highly unlikely that a species otherwise restricted to Indonesia should arrive unaided on the opposite side of the Pacific Ocean and only in a solitary suburban location. Undoubtedly, a breach of biosecurity had occurred, yet it seems improbable that the "usual route" was responsible; that is, the legal (or otherwise) importation of infested host plants (large coconut or sago palms in this case). Scientists studying the situation suspect the demand for the weevil larvae as food led to illegal import of live weevils as a gastronomic souvenir. Perhaps the Californian dietary interest in locavory (eating local produce) could extend to include palm grubs? More seriously, can Californians imagine life without native palms dotting the landscape, and no palms in Palm Springs and no dates from the Coachella Valley?

now harvest intensively for commercial sale (mopane are available in certain urban supermarkets) and this increased demand has degraded a common property resource to an over-exploited and unsustainable "free-for-all" resource. Demand, especially from urban South Africa, has led to forest damage and localized moth extinction, including in Botswana. An optimistic sign of sustainability is a trial in Kruger National Park, where local people from Limpopo Province are supervised while they collect mopane worms within the park during a short, pre-Christmas season.

In the Philippines, June beetles (melolonthine scarabs), weaver ants (*Oecophylla smaragdina*) (see **Plate 6b**), mole crickets, palmworms (weevil larvae; Box 1.5, see also **Plate 2a**) and locusts are eaten in some regions. Locusts form an important dietary supplement during outbreaks, which apparently have become less common since the widespread use of insecticides. Various species of grasshoppers and locusts were eaten commonly by native tribes in western North America prior to the arrival of Europeans. The number and identity of species used have been poorly documented, but species of *Melanoplus* were among those consumed. Harvesting involved driving grasshoppers into a pit in the ground using fire or advancing people, or herding them into a bed of coals. Today, people in Central America, especially Mexico, harvest, sell, cook and consume grasshoppers.

Australian Aborigines use (or once used) a wide range of insect foods, especially moth larvae. The caterpillars of wood or ghost moths (Cossidae and Hepialidae) (Fig. 1.3, see also **Plate 2c**) are called witchety grubs, from the Aboriginal word "witjuti" for the *Acacia* species (wattles), on the roots and stems of which the grubs feed. Witchety grubs, which are regarded as a delicacy, contain 7–9% protein, 14–38% fat and 7–16% sugars, as well as being good sources of iron and calcium. Adults of the bogong moth, *Agrotis infusa* (Noctuidae), formed another important Aboriginal food, once collected in their millions from aestivating sites in narrow caves and crevices on mountain summits in southeastern Australia. Moths cooked in hot ashes provided a rich source of dietary fat.

Aboriginal people living in central and northern Australia eat the contents of the apple-sized galls of *Cystococcus pomiformis* (Hemiptera: Eriococcidae), commonly called bush coconuts or bloodwood apples (see **Plate 4f,g**). These galls occur only on bloodwood eucalypts (*Corymbia* species) and can be very abundant after a favourable growing season. Each mature gall

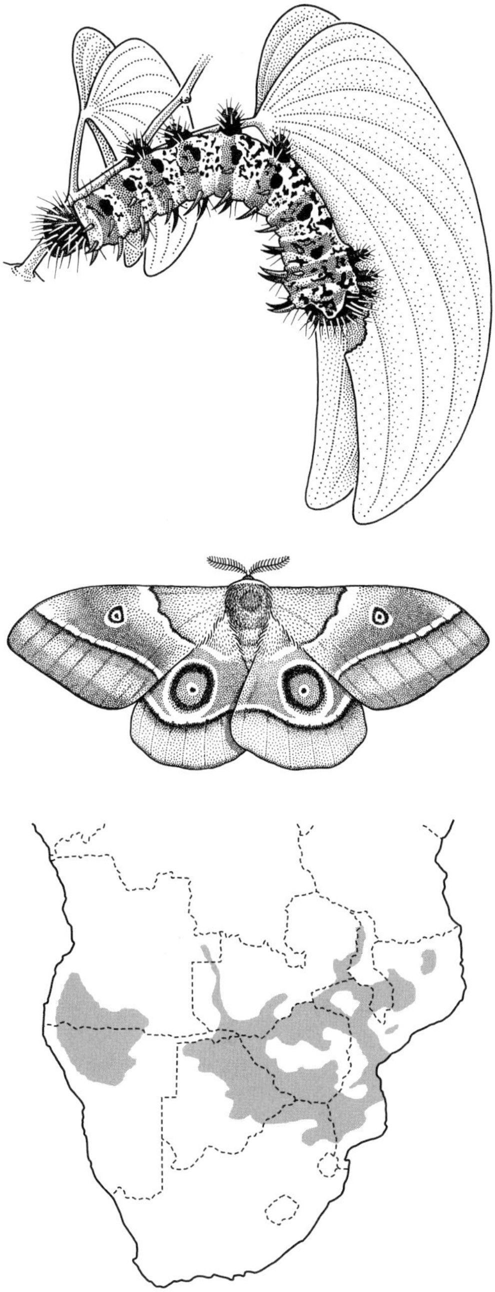

Fig. 1.2 Mopane moths and host tree. (a) Larva of *G. belina* on mopane (*Colophospermum mopane*) leaves. (b) Adult of *Gonimbrasia belina*. (c) Distribution of mopane woodland in southern Africa. (After photographs by R.G. Oberprieler and map from van Voortthuizen 1976.)

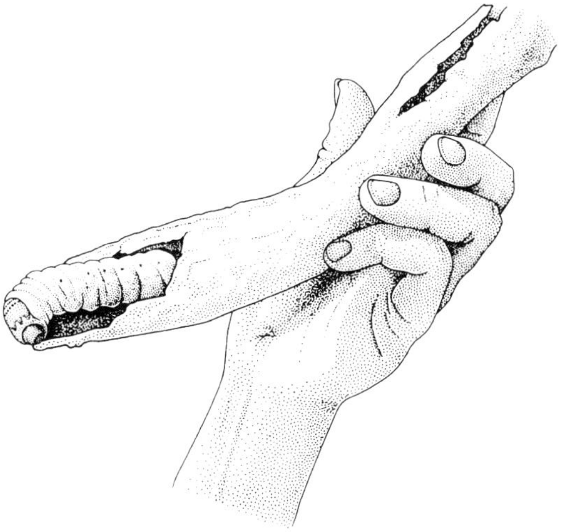

Fig. 1.3 A delicacy of the Australian Aborigines – a witchety (or witjuti) grub, a caterpillar of a wood moth (Lepidoptera: Cossidae) that feeds on the roots and stems of witjuti bushes (certain *Acacia* species). (After Cherikoff & Isaacs 1989.)

contains a single adult female, up to 4 cm long, which is attached by her mouth area to the base of the inner gall and has part of her abdomen plugging a hole in the gall apex. The inner wall of the gall is lined with white edible flesh, about 1 cm thick, which serves as the feeding site for the male offspring of the female. Aborigines relish the watery female insect and her nutty-flavoured nymphs, then scrape out and consume the white coconut-like flesh of the inner gall.

A favourite source of sugar for Australian Aboriginals living in arid regions comes from species of *Melophorus* and *Camponotus* (Formicidae), popularly known as honeypot ants (**Plate 2g–h**). Specialized workers (called repletes) store nectar, fed to them by other workers, in their huge distended crops (see Fig. 2.4). Repletes serve as food reservoirs for the ant colony and regurgitate part of their crop contents when solicited by another ant. Aborigines dig repletes out from their underground nests, an activity most frequently undertaken by women, who may excavate pits to a depth of a metre or more in search of these sweet rewards. Individual nests rarely supply more than 100 g of a honey that is essentially similar in composition to commercial honey. Honeypot ants in the western United States and Mexico belong to a different genus, *Myrmecocystus*. The repletes, a highly valued food, are collected by the rural people of Mexico, a difficult process in the hard soil of the stony ridges where the ants nest.

Perhaps the general western rejection of entomophagy is only an issue of marketing being required to counter a popular perception that insect food is for the poor and protein-deprived of the developing world. In reality, certain sub-Saharan Africans apparently prefer caterpillars to beef. Ant grubs (so called "ant eggs") and eggs of water boatmen (Corixidae) and backswimmers (Notonectidae) are much sought after in Mexican gastronomy as "caviar". In parts of Asia, a diverse range of insects can be purchased (see **Plate 2d–f**). Traditionally, desirable water beetles for human consumption are valuable enough to be farmed in Guangdong. The culinary culmination may be the meat of the giant water bug *Lethocerus indicus* or the Thai and Laotian *mangda* sauces made with the flavours extracted from the male abdominal glands, for which high prices are paid. Even in the urban United States some insects may become popular as a food novelty. The millions of 17-year cicadas that periodically plague northeastern states are edible. Newly hatched cicadas, called tenerals, are best for eating because their soft body cuticle means that they can be consumed without first removing the legs and wings. These tasty morsels can be marinated or dipped in batter and then deep-fried, boiled and spiced, roasted and ground, or stir-fried with favourite seasonings.

Large-scale harvest or mass production of insects for human consumption brings some practical and other problems. The small size of most insects presents difficulties in collection or rearing and in processing for sale. The unpredictability of many wild populations needs to be overcome by the development of culture techniques, especially as over-harvesting from the wild could threaten the viability of some insect populations. Another problem is that not all insect species are safe to eat. Warningly coloured insects are often distasteful or toxic (Chapter 14) and some people can develop allergies to insect material (section 15.6.3). However, several advantages derive from eating insects. The encouragement of entomophagy in many rural societies, particularly those with a history of insect use, may help diversify peoples' diets. By incorporating mass-harvesting of pest insects into control programmes, the use of pesticides can be reduced. Furthermore, if carefully regulated, cultivating insects for protein should be less environmentally damaging than cattle ranching, which devastates forests and native grasslands. Insect farming (the rearing of mini-livestock) is compatible with low-input sustainable agriculture and most insects have a high food conversion efficiency compared with conventional meat animals. However, the unregulated harvesting of wild insects can, and is, causing conservation concerns, especially in parts of Asia and Africa, where populations of some edible insects are threatened by over-collection as well as by habitat loss.

1.8.2 Insects as feed for domesticated animals

Although many people do not relish the prospect of eating insects, the concept of insects as a protein source for domesticated animals is quite acceptable. The nutritive significance of insects as feed for fish, poultry, pigs, and farm-grown mink certainly is recognized in China, where feeding trials have shown that insect-derived diets can be cost-effective alternatives to more conventional fishmeal diets. The insects involved are primarily the pupae of silkworms (*Bombyx mori*), the larvae and pupae of house flies (*Musca domestica*), and the larvae of mealworms (*Tenebrio molitor*). The same or related insects are being used or investigated elsewhere, particularly as poultry or fish feedstock. Silkworm pupae, a by-product of the silk industry, provide a high-protein supplement for chickens. In India, poultry are fed the meal that remains after the oil has been extracted from the pupae. Fly larvae fed to chickens can recycle animal manure, and development of insect recycling systems for converting organic wastes into feed supplements is ongoing (see Box. 9.1).

Clearly, insects have the potential to form part of the nutritional base of people and their domesticated animals. Further research is needed, and a database with accurate identifications is required to handle biological information. We must know which species we are dealing with in order to make use of information gathered elsewhere on the same or related insects. Data on the nutritional value, seasonal occurrence, host plants, or other dietary requirements, and rearing or collecting methods must be collated for all actual or potential food insects. Opportunities for insect food enterprises are numerous, given the immense diversity of insects.

FURTHER READING

Basset, Y., Cizek, L., Cuénoud, P. *et al.* (2012) Arthropod diversity in a tropical forest. *Science* **338**, 1481–4.

Boppré, M. & Vane-Wright, R.I. (2012) The butterfly house industry: conservation risks and education opportunities. *Conservation and Society* **10**, 285–303.

Bossart, J.L. & Carlton, C.E. (2002) Insect conservation in America. *American Entomologist* **40**(2), 82–91.

Brock, R.L. (2006) Insect fads in Japan and collecting pressure on New Zealand insects. *The Weta* **32**, 7–15.

Cardoso, P., Erwin, T.L., Borges, P.A.V. & New, T.R. (2011) The seven impediments in invertebrate conservation and how to overcome them. *Biological Conservation* **144**, 2647–55.

DeFoliart, G.R. (1989) The human use of insects as food and as animal feed. *Bulletin of the Entomological Society of America* **35**, 22–35.

DeFoliart, G.R. (1999) Insects as food; why the western attitude is important. *Annual Review of Entomology* **44**, 21–50.

DeFoliart, G.R. (2012) The human use of insects as a food resource: a bibliographic account in progress. [Available online at http://www. food-insects.com]

Erwin, T.L. (1982) Tropical forests: their richness in Coleoptera and other arthropod species. *The Coleopterists Bulletin* **36**, 74–5.

Foottit, R.G. & Adler, P.H. (eds.) (2009) *Insect Biodiversity: Science and Society*. Wiley-Blackwell, Chichester.

Gallai, N., Salles, J.-M., Settele, J. & Vaissière, B.E. (2009) Economic valuation of the vulnerability of world agriculture confronted with pollinator decline. *Ecological Economics* **68**, 810–21.

Goka, K., Kojima, H. & Okabe, K. (2004) Biological invasion caused by commercialization of stag beetles in Japan. *Global Environmental Research* **8**, 67–74.

International Commission of Zoological Nomenclature (1999) *International Code of Zoological Nomenclature*, 4th edn. International Trust for Zoological Nomenclature, London. [Available online at http://iczn.org/code]

Kawahara, A.Y. (2007) Thirty-foot of telescopic nets, bug-collecting video games, and beetle pets: entomology in modern Japan. *American Entomologist* **53**, 160–72.

Lemelin, R.H. (ed) (2012) *The Management of Insects in Recreation and Tourism*. Cambridge University Press, Cambridge.

Lockwood, J.A. (2009) *Six-legged Soldiers: Using Insects as Weapons of War*. Oxford University Press, New York.

Losey, J.E. & Vaughan, M. (2006) The economic value of ecological services provided by insects. *BioScience* **56**, 311–23.

New, T.R. (2009) *Insect Species Conservation*. Cambridge University Press, Cambridge.

New, T.R. (2010) *Beetles in Conservation*. Wiley-Blackwell, Chichester.

New, T.R. (ed.) (2012) *Insect Conservation: Past, Present and Prospects*. Springer, Dordrecht, New York.

New, T.R. (2012) *Hymenoptera and Conservation*. Wiley-Blackwell, Chichester.

New, T.R. & Samways, M.J. (2014) Insect conservation in the southern temperate zones: an overview. *Austral Entomology* **53**, 26–31.

Novotny, V., Drozd, P., Miller, S.E. *et al.* (2006) Why are there so many species of herbivorous insects in tropical rainforests? *Science* **313**, 1115–8.

Pech, P., Fric, Z. & Konvicka, M. (2007) Species-specificity of the *Phengaris* (*Maculinea*)–*Myrmica* host system: fact or myth? (Lepidoptera: Lycaenidae; Hymenoptera: Formicidae). *Sociobiology* **50**, 983–1003.

Samways, M.J. (2005) *Insect Diversity Conservation*. Cambridge University Press, Cambridge.

Samways, M.J., McGeoch, M.A. & New, T.R. (2010) *Insect Conservation: A Handbook of Approaches and Methods (Techniques in Ecology and Conservation)*. Oxford University Press, Oxford.

Sands, D.P.A. & New, T.R. (2013) *Conservation of the Richmond Birdwing Butterfly in Australia*. Springer Dordrecht, New York.

Saul-Gershenz, L. (2009) Insect zoos. In: *Encyclopedia of Insects*, 2nd edn (eds *V.H. Resh & R.T. Cardé*), pp. 516–23. Elsevier, San Diego, CA.

Speight, M.R., Hunter, M.D. & Watt, A.D. (2008) *Ecology of Insects. Concepts and Applications*, 2nd edn. Wiley-Blackwell, Chichester.

Thomas, J.A., Simcox, D.J. & Clarke, R.T. (2009) Successful conservation of a threatened *Maculinea* butterfly. *Science* **325**, 80–3.

Tsutsui, N.D. & Suarez, A.V. (2003) The colony structure and population biology of invasive ants. *Conservation Biology* **17**, 48–58.

van Huis, A. (2013) Potential of insects as food and feed in assuring food security. *Annual Review of Entomology* **58**, 563–83.

van Huis, A., van Itterbeeck, J., Klunder, H. *et al.* (2013) Edible Insects: Future Prospects for Food and Feed Security. *FAO Forestry Paper 171*, 187 pp. Food and Agriculture Organization of the United Nations, Rome.

Wagner, D.L. & Van Driesche, R.G. (2010) Threats posed to rare or endangered insects by invasions of nonnative species. *Annual Review of Entomology* **55**, 547–68.

Wetterer, J.K. (2013) Exotic spread of *Solenopsis invicta* Buren (Hymenoptera: Formicidae) beyond North America. *Sociobiology* **60**, 50–5.

Chapter 2

EXTERNAL ANATOMY

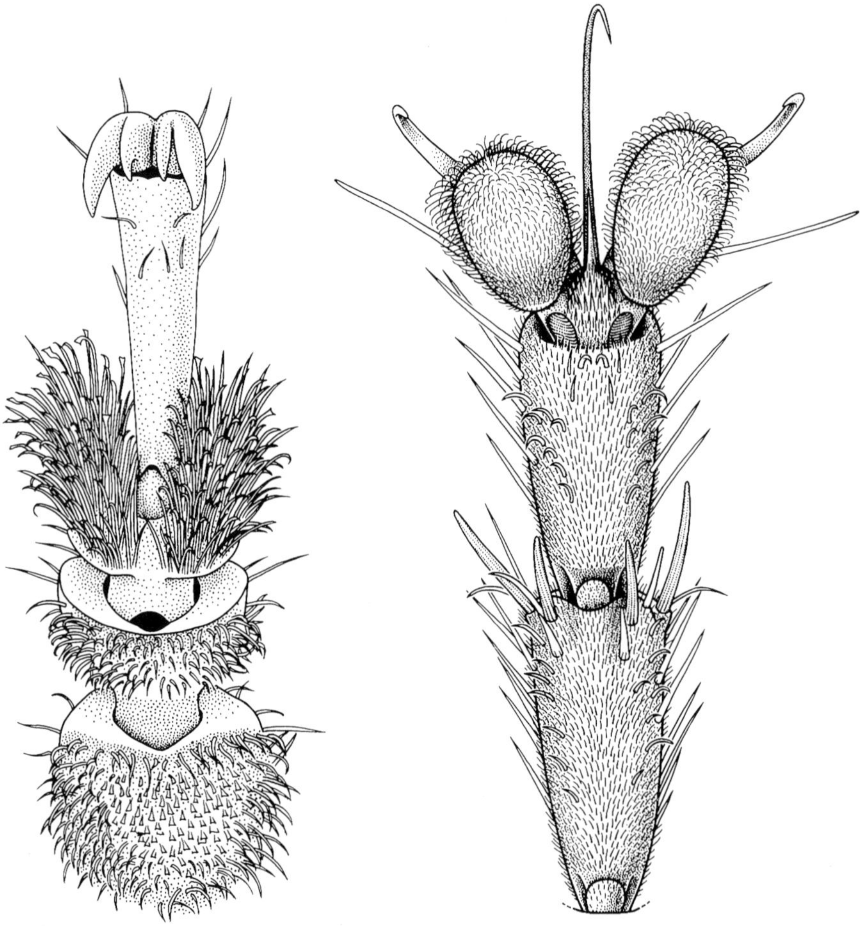

"Feet" of leaf beetle (left) and bush fly (right). (From scanning electron micrographs by C.A.M. Reid and A.C. Stewart.)

Insects are segmented invertebrates that possess the articulated external skeleton (exoskeleton) characteristic of all arthropods. Groups are differentiated by various modifications of the exoskeleton and the appendages – for example, the Hexapoda to which the Insecta belong (section 7.2) is characterized by having six-legged adults. Many anatomical features of the appendages, especially of the mouthparts, legs, wings and abdominal apex, are important in recognizing the higher groups within the hexapods, including insect orders, families and genera. Differences between species frequently are indicated by less obvious anatomical differences. Furthermore, the biomechanical analysis of morphology (e.g. studying how insects fly or feed) depends on a thorough knowledge of structural features. Clearly, an understanding of external anatomy is necessary in order to interpret and appreciate the functions of the various insect designs and to allow identification of insects and their hexapod relatives. In this chapter, we describe and discuss the cuticle, body segmentation and the structure of the head, thorax and abdomen and their appendages.

Knowledge of the following basic classification and terms is fundamental to the information that follows in this chapter. Adult insects normally have wings (most of the Pterygota), the structure of which may diagnose orders, but there is a group of primitively wingless insects (the "apterygotes") (see section 7.4.1 and Taxobox 2 for defining features). Within the Insecta, three major patterns of development can be recognized (section 6.2). Apterygotes (and non-insect hexapods) develop to adulthood with little change in body form (**ametaboly**), except for sexual maturation through development of gonads and genitalia. All other insects either have a gradual change in body form (**hemimetaboly**), with external wing buds getting larger at each moult, or an abrupt change from a wingless immature insect to a winged adult stage via a pupal stage (**holometaboly**). Immature stages of hemimetabolous insects are generally called **nymphs**, whereas those of holometabolous insects are referred to as **larvae**.

Anatomical structures of different taxa are **homologous** if they share an evolutionary origin, i.e. if the genetic basis is inherited from an ancestor common to them both. For instance, the wings of all insects are believed to be homologous; this means that wings (but not necessarily flight; see section 8.4) originated once. Homology of structures generally is inferred by comparison of similarity in **ontogeny** (development from egg to adult), composition (size and detailed appearance) and position (on the same segment and same relative location on that segment). The homology of insect wings is demonstrated by similarities in venation and articulation – the wings of all insects can be derived from the same basic pattern or ground-plan (as explained in section 2.4.2). Sometimes association with other structures of known homologies is helpful in establishing the homology of a structure of uncertain origin. Another sort of homology, called **serial homology**, refers to corresponding structures on different segments of an individual insect. Thus, the appendages of each body segment are serially homologous, although in living insects those on the head (antennae and mouthparts) are very different in appearance from those on the thorax (walking legs) and abdomen (genitalia and cerci). The way in which molecular developmental studies are confirming these serial homologies is described in Box 6.1.

2.1 THE CUTICLE

The cuticle is a key contributor to the success of the Insecta. This inert layer provides the strong **exoskeleton** of body and limbs, the **apodemes** (internal supports and muscle attachments) and wings, and acts as a barrier between living tissues and the environment. Internally, cuticle lines the tracheal tubes (section 3.5), some gland ducts and the foregut and hindgut of the digestive tract. Cuticle may range from rigid and armour-like, as in most adult beetles, to thin and flexible, as in many larvae. Restriction of water loss is a critical function of cuticle and is vital to the success of insects on land. The mechanical strength of the exoskeleton derives both from the chemical and physical characteristics of the cuticle (described below) and from the shape of various parts (plates, appendage segments, etc.) that can be curved, corrugated or tubular, often with stiffening flanges, ribs or other reinforcements.

The Insects: An Outline of Entomology, Fifth Edition. P.J. Gullan and P.S. Cranston.

Companion Website: www.wiley.com/go/gullan/insects

The cuticle is thin, but its structure is complex and varies among different taxonomic groups, as well as through the development of an individual insect (e.g. larva versus pupa versus adult) and on different parts of its body. A single layer of cells, the **epidermis**, lies beneath and secretes the cuticle, which consists of a thicker **procuticle** overlaid with thin **epicuticle** (Fig. 2.1). The epidermis and cuticle together form an **integument** – the outer covering of the living tissues of an insect. The epidermis is closely associated with

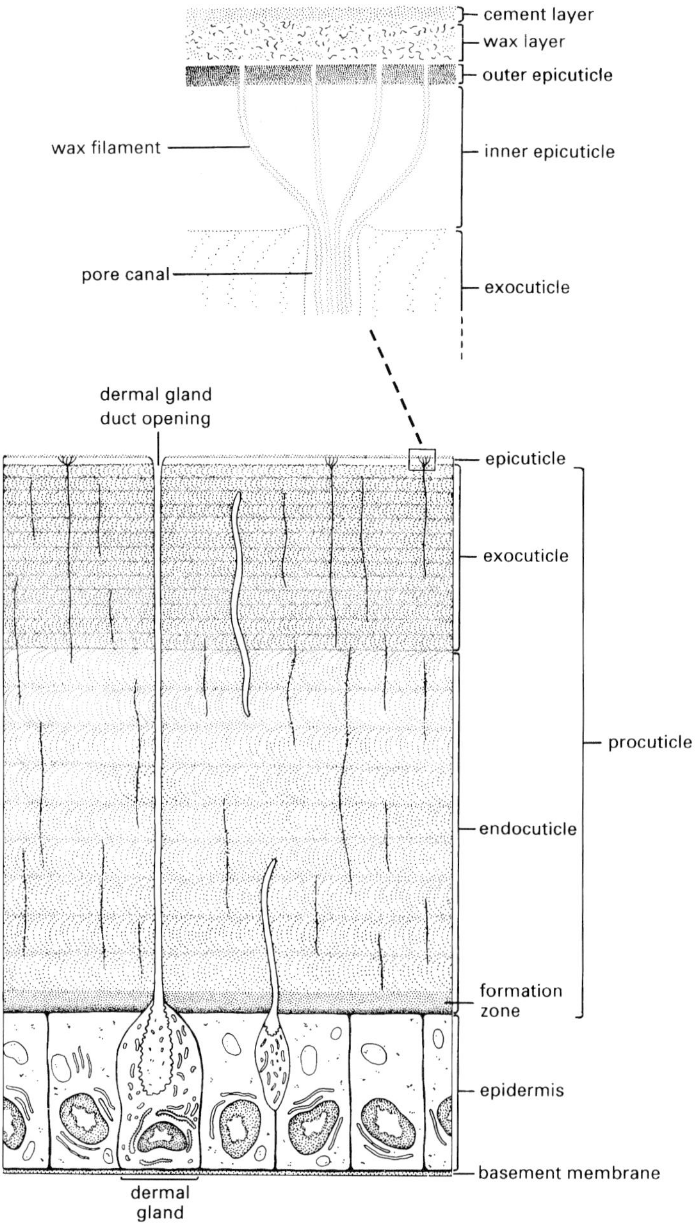

Fig. 2.1 The general structure of insect cuticle; the enlargement above shows details of the epicuticle. (After Hepburn 1985; Hadley 1986; Binnington 1993.)

moulting – the events and processes leading up to and including **ecdysis**, i.e. the shedding of the old cuticle (section 6.3).

The epicuticle ranges from 1 to 4 μm in thickness, and typically consists (from internal to external) of: an **inner epicuticle**; an **outer epicuticle** (sometimes called the cuticulin layer); a lipid or wax layer; and a variably discrete cement layer. The chemistry of the epicuticle and its outer layers is vital in preventing dehydration, a function derived from water-repelling (hydrophobic) lipids, especially hydrocarbons. These compounds include free and protein-bound lipids, and the outermost waxy coatings give a bloom to the external surface of some insects. Other cuticular patterns, such as light reflectivity, are produced by various kinds of epicuticular surface microsculpturing, such as close-packed, regular or irregular tubercles, ridges or tiny hairs. Lipid composition can vary, and waxiness can increase seasonally or under dry conditions. Besides being water retentive, surface waxes may deter predation, provide patterns for mimicry or camouflage, repel excess rainwater, reflect solar and ultraviolet radiation, or give species-specific olfactory cues (see sections 4.3.2 and 7.1.2).

The epicuticle is inextensible and unsupportive. Instead, support is given by the underlying chitinous cuticle, which is known as procuticle when it is first secreted. This differentiates into a thicker **endocuticle** covered by a thinner **exocuticle**, due to **sclerotization** of the latter. The procuticle is from 10 μm to 0.5 mm thick and consists primarily of chitin complexed with protein. This contrasts with the overlying epicuticle, which lacks chitin. There is a diversity of cuticular proteins, from 10 to more than 100 kinds in any particular type of cuticle, with hard cuticles having different proteins to flexible cuticles. Different insect species have different proteins in their cuticle. This variation in protein composition contributes to the specific characteristics of different cuticles.

Chitin is found as a supporting element in fungal cell walls as well as in arthropod exoskeletons. It is an unbranched polymer of high molecular weight – an amino polysaccharide predominantly composed of β-(1–4)-linked units of *N*-acetyl-D-glucosamine (Fig. 2.2). Chitin molecules are grouped into bundles and assembled into flexible microfibrils that are embedded in, and intimately linked to, a protein matrix, giving great tensile strength. The commonest arrangement of chitin microfibrils is in a sheet, in which the microfibrils are in parallel. In the exocuticle, each successive sheet lies in the same plane but may be orientated at a slight angle relative to the previous sheet, such that a thickness of many sheets produces a helicoid arrangement, which in sectioned cuticle appears as alternating light and dark bands (lamellae). Thus, the parabolic patterns and lamellar arrangement, visible so clearly in sectioned cuticle, represent an optical artefact resulting from microfibrillar orientation (Fig. 2.3). In the endocuticle, alternate stacked or helicoid arrangements of microfibrillar sheets may occur, often giving rise to thicker lamellae than in the exocuticle. Different

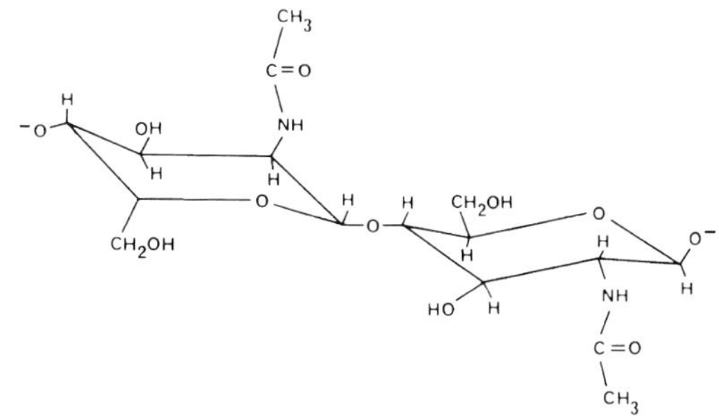

Fig. 2.2 Structure of part of a chitin chain, showing two linked units of *N*-acetyl-D-glucosamine. (After Cohen 1991.)

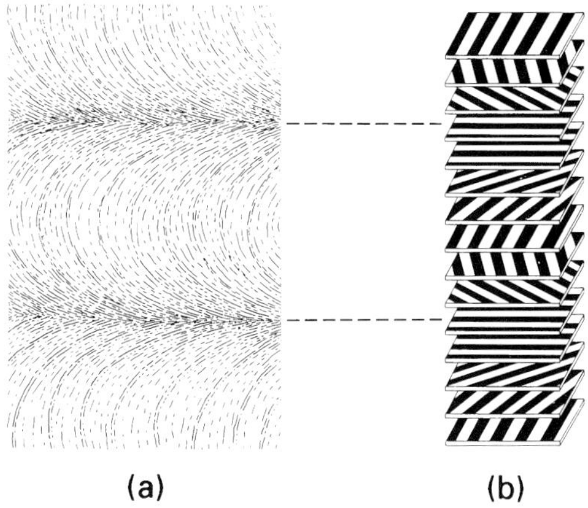

Fig. 2.3 The ultrastructure of cuticle (from a transmission electron micrograph). (a) The arrangement of chitin microfibrils in a helicoidal array produces characteristic (though artefactual) parabolic patterns. (b) Diagram of how the rotation of microfibrils produces a lamellar effect owing to microfibrils being either aligned or non-aligned to the plane of sectioning. (After Filshie 1982.)

arrangements may be laid down during darkness compared with during daylight, allowing precise age determination in many adult insects.

Much of the strength of cuticle comes from extensive hydrogen bonding of adjacent chitin chains. Additional stiffening comes from sclerotization, an irreversible process that darkens the exocuticle and results in the proteins becoming water-insoluble. Sclerotization may result from linkages of adjacent protein chains by phenolic bridges (quinone tanning) or from controlled dehydration of the chains, or both mechanisms acting together. Only exocuticle becomes sclerotized. The deposition of pigment in the cuticle, including deposition of melanin, may be associated with quinones, but is additional to sclerotization and not necessarily associated with it. The extreme hardness of the cutting edge of some insect mandibles is correlated with the presence of zinc and/or manganese in the cuticle.

In contrast to the solid cuticle typical of sclerites and mouthparts such as mandibles, softer, plastic, highly flexible or truly elastic cuticles occur in insects in varying locations and proportions. Where elastic or spring-like movement occurs, such as in wing ligaments or for the jump of a flea, **resilin** – a "rubber-like" protein – is present. The coiled polypeptide chains of this protein function as a mechanical spring under tension or compression, or in bending.

In soft-bodied larvae and in the membranes between segments, the cuticle must be tough, but also flexible and capable of extension. This "soft" cuticle, sometimes termed **arthrodial membrane**, is evident in gravid females, for example in the ovipositing migratory locust, *Locusta migratoria* (Orthoptera: Acrididae), in which intersegmental membranes may be expanded up to 20-fold for oviposition. Similarly, the gross abdominal dilation of gravid queen termites (see **Plate 6c**), bees and ants is possible through expansion of the unsclerotized cuticle. In these insects, the overlying unstretchable epicuticle expands by unfolding from an originally highly folded state, and some new epicuticle is formed. An extreme example of the distensibility of arthrodial membrane is seen in honeypot ants (Fig. 2.4; see also **Plate 2h** and section 12.2.3). In *Rhodnius* nymphs (Hemiptera: Reduviidae), changes in molecular structure of the cuticle allow actual stretching of the abdominal membrane to occur in response to intake of a large fluid volume during feeding.

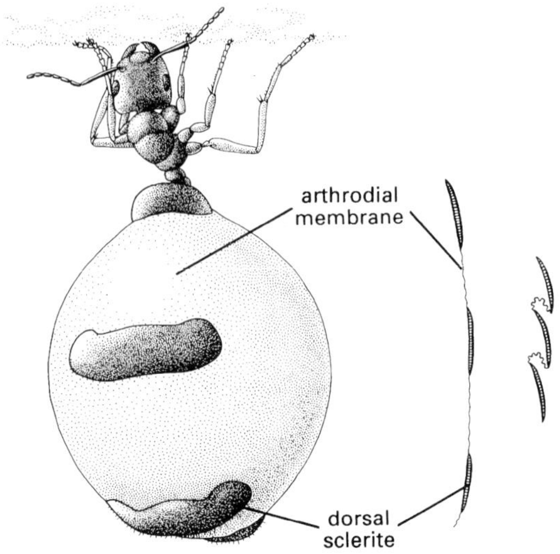

Fig. 2.4 A specialized worker, or replete, of the honeypot ant, *Camponotus inflatus* (Hymenoptera: Formicidae), which holds honey in its distensible abdomen and acts as a food store for the colony. The arthrodial membrane between tergal plates is depicted to the right in its unfolded and folded conditions. (After Hadley 1986; Devitt 1989.)

Cuticular structural components, waxes, cements, pheromones (Chapter 4) and defensive and other compounds are products either of the epidermis, which is a near-continuous, single-celled layer beneath the cuticle, or of secretory cells associated with the epidermis. Many of these compounds are secreted to the outside of the insect epicuticle. Numerous fine **pore canals** traverse the procuticle and then branch into numerous finer **wax canals** (containing wax filaments) within the epicuticle (see enlargement in Fig. 2.1); this system transports lipids (waxes) from the epidermis to the epicuticular surface. The wax canals may also have a structural role within the epicuticle. **Dermal glands** (exocrine glands) associated with the epidermis may produce cement and/or wax or other products, which are transported from the secretory cells via ducts to the surface of the cuticle or to a reservoir that opens at the surface. Wax-secreting glands are particularly well developed in mealybugs and other scale insects (Fig. 2.5). Ants also have an impressive number of exocrine glands, with 20 different types identified from the legs alone.

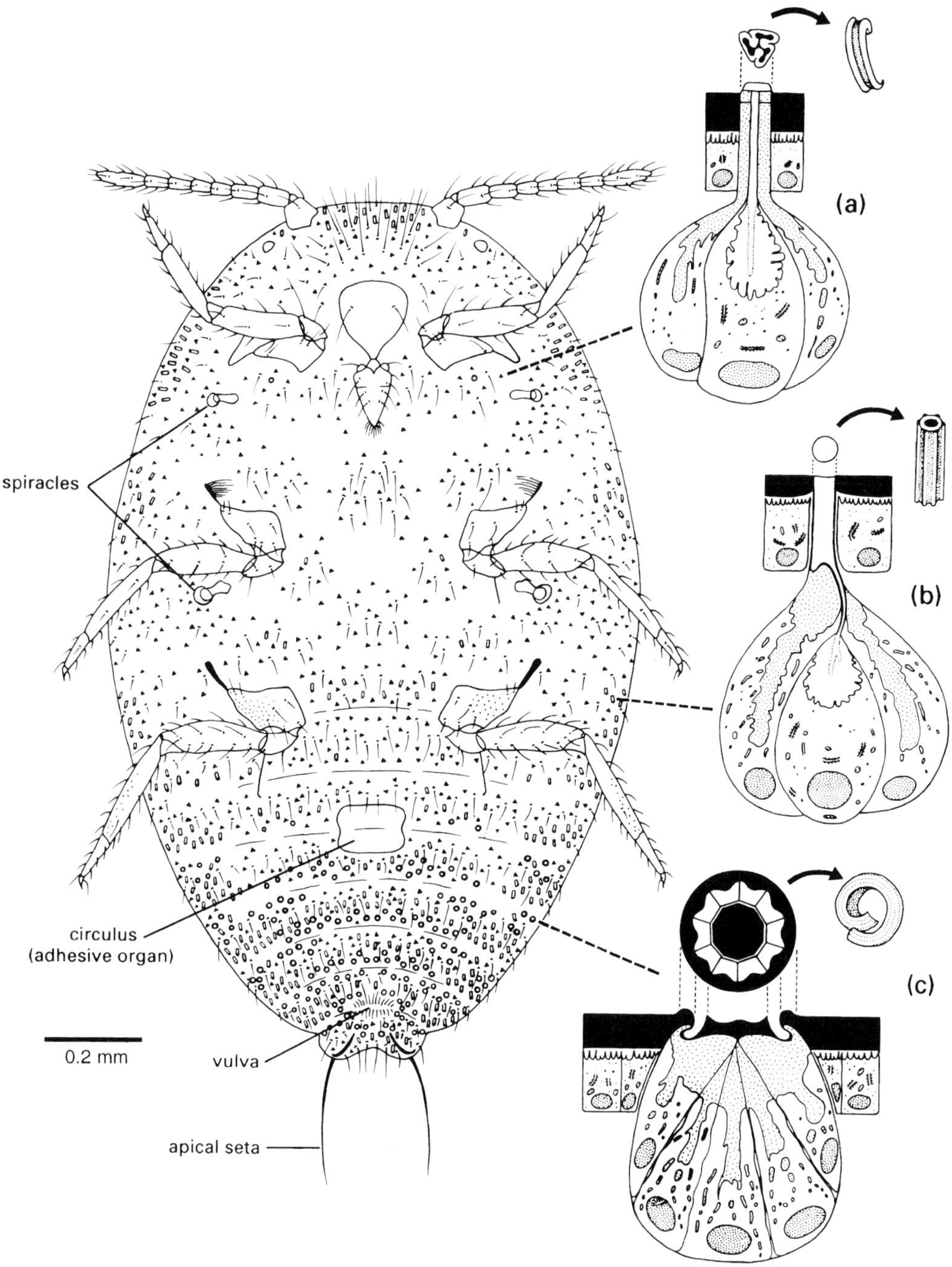

Fig. 2.5 The cuticular pores and ducts on the venter of an adult female of the citrus mealybug, *Planococcus citri* (Hemiptera: Pseudococcidae). Enlargements depict the ultrastructure of the wax glands and the various wax secretions (arrowed) associated with three types of cuticular structure: (a) a trilocular pore; (b) a tubular duct; and (c) a multilocular pore. Curled filaments of wax from the trilocular pores form a protective body-covering and prevent contamination with their own sugary excreta, or honeydew; long, hollow, and shorter curled filaments from the tubular ducts and multilocular pores, respectively, form the ovisac. (After Foldi 1983; Cox 1987.)

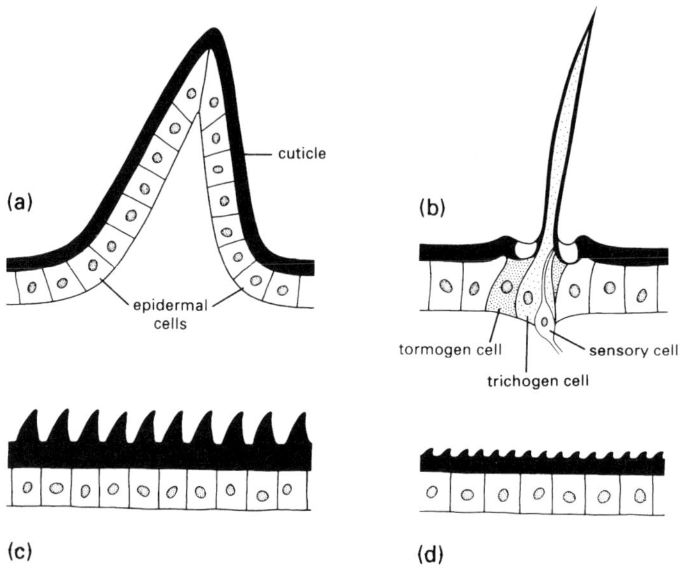

Fig. 2.6 The four basic types of cuticular protuberances: (a) a multicellular spine; (b) a seta, or trichoid sensillum; (c) acanthae; and (d) microtrichia. (After Richards & Richards 1979.)

Insects are well endowed with cuticular extensions, varying from fine and hair-like to robust and spine-like. Four basic types of protuberance (Fig. 2.6), all with sclerotized cuticle, can be recognized:

1 **spines** are multicellular with undifferentiated epidermal cells;

2 **setae**, also called **hairs**, **macrotrichia** or **trichoid sensilla**, are multicellular with specialized cells;

3 **acanthae** are unicellular in origin;

4 **microtrichia** are subcellular, with several to many extensions per cell.

Setae sense much of the insect's tactile environment. Large setae may be called bristles or chaetae, with the most modified being **scales**, the flattened setae found on butterflies and moths (Lepidoptera) and sporadically elsewhere in the insects. Three separate cells form each seta, one for hair formation (**trichogen cell**), one for socket formation (**tormogen cell**) and one sensory cell (see Fig. 4.1).

There is no such cellular differentiation in multicellular spines, unicellular acanthae, or subcellular microtrichia. The functions of these types of protuberances are diverse and sometimes debatable, but their sensory function appears limited. The production of pattern, including colour, may be significant for some of the microscopic projections. Spines are immovable, but if they are articulated, then they are called **spurs**. Both spines and spurs may bear unicellular or subcellular processes.

2.1.1 Colour production

The diverse colours of insects are produced by the interaction of light with the cuticle and/or underlying cells or fluid by two different mechanisms. Physical (structural) colours result from light scattering, interference and diffraction, whereas pigmentary colours are due to the absorption of visible light by a range of chemicals. Often both mechanisms occur together to produce a colour different from that produced by either mechanism alone.

All physical colours derive from the cuticle and its protuberances. **Interference** colours, such as iridescence and ultraviolet, are produced by refraction from varyingly spaced, close reflective layers produced by microfibrillar orientation within the exocuticle, or, in some beetles, the epicuticle, and by diffraction from regularly textured surfaces such as on many scales. Colours produced by light scattering depend on the size of surface irregularities relative to the wavelength of light. Thus, whites are produced by structures larger than the wavelength of light, such that all light is reflected, whereas blues are produced by irregularities that reflect only short wavelengths. The black colour on the wings of some butterflies, such as *Papilio ulysses* (Lepidoptera: Papilionidae), is produced by the absorption of most light by a combination of light-absorbing pigments and specially structured wing scales that prevent light from being scattered or reflected.

Insect pigments are produced in three ways:
1 by the insect's own metabolism;
2 by sequestering from a plant source;
3 by microbial endosymbionts (rarely).

Pigments may be located in the cuticle, epidermis, haemolymph or fat body. Cuticular darkening is the most ubiquitous insect colour. This may be due to either sclerotization (unrelated to pigmentation) or the exocuticular deposition of melanins, a group of polymers that may give a black, brown, yellow or red colour. Carotenoids, ommochromes, papiliochromes and pteridines (pterins) mostly produce yellows to reds, flavonoids give yellow, and tetrapyrroles (including breakdown products of porphyrins such as chlorophyll and haemoglobin) create reds, blues and greens. Quinone pigments occur in scale insects as red and yellow anthraquinones (e.g. carmine from cochineal insects), and in aphids as yellow to red to dark blue-green or black aphins.

Colours have an array of functions in addition to the obvious roles of colour patterns in sexual and defensive display. For example, the ommochromes are the main visual pigments of insect eyes, whereas black melanin, an effective screen for possibly harmful light rays, can convert light energy into heat and may act as a sink for free radicals that could otherwise damage cells. The red haemoglobins, which are widespread respiratory pigments in vertebrates, occur in a few insects, notably in some midge larvae and a few aquatic bugs, in which they have a similar respiratory function.

2.2 SEGMENTATION AND TAGMOSIS

Metameric segmentation, so distinctive in annelids, is visible only in some unsclerotized larvae (Fig. 2.7a). The segmentation seen in the sclerotized adult or nymphal insect is not directly homologous with that of larval insects, as sclerotization extends beyond each primary segment (Fig. 2.7b,c). Each apparent segment represents an area of sclerotization that commences in front of the fold that demarcates the primary segment and extends almost to the rear of that segment, leaving an unsclerotized area of the primary segment, the **conjunctival** or **intersegmental membrane**. This **secondary segmentation** means that the muscles, which are always inserted on the folds, are attached to solid rather than to soft cuticle. The apparent segments of adult insects, such as on the abdomen, are secondary in origin, but we refer to them simply as segments throughout this book.

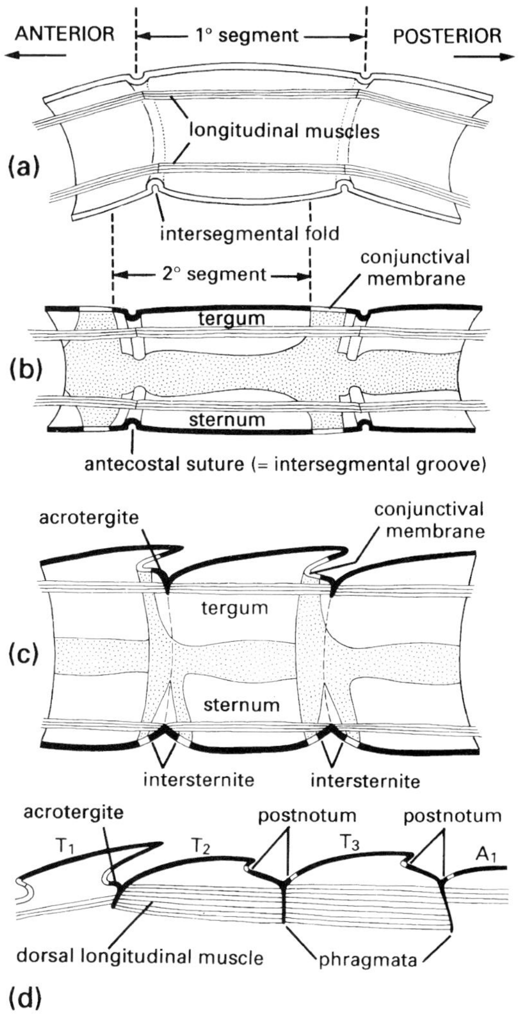

Fig. 2.7 Types of body segmentation. (a) Primary segmentation, as seen in soft-bodied larvae of some insects. (b) Simple secondary segmentation. (c) More-derived secondary segmentation. (d) Longitudinal section of dorsum of the thorax of winged insects, in which the acrotergites of the second and third segments have enlarged to become the postnota. (After Snodgrass 1935.)

In adult and nymphal insects, and hexapods in general, one of the most striking external features is the amalgamation of segments into functional units. This process of **tagmosis** has given rise to the familiar **tagmata** (regions) of **head**, **thorax** and **abdomen**. In this process, the 20 original segments have been divided into an embryologically detectable six-segmented head, three-segmented thorax and 11-segmented abdomen (plus primitively the telson), although varying degrees of fusion mean that the full complement is never visible.

An understanding of orientation is needed to discuss external morphology in more detail. The bilaterally symmetrical body may be described according to three axes:

1 **longitudinal**, or **anterior** to **posterior**, also termed **cephalic** (head) to **caudal** (tail);

2 **dorsoventral**, or **dorsal** (upper) to **ventral** (lower);

3 **transverse**, or **lateral** (outer) through the longitudinal axis to the opposite lateral (Fig. 2.8).

For appendages, such as legs or wings, **proximal** or **basal** refers to near the body, whereas **distal** or **apical** means distant from the body. In addition, structures are **mesal** or **medial** if they are nearer to the midline, or lateral if they are closer to the body margin, relative to other structures.

Four principal regions of the body surface can be recognized: the **dorsum** or upper surface; the **venter** or lower surface; and the two lateral pleura (singular: **pleuron**), separating the dorsum from the venter and

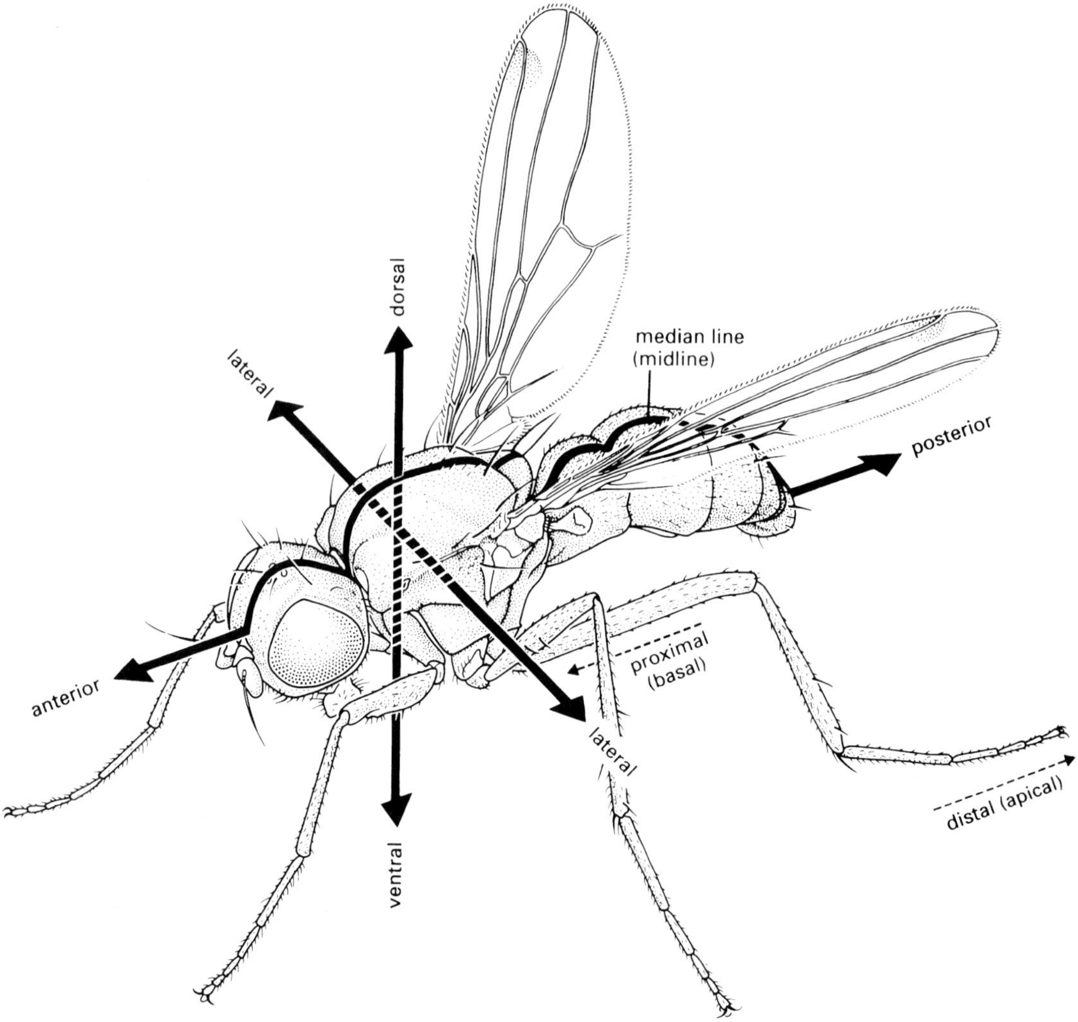

Fig. 2.8 The major body axes and the relationship of parts of the appendages to the body, shown for a sepsid fly. (After McAlpine 1987.)

bearing limb bases, if these are present. Sclerotization that takes place in defined areas gives rise to plates called **sclerites**. The major segmental sclerites are the **tergum** (the dorsal plate; plural: terga), the **sternum** (the ventral plate; plural: sterna) and the pleuron (the side plate). If a sclerite is a subdivision of the tergum, sternum or pleuron, the diminutive terms **tergite**, **sternite** and **pleurite** may be applied.

The abdominal pleura are often at least partly membranous, but on the thorax they are sclerotized and usually linked to the tergum and sternum of each segment. This fusion forms a box, which contains the leg muscle insertions and, in winged insects, the flight muscles. With the exception of some larvae, the head sclerites are fused into a rigid capsule. In larvae (but not nymphs), the thorax and abdomen may remain membranous, tagmosis may be less apparent (such as in most wasp larvae and fly maggots), and the terga, sterna and pleura are rarely distinct.

2.3 THE HEAD

The rigid cranial capsule has two openings, one posteriorly through the **occipital foramen** to the prothorax, the other to the mouthparts. Typically, the mouthparts are directed ventrally (**hypognathous**), although sometimes anteriorly (**prognathous**) as in many beetles, or posteriorly (**opisthognathous**) as in, for example, aphids, cicadas and leafhoppers. Several regions can be recognized on the head (Fig. 2.9): the posterior horseshoe-shaped **posterior cranium** (dorsally the **occiput**) contacts the **vertex** dorsally and the genae (singular: **gena**) laterally; the vertex abuts the

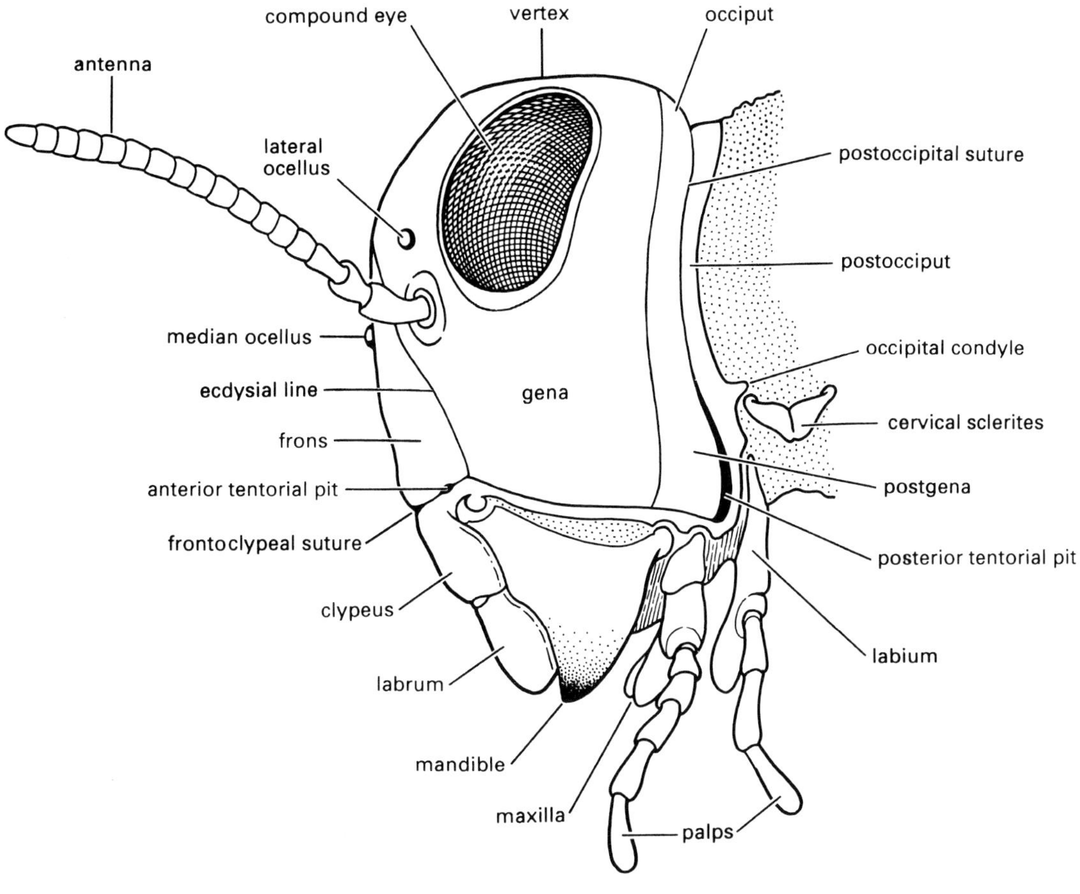

Fig. 2.9 Lateral view of the head of a generalized pterygote insect. (After Snodgrass 1935.)

frons anteriorly and more anteriorly lies the **clypeus**, both of which may be fused into a **frontoclypeus**. In adult and nymphal insects, paired **compound eyes** lie more or less dorsolaterally between the vertex and genae, with a pair of sensory **antennae** placed more medially. In many insects, three light-sensitive "simple" eyes, or ocelli (singular: **ocellus**), are situated on the anterior vertex, typically arranged in a triangle, and many larvae have stemmatal eyes.

The head regions are often somewhat weakly delimited, with some indications of their extent coming from **sutures** (external grooves or lines on the head). Three sorts of suture may be recognized:

1 remnants of original segmentation, generally restricted to the **postoccipital suture**;
2 **ecdysial lines** of weakness where the head capsule of the immature insect splits at moulting (section 6.3), including an often prominent inverted "Y", or **epicranial suture**, on the vertex (Fig. 2.10); the frons is delimited by the arms (also called **frontal sutures**) of this "Y";
3 grooves that reflect the underlying internal skeletal ridges, such as the **frontoclypeal** or **epistomal suture**, which often delimits the frons from the more anterior clypeus.

The head endoskeleton consists of several invaginated ridges and arms (**apophyses**, or elongate apodemes), the most important of which are the two pairs of **tentorial arms**, one pair being posterior, the other anterior, sometimes with an additional dorsal component. Some of these arms may be absent or, in pterygotes, fused to form the **tentorium**, an endoskeletal strut. Pits are discernible on the surface of the cranium at the points where the tentorial arms invaginate. These pits and the sutures may provide prominent landmarks on the head but usually they bear little or no association with the segments.

The segmental origin of the head is most clearly demonstrated by the mouthparts (section 2.3.1). From anterior to posterior, there are six fused head segments:

1 preantennal (or ocular);
2 antennal, with each antenna equivalent to an entire leg;
3 labral (previously sometimes called the intercalary segment; see section 2.3.1);
4 mandibular;
5 maxillary;
6 labial.

The neck is mainly derived from the first part of the thorax and is not a segment.

2.3.1 Mouthparts

The mouthparts are formed from appendages of head segments 3–6. In omnivorous insects, such as cockroaches, crickets and earwigs, the mouthparts are of a biting and chewing type (**mandibulate**) and resemble the probable basic design of ancestral pterygote insects more closely than do the mouthparts of the majority of modern insects. Extreme modifications of basic mouthpart structure, correlated with feeding specializations, occur in most Lepidoptera, Diptera, Hymenoptera, Hemiptera and a number of the smaller orders. Here, we first discuss basic mandibulate mouthparts, as exemplified by the European earwig, *Forficula auricularia* (Dermaptera: Forficulidae) (Fig. 2.10), and then describe some of the more common modifications associated with more specialized diets.

There are five basic components of the mouthparts:

1 **labrum**, or "upper lip", with a ventral surface called the **epipharynx**;
2 **hypopharynx**, a tongue-like structure;
3 **mandibles**, or jaws;
4 maxillae (singular: **maxilla**);
5 **labium**, or "lower lip" (Fig. 2.10).

The labrum forms the roof of the preoral cavity and mouth (see Fig. 3.14) and covers the base of the mandibles. Until recently, the labrum generally was considered to be associated with head segment 1. However, recent studies of the embryology, gene expression and nerve supply of the labrum show that it is innervated by the tritocerebrum of the brain (the fused ganglia of the third head segment) and is formed from fusion of parts of a pair of ancestral appendages on head segment 3. Projecting forwards from the back of the preoral cavity is the hypopharynx, a lobe of uncertain origin, but perhaps associated with the mandibular segment; in apterygotes, earwigs and nymphal mayflies the hypopharynx bears a pair of lateral lobes, the superlinguae (singular: **superlingua**) (Fig. 2.10). It divides the cavity into a dorsal food pouch, or

Fig. 2.10 Frontal view of the head and dissected mouthparts of an adult European earwig, *Forficula auricularia* (Dermaptera: Forficulidae). Note that the head is prognathous and thus a gular plate, or gula, occurs in the ventral neck region.

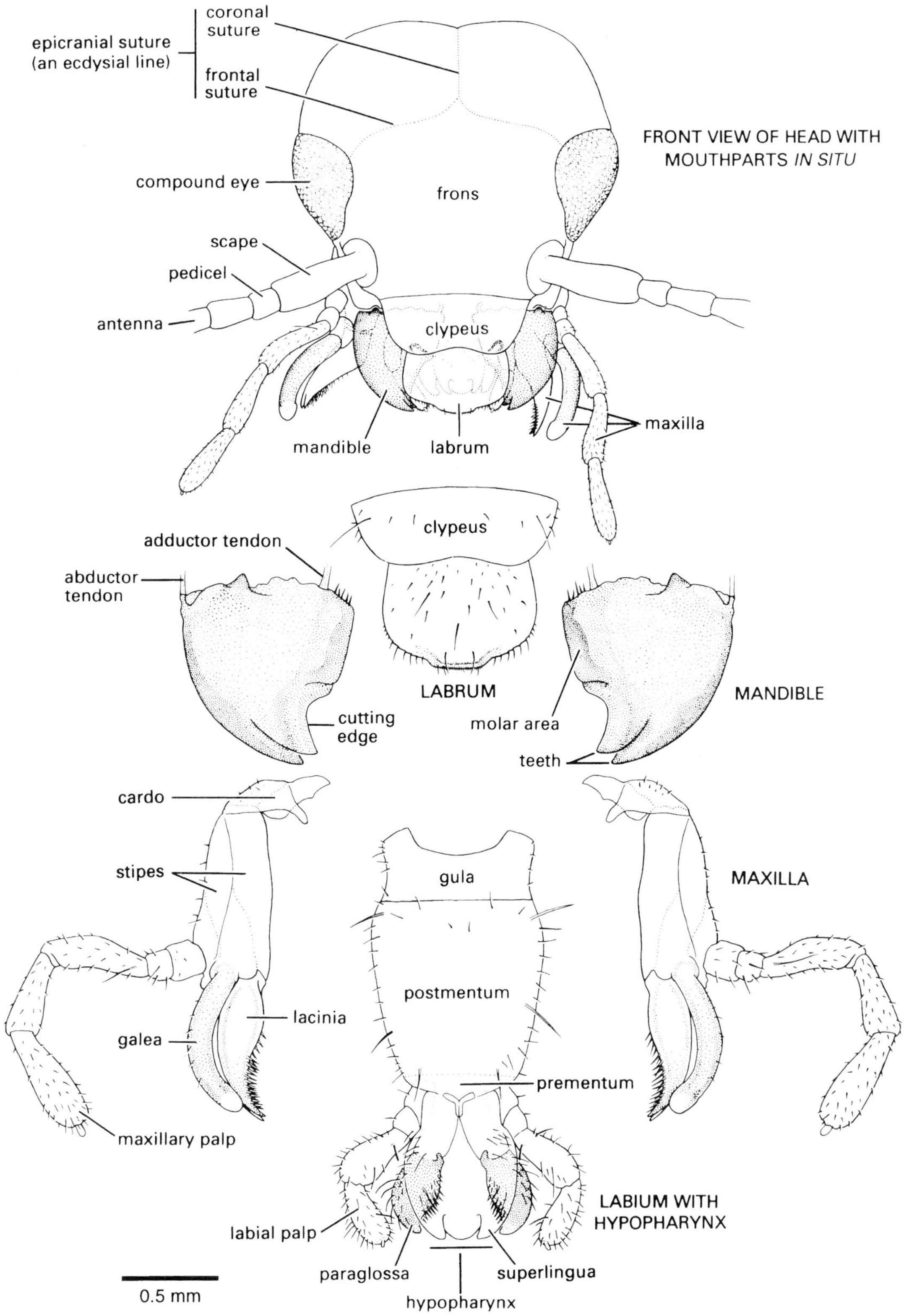
epicranial suture
(an ecdysial line)
coronal suture
frontal suture
FRONT VIEW OF HEAD WITH MOUTHPARTS *IN SITU*
compound eye
frons
scape
pedicel
antenna
clypeus
mandible
labrum
maxilla
adductor tendon
abductor tendon
clypeus
LABRUM
MANDIBLE
cutting edge
molar area
teeth
cardo
stipes
gula
MAXILLA
postmentum
lacinia
galea
prementum
maxillary palp
LABIUM WITH HYPOPHARYNX
labial palp
paraglossa
superlingua
hypopharynx
0.5 mm

cibarium, and a ventral **salivarium** into which the salivary duct opens (see Fig. 3.14). The mandibles, maxillae and labium are the paired appendages of segments 4–6 and are highly variable in structure among insect orders; their serial homology with walking legs is more apparent than for the labrum and hypopharynx.

The mandibles cut and crush food and may be used for defence. Generally, they have an apical cutting edge and the more basal molar area grinds the food. They can be extremely hard (approximately 3 on Moh's scale of mineral hardness, or an indentation hardness of about 30 kg mm^{-2}) and thus many termites and beetles have no physical difficulty in boring through foils made from common metals such as copper, lead, tin and zinc. Behind the mandibles lie the maxillae, each consisting of a basal part composed of the proximal **cardo** and the more distal **stipes** and, attached to the stipes, two lobes – the mesal **lacinia** and the lateral **galea** – and a lateral, segmented **maxillary palp**, or palpus (plural: palps or palpi). Functionally, the maxillae assist the mandibles in processing food; the pointed and sclerotized lacinae hold and macerate the food, whereas the galeae and palps bear sensory setae (mechanoreceptors) and chemoreceptors that sample items before ingestion. The appendages of the sixth segment of the head are fused with the sternum to form the labium, which is believed to be homologous to the second maxillae of Crustacea. In prognathous insects, such as the earwig, the labium attaches to the ventral surface of the head via a ventromedial sclerotized plate called the **gula** (Fig. 2.10). There are two main parts to the labium: the proximal **postmentum**, closely connected to the posteroventral surface of the head and sometimes subdivided into a submentum and mentum; and the free distal **prementum**, typically bearing a pair of **labial palps** lateral to two pairs of lobes, the mesal glossae (singular: **glossa**) and the more lateral paraglossae (singular: **paraglossa**). The glossae and paraglossae, including sometimes the distal part of the prementum to which they attach, are known collectively as the **ligula**; the lobes may be variously fused or reduced as in *Forficula* (Fig. 2.10), in which the glossae are absent. The prementum with its lobes forms the floor of the preoral cavity (functionally a "lower lip"), whereas the labial palps have a sensory function, similar to that of the maxillary palps.

During insect evolution, an array of different mouthpart types have been derived from the basic design described above. Often, feeding structures are characteristic of all members of a genus, family or order of insects, so that knowledge of mouthparts is useful for both taxonomic classification and identification, and for ecological generalization (section 10.6). Generally, mouthpart structure is categorized according to feeding method, but mandibles and other components may function in defensive combat or even male–male sexual contests, as in the enlarged mandibles on certain male beetles (Lucanidae). Insect mouthparts have diversified in different orders, with feeding methods that include lapping, suctorial feeding, biting, or piercing combined with sucking, and filter feeding, in addition to the basic chewing mode.

The mouthparts of bees are of a chewing and lapping-sucking type. Lapping is a mode of feeding in which liquid or semi-liquid food adhering to a protrusible organ, or "tongue", is transferred from substrate to mouth. In the honey bee, *Apis mellifera* (Hymenoptera: Apidae), the elongate and fused labial glossae form a hairy tongue, which is surrounded by the maxillary galeae and the labial palps to form a tubular proboscis containing a food canal (Fig. 2.11). In feeding, the

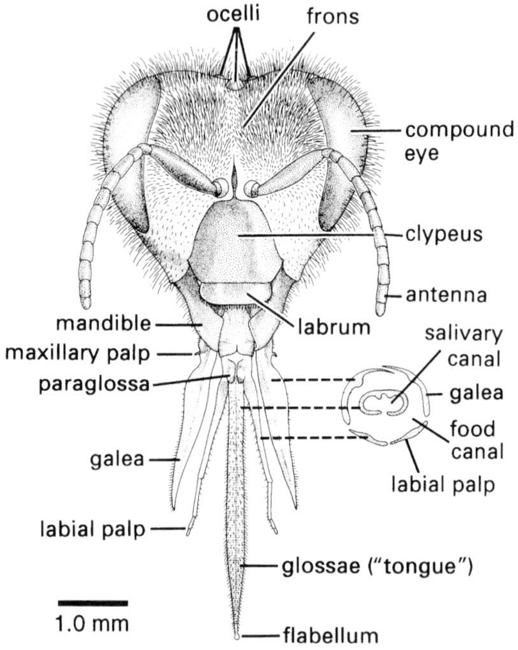

Fig. 2.11 Frontal view of the head of a worker honey bee, *Apis mellifera* (Hymenoptera: Apidae), with transverse section of proboscis showing how the "tongue" (fused labial glossae) is enclosed within the sucking tube formed from the maxillary galae and labial palps. (Inset after Wigglesworth 1964.)

tongue is dipped into the nectar or honey, which adheres to the hairs, and then is retracted so that adhering liquid is carried into the space between the galeae and labial palps. This back-and-forth glossal movement occurs repeatedly. Movement of liquid to the mouth apparently results from the action of the cibarial pump, facilitated by each retraction of the tongue pushing liquid up the food canal. The maxillary laciniae and palps are rudimentary and the paraglossae embrace the base of the tongue, directing saliva from the dorsal salivary orifice around into a ventral channel from whence it is transported to the **flabellum**, a small lobe at the glossal tip; saliva may dissolve solid or semi-solid sugar. The sclerotized, spoon-shaped mandibles lie at the base of the proboscis and have a variety of functions, including the manipulation of wax and plant resins for nest construction, the feeding of larvae and the queen, grooming, fighting and the removal of nest debris, including dead bees.

Most adult Lepidoptera, and some adult flies, obtain their food solely by sucking up liquids using suctorial (**haustellate**) mouthparts that form a proboscis or rostrum (Fig. 2.12, Fig. 2.13 and Fig. 2.14). Pumping of the liquid food is achieved by muscles of the cibarium and/or pharynx. The proboscis of moths and butterflies, formed from the greatly elongated maxillary galeae, is extended (Fig. 2.12a) by increases in haemolymph ("blood") pressure. It is loosely coiled by the inherent elasticity of the cuticle, but tight coiling requires contraction of intrinsic muscles (Fig. 2.12b). A cross-section of the proboscis (Fig. 2.12c) shows how the food canal, which opens basally into the cibarial pump, is formed by apposition and interlocking of the two galeae. The proboscis of most adult lepidopterans is very obvious (see **Plate 3a** and **Plate 5g**), and that of some male hawkmoths (Sphingidae), such as *Xanthopan morgani*, can attain great length (see Fig. 11.7).

A few moths, and many flies, combine sucking with piercing or biting. For example, moths that pierce fruit and exceptionally suck blood (some species of Noctuidae) have spines and hooks at the tip of their proboscis that are rasped against the skin of either ungulate mammals or fruit. For at least some moths, penetration is effected by the alternate protraction and retraction of the two galeae that slide along each other.

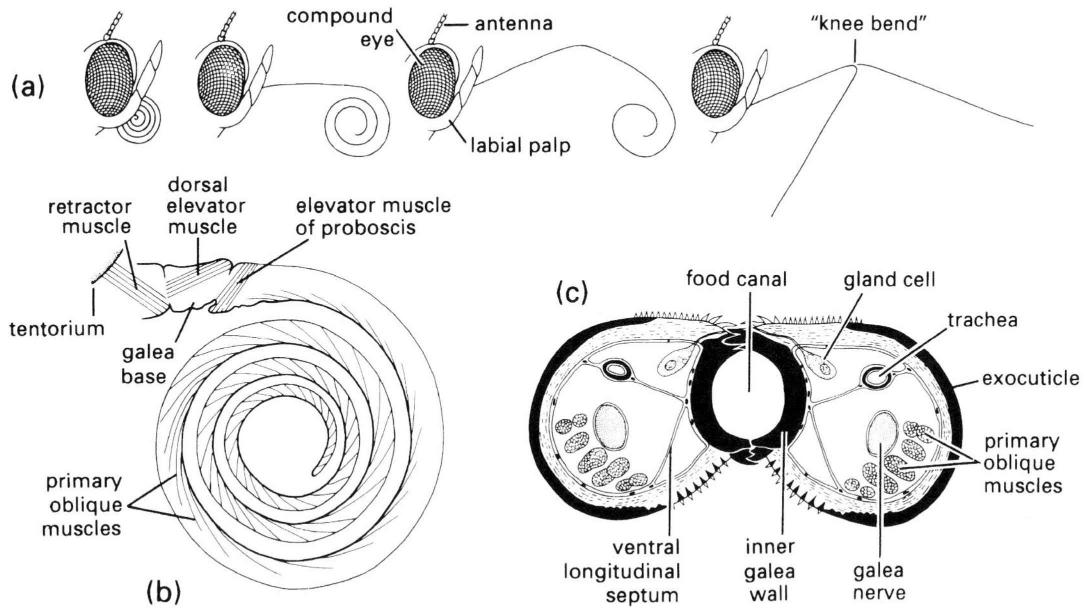

Fig. 2.12 Mouthparts of a white butterfly, *Pieris* sp. (Lepidoptera: Pieridae). (a) Positions of the proboscis showing, from left to right, at rest, with proximal region uncoiling, with distal region uncoiling, and fully extended with tip in two of many possible different positions due to flexing at the "knee bend". (b) Lateral view of proboscis musculature. (c) Transverse section of the proboscis in the proximal region. (After Eastham & Eassa 1955.)

All dipterans typically have a tubular sucking organ, the proboscis, comprising elongate mouthparts (usually including the labrum) (Fig. 2.13a). A biting-and-sucking type of proboscis appears to be a primitive dipteran feature. Although biting functions have been lost and regained with modifications more than once, blood feeding is frequent, and leads to the importance of the Diptera as vectors of disease. Blood-feeding flies have a variety of skin-penetration and feeding mechanisms. In the horse flies (Brachycera: Tabanidae), the labium of the adult fly forms a non-piercing sheath for the other mouthparts, which together contribute to the piercing structure. In contrast, the biting calyptrate dipterans (Brachycera: Calyptratae, e.g. stable flies and tsetse flies) lack mandibles and maxillae and the principal piercing organ is the highly modified labium.

The blood-feeding female nematocerans – Culicidae (mosquitoes), Ceratopogonidae (biting midges), Psychodidae: Phlebotominae (sand flies) and Simuliidae (black flies) – have generally similar mouthparts, but differ in proboscis length, allowing penetration of the host to different depths. Mosquitoes can probe deep in search of capillaries, but other blood-feeding nematocerans operate more superficially, where a pool of blood is induced in the wound. The labium, which ends in two sensory **labella** (singular: labellum), forms a protective sheath for the functional mouthparts (Fig. 2.13a). Enclosed are serrate-edged, cutting mandibles and maxillary lacinia, the curled labrum-epipharynx and the hypopharynx, all of which are often termed stylets (Fig. 2.13b). When feeding, the labrum, mandibles and laciniae act as a single unit driven through the skin of the host. The flexible labium remains bowed outside the wound. Saliva, which may contain anticoagulant, is injected through a salivary duct that runs the length of the sharply pointed and often toothed hypopharynx. Blood is transported up a food canal formed from the curled labrum sealed by either the paired mandibles or the hypopharynx. Capillary blood can flow unaided, but blood must be sucked or pumped from a pool via a pumping action from two muscular pumps: the cibarial located at the base of the food canal, and the pharyngeal in the pharynx between the cibarium and midgut.

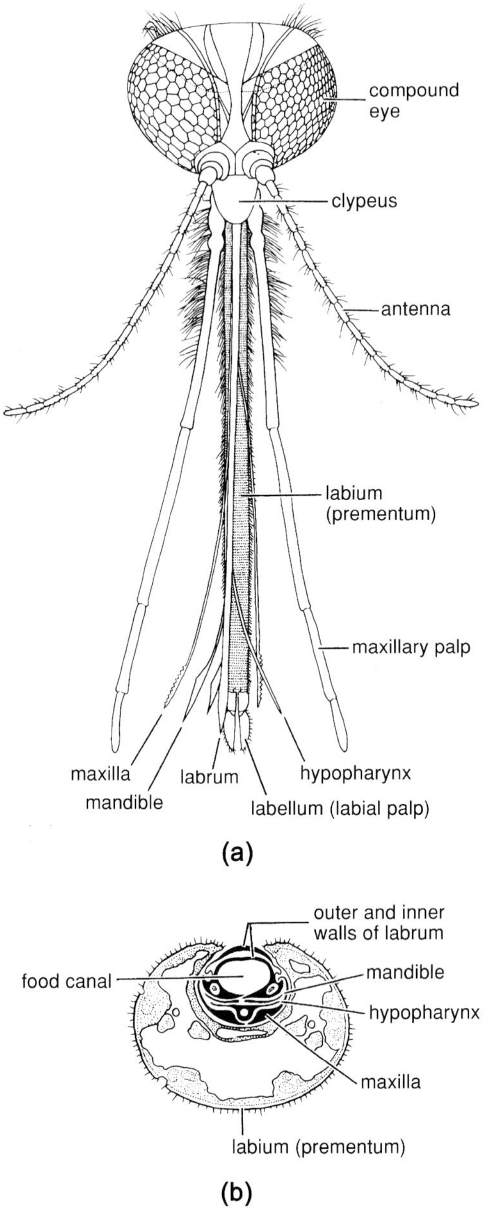

Fig. 2.13 Mouthparts of female mosquito in: (a) frontal view; (b) transverse section. ((a) After Freeman & Bracegirdle 1971; (b) after Jobling 1976.)

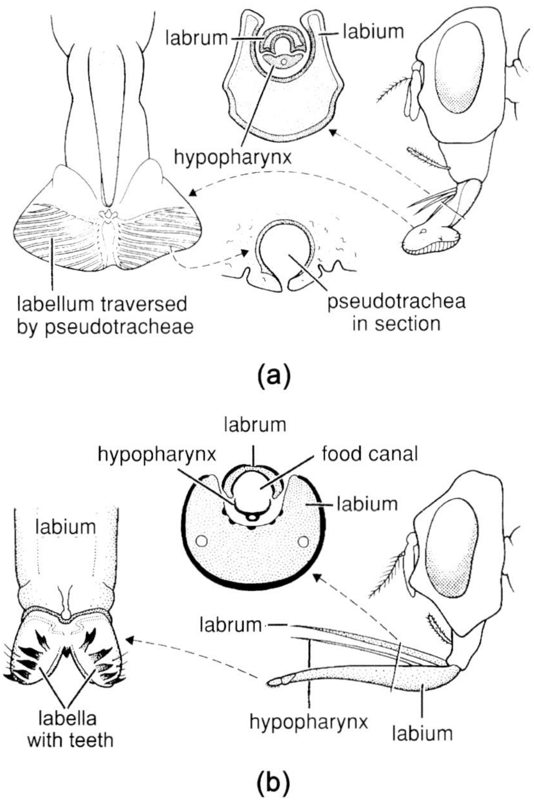

Fig. 2.14 Mouthparts of adult Diptera. (a) House fly, *Musca* (Muscidae). (b) Stable fly, *Stomoxys* (Muscidae). (After Wigglesworth 1964.)

Many mouthparts are lost in brachyceran flies, and the remaining mouthparts are modified for lapping food using pseudotracheae of the labella as "sponges", as in a house fly (Muscidae: *Musca*) (Fig. 2.14a). With neither mandibles nor maxillary lacinia to make a wound, blood-feeding cyclorrhaphans often use modified labella in which the inner surfaces are adorned with sharp teeth, as in stable flies (Muscidae: *Stomoxys*) (Fig. 2.14b). Through muscular contraction and relaxation, the labellar lobes dilate and contract repeatedly, creating an often painful rasping of the labellar teeth to create a pool of blood. The hypopharynx applies saliva, which is dissipated via the labellar pseudotracheae. Uptake of blood is via capillary action through "food furrows" lying dorsal to the pseudotracheae, with the aid of three pumps operating synchronously to produce continuous suction from labella to pharynx. A prelabral pump produces the contractions in the labella, with a more proximal labral pump linked via a feeding tube to the cibarial pump.

The mouthparts of adult flies and how they are used in feeding have implications for disease transmission. Shallow-feeding species such as black flies are more involved in transmission of microfilariae, such as those of *Onchocerca* (section 15.3.6), which aggregate just beneath the skin, whereas deeper feeders such as mosquitoes transmit pathogens that circulate in the blood, as in malaria (section 15.3.1). The transmission from fly to host is aided by the introduction of saliva into the wound, and many parasites aggregate in the salivary glands or ducts. Filariae, in contrast, are too large to enter the wound through this route, and leave the insect host by rupturing the labium or labella during feeding.

Other mouthpart modifications for piercing and sucking are seen in the true bugs (Hemiptera), thrips (Thysanoptera), fleas (Siphonaptera) and sucking lice (Psocodea: Anoplura). In each of these orders, different mouthpart components form needle-like stylets capable of piercing the plant or animal tissues upon which the insect feeds. Bugs have extremely long, thin, paired mandibular and maxillary stylets, which fit together to form a flexible stylet-bundle containing a food canal and a salivary canal (Taxobox 20). Thrips have three stylets – paired maxillary stylets (laciniae) plus the left mandibular one (Fig. 2.15). Sucking lice have three stylets – the hypopharyngeal (dorsal), the salivary (median) and the labial (ventral) – lying in a ventral sac on the head and opening at a small eversible proboscis armed with internal teeth that grip the host during blood-feeding (Fig. 2.16). Fleas possess a single stylet derived from the epipharynx, and the laciniae of the maxillae form two long cutting blades that are ensheathed by the labial palps (Fig. 2.17). The Hemiptera and the Thysanoptera are sister groups and belong to the same assemblage as the Psocodea (see Fig. 7.5), with the lice originating from a psocid-like ancestor with mouthparts of a more

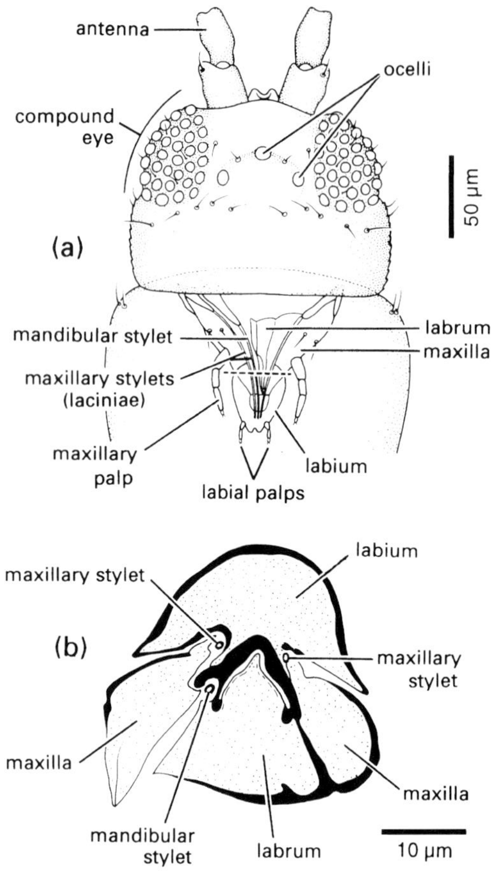

Fig. 2.15 Head and mouthparts of a thrips, *Thrips australis* (Thysanoptera: Thripidae). (a) Dorsal view of head showing mouthparts through prothorax. (b) Transverse section through proboscis. The plane of the transverse section is indicated by the dashed line in (a). (After Matsuda 1965; CSIRO 1970.)

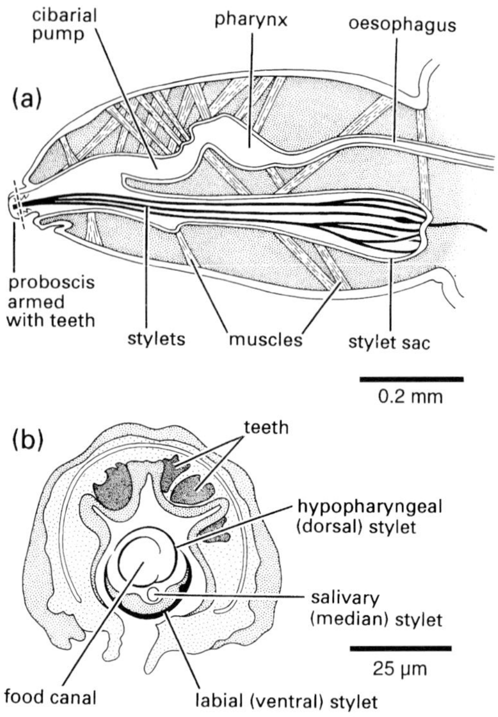

Fig. 2.16 Head and mouthparts of a sucking louse, *Pediculus* (Psocodea: Anoplura: Pediculidae). (a) Longitudinal section of head (nervous system omitted). (b) Transverse section through eversible proboscis. The plane of the transverse section is indicated by the dashed line in (a). (After Snodgrass 1935.)

generalized, mandibulate type. The Siphonaptera are distant relatives of the other three taxa; thus similarities in mouthpart structure among these orders result largely from parallel or, in the case of fleas, convergent evolution.

Slightly different piercing mouthparts are found in antlions and the predatory larvae of other lacewings (Neuroptera). The stylet-like mandible and maxilla on each side of the head fit together to form a sucking tube (see Fig. 13.2c), and in some families (Chrysopidae, Myrmeleontidae and Osmylidae) there is also a narrow poison channel. Generally, labial palps are present, maxillary palps are absent, and the labrum is reduced. Prey is seized by the pointed mandibles and maxillae, which are inserted into the victim; its body contents are digested extra-orally and sucked up by pumping of the cibarium.

A unique modification of the labium for prey capture occurs in nymphal damselflies and dragonflies (Odonata). These predators catch other aquatic organisms by extending their folded labium (or "mask") rapidly and seizing mobile prey using prehensile apical hooks on modified labial palps (see Fig. 13.4). The labium is hinged between the prementum and postmentum and, when folded, covers most of the underside of the head. Labial extension involves the sudden release of energy, produced by increases in blood pressure brought about by the contraction of thoracic and abdominal muscles, and stored elastically in a cuticular

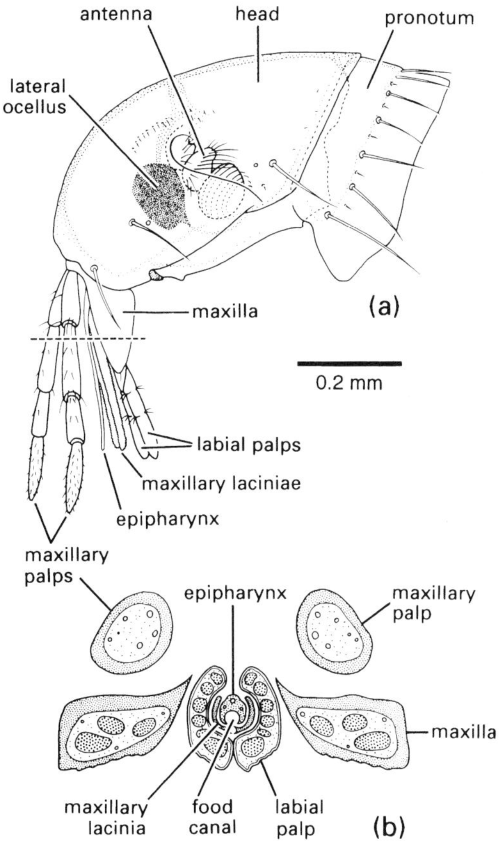

Fig. 2.17 Head and mouthparts of a human flea, *Pulex irritans* (Siphonaptera: Pulicidae). (a) Lateral view of head. (b) Transverse section through mouthparts. The plane of the transverse section is indicated by the dashed line in (a). (After Snodgrass 1946; Herms & James 1961.)

click mechanism at the prementum–postmentum joint. As the click mechanism is disengaged, the elevated hydraulic pressure shoots the labium rapidly forwards. Labial retraction then brings the captured prey to the other mouthparts for maceration.

Filter feeding in aquatic insects has been studied best in larval mosquitoes (Diptera: Culicidae), black flies (Diptera: Simuliidae) and net-spinning caddisflies (Trichoptera: many Hydropsychoidea and Philopotamoidea), which obtain their food by filtering particles (including bacteria, microscopic algae and detritus) from the water in which they live. The mouthparts of the dipteran larvae have an array of setal "brushes" and/or "fans", which generate feeding currents or trap particulate matter and then move it to the mouth. In contrast, the caddisflies spin silk nets that filter particulate matter from flowing water and then use their mouthpart brushes to remove particles from the nets. Thus, insect mouthparts are modified for filter feeding chiefly by the elaboration of setae. In mosquito larvae, the lateral palatal brushes on the labrum generate the feeding currents (Fig. 2.18); they beat actively, causing particle-rich surface water to flow towards the mouthparts, where setae on the mandibles and maxillae help to move particles into the pharynx, where food boluses form at intervals.

In some adult insects, such as mayflies (Ephemeroptera), some Diptera (warble flies), a few moths (Lepidoptera) and male scale insects (Hemiptera: Coccoidea), mouthparts are greatly reduced and nonfunctional. Atrophied mouthparts correlate with short adult lifespan.

2.3.2 Cephalic sensory structures

The most obvious sensory structures of insects are on the head. Most adults and many nymphs have **compound eyes** dorsolaterally on the head (probably derived from segment 1 of the head) and three **ocelli** on the vertex of the head. The median, or anterior, ocellus lies on the frons and is formed from a fused pair of ocelli; the two lateral ocelli are located more posteriorly on the head. The only visual structures of larval insects are **stemmata** (singular: stemma), or simple eyes, positioned laterally on the head, either singly or in clusters. The structure and functioning of these three types of visual organs are described in detail in section 4.4.

Antennae are mobile, segmented, paired appendages. Primitively, they appear to be eight-segmented in nymphs and adults, but often there are numerous subdivisions, sometimes called **antennomeres**. The entire antenna typically has three main divisions (Fig. 2.19a): the first segment, or **scape**, generally is larger than the other segments and is the basal stalk; the second segment, or **pedicel**, nearly always contains a sensory organ known as **Johnston's organ**, which responds to movement of the distal part of the antenna relative to the pedicel; the remainder of the antenna, called the **flagellum**, is often filamentous and multisegmented (with many **flagellomeres**), but may be reduced or variously modified (Fig. 2.19b–i).

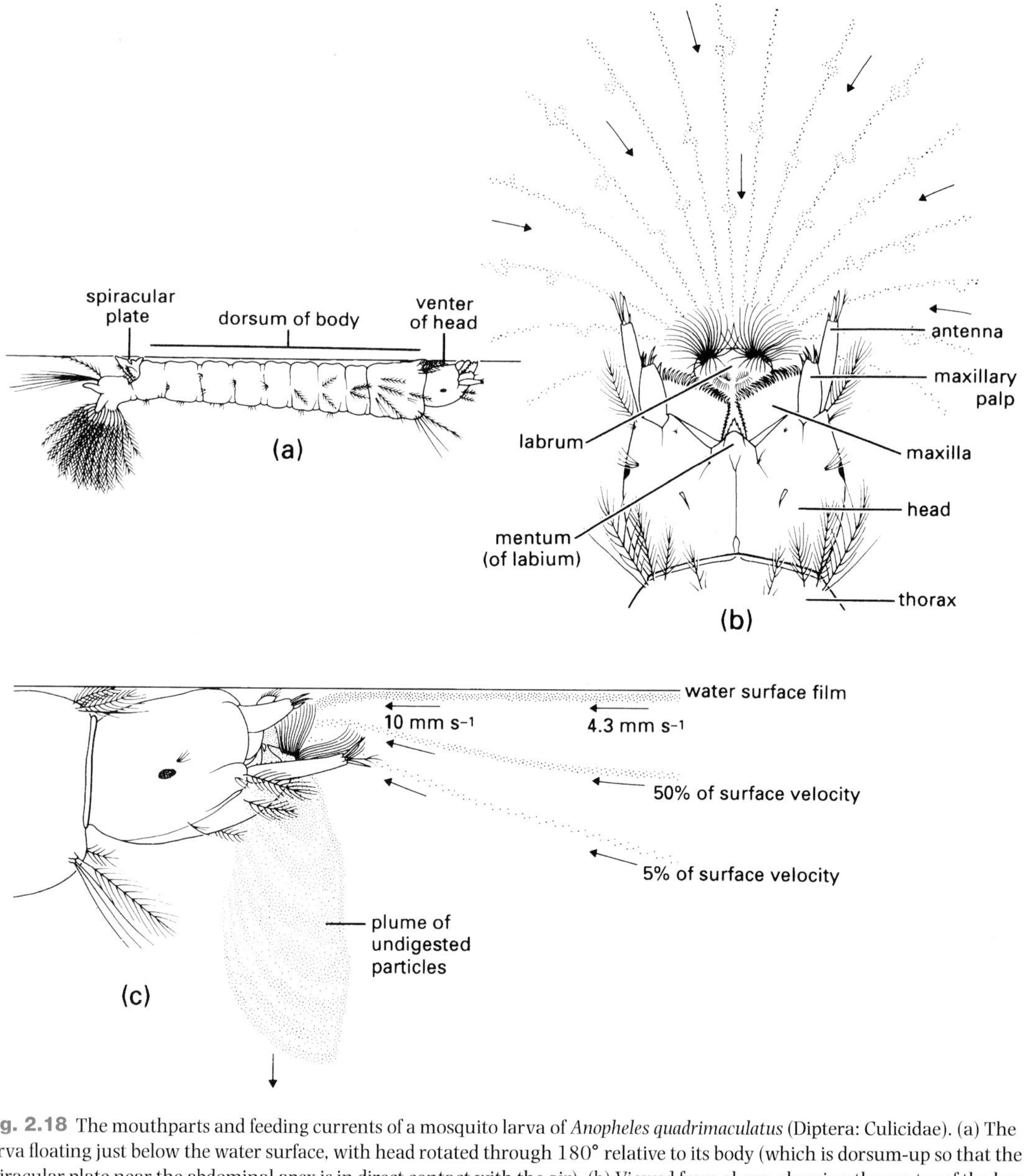

Fig. 2.18 The mouthparts and feeding currents of a mosquito larva of *Anopheles quadrimaculatus* (Diptera: Culicidae). (a) The larva floating just below the water surface, with head rotated through 180° relative to its body (which is dorsum-up so that the spiracular plate near the abdominal apex is in direct contact with the air). (b) Viewed from above, showing the venter of the head and the feeding current generated by setal brushes on the labrum (direction of water movement and paths taken by surface particles are indicated by arrows and dotted lines, respectively). (c) Lateral view showing the particle-rich water being drawn into the preoral cavity between the mandibles and maxillae, and its downward expulsion as the outward current. ((b,c) After Merritt *et al.* 1992.)

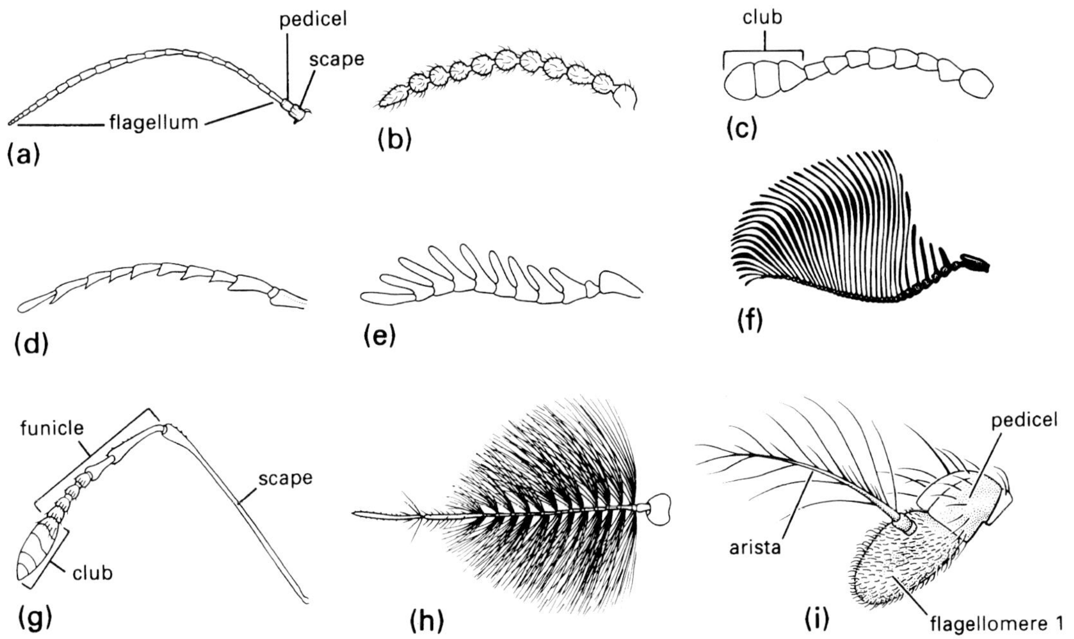

Fig. 2.19 Some types of insect antennae: (a) filiform – linear and slender; (b) moniliform – like a string of beads; (c) clavate or capitate – distinctly clubbed; (d) serrate – saw-like; (e) pectinate – comb-like; (f) flabellate – fan-shaped; (g) geniculate – elbowed; (h) plumose – bearing whorls of setae; and (i) aristate – with enlarged third segment bearing a bristle.

The antennae are reduced or almost absent in some larval insects.

Numerous sensory organs, or sensilla (singular: **sensillum**), in the form of hairs, pegs, pits or cones, occur on antennae and function as chemoreceptors, mechanoreceptors, thermoreceptors and hygroreceptors (Chapter 4). Antennae of male insects may be more elaborate than those of the corresponding females, increasing the surface area available for detecting female sex pheromones (section 4.3.2).

The mouthparts, other than the mandibles, are well endowed with chemoreceptors and tactile setae. These sensilla are described in detail in Chapter 4.

2.4 THE THORAX

The thorax is composed of three segments: the first or **prothorax**; the second or **mesothorax**; and the third or **metathorax**. Primitively, and in apterygotes (bristletails and silverfish) and immature insects, these segments are similar in size and structural complexity. In most winged insects, the mesothorax and metathorax are enlarged relative to the prothorax and form a **pterothorax**, bearing the wings and associated musculature. Wings occur only on the second and third segments in extant insects, although some fossils have prothoracic winglets (see Fig. 8.3) and homeotic mutants may develop prothoracic wings or wing buds. Almost all nymphal and adult insects have three pairs of thoracic legs – one pair per segment. Typically, the legs are used for walking, although various other functions and associated modifications occur (section 2.4.1). Openings (**spiracles**) of the gas-exchange, or tracheal, system (section 3.5) are present laterally on the second and third thoracic segments with at most one pair per segment. However, a secondary condition in some insects is for the mesothoracic spiracles to open on the prothorax.

The tergal plates of the thorax are simple structures in apterygotes and in many immature insects, but are variously modified in winged adults. Thoracic terga are called nota (singular: **notum**), to distinguish them from the abdominal terga. The **pronotum** of the prothorax may be simple in structure and small in comparison with the other nota, but in beetles,

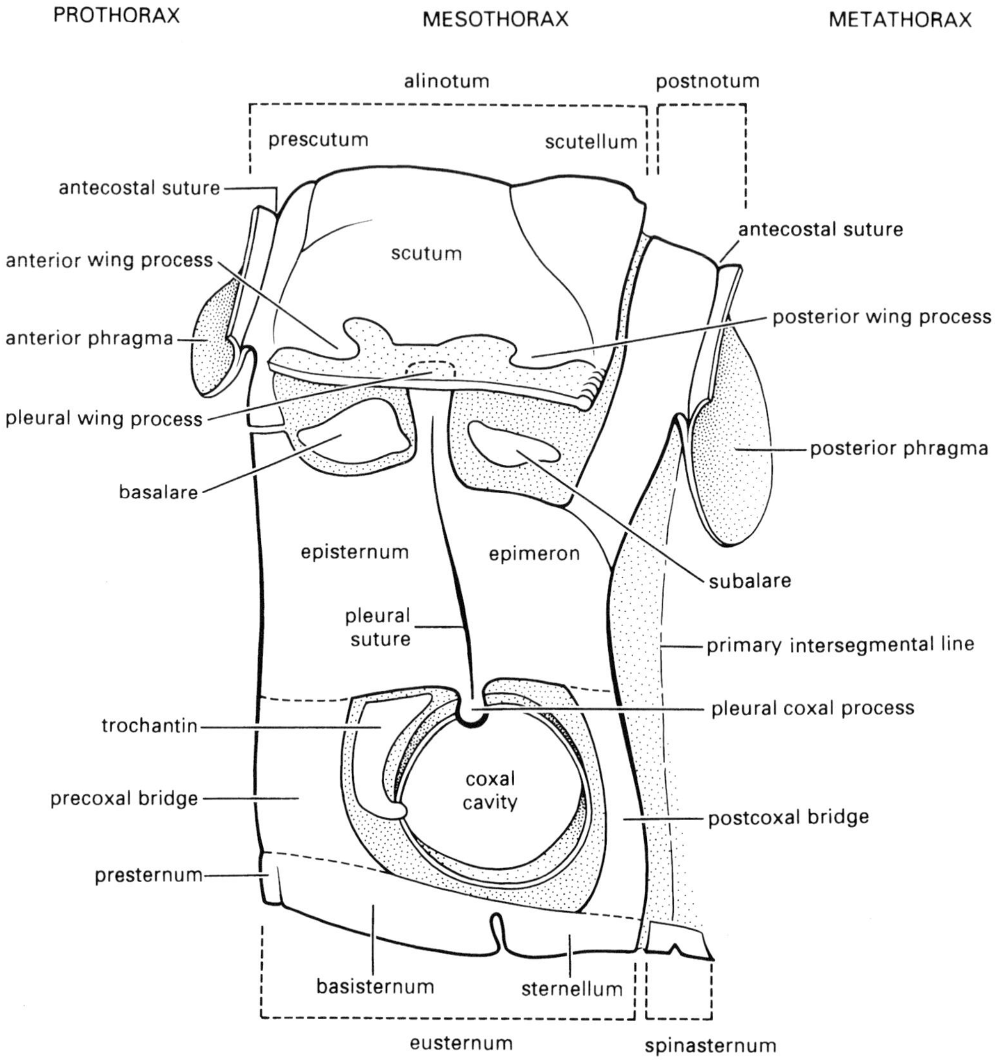

Fig. 2.20 Diagrammatic lateral view of a wing-bearing thoracic segment, showing the typical sclerites and their subdivisions. (After Snodgrass 1935.)

mantids, many bugs and some Orthoptera, the pronotum is expanded, and in cockroaches it forms a shield that covers part of the head and mesothorax. The pterothoracic nota each have two main divisions – the anterior wing-bearing **alinotum** and the posterior phragma-bearing **postnotum** (Fig. 2.20). Phragmata (singular: **phragma**) are plate-like apodemes that extend inwards below the **antecostal sutures**, marking the primary intersegmental folds between segments; phragmata provide attachment for the longitudinal flight muscles (Fig. 2.7d). Each alinotum (sometimes confusingly referred to as a "notum") may be traversed by sutures that mark the position of internal strengthening ridges and commonly divide the plate into three areas – the anterior **prescutum**, the **scutum**, and the smaller posterior **scutellum**.

The lateral pleural sclerites may be derived from the subcoxal segment of the ancestral insect leg (see Fig. 8.5a). These sclerites can be separate, as in silverfish, or fused into an almost continuous sclerotic area, as in most winged insects. In the pterothorax, the pleuron is divided into two main areas – the anterior **episternum** and the posterior **epimeron** – by an internal **pleural ridge**, which is visible externally as the **pleural suture** (Fig. 2.20). The pleural ridge runs from the **pleural coxal process** (which articulates with the coxa) to the **pleural wing process** (which articulates with the wing), providing reinforcement for these articulation points. The **epipleurites** are small sclerites beneath the wing and consist of the **basalaria** anterior to the pleural wing process and the posterior **subalaria**, but are often reduced to just one basalare and one subalare, which serve as attachment points for some direct flight muscles. The **trochantin** is the small sclerite anterior to the coxa.

The degree of ventral sclerotization on the thorax varies greatly in different insects. Sternal plates, if present, are typically two per segment: the **eusternum**, and the following intersegmental sclerite or **intersternite** (Fig. 2.7c), commonly called the **spinasternum** (Fig. 2.20) because it usually has an internal apodeme called the **spina** (except for the metasternum, which never has a spinasternum). The eusterna of the prothorax and mesothorax may fuse with the spinasterna of their segment. Each eusternum may be simple or divided into separate sclerites – typically the **presternum**, **basisternum** and **sternellum**. The eusternum may be fused laterally with one of the pleural sclerites and is then called the **laterosternite**. Fusion of the sternal and pleural plates may form **precoxal** and **postcoxal bridges** (Fig. 2.20).

2.4.1 Legs

In most adult and nymphal insects, segmented **fore**, mid and **hind** legs occur on the prothorax, mesothorax and metathorax, respectively. Typically, each leg has six segments (Fig. 2.21) and these are, from proximal to distal: **coxa**, **trochanter**, **femur**, **tibia**, **tarsus** and **pretarsus** (or more correctly **post-tarsus**) with claws. Additional segments – the prefemur, patella and basitarsus (see Fig. 8.5a) – are recognized in some fossil insects and other arthropods, such as arachnids, and one or more of these segments are evident in some Ephemeroptera and Odonata. Primitively, two further segments lie proximal to the coxa, and in extant insects one of these, the epicoxa, is associated with the wing articulation or the tergum, and the other, the subcoxa, with the pleuron (Fig. 8.5a).

The tarsus is subdivided into five or fewer components, giving the impression of segmentation; but, because there is only one tarsal muscle, **tarsomere**

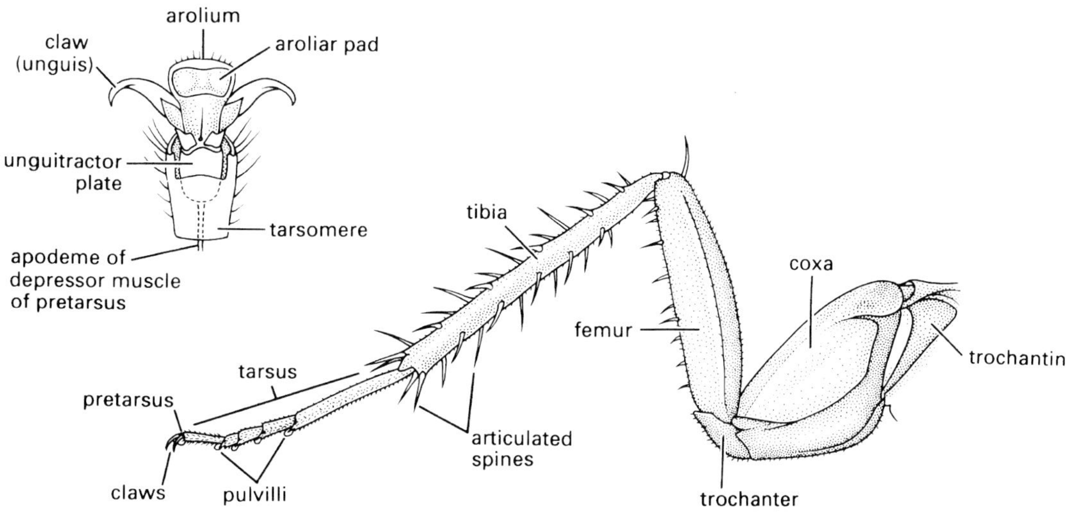

Fig. 2.21 The hind leg of a cockroach, *Periplaneta americana* (Blattodea: Blattidae), with enlargement showing ventral surface of pretarsus and last tarsomere. (After Cornwell 1968; enlargement after Snodgrass 1935.)

is a more appropriate term for each "pseudosegment". The first tarsomere sometimes is called the basitarsus, but should not be confused with the segment called the basitarsus in certain fossil insects. The underside of the tarsomeres may have ventral pads, **pulvilli**, also called **euplantulae**, which assist in adhesion to surfaces. The surface of each pad is either setose (hairy) or smooth. Terminally on the leg, the small pretarsus (shown in the enlargement in Fig. 2.21) bears a pair of lateral **claws** (also called **ungues**) and usually a median lobe, the **arolium**. In Diptera, there may be a central spine-like or pad-like **empodium** (plural: empodia), which is not the same as the arolium, and a pair of lateral pulvilli (as shown for the bush fly, *Musca vetustissima*, depicted on the right side of the vignette at the beginning of this chapter). These structures allow flies to walk on walls and ceilings. The pretarsus of Hemiptera may bear a variety of structures, some of which appear to be pulvilli, whereas others have been called empodia or arolia, but the homologies are uncertain. In some beetles, such as Coccinellidae, Chrysomelidae and Curculionidae, the ventral surface of some tarsomeres is clothed with adhesive setae that facilitate climbing. The left side of the vignette at the beginning of this chapter shows the underside of the tarsus of the leaf beetle *Rhyparida* (Chrysomelidae).

Generally, the femur and tibia are the longest leg segments, but variations in the lengths and robustness of each segment relate to their functions. For example, walking (**gressorial**) and running (**cursorial**) insects usually have well-developed femora and tibiae on all legs, whereas jumping (**saltatorial**) insects such as grasshoppers have disproportionately developed hind femora and tibiae. In aquatic beetles (Coleoptera) and bugs (Hemiptera), the tibiae and/or tarsi of one or more pairs of legs usually are modified for swimming (**natatorial**), with fringes of long, slender hairs. Many ground-dwelling insects, such as mole crickets (Orthoptera: Gryllotalpidae), nymphal cicadas (Hemiptera: Cicadidae), and scarab beetles (Scarabaeidae), have the tibiae of the fore legs enlarged and modified for digging (**fossorial**) (see Fig. 9.2), whereas the fore legs of some predatory insects, such as mantispid lacewings (Neuroptera) and mantids (Mantodea), are specialized for seizing prey (**raptorial**) (see Fig. 13.3). The tibia and basal tarsomere of each hind leg of honey bees are modified for the collection and carriage of pollen (see Fig. 12.4).

These "typical" thoracic legs are a distinctive feature of insects, whereas abdominal legs are confined to the immature stages of holometabolous insects. There have been conflicting views on whether the thoracic legs on the thorax of immature stages of the Holometabola are: (i) developmentally identical (serially homologous) to those of the abdomen; and/or (ii) homologous with those of the adult. Detailed study of musculature and innervation shows similarity of development of thoracic legs throughout all stages of insects with ametaboly (without metamorphosis, as in silverfish) and hemimetaboly (partial metamorphosis and no pupal stage) and in adult Holometabola, having identical innervation through the lateral nerves. Moreover, the oldest known larva (from the Upper Carboniferous) has thoracic and abdominal legs/leglets, each with a pair of claws, as in the legs of nymphs and adults. Although larval legs appear similar to those of adults and nymphs, the term **proleg** is used for the larval leg. Prolegs on the abdomen, especially on caterpillars, usually are lobe-like and each bears an apical circle or band of small sclerotized hooks, or **crochets**. The thoracic prolegs may possess the same number of segments as the adult leg, but the number is more often reduced, apparently through fusion. In other cases, the thoracic prolegs, like those of the abdomen, are unsegmented outgrowths of the body wall, often bearing apical hooks.

2.4.2 Wings

Wings are developed fully only in the adult, or exceptionally in the subimago, the penultimate stage of Ephemeroptera. Typically, functional wings are flap-like cuticular projections supported by tubular, sclerotized **veins**. The major veins are longitudinal, running from the wing base towards the tip, and are more concentrated at the anterior margin. Additional supporting **cross-veins** are transverse struts, which join the longitudinal veins to give a more complex structure. The major veins usually contain tracheae, blood vessels and nerve fibres, with the intervening membranous areas comprising the closely appressed dorsal and ventral cuticular surfaces. Generally, the major veins are alternately "convex" and "concave" in relation to the surface plane of the wing, especially near the wing attachment; this configuration is described by plus (+) and minus (−) signs. Most veins lie in an anterior area of the wing called the **remigium** (Fig. 2.22), which, powered by the thoracic flight muscles, is responsible for most of the movements of flight.

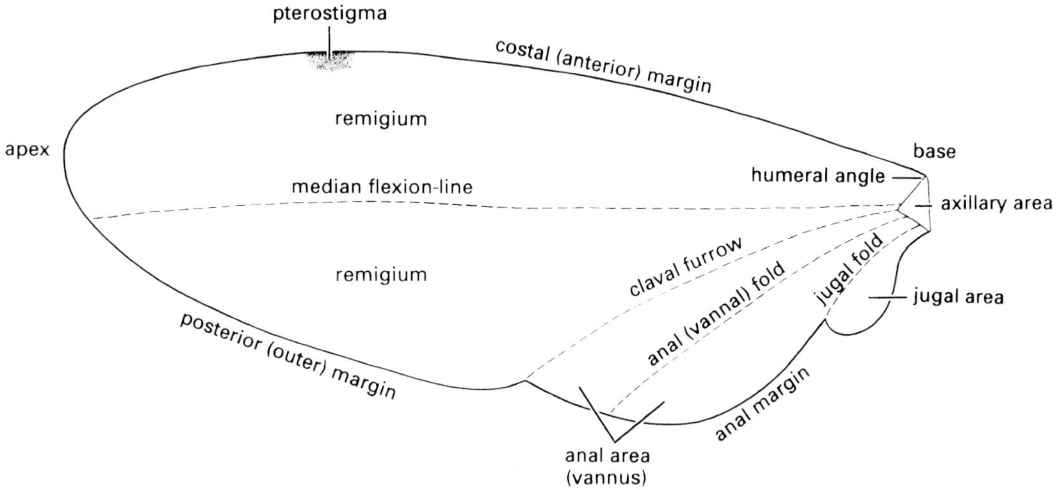

Fig. 2.22 Nomenclature for the main areas, folds and margins of a generalized insect wing.

The area of wing posterior to the remigium is sometimes called the **clavus**, but more often two areas are recognized: an anterior **anal area** (or **vannus**) and a posterior **jugal area** (or **jugum**). Wing areas are delimited and subdivided by **fold-lines**, along which the wing can be folded, and **flexion-lines**, at which the wing flexes during flight. The fundamental distinction between these two types of lines is often blurred, as fold-lines may permit some flexion, and vice versa. The **claval furrow** (a flexion-line) and the **jugal fold** (or fold-line) are nearly constant in position in different insect groups, but the **median flexion-line** and the **anal** (or vannal) **fold** (or fold-line) form variable and unsatisfactory area boundaries. Wing folding may be very complicated; transverse folding occurs in the hind wings of Coleoptera and Dermaptera, and in some insects the enlarged anal area may be folded like a fan.

The fore and hind wings of insects in many orders are coupled together, which improves the aerodynamic efficiency of flight. The commonest coupling mechanism (seen clearly in Hymenoptera and some Trichoptera) is a row of small hooks, or **hamuli**, along the anterior margin of the hind wing that engages with a fold along the posterior margin of the fore wing (**hamulate coupling**). In some other insects (e.g. Mecoptera, Lepidoptera and some Trichoptera), a jugal lobe of the fore wing overlaps the anterior hind wing (**jugate coupling**), or the margins of the fore and hind wing overlap broadly (**amplexiform coupling**), or one or more hind-wing bristles (the **frenulum**) hook under a retaining structure (the **retinaculum**) on the fore wing (**frenate coupling**). The mechanics of flight are described in section 3.1.4 and the evolution of wings is covered in section 8.4.

All winged insects share the same basic wing venation, comprising eight veins, named from anterior to posterior of the wing as: **precosta** (PC), **costa** (C), **subcosta** (Sc), **radius** (R), **media** (M), **cubitus** (Cu), **anal** (A) and **jugal** (J). Primitively, each vein has an anterior convex (+) **sector** (a branch with all of its subdivisions) and a posterior concave (−) sector. In almost all extant insects, the precosta is fused with the costa and the jugal vein is rarely apparent. The wing nomenclatural system presented in Fig. 2.23 is that of Kukalová-Peck and is based on detailed comparative studies of fossil and living insects. This system can be applied to the venation of all insect orders, although as yet it has not been widely applied because the various schemes devised for each insect order have a long history of use and there is a reluctance to discard familiar systems. Thus, in most textbooks, the same vein may be referred to by different names in different insect orders because the structural homologies were not recognized correctly in early studies. For example, until 1991, the venational scheme for Coleoptera labelled the radius posterior (RP) as the media (M) and the media posterior (MP) as the cubitus (Cu). Correct interpretation of venational homologies is essential for phylogenetic studies, and

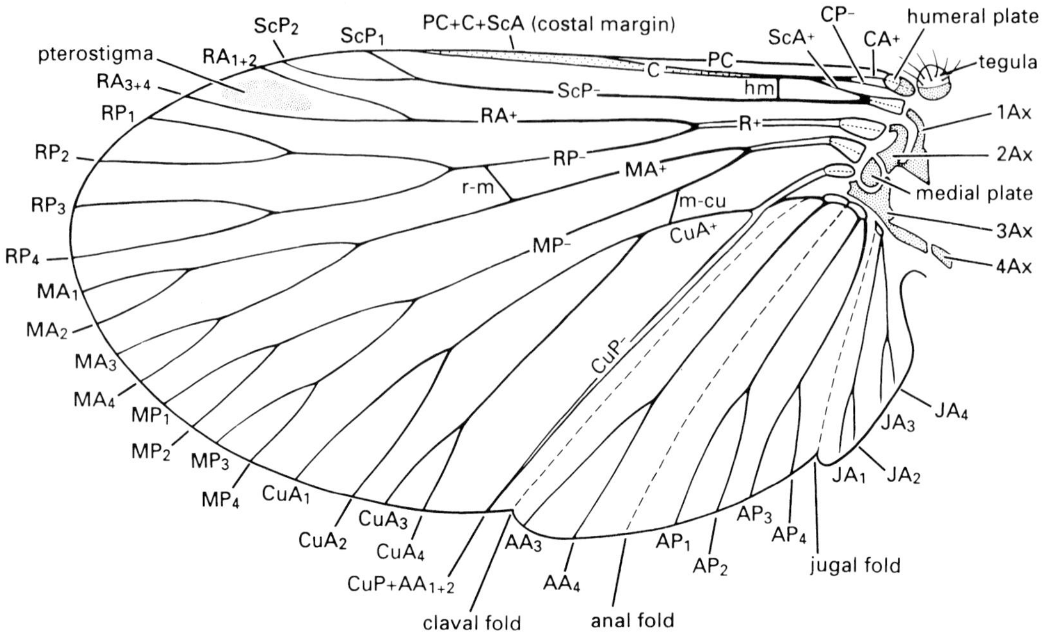

Fig. 2.23 A generalized wing of a neopteran insect (any living, winged insect, other than Ephemeroptera and Odonata), showing the articulation and the Kukalová-Peck nomenclatural scheme of wing venation. Notation is as follows: AA, anal anterior; AP, anal posterior; Ax, axillary sclerite; C, costa; CA, costa anterior; CP, costa posterior; CuA, cubitus anterior; CuP, cubitus posterior; hm, humeral vein; JA, jugal anterior; MA, media anterior; m-cu, cross-vein between medial and cubital areas; MP, media posterior; PC, precosta; R, radius; RA, radius anterior; r-m, cross-vein between radial and median areas; RP, radius posterior; ScA, subcosta anterior; ScP, subcosta posterior. Branches of the anterior and posterior sector of each vein are numbered, e.g. CuA_{1-4}. (After CSIRO 1991.)

the establishment of a single, universally applied scheme is essential.

Cells are areas of the wing delimited by veins, and may be **open** (extending to the wing margin) or **closed** (surrounded by veins). They are named usually according to the longitudinal veins or vein branches that they lie behind, except that certain cells are known by special names, such as the discal cell in Lepidoptera (Fig. 2.24a) and the triangle in Odonata (Fig. 2.24b). The **pterostigma** is an opaque or pigmented spot anteriorly near the apex of the wing (Fig. 2.22, Fig. 2.23 and Fig. 2.24b).

Wing venation patterns are consistent within groups (especially families and orders) but often differ between groups and, together with folds or pleats, provide major features used in insect classification and identification. Relative to the basic scheme outlined above, venation may be greatly reduced by loss or postulated fusion of veins, or increased in complexity by numerous cross-veins or substantial terminal branching. Other features that may be diagnostic of the wings of different insect groups are pigment patterns and colours, hairs and scales. Scales occur on the wings of Lepidoptera, many Trichoptera, and a few psocids (Psocodea) and flies, and may be highly coloured and have various functions, including waterproofing. Shedding of scales can allow escape from predators. Hairs consist of small microtrichia, either scattered or grouped, and larger macrotrichia, typically on the veins.

Usually, two pairs of functional wings lie dorsolaterally as **fore wings** on the mesothorax and as **hind wings** on the metathorax; typically the wings are membranous and transparent. However, from this basic pattern are derived many other conditions, often involving variation in the relative size, shape and degree of sclerotization of the fore and hind wings. Examples of fore-wing modification include: the thickened, leathery fore wings of Blattodea, Dermaptera

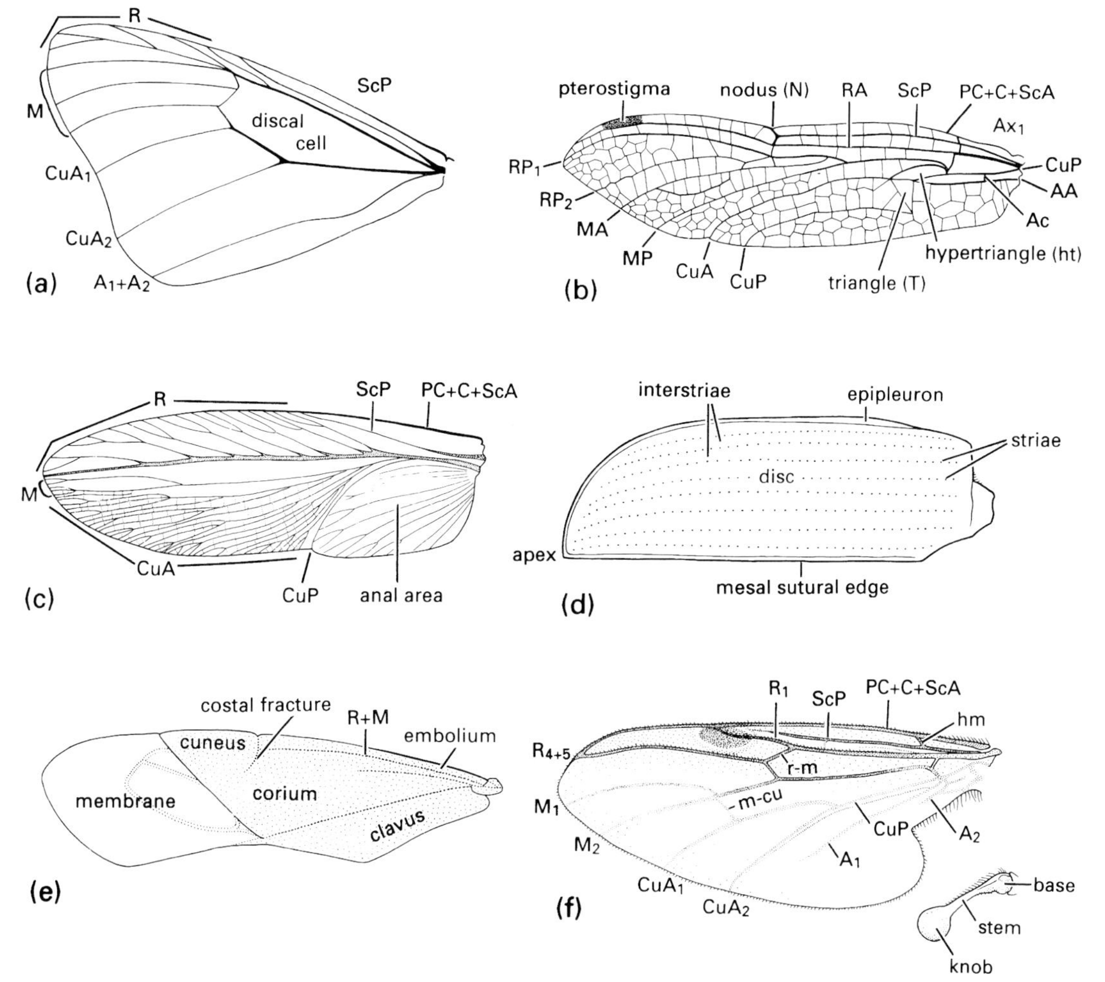

Fig. 2.24 The left wings of a range of insects showing some of the major wing modifications: (a) fore wing of a butterfly of *Danaus* (Lepidoptera: Nymphalidae); (b) fore wing of a dragonfly of *Urothemis* (Odonata: Anisoptera: Libellulidae); (c) fore wing or tegmen of a cockroach of *Periplaneta* (Blattodea: Blattidae); (d) fore wing or elytron of a beetle of *Anomala* (Coleoptera: Scarabaeidae); (e) fore wing or hemelytron of a mirid bug (Hemiptera: Heteroptera: Miridae), showing three wing areas – the membrane, corium and clavus; (f) fore wing and haltere of a fly of *Bibio* (Diptera: Bibionidae). Nomenclatural scheme of venation consistent with that depicted in Fig. 2.23; that of (b) after J.W.H. Trueman, unpublished. ((a–d) After Youdeowei 1977; (f) after McAlpine 1981.)

and Orthoptera, which are called tegmina (singular: **tegmen**; Fig. 2.24c); the hardened fore wings of Coleoptera that form protective wing cases or elytra (singular: **elytron**; Fig. 2.24d); and the hemelytra (singular: **hemelytron**) of heteropteran Hemiptera, with the basal part thickened and the apical part membranous (Fig. 2.24e). Typically, the heteropteran hemelytron is divided into three wing areas: the membrane, **corium** and **clavus**. Sometimes the corium is divided further, with the **embolium** anterior to R+M, and the **cuneus** distal to a **costal fracture**. In Diptera the hind wings are modified as stabilizers (**halteres**) (Fig. 2.24f) and do not function as wings, whereas in male Strepsiptera the fore wings form halteres and the hind wings are used in flight (Taxobox 23). In male scale insects, the fore wings have highly reduced venation and the hind wings form hamulohalteres (different in structure to the halteres) or are lost completely.

Small insects confront different aerodynamic challenges compared with larger insects and their wing area often is expanded to aid wind dispersal. Thrips (Thysanoptera), for example, have very slender wings but have a fringe of long setae or cilia to extend the wing area (Taxobox 19). In termites (Blattodea: Termitoidae) and ants (Hymenoptera: Formicidae), the winged reproductives, or alates, have large **deciduous** wings that are shed after the nuptial flight. Some insects are wingless, or **apterous**, either primitively as in silverfish (Zygentoma) and bristletails (Archaeognatha), which diverged from other insect lineages prior to the origin of wings, or secondarily as in all parasitic lice (Psocodea) and fleas (Siphonaptera), which evolved from winged ancestors. Secondary partial wing reduction occurs in a number of short-winged, or **brachypterous**, insects.

In all winged insects (Pterygota), a triangular area at the wing base, the **axillary area** (Fig. 2.22), contains the movable **articular sclerites** via which the wing articulates on the thorax. These sclerites are derived, by reduction and fusion, from a band of articular sclerites in the ancestral wing. Three different types of wing articulation among living Pterygota result from unique patterns of fusion and reduction of the articular sclerites. In Neoptera (all living winged insects except the Ephemeroptera and Odonata), the articular sclerites consist of the **humeral plate**, the **tegula**, and usually three, rarely four, **axillary sclerites** (1Ax, 2Ax, 3Ax and 4Ax) (Fig. 2.23). The Ephemeroptera and Odonata each has a different configuration of these sclerites compared with the Neoptera (literally meaning "new wing"). Odonate and ephemeropteran adults cannot fold their wings back along the abdomen, as can neopterans. In Neoptera, the wing articulates via the articular sclerites with the anterior and posterior wing processes dorsally, and ventrally with the pleural wing processes and two small pleural sclerites (the basalare and subalare) (Fig. 2.20).

2.5 THE ABDOMEN

Primitively, the insect abdomen is 11-segmented, although segment 1 may be reduced or incorporated into the thorax (as in many Hymenoptera) and the terminal segments usually are variously modified and/or diminished (Fig. 2.25a). Generally, at least the first seven abdominal segments of adults (the **pregenital segments**) are similar in structure and lack appendages. However, apterygotes (bristletails and silverfish) and many immature aquatic insects have abdominal appendages. Apterygotes possess a pair of **styles** – rudimentary appendages that are serially homologous with the distal part of the thoracic legs – and, mesally, one or two pairs of **protrusible** (or exsertile) **vesicles** on at least some abdominal segments. These vesicles are derived from the coxal and trochanteral **endites** (inner annulated lobes) of the ancestral abdominal appendages (see Fig. 8.5b). Aquatic larvae and nymphs may have gills laterally on some to most abdominal segments (Chapter 10). Some of these may be serially homologous with thoracic wings (e.g. the plate gills of mayfly nymphs) or with other leg derivatives. Spiracles typically are present on segments 1–8, but reductions in number occur frequently in association with modifications of the tracheal system (section 3.5), especially in immature insects, and with specializations of the terminal segments in adults.

2.5.1 Terminalia

The anal-genital part of the abdomen, known as the **terminalia**, consists generally of segments 8 or 9 to the abdominal apex. Segments 8 and 9 bear the genitalia; segment 10 is visible as a complete segment in many "lower" insects but always lacks appendages; and the small segment 11 is represented by a dorsal epiproct and pair of ventral paraprocts derived from the sternum (Fig. 2.25b). A pair of appendages, the cerci (singular: **cercus**), articulates laterally on segment 11; typically these are annulated and filamentous, but have been modified (e.g. the forceps of earwigs) or reduced in different insect orders. An annulated caudal filament, the median **appendix dorsalis**, arises from the tip of the epiproct in apterygotes, most mayflies (Ephemeroptera) and a few fossil Insects. A similar structure in nymphal stoneflies (Plecoptera) is of uncertain homology. These terminal abdominal segments have excretory and sensory functions in all insects, but in adults there is an additional reproductive function.

The organs concerned specifically with mating and the deposition of eggs are known collectively as the **external genitalia**, although they may be largely internal. The components of the external genitalia of

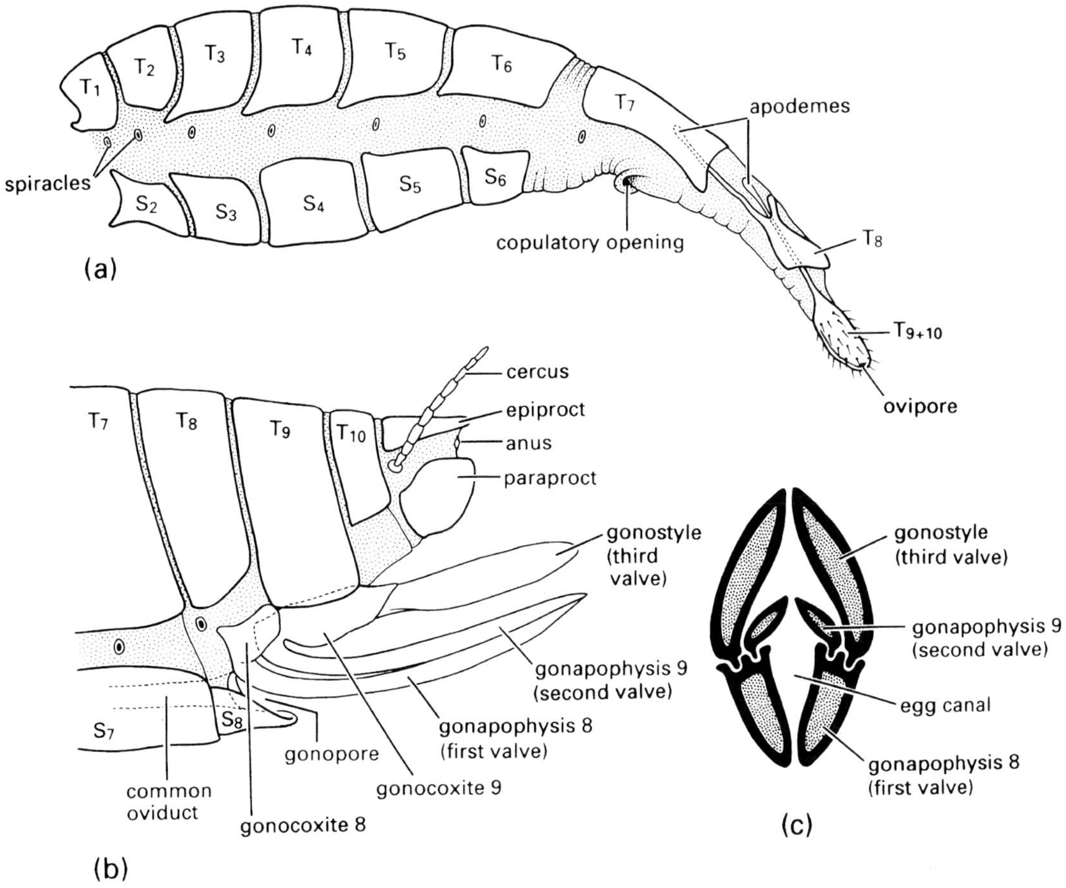

Fig. 2.25 The female abdomen and ovipositor. (a) Lateral view of the abdomen of an adult tussock moth (Lepidoptera: Lymantriidae), showing the substitutional ovipositor formed from the extensible terminal segments. (b) Lateral view of a generalized orthopteroid ovipositor composed of appendages of segments 8 and 9. (c) Transverse section through the ovipositor of a katydid (Orthoptera: Tettigoniidae). $T_1 - T_{10}$, terga of first to tenth segments; $S_2 - S_8$, sterna of second to eighth segments. ((a) After Eidmann 1929; (b) after Snodgrass 1935; (c) after Richards & Davies 1959.)

insects are very diverse in form and often have considerable taxonomic value, particularly amongst species that appear structurally similar in other respects. The male external genitalia have been used widely to aid in distinguishing species, whereas the female external genitalia may be simpler and less varied. The diversity and species-specificity of genitalic structures are discussed in section 5.5.

The terminalia of adult female insects include internal structures for receiving the male copulatory organ and his spermatozoa (sections 5.4 and 5.6) and external structures used for oviposition (egg-laying; section 5.8). Most female insects have an egg-laying tube, or **ovipositor**; it is absent in termites (Temitoidae), parasitic lice (Psocodea), female scale insects (Coccoidea), many Plecoptera and most Ephemeroptera. Ovipositors take two forms:

1 true, or **appendicular ovipositor**, formed from appendages of abdominal segments 8 and 9 (Fig. 2.25b);

2 **substitutional ovipositor**, composed of extensible posterior abdominal segments (Fig. 2.25a).

Substitutional ovipositors include a variable number of the terminal segments and clearly have been

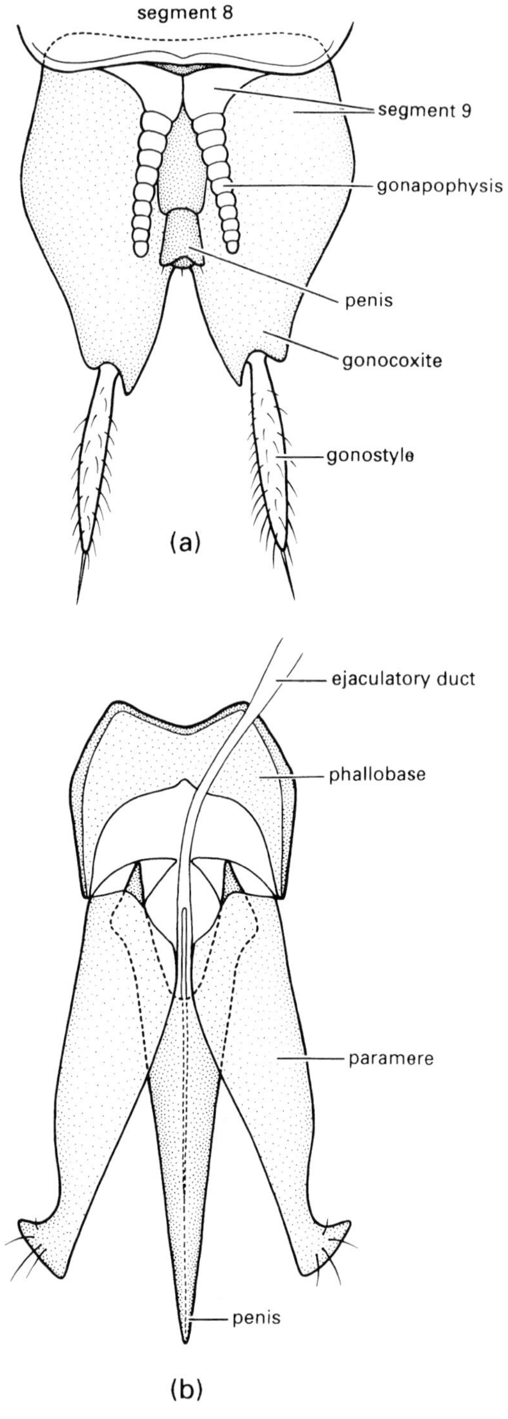

Fig. 2.26 (*left*) Male external genitalia. (a) Abdominal segment 9 of the bristletail *Machilis variabilis* (Archaeognatha: Machilidae). (b) Aedeagus of a click beetle (Coleoptera: Elateridae). ((a) After Snodgrass 1957.)

derived convergently several times, even within some orders. They occur in many insects, including most Lepidoptera, Coleoptera and Diptera. In these insects, the terminalia are telescopic and can be extended as a slender tube, manipulated by muscles attached to apodemes of the modified terga (Fig. 2.25a) and/or sterna.

Appendicular ovipositors represent the primitive condition for female insects and are present in Archaeognatha, Zygentoma, many Odonata, Orthoptera, some Hemiptera, some Thysanoptera and Hymenoptera. In some Hymenoptera, the ovipositor is modified as a poison-injecting sting (see Fig. 14.11) and the eggs are ejected at the base of the sting. In all other cases, the eggs pass down a canal in the shaft of the ovipositor (section 5.8). The shaft is composed of three pairs of **valves** (Fig. 2.25b,c) supported on two pairs of **valvifers** – the coxae plus trochanters, or **gonocoxites**, of segments 8 and 9 (Fig. 2.25b). The gonocoxites of segment 8 have a pair of trochanteral endites (inner lobe from each trochanter), or **gonapophyses**, which form the first valves, whereas the gonocoxites of segment 9 have a pair of gonapophyses (the second valves) plus a pair of **gonostyles** (the third valves) derived from the distal part of the appendages of segment 9 (and homologous with the styles of the apterygotes mentioned above). In each half of the ovipositor, the second valve slides in a tongue-and-groove fashion against the first valve (Fig. 2.25c), whereas the third valve generally forms a sheath for the other valves.

The external genitalia of male insects include an organ for transferring the spermatozoa (either packaged in a **spermatophore** or free in fluid) to the female, and often involve structures that grasp and hold the partner during mating. Numerous terms are applied to the various components in different insect groups, and homologies are difficult to establish. Males of Archaeognatha, Zygentoma and Ephemeroptera have relatively simple genitalia, consisting of gonocoxites, gonostyles and sometimes gonapophyses on segment 9 (and also on segment 8 in Archaeognatha), as in the female, except with a median **penis** (**phallus**) or, if paired or bilobed, penes, on segment 9 (Fig. 2.26a). The penis (or penes) might be derived from the fused inner lobes (endites) of either the ancestral coxae or trochanters of segment 9. In the orthopteroid orders, the gonocoxites are reduced or absent, although gonostyles may be present (called styles), and there is a median penis with a lobe called a **phallomere** on each side of it. The evolutionary fate of the gonapophyses and the origin of the phallomeres are uncertain. In the "higher" insects – the hemipteroids and the holometabolous orders – the homologies and terminology of the male structures are even more confusing if one tries to compare the terminalia of different orders. The whole copulatory organ of higher insects generally is known as the **aedeagus** (edeagus) and, in addition to insemination, it may clasp and provide sensory stimulation to the female. Typically, there is a median tubular penis (sometimes the term "aedeagus" is restricted to this lobe), which often has an inner tube, the **endophallus**, which is everted during insemination (see Fig. 5.4b). The ejaculatory duct opens at the gonopore, either at the tip of the penis or the endophallus. Lateral to the penis is a pair of lobes or **parameres**, which may have a clasping and/or sensory function. Their origin is uncertain; they may be homologous with the gonocoxites and gonostyles of lower insects, with the phallomeres of orthopteroid insects, or may be derived *de novo*, perhaps even from segment 10. This trilobed type of aedeagus is well exemplified in many beetles (Fig. 2.26b), but modifications are too numerous to describe here.

Much variation in male external genitalia correlates with mating position, which is very variable between and sometimes within orders. Mating positions include end-to-end, side-by-side, male below with his dorsum up, male on top with female dorsum up, and even venter-to-venter. In some insects, torsion of the terminal segments may take place post-metamorphosis or just prior to or during copulation, and asymmetries of male clasping structures occur in many insects. Copulation and associated behaviours are discussed in more detail in Chapter 5.

FURTHER READING

There are many excellent, recent articles that take a more genetic or developmental approach to understanding insect structure. The following sources provide basic morphological and functional information.

Andersen, S.O. (2010) Insect cuticular sclerotization: a review. *Insect Biochemistry and Molecular Biology* **40**, 166–78.

Beutel, R.G., Friedrich, F., Ge, S.-Q. & Yang, Z.-K. (2014) *Insect Morphology and Phylogeny*. Walter De Gruyter Inc., Berlin.

Binnington, K. & Retnakaran, A. (eds.) (1991) *Physiology of the Insect Epidermis*. CSIRO Publications, Melbourne.

Chapman, R.F. (2013) *The Insects. Structure and Function*, 5th edn. (eds. S.J. Simpson & A.E. Douglas), Cambridge University Press, Cambridge.

Krenn, H.W. & Aspöck, H. (2012) Form, function and evolution of the mouthparts of blood-feeding Arthropoda. *Arthropod Structure and Development* **41**, 101–18.

Krenn, H.W., Plant, J.D. & Szucsich, N.U. (2005) Mouthparts of flower-visiting insects. *Arthropod Structure and Development* **34**, 1–40.

Lawrence, J.F., Nielsen, E.S. & Mackerras, I.M. (1991) Skeletal anatomy and key to orders. In: *The Insects of Australia*, 2nd edn. (CSIRO), pp. 3–32. Melbourne University Press, Carlton.

Nichols, S.W. (1989) *The Torre-Bueno Glossary of Entomology*, 2nd edn. The New York Entomological Society in co-operation with the American Museum of Natural History, New York.

Resh, V.H. & Cardé, R.T. (eds.) (2009) *Encyclopedia of Insects*, 2nd edn. Elsevier, San Diego, CA. [Particularly see articles on anatomy; head; thorax; abdomen and genitalia; mouthparts; wings.]

Richards, A.G. & Richards, P.A. (1979) The cuticular protuberances of insects. *International Journal of Insect Morphology and Embryology* **8**, 143–57.

Snodgrass, R.E. (1935) *Principles of Insect Morphology*. McGraw-Hill, New York. [This classic book remains a valuable reference, despite its age.]

Wootton, R.J. (1992) Functional morphology of insect wings. *Annual Review of Entomology* **37**, 113–40.

Chapter 3

INTERNAL ANATOMY AND PHYSIOLOGY

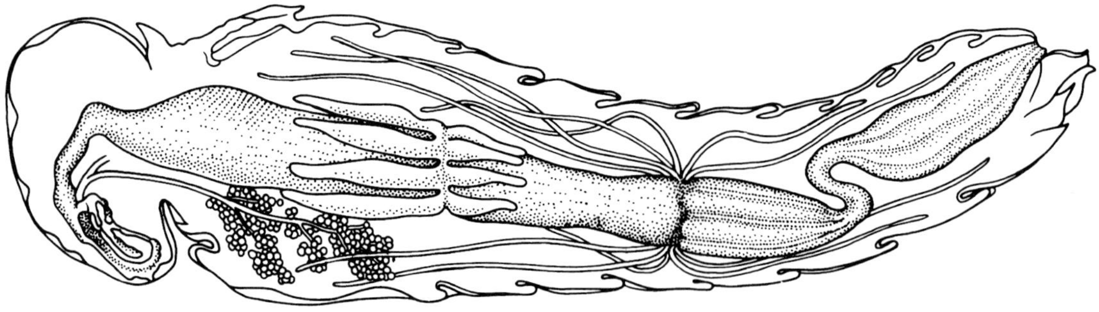

Internal structures of a locust. (After Uvarov 1966.)

The dissected open body of an insect is a complex and compact masterpiece of functional design. The "insides" of two omnivorous insects, a cockroach and a cricket, which have relatively unspecialized digestive and reproductive systems are shown in Fig. 3.1. The digestive system, which includes salivary glands as well as an elongate gut, consists of three main sections. These function in storage, biochemical breakdown, absorption and excretion. Each gut section has more than one physiological role and this may be reflected in local structural modifications, such as thickening of the gut wall or diverticula (extensions) from the main lumen. The reproductive systems depicted in Fig. 3.1 exemplify the female and male organs of many insects. These may be dominated in males by very visible accessory glands, especially as the testes of many adult insects are degenerate or absent. This is because the spermatozoa are produced in the pupal or penultimate stage and stored. In gravid female insects, the body cavity may be filled with eggs at various stages of development, thereby obscuring other internal organs. Likewise, the internal structures (except the gut) of a well-fed, late-stage caterpillar may be hidden within the mass of fat body tissue.

The insect body cavity, called the **haemocoel** (hemocoel) and filled with fluid **haemolymph** (hemolymph), is lined with endoderm and ectoderm. It is not a true coelom, which is defined as a mesoderm-lined cavity. Haemolymph (so-called because it combines many roles of vertebrate blood (haem/hem and lymph) bathes all internal organs, delivers nutrients, removes metabolites, and performs immune functions. Unlike vertebrate blood, haemolymph rarely has respiratory pigments and therefore has little or no role in gas exchange. In insects, this function is performed by the **tracheal system**, a ramification of air-filled tubes (**tracheae**), which sends fine branches throughout the body. Gas entry to, and exit from, tracheae is controlled by sphincter-like structures called **spiracles** that open through the body wall. Non-gaseous wastes are filtered from the haemolymph by filamentous **Malpighian tubules** (named after their discoverer), which have free ends distributed through the haemocoel. Their contents are emptied into the gut from which, after further modification, wastes are eliminated eventually via the anus.

All motor, sensory and physiological processes in insects are controlled by the nervous system in conjunction with hormones (chemical messengers). The brain and ventral nerve cord are readily visible in dissected insects, but most endocrine centres, neurosecretion sites, numerous nerve fibres, muscles and other tissues cannot be seen by the unaided eye.

This chapter describes insect internal structures and their functions. Topics covered are the muscles and locomotion (walking, swimming and flight), the nervous system and co-ordination, endocrine centres and hormones, the haemolymph and its circulation, the tracheal system and gas exchange, the gut and digestion, the fat body, nutrition and microorganisms, the excretory system and waste disposal, and lastly the reproductive organs and gametogenesis. Boxes deal with four special topics, namely neuropeptide research (Box 3.1), tracheal hypertrophy in mealworms at low oxygen concentrations (Box 3.2), the filter chamber of Hemiptera (Box 3.3), and insect cryptonephric systems (Box 3.4). A full account of insect physiology cannot be provided in one chapter. For more comprehensive overviews of specific topics, we direct readers to Chapman (2013) and to relevant chapters in the *Encyclopedia of Insects* (Resh & Cardé 2009). Most recent advances in insect physiology involve molecular biology; an overview of applications of molecular science to insects can be obtained from a series of volumes edited by Gilbert *et al.* (2005) and Gilbert (2012), and other sources listed in the Further reading section.

3.1 MUSCLES AND LOCOMOTION

As stated in section 1.3.4, much of the success of insects relates to their ability to sense, interpret and move around their environment. Although the origin of flight, at least 350 million years ago, was a major innovation, terrestrial and aquatic locomotion also is well developed. Power for movement originates from muscles operating against a skeletal system, either the rigid cuticular exoskeleton or, in soft-bodied larvae, a hydrostatic skeleton.

The Insects: An Outline of Entomology, Fifth Edition. P.J. Gullan and P.S. Cranston.

Companion Website: www.wiley.com/go/gullan/insects

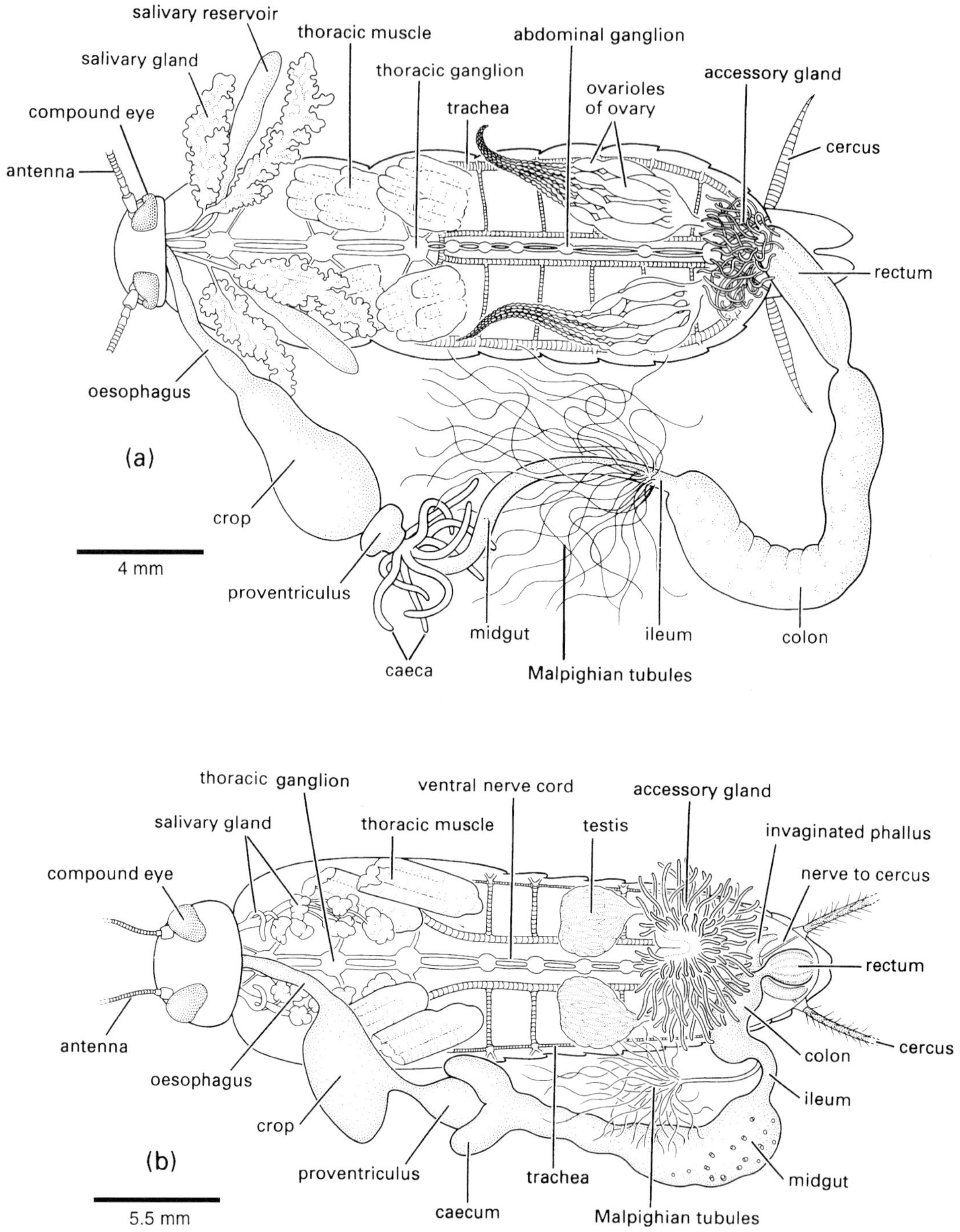

Fig. 3.1 Dissections of: (a) a female American cockroach, *Periplaneta americana* (Blattodea: Blattidae); and (b) a male black field cricket, *Teleogryllus commodus* (Orthoptera: Gryllidae). The fat body and most of the tracheae have been removed; most details of the nervous system are not shown.

3.1.1 Muscles

Vertebrates and many non-hexapod invertebrates have three kinds of muscles: striated, smooth and cardiac. Insects possess only **striated muscles**, so-called because of the overlapping thicker myosin and thinner actin filaments that give the appearance of cross-banding under a microscope. Each striated muscle fibre comprises many cells, with a common plasma membrane and **sarcolemma**, or outer sheath. The sarcolemma is invaginated, but not broken, where an oxygen-supplying tracheole (section 3.5, Fig. 3.10b) contacts the muscle fibre. Contractile **myofibrils** run the length of the fibre, arranged in sheets or cylinders. When viewed under high magnification, a myofibril can be seen to comprise a thin actin filament sandwiched between a pair of thicker myosin filaments. Muscle contraction involves the sliding of filaments past each other, stimulated by nerve impulses. Innervation comes from one to three motor axons per bundle of fibres, each separately tracheated and referred to as one muscle unit, with several units grouped in a functional muscle.

There are several different muscle types. The most important division is between those that respond synchronously, with one contraction cycle per impulse, and fibrillar muscles that contract asynchronously, with multiple contractions per impulse. Asynchronous muscles include some associated with flight (see below) and the tymbal of cicadas (section 4.1.4).

The basic mode of action does not differ between muscles of insects and vertebrates, although insects can produce prodigious muscular feats for their size, such as the leap of a flea or the repetitive stridulation of the cicada tympanum. Reduced body size benefits insects because of the relationship between: (i) power, which is proportional to muscle cross-section and decreases with reduction in size by the square root; and (ii) the body mass, which decreases with reduction in size by the cube root. Thus, the power : mass ratio increases as body size decreases.

3.1.2 Muscle attachments

The muscles of vertebrates work against an internal skeleton, but the muscles of insects must attach to the inner surface of an external skeleton. As musculature is mesodermal and the exoskeleton is of ectodermal origin, fusion must take place. This occurs by the growth of **tonofibrillae**, which are fine connecting fibrils that link the epidermal end of the muscle to the epidermal layer (Fig. 3.2a,b). At each moult, tonofibrillae are discarded along with the cuticle, and therefore must be regrown.

At the site of tonofibrillar attachment, the inner cuticle often is strengthened through ridges or **apodemes**, which, when elongated into arms, are termed **apophyses** (Fig. 3.2c). These muscle attachment sites, particularly the long, slender apodemes for

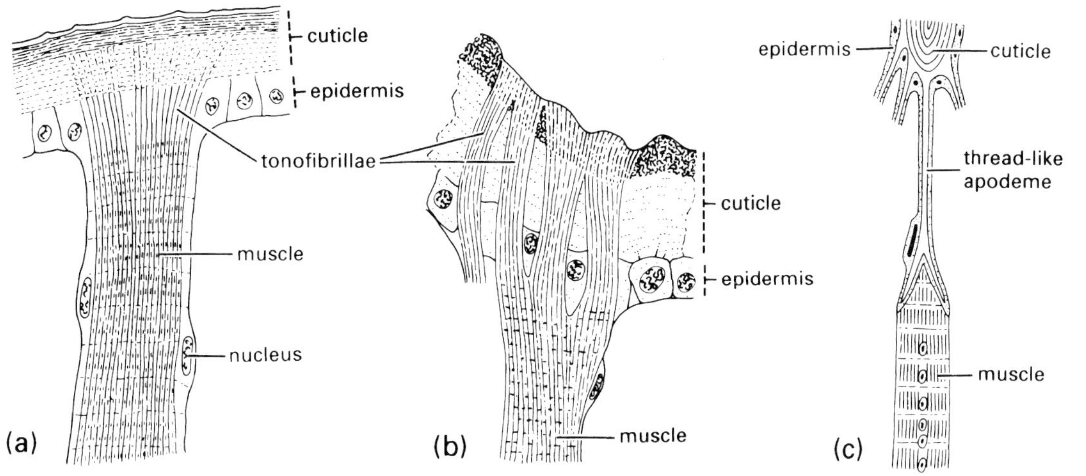

Fig. 3.2 Muscle attachments to body wall: (a) tonofibrillae traversing the epidermis from the muscle to the cuticle (b) a muscle attachment in an adult beetle of *Chrysobothrus femorata* (Coleoptera: Buprestidae) (c) a multicellular apodeme with a muscle attached to one of its thread-like, cuticular "tendons" or apophyses. (After Snodgrass 1935.)

individual muscle attachments, often include resilin to give an elasticity that resembles that of vertebrate tendons.

Some insects, including soft-bodied larvae, have mainly thin, flexible cuticle without the rigidity to anchor muscles unless given additional strength. The body contents form a **hydrostatic skeleton**, with turgidity maintained by criss-crossed body wall "turgor" muscles that continuously contract against the incompressible fluid of the haemocoel, giving a strengthened foundation for other muscles. If the larval body wall is perforated, the fluid leaks, the haemocoel becomes compressible, and the turgor muscles cause the larva to become flaccid.

3.1.3 Crawling, wriggling, swimming and walking

Soft-bodied larvae with hydrostatic skeletons move by crawling. Muscular contraction in one part of the body gives equivalent extension in a relaxed part elsewhere on the body. In apodous (legless) larvae, such as dipteran "maggots", waves of contractions and relaxation run from head to tail. Bands of adhesive hooks or tubercles successively grip and detach from the substrate to provide a forward motion, aided in some maggots by use of their mouth hooks to grip the substrate. In water, lateral waves of contraction against the hydrostatic skeleton can give a sinuous, snake-like, swimming motion, with anterior-to-posterior waves giving an undulating motion.

Larvae with thoracic legs and abdominal prolegs, like caterpillars, develop posterior-to-anterior waves of turgor muscle contraction, with as many as three waves visible simultaneously. Locomotor muscles operate in cycles of successive detachment of the thoracic legs, reaching forwards and grasping the substrate. These cycles occur in concert with inflation, deflation and forward movement of the posterior prolegs.

Insects with hard exoskeletons can contract and relax pairs of agonistic and antagonistic muscles that attach to the cuticle. Compared to crustaceans and myriapods, insects have fewer legs (six) and these are located more ventrally and closer together on the thorax, allowing concentration of locomotor muscles (both flying and walking) into the thorax, and providing more control and greater efficiency. Motion with six legs at low to moderate speed allows continuous contact with the ground by a tripod of fore and hind legs on one side and mid leg on the opposite side thrusting rearwards (retraction), whilst each opposite leg is moved forwards (protraction) (Fig. 3.3). The centre of gravity of the slow-moving insect always lies within this tripod, giving great stability. Motion is imparted through thoracic muscles acting on the leg bases, with transmission via internal leg muscles through the leg to extend or flex the leg. Anchorage to the substrate, needed to provide a lever to propel the body, is through pointed claws and adhesive pads (the arolium or, in flies and some beetles, pulvilli). Claws, such as those illustrated in the vignette at the beginning of Chapter 2, can obtain purchase on the slightest roughness in a surface, and the pads of some insects can adhere to perfectly smooth surfaces through the application of lubricants to the tips of numerous fine hairs and by the action of close-range molecular forces between the hairs and the substrate.

When faster motion is required, there are several alternatives: increasing the frequency of the leg movement by shortening the retraction period; increasing the stride length; altering the triangulation basis of support to adopt quadrupedy (use of four legs); or even hind-leg bipedality, with the other legs held above the substrate. At high speeds, even those insects that maintain triangulation are very unstable and may have no legs in contact with the substrate at intervals. This instability at speed seems to cause no difficulty for cockroaches, which when filmed with high-speed video cameras have been shown to maintain speeds of up to $1\,\mathrm{m\,s^{-1}}$ whilst twisting and turning up to 25 times per second. This motion was maintained by sensory information received from one antenna whose tip maintained contact with an experimentally provided wall, even when it had a zig-zagging surface.

Many insects jump, some prodigiously, usually using modified hind legs. In orthopterans, flea beetles (Chrysomelidae: Alticini) and a range of weevils (Curculionidae), an enlarged hind femur contains large muscles whose slow contraction produces energy that is stored by either distortion of the femoro-tibial joint or in some spring-like sclerotization, for example the hind-leg tibial extension tendon. In fleas, the energy is produced by the trochanter levator muscle raising the femur, and is stored by compression of an elastic resilin pad in the coxa. In all these jumpers, release of tension is sudden, resulting in propulsion of the insect into the air – usually in an uncontrolled manner, although fleas can attain their hosts with some control over the leap. The main benefit for

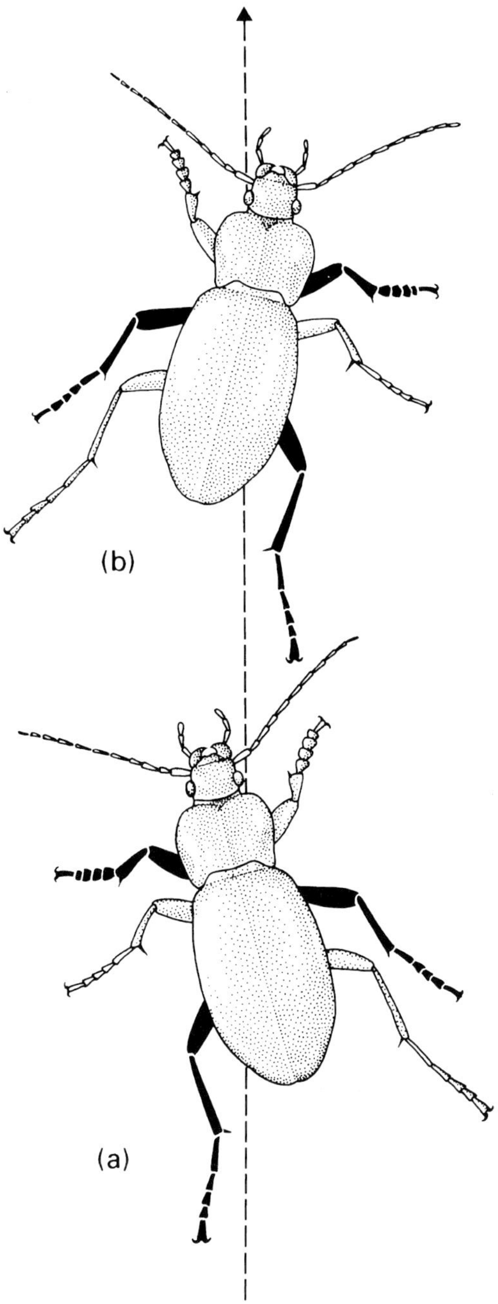

Fig. 3.3 A ground beetle (Coleoptera: Carabidae: *Carabus*) walking in the direction of the dashed line. The three blackened legs are those in contact with the ground in the two positions illustrated – (a) is followed by (b). (After Wigglesworth 1972.)

flighted jumping insects may be to get into the air before opening the wings, so as to avoid damage from the surrounding substrate.

In swimming, contact with the water is maintained during protraction, so the insect must impart more thrust to the rowing motion than to the recovery stroke in order to progress. This is achieved by expanding the effective leg area during retraction by extending fringes of hairs and spines (see Fig. 10.7) that collapse onto the folded leg during the recovery stroke. We have seen already how some insect larvae swim using contractions against their hydrostatic skeleton. Others, including many nymphs and the larvae of caddisflies, can walk underwater and, particularly in running waters, do not swim routinely.

The surface film of water can support some specialist insects, most of which have hydrofuge (water-repelling) cuticles or hair fringes, and some, such as gerrid water-striders (see Fig. 5.7), move by rowing with hair-fringed legs.

3.1.4 Flight

The development of flight allowed insects much greater mobility, which helped in food and mate location and provided much improved powers of dispersal. Importantly, flight opened up many new environments for exploitation. Plant microhabitats such as flowers and foliage are more easily accessed by winged insects than by those without flight.

Fully developed, functional, flying wings occur only in adult insects, although in nymphs the developing wings are visible as wing buds in all but the earliest instars. Usually, two pairs of functional wings arise dorsolaterally, as fore wings on the second and hind wings on the third thoracic segment. Some wing variations are described in section 2.4.2.

To fly, the forces of weight (gravity) and drag (air resistance to movement) must be overcome. In gliding flight, in which the wings are held rigidly outstretched, these forces are overcome through the use of passive air movements – known as the relative wind. The insect attains lift by adjusting the angle of the leading edge of the wing when orientated into the wind. As this angle (the attack angle) increases, so lift increases, until stalling occurs, i.e. when lift is catastrophically lost. In contrast to aircraft, nearly all of which stall at around 20°, the attack angle of insects can be raised to more than 30°, even to as high as 50°, giving great manoeuvrability. Enhanced lift and reduced drag can be attained with wing scales and hairs, which affect the boundary layer across the wing surface.

Most insects can glide, and dragonflies (Odonata) and some grasshoppers (Orthoptera), notably locusts, glide extensively. However, most winged insects fly by beating their wings. Examination of wing beat is difficult because the frequency of even a large, slow-flying butterfly may be five times a second (5 Hz), a bee may beat its wings at 180 Hz, and some midges emit an audible buzz with their wing-beat frequency of greater than 1000 Hz. However, through the use of slowed-down, high-speed cine film, the insect wing beat can be slowed down from faster than the eye can see until a single beat can be analysed. This reveals that a single beat comprises three interlinked movements. The first is a cycle of downward, forward motion followed by an upward and backward motion. Second, during the cycle each wing is rotated around its base. The third component occurs as various parts of the wing flex in response to local variations in air pressure. Unlike gliding, in which the relative wind derives from passive air movement, in true flight the relative wind is produced by the moving wings. The flying insect makes constant adjustments, so that during a wing beat, the air ahead of the insect is thrown backwards and downwards, impelling the insect upwards (lift) and forwards (thrust). In climbing, the emergent air is directed more downwards, reducing thrust but increasing lift. In turning, the wing on the inside of the turn is reduced in power by a decrease in the amplitude of the beat.

Despite the elegance and intricacy of detail of insect flight, the mechanisms responsible for beating the wings are straightforward. The thorax of the wing-bearing segments can be envisaged as a box with the sides (pleura) and base (sternum) rigidly fused, and the wings connected where the rigid tergum is attached to the pleura by flexible membranes. This membranous attachment and the wing hinge are composed of resilin (section 2.1), which gives crucial elasticity to the thoracic box. Flying insects have one of two kinds of arrangements of muscles powering their flight:

1 **direct flight muscles** connected to the wings, or
2 an **indirect system** in which there is no muscle-to-wing connection, but rather muscle action deforms the thoracic box in order to move the wings.

Some groups such as Odonata and Blattodea appear to use direct flight muscles to varying degrees, although at least some recovery muscles may be indirect. Others use indirect muscles for flight, with direct muscles providing wing orientation rather than power production.

Direct flight muscles produce the upward stroke by contraction of muscles attached to the wing base inside the pivotal point (Fig. 3.4a). The downward wing stroke is produced through contraction of muscles that extend from the sternum to the wing base outside the pivot point (Fig. 3.4b). In contrast, **indirect flight muscles** are attached to the tergum and sternum. Contraction causes the tergum, and with it the very base of the wing, to be pulled down. This movement levers the outer, main part of the wing in an upward stroke (Fig. 3.4c). The down beat is powered by contraction of the second set of muscles, which run from front to back of the thorax, thereby deforming the box and lifting the tergum (Fig. 3.4d). At each stage in the cycle, when the flight muscles relax, energy is conserved because the elasticity of the thorax restores its shape.

It seems that primitively the four wings may be controlled independently, with small variations in timing and rate allowing alteration in the direction of flight. However, excessive variation impedes controlled flight and the beat of all wings is usually harmonized, as in butterflies, bugs and bees, for example, by locking the fore and hind wings together, and also by neural control. For insects with slower wing-beat frequencies (< 100 Hz), such as dragonflies, one nerve impulse for each beat can be maintained by **synchronous muscles**. However, in faster-beating wings, which may attain a frequency of 100 Hz to over 1000 Hz, one impulse per beat is impossible and **asynchronous muscles** are required. In these insects, the wing is constructed such that only two wing positions are stable – fully up and fully down. As the wing moves from one extreme to the alternate one, it passes through an intermediate, unstable position. As it passes this unstable ("click") point, thoracic elasticity snaps the wing through to the alternate stable position. Insects with this asynchronous mechanism have peculiar fibrillar flight muscles with the property that on sudden release of muscle tension, as at the click point, the next muscle contraction is induced. Thus, muscles can oscillate, contracting at a much higher frequency than the nerve impulses, which need be only periodic to maintain the insect in flight. Harmonization of the wing beat on each side is maintained through the rigidity of the thorax – as the tergum is depressed or relaxed, what happens to one wing must happen identically to the other. However, insects with indirect

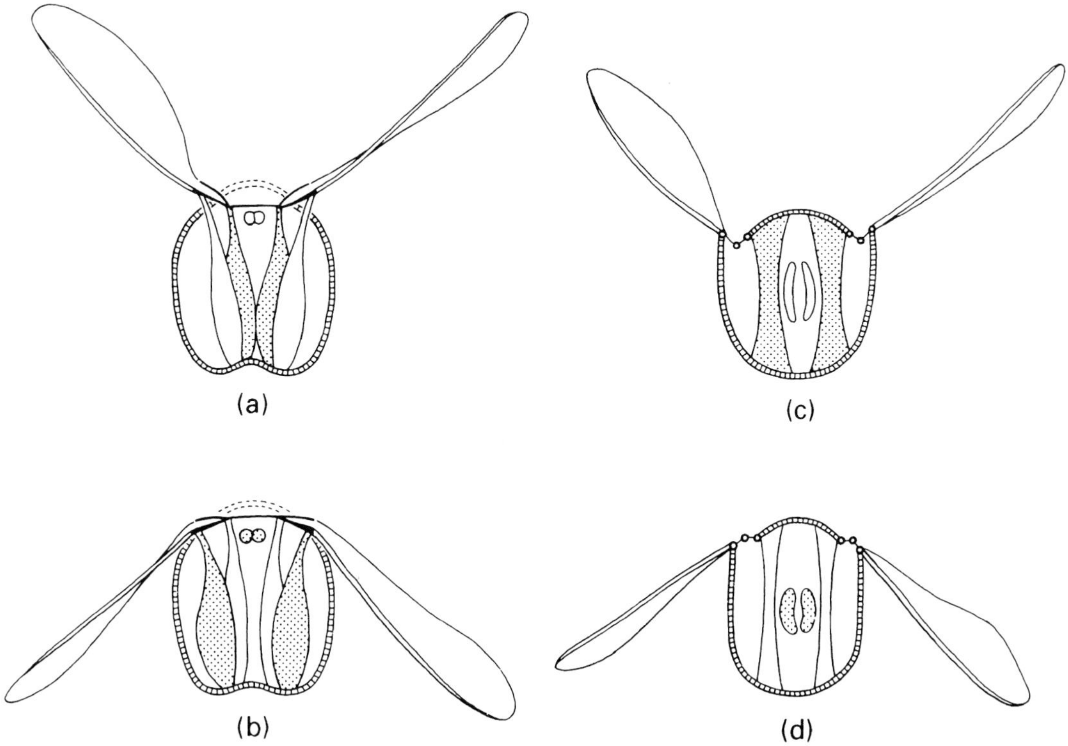

Fig. 3.4 Direct flight mechanisms: thorax during (a) upstroke and (b) downstroke of the wings. Indirect flight mechanisms: thorax during (c) upstroke and (d) downstroke of the wings. Stippled muscles are those contracting in each illustration. (After Blaney 1976.)

flight muscles retain direct muscles, which are are used in making fine adjustments in wing orientation during flight.

Direction and any deviations from the flight course, perhaps caused by air movements, are sensed predominantly through an insect's eyes and antennae. However, the true flies (Diptera) have extremely sophisticated sensory equipment, with their hind wings modified as balancing organs. These halteres, which each comprise a base, stem and apical knob (see Fig. 2.24f), beat in time but out of phase with the fore wings. The knob, which is heavier than the rest of the organ, tends to keep the halteres beating in one plane. When the fly alters direction, whether voluntarily or otherwise, the haltere is twisted. The stem, which is richly endowed with sensilla, detects this movement, and the fly can respond accordingly.

Initiation of flight, for whatever reason, may involve the legs springing the insect into the air. Loss of tarsal contact with the ground causes neural firing of the direct flight muscles. In flies, flight activity originates in contraction of a mid-leg muscle, which both propels the leg downwards (and the fly upwards) and simultaneously pulls the tergum downwards to inaugurate flight. The legs are also important when landing because there is no gradual braking by running forwards – all the shock is taken on the outstretched legs, endowed with pads, spines and claws for adhesion.

3.2 THE NERVOUS SYSTEM AND CO-ORDINATION

The complex nervous system of insects integrates a diverse array of external sensory and internal physiological information and generates some of the behaviours discussed in Chapter 4. In common with other animals, the basic component is the nerve cell, or **neuron** (neurone), composed of a cell body with two projections (fibres): the **dendrite**, which receives stimuli; and the **axon**, which transmits information, either to another neuron or to an effector organ such as a muscle. Insect neurons release a variety of chemicals at **synapses** to either stimulate or inhibit effector neurons or muscles. Important neurotransmitters include acetylcholine and catecholamines such as dopamine, as in vertebrates. More insect-specific neuromuscular transmitters are L-glutamate (stimulatory) and gamma-aminobutyric acid (GABA) (inhibition), with muscle activity being modulated by neurochemicals including octopamine, serotonin and proctolin.

Neurons (Fig. 3.5) are of at least four types:

1 **sensory neurons** receive stimuli from the insect's environment and transmit them to the central nervous system (see below);
2 **interneurons** (or association neurons) receive information from and transmit it to other neurons;
3 **motor neurons** receive information from interneurons and transmit it to muscles;
4 **neurosecretory cells** (neuroendocrine cells), as in section 3.3.1.

The cell bodies of interneurons and motor neurons are aggregated, with the fibres interconnecting all types of nerve cells to form nerve centres called **ganglia**. Simple reflex behaviour has been well studied in insects (described further in section 4.5), but insect behaviour can be complex, involving integration of neural information within the ganglia.

The **central nervous system** (CNS) (Fig. 3.6) is the principal division of the nervous system, and consists of series of ganglia joined by paired longitudinal nerve cords called **connectives**. Ancestrally, there was a pair of ganglia per body segment, but usually the two ganglia of each thoracic and abdominal segment are now fused into a single structure and the ganglia of all head segments are coalesced to form two ganglionic centres – the **brain** and the **suboesophageal** (subesophageal) **ganglion** (seen in Fig. 3.7). The chain of thoracic and abdominal ganglia found on the floor of the body cavity is called the **ventral nerve cord**. The brain, or the dorsal ganglionic centre of the head, is composed of three pairs of fused ganglia (from the first three head segments):

1 the **protocerebrum**, associated with the eyes and thus bearing the optic lobes;
2 the **deutocerebrum**, innervating the antennae;
3 the **tritocerebrum**, concerned with handling the signals that arrive from the body.

Coalesced ganglia of the three mouthpart-bearing segments (mandibular, maxillary and labial) form the suboesophageal ganglion, with nerves emerging that innervate the mouthparts.

The **visceral** (or **sympathetic**) **nervous system** consists of three subsystems: the stomodeal (or stomatogastric) nervous system (which includes the frontal

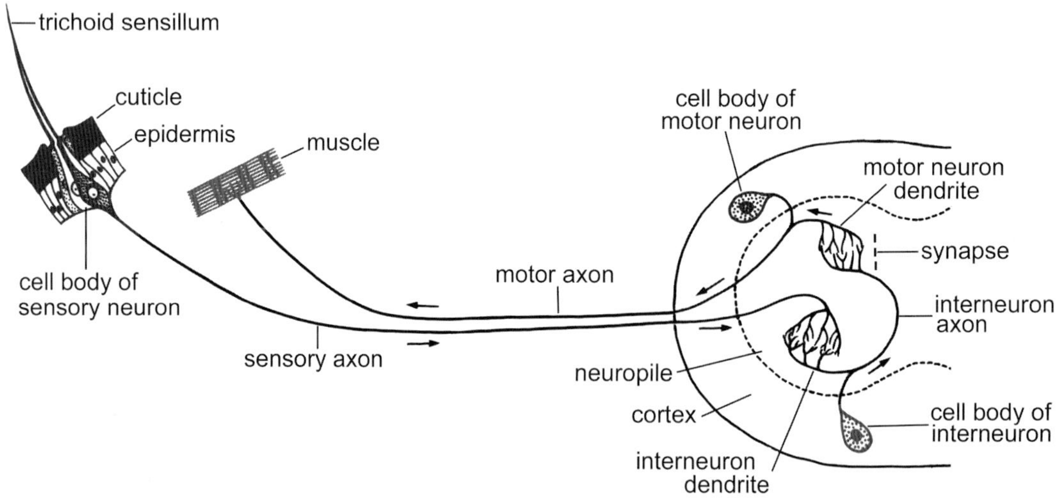

Fig. 3.5 Diagram of a simple reflex mechanism of an insect. The arrows show the paths of nerve impulses along nerve fibres (axons and dendrites). The ganglion, with its outer cortex and inner neuropile, is shown on the right. (After various sources.)

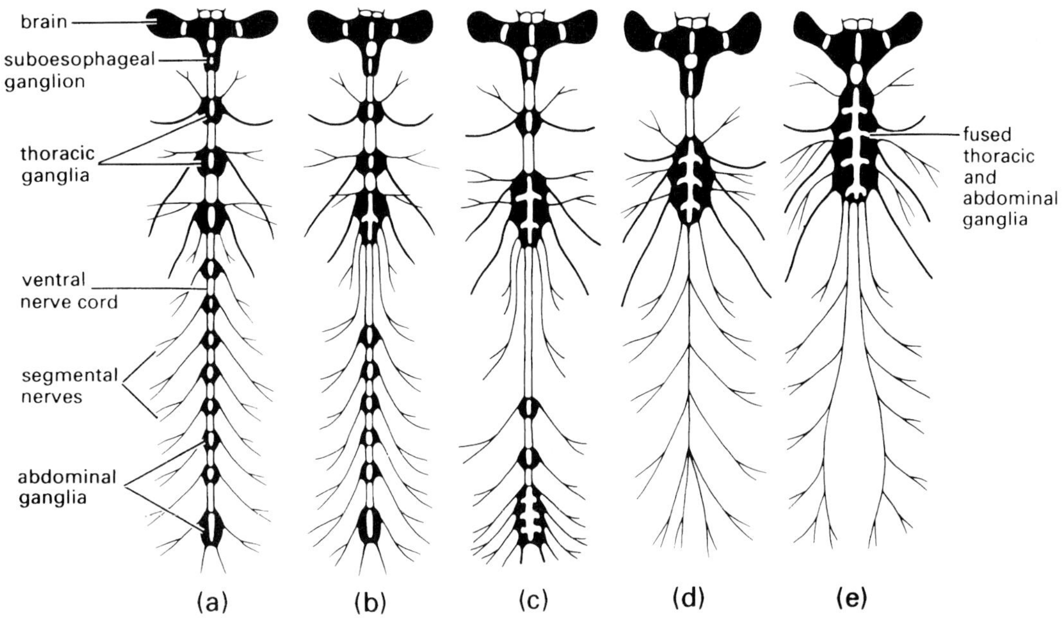

Fig. 3.6 The central nervous system of various insects showing the diversity of arrangement of ganglia in the ventral nerve cord. Varying degrees of fusion of ganglia occur from the least to the most specialized: (a) three separate thoracic and eight abdominal ganglia, as in *Dictyopterus* (Coleoptera: Lycidae) and *Pulex* (Siphonaptera: Pulicidae); (b) three thoracic and six abdominal ganglia, as in *Blatta* (Blattodea: Blattidae) and *Chironomus* (Diptera: Chironomidae); (c) two thoracic and considerable abdominal fusion of ganglia, as in *Crabro* and *Eucera* (Hymenoptera: Crabronidae and Anthophoridae); (d) highly fused, with one thoracic and no abdominal ganglia, as in *Musca*, *Calliphora* and *Lucilia* (Diptera: Muscidae and Calliphoridae); (e) extreme fusion, with no separate suboesophageal ganglion, as in *Hydrometra* (Hemiptera: Hydrometridae) and *Rhizotrogus* (Scarabaeidae). (After Horridge 1965.)

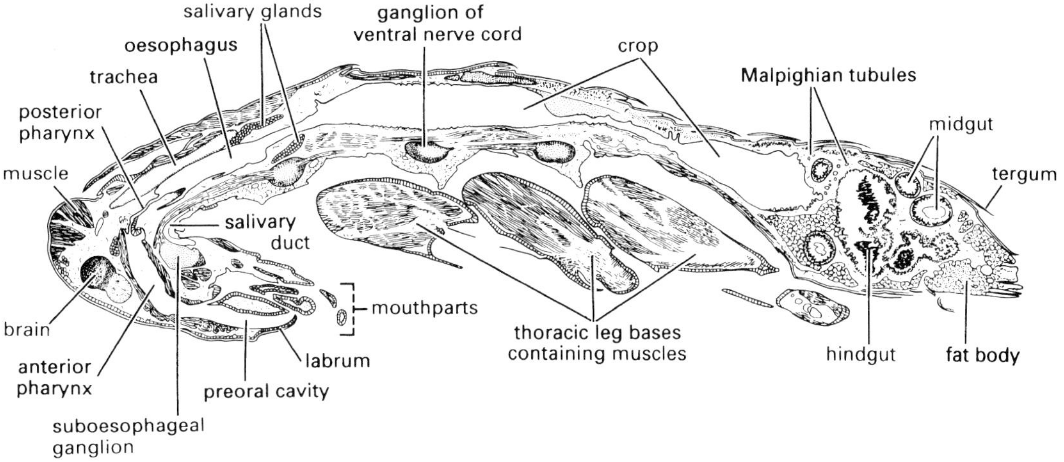

Fig. 3.7 Mediolongitudinal section of an immature cockroach of *Periplaneta americana* (Blattodea: Blattidae) showing internal organs and tissues.

ganglion); the ventral visceral nervous system; and the caudal visceral nervous system. Together, the nerves and ganglia of these subsystems innervate the anterior and posterior gut, several endocrine organs (corpora cardiaca and corpora allata), the reproductive organs, and the tracheal system including the spiracles.

The **peripheral nervous system** consists of all of the motor neuron axons that radiate to the muscles from the ganglia of the CNS and stomodeal nervous system plus the sensory neurons of the cuticular sensory structures (the sense organs) that receive mechanical, chemical, thermal or visual stimuli from an insect's environment. Insect sensory systems are discussed in detail in Chapter 4.

3.3 THE ENDOCRINE SYSTEM AND THE FUNCTION OF HORMONES

Hormones are chemicals produced within an organism's body and transported, generally in body fluids, away from their point of synthesis to sites where they influence many physiological processes, despite being present in extremely small quantities. Insect hormones have been studied in detail in few species, but similar patterns of production and function are likely to apply widely. The actions and interrelationships of these chemical messengers are varied and complex, but the role of hormones in the moulting process is of overwhelming importance and will be discussed more fully in this context in section 6.3. Here, we provide an overview of the endocrine centres and the hormones that they export.

Historically, the implication of hormones in the processes of moulting and metamorphosis resulted from simple but elegant experiments. These utilized techniques that removed the influence of the brain (**decapitation**), isolated the haemolymph of different parts of the body (**ligation**), or artificially connected the haemolymph of two or more insects by joining their bodies. Ligation and decapitation of insects enabled researchers to localize the sites of control of developmental and reproductive processes, and to show that substances are released that affect tissues at sites distant from the point of release. In addition, critical developmental periods for the action of these controlling substances have been identified. The blood-sucking bug *Rhodnius prolixus* (Hemiptera: Reduviidae) and various moths and flies were the principal experimental insects. More-refined technologies allowed microsurgical removal or transplant of various tissues, haemolymph transfusion, hormone extraction and purification, and radioactive labelling of hormone extracts. Today, molecular biological (Box 3.1) and advanced chemical analytical techniques allow hormone isolation, characterization and manipulation.

3.3.1 Endocrine centres

Insect hormones are produced by neuronal, neuroglandular or glandular centres (Fig. 3.8). Hormonal production by some organs, such as the ovaries, is secondary to their main function, but several tissues and organs are specialized for an endocrine role.

Neurosecretory cells

Neurosecretory cells (NSC) (also called neuroendocrine cells) are modified neurons found throughout

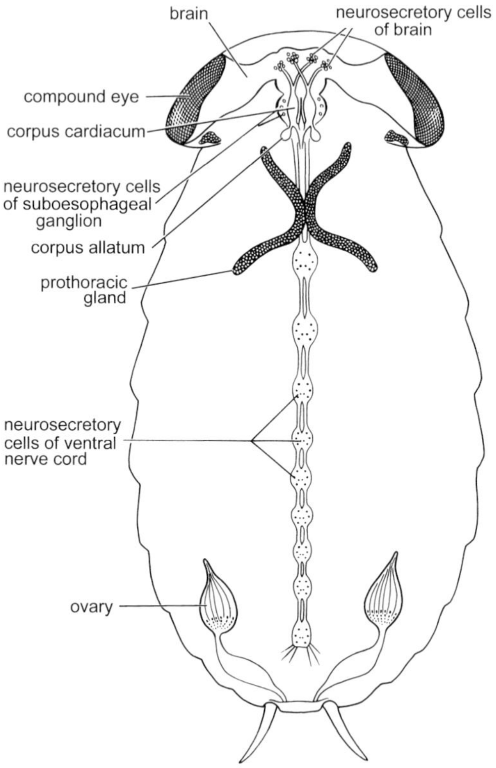

Fig. 3.8 The main endocrine centres in a generalized insect. (After Novak 1975.)

Box 3.1 Molecular genetic techniques and their application to neuropeptide research

Molecular biology is essentially a technical "toolbox" for the isolation, analysis and manipulation of DNA and its RNA and protein products. Molecular genetics is concerned primarily with the nucleic acids, whereas research on the proteins and their constituent amino acids involves chemistry. Thus, genetics and chemistry are integral to molecular biology. Molecular biological tools include:

- a means of cutting DNA at specific sites using restriction enzymes, and of rejoining naked ends of cut fragments using ligase enzymes;
- techniques that produce numerous identical copies by repeated cycles of amplification of a segment of DNA (e.g. polymerase chain reaction, PCR);
- methods for rapid sequencing of nucleotides of DNA or RNA, and amino acids of proteins;
- the ability to synthesize specific sequences of DNA, RNA or proteins;
- the ability to search a genome for a specific nucleotide sequence using probes of nucleic acid segments complementary to the sequence being sought;
- site-directed mutation of specific DNA segments *in vitro*;
- genetic engineering – the isolation and transfer of intact genes into other organisms, with subsequent stable transmission and gene expression;
- cytochemical techniques to observe gene transcription;
- immunochemical and histochemical techniques to identify how, when and where a specific gene product functions;
- relatively rapid and inexpensive sequencing and assembly of the complete genome, the totality of DNA in an organism (genomics);
- the ability to silence (knockdown) genes of interest *in vivo* by "reverse genetics" using **RNA interference** (RNAi);
- the ability to induce overexpression or ecotopic expression *in vivo* of specific genes.

Insect peptide hormones used to be difficult to study because of the minute quantities produced by individual insects and their structural complexity and occasional instability. The realization that these proteins play crucial roles in insect physiology (see Table 3.1), and the recent availability of some of the techniques listed above, have led to an explosion of studies of neuropeptide signalling systems. Nevertheless, although neuropeptide amino acid sequences provide an entry to the powerful capabilities of molecular genetics, there remain many "orphaned" neuropeptide systems in which functions and receptors remain unknown. This applies even in the very well-studied *Drosophila* genome. Although signalling systems seem to be highly conserved through evolutionary time, presumably due to their centrality to insect physiology, some neuropeptides and receptors appear to have been lost even at an ordinal level. With increasing availability of whole genomes across a greater diversity of insects, and the use of reverse genetics to "silence" genes, there is scope for wider understanding of these pervasive signalling chemicals. Not least is the future potential for neuropeptide research in control of insect pests, discussed in section 16.4.3.

the nervous system (within the CNS, peripheral nervous system and stomodeal nervous system), but they occur in major groups in the brain. These cells produce most of the known insect hormones, the notable exceptions being the production by non-neural tissues of ecdysteroids and juvenile hormones. However, the synthesis and release of the latter hormones are regulated by neurohormones from NSC.

Corpora cardiaca

The **corpora cardiaca** (singular: corpus cardiacum) are a pair of neuroglandular bodies located on either side of the aorta and behind the brain. As well as producing their own neurohormones (such as adipokinetic hormone, AKH), they store and release neurohormones, including **prothoracicotropic hormone** (PTTH, formerly called brain hormone or ecdysiotropin), originating from the NSC of the brain. PTTH stimulates the secretory activity of the prothoracic glands.

Prothoracic glands

The **prothoracic glands** are diffuse, paired glands, generally located in the thorax or the back of the head. In cyclorrhaphous Diptera they are part of the ring gland, which also contains the corpora cardiaca and corpora allata. The prothoracic glands secrete an ecdysteroid, usually ecdysone (sometimes called moulting hormone), which, after hydroxylation, elicits the moulting process of the epidermis (section 6.3). In most insects, the prothoracic glands degenerate in the adult, but they are retained in bristletails (Archaeognatha) and silverfish (Zygentoma), which continue to moult as adults.

Corpora allata

The **corpora allata** (singular: corpus allatum) are small, discrete, paired glandular bodies derived from the epithelium and located on either side of the foregut. In some insects they fuse to form a single gland. Their function is to secrete juvenile hormone (JH), which has regulatory roles in both metamorphosis and reproduction. In Lepidoptera, the corpora allata also store and release PTTH.

Inka cells

These endocrine cells are a major component of the epitracheal glands, which are paired structures attached to tracheal trunks near each spiracle, and are found in the prothoracic and abdominal segments of Lepidoptera, Diptera, and some Coleoptera and Hymenoptera. In other Holometabola, including most beetles and bees, and all hemimetabolous insects thus far examined, numerous small **Inka cells** are dispersed throughout the tracheal system. Inka cells produce and release **pre-ecdysis** and **ecdysis triggering hormones** (PETH and ETH), which are peptides that activate the ecdysis sequence by acting on receptors in the CNS. The ecdysis sequence consists of pre-ecdysis, ecdysis and post-ecdysis behaviours, and involves specific contractions of skeletal muscles, which lead to movements that facilitate the splitting and shedding of the old cuticle.

3.3.2 Hormones

Three hormones or hormone types are integral to growth and reproductive functions in insects. These are the ecdysteroids, the juvenile hormones and the neurohormones (also called neuropeptides).

Ecdysteroid is a general term applied to any steroid with moult-promoting activity. All ecdysteroids are derived from sterols, such as cholesterol, which insects cannot synthesize *de novo* and must obtain from their diet. Ecdysteroids occur in all insects and form a large group of compounds, of which ecdysone and 20-hydroxyecdysone are the most common members. **Ecdysone** (also called a-ecdysone) is released from the prothoracic glands into the haemolymph and usually is converted to the more active hormone 20-hydroxyecdysone in several peripheral tissues, especially the fat body. The 20-hydroxyecdysone (referred to as ecdysterone or b-ecdysone in older literature) is the most widespread and physiologically important ecdysteroid in insects. The action of ecdysteroids in eliciting moulting is well studied, and functions similarly in different insects. Ecdysteroids are produced also by the ovary of the adult female insect and may be involved in ovarian maturation (e.g. yolk deposition) or be packaged in the eggs to be metabolized during the formation of embryonic cuticle.

Juvenile hormones form a family of related sesquiterpenoid compounds, so that the symbol JH may denote one or a mixture of hormones, including JH-I, JH-II, JH-III and JH-0. The occurrence of mixed-JH-producing insects (such as the tobacco hornworm, *Manduca sexta*) adds to the complexity of unravelling the functions of the homologous JHs. These hormones are signalling molecules, and act via lipid activation of proteins that play a diversity of roles in development and physiology. Lipid-based signalling systems are known to have diverse modes of action and generally do not require high-affinity binding to receptor sites. Insect JHs have two major roles – the control of metamorphosis and the regulation of reproductive development. Larval characteristics are maintained and metamorphosis is inhibited by JH; adult development requires a moult in the absence of JH (see section 6.3 for details). Thus, JH controls the degree and direction of differentiation at each moult. In the adult female insect, JH stimulates the deposition of yolk in the eggs and affects accessory gland activity and pheromone production (section 5.11).

Neurohormones, the largest class of insect hormones, are peptides (small proteins) with the alternative name of **neuropeptides**. These protein messengers are the master regulators of all insect physiological processes, including development, homeostasis, metabolism and reproduction, as well as the secretion of the JHs and ecdysteroids. Over a hundred neuropeptides have been recognized, many existing in multiple forms encoded by the same gene but resulting from gene duplication, mutation and selection giving rise to closely related signalling systems. Some important physiological processes controlled by neurohormones in some or all insects are summarized in Table 3.1. The diversity and vital co-ordinating roles of these small molecules are characterized increasingly via peptide molecular biology (Box 3.1) combined with the availability of the complete *Drosophila* genome. Neuropeptides regulate by both inhibitory and stimulatory signals, and reach action sites (receptors) via nerve axons or the haemolymph. Others control indirectly via their action on other endocrine glands (corpora allata and prothoracic glands). Receptors are high-affinity binding sites located in the plasma membrane of the target cells, with most classified as G protein-coupled receptors (GPCRs). Exceptions include prothoracicotropic hormone (PTTH) and insulin-like peptides that activate by binding to a receptor tyrosine kinase. Some ligand-receptor pairs are very specific in responding to a single neuropeptide type, whereas other receptors respond to several types of ligand.

3.4 THE CIRCULATORY SYSTEM

Haemolymph, the insect body fluid (with properties and functions as described in section 3.4.1), circulates freely around the internal organs. The pattern of flow is regular between compartments and appendages, assisted by muscular contractions of parts of the body, especially the peristaltic contractions of a longitudinal **dorsal vessel**, part of which is sometimes called the heart. Haemolymph does not directly contact the cells because the internal organs and the epidermis are covered in a basement membrane, which may regulate the exchange of materials. This open circulatory system has only a few vessels and compartments to direct haemolymph movement, in contrast to the closed network of blood-conducting vessels seen in vertebrates.

3.4.1 Haemolymph

The volume of the haemolymph may be substantial (20–40% of body weight) in soft-bodied larvae, which use the body fluid as a hydrostatic skeleton, but is less than 20% of body weight in most nymphs and adults. Haemolymph is a watery fluid containing ions, molecules and cells. It is often clear and colourless, but may be variously pigmented yellow, green or blue, or rarely, in the immature stages of a few aquatic and endoparasitic flies, red owing to the presence of haemoglobin. All chemical exchanges between insect tissues are mediated via the haemolymph – hormones are transported, nutrients are distributed from the gut, and wastes are removed to the excretory organs. However, insect haemolymph rarely contains respiratory pigments and hence has a very low oxygen-carrying capacity. Local changes in haemolymph pressure are important in ventilation of the tracheal system (section 3.5.1), in thermoregulation (section 4.2.2), and at moulting to aid splitting of the old and expansion of the new cuticle. The haemolymph serves also as a water reserve, as its main constituent, **plasma**, is an aqueous solution of inorganic ions, lipids, sugars (mainly trehalose), amino acids, proteins, organic acids and other compounds. High concentrations of amino acids and organic phosphates characterize insect haemolymph, which also is the site of deposition of molecules associated with cold protection (section 6.6.1). Haemolymph proteins include those that act in storage (hexamerins) and those that transport lipids (lipophorin) or complex with iron (ferritin) or juvenile hormone (JH-binding protein). Hexamerins are large proteins that are synthesized in the fat body and occur in the haemolymph at very high concentrations in many insects, and are believed to act as a source of energy and amino acids during non-feeding periods, such as pupation. However, their functions appear to be much more diverse as, at least in some insects, they may have roles in sclerotization of insect cuticle and in immune responses, serve as carriers for juvenile hormone and ecdysteroids, and also be involved in caste formation in termites via JH regulation. Hexamerins evolved from copper-containing haemocyanins (that function as respiratory pigments in many arthropods) and some insects, such as stoneflies (Plecoptera), have haemocyanins that function to transport oxygen in the haemolymph. In stoneflies, this pigment-based oxygen transport system occurs in conjunction with a tracheal system that takes oxygen directly to the tissues. Haemocyanins probably were the ancestral method for oxygen transport in insects (section 8.3), but subsquently were widely lost.

The blood cells, or **haemocytes** (hemocytes), are of several types (mainly plasmatocytes, granulocytes,

Table 3.1 Examples of some important insect physiological processes mediated by neuropeptides; note that usually only one, of often multiple, function of each neuropeptide is listed. (After Keeley & Hayes 1987; Holman *et al.* 1990; Gäde *et al.* 1997; Altstein 2003.)

Neuropeptide	Action
Growth and development	
Allatostatins and allatotropins	Induce/regulate juvenile hormone (JH) production
Bursicon	Controls cuticular sclerotization
Corazonin	Initiates ecdysis
Crustacean cardioactive peptide (CCAP)	Switches on ecdysis behaviour
Diapause hormone (DH)	Causes dormancy in silkworm eggs
Pre-ecdysis triggering hormone (PETH)	Stimulates pre-ecdysis behaviour
Ecdysis triggering hormone (ETH)	Initiates behaviour at ecdysis
Eclosion hormone (EH)	Controls events at ecdysis
JH esterase inducing factor	Stimulates JH degradative enzyme
Prothoracicotropic hormone (PTTH)	Induces ecdysteroid secretion from prothoracic gland
Puparium tanning factor	Accelerates fly puparium tanning
Reproduction	
Antigonadotropin	Suppresses oocyte development
Ovarian ecdysteroidogenic hormone (OEH = EDNH)	Stimulates ovarian ecdysteroid production
Ovary maturing peptide (OMP)	Stimulates egg development
Oviposition peptides	Stimulate egg deposition
Prothoracicotropic hormone (PTTH)	Affects egg development
Pheromone biosynthesis activating neuropeptide (PBAN)	Regulates pheromone production
Homeostasis	
Metabolic peptides (= AKH/RPCH family)	
Adipokinetic hormone (AKH)	Releases lipid from fat body
Hyperglycaemic hormone	Releases carbohydrate from fat body
Hypoglycaemic hormone	Enhances carbohydrate uptake
Protein synthesis factors	Enhance fat body protein synthesis
Diuretic and antidiuretic peptides	
Antidiuretic peptide (ADP)	Suppresses water excretion
Diuretic peptide (DP)	Enhances water excretion
Chloride-transport stimulating hormone	Stimulates Cl^- absorption (rectum)
Ion-transport peptide (ITP)	Stimulates Cl^- absorption (ileum)
Myotropic peptides	
Cardiopeptides	Increase heartbeat rate
Kinin family (e.g. leukokinins and myosuppressins)	Regulate gut contraction
Proctolin	Modifies excitation response of some muscles
Corazonin	Cardiostimulatory
Chromatotropic peptides	
Melanization and reddish colouration hormone (MRCH)	Induces darkening
Pigment-dispersing hormone (PDH)	Disperses pigment
Corazonin	Causes dark pigmentation in locusts

prohaemocytes and oenocytoids) and all are nucleate. They have four basic functions:

1 phagocytosis – the ingestion of small particles and substances such as metabolites;
2 encapsulation of parasites and other large foreign materials;
3 haemolymph coagulation;
4 storage and distribution of nutrients.

The haemocoel contains two additional types of cells. **Nephrocytes** (sometimes called pericardial cells) generally occur on or near the dorsal vessel and regulate haemolymph composition by sieving/filtering certain substances and metabolizing them for use or excretion elsewhere. **Oenocytes** are of epidermal origin but may occur in the haemocoel, fat body or epidermis. Although their functions are unclear in most insects, they have many roles and are involved in lipid metabolism, including cuticle lipid (hydrocarbon) synthesis, as well as detoxification and developmental signalling in some insects and, in some chironomids, they produce haemoglobins.

3.4.2 Circulation

Circulation in insects is maintained mostly by a system of muscular pumps moving haemolymph through compartments separated by fibromuscular septa or membranes. The main pump is the pulsatile dorsal vessel. The anterior part may be called the aorta and the posterior part may be called the heart, but the two terms are inconsistently applied. The dorsal vessel is a simple tube, generally composed of one layer of myocardial cells and with segmentally arranged openings, or **ostia**. The lateral ostia typically permit the one-way flow of haemolymph into the dorsal vessel as a result of valves that prevent backflow. In many insects there also are more ventral ostia that permit haemolymph to flow out of the dorsal vessel, probably to supply adjacent active muscles. There may be up to three pairs of thoracic ostia and nine pairs of abdominal ostia, although there is an evolutionary tendency towards reduction in the number of ostia. The dorsal vessel lies in a compartment, the **pericardial sinus**, above a **dorsal diaphragm** (a fibromuscular septum – a separating membrane) formed of connective tissue and segmental pairs of **alary muscles**. The alary muscles support the dorsal vessel but their contractions do not affect heartbeat. Haemolymph enters the pericardial sinus via segmental openings in the diaphragm and/or at the posterior border, and then moves into the dorsal vessel via the ostia during a muscular relaxation phase. Waves of contraction, which normally start at the posterior end of the body, pump the haemolymph forwards in the dorsal vessel and out via the aorta into the head. Next, the appendages of the head and thorax are supplied with haemolymph as it circulates posteroventrally and eventually returns to the pericardial sinus and the dorsal vessel. A generalized pattern of haemolymph circulation in the body is shown in Fig. 3.9a; however, in adult insects there also may be a periodic reversal of haemolymph flow in the dorsal vessel (from thorax posteriorly) as part of normal circulatory regulation.

Another important component of the circulation of many insects is the **ventral diaphragm** (Fig. 3.9b) – a fibromuscular septum that lies in the floor of the body cavity and is associated with the ventral nerve cord. Circulation of the haemolymph is aided by active peristaltic contractions of the ventral diaphragm, which direct the haemolymph backwards and laterally in the **perineural sinus** below the diaphragm. Haemolymph flow from the thorax to the abdomen also may be dependent, at least partially, on expansion of the abdomen, thus "sucking" haemolymph posteriorly. Haemolymph movements are especially important in insects that use the circulation in thermoregulation (e.g. some Odonata, Diptera, Lepidoptera and Hymenoptera). The diaphragm may facilitate rapid exchange of chemicals between the ventral nerve cord and the haemolymph, either by actively moving the haemolymph and/or moving the cord itself.

Haemolymph generally is circulated to appendages unidirectionally by various tubes, septa, valves and pumps (Fig. 3.9c). The muscular pumps are termed **accessory pulsatile organs** and occur at the antennal and wing bases, and sometimes in the legs. Furthermore, the antennal pulsatile organs may release neurohormones that are carried to the antennal lumen to influence the sensory neurons. Wings have a circulation, although apparent often only in the young adult. At least in some Lepidoptera, circulation in the wing occurs by the reciprocal movement of haemolymph (in the wing vein sinuses) and air (within the elastic wing tracheae) into and from the wing, brought about by pulsatile organ activity, reversals of heartbeat, and tracheal volume changes.

The insect circulatory system displays an impressive synchronization between the activities of the dorsal

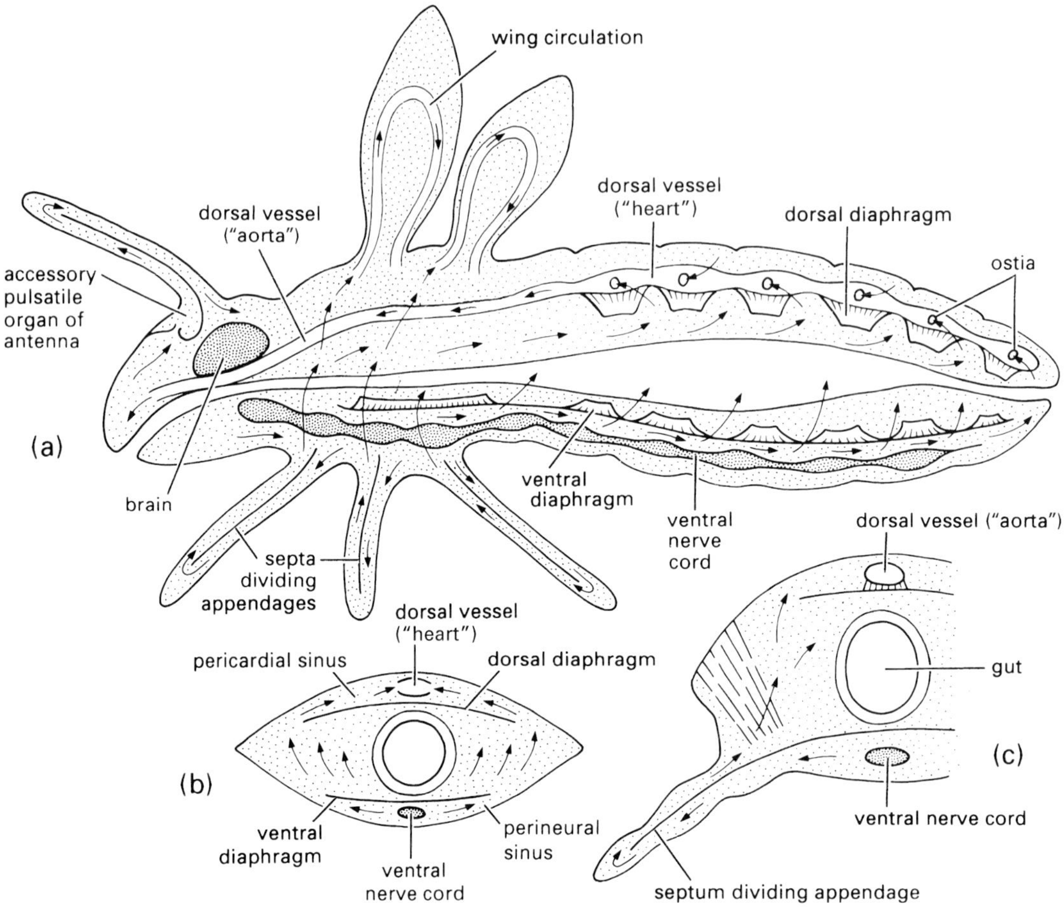

Fig. 3.9 Schematic diagram of a well-developed circulatory system: (a) longitudinal section through body; (b) transverse section of the abdomen; (c) transverse section of the thorax. Arrows indicate directions of haemolymph flow. (After Wigglesworth 1972.)

vessel, fibromuscular diaphragms and accessory pumps, mediated by both nervous and neurohormonal regulation. The physiological regulation of many body functions by the neurosecretory system occurs via neurohormones transported in the haemolymph.

3.4.3 Protection and defence by the haemolymph

Haemolymph provides various kinds of protection and defence from: (i) physical injury; (ii) the entry of disease organisms, parasites or other foreign substances; and sometimes (iii) the actions of predators. In some insects, the haemolymph contains malodorous or distasteful chemicals, which are deterrent to predators (Chapter 14). Injury to the integument elicits a wound-healing process that involves haemocytes and plasma coagulation. A haemolymph clot seals the wound and reduces further haemolymph loss and bacterial entry. If disease organisms or particles enter an insect's body, then immune responses are invoked. These include the cellular defence mechanisms of phagocytosis, encapsulation and nodule formation mediated by the haemocytes, as well as the actions of humoral factors such as enzymes or other proteins (e.g. lysozymes, prophenoloxidase, lectins and peptides).

The immune system of insects bears little resemblance to the complex immunoglobulin-based vertebrate system, yet insects infected sublethally with bacteria can acquire resistance against subsequent infection. Haemocytes are involved in phagocytosing bacteria but, in addition, immunity proteins with antibacterial activity appear in the haemolymph after a primary infection. For example, lytic peptides called cecropins, which disrupt the cell membranes of bacteria and other pathogens, have been isolated from certain moths. Furthermore, some neuropeptides may participate in cell-mediated immune responses by exchanging signals between the neuroendocrine system and the immune system, as well as influencing the behaviour of cells involved in immune reactions.

The enzyme phenoloxidase (PO) plays a central role in insect immunochemistry. It is produced as an inactive zymogen (proPO), mostly in haematocytes (oenocytoids). PO produces indoles, which are polymerized to melanin (used in encapsulation of foreign material), and the enzymatic reactions involved produce quinones, diphenols, superoxide, hydrogen peroxide and reactive nitrogen intermediates. These help defend against bacteria, fungi and viruses. Activation and inhibition of phenoloxidase involves zymogens (enzyme precursors), inhibitor enzymes, and signalling molecules in a diversity of cell types. Phenoloxidase-based immunity is metabolically expensive and has fitness costs.

The insect immune response to ingestion of pathogens is located in the alimentary system, especially the mid-gut, where two responses are induced: reactive oxygen species (ROS) and antimicrobial peptides (AMP) are produced. Management of ROS is needed to protect the gut from self-harm from peroxide and hypochlorous acids. Likewise, AMPs have to be regulated to safeguard beneficial resident microbiota, while responding rapidly and effectively against pathogenic intruders.

3.5 THE TRACHEAL SYSTEM AND GAS EXCHANGE

In common with all aerobic animals, insects must obtain oxygen from their environment and eliminate carbon dioxide respired by their cells. This **gas exchange** must be distinguished from **respiration**, which refers strictly to oxygen-consuming, cellular metabolic processes. Gas exchange occurs largely through internal air-filled tracheae – tubes that branch and ramify throughout the body (Fig. 3.10). The finest branches contact all internal organs and tissues, and are especially numerous in tissues with high oxygen requirements. Air usually enters the tracheae via **spiracles**, which are muscle-controlled openings positioned laterally on the body, primitively with one pair per post-cephalic segment (but not now on the prothorax or posterior abdomen). No extant insect has more than 10 pairs (two thoracic and eight abdominal) (Fig. 3.11a), most have eight or nine, and some have one (Fig. 3.11c), two or none (Fig. 3.11d–f). Typically, spiracles (Fig. 3.10a) have a chamber, or **atrium**, with an opening-and-closing mechanism, or **valve**, either projecting externally or at the inner end of the atrium. In the latter type, a filter apparatus sometimes protects the outer opening. Each spiracle may be set in a sclerotized cuticular plate called a **peritreme**.

The tracheae are invaginations of the epidermis and thus their lining is continuous with the body cuticle. The characteristic ringed appearance of the tracheae seen in tissue sections (as in Fig. 3.7) is due to the spiral ridges or thickenings of the cuticular lining, the **taenidia**, which allow the tracheae to be flexible but to resist compression (analogous to the function of the ringed hose of a vacuum cleaner). The cuticular linings of the tracheae are shed with the rest of the exoskeleton when the insect moults. Usually, even the linings of the finest branches of the tracheal system are shed at ecdysis, but linings of the fluid-filled blind endings, the **tracheoles**, may or may not be shed. Tracheoles are less than 1 μm in diameter and closely contact the respiring tissues (Fig. 3.10b), sometimes indenting into the cells that they supply. However, the tracheae that supply oxygen to the ovaries of many insects have very few tracheoles, the taenidia are weak or absent, and the tracheal surface is evaginated as tubular spirals projecting into the haemolymph. These aptly named **aeriferous tracheae** have a highly permeable surface that allows direct aeration of the surrounding haemolymph from tracheae that may exceed 50 μm in diameter.

In terrestrial and many aquatic insects, the tracheae open to the exterior via the spiracles (an **open tracheal system**) (Fig. 3.11a–c). In contrast, in some aquatic and many endoparasitic larvae, spiracles are absent (a **closed tracheal system**) and the tracheae divide peripherally to form a network. This covers the general body surface (allowing cutaneous

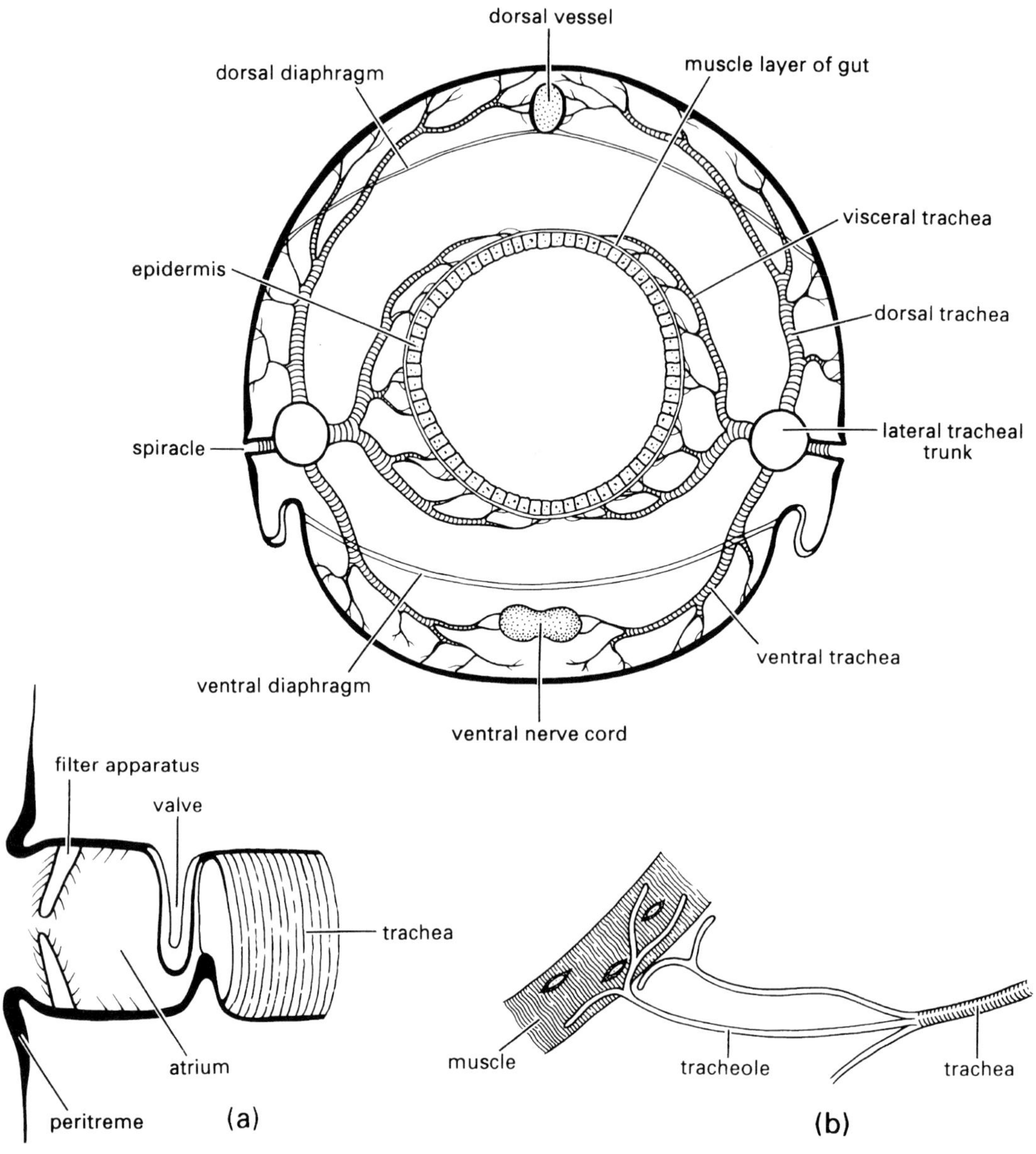

Fig. 3.10 Schematic diagram of a generalized tracheal system seen in a transverse section of the body at the level of a pair of abdominal spiracles. Enlargements show: (a) an atriate spiracle with closing valve at inner end of atrium; (b) tracheoles running to a muscle fibre. (After Snodgrass 1935.)

gas exchange) (Fig. 3.11d) or lies within specialized filaments or lamellae (tracheal gills) (Fig. 3.11e,f). Some aquatic insects with an open tracheal system carry **gas gills** with them (e.g. bubbles of air); these may be temporary or permanent (section 10.3.4).

The volume of the tracheal system ranges between 5% and 50% of the body volume, depending on species and stage of development. The more active the insect, the more extensive is the tracheal system. The tracheal system can expand during an insect's

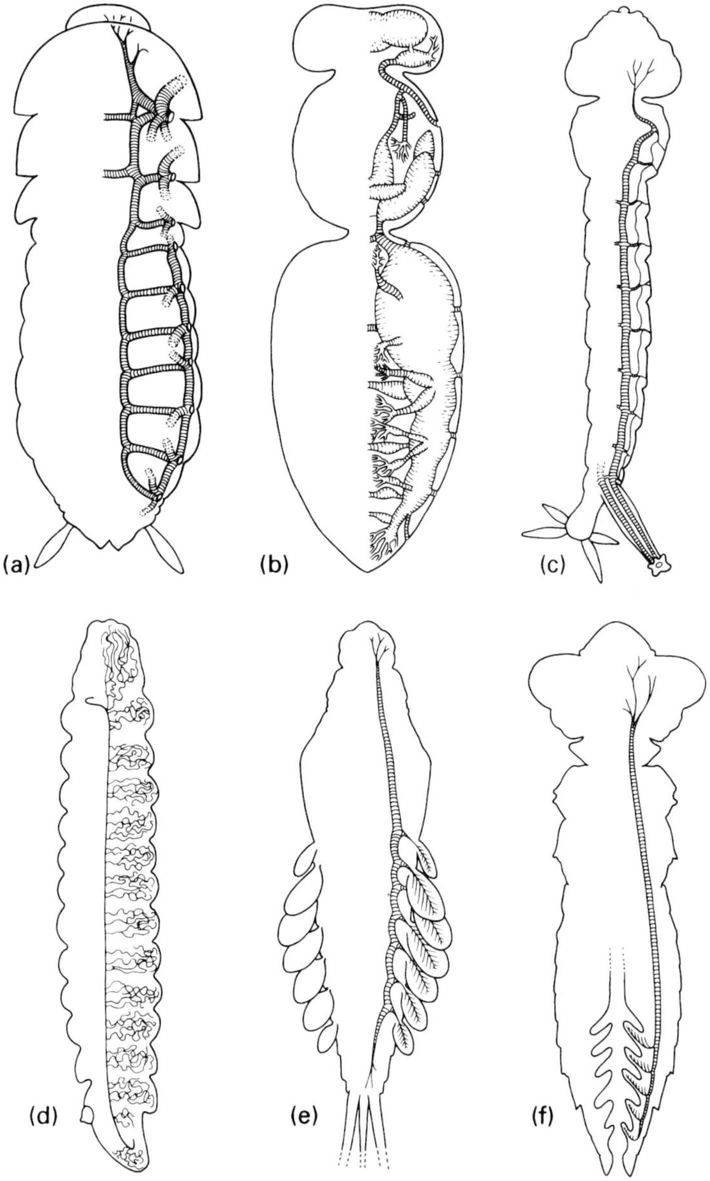

Fig. 3.11 Some basic variations in the open (a–c) and closed (d–f) tracheal systems of insects. (a) Simple tracheae with valved spiracles, as in cockroaches. (b) Tracheae with mechanically ventilated air sacs, as in honey bees. (c) Metapneustic system with only the terminal spiracles functional, as in mosquito larvae. (d) Entirely closed tracheal system with cutaneous gas exchange, as in most endoparasitic larvae. (e) Closed tracheal system with abdominal tracheal gills, as in mayfly nymphs. (f) Closed tracheal system with rectal tracheal gills, as in dragonfly nymphs. (After Wigglesworth 1972; details in (a) after Richards & Davies 1977, (b) after Snodgrass 1956, (c) after Snodgrass 1935, (d) after Wigglesworth 1972.)

growth in response to oxygen needs of all or part of the insect (Box 3.2). For example, in *Drosophila* larvae, tracheal cells are extremely sensitive to oxygen levels, and modulate the expression of hypoxia-induced proteins, which leads to new trachea being grown in the direction of the oxygen-deprived tissue. Also, the larger the insect, the greater the fraction of the body devoted to tracheal volume. Parts of tracheae may be dilated or enlarged to increase the reservoir of air, and the dilations may form **air sacs** (Fig. 3.11b), which are collapsible as the taenidia of the cuticular lining are reduced or absent. Sometimes the tracheal volume may decrease within a developmental stage, as air sacs are occluded by growing tissues. Air sacs reach their

Box 3.2 Tracheal hypertrophy in mealworms at low oxygen concentrations

Resistance to diffusion of gases in insect tracheal systems arises from the spiracular valves, when they are partially or fully closed, the tracheae, and the cytoplasm supplied by the tracheoles at the end of the tracheae. Air-filled tracheae will have a much lower resistance per unit length than the watery cytoplasm because oxygen diffuses several orders of magnitude faster in air than it does in cytoplasm for the same gradient of oxygen partial pressure. Until recently, the tracheal system was believed to provide more than sufficient oxygen (at least in non-flying insects that lack air sacs), with the tracheae offering trivial resistance to the passage of oxygen. Experiments on mealworm larvae, *Tenebrio molitor* (Coleoptera: Tenebrionidae), that were reared in different levels of oxygen (all at the same total gas pressure) showed that the main tracheae that supply oxygen to the tissues in the larvae hypertrophy (increase in size) at lower oxygen levels. The dorsal (D), ventral (V) and visceral (or gut, G) tracheae were affected, but not the lateral longitudinal tracheae that interconnect the spiracles (the four tracheal categories are illustrated in an inset on the graph). The dorsal tracheae supply the dorsal vessel and dorsal musculature, the ventral tracheae supply the nerve cord and ventral musculature, whereas the visceral tracheae supply the gut, fat body and gonads. The graph shows that the cross-sectional areas of the tracheae were greater when the larvae were reared in 10.5% oxygen than when they were reared in 21% oxygen (as in normal air) (After Loudon 1989). Each point on the graph is for a single larva and is the average of the summed areas of the dorsal, ventral and visceral tracheae for six pairs of abdominal spiracles. This hypertrophy appears to be inconsistent with the widely accepted hypothesis that tracheae contribute an insignificant resistance to net oxygen movement in insect tracheal systems. Alternatively, hypertrophy may simply increase the amount of air (and thus oxygen) that can be stored in the tracheal system, rather than reduce resistance to air flow. This might be particularly important for mealworms because they normally live in a dry environment and may minimize the opening of their spiracles. Whatever the explanation, the observations suggest that some adjustment can be made to the size of the tracheae in mealworms (and perhaps other insects) to match the requirements of the respiring tissues.

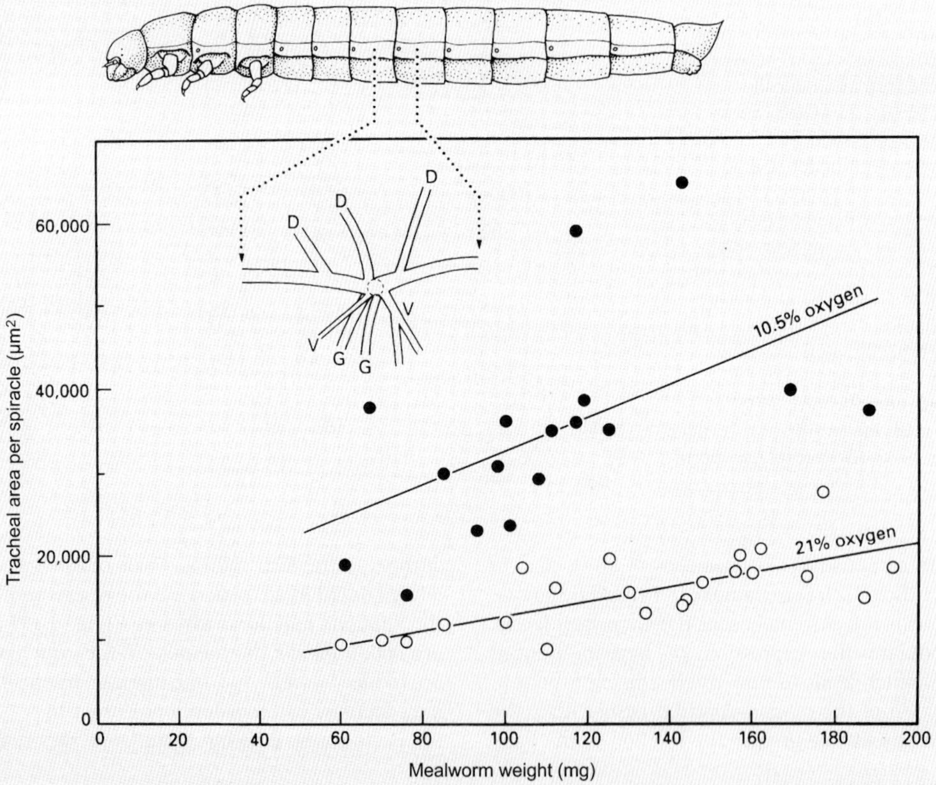

greatest development in very active flying insects, such as bees and cyclorrhaphous Diptera. They may assist flight by increasing buoyancy, but their main function is in ventilation of the tracheal system.

3.5.1 Diffusion and ventilation

Oxygen enters the spiracle, passes through the length of the tracheae to the tracheoles and into the target cells, by a combination of ventilation and diffusion along a concentration gradient, from high in the external air to low in the tissue. Net movement of oxygen molecules in the tracheae is inward, but for carbon dioxide and (in terrestrial insects) water vapour molecules it is outward. Hence, gas exchange in most terrestrial insects is a compromise between securing sufficient oxygen and reducing water loss via the spiracles. Some insect species exchange gases continuously (the spiracles always open), others cyclically, and yet others discontinuously. During periods of inactivity, the spiracles in many adult and diapausing pupal insects are mostly shut, opening only periodically, so that gases are exchanged discontinuously in a three-phase cycle referred to as **discontinuous gas exchange**. When the spiracles are closed, cellular respiration in the body causes oxygen to be used up, and the partial pressure of oxygen within the tracheal system drops until a certain threshold level at which the spiracles rapidly open and close at high frequency (the flutter phase). Some atmospheric oxygen enters the tracheae during the flutter phase, but carbon dioxide continues to build up until eventually it triggers an open spiracular phase, during which tracheal oxygen, carbon dioxide and water are exchanged with the outside air, either by diffusion or active ventilation. Comparative study of many insects living in different environments suggests that discontinuous gas exchange evolved to reduce tracheal water loss, although this may not be the only explanation, as some insects in dry areas exchange gases continuously. In insects of xeric environments, the spiracles may be small with deep atria or have a mesh of cuticular projections in the orifice. Insects in low-oxygen environments, such as underground nests, may use discontinuous gas exchange as a method for increasing tracheal concentration gradients to facilitate uptake of oxygen and emission of carbon dioxide.

In insects without air sacs, such as most holometabolous larvae, diffusion appears to be the primary mechanism for the movement of gases in the tracheae, and is always the sole mode of gas exchange at the tissues. The efficiency of diffusion is related to the distance of diffusion and probably to the diameter of the tracheae (Box 3.2). Rapid cycles of tracheal compression and expansion have been observed in the head and thorax of some insects using X-ray videoing. Movements of the haemolymph and body could not explain these cycles, which appear to be a distinct mechanism of gas exchange in insects. In addition, large or dilated tracheae may serve as an oxygen reserve when the spiracles are closed. In very active insects, especially large ones, active pumping movements of the thorax and/or abdomen **ventilate** (pump air through) the outer parts of the tracheal system, and thus the diffusion pathway to the tissues is reduced. Rhythmic thoracic movements and/or dorsoventral flattening or telescoping of the abdomen expels air, via the spiracles, from extensible or some partially compressible tracheae, or from air sacs. Co-ordinated opening and closing of the spiracles usually accompanies ventilatory movements and provides the basis for the unidirectional airflow that occurs in the main tracheae of larger insects. Anterior spiracles open during inspiration and posterior ones open during expiration. The presence of air sacs, especially if large or extensive, facilitates ventilation by increasing the volume of tidal air that can be changed as a result of ventilatory movements. If the main tracheal branches are strongly ventilated, diffusion appears sufficient to oxygenate even the most actively respiring tissues, such as flight muscles. However, the design of the gas-exchange system of insects places an upper limit on size because if oxygen has to diffuse over a considerable distance, the requirements of a very large and active insect either could not be met, even with ventilatory movements and compression and expansion of tracheae, or would result in substantial loss of water through the spiracles. Interestingly, many large insects are long and thin, thereby minimizing the diffusion distance from the spiracle along the trachea to any internal organ.

3.6 THE GUT, DIGESTION AND NUTRITION

Insects of different groups consume an astonishing variety of foods, including watery xylem sap (e.g. nymphs of spittle bugs and cicadas), vertebrate blood (e.g. bed bugs and female mosquitoes), dry wood (e.g. some termites), bacteria and algae (e.g. black fly and many caddisfly larvae), and the internal tissues of other insects (e.g. endoparasitic wasp larvae). The diverse range of mouthpart types (section 2.3.1) correlates with the diets of different insects, but gut structure and function also reflect the mechanical properties and the nutrient composition of the food eaten. Four major feeding specializations can be identified, depending on whether the food is solid or liquid, or of plant or animal origin (Fig. 3.12). Some insect species clearly fall into a single category, but others with generalized diets may fall between two or more of them, and most holometabolous insects will occupy different categories at different stages of their life (e.g. moths and butterflies switch from solid-plant as larvae to liquid-plant as adults). Gut morphology and physiology relate to these dietary differences in the following ways. Insects that take solid food typically have a wide, straight, short gut with strong musculature and obvious protection from abrasion (especially in the midgut, which has no cuticular lining). These features are most obvious in solid-feeders with rapid throughput of food, as in plant-feeding caterpillars. In contrast, insects feeding on blood, sap or nectar usually have long, narrow, convoluted guts to allow maximal contact with the liquid food; here, protection from abrasion is unnecessary. The most obvious gut specialization of liquid-feeders is a mechanism for removing excess water to concentrate nutrient substances prior to digestion, as seen in hemipterans (Box 3.3). From a nutritional viewpoint, most plant-feeding insects need

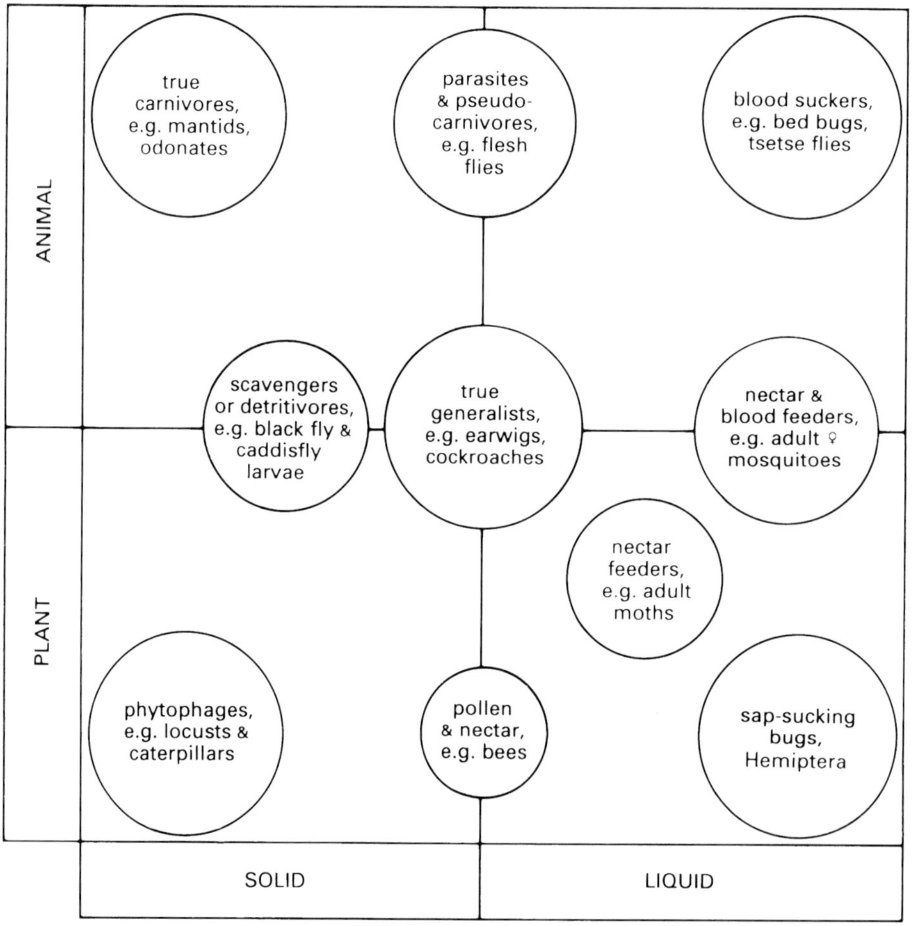

Fig. 3.12 The four major categories of insect-feeding specialization. Many insects are typical of one category, but others cross two categories (or more, as in generalist cockroaches). (After Dow 1986.)

Box 3.3 The filter chamber of Hemiptera

Most Hemiptera have an unusual arrangement of the midgut that is related to their habit of feeding on plant fluids. An anterior and a posterior part of the gut (typically involving the midgut) are in intimate contact to allow concentration of the liquid food. This **filter chamber** allows excess water and relatively small molecules, such as simple sugars, to be passed quickly and directly from the anterior gut to the hindgut, thereby short-circuiting the main absorptive portion of the midgut. Thus, the digestive region is not diluted by water nor congested by superabundant food

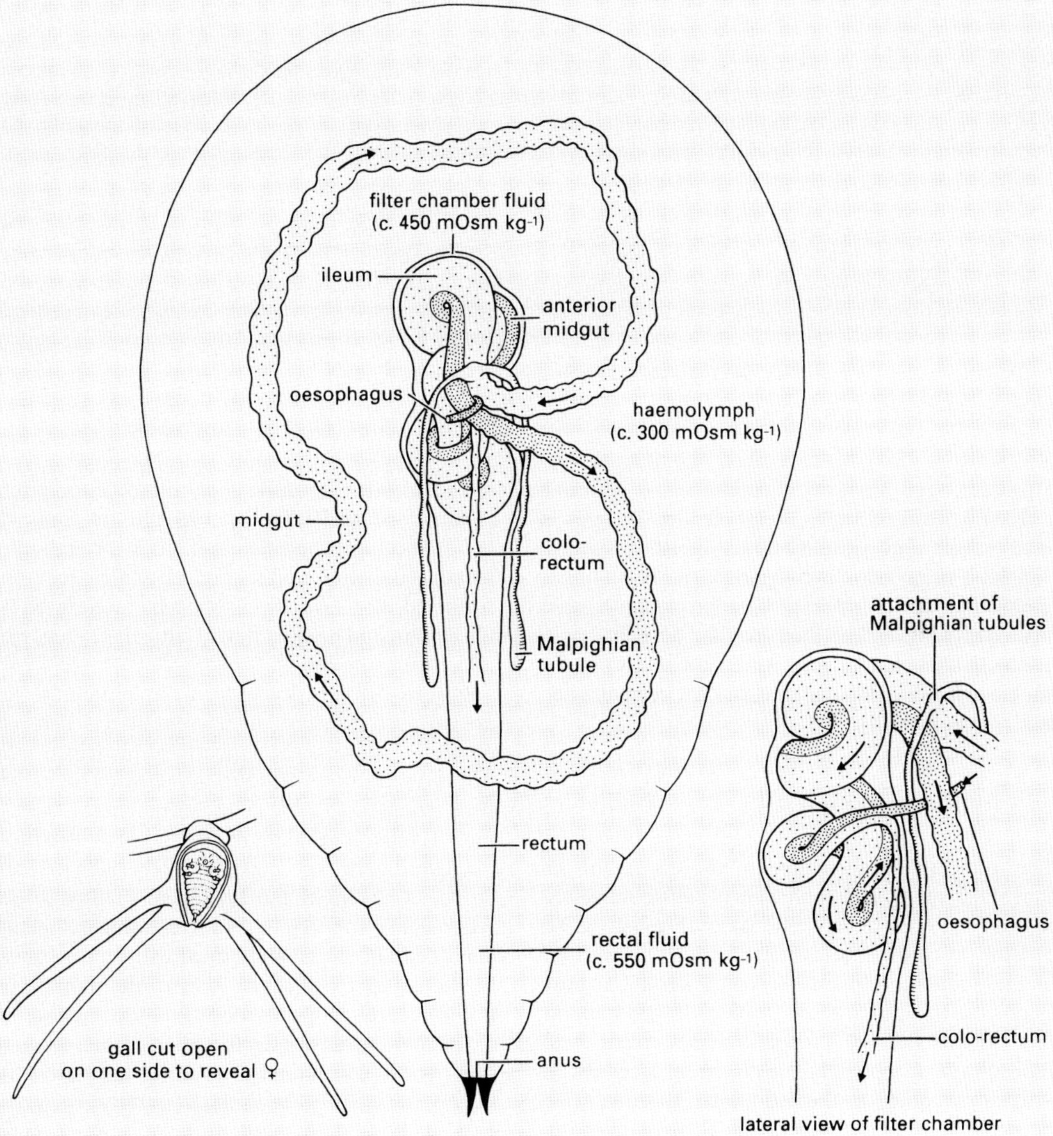

molecules. Well-developed filter chambers are characteristic of cicadas and spittle bugs, which feed on xylem (sap that is rich in ions, low in organic compounds, and with low osmotic pressure), and leafhoppers and coccoids, which feed on phloem (sap that is rich in nutrients, especially sugars, and with high osmotic pressure). The gut physiology of such sap suckers has been rather poorly studied because accurate recording of gut fluid composition and osmotic pressure depends on the technically difficult task of taking readings from an intact gut.

Adult female coccoids of gall-inducing *Apiomorpha* species (Eriococcidae) (section 11.2.5) tap the vascular tissue of the gall wall to obtain phloem sap. Some species have a highly developed filter chamber formed from loops of the anterior midgut and anterior hindgut enclosed within the membranous rectum. Depicted here is the gut of an adult female of *A. munita* viewed from the ventral side of the body. The thread-like sucking mouthparts (see Fig. 11.4c) in series with the cibarial pump connect to a short oesophagus, which can be seen here in both the main drawing and the enlarged lateral view of the filter chamber. The oesophagus terminates at the anterior midgut, which coils upon itself as three loops of the filter chamber. It emerges ventrally and forms a large midgut loop lying free in the haemolymph. Absorption of nutrients occurs in this free loop. The Malpighian tubules enter the gut at the beginning of the ileum, before the ileum enters the filter chamber where it is closely apposed to the much narrower anterior midgut. Within the irregular spiral of the filter chamber, the fluids in the two tubes move in opposite directions (as indicated by the arrows).

The filter chamber of these coccoids apparently transports sugar (perhaps by active pumps) and water (passively) from the anterior midgut to the ileum and then via the narrow colo-rectum to the rectum, from which it is eliminated as honeydew. In *A. munita*, other than water, the honeydew is mostly sugar (accounting for 80% of the total osmotic pressure of about 550 mOsm kg^{-1}*). Remarkably, the osmotic pressure of the haemolymph (about 300 mOsm kg^{-1}) is much lower than that within the filter chamber (about 450 mOsm kg^{-1}) and rectum. Maintenance of this large osmotic difference may be facilitated by the impermeability of the rectal wall.

*Osmolarity values are from unpublished data of P.D. Cooper and A.T. Marshall.

to process large amounts of food because nutrient levels in leaves and stems are often low. The gut is usually short and without storage areas, as food is available continuously. By comparison, a diet of animal tissue is nutrient-rich and, at least for predators, well balanced. However, the food may be available only intermittently (such as when a predator captures prey or a blood meal is obtained) and the gut normally has large storage capacity.

3.6.1 Structure of the gut

There are three main regions to the insect gut (or alimentary canal), with sphincters (valves) controlling food–fluid movement between regions (Fig. 3.13). The **foregut (stomodeum)** is concerned with ingestion, storage, grinding and transport of food to the next region, the **midgut (mesenteron)**. In the midgut, digestive enzymes are produced and secreted, and absorption of the products of digestion occurs. The material remaining in the gut lumen, together with urine from the Malpighian tubules, then enters the **hindgut (proctodeum)**, where absorption of water, salts and other valuable molecules occurs, prior to elimination of the faeces through the anus. The gut epithelium is one cell-layer thick throughout the length of the canal, and rests on a basement membrane surrounded by a variably developed muscle layer. Both the foregut and hindgut have a cuticular lining, which is lacking in the midgut.

Regions of the gut display local specializations, which are variously developed in different insects, depending on diet. Typically, the foregut is subdivided into a **pharynx**, an **oesophagus** (esophagus) and a **crop** (food storage area), and in insects that ingest solid food there is often a grinding organ, the **proventriculus** (or gizzard). The proventriculus is especially well developed in Orthoptera and Blattodea, such as crickets, cockroaches and termites, in which the epithelium is longitudinally ridged and armed with spines or teeth. At the anterior end of the foregut, the mouth opens into a preoral cavity, which is bounded by the bases of the mouthparts and often divided into an upper area, or **cibarium**, and a lower part, or **salivarium** (Fig. 3.14a). The paired labial or **salivary glands** vary in size and arrangement from simple elongated tubes to complex branched or lobed structures.

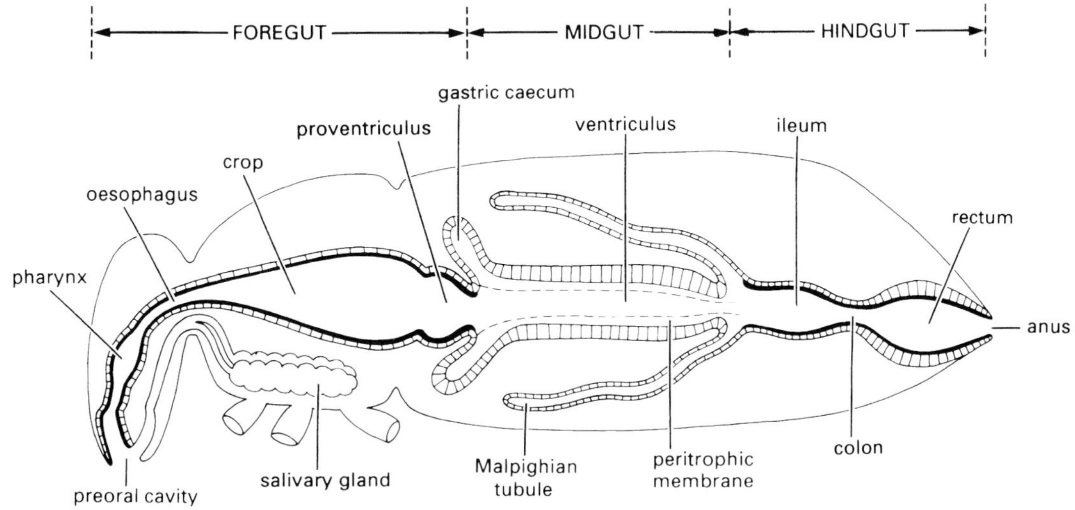

Fig. 3.13 Generalized insect alimentary canal showing division into three regions. The cuticular lining of the foregut and hindgut are indicated by thicker black lines. (After Dow 1986.)

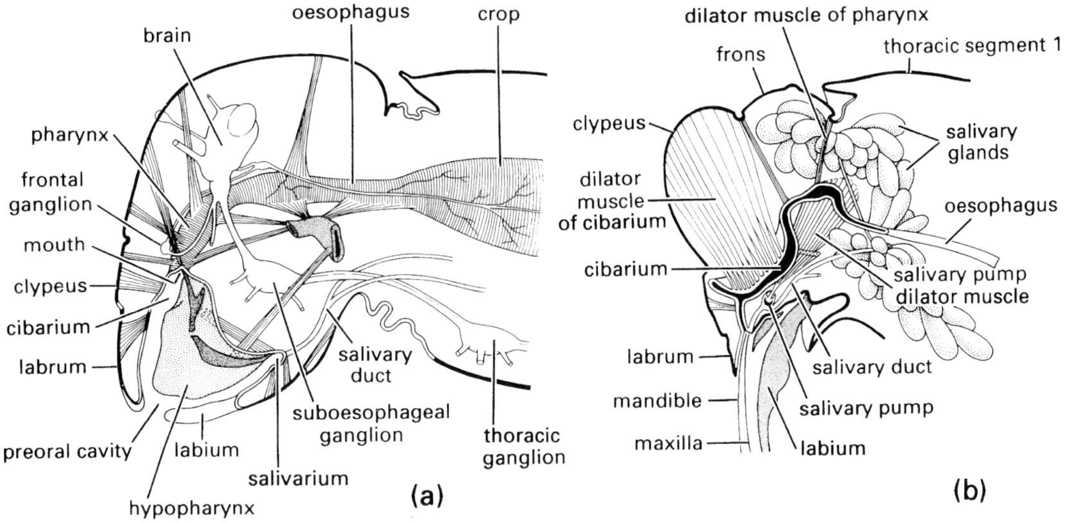

Fig. 3.14 Preoral and anterior foregut morphology in: (a) a generalized orthopteroid insect; and (b) a xylem-feeding cicada. Musculature of the mouthparts and the (a) pharyngeal or (b) cibarial pump are indicated but not fully labelled. Contraction of the respective dilator muscles causes dilation of the pharynx or cibarium and fluid is drawn into the pump chamber. Relaxation of these muscles results in elastic return of the pharynx or cibarial walls and the expulsion of food upwards into the oesophagus. (After Snodgrass 1935.)

Complicated glands occur in many Hemiptera, which produce two types of saliva (see section 3.6.2). In Lepidoptera, the labial glands produce silk, whereas mandibular glands secrete the saliva. Several types of secretory cell may occur in the salivary glands of one insect. The secretions from these cells are transported along cuticular ducts and emptied into the ventral part of the preoral cavity. In insects that store meals in their foregut, the crop may take up the greater portion of the food and often is capable of extreme distension, with a posterior sphincter controlling food retention. The crop may be an enlargement of part of the tubular gut (Fig. 3.7) or a lateral diverticulum.

The generalized midgut has two main areas – the tubular **ventriculus** and blind-ending lateral diverticula called **caeca** (ceca). Most cells of the midgut are structurally similar, being columnar and with microvilli (finger-like protrusions) covering the inner surface. The distinction between the almost indiscernible foregut epithelium and the thickened epithelium of the midgut usually is visible in histological sections (Fig. 3.15). The midgut epithelium mostly is separated from the food by a thin sheath called the **peritrophic matrix** (or **membrane** or **envelope**), which consists of a network of chitin fibrils in a protein–glycoprotein matrix. Some of these proteins, called peritrophins, may have evolved from gastrointestinal mucus proteins by acquiring the ability to bind chitin. The peritrophic membrane either is delaminated from the whole midgut or is produced by cells in the anterior region of the midgut. Exceptionally, Hemiptera and Thysanoptera lack a peritrophic membrane, as do just the adults of several other orders, such as Lepidoptera.

Typically, the beginning of the hindgut is defined by the entry point of the Malpighian tubules, often into a distinct **pylorus** forming a muscular pyloric sphincter, followed by the ileum, colon and rectum. The main functions of the hindgut are the absorption of water, salts and other useful substances from the faeces and urine. A detailed discussion of the structure and function of the hindgut is presented in section 3.7.1.

3.6.2 Saliva and food ingestion

Salivary secretions dilute the ingested food and adjust its pH and ionic content. The saliva often contains

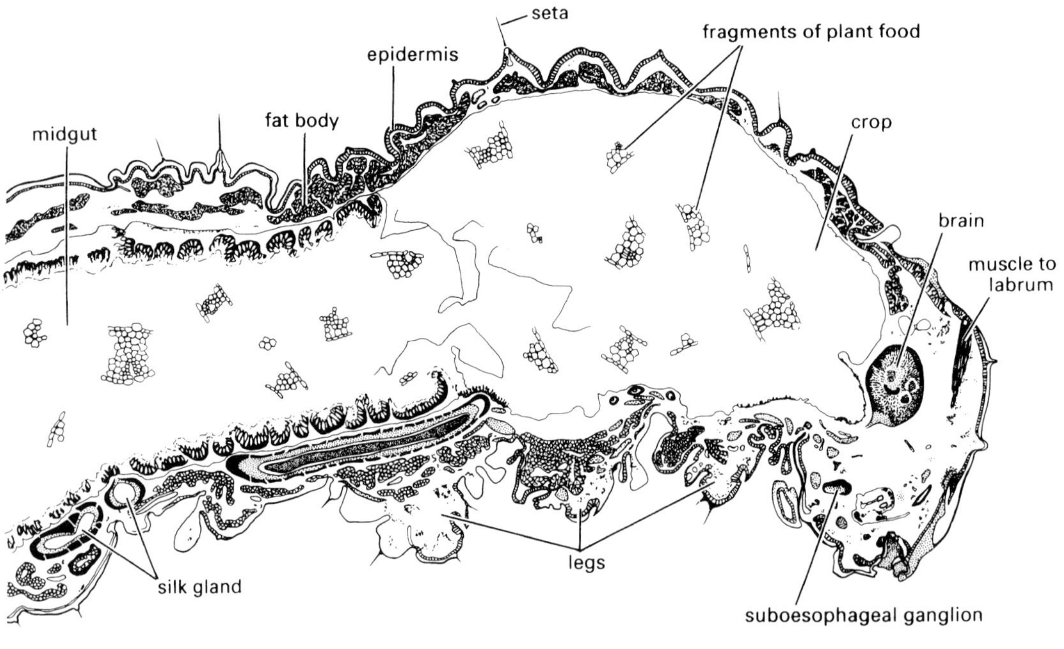

Fig. 3.15 Longitudinal section through the anterior body of a caterpillar of the small white, small cabbage white or cabbage white butterfly, *Pieris rapae* (Lepidoptera: Pieridae). Note the thickened epidermal layer lining the midgut.

digestive enzymes and in blood-feeding insects, anticoagulants and thinning agents are also present. In insects with extra-intestinal digestion, such as predatory Hemiptera, digestive enzymes are exported into the food and the resulting liquid is ingested. Most Hemiptera produce an alkaline watery saliva that is a vehicle for enzymes (either digestive or lytic), and a proteinaceous solidifying saliva that either forms a complete sheath around the mouthparts (stylets) as they pierce and penetrate the food or just a securing flange at the point of entry (section 11.2.4, Fig. 11.4c). Stylet-sheath feeding is characteristic of many phloem- and xylem-feeding Hemiptera, such as aphids, scale insects (coccoids) and spittle bugs, which leave visible tracks formed of exuded solidifying saliva in the plant tissue on which they have fed. The stylet-sheath track can be either intercellular or intracellular, depending on the hemipteran species. The sheath may function to guide the stylets, prevent loss of fluid from damaged cells, and/or absorb necrosis-inducing compounds to reduce defensive reaction by the plant. By comparison, certain leafhoppers (the Typhlocybinae) and many Heteroptera feed by moving the stylets so as to rupture plant cells, and these bugs mostly feed intracellularly on leaf or stem parenchyma cells rather than in vascular tissue. This cell-rupture strategy includes a macerate-and-flush technique in which salivary enzymes dissolve cell walls, as in Miridae, and a lacerate-and-flush method in which stylet movement damages plant cells, as in Lygaeidae. The released plant fluids are "flushed out" with saliva and ingested by sucking. A third feeding strategy, osmotic-pump feeding, occurs in Coreidae and involves increasing the osmotic concentration in the intercellular spaces to acquire ("pump") plant-cell contents without mechanically damaging the cell membranes. Coreid feeding causes collapsed cells and watery lesions.

In fluid-feeding insects, prominent dilator muscles attach to the walls of the pharynx and/or the preoral cavity (cibarium) to form a pump (Fig. 3.14b), although most other insects have some sort of pharyngeal pump (Fig. 3.14a) for drinking and for air intake to facilitate cuticle expansion during a moult.

3.6.3 Digestion of food

Most digestion occurs in the midgut, where the epithelial cells produce and secrete digestive enzymes and also absorb the resultant food breakdown products. Insect food consists principally of polymers of carbohydrates and proteins, which are digested by enzymatically breaking these large molecules into small monomers. The midgut pH usually is 6.0–7.5, although very alkaline values (pH 9–12) occur in many plant-feeding insects that extract hemicelluloses from plant cell walls, and very low pH values occur in many Diptera. High pH may prevent or reduce the binding of dietary tannins to food proteins, thereby increasing the digestibility of ingested plants. In some insects, gut lumenal surfactants (detergents) may have an important role in preventing the formation of tannin–protein complexes, particularly in insects with near-neutral gut pH.

In most insects, the midgut epithelium is separated from the food bolus by the peritrophic matrix (also called the peritrophic membrane or envelope; see section 3.6.1), which constitutes a very efficient high-flux sieve. It is perforated by pores, which allow passage of small molecules while restricting large molecules, bacteria and food particles from directly accessing the midgut cells. The peritrophic membrane also may protect herbivorous insects from ingested allelochemicals such as tannins (section 11.2). In some insects, all or most midgut digestion occurs inside the peritrophic membrane in the **endoperitrophic space**. In others, only initial digestion occurs there and smaller food molecules then diffuse out into the **ectoperitrophic space**, where further digestion takes place (Fig. 3.16). A final phase of digestion usually occurs on the surface of the midgut microvilli, where certain enzymes are either trapped in a mucopolysaccharide coating or bound to the cell membrane. Thus, the peritrophic membrane forms a permeability barrier and helps to compartmentalize the phases of digestion, in addition to providing mechanical protection of the midgut cells, which was once believed to be its principal function. Fluid containing partially digested food molecules and digestive enzymes is thought to circulate through the midgut in a posterior direction in the endoperitrophic space and forwards in the ectoperitrophic space, as indicated in Fig. 3.16. This endo–ectoperitrophic circulation may facilitate digestion by moving food molecules to sites of final digestion and absorption, and/or by conserving digestive enzymes, which are removed from the food bolus before it passes to the hindgut.

Unusually, both Hemiptera and Thysanoptera (but not Psocodea), which lack a peritrophic membrane, have an extracellular lipoprotein membrane, the

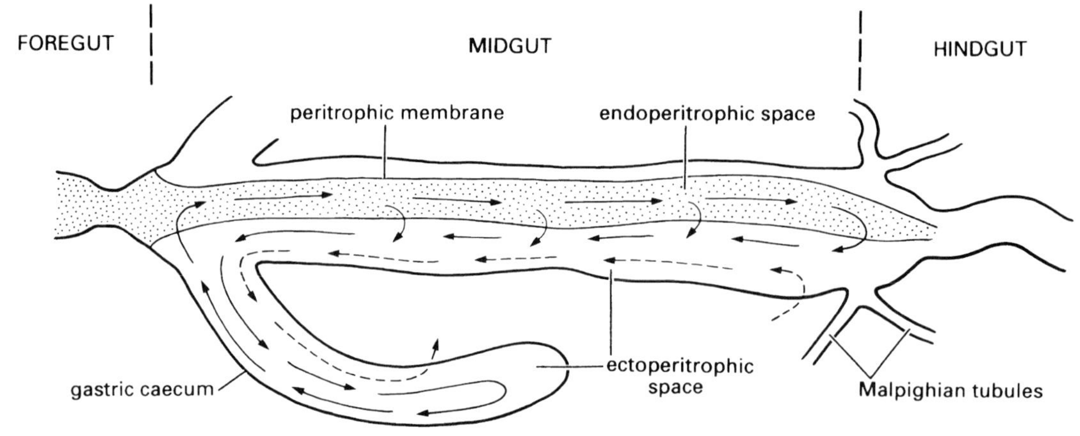

Fig. 3.16 Generalized scheme of the endo–ectoperitrophic circulation of digestive enzymes in the midgut. (After Terra & Ferreira 1981.)

perimicrovillar membrane (PPM), ensheathing the microvilli of the midgut cells and forming a closed space. The PPM and the associated perimicrovillar space may function to improve absorption of organic molecules, such as amino acids, by generating a concentration gradient between the liquid contents of the gut lumen and the fluid in the perimicrovillar space adjacent to the midgut cells. The PPM contains membrane-bound alpha-glucosidase, which breaks down sucrose in the ingested plant sap.

3.6.4 The fat body

In many insects, especially in holometabolous larvae, fat body tissue is a conspicuous internal component (Fig. 3.7 and Fig. 3.15), typically forming a pale tissue comprising loose sheets, ribbons or lobes of cells lying in the haemocoel. The structure of this organ is ill-defined and taxonomically variable, but often caterpillars and other larvae have a peripheral fat body beneath the cuticle and a central layer around the gut. The fat body has multiple metabolic functions, including: the metabolism of carbohydrates, lipids and nitrogenous compounds; the storage of glycogen, fat and protein; the synthesis and regulation of blood sugar; the synthesis of major haemolymph proteins (such as haemoglobins, vitellogenins for yolk formation, and storage proteins); and as an endocrine organ producing growth factors and hydroxylating the ecdysone synthesized in the prothoracic glands to 20-hydroxyecdysone (the active moulting hormone; section 3.3.2). Fat body cells can switch their activities in response to nutritional and hormonal signals in order to supply the requirements of insect growth, metamorphosis and reproduction. For example, specific storage proteins are synthesized by the fat body during the final larval instar of holometabolous insects, and accumulate in the haemolymph to be used during metamorphosis as a source of amino acids for the synthesis of proteins during pupation. **Calliphorin**, a haemolymph storage protein synthesized in the fat body of larval blow flies (Diptera: Calliphoridae: *Calliphora*), may form about 75% (about 7 mg) of the haemolymph protein of a late-instar maggot; the amount of calliphorin falls to around 3 mg at the time of pupariation and to 0.03 mg after emergence of the adult fly. The production and deposition of proteins specifically for amino acid storage is a feature that insects share with seed plants but not with vertebrates. Humans, for example, excrete any dietary amino acids that are in excess of immediate needs.

The principal cell type found in the fat body is the **trophocyte** (or adipocyte), which is responsible for most of the above metabolic and storage functions. Visible differences in the extent of the fat body in different individuals of the same insect species reflect the amount of material stored in the trophocytes; little body fat indicates either active tissue construction or starvation. Two other cell types – urocytes and bacteriocytes (also called mycetocytes) – may occur in the fat body of some insect groups. **Urocytes** temporarily store spherules of urates, including uric acid, one of the nitrogenous waste products of insects. Amongst studied cockroaches, rather than being permanent stores of excreted waste uric acid (storage excretion), urocytes recycle urate nitrogen, perhaps with assistance of mycetocyte bacteria. **Bacteriocytes** (mycetocytes) contain symbiotic microorganisms, and are either scattered through the fat body of cockroaches or are contained within special organs, sometimes surrounded by the fat body. These symbionts are important in insect nutrition.

3.6.5 Nutrition and microorganisms

Broadly defined, nutrition concerns the nature and processing of foods needed to meet the requirements for growth and development, involving feeding behaviour (Chapter 2) and digestion. Insects often have unusual or restricted diets. Sometimes, although only one or a few foods are eaten, the diet provides a complete range of the chemicals essential to metabolism. In these cases, monophagy is a specialization without nutritional limitations. In other cases, a restricted diet may require utilization of microorganisms in digesting or supplementing the directly available nutrients. In particular, insects cannot synthesize sterols (required for synthesis of moulting hormone) or carotenoids (used in visual pigments), and these must come from the diet or from microorganisms.

Insects may harbour extracellular or intracellular microorganisms, which are referred to as **symbionts** because they are dependent on their insect hosts. These microorganisms contribute to the nutrition of their hosts by functioning in sterol, vitamin, carbohydrate or amino acid synthesis and/or metabolism. Symbiotic microorganisms may be bacteria or bacteroids, yeasts or other unicellular fungi, or protists. Historically, studies on their function were hampered by difficulties in removing them (e.g. with antibiotics, to produce aposymbiotic hosts) without harming the host insect, and also in culturing the microorganisms outside the host. The diets of their hosts provided some clues as to the functions of these microorganisms. Insect hosts include many sap-sucking hemipterans (such as aphids, psylloids, whiteflies, scale insects, thrips, leafhoppers and cicadas) and sap- and blood-sucking heteropterans (Hemiptera), parasitic lice (Psocodea), some wood-feeding insects (such as termites and some longicorn beetles and weevils), many seed- or grain-feeding insects (certain beetles), and some omnivorous insects (such as cockroaches, some termites and some ants). Predatory insects never seem to contain such symbionts. That microorganisms are required by insects on suboptimal diets has been confirmed by modern studies showing, for example, that critical dietary shortfall in certain essential amino acids in aposymbiotic aphids is compensated for by production by *Buchnera* symbionts (Gammaproteobacteria). Similarly, the nutritionally poor xylem-sap diet of sharpshooters (Hemiptera: Cicadellidae) is augmented by the biosyntheses of its two main symbionts – *Sulcia muelleri* (Bacteroidetes), which has genes for the synthesis of essential amino acids, and *Baumannia cicadellinicola* (Gammaproteobacteria), which provides its host with cofactors including B-vitamins. In some termites, spirochaete bacteria provide much of a colony's nitrogen, carbon and energy requirements via acetogenesis and nitrogen fixation.

Extracellular symbionts may be free in the gut lumen or housed in diverticula or pockets of the midgut or hindgut. For example, termite hindguts contain a veritable fermenter, comprising many bacteria, fungi and protists, including flagellates. These assist in the degradation of the otherwise refractory dietary lignocellulose, and in the fixation of atmospheric nitrogen. The process involves generation of methane, and calculations suggest that tropical termites' symbiont-assisted cellulose digestion produces a significant proportion of the world's methane (a greenhouse gas).

Transmission of extracellular symbionts from one individual insect to another involves oral uptake, often by the offspring. Microorganisms may be acquired from the anus or the excreta of other individuals, or eaten at a specific time, as in some bugs, in which the newly hatched young eat the contents of special symbiont-containing capsules deposited with the eggs.

Intracellular symbionts (**endosymbionts**) may occur in as many as 70% of all insect species. Endosymbionts probably mostly have a mutualistic association with their host insect, but some appear parasitic on their host. Examples of the latter include *Wolbachia* (section 5.10.4, although even this bacterium may sometimes have nutritional benefits for the host insect, as in the bed bug *Cimex lectularius*), *Spiroplasma* and microsporidia. Endosymbionts may be housed in the gut epithelium, as in lygaeid bugs and some weevils; however, most insects with intracellular microorganisms house them in symbiont-containing cells called bacteriocytes or mycetocytes. These cells are in the body cavity, usually associated with the fat body or the gonads, and often in special aggregations, forming an organ called a **bacteriome** or mycetome. In such insects, the symbionts are transferred to the ovary and then to the eggs or embryos prior to oviposition or parturition – a process referred to as **vertical** or **transovarial transmission**. Lacking evidence for lateral transfer (to an unrelated host), this method of transmission, which is found in many Hemiptera and cockroaches, indicates a very close association or coevolution of the insects and their microorganisms. Evidence of benefits of endosymbionts to their hosts is accruing rapidly due to genome sequencing and studies of gene expression. For example, as mentioned above, the provision of the otherwise dietarily scarce essential amino acids to aphids by their bacteriocyte-associated *Buchnera* symbiont is well substantiated. Of interest for further research is the suggestion that aphid biotypes with *Buchnera* bacteriocytes show enhanced ability to transmit certain plant viruses of the genus *Luteovirus* relative to antibiotic-treated, symbiont-free individuals. The relationship between bacteriocyte endosymbionts and their phloem-feeding host insects is a very tight phylogenetic association (cf. *Wolbachia* infections, section 5.10.4), suggesting a very long association with co-diversification.

Some insects that maintain fungi essential to their diet cultivate them externally to their body as a means of converting woody substances to an assimilable form. Examples are the fungus gardens of some ants (Formicidae) and termites (Termitidae) (section 9.5.2 and section 9.5.3), and the fungi transmitted by certain timber pests, namely, wood wasps (Hymenoptera: Siricidae) and ambrosia beetles (Coleoptera: Scolytinae).

The genomes of *Drosophila melanogaster*, *Anopheles gambiae* and *Bombyx morii* were found to lack cellulase genes, and thus endogenous cellulases were presumed to be absent from all insects. However, careful experiments on cellulose-ingesting insects show that at least some cockroaches and termites, crickets, longicorn beetles and the mustard leaf beetle, *Phaedon cochleariae* (Chrysomelidae), are capable of expressing endogenous cellulases and do not necessarily rely solely on symbionts.

3.7 THE EXCRETORY SYSTEM AND WASTE DISPOSAL

Excretion – the removal from the body of the waste products of metabolism, especially nitrogenous compounds – is essential. It differs from defecation in that excretory wastes have been metabolized in cells of the body rather than simply passing directly from the mouth to the anus (sometimes essentially unchanged chemically). Of course, insect faeces, either in liquid form or packaged in pellets and known as **frass**, contain both undigested food and metabolic excretions. Aquatic insects eliminate dilute wastes from their anus directly into the water, and so their faecal material is flushed away. In comparison, terrestrial insects generally must conserve water. This requires efficient waste disposal in a concentrated or even dry form, while simultaneously avoiding the potentially toxic effects of nitrogen. Furthermore, both terrestrial and aquatic insects must conserve ions, such as sodium (Na^+), potassium (K^+) and chloride (Cl^-), which may be limiting in their food or, in aquatic insects, lost into the water by diffusion. Production of insect urine or frass therefore results from two intimately related processes: excretion and **osmoregulation** – the maintenance of a favourable body-fluid composition (osmotic and ionic homeostasis). The system responsible for excretion and osmoregulation is referred to loosely as the excretory system, and its activities are performed largely by the Malpighian tubules and hindgut as outlined below. However, the haemolymph composition in freshwater insects must be regulated in response to the constant loss of salts (as ions) to the surrounding water, and ionic regulation involves both the typical excretory system and special cells, called **chloride cells**, which usually are associated with the hindgut. Chloride cells are capable of absorbing inorganic ions from very dilute solutions, and are best studied in aquatic bugs and nymphal dragonflies and damselflies.

3.7.1 The Malpighian tubules and rectum

The main organs of excretion and osmoregulation in insects are the Malpighian tubules and the rectum and/or ileum, which all act together (Fig. 3.17). Malpighian tubules are outgrowths of the alimentary canal and consist of long, thin tubules (Fig. 3.1) formed of a single layer of cells surrounding a blind-ending lumen. They range in number from as few as two in most scale insects (coccoids) to over 200 in large locusts. Generally, they are free, waving around in the haemolymph, where they filter out solutes. Aphids are the only insects that lack Malpighian tubules, although they are reduced to papillae in Strepsiptera. The vignette at the beginning of this chapter shows the gut of *Locusta*, but with only a few of the many

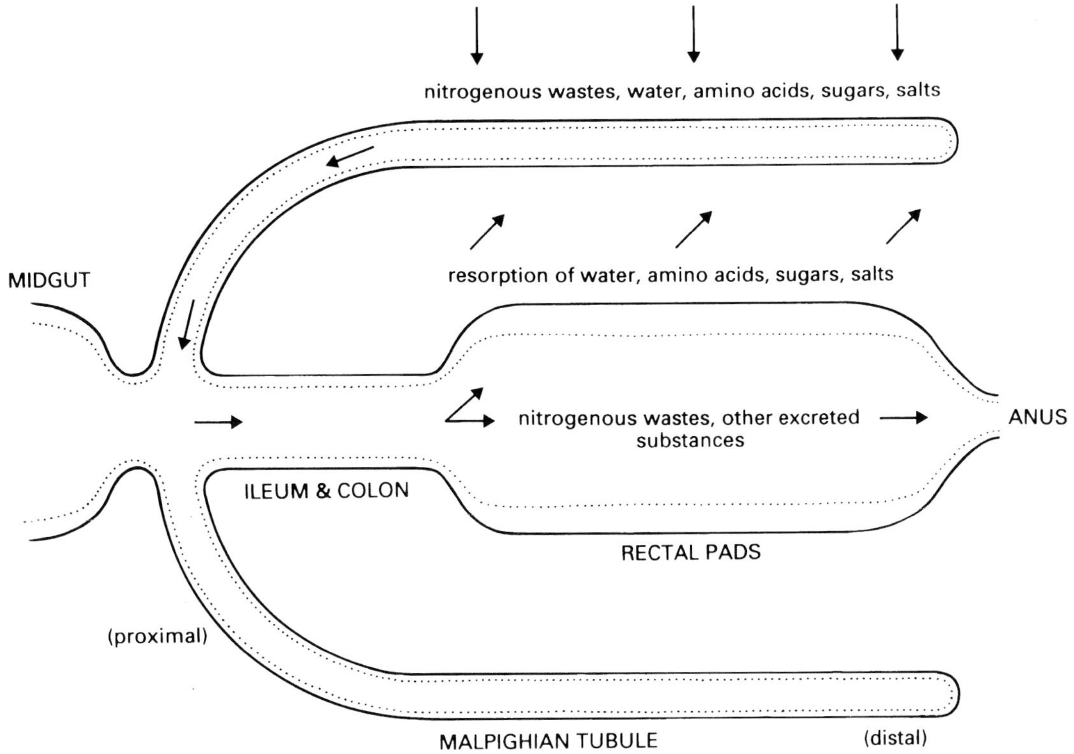

Fig. 3.17 Schematic diagram of a generalized excretory system showing the path of elimination of wastes. (After Daly *et al.* 1978.)

Malpighian tubules depicted. Similar structures are believed to have arisen convergently in different arthropod groups, such as myriapods and arachnids, in response to the physiological stresses of life on dry land. Insect Malpighian tubules are considered to belong to the hindgut and to be ectodermal in origin. Their position marks the junction of the midgut and the cuticle-lined hindgut.

The anterior hindgut is called the **ileum**, the generally narrower middle portion is the **colon**, and the expanded posterior section is the **rectum** (Fig. 3.13). In many terrestrial insects, the rectum is the only site of water and solute resorption from the excreta, but in other insects, for example the desert locust *Schistocerca gregaria* (Orthoptera: Acrididae), the ileum makes some contribution to osmoregulation. In a few insects, such as the cockroach *Periplaneta americana* (Blattodea: Blattidae), even the colon may be a potential site of some fluid absorption. The resorptive role of the rectum (and sometimes the anterior hindgut) is indicated by its anatomy. In most insects, specific parts of the rectal epithelium are thickened to form **rectal pads** or papillae, composed of aggregations of columnar cells. Typically, there are six pads arranged longitudinally, but there may be fewer pads or many papillate ones.

The general picture of insect excretory processes outlined here is applicable to most freshwater species and to the adults of many terrestrial species. The Malpighian tubules produce a filtrate (the primary urine) that is isosmotic but ionically dissimilar to the haemolymph, and then the hindgut, especially the rectum, selectively resorbs water and certain solutes but eliminates others (Fig. 3.17). Details of Malpighian tubule and rectal structure and of filtration and absorption mechanisms differ between taxa, in relation to both taxonomic position and dietary composition (Box 3.4 gives an example of one type of specialization – cryptonephric systems), but the excretory system of the desert locust *S. gregaria* (Fig. 3.18) exemplifies the general structure

Box 3.4 Cryptonephric systems*

Many larval and adult Coleoptera, larval Lepidoptera and some larval Symphyta have a modified arrangement of the excretory system that is involved either with efficient dehydration of faeces before their elimination (in beetles) or ionic regulation (in plant-feeding caterpillars). These insects have a **cryptonephric system** in which the distal ends of the Malpighian tubules are held in contact with the rectal wall by the perinephric membrane. Such an arrangement allows some beetles that live on a very dry diet, such as stored grain or dry carcasses, to be extraordinarily efficient in their conservation of water. Water even may be extracted from the humid air in the rectum. In the cryptonephric system of the mealworm, *Tenebrio molitor* (Coleoptera: Tenebrionidae), shown here, ions (principally of potassium chloride, KCl) are transported into, and concentrated in, the six Malpighian tubules, creating an osmotic gradient that draws water from the surrounding perirectal space and the rectal lumen. The tubule fluid is then transported forwards to the free portion of each tubule, from which it is passed to the haemolymph or recycled in the rectum.

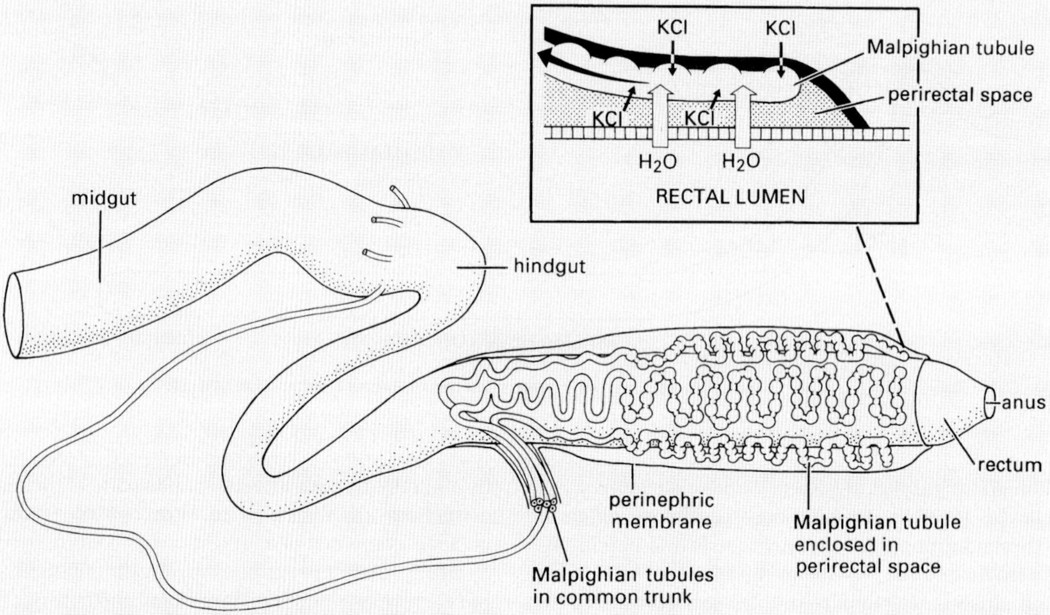

*After Grimstone *et al.* 1968; Bradley 1985.

and principles of insect excretion. The Malpighian tubules of the locust produce an isosmotic filtrate of the haemolymph, which is high in K^+, low in Na^+, and has Cl^- as the major anion. The active transport of ions, especially K^+, into the tubule lumen generates an osmotic pressure gradient so that water passively follows (Fig. 3.18a). Sugars and most amino acids also are filtered passively from the haemolymph (probably via junctions between the tubule cells), whereas the amino acid proline (later used as an energy source by the rectal cells) and non-metabolizable and toxic organic compounds are transported actively into the tubule lumen. Sugars, such as sucrose and trehalose, are resorbed from the lumen and returned to the haemolymph. The continuous secretory activity of each Malpighian tubule leads to a flow of primary

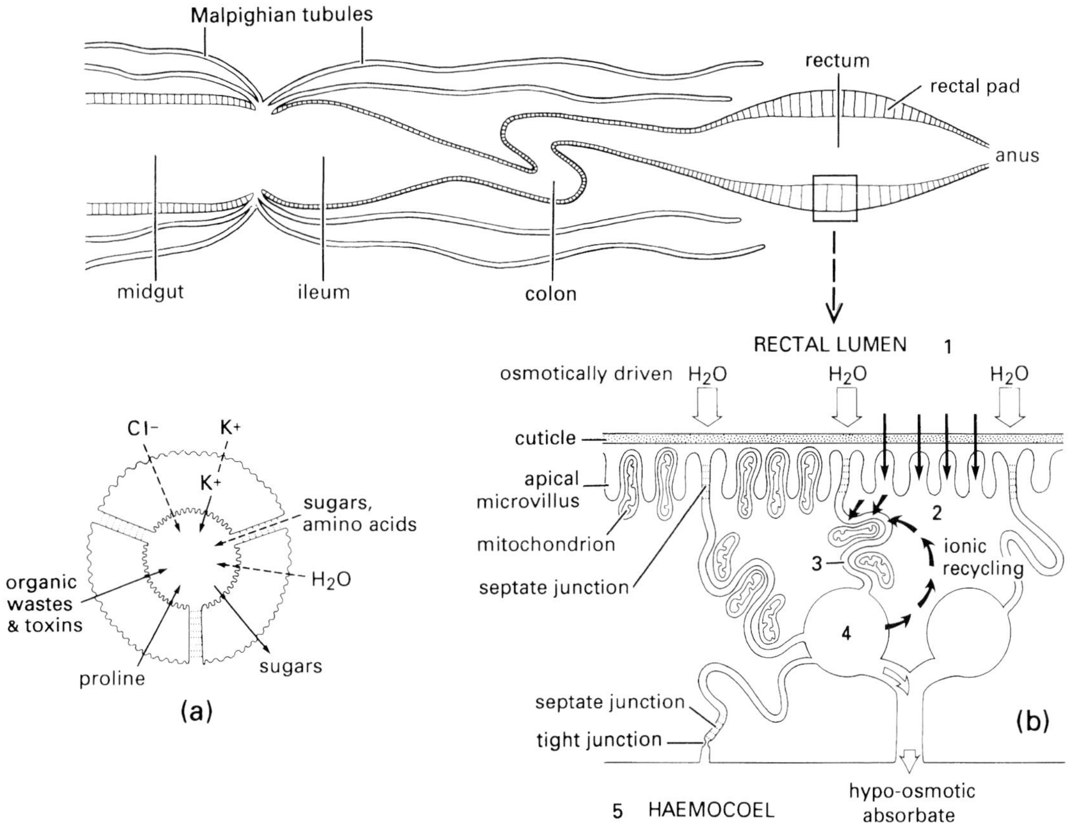

Fig. 3.18 Schematic diagram of the organs in the excretory system of the desert locust, *Schistocerca gregaria* (Orthoptera: Acrididae). Only a few of the > 100 Malpighian tubules are drawn. (a) Transverse section of one Malpighian tubule showing probable transport of ions, water and other substances between the surrounding haemolymph and the tubule lumen; active processes are indicated by solid arrows and passive processes by dashed arrows. (b) Diagram illustrating the movements of solutes and water in the rectal pad cells during fluid resorption from the rectal lumen. Pathways of water movement are represented by open arrows, and solute movements by black arrows. Ions are actively transported from the rectal lumen (compartment 1) to the adjacent cell cytoplasm (compartment 2) and then to the intercellular spaces (compartment 3). Mitochondria are positioned so as to provide the energy for this active ion transport. Fluid in the spaces is hyperosmotic (higher ion concentration) to the rectal lumen and draws water by osmosis from the lumen via the septate junctions between the cells. Water thus moves from compartment 1 to 3 to 4 and finally to 5, the haemolymph in the haemocoel. (After Bradley 1985.)

urine from its lumen towards and into the gut. In the rectum, the urine is modified by removal of solutes and water to maintain fluid and ionic homeostasis of the locust's body (Fig. 3.18b). Specialized cells in the rectal pads carry out active recovery of Cl^- under hormonal stimulation. This pumping of Cl^- generates electrical and osmotic gradients that lead to some resorption of other ions, water, amino acids and acetate.

3.7.2 Nitrogen excretion

Many predatory, blood-feeding and even plant-feeding insects ingest nitrogen, particularly certain amino acids, far in excess of requirements. Most insects excrete nitrogenous metabolic wastes at some or all stages of their life, although some nitrogen is stored in the fat body or as proteins in the haemolymph in some

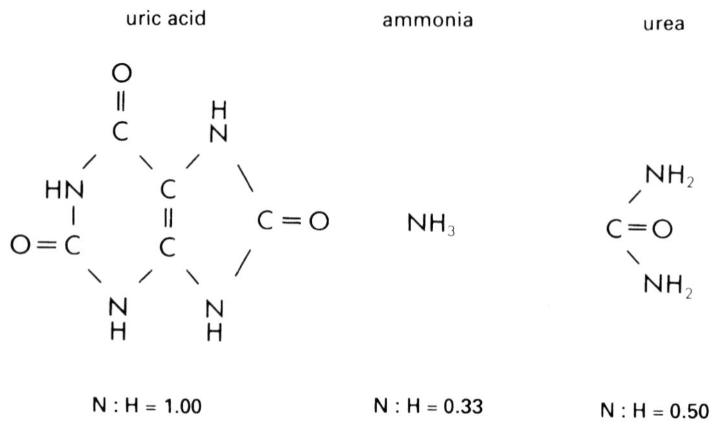

Fig. 3.19 Molecules of the three common nitrogenous excretory products. The high N : H ratio of uric acid relative to both ammonia and urea means that less water is used for uric acid synthesis (as hydrogen atoms are derived ultimately from water).

insects. Many aquatic insects and some flesh-feeding flies excrete large amounts of ammonia, whereas in terrestrial insects wastes generally consist of **uric acid** and/or certain of its salts (urates), often in combination with **urea**, pteridines, certain amino acids and/or relatives of uric acid, such as hypoxanthine, allantoin and allantoic acid. Amongst these waste compounds, ammonia is relatively toxic and usually must be excreted as a dilute solution, or else rapidly volatilized from the cuticle or faeces (as in cockroaches). Urea is less toxic but is more soluble, requiring a lot of water for its elimination. Uric acid and urates require less water for their synthesis than either ammonia or urea (Fig. 3.19), are non-toxic and, having low solubility in water (at least in acidic conditions), can be excreted essentially dry, without causing osmotic problems. Waste dilution can be achieved easily by aquatic insects, but water conservation is essential for terrestrial insects and uric acid excretion (**uricotelism**) is highly advantageous.

Historically, deposition of urates in specific cells of the fat body (section 3.6.4) was viewed as "excretion" by storage of uric acid. However, in at least some insects such as cockroaches, it is now known to constitute a metabolic store for recycling by the insect with the assistance of symbiotic microorganisms. Cockroaches, including *P. americana*, do not excrete uric acid in the faeces even if fed a high-nitrogen diet but do produce large quantities of internally stored urates. The cockroach fat body houses endosymbiont bacteria (*Blattabacterium*) capable of synthesizing amino acids from ammonia, urea and glutamate, as determined from genome analysis.

By-products of feeding and metabolism need not be excreted as waste – for example, the antifeedant defensive compounds of plants may be sequestered directly or may form the biochemical base for synthesis of chemicals used in communication (Chapter 4), including in warning and defence. White-pigmented uric-acid derivatives colour the epidermis of some insects and provide the white in the wing scales of certain butterflies (Lepidoptera: Pieridae).

3.8 REPRODUCTIVE ORGANS

The reproductive organs of insects exhibit an incredible variety of forms, but there is a basic design and function to each component so that even the most aberrant reproductive system can be understood in terms of a generalized plan. Individual components of the reproductive system can vary in shape (e.g. of gonads and accessory glands), position (e.g. of the attachment of accessory glands) and number (e.g. of ovarian or testicular tubes, or sperm storage organs) between different insect groups, and sometimes even between different species within a genus. Knowledge of the homology of the components assists in interpreting structure and function in different insects. Generalized male and female systems are depicted in Fig. 3.20, and a comparison of the corresponding reproductive structures of male and female insects is provided in Table 3.2. Many other aspects of reproduction, including copulation and regulation of physiological processes, are discussed in detail in Chapter 5.

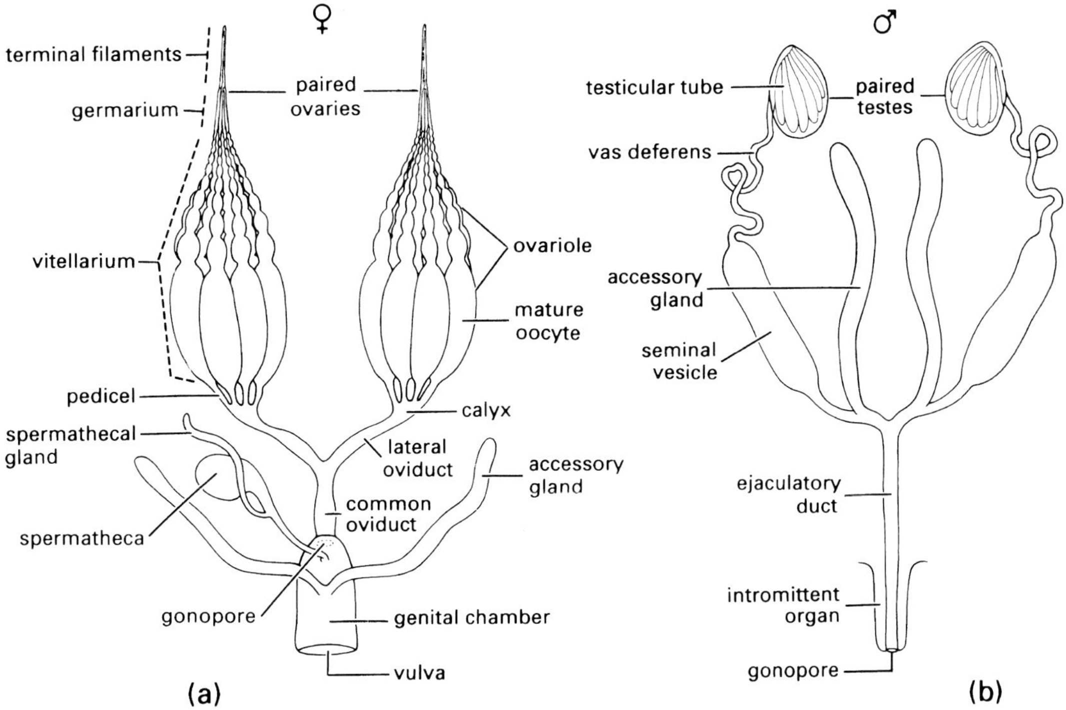

Fig. 3.20 Comparison of generalized: (a) female reproductive system; and (b) male reproductive system. (After Snodgrass 1935.)

Table 3.2 The corresponding female and male reproductive organs of insects.

Female reproductive organs	Male reproductive organs
Paired ovaries composed of ovarioles (ovarian tubes)	Paired testes composed of follicles (testicular tubes)
Paired oviducts (ducts leading from ovaries)	Paired vasa deferentia (ducts leading from testes)
Egg calyces (if present, reception of eggs)	Seminal vesicles (sperm storage)
Common (median) oviduct and vagina	Median ejaculatory duct
Accessory glands (ectodermal origin: colleterial or cement glands)	Accessory glands (two types): (i) ectodermal origin (ii) mesodermal origin
Bursa copulatrix (copulatory pouch) and spermatheca (sperm storage)	No equivalent
Ovipositor (if present)	Genitalia (if present): aedeagus and associated structures

3.8.1 The female reproductive system

The main functions of the female reproductive system are egg production, including the provision of a protective coating in many insects, and the storage of the male's spermatozoa until the eggs are ready to be fertilized. Transport of the spermatozoa to the female's storage organ and their subsequent controlled release requires movement of the spermatozoa, which in some species is mediated by muscular contractions of parts of the female reproductive tract.

The basic components of the female reproductive system (Fig. 3.20a) are paired **ovaries**, which empty their mature **oocytes** (eggs) via the calyces (singular: **calyx**) into the **lateral oviducts**, which unite to form the **common** (or **median**) **oviduct**. The **gonopore** (opening) of the common oviduct usually is concealed in an inflection of the body wall that typically forms a cavity, the **genital chamber**. This chamber serves as a copulatory pouch during mating, and thus often is known as the **bursa copulatrix**. Its external opening is the **vulva**. In many insects, the vulva is narrow and the genital chamber becomes an enclosed pouch or tube, referred to as the **vagina**. Two sorts of ectodermal glands open into the genital chamber. The first is the **spermatheca**, which stores spermatozoa until needed for egg fertilization. Typically, the spermatheca is single, generally sac-like with a slender duct, and often has a diverticulum that forms a tubular **spermathecal gland**. The gland or glandular cells within the storage part of the spermatheca provide nourishment to the contained spermatozoa. The second type of ectodermal gland, known collectively as accessory glands, opens more posteriorly in the genital chamber and has a variety of functions, depending on the species (see section 5.8).

Each ovary is composed of a cluster of ovarian or egg tubes, the **ovarioles**, each consisting of a terminal filament, a **germarium** (in which mitosis gives rise to primary oocytes), a **vitellarium** (in which oocytes grow by deposition of yolk in a process known as vitellogenesis; section 5.11.1) and a **pedicel** (or stalk). An ovariole contains a series of developing oocytes, each surrounded by a layer of follicle cells forming an epithelium (the oocyte and its epithelium is termed a follicle); the youngest oocytes occur near the apical germarium and the most mature near the pedicel.

Three different types of ovariole are recognized, based on the manner in which the oocytes are nourished. A **panoistic ovariole** lacks specialized nutritive cells so that it contains only a string of follicles, with the oocytes obtaining nutrients from the haemolymph via the follicular epithelium. Ovarioles of the other two types contain trophocytes (nurse cells) that contribute to the nutrition of the developing oocytes. In a **telotrophic** (or **acrotrophic**) **ovariole**, the trophocytes are confined to the germarium and remain connected to the oocytes by cytoplasmic strands as the oocytes move down the ovariole. In a **polytrophic ovariole**, a number of trophocytes are connected to each oocyte and move down the ovariole with it, providing nutrients until depleted; thus individual oocytes alternate with groups of successively smaller trophocytes. Different suborders or orders of insects usually have only one of these three ovariole types.

Accessory glands of the female reproductive tract often are referred to as **colleterial** (or **cement**) **glands** because in most insect orders their secretions surround and protect the eggs or cement them to the substrate (section 5.8). In other insects, the accessory glands may function as **poison glands** (as in many Hymenoptera) or as "**milk**" **glands** in the few insects (e.g. tsetse flies, *Glossina* spp.) that exhibit adenotrophic viviparity (section 5.9). Accessory glands of a variety of forms and functions appear to have been derived independently in different orders and even may be non-homologous within an order, as in Coleoptera.

3.8.2 The male reproductive system

The main functions of the male reproductive system are the production and storage of spermatozoa, and their transport in a viable state to the reproductive tract of the female. Morphologically, the male tract consists of paired **testes**, each containing a series of testicular tubes or follicles (in which spermatozoa are produced), which open separately into the mesodermally derived sperm duct or **vas deferens**, which usually expands posteriorly to form a sperm-storage organ, or **seminal vesicle** (Fig. 3.20b). Typically, tubular, paired **accessory glands** are formed as diverticula of the vasa deferentia, but sometimes the vasa deferentia themselves are glandular and fulfil the functions of accessory glands (see below). The paired vasa deferentia unite where they lead into the ectodermally derived **ejaculatory duct** – the tube that transports the semen or the sperm package to the gonopore. In a few insects, particularly certain flies, the accessory glands consist of an enlarged glandular part of the ejaculatory duct.

Thus, the accessory glands of male insects can be classified into two types according to their mesodermal or ectodermal derivation. Almost all are mesodermal in origin, and those apparently of ectodermal origin have been poorly studied. Furthermore, the mesodermal structures of the male tract frequently differ morphologically from the basic paired sacs or tubes described above. For example, in male cockroaches and many other orthopteroids, the seminal vesicles and the numerous accessory gland tubules (Fig. 3.1) are clustered into a single median structure called the **mushroom body**. Secretions of the male accessory glands form the **spermatophore** (the package that surrounds the spermatozoa of many insects), contribute to the seminal fluid, which nourishes the spermatozoa during transport to the female, are involved in activation (induction of motility) of the spermatozoa, and may alter female behaviour (induce non-receptivity to further males and/or stimulate oviposition; see section 5.4, section 5.11 and Box 5.6).

FURTHER READING

Alstein, M. (2003) Neuropeptides. In: *Encyclopedia of Insects* (eds V.H. Resh & R.T. Cardé), pp. 782–5. Academic Press, Amsterdam.

Bourtzis, K. & Miller, T.A. (eds.) (2003, 2006, 2008) *Insect Symbiosis*. Vols **1–3**. CRC Press, Boca Raton, FL.

Caers, J., Verlinden, H., Zels, S. *et al.* (2012) More than two decades of research on insect neuropeptide GPCRs: an overview. *Frontiers in Endocrinology* **3**, 151. doi: 10.3389/fendo.2012.00151

Chapman, R.F. (2013) *The Insects. Structure and Function*, 5th edn. (eds. S.J. Simpson & A.E. Douglas), Cambridge University Press, Cambridge.

Chown, S.L., Gibbs, A.G., Hetz, S.K. *et al.* (2006) Discontinuous gas exchange in insects: a clarification of hypotheses and approaches. *Physiological and Biochemical Zoology* **79**, 333–43.

Davey, K.G. (1985) The male reproductive tract/The female reproductive tract. In: *Comprehensive Insect Physiology, Biochemistry, and Pharmacology*, Vol. **1**: *Embryogenesis and Reproduction* (eds.G.A. Kerkut & L.I. Gilbert), pp. 1–14, 15–36. Pergamon Press, Oxford.

Dillon, R.J. & Dillon, V.M. (2004) The gut bacteria of insects: nonpathogenic interactions. *Annual Review of Entomology* **49**, 71–92.

Gilbert, L.I. (ed.) (2012) *Insect Molecular Biology and Biochemistry*. Academic Press, London.

Gilbert, L.I., Iatrou, K. & Gill, S.S. (eds.) (2005) *Comprehensive Molecular Insect Science*. Vols **1–6**. Elsevier, Oxford.

González-Santoyo, I. & Córdoba-Aguilar, A. (2012) Phenoloxidase: a key component of the insect immune system. *Entomologia Experimentalis et Applicata* **142**, 1–16.

Harrison, F.W. & Locke, M. (eds.) (1998) *Microscopic Anatomy of Invertebrates*, Vol. **11B**: Insecta. Wiley–Liss, New York.

Hegedus, D., Erlandson, E., Gillott, C. & Toprak, U. (2009) New insights into peritrophic matrix synthesis, architecture, and function. *Annual Review of Entomology* **54**, 285–302.

Hoy, M. (2003) *Insect Molecular Genetics: An Introduction to Principles and Applications*, 2nd edn. Academic Press, San Diego, CA.

Klowden, M.J. (2013) *Physiological Systems in Insects*, 3rd edn. Academic Press, London, San Diego.

Nation, J.L. (2008) *Insect Physiology and Biochemistry*, 2nd edn. CRC Press, Boca Raton, FL.

Nijhout, H.F. (2013) Arthropod developmental endocrinology. In: *Arthropod Biology and Evolution* (eds A. Minelli, G. Boxshall & G. Fusco), pp. 123–48. Springer-Verlag, Berlin, Heidelberg.

Niven, J.E., Graham, C.M. & Burrows, M. (2008) Diversity and evolution of the insect ventral nerve cord. *Annual Review of Entomology* **53**, 253–71.

Ponton, F., Wilson, K., Holmes, A.J. *et al.* (2013) Integrating nutrition and immunology: a new frontier. *Journal of Insect Physiology* **59**, 130–7.

Resh, V.H. & Cardé, R.T. (eds.) (2009) *Encyclopedia of Insects*, 2nd edn. Elsevier, San Diego, CA. [Particularly see articles on accessory glands; brain and optic lobes; circulatory system; digestion; digestive system; excretion; fat body; flight; hemolymph; immunology; muscle system; neuropeptides; nutrition; respiratory system; salivary glands; symbionts aiding digestion; tracheal system; walking and jumping; water and ion balance, hormonal control of.]

Schmid-Hempel, P. (2005) Evolutionary ecology of insect immune defenses. *Annual Review of Entomology* **50**, 529–51.

Sharma, A., Khan, A.N., Subrahmanyam, S. *et al.* (2013) Salivary proteins of plant-feeding hemipteroids – implication in phytophagy. *Bulletin of Entomological Research*. doi:10.1017/S0007485313000618.

Terra, W.R. (1990) Evolution of digestive systems of insects. *Annual Review of Entomology* **35**, 181–200.

White, C.R., Blackburn, T.M., Terblanche, J.S. *et al.* (2007). Evolutionary responses of discontinuous gas exchange in insects. *Proceedings of the National Academy of Sciences* **104**, 8357–61.

White, J.A., Giorgini, M., Strand, M.R. & Pennacchio, F. (2013) Arthropod endosymbiosis and evolution. In: *Arthropod Biology and Evolution* (eds A. Minelli, G. Boxshall & G. Fusco), pp. 441–77. Springer-Verlag, Berlin, Heidelberg.

Chapter 4

SENSORY SYSTEMS AND BEHAVIOUR

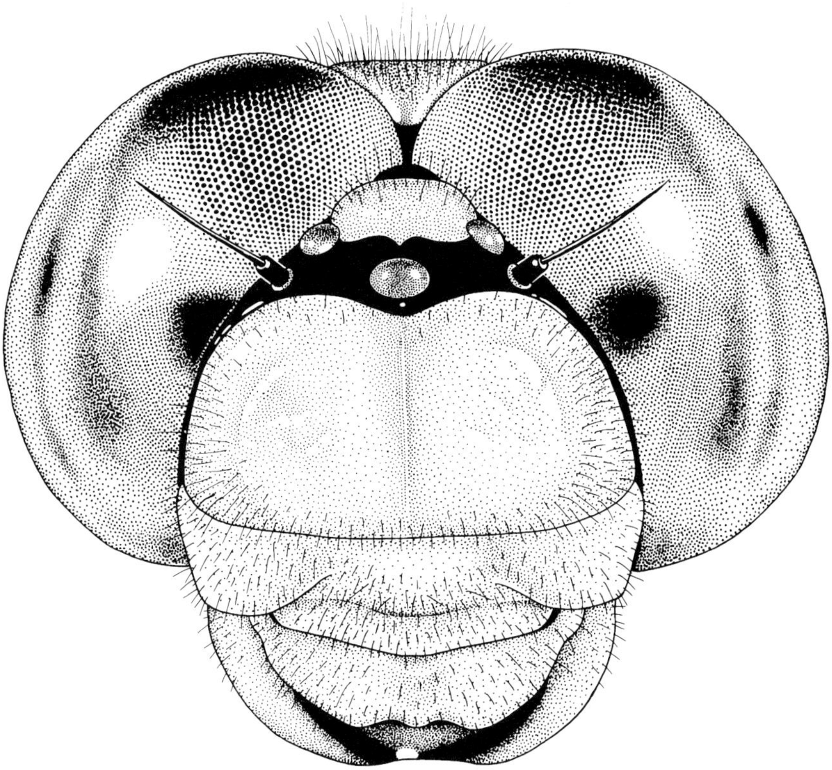

Head of a dragonfly showing enormous compound eyes. (After Blaney 1976.)

The success of insects derives, at least in part, from their ability to sense and interpret their surroundings and to discriminate on a fine scale. Insects respond selectively to cues from a heterogeneous environment, differentiate between hosts (both plant and animal), and respond to microclimatic factors, such as humidity, temperature and airflow.

Sensory complexity allows for both simple and complex behaviours of insects. For example, to control flight, the aerial environment must be sensed and appropriate responses made. Because much insect activity is nocturnal, orientation and navigation cannot rely solely on conventional visual cues, and in many night-active species, odours and sounds play a major role in communication. The range of sensory information used by insects differs from that of humans. We rely heavily on visual information, and although many insects have well-developed vision, most insects make greater use of olfaction and hearing than humans do.

The insect is isolated from its external surroundings by a relatively inflexible, insensitive and impermeable cuticular barrier. The answer to the enigma of how this armoured insect can perceive its immediate environment lies in frequent and abundant cuticular structures that detect external stimuli. Sensory organs (sensilla, singular: **sensillum**) protrude from the cuticle, or sometimes lie within or beneath it. Specialized cells detect stimuli that may be categorized as mechanical, thermal, chemical and visual. Other cells (the neurons) transmit messages to the central nervous system (section 3.2), where they are integrated. The nervous system instigates and controls appropriate behaviours, such as posture, movement and feeding, as well as behaviours associated with mating and oviposition.

This chapter surveys these sensory systems and presents selected behaviours that are elicited or modified by environmental stimuli. The means of detection and, where relevant, production of these stimuli are treated in the following sequence: touch, position, sound, temperature, chemicals (with particular emphasis on communication chemicals, called pheromones) and light. The chapter concludes with a section that relates some aspects of insect behaviour to the preceding discussion on stimuli. Boxes cover four special topics: aural location of a host by a parasitoid fly (Box 4.1); reception of communication molecules (Box 4.2); the electroantennogram (Box 4.3); and biological clocks (Box 4.4).

4.1 MECHANICAL STIMULI

The stimuli grouped here are those associated with distortion caused by mechanical movement, as a result of the environment itself, the insect in relation to the environment, or internal forces derived from the muscles. The mechanical stimuli sensed include touch, body stretching and stress, position, pressure, gravity and vibrations, including pressure changes of the air and substrate involved in sound transmission and hearing.

4.1.1 Tactile mechanoreception

Insects are covered externally with cuticular projections, called microtrichia if many arise from one cell, or hairs, bristles, setae or macrotrichia if they are of multicellular origin. Most flexible projections arise from an innervated socket. These are sensilla, termed **trichoid sensilla** (literally, hair-like little sense organs), and develop from epidermal cells that switch from cuticle production. Three cells are involved (Fig. 4.1):

1 a **trichogen cell**, which grows the conical hair;
2 a **tormogen cell**, which grows the socket;
3 a **sensory neuron**, or nerve cell, which grows a **dendrite** into the hair and an **axon** that winds inwards to link with other axons to form a nerve connected to the central nervous system.

Fully developed trichoid sensilla fulfil tactile functions. As touch sensilla they respond to the movement of the hair by firing impulses from the dendrite at a frequency related to the extent of the deflection. Touch sensilla are stimulated only during actual movement of the hair. Each hair can differ in sensitivity, with some being so sensitive that they respond to vibrations of air particles caused by noise (section 4.1.3).

The Insects: An Outline of Entomology, Fifth Edition. P.J. Gullan and P.S. Cranston.

Companion Website: www.wiley.com/go/gullan/insects

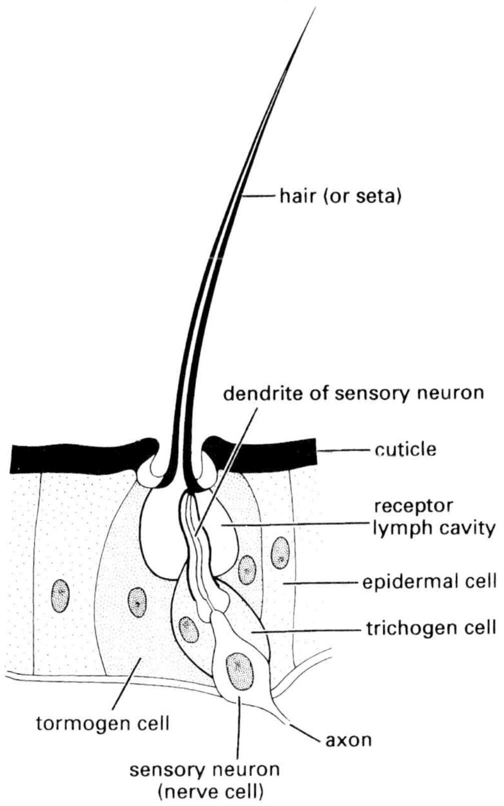

Fig. 4.1 Longitudinal section of a trichoid sensillum showing the arrangement of the three associated cells. (After Chapman 1991.)

4.1.2 Position mechanoreception (proprioceptors)

Insects require continuous knowledge of the relative position of their body parts, such as limbs or head, and need to detect how the orientation of the body relates to gravity. This information is conveyed by **proprioceptors** (self-perception receptors), of which three types are described here. One type of trichoid sensillum gives a continuous sensory output at a frequency that varies with the position of the hair. Sensilla often form a bed of grouped small hairs, a **hair plate**, at joints or at the neck, in contact with the cuticle of an adjacent body part (Fig. 4.2a). The degree of flexion of the joint gives a variable stimulus to the sensilla, thereby allowing monitoring of the relative positions of different parts of the body.

The second type, stretch receptors, comprise internal proprioceptors associated with muscles such as those of the abdominal and gut walls. Alteration of the length of the muscle fibre is detected by multiple-inserted neuron endings, producing variation in the rate of firing of the nerve cell. Stretch receptors monitor body functions such as abdominal or gut distension, or ventilation rate.

The third type, stress detectors on the cuticle, function via stress receptors called **campaniform sensilla**. Each sensillum comprises a central cap or peg surrounded by a raised circle of cuticle and with a single neuron per sensillum (Fig. 4.2b). These sensilla are located on joints, such as those of legs and wings, and other places liable to distortion. Locations include the haltere (the knob-like modified hind wing of Diptera), at the base of which there are dorsal and ventral groups of campaniform sensilla that respond to distortions created during flight.

4.1.3 Sound reception

Sound is a pressure fluctuation transmitted in a wave form via movement of the air or the substrate, including water. Sound and hearing are terms often applied to the quite limited range of frequencies of airborne vibration that humans perceive with their ears, usually in adults this is from 20 to 20,000 Hz (1 hertz (Hz) is a frequency of one cycle per second). Such a definition of sound is restrictive, particularly as amongst insects some receive vibrations ranging from as low as 1–2 Hz to ultrasound frequencies perhaps as high as 100 kHz. Specialized emission and reception across this range of frequencies of vibration are considered here. The reception of these frequencies involves a variety of organs, none of which resemble the ears of mammals.

An important role of insect sound is in intraspecific acoustic communication. For example, courtship in most orthopterans is acoustic, with males producing species-specific sounds ("songs") that the predominantly non-singing females detect and upon which they base their choice of mate. Hearing also allows detection of predators, such as insectivorous bats, which use ultrasound in hunting. It is probable that each species of insect detects sound within one or two relatively narrow ranges of frequencies that relate to these functions.

The insect mechanoreceptive communication system can be viewed as a continuum from substrate

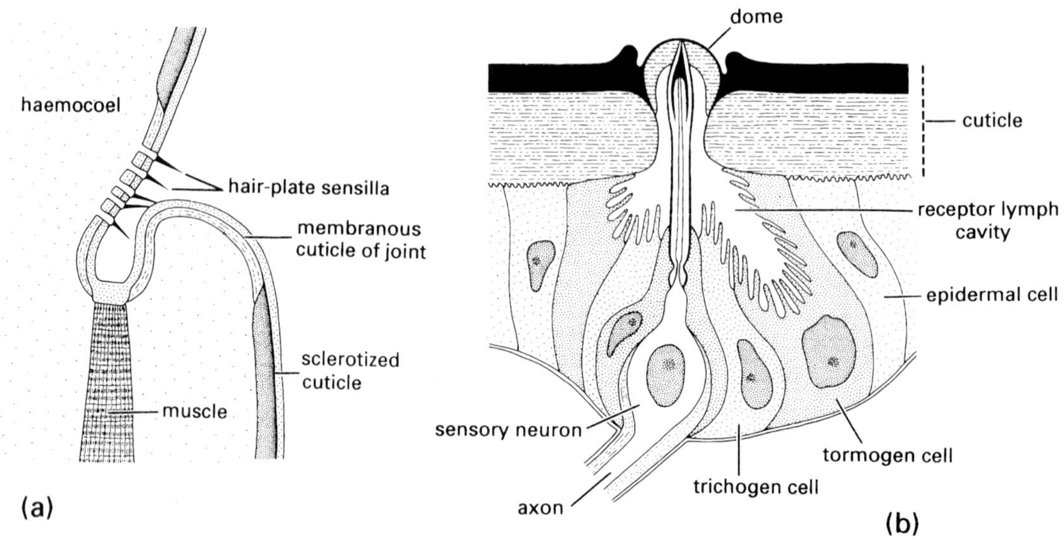

Fig. 4.2 Proprioceptors: (a) sensilla of a hair plate located at a joint, showing how the hairs are stimulated by contacting adjacent cuticle; (b) campaniform sensillum on the haltere of a fly. ((a) After Chapman 1982; (b) after Snodgrass 1935; McIver 1985.)

vibration reception, grading through the reception of very near airborne vibration, to hearing of more distant sound using thin cuticular membranes called tympana (singular: **tympanum**; adjective: tympanal). Substrate signalling probably appeared first in insect evolution; the sensory organs used to detect substrate vibrations appear to have been co-opted and modified many times in different insect groups to allow for reception of airborne sound at considerable distance and at a range of frequencies.

Non-tympanal vibration reception

Two types of vibration or sound reception that do not involve tympana (see next subsection for tympanal reception) are the detection of substrate-borne signals and the ability to perceive the relatively large translational movements of the surrounding medium (air or water) that occur very close to a sound. The latter, referred to as near-field sound, is detected by either sensory hairs or specialized sensory organs.

A simple form of sound reception occurs in species that have very sensitive, elongate, trichoid sensilla that respond to vibrations produced by a near-field sound. For example, caterpillars of the noctuid moth *Barathra brassicae* have thoracic hairs about 0.5 mm long that respond optimally to vibrations of 150 Hz. Although in air this system is effective only for local sounds, caterpillars can respond to the vibrations caused by the audible approach of parasitic wasps.

The cerci of many insects, especially crickets, are clothed in long, fine trichoid sensilla (filiform setae or hairs) that sense air currents, which can convey information about the approach of predatory or parasitic insects or a potential mate. The direction of approach of another animal is indicated by which hairs are deflected; the sensory neuron of each hair is tuned to respond to movement in a particular direction. The dynamics (the time-varying pattern) of air movement gives information on the nature of the stimulus (and thus on what type of animal is approaching), and is indicated by the properties of the mechanosensory hairs. The length of each hair determines the response of its sensory neuron to the stimulus: neurons that innervate short hairs are most sensitive to high-intensity, high-frequency stimuli, whereas long hairs are more sensitive to low-intensity, low-frequency stimuli. The responses of many sensory neurons innervating different hairs on the cerci are integrated in the central nervous system to allow the insect to behave appropriately to detected air movement.

For low-frequency sounds in water (a medium more viscous than air), longer distance transmission is possible. Although few aquatic insects communicate through underwater sounds, "drumming" sounds are produced by some aquatic larvae to assert territory, and noises are produced by underwater diving hemipterans such as corixids and nepids.

Many insects can detect vibrations transmitted through a substrate at a solid–air or solid–water boundary or along a water–air surface. The perception of substrate vibrations is particularly important for ground-dwelling insects, especially nocturnal species, and social insects living in dark nests. Some insects living on plant surfaces, such as sawflies (Hymenoptera: Pergidae), communicate with each other by tapping the stem. Various plant-feeding bugs (Hemiptera), such as leafhoppers, planthoppers and pentatomids, produce vibratory signals that are transmitted through the host plant. Water-striders (Hemiptera: Gerridae; see Box 10.2) create pulsed waves across the water surface film to communicate in courtship and aggression. Moreover, they can detect the vibrations produced by the struggles of prey that fall onto the water surface. Whirligig beetles (Coleoptera: Gyrinidae; see Fig. 10.7) can navigate using a form of echolocation: waves that move on the water surface ahead of them are reflected from obstacles and are sensed by their antennae in time to take evasive action.

The specialized sensory organs that receive vibrations are subcuticular mechanoreceptors called **chordotonal organs**. A chordotonal organ consists of one to many **scolopidia**, each of which consists of three linearly arranged cells: a sub-tympanal **cap cell** placed on top of a sheath cell (**scolopale cell**), which envelops the end of a nerve cell dendrite (Fig. 4.3). All adult insects, and many larvae, have a particular chordotonal organ, **Johnston's organ**, lying within the pedicel, the second antennal segment. The primary function is to sense movements of the antennal flagellum relative to the rest of the body, as in the detection of flight speed by air movement. Additionally, it functions in hearing in some insects. In male mosquitoes (Culicidae) and midges (Chironomidae), many scolopidia are contained in the swollen pedicel. These scolopidia are attached at one end to the pedicel wall and at the other, sensory, end to the base of the third antennal segment. This greatly modified Johnston's organ is the male receptor for the female wing tone (see section 4.1.4), as shown when males are rendered unreceptive to the sound of the female by amputation of the terminal flagellum or arista of the antenna.

Detection of substrate vibration involves the **subgenual organ**, a chordotonal organ located in the

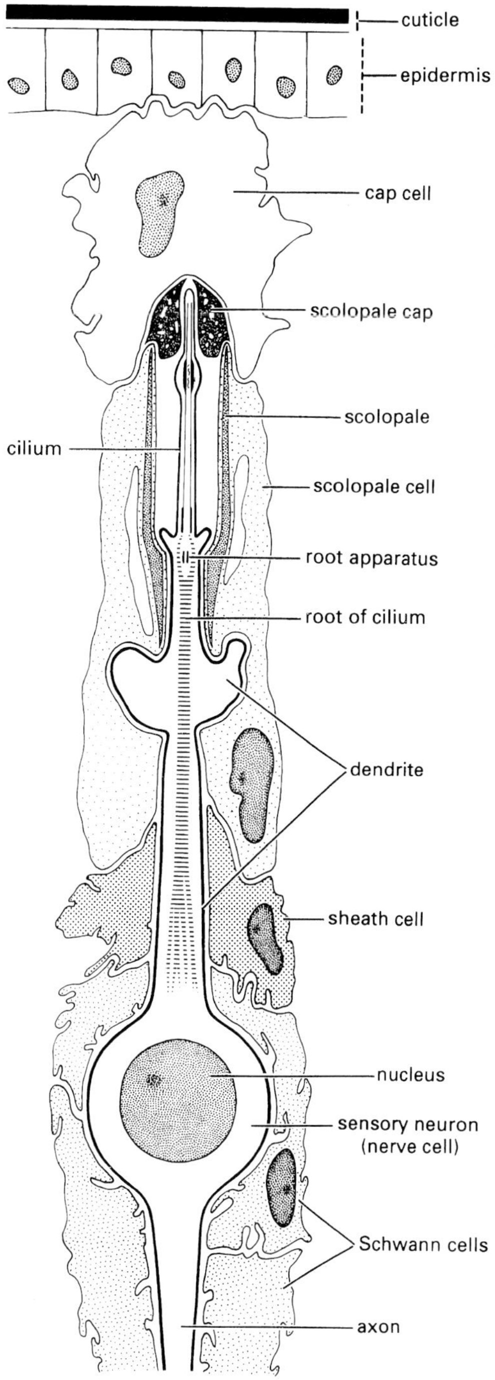

Fig. 4.3 Longitudinal section of a scolopidium, the basic unit of a chordotonal organ. (After Gray 1960.)

proximal tibia of each leg. Subgenual organs are found in most insects, but not in the Coleoptera and Diptera. The subgenual organ consists of a semi-circle of many sensory cells lying in the haemocoel, connected at one end to the inner cuticle of the tibia, and at the other end to the trachea. There are subgenual organs within all legs: the organs of each pair of legs may respond specifically to substrate-borne sounds of differing frequencies. Vibration reception may involve either direct transfer of low-frequency substrate vibrations to the legs, or there may be more complex amplification and transfer. Airborne vibrations can be detected if they cause vibration of the substrate and hence of the legs.

Tympanal reception

The most elaborate sound reception system in insects involves a specific receptor structure, the **tympanum**. This membrane responds to distant sounds transmitted by airborne vibration. Tympanal membranes are linked to chordotonal organs and are associated with air-filled sacs, especially modified trachea, that enhance sound reception. Tympanal organs acting as "ears" have evolved in at least seven orders of insects in a diversity of locations:

- the ventral thorax between the metathoracic legs of mantids;
- the thorax of many noctuid moths;
- the prothoracic legs of many orthopterans;
- the abdomen of other orthopterans, cicadas, and some moths and beetles;
- the wing bases of certain moths and the radial vein of some lacewings;
- under the elytra in tiger beetles, which only hear when in flight;
- the prosternum of some flies (Box 4.1);
- the cervical membranes of a few scarab beetles;
- the mouthparts of a few moths.

Variation in the location of these organs and their occurrence in distantly related insect groups indicates that tympanal hearing has evolved several times in insects. Neuroanatomical studies suggest that all insect tympanal organs evolved from proprioceptors, and the wide distribution of proprioceptors throughout the insect cuticle must account for the variety of positions of tympanal organs.

Tympanal sound reception is particularly well developed in orthopterans, especially the crickets and katydids. In most of these ensiferan Orthoptera, the tympanal organs are located in the tibia of each fore leg (Fig. 4.4; see also Fig. 9.2a). Behind the paired tympanal membranes lies an acoustic trachea that runs from a prothoracic spiracle down each leg to the tympanal organ (Fig. 4.4a).

Katydids (bushcrickets) and crickets have similar hearing systems. The system in crickets appears to be less specialized because their acoustic tracheae remain connected to the ventilatory spiracles of the prothorax. The acoustic tracheae of katydids form a system completely isolated from the ventilatory tracheae, opening via a separate pair of acoustic spiracles. In many katydids, the tibial base has two separated longitudinal slits, each of which leads into a tympanic chamber (Fig. 4.4b). The acoustic trachea, which lies centrally in the leg, is divided in half by a membrane at this point, such that one half closely connects with the anterior and the other half with the posterior tympanal membrane. The primary route of sound to the tympanal organ is usually from the acoustic spiracle and along the acoustic trachea to the tibia. The change in cross-sectional area from the enlargement of the trachea behind each spiracle (sometimes called a tracheal vesicle) to the tympanal organ in the tibia functions as a horn in amplifying the sound. Although the slits of the tympanic chambers allow the entry of sound, their exact function is debatable. They may allow directional hearing, because very small differences in the time of arrival of sound waves at the tympanum can be detected by pressure differences across the membrane.

Entry of sound to the tympanal organs by air- and substrate-borne signals cause the tympanal membranes to vibrate. Vibrations are sensed by three chordotonal organs: the **subgenual organ**, the **intermediate organ** and the **crista acustica** (Fig. 4.4c). The subgenual organs, which have a form and function similar to those in non-orthopteroid insects, are present on all legs, but the crista acustica and intermediate organs are found only on the fore legs in conjunction with the tympana. This implies that the tibial hearing organ is a serial homologue of the proprioceptor units of the mid and hind legs. The crista acustica consists of a row of up to 60 scolopidial cells attached to the acoustic trachea and is the main sensory organ for airborne sound in the 5–50 kHz range. The intermediate organ, which consists of 10–20 scolopidial cells, is posterior to the subgenual organ and is virtually continuous with the crista acustica. The role of the intermediate organ is uncertain but it may respond to airborne sound of frequencies from

2 to 14 kHz. Each of the three chordotonal organs is innervated separately, but the neuronal connections between the three imply that signals from the different receptors are integrated.

Hearing insects can locate a point source of sound, but exactly how they do so varies between taxa. Localization of sound directionality clearly depends upon detection of differences in the sound received by one tympanum relative to another, or in some orthopterans by a tympanum within a single leg. Sound reception varies with the orientation of the body relative to the sound source, allowing some precision in locating the source. The unusual means of sound reception and sensitivity of detection of direction of sound source shown by ormiine flies is discussed in Box 4.1.

Night activity is common, as shown by the abundance and diversity of insects attracted to artificial light, especially at the ultraviolet end of the spectrum, and on moonless nights. Night flight allows avoidance of visual-hunting predators, but exposes the insect to specialist nocturnal predators – the insectivorous bats (Microchiroptera). These bats employ a biological sonar system, using ultrasonic frequencies that range (according to species) from 8 to 215 kHz, for navigating and for detecting and locating prey, predominantly flying insects.

Although bat predation on insects occurs in the darkness of night and high above a human observer, it is evident that a range of insect taxa can detect bat ultrasounds and take appropriate evasive action. The behavioural response to ultrasound, known as the acoustic startle response, involves very rapid and co-ordinated muscle contractions. This leads to reactions such as "freezing", unpredictable deviation in flight, or rapid cessation of flight and plummeting towards the ground. Instigation of these reactions, which assist in escape from predation, obviously requires that the insect hears the ultrasound produced by the bat. Physiological experiments show that the response can be within a few milliseconds of the emission of such a sound, which would precede the detection of the prey by a bat.

Certain insects belonging to at least six orders detect and respond to ultrasound: praying mantids (Mantodea), locusts, katydids and crickets (Orthoptera), lacewings (Neuroptera), moths (Lepidoptera), beetles

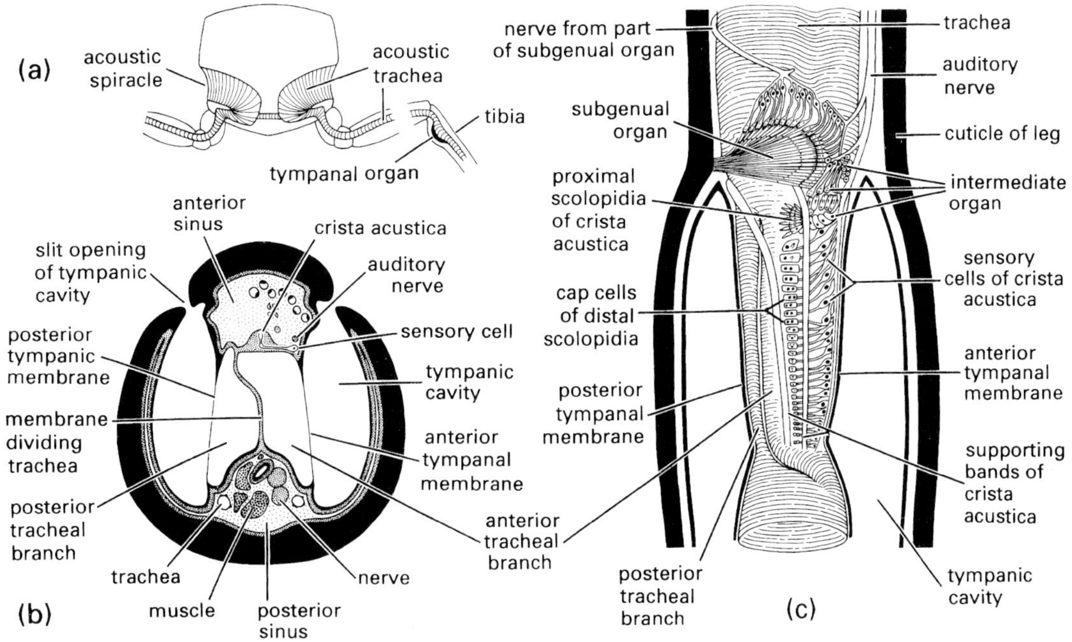

Fig. 4.4 Tympanal organs of a katydid, *Decticus* (Orthoptera: Tettigoniidae): (a) transverse section through the fore legs and prothorax to show the acoustic spiracles and tracheae; (b) transverse section through the base of the fore tibia; (c) longitudinal breakaway view of the fore tibia. (After Schwabe 1906, in Michelsen & Larsen 1985.)

Box 4.1 Aural location of host by a parasitoid fly

Parasitoid insects predominantly use chemical and visual cues to track down hosts (section 13.1), upon which the development of their immature stage depends. Locating a host from afar by orientation towards a sound that is specific for that host is rather unusual behaviour. Although close-up low-frequency air movements produced by prospective hosts can be detected, for example by fleas and some blood-feeding flies (section 4.1.3), host location by distant sound is developed best in flies of the tribe Ormiini (Diptera: Tachinidae). The hosts are male crickets, for example of the genus *Gryllus*, and katydids, whose mate-attracting songs (chirps) range in frequency from 2 to 7 kHz. Under the cover of darkness, the female *Ormia* locates the calling host insect, on or near which she deposits first-instar larvae (larviposits). The larvae burrow into the host, in which they develop by feeding on selected tissues for 7–10 days, after which the third-instar larvae emerge from the dying host and pupariate in the ground.

Location of a calling host is a complex process compared with simply being able to detect its presence by hearing the call, as will be understood by anyone who has tried to locate a calling cricket or katydid. Directional hearing is a prerequisite to be able to orientate towards and localize the source of the sound. In most animals with directional hearing, the two receptors ("ears") are separated by a distance greater than the wavelength of the sound, such that the differences (e.g. in intensity and timing) between the sounds received by each "ear" are large enough to be detected and converted by the receptor and nervous system. However, in small animals, such as the house-fly-sized ormiine female, with a hearing system spanning less than 1.5 mm, the "ears" are too close together to create interaural differences in intensity and timing. A very different approach to sound detection is required.

As in other hearing insects, the reception system consists of a flexible tympanal membrane, an air sac apposed to the tympanum, and a chordotonal organ linked to the tympanum (section 4.1.3). Uniquely amongst hearing insects, the ormiine paired tympanal membranes are located on the prosternum, ventral to the neck (cervix), facing forwards and are somewhat obscured by the head (as illustrated here in the side view of a female fly of *Ormia*; after Robert *et al.* 1994). On the inner surface of these thin (1 μm) membranes are attached a pair of auditory sense organs, the bulbae acusticae – chordotonal organs comprising many scolopidia (section 4.1.3). The bulbae are located within an unpartitioned prosternal chamber, which is enlarged by relocation of the anterior musculature and connected to the external environment by tracheae. A sagittal view of this hearing organ is shown in the diagram to the right of the fly. The structures are sexually dimorphic, with strongest development being in the host-seeking female.

What is anatomically unique amongst hearing animals, including all other insects studied, is that in this fly there is no separation of the "ears" – the auditory chamber that contains the sensory organs is undivided. Furthermore,

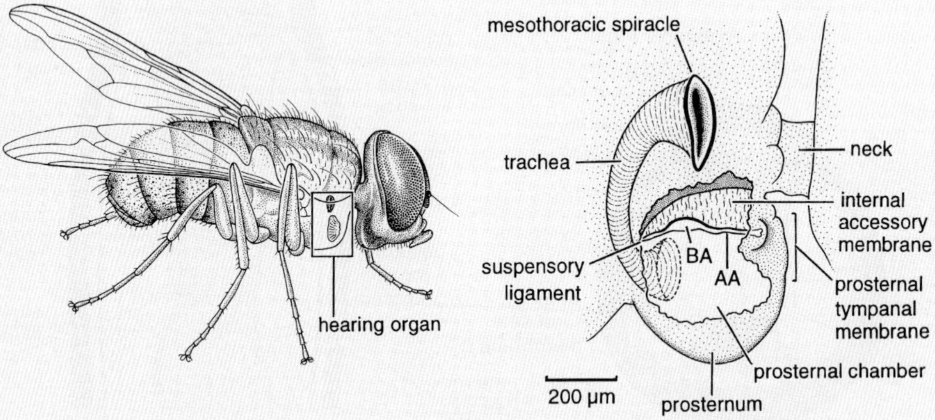

AA = auditory apodeme   BA = bulbae acusticae

the tympana virtually abut, such that the difference in arrival time of sound at each ear is <1–2 microseconds. The answer to the physical dilemma is revealed by close examination, which shows that the two tympana actually are joined by a cuticular structure that functions to connect the ears. This mechanical intra-aural coupling involves the connecting cuticle acting as a flexible lever that pivots about a fulcrum and functions to increase the time lag between the nearer-to-noise (ipsilateral) tympanum and the further-from-noise (contralateral) tympanum by about 20-fold. The ipsilateral tympanic membrane is first to be excited to vibrate by incoming sound, slightly before the contralateral one, with the connecting cuticle then commencing to vibrate. In a complex manner involving some damping and cancellation of vibrations, the ipsilateral tympanum produces most vibrations.

This magnification of interaural differences allows for very sensitive directionality in sound reception. This novel design discovered in ormiine hearing has possible applications in human hearing-aid technology.

(Coleoptera), and some parasitoid flies (Diptera). Tympanal organs occur in different sites amongst these insects, showing that ultrasound reception has several independent origins amongst these orders. Orthoptera are major acoustic communicators that use sound in intraspecific sexual signalling. Evidently, hearing ability arose early in orthopteran evolution, probably at least some 200 mya, long before bats evolved (perhaps some 50 Ma in the Eocene, from which the oldest fossil comes). The orthopteran ability to hear bat ultrasounds thus is an **exaptation** – in this case, a morphological–physiological predisposition that has been modified to permit sensitivity to ultrasound. The crickets, katydids and acridid grasshoppers that communicate intraspecifically and also hear ultrasound have sensitivity to high- and low-frequency sound – and perhaps limit their discrimination to only two discrete frequencies. The ultrasound elicits aversion; the low-frequency sound (under suitable conditions) elicits attraction.

In contrast, the tympanal hearing that has arisen independently in several other insects appears to be receptive specifically to ultrasound. The two receptors of a "hearing" noctuoid moth, though differing in threshold, are tuned to the same ultrasonic frequency, and it has been demonstrated experimentally that the moths show behavioural (startle) and physiological (neural) response to bat sonic frequencies. The female of the parasitic tachinid fly *Ormia* (Box 4.1) locates its orthopteran host by tracking its mating calls. The structure and function of the "ear" is sexually dimorphic: the tympanic area of the female fly is larger, and is sensitive to the 5 kHz frequency of the cricket host and also to the ultrasounds made by insectivorous bats. In contrast, the male fly has a smaller tympanic area that responds optimally at 10 kHz and extending up to ultrasound. This suggests that the acoustic response originally was present in both sexes and was used to detect and avoid bats, with sensitivity to cricket calls a later modification only in the female.

At least in these cases, and probably in other groups in which tympanal hearing is limited in taxonomic range and complexity, ultrasound reception appears to have coevolved with the sounds made by the bats that seek to eat them.

4.1.4 Sound production

The most common method of sound production by insects is **stridulation**, in which one specialized body part, the **scraper**, is rubbed against another, the **file**. The file is a series of teeth, ridges or pegs, which vibrate through contact with a ridged or plectrum-like scraper. The file itself makes little noise, and so this has to be amplified to generate airborne sound. The horn-shaped burrow of the mole cricket is an excellent sound enhancer (Fig. 4.5). Other insects show many modifications of the body, particularly of wings and internal air sacs of the tracheal system, to provide amplification and resonance.

Sound production by stridulation occurs in some species of many orders of insects, but the Orthoptera show the most elaboration and diversity. All stridulating orthopterans enhance their sounds using the tegmina (the modified fore wings). The file of katydids and crickets is formed from a basal vein of one or both tegmina, and rasps against a scraper on the other wing. Grasshoppers and locusts (Acrididae) rasp a file on the hind femora against a similar scraper on the tegmen.

Many insects lack the body size, power or sophistication to produce high-frequency airborne sounds,

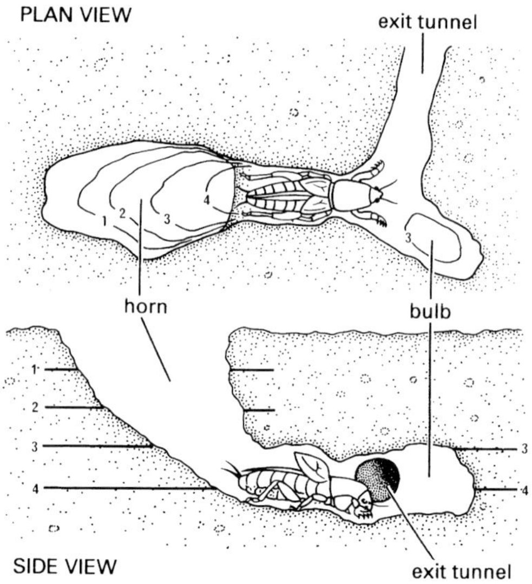

Fig. 4.5 The singing burrow of a mole cricket, *Scapteriscus acletus* (Orthoptera: Gryllotalpidae), in which the singing male sits with his head in the bulb and tegmina raised across the throat of the horn. The depth markers in the burrow are in centimetres, with the burrow being just over 4 cm deep. (After Bennet-Clark 1989.)

but some can generate low-frequency sound that is transmitted by vibration of the substrate (such as wood, soil or a host plant), which is a denser medium. Insect communication via substrate vibration has been recorded in most orders of winged (pterygote) insects and is the most common modality in at least 10 orders, but it has been poorly studied in most groups. Examples of insects that communicate with each other via substrate vibration are psylloids and treehoppers (Hemiptera), some stoneflies (Plecoptera) and heelwalkers (Mantophasmatodea). Such vibrational signals can lead a male to a nearby receptive female, although pheromones may be used for long-range attraction. Substrate vibrations are also a by-product of airborne sound production, as in acoustic signalling insects such as some katydids, whose whole body vibrates whilst producing audible airborne stridulatory sounds. Body vibrations, which are transferred through the legs to the substrate (plant or ground), are of low frequencies of 1–5000 Hz. Substrate vibrations can be detected by the female and appear to be used in closer-range localization of the calling male, in contrast to the airborne signal used at greater distance.

A second method of sound production involves alternate muscular distortion and relaxation of a specialized area of elastic cuticle, the **tymbal** (or timbal), to give individual clicks or variably modulated pulses of sound. Tymbal sound produced by cicadas is very audible to us, but many other hemipterans and some moths also produce sounds from a tymbal. In the cicadas, only the males have these paired tymbals, which are located dorsolaterally, one on each side, on the first abdominal segment. The tymbal membrane is supported by a number of ribs. A strong tymbal muscle distorts the membrane and ribs to produce a sound; on relaxation, the elastic tymbal returns to rest. High-frequency sounds are produced by the tymbal muscle contracting asynchronously, with many contractions per nerve impulse (section 3.1.1). A group of chordonotal sensilla is present and a smaller tensor muscle controls the shape of the tymbal, thereby allowing alteration of the acoustic property. The noise of one or more clicks is emitted as the tymbal distorts, and further sounds may be produced during the elastic return on relaxation. The first abdominal segment contains air sacs – modified tracheae – tuned to resonate at, or close to, the natural frequency of tymbal vibration.

The calls of cicadas generally are in the range of 3–16 kHz, usually of high intensity, carrying as far as 1 km, even in dense forest. Sound is received by both sexes via tympanic membranes that lie ventral to the position of the male tymbal on the first abdominal segment. Cicada calls are species-specific – studies in New Zealand and North America have shown specificity of duration and cadence of introductory cueing phases inducing timed responses from a prospective mate. Interestingly however, song structures are very homoplasious, with similar songs found in distantly related taxa, but closely related taxa differing markedly in their song.

In other sound-producing hemipterans, both sexes may possess tymbals but because they lack abdominal air sacs, the sound is very damped compared with that of cicadas. The sounds produced by *Nilaparvata lugens* (the brown planthopper; Delphacidae) and other non-cicadan hemipterans are transmitted by vibration of the substrate, and are specifically associated with mating.

Many moths produce ultrasound, mainly using either tymbals on the thorax or abdomen, or a file and scraper on the wings, legs or genitalia. Ultrasonic songs of moths are used primarily for sexual communication, particularly in mate attraction and courtship, and range from loud to very low in intensity. Some tiger moths (Arctiidae) produce ultrasound using metathoracic tymbals and can hear the ultrasound produced by predatory bats. Aside from intraspecific communication between these tiger moths, the evolution of sound production and ultrasound hearing exemplifies an "arms war" in the bat–moth predation interactions. Thus, the high-frequency clicking sounds that some arctiids produce cause bats to veer away from attack, and have the following (not mutually exclusive) roles:

- interference with bat sonar systems;
- aural mimicry of a bat to delude the predator about the presence of a prey item;
- warning of distastefulness (aposematism; see section 14.4).

The humming or buzzing sound characteristic of swarming mosquitoes, gnats and midges is a flight tone produced by the frequency of wing beat. This tone, which can be virtually species-specific, differs between the sexes: the male produces a higher tone than the female. The tone also varies with age and ambient temperature for both sexes. Males are able to recognize females of their own species by wing-beat frequency. Male insects that form nuptial (mating) swarms recognize the swarm site by species-specific environmental markers rather than by audible cues (section 5.1); they are insensitive to the wing tone of males of their species. Neither can the male detect the wing tone of immature females – the Johnson's organ in his antenna responds only to the wing tone of physiologically receptive females.

4.2 THERMAL STIMULI

4.2.1 Thermoreception

Insects evidently detect variation in temperature, as seen by their behaviour (section 4.2.2), yet the nature and location of receptors is poorly known. Most studied insects sense temperature with their antennae – amputation leads to a thermal response different from that of an intact insect. Antennal temperature receptors are few (presumably ambient temperature is similar at all points along the antenna), are exposed or concealed in pits, and may be associated with humidity receptors in the same sensillum. In leafcutter ants (*Atta* species), thermo-sensitive peg-in-pit sensilla coeloconica (**coeloconic sensilla**) are clustered in the apical antennal flagellomere and respond both to changes in air temperature and to radiant heat. In the cockroach *Periplaneta americana*, the arolium and pulvilli of the tarsi bear temperature receptors, and thermoreceptors have been found on the legs of certain other insects. Central sensors must exist to detect internal temperature, but there is little experimental evidence. A small set of warmth-activated neurons have been identified in the brain of *Drosophila* and appear to allow the flies to avoid non-preferred temperatures. A similar mechanism is likely to occur in other insects, especially those using warmth sensing to detect vertebrate hosts or specific microhabitats. In the large saturniid moth, *Hyalophora cecropia*, the thoracic neural ganglia have a role in instigating temperature-dependent flight muscle activity.

An extreme form of temperature detection is illustrated in jewel beetles (Buprestidae) belonging to the largely Holarctic genus *Melanophila* and also in *Merimna atrata* (the "fire-beetle" of Australia). These day-flying beetles can detect and orientate towards large-scale forest fires, where they oviposit in still-smouldering pine trunks. Adults of *Melanophila* eat insects killed by fire, and their larvae develop as pioneering colonists boring into trees killed by fire. Detection and orientation in *Melanophila* and *Merimna* to distant fires is achieved neither by smell nor sight but by response to infrared (IR) radiation detected by sensilla located in pit organs. In *Melanophila* these pit organs lie next to the coxal cavities of the mesothoracic legs, which are exposed when the beetle is in flight. In *Merimna* the receptor organs lie on the posterolateral abdomen. Within the pits, each of the 50–100 small spherical sensilla respond with sub-nanometre-scale, heat-induced expansion of fluid contained in the sphere, which acts as a "pressure vessel" in which expansion is converted rapidly to a mechanoreceptive dendrite signal. Such hypersensitive receptors allow a flying adult buprestid to locate a source of IR (indicating fire) perhaps from a distance of 100 km or more. The sensory complex so closely resembles the hearing system of noctuid moths that it can be said that fire-beetles "hear" fire. Calculations of the strength of a remote signal suggest it is so weak as to be difficult to distinguish from thermal background. Evidently though, the system does work, perhaps by summation of multiple sensors, and it is a feat of some interest to technologists.

Individuals of the North American western conifer seed bug, *Leptoglossus occidentalis* (Hemiptera:

Coreidae), which feed by sucking the contents of conifer seeds, are attracted by IR radiation from seed cones, which can be up to 15°C warmer than surrounding needles. These bugs have IR receptor sites on their ventral abdomen; occlusion of these receptors impairs their response to IR. In this system, the warmth generated by the cones is harmful to the plant because it leads to seed herbivory. In other plants, the flowers, inflorescences or cones may produce heat that lures pollinators to the plant's reproductive parts. An Australian cycad, *Macrozamia lucida*, undergoes daily changes in cone thermogenesis and production of volatiles that drive pollen-laded thrips (Thysanoptera: *Cycadothrips*) from male cones and attract them to female cones for pollination. However, in this push–pull pollination system, the thrips may be responding to the heat-released plant volatiles rather than to temperature *per se*.

4.2.2 Thermoregulation

Insects are **poikilothermic**, that is they lack the means to maintain homeothermy – a constant body temperature that is independent of fluctuations in their surroundings. Although the temperature of an inactive insect tends to track ambient, many insects can alter their temperature, both upwards and downwards, even if only for a short time. The temperature of an insect can be varied from ambient either behaviourally using external heat (ectothermy) or by physiological mechanisms (endothermy). Endothermy relies on internally generated heat, predominantly from metabolism associated with flight. Some 94% of flight energy is generated as heat (only 6% is directed to mechanical force on the wings), thus flight demands much energy and also produces a great deal of heat.

Understanding thermoregulation requires some appreciation of the relationship between heat and mass (or volume). In small insects, heat generated is rapidly dissipated. In an environment at 10°C, a 100 mg bumble bee with a body temperature of 40°C experiences a temperature drop of 1°C per second in the absence of any further heat generation. The larger the body, the slower is this heat loss – which is one factor enabling larger organisms to be homeothermic, with the greater mass buffering against heat loss. Another consequence of the mass–heat relationship is that a small insect can warm up quickly from an external heat source, even one as restricted as a light fleck. Clearly, with insects showing a 500,000-fold variation in mass and 1000-fold variation in metabolic rate, there is scope for a range of variants on thermoregulatory physiologies and behaviours. We review the conventional range of thermoregulatory strategies below, and discuss tolerance of extreme temperature in section 6.6.1 and section 6.6.2.

Behavioural thermoregulation (ectothermy)

The extent to which radiant energy (either solar or substrate) influences body temperature is related to the aspect that a diurnal insect adopts. Basking, by which many insects maximize heat uptake, involves both posture and orientation relative to the heat source. The setae of some "furry" caterpillars, such as gypsy moth larvae (Lymantriidae), serve to insulate the body against convective heat loss while not impairing radiant heat uptake. Wing position and orientation may enhance heat absorption or, alternatively, provide shading from excessive solar radiation. Cooling may include shade-seeking behaviour, such as seeking cooler environmental microhabitats or altered orientation on plants. Many desert insects avoid temperature extremes by burrowing, whereas others living in exposed places avoid excessive heating by "stilting", i.e. raising themselves on extended legs to elevate most of the body out of the narrow boundary layer close to the ground. Conduction of heat from the substrate is thus reduced, and convection enhanced in the cooler moving air above the boundary layer.

There is a complex (and disputed) relationship between temperature regulation and insect colour and surface sculpturing. Amongst some desert beetles (Tenebrionidae), black species become active earlier in the day at lower ambient temperatures than do pale ones, which in turn can remain active longer during hotter times. White paint applied to black tenebrionid beetles results in substantial body-temperature changes: black beetles warm up more rapidly at a given ambient temperature and overheat more quickly compared with ones painted white, which reflect heat. These physiological differences correlate with differences in thermal ecology between dark and pale species. Further evidence of the role of colour comes from the beclouded cicada (Hemiptera: *Cacama valvata*) in which basking involves directing the dark dorsal surface towards the sun, in contrast to cooling, where only the pale ventral surface is exposed.

For insects in aquatic habitats, body temperature must follow water temperature. They have little or no ability to regulate body temperature beyond seeking microclimatic differences within a water body.

Physiological thermoregulation (endothermy)

Some insects can be endothermic because the thoracic flight muscles have a very high metabolic rate and produce a lot of heat. During flight, the thorax can be maintained at a relatively constant high temperature. Regulation may involve clothing the thorax with insulating scales or hairs, but insulation must be balanced by dissipation of any excess heat generated during flight. Some butterflies and locusts alternate heat-producing flight with gliding, which allows cooling; but many insects must fly continuously and cannot glide. Bees and many moths prevent thoracic overheating in flight by increasing the heart rate and circulating haemolymph from the thorax to the poorly insulated abdomen, where radiation and convection dissipate heat. At least in some bumble bees (*Bombus*) and carpenter bees (*Xylocopa*), a counter-current system that normally prevents heat loss is bypassed during flight to enhance abdominal heat loss.

Insects that produce elevated temperatures during flight often require a warm thorax before they can take off. When ambient temperatures are low, these insects use the flight muscles to generate heat prior to switching them for use in flight. Mechanisms differ according to whether the flight muscles are synchronous or asynchronous (section 3.1.4). Insects with synchronous flight muscles warm up by contracting antagonistic muscle pairs synchronously and/or synergistic muscles alternately. This activity generally produces some wing vibration, as seen for example in odonates prior to flight. Asynchronous flight muscles are warmed by operating the flight muscles whilst the wings are uncoupled, or the thoracic box is held rigid by accessory muscles to prevent wing movement. Usually, no wing movement is seen, though ventilatory pumping movements of the abdomen may be visible. When the thorax is warm but the insect is sedentary (e.g. whilst feeding), many insects maintain temperature by shivering, which may be prolonged. In contrast, foraging honey bees may cool off during rest, and must then warm up before take-off.

4.3 CHEMICAL STIMULI

In comparison with vertebrates, insects show a more profound use of chemicals in communication, particularly with other individuals of their own species. Insects produce chemicals for many purposes and their perception of the external environment is through specific chemoreceptors.

4.3.1 Chemoreception

The chemical senses may be divided into **taste**, for detection of aqueous chemicals, and **smell**, for airborne ones – but the distinction is relative. Alternative terms are contact (taste, gustatory) and distant (smell, olfactory) chemoreception. For aquatic insects, all chemicals sensed are in aqueous solution, and strictly all chemoreception should be termed "taste". However, if an aquatic insect has a chemoreceptor that is structurally and functionally equivalent to one in a terrestrial insect that is olfactory, then the aquatic insect is said to "smell" the chemical.

Chemosensors trap chemical molecules, which are transferred to a site for recognition, where they specifically depolarize a membrane and stimulate a nerve impulse. Effective trapping involves localization of the chemoreceptors. Thus, many contact (taste) receptors occur on the mouthparts, such as the labella of higher Diptera (see Fig. 2.14a) where salt and sugar receptors occur, and on the ovipositor, to assist with identification of suitable oviposition sites. The antennae, which often are forward-directed and prominent, are first to encounter sensory stimuli and are endowed with many distant chemoreceptors, some contact chemoreceptors, and many mechanoreceptors. The legs, particularly the tarsi, which are in contact with the substrate, also have many chemoreceptors. In butterflies, stimulation of the tarsi by sugar solutions evokes an automatic extension of the proboscis. In blow flies, a complex sequence of stereotyped feeding behaviours is induced when a tarsal chemoreceptor is stimulated with sucrose. The proboscis starts to extend, and following sucrose stimulation of the chemoreceptors on the labellum, further proboscis extension occurs and the labellar lobes open. With more sugar stimulus, the source is sucked until stimulation of the mouthparts ceases. When this happens, a predictable pattern of search for further food follows.

Insect chemoreceptors are sensilla with one or more pores (holes). Two classes of sensilla can be defined based on their ultrastructure: **uniporous**, with one pore, and **multiporous**, with several to many pores. Uniporous sensilla range in appearance from hairs to pegs, plates, or simply pores in a cuticular depression, but all have relatively thick walls and a simple permeable pore, which may be apical or central. The hair or peg contains a chamber, which is in basal contact with a dendritic chamber that lies beneath the cuticle. The outer chamber may extrude a viscous liquid, presumed to entrap and transfer chemicals to the dendrites. It is assumed that these uniporous chemoreceptors predominantly detect chemicals by contact, although there is evidence for some olfactory function. Gustatory (contact) neurons are best classified according to their function and, thus, in relation to feeding. There are cells whose activity in response to chemical stimulation either is to enhance or reduce feeding. These receptors are called phagostimulatory or phagodeterrent.

The major olfactory role comes from multiporous sensilla (Box 4.2), which are hair- or peg-like setae, with many round pores or slits in the thin walls, leading into a chamber known as the **pore kettle**. This is richly endowed with pore tubules, which run inwards to meet multibranched dendrites. Development of an electroantennogram (Box 4.3) allowed revelation of the specificity of chemoreception by the antenna. Used in conjunction with the scanning electron microscope, micro-electrophysiology and modern molecular techniques have extended our understanding of the ability of insects to detect and respond to very weak chemical signals (Box 4.2). Great sensitivity is achieved by spreading very many receptors over as great an area as possible, and allowing the maximum volume of air to flow across the receptors. Thus, the antennae of many male moths are large, and frequently the surface area is enlarged by pectinations that form a sieve-like basket (Fig. 4.6). Each antenna of the male silkworm moth (Bombycidae: *Bombyx mori*) has some 17,000 sensilla of different sizes and several ultrastructural morphologies. Sensilla respond specifically to sex-signalling chemicals produced by the female (sex pheromones; section 4.3.2). As each sensillum has up to 3000 pores, each 10–15 nm in diameter, there are some 45 million pores per moth. Calculations concerning the silkworm moth suggest that just a few molecules could stimulate a nerve impulse above the background rate, and behavioural change may be elicited by less than a hundred molecules.

Some chemicals repel rather than attract insects. For example, DEET (*N*, *N*-diethyl-3-methylbenzamide) effectively repels mosquitoes, which avoid the chemical after detecting it using specific olfactory

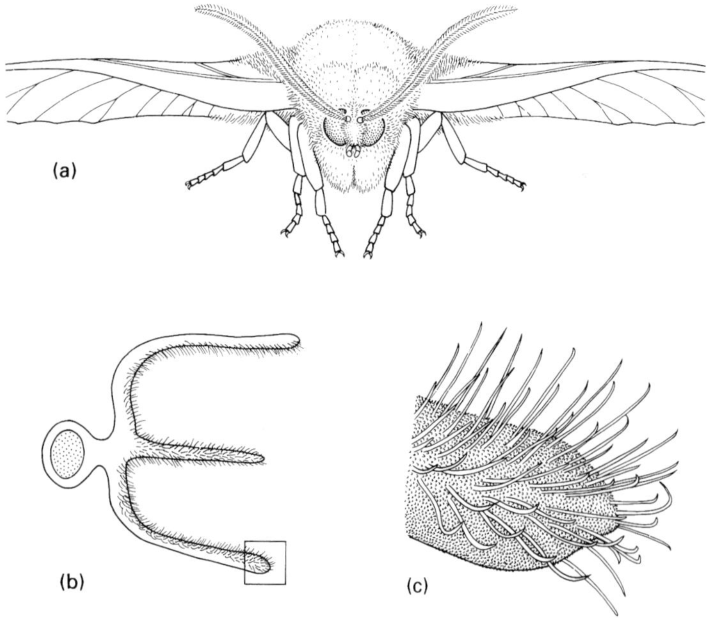

Fig. 4.6 The antennae of a male moth of *Trictena atripalpis* (Lepidoptera: Hepialidae): (a) anterior view of head showing tripectinate antennae of this species; (b) cross-section through an antenna showing the three branches; (c) enlargement of tip of the outer branch of one pectination showing olfactory sensilla.

Box 4.2 Reception of communication molecules

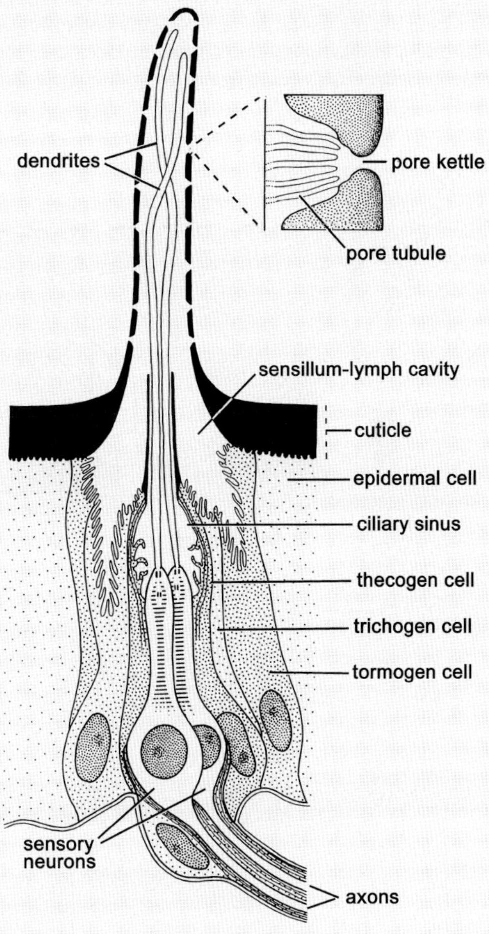

Pheromones, and indeed all signalling chemicals (semiochemicals), must be detectable in even the smallest quantities. For example, the moth approaching a pheromone source portrayed in Fig. 4.7 must detect an initially weak signal and then respond appropriately by orientating towards it, distinguishing abrupt changes in concentration ranging from zero to short-lived concentrated puffs. This involves a physiological ability to monitor continuously and respond to aerial pheromone levels in a process involving extra- and intracellular events.

Ultrastructural studies of *Drosophila melanogaster* and several species of moth allow identification of several types of chemoreceptive (olfactory) sensilla, namely: sensilla basiconica, sensilla trichodea and sensilla coeloconica. These sensillar types are widely distributed across insect taxa and structures, but most often are concentrated on the antennae. Each sensillar type has from two to multiple subtypes, which differ in their sensitivity and tuning to different communication chemicals. The structure of a generalized multiporous olfactory sensillum shown in the accompanying illustration follows Birch and Haynes (1982) and Zacharuk (1985).

To be detected, the chemical must arrive at a pore of an olfactory sensillum. In a multiporous sensillum, it enters a pore kettle and contacts and crosses the cuticular lining of a pore tubule. Because pheromones (and other semiochemicals) largely are hydrophobic (lipophilic) compounds, they must be made soluble in order to reach the receptors. This role falls to odorant-binding proteins (OBP) produced in the tormogen and trichogen cells (Fig. 4.1), from which they are secreted into the sensillum-lymph cavity that surrounds the dendrite of the receptor. Specific OBPs bind the semiochemical into a soluble ligand (OBP–pheromone complex) that is protected as it diffuses through the lymph to the dendrite surface. Here, interaction with negatively charged sites transforms the complex, releasing the pheromone to the binding site of the appropriate olfactory receptors located on the dendrite of the neuron, and triggering a cascade of neural activity leading to the appropriate behaviour.

Much research has involved detection of pheromones because of their use in pest management (see section 16.9), but the principles revealed apparently apply to semiochemical reception across a range of organs and taxa. Thus, experiments with the electroantennogram (Box 4.3) show highly specific responses to particular semiochemicals, and failure to respond even to "trivially" modified compounds. Studied OBPs appear to be one-to-one matched with each semiochemical, but insects apparently respond to more chemical cues than there are OBPs yet revealed. Additionally, olfactory receptors on the dendrite surface seemingly may be less specific, being triggered by a range of unrelated ligands. Furthermore, the model above does not address the frequently observed synergistic effects, in which a cocktail of chemicals provokes a stronger response than does any component alone. How insects are so spectacularly sensitive to so many specific chemicals, alone or in combination, is an active research area, with microphysiology and molecular tools providing many new insights.

Box 4.3 The electroantennogram

Electrophysiology is the study of the electrical properties of biological material, such as all types of nerve cells, including the peripheral sensory receptors of insects. Insect antennae bear a large number of sensilla and are the major site of olfaction in most insects. Electrical recordings can be made from either individual sensilla on the antenna (single-cell recordings) or from the whole antenna (electroantennogram) (as described by Rumbo 1989).

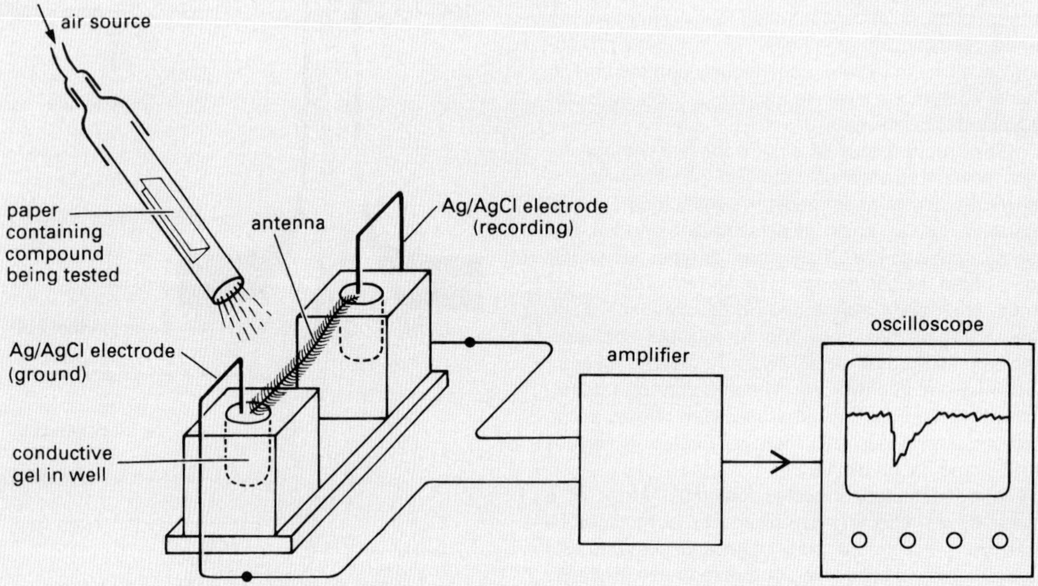

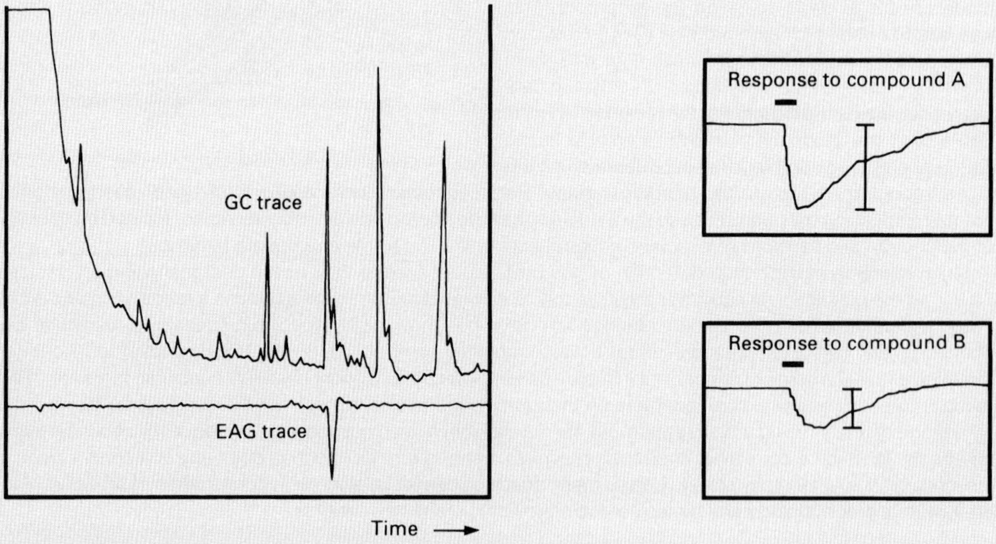

The electroantennogram (EAG) technique measures the total response of insect antennal receptor cells to particular stimuli. Recordings can be made using the antenna either excised, or attached to an isolated head or to the whole insect. In the illustrated example, the effects of a particular biologically active compound (a pheromone) blown across the isolated antenna of a male moth are being assessed. The recording electrode, connected to the apex of the antenna, detects the electrical response, which is amplified and visualized as a trace as in the EAG set-up illustrated in the upper drawing. Antennal receptors are very sensitive and specifically perceive particular odours, such as the sex pheromone of potential conspecific partners or volatile chemicals released by the insect's host. Different compounds usually elicit different EAG responses from the same antenna, as depicted in the two traces on the lower right.

This elegant and simple technique has been used extensively in pheromone identification studies as a quick method of bioassaying compounds for activity. For example, the antennal responses of a male moth to the natural sex pheromone obtained from conspecific female moths are compared with responses to synthetic pheromone components or mixtures. Clean air is blown continuously over the antenna at a constant rate and the samples to be tested are introduced into the air stream, and the EAG response is observed. The same samples can be passed through a gas chromatograph (GC) (which can be interfaced with a mass spectrometer to determine molecular structure of the compounds being tested). Thus, the biological response from the antenna can be related directly to the chemical separation (seen as peaks in the GC trace), as illustrated here in the graph on the lower left (after Struble & Arn 1984).

In addition to lepidopteran species, EAG data have been collected for cockroaches, beetles, flies, bees and other insects, to measure antennal responses to a range of volatile chemicals affecting host attraction, mating, oviposition and other behaviours. EAG information is valuable in assessing the impact of RNA interference (section 16.4.4) to identify (by "knockdown") target genes associated with particular senses and sensilla.

receptor neurons in short trichoid sensilla on their antennae. The presence of DEET induces avoidance in sugar-seeking female and male mosquitoes, and deters females from landing when in search of a blood meal.

4.3.2 Semiochemicals: pheromones

Many insect behaviours rely on smell. Chemical odours, termed **semiochemicals** (from the Greek word *semion* – signal), are especially important in both interspecific and intraspecific communication. The latter is particularly highly developed in insects, and involves the use of chemicals called **pheromones**. When first recognized in the 1950s, pheromones were defined as: "substances that are secreted to the outside by one individual and received by a second individual of the same species in which they release a specific reaction, for example a definite behaviour or developmental process". This definition remains valid today, despite the discovery of a hidden complexity of pheromone cocktails.

Pheromones are predominantly volatile, but sometimes are liquid contact chemicals. All are produced by exocrine glands (those that secrete to the outside of the body) derived from epidermal cells. Pheromones may be released over the surface of the cuticle or from specific dermal structures. Scent organs may be located almost anywhere on the body. Sexual scent glands on female Lepidoptera lie in eversible sacs or pouches between the eighth and ninth abdominal segments, the organs are mandibular in the female honey bee, they are located on the swollen hind tibiae of female aphids, and in the last abdominal tergite in female blaberid cockroaches.

Cuticular hydrocarbons are complex mixtures that serve the multiple functions of waterproofing and chemical communication. They are produced from fatty acids, synthesized in the ectodermally derived oenocytes and perhaps elsewhere, and shuttled from the haemolymph to the outer epicuticle by lipophorin (a transporter lipoprotein). The cuticular hydrocarbon profile of any one species can be highly complex, with up to a hundred different compounds sometimes present. Generally, the *n*-alkanes (saturated hydrocarbons) are involved in controlling water movement across the cuticle, whereas unsaturated hydrocarbons (such as alkenes) and the methyl-branched alkanes are typically involved in communication. These molecules induce various behaviours in solitary and social insects, and can be involved in species, gender, nest-mate and caste recognition, chemical mimicry, and dominance and fertility cues, or act as sex or primer pheromones or even cues for parasitoids. The hydrocarbons that constitute the queen pheromone of *Lasius* ants are evolutionarily conserved compared with the "signature mixtures" of hydrocarbons involved in nest-mate recognition in ants. Furthermore, the worker response to queen pheromone in *Lasius* appears to be innate, whereas ants learn their colony's hydrocarbon profile. Hydrocarbon profiles of insects can provide chemotaxonomic characters for species delimitation (see section 7.1.2 and Box 7.2).

Classification of pheromones by chemical structure reveals that many naturally occurring compounds (such as host odours) and pre-existing metabolites (such as cuticular lipids) have been co-opted by insects directly or to serve in the biochemical synthesis of a wide variety of compounds that function in communication. Chemical classification is less valuable to many entomologists than the behaviours that the chemicals elicit. Numerous insect behaviours are governed by chemicals, but pheromones that release specific behaviours can be distinguished from those that prime long-term, irreversible physiological changes. Thus, the stereotyped sexual behaviour of a male moth is released by the female-emitted sex pheromone, whereas the crowding pheromone of locusts primes maturation of gregarious phase individuals (section 6.10.5). Here, further classification of pheromones is based on five categories of behaviour associated with sex, aggregation, spacing, trail forming, and alarm.

Sex pheromones

Male and female conspecific insects often communicate with each other using chemical sex pheromones. Mate location and courtship may involve chemicals in two stages, with sex attraction pheromones acting at a distance, followed by close-up courtship pheromones employed prior to mating. The sex pheromones involved in attraction often differ from those used in courtship. Production and release of sex attractant pheromones tends to be restricted to the female, although there are lepidopterans and scorpionflies in which males are the releasers of distance attractants that lure females. The producer releases volatile pheromones that stimulate characteristic behaviour in those members of the opposite sex within range of the odour plume. An aroused recipient raises their antennae, orientates towards the source, and walks or flies upwind towards the source, often making a zig-zag track (Fig. 4.7) based on the ability to respond rapidly to minor changes in pheromone concentration by direction change (Box 4.2). Each successive action appears to depend upon an increase in concentration of this airborne pheromone. As the insect approaches the source, cues such as sound and vision may be involved in close-up courtship behaviour.

Courtship (section 5.2), which involves co-ordination of the two sexes, may require close-up chemical

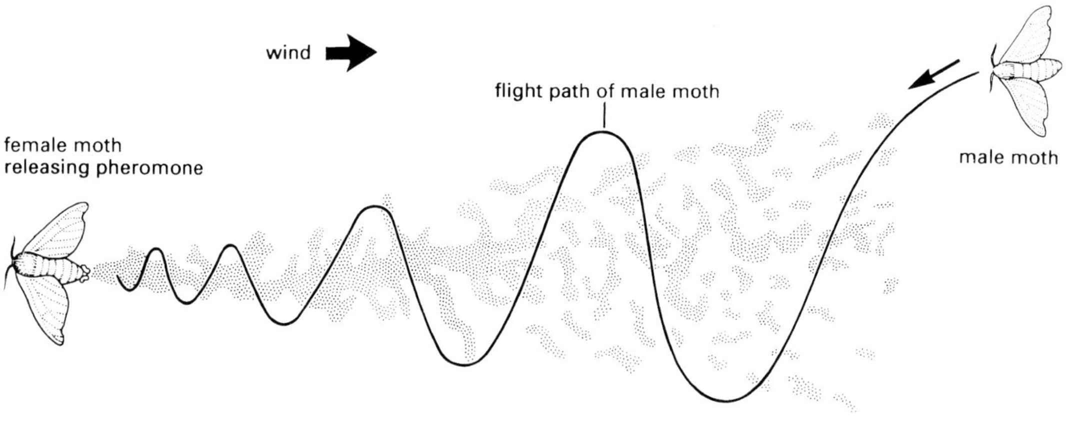

Fig. 4.7 Location of pheromone-emitting female by male moth tacking upwind. The pheromone trail forms a somewhat discontinuous plume because of turbulence, intermittent release, and other factors. (After Haynes & Birch 1985.)

Fig. 4.8 A pair of queen butterflies, *Danaus gilippus* (Lepidoptera: Nymphalidae: Danainae), showing aerial "hairpencilling" by the male. The male (above) has splayed hairpencils (at his abdominal apex) and is applying pheromone to the female (below). (After Brower *et al.* 1965.)

stimulation of the partner with a courtship pheromone. This pheromone may be simply a high concentration of the attractant pheromone, but "aphrodisiac" chemicals do exist, as seen in the queen butterfly (Nymphalidae: *Danaus gilippus*). Males of this species, as with some other lepidopterans, have extrusible abdominal hairpencils (brushes), which produce a pheromone that is dusted directly onto the antennae of the female, whilst both are in flight (Fig. 4.8). The effect of this pheromone is to placate a natural escape reaction of the female, which alights, folds her wings, and allows copulation. In *D. gilippus*, this male courtship pheromone, a pyrrolizidine alkaloid called danaidone, is essential to successful courtship. However, the butterfly cannot synthesize it without acquiring the chemical precursor by feeding on selected plants as an adult. In the arctiid moth, *Creatonotus gangis*, the precursor of the male courtship pheromone likewise cannot be synthesized by the moth, but is sequestered by the larva in the form of a toxic alkaloid from the host plant. The larva uses the chemical in its defence and at metamorphosis the toxins are transferred to the adult. Both sexes use them as defensive compounds, with the male additionally converting them to his pheromone. This he emits from inflatable abdominal tubes, called **coremata**, whose development is regulated by the alkaloid pheromone precursor.

A spectacular example of deceitful sexual signalling occurs in bolas spiders, which do not build a web, but whirl a single thread, terminating in a sticky globule, towards their moth prey (like gauchos using a bolas to hobble cattle). The spiders lure male moths to within reach of the bolas using synthetic lures of female moth sex-attractant pheromone cocktails. The proportions of the components vary according to the abundance of particular moth species available as prey. Similar principles are applied by humans to control pest insects, using lures containing synthetic sex pheromones or other attractants (section 16.9). Certain chemical compounds (e.g. methyl eugenol), which either occur naturally in plants or can be synthesized in the laboratory, are used to lure male fruit flies (Tephritidae) for pest management purposes. These male lures, sometimes called **parapheromones**, probably because the compounds may be used by the flies as a component in the synthesis of their sex pheromones and can improve mating success, perhaps by enhancing the male's sexual signals.

Sex pheromones, once thought to be unique, species-specific chemicals, often are chemical blends. The same chemical (e.g. a particular 14-carbon chain alcohol) may be present in a range of related and unrelated species, but is blended in different proportions with other chemicals. Each component may elicit only part of the sex attraction behaviour, or a partial or complete mixture may be required. Often the blend produces a greater response than does any individual component, a **synergism** that is widespread in insects that produce pheromone mixtures. Chemical structural similarity of pheromones may indicate systematic relationship amongst the producers. However, obvious anomalies arise when identical or very similar pheromones are synthesized from chemicals derived from identical diets by unrelated insects.

Even if individual components are shared by many species, the mixture of pheromones is very often species-specific. It is evident that pheromones, and the stereotyped behaviours that they evoke, are highly significant in maintenance of reproductive isolation between species. The species-specificity of sex pheromones avoids cross-species mating before males and females even come into contact.

Aggregation pheromones

The release of an aggregation pheromone causes conspecific insects of both sexes to crowd around the source of the pheromone. Aggregation may lead to increased likelihood of mating but, in contrast to many sex pheromones, both sexes may produce and respond to aggregation pheromones. The potential benefits provided by aggregation include security from predation, maximum utilization of scarce food, overcoming of host resistance, and cohesion of social insects, as well as increasing the chances of mating.

Aggregation pheromones are known in six insect orders, including cockroaches, but their presence and mode of action has been studied in most detail in Coleoptera, particularly in economically damaging species such as stored-grain beetles (from several families) and timber and bark beetles (Curculionidae: Scolytinae). A well-researched example of a complex suite of aggregation pheromones is provided by the Californian western pine beetle, *Dendroctonus brevicomis* (Scolytinae), which attacks ponderosa pine (*Pinus ponderosa*). On arrival at a new tree, colonizing females release the pheromone *exo*-brevicomin from frass, augmented by myrcene, a terpene originating from the damaged pine tree. Both sexes of western pine beetle are attracted by this mixture, and newly arrived males then add to the chemical mix by releasing another pheromone, frontalin. The cumulative lure of frontalin, *exo*-brevicomin and myrcene is synergistic, i.e. greater than that induced by any of the pheromones alone. The aggregation of many pine beetles overwhelms the tree's defensive secretion of resins.

Spacing pheromones

There is a limit to the number of western pine beetles (*D. brevicomis*; see above) that attack a single tree. Cessation is assisted by reduction in the production of attractant aggregation pheromones, but also deterrent chemicals are produced. After the beetles mate on the tree, both sexes produce "anti-aggregation" pheromones called verbenone and *trans*-verbenol. These deter further beetles from landing close by, encouraging spacing-out of new colonists. Therefore, when the resource is saturated, further arrivals are repelled.

Such semiochemicals, called spacing, epideictic or dispersion pheromones, may effect appropriate spacing on food resources, as with some phytophagous insects. Several species of tephritid flies lay eggs singly in fruit where the solitary larva is to develop. Spacing occurs because the ovipositing female deposits an oviposition-deterrent pheromone on the fruit on which she has laid an egg, thereby deterring subsequent oviposition. Social insects utilize pheromones for many purposes, including in regulation of spacing between colonies. Spacer pheromones of colony-specific odours may ensure an even dispersal of colonies of conspecifics, as in African weaver ants (Formicidae: *Oecophylla longinoda*).

Trail-marking pheromones

Many social insects use pheromones to mark their trails, particularly those leading to food and the nest. Trail-marking pheromones are volatile and short-lived chemicals that evaporate within days unless reinforced (perhaps as a response to a food resource that is longer lasting than usual). Trail-marking pheromones in ants mostly are excreted by the poison gland, Dufour's gland or other abdominal glands. These pheromones need not be species-specific, as several species share some common chemicals. Dufour's gland secretions of some ant species may be more species-specific chemical mixtures, and are associated with marking of territory and pioneering trails. In some ant species, trail-marking pheromones are released from exocrine glands on the hind legs. Ant trails appear to be non-polar, i.e. the direction to the nest or food resource cannot be determined by the trail odour.

In contrast to trails laid on the ground, an airborne trail – an odour plume – has directionality because of increasing concentration of the odour towards the source. An insect may rely upon angling the flight path relative to the direction of the wind that brings the odour, resulting in a zig-zag upwind flight towards the source. Each direction change is made where the odour diminishes at the edge of the plume (Fig. 4.7).

Alarm pheromones

About two centuries ago it was recognized that workers of honey bees (*Apis mellifera*) were alarmed by a freshly extracted sting. Many aggregating insects produce chemical releasers of alarm behaviour – alarm pheromones – that characterize most social insects (termites and eusocial hymenopterans). Alarm pheromones are known also in several hemipterans, including subsocial treehoppers (Membracidae), aphids (Aphididae) and some other true bugs. These pheromones are volatile, non-persistent compounds that disperse readily throughout the aggregation. Alarm is provoked by the presence of a predator, or in many social insects, a threat to the nest. The behaviour elicited may be rapid dispersal, such as in hemipterans that drop from the host plant, or escape from an unwinnable conflict with a large predator, as in poorly defended ants living in small colonies. The alarm behaviour of many eusocial insects is most familiar to us when disturbance of a nest induces many ants, bees or wasps to aggressively defend their nest. Alarm pheromones attract aggressive workers and these recruits attack the cause of the disturbance by biting, stinging or firing repellent chemicals. Emission of more alarm pheromone mobilizes further defenders. Alarm pheromone may be daubed over an intruder to aid in directing the attack.

Alarm pheromones may have been derived over evolutionary time from chemicals used as general anti-predator devices (allomones; section 4.3.3), utilizing glands co-opted from many different parts of the body to produce these substances. For example, hymenopterans commonly produce alarm pheromones from mandibular glands and also from poison glands, metapleural glands, the sting shaft and even the anal area. All these glands also may produce defensive chemicals.

4.3.3 Semiochemicals: kairomones, allomones and synomones

Communication chemicals (semiochemicals) may function between individuals of the same species (pheromones) or between different species (**allelochemicals**). Interspecific semiochemicals may be grouped according to the benefits they provide to the producer and receiver. Those that benefit the receiver but disadvantage the producer are known as **kairomones**. **Allomones** benefit the producer by modifying the behaviour of the receiver while having a neutral effect on the receiver. **Synomones** benefit both the producer and the receiver. This terminology has to be applied in the context of the specific behaviour induced in the recipient, as seen in the examples discussed below. A particular chemical can act as an intraspecific pheromone and may also fulfil all three categories of interspecific communication, depending on circumstances. The use of the same chemical for two or more functions in different contexts is referred to as semiochemical parsimony.

Kairomones

Myrcene, the terpene produced by a ponderosa pine when it is damaged by the western pine beetle (section 4.3.2), acts as a synergist with aggregation pheromones to lure more beetles. Thus, myrcene and other terpenes produced by damaged conifers can be kairomones, disadvantaging the producer by luring damaging timber beetles. A kairomone need not indicate insect attack: elm bark beetles (Curculionidae: Scolytinae: *Scolytus* spp.) respond to α-cubebene, a product of the Dutch elm disease fungus *Ceratocystis ulmi* from a weakened or dead elm tree (*Ulmus*). Elm beetles themselves inoculate previously healthy elms with the fungus, but pheromone-induced aggregations of beetles form only when the kairomone (fungal α-cubenene) indicates suitability for colonization. Host-plant detection by phytophagous insects also involves reception of plant chemicals, which act as kairomones.

Insects produce many communication chemicals, with clear benefits. However, these semiochemicals also may act as kairomones if other insects recognize them. In "hijacking" the chemical messenger for their own use, specialist parasitoids (Chapter 13) use chemicals emitted by the host, or plants attacked by the host, to locate a suitable site for development of its offspring.

Allomones

Allomones are chemicals that benefit the producer but have neutral effects on the recipient. For example, defensive and/or repellent chemicals are allomones that advertise distastefulness and protect the producer from lethal experiment by prospective predators. The effect on a potential predator is considered to be neutral, as it is warned against wasting energy in seeking a distasteful meal.

The worldwide beetle family Lycidae has many distasteful and warning-coloured (aposematic) members, including species of *Metriorrhynchus* that are protected by odorous alkylpyrazine allomones. In Australia, several distantly related beetle families include many mimics that are modelled visually on *Metriorrhynchus*. Some mimics are remarkably convergent in colour and distasteful chemicals, and possess nearly identical alkylpyrazines. Others share the allomones but differ in distasteful chemicals, whereas some have the warning chemical but appear to lack distastefulness. Other insect mimicry complexes involve allomones. Mimicry and insect defences in general are considered further in Chapter 14.

Some defensive allomones also can act as sex pheromones. Examples include chemicals from the defensive glands of various bugs (Heteroptera), grasshoppers (Acrididae) and beetles (Staphylinidae), as well as plant-derived toxins used by some Lepidoptera (section 4.3.2). Many female ants, bees and wasps have exploited the secretions of the glands associated with their sting – the poison (or venom) gland and Dufour's gland – as male attractants and releasers of male sexual activity.

A novel use of allomones occurs in certain orchids, whose flowers produce similar odours to female sex pheromone of the wasp or bee species that acts as their specific pollinator. Male wasps or bees are deceived by this chemical mimicry and also by the colour and shape of the flower, with which they attempt to copulate (**pseudocopulation**; section 11.3.1). Thus, the orchid's odour acts as an allomone beneficial to the plant by attracting its specific pollinator, whereas the effect on the male insects is near neutral – at most they waste time and effort.

Synomones

Terpenes produced by damaged pines are kairomones for pest beetles, but if identical chemicals are used by beneficial parasitoids to locate and attack the bark beetles, the terpenes are acting as synomones (by benefiting both the producer and the receiver). Thus, α-pinene and myrcene produced by damaged pines are kairomones for species of *Dendroctonus* but synomones for the pteromalid hymenopterans that parasitize these timber beetles. In like manner, α-cubebene produced by Dutch elm fungus is a synomone for the braconid hymenopteran parasitoids of elm bark beetles (for which it is a kairomone).

An insect parasitoid may respond to host-plant odour directly, like the phytophage it seeks to parasitize, but this means of searching cannot guarantee the parasitoid that the phytophage host is actually present. There is a greater chance of success for the parasitoid if it can identify and respond to the specific plant chemical defences that the phytophage provokes. If an insect-damaged host plant produces a repellent odour, such as a volatile terpenoid, then the chemical can act as:

- an allomone that deters non-specialist phytophages;
- a kairomone that attracts a specialist phytophage;
- a synomone that lures the parasitoid of the phytophage.

Of course, phytophagous, parasitic and predatory insects rely on more than odours to locate potential hosts or prey, and visual discrimination is implicated in resource location (section 4.4).

4.3.4 Carbon dioxide as a sensory cue

Carbon dioxide (CO_2) is important to the biology and behaviour of many insects, which can detect and measure the concentration of this environmental chemical using specialized receptor cells. Sensory structures that detect atmospheric CO_2 have been identified on the antennae and the mouthparts of insects. They occur on the labial palps of adult Lepidoptera, the maxillary palps of larval Lepidoptera and adult blood-feeding nematoceran Diptera (mosquitoes and biting midges), the antennae of some brachyceran Diptera (the stable fly *Stomoxys calcitrans* and Queensland fruit fly *Bactrocera tryoni*), Hymenoptera (honey bees and *Atta* ants) and termites, and on the antennae and/or maxillary palps of adult and larval Coleoptera. Known sensory structures mostly are clusters of sensilla, which may be aggregated into distinct sensory organs, often recessed in capsules or pits, such as the pit organ on the apical segment of the labial palps of adult butterflies and moths and which contains several CO_2 detecting sensory cones (sensilla). Each sensillum is thin-walled and has wall pores and branched or lamellated dendrites with an increased distal surface area. In most studied insects, each sensillum contains one receptor cell (RC) that differs physiologically to the typical odorant receptor cell (Box 4.2), which measures the rate of odorant molecules adsorbed irreversibly by the sensillum. In contrast, adsorption of CO_2 to its RCs is reversible and these RCs have a bidirectional response to change in CO_2 concentrations, such that both increases and decreases can be detected, and CO_2 RCs can signal the background levels of CO_2 continually over a broad range of concentrations without sensory adaptation. The detection of CO_2 levels, or gradients in CO_2 concentration, has been implicated in many insect activities, including allowing or assisting:

- adult insects such as butterflies or moths to locate healthy host plants for oviposition (plants release or uptake CO_2 depending on their photosynthetic activity and time of day) or adult tephritid fruit flies to home in on damaged fruit (which release CO_2 from wounds);

- foraging insects, such as lepidopteran or coleopteran larvae, to locate or select roots, fruits or flowers for feeding (these plant tissues are sources of CO_2);
- certain fruit-feeding flies such as *Drosophila* to avoid unripe fruits (which emit more CO_2 than do ripe ones);
- blood-feeding insects such as mosquitoes to detect vertebrate hosts; and
- some social insects to regulate levels of CO_2 in their nests by fanning behaviour at the nest entrance (as in bees) or by altering nest architecture (ants and termites).

Global atmospheric CO_2 concentration has increased steadily from about 280 parts per million (ppm) prior to the Industrial Revolution of the late 18th to early 19th centuries to reach 400 ppm in 2013. A much-elevated level of atmospheric CO_2 will affect the physiology of the CO_2-sensing systems of insects (and other organisms), with concomitant effects on behaviour and reproduction.

4.4 INSECT VISION

Excepting a few blind subterranean and endoparasitic species, most insects have some sight, and many possess highly developed visual systems. The basic components needed for vision are a lens to focus light onto **photoreceptors** – cells containing light-sensitive molecules – and a nervous system complex enough to process visual information. In insect eyes, the photoreceptive structure is the **rhabdom**, comprising several adjacent **retinula** (or nerve) **cells** and consisting of close-packed **microvilli** containing visual pigment. Light falling onto the rhabdom changes the configuration of the visual pigment, triggering a change of electrical potential across the cell membrane. This signal is then transmitted via chemical synapses to nerve cells in the brain. Comparison of the visual systems of different kinds of insect eyes involves two main considerations: (i) their resolving power for images, i.e. the amount of fine detail that can be resolved; and (ii) their light sensitivity, i.e. the minimum ambient light level at which the insect can still see. Eyes of different kinds and in different insects vary widely in resolving power and light sensitivity and thus in details of function.

The compound eyes are the most obvious and familiar visual organs of insects, but there are three other means by which an insect may perceive light: dermal detection, stemmata and ocelli. The dragonfly head depicted in the vignette at the beginning of this chapter is dominated by its huge compound eyes with the three ocelli and paired antennae in the centre.

4.4.1 Dermal detection

In insects able to detect light through their body surface, there are sensory receptors below the body cuticle but no optical system with focusing structures. Evidence for this general response to light comes from the persistence of photic responses after covering all visual organs, for example in cockroaches and certain beetle and lepidopteran larvae. Some blind cave insects, with no recognizable visual organs, respond to light, as do decapitated cockroaches. In most cases, the sensitive cells and their connection with the central nervous system have yet to be discovered. However, within the brain itself, aphids have light-sensitive cells that detect changes in day length – an environmental cue that controls the mode of reproduction (i.e. either sexual or parthenogenetic). The setting of the biological clock (Box 4.4) relies upon the ability to detect photoperiod.

4.4.2 Stemmata

The only visual organs of larval holometabolous insects are stemmata, sometimes called larval ocelli (Fig. 4.9a). These organs are located on the head, and vary from a single pigmented spot on each side to six or more larger stemmata, each with numerous photoreceptors and associated nerve cells. Larval hymenopterans (except symphytans) and fleas have no stemmata. In the simplest stemma, a cuticular lens overlies a crystalline body secreted by several cells. Light is focused by the lens onto a single rhabdom. Each stemma points in a different direction so that the insect sees only a few points in space, according to the number of stemmata. Some caterpillars increase the field of view and fill in the gaps between the direction of view of adjacent stemmata by scanning movements of the head. Other larvae, such as those of sawflies and tiger beetles, possess more sophisticated stemmata. They consist of a two-layered lens that forms an image on an extended retina composed of many rhabdoms, each receiving light from a different part of the image. In general, stemmata seem designed for high light sensitivity, with resolving power being relatively low.

Box 4.4 Biological clocks

Seasonal changes in environmental conditions allow insects to adjust their life histories to optimize the use of suitable conditions and to minimize the impact of unsuitable ones (e.g. through diapause; section 6.5). Similar physical fluctuations on a daily scale encourage a diurnal (daily) cycle of activity and quiescence. Nocturnal insects are active at night; diurnal ones by day; and in crepuscular insects activity occurs at dusk and dawn when light intensities are transitional. The external physical environment, such as light–dark or temperature, controls some daily activity patterns, called exogenous rhythms. However, many other periodic activities are internally driven endogenous rhythms that have a clock-like or calendar-like frequency irrespective of external conditions. Endogenous periodicity is frequently about 24 h (circadian), but lunar and tidal periodicities govern the emergence of adult aquatic midges from large lakes and the marine intertidal zones, respectively. This unlearned, once-in-a-lifetime rhythm that allows synchronization of eclosion demonstrates the innate ability of insects to measure passing time.

Experimentation is required to discriminate between exogenous and endogenous rhythms. This involves observing what happens to rhythmic behaviour when external environmental cues are altered, removed or made invariant. Such experiments show that inception (setting) of endogenous rhythms is found to be day length, with the clock then free-running, without daily reinforcement by the light–dark cycle, often for a considerable period. Thus, if nocturnal cockroaches that become active at dusk are kept at constant temperature in constant light or dark, they will maintain the dusk commencement of their activities at a circadian rhythm of 23–25 h. Rhythmic activities of other insects may require an occasional clock-setting (such as darkness) to prevent the circadian rhythm drifting, either through adaptation to an exogenous rhythm or into arrhythmia.

Biological clocks allow solar orientation – the use of the sun's elevation above the horizon as a compass – provided that there is a means of assessing (and compensating for) the passage of time. Some ants and honey bees use a "light-compass", finding direction from the sun's elevation and using the biological clock to compensate for the sun's movement across the sky. Evidence for this came from an elegant experiment with honey bees trained to forage in the late afternoon at a feeding table (F) placed 180 m NW of their hive (H), as depicted in the left-hand figure (after Lindauer 1960). Overnight, the hive was moved to a new location to prevent use of familiar landmarks in foraging, and a selection of four feeding tables (F_{1-4}) was provided at 180 m, NW, SW, SE and NE from the hive. In the morning, despite the sun being at a very different angle to that during the afternoon training, 15 of the 19 bees were able to locate the NW table (as depicted in the right-hand figure). The honey bee "dance language"

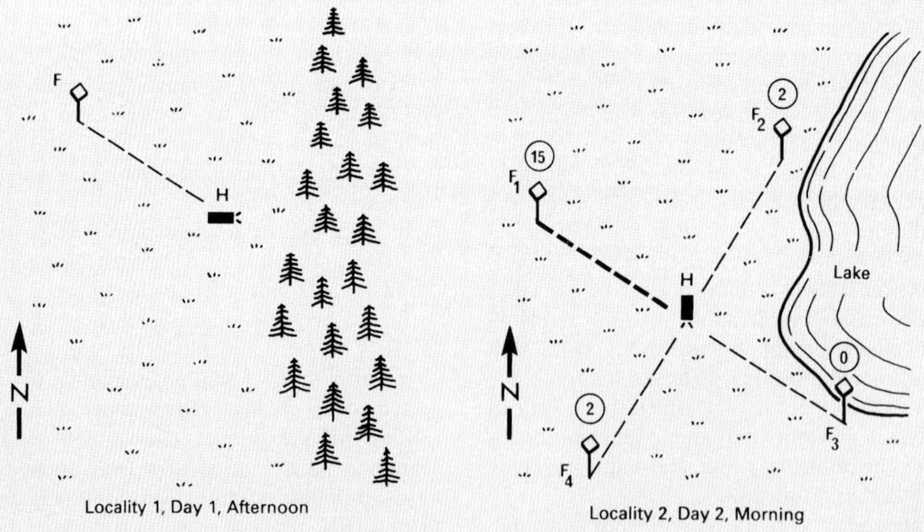

that communicates direction and distance of food to other workers (see Box 12.1) depends upon their capacity to calculate direction from the sun.

The circadian pacemaker (oscillator) that controls the rhythm is located in the brain; it is not an external photoperiod receptor. Experimental evidence shows that in cockroaches, beetles and crickets a pacemaker lies in the optic lobes, whereas in some silkworms it lies in the cerebral lobes of the brain. In the well-studied *Drosophila*, a major oscillator site appears to be located between the lateral protocerebellum and the medulla of the optic lobe. However, visualization of the sites of activity of the *period* gene (a clock gene) is not localized, and there is increasing evidence of multiple pacemaker centres located throughout the tissues. Whether they communicate with each other or run independently is not yet clear.

4.4.3 Ocelli

Many adult insects, as well as some nymphs, have dorsal ocelli in addition to compound eyes. These ocelli are unrelated embryologically to the stemmata. Typically, three small ocelli lie in a triangle on top of the head. The cuticle covering an ocellus is transparent and may be curved as a lens. It overlies transparent epidermal cells, so that light passes through to an extended retina made up of many rhabdoms (Fig. 4.9b). Individual groups of retinula cells that contribute to one rhabdom or the complete retina are surrounded by pigment cells or by a reflective layer. The focal plane of the ocellar lens often lies below the rhabdoms so that the retina receives a blurred image. However, the median ocellus of some insects (e.g. dragonflies) does focus. The axons of the ocellar retinula cells converge onto only a few neurons that connect the ocelli to the brain. In the ocellus of the dragonfly *Sympetrum*, some 675 receptor cells converge onto one large neuron, two medium-sized neurons, and a few small ones in the ocellar nerve.

The ocelli thus integrate light over a large visual field, both optically and neurally. They are very sensitive to low light intensities and to subtle changes in light, but do not provide high-resolution vision. They appear to function as "horizon detectors" for control of roll and pitch movements in flight, and to register cyclical changes in light intensity that correlate with diurnal behavioural rhythms.

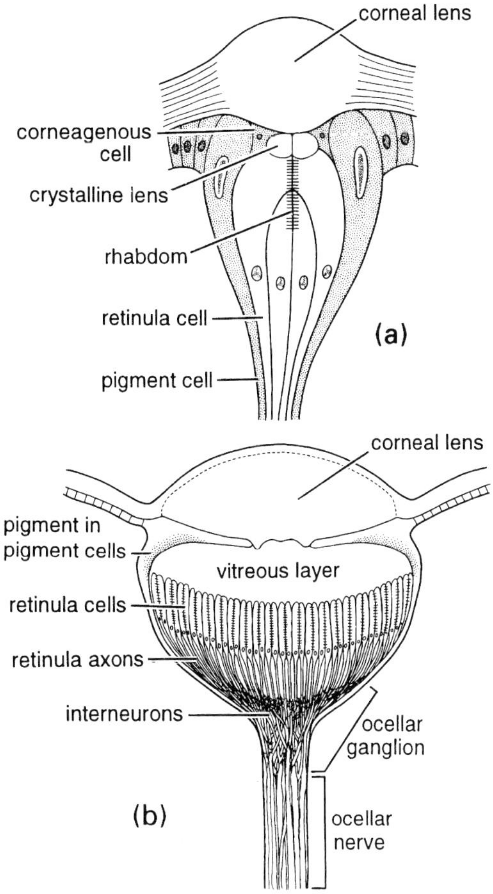

Fig. 4.9 Longitudinal sections through simple eyes: (a) a simple stemma of a lepidopteran larva; (b) a light-adapted median ocellus of a locust. ((a) After Snodgrass 1935; (b) after Wilson 1978.)

4.4.4 Compound eyes

The most sophisticated insect visual organ is the compound eye. Virtually all adult insects and nymphs have a pair of large, prominent compound eyes, which often cover nearly 360 degrees of visual space.

The compound eye is based on the repetition of many individual units called **ommatidia** (Fig. 4.10). Each ommatidium resembles a simple stemma: it has a cuticular lens overlying a crystalline cone, which directs and focuses light onto eight (or maybe 6–10) elongate retinula cells (see the transverse section in Fig. 4.10b). The retinula cells are clustered around the longitudinal axis of each ommatidium and each contributes a rhabdomere to the rhabdom at the centre of the ommatidium. Each cluster of retinula cells is surrounded by a ring of light-absorbing pigment cells, which optically isolates an ommatidium from its neighbours.

The corneal lens and crystalline cone of each ommatidium focus light onto the distal tip of the rhabdom from a region about 2–5 degrees across. The field of view of each ommatidium differs from that of its neighbours, and together the array of all the ommatidia provides the insect with a panoramic image of the world. Thus, the actual image formed by the compound eye is of a series of apposed points of light of different intensities, hence the name **apposition eye**.

The light sensitivity of apposition eyes is limited severely by the small diameter of facet lenses. Crepuscular and nocturnal insects, such as moths and some beetles, overcome this limitation with a modified optical design of compound eyes, called **optical superposition eyes**. In these, ommatidia are not isolated optically from each other by pigment cells. Instead, the retina is separated by a wide, clear zone from the corneal facet lenses, and many lenses co-operate to focus light on an individual rhabdom (light from many lenses super-imposes on the retina). The light sensitivity of these eyes is thus greatly enhanced. In some optical superposition eyes, screening pigment moves into the clear zone during light adaptation, and by this means the ommatidia become isolated optically, as in the apposition eye. At low light levels, the screening pigment moves again towards the outer surface of the eye so as to open up the clear zone to enable optical superposition to occur.

Because the light arriving at a rhabdom has passed through many facet lenses, blurring is an issue in optical superposition eyes and resolution generally is not as good as in apposition eyes. However, high light sensitivity is much more important than good resolving power in crepuscular and nocturnal insects whose main concern is to see anything at all. In the eyes of some insects, photon-capture is increased even further by a mirror-like tapetum of small tracheae at the base of the retinula cells. This reflects light that has passed unabsorbed through a rhabdom, allowing it a second pass. Light reflecting from the tapetum produces the bright eye shine seen when an insect

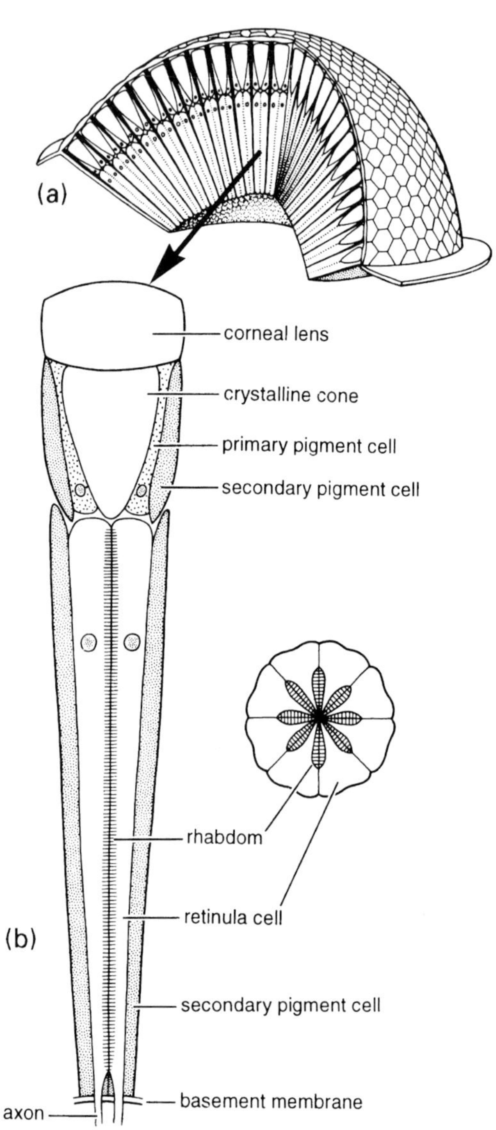

Fig. 4.10 Details of the compound eye: (a) a cutaway view showing the arrangement of the ommatidia and the facets; (b) a single ommatidium with an enlargement of a transverse section. (After CSIRO 1970; Rossel 1989.)

with an optical superposition eye is illuminated in a flashlight or car headlight beam at night.

In comparison with a vertebrate eye, the resolving power of insect compound eyes is rather unimpressive. However, for the purpose of flight control, navigation, prey capture, predator avoidance and mate-finding they obviously do a splendid job. Bees can memorize quite sophisticated shapes and patterns, and flies and odonates hunt down prey insects or mates in extremely fast, aerobatic flight. Insects in general are exquisitely sensitive to image motion, which provides them with useful cues for avoiding obstacles and for landing, and for distance judgment. Insects, however, cannot easily use binocular vision for the perception of distance because their eyes are so close together and their resolution is quite poor. A notable exception is the praying mantid, which is the only insect known to make use of binocular disparity to localize prey.

Within each ommatidium, most studied insects possess several classes of retinula cells that differ in their spectral sensitivities. This feature means that each retinula cell responds best to light of a different wavelength. Variations in the molecular structure of visual pigments are responsible for these differences in spectral sensitivity and are a prerequisite for the colour vision of flower visitors such as bees and butterflies. Many insects have three colour-receptor types (blue, green and ultraviolet), but some insects are pentachromats, with five classes of receptors of differing spectral sensitivities, as compared with humans, who are di- or trichromats. Most insects can perceive ultraviolet light (which is invisible to us), allowing them to see distinctive alluring flower patterns that are visible only in the ultraviolet.

Light emanating from the sky and reflected light from water surfaces or shiny leaves is polarized, i.e. it has greater vibration in some planes than in others. Many insects can detect the plane of polarization of light, and they utilize this in navigation, as a compass or as an indicator of water surfaces. The pattern of polarized skylight changes its position as the sun moves across the sky, so that insects can use small patches of clear sky to infer the position of the sun, even when it is not visible. In like manner, an African dung beetle has been shown to orientate its dung rolling using polarized moonlight. Even in the absence of the moon, these nocturnal beetles can maintain a straight line in dung rolling using the orientation of the Milky Way to orientate. The microvillar organization of the insect rhabdomere makes insect photoreceptors inherently sensitive to the plane of polarization of light, unless sensitivity is reduced by scrambling the alignment of microvilli. Insects with well developed navigational abilities often have a specialized region of retina in the dorsal visual field, the dorsal rim, in which retinula cells are highly sensitive to the plane of polarization of light. Ocelli and stemmata also may be involved in the detection of polarized light.

4.4.5 Light production

The most spectacular visual displays of insects involve light production, or **bioluminescence**. Some insects co-opt symbiotic luminescent bacteria or fungi, but self-luminescence is found in a few Collembola, one hemipteran (the fulgorid lantern bug), one genus of cockroaches (Blaberidae: *Lucihormetica*), a few dipteran fungus gnats, and a diverse group amongst several families of coleopterans. The luminescent beetles are members of the Elateridae (click beetles), Phengodidae and notably the Lampyridae (about 2000 species, which are all luminescent, at least as larvae), and are commonly given colloquial names including fireflies, glow worms and lightning bugs. Either sex and any or all life-history stages may glow, using one to many luminescent organs, which may be located nearly anywhere on the body. Light emitted may be white, yellow, red or green.

Light emission in the lampyrid firefly *Photinus pyralis* may be typical of luminescent Coleoptera. The enzyme luciferase oxidizes a substrate, luciferin, in the presence of an energy source of adenosine triphosphate (ATP) and oxygen, to produce oxyluciferin, CO_2 and light. Variation in ATP release controls the rate of flashing, and differences in pH may allow variation in the frequency (colour) of light emitted.

In lampyrid larvae, light emission serves as a warning of distastefulness, but the principal role of bioluminescence in adults of nocturnal taxa is courtship signalling. This involves either a continuous light signal (a glow) as found in larva-like females of some groups, or short, intermittent light signals (flashes) as found in species of *Photinus* and *Photuris*. There is species-specific variation in the duration, number and rate of flashes in a pattern, and in the frequency of repetition of the pattern (Fig. 4.11). Generally, a mobile male advertises his presence by instigating the signalling with one or more flashes and a sedentary female indicates her location with a flash in response. As with all communication systems, there

Fig. 4.11 The flash patterns of males of nine of *Photinus* firefly species (Coleoptera: Lampyridae), each of which generates a distinctive pattern of signals in order to elicit a response from their conspecific females. (After Lloyd 1966.)

is scope for abuse, for example that involving luring of prey by a carnivorous female lampyrid of *Photurus* (section 13.1.2).

Bioluminescence is involved in both luring prey and mate-finding in Australian and New Zealand cave-dwelling *Arachnocampa* fungus gnats (Diptera: Mycetophilidae). Their luminescent displays in the dark zone of caves have become tourist attractions in some places. All developmental stages of these flies use a reflector to concentrate light that they produce from modified Malpighian tubules. In the dark zone of a cave, the larval light lures prey, particularly small flies, onto a sticky thread suspended by the larva from the cave ceiling. The flying adult male locates the luminescent female while she is still in the pharate state, and waits for the opportunity to mate upon her emergence.

4.5 INSECT BEHAVIOUR

Many of the insect behaviours mentioned in this chapter appear very complex, but behaviourists attempt to reduce them to simpler components. Thus, individual **reflexes** (simple responses to simple stimuli) can be identified, such as the flight response when the legs lose contact with the ground, and cessation of flight when contact is regained. Some extremely rapid reflex actions, such as the feeding lunge of odonate nymphs, or some "escape reactions" of many insects, depend upon a reflex involving **giant axons** that conduct impulses rapidly from the sense organs to the muscles. The integration of multiple reflexes associated with movement of the insect can be divided into:

- **kinesis** (plural: kineses), in which unorientated action varies according to stimulus intensity;

- **taxis** (plural: taxes), in which movement is directly towards or away from the stimulus.

Kineses include akinesis, which is unstimulated lack of movement, orthokinesis, in which speed depends upon stimulus intensity, and klinokinesis, which is a "random walk" with course changes (turns) being made when unfavourable stimuli are perceived and with the frequency of turns depending on the intensity of the stimulus. Increased exposure to unfavourable stimuli leads to increased tolerance (**acclimation**), so that random walking and acclimation will lead the insect to a favourable environment. The male response to the plume of sex attractant (Fig. 4.7) is an example of klinokinesis to a chemical stimulus. Ortho- and klinokineses are effective responses to diffuse stimuli, such as temperature or humidity, but different, more efficient responses are seen when an insect is confronted by less diffuse, gradient or point-source stimuli.

Kineses and taxes can be defined with respect to the type of stimulus eliciting a response. Appropriate prefixes include: anemo- for air currents, astro- for solar, lunar or astral (including polarized light), chemo- for taste and odour, geo- for gravity, hygro- for moisture, phono- for sound, photo- for light, rheo- for water current, and thermo- for temperature. Orientation and movement may be positive or negative with respect to the stimulus source. For example, resistance to gravity is termed negative geotaxis, attraction to light is termed positive phototaxis, and repulsion from moisture is termed negative hygrotaxis.

In klinotaxic behaviour, an insect moves relative to a gradient (or cline) of stimulus intensity, such as a light source or a sound emission. The strength of the stimulus is compared on each side of the body by moving the receptors from side to side (as in head waving of ants when they follow an odour trail), or by detection of differences in stimulus intensity between the two sides of the body using paired receptors, such as maxillary or antennal chemosensors. The tympanal organs detect the direction of a sound source by comparing differences in intensity between the two organs. Orientation with respect to a constant angle of light is termed menotaxis and includes the "light-compass" referred to in Box 4.4. Visual fixation of an object, such as prey, is termed telotaxis.

Often, the relationship between stimulus and behavioural response is complex, as a **threshold** intensity may be required before an action ensues. A particular stimulatory intensity is termed a **releaser** for a particular behaviour. Furthermore, complex behaviour elicited by a single stimulus may comprise several sequential steps, each of which may have a higher threshold, requiring an increased stimulus. As described in section 4.3.2, a male moth responds to a low-level sex pheromone stimulus by raising the antennae; at higher levels he orientates towards the source, and at an even higher threshold, flight is initiated. Increasing concentration encourages continued flight, and a second, higher threshold may be required before courtship ensues. In other behaviours, several different stimuli are involved, such as for courtship through to mating. This sequence can be seen as a long chain reaction of stimulus, action, new stimulus, next action, and so on, with each successive behavioural stage depending upon the occurrence of an appropriate new stimulus. An inappropriate stimulus during a chain reaction (such as the presentation of food while courting) is not likely to elicit the usual response.

Most insect behaviours are considered to be **innate**, i.e. they are programmed genetically to arise stereotypically upon first exposure to the appropriate stimulus. However, other behaviours are environmentally and physiologically modified. For example, virgins and mated females may respond in very different ways to identical stimuli, and immature insects often respond to different stimuli compared with conspecific adults. Furthermore, experimental evidence shows that learning can modify innate behaviour. Learning is defined as the acquisition of neuronal representations of new information (such as spatial changes to the environment, or new visual or olfactory characteristics), but its occurrence can be assessed only indirectly by its possible effect on behaviour. By experimental teaching (using training and reward), bees and ants can learn to fly or walk through a maze, and butterflies can be induced to alter their favourite flower colour. Fruit fly larvae (*Drosophila* species) can be trained to associate an odour with an unpleasant stimulus, as demonstrated by future avoidance of that learned odour. These examples of associative learning derive largely from laboratory studies. However, study of natural behaviour (ethology) is more relevant to understanding the role played in the evolutionary success of insects' behavioural plasticity, including the ability to modify behaviour through learning. In pioneering ethological studies, Niko Tinbergen showed that a digger wasp (Crabronidae: *Philanthus* *triangulum*) can learn the location of its chosen nest site by making a short flight to memorize elements of the local terrain. Adjustment of prominent landscape features around the nest misleads the homing wasp. However, as wasps identify landmark relationships rather than individual features, the confusion may be only temporary. Closely related *Bembix* digger wasps (Sphecidae) learn nest locations through more distant and subtle markers, including the appearance of the horizon, and are not tricked by investigator ethologists moving local small-scale landmarks.

Much research shows evidence of learning in insects, and genetically based individual variation in learning has been recorded in a few species. The widespread notion that insects have little learning ability clearly is wrong.

FURTHER READING

Blomquist, G.J. & Bagnères, A.-G. (eds) (2010) *Insect Hydrocarbons: Biology, Biochemistry, and Chemical Ecology*. Cambridge University Press, Cambridge.

Chapman, R.F. (2003) Contact chemoreception in feeding by phytophagous insects. *Annual Review of Entomology* **48**, 455–84.

Chapman, R.F. (2013) *The Insects. Structure and Function*, 5th edn (eds S.J. Simpson & A.E. Douglas). Cambridge University Press, Cambridge.

Cocroft, R.B. & Rodríguez, R.L. (2005) The behavioral ecology of insect vibrational communication. *BioScience* **55**, 323–34.

Connor, W.E. & Corcoran, A.J. (2012) Sound strategies: the 65-million-year-old battle between bats and insects. *Annual Review of Entomology* **57**, 21–39.

Drijfhout, F.P. (2010) Cuticular hydrocarbons: a new tool in forensic entomology. In: *Current Concepts in Forensic Entomology* (eds J. Amendt, C.P. Campobasso, M.L. Goff & M. Grassberger), pp. 179–203.Springer, Dordrecht, Heidelberg, London, New York.

Drosopoulos, S. & Claridge, M.F. (eds) (2006) *Insect Sounds and Communication: Physiology, Behaviour, Ecology and Evolution*. CRC Press, Boca Raton, FL.

Dukas, R. (2008) Evolutionary biology of insect learning. *Annual Review of Entomology* **53**, 145–60.

Gitau, C.W., Bashford, R., Carnegie, A.J. & Gurr, G.M. (2013) A review of semiochemicals associated with bark beetles (Coleoptera: Curculionidae: Scolytinae) pests of coniferous trees: a focus on beetle interactions with other pests and associates. *Forest Ecology and Management* **297**, 1–14.

Greenfield, M.D. (2002) *Signalers and Receivers: Mechanisms and Evolution of Arthropod Communication*. Oxford University Press, New York.

Guerenstein, P.G. & Hildebrand, J.G. (2008) Roles and effects of environmental carbon dioxide in insect life. *Annual Review of Entomology* **53**, 161–78.

Heinrich, B. (1993) *The Hot-blooded Insects: Strategies and Mechanisms of Thermoregulation*. Harvard University Press, Cambridge, MA.

Holman, L., Lanfear, R. & d'Ettorre, P. (2013) The evolution of queen pheromones in the ant genus *Lasius*. *Journal of Evolutionary Biology* **26**, 1549–58.

Howse, P., Stevens, I. & Jones, O. (1998) *Insect Pheromones and their Use in Pest Management*. Chapman & Hall, London.

Hoy, R.R. & Robert, D. (1996) Tympanal hearing in insects. *Annual Review of Entomology* **41**, 433–50.

Kocher, S.D. & Grozinger, C.M. (2011) Cooperation, conflict, and the evolution of queen pheromones. *Journal of Chemical Ecology* **37**, 1263–75.

Leal, W.S. (2005) Pheromone reception. *Topics in Current Chemistry* **240**, 1–36.

Nakano, R., Ishikawa, Y., Tatsuki, S. *et al.* (2009) Private ultrasonic whispering in moths. *Communicative and Integrative Biology* **2**, 123–6.

Resh, V.H. & Cardé, R.T. (eds) (2009) *Encyclopedia of Insects*, 2nd edn. Elsevier, San Diego, CA. [Particularly see articles on: bioluminescence; chemoreception; circadian rhythms; eyes and vision; hearing; magnetic sense; mechanoreception; ocelli and stemmata; orientation; pheromones; vibrational communication.]

Robert, D. & Hoy, R.R. (2007) Auditory systems in insects. In: *Invertebrate Neurobiology* (eds G. North & R.J. Greenspan), pp. 155–84. Cold Spring Harbor Laboratory Press, New York.

Stumpner, A. & von Helversen, D. (2001) Evolution and function of auditory systems in insects. *Naturwissenschaften* **88**, 159–70.

Syed, Z. & Leal, W.S. (2008) Mosquitoes smell and avoid the insect repellent DEET. *Proceedings of the National Academy of Sciences* **105**, 13598–603.

Wajnberg, E. & Colazza, S. (eds) (2013) *Chemical Ecology of Insect Parasitoids*. Wiley-Blackwell, Chichester.

Wyatt, T.D. (2003) *Pheromones and Animal Behavior: Communication by Smell and Taste*. Cambrige University Press, Cambridge.

Chapter 5

REPRODUCTION

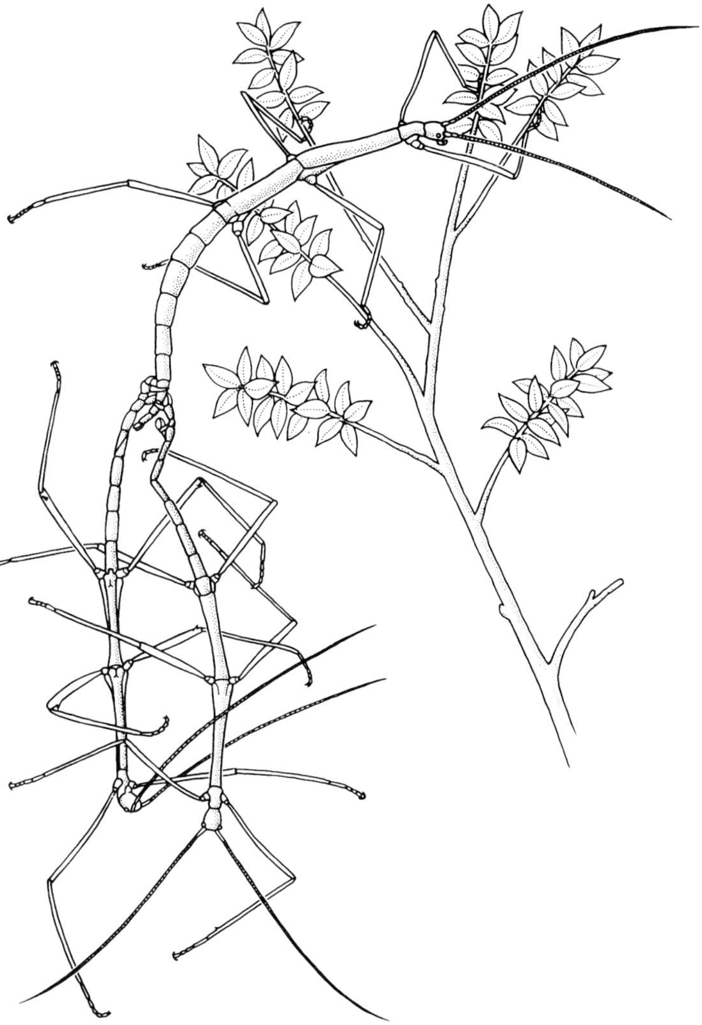

Two male stick-insects fighting over a female. (After Sivinski 1978.)

Most insects are sexual, and thus mature males and females must be present at the same time and place in order for reproduction to occur. As insects are generally short-lived, their life history, behaviour and reproductive condition must be synchronized. This requires finely tuned and complex physiological responses to the external environment. Furthermore, reproduction also depends on monitoring of internal physiological stimuli, and the neuroendocrine system plays a key regulatory role in this. Mating and egg production are known to be controlled by a series of hormonal and behavioural changes, yet there is much still to learn about the control and regulation of insect reproduction, particularly compared with knowledge of vertebrate reproduction.

These complex behavioural and regulatory systems are highly successful. For example, pest insects often increase in abundance rapidly. A combination of short generation time, high fecundity and population synchronization to environmental cues allows many insect populations to respond extremely quickly to appropriate environmental conditions, such as a crop monoculture or release from a controlling predator. In these situations, temporary or obligatory loss of males (**parthenogenesis**) has proved to be an effective means by which some insects rapidly exploit temporarily (or seasonally) abundant resources.

This chapter examines the different mechanisms associated with courtship and mating, avoidance of interspecies mating, ensuring paternity, and determination of the sex of offspring. We also examine the elimination of sex, and describe some extreme cases in which the adult stage has been dispensed with altogether. These observations relate to hypotheses concerning sexual selection, including those linked to why insects have such remarkable diversity of genitalic structures. The concluding summary of the physiological control of reproduction emphasizes the extreme complexity and sophistication of mating and oviposition in insects. Eight boxes cover special topics, namely courtship and mating in Mecoptera (Box 5.1), mating in katydids and crickets (Box 5.2), cannibalistic mating in mantids (Box 5.3), puddling and gifts in Lepidoptera (Box 5.4), sperm precedence (Box 5.5), control of mating and oviposition in a blow fly (Box 5.6), mothers' choices of oviposition sites (Box 5.7), and the egg-tending fathers of giant water bugs (Box 5.8).

5.1 BRINGING THE SEXES TOGETHER

Insects often are at their most conspicuous when synchronizing the time and place for mating. The flashing lights of fireflies, the singing of crickets and the cacophony of cicadas are spectacular examples. However, there is a wealth of less ostentatious behaviour, which is of equal significance in bringing the sexes together and signalling readiness to mate to other members of the species. All signals are species-specific, serving to attract conspecific individuals of the opposite sex. However, abuse of these communication systems can take place, as when females of a predatory species of firefly lure males of another species to their death by emulating the flashing signal of that species (section 13.1.2).

Swarming is a characteristic and perhaps fundamental behaviour of insects, as it occurs widely in many insect groups, from mayflies and odonates to many flies and butterflies. Swarming sites are identified by visual markers (Fig. 5.1) and are usually species-specific, although mixed-species **swarms** have been reported, especially in the tropics and subtropics. Swarms are predominantly of the male sex only, though female-only swarms do occur. Swarms are most evident when many individuals are involved, such as when midge swarms are so dense that they can be mistaken for smoke from burning buildings, but small swarms may be more significant in evolution. A single male insect holding station over a spot is a swarm of one – he awaits the arrival of a receptive female that has responded in like manner to visual cues that identify the site. The precision of swarm sites allows more effective mate finding than does searching, particularly when individuals are rare or dispersed and at low density. The formation of a swarm allows insects of differing genotypes to meet and outbreed. This is of particular importance if larval development sites are patchy and locally dispersed; inbreeding would occur if adults did not disperse.

In addition to aerial aggregations, some male insects form substrate-based aggregations where they may defend a territory against conspecific males and/or court arriving females. Species in which males hold territories that contain no resources (such as oviposition substrates) of importance to females and exhibit male–male aggression plus courtship of females are

The Insects: An Outline of Entomology, Fifth Edition. P.J. Gullan and P.S. Cranston.

Companion Website: www.wiley.com/go/gullan/insects

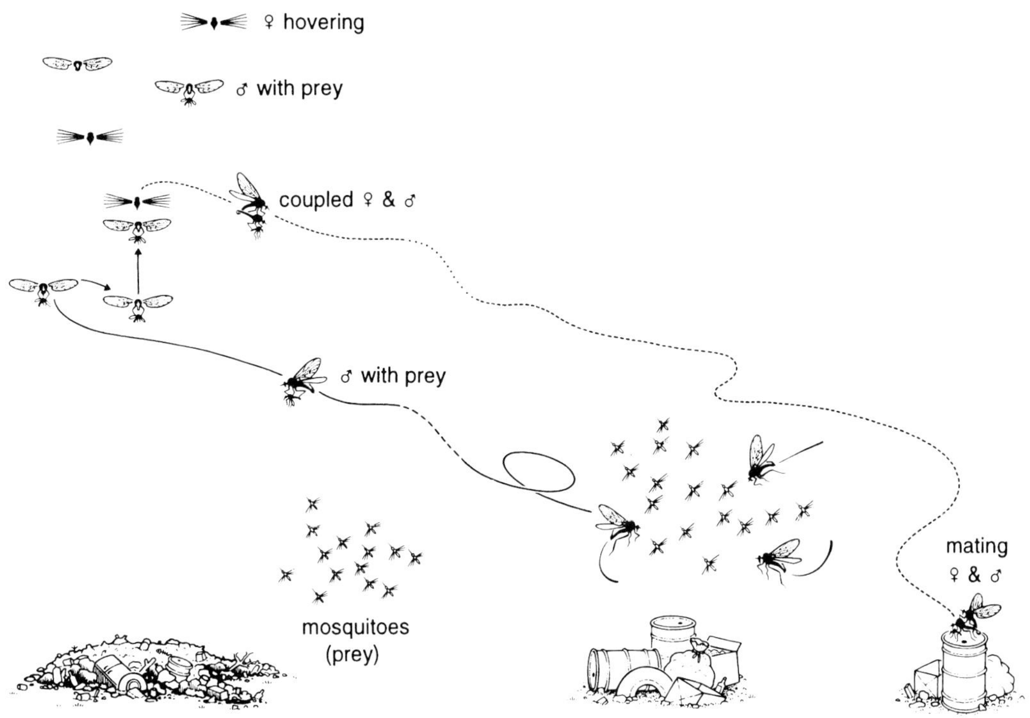

Fig. 5.1 Males of the Arctic fly, *Rhamphomyia nigrita* (Diptera: Empididae), hunt for prey in swarms of *Aedes* mosquitoes (lower mid-right of drawing) and carry the prey to a specific visual marker of the swarm site (left-hand side of drawing). Swarms of both the empidids and the mosquitoes form near conspicuous landmarks, including refuse heaps or oil drums, which are common in parts of the tundra. Within the mating swarm (upper left), a male empidid rises towards a female hovering above, they pair, and the prey is transferred to the female; the mating pair alights (lower far right) and the female feeds as they copulate. Females appear to obtain food only via males and, as individual prey items are small, must mate repeatedly in order to obtain sufficient nutrients to develop a batch of eggs. (After Downes 1970).

said to have a **lek** mating system. Lek behaviour is common in fruit flies of the families Drosophilidae and Tephritidae. Polyphagous fruit flies are more likely to have a lek mating system than are monophagous species, because in monophagous species males can expect to encounter females at the particular fruit that serves as the oviposition site.

Insects that form aerial or substrate-based mating aggregations often do so on hilltops, although some swarming insects aggregate above a water surface or use landmarks such as bushes or cattle. Most species probably use visual cues to locate an aggregation site, although uphill wind currents may guide insects to hilltops.

In other insects, the sexes may meet via attraction to a common resource and the meeting site might not be visually located. For species whose larval development medium is discrete, such as rotting fruit, animal dung or a specific host plant or vertebrate host, the best place for adult males and females to meet and court is at that resource. For example, the olfactory receptors by which the female dung fly finds a fresh pile of dung (the larval development site) can be employed by both sexes to facilitate meeting.

Another odoriferous communication involves one or both sexes producing and emitting a **pheromone**, which is a chemical, or a mixture of chemicals, perceptible to another member of the species (section 4.3.2). Substances emitted with the intention of altering the sexual behaviour of the recipient are termed sex pheromones. Generally, these are produced by the female and announce her presence and sexual availability to conspecific males. Recipient males that detect the odour plume become aroused and orientate from downwind towards the source. Many insects have species-specific sex pheromones; the diversity and specificity of which are important in maintaining the reproductive isolation of a species.

When the sexes are in proximity, mating in some species takes place without delay. For example, when a conspecific female arrives at a swarm of male flies, a nearby male, recognizing her by the particular sound of her wing-beat frequency, immediately copulates with her. However, more elaborate and specialized close-range behaviours, termed courtship, are commonplace.

5.2 COURTSHIP

Although the long-range attraction mechanisms discussed above reduce the number of species present at a prospective mating site, generally there remains an excess of potential partners. Further discrimination among species and conspecific individuals usually takes place. Courtship is the close-range, intersexual behaviour that induces sexual receptivity prior to (and often during) mating, and acts as a mechanism for species recognition. During courtship, one or both sexes seek to facilitate insemination and fertilization by influencing the other's behaviour.

Courtship may include visual displays, predominantly by males, including movements of adorned parts of the body such as antennae, eyestalks and "picture" wings, and ritualized movements ("dancing"). Tactile stimulation such as rubbing and stroking often occurs later in courtship, often immediately prior to mating, and may continue during copulation. Antennae, palps, head horns, external genitalia and legs are used in tactile stimulation. Acoustic courtship and mating recognition systems are common in many insects (e.g. Hemiptera, Orthoptera and Plecoptera). Insects such as crickets, which use long-range calling, may have different calls for use in close-range courtship. Others, such as fruit flies (*Drosophila*), have no long-distance call and sing (by wing vibration) only in close-up courtship. In some predatory insects, including empidid flies and mecopterans, the male courts a prospective mate by offering a prey item as a nuptial gift (Fig. 5.1; Box 5.1).

If the sequence of display proceeds correctly, courtship grades into mating. Sometimes, the sequence need not be completed before copulation commences. In other instances, courtship must be prolonged and repeated. It may be unsuccessful if one sex fails to respond or makes inappropriate responses. Generally, members of different species differ in some elements of their courtships, and interspecies matings do not occur. The great specificity and complexity of insect courtship behaviours can be interpreted in terms of mate location, synchronization of behaviour and species recognition, and viewed as having evolved as pre-mating isolating mechanisms. Important as this view is, there is equally compelling evidence that courtship is an extension of a wider phenomenon of competitive communication and involves sexual selection.

5.3 SEXUAL SELECTION

Many insects are sexually dimorphic, usually with the male adorned with secondary sexual characteristics, some of which have been noted above in relation to courtship display. In many mating systems, courtship can be viewed as intraspecific competition for mates, with certain male behaviours inducing female response in ways that can increase the mating success of particular males. Because females differ in their responsiveness to male stimuli, females can be said to choose between mates, and courtship thus is competitive. Female choice might involve no more than selection of the winners of male–male interactions, or may be as subtle as discrimination between the sperm of different males (section 5.7). All elements of communication associated with gaining fertilization of the female, from long-distance sexual calling through to insemination, are seen as competitive courtship between males. By this reasoning, members of a species avoid hybrid matings because of a specific-mate recognition system that evolved under the direction of female choice, rather than as a mechanism to promote species cohesion.

Understanding sexual dimorphism in insects, such as in staghorn beetles, song in orthopterans and cicadas, and wing colour in butterflies and odonates, helped Charles Darwin to recognize the operation of sexual selection – the elaboration of features associated with sexual competition rather than directly with survival. Since Darwin's day, studies of sexual selection often have featured insects because of their short generation time, facility of manipulation in the laboratory, and relative ease of observation

Box 5.1 Courtship and mating in Mecoptera

Sexual behaviour has been well studied in hangingflies (Bittacidae) of the North American *Hylobittacus* (*Bittacus*) *apicalis* and *Bittacus* species and the Australian *Harpobittacus* species, and in the Mexican *Panorpa* scorpionflies (Panorpidae). Adult males hunt for arthropod prey, such as caterpillars, bugs, flies and katydids. These same food items may be presented to a female as a nuptial offering to be consumed during copulation. Females are attracted by a sex pheromone emitted from one or more eversible vesicles or pouches near the end of the male's abdomen as he hangs in the foliage using prehensile fore tarsi.

Courting and mating in Mecoptera are exemplified by the sexual interactions in *Harpobittacus australis* (Bittacidae). The female closely approaches the "calling" male; he then ceases pheromone emission by retracting the abdominal vesicles. Usually, the female probes the offered prey briefly, presumably testing its quality, while the male touches or rubs her abdomen and seeks her genitalia with his own. If the female rejects the nuptial gift, she refuses to copulate. However, if the prey is suitable, the genitalia of the pair couple and the male temporarily withdraws the prey with his hind legs. The female lowers herself until she hangs head downwards, suspended by her terminalia. The male then surrenders the nuptial offering (in the illustration, a caterpillar) to the female, which feeds as copulation proceeds. During this stage, the male frequently supports the female by holding either her legs or the prey that she is feeding on. The derivation of the common name "hangingflies" is obvious!

Detailed field observations and manipulative experiments have demonstrated female choice of male partners in species of Bittacidae. Both sexes mate several times per day with different partners. Females discriminate against males that provide small or unsuitable prey, either by rejection or by copulating only for a short time, which is insufficient to pass the complete ejaculate. Given an acceptable nuptial gift, the duration of copulation correlates with the size of the offering. Each copulation in field populations of *Ha. australis* lasts from 1 to a maximum of about 17 minutes for prey from 3 to 14 mm long. In the larger *Hy. apicalis*, copulations involving prey of the size of houseflies or larger (19–55 mm^2; length × width) last from 20 to 31 minutes, resulting in maximal sperm transfer, increased oviposition, and the induction of a refractory period (female non-receptivity to other males) of several hours. Copulations that last less than 20 minutes reduce or eliminate male fertilization success. (Data from Thornhill 1976; Alcock 1979.)

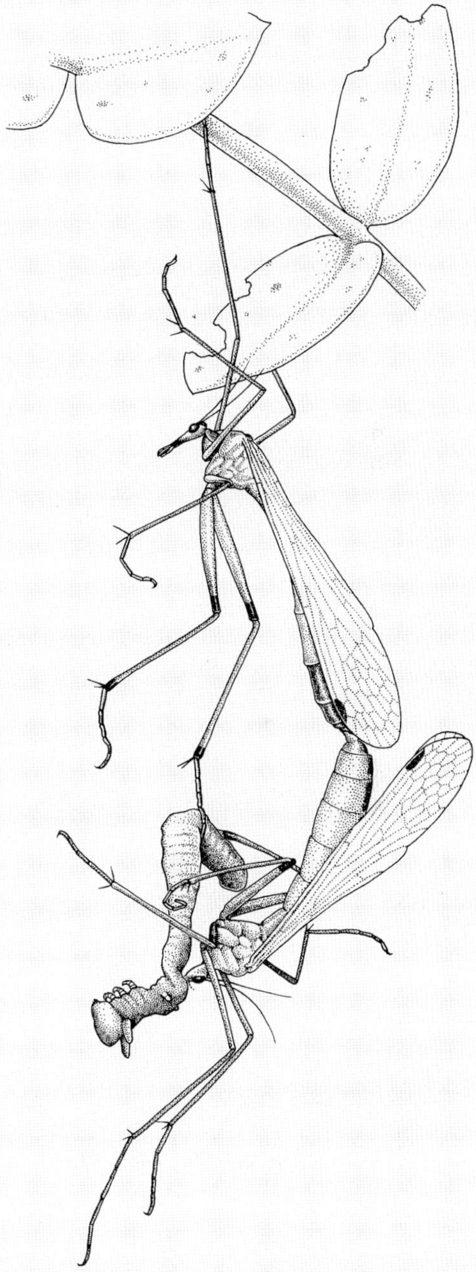

in the field. For example, dung beetles belonging to the large and diverse genus *Onthophagus* may display elaborate horns that vary in size between individuals and in position on the body between species. Large horns are restricted nearly exclusively to males, with only one species known in which the female has better developed protuberances than conspecific males. Studies show that females preferentially select males with larger horns as mates. Males size each other up and may fight, but there is no lek. Benefits to the female come from long-horned males' better defensive capabilities against intruders seeking to oust the resident from the resource-rich nest site, provisioned with dung, his mate and their young (see Fig. 9.6). However, the system is more complicated, at least in the North American *Onthophagus taurus*. In this dung beetle, male horn size is dimorphic, with insects greater than a certain threshold size having large horns, and those below a certain size having only minimal horns (Fig. 5.2). The nimble small-horned males attain some mating success through sneakily circumventing the large-horned but clumsy male defending the tunnel entrance, either by evasion or by digging a side tunnel to access the female.

Darwin could not understand why the size and location of horns varied, but now elegant comparative studies have shown that large horns bear a developmental cost. Organs located close to a large horn are diminished in size – evidently, resources are reallocated during development so that either eyes, antennae or wings apparently "pay" for being close to a male's large horn. Regular-sized adjacent organs are developed in females of the same species with smaller horns and male conspecifics with weakly developed horns. Exceptionally, the species with the female having long horns on the head and thorax commensurately has reduced adjacent organs, and a sex reversal in defensive roles is assumed to have taken place. The different locations of the horns appear to be explained by selective sacrifice of adjacent organs according to species behaviour. Thus, nocturnal species that require good eyes have their horns located elsewhere than on the head; those requiring flight to locate dispersed dung have horns on the head where they interfere with eye or antennal size, but do not compromise the wings. Presumably, the upper limit to horn elaboration either is the burden of ever-increasing deleterious effects on adjacent vital functions, or an upper limit on the volume of new cuticle that can develop sub-epidermally in the pharate pupa within the final-instar larva, under juvenile hormonal control.

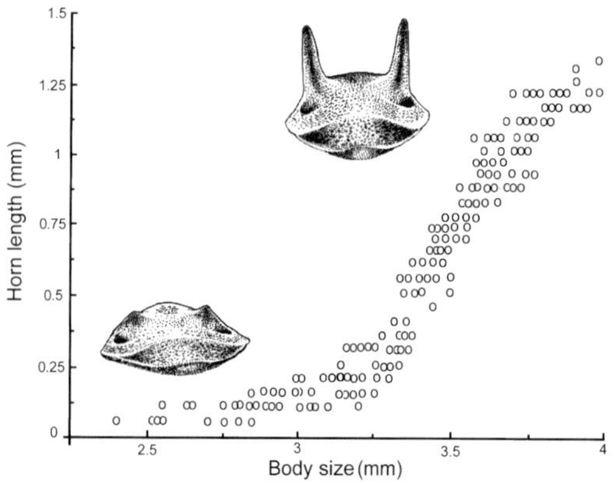

Fig. 5.2 Relationship between length of horn and body size (thorax width) of male scarabs of *Onthophagus taurus*. (After Moczek & Emlen 2000; with beetle heads drawn by S.L. Thrasher.)

Size alone may be important in female choice: in some stick-insects (also called walking sticks), larger males often monopolize females. Males fight over their females by boxing at each other with their legs while grasping the female's abdomen with their claspers (as shown for *Diapheromera velii* in the vignette at the beginning of this chapter). Ornaments used in male-to-male combat include the extraordinary "antlers" of *Phytalmia* (Tephritidae) (Fig. 5.3) and the eyestalks of a few other flies (such as Diopsidae), which are used in competition for access to the oviposition site visited by females. Furthermore, in studied species of diopsid (stalk-eyed flies), female mate choice is based on eyestalk length up to a dimension of eye separation that can surpass the body length. Cases such as these provide evidence for two apparently alternative but probably non-exclusive explanations for male adornments: "sexy sons" or "good genes". If the female choice commences arbitrarily for any particular adornment, their selection alone will drive the increased frequency and development of the elaboration in male offspring in ensuing generations (the "sexy sons"), despite any countering selection against conventional unfitness. Alternatively, females may choose mates that can demonstrate their fitness by carrying around apparently deleterious elaborations, thereby indicating a superior genetic background ("good genes"). Darwin's interpretation of the enigma

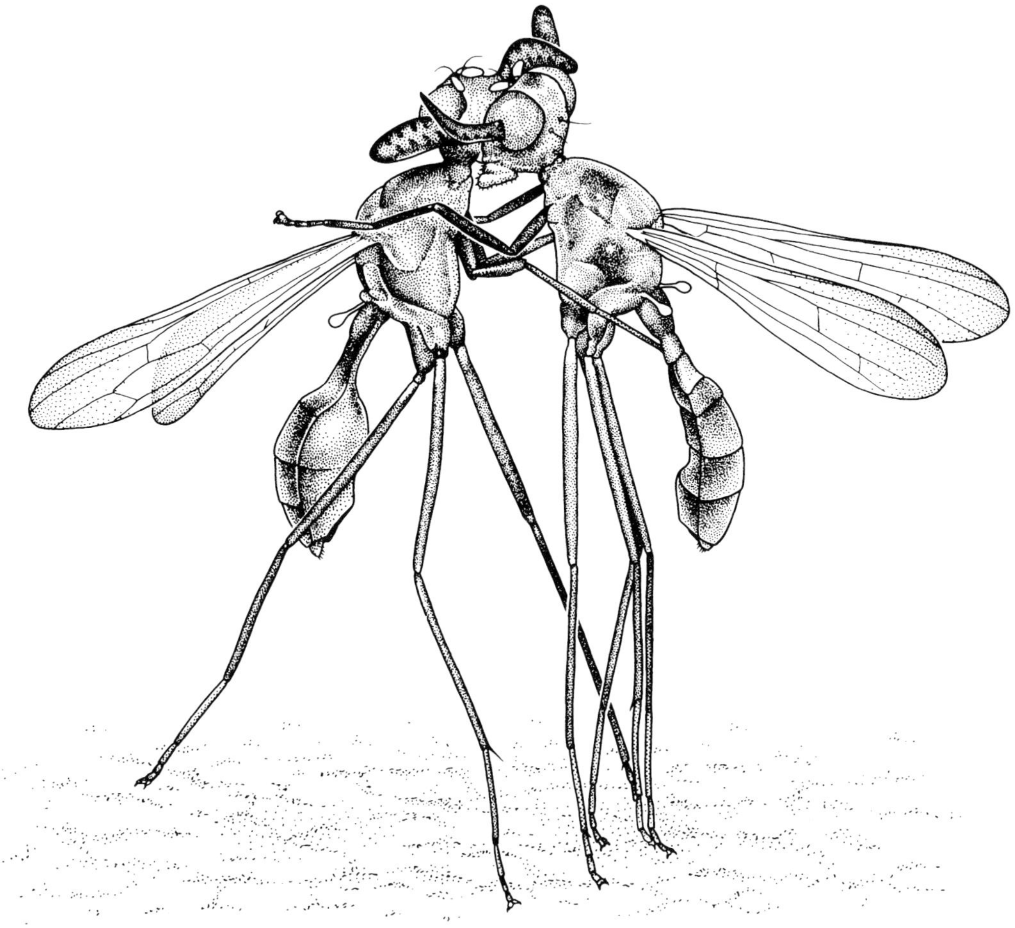

Fig. 5.3 Two males of *Phytalmia mouldsi* (Diptera: Tephritidae) fighting over access to the oviposition site at the larval substrate visited by females. These tropical rainforest flies thus have a resource-defence mating system. (After Dodson 1989, 1997.)

of female choice certainly is substantiated, not least by studies of insects.

5.4 COPULATION

The evolution of male external genitalia made it possible for insects to transfer sperm directly from male to female during copulation. All pterygote insects were freed from reliance on indirect methods, such as the male depositing a **spermatophore** (sperm packet) for the female to pick up from the substrate, as in Collembola, Diplura and apterygote insects. In pterygotes, copulation (sometimes referred to as mating) involves the physical apposition of male and female genitalia, usually followed by insemination – the transfer of sperm via the insertion of part of the male's **aedeagus** (edeagus), the penis, into the reproductive tract of the female. In males of many species, the extrusion of the aedeagus during copulation is a two-stage process. The complete aedeagus is extended from the abdomen, then the intromittent organ is everted or extended to produce an expanded, often elongate, structure (variably called the **endophallus**, flagellum or vesica) capable of depositing semen deep within the female's reproductive tract (Fig. 5.4). In many insects, the male terminalia have specially modified claspers, which lock with specific parts of the female terminalia

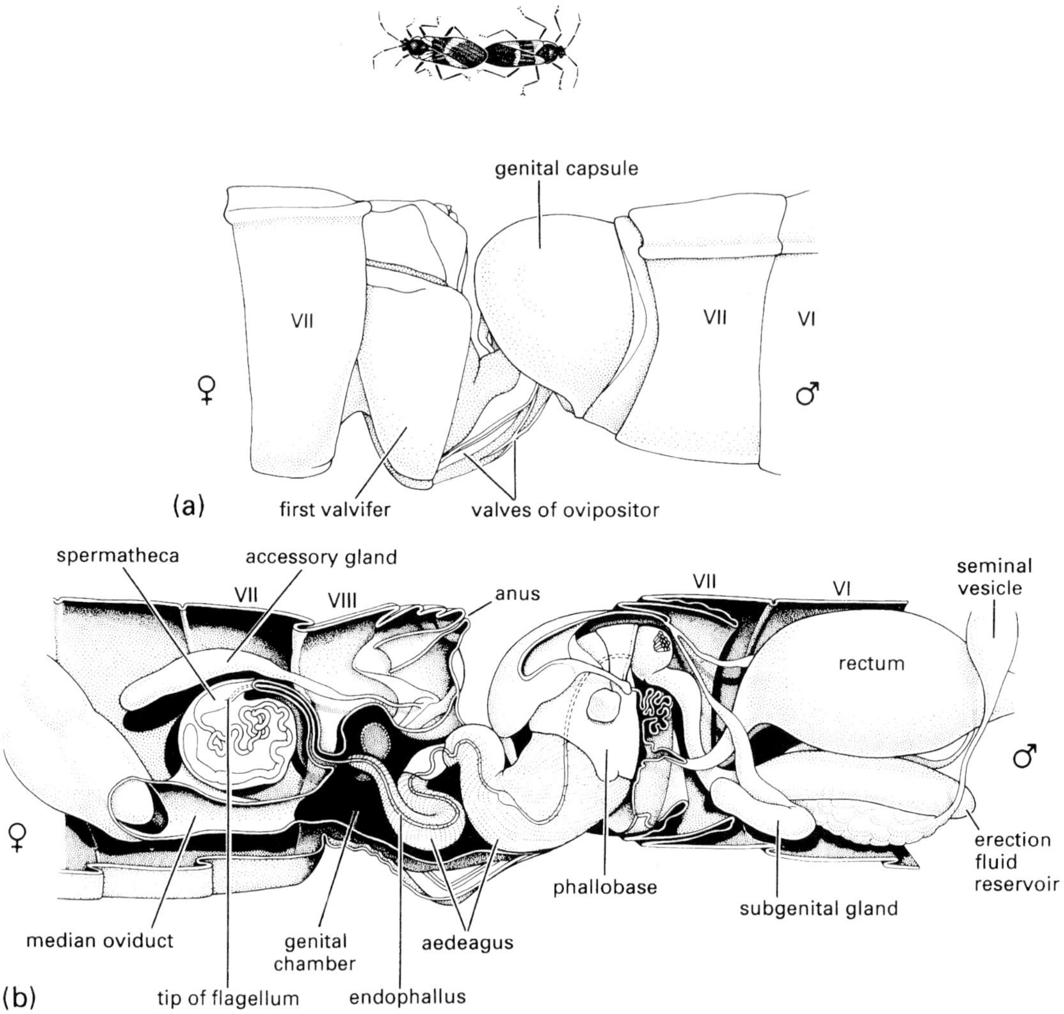

Fig. 5.4 Posterior ends of a pair of copulating milkweed bugs, *Oncopeltus fasciatus* (Hemiptera: Lygaeidae). Mating commences with the pair facing in the same direction, then the male rotates his eighth abdominal segment (90°) and genital capsule (180°), erects the aedeagus and gains entry to the female's genital chamber, before he swings around to face in the opposite direction. The bugs may copulate for several hours, during which they walk around with the female leading and the male walking backwards. (a) Lateral view of the terminal segments, showing the valves of the female's ovipositor in the male genital chamber; (b) longitudinal section showing internal structures of the reproductive system, with the tip of the male's aedeagus in the female's spermatheca. (After Bonhag & Wick 1953.)

to maintain the connection of their genitalia during sperm transfer.

This mechanistic definition of copulation ignores the sensory stimulation that is a vital part of the copulatory act in insects, as it is in other animals. In over a third of all insect species surveyed, the male indulges in copulatory courtship – behaviour that appears to stimulate the female during mating. The male may stroke, tap or bite the body or legs of the female, wave antennae, produce sounds, or thrust or vibrate parts of his genitalia.

Sperm are received by the female insect in a copulatory pouch (genital chamber, vagina or **bursa copulatrix**) or directly into a **spermatheca** or its duct (as in *Oncopeltus*; Fig. 5.4). A spermatophore is the means of sperm transfer in most orders of insects; only some Heteroptera, Coleoptera, Diptera and Hymenoptera deposit unpackaged sperm. Sperm transfer requires lubrication, which is obtained from the seminal fluids and, in insects that use a spermatophore, packaging of sperm. Secretions of the male reproductive tract serve both of these functions, as well as sometimes facilitating the final maturation of sperm, supplying energy for sperm maintenance, regulating female physiology and, in a few species, providing nourishment to the female (section 5.4.1; Box 5.2). The male seminal fluid secretions (proteins and other molecules) are the product of the accessory glands, seminal vesicles, ejaculatory duct and bulb, and testes. The seminal fluid chemicals may elicit various post-mating physiological and behavioural responses in the female by entering her haemolymph and acting on her nervous and/or endocrine system. Two major effects are: (i) induction or promotion of oviposition (egg-laying) by increasing ovulation or egg-laying rates; and/or (ii) repression of sexual receptivity to reduce her likelihood of re-mating.

Box 5.2 Mating in katydids and crickets

During copulation, the males of many species of katydids (Orthoptera: Tettigoniidae) and some crickets (Orthoptera: Gryllidae) transfer elaborate spermatophores, which are attached externally to the female's genitalia (see **Plate 3b**). Each spermatophore consists of a large, proteinaceous, sperm-free portion, the **spermatophylax**, which is eaten by the female after mating, and a sperm ampulla, eaten after the spermatophylax has been consumed and the sperm have been transferred to the female. The illustration on the upper left shows a recently mated female Mormon cricket, *Anabrus simplex*, with a spermatophore attached to her gonopore; in the illustration on the upper right, the female is consuming the spermatophylax of the spermatophore (after Gwynne 1981). The schematic illustration underneath depicts the posterior of a female Mormon cricket, showing the two parts of the spermatophore: the spermatophylax (cross-hatched) and the sperm ampulla (stippled) (after Gwynne 1990). During consumption of the spermatophylax, sperm are transferred from the ampulla, and substances are ingested or transferred that "turn off" female receptivity to further males. Insemination also stimulates oviposition by the female, thereby increasing the probability that the male supplying the spermatophore will fertilize the eggs.

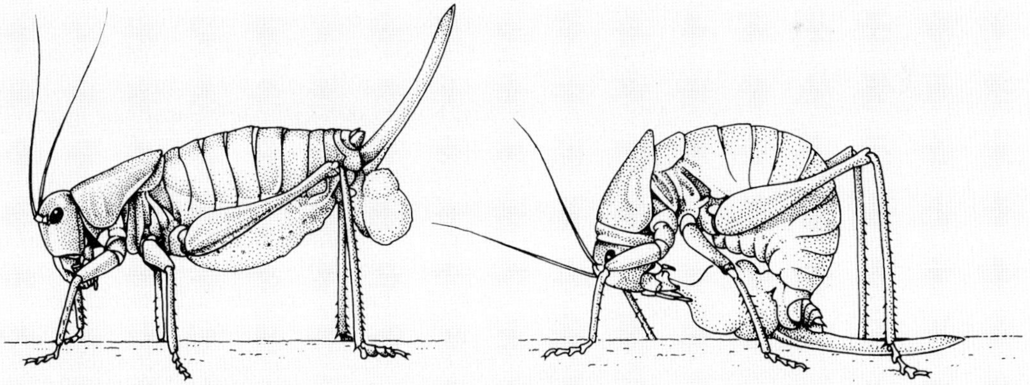

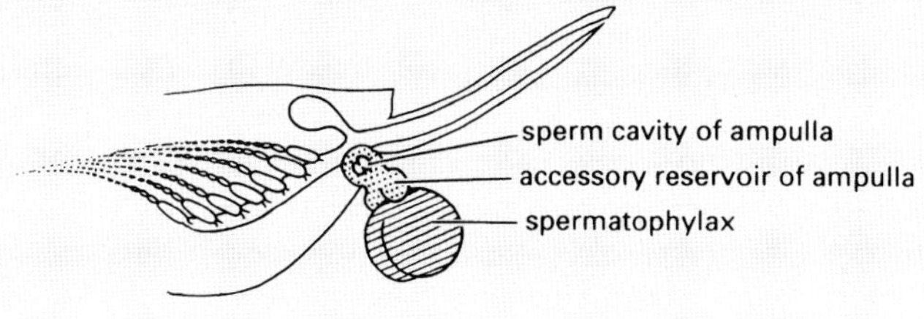

There are two main hypotheses for the adaptive significance to the male of this form of nuptial feeding. The spermatophylax may be a form of parental investment in which nutrients from the male increase the number or size of the eggs sired by that male. Alternatively, viewed in the context of sexual conflict, the "candymaker" hypothesis suggests that the spermatophylax is a "sensory trap" that lures the female into superfluous and/or prolonged matings that have a couple of possible benefits to the male. In particular, the spermatophylax may serve as a sperm-protection device by preventing the ampulla from being removed until after the complete ejaculate has been transferred, which may not be in the female's interests. Such prolonged insemination also may allow transfer of male secretions that manipulate the re-mating or oviposition behaviour of the female to the benefit of the male's reproductive success. Of course, the spermatophylax may serve the purposes of both offspring nutritional investment and male mating effort, and there is evidence from different species to support each hypothesis. Experimental alteration of the size of the spermatophylax has demonstrated that females take longer to eat larger ones, but in some katydid species, the spermatophylax is larger than is needed to allow complete insemination and, in this case, the nutritional bonus to the female may benefit the male's offspring. However, recent chemical analysis of the spermatophylax of a *Gryllodes* cricket showed that its free amino acid composition was highly imbalanced, being low in essential amino acids and high in phagostimulatory amino acids that probably increase the attractiveness of a low-value food. The latter data lend support to the candymaker hypothesis. Nevertheless, some female orthopterans may benefit directly from consumption of the spermatophylax, as shown for the females of a European katydid able to route male-derived nutrients to their own metabolism within a few hours after spermatophylax consumption. The function of the spermatophylax no doubt varies among genera, although phylogenetic analysis suggests that the ancestral condition within the Tettigoniidae was to possess a small spermatophylax that protected the ejaculate.

5.4.1 Nuptial feeding and other "gifts"

Feeding of the female by the male before, during or after copulation has evolved independently in several different insect groups. From the female's perspective, feeding takes one of three forms:

1 receipt of food collected, captured or regurgitated by the male (Box 5.1); or

2 obtaining attractive chemicals (often a form of nourishment) from a glandular product (including the spermatophore) of the male (Box 5.2); or

3 cannibalization of males during or after copulation (Box 5.3).

There is controversy concerning whether, and how much, the female typically benefits from such male-proffered gifts. In some instances, nuptial gifts may exploit the sensory preferences of the female and provide little nutritional benefit, while luring the female to accept larger ejaculates or extra copulations, and thus allow male control of insemination. A meta-analysis showed that female lifetime egg and offspring production increased with mating rate in groups of insects regardless of whether they used nuptial feeding or not, but that egg production increased to a larger extent in insects using nuptial gifts, and increased mating rate tended to increase female longevity of insects with nuptial feeding but decrease longevity of species not using nuptial feeding. Thus, there appears to be no negative effect of polyandry (mating of females with multiple males) on the female reproductive fitness of insects with nuptial feeding.

From the male's perspective, nuptial feeding may represent parental investment (provided that the male can be sure of his paternity), as it may increase the number or survival of the male's offspring indirectly via nutritional benefits to the female. Alternatively, courtship feeding may increase the male's fertilization success by preventing the female from interfering with sperm transfer and by inducing longer post-copulatory sexual refractory periods in females. These two hypotheses concerning the function of nuptial feeding are not necessarily mutually exclusive; their explanatory value appears to vary between insect groups and may depend, at least partly, on the nutritional status of the female at the time of mating. The mating processes of Mecoptera (Box 5.1), Orthoptera (Box 5.2) and Mantodea (Box 5.3) exemplify the three nuptial feeding types seen in insects.

In some other insect orders, such as the Lepidoptera and Coleoptera, the female sometimes acquires metabolically essential substances or defensive chemicals

Box 5.3 Cannibalistic mating in mantids

The sex life of mantids (Mantodea) is the subject of some controversy, partly as a consequence of behavioural observations made under unnatural conditions in the laboratory. For example, there are many reports of the male being eaten by the generally larger female before, during or after mating. Males decapitated by females are even known to copulate more vigorously because of the loss of the suboesophageal ganglion that normally inhibits copulatory movements. Sexual cannibalism has been attributed to food deprivation in confinement, but female mantids of at least some species may indeed eat their partners in the wild.

Courtship displays may be complex or absent, depending on species, but generally the female attracts the male via sex pheromones and visual cues. Typically, the male approaches the female cautiously, arresting movement if she turns her head towards him, and then he leaps onto her back from beyond her strike reach. Once mounted, he crouches to elude his partner's grasp. Copulation usually lasts at least half an hour and may continue for several hours, during which sperm are transferred from the male to the female in a spermatophore. After mating, the male retreats hastily. If the male were in no danger of becoming the female's meal, his distinctive behaviour in the presence of the female would be inexplicable. Furthermore, suggestions of gains in reproductive fitness of the male via indirect nutritional benefits to his offspring are negated by the obvious unwillingness of the male to participate in the ultimate nuptial sacrifice – his own life!

Whereas there is no evidence yet for an increase in male reproductive success as a result of sexual cannibalism, females that obtain an extra meal by eating their mate may gain a selective advantage, especially if food is limiting. This hypothesis is supported by a positive relationship between female mantid hunger and the probability of sexual cannibalism. In experiments with captive females of the Asian mantid *Hierodula membranacea* that were fed different quantities of food, the frequency of sexual cannibalism was higher for females of poorer nutritional condition and, among the females on the poorest diet, those that ate their mates produced significantly larger oothecae (egg packages) and hence more offspring. The cannibalized males would be making a parental investment only if their sperm fertilize the eggs that they have nourished. The crucial data on sperm competition in mantids are not available, and so currently the advantages of this form of nuptial feeding are attributed entirely to the female.

Box 5.4 Puddling and gifts in Lepidoptera

Anyone who has visited a tropical rainforest will have seen drinking clusters of perhaps hundreds of newly eclosed male butterflies, attracted particularly to mud, urine, faeces and human sweat (see **Plate 3a**). Male butterflies and moths frequently drink at pools of liquid, a behaviour known as **puddling**. It can involve copious quantities of liquid being ingested orally and expelled anally. Puddling allows the uptake of minerals, such as sodium, which are deficient in the larval (caterpillar) folivore diet, or the acquisition of nitrogen, depending on the substrate. Temperate butterflies mostly appear to seek salts, whereas some tropical butterflies are more interested in protein (nitrogen) sources. The sex bias in puddling occurs because the male uses the sodium (or perhaps nitrogen in some species) obtained by puddling as a nuptial gift for his mate. In the moth *Gluphisia septentrionis* (Notodontidae), the sodium gift amounts to more than half of the puddler's total body sodium and appears to be transferred to the female via his spermatophore (Smedley & Eisner 1996). The female then apportions much of this sodium to her eggs, which contain several times more sodium than eggs sired by males that have been experimentally prevented from puddling. Such paternal investment in the offspring is of obvious advantage to them in supplying an ion important to body function.

In some other lepidopteran species, such "salted" gifts may function to increase the male's reproductive fitness not only by improving the quality of his offspring but also by increasing the total number of eggs that he can fertilize, assuming that he re-mates. In the skipper butterfly, *Thymelicus lineola* (Hesperiidae), females usually mate only once, and male-donated sodium appears essential for both their fecundity and longevity (Pivnick & McNeil 1987). These skipper males mate many times and can produce spermatophores without access to sodium from puddling, but after their first mating, they father fewer viable eggs compared with re-mating males that have been allowed to puddle. This raises the question of whether females, which should be selective in the choice of their sole partner, can discriminate between males based on their sodium load. If they can, then sexual selection via female choice also may have selected for male puddling.

In other studies, copulating male lepidopterans have been shown to donate a diversity of nutrients, including zinc, phosphorus, lipids and amino acids, to their partners. Thus, paternal contribution of chemicals to offspring may be widespread within the Lepidoptera.

from the male during copulation, but oral uptake by the female usually does not occur. The chemicals are transferred with the ejaculate of the male. Such nuptial gifts may function solely as a form of parental investment (as may be the case in puddling; Box 5.4), but may also be a form of male mating effort (see Box 14.3).

5.5 DIVERSITY IN GENITALIC MORPHOLOGY

The components of the terminalia of insects are very diverse in structure and frequently exhibit species-specific morphology (Fig. 5.5), even in otherwise similar species. Variations in external features of the male genitalia often allow differentiation of species, whereas external structures in the female usually are simpler and less varied. Conversely, the internal genitalia of female insects often show greater diagnostic variability than the internal structures of the males. However, recent development of techniques to evert the endophallus of the male aedeagus allows increasing demonstration of the species-specific shapes of these male internal structures. In general, external genitalia of both sexes are much more sclerotized than the internal genitalia, although parts of the reproductive tract are lined with cuticle. Increasingly, characteristics of insect internal genitalia and even soft tissues are recognized as allowing species delineation and providing evidence of phylogenetic relationships.

Observations that genitalia frequently are complex and species-specific in form, sometimes appearing to correspond tightly between the sexes, led to formulation of the "lock-and-key" hypothesis as an explanation of this phenomenon. Species-specific male genitalia (the "keys") were believed to fit only the conspecific female genitalia (the "locks"), thus preventing interspecific mating or fertilization. For example, in some katydids, interspecific copulations are unsuccessful in transmitting spermatophores because the specific structure of the male claspers (modified cerci) fails to fit the subgenital plate of the "wrong" female. The lock-and-key hypothesis was postulated first in 1844 and has been the subject of controversy ever since.

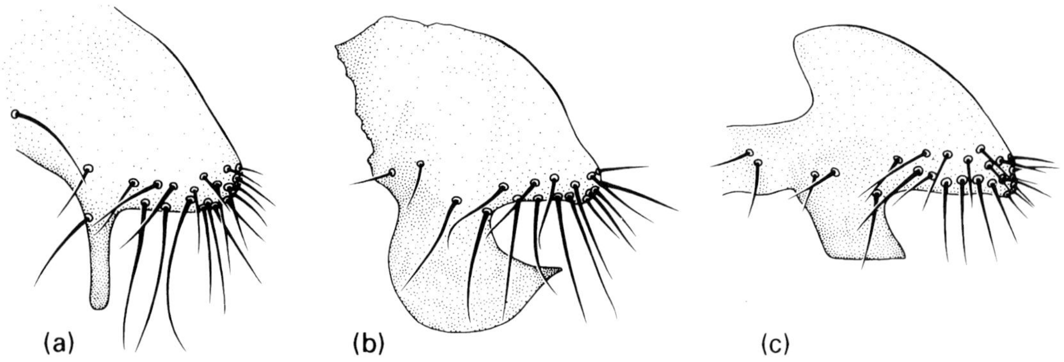

Fig. 5.5 Species-specificity in part of the male genitalia of three sibling species of *Drosophila* (Diptera: Drosophilidae). The epandrial processes of tergite 9 in: (a) *D. mauritiana*; (b) *D. simulans*; (c) *D. melanogaster*. (After Coyne 1983.)

In many (but not all) insects, mechanical exclusion of "incorrect" male genitalia by the female is seen as unlikely for several reasons:

1 morphological correlation between conspecific male and female parts may be poor;

2 interspecific, intergeneric and even interfamilial hybrids can be induced;

3 amputation experiments have demonstrated that male insects do not need all parts of the genitalia to inseminate conspecific females successfully.

Some support for the lock-and-key hypothesis comes from studies of certain noctuid moths in which structural correspondence in the internal genitalia of the male and female is thought to indicate their function as a post-copulatory but prezygotic isolating mechanism. Laboratory experiments involving interspecific matings support a lock-and-key function for the internal structures of other noctuid moths. Interspecific copulation can occur, although without a precise fit of the male's vesica (the flexible tube everted from the aedeagus during insemination) into the female's bursa (genital pouch), the sperm may be discharged from the spermatophore to the cavity of the bursa, instead of into the duct that leads to the spermatheca, resulting in fertilization failure. In conspecific pairings, the spermatophore is positioned so that its opening lies opposite that of the duct (Fig. 5.6).

In species of Japanese ground beetles of the genus *Carabus* (subgenus *Ohomopterus*: Carabidae), the male's copulatory piece (a part of the endophallus) is a precise fit for the vaginal appendix of the conspecific female. During copulation, the male everts his endophallus in the female's vagina and the copulatory piece is inserted into the vaginal appendix. Closely related parapatric species (which have immediately adjacent ranges) are of similar size and external appearance, but their copulatory piece and vaginal appendix are very different in shape. Although hybrids occur in areas of overlap of species, matings between different species of beetles have been observed to result in broken copulatory pieces and ruptured vaginal membranes, as well as reduced fertilization rates compared with conspecific pairings. Thus, the genital lock-and-key appears to select strongly against hybrid matings.

Mechanical reproductive isolation is not the only available explanation of species-specific genital morphology. Five other hypotheses have been advanced: pleiotropy; genitalic recognition; female choice; intersexual conflict; and male–male competition. The first two of these are further attempts to account for reproductive isolation of different species, whereas the last three are concerned with sexual selection, a topic that is addressed in more detail in sections 5.3 and 5.7.

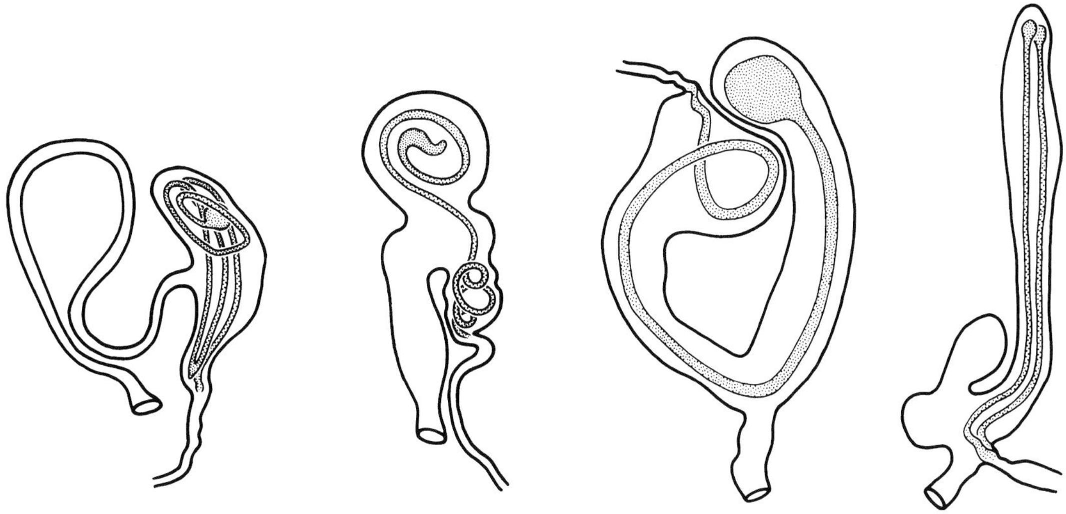

Fig. 5.6 Spermatophores lying within the bursae of the female reproductive tracts of moth species from four different genera (Lepidoptera: Noctuidae). The sperm leave via the narrow end of each spermatophore, which has been deposited so that its opening lies opposite the "seminal duct" leading to the spermatheca (not drawn). The bursa on the far right contains two spermatophores, indicating that the female has re-mated. (After Williams 1941; Eberhard 1985.)

The pleiotropy hypothesis explains genitalic differences between species as chance effects of genes that primarily code for other vital characteristics of the organism. This idea fails to explain why genitalia should be more affected than other parts of the body. Nor can pleiotropy explain genital morphology in groups (such as the Odonata) in which organs other than the primary male genitalia have an intromittent function (like those on the anterior abdomen in odonates). Such secondary genitalia consistently become subject to the postulated pleiotropic effects, whereas the primary genitalia do not, a result inexplicable by the pleiotropy hypothesis.

The hypothesis of genitalic recognition involves reproductive isolation of species via female sensory discrimination between different males based upon genitalic structures, both internal and external. The female thus responds only to the appropriate genital stimulation of a conspecific male and never to that of any male of another species.

In contrast, the female-choice hypothesis involves female sexual discrimination amongst conspecific males based on qualities that can vary intraspecifically and for which the female shows preference. This idea has nothing to do with the origin of reproductive isolation, although female choice may lead to reproductive isolation or speciation as a by-product. The female-choice hypothesis predicts diverse genitalic morphology in taxa with promiscuous females, and uniform genitalia in strictly monogamous taxa. This prediction seems to be fulfilled in some insects. For example, in Neotropical butterflies of the genus *Heliconius*, species in which females mate more than once are more likely to have species-specific male genitalia than species in which females mate only once. The greatest reduction in external genitalia (to near absence) occurs in termites, which, as might be predicted, form monogamous pairs.

Variation in genitalic and other body morphology also may result from intersexual conflict over control of fertilization. According to this hypothesis, females evolve barriers to successful fertilization in order to control mate choice, whereas males evolve mechanisms to overcome these barriers. For example, in many species of water-striders (Gerridae), males possess complex genital processes and modified appendages (Fig. 5.7) for grasping females, which in turn exhibit behaviours or morphological traits (e.g, abdominal spines) for dislodging males. Female water-striders of *Gerris gracilicornis* have a shield over their vulvar opening that prevents mounted males from forcing intromission.

Another example is the long spermathecal tube of some female crickets (Gryllinae), fleas (Ceratophyllinae), flies (e.g. Tephritidae) and beetles (e.g. Chrysomelidae), which corresponds to a long spermatophore tube in the male, suggesting an evolutionary contest over control of sperm placement in the spermatheca. In the cowpea seed beetle, *Callosobruchus maculatus* (Chrysomelidae: Bruchinae), spines on the male's intromittent organ wound the genital tract of the female during copulation, either to reduce re-mating and/or increase female oviposition rate, both of which would increase his fertilization success. The female responds by kicking to dislodge the male, thus shortening copulation time, reducing genital damage and presumably maintaining some control over fertilization. It is also possible that **traumatic insemination** (known in Cimicoidea, including bed bugs *Cimex lectularius*, and in a few species of Miridae and Nabidae), in which the male inseminates the female via the haemocoel by piercing her body wall with his aedeagus, evolved as a mechanism for the male to short-circuit the normal insemination pathway controlled by the female. A form of traumatic insemination involving male copulatory wounding of the female's body wall near the genital opening and sperm transfer into the genital tract through the wound has been recorded in the *Drosophila bipectinata* complex. Such examples of apparent intersexual conflict could be viewed as male attempts to circumvent female choice.

Another possibility is that species-specific elaborations of male genitalia may result from interactions between conspecific males vying for inseminations. Selection may act on male genitalic clasping structures to prevent usurpation of the female during copulation, or act on the intromittent organ itself to produce structures that can remove or displace the sperm of other males (section 5.7). However, although sperm displacement has been documented in a few insects, this phenomenon is unlikely to be a general explanation of male genitalic diversity because the penis of male insects often cannot reach the sperm-storage organ(s) of the female or, if the spermathecal ducts are long and narrow, sperm flushing should be impeded.

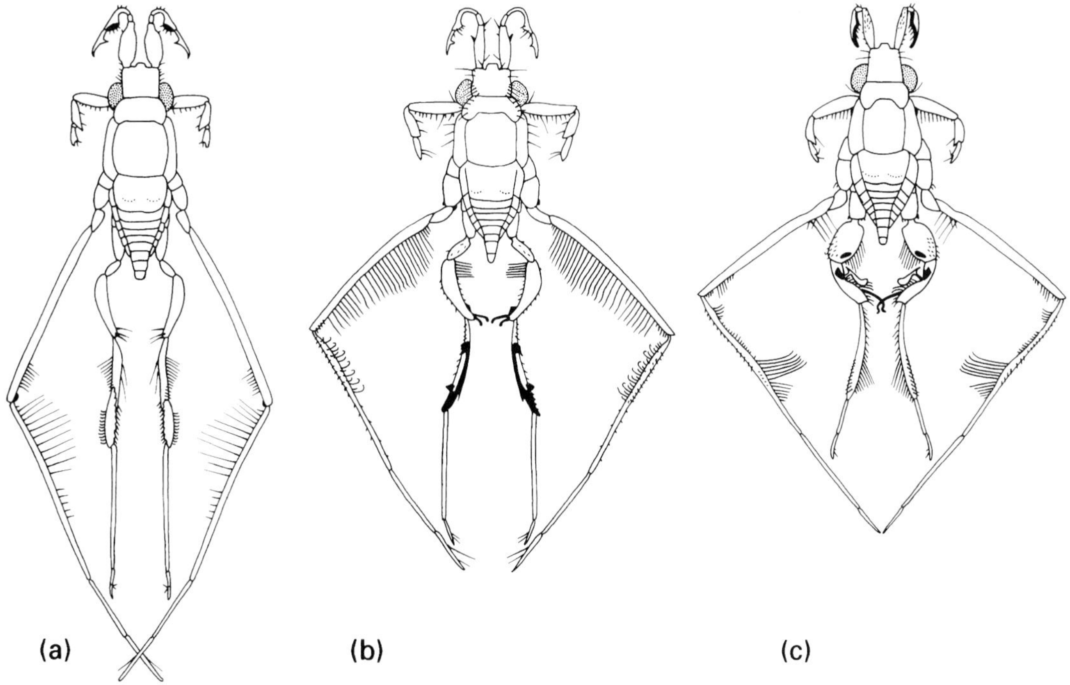

Fig. 5.7 Males of three species of the water-strider genus *Rheumatobates*, showing species-specific antennal and leg modifications (mostly flexible setae). These non-genitalic male structures are specialized for contact with the female during mating, when the male rides on her back. Females of all species have a similar body form. (a) *R. trulliger*; (b) *R. rileyi*; (c) *R. bergrothi*. (After Hungerford 1954.)

Functional generalizations about the species-specific morphology of insect genitalia are controversial because different explanations no doubt apply in different groups. For example, male–male competition (via sperm removal and displacement; Box 5.5) may be important in accounting for the shape of odonate penes, but appears irrelevant as an explanation in noctuid moths. Female choice, intersexual conflict and male–male competition may have little selective effect on genitalic structures of insect species in which the female mates with only one male (as in termites). In such species, sexual selection may affect features that determine which male is chosen as a partner, but not how the male's genitalia are shaped. Furthermore, both mechanical and sensory lock-and-key mechanisms will be unnecessary if isolating mechanisms, such as courtship behaviour or seasonal or ecological differences, are well developed. Therefore, we might predict morphological constancy (or a high level of similarity, allowing for some pleiotropy) in genitalic structures among species in a genus that has species-specific pre-copulatory displays involving non-genital structures followed by a single insemination of each female.

5.6 SPERM STORAGE, FERTILIZATION AND SEX DETERMINATION

Many female insects store the sperm that they receive from one or more males in their sperm storage organ, or spermatheca. Females of most insect orders have a single spermatheca, but some flies are notable in having more, often two or three. Sometimes sperm remain viable in the spermatheca for a considerable time, even three or more years in the case of honey bees. During storage, secretions from the female's spermathecal gland maintain the viability of sperm.

Eggs are fertilized as they pass down the median oviduct and vagina. The sperm enter the egg via one or more **micropyles**, which are narrow canals that pass through the eggshell. The micropyle or micropylar area is orientated towards the opening of the spermatheca during egg passage, facilitating sperm entry. In many insects, the release of sperm from the spermatheca appears to be controlled very precisely in timing and number. In queen honey bees, as few as 20 sperm per egg may be released, suggesting extraordinary economy of use.

The fertilized eggs of most insects give rise to both males and females, with the sex being dependent upon specific determining mechanisms, which are predominantly genetic. Most insects are **diploid**, i.e. having one set of chromosomes from each parent. The most common mechanism is for sex of the offspring to be determined by the inheritance of sex chromosomes (X-chromosomes; heterochromosomes), which are differentiated from the remaining autosomes. Individuals are thus allocated to sex according to the presence of one (X0) or two (XX) sex chromosomes. Although XX is usually female and X0 usually male, this allocation varies within and between taxonomic groups. Mechanisms involving multiple sex chromosomes also occur, and there are related observations of complex fusions between sex chromosomes and autosomes. *Drosophila* (Diptera) has an XY sex-determination system, in which females (the homogametic sex) have two of the same kind of sex chromosome (XX) and males (the heterogametic sex) have two kinds of sex chromosome (XY). Lepidoptera have a ZW sex-determination system, in which females have two different kinds of chromosomes (ZW) and males have two of the same kind of chromosome (ZZ). Frequently, we cannot recognize sex chromosomes, particularly as gender is known to be determined by single genes in certain insects, such as in many hymenopterans and some mosquitoes and midges. For example, a gene called *complementary sex determiner* (*csd*) has been cloned in honey bees (*Apis mellifera*) and shown to be the primary switch in the sex-determination cascade of that species. There are many variations of *csd* in honey bees and as long as two versions are inherited, the bee becomes female, whereas an unfertilized (haploid) egg, with a single copy of *csd*, becomes male. Additional complications with the determination of sex arise with the interaction of both the internal and external environment on the genome (epigenetic factors). Furthermore, great variation is seen in sex ratios at birth. Although the ratio is often one male to one female, there are many deviations, ranging from 100% of one sex to 100% of the other.

In **haplodiploidy** (male haploidy), the male sex has only one set of chromosomes. This arises either through his development from an unfertilized egg (containing half of the female chromosome complement following meiosis), called **arrhenotoky** (section 5.10.1), or from a fertilized egg in which the paternal set of chromosomes is inactivated and eliminated, called **paternal genome elimination** (as in many male scale insects). Arrhenotoky is exemplified by honey bees, in which females (queens and workers) develop from fertilized eggs, whereas males (drones) develop from unfertilized eggs. However, sex may be determined by a single gene (e.g. the *csd* locus, which is well studied in honey bees, as noted above) that is heterozygous in females and hemizygous in (haploid) males. Female insects control the sex of offspring through their ability to store sperm and control fertilization of eggs. Evidence points to a precise control of sperm release from storage, but very little is known about this process in most insects. The presence of an egg in the genital chamber may stimulate contractions of the spermathecal walls, leading to sperm release.

5.7 SPERM COMPETITION

Multiple matings are common in many insect species. The occurrence of re-mating under natural conditions can be determined by observing the mating behaviour of individual females, or by dissection to establish the amount of ejaculate or the number of spermatophores present in the female's sperm-storage organs. Some of the best documentation of re-mating comes from studies of many Lepidoptera, in which part of each spermatophore persists in the bursa copulatrix of the female throughout her life (Fig. 5.6). These studies show that re-mating occurs, to some extent, in almost all species of Lepidoptera for which adequate field data are available.

The combination of internal fertilization, sperm storage, multiple mating by females, and the overlap within a female of ejaculates from different males leads to a phenomenon known as **sperm competition**. This occurs within the reproductive tract of the female at the time of oviposition, when sperm from two or more males compete to fertilize the eggs. Both physiological and behavioural mechanisms determine the outcome of sperm competition. Thus, events inside the female's reproductive tract, combined with various attributes of mating behaviour, determine which sperm will succeed in reaching the eggs. It is important to realize that male reproductive fitness is measured in terms of the number of eggs fertilized or offspring fathered and not simply the number of copulations achieved, although these measures sometimes are correlated. Often there may be a trade-off between the number of copulations that a male can secure and the number of eggs that he will fertilize at each mating. A high copulation frequency is generally associated with low time or energy investment per copulation and also with low certainty of paternity. At the other extreme, males that exhibit substantial parental investment, such as feeding their mates (Box 5.1, Box 5.2 and Box 5.3), and other adaptations that more directly increase certainty of paternity, will inseminate fewer females over a given period.

There are two main types of sexually selected adaptations in males that increase certainty of paternity. The first strategy involves mechanisms by which males can ensure that females use their sperm preferentially. Such **sperm precedence** is achieved usually by displacing the ejaculate of males that have mated previously with the female (Box 5.5). The second strategy is to reduce the effectiveness or occurrence of subsequent inseminations by other males. Various mechanisms appear to achieve this result, including mating plugs, use of male-derived secretions that "switch off" female receptivity (see the introduction to section 5.4 and Box 5.6), prolonged copulation (Fig. 5.8), guarding of females, and improved structures for gripping the female during copulation to prevent "take-over" by other males. A significant selective advantage would accrue to any male that could both achieve sperm precedence and prevent other males from successfully inseminating the female until his sperm had fertilized at least some of her eggs.

Box 5.5 Sperm precedence

The penis or aedeagus of a male insect may be modified to facilitate placement of his own sperm in a strategic position within the spermatheca of the female, or even to remove a rival's sperm. Sperm displacement of the former type, called stratification, involves pushing previously deposited sperm to the back of a spermatheca, and occurs in systems in which a "last-in-first-out" principle operates (i.e. the most recently deposited sperm are the first to be used when the eggs are fertilized). Last-male sperm precedence occurs in many insect species; in others there is either first-male precedence or no precedence (because of sperm mixing). In some dragonflies, males appear to use inflatable lobes on the penis to reposition rival sperm. Such sperm-packing enables the copulating male to place his sperm closest to the oviduct. However, stratification of sperm from separate inseminations may occur in the absence of any deliberate repositioning, by virtue of the tubular design of the storage organs.

A second strategy of sperm displacement is removal, which can be achieved either by direct scooping out of existing sperm prior to depositing an ejaculate or, indirectly, by flushing out a previous ejaculate with a subsequent one. An unusually long penis that can reach well into the spermathecal duct may facilitate flushing of a rival's sperm from the spermatheca. A number of structural and behavioural attributes of male insects can be interpreted as devices to facilitate this form of sperm precedence, but some of the best-known examples come from odonates.

Copulation in Odonata involves the female placing the tip of her abdomen against the underside of the anterior abdomen of the male, where his sperm are stored in a reservoir of his secondary genitalia. In some dragonflies and most damselflies, such as the pair of copulating *Calopteryx* damselflies (Calopterygidae) illustrated in the wheel position (after Zanetti 1975), the male spends the greater proportion of the copulation time physically removing the sperm of other males from the female's sperm-storage organs (spermathecae and bursa copulatrix). Only at the last minute does he introduce his own. In these species, the male's penis is structurally complex, sometimes with an extensible head used as a scraper and a flange to trap the sperm, plus lateral horns or hook-like distal appendages with recurved spines to remove rival sperm (inset to figure; after Waage 1986). A male's ejaculate may be lost if another male mates with the female before she oviposits. Thus, it is not surprising that male odonates guard their mates, which explains why they are so frequently seen as pairs flying in tandem.

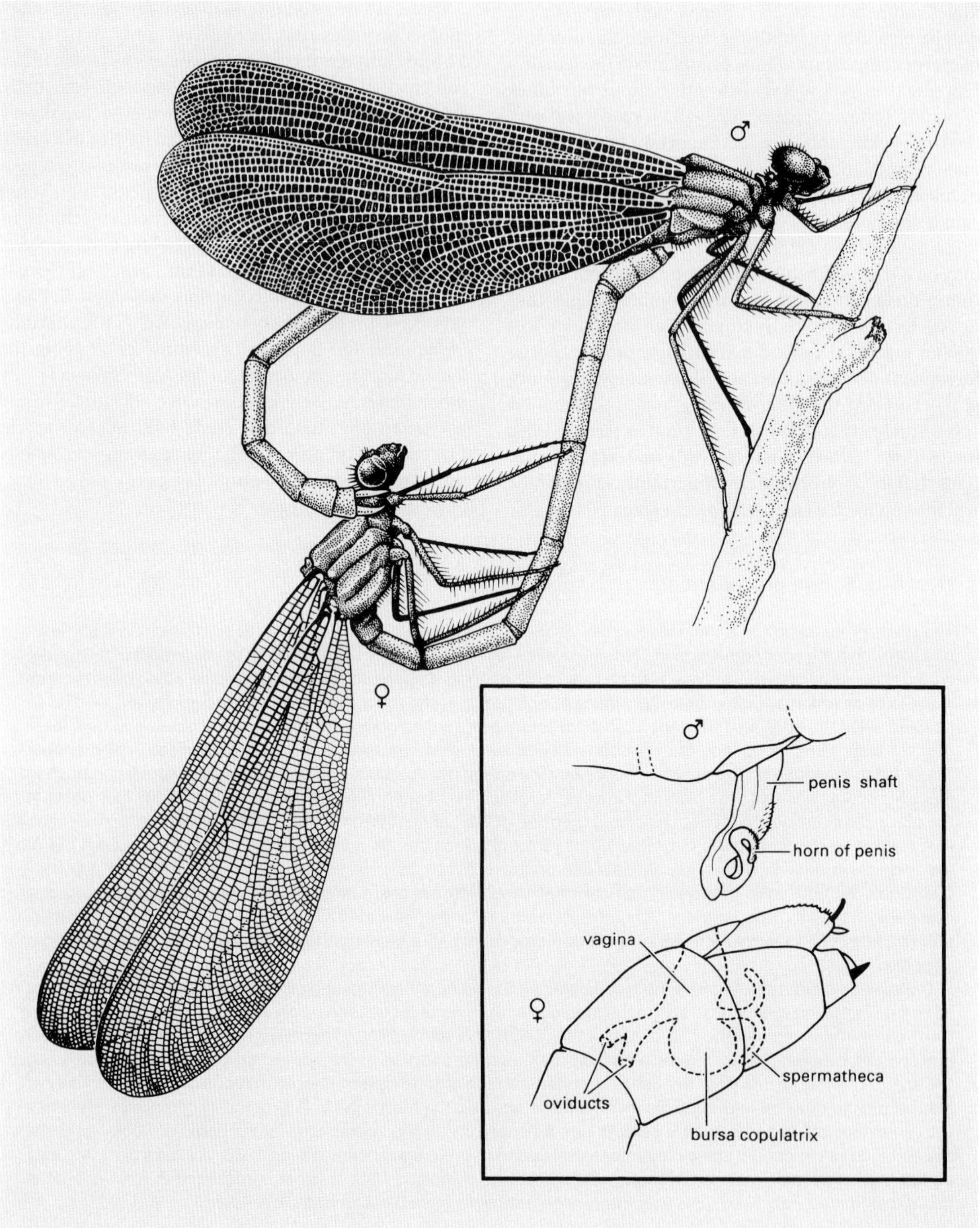
♂
♀
♂
penis shaft
horn of penis
vagina
♀
spermatheca
oviducts
bursa copulatrix

Box 5.6 Control of mating and oviposition in a blow fly

The sheep blow fly, *Lucilia cuprina* (Diptera: Calliphoridae), costs the Australian sheep industry many millions of dollars annually, through losses caused by myiases or fly "strikes". This pestiferous fly may have been introduced to Australia from Africa in the late 19th century. The reproductive behaviour of *L. cuprina* has been studied in some detail because of its relevance to a control programme for this pest. Ovarian development and reproductive behaviour of the adult female are highly stereotyped and are readily manipulated via precise feeding of protein. Most females are **anautogenous**, i.e. they require a protein meal in order to develop their eggs, and usually mate after feeding and before their oocytes have reached early vitellogenesis. After their first mating, females reject further mating attempts for several days. The "switch-off" is activated by a peptide produced in the accessory glands of the male and transferred to the female during mating. Mating also stimulates oviposition; virgin females rarely lay eggs, whereas mated females readily do so. The eggs of each fly are laid in a single mass of a few hundred (illustration at top right of figure), and then a new ovarian cycle commences with another batch of synchronously developing oocytes. Females may lay one to four egg masses before re-mating.

Unreceptive females respond to male mating attempts by curling their abdomen under their body (illustration at top left), by kicking at the males (illustration at top centre), or by actively avoiding them. Receptivity gradually returns to previously mated females, in contrast to their gradually diminishing tendency to lay. If re-mated, such non-laying females resume laying. Neither the size of the female's sperm store nor the mechanical stimulation of copulation can explain these changes in female behaviour. Experimentally, it has been demonstrated that the female mating refractory period and readiness to lay are related to the amount of male accessory gland substance deposited in the female's bursa copulatrix during copulation. If a male repeatedly mates during one day (a multiply-mated male), less gland material is transferred at each successive copulation. Thus, if one male is mated, during one day, to a succession of females, which are later tested at intervals for their receptivity and readiness to lay, then the proportion of females either unreceptive or laying is inversely related to the number of females with which the male had previously mated. The graph on the left shows the percentage of females unreceptive to further mating

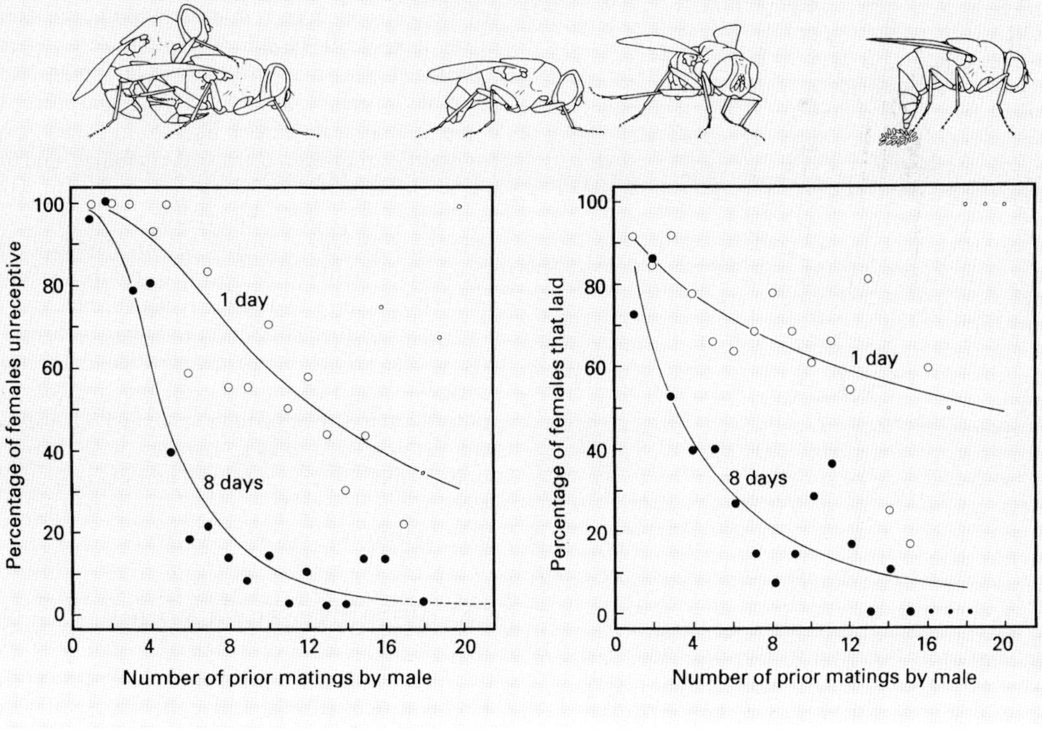

when tested 1 day or 8 days after having mated with multiply-mated males. The percentage unreceptive values are based on up to 29 tests of different females. The graph on the right shows the percentage of females that laid eggs during 6 h of access to oviposition substrate presented 1 day or 8 days after mating with multiply-mated males. The percentage laid values are based on tests of up to 15 females. These two plots represent data from different groups of 30 males; samples of female flies numbering less than five are represented by smaller symbols. (After Bartell *et al.* 1969; Barton Browne *et al.* 1990; Smith *et al.* 1990.)

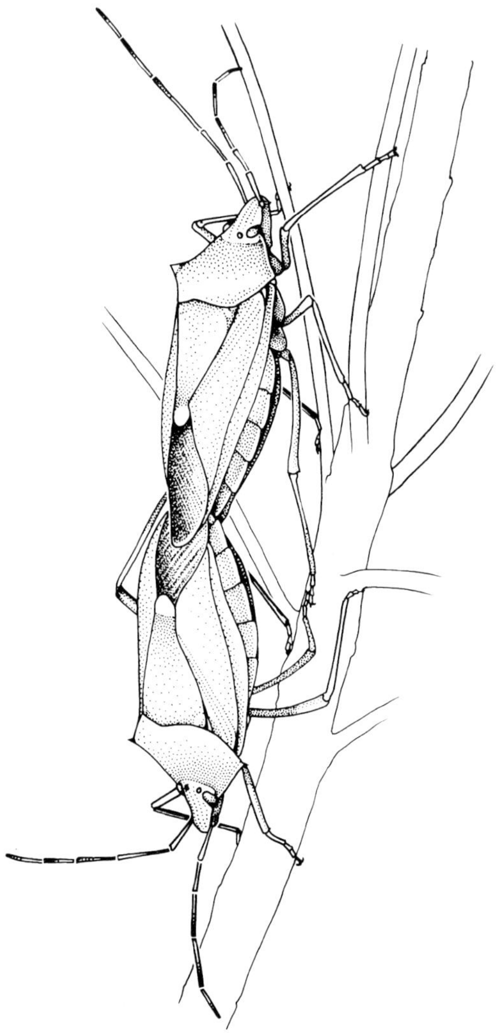

Fig. 5.8 A copulating pair of stink or shield bugs of the genus *Poecilometis* (Hemiptera: Pentatomidae). Many heteropteran bugs engage in prolonged copulation, which prevents other males from inseminating the female until either she becomes non-receptive to further males or she lays the eggs fertilized by the "guarding" male.

The factors that determine the outcome of sperm competition are not totally under male control. Female choice is a complicating influence, as shown in the above discussions on sexual selection and on morphology of genitalic structures. Female choice of sexual partners may be two-fold. First, there is good evidence that the females of many species choose among potential mating partners. For example, females of many mecopteran species mate selectively with males that provide food of a certain minimum size and quality (Box 5.1). In some insects, such as a few beetles and some moth and katydid species, females have been shown to prefer larger males as mating partners. Second, subsequent to copulation, the female might discriminate between partners as to which sperm will be used. One idea is that variation in the stimuli of the male genitalia induces the female to use one male's sperm in preference to those of another, based upon an "internal courtship". Differential sperm use is possible because females have control over sperm transport to storage, sperm maintenance and also use at oviposition. Research on *Drosophila* in which transgenic males either had red or green fluorescent labels on their sperm allowed *in vivo* visualization of interactions between the sperm of competing males and between the female's reproductive tract and the ejaculates. This labelling technique has been applied also to the sperm of two closely related *Drosophila* species to allow visualization of interspecific sperm competition in females doubly-mated with conspecific and heterospecific males. Females were found to control which sperm were stored and where they were stored to bias against the heterospecific sperm in favour of the conspecific sperm for fertilization.

5.8 OVIPARITY (EGG-LAYING)

The vast majority of female insects are **oviparous**, i.e. they lay eggs. Generally, ovulation – expulsion of eggs from the ovary into the oviducts – is followed rapidly by fertilization and then oviposition. The physiological

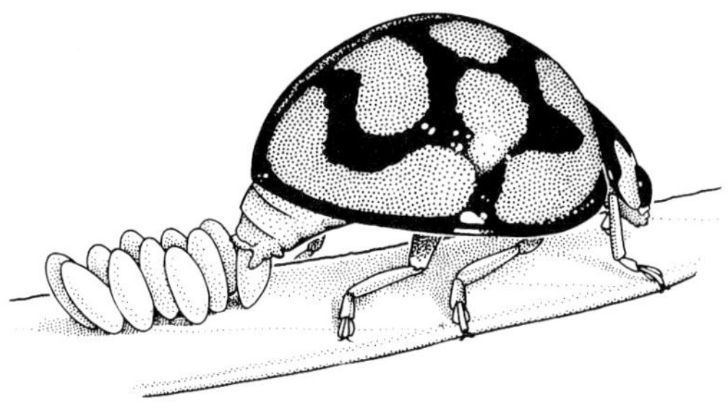

Fig. 5.9 Oviposition by a South African ladybird beetle, *Cheilomenes lunata* (Coleoptera: Coccinellidae). The eggs adhere to the leaf surface because of a sticky secretion applied to each egg. (After Blaney 1976.)

mechanism(s) controlling ovulation is(are) poorly known in insects, whereas oviposition appears to be under both hormonal and neural control, and both processes may be stimulated by peptides in the male's seminal fluid. Oviposition, the process of the egg passing from the external genital opening or vulva to the outside of the female (Fig. 5.9), may be associated with behaviours such as digging or probing into an egg-laying site, but often the eggs are simply dropped to the ground or into water. Usually, the eggs are deposited on or near the food required by the offspring upon hatching (Box 5.7). Care of eggs after laying often is lacking or minimal, but social insects (Chapter 12) have highly developed care, and certain aquatic insects show very unusual paternal care (Box 5.8).

An insect egg within the female's ovary is complete when an oocyte becomes covered with an outer protective coating, the eggshell, comprising the **vitelline membrane** and the **chorion**. The chorion may be composed of any or all of the following layers: wax layer; innermost chorion; endochorion; and exochorion (Fig. 5.10). Ovarian follicle cells produce the eggshell and the surface sculpturing of the chorion usually reflects the outline of these cells. Typically, the eggs are yolk-rich and thus large relative to the size of the adult insect; egg cells range in length from 0.2 mm to about 13 mm. Embryonic development within the egg begins after egg activation (section 6.2.1), which usually is stimulated by entry of the sperm.

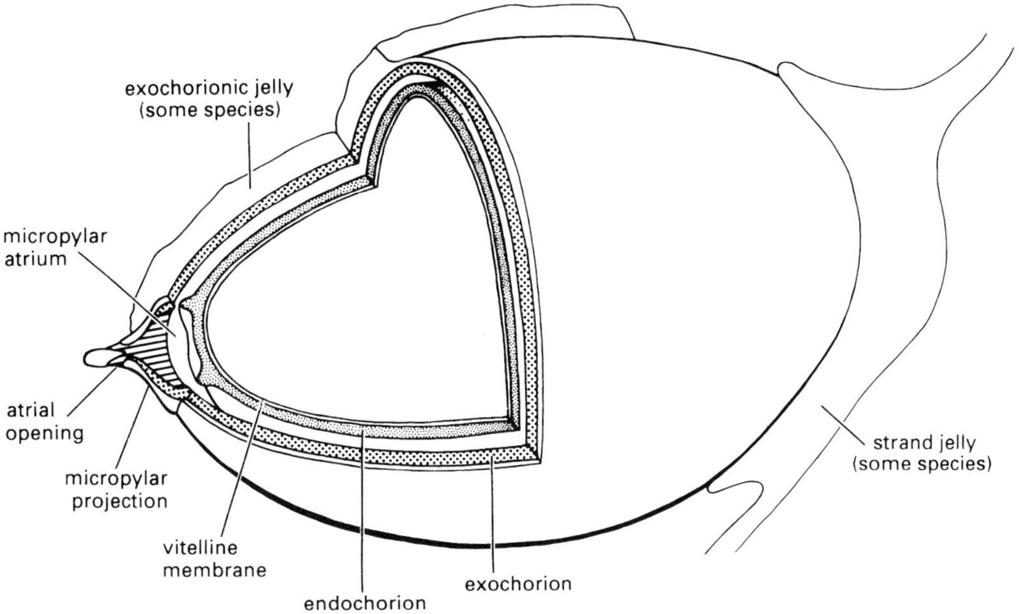

Fig. 5.10 The generalized structure of a libelluloid dragonfly egg (Odonata: Corduliidae, Libellulidae). Libelluloid dragonflies oviposit into freshwater but always exophytically (i.e. outside of plant tissues). The endochorionic and exochorionic layers of the eggshell are separated by a distinct gap in some species. A gelatinous matrix may be present on the exochorion or as connecting strands between eggs. (After Trueman 1991.)

Box 5.7 Does mother know best?

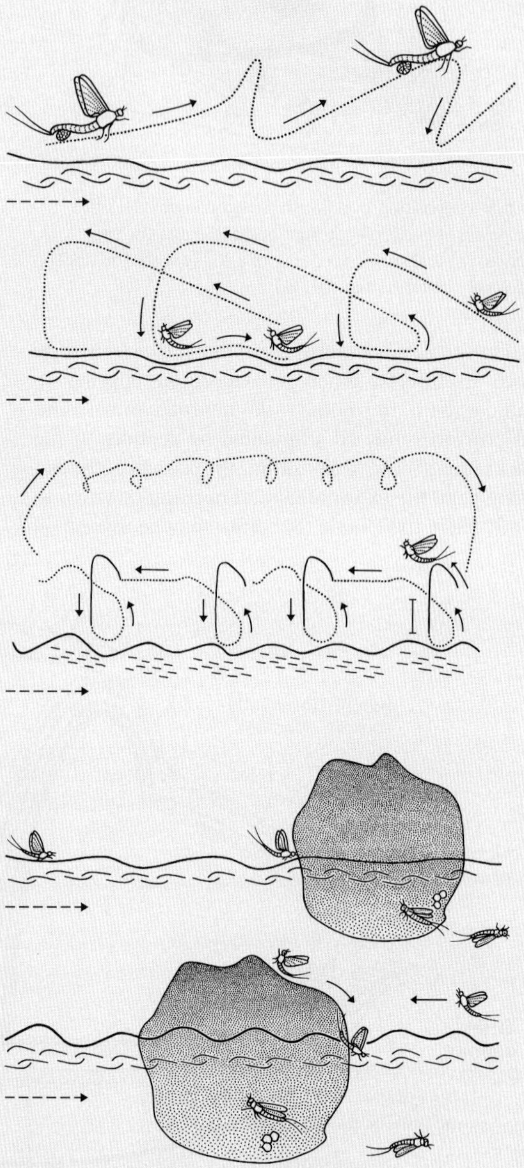

Among hemimetabolous insects, in which the immature stages (nymphs) resemble small adults and generally behave in a similar manner to the adults, it might be expected that oviposition in the place that has sustained the parents is a suitable strategy for the next generation. For example, the best way to ensure a suitable host plant for the next generation is to live on it as an adult and to oviposit right there. Furthermore, mobile nymphs can "rectify" (within limits) any oviposition errors made by their mother by relocation. Certainly, many terrestrial hemimetabolous insects do little more than the above. However, those that exhibit sociality and parental care show enhanced site selection, including the extraordinary parenting behaviour of giant water bugs (Box 5.8).

In **holometabolous** insects, in which immature stages (larvae) very often develop on a different medium to the adults, the simple hemimetabolous strategy often will fail. Instead, the gravid female must seek the appropriate larval habitat – one that she may have left as a larva prior to pupation and the onset of her adult life. There are many examples in this book of a female's ability to lay eggs appropriately, so that the next generation is able to survive and thrive. This has been termed the preference–performance hypothesis (female preference leads to immature performance) or, more popularly, "mother knows best". However, it is not universal – for example, Australian Richmond birdwing butterflies (*Ornithoptera richmondia*) will oviposit onto a toxic *Aristolochia* vine, but this is a non-native, alien host plant closely related to the "correct" host vine (section 1.7). Perhaps the strongest demonstrated negation of the universality of the hypothesis is found in a vine weevil (*Otiorhynchus sulcatus*) feeding on raspberry (*Rubus idaeus*), with the adult feeding above ground and its larva feeding on the roots. Foliage biomass is unrelated to root biomass, and thus mother weevils are unable to distinguish between larva-infested plants and clean ones. Indeed, mothers tend more to lay eggs on plants with smaller root systems, surely leading to reduced performance by their larvae.

Much study has focused on the performance of immature stages of aquatic insects in relation to the oviposition behaviour of the aerial adult. Thus, the desiccation-resistant eggs of *Aedes* mosquitoes are laid into dry locations to await the seasonal filling with rainwater (section 6.5). Some female aquatic insects detect the presence of predators, such as odonate nymphs, and avoid laying eggs into such locations. Despite this evidence for selective oviposition, many aquatic insects superficially seem rather unselective in site selection, beyond the rather poorly substantiated "upstream" compensatory flights to counter downstream larval/nymphal drift.

Mayflies (Ephemeroptera) are foremost amongst insects that lay their eggs in a manner that seems random or even counter-productive with respect to immature survivorship. Their oviposition behaviours can be categorized into five functional groups: splashers; bombers; dippers; floaters; and landers (illustrated here; after Encalada & Peckarsky 2007; Peckarsky *et al.* 2012). Stream-dwelling mayflies are highly susceptible to predation, including by vertebrates such as trout (*Salmo* species), yet oviposit and attain very high larval densities in the riskiest streams. It seems that the preferred oviposition site for *Baetis* mayfly females is associated with large rocks protruding from fast, relatively cool and well-oxygenated streams, where they land and lay beneath the surface – but these also are the waters preferred by trout. Thus, a complex trade-off exists between heightened exposure to predation by trout of both the female during egg-laying and her developing nymphs, countered by enhanced hatching success for her well-located eggs. Mother indeed may know best!

Box 5.8 Egg-tending fathers – the giant water bugs

Care of eggs by adult insects is common in those that show sociality (Chapter 12), but tending solely by male insects is very unusual. This behaviour is known best in the giant water bugs, the Nepoidea, comprising the families Belostomatidae and Nepidae, whose common names – giant water bugs, water scorpions, toe biters – reflect their size and behaviours. These are predators, amongst which the largest species specialize in vertebrate prey such as tadpoles and small fish, which they capture with raptorial forelegs and piercing mouthparts. Evolutionary attainment of the large adult body size necessary for feeding on these large items is inhibited by the fixed number of five nymphal instars in Heteroptera and the limited size increase at each moult (Dyar's rule; section 6.9.1). These phylogenetic (evolutionarily inherited) constraints have been overcome in intriguing ways – by the commencement of development at a large size via oviposition of large eggs, and in one family, with specialized paternal protection of the eggs.

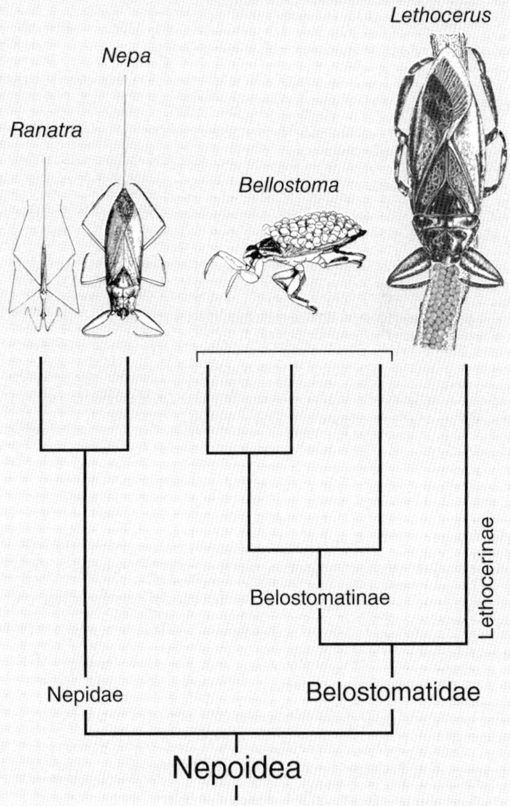

Egg-tending in the subfamily Belostomatinae involves the males "back-brooding" – carrying the eggs on their backs, in a behaviour shared by over a hundred species in five genera. The male mates repeatedly with a female, perhaps up to a hundred times, thus guaranteeing that the eggs she deposits on his back are his alone, which encourages his subsequent tending behaviour. Active male-tending behaviour, called "brood-pumping", involves underwater undulating "press-ups" by the anchored male, creating water currents across the eggs. This is an identical, but slowed-down, form of the pumping display used in courtship. Males of other taxa "surface-brood", with the back (and thus eggs) held horizontally at the water surface such that the interstices of the eggs are wet and the apices aerial. This position, which is unique to brooding males, exposes the males to higher levels of predation. A third behaviour, "brood-stroking", involves the submerged male sweeping and circulating water over the egg pad. Tending by the male results in >95% successful emergence, in contrast to death of all eggs if removed from the male, whether aerial or submerged.

Members of the Lethocerinae, sister group to the Belostomatinae, show related behaviours that help us to understand the origins of aspects of these paternal egg defences. Following courtship, which involves display pumping as in Belostomatinae, the pair copulate frequently between bouts of laying in which eggs are placed on a stem or other projection above the surface of a pond or lake. After completion of egg-laying, the female leaves the male to attend the eggs and she swims away and plays no further role. The "emergent brooding" male tends the aerial eggs for a few days to a week until they hatch. His roles include periodically submerging himself to absorb and drink water that he regurgitates over the eggs, shielding the eggs, and display posturing against airborne threats. Unattended eggs die from desiccation; those immersed by rising water are abandoned and drown.

Insect eggs have a well-developed chorion that enables gas exchange between the external environment and the developing embryo (see section 5.8). The problem with a large egg relative to a smaller one is that the surface-area increase of the sphere is much less than the increase in volume. Because oxygen is scarce in water and diffuses much more slowly than in air (section 10.3), the increased-sized egg hits a limit of the ability for oxygen diffusion from water to egg. For such an egg in a terrestrial environment, gas exchange is easy, but desiccation through loss of water becomes an issue. Although terrestrial insects use waxes around the chorion to avoid desiccation, the long aquatic history of the Nepoidea means that any such mechanism has been lost and is unavailable, providing an another example of phylogenetic inertia.

In the phylogeny of Nepoidea (shown here in reduced form from Smith 1997), a stepwise pattern of acquisition of paternal care can be seen. In the sister family to Belostomatidae, the Nepidae (the waterscorpions), all eggs, including the largest, develop immersed. Gas exchange is facilitated by expansion of the chorion surface area into either a crown or two long horns: the eggs never are brooded. No such chorionic elaboration evolved in Belostomatidae: the requirement by large eggs for oxygen with the need to avoid drowning or desiccation could have been fulfilled by oviposition on a wave-swept rock – although this strategy is unknown in any extant taxa. Two alternatives developed – avoidance of submersion and drowning by egg-laying on emergent structures (Lethocerinae), or, perhaps in the absence of any other suitable substrate, egg-laying onto the back of the attendant mate (Belostomatinae). In Lethocerinae, watering behaviours of the males counter the desiccation problems encountered during emergent brooding of aerial eggs; in Belostomatinae, the pre-existing male courtship pumping behaviour is a pre-adaptation for the oxygenating movements of the back-brooding male. Surface-brooding and brood-stroking are seen as more-derived male-tending behaviours.

The traits of large eggs and male-brooding behaviour appeared together, and the traits of large eggs and egg respiratory horns also appeared together, because the first was impossible without the second. Thus, large body size in Nepoidea must have evolved twice. Paternal care and egg respiratory horns are different adaptations that facilitate gas exchange, and thus survival of large eggs.

The eggshell has a number of important functions. Its design allows for selective entry of the sperm at the time of fertilization (section 5.6). Its elasticity facilitates oviposition, especially for species in which the eggs are compressed during passage down a narrow egg-laying tube, as described below. Its structure and composition afford the embryo protection from deleterious conditions such as unfavourable humidity and temperature, and microbial infection, while also allowing for the exchange of oxygen and carbon dioxide between the inside and outside of the egg.

The differences seen in composition and complexity of layering of the eggshell in different insect groups generally are correlated with the environmental conditions encountered at the site of oviposition. In parasitic wasps, the eggshell is usually thin and relatively homogeneous, allowing flexibility during passage down the narrow ovipositor, and because the embryo develops within host tissues where desiccation is not a hazard, the wax layer of the eggshell is absent. In contrast, many insects lay their eggs in dry places and here the problem of avoiding water loss while obtaining oxygen is often acute because of the high surface area to volume ratio of most eggs. The majority of terrestrial eggs, therefore, have a hydrofuge (water repellent) waxy chorion that contains a meshwork holding a layer of gas in contact with the outside atmosphere via narrow holes, or aeropyles.

The females of many insects (e.g. Zygentoma, many Odonata, all Orthoptera (see **Plate 1d**), some Hemiptera, some Thysanoptera and most Hymenoptera) have appendages of the eighth and ninth abdominal segments modified to form an egg-laying organ or **ovipositor** (section 2.5.1). In other insects (e.g. many Lepidoptera, Coleoptera and Diptera), it is the posterior segments rather than appendages of the female's abdomen that function as an ovipositor (a "substitutional" ovipositor). Often, these segments can be protracted into a telescopic tube in which the opening of the egg passage is close to the distal end. The ovipositor or the modified end of the abdomen enables the insect to insert its eggs into particular sites, such as into crevices, soil, plant tissues or, in the case of many parasitic species, into an arthropod host (see **Plate 6f** and **Plate 8d**). Other insects, such as termites, parasitic lice and many Plecoptera, lack an egg-laying organ and eggs simply are deposited on a surface.

In certain Hymenoptera (ants, bees and some wasps), the ovipositor has lost its egg-laying function and is used as a poison-injecting sting. The stinging Hymenoptera eject the eggs from the opening of the genital chamber at the base of the modified ovipositor. However, in most wasps, the eggs pass down the canal of the ovipositor shaft, even if the shaft is very narrow (Fig. 5.11). In some parasitic wasps with very slender ovipositors, the eggs are extremely compressed and stretched as they move through the narrow canal of the shaft.

The valves of an insect ovipositor usually are held together by interlocking tongue-and-groove joints, which prevent lateral movement but allow the valves to slide back and forth on one another. Such movement, and sometimes also the presence of serrations on the tip of the ovipositor, is responsible for the piercing action of the ovipositor into an egg-laying site. Movement of eggs down the ovipositor tube is possible because of many posteriorly directed "scales" (microsculpturing) located on the inside surface of the valves. Ovipositor scales vary in shape (from plate-like to spine-like) and in arrangement among insect groups, and are seen best under a scanning electron microscope.

The scales found in the conspicuous ovipositors of crickets and katydids (Orthoptera: Gryllidae and Tettigoniidae) exemplify these variations. The ovipositor of the field cricket *Teleogryllus commodus* (Fig. 5.12)

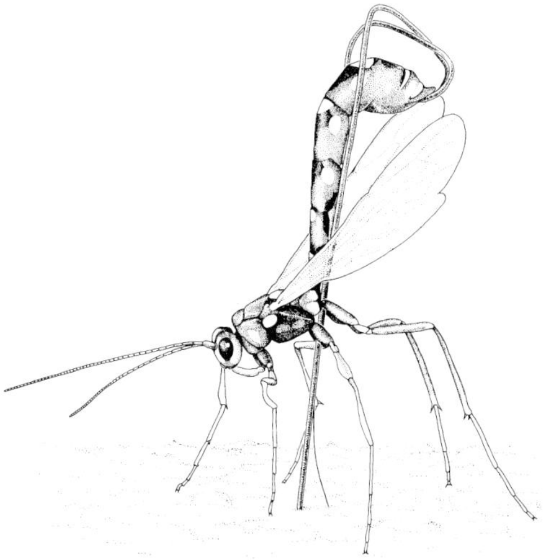

Fig. 5.11 A female of the parasitic wasp *Megarhyssa nortoni* (Hymenoptera: Ichneumonidae) probing a pine log with her very long ovipositor in search of a larva of the sirex wood wasp, *Sirex noctilio* (Hymenoptera: Siricidae). If a larva is located, she stings and paralyzes it before laying an egg on it.

possesses overlapping plate-like scales and scattered, short sensilla along the length of the egg canal. These sensilla may provide information on the position of the egg as it moves down the canal, whereas a group of larger sensilla at the apex of each dorsal valve presumably signals that the egg has been expelled. In addition, in *T. commodus* and some other insects, there are scales on the outer surface of the ovipositor tip, which are orientated in the opposite direction to those on the inner surface. These are thought to assist with penetration of the substrate and in holding the ovipositor in position during egg-laying.

In addition to the eggshell, many eggs are provided with a proteinaceous secretion or cement that coats them and fastens them to a substrate, such as a vertebrate hair in the case of sucking lice, or a plant surface in the case of many beetles (Fig. 5.9). Colleterial glands, accessory glands of the female reproductive tract, produce such secretions. In other insects, groups of thin-shelled eggs are enclosed in an **ootheca**, which protects the developing embryos from desiccation. The colleterial glands produce the tanned, purse-like ootheca of cockroaches (Taxobox 15) and the frothy ootheca of mantids (see **Plate 3e**), whereas the foamy ootheca that surrounds locust and other orthopteran eggs in the soil is formed

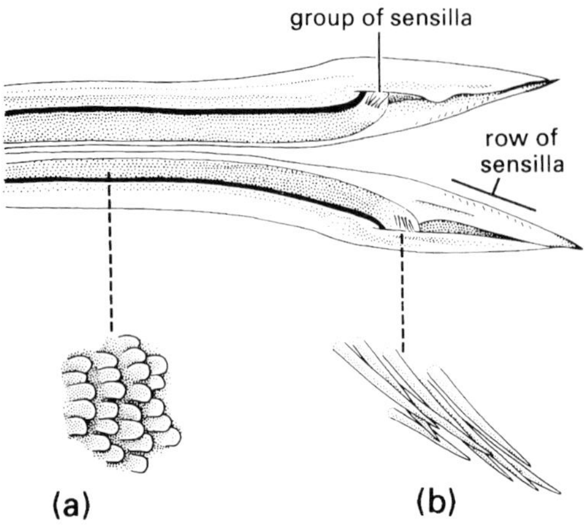

Fig. 5.12 Tip of the ovipositor of a female of the black field cricket, *Teleogryllus commodus* (Orthoptera: Gryllidae), split open to reveal the inside surface of the two halves of the ovipositor. Enlargements show: (a) posteriorly directed ovipositor scales; (b) distal group of sensilla. (After Austin & Browning 1981.)

from the accessory glands in conjunction with other parts of the reproductive tract.

5.9 OVOVIVIPARITY AND VIVIPARITY

Most insects are oviparous, with the act of laying coinciding with initiation of egg development. However, some species are viviparous, with initiation of egg development taking place within the mother, which thus gives birth to live young. The life cycle is shortened by retention of eggs and even of developing young within the mother. Four main types of viviparity are observed in different insect groups, with many of the specializations prevalent in various higher dipterans.

1 **Ovoviviparity**, in which fertilized eggs containing yolk and enclosed in some form of eggshell are incubated inside the reproductive tract of the female. This occurs in some cockroaches (Blattidae), some aphids and scale insects (Hemiptera), a few beetles (Coleoptera) and thrips (Thysanoptera), and some flies (Muscidae, Calliphoridae and Tachinidae). The fully developed eggs hatch immediately after being laid or just prior to ejection from the female's reproductive tract.

2 **Pseudoplacental viviparity** occurs when a yolk-deficient egg develops in the genital tract of the female. The mother provides a special placenta-like tissue, through which nutrients are transferred to developing embryos. There is no oral feeding and larvae are laid upon hatching. This form of viviparity occurs in many aphids (Hemiptera), some earwigs (Dermaptera), a few psocids (Psocodea), and in polyctenid bugs (Hemiptera).

3 **Haemocoelous viviparity** involves embryos developing free in the female's haemolymph, with nutrients being taken up by osmosis. This form of internal parasitism occurs only in Strepsiptera, in which the larvae exit through a brood canal (Taxobox 23), and in some gall midges (Diptera: Cecidomyiidae), where the larvae may consume the mother (as in paedogenetic development; section 5.10).

4 **Adenotrophic viviparity** occurs when a poorly developed larva hatches and feeds orally from accessory ("milk") gland secretions within the "uterus" of the mother's reproductive system. The full-grown larva is deposited and pupariates immediately. The dipteran "pupiparan" families, namely the Glossinidae (tsetse flies), Hippoboscidae (louse or wallaby flies, keds), and Nycteribidae and Streblidae (bat flies), have adenotrophic viviparity.

5.10 OTHER MODES OF REPRODUCTION

Sexual reproduction (**amphimixis**) with separate male and female individuals (**gonochorism**) is the usual mode of reproduction in insects, and **diplodiploidy**, in which males as well as females are diploid, occurs as the ancestral system in almost all insect orders. However, other modes of reproduction are not uncommon. Various types of asexual reproduction occur in many insect groups; development from unfertilized eggs is a widespread phenomenon, whereas the production of multiple embryos from a single egg is rare. Some species exhibit alternating sexual and asexual reproduction, depending on season or food availability. A few species possess both male and female reproductive systems in one individual (**hermaphroditism**), but self-fertilization has been established for species in just one genus.

5.10.1 Parthenogenesis, paedogenesis (pedogenesis) and neoteny

Some, or a few, representatives of virtually every insect order have dispensed with mating, with females producing viable eggs even though unfertilized. In other groups, notably the Hymenoptera, mating occurs but the sperm need not be used in fertilizing all the eggs. Development from unfertilized eggs is called **parthenogenesis**, which in some species may be **obligatory**, but in many others is **facultative**. The female may produce parthenogenetically only female eggs (**thelytokous parthenogenesis** or **thelytoky**), only male eggs (**arrhenotokous parthenogenesis** or **arrhenotoky**), or eggs of both sexes (**amphitokous** or **deuterotokous parthenogenesis**, also known as **amphitoky** or **deuterotoky**). The largest insect group showing arrhenotoky is the Hymenoptera, but some species also display thelytoky. Where thelytoky occurs in Hymenoptera, it either has a nuclear genetic basis (as in eusocial species such as *Apis mellifera capensis* and many ants), or is caused by parthenogenesis-inducing bacteria such as *Wolbachia* (section 5.10.4) (as occurs in many solitary wasps). Within the Hemiptera, aphids display thelytoky and most whiteflies are arrhenotokous. Certain Diptera and a few Coleoptera are thelytokous, and Thysanoptera display all three types of parthenogenesis. Facultative parthenogenesis, and variation in sex of egg produced, may be a response to fluctuations in environmental conditions, as occurs in aphids that vary the sex of their offspring and mix parthenogenetic and sexual cycles according to season.

Some insects abbreviate their life cycles by loss of the adult stage, or even loss of both adult and pupal stages. In this precocious stage, reproduction is almost exclusively by parthenogenesis. **Larval paedogenesis**, the production of young by the larval insect, has arisen at least three times in the gall midges (Diptera: Cecidomyiidae) and once in the Coleoptera (*Micromalthus debilis*). In some gall midges, in an extreme case of haemocoelous viviparity, the precocially developed eggs hatch internally and the larvae may consume the body of the mother-larva before leaving to feed on the surrounding fungal medium. In the well-studied gall midge *Heteropeza pygmaea*, eggs develop into female larvae, which may metamorphose to female adults or produce more larvae paedogenetically. These larvae, in turn, may be males, females or a mixture of both sexes. Female larvae may become adult females or repeat the larval paedogenetic cycle, whereas male larvae must develop to adulthood.

In **pupal paedogenesis**, which sporadically occurs in gall midges, embryos are formed in the haemocoel of a paedogenetic mother-pupa, termed a hemipupa as it differs morphologically from the "normal" pupa. This production of live young in pupal paedogenetic insects also destroys the mother-pupa from within, either by larval perforation of the cuticle or by the eating of the mother by the offspring. Paedogenesis appears to have evolved to allow maximum use of locally abundant but ephemeral larval habitats, such as a mushroom fruiting body. When a gravid female detects an oviposition site, eggs are deposited, and the larval population builds up rapidly through paedogenetic development. Adults are developed only in response to conditions adverse to larvae, such as food depletion and overcrowding. Adults may be female only, or males may occur in some species under specific conditions.

In true paedogenetic taxa, there are no reproductive adaptations beyond precocious egg development. In contrast, in **neoteny**, a non-terminal instar develops reproductive features of the adult, including the ability to locate a mate, copulate and deposit eggs (or larvae) in a conventional manner. For example, the scale insects (Hemiptera: Coccoidea) appear to have neotenous females. Whereas a moult to the winged adult male follows the final immature instar, development of the reproductive female involves omission of one or more instars relative to the male. In appearance, the female is a sedentary nymph-like or larviform instar, resembling a larger version of the previous (second or third) instar in all but the presence of a vulva and developing eggs. Neoteny also occurs in all female members of the order Strepsiptera; in these insects female development ceases at the puparium stage. In some other insects (e.g. marine midges; Chironomidae), the adult appears larva-like, but this is evidently not due to neoteny because complete metamorphic development is retained, including a pupal instar. Their larviform appearance therefore results from suppression of adult features, rather than the paedogenetic acquisition of reproductive ability in the larval stage.

5.10.2 Hermaphroditism

Several species of *Icerya* (Hemiptera: Monophlebidae) that have been studied cytologically have gynomonoecious hermaphrodites. These are female-like but possess an ovotestis (a gonad that is part testis, part ovary). In these species, occasional males arise from unfertilized eggs and are apparently functional, but normally self-fertilization is assured by production of male gametes prior to female gametes in the body of one individual (protandry of the hermaphrodite).

The coexistence of both hermaphrodites and males is a rare condition called **androdioecy**. Without doubt, hermaphroditism greatly assists the spread of the pestiferous cottony-cushion scale, *I. purchasi* (see Box 16.3), as single nymphs of this and other hermaphroditic *Icerya* species can initiate new infestations if dispersed or accidentally transported to new plants. Furthermore, all iceryines are arrhenotokous, with unfertilized eggs developing into males and fertilized eggs into females.

5.10.3 Polyembryony

This form of asexual reproduction involves the production of two or more embryos from one egg by subdivision (fission). It is restricted predominantly to parasitic insects; it occurs in at least one strepsipteran and representatives of four wasp families, especially the Encyrtidae. It appears to have arisen independently within each wasp family. In these parasitic wasps, the number of larvae produced from a single egg varies in different genera but is influenced by the size of the host, with from fewer than 10 to several hundred, and in *Copidosoma* (Encyrtidae) more than 1000 embryos, arising from one small, yolkless egg. Nutrition for a large number of developing embryos obviously cannot be supplied by the original egg and is acquired from the host's haemolymph through a specialized enveloping membrane called the **trophamnion**. Typically, the embryos develop into larvae when the host moults to its final instar, and these larvae consume the host insect before pupating and emerging as adult wasps.

5.10.4 Reproductive effects of endosymbionts

Wolbachia, an intracellular bacterium (Proteobacteria: Rickettsiales) discovered first infecting the ovaries of *Culex pipiens* mosquitoes, causes some inter-populational (intraspecific) matings to produce inviable embryos. It has been shown that such crosses, in which embryos abort before hatching, can be returned to viability after treatment of the parents with antibiotic – thus implicating the microorganism in the sterility. This phenomenon, termed **cytoplasmic** or **reproductive incompatibility**, now has been demonstrated in a very wide range of invertebrates that host many "strains" of *Wolbachia*. Surveys have suggested that up to 76% of insect species may be infected. *Wolbachia* is transferred vertically (inherited by offspring from the mother via the egg), and causes several different but related effects. Specific effects include the following:

- Cytoplasmic (reproductive) incompatibility, with directionality varying according to whether one, the other, or both sexes of the partners are infected, and with which strain. Unidirectional incompatibility typically involves an infected male and an uninfected female, with the reciprocal cross (uninfected male with infected female) being compatible (producing viable offspring). Bidirectional incompatibility usually involves both partners being infected with different strains of *Wolbachia* and no viable offspring are produced from any mating.
- Parthenogenesis, or sex-ratio bias towards the diploid sex (usually female) in insects with haplodiploid genetic systems (section 5.6, section 12.2 and section 12.4.1). In the parasitic wasps, this involves infected females that produce only fertile female offspring. The mechanism is usually gamete duplication, involving disruption of meiotic chromosomal segregation such that the nucleus of an unfertilized, *Wolbachia*-infected egg contains two sets of identical chromosomes (diploidy), producing a female. Normal sex ratios are restored by treatment of parents with antibiotics, or by development at elevated temperature, to which *Wolbachia* is sensitive.
- Feminization, the conversion of genetic males into functional females, perhaps caused by specific inhibitions of male-determiner genes. This effect has been studied in terrestrial isopods and a few insects (one species each from the Lepidoptera and Hemiptera), but may be more widespread in other arthropods. In the butterfly *Eurema hecabe* (Pieridae), feminizing *Wolbachia* endosymbionts have been shown to act continuously on genetic males during larval development, leading to female phenotypic expression; antibiotic treatment of larvae leads to intersexual development. Lepidopteran sex determination is thought to be complete in early embryogenesis, and thus the *Wolbachia*-induced feminization apparently does not target embryonic sex determination.
- Male killing, usually during early embryogenesis, and possibly as a result of lethal feminization (see below).

These strategies of *Wolbachia* can be viewed as reproductive parasitism (see also section 3.6.5), in which the bacterium manipulates its host into producing an imbalance of female offspring (this being the sex responsible for the vertical transmission of the infection), compared with uninfected hosts. Only in a very few cases have infections been shown to benefit the insect host, primarily via enhanced fecundity or nutritional contribution. Certainly, with evidence derived from phylogenies of *Wolbachia* and their hosts, *Wolbachia* often has been transferred horizontally between unrelated hosts, and no coevolution is apparent.

Although *Wolbachia* is now the best-studied sex-ratio modifying organism, there are other somewhat similar cytoplasm-dwelling organisms (such as *Cardinium* bacteria in the Bacteroidetes), with the most extreme sex-ratio distorters being known as male-killers. This phenomenon of male lethality has been found in at least five orders of insects, and is associated with a range of maternally inherited, symbiotic–infectious causative organisms, from bacteria to viruses and microsporidia. Each acquisition seems to be independent, and others are suspected to exist. Certainly, if parthenogenesis often involves such associations, many such interactions remain to be discovered. Furthermore, much remains to be learned about the effects of insect age, re-mating frequency, and temperature on the expression and transmission of *Wolbachia*. Even less is known about *Cardinium*, which occurs most commonly in haplodiploid insects (for example, certain parasitic wasps and some armoured scale insects) and has been implicated in the induction of parthenogenesis. There is an intriguing case involving the parasitic wasp *Asobara tabida* (Braconidae), in which the elimination of *Wolbachia* by antibiotics causes the inhibition of egg production, thus rendering the wasps infertile. Such obligatory infection with *Wolbachia* also occurs in filarial nematodes (section 15.3.6) and in the bed bug *Cimex lectularius*.

5.11 PHYSIOLOGICAL CONTROL OF REPRODUCTION

The initiation and termination of some reproductive events often depend on environmental factors, such as temperature, humidity, photoperiod or availability of food or a suitable egg-laying site. Such external influences may be modified by internal factors such as nutritional condition and the state of maturation of the oocytes. Copulation may trigger oocyte development, oviposition and/or inhibition of sexual receptivity in the female via enzymes or peptides transferred to the female reproductive tract in male accessory gland secretions (Box 5.6). Fertilization following mating normally triggers embryogenesis via egg activation (Chapter 6). Regulation of reproduction is complex and involves sensory receptors, neuronal transmission and integration of messages in the brain, as well as chemical messengers (hormones) transported in the haemolymph or via the nerve axons to target tissues or to other endocrine glands. Certain parts of the nervous system, particularly neurosecretory cells in the brain, produce neurohormones or neuropeptides (proteinaceous messengers) and also control the synthesis of two groups of insect hormones – the ecdysteroids and the juvenile hormones (JHs). More detailed discussions of the regulation and functions of all of these hormones are provided in Chapters 3 and 6. Neuropeptides, steroid hormones and JH all play essential roles in the regulation of reproduction, as summarized in Fig. 5.13.

JHs and/or ecdysteroids are essential to reproduction, with JH triggering the functioning of organs such as the ovary, accessory glands and fat body, whereas ecdysteroids influence morphogenesis as well as gonad functions. Neuropeptides act in many stages of reproduction (see Table 3.1), regulating endocrine function (via the corpora allata and prothoracic glands), and directly influencing reproductive events, especially ovulation and oviposition or larviposition.

5.11.1 Vitellogenesis and its regulation

In the ovary, both nurse cells (or trophocytes) and ovarian follicle cells are associated with the oocytes (section 3.8.1). These cells pass nutrients to the growing oocytes. The relatively slow period of oocyte growth is followed by a period of rapid yolk deposition, or **vitellogenesis**, which mostly occurs in the terminal oocyte of each ovariole and leads to the production of fully developed eggs. Vitellogenins, specific female lipoglycoproteins, are synthesized mostly in the fat body under the control of JH. After migration via the haemolymph and entry to the oocytes by endocytosis, these proteins are called vitellins and their chemical structure may differ slightly from that of vitellogenins. Lipid bodies – mostly triglycerides from the follicle cells, nurse cells or fat body – also are deposited in the growing oocyte.

Vitellogenesis has been a favoured area of insect hormone research because it is amenable to experimental manipulation with artificially supplied hormones, and analysis is facilitated by the large

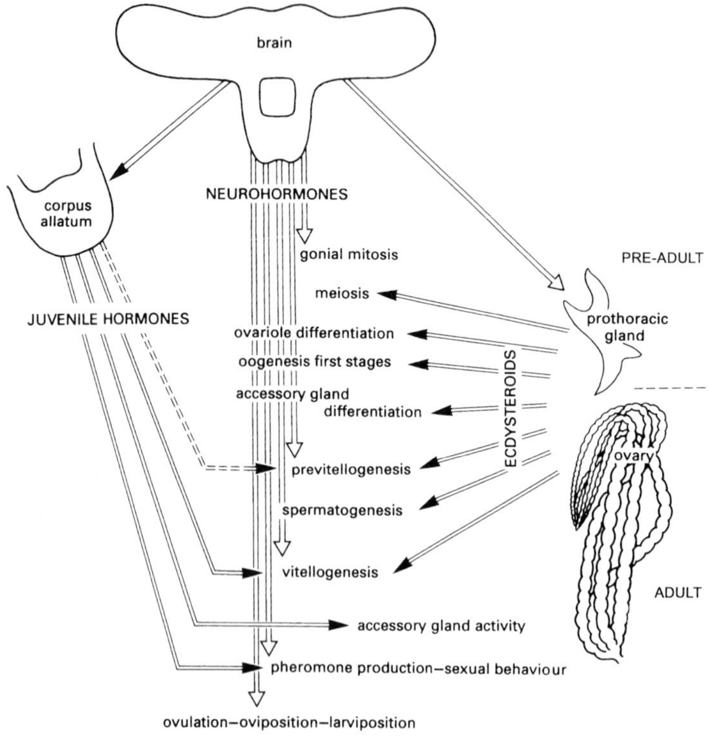

Fig. 5.13 A schematic diagram of the hormonal regulation of reproductive events in insects. The transition from ecdysteroid production by the pre-adult prothoracic gland to the adult ovary varies between taxa. (After Raabe 1986.)

amounts of vitellogenins produced during egg growth. Vitellogenesis is regulated differently between insect groups, with JH from the corpora allata, ecdysteroids from the prothoracic glands or the ovary, and brain neurohormones (neuropeptides such as ovarian ecdysteroidogenic hormone, OEH) inducing or stimulating vitellogenin synthesis to varying degrees in different insect species (Fig. 5.13).

In some flies (specifically *Aedes aegypti*, a mosquito, and *Neobellieria bullata*, a flesh fly), egg development in ovarian follicles in the previtellogenic stage can be inhibited by antigonadotropins. This seems to be restricted to insects with oocytes that undergo vitellogenesis in several ovarian cycles. The peptides responsible for this suppression were known as oostatic hormones but are now termed trypsin-modulating oostatic factors (TMOF). These factors are synthesized in the ovary or associated neurosecretory tissues. Their action depends upon inhibiting proteolytic enzyme synthesis ("trypsin-modulation") and thus inhibiting blood digestion in the midgut, which in turn prevents the ovary from accumulating vitellogenin from the haemolymph, thus constraining ovarian development. These physiological effects have led to consideration of the use of such peptides as candidates in the control of haematophagous insects.

Vitellogenesis is contolled by JH in most studied insects, but increasingly roles are being recognized for ecdysteroids and neuropeptides, a group of proteins for which reproductive regulation is but one of a diverse array of functions (see Table 3.1).

FURTHER READING

Austin, A.D. & Browning, T.O. (1981) A mechanism for movement of eggs along insect ovipositors. *International Journal of Insect Morphology and Embryology* **10**, 93–108.

Avila, F.W., Sirot, L.K., LaFlamme, B.A., Rubinstein, C.D. & Wolfner, M.F. (2011) Insect seminal fluid proteins: identification and function. *Annual Review of Entomology* **56**, 21–40.

Chapman, R.F. (2013) *The Insects. Structure and Function*, 5th edn. (eds. S.J. Simpson & A.E. Douglas), Cambridge University Press, Cambridge. [Particularly see chapters 12 and 13.]

Chippindale, A.K. (2013) Evolution: sperm, cryptic choice, and the origin of species. *Current Biology* **23**(19), R885–7.
Choe, J.C. & Crespi, B.J. (eds.) (1997) *The Evolution of Mating Systems in Insects and Arachnids*. Cambridge University Press, Cambridge.
Clark, K.E., Hartley, S.E. & Johnson, S.N. (2011) Does mother know best? The preference–performance hypothesis and parent–offspring conflict in aboveground–belowground herbivore life cycles. *Ecological Entomology* **36**, 117–24.
Eberhard, W.G. (1994) Evidence for widespread courtship during copulation in 131 species of insects and spiders, and implications for cryptic female choice. *Evolution* **48**, 711–33.
Eberhard, W.G. (2004) Male–female conflict and genitalia: failure to confirm predictions in insects and spiders. *Biological Reviews* **79**, 121–86.
Emlen, D.F.J. (2001) Costs and diversification of exaggerated animal structures. *Science* **291**, 1534–6.
Emlen, D.F. (2008) The evolution of animal weapons. *Annual Review of Ecology, Evolution and Systematics* **39**, 387–413.
Encalada, A.C. & Peckarsky, B.L. (2007) A comparative study of the costs of alternative mayfly oviposition behaviors. *Behavioral Ecology and Sociobiology* **61**, 1437–48.
Gwynne, D.T. (2008) Sexual conflict over nuptial gifts in insects. *Annual Review of Entomology* **53**, 83–101.
Han, C.S. & Jablonski, P.G. (2009) Female genitalia concealment promotes intimate male courtship in a water strider. *PLoS ONE* **4**(6), e5793. doi:10.1371/journal.pone.0005793.
Judson, O. (2002) *Dr Tatiana's Advice to All Creation. The Definitive Guide to the Evolutionary Biology of Sex*. Metropolitan Books, Henry Holt & Co., New York.
Mikkola, K. (1992) Evidence for lock-and-key mechanisms in the internal genitalia of the *Apamea* moths (Lepidoptera, Noctuidae). *Systematic Entomology* **17**, 145–53.
Normark, B.B. (2003) The evolution of alternative genetic systems in insects. *Annual Review of Entomology* **48**, 397–423.
Peckarsky, B.L., Encalada, A.C. & Macintosh, A.R. (2012) Why do vulnerable mayflies thrive in trout streams? *American Entomologist* **57**, 152–64.
Rabeling, C. & Kronauer, D.J.C. (2013) Thelytokous parthenogenesis in eusocial Hymenoptera. *Annual Review of Entomology* **58**, 273–92.
Resh, V.H. & Cardé, R.T. (eds.) (2009) *Encyclopedia of Insects*, 2nd edn. Elsevier, San Diego, CA. [Particularly see articles on mating behaviors; parthenogenesis; polyembryony; puddling behavior; and the four chapters on reproduction; sex determination; sexual selection; *Wolbachia*.]
Ross, L., Penn, I. & Shuker, D.M. (2010) Genomic conflict in scale insects: the causes and consequences of bizarre genetic systems. *Biological Reviews* **85**, 807–28.
Saridaki, A. & Bourtzis, K. (2010) *Wolbachia*: more than just a bug in insect genitals. *Current Opinion in Microbiology* **13**, 67–72.
Simmons, L.W. (2001) *Sperm Competition and its Evolutionary Consequences in the Insects*. Princeton University Press, Princeton, NJ.
Simmons, L.W. (2014) Sexual selection and genital evolution. *Austral Entomology* **53**, 1–17.
Siva-Jothy, M.T. (2006) Trauma, disease and collateral damage: conflict in mirids. *Philosophical Transactions of the Royal Society B* **361**, 269–75.
Sota, T. & Kubota, K. (1998) Genital lock-and-key as a selective agent against hybridization. *Evolution* **52**, 1507–13.
Werren, J.H., Baldo, L. & Clark, M.E. (2008) *Wolbachia*: master manipulators of invertebrate biology. *Nature Reviews Microbiology* **6**, 741–51.
Wilder, S.M., Rypstra, A.L. & Elgar, M.A. (2009) The importance of ecological and phylogenetic conditions for the occurrence and frequency of sexual cannibalism. *Annual Review of Entomology* **40**, 21–39.
Vahed, K. (2007) All that glitters is not gold: sensory bias, sexual conflict and nuptial feeding in insects and spiders. *Ethology* **113**, 105–27.

Chapter 6

INSECT DEVELOPMENT AND LIFE HISTORIES

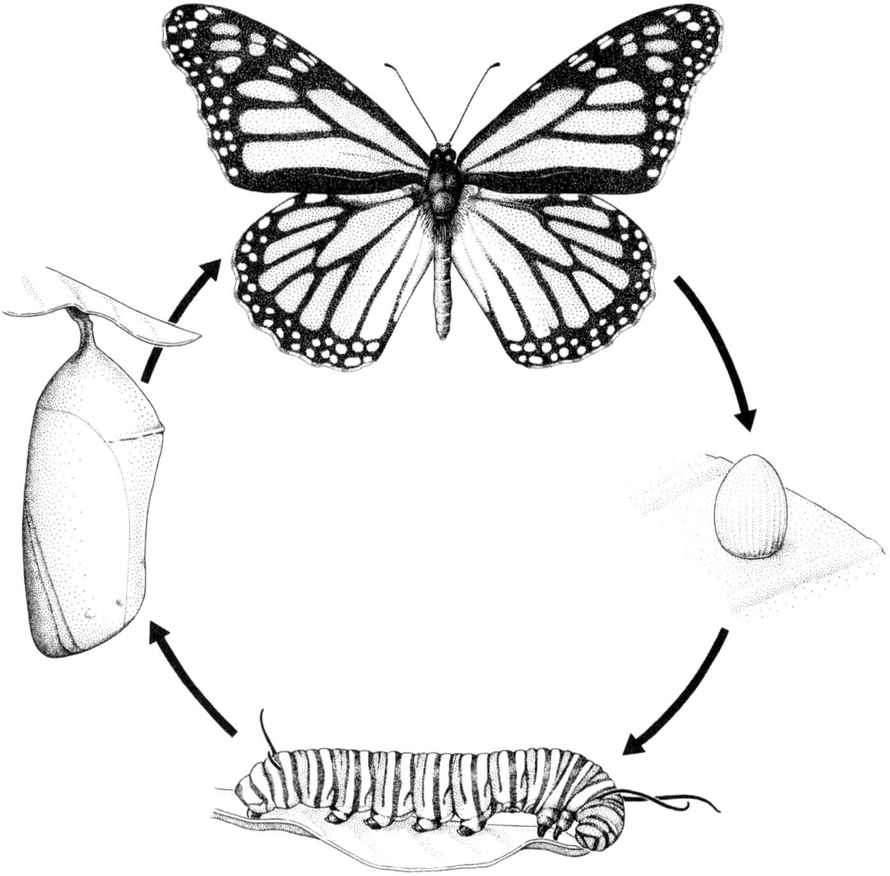

Life-cycle phases of the monarch or wanderer butterfly, *Danaus plexippus*. (After photographs by P.J. Gullan.)

In this chapter we discuss the pattern of growth from egg to adult – the ontogeny – and life histories of insects. The various growth phases from the egg, through immature development to the emergence of the adult are considered. Molecular insights into the embryological development of insects are considered in Box 6.1. We discuss the different kinds of metamorphosis, and suggest that complete metamorphosis reduces competition between conspecific juveniles and adults, by providing a clear ecological differentiation between immature and adult stages. Amongst the different aspects of life histories covered are voltinism, resting stages, the coexistence of different forms within a single species, migration, age determination, allometry, and genetic and environmental effects on development. We include a box on a method of calculating physiological age, or day-degrees (degree-days) (Box 6.2). The influence of environmental factors, namely temperature, photoperiod, humidity, toxins and biotic interactions, upon life-history traits is vital to any applied entomological research. Likewise, knowledge of the process and hormonal regulation of moulting is fundamental to insect control.

Insect life-history characteristics are very diverse, and the variability and range of strategies seen in many higher taxa imply that these traits are highly adaptive. For example, diverse environmental factors trigger termination of egg dormancy in different species of *Aedes*, even though the species in this genus are closely related. However, phylogenetic constraint, such as the restrained instar number of Nepoidea (see Box 5.8), undoubtedly plays a role in life-history evolution in insects.

Thus, this chapter describes the patterns of insect growth and development, and explores the various environmental, including hormonal, influences on growth, development and life cycles. We do not attempt to explain why different insects look the way that they do, i.e. how their morphology evolved to produce the observable differences in body form among species and higher taxa. However, evolutionary developmental biology (**evo-devo**) has provided a mechanistic framework to help explain the evolution of morphological diversity. Research on *Drosophila* species, for which both the genome and the pattern of development are relatively well understood, has shown that: "(i) form evolves largely by altering the expression of functionally conserved proteins; and (ii) such changes largely occur through mutations in the *cis*-regulatory sequences of pleiotropic developmental regulatory loci and of target genes within the vast networks they control." (Carroll, 2008: 3). For example, the Hox proteins of arthropods have highly conserved sequences across the various arthropod groups, but their expression varies substantially among major taxa (Box 6.1). At the level of species, differences in morphology (such as the differing patterns of pigmentation on the wings and abdomen of adult flies of closely related *Drosophila* species) can be explained by changes in the **regulatory sequences** that control the expression of the genes that code for pigmentation. More detailed information on the genetics of morphological evolution are beyond the scope of an entomology textbook, and we refer interested readers to Box 6.1 and to articles listed in the Further reading.

6.1 GROWTH

Insect growth is discontinuous, at least for the sclerotized cuticular parts of the body, because the rigid cuticle limits expansion. Size increase is by **moulting** – the periodic formation of new cuticle of greater surface area, and subsequent **ecdysis**, the shedding of the old cuticle. Thus, for sclerite-bearing body segments and appendages, increases in body dimensions are confined to the postmoult period immediately after moulting, before the cuticle stiffens and hardens (section 2.1). Hence, the sclerotized head capsule of a beetle or moth larva increases in dimensions in a saltatory manner (in major increments) during development, whereas the membranous nature of body cuticle allows the larval body to grow more or less continuously.

Studies concerning insect development involve two components of growth. The first, the **moult increment**, is the increment in size occurring between one **instar** (growth stage, or the form of the insect between two successive moults) and the next. Generally, increase in size is measured as the increase in a single dimension (length or width) of some sclerotized body part, rather than a weight increment, which

The Insects: An Outline of Entomology, Fifth Edition. P.J. Gullan and P.S. Cranston.

Companion Website: www.wiley.com/go/gullan/insects

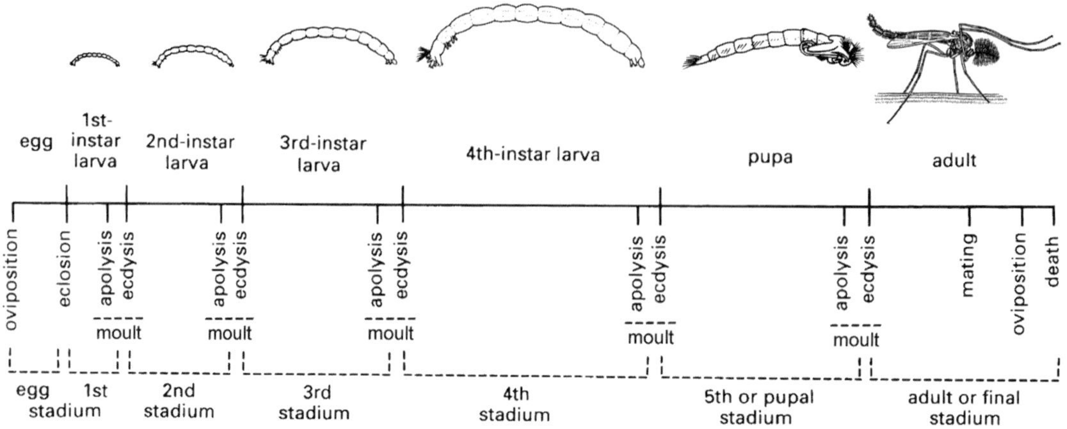

Fig. 6.1 Schematic drawing of the life cycle of a non-biting midge (Diptera: Chironomidae, *Chironomus*), showing the various events and stages of insect development.

may be misleading because of variability in food or water intake. The second component of growth is the **intermoult period** or interval, better known as the **stadium** or instar duration, which is defined as the time between two successive moults, or more precisely between successive ecdyses (Fig. 6.1 and section 6.3). The magnitude of both moult increments and intermoult periods may be affected by food supply, temperature, larval density and physical damage (such as loss of appendages) (section 6.10), and may differ between the sexes of a species.

In collembolans, diplurans and apterygote insects, growth is **indeterminate** – the animals continue to moult until they die. There is no definitive terminal moult in such animals, but they do not continue to increase in size throughout their adult life. In the vast majority of insects, growth is **determinate**, as there is a distinctive instar that marks the cessation of growth and moulting. All insects with determinate growth become reproductively mature in this final instar, called the adult or imaginal instar. A reproductively mature individual is called an adult or **imago** (plural: imagines or imagos). In most insect orders it is fully winged, although secondary wing loss has occurred independently in the adults of several groups, such as lice, fleas and certain parasitic flies, and in the adult females of all scale insects (Hemiptera: Coccoidea). In just one order of insects, the Ephemeroptera or mayflies, a subimaginal instar immediately precedes the final or imaginal instar. This **subimago**, although capable of flight, only rarely is reproductive. In the few mayfly groups in which the female mates as a subimago, she dies without moulting to an imago, so that the subimaginal instar actually is the final growth stage.

In some pterygote taxa, the total number of pre-adult growth stages or instars may vary within a species, depending on environmental conditions, such as developmental temperature, diet and larval density. In many other species, the total number of instars (although not necessarily final adult size) is determined genetically, and is constant regardless of environmental conditions.

6.2 LIFE-HISTORY PATTERNS AND PHASES

Growth is an important part of an individual's **ontogeny**, the developmental history of that organism from egg to adult. Equally significant are the changes, both subtle and dramatic, that take place in body form as insects moult and grow larger. Changes in form (morphology) during ontogeny affect internal and external structures, but only external changes are apparent at each moult. Three broad patterns of developmental morphological change during ontogeny can be distinguished, based on how much external alteration occurs in post-embryonic development.

In the basic (ancestral) pattern, **ametaboly**, the hatchling emerges from the egg in a form essentially resembling a miniature adult, lacking only genitalia. This pattern is retained by the primitively wingless orders, the Archaeognatha (bristletails; Taxobox 2) and Zygentoma (silverfish; Taxobox 3), in which adults continue to moult after sexual maturity. In contrast, all pterygote (winged) insects (section 7.4.2) undergo a change in form, a **metamorphosis**, between the immature phase of development and the winged or secondarily wingless (apterous) adult or imaginal phase. Development of pterygotes can be subdivided into **hemimetaboly** (partial or incomplete metamorphosis; Fig. 6.2) and **holometaboly**

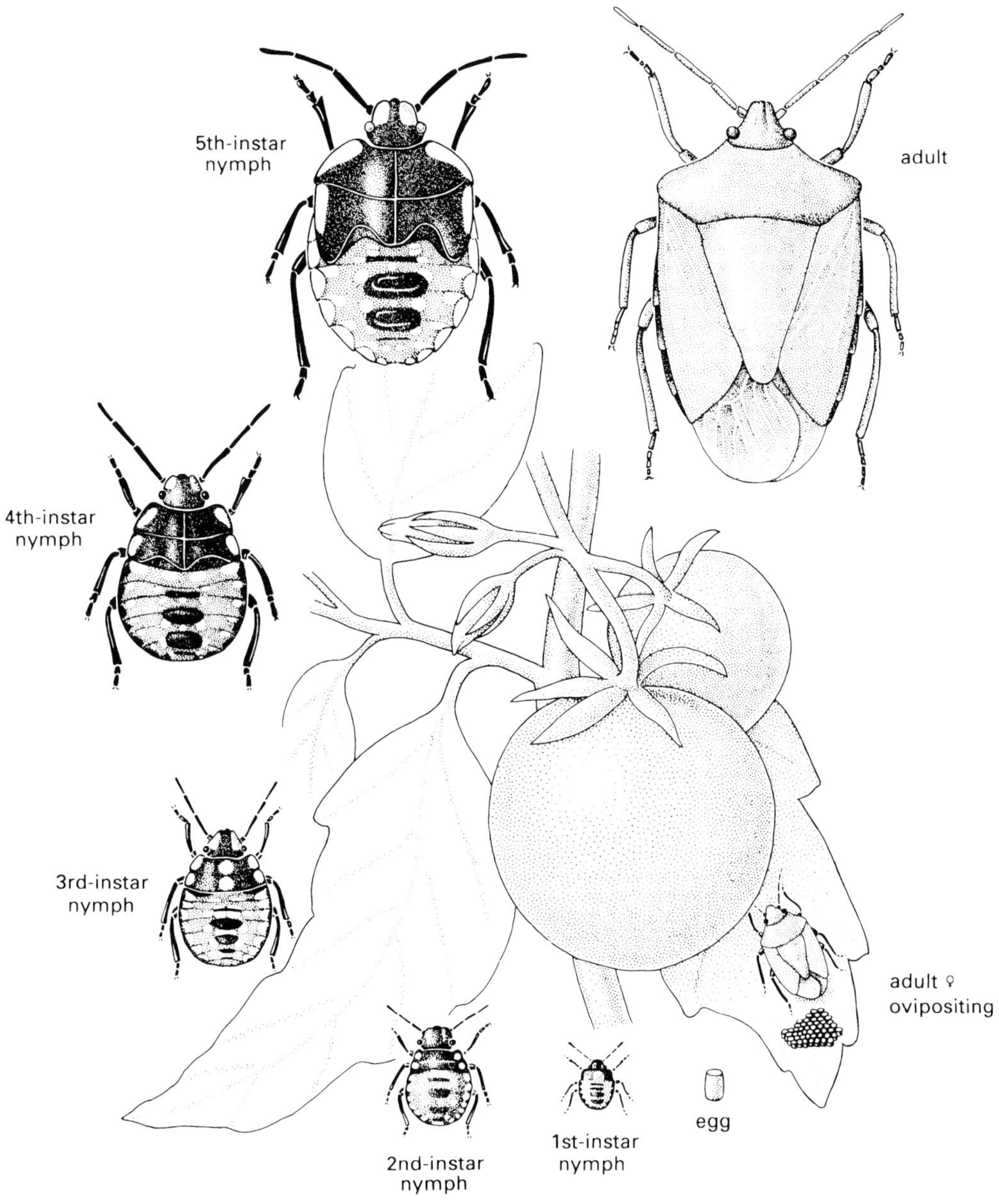

Fig. 6.2 The life cycle of a hemimetabolous insect, the southern green stink bug or green vegetable bug, *Nezara viridula* (Hemiptera: Pentatomidae), showing the eggs, nymphs of the five instars, and the adult bug on a tomato plant. This cosmopolitan and polyphagous bug is an important world pest of food and fibre crops. (After Hely *et al.* 1982.)

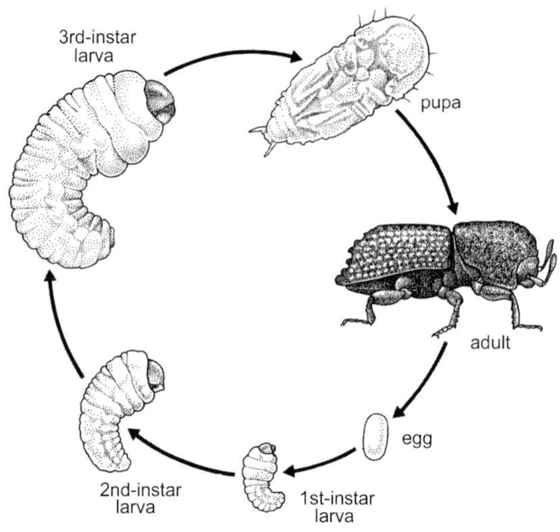

Fig. 6.3 The life cycle of a holometabolous insect, a bark beetle, *Ips grandicollis* (Coleoptera: Curculionidae: Scolytinae), showing the egg, the three larval instars, the pupa, and the adult beetle. (After Johnson & Lyon 1991.)

(complete metamorphosis; Fig. 6.3 and the vignette at the beginning of this chapter, which shows the life-cycle phases of the monarch butterfly).

In all but the youngest instars, developing wings are visible in external sheaths on the dorsal surface of nymphs of hemimetabolous insects. The term **exopterygote** has been applied to this type of "external" wing growth. Orders with hemimetabolous and exopterygote development once were grouped into "Hemimetabola" (also called Exopterygota), but this is a grade of organization rather than a monophyletic group, or **clade** (section 7.1.1). In contrast, pterygote orders displaying holometabolous development share the evolutionary innovation of a resting stage or **pupal instar** in which development of the major structural differences between immature (larval) and adult stages is concentrated. All insects in the clade called the Endopterygota or Holometabola share this derived pattern of development. In many Holometabola, expression of adult features is retarded until the pupal stage; however, in *Drosophila* and other derived taxa, uniquely adult structures including wings arise internally in larvae as groups of undifferentiated cells (or primordia) called **imaginal discs** (or **imaginal buds**) (Fig. 6.4). The term **endopterygote** reflects the wing development in invaginated pockets of the integument, with eversion only at the larval–pupal moult.

Holometaboly allows the immature and adult stages of an insect to specialize in different resources, undoubtedly contributing to the successful radiation of the group (see section 8.5).

6.2.1 Embryonic phase

The egg stage begins as soon as the female deposits the mature egg. For practical reasons, the age of an egg is estimated from the time of its deposition even though the egg existed before oviposition. The beginning of the egg stage, however, need not mark the commencement of an individual insect's ontogeny, which actually begins when embryonic development within the egg is triggered by **activation**. This trigger usually results from fertilization in sexually reproducing insects, but in parthenogenetic species appears to be induced by various events at oviposition, including the entry of oxygen to the egg or mechanical distortion.

Following activation of the insect egg cell, the **zygote** nucleus subdivides by mitotic division to produce many daughter nuclei, giving rise to a **syncytium**. These nuclei and their surrounding cytoplasm, called cleavage **energids**, migrate to the egg periphery where the membrane infolds, leading to cellularization of the superficial layer to form the one-cell thick blastoderm. This distinctive superficial cleavage during early embryogenesis in insects is the result of the large amount of yolk in the egg. The blastoderm usually gives rise to all the cells of the larval body, whereas the central yolky part of the egg provides the nutrition for the developing embryo and will be used up by the time of **eclosion**, or emergence from the egg.

Regional differentiation of the blastoderm leads to the formation of the germ **anlage** or germ disc (Fig. 6.5a), which is the first sign of the developing embryo, whereas the remainder of the blastoderm becomes a thin membrane, the **serosa**, or embryonic cover. Next, the germ anlage develops an infolding in a process called gastrulation (Fig. 6.5b) and sinks into the yolk, forming a two-layered embryo containing the

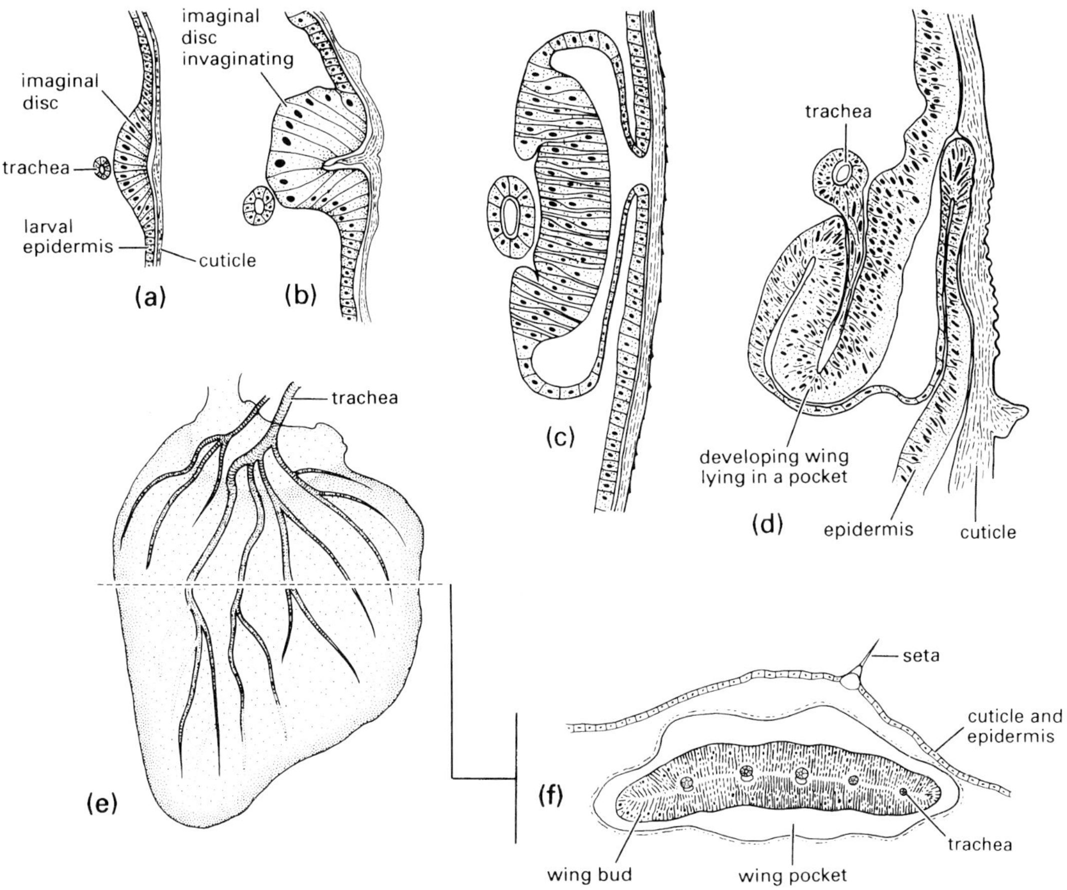

Fig. 6.4 Stages in the development of the wings of the small white, small cabbage white, or cabbage white butterfly, *Pieris rapae* (Lepidoptera: Pieridae). A wing imaginal disc in: (a) a first-instar larva; (b) a second-instar larva; (c) a third-instar larva; and (d) a fourth-instar larva. The wing bud, as it appears: (e) if dissected out of the wing pocket; or (f) cut in cross-section in a fifth-instar larva. ((a–e) After Mercer 1900.)

amniotic cavity (Fig. 6.5c). After gastrulation, the germ anlage becomes the **germ band**, which externally is characterized by segmental organization (commencing in Fig. 6.5d with the formation of the protocephalon). The germ band essentially forms the ventral regions of the future body, which progressively differentiates with the head, body segments and appendages becoming increasingly well defined (Fig. 6.5e–g). At this time, the embryo undergoes movement called **katatrepsis**, which brings it into its final position in the egg. Later, near the end of embryogenesis (Fig. 6.5h,i), the edges of the germ band grow over the remaining yolk and fuse mid-dorsally to form the lateral and dorsal parts of the insect – a process called **dorsal closure**.

In the well-studied *Drosophila*, the complete embryo is large, extending across the egg, and becomes segmented at the cellularization stage, termed "long germ" (as in most studied Diptera, Coleoptera and Hymenoptera). This long-germ band contains the primordia (rudiments) of all segments that will form the embryo. In contrast, in "short-germ band" insects (such as locusts and the beetle *Tribolium*), the embryo derives from only a small region of the blastoderm, and the germ band mostly forms only the anterior

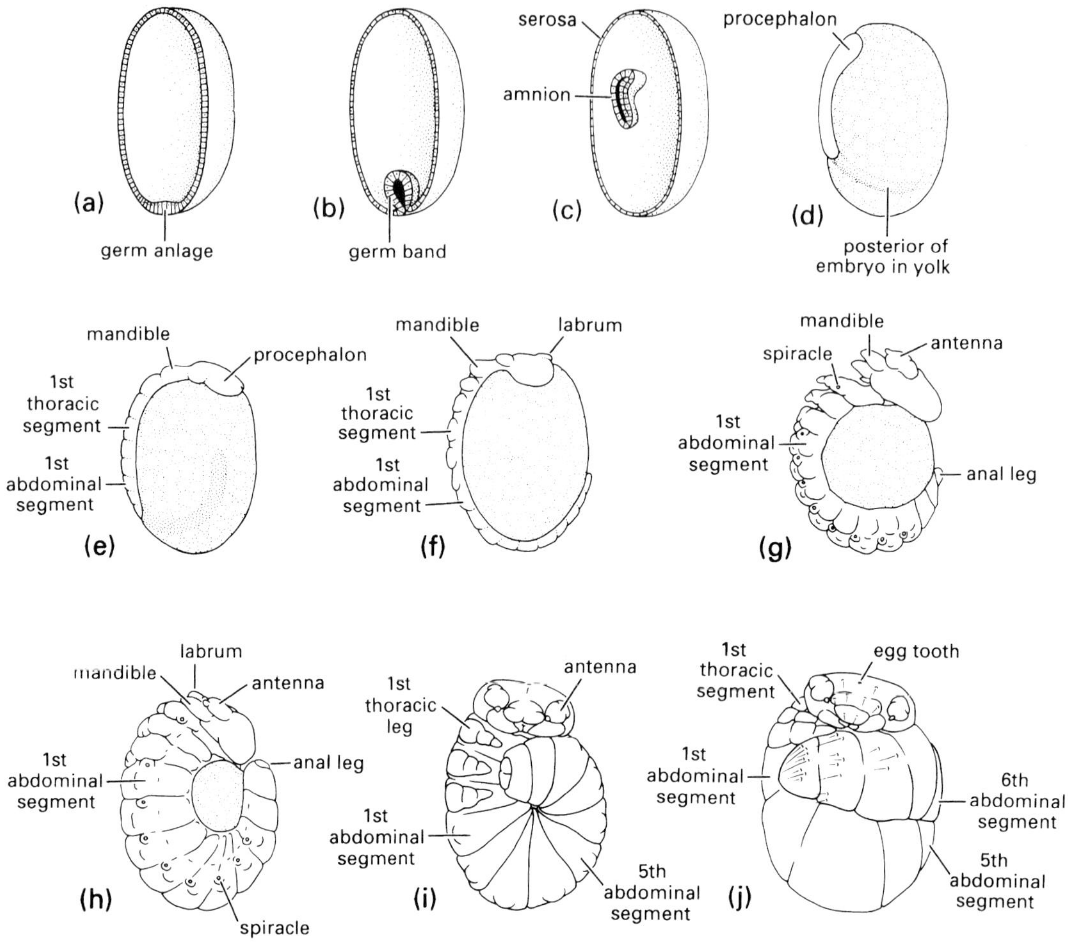

Fig. 6.5 Embryonic development of the scorpionfly *Panorpodes paradoxa* (Mecoptera: Panorpodidae): (a–c) schematic drawings of egg halves from which yolk has been removed to show position of embryo; (d–j) gross morphology of developing embryos at various ages. Age from oviposition: (a) 32 h; (b) 2 days; (c) 7 days; (d) 12 days; (e) 16 days; (f) 19 days; (g) 23 days; (h) 25 days; (i) 25–26 days; (j) full grown at 32 days. (After Suzuki 1985.)

parts of the head with the posterior segments added post-cellularization, during subsequent growth. In "intermediate-germ band" insects (such as damselflies and crickets), the germ band forms from two ventrolateral cell aggregations, which fuse ventrally to become primordia of the head and thorax, with abdominal segments budded off later, as in short-germ band insects. In the developing long-germ embryo, the syncytial phase is followed by cell membrane intrusion to form the blastoderm phase.

Functional specialization of cells and tissues occurs during the latter period of embryonic development, so that by the time of hatching (Fig. 6.5j), the embryo is a tiny proto-insect packed into an eggshell. In ametabolous and hemimetabolous insects, this stage may be recognized as a **pronymph** – a special hatching stage (section 8.5). Molecular developmental processes involved in organizing the polarity and differentiation of areas of the body, including segmentation, are summarized in Box 6.1.

Box 6.1 Molecular insights into insect development

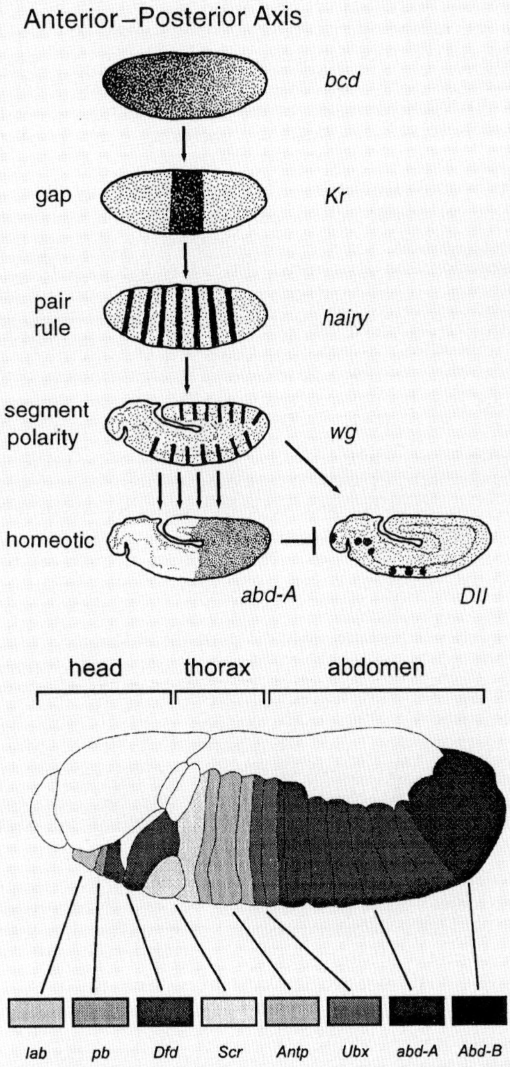

The formation of segments in the early embryo of *Drosophila* is understood better than almost any other complex developmental process. Segmentation is controlled by a hierarchy of transcription factors, which are proteins that bind to DNA and enhance or repress production of specific messages. In the absence of a message, the coded protein is not produced. Thus, transcription factors act as molecular switches, turning on and off the production of specific proteins. In addition to controlling genes below them in the hierarchy, many transcription factors also act on other genes at the same level, as well as regulating their own concentrations. Mechanisms and processes observed in *Drosophila* have had much wider relevance, even extending to vertebrate development, and cloning of human genes. Nevertheless, *Drosophila* is a highly derived fly, showing some quite novel changes in transcription compared to other insects. In this regard, the red flour beetle *Tribolium castaneum*, including studies using induced loss of function by RNA interference (RNAi), is a popular emerging model for developmental studies in holometabolous insects.

During oogenesis (section 6.2.1) in *Drosophila*, the anterior–posterior and dorsal–ventral axes are established by localization of maternal messenger RNAs (mRNAs) or proteins at specific positions within the egg. For example, the mRNAs from the *bicoid* (*bcd*) and *nanos* genes become localized at anterior and posterior ends of the egg, respectively. At oviposition, these messages are translated and proteins are produced that establish concentration gradients by diffusion from each end of the egg. These protein gradients differentially activate or inhibit zygotic genes lower in the segmentation hierarchy – as shown in the upper figure (after Nagy 1998), with zygotic gene hierarchy on the left and representative genes on the right – as a result of their differential thresholds of action. The first class of zygotic genes to be activated is the gap genes, for example *Kruppel* (*Kr*), which divide the embryo into broad, slightly overlapping zones from anterior to posterior. The maternal and gap proteins establish a complex of overlapping protein gradients that provide a chemical framework that controls the periodic (alternate segmental) expression of the pair-rule genes. For example, the pair-rule protein hairy is expressed in seven stripes along the length of the embryo while it is still in the syncytial stage. The pair-rule proteins, in addition to the proteins produced by genes higher in the hierarchy, then act to regulate the segment polarity genes, which are expressed with segmental periodicity and represent the final step in the determination of segmentation. Because there are many members of the various classes of segmentation genes, each row of cells in the anterior–posterior axis must contain a unique combination and concentration of the transcription factors that inform cells of their position along the anterior–posterior axis.

Once the segmentation process is complete, each developing segment is given its unique identity by the homeotic genes. Although these genes were first discovered in *Drosophila* it has since been established that they are very ancient, and some form or remnant of a cluster is found in all multicellular animals. When this was realized, it was agreed that this group of genes would be called the Hox genes, although both terms, homeotic and Hox, are still in use for the same group of genes. In many organisms, these genes form a single cluster on one chromosome, although in *Drosophila* they are organized into two clusters, an anteriorly expressed Antennapedia complex (Antp-C) and a posteriorly expressed Bithorax complex (Bx-C). The composition of these clusters in *Drosophila* is as follows (from anterior to posterior): (Antp-C) – *labial* (*lab*), *proboscipedia* (*pb*), *Deformed* (*Dfd*), *Sex combs reduced* (*Scr*), *Antennapedia* (*Antp*); (Bx-C) – *Ultrabithorax* (*Ubx*), *abdominal-A* (*abd-A*) and *Abdominal-B* (*Abd-B*), as illustrated in the lower figure of a *Drosophila* embryo (after Carroll 1995; Purugganan 1998). The evolutionary conservation of the Hox genes is remarkable, for not only are they conserved in their primary structure, but they follow the same order on the chromosome, and their temporal order of expression and anterior border of expression along the body correspond to their chromosomal position. In the lower figure, the anterior zone of expression of each gene and the zone of strongest expression is shown (for each gene there is a zone of weaker expression posteriorly); as each gene switches on, protein production from the gene anterior to it is repressed.

The zone of expression of a particular Hox gene may be morphologically very different in different organisms, so it is evident that Hox gene activities demarcate relative positions but not particular morphological structures. A single Hox gene may regulate directly or indirectly many targets; for example, *Ultrabithorax* regulates some 85–170 genes. These downstream genes may operate at different times and also have multiple effects (pleiotropy); for example, *wingless* in *Drosophila* is involved successively in segmentation (embryo), Malpighian tubule formation (larva), and leg and wing development (larva–pupa).

Boundaries of transcription factor expression are important locations for the development of distinct morphological structures, such as limbs, tracheae and salivary glands. Studies of the development of legs and wings have revealed something about the processes involved. Limbs arise at the intersection between expression of wingless, engrailed and decapentaplegic (dpp), a protein that helps to inform cells of their position in the dorsal–ventral axis. Under the influence of the unique mosaic of gradients created by these gene products, limb primordial cells are stimulated to express the gene *distal-less* (*Dll*), which is required for proximodistal limb growth. As potential limb primordial cells (anlage) are present on all segments, as are limb-inducing protein gradients, prevention of limb growth on inappropriate segments (i.e. the *Drosophila* abdomen) must involve repression of *Dll* expression on such segments. In Lepidoptera, in which larval prolegs typically are found on the third to sixth abdominal segments, homeotic gene expression is fundamentally similar to that of *Drosophila*. In the early lepidopteran embryo, *Dll* and *Antp* are expressed in the thorax, as in *Drosophila*, with *abd-A* expression dominant in abdominal segments including 3–6, which are prospective for proleg development. Then, a dramatic change occurs, with abd-A protein repressed in the abdominal proleg cell anlagen, followed by activation of *Dll* and upregulation of *Antp* expression as the anlagen enlarge. Two genes of the Bithorax complex (Bx-C), *Ubx* and *abd-A*, repress *Dll* expression (and hence prevent limb formation) in the abdomen of *Drosophila*. Therefore, expression of prolegs in the caterpillar abdomen results from repression of Bx-C proteins thus derepressing *Dll* and *Antp* and thereby permitting their expression in selected target cells, with the result that prolegs develop.

A somewhat similar condition exists with respect to wings, in that the default condition is presence on all thoracic and abdominal segments, with Hox gene repression reducing the number from this default condition. In the prothorax, the homeotic gene *Scr* has been shown to repress wing development. Other effects of *Scr* expression in the posterior head, labial segment and prothorax appear homologous across many insects, including ventral migration and fusion of the labial lobes, specification of labial palps, and development of sex combs on male prothoracic legs. Experimental mutational damage to *Scr* expression leads, amongst other deformities, to appearance of wing primordia from a group of cells located just dorsal to the prothoracic leg base. These mutant prothoracic wing anlagen are situated very close to the site predicted by Kukalová-Peck from palaeontological evidence (section 8.4, Fig. 8.5b). Furthermore, the apparent default condition (lack of repression of wing expression) would produce an insect resembling the hypothesized "protopterygote", with winglets present on all segments.

Regarding the variations in wing expression seen in the pterygotes, *Ubx* activity differs in *Drosophila* between the meso- and metathoracic imaginal discs: the anterior produces a wing; the posterior produces a haltere. *Ubx* is unexpressed in the wing (mesothoracic) imaginal disc but is strongly expressed in the metathoracic disc, where its activity suppresses wing and enhances haltere formation. However, in some studied non-dipterans, *Ubx* is expressed as in *Drosophila* – not in the fore wing but strongly in the hind-wing imaginal disc – despite the elaboration of a complete hind wing as in butterflies or beetles. Thus, very different wing morphologies seem to result from variation in "downstream" response to wing-pattern genes regulated by *Ubx*, rather than from homeotic control.

Clearly, much is yet to be learnt concerning the multiplicity of morphological outcomes from the interaction between Hox genes and their downstream interactions with a wide range of genes. It is tempting to relate major variation in Hox pathways with morphological disparities associated with high-level taxonomic rank (e.g. animal classes), more subtle changes in Hox regulation with intermediate taxonomic levels (e.g. orders/suborders), and changes in downstream regulatory/functional genes perhaps with suborder/family rank. Notwithstanding progress in the case of the Strepsiptera (section 7.4.2), such simplistic relationships between a few well-understood major developmental features and taxonomic radiations may not lead to great insight into insect macroevolution in the immediate future. Estimated phylogenies from other sources of data will be necessary to help interpret the evolutionary significance of homeotic changes for some time to come.

6.2.2 Larval or nymphal phase

Hatching from the egg may be by a pronymph, nymph or larva. Eclosion conventionally marks the beginning of the first stadium, when the young insect is said to be in its first instar (Fig. 6.1). This stage ends at the first ecdysis, when the old cuticle is cast to reveal the insect in its second instar. Third and often subsequent instars generally follow. Thus, the development of the immature insect is characterized by repeated moults separated by periods of feeding, with hemimetabolous insects generally undergoing more moults to reach adulthood than do holometabolous insects.

All immature holometabolous insects are called **larvae**. Immature terrestrial insects with hemimetabolous development, such as cockroaches and termites (Blattodea), grasshoppers and crickets (Orthoptera), mantids (Mantodea) and bugs (Hemiptera), always are called **nymphs**. However, immature individuals of aquatic hemimetabolous insects (Odonata, Ephemeroptera and Plecoptera), although possessing external wing pads at least in later instars, also are frequently, but incorrectly, referred to as larvae (or sometimes naiads). True larvae look very different from the final adult form in every instar, whereas nymphs more closely approach the adult appearance at each successive moult. Larval diets and lifestyles are very different from those of their adults. In contrast, nymphs often eat the same food and coexist with the adults of their species. Competition thus is rare between larvae and their adults but is likely to be prevalent between nymphs and their adults.

The great variety of endopterygote larvae can be classified into a few functional rather than phylogenetic types. Often the same larval type occurs convergently in unrelated orders. The three most common forms are the polypod, oligopod and apod larvae (Fig. 6.6). Lepidopteran caterpillars (Fig. 6.6a,b) are characteristic **polypod** larvae, with cylindrical bodies with short thoracic legs and abdominal prolegs (pseudopods). Symphytan Hymenoptera (sawflies; Fig. 6.6c) and most Mecoptera also have polypod larvae. Such larvae are rather inactive and are mostly phytophagous. **Oligopod** larvae (Fig. 6.6d–f) lack abdominal prolegs but have functional thoracic legs and frequently prognathous mouthparts. Many are active predators, but others are slow-moving detritivores living in soil or are phytophages. This larval type occurs in at least some members of most orders of insects but not in the Lepidoptera, Mecoptera, Siphonaptera, Diptera or Strepsiptera. **Apod** larvae (Fig. 6.6g–i) lack true legs, are worm-like or maggot-like, and live in soil, mud, dung, decaying plant or animal matter, or within the bodies of other organisms as parasitoids (Chapter 13). The Siphonaptera, aculeate Hymenoptera, nematoceran Diptera, and many Coleoptera, typically have apod larvae with a well-developed head. In the maggots of brachyceran Diptera, the mouth hooks may be the only obvious evidence of the cephalic region. The grub-like apod larvae of some parasitic and gall-inducing wasps and flies are greatly reduced in external structure, and are difficult to identify to order level even by a specialist entomologist. Furthermore, the early-instar larvae of some parasitic wasps resemble a naked embryo, but change into typical apod larvae in later instars.

A major change in form during the larval phase, such as different larval types in different instars, is called larval **heteromorphosis** (or **hypermetamorphosis**). In the Strepsiptera and certain beetles this

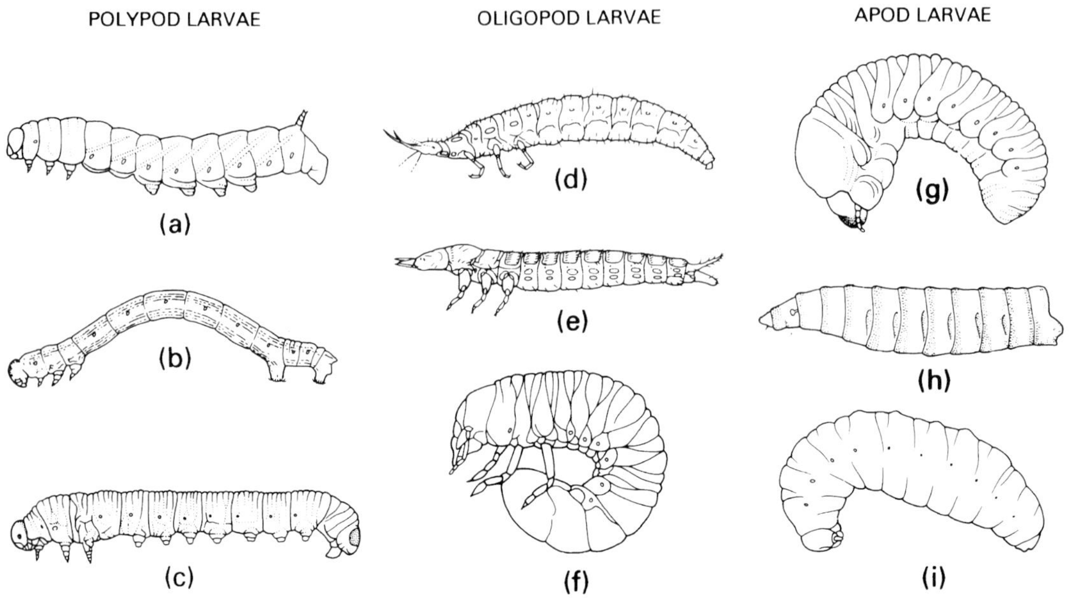

Fig. 6.6 Examples of larval types. Polypod larvae: (a) Lepidoptera: Sphingidae; (b) Lepidoptera: Geometridae; (c) Hymenoptera: Diprionidae. Oligopod larvae: (d) Neuroptera: Osmylidae; (e) Coleoptera: Carabidae; (f) Coleoptera: Scarabaeidae. Apod larvae: (g) Coleoptera: Scolytinae; (h) Diptera: Calliphoridae; (i) Hymenoptera: Vespidae. ((a,e–g) After Chu 1949; (b,c) after Borror *et al.* 1989; (h) after Ferrar 1987; (i) after CSIRO 1970.)

involves an active first-instar larva, or **triungulin**, followed by several grub-like, inactive, sometimes legless, later-instar larvae. This developmental phenomenon occurs most commonly in parasitic insects in which a mobile first instar is necessary for host location and entry. Larval heteromorphosis and diverse larval types are typical of many parasitic wasps.

6.2.3 Metamorphosis

All pterygote insects undergo varying degrees of transformation from the immature to the adult phase of their life history. Some exopterygotes, such as cockroaches, show only slight morphological changes during post-embryonic development, whereas the body is substantially reconstructed at metamorphosis in many endopterygotes. The evolution of metamorphosis is discussed in section 8.5.

Orders belonging to the Holometabola (= Endopterygota) have a metamorphosis involving a pupal stadium, during which adult structures are elaborated from certain larval structures and from imaginal discs (e.g. Fig. 6.4). In some holometabolous insects, such as *Drosophila*, most larval tissues are destroyed at metamorphosis, and the pupal and adult structures are formed largely from imaginal discs. Alterations in body shape, which are the essence of metamorphosis, are brought about by differential growth of various body parts. Organs that will function in the adult but that were undeveloped in the larva grow at a faster rate than the body average. The accelerated growth of wing pads is the most obvious example, but legs, genitalia, gonads and other internal organs may increase considerably in size and complexity.

One trigger for onset of complete metamorphosis is attainment of a particular body size (the critical mass), which programmes the brain for metamorphosis by altering hormone levels, as discussed in section 6.3.

The moult into the pupal instar is called **pupation**, or the larval–pupal moult. Many insects survive conditions unfavourable for development in the "resting", non-feeding pupal stage. However, what looks like a pupa actually often is a fully developed adult within the pupal cuticle, referred to as a **pharate** (cloaked) adult. Typically, a protective cell or cocoon surrounds

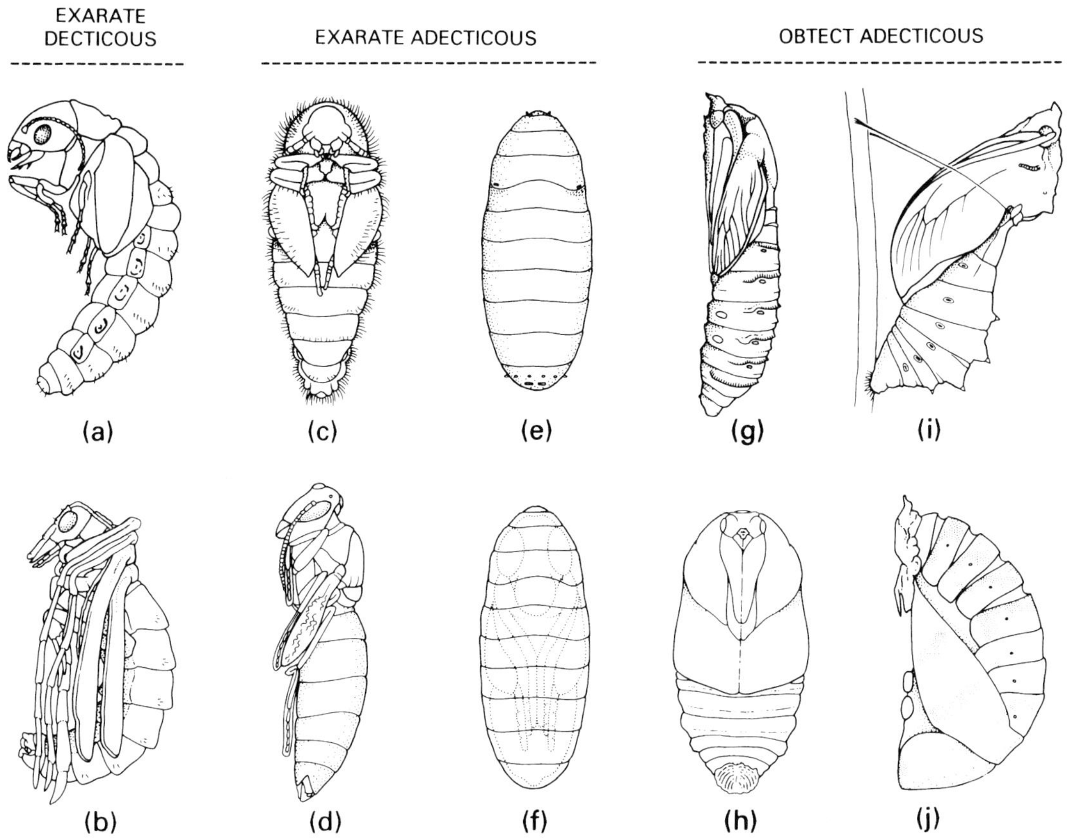

Fig. 6.7 Examples of pupal types. Exarate decticous pupae: (a) Megaloptera: Sialidae; (b) Mecoptera: Bittacidae. Exarate adecticous pupae: (c) Coleoptera: Dermestidae; (d) Hymenoptera: Vespidae; (e,f) Diptera: Calliphoridae, puparium and pupa within. Obtect adecticous pupae: (g) Lepidoptera: Cossidae; (h) Lepidoptera: Saturniidae; (i) Lepidoptera: Papilionidae, chrysalis; (j) Coleoptera: Coccinellidae. ((a) After Evans 1978; (b,c,e,g) after CSIRO 1970; (d) after Chu 1949; (h) after Common 1990; (i) after Common & Waterhouse 1972; (j) after Palmer 1914.)

the pupa and then, prior to emergence, the pharate adult. Only certain Coleoptera, Diptera, Lepidoptera and Hymenoptera have unprotected pupae.

Several pupal types (Fig. 6.7) are recognized, and these appear to have arisen convergently in different orders. Most pupae are **exarate** (Fig. 6.7a–d) – their appendages (e.g. legs, wings, mouthparts and antennae) are not closely appressed to the body; the remaining pupae are **obtect** (Fig. 6.7g–j) – their appendages are cemented to the body and the cuticle is often heavily sclerotized (as in almost all Lepidoptera). Exarate pupae can have articulated mandibles (**decticous**), which the pharate adult uses to cut through the cocoon, or the mandibles can be non-articulated (**adecticous**), in which case the adult usually first sheds the pupal cuticle and then uses its mandibles and legs to escape the cocoon or cell. In some cyclorrhaphous Diptera (the Schizophora), the adecticous exarate pupa is enclosed in a **puparium** (Fig. 6.7e,f) – the sclerotized cuticle of the last larval instar. Escape from the puparium is facilitated by eversion of a membranous sac on the head of the emerging adult, the **ptilinum**. Insects with obtect pupae may lack a cocoon, as in coccinellid beetles and most nematocerous and orthorrhaphous Diptera. If a cocoon is present, as in most Lepidoptera, emergence from the cocoon is either by the pupa using backwardly directed abdominal spines or a projection on the head, or an adult emerges from the pupal cuticle before escaping the cocoon, sometimes helped by a fluid that dissolves the cocoon's silk.

6.2.4 Imaginal or adult phase

Except for the mayflies, pterygote insects do not moult again once the adult phase is reached. The adult, or imaginal, stage has a reproductive role, and is often the dispersive stage in insects with relatively sedentary larvae. The imago that emerges (ecloses) from the cuticle of the previous instar may be capable of reproduction almost immediately or a period of maturation may precede sperm transfer or oviposition. Depending on species and food availability, there may be from one to several reproductive cycles in the adult stadium. The adults of certain species, such as some mayflies, midges and male scale insects, are very short-lived. These insects have reduced or no mouthparts, and fly for only a few hours or at the most a day or two – they simply mate and die. Most adult insects live at least a few weeks, often a few months and sometimes for several years; termite reproductives and queen ants and bees are particularly long-lived. The evolution of eusociality (section 12.4) is associated with a 100-fold increase in adult lifespan, based on a comparison of the mean average longevity of ant, termite and honeybee queens with that of adult solitary insects from eight orders.

Adult life begins at eclosion from the pupal or last-nymphal cuticle. Metamorphosis, however, may have been complete for some hours, days or weeks previously, and the pharate adult may have rested within the pupal cuticle until the appropriate environmental trigger for emergence. Changes in temperature or light, and perhaps chemical signals, may synchronize adult emergence in most species.

Our understanding of hormonal control of emergence derives substantially from studies of the tobacco hornworm, *Manduca sexta* (Sphingidae), notably by James Truman, Lynn Riddiford and colleagues. Despite the huge range of insect body plans, highly conserved systems of eclosion control in other taxa resemble that found for *M. sexta*. At least six hormones are involved in eclosion. A few days pre-eclosion, a rise in the neuropeptide **corazonin** and decline in ecdysteroid level initiates physiological and behavioural preparation for ecdysis, including the release of neuropeptide hormones and transmitters. **Ecdysis triggering hormones** (PETH and ETH), secreted from **Inka cells** clustered in epitracheal glands and/or dispersed throughout the tracheal system, and **eclosion hormones** (EH), from neurosecretory cells in the brain, stimulate pre-eclosion behaviour, such as seeking a site suitable for ecdysis and inducing contractions that aid extrication from the old cuticle. PETH (pre-edysis triggering hormone) is released first and ETH and EH then stimulate each other's release, forming a positive feedback loop. The build-up of EH releases **crustacean cardioactive peptide** (CCAP) from cells in the ventral nerve cord. CCAP induces a switch from pre-eclosion to eclosion behaviours, including body contractions and wing-base movements, and accelerates heartbeat. EH appears also to permit the release of further neurohormones – **cardiopeptides** and **bursicon** – that are involved in wing expansion and cuticle hardening after ecdysis. Cardiopeptides stimulate the heart, pumping haemolymph into the thorax and thus into the wings. Bursicon induces a brief increase in cuticle plasticity, allowing wing expansion, followed by sclerotization of the cuticle.

The newly emerged, or **teneral**, adult has soft cuticle, allowing expansion of the body surface by swallowing air, by taking air into the tracheal sacs, and by locally increasing haemolymph pressure by muscular activity. The wings normally hang down (Fig. 6.8), which aids their inflation. Pigment deposition in the cuticle and epidermal cells occurs just before or after emergence, and is either linked to, or followed by, sclerotization of the body cuticle under the influence of the neurohormone bursicon. A newly moulted insect is often white (see **Plate 3d**).

Following emergence from the pupal cuticle, many holometabolous insects void a faecal fluid called the **meconium**. This represents the metabolic wastes that have accumulated during the pupal stadium. Sometimes, the teneral adult retains the meconium in the rectum until sclerotization is complete, thus aiding increase in body size.

Reproduction is the main function of adult life, and the length of the imaginal stadium, at least in the female, is related to the duration of egg production. Reproduction is discussed in detail in Chapter 5. Senescence correlates with termination of reproduction, and death may be predetermined in the ontogeny of an insect. Females may die after egg deposition and males may die after mating. An extended post-reproductive life is important in distasteful, aposematic insects as it allows predators to learn distastefulness from expendable prey individuals (section 14.4).

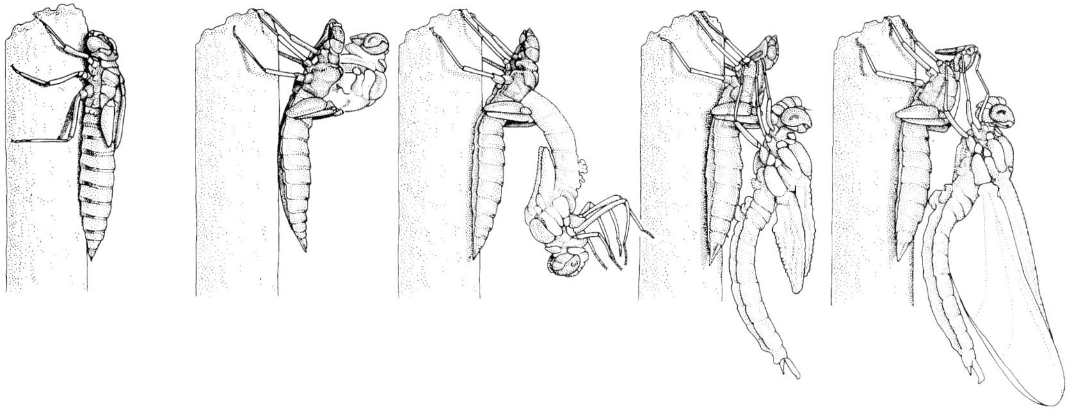

Fig. 6.8 The nymphal–imaginal moult of a male dragonfly of *Aeshna cyanea* (Odonata: Aeshnidae). The final-instar nymph climbs out of the water prior to the shedding of its cuticle. The old cuticle splits mid-dorsally, the teneral adult frees itself, swallows air and must wait many hours for its wings to expand and dry. (After Blaney 1976.)

6.3 PROCESS AND CONTROL OF MOULTING

For practical reasons, an instar is defined as extending from ecdysis to ecdysis (Fig. 6.1), since shedding of the old cuticle is an obvious event. However, in terms of morphology and physiology, a new instar comes into existence at the time of **apolysis**, when the epidermis separates from the cuticle of the previous stage. Apolysis is difficult to detect in most insects but knowledge of its occurrence may be important because many insects spend a substantial period in the pharate state (cloaked within the cuticle of the previous instar), awaiting conditions favourable for emergence as the next stage. Insects often survive adverse conditions as pharate pupae or pharate adults (e.g. some diapausing adult moths) because in this state the double cuticular layer restricts water loss during a developmental period during which metabolism is reduced and requirements for gaseous exchange are minimal.

Moulting is a complex process, involving hormonal, behavioural, epidermal and cuticular changes that lead up to the shedding of the old cuticle. The epidermal cells are actively involved in moulting – they are responsible for partial breakdown of the old cuticle and formation of the new cuticle. The moult commences with the retraction of the epidermal cells from the inner surface of the old cuticle, usually in an antero-posterior direction. This separation is incomplete because muscles and sensory nerves do retain connection with the old cuticle for some time. Apolysis either is correlated with or followed by mitotic division of the epidermal cells, leading to increases in the volume and surface area of the epidermis. The subcuticular or **apolysial space** formed after apolysis fills with the secreted but inactive moulting fluid. The chitinolytic and proteolytic enzymes of the moulting fluid are not activated until the epidermal cells have laid down the protective outer layer of a new cuticle. Then the inner part of the old cuticle (the **endocuticle**) is lysed and presumably resorbed, while the new pharate cuticle continues to be deposited as an undifferentiated **procuticle**. Ecdysis commences with the remnants of the old cuticle splitting along the dorsal midline as a result of increase in haemolymph pressure. The cast cuticle consists of the indigestible protein, lipid and chitin of the old **epicuticle** and **exocuticle**. Once free of the constraints of this previous "skin", the newly ecdysed insect expands the new cuticle. This is attained by swallowing air or water and/or by increasing haemolymph pressure in different body parts to smooth out the wrinkled and folded epicuticle and stretch the procuticle. After cuticular expansion, some or much of the body surface may become sclerotized by the chemical stiffening and darkening of the procuticle to form exocuticle (section 2.1). However, in larval insects, most of the body cuticle remains membranous and exocuticle is confined to the head capsule. Following ecdysis, more proteins and chitin are secreted from the epidermal cells, thus adding to the inner part of the procuticle, that is, the endocuticle, which may continue to be deposited well into the intermoult period. Sometimes, the endocuticle is partially sclerotized during the stadium, and frequently the outer surface of the cuticle is covered in wax secretions. Finally, the stadium draws to an end and apolysis is initiated once again.

The above events are controlled by hormones acting on the epidermal cells to cause the cuticular changes and on the nervous system to co-ordinate ecdysis behaviours. Hormonal regulation of moulting has been studied most thoroughly at metamorphosis, when endocrine influences on moulting *per se* are difficult to separate from those involved in the control of morphological change. A classical simplified view of the hormonal regulation of moulting and metamorphosis is presented schematically in Fig. 6.9. The role of endocrine centres and their hormones are detailed

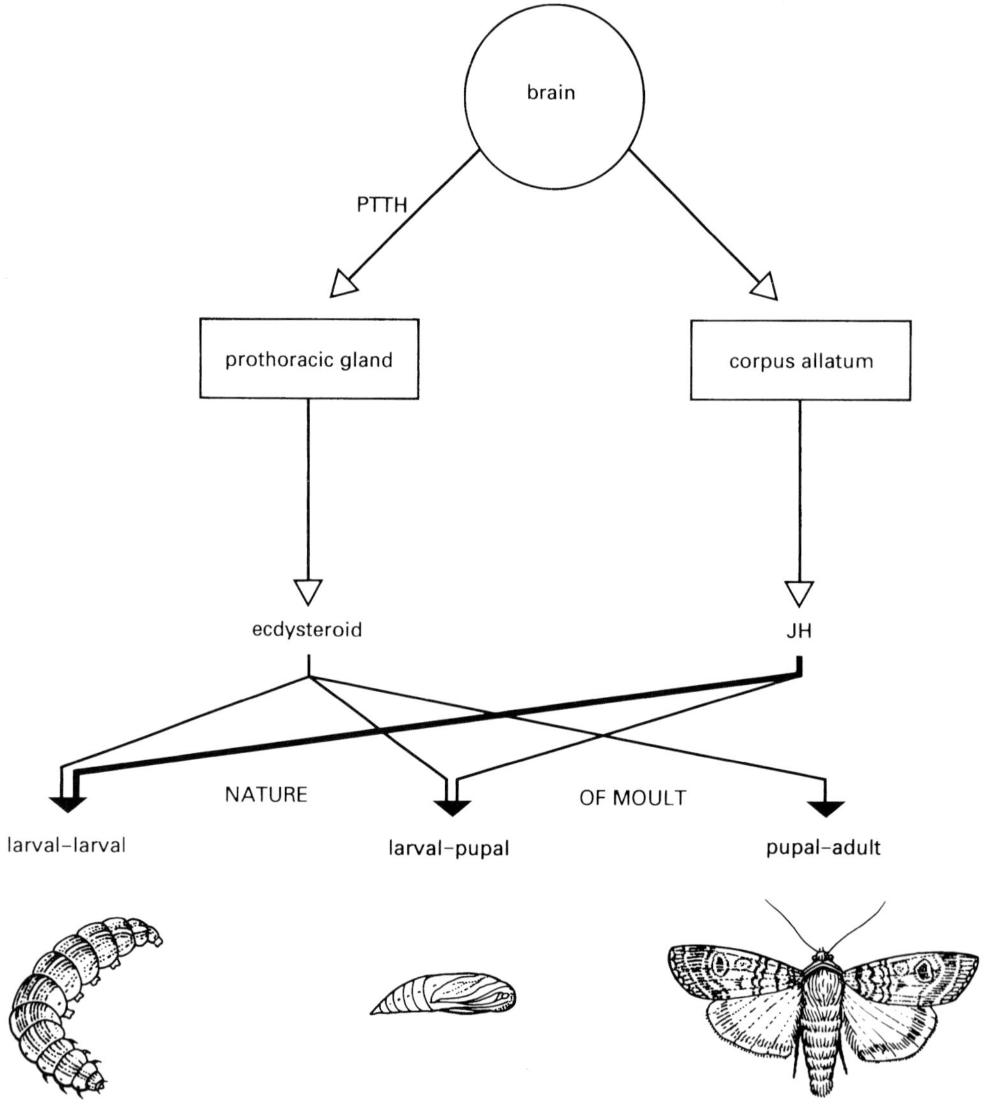

Fig. 6.9 Schematic diagram of the classical view of endocrine control of the epidermal processes that occur in moulting and metamorphosis in a holometabolous insect. This scheme simplifies the complexity of ecdysteroid and JH secretion, and does not indicate the influence of neuropeptides such as eclosion hormone. JH, juvenile hormone; PTTH, prothoracicotropic hormone. (After Richards 1981.)

in section 3.3. Three major types of hormones control moulting and metamorphosis:

1 neuropeptides, including **prothoracicotropic hormone** (PTTH), **ecdysis-triggering hormones** (PETH, ETH) and eclosion hormone (EH);

2 **ecdysteroids**;

3 **juvenile hormone** (JH), which may occur in several different forms even in the same insect.

Neurosecretory cells in the brain secrete PTTH, which passes down nerve axons to the corpora cardiaca (or to the corpora allata in Lepidoptera), paired neuroglandular bodies that store and release PTTH into the haemolymph. The PTTH initiates each moult by stimulating ecdysteroid synthesis and secretion by the prothoracic or moulting glands. Ecdysteroid release initiates changes in epidermal cells that lead to the production of new cuticle. The characteristics of the moult are regulated by JH from the corpora allata; JH inhibits the expression of adult features so that a high haemolymph level (titre) of JH is associated with a larval–larval moult, and a lower titre with a larval–pupal moult; JH is absent at the pupal–adult moult.

Ecdysis is mediated by ETH and EH. EH is important at every moult in the life history of all studied insects. This neuropeptide acts on a steroid-primed central nervous system to trigger co-ordinated motor activities that allow escape from the old cuticle. Eclosion hormone derives its name from the pupal–adult ecdysis, or eclosion, for which its importance was discovered first and before its wider role was realized. Indeed, the association of EH with moulting appears to be ancient, as crustaceans have EH homologues. In the well-studied tobacco hornworm, *Manduca sexta* (section 6.2.4), ETHs are as important to ecdysis as EH, with PETH initiating the pre-ecdysis behaviour that loosens muscle attachments from the old cuticle, and ETH stimulating release of EH from the brain. Bursicon, another neuropeptide, controls sclerotization of the exocuticle and post-moult deposition of endocuticle in many insects.

The relationship between this hormonal environment and the epidermal activities that control moulting and cuticular deposition in *M. sexta* are presented in Fig. 6.10. A correlation exists between the ecdysteroid and JH titres and the cuticular changes

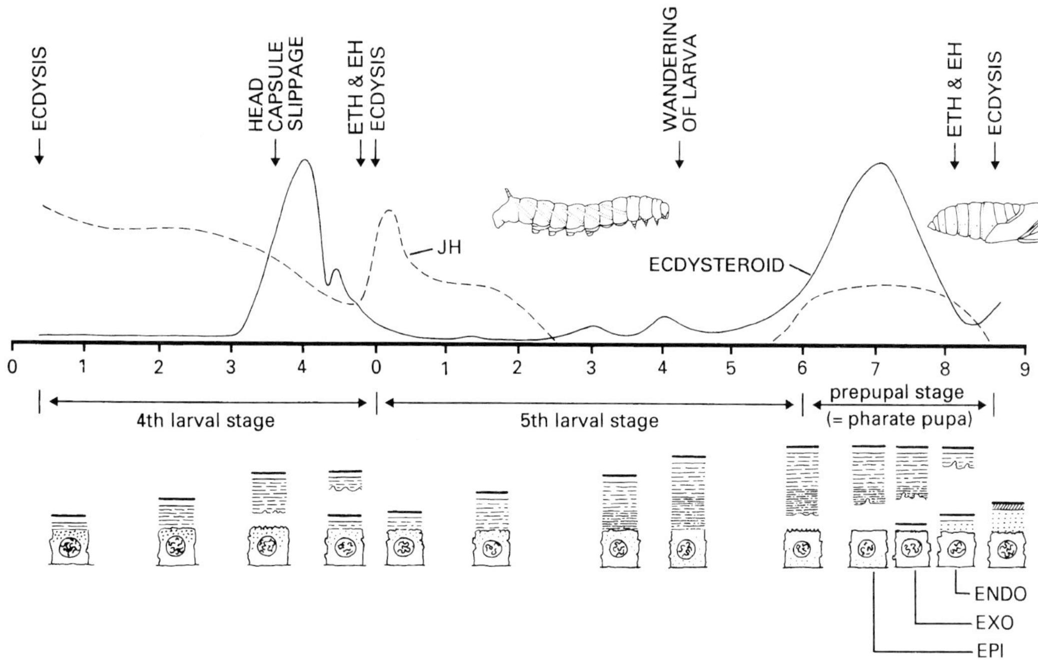

Fig. 6.10 Diagrammatic view of the changing activities of the epidermis during the fourth and fifth larval instars and prepupal (= pharate pupal) development in the tobacco hornworm, *Manduca sexta* (Lepidoptera: Sphingidae), in relation to the hormonal environment. The dots in the epidermal cells represent granules of the blue pigment insecticyanin. ETH, ecdysis triggering hormone; EH, eclosion hormone; JH, juvenile hormone; EPI, EXO, ENDO, deposition of pupal epicuticle, exocuticle and endocuticle, respectively. The numbers on the *x*-axis represent days. (After Riddiford 1991.)

that occur in the last two larval instars and in prepupal development. Thus, during the moult at the end of the fourth larval instar, the epidermis responds to the surge of ecdysteroid by halting synthesis of endocuticle and the blue pigment insecticyanin. A new epicuticle is synthesized, much of the old cuticle is digested, and production of endocuticle and insecticyanin resumes by the time of ecdysis. In the final (fifth) larval instar (but not earlier), JH inhibits the secretion of PTTH and ecdysteroids, and thus the level of JH must decline to zero before the ecdysteroid level can rise. When ecdysteroid initiates the next moult, the epidermal cells produce a stiffer cuticle with thinner lamellae (the pupal cuticle). This larval–pupal moult is distinctive also by actions of genes, including *broad* and *krüppel*, which are linked to development in holometabolans.

Decline in ecdysteroid level towards the end of each moult seems essential for, and may be the physiological trigger causing, ecdysis to occur. A cascade of small-peptide hormones is released after formation of the new cuticle and reduction of the ecdysteroid below a threshold level (see section 6.2.4). Apolysis at the end of the fifth larval instar thus marks the beginning of a prepupal period when the developing pupa is pharate within the larval cuticle. Differentiated exocuticle and endocuticle appear at this larval–pupal moult. During larval life, the epidermal cells covering most of the body do not produce exocuticle. The caterpillar's soft and flexible cuticle allows the considerable growth seen within an instar, especially the last larval instar, as a result of voracious feeding.

In the tobacco hornworm, onset of metamorphosis involves attainment of a critical mass by the final-instar larva. This causes the reduction (to zero) in amount of circulating JH by reduced corpora allata activity and enzymatic degradation of JH in the haemolymph. This reduction in JH initiates a subsequent cessation of feeding caused by rising ecdysteroid levels; however, growth does not cease immediately that the critical mass is attained, but continues until sometime during the next 24 h after JH levels reach zero, when a photoperiod "gate" opens. At this time, the previously suppressed PTTH is expressed, triggering the burst of ecdysteroid (as ecdysone) that stimulates behavioural changes and induces commencement of a moult to pupal development (schematically presented in Fig. 6.11). The final body size at metamorphosis evidently depends upon how much feeding (and growth) takes place between attainment of the critical mass and the delayed onset of PTTH secretion, and ultimately affects the size of the emerged adult.

6.4 VOLTINISM

Insects are short-lived creatures, whose lives can be measured by their **voltinism** – the numbers of generations per year. Most insects take a year or less to develop, with either one generation per year (**univoltine** insects), or two (**bivoltine** insects), or more than two (**multivoltine**, or polyvoltine, insects). Generation times in excess of one year (**semivoltine** insects) are found, for example amongst some inhabitants of the polar extremes, where suitable conditions for development may exist for only a few weeks in each year. Large insects that rely upon nutritionally poor diets also develop slowly over many years. For example, periodical cicadas feeding on sap from tree roots may take either 13 or 17 years to mature, and beetles that develop within dead wood have been known to emerge after more than 20 years of development.

Most insects do not develop continuously throughout the year, but arrest their development during unfavourable times by quiescence or diapause (section 6.5). Many univoltine and some bivoltine species enter diapause at some stage, awaiting suitable conditions before completing their life cycle. For some univoltine insects, many social insects, and others that take longer than a year to develop, adult longevity may extend to several years. In contrast, the adult life of multivoltine insects may be as little as a single evening for many Ephemeroptera, or even a few hours at low tide for marine midges such as *Clunio* (Diptera: Chironomidae).

Multivoltine insects tend to be small and fast-developing, using resources that are more evenly available throughout the year. Univoltinism is common amongst temperate insects, particularly those that use resources that are seasonally restricted. These might include insects whose aquatic immature stages rely on spring algal bloom, or phytophagous insects using short-lived annual plants. Bivoltine insects include those that develop slowly on evenly spread resources, and those that track a bimodally distributed factor, such as spring and autumn/fall temperature. Some species have fixed voltinism patterns, whereas others may vary with geography, particularly in insects with broad latitudinal or elevational ranges.

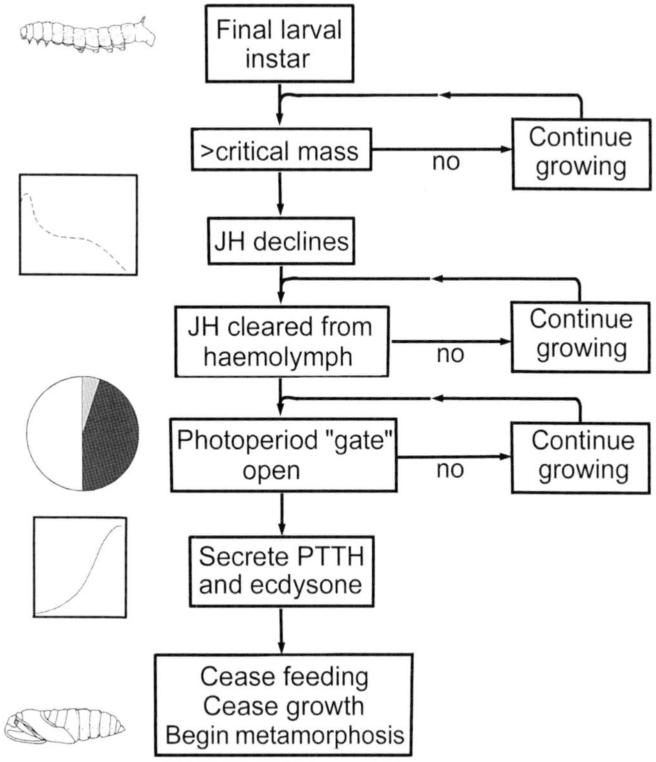

Fig. 6.11 A flow chart of the events prior to metamorphosis that determine body size in the tobacco hornworm, *Manduca sexta* (Lepidoptera: Sphingidae). During the last larval instar, there are three physiological decision points. The final size of the insect is determined by the amount of growth that occurs in the intervals between these three conditional events. JH, juvenile hormone; PTTH, prothoracicotropic hormone. (After Nijhout *et al.* 2006.)

6.5 DIAPAUSE

The developmental progression from egg to adult often is interrupted by a period of dormancy. This occurs particularly in temperate areas when environmental conditions become unsuitable, such as in seasonal extremes of high or low temperatures, or drought. Dormancy may occur in summer (**aestivation** (estivation)) or in winter (**hibernation**), and may involve either quiescence or diapause. **Quiescence** is a halted or slowed development as a direct response to unfavourable conditions, with development resuming immediately favourable conditions return. In contrast, **diapause** involves arrested development combined with adaptive physiological changes, with development recommencing not necessarily on return of suitable conditions, but only following particular physiological stimuli. Distinguishing between quiescence and diapause requires detailed study.

Diapause at a fixed time, regardless of varied environmental conditions, is termed **obligatory**. Univoltine insects (those with one generation per year) often have obligatory diapause to extend an essentially short life cycle to one full year. Diapause that is optional is termed **facultative**, and this occurs widely in insects, including many bi- or multivoltine insects in which diapause occurs only in the generation that must survive the unfavourable conditions. Facultative diapause can be induced by food. Thus, when prey populations of summer aphids are low, the ladybird beetles *Hippodamia convergens* and *Semidalia unidecimnotata* aestivate, but if aphids remain at high densities (as in irrigated crops), the predators continue to develop without diapause.

Diapause can last from days to months or in rare cases years, and can occur in any life-history stage from egg to adult. The diapausing stage predominantly is fixed within any species and can vary between close relatives. Egg and/or pupal diapause is common, probably because these stages are relatively closed systems, with only gases being exchanged during embryogenesis and metamorphosis, respectively, allowing better survival during environmental stress. In the adult stage, reproductive diapause describes the cessation or suspension of reproduction in mature insects. In this state, metabolism may be redirected to migratory flight (section 6.7), production of cryoprotectants (section 6.6.1), or simply reduced during conditions inclement for the survival of adult (and/or immature) stages. Reproduction commences post-migration or when conditions for successful oviposition and immature-stage development return.

Much research on diapause has been carried out in Japan in relation to silk production from cultured silkworms (*Bombyx mori*). Optimal silk production comes from the generation with egg diapause, but this is in conflict with commercial needs for continuous production from individuals reared from non-diapausing eggs. The complex mechanisms that promote and break diapause in this species are now well understood. However, these mechanisms may not apply generally, and as the example of *Aedes* (see below) indicates, several different mechanisms may be at play in different, even closely related, insects, and much remains to be discovered.

Major environmental cues that induce and/or terminate diapause are photoperiod, temperature, food quality, moisture, pH and chemicals including oxygen, urea and plant secondary compounds. Identification of the contribution of each may be difficult, as for example in species of the mosquito genus *Aedes* that lay diapausing eggs into seasonally dry pools or containers. Flooding of the oviposition site at any time may terminate embryonic diapause in some *Aedes* species. In other species, many successive inundations may be required to break diapause, with the cues apparently including chemical changes such as lowering of pH by microbial decomposition of pond detritus. Furthermore, one environmental cue may enhance or override a previous one. For example, if an appropriate diapause-terminating cue of inundation occurs while the photoperiod and/or temperature is "wrong", then either diapause continues, or just a few eggs hatch.

Photoperiod is significant in diapause because changed day length predicts much about future seasonal environmental conditions. Photoperiod increases as summer heat approaches, and diminishes towards winter cold (section 6.10.2). Insects can detect day-length or night-length changes (photoperiodic stimuli), sometimes with extreme accuracy, through brain photoreceptors rather than compound eyes or ocelli. The insect brain also codes for diapause: transplant of a pupal brain of a diapausing moth into a non-diapausing pupa induces diapause in the recipient. The reciprocal operation causes resumption of development in a diapausing recipient. This programming may long precede the diapause and even span a generation, such that maternal conditions can govern the diapause in the developing stages of her offspring.

Many studies have shown endocrine control of diapause, but substantial variation in mechanisms for the regulation of diapause reflects the multiple, independent evolution of this phenomenon. Generally, in diapausing larvae, the production of ecdysteroid moulting hormone from the prothoracic gland ceases, and JH plays a role in termination of diapause. Resumption of ecdysteroid secretion from the prothoracic glands (under the influence of PTTH) appears essential for the termination of pupal diapause. JH is important in diapause regulation in adult insects but, as with the immature stages, may not be the only regulator. In larvae, pupae and adults of *Bombyx mori*, complex antagonistic interactions occur between a **diapause hormone** (DH), originating from paired neurosecretory cells in the suboesophageal ganglion, and JH from the corpora allata. The adult female produces diapause eggs when the ovariole is under the influence of DH, whereas in the absence of this hormone and in the presence of JH, non-diapause eggs are produced. In moths of *Helicoverpa* and *Heliothis* species, pupal diapause can be terminated experimentally either by ecdysteroid or by DH, but the action of DH requires temperatures above a certain threshold. It is postulated that diapause termination of these pupae results from DH and PTTH working together to bring about renewed development associated with diapause termination, with both hormones probably acting on the prothoracic gland.

6.6 DEALING WITH ENVIRONMENTAL EXTREMES

The most obvious environmental variables that confront an insect are seasonal fluctuations in temperature and humidity. The extremes of temperature and humidity experienced by insects in their natural environments span the range of conditions encountered by terrestrial organisms. For reasons of human interest in cryobiology (revivable preservation), responses to extremes of cold and desiccation have been studied better than those to high temperatures alone.

The options available for evading extremes are behavioural avoidance, such as by burrowing into soil of a more equable temperature, migration (section 6.7), diapause (section 6.5), and *in situ* tolerance/survival in a highly altered physiological condition, the topic of the following sections.

6.6.1 Cold

Biologists have long been interested in the occurrence of insects at the extremes of the Earth, in surprising diversity and sometimes in large numbers. Holometabolous insects are abundant in refugial sites within 3° of the North Pole. Fewer insects, notably a chironomid midge and some penguin and seal lice, are found on the Antarctic proper. Freezing, high elevations, including glaciers, sustain resident insects, such as the Himalayan *Diamesa* glacier midge (Diptera: Chironomidae), which sets a record for cold activity, being active at an air temperature of −16°C. Snowfields also support seasonally cold-active insects such as grylloblattids, *Chionea* (Diptera: Tipulidae) and the "snowfleas", *Boreus* spp. (Mecoptera). Low-temperature environments pose physiological problems that resemble dehydration in terms of the reduction of available water, but clearly also include the need to avoid freezing of body fluids. Expansion and ice-crystal formation typically kill mammalian cells and tissues, but perhaps some insect cells can tolerate freezing. Insects may possess one or several of a suite of mechanisms – collectively termed **cryoprotection** – that allows survival of cold extremes. These mechanisms may apply in any life-history stage, from resistant eggs to adults. Although they form a continuum, the following categories can aid understanding.

Freeze tolerance

Freeze-tolerant insects include some of the most cold-hardy species, mainly occurring in Arctic, sub-Arctic and Antarctic locations, which experience the most extreme winter temperatures (e.g. −40 to −80°C). Protection is provided by seasonal production of ice-nucleating agents (INA) under the induction of falling temperatures and prior to onset of severe cold. These proteins, lipoproteins and/or endogenous crystalline substances such as urates, act as sites where (safe) freezing is encouraged outside cells, such as in the haemolymph, gut or Malpighian tubules. Controlled and gentle extra-cellular ice formation acts also to gradually dehydrate cell contents, and thus avoid freezing. In addition, substances such as glycerol and/or related polyols, and sugars including sorbitol and trehalose, allow supercooling (i.e. remaining liquid at subzero temperature without ice formation) and also protect tissues and cells prior to full INA activation and after freezing. Antifreeze proteins may also be produced; these fulfil some of the same protective roles, especially during freezing conditions in autumn/fall and during the spring thaw, outside the core deep-winter freeze. Onset of internal freezing often requires body contact with external ice to trigger ice nucleation, and may take place with little or no internal supercooling. Freeze tolerance does not guarantee survival, which depends not only on the actual minimum temperature experienced but also upon acclimation before cold onset, the rapidity of onset of extreme cold, and perhaps also the range and fluctuation in temperatures experienced during thawing. In the well-studied galling tephritid fly *Eurosta solidaginis*, all these mechanisms have been demonstrated, plus tolerance of cell freezing, at least in fat body cells.

Freeze avoidance

Freeze avoidance describes both a survival strategy and a species' physiological ability to survive low temperatures without internal freezing. In this definition, insects that avoid freezing by supercooling can survive extended periods in the supercooled state, and show high mortality below the supercooling point but little above it, and are freeze avoiders. Mechanisms for encouraging supercooling include evacuation of the digestive system to remove the promoters of ice nucleation, plus pre-winter synthesis of polyols and antifreeze agents. In these insects, cold hardiness (potential to survive cold) can be calculated readily by comparison of the supercooling point (below which death occurs) and the lowest temperature experienced by the insect. Freeze avoidance has been studied in the autumnal moth, *Epirrita autumnata* (Geometridae) and goldenrod gall moth, *Epiblema scudderiana* (Tortricidae).

Chill tolerance

Chill-tolerant species occur mainly from temperate areas polewards, where insects survive frequent encounters with subzero temperatures. This category contains species with extensive supercooling ability (see above) and cold tolerance, but is distinguished from these by mortality that is dependent on duration of cold exposure and low temperature (above the supercooling point), i.e. the longer and the colder the freezing spell, the more deaths are attributable to freezing-induced cellular and tissue damage. A notable ecological grouping that demonstrates high chill tolerance are species that survive extreme cold (lower than supercooling point) by relying on snow cover, which provides "milder" conditions where chill tolerance permits survival. Examples of studied chill-tolerant species include the beech weevil, *Rhynchaenus fagi* (Curculionidae), in the United Kingdom, and the bertha armyworm, *Mamestra configurata* (Noctuidae), in Canada.

Chill susceptibility

Chill-susceptible species lack cold hardiness, and although they may supercool, death is rapid on exposure to subzero temperatures. Such temperate insects tend to vary in summer abundances according to the severity of the preceding winter. Thus, several studied European pest aphids (Hemiptera: Aphididae: *Myzus persicae*, *Sitobion avenae* and *Rhopalosiphum padi*) can supercool to −24°C (adults) or −27°C (nymphs) yet show high mortality when held at subzero temperatures for just a minute or two. Eggs show much greater cold hardiness than nymphs or adults. As overwintering eggs are produced only by sexual (**holocyclic**) species or clones, aphids with this life cycle predominate at increasingly high latitudes in comparison with those in which overwintering is in a nymphal or adult stage (**anholocyclic** species or clones).

Opportunistic survival

Opportunistic survival is observed in insects living in stable, warm climates in which cold hardiness is little developed. Even though supercooling is possible, in species that lack avoidance of cold through diapause or quiescence (section 6.5), mortality occurs when an irreversible lower threshold for metabolism is reached. Survival of predictable or sporadic cold episodes for these species depends upon exploitation of favourable sites, for example by migration (section 6.7) or by local opportunistic selection of appropriate microhabitats.

Clearly, low-temperature tolerance is acquired convergently, with a range of different mechanisms and chemistries involved in different groups. A unifying feature may be that the mechanisms for cryoprotection are rather similar to those shown for avoidance of dehydration, which may be preadaptive for cold tolerance. Although each of the above categories contains a few unrelated species, amongst the terrestrial bembidiine Carabidae (Coleoptera), the Arctic and sub-Arctic regions contain a radiation of cold-tolerant species. A preadaptation to aptery (wing loss) has been suggested for these beetles, as it is too cold to warm flight muscles. Nonetheless, the summer Arctic is plagued by actively flying (and human-biting) flies, which warm themselves by their resting orientation towards the sun.

6.6.2 Heat

The hottest inhabited terrestrial environment – vents in thermally active areas – support a few specialist insects. For example, the hottest waters in thermal springs of Yellowstone National Park are too hot for us to touch, but by selection of slightly cooler microhabitats amongst the cyanobacteria/blue-green algal mats, a brine fly, *Ephydra bruesi* (Ephydridae), can survive at 43°C. At least some other species of ephydrids, stratiomyids and chironomid larvae (all Diptera) tolerate nearly 50°C in Iceland, New Zealand, South America and perhaps other sites where volcanism provides hot-water springs. The other aquatic temperature-tolerant taxa are found principally amongst the Odonata and Coleoptera.

High temperatures tend to kill cells by denaturing proteins, altering membrane and enzyme structures and properties, and by loss of water (dehydration). Inherently, the stability of non-covalent bonds that determine the complex structure of proteins determines the upper limits, but below this threshold there are many different but interrelated temperature-dependent biochemical reactions. **Acclimation**, in which a gradual exposure to increasing (or decreasing) temperatures takes place, certainly provides a greater disposition to survival at extreme temperatures compared with instantaneous exposure. Acclimation conditioning should be considered when comparing effects of temperature on insects.

Options of dealing with high air temperatures include behaviours such as use of a burrow during the hottest times. This activity takes advantage of the buffering of soils, including desert sands, against temperature extremes so that near-stable temperatures occur within a few centimetres of the fluctuations of the exposed surface. Overwintering pupation of temperate insects frequently takes place in a burrow made by a late-instar larva, and in hot, arid areas night-active insects such as predatory carabid beetles may pass the extremes of the day in burrows. Arid-zone ants, including Saharan *Cataglyphis*, Australian *Melophorus* and Namibian *Ocymyrmex*, show several behavioural features to maximize their ability to survive in some of the hottest places on Earth. Long legs hold the body in the cooler air above the substrate, they can run as fast as 1 ms^{-1} and are good navigators, so as to allow rapid return to the burrow. Tolerance of high temperature is an advantage to *Cataglyphis* because they scavenge upon insects that have died from heat stress. However, *Cataglyphis bombycina* suffers predation from a lizard that also has a high temperature tolerance, and predator avoidance restricts the aboveground activity of *Cataglyphis* to a very narrow temperature band, between that at which the lizard ceases activity and its own upper lethal thermal threshold. *Cataglyphis* minimizes exposure to high temperatures using the strategies outlined above, as well as thermal respite activity – climbing and pausing on grass stems above the desert substrate, which may exceed 46°C. Physiologically, *Cataglyphis* may be amongst the most thermally tolerant land animals because they can accumulate high levels of "heat-shock proteins" in advance of their departure to forage from their (cool) burrow to the ambient external heat. The few minutes duration of the foraging frenzy is too short for synthesis of these protective proteins after exposure to the heat.

These "heat-shock proteins" (abbreviated as "hsp") may be best termed stress-induced proteins when they are involved in temperature-related activities, as at least some of the suite can be induced also by desiccation and cold. Their function at higher temperatures appears to be to act as molecular chaperones assisting in protein folding. In cold conditions, protein folding is not the problem, but rather it is loss of membrane fluidity, which can be restored by fatty acid changes and by denaturing of membrane phospholipids, perhaps also under some control of stress proteins.

The most remarkable specialization to extreme conditions involves a larval chironomid midge, *Polypedilum vanderplanki*, which lives in Africa in temporary pools on granite outcrops, such as natural depressions and those formed by grinding grain. As the pools dry out seasonally, larvae that have failed to complete development lose water until they are almost completely dehydrated. In this condition, termed **cryptobiosis** (alive but with all metabolism ceased) or **anhydrobiosis** (alive without water), larvae tolerate environmental extremes, including artificially imposed temperatures in dry air from more than 100°C for 3 h down to −270°C for 77 h. In this state, larvae can survive in a vacuum, under high pressure (1.2 GPa), in 100% ethanol for a week, and can tolerate high levels of irradiation, such that larvae were sent into space for research purposes. On revival, by wetting, the larvae rapidly are restored to the hydrated active state, resume feeding and continue development either until the onset of another cycle of desiccation, or to pupation and adult emergence.

The molecular biochemistry of this phenomenon involves desiccation slow enough to allow production of trehalose by upregulation of synthesis pathways. Larvae that desiccate too rapidly (6 h) do not revive, whereas those that take 48 h to desiccate show 100% revival. Trehalose production takes place in fat cells, triggered by onset of desiccation, and the sugar is transported in the haemolymph and hence into all somatic cells. But this is not enough to protect the larvae; in addition, genes are upregulated to produce enzymes that "mop-up" the free oxygen radicals produced by oxidative stress. Late Embryo Abundant (LEA) proteins are produced that prevent proteins aggregating as water is lost, and also to provide a framework to support the trehalose alone and complexed with proteins. As further water is lost, a process of vitrification takes place to produce a glassy, fully desiccated larva.

During the drying phase, trehalose degrading enzymes are upregulated, but they are only activated during rehydration. Likewise DNA repair enzymes are produced prior to completion of desiccation, but also are activated only during revival. Evidently, during the vitrified stage, DNA is not truly intact but is highly protected, with necessary repair when anhydrobiosis ends.

6.6.3 Aridity

In terrestrial environments, temperature and humidity are intimately linked, and responses to high temperatures are inseparable from concomitant water stress. Although free water may be unavailable in the arid tropics for long periods, many insects are active year-round in places such as the Namib Desert, an essentially rain-free desert in southwest Africa. This desert has provided a research environment for the study of water relations in arid-zone insects ever since the discovery of "fog basking" amongst some tenebrionid beetles. The cold oceanic current that abuts the hot Namib Desert produces daily fog that sweeps inland. This provides a source of aerial moisture that can be precipitated onto the bodies of beetles that present a head-down stance on the slip face of sand dunes, facing the fog-laden wind. The precipitated moisture then runs to the mouth of the beetle. Such atmospheric water gathering is just one of a range of insect behaviours and morphologies that allow survival under these stressful conditions. Two different strategies exemplified by different beetles can be compared and contrasted: in detritivorous tenebrionids and in predatory carabids, both of which have many aridity-tolerant species.

The greatest water loss by most insects occurs via evaporation from the cuticle, with lesser amounts lost through gas exchange at the spiracles and through excretion. Some arid-zone beetles have reduced their water loss 100-fold by one or more strategies, including extreme reduction in evaporative water loss through the cuticle (section 2.1), reduction in spiracular water loss, reduction in metabolism, and extreme reduction of excretory loss. In the studied arid-zone species of tenebrionids and carabids, cuticular water permeability is reduced to almost zero, such that water loss is virtually a function of metabolic rate alone – i.e. loss is by the gas-exchange pathway, and is predominantly related to variation in the local humidity around the spiracles. Enclosure of the spiracles in a humid subelytral space is an important mechanism for reduction of such losses. Observation of unusually low levels of sodium in the haemolymph of studied tenebrionids compared with levels in arid-zone carabids (and most other insects) implies reduced sodium pump activity, reduced sodium gradient across cell membranes, a concomitantly inferred reduction in metabolic rate, and reduced respiratory water loss. Uric acid precipitation when water is reabsorbed from the rectum allows the excretion of virtually dry urine (section 3.7.2), which, with retention of free amino acids, minimizes loss of everything except the nitrogenous wastes. All these mechanisms allow the survival of a tenebrionid beetle in an arid environment with seasonal food and water shortage. In contrast, desert carabids include species that maintain a high sodium pump activity and sodium gradient across cell membranes, implying a high metabolic rate. They also excrete more-dilute urine, and appear less able to conserve free amino acids. Behaviourally, carabids are active predators, requiring a high metabolic rate for pursuit, which would incur greater rates of water loss. This may be compensated for by the higher water content of their prey, compared with the desiccated detritus that forms the tenebrionid diet.

To test if these distinctions are different "adaptive" strategies, or if tenebrionids differ more generally from carabids in their physiology, irrespective of any arid tolerance, will require wider sampling of taxa, and some appropriate tests to determine whether the observed physiological differences are correlated with taxonomic relationships (i.e. are preadaptive for life in low-humidity environments) or ecology of the species.

6.7 MIGRATION

Diapause, as described in section 6.5, allows an insect to track its resources in time – when conditions become inclement, development ceases until diapause breaks. An alternative to shutdown is to track resources in space by directed movement. The term **migration** once was restricted to the to-and-fro major movements of vertebrates, such as wildebeest, salmonid fish and migratory birds including swallows, shorebirds and maritime terns. However, there are good reasons to expand this to include organisms that fulfil some or all of the following criteria, in and around specific phases of movement:

- persistent movement away from an original home range;
- relatively straight movement in comparison with station-tending or zig-zagging within a home range;
- undistracted by (unresponsive to) stimuli from home range;
- distinctive pre- and post-movement behaviours;
- reallocation of energy within the body.

All migrations in this wider sense are attempts to provide a homogeneous suitable environment despite temporal fluctuations in a single home range. Criteria such as length of distance travelled, geographical area in which migration occurs, and whether or not the outward-bound individual undertakes a return are unimportant to this definition. Furthermore, thinning out of a population (dispersal) or advance across a similar habitat (range extension) are not migration. According to this definition, seasonal movements from the upper mountain slopes of the Sierra Nevada down to California's Central Valley by the convergent ladybird beetle (*Hippodamia convergens*) is as much a migratory activity as is a transcontinental movement of a monarch butterfly (*Danaus plexippus*).

Pre-migration behaviours in insects include redirecting metabolism to energy storage, cessation of reproduction, and production of wings in polymorphic species in which winged and wingless forms coexist (polyphenism; section 6.8.2). Feeding and reproduction are resumed post-migration. Some responses are under hormonal control, whereas others are environmentally induced. Evidently, pre-migration changes must anticipate the altered environmental conditions that migration has evolved to avoid. As with induction of diapause (section 6.5), the principal cue is change in day length (photoperiod). A strong linkage exists between the several cues for onset and termination of reproductive diapause and induction and cessation of migratory response in studied species, including monarch butterflies and milkweed bugs (*Oncopeltus fasciatus*). Individuals of both species migrate south from their extensive range associated with North American host milkweed plants (Apocynaceae). At least in the migrant generation of monarchs, a magnetic compass complements solar navigation in deriving the bearings towards the overwintering site. Shortening day length induces a reproductive diapause in which flight inhibition is removed and energy is transferred to flight instead of reproduction. The overwintering generation of both species is in diapause, which ends with a two- (or more) stage migration from south to north that essentially tracks the sequential development of subtropical to temperate annual milkweeds as far as southern Canada. The first flight in early spring from the overwintering area is short, with both reproduction and flight effort occurring during short-length days, but the next generation extends far northwards in longer days, either as individuals or by consecutive generations. Few if any of the returning individuals are the original outward migrants. In the milkweed bugs, there is a circadian rhythm (see Box 4.4), with oviposition and migration temporally segregated in the middle of the day, and mating and feeding concentrated at the end of the daylight period. Both milkweed bugs and monarch butterflies have non-migratory multivoltine relatives that remain in the tropics. Therefore, it seems that the ability to diapause and thus escape south in the autumn/fall has allowed just these two species to invade summer milkweed stands of the temperate region.

It is a common observation that insects living in "temporary" habitats of limited duration have a higher proportion of flighted species, and within polymorphic taxa, more flighted individuals. In longer-lasting habitats, loss of flight ability, either permanently or temporarily, is more common. Thus, amongst European water-striders (Hemiptera: Gerridae), species associated with small ephemeral water bodies are winged and regularly migrate to seek new water bodies; those associated with large lakes tend to winglessness and sedentary life histories. Evidently, flightedness relates to the tendency (and ability) to migrate in locusts, as exemplified in *Chortoicetes terminifera* (the Australian migratory locust) and *Locusta migratoria* (Orthoptera: Acrididae), which demonstrate adaptive migration to exploit transient favourable conditions in arid regions (see section 6.10.5 for *L. migratoria* behaviour).

Although such massed movements described above are very conspicuous, even the "passive dispersal" of small and lightweight insects can fulfil many of the criteria of migration. Thus, even reliance upon wind (or water) currents for movement may involve the insect being capable of any or all of the following:

- changing their behaviour to enable them to embark, such as young scale insects crawling to a leaf apex and adopting a posture there to enhance the chances of extended aerial movement;
- being in appropriate physiological and developmental condition for the journey, as in the flighted stage of otherwise apterous aphids;
- sensing appropriate environmental cues to depart, such as seasonal failure of the host plant of many aphids;
- recognizing environmental cues on arrival, such as odours or colours of a new host plant, and making controlled departure from the current environment.

Naturally, embarkation on such journeys does not always bring success and there are many strandings of migratory insects in unsuitable habitat, such as on ice fields and in open oceans. Nonetheless, clearly some fecund insects make use of predictable meteorological conditions to make long journeys in a consistent direction, depart from the air current and establish in a suitable, novel habitat. Aphids are a prime example, but certain thrips and scale insects and other agriculturally damaging pests can locate new host plants in this way.

6.8 POLYMORPHISM AND POLYPHENISM

The existence of several generations per year often is associated with morphological change between generations. Similar variation may occur contemporaneously within a population, such as the existence simultaneously of both winged and flightless forms ("**morphs**"). Sexual differences between males and females, and the existence of strong differentiation in social insects such as ants and bees, are further obvious examples of the phenomenon. The term **polymorphism** encompasses all such discontinuities, which occur in the same life-history phase at a frequency greater than might be expected from repeated mutation alone. It is defined as the simultaneous or recurrent occurrence of distinct morphological differences, reflecting and often including physiological, behavioural and/or ecological differences among conspecific individuals.

6.8.1 Genetic polymorphism

The distinction between the sexes is an example of a particular polymorphism, namely sexual dimorphism, which in insects is almost totally under genetic determination. Environmental factors may affect sexual expression, as in castes of some social insects or in feminization of genetically male insects by mermithid nematode infections. Aside from the dimorphism of the sexes, different genotypes may co-occur within a single species, maintained by natural selection at specific frequencies that vary from place to place and time to time throughout the range. For example, adults of some gerrid bugs are fully winged and capable of flight, whereas other coexisting individuals of the same species are brachypterous and cannot fly. Intermediates are at a selective disadvantage and the two genetically determined morphs coexist in a balanced polymorphism. Some of the most complex, genetically based, polymorphisms have been discovered in butterflies that mimic chemically protected butterflies of another species (the **model**) for purposes of defence from predators (section 14.5). Some butterfly species may mimic more than one model and, in these species, the accuracy of the several distinct **mimicry** patterns is maintained because inappropriate intermediates are not recognized by predators as being distasteful and are eaten. Mimetic polymorphism predominantly is restricted to the females, with the males generally monomorphic and non-mimetic. The basis for the switching between the different mimetic morphs is relatively simple Mendelian genetics, which may involve relatively few genes or supergenes.

It is a common observation that some individual species with a wide range of latitudinal distributions show different life-history strategies according to location. For example, populations living at high latitudes (nearer the pole) or high elevation may be univoltine, with a long dormant period, whereas populations nearer the equator or lower in elevation may be multivoltine, and develop continuously, without dormancy. Dormancy is environmentally induced (section 6.5 and section 6.10.2), but the ability of the insect to recognize and respond to these cues is programmed genetically. In addition, at least some geographical variation in life histories is the result of genetic polymorphism.

6.8.2 Environmental polymorphism, or polyphenism

A phenotypic difference between generations that lacks a genetic basis and is determined entirely by the environment often is termed **polyphenism**. The expression of a particular phenotype depends upon one or more genes that are triggered by an environmental cue. An example is the temperate to tropical Old World butterfly *Eurema hecabe* (Lepidoptera: Pieridae), which shows a seasonal change in wing colour between summer and autumn/fall morphs. Photoperiod induces morph change, with a dark-winged summer morph induced by a long day of greater than 13 h. A short day of less than 12 h induces the paler-winged autumn/fall morph, particularly at temperatures of under 20°C, with temperature affecting males more than females. A second example is the colour polyphenism of the caterpillars of the American peppered moth *Biston betularia cognataria* (Lepidoptera: Geometridae), which are generalist herbivores with a body colour that varies to match the green or brown twigs of their host plants. Experiments have shown that the caterpillars change the pigments in their epidermal cells in response primarily to their visual experience, rather than to their diet. This larval colour polyphenism is not related to the genetic polymorphism for melanic forms seen in adult moths (see Box 14.1). In another genus of geometrid moths, caterpillars of *Nemoria arizonaria* of the spring generation feed on and develop to resemble oak catkins (flowers), whereas the summer-generation caterpillars feed on leaves and resemble oak twigs. These catkin and twig morphs are induced solely by the larval diet.

Amongst the most complex polyphenisms are those seen in the aphids (Hemiptera: Aphidoidea). Within parthenogenetic lineages (i.e. those in which there is absolute genetic identity), the females may show up to eight distinct phenotypes, in addition to polymorphisms in sexual forms. These female aphids may vary in morphology, physiology, fecundity, offspring timing and size, development time, longevity, and host-plant choice and utilization. Environmental cues responsible for alternative morphs are similar to those that govern diapause and migration in many insects (section 6.5 and section 6.7), including photoperiod, temperature and maternal effects, such as elapsed time (rather than number of generations) since the winged founding mother. Overcrowding triggers many aphid species to produce a winged dispersive phase. Crowding also is responsible for one of the most dramatic examples of polyphenism, the phase transformation from the solitary young locusts (hoppers) to the gregarious phase (section 6.10.5). Studies on the physiological mechanisms that link environmental cues to these phenotype changes have implicated JH in many aphid morph shifts.

If aphids show the greatest number of polyphenisms, the social insects are a close second, and undoubtedly have a greater degree of morphological differentiation between morphs, termed **castes**. This is discussed in more detail in Chapter 12; suffice it to say that maintenance of the phenotypic differences between castes as different as queens, workers and soldiers includes physiological mechanisms such as pheromones transferred with food, olfactory and tactile stimuli, and endocrine control including JH and ecdysone. Superimposed on these polyphenisms are the dimorphic differences between the sexes, which impose some limits on variation.

6.9 AGE-GRADING

Identification of the growth stages or ages of insects in a population is important in ecological or applied entomology. Information on the proportion of a population in different developmental stages and the proportion of the adult population at reproductive maturity can be used to construct time-specific life-tables or budgets to determine factors that cause and regulate fluctuations in population size and dispersal rate, and to monitor fecundity and mortality factors in the population. Such data are integral to predictions of pest outbreaks as a result of climate, and to the construction of models of population response to the introduction of a control programme.

Many different techniques have been proposed for estimating either the growth stage or the age of insects. Some provide an estimate of chronological (calendar) age within a stadium, whereas most estimate either instar number or relative age within a stadium, in which case the term **age-grading** is used in place of age determination.

6.9.1 Age-grading of immature insects

For many population studies it is important to know the number of larval or nymphal instars in a species and to be able to recognize the instar to which any immature individual belongs. Generally, such information is available or its acquisition is feasible for species with a constant and relatively small number of immature instars, especially those with a lifespan of a few months or less. However, it is more difficult to obtain such data for species with either many or a variable number of instars, or with overlapping generations. The latter situation may occur in species with many asynchronous generations per year or in species with a life cycle of longer than one year. In some species, there are readily discernible qualitative (e.g. colour) or meristic (e.g. antennal segment number) differences between consecutive immature instars. More frequently, the only obvious difference between successive larval or nymphal instars is the increase in size that occurs after each moult (the moult increment). Thus, it should be possible to determine the actual number of instars in the life history of a species from a frequency histogram of measurements of a sclerotized body part (Fig. 6.12).

Entomologists have sought to quantify this size progression for a range of insects. One of the earliest attempts was that of H.G. Dyar, who in 1890 established a "rule" from observations on the caterpillars of 28 species of Lepidoptera. Dyar's measurements

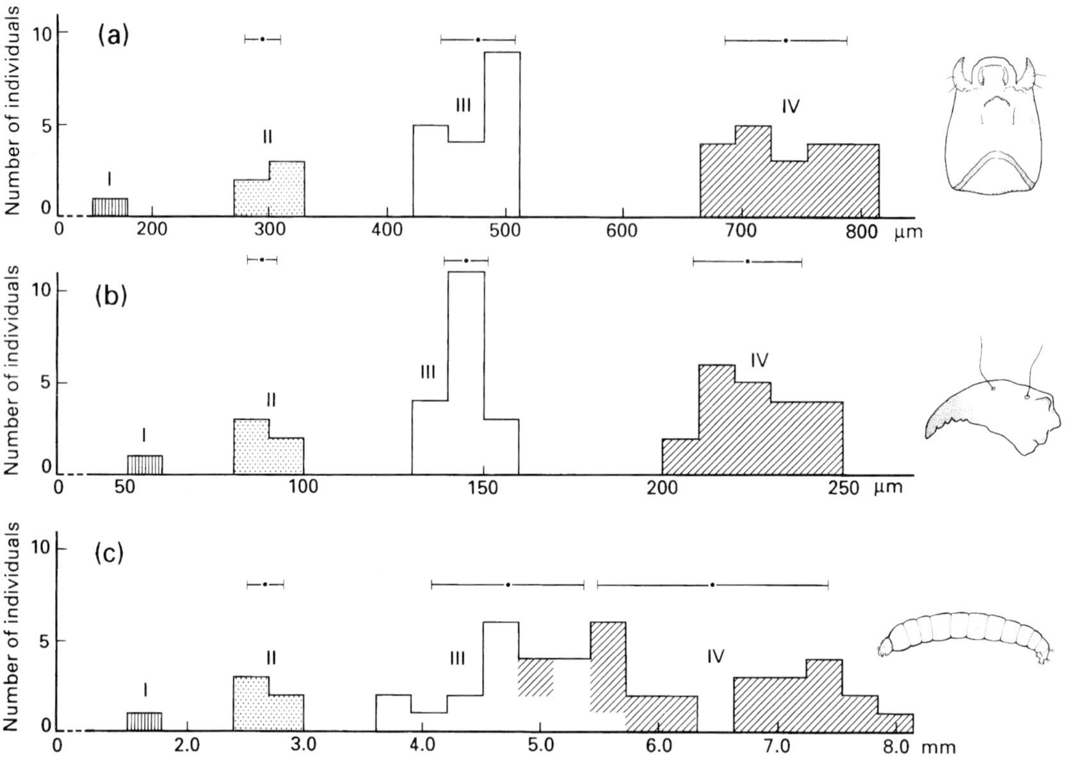

Fig. 6.12 Growth and development in a marine midge, *Telmatogeton* (Diptera: Chironomidae), showing increases in: (a) head capsule length; (b) mandible length; and (c) body length, between the four larval instars (I–IV). The dots and horizontal lines above each histogram represent the means and standard deviations of measurements for each instar. Note that the lengths of the sclerotized head and mandible fall into discrete size classes representing each instar, whereas body length is an unreliable indicator of instar number, especially for separating the third- and fourth-instar larvae.

showed that the width of the head capsule increased in a regular linear progression in successive instars by a ratio (range 1.3–1.7) that was constant for a given species. **Dyar's rule** states that:

post-moult size/pre-moult size (or moult increment) = constant

Thus, if logarithms of measurements of some sclerotized body part in different instars are plotted against the instar number, a straight line should result; any deviation from a straight line indicates a missing instar. In practice, however, there are many departures from Dyar's rule, as the progression factor is not always constant, especially in field populations subject to variable conditions of food and temperature during growth.

A related empirical "law" of growth is Przibram's rule, which states that an insect's weight is doubled during each instar, and at each moult all linear dimensions are increased by a ratio of 1.26. The growth of most insects shows no general agreement with this rule, which assumes that the dimensions of a part of the insect body should increase at each moult by the same ratio as the body as a whole. In reality, growth in most insects is **allometric**, i.e. the parts grow at rates peculiar to themselves, and often are very different from the growth rate of the body as a whole. The horned adornments on the head and thorax of *Onthophagus* dung beetles discussed in section 5.3 exemplify some trade-offs associated with allometric growth.

6.9.2 Age-grading of adult insects

The age of an adult insect is not determined easily. However, adult age is of great significance, particularly in the insect vectors of disease. For instance, it is crucial to epidemiology that the age (longevity) of an adult female mosquito be known, as this relates to the number of blood meals taken and therefore the number of opportunities for pathogen transmission (e.g. see section 15.3.1). Most techniques for assessing the age of adult insects estimate relative (not chronological) age and hence age-grading is the appropriate term.

Four general categories of age assessment have been used or proposed, relating to:

1 age-related changes in physiology and morphology of the reproductive system;

2 changes in somatic structures;

3 external wear and tear;

4 changes in gene expression (transcriptional profiling).

The third approach has proved unreliable, but the first two methods have wide applicability, and the fourth is in the development phase.

In the first method, age is graded according to reproductive physiology in a technique applicable only to females. Examination of an ovary of a **parous** insect (one that has laid at least one egg) shows that evidence remains after each egg is laid (or even resorbed), in the form of a **follicular relic** that denotes an irreversible change in the epithelium. The deposition of each egg, together with contraction of the previously distended membrane, leaves one follicular relic per egg. The actual shape and form of the follicular relic varies between species, but one or more residual dilations of the lumen, with or without pigment or granules, is common in the Diptera. Females that have no follicular relic have not developed an egg and are termed **nulliparous**.

Counting follicular relics can give a comparative measure of the physiological age of a female insect, for example allowing discrimination of parous from nulliparous individuals, and often allowing further segregation within parous individuals according to the number of ovipositions. The chronological age can be calculated if the time between successive ovipositions (the **ovarian cycle**) is known. However, in many medically significant flies in which there is one ovarian cycle per blood meal, physiological age (number of cycles) is more significant than the precise chronological age.

The second generally applicable method of age determination has a more direct relationship with chronology, and most of the somatic features that allow age estimation are present in both sexes. Estimates of age can be made from measures of cuticle growth, fluorescent pigments, fat body size, cuticular hardness or composition and, in females only, colour and/or patterning of the abdomen. Cuticular growth estimates of age rely upon there being a daily rhythm of deposition of the endocuticle. In hemimetabolous insects, cuticular layers are more reliable, whereas in holometabolous ones, the apodemes (internal skeletal projections upon which muscles attach) are more dependable. The daily layers are most distinctive when the temperature for cuticle formation is not attained for part of each day. This use of growth rings is confounded by development temperatures too cold for deposition, or too high for the daily cycle of deposition and cessation. A further drawback to the technique is that deposition ceases after a certain age is attained, perhaps only 10–15 days after eclosion. Physiological age can be determined by measuring the pigments that accumulate in the ageing cells of many animals, including insects. These pigments fluoresce and can be studied by fluorescence microscopy. Lipofuscin from post-mitotic cells in most body tissues, and pteridine eye pigments, have been measured in this way, especially in flies. Near-infrared reflectance spectroscopy (NIRS) is a non-destructive method used to measure the near-infrared energy absorbed at specific wavelengths by biological material. It has not been widely used, but allows the relative age of young and old females of *Anopheles* mosquitoes to be predicted with about 80% accuracy.

A new approach, using transcriptional profiles, has been developed to age female mosquitoes. Assay of genes that display age-related transcription patterns have been shown to allow age prediction of females of *Aedes aegypti* and *Anopheles gambiae*.

6.10 ENVIRONMENTAL EFFECTS ON DEVELOPMENT

The rate or manner of insect development or growth may depend upon a number of factors. These include the type and amount of food, the amount of moisture (for terrestrial species) and heat (measured as temperature), or the presence of environmental signals (e.g. photoperiod), mutagens and toxins, or other organisms, either predators or competitors. Two or more of these factors may interact to complicate interpretation of growth characteristics and patterns.

6.10.1 Temperature

Most insects are **poikilothermic**, that is with body temperature more or less directly varying with environmental temperature, and thus heat drives the rate of growth and development when food is unlimited. A rise in temperature, within a favourable range, will speed up the metabolism of an insect ,and consequently increase its rate of development. Each species and each stage in the life history may develop at its own rate in relation to temperature. Thus, **physiological time**, a measure of the amount of heat required over time for an insect to complete development or a stage of development, is more meaningful as a measure of development time than age in calendar time. Knowledge of temperature–development relationships and the use of physiological time allow comparison of the life cycles and/or fecundity of pest species in the same system (Fig. 6.13), and prediction of the larval feeding periods, generation length and time of adult emergence under variable temperature conditions that exist in the field. Such predictions are especially important for pest insects, as control measures must be timed carefully in order to be effective.

Physiological time is the cumulative product of total development time (in hours or days) multiplied by the temperature (in degrees) above the **developmental (or growth) threshold**, or the temperature below which no development occurs. Thus, physiological time is commonly expressed as **day-degrees** (also degree-days) (D°) or hour-degrees (h°). Normally, physiological time is estimated for a species by rearing a number of individuals of the life-history stage(s) of interest under different constant temperatures in several identical growth cabinets. The developmental threshold is estimated by the linear regression x-axis method, as outlined in Box 6.2, although more

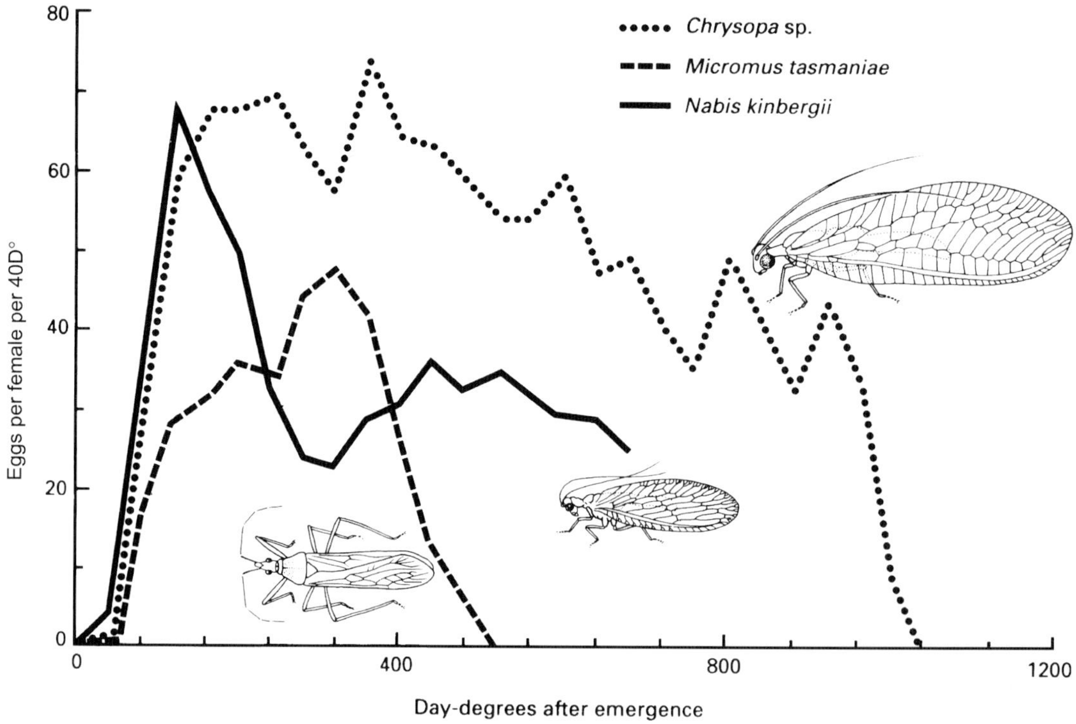

Fig. 6.13 Age-specific oviposition rates of three predators of cotton pests, *Chrysopa* sp. (Neuroptera: Chrysopidae), *Micromus tasmaniae* (Neuroptera: Hemerobiidae) and *Nabis kinbergii* (Hemiptera: Nabidae), based on physiological time above respective development thresholds of 10.5°C, −2.9°C and 11.3°C. D°, day-degrees. (After Samson & Blood 1979.)

Box 6.2 Calculation of day-degrees

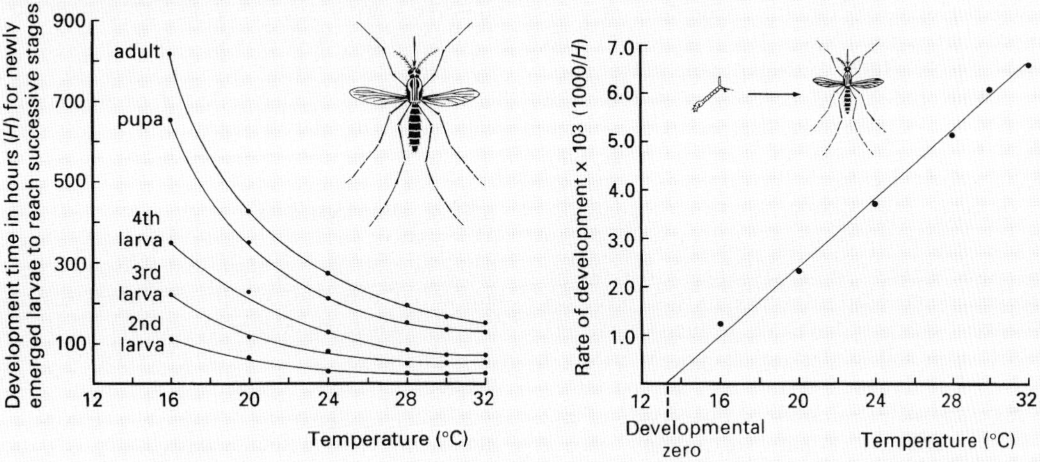

Day-degrees (or degree-days) can be estimated simply (after Daly *et al.* 1978), as exemplified by data on the relationship between temperature and development in the yellow-fever mosquito, *Aedes aegypti* (Diptera: Culicidae) (after Bar-Zeev 1958).

1 In the laboratory, establish the average time required for each stage to develop at different constant temperatures. The graph on the left of the figure shows the time in hours (H) for newly hatched larvae of *Ae. aegypti* to reach successive stages of development when incubated at various temperatures.

2 Plot the reciprocal of development time ($1/H$), the development rate, against temperature to obtain a sigmoid curve with the middle part of the curve approximately linear. The graph on the right shows the linear part of this relationship for the total development of *Ae. aegypti* from the newly hatched larva to the adult stage. A straight line would not be obtained if extreme development temperatures (e.g. higher than 32°C or lower than 16°C) had been included.

3 Fit a linear regression line to the points and calculate the slope of this line. The slope represents the amount in hours by which development rates are increased for each 1 degree of increased temperature. Hence, the reciprocal of the slope gives the number of hour-degrees, above threshold, required to complete development.

4 To estimate the developmental threshold, the regression line is projected to the X-axis (abscissa) to give the developmental zero, which in the case of *Ae. aegypti* is 13.3°C. This zero value may differ slightly from the actual developmental threshold determined experimentally, probably because at low (or high) temperatures the temperature–development relationship is rarely linear. For *Ae. aegypti*, the developmental threshold actually lies between 9 and 10°C.

5 The equation of the regression line is $1/H = k(T^{\circ} - T^{t})$, where H is the development period, T° is temperature, T^{t} is development threshold temperature, and k is the slope of the line.

Thus, the physiological time for development is $H(T^{\circ} - T^{t}) = 1/k$ hour-degrees, or $H(T^{\circ} - T^{t})/24 = 1/k = K$ day-degrees, where K is the thermal constant, or K-value.

By inserting the values of H, T° and T^{t} for the data from *Ae. aegypti* in the equation given above, the value of K can be calculated for each of the experimental temperatures from 14 to 36°C:

Temperature (°C)	14	16	20	24	28	30	32	34	36
K	1008	2211	2834	2921	2866	2755	2861	3415	3882

Thus, the K-value for *Ae. aegypti* is approximately independent of temperature, except at extremes (14 and 34–36°C), and averages about 2740 hour-degrees (or degree-hours) or 114 day-degrees (or degree-days) between 16 and 32°C.

accurate threshold estimates can be obtained by more time-consuming methods.

In practice, the application of laboratory-estimated physiological time to natural populations may be complicated by several factors. Under fluctuating temperatures, especially if the insects experience extremes, growth may be retarded or accelerated compared with the same number of day-degrees under constant temperatures. Furthermore, the temperatures actually experienced by the insects, in their often sheltered microhabitats on plants or in soil or litter, may be several degrees different from the temperatures recorded at a meteorological station even just a few metres away. Insects may select microhabitats that ameliorate cold night conditions or reduce or increase daytime heat. Thus, predictions of insect life-cycle events based on extrapolation from laboratory to field temperature records may be inaccurate. For these reasons, the laboratory estimates of physiological time should be corroborated by calculating the hour-degrees or day-degrees required for development under more natural conditions, but using the laboratory-estimated developmental threshold, as follows.

1 Place newly laid eggs or newly hatched larvae in their appropriate field habitat and record temperature each hour (or calculate a daily average – a less accurate method).

2 Estimate the time for completion of each instar by discarding all temperature readings below the developmental threshold of the instar and subtracting the developmental threshold from all other readings to determine the effective temperature for each hour (or simply subtract the development threshold temperature from the daily average temperature). Sum the degrees of effective temperature for each hour from the beginning to the end of the stadium. This procedure is called thermal summation.

3 Compare the field-estimated number of hour-degrees (or day-degrees) for each instar with that predicted from the laboratory data. If there are discrepancies, then microhabitat and/or fluctuating temperatures may be influencing insect development, or the developmental zero read from the graph may be a poor estimate of the developmental threshold.

Another problem with laboratory estimation of physiological time is that insect populations maintained for lengthy periods under laboratory conditions frequently undergo acclimation to constant conditions, or even genetic change in response to the altered environment or as a result of population reductions that produce genetic "bottle-necks". Therefore, insects maintained in rearing cages may exhibit different temperature–development relationships from individuals of the same species in wild populations.

For all of the above reasons, any formula or model that purports to predict insect response to environmental conditions must be tested carefully for its fit with natural population responses.

6.10.2 Photoperiod

Many insects, perhaps most, do not develop continuously all year round, but avoid some seasonally adverse conditions by a resting period (section 6.5) or migration (section 6.7). Summer dormancy (aestivation) and winter dormancy (hibernation) provide two examples of avoidance of seasonal extremes. The most predictable environmental indicator of changing seasons is **photoperiod** – the length of the daily light phase or, more simply, day length. Near the equator, although sunrise to sunset of the longest day may be only a few minutes longer than on the shortest day, if the period of twilight is included then total day length shows more marked seasonal change. The photoperiod response is to duration rather than intensity and there is a critical threshold intensity of light below which the insect does not respond; this threshold is often as dim as twilight, but rarely as low as bright moonlight. Many insects appear to measure the duration of the light phase in the 24 h period, and some have been shown experimentally to measure the duration of dark. Others recognize long days by light falling within the "dark" half of the day.

Most insects can be described as "long-day" species, with growth and reproduction in summer, and with dormancy commencing with decreasing day length. Others show the reverse pattern, with "short-day" (often autumn/fall and spring) activity and summer aestivation. In some species, the life-history stage in which photoperiod is assessed is in advance of the stage that reacts, as is the case when the photoperiodic response of the maternal generation of silkworms affects the eggs of the next generation.

The ability of insects to recognize seasonal photoperiod and other environmental cues requires some means of measuring time between the cue and the subsequent onset or cessation of diapause. This is achieved through a "biological clock" (see Box 4.4), which may be driven by internal (endogenous) or external (exogenous) daily cycles, called **circadian rhythms**. Interactions between the short time periodicity of circadian rhythms and longer-term seasonal rhythms, such as photoperiod recognition, are complex and diverse, and have probably evolved many times within the insects.

6.10.3 Humidity

The high surface area : volume ratio of insects means that loss of body water is a serious hazard in a terrestrial environment, especially a dry one. Low moisture content of the air can affect the physiology and thus the development, longevity and oviposition of many insects. Air holds more water vapour at high than at low temperatures. The relative humidity (RH) at a particular temperature is the ratio of actual water vapour present to that necessary for saturation of the air at that temperature. At low relative humidity, development may be retarded, for example in many pests of stored products; but at high relative humidity or in saturated air (100% RH), insects or their eggs may drown or be infected more readily by pathogens. The fact that stadia may be greatly lengthened by unfavourable humidity has serious implications for estimates of development times, whether calendar or physiological time is used. The complicating effects of low, and sometimes even high, air moisture levels should be taken into account when gathering such data.

6.10.4 Mutagens and toxins

Stressful conditions induced by toxic or mutagenic chemicals may affect insect growth and form to varying degrees, ranging from death at one extreme to slight phenotypic modifications at the other end of the spectrum. Some life-history stages may be more sensitive to mutagens or toxins than others, and phenotypic effects may not be measured easily by crude estimates of stress, such as percentage survival. One measure of the amount of genetic or environmental stress experienced by insects during development is the incidence of **fluctuating asymmetry**, or the quantitative differences between the left and right sides of each individual in a sample of the population. Insects should be bilaterally symmetrical, with left and right halves of their bodies as mirror images – except for obvious differences in structures such as the genitalia of some male insects. Under stressful (unstable) developmental conditions, the degree of asymmetry tends to increase. However, the underlying developmental genetics is poorly understood, and the technique is unreliable, with doubts having been raised about interpretation, such as variation in asymmetry response between different organ systems measured.

6.10.5 Biotic effects

In most insect orders, adult size has a strong genetic component, and growth is strongly determinate. In many Lepidoptera, for example, final adult size is relatively constant within a species; reduction in food quality or availability delays caterpillar growth rather than causing reduced final adult size, although there are exceptions. In contrast, in flies that have limited or ephemeral larval resources, such as a dung pat/pad or temporary pool, cessation of larval growth would result in death as the habitat shrinks. Thus, larval crowding and/or limitation of food supply tend to shorten development time and reduce final adult size. In some mosquitoes and midges, success in short-lived pool habitats is attained by a small proportion of the larval population developing with extreme rapidity relative to their slower-growing siblings. In paedogenetic gall midges (section 5.10.1), crowding with reduced food supply terminates larva-only reproductive cycles and induces the production of adults, allowing dispersal to more favourable habitats.

Food quality appears important in all these cases, but there may be related effects, for example as a result of crowding. Clearly, it can be difficult to segregate out food effects from other potentially limiting factors. In the California red scale, *Aonidiella aurantii* (Hemiptera: Diaspididae), development and reproduction on orange trees is fastest on fruit, intermediate on twigs, and slowest on leaves. Although these differences may reflect differing nutritional status, a microclimatic explanation cannot be excluded, as fruit may retain heat longer than do stems and leaves, and such slight temperature differences might affect the development of the insects.

The effects of crowding on development are well understood in some insects, as in locusts in which two extreme phases, termed **solitary** and **gregarious** (Fig. 6.14), differ in morphometrics, colour and behaviour. At low densities, locusts develop into the solitary phase, with a characteristic uniform-coloured

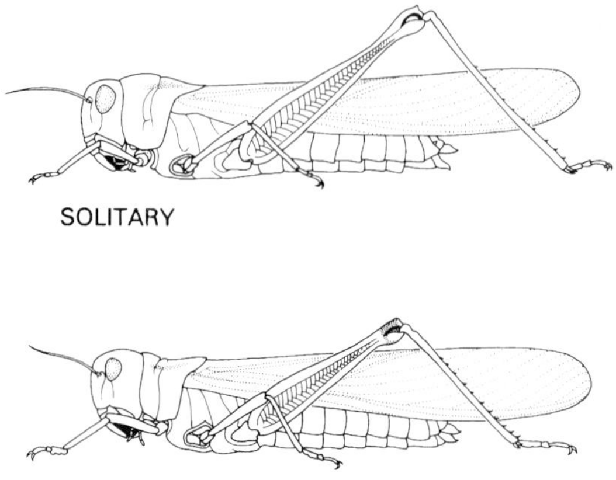

Fig. 6.14 Solitary and gregarious females of the migratory locust, *Locusta migratoria* (Orthoptera: Acrididae). The solitaria adults have a pronounced pronotal crest and the femora are larger relative to the body and wing than in the gregaria adults. Intermediate morphologies occur in the transiens (transient stage), during the transformation from solitaria to gregaria or the reverse.

"hopper" (nymph), and large-sized adult with large hind femora. As densities increase, induced in nature by high survivorship of eggs and young nymphs under favourable climatic conditions, graded changes occur and a darker-striped nymph develops into a smaller locust with shorter hind femora. The most conspicuous difference is behavioural, with more solitary individuals shunning each other's company but making concerted nocturnal migratory movements that result eventually in aggregations in one or a few places of gregarious individuals, which tend to form enormous and mobile swarms. The behavioural shift is induced by crowding, as can be shown by splitting a single locust egg pod into two: rearing the offspring at low densities induces solitary locusts, whereas their siblings reared under crowded conditions develop into gregarious locusts. The response to high population density results from the integration of several cues, including the sight, odour and touch of conspecifics (perhaps mainly via cuticular hydrocarbons acting as contact primer pheromones and detected by the antennae), which lead to endocrine and neuroendocrine (ecdysteroid) changes associated with developmental transformation.

Under certain circumstances, biotic effects can override growth factors. Across much of the eastern United States, 13- and 17-year periodical cicadas (*Magicicada* spp.) emerge highly synchronously. At any given time, nymphal cicadas are of various sizes and in different instars, according to the nutrition they have obtained from feeding on the xylem sap from roots of a variety of trees. Whatever their growth condition, after the elapse of 13 or 17 years since the previous emergence and egg-laying, the final moult of all nymphs prepares them for synchronous emergence as adults. In a very clever experiment, host plants were induced to flush new foliage and flowers twice in one year, inducing adult cicada emergence one year early compared to controls on the roots of single-flushing trees. This implies that synchronized timing for cicadas depends on an ability to "count off" annual events – the predictable flush of sap with the passing of each spring once a year (except when experimenters manipulate it!).

FURTHER READING

Binnington, K. & Retnakaran, A. (eds) (1991) *Physiology of the Insect Epidermis*. CSIRO Publications, Melbourne.

Carroll, S.B. (2008) Evo-devo and an expanding evolutionary synthesis: a genetic theory of morphological evolution. *Cell* **134**, 25–36.

Chown, S.L. & Terblanche, J.S. (2006) Physiological diversity in insects: ecological and evolutionary contexts. *Advances in Insect Physiology* **33**, 50–152.

Daly, H.V. (1985) Insect morphometrics. *Annual Review of Entomology* **30**, 415–38.

Danks, H.V. (ed.) (1994) *Insect Life Cycle Polymorphism: Theory, Evolution and Ecological Consequences for Seasonality and Diapause Control*. Kluwer Academic, Dordrecht.

Dingle, H. (2002) Hormonal mediation of insect life histories. In: *Hormones, Brain and Behavior*, Vol. **3** (eds D.W. Pfaff, A.P. Arnold, S.E. Fahrbach, A.M. Etgen & R.T. Rubin), pp. 237–79. Academic Press, San Diego, CA.

Gilbert, L.I. (ed.) (2009) *Insect Development: Morphogenesis, Molting and Metamorphosis*. Academic Press, London.

Hayes, E.J. & Wall, R. (1999) Age-grading adult insects: a review of techniques. *Physiological Entomology* **24**, 1–10.

Heffer, A. & Pick, L. (2013) Conservation and variation in *Hox* genes: how insect models pioneered the evo-devo field. *Annual Review of Entomology* **58**, 161–79.

Heming, B.-S. (2003) *Insect Development and Evolution*. Cornell University Press, Ithaca, NY.

Karban, R., Black, C.A. & Weinbaum, S.A. (2000) How 17-year cicadas keep track of time. *Ecology Letters* **3**, 253–6.

Minelli, A. & Fusco, G. (2013) Arthropod post-embryonic development. In: Arthropod Biology *and Evolution* (eds *A. Minelli, G. Boxshall & G. Fusco)*, pp. 91–122. Springer-Verlag, Berlin.

Nijhout, H.F., Davidowitz, G. & Roff, D.A. (2006) A quantitative analysis of the mechanism that controls body size in *Manduca sexta*. *Journal of Biology* **5**, 1–16.

Resh, V.H. & Cardé, R.T. (eds) (2009) *Encyclopedia of Insects*, 2nd edn. Elsevier, San Diego, CA. [Particularly see articles on: aestivation; cold/heat protection; development, hormonal control of; diapause; embryogenesis; greenhouse gases, global warming, and insects; growth, individual; imaginal discs; metamorphosis; migration; molting; segmentation; thermoregulation.]

Simpson, S.J., Sword, G.A. & De Loof, A. (2005) Advances, controversies and consensus in locust phase polymorphism research. *Journal of Orthoptera Research* **14**, 213–22.

Simpson, S.J., Sword, G.A. & Lo, N. (2011) Polyphenism in insects. *Current Biology* **21**, R738–R749. doi:10.1016/j.cub.2011.06.006.

Stansbury, M.S. & Moczek, A.P. (2013) The evolvability of arthropods. In: Arthropod Biology *and Evolution* (eds *A. Minelli, G. Boxshall & G. Fusco)*, pp. 479–93. Springer-Verlag, Berlin.

Whitman, D.W. & Ananthakrishnan, T.N. (eds) (2009) *Phenotypic Plasticity of Insects: Mechanisms and Consequences*. Science Publishers, Enfield, NH.

Chapter 7

INSECT SYSTEMATICS: PHYLOGENY AND CLASSIFICATION

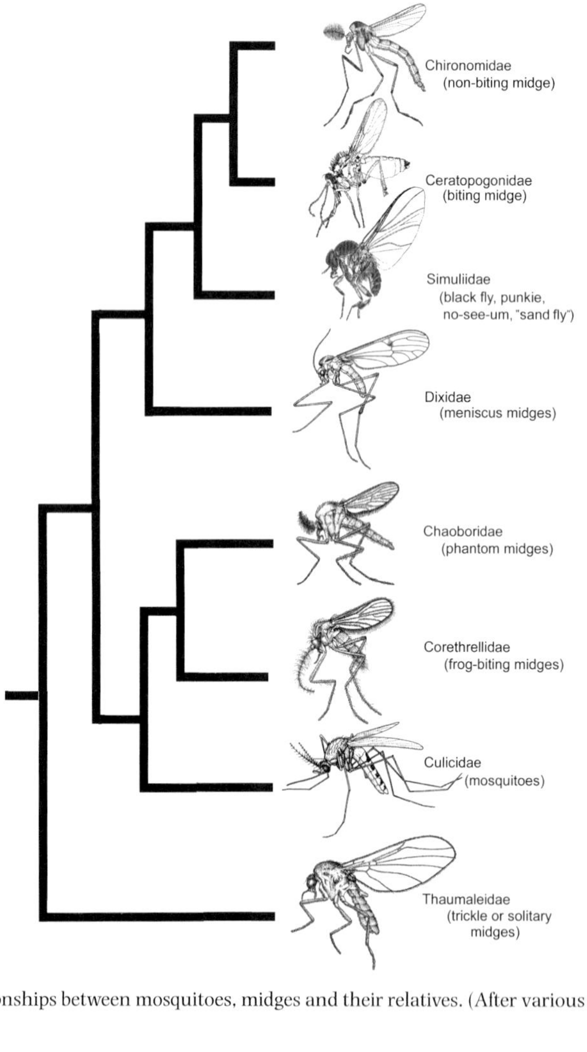

Tree showing proposed relationships between mosquitoes, midges and their relatives. (After various sources.)

It is tempting to think that we know every organism in the living world because so many guides are produced to identity and classify organisms such as birds, mammals, butterflies and flowers. However, such treatments vary, perhaps concerning the taxonomic status of a geographical race of bird, or of the family to which a species of flowering plant belongs. Scientists are not being perverse: differences reflect uncertainty, and an agreed stable classification may be elusive. Changes arise from continuing acquisition of knowledge concerning relationships, increasingly this is through the addition of molecular data to previous anatomical studies. This is especially true for insects, with new data and changing ideas on evolution leading to a dynamic classification, even at the level of insect orders. Knowledge of insects changes as new species are being discovered, particularly in the tropics, leading to revised estimates of the evolution and biodiversity of global insects (see section 1.3).

The study of the kinds and diversity of organisms and their inter-relationships – **systematics** – encompasses two more-narrowly defined but highly interdependent fields. The first is **taxonomy**, which is the science and practice of classification. It includes recognizing, describing and naming species, and classifying them into a ranked and named system (e.g. of genera, families, etc.) that, in modern systematics, aims to reflect their evolutionary history. The second field is **phylogenetics** – the study of evolutionary relatedness of **taxa** (groups) – and it provides information essential for constructing a natural (evolutionary) classification of organisms. Taxonomy includes some time-consuming activities, including exhaustive library searches and specimen study, curation of collections, measurements of features from specimens, and sorting of perhaps thousands of individuals into morphologically distinctive and coherent groups (which are first approximations to species), and perhaps hundreds of species into higher groupings. These essential tasks require considerable skill and are fundamental to the wider science of systematics, which involves the investigation of the origin, diversification and distribution (**biogeography**), both historical and current, of organisms. Modern systematics has become an exciting and controversial field of research, due largely to the accumulation of increasing amounts of nucleotide sequence data and the application of explicit analytical methods to both morphological and DNA data, and partly to increasing interest in the documentation and preservation of biological diversity.

Taxonomy provides the database for systematics. The collection of these data and their interpretation have been the subject of challenging debate. Similarly, the unravelling of evolutionary history, phylogenetics, is a stimulating and contentious area of biology, particularly regarding the insects. Entomologists are prominent participants in the vital biological enterprise of systematics. In this chapter, the methods of systematics are reviewed briefly, followed by details of the current ideas on a classification based on the postulated evolutionary relationships within the Hexapoda, of which the Insecta forms the largest group. Two examples of how entomologists use multiple data sources to recognize insect species are covered in Boxes 7.1 and 7.2. Box 7.3 describes the use of DNA barcoding in species discovery.

7.1 SYSTEMATICS

Systematics, whether based on morphological or molecular data, depends upon the study and interpretation of characters and their states. A character is any observable feature of a taxon, which may differentiate it from other taxa, and the different conditions of a character are called its states. Characters vary in the number of states recognized, and are either binary – having two states, such as presence or absence of wings – or multistate – having more than two states, such as the three states, "digitiform", "hooked" or "rhomboid", for the shape of the epandrial process of the genitalia of *Drosophila* shown in Fig. 5.5. An attribute is the possession of a particular state of a character; thus a digitiform epandrial process is an attribute of *D. mauritiana* (Fig. 5.5a), whereas a hooked process is an attribute of *D. similans* (Fig. 5.5b) and a rhomboid one is an attribute of *D. melanogaster* (Fig. 5.5c). The choice of characters and their states depends on their intended use. A diagnostic character state can define a taxon and distinguish it from relatives; ideally it should be unambiguous and, if possible, unique to the taxon. Character states should not be too

The Insects: An Outline of Entomology, Fifth Edition. P.J. Gullan and P.S. Cranston.

Companion Website: www.wiley.com/go/gullan/insects

variable within a taxon if they are to be used for the purposes of diagnosis, classification and identification. Characters showing variation due to environmental effects are less reliable for use in systematics than those under strong genetic control. For example, in some insects, size-related features may vary depending on the nutrition available to the individual developing insect. Characters can be classified according to their precision of measurement. A qualitative character has discrete (clearly distinguishable) states, such as the shapes of the epandrial process (described above). A quantitative character has states with values that can be counted or measured, and these can be further distinguished as meristic (countable) traits (e.g. number of antennal segments, or number of setae on a wing vein) versus continuous quantitative, in which the measurements of a continuously varying trait (e.g. length or width of a structure) can be divided into states arbitrarily or by statistical gap coding.

Although the various insect groups (taxa), especially the orders, are fairly well defined based on morphological characters, the phylogenetic relationships among insect taxa are a matter of much conjecture, even at the level of orders. For example, the order Strepsiptera is a discrete group that is recognized easily by its parasitoid lifestyle and the adult male having the fore wings modified as balancing organs (Taxobox 23), yet the identity of its closest relative has been contentious. Stoneflies (Plecoptera) and mayflies (Ephemeroptera) somewhat resemble each other, but this resemblance is superficial and misleading as an indication of relationship. The stoneflies are more closely related to the cockroaches, termites, mantids, earwigs, grasshoppers, crickets and their allies, than to mayflies. Resemblance may not indicate evolutionary relationships. Similarity may derive from being related, but equally it can arise through **homoplasy**, meaning convergent or parallel evolution of structures either by chance or by selection for similar functions. Only similarity as a result of common ancestry (**homology**) provides information regarding phylogeny. Two criteria for homology are:

1 similarity in outward appearance, development, composition and position of features; and

2 conjunction, i.e. two homologous features (character states) cannot occur simultaneously in the same organism.

A test for homology is congruence (correspondence) with other homologies. Thus, if a character state in one species is postulated to be homologous to the state of the same character in another species, then this hypothesis of homology (and shared ancestry) would be better supported if shared states of many more characters were identified in the two species.

In segmented organisms such as insects (section 2.2), features may be repeated on successive segments. For example, each thoracic segment has a pair of legs, and the abdominal segments each have a pair of spiracles. **Serial homology** refers to the correspondence of an identically derived feature of one segment with the similar feature on another segment (Chapter 2).

Morphology (mostly from external anatomy, but increasingly from internal structures using tomography techniques) provides data upon which insect taxa are described, relationships reconstructed, and classifications proposed. Morphological characters can be derived from all parts of the insect body, including internal tissues and genitalia. The genitalia, especially of male insects, are often complex structures with component parts diverging separately, which provide ideal characters for phylogenetic analysis. Some of the ambiguity and lack of clarity regarding insect relationships was blamed on inherent deficiencies in the phylogenetic information provided by morphology. After investigations of the utility of chromosomes and then differences in electrophoretic mobility of proteins, molecular sequence data from the mitochondrial and the nuclear genomes have become standard for solving many unanswered questions, including those concerning both higher relationships among insect groups and recognition of species. However, molecular data are not foolproof; as with all data sources, the signal can be obscured by homoplasy, and there are other problems (discussed below). Nevertheless, with appropriate choice of taxa and genes, molecules do help resolve certain phylogenetic questions that morphology has been unable to answer. This is particularly true for deeper relationships (family or ordinal levels) where, using morphology, it may be difficult or impossible to recognize homologous character states in the taxa being compared, due to independent evolution of unique structures within each lineage. Another source of useful data for inferring the phylogenies of some insect groups derives from the DNA of their specialist bacterial symbionts. For example, the primary endosymbionts (but not the secondary endosymbionts) of aphids, mealybugs and psyllids co-speciate with their hosts, and bacterial relationships can be used (with caution) to estimate host relationships. Evidently, the preferred approach to estimating phylogenies is a holistic one, using data from as many sources as possible, and retaining an awareness that not all similarities are equally informative in revealing phylogenetic patterns.

7.1.1 Phylogenetic methods

The various methods that attempt to recover the pattern produced by evolutionary history rely on observations on living and fossil organisms. Historically, relationships of insect taxa were based on estimates of overall similarity (**phenetics**) derived principally from morphology, and taxonomic rank was governed by degree of difference. It is recognized now that analyses of overall similarity are unlikely to recover the pattern of evolution and thus phenetic classifications are considered artificial. Their use in phylogeny largely has been abandoned, except perhaps for organisms such as viruses and bacteria, which exhibit reticulate evolution (characterized by intercrossing between lineages, producing a network of relationships). However, phenetic methods are useful in DNA barcoding (section 18.3.3) in which identification of an unknown species is often possible based on comparing the nucleotide sequences of one of its genes with those in a database of identified species of the group, although this process does have its problems. Alternative methods in phylogenetics are based on the premise that the pattern produced by evolutionary processes can be estimated, and, furthermore, ought to be reflected in the classification. Popular methods in current use for all types of data include cladistics (maximum parsimony, MP), maximum likelihood (ML) and Bayesian inference (BI). A detailed discussion of the methods of phylogenetic inference is beyond the scope of an entomology textbook and readers should consult other sources for in-depth information. Here, the basic principles and terms are explained, followed by a section on molecular phylogenetics that considers the use of genetic markers and problems with estimating relationships of insects using nucleotide sequence data.

Phylogenetic trees, also called dendrograms, are the branching diagrams that depict purported relationships or resemblances among taxa. Different kinds of trees emphasize different components of the evolutionary process. **Cladograms** depict only the branching pattern of ancestor–descendant relationships, and branch lengths have no meaning. **Phylograms** show the branching pattern and the number of character state changes are represented by differences in the branch lengths. **Chronograms** explicitly represent time through branch lengths. Evolution is directional (through time), and thus phylogenetic trees usually are drawn with a root; unrooted trees must be interpreted with extreme care. In modern systematics, phylogenetic trees are the basis for erecting new or revised classifications, although there are no universally accepted rules for how to convert the topology of a tree into a ranked classification. Here, we use cladograms to explain some terms that also apply more widely to phylogenetic trees derived by other analytical methods.

The cladistic method (**cladistics**) seeks patterns of special similarity based only on shared, evolutionarily novel features (**synapomorphies**). Synapomorphies are contrasted with shared ancestral features (**plesiomorphies** or **symplesiomorphies**), which do not indicate closeness of relationship. The terms apomorphy and plesiomorphy are relative terms because an apomorphy at one level in the taxonomic hierarchy, for example the ordinal apomorphy of all beetles having fore wings in the form of protective sheaths (the elytra), becomes a plesiomorphy if we consider wing characters among families of beetles. Furthermore, features that are unique to a particular group (**autapomorphies**) but unknown outside the group do not indicate inter-group relationships, although they are very useful for diagnosing the group. Construction of a cladogram (Fig. 7.1), a treelike diagram portraying the phylogenetic branching pattern, is fundamental to cladistics. Cladograms are constructed so that character state changes across the tree are minimized, based on the principle of parsimony (i.e. a simple explanation is preferred to a more complex one). The tree with the fewest changes (the "shortest" tree) is considered to have the optimal topology (tree shape). It is important to remember that parsimony is a rule for evaluating hypotheses, not a description of evolution.

From a cladogram, **monophyletic** groups, or **clades**, their relationships to each other, and a classification, can be inferred directly. Only the branching pattern of relationships is considered. **Sister groups** are taxa that are each other's closest relatives; they arise from the same **node** (branching point) on a tree. A monophyletic group contains a hypothetical ancestor and *all* of its descendants (Fig. 7.1a). Further groupings can be identified from Fig. 7.1: a **paraphyletic** group lacks a clade from amongst the descendants of a common ancestor, and often is created by the recognition (and removal) of a derived subgroup; **polyphyletic** groups fail to include two or more clades from amongst the descendants of a common ancestor (e.g. A and D in Fig. 7.1c). Thus, when we recognize the monophyletic Pterygota (winged plus secondarily apterous insects), two other extant

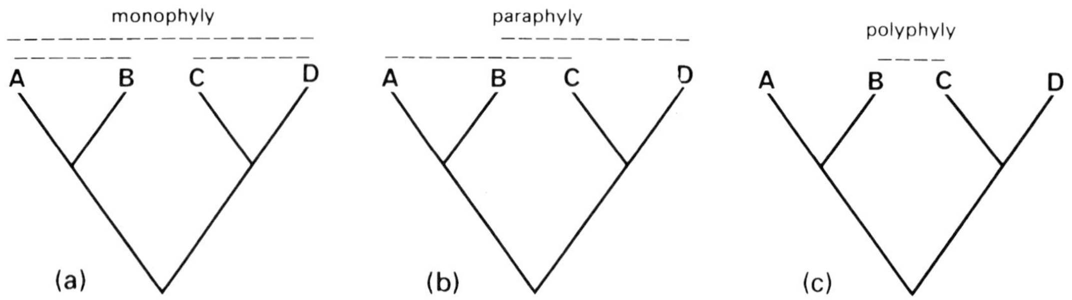

Fig. 7.1 A cladogram showing the relationships of four species, A, B, C and D, and examples of: (a) the three monophyletic groups; (b) two of the four possible (ABC, ABD, ACD, BCD) paraphyletic groups; and (c) one of the four possible (AC, AD, BC and BD) polyphyletic groups that could be recognized based on this cladogram.

orders (Archaeognatha and Zygentoma) form a grade of primitively wingless insects (Fig. 7.2). If this grade is treated as a named group ("Apterygota"), it would be paraphyletic. A group of flying insects with fully developed wings restricted to the mesothorax (true flies, male scale insects and a few mayflies) would be polyphyletic. Paraphyletic groups should be avoided if possible because their only defining features are ancestral ones shared with other indirect relatives. Thus, the absence of wings in the paraphyletic apterygotes is an ancestral feature shared by many other invertebrates. The mixed ancestry of polyphyletic groups means that they are biologically uninformative, and such artificial taxa should never be included in any classification.

"Tree thinking" (interpreting and using phylogenies) is fundamental to all of biology, and yet misinterpretations of trees abound. The most common error is to read trees as ladders of evolutionary progress, with the often species-poor sister group erroneously referred to as the "basal" taxon and misinterpreted as displaying characteristics found in the common ancestor. All extant species are mixes of ancestral and derived features, and there is no *a priori* reason to assume that a species-poor lineage retains more plesiomorphies than its species-rich sister lineage. Trees have basal nodes, but they do *not* have basal taxa. Readers should refer to papers by Gregory (2008) and Omland *et al.* (2008) for more information on this important topic.

Molecular phylogenetics

Pioneers in the field of molecular systematics used chromosome structure, or the chemistry of molecules such as enzymes, carbohydrates and proteins, to access the genetic basis for evolution. Although these techniques are largely superceded, understanding the banding patterns in "giant" chromosomes remains important in locating gene function in some medically important flies, and comparative patterns of enzyme polymorphism (isoenzymes) remains valuable in population genetics. The field of phylogenetic systematics has been revolutionized by ever-more automated and cheaper techniques that access the genetic code of life directly, via the nucleotides (bases) of DNA and RNA, and the amino acids of proteins that are coded for by the genes. Obtaining genetic data requires extraction of DNA, use of the polymerase chain reaction (PCR) to amplify DNA, and a range of procedures to sequence, that is, to determine the order of the nucleotide bases – adenine, guanine (purines), cytosine and thymine (pyrimidines) – that make up the selected section of DNA. Generally, comparable sequences of 300–1000 nucleotides are sought from each of several genes (the more the better), for comparisons across a range of taxa of interest.

Choice of genes for phylogenetic study involves selection of those with an appropriate substitution (mutation) rate of molecular evolution for the question at hand. Closely related (recently diverged) taxa may be near identical in slower-evolving genes that provide little or no phylogenetic information, but should differ more in fast-evolving genes. From an ever-increasing database, appropriate genes ("markers") can be selected with succesful previous history of use, and a "cook book" method followed. Already tried and tested primers (nucleic acid strands that start DNA replication at a chosen place on a selected gene) can be selected appropriately. For insect molecular phylogenetics, a suite of genes typically includes one, some or all of: the mitochondrial genes *16S*, *COI* and *COII*; the nuclear small subunit rRNA (*18S*); part of the large subunit rRNA (*28S*); and progressively more nuclear protein-coding genes such as elongation factor 1 alpha ($EF-1\alpha$), histone 3 (*H3*), wingless (*wg*) and rudimentary *CAD*. Very slowly mutating (highly conserved) genes are needed to infer older branching patterns, of 100 million years and more.

Revolutions in techniques for acquisition of genetic data have lead to new sources of data for taxonomic and phylogenetic study. Since 2007, complete genomes have been sequenced for almost 50 insect species, including fruit flies (more than 20 *Drosophila* species), the pea aphid (*Acyrthosiphum pisum*), the red flour beetle (*Tribolium castaneum*) and several lepidopteran (including *Bombyx mori* and *Danaus plexippus*), mosquito (Culicidae), ant (Formicidae) and bee (Apidae) species. As well as insect whole genomes, many more whole mitochondrial genomes (mitogenomes) are available for insects, including over 20 species of Lepidoptera, and almost 50 species of Orthoptera. Mitogenomes and transcriptome data are increasingly used in phylogenetic studies within and among orders of insects. The transcriptome is the set of all transcribed RNA molecules (including mRNA, rRNA, tRNA and non-coding RNA) from one or a population of cells of an organism.

Particular sequences of nucleotide bases (haplotypes) allow examination of genetic variation between individuals in a population. However, in molecular phylogenetics, the haplotype is used more as characteristic of a species (or to represent a higher taxon) to be compared among other taxa. The first procedure in such an analysis is to "align" comparable (homologous) sequences in a species-by-nucleotide matrix – rows being the sampled taxa, columns being the nucleotide identified at a particular position (site) in the gene, reading from where the primer commenced. This matrix is quite comparable with one scored for morphology, in which each column is one character, and the different nucleotides are the various "states" at the site. Many characters will be invariate, unchanged amongst all taxa studied, whereas others will show variation due to substitution (mutation) at a site. Some sites are more prone to substitution than others, which are more constrained. For example, each amino acid is coded for by a triplet of nucleotides, but the state of the third nucleotide in each triplet is freer to vary (compared to the first and second positions) without affecting the resultant amino acid.

The aligned species-by-nucleotide matrix, consisting of the ingroup (of taxa under study) and one or more outgroups (more distantly removed taxa), can be analysed by one or more of a suite of methods. Some molecular phylogeneticists argue that parsimony – minimizing the number of character-state changes across the tree, with the same weight being allocated to each substitution observed in a column (character) – is the only justifiable approach, given our uncertainty about how molecular evolution takes place. However, increasingly, analyses are based on more-complex models, involving application of weights to different kinds of mutations – for example, a transition between one purine and the other (adenine ↔ guanine) or between two pyrimidines (cytosine ↔ thymine) is more "likely" to have occurred than a transversion between a purine and a chemically more dissimilar pyrimidine (and vice versa). Most systematists examine results (phylogenetic trees) based on both the simplest model (parsimony) and ever-more complex "likelihood" models, including Bayesian statistical programs, requiring one or a suite of powerful computers running over days or weeks. Each model and analysis method should provide estimates of confidence (support) for each relationship portrayed to allow for critical assessment of all hypothesized relationships.

The procedures described briefly above are implemented widely, covering much of the immense diversity of the hexapods, and generating results (or hypotheses of relationships) that are published in numerous journal articles each year. The reader may be excused from asking why, if this is the case, do so many unresolved or contradictory ideas on insect relationships exist? Today we may seem to know less about some relationships than we did previously based on either morphology or earlier molecular work. Molecular data, which can provide many thousands of characters and more than adequately variable character states (mutations), has yet to deliver its promise of a full understanding of the evolution of the insects.

There are many problems encountered with molecular phylogenetics that were unforeseen by early practitioners, some of these are outlined here.

1 It is relatively straightforward to assess the morphology of a group of insects, including using historical specimens, but it much less easy to obtain the same diversity of appropriately preserved species for molecular study. Collection requires specialist knowledge and techniques, with increasing legal impediments for genetic collections.

2 Even with a good sample of the diversity of material, the procedures for DNA extraction and sequencing can be unsuccessful for some individual specimens or even broader groups, or for some genes, for a number of reasons, including poor preservation causing nucleases to degrade DNA or rearrangements in the DNA of some species, resulting in failure of primers (see point **8**).

3 Given sequences, the procedure of alignment – constructing columns of homologous nucleotides – becomes non-trivial as more distantly related taxa are included, because the gene sections sequenced may differ in length. One or more sequences with respect to others may be longer or shorter due to insertions or deletions (indels) of some to many nucleotides in one or more places. It is problematic as to how to align such sequences by insertion of "gaps" (absences of nucleotides) and how to "weight" such inferred changes relative to regular nucleotide substitutions. Although nuclear ribosomal genes (*18S* and *28S*) have been commonly chosen, especially for deeper divergences, these genes may present major alignment problems compared to protein-encoding nuclear genes.

4 There are only four nucleotides amongst which substitutions can take place, so the higher the mutation rate, the more likely that a second (or further) mutation may take place that could revert a nucleotide either to its original or another condition (a so-called "multiple hit"). The existence and history of the substitions that resulted in an identical end state cannot be recognized: an adenine is an adenine whether or not it has mutated via a guanine and back.

5 There is variation in the phylogenetic information between different parts of a gene – there may be regions, said to be hypervariable, with a high concentration of substitutions, and therefore many multiple hits may be interspersed (unrecognized) within sections that are near impossible to align.

6 There are differences in propensities of sites to mutate – not only in the 1st and 2nd nucleotides relative to the 3rd positions (above), but related to the general and specific (secondary and tertiary) structure of the protein or RNA molecule from which the sequence came.

7 The evolutionary signal derived from phylogenetic analysis of one gene provides an insight into the evolution of that gene, but this need not be the "true" history of the organisms. Perhaps there is no "true" history derivable from genes: many insects diversified very rapidly, but speciation is a drawn-out process. Population-level processes of gene flow, genetic drift, selection and mutation take place at much shorter time-scales and can create multiple histories of genes. Genes can duplicate within a lineage, with each copy (paralogue/paralog) subjected subsequently to different substitutions. Only similar (homologous) copies of genes should be compared in phylogenetic study of the organisms.

8 In contrast to population genetics, a major challenge to phylogenetics is to obtain primers that can amplify the targeted locus from a diversity of organisms. But primer sites diverge with time, reducing the "cross-reactivity" of primers between distant taxa. Primers with better ability across distant taxa (incorporating degenerate bases) may lack specificity for the study group, increasing the propensity of non-specific priming of non-target sequences.

These, and other problems with genetic data, do not imply that such studies cannot untangle insect evolutionary history, but may explain conflicting results from molecular phylogenetic studies. Some problems can be addressed, for example, by better sampling, improved models for alignment and for site-specific variation in substitution rates according to better understanding of molecular structures. As massively more molecular data become available, especially from whole genomes, computation is challenged by large matrices, including those with missing data (lacking genes or taxa). Perhaps phylogenetic analyses will converge on consistent relationships among the groups of insects that interest us. Our approach to portraying and discussing the evolutionary relationships in this chapter is conservative, showing only strongly supported groupings from molecular studies where there is a congruent morphological basis.

7.1.2 Taxonomy and classification

Current classifications of insects are based on a combination of ideas for how to recognize the rank and relationships of groups, with most orders being based on groups (taxa) with distinctive morphology. It does not follow that these groups are monophyletic, for instance the traditionally defined Blattodea and Psocoptera were each paraphyletic (see below in the discussion of each order in section 7.4.2). However, it is unlikely that any higher-level groups are polyphyletic.

In many cases, the present groupings coincide with the earliest colloquial observations on insects, for example the term "beetles" for Coleoptera. However, in other cases, such old colloquial names cover disparate modern groupings, as with the old term "flies", now seen to encompass unrelated orders from mayflies (Ephemeroptera) to true flies (Diptera). Refinements continue as classification is found to be out of step with our developing understanding of the evolution of the Hexapoda. Thus, classifications increasingly combine traditional views with current ideas on phylogeny.

Difficulties with attaining a comprehensive, coherent classification of the insects arise when phylogeny is obscured by complex evolutionary diversifications, and this is true at all taxonomic levels (see Table 1.1 for the taxonomic categories). These include radiations associated with adoption of specialized plant or animal feeding (phytophagy and parasitism; section 8.6) and radiations from a single founder on isolated islands (section 8.7). Sometimes, the evolution of reproductive isolation of closely related taxa is not accompanied by obvious (to humans) morphological differences among the entities, and it is a challenge to delimit species. Difficulties may arise also because of conflicting evidence from immature and adult insects, but, above all, problems derive from the immense number of species (section 1.3.2).

Scientists who study the taxonomy of insects – i.e. describe, name and classify them – face a daunting task. Virtually all the world's vertebrates are described, their past and present distributions verified and their behaviours and ecologies studied at some level. In contrast, perhaps only 5–20% of the estimated number of insect species have been described formally, let alone studied biologically. The disproportionate allocation of taxonomic resources is exemplified by Q.D. Wheeler's report for the United States of seven described mammal species per mammal taxonomist in contrast to 425 described insects per insect taxonomist. These ratios, which probably have worldwide application, become even more alarming if we include estimates of undescribed species. Few mammal species are unnamed, but estimates of global insect diversity may involve undescribed millions of species.

New species and other taxa of insects (and other organisms) are named according to a set of rules developed by international agreement and revised as practices and technologies change. For all animals, the International Code of Zoological Nomenclature (ICZN) regulates names in the species, genus and family groups (i.e. from infraspecific to superfamily level). The standard use of a unique binomen (to names) for each species (section 1.4) assures that people everywhere can communicate clearly. Unique and universal names for taxa result from making new names available through permanent publication, with the appropriate designation of type specimens (that serve as the reference for names), and by recognizing the priority of names (i.e, the oldest available name is the valid name of a taxon). Today, much controversy is centred upon what constitutes a "published" work for the purposes of nomenclature, principally due to the difficulty of ensuring permanence and long-term accessibility of electronic publications.

When a new insect species is recognized and a new name published, it must be accompanied by a published description of the appropriate life stages (almost always including the adult) in sufficient detail that the species can be distinguished from its close relatives. Features diagnostic of the new species must be explained, and any variations in appearance or habits described or discussed. Nucleotide sequences may be available for one or more genes of the new species and, sometimes, diagnostic sequences form part of the species description. Sequences should be deposited in online databases, such as GenBank. Indeed, comparison of sequences from related specimens and populations often leads to the recognition of new species that were previously "cryptic" due to similar morphology, as occurred with *Gonipterus* weevils (Box 7.1). Species delimitation remains one of the challenging tasks of taxonomy and new character systems are being explored in many insect groups.

How entomologists recognize insect species

Although there are a myriad of ideas on what constitutes a species, most species concepts centre on sexually reproducing organisms and can be applied to most insects. Different species concepts adopt different properties, such as phenetic distinctness, diagnosability or reproductive incompatibility, as their species criterion. The use of various defining properties for species has led to diasagreements over species concepts, but the actual practice of deciding how many species exist among a sample of related or similar-looking specimens generally is less contentious. Systematists mostly use recognizable discontinuities in the distribution of features (character states) and/or in the distribution of individuals in time and space. The recognition of species is a specialist taxonomic activity, whereas identification can be performed by anyone using the appropriate identification tools, such as dichotomous keys, illustrated computer interactive keys, a reference database, or, for a few pest insects, a DNA probe. Insect systematists use a variety of data sources – morphology, molecules (e.g. DNA, RNA, cuticular hydrocarbons, neuropeptides, etc.), karyotypes, behaviour, ecology and distribution – to sort individuals into groups (Box 7.1 and Box 7.2). The name **integrative taxonomy**

Box 7.1 *Gonipterus* weevils – recognition of a species complex

A eucalypt snout weevil of the genus *Gonipterus*, believed to be the eastern Australian species *G. scutellatus*, has been causing damage in plantations in Western Australia, New Zealand, Western Europe, North and South America, and Africa. Leaf feeding by these weevils causes severe damage to the foliage (seen here for *G. platensis*, mostly after photographs by M. Matsuki), with adults (see **Plate 8f**) chewing the leaf edges and larvae first grazing the leaf surface (shown bottom middle, with a frass filament from the anus) and later chewing the whole leaf. In its native range, the weevil is controlled by a tiny mymarid wasp, *Anaphes nitens*, which parasitizes weevil eggs (laid in pods on the leaves, as seen in the illustration), but the wasp's introduction as a biological control agent elsewhere has met with mixed success. Incomplete control often indicates a taxonomic problem that results in erroneous association of the target pest species with its natural enemy. In this case, the identity of the pest weevil had been problematic for almost 100 years, a number of species names had been synonymized under *G. scutellatus*, and variation had been observed in the complex structures of the male genitalia. In order to determine whether the observed morphological differences had a genetic basis, a collaborative study obtained and analysed sequences of *CO1* (section 7.1) from many populations across Australia, plus a few from overseas populations. The outcome was that clusters of males based on similarities of the internal sclerite(s) of the aedeagus correlated well with haplotype clusters from the *CO1* analyses. Five (of 10 clusters) could be associated with already named species, others were undescribed and showed some regional and host specificity. The overseas pests were shown to belong to three different species, according to region: *G. platensis* (New Zealand, America, western Europe), *G. pulverulentus* (eastern South America) and an undescribed species (Africa, France, Italy). The problem species in Western Australian plantations turned out to be *G. platensis*, likely acquired recently from its home range in Tasmania. Integrating the morphology of the *Gonipterus scutellatus* complex with data on species limits derived from haplotype analysis and geographic distribution allows better understanding of the relationships amongst the weevils and their host eucalypt species (many are used in commerce), and the variable efficacy of the mymarid wasps in biological control programmes. In the latter case, each *Gonipterus* species may have its own specific mymarid parasitoid (as yet undetermined), but failure of biological control also seems to be associated with poor wasp performance in seasons when temperatures are below 10°C.

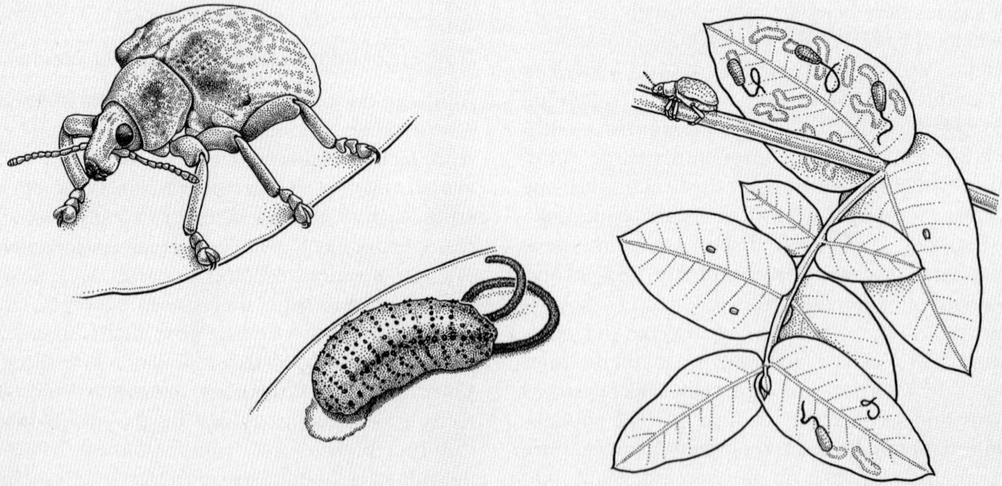

Box 7.2 Integrative taxonomy of woodroaches

The woodroaches (Blattodea: *Cryptocercus*) of eastern North America provide a good example of an insect group that has been studied taxonomically by integrating a variety of data sources. All populations within the *Cryptocercus punctulatus* species complex in the Appalachian Mountains are morphologically, behaviourally and ecologically similar, but have been described as four species based on unique male diploid chromosome numbers ($2n = 37$, 39, 43 or 45) and some diagnostic bases from two rRNA genes. Consistent morphological differences in the reproductive structures, especially of adult females, also support these species. However, there appear to be five taxa (as shown in the tree), based on evidence from recent analysis of data from one nuclear and two mitochondrial genes, and supported by cuticular hydrocarbon data (see section 4.3.2 for information on cuticular hydrocarbons). The latter two datasets differ from the karyotype information in indicating two distinct and not closely related clades (or taxa) with $2n = 43$, indicating parallel evolution of that karyotype, perhaps by convergent reduction of chromosome number. Woodroaches with $2n = 43$ (currently *C. punctulatus sensu stricto*) occur over a wide geographic range compared with the other species, and variation in reproductive structures has not been examined across the range. Not all groups recognized by the hydrocarbon data are fully concordant with those delimited by other evidence, since some woodroaches in the two clades that are distinct genetically and either $2n = 43$ or $2n = 45$ have similar hydrocarbons, plus there are two distinct hydrocarbon groups within $2n = 45$ woodroaches. Thus, hydrocarbon information fails to support two of the putative *Cryptocercus* species but fully supports other species. At least in *Cryptocercus*, it seems that differentiation of cuticular hydrocarbons can occur with minimal karyotype and genetic change, and changes in genes coding for cuticular hydrocarbons do not necessarily accompany changes in other genes or chromosome number.

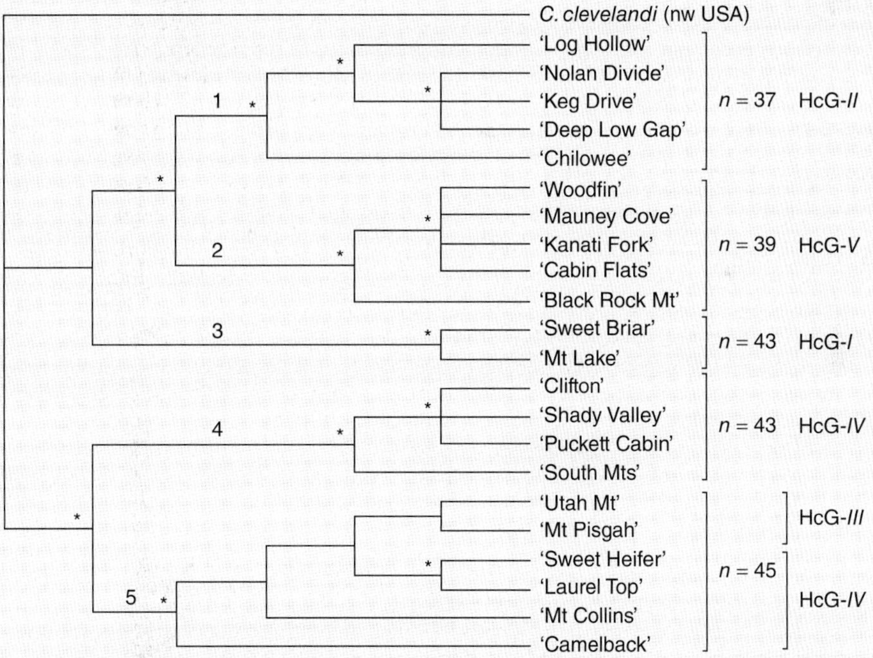

Strict consenus tree of the *Cryptocercus punctulatus* species complex from 22 locations, based on combined datasets of mtDNA and nuclear DNA. Nodes with strong support are indicated by an asterisk, and the five putative species are numbered 1–5. Each terminal is labelled with the collection location name, the male diploid chromosome number, and the hydrocarbon group (HcG). (After Everaerts *et al.* 2008.)

has been applied to taxonomic approaches that utilize information from multiple disciplines or fields of study.

Traditionally, differences in external appearance (e.g. Fig. 5.7), often coupled with internal genitalic features (e.g. Fig. 5.5), were used almost exclusively for discriminating insect species. In recent decades, such species-specific variation in morphology is frequently supported by other data, especially nucleotide sequences from mitochondrial or nuclear genes. In insect groups with morphologically similar species, particularly sibling or cryptic species complexes (e.g. the *Anopheles gambiae* complex discussed in Box 15.2), non-morphological data are important to species recognition and subsequent identification.

DNA barcoding has been promoted as a molecular diagnostic tool for species-level identification (see section 18.3.3) and discovery (Box 7.3). Although the standard barcode is a region of mitochondrial DNA, nucleotide sequences from other genes can be used. Two criteria should be used for species recognition using sequence data: (i) reciprocal monophyly, meaning that all members of a group must share a more recent common ancestor with each other than with any member of another group on a phylogenetic tree; and (ii) a genetic distance cut-off, such that interspecific distances are greater than intraspecific distances by a factor that is different for each gene (e.g. 10 times the average intraspecific distance has been proposed for species delimitation using *COI* data). In practice, reliable species recognition requires thorough taxon and specimen sampling to avoid underestimating intraspecific variation through poor geographic

Box 7.3 DNA barcoding and species discovery

Accurate and reliably repeatable, and above all cheap, methods for sequencing of DNA allow the acquisition of massive amounts of genomic information. This information-rich source of data challenges analysis, and advances in the technical methods must keep pace with the increasing ease of obtaining whole-organismal genomes. However, valuable earlier approaches that generate data pertaining to species discrimination remain useful, notably the so-called "barcoding" of organisms. The name relates by analogy to the unique universal product codes affixed to most retail items that we purchase.

For most insects, **DNA barcoding** is based on part of the mitochondrial gene, *cytochrome c oxidase* subunit I (*COI* or *cox*1), which shows variations (mutations) at a frequency that allows clustering of individuals from the same species and provides increasing distinction from more distantly related species. Advocates for barcoding argued for a standardized use of this particular gene region for investigating all species. In an ideal world, all taxa would have their own distinctive signature of unique sequence(s), allowing identification of all life from little more than 500 base pairs of DNA.

Since 2003 when the formal DNA barcoding system was proposed, there has been much progress, especially in identifying morphologically cryptic species and resolving complexes of very similar, but often biologically different, species. Major projects to barcode large sections of biodiversity have been funded and undertaken, with variable progress towards the goal. Problems arrive with organisms that are difficult to survey widely, and thus the library against which new barcode sequences are compared is slow to build for some taxa, especially insects. In morphological studies it is possible to survey all recorded diversity through specimens in repositories, although this can be complicated logistically, and is susceptible to underfunded curation of many collections. Although these museum specimens, if relatively fresh and appropriately preserved, can retain DNA – many do not, or the DNA is too degraded for systematic use. Although DNA can be amplified from small parts of a preserved specimen, many museums are reluctant to allow valuable old and rare specimens, especially types, to be used in molecular studies.

A notable success of DNA barcoding involves the integration of molecular data in associating phytophagous caterpillars, pupal and adult lepidopterans and their parasitoids in the Area de Conservación Guanacaste (ACG) in northwestern Costa Rica. By 2011, about 5000 species of caterpillar had been reared, their DNA barcodes recorded, host plants identified, and associated parasitoids integrated into a barcoding scheme. The 30+-year-long project, famed for its use of **parataxonomists**, has enormously extended the known diversity of the system, with cryptic species, novel associations and tight coevolution uncovered in all studied insects.

sampling, or overestimating interspecific distance through omission of species. Molecular data can provide clues to possible species segregates and thus allow the re-interpretation of existing morphological data by highlighting phylogenetically informative anatomical characters.

Chemotaxonomy is another tool for species determination; the chemical profiles of different insects can be compared using statistical methods such as principal component analysis (PCA). Hydrocarbons occur in the lipid layer of the insect cuticle (section 2.1 and 4.3.2), are easy to extract and usually chemically stable, and the hydrocarbon blend can be species specific (see Box 7.2). For example, the cuticular hydrocarbon profiles of two cryptic species of *Macrolophus* (Hemiptera: Miridae) provide a reliable phenotypic character for distinguishing adults of the two species, and these species-specific patterns are not altered by diet. In termites, which can be difficult to identify by morphology (particularly from workers), there is some convincing evidence that hydrocarbon profiles can differentiate species. In a few studies that report more than one hydrocarbon phenotype for a named termite species, other sources of evidence (such as intercolony agonistic experiments and genetic data) suggest that these phenotypes are distinct taxonomically. However, data must be interpreted with care because the environment may influence hydrocarbon composition, as found for Argentine ants fed on different diets. Microorganisms (such as pathogenic fungi) living on an insect's cuticle also can influence hydrocarbon profiles by using the hydrocarbons as an energy source. A detailed behavioural study of an African *Macrotermes* species showed that although the level of mortality due to aggressive encounters increased with differences in cuticular hydrocarbons between colonies, termites displayed lower aggression to neighbouring colonies than to more distant ones regardless of hydrocarbon phenotype. However, different hydrocarbons have different physiological or behavioural functions (waterproofing versus communication; see section 4.3.2) and some are likely to be more environmentally stable than others. Clearly, more studies are needed of different types of insects using multiple data sources, including behavioural bioassys, to test for concordance between hydrocarbon phenotypes and taxonomic groups.

Despite taxonomic problems at the species and genus levels due to immense insect diversity and phylogenetic controversies at all taxonomic levels, we are moving towards a consensus view on many of the internal relationships of Insecta and their wider grouping, the Hexapoda. These are discussed below.

7.2 THE EXTANT HEXAPODA

The Hexapoda (usually given the rank of superclass) contains all six-legged arthropods. The closest relatives of hexapods were considered to be the myriapods (centipedes, millipedes and their allies), but molecular sequence and developmental data plus some morphology (especially of the compound eye and nervous system) suggest a more recent shared ancestry for hexapods and crustaceans (see Box 8.1).

Diagnostic features of the Hexapoda include the possession of a unique **tagmosis** (section 2.2), which is the specialization of successive body segments that more or less unite to form sections or tagmata, namely the head, thorax and abdomen. The head is composed of a pregnathal region (usually considered to be three segments) and three gnathal segments bearing mandibles, maxillae and labium, respectively; the eyes are variously developed and may be lacking. The thorax comprises three segments, each of which bears one pair of legs, and each thoracic leg has a maximum of six segments in extant forms, but was ancestrally 11-segmented, with up to five **exites** (outer appendages of the leg), a coxal **endite** (an inner appendage of the leg) and two terminal claws. The abdomen originally had 11 segments plus a telson or some homologous structure; if abdominal limbs are present, they are smaller and weaker than those on the thorax, and ancestrally were present on all except the tenth segment.

The earliest branches in the hexapod phylogeny undoubtedly involve organisms whose ancestors were terrestrial (non-aquatic) and wingless. However, any combined grouping of these taxa is not monophyletic, being based on evident symplesiomorphies (e.g. lack of wings, or the occurrence of moults after sexual maturity) or otherwise doubtfully derived character states. Therefore, a grouping of apterygote (primitively wingless) hexapods is a grade of organization, not a clade. Included orders are Protura, Collembola, Diplura, Archaeognatha and Zygentoma. The Insecta proper comprise Archaeognatha, Zygentoma and the huge radiation of Pterygota (the primarily winged hexapods).

Relationships are uncertain among the component taxa that branch from nodes near the base of the phylogenetic tree of the Hexapoda, although the cladogram shown in Fig. 7.2 and the classification presented in the following sections reflect our current synthetic view. Traditionally, Collembola, Protura and Diplura were grouped as "Entognatha", based primarily on resemblance in mouthpart morphology. Entognathan mouthparts are enclosed in folds of the head that form a gnathal pouch, in contrast to mouthparts of the Insecta (Archaeognatha + Zygentoma + Pterygota), which are exposed (ectognathous). However, the Diplura has sometimes been placed as sister to the Insecta, thus rendering the Entognatha paraphyletic. Also, the Collembola and Protura have sometimes been grouped as the Ellipura based on certain shared morphological features, including an apparently advanced form of entognathy. Some molecular evidence supports the monophyly of the Entognatha, and analyses of data from rRNA genes suggest that Diplura is sister to Protura, in a group named the Nonoculata ("no eyes") due to the absence of even simple eyes; in contrast, many Collembola have clustered simple eyes. Data from protein sequences of certain nuclear genes suggest that Diplura is sister to Collembola, with Protura sister to these two plus all insects. Recent transcriptome data also fails to resolve interordinal relationships of Entognatha, although Diplura may be sister to the insects rather than the other entognaths, as proposed earlier. We have used the latter classification here and treat the Entognatha as an informal group, but further gene and taxon sampling is required to confirm one of the several purported relationships.

7.3 INFORMAL GROUP ENTOGNATHA: COLLEMBOLA (SPRINGTAILS), DIPLURA (DIPLURANS) AND PROTURA (PROTURANS)

7.3.1 Order Collembola (springtails) (see also Taxobox 1)

Collembolans are minute to small, soft bodied, and often with rudimentary eyes or ocelli. The antennae are four- to six-segmented. The mouthparts are entognathous, consisting predominantly of elongate maxillae and mandibles enclosed by lateral folds of head, and lacking maxillary and labial palps. The legs are four-segmented. The abdomen is six-segmented with a sucker-like ventral tube or collophore, a retaining hook and a furcula (forked jumping organ) on segments 1, 3 and 4, respectively. A gonopore is present on segment 5; the anus on segment 6. Cerci are absent. Larval development is epimorphic, that is, with segment number constant through development. Collembola either form the sister group to Protura in a grouping called Ellipura or form the sister group to the Diplura + Protura (the Nonoculata).

7.3.2 Order Diplura (diplurans) (see also Taxobox 1)

Diplurans are small to medium sized, mostly unpigmented, possess long, moniliform antennae (like a string of beads), but lack eyes. The mouthparts are entognathous, with tips of well-developed mandibles and maxillae protruding from the mouth cavity, and maxillary and labial palps reduced. The thorax is poorly differentiated from the 10-segmented abdomen. The legs are five-segmented and some abdominal segments have small styles and protrusible vesicles. A gonopore lies between segments 8 and 9, and the anus is terminal. Cerci are slender to forceps-shaped. The tracheal system is relatively well developed, whereas it is absent or poorly developed in other entognath groups. Larval development is epimorphic, with segment number constant through development. Diplura may be the sister group to Insecta, although other hypotheses suggest a sister relationship to the Protura or to Collembola + Protura within the Entognatha.

7.3.3 Order Protura (proturans) (see also Taxobox 1)

Proturans are small, delicate, elongate, mostly unpigmented hexapods, lacking eyes and antennae, with entognathous mouthparts consisting of slender mandibles and maxillae that slightly protrude from the mouth cavity. Maxillary and labial palps are present. The thorax is poorly differentiated from the 12-segmented abdomen. Legs are five-segmented. A gonopore lies between segments 11 and 12, and the anus is terminal. Cerci are absent. Larval development is anamorphic, that is, with segments added posteriorly during development. Protura either is sister to Collembola, forming Ellipura in a weakly supported relationship based on some morphological features, or might be sister to the Diplura in a group called Nonoculata, based on ribosomal RNA gene data, the absence of eyes, and Malpighian tubules either reduced to short papillae or absent.

7.4 CLASS INSECTA (TRUE INSECTS)

Insects range from minute to large (0.2 mm to >30 cm long), with very variable appearance. Adult insects typically have ocelli and compound eyes, and the mouthparts are exposed (ectognathous), with the maxillary and labial palps usually well developed. The thorax may be weakly developed in immature stages but is distinct and often highly developed in winged adults, associated with the sclerites and musculature required for flight; it is weakly developed in wingless taxa. Thoracic legs each have six segments (or podites): coxa, trochanter, femur, tibia, tarsus and pretarsus. The abdomen is primitively 11-segmented, with the gonopore nearly always on segment 8 in the female and segment 9 in the male. Cerci are primitively present. Gas exchange is predominantly tracheal, with spiracles present on both the thorax and abdomen, but may be variably reduced or absent as in some immature stages. Larval/nymphal development is epimorphic, that is, with the number of body segments constant during development.

The orders of insects traditionally have been divided into two groups. Monocondylia is represented by just one small order, Archaeognatha, in which each mandible has a single posterior articulation with the head. Dicondylia, which contains all of the other orders and the overwhelming majority of species, has mandibles characterized by a secondary anterior articulation in addition to the primary posterior one. The traditional groupings of "Apterygota" for the primitively wingless hexapods and "Thysanura" for the primitively wingless taxa Archaeognatha + Zygentoma are both paraphyletic according to most modern analyses (Fig. 7.2).

The traditional classification of Insecta recognized at least 30 orders. However, recent studies, especially using molecular data, have shown that two of the traditional orders (Blattodea and Psocoptera) are each paraphyletic. In each case, another group nested within had been accorded order status due to its possession of diagnostic autapomorphic features. The requirement for monophyly of orders means that only 28 orders of insects are recognized here, with the Isoptera (termites) subsumed into the Blattodea, and the Phthiraptera (parasitic lice) plus Psocoptera forming the order Psocodea. Evidence for these relationships is discussed below. Although a new ordinal-level classification is used for these groups, separate taxon boxes have been provided for the termites and parasitic lice due to the distinctive biology and morphology of each group. Hypotheses of relationships for all insect orders are summarized in Fig. 7.2, with uncertain associations or alternate hypotheses shown by dashed lines.

7.4.1 Apterygote Insecta (former Thysanura *sensu lato*)

The two extant orders of primitively wingless insects, Archaeognatha and Zygentoma, almost certainly are not sister groups, based on most analyses of morphological and molecular data. Thus, the traditional grouping of Thysanura *sensu lato* (meaning in the broad sense in which the name was first used for apterous insects with "bristle tails") should *not* be used. The taxonomic placement of the few, very old wingless fossils (from the Devonian) is uncertain.

Order Archaeognatha (Microcoryphia; archaeognathans or bristletails) (see also Taxobox 2)

Archaeognathans are medium sized, elongate cylindrical, and primitively wingless ("apterygotes"). The head bears three ocelli and large compound eyes that are in contact medially. The antennae are multisegmented. The mouthparts project ventrally, can be partially retracted into the head, and include elongate mandibles with a single condyle (articulation point) each and elongate seven-segmented maxillary palps. Often a coxal style occurs on coxae of legs 2 and 3, or 3 alone. Tarsi are two- or three-segmented. The abdomen continues in an even contour from the humped thorax, and bears ventral muscle-containing styles (representing reduced limbs) on segments 2–9, and generally one or two pairs of eversible vesicles medial to the styles on segments 1–7. Cerci are multisegmented and shorter than the median caudal appendage. Development occurs without change in body form.

The two extant families of Archaeognatha, Machilidae and Meinertellidae, form an undoubted monophyletic group. The order is the oldest of the extant Insecta, with putative fossils from at least the

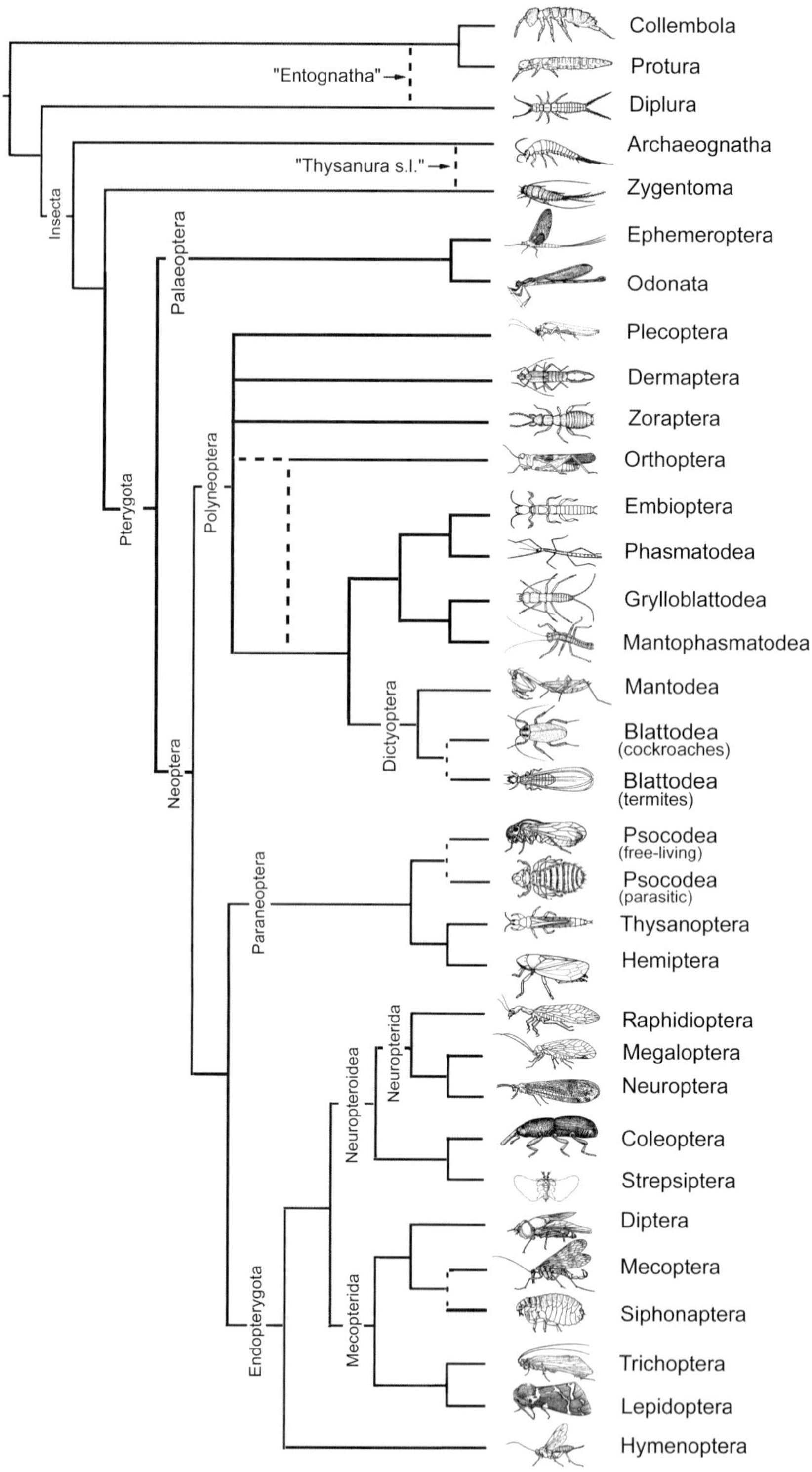
Collembola
Protura
"Entognatha"
Diplura
Archaeognatha
"Thysanura s.l."
Zygentoma
Insecta
Ephemeroptera
Palaeoptera
Odonata
Plecoptera
Dermaptera
Zoraptera
Orthoptera
Pterygota
Polyneoptera
Embioptera
Phasmatodea
Grylloblattodea
Mantophasmatodea
Mantodea
Dictyoptera
Blattodea
(cockroaches)
Blattodea
(termites)
Neoptera
Psocodea
(free-living)
Psocodea
(parasitic)
Paraneoptera
Thysanoptera
Hemiptera
Raphidioptera
Neuropterida
Megaloptera
Neuroptera
Neuropteroidea
Coleoptera
Strepsiptera
Diptera
Mecoptera
Siphonaptera
Endopterygota
Mecopterida
Trichoptera
Lepidoptera
Hymenoptera

Fig. 7.2 Cladogram of postulated relationships of extant hexapods, based on combined morphological and nucleotide sequence data. Dashed lines indicate uncertain relationships or alternative hypotheses. Thysanura *sensu lato* refers to Thysanura in the broad sense. An expanded concept is depicted for each of two orders – Blattodea (including termites) and Psocodea (former Psocoptera and Phthiraptera) – but intraordinal relationships are shown simplified (see Fig. 7.4 and Fig. 7.5 for full details). (Data from several sources.)

Carboniferous and perhaps earlier, and is sister group to Zygentoma + Pterygota (Fig. 7.2). The fossil taxon Monura used to be considered an order of apterygotes related to Archaeognatha, but these monuran fossils (genus *Dasyleptus*) appear to be immature bristletails and the group usually is considered to be sister to all other archaeognathans.

Order Zygentoma (silverfish) (see also Taxobox 3)
Zygentomans are medium sized, dorsoventrally flattened, and primitively wingless ("apterygotes"). Eyes and ocelli are present, reduced or absent, the antennae are multisegmented. The mouthparts are ventrally to slightly forward projecting, and include a special form of double-articulated (dicondylous) mandibles, and five-segmented maxillary palps. The abdomen continues the even contour of the thorax, and includes ventral muscle-containing styles (representing reduced limbs) on at least segments 7–9, sometimes on 2–9, and with eversible vesicles medial to the styles on some segments. Cerci are multisegmented and subequal to the length of the median caudal appendage. Development occurs without change in body form.

There are five extant families and the order probably dates from at least the Carboniferous. Zygentoma is the sister group of the Pterygota (Fig. 7.2) in a group called Dicondylia because of the presence of two articulation points on the base of the mandibles.

7.4.2 Pterygota

Pterygota, sometimes treated as an infraclass, comprises the winged or secondarily wingless (apterous) insects, with thoracic segments of adults usually large, and with the meso- and metathorax variably united to form a pterothorax that bears the wings. The lateral regions of the thorax are well developed. Abdominal segments number 11 or fewer, and lack the styles and vesicular appendages found on apterygote insects. Most Ephemeroptera have a median terminal filament, but other insects do not. The spiracles primarily have a muscular closing apparatus. Mating is by copulation. Metamorphosis is hemi- to holometabolous, with no adult ecdysis, except for the subimago (subadult) stage in Ephemeroptera.

There has been decades of debate concerning the relationships of the Odonata (damselflies and dragonflies) and Ephemeroptera (mayflies) to the rest of the pterygotes, the Neoptera ("new wings"). All three possible arrangements (Fig. 7.3) of these three taxa have been suggested, based on different evidence. The hypothesis of Ephemeroptera as sister to Neoptera (a grouping called Chiastomyaria) has had some support from DNA data. A group consisting of the Odonata and Neoptera (called the Metapterygota) has also found favour, and is characterized morphologically by features such as the loss of moulting in the adult stage, absence of the caudal filament of the abdomen and possession of a strong anterior mandibular articulation. Nonetheless, we accept Ephemeroptera as sister to Odonata, in a monophyletic group called Palaeoptera (as explained below), giving a higher classification of the Pterygota into two divisions – Palaeoptera and Neoptera.

Division Palaeoptera

Insects in this major group have wings that cannot be folded against the body at rest because the wing articulation with the thorax is via axillary plates that are fused with veins. This condition has been termed "palaeopteran" or "palaeopterous" ("old wings"). The two living orders with such wings typically have triadic

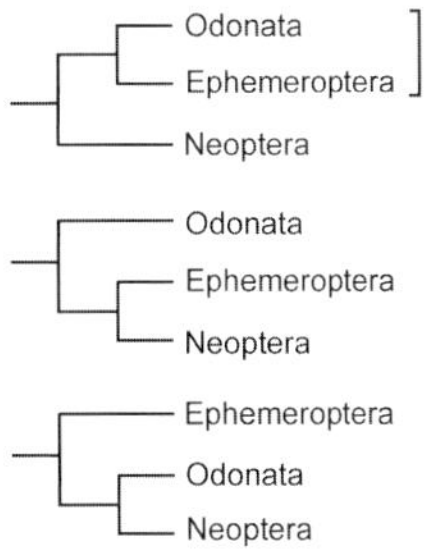

Fig. 7.3 The three possible relationships of Ephemeroptera, Odonata and Neoptera. The top tree shows a monophyletic Palaeoptera, as indicated by the vertical line on the top right.

veins (paired main veins with intercalated longitudinal veins of opposite convexity/concavity to the adjacent main veins) and a network of cross-veins (illustrated in Taxobox 4 and Taxobox 5). This type of wing venation and articulation, certain structures of the head, as well as recent, rigorous re-analyses of gene sequence data, all indicate that Odonata and Ephemeroptera form a monophyletic group. The Palaeoptera also includes some extinct orders (section 8.2.1), but their exact relationships to Odonata and Ephemeroptera are unclear. Although current data suggest that the palaeopterous condition is a synapomorphy of the Palaeoptera, somewhat different sets of wing-vein fusions may be responsible for this condition amongst different orders of the Palaeoptera. Also, it is most likely that a foldable wing was the ancestral condition for pterygotes (see section 8.4) and that the non-foldable wing bases of Ephemeroptera and Odonata were derived secondarily, as well as possibly independently. Thus, the name "old wings" may be a misnomer.

Order Ephemeroptera (mayflies) (see also Taxobox 4)

Ephemeroptera has a fossil record dating back to the Permian and is represented today by a few thousand species. In addition to their "palaeopteran" wing features, mayflies display a number of unique characteristics, including the non-functional, strongly reduced adult mouthparts, the presence of just one axillary plate in the wing articulation, a hypertrophied costal brace, and male fore legs modified for grasping the female during copulatory flight. Retention of a subimago (subadult stage) is unique. Nymphs (naiads) are aquatic, and the mandible articulation, which is intermediate between monocondyly and the dicondylous ball-and-socket joint of all higher Insecta, may be diagnostic.

Historic contraction of ephemeropteran diversity and remnant high levels of homoplasy render phylogenetic reconstruction difficult. Ephemeroptera was divided into two suborders: Schistonota (with nymphal fore-wing pads separate from each other for over half their length) and Pannota ("fused back" – with more extensively fused fore-wing pads). Combined morphology and molecular analyses suggest that Pannota is monophyletic, but that Schistonota is paraphyletic. Carapacea (Baetiscidae + Prosopistomatidae, possessing a notal shield or carapace), Furcatergalia (the pannote families plus some other families such as Leptophlebiidae), Fossoriae (the burrowing mayflies), and superfamilies Caenoidea and Ephemerelloidea all are monophyletic. Other previously recognized lineages, including Pisciforma (the minnow-like mayflies), Setisura (the flat-headed mayflies) and Ephemeroidea, plus families Ameletopsidae and Coloburiscidae, were found to be non-monophyletic.

Order Odonata (damselflies and dragonflies) (see also Taxobox 5)

Odonates have "palaeopteran" wings as well as many additional unique features, including the presence of two axillary plates (humeral and posterior axillary) in the wing articulation and many features associated with specialized copulatory behaviour, including possession of secondary copulatory apparatus on ventral segments 2–3 of the male and the formation of a tandem wheel during copulation (see Box 5.5). The immature stages are aquatic and possess a highly modified prehensile labium for catching prey (see Fig. 13.4).

Odonatologists (those who study odonates) traditionally recognized three groups generally ranked as suborders: Zygoptera (damselflies), Anisozygoptera (fossil taxa plus one extant genus *Epiophlebia* with two species in family Epiophlebiidae) and Anisoptera (dragonflies), but the extant Anisozygoptera now usually are included with Anisoptera in the suborder Epiprocta. Assessment of the monophyly or paraphyly of each suborder has relied very much on interpretation of the very complex wing venation, including that of many fossils. Interpretation of wing venation within the odonates and between them and other insects has been prejudiced by prior ideas about relationships. Thus, the Comstock and Needham naming system for wing veins implies that the common ancestor of modern Odonata was anisopteran, and the venation of zygopterans is reduced. In contrast, the Tillyard naming system implies that Zygoptera is a grade (is paraphyletic) to Anisozygoptera, which itself is a grade on the way to a monophyletic Anisoptera. The recent consensus for extant odonates, based on morphological and molecular data, has both Zygoptera and Epiprocta monophyletic, and Anisoptera as the monophyletic sister group to the Epiophlebiidae.

Zygoptera contains three broad superfamilial groupings, the Coenagrionoidea, Lestoidea and Calopterygoidea, but interrelationships are uncertain. Relationships among major lineages within Anisoptera also are controversial and there is no current consensus. Likewise, the positions of many fossil taxa are contentious but are important to understanding the evolution of the odonate wing.

Division Neoptera

Neopteran ("new wing") insects diagnostically have wings capable of being folded back against their abdomen when at rest, with wing articulation that derives from separate movable sclerites in the wing base, and wing venation with none to few triadic veins and mostly lacking anastomosing (joining) cross-veins (see Fig. 2.23).

The phylogeny (and hence classification) of the neopteran orders remains subject to debate, mainly concerning: (i) the placement of many extinct orders described only from fossils of variably adequate preservation; (ii) the relationships among the Polyneoptera; and (iii) relationships of some groups of Holometabola (the endopterygote orders).

Here, we summarize the most recent research findings, based on both morphology and molecules. No single or combined data set provides unambiguous resolution of insect order-level phylogeny, and there are several areas of controversy. Some questions arise from inadequate data (insufficient or inappropriate taxon sampling) and character conflict within existing data (support for more than one relationship). In the absence of a robust phylogeny, ranking is somewhat subjective and "informal" ranks abound.

A group of 10 orders termed the Polyneoptera is sister to the remaining Neoptera, which can be divided readily into two groups, namely Paraneoptera (the hemipteroids) and Holometabola (= Endopterygota) (Fig. 7.2). Monophyly of Holometabola is undisputed, but some molecular data have suggested paraphyly of Paraneoptera (with Psocodea sister to Holometabola). These three clades – Polyneoptera, Paraneoptera and Holometabola – may be given the rank of subdivision. Polyneoptera and Paraneoptera exhibit hemimetabolous development, in contrast to the complete metamorphosis of Holometabola, although some groups within the Paraneoptera have a convergent form of holometaboly, with one or more quiescent pupal-like stages.

Subdivision Polyneoptera (or Orthopteroid–Plecopteroid assemblage)

This grouping comprises the 10 orders Plecoptera, Mantodea, Blattodea (including the fomer Isoptera), Grylloblattodea, Mantophasmatodea, Orthoptera, Phasmatodea, Embioptera, Dermaptera and Zoraptera.

Some branching events amongst the polyneopteran orders are becoming better understood, but deeper relationships remain poorly resolved, and often contradictory between those suggested by morphology and those from molecular data, or between data from different genes. There is support for monophyly of the Polyneoptera, based on the shared presence of an expanded anal area in the hind wing of winged groups (except the small-bodied Zoraptera and Embioptera), tarsal euplantulae (lacking only in Zoraptera) and recent analyses of nucleotide sequences. Monophyly has been uncertain in the past due to the diversity of morphology among the 10 included orders, the lack of convincing morphological synapomorphies for the whole group, and several contradictory or poorly resolved analyses based on nucleotide sequences. In contrast, recent independent analyses of data from either nuclear protein-coding genes or transcriptomes (ESTs) found strong support for monophyly of the Polyneoptera. Within Polyneoptera, two groupings appear robust, based on both morphological and molecular data. The first is the Dictyoptera (Fig. 7.4), comprising Blattodea (cockroaches and termites) and Mantodea (mantids). All groups within Dictyoptera share distinctive features of the head skeleton (perforated tentorium), mouthparts (paraglossal musculature), digestive system (toothed proventriculus) and female genitalia (shortened ovipositor above a large subgenital plate), which support monophyly. This is substantiated by nearly all analyses based on nucleotide sequences. The second robust clade, sometimes called Xenonomia, is the Grylloblattodea (the ice crawlers or rock crawlers; now apterous, but with winged fossils) and the recently established order Mantophasmatodea. Relationships of each of Plecoptera (stoneflies), Dermaptera (earwigs), Orthoptera (crickets, katydids, grasshoppers, locusts, etc.), Embioptera (webspinners) and Zoraptera (zorapterans) to other orders are uncertain, and the position of Zoraptera has been especially problematic. Sometimes Phasmatodea (stick-insects or phasmids) and Orthoptera have been treated as sisters in a grouping called Orthopterida, but recent molecular evidence suggests a sister-group relationship between Phasmatodea and Embioptera.

Order Plecoptera (stoneflies) (see also Taxobox 6)

Plecoptera are mandibulate in the adult, with filiform antennae, bulging compound eyes, two to three ocelli and subequal thoracic segments. The fore and hind wings are membranous and similar except that the hind wings are broader; aptery and brachyptery are frequent. The abdomen is 10-segmented, with

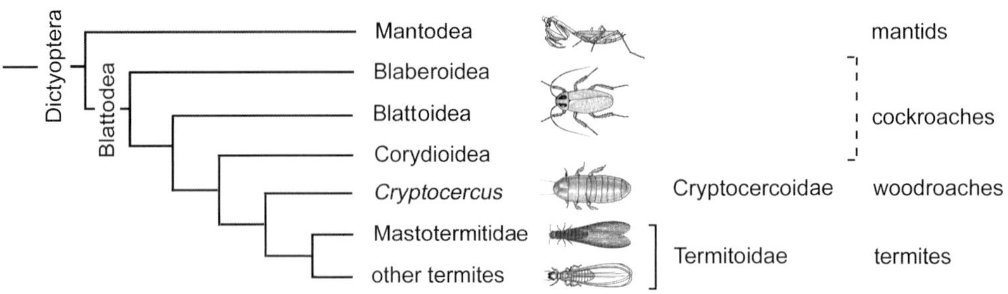

Fig. 7.4 Cladogram of postulated relationships within Dictyoptera, based on combined morphological and nucleotide sequence data. The revised concept of order Blattodea includes the termites, which are given the rank of epifamily (-oidae) as Termitoidae; similarly the woodroaches (Cryptocercoidae: Cryptocercidae). (Data from several sources, including Djernæs *et al.* 2012, with classification of Beccaloni & Eggleton 2013.)

remnants of segments 11 and 12 present, including cerci. Nymphs are aquatic.

Monophyly of the order is supported by a few morphological characters, including in the adult the looping and partial fusion of gonads and male seminal vesicles, and the absence of an ovipositor. In nymphs, the presence of strong, oblique, ventro-longitudinal muscles running intersegmentally and allowing lateral undulating swimming, and the probably widespread "cercus heart", an accessory circulatory organ associated with posterior abdominal gills, support the monophyly of the order. Nymphal plecopteran gills may occur on almost any part of the body, or may be absent. This varied distribution causes problems of homology of gills between families, and between those of Plecoptera and other orders. Whether Plecoptera are ancestrally aquatic or terrestrial is debatable. Plecoptera is one of the oldest lineages of Polyneoptera, but its uncertain relationships are portrayed here as unresolved (Fig. 7.2). It may be sister to Dermaptera, as suggested by some rDNA, nuclear-protein coding DNA and mitochondrial genome data, or sister to some or all other Polyneoptera, except Dermaptera and Zoraptera. The earliest stonefly fossils are from the Early Permian and include terrestrial nymphs, but modern families do not appear until the Mesozoic.

Internal relationships have been proposed as two predominantly vicariant suborders, the austral (southern hemisphere) Antarctoperlaria and the northern Arctoperlaria. The monophyly of Antarctoperlaria is argued based on the unique sternal depressor muscle of the fore trochanter, lack of the usual tergal depressor of the fore trochanter, and presence of floriform chloride cells, which may have an additional sensory function. Included taxa are the large-sized Eustheniidae and Diamphipnoidae, the Gripopterygidae and Austroperlidae – all southern hemisphere families. Some nucleotide sequence studies support this clade.

The sister group Arctoperlaria lacks defining morphology, but is united by a variety of mechanisms linked to drumming sound associated with mate finding. Component families include Capniidae, Leuctridae, Nemouridae (including Notonemouridae), Perlidae, Chloroperlidae, Pteronarcyidae and several smaller families, and the suborder is essentially northern hemisphere, with a lesser radiation of Notonemouridae into the southern hemisphere. Nucleotide sequence data appear to support the monophyly of Arctoperlaria and Antarctoperlaria. Relationships amongst extant Plecoptera have been used in hypothesizing origins of wings from "thoracic gills", and in tracing the possible development of aerial flight from surface flapping with legs trailing on the water surface, and forms of gliding. Current views of the phylogeny suggest these traits are secondary and reductional.

Order Dermaptera (earwigs) (see also Taxobox 7)

Adult earwigs are elongate and dorsoventrally flattened, with mandibulate, forward-projecting mouthparts, compound eyes ranging from large to absent, no ocelli, and short annulate antennae. The tarsi are three segmented with a short second tarsomere. Many species are apterous or, if winged, the fore wings are small, leathery and smooth, forming unveined tegmina, and the hind wings are large, membranous, semi-circular and dominated by an anal fan of radiating vein branches connected by cross-veins.

Traditionally, three suborders of earwigs have been recognized, with the majority of species in suborder Forficulina. Five species commensal or semi-parasitic on bats in Southeast Asia have been placed in their own family (Arixeniidae) and suborder (Arixeniina). Similarly, 11 species semi-parasitic on African rodents have been placed in their own family (Hemimeridae) and suborder (Hemimerina). Earwigs in both of these epizoic groups are blind, apterous and exhibit pseudo-placental viviparity. Recent morphological and molecular studies show derivation of both groups from within the superfamily Forficuloidea of Forficulina, rendering the latter paraphyletic and overturning the old suborder classification in favour of all extant earwigs being in one suborder, Neodermaptera. Only some of the 11 extant neodermapteran families appear to be supported by synapomorphies. Other families, such as Pygidicranidae and Spongiphoridae, may be paraphyletic, as much weight has been placed on plesiomorphies, especially of the penis specifically and genitalia more generally, or homoplasies (convergences) in furcula form and wing reduction.

Different data sources have suggested a dermapteran sister-group relationship to Dictyoptera, Embioptera or Zoraptera. Although we consider the relationship of Dermaptera is best considered as unresolved, a tentative relationship as sister to Plecoptera finds some support.

Order Zoraptera (zorapterans or angel insects) (see also Taxobox 8)

Zoraptera is one of the smallest and probably the least known pterygote order. Zorapterans are small, rather termite-like insects, with simple morphology. They have biting, generalized mouthparts, including five-segmented maxillary palps and three-segmented labial palps, and expanded hind femora bearing stout ventral spines. Sometimes both sexes are apterous, and in alate forms the hind wings are smaller than the fore wings; the wings are shed as in ants and termites. Wing venation is highly specialized and reduced.

Traditionally, the order contains only one family (Zorotypidae) and one extant genus (*Zorotypus*); a proposal to divide it into several genera of uncertain monophyly, delimited predominantly on wing venation, is not widely accepted. The phylogenetic position of Zoraptera, based on data from both morphology and nucleotide sequences, has been controversial, with many different placements suggested. Currently, only four sister-group hypotheses are considered plausible, based on recent phylogenetic analyses; these are as sister to Dermaptera, Dictyoptera, Embioptera or all the rest of Polyneoptera. Molecular data (rDNA and nuclear protein-coding) support Zoraptera as a close relative of Dictyoptera, even though ribosomal DNA (especially *18S*) of Zoraptera has been shown to be unusual. Another suggested relationship is Zoraptera + Embioptera, supported by shared states of wing-base structure and hind-leg musculature, and several reduction or loss features. Recent analyses of transcriptome data suggest either Zoraptera as sister to all other Polyneoptera or just to Dermaptera, and reject Zoraptera as sister to either Embioptera or Dictyoptera.

Order Orthoptera (grasshoppers, locusts, katydids and crickets) (see also Taxobox 9)

Orthopterans are medium-sized to large insects, with hind legs enlarged for jumping (saltation). The compound eyes are well developed, the antennae are elongate and multisegmented, and the prothorax is large with a shield-like pronotum curving downwards laterally. The fore wings form narrow, leathery tegmina, and the hind wings are broad, with numerous longitudinal and cross-veins and folded beneath the tegmina by pleating; aptery and brachyptery are frequent. The abdomen has eight or nine annular visible segments, with the two or three terminal segments reduced, and one-segmented cerci. The ovipositor is well developed, formed from highly modified abdominal appendages.

The Orthoptera is one of the oldest extant insect orders, with a fossil record dating back nearly 300 million years. Several morphological features and some molecular data have suggested that the Orthoptera is closely related to Phasmatodea, to the extent that some entomologists united the orders in the past. However, molecular evidence, different wing-bud development, egg morphology and lack of auditory organs in phasmatids strongly support distinction. The placement of Orthoptera has not been resolved by molecular data, which suggest that Orthoptera is sister to various different groupings within the Polyneoptera. It may be sister to all of Polyneoptera except Dermaptera, Plecoptera and Zoraptera. Two of the alternative positions for Orthoptera are shown by dashed lines in Fig. 7.2, and much further study is required.

The division of Orthoptera into two monophyletic suborders, Caelifera (grasshoppers and locusts – predominantly day-active, fast-moving, visually acute, terrestrial herbivores) and Ensifera (katydids and crickets – often night-active, camouflaged or mimetic, predators, omnivores or phytophages), is supported on morphological and molecular evidence. Relationships of major groupings within Ensifera vary among studies. For Caelifera, the seven or eight superfamilies sometimes are divided into four major groups, namely Tridactyloidea, Tetrigoidea, Eumastacoidea, and the "higher Caelifera", containing acridoid grasshoppers (Acridoidea) plus several less-speciose superfamilies. Tridactyloidea may be sister to the rest of the Caelifera, but internal relationships are not resolved.

Order Embioptera (= Embiidina, Embiodea) (embiopterans or webspinners) (see also Taxobox 10)

Embiopterans have an elongate, cylindrical body, somewhat flattened in the male. The head has kidney-shaped compound eyes that are larger in males than in females, and lacks ocelli. The antennae are multisegmented, and the mandibulate mouthparts project forwards (prognathy). All females and some males are apterous; wings, if present, are characteristically soft and flexible, with blood sinus veins stiffened for flight by blood pressure. The legs are short, with three-segmented tarsi, and the basal segment of each fore tarsus is swollen because it contains silk glands. The hind femora are swollen by strong tibial muscles. The abdomen is 10-segmented, with rudiments of segment 11 and with two-segmented cerci. The female external genitalia are simple (a rudimentary ovipositor) and those of males are complex and asymmetrical.

Several ordinal names have been used for webspinners but Embioptera is preferred because this name has the widest usage in published work and its ending matches the names of some related orders. Most of the rules of nomenclature do not apply to names above the family group and thus there is no name priority at ordinal level. The Embioptera is undoubtedly monophyletic, based above all on the ability to produce silk from unicellular glands in the anterior basal tarsus, and to spin the silk – pull and shear it to form sheets – to construct silken domiciles. Based on morphology, suggested sister-group relationships of Embioptera are to Dermaptera, Phasmatodea, Plecoptera or Zoraptera, but recent assessment of data from head morphology and nucleotide sequences favours Embioptera + Phasmatodea. Historically, the classification of embiopterans has overemphasized the male genitalia. Recent morphological and molecular analyses, including representatives of most of the described higher taxa of extant Embioptera, suggest that at least nine of the 13 recognized families are monophyletic.

Order Phasmatodea (phasmids, stick-insects or walking sticks) (see also Taxobox 11)

Phasmatodea exhibit body shapes that are variations on elongate cylindrical and stick-like or flattened, or often leaf-like. The mouthparts are mandibulate. The compound eyes are relatively small and placed anterolaterally, with ocelli only in winged species, and often only in males. The wings, if present, are functional in males, but often reduced in females, and many species are apterous in both sexes. Fore wings form short leathery tegmina, whereas the hind wings are broad, with a network of numerous cross-veins and with the anterior margin toughened to protect the folded wing. The legs are elongate, slender, and adapted for walking, with five-segmented tarsi. The abdomen is 11-segmented, with segment 11 often forming a concealed supra-anal plate in males or a more obvious segment in females.

Traditionally, Phasmatodea was considered as sister to Orthoptera within the orthopteroid assemblage. Evidence from morphology in support of this grouping comes from some genitalic and wing features and a limited neurophysiological study. Phasmatodea are distinguished from the Orthoptera by their body shape, asymmetrical male genitalia, proventricular structure, and lack of rotation of nymphal wing pads during development. Evidence for a sister-group relationship of Phasmatodea to Embioptera comes from recent analyses of morphological (especially head) and molecular (including transcriptome) data. Further data and broader taxon sampling are needed to resolve the issue. Also, there is no workable family or subfamily classification of Phasmatodea, and recent phylogenetic studies have shown non-monophyly of most higher groups. The only certainty in internal relationships is that plesiomorphic western North American *Timema* (suborder Timematodea) is sister to the remaining extant phasmids of suborder Euphasmatodea (or Euphasmida).

Order Grylloblattodea (= Grylloblattaria, Notoptera) (grylloblattids, ice crawlers or rock crawlers) (see also Taxobox 12)

Grylloblattids are moderate-sized, soft-bodied insects, with anteriorly projecting mandibulate mouthparts and the compound eyes are either reduced or absent. The antennae are multisegmented and the mouthparts mandibulate. The quadrate prothorax is larger than the meso- or metathorax, and wings are absent. The legs have large coxae and five-segmented tarsi. Ten abdominal segments are visible, with rudiments of segment 11, including five- to nine-segmented cerci. The female has a short ovipositor, and the male genitalia are asymmetrical.

Several ordinal names have been used for these insects but Grylloblattodea is preferred because this name has the widest usage in published work and its ending matches the names of some related orders. Most of the rules of nomenclature do not apply to names above the family group and thus there is no name priority at ordinal level. Initially, the phylogenetic placement of Grylloblattodea also was controversial, generally being argued to be relictual, either "bridging the cockroaches and orthopterans" or "primitive amongst orthopteroids". The antennal musculature resembles that of mantids and embiids, mandibular musculature resembles Dictyoptera, and the maxillary muscles those of Dermaptera. Grylloblattids resemble orthopteroids embryologically and based on sperm ultrastructure. Molecular phylogenetic study emphasizing grylloblattids strongly supports a sister-group relationship to the Mantophasmatodea. Morphological evidence for the latter relationship includes several putative synapomorphies of the head, and shared loss of ocelli and wings.

Grylloblattodea is sometimes claimed to have a diverse fossil record, but allocation of most fossils to this order is dubious, although several Permian and Jurassic fossils of winged taxa may represent stem-group Grylloblattodea or distinct but related lineages. The species diversity of extant grylloblattids is very low, and they occur only in western North America and eastern Asia.

Order Mantophasmatodea (heelwalkers) (see also Taxobox 13)

Mantophasmatodea is the most recently recognized order, comprising three extant families from sub-Saharan Africa, as well as Baltic amber specimens and a recently described representative from the Middle Jurassic. Mantophasmatodeans all are apterous, without even wing rudiments. The head is hypognathous, with generalized mouthparts and long, slender, multisegmented antennae. Coxae are not enlarged, the fore and mid femora are broadened and have bristles or spines ventrally; hind legs are elongate; tarsi are five-segmented, with euplanulae on the basal four; the ariolum is very large and the distal tarsomere is held off the substrate in life. Male cerci are prominent, clasping and not differentially articulated with tergite 10; female cerci are short and one-segmented. A distinct short ovipositor projects beyond a short subgenital lobe, lacking any protective operculum (plate below ovipositor) as seen in phasmids.

Based on morphology, placement of the new order was difficult, but relationships with phasmids (Phasmatodea) and/or ice crawlers (Grylloblattodea) were suggested. Nucleotide sequence data have justified the rank of order, and data from ribosomal and nuclear protein-coding genes strongly support a sister-group relationship to Grylloblattodea, whereas mitochondrial genes suggest a sister relationship to Phasmatodea. A grouping of Mantophasmatodea and Grylloblattodea variously has been called Notoptera or Xenonomia by different authors. The presence of a convincing fossil mantophasmatodean (*Juramantophasma*) from the Jurassic (165 mya) of China suggests that the order is at least early Mesozoic in origin and also was once more widely distributed than its current African distribution.

Monophyly of each of the extant families, Austrophasmatidae (South African except for *Striatophasma* from Namibia) and Mantophasmatidae (from Namibia), has been supported by mitochondrial and neuropeptide sequences. The sole representative of Tanziophasmatidae (a single male specimen from Tanzania) either is put in its own genus *Tanziophasma* or sometimes treated as a species of *Mantophasma* (in Mantophasmatidae), but no living specimens are available for DNA analysis.

Order Mantodea (mantids, mantises or praying mantids) (see also Taxobox 14)

Mantodea are predatory, with males generally smaller than females. The small, triangular head is mobile, with slender antennae, large, widely separated eyes, and mandibulate mouthparts. The prothorax is narrow and elongate, with the meso- and metathorax shorter. The fore wings form leathery tegmina with a reduced anal area; the hind wings are broad and membranous, with long unbranched veins and many cross-veins, but often are reduced or absent. The fore legs are raptorial, whereas the mid and hind legs are elongate for walking. The abdomen has a visible 10th segment, bearing variably segmented cerci. The ovipositor predominantly is internal, and the external male genitalia are asymmetrical.

Mantodea forms the sister group to Blattodea (including the termites) (Fig. 7.4), and shares many features with Blattodea, such as strong direct flight muscles and weak indirect (longitudinal) flight muscles, asymmetrical male genitalia and multisegmented cerci. Derived features of Mantodea relative to Blattodea involve modifications associated with predation, including leg morphology, an elongate prothorax, and features associated with visual predation, namely the mobile head with large, separated eyes. The current family-level classification of Mantodea is unnatural, with many of the up to 15 recognized families probably paraphyletic. A phylogeny based on multiple genes and one or more representatives of all families found that relationships reflected biogeography more than the traditional classification, due to morphological convergences confounding delimitation of higher groups.

Order Blattodea (cockroaches and termites) (see also Taxobox 15 and Taxobox 16)

Termites are considered part of Blattodea but are discussed separately below. Cockroaches are dorsoventrally flattened insects with filiform, multisegmented antennae and mandibulate, ventrally projecting mouthparts. The prothorax has an enlarged, shield-like pronotum that often covers the head; the meso- and metathorax are rectangular and subequal. The fore wings are sclerotized tegmina protecting membranous hind wings folded fan-like beneath. Hind wings often may be reduced or absent and, if present, characteristically have many vein branches and a large anal lobe. The legs may be spiny and the tarsi are five-segmented. The abdomen has 10 visible segments in the male, with a subgenital plate (sternum 9) usually bearing one or two styles and concealing well-developed asymmetrical genitalia and the reduced 11th segment. Cerci have one or usually many segments; the female ovipositor valves are small, concealed beneath tergum 10.

Although Blattodea was long considered as an order (and hence monophyletic), convincing evidence shows that the termites arose from within the cockroaches, and the "order" thus is paraphyletic if termites are excluded from it. The sister group of the termites is the wingless *Cryptocercus* woodroaches of North America and eastern Asia, which undoubtedly are cockroaches (Fig. 7.4). Other internal relationships of the Blattodea are not well understood, with apparent conflict between morphology and molecular data. Usually, eight families of cockroaches and nine families of termites are recognized. Ectobiidae (formerly Blatellidae) and Blaberidae (the largest families) may be sister groups, or Blaberidae may render Ectobiidae paraphyletic. The many early fossils allocated to Blattodea that possess a well-developed ovipositor are considered best as belonging to a blattoid stem-group, that is, from prior to the ordinal diversification of the Dictyoptera.

Epifamily Termitoidae (former order Isoptera; termites, "white ants") (see also Taxobox 16)

Termites form a distinctive clade within Blattodea due to their eusociality and polymorphic caste system of reproductives, workers and soldiers. Mouthparts are blattoid and mandibulate. Antennae are long and multisegmented. The fore and hind wings generally are similar, membranous and with restricted venation; but *Mastotermes* (Mastotermitidae; with one extant species, *M. darwiniensis*), with complex wing venation and a broad hind-wing anal lobe, is exceptional. The male has no intromittent organ, in contrast to the complex, asymmetrical genitalia of Blattodea and Mantodea. Females have no ovipositor, except *Mastotermes*, which have a reduced blattoid-type ovipositor.

The termites comprise a morphologically derived group within Dictyoptera. A long-held view that Mastotermitidae is the sister group of the rest of the termites is upheld by all studies – the distinctive features mentioned above evidently are plesiomorphies. Recent studies that included structure of the proventriculus and nucleotide sequence data demonstrate that termites arose from within the cockroaches, thereby rendering Blattodea paraphyletic if termites are maintained at ordinal rank as the Isoptera (Fig. 7.4). Thus, the termite clade has been lowered in rank to epifamily (a category between superfamily and family) to reduce disruption to the current classification – the names of all cockroach and termite families are maintained, as proposed by several termite researchers. Given that overwhelming evidence puts the genus *Cryptocercus* as the sister group to termites, the alternative hypothesis of an independent origin (hence convergence) of the semisociality (parental care and transfer of symbiotic gut flagellates between generations) of *Cryptocercus* and the sociality of termites (section 12.4.2) no longer is supported. There is still no consensus on the number of termite families or on relationships among the families of so-called "lower termites" (non-Termitidae). Furthermore, Rhinotermitidae is paraphyletic with respect to Termitidae.

Subdivision Paraneoptera (Acercaria, or Hemipteroid assemblage)

This subdivision comprises the orders Psocodea (composed of former orders Psocoptera and Phthiraptera), Thysanoptera and Hemiptera. This group is defined by derived features of the mouthparts, including the slender, elongate maxillary lacinia separated from the stipes and a swollen postclypeus containing an enlarged cibarium (sucking pump), and the reduction in tarsomere number to three or fewer, Malpighian tubules to four and abdominal ganglia to a single complex. The monophyly of Paraneoptera is supported by morphology and ribosomal DNA, although nuclear protein-coding genes analysed alone and some transcriptome data recover paraphyly of Paraneoptera, with Psocodea as sister to Holometabola.

Within Paraneoptera, the monophyletic order Psocodea (formerly treated as a superorder) contains "Phthiraptera" (parasitic lice) and "Psocoptera" (bark lice and book lice). Phthiraptera arose from within Psocoptera, rendering that group paraphyletic if Phthiraptera is maintained at ordinal rank. Although sperm morphology and some molecular sequence data imply that Hemiptera is sister to Psocodea + Thysanoptera, a grouping of Thysanoptera + Hemiptera (called superorder Condylognatha) is supported by molecular data and derived head structures including the stylet mouthparts, midgut structure and function, features of the fore-wing base and the presence of a sclerotized ring between antennal flagellomeres. Condylognatha thus forms the sister group to Psocodea (Fig. 7.5).

Order Psocodea (bark lice, book lice, chewing lice and sucking lice) (see also Taxobox 17 and Taxobox 18)

The use of the order name Psocodea is advocated for seven suborders that comprised the former orders "Psocoptera" (non-parasitic lice, or bark lice and book lice) and "Phthiraptera" (parasitic lice, or chewing lice and sucking lice) (Fig. 7.5). Non-parasitic lice are small cryptic insects, with a large, mobile head, a bulbous postclypeus, and membranous wings held roof-like over the abdomen, except for a few groups (e.g. the book lice family Liposcelididae) that have dorsoventrally flattened bodies, a prognathous head and are often wingless. Parasitic lice are wingless obligate ectoparasites of birds and mammals, with dorsoventrally flattened bodies.

Phylogenetic analysis of both morphological and molecular data showed that the traditionally recognized order Psocoptera was rendered paraphyletic by a sister-group relationship of Liposcelididae (book lice) to part of the traditionally recognized order Phthiraptera. A revised classification treats "Psocoptera" plus "Phthiraptera" as a single order. Among the three suborders of non-parasitic lice, Troctomorpha (bark and book lice), Trogiomorpha (bark lice) and Psocomorpha (bark lice), there is support for the monophyly of Psocomorpha and probably Trogiomorpha. Relationships among the parasitic lice are reasonably resolved, with all four morphologically defined suborders (Anoplura, Amblycera, Ischnocera and Rhyncophthirina) probably monophyletic. Traditionally, the group consisting of Amblycera, Ischnocera and Rhyncophthirina had been treated as a monophyletic Mallophaga (biting and chewing lice) based

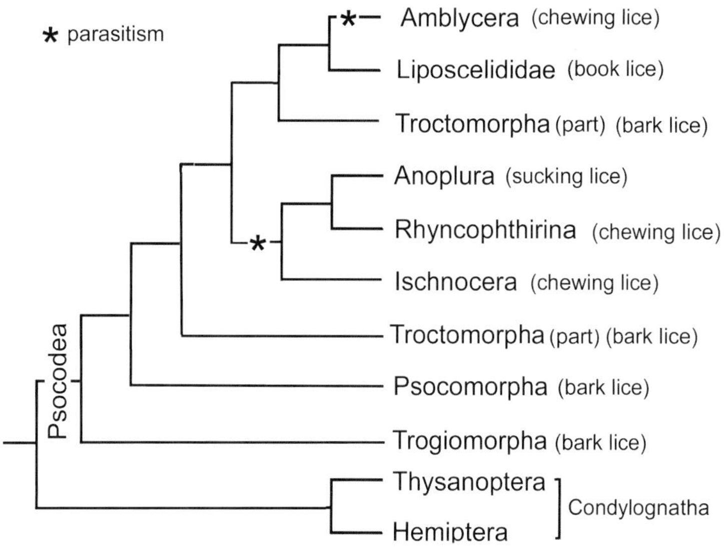

Fig. 7.5 Cladogram of postulated relationships among suborders of Psocodea, with Condylognatha as the sister group. The depicted hypothesis of relationships suggests two origins of parasitism in the order. (After Johnson *et al.* 2004; Yoshizawa & Johnson 2010).

on their feeding mode and morphology, in contrast to the piercing and blood-feeding Anoplura. However, *18S* rDNA data alone and combined evidence from multiple genes suggest that the Liposcelididae (book lice; Troctomorpha) is sister to just the Amblycera, rather than to all parasitic lice; thus parasitic lice are not monophyletic, and parasitism either arose twice or has been lost secondarily in some groups (Fig. 7.5). The fossil record of parasitic lice is very poor, but the recent description of an amber fossil member of the Liposcelididae from about 100 Ma suggests that the parasitic lice must have diverged from the rest of the Psocodea at least by the mid Cretaceous, and thus hosts of parasitic lice may have included the early mammals, early birds and perhaps certain dinosaurs. At lower taxonomic levels (genera and species), robust estimates of relationship are needed to estimate evolutionary interactions between parasitic lice and their bird and mammal hosts. For certain groups of lice, such as the sucking lice that parasitize primates, phylogenies have demonstrated that the lice co-speciate with their hosts (see section 13.3.3).

Order Thysanoptera (thrips) (see also Taxobox 19)
The development of Thysanoptera includes two or three pupal stages in a form of holometaboly convergent with that of the Holometabola(= Endopterygota). The thrips head is elongate, and the mouthparts are unique in that the maxillary laciniae form grooved stylets, the right mandible is atrophied, but the left mandible forms a stylet; all three stylets together form the feeding apparatus. The tarsi are each one- or two-segmented, and the pretarsus has an apical protrusible adhesive ariolum (bladder or vesicle). The narrow wings have a fringe of long marginal setae, called cilia. Reproduction in thrips is haplodiploid.

Recent molecular evidence strongly supports a traditional morphological division of the Thysanoptera into two suborders: Tubulifera, containing the single speciose family Phlaeothripidae, and the Terebrantia. Terebrantia includes one speciose family, Thripidae, which is monophyletic, and seven smaller families, of which either Merothripidae or Aelothripidae may be sister to the rest of Terebrantia. Phylogenies have been generated at lower taxonomic levels concerning aspects of the evolution of sociality, especially the origins of gall-inducing thrips and of "soldier" castes in Australian gall-inducing Thripidae.

Order Hemiptera (bugs, moss bugs, cicadas, leafhoppers, planthoppers, spittle bugs, treehoppers, aphids, jumping plant lice, scale insects and whiteflies) (see also Box 10.2 and Taxobox 20)
Hemiptera, the largest non-endopterygote order, has diagnostic mouthparts, with mandibles and maxillae modified as needle-like stylets, lying in a beak-like, grooved labium, collectively forming a rostrum or proboscis. Within this, the stylet bundle contains two canals, one delivering saliva and the other uptaking fluid. Hemiptera lack maxillary and labial palps. The prothorax and mesothorax usually are large, and the metathorax small. Venation of both pairs of wings can be reduced; some species are apterous, and male scale insects have only one pair of wings (the hind wings are hamulohalteres). Legs often have complex pretarsal adhesive structures. Cerci are lacking.

Hemiptera and Thysanoptera are sisters in a grouping called Condylognatha within Paraneoptera (Fig. 7.2). Hemiptera once was divided into two groups, Heteroptera (true bugs) and "Homoptera" (cicadas, leafhoppers, planthoppers, spittle bugs, aphids, psylloids, scale insects and whiteflies), treated as either suborders or as orders. All "homopterans" are terrestrial plant feeders, and many share a common biology of producing honeydew and being ant-attended. Although recognized by features such as wings (if present) usually held roof-like over the abdomen and fore wings either membranous or in the form of tegmina of uniform texture, "Homoptera" represents a paraphyletic grade rather than a clade and the name should not be used (Fig. 7.6). This view is supported strongly by phylogenetic analyses of morphological data and of multi-locus nucleotide sequences (nuclear and mitochondrial protein-coding and ribosomal genes), which suggest the relationships shown in Fig. 7.6. However, available mitochondrial genome sequences suggest different patterns of relationship, but suffer from unbalanced taxon sampling and problematic features of sternorrhynchan mitogenomes.

The rank of hemipteran clades has been much disputed and names abound. The four monophyletic taxa, Fulgoromorpha, Cicadomorpha, Coleorrhyncha and Heteroptera (sometimes collectively termed the Euhemiptera), each may be treated as a suborder, or Fulgoromorpha + Cicadomorpha may be united into suborder Auchenorrhyncha. Also, sometimes Coleorrhyncha and Heteroptera are treated as one suborder,

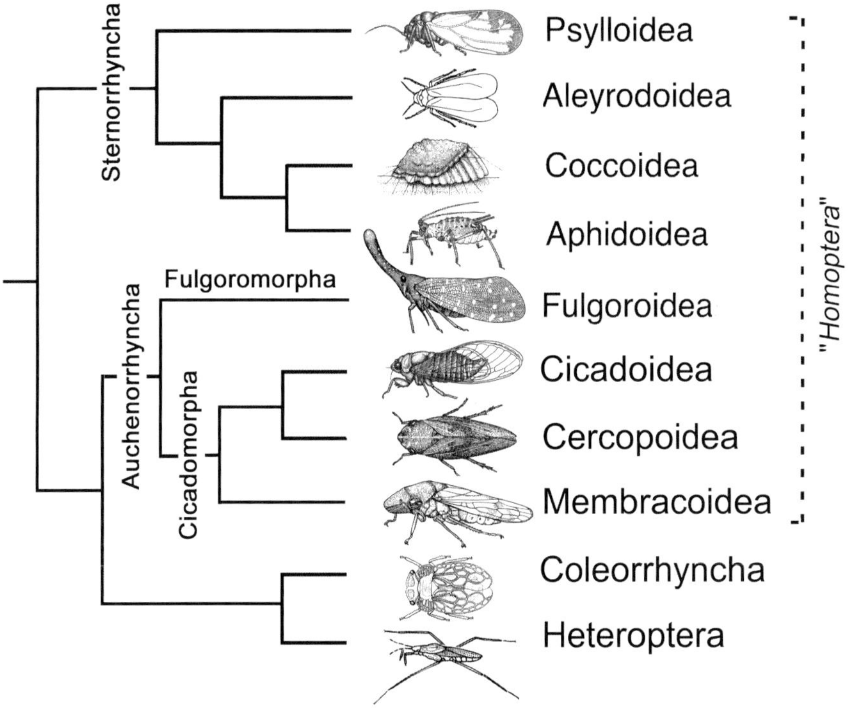

Fig. 7.6 Cladogram of postulated relationships within Hemiptera, based on combined morphological and nucleotide sequence data. The dashed line and italicized name indicate the paraphyly of Homoptera. (After Cryan & Urban 2012.)

Prosorrhyncha. The sister group to Euhemiptera is suborder Sternorrhyncha, which contains the aphids (Aphidoidea), jumping plant lice (Psylloidea), scale insects (Coccoidea) and whiteflies (Aleyrodoidea). Sternorrhynchans are characterized principally by their possession of a particular kind of gut filter chamber, a rostrum that appears to arise between the bases of their front legs, a one- or two-segmented tarsus and, if winged, by absence of the vannus and vannal fold in the hind wings. Among Euhemiptera, the traditional grouping Auchenorrhyncha was suspected to be paraphyletic, but a recent multi-gene analysis supports monophyly. Auchenorrhyncha contains the Fulgoromorpha (= Archaeorrhyncha; planthoppers) and Cicadomorpha (= Clypeorrhyncha; cicadas, leafhoppers, treehoppers and spittle bugs) and is defined morphologically by shared possession of a tymbal acoustic system, an aristate antennal flagellum, a labium arising from the posteroventral region of the head, male lateral ejaculatory ducts, and structures associated with wing coupling and the fore-wing base, and by shared absence of a sclerotized gula. Many species of Auchenorrhyncha also share endosymbiont bacteria (*Sulcia*) of phylum Bacteroidetes, whereas the diverse endosymbiont groups of Heteroptera and Coleorrhyncha belong to the phylum Proteobacteria. These bacteria provide essential compounds that are not available in the diet (section 3.6.5).

Heteroptera (true bugs, including assassin bugs, back-swimmers, lace bugs, stink bugs, waterstriders and others) is the sister group of the Coleorrhyncha, containing only one family, Peloridiidae or moss bugs. Although small, cryptic and rarely collected, moss bugs have generated considerable phylogenetic interest due to their combination of ancestral and derived hemipteran features, and their exclusively "relictual" Gondwanan distribution. Heteropteran diversity is distributed amongst about 75 families, forming the largest hemipteran clade. Heteroptera is diagnosed most easily by the presence of metapleural scent glands, and monophyly is undisputed. Seven infraorders (the rank between suborder and superfamily) are recognized:

Cimicomorpha, Dipsocoromorpha, Enicocephalomorpha, Gerromorpha, Leptopodomorpha, Nepomorpha and Pentatomomorpha, with some molecular and morphological data supporting Nepomorpha (true water bugs) as the sister group of the rest.

Subdivision Holometabola (= Endopterygota)

Endopterygota comprise insects with holometabolous development, in which immature (larval) instars are very different from their respective adults. The adult wings and genitalia are internalized in their pre-adult expression, developing in imaginal discs that are evaginated at the penultimate moult. Larvae lack true ocelli. The "resting stage" or pupa is non-feeding and precedes the adult (imago), which may persist for some time as a pharate ("cloaked" in pupal cuticle) adult (see section 6.2.3 and section 6.2.4). Unique derived features of endopterygotes are less evident in adults than in earlier developmental stages, but the clade is recovered consistently from all phylogenetic analyses. Recent analyses of nuclear protein-coding genes and transcriptome data have supplemented earlier studies based on ribosomal and mitochondrial DNA and successfuly confirmed most ordinal relationships within Holometabola.

One of the strongest groups among endopterygotes is Amphiesmenoptera, the sister-group relationship between the Trichoptera (caddisflies) and Lepidoptera (butterflies and moths). A plausible scenario of an ancestral amphiesmenopteran taxon envisages a larva living in damp soil amongst liverworts and mosses, followed by radiation into water (Trichoptera) or into terrestrial plant feeding (Lepidoptera).

Antliophora, a second well-supported relationship, unites Diptera (true flies), Mecoptera (scorpionflies, hangingflies and snowfleas) and Siphonaptera (fleas). Fleas once were considered to be sister to Diptera, but nucleotide sequence evidence supports a closer relationship to mecopterans (Fig. 7.7). Longstanding morphological evidence, and all recent molecular data, support a sister group relationship between Antliophora and Amphiesmenoptera – a combined group referred to as Mecopterida (or Panorpida).

Another strongly supported relationship is between three orders – Neuroptera, Megaloptera and Raphidioptera – together called Neuropterida, and sometimes treated as a group of ordinal rank, which shows a sister-group relationship to Coleoptera + Strepsiptera (see below). Morphology and all molecular data support a close relationship of Coleoptera with Neuropterida. The higher group comprising

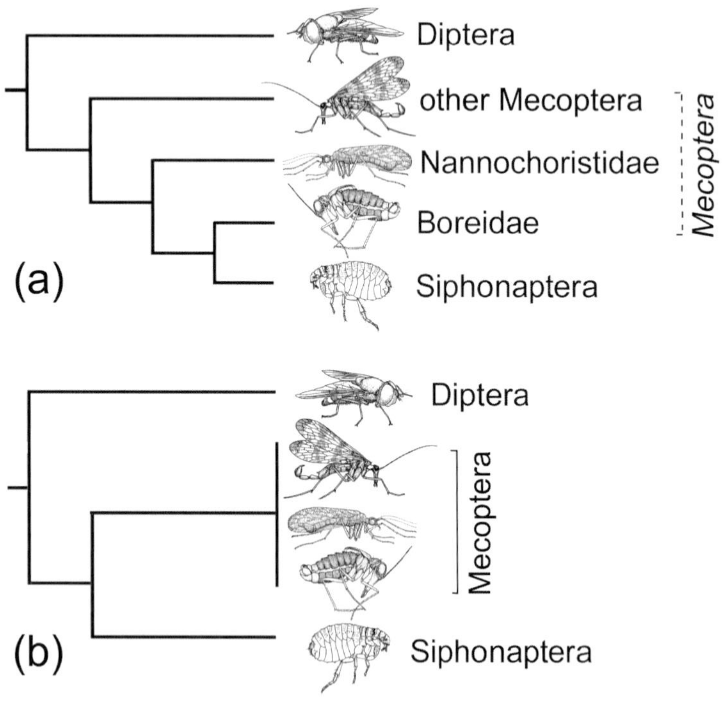

Fig. 7.7 Two competing hypotheses for the relationships of Antliophora. (a) Based on nucleotide sequence data, including ribosomal genes, supported by morphology. (After Whiting 2002.) (b) Based on nucleotide sequence data from single-copy protein-coding genes. (After Wiegmann *et al.* 2009.). The order Mecoptera is paraphyletic in (a), as indicated by its italicized name and the dashed line, whereas it is monophyletic in (b), as indicated by the solid line.

Neuropterida, Coleoptera and Strepsiptera is called Neuropteroidea.

Strepsiptera has been phylogenetically enigmatic, but resemblance of their first-instar larvae (called triungula) and adult males to those of certain Coleoptera, notably parasitic Ripiphoridae, and some wing-base features, have been cited as indicative of a close relationship. Recent phylogenetic studies of nuclear protein-coding genes and whole genomes support the Strepsiptera either as the sister group of Coleoptera or as a group within Coleoptera. A relationship of Strepsiptera with Diptera, as suggested by some earlier molecular evidence and by haltere development, has been refuted. Strepsiptera has undergone much morphological and molecular evolution, and is highly derived relative to other taxa. Such long-isolated evolution of the genome can create a problem known as "long-branch attraction", in which nucleotide sequences may converge by chance mutations alone with those of an unrelated taxon with a similarly long independent evolution.

The relationship of one major order of endopterygotes, Hymenoptera, is as sister to all other orders of Holometamola, as shown by recent analyses of all data sources.

Relationships within Holometabola are summarized in Fig. 7.2.

Orders Neuroptera (lacewings, owlflies, antlions), and Megaloptera (alderflies, dobsonflies, fishflies) and Raphidioptera (snakeflies) (see also Box 10.4 and Taxobox 21)

Neuropterida comprises three minor (species-poor) orders, whose adults have multisegmented antennae, large, separated eyes, and mandibulate mouthparts. The prothorax may be larger than either the meso- or metathorax, which are about equal in size. Legs sometimes are modified for predation. The fore and hind wings are quite similar in shape and venation, with folded wings often extending beyond the abdomen. The abdomen lacks cerci, and members of the three groups share the synapomorphic fusion of the third ovipositor valvulae.

Many adult neuropterans are predators, and have wings typically characterized by numerous cross-veins and "twigging" at the ends of veins. Neuropteran larvae usually are active predators, with slender, elongate mandibles and maxillae combined to form piercing and sucking mouthparts. Megalopterans are predatory only in the aquatic larval stage; although adults have strong mandibles, they are not used in feeding. Adults closely resemble neuropterans, except for the presence of an anal fold in the hind wing. Raphidiopterans are terrestrial predators as adults and as larvae. The adult is mantid-like, with an elongate prothorax, and the head is mobile and used to strike, snake-like, at prey. The larval head is large and forwardly directed.

Megaloptera, Raphidioptera and Neuroptera may be treated as separate orders, united in Neuropterida, or Raphidioptera may be included in Megaloptera. Neuropterida undoubtedly is monophyletic, with support from morphology (e.g. structure of wing-base sclerites and ovipositor) and from nucleotide sequences. Molecular data, including transcriptomes, also support the long-held view that Neuropterida forms a sister group to Coleoptera (now Coleoptera + Strepsiptera). Neuroptera and Raphidioptera are each monophyletic, but megalopteran monophyly has been questioned. Likely internal relationships are either: (i) Megaloptera rendered paraphyletic by Raphidioptera, as suggested by combined nuclear and mitochondrial genes; or (ii) Megaloptera and Neuroptera as sister groups, with these two sister to Raphidioptera, as suggested by transcriptomes, some analyses of whole mitochondrial genomes and some larval features. The position of Raphidioptera in molecular analyses is affected by rate variation, unless such variation is explicitly compensated for.

Order Coleoptera (beetles) (see also Box 10.3 and Taxobox 22)

The major shared derived feature of Coleoptera is the development of the fore wings as sclerotized rigid elytra, which extend to cover some or many of the abdominal segments, and beneath which the propulsive hind wings are elaborately folded when at rest.

Coleoptera is the sister-group either to Strepsiptera or (if Strepsiptera is a highly derived beetle taxon) to Neuropterida. Within Coleoptera, four modern lineages (treated as monophyletic suborders) are recognized: Archostemata, Adephaga, Myxophaga and Polyphaga. Relationships among the suborders are uncertain. Archostemata includes the small Recent families Crowsoniellidae, Cupedidae, Jurodidae, Micromalthidae and Ommatidae, and might be the sister group to the remaining extant Coleoptera. The few known larvae are wood-miners with a sclerotized ligula and a large mola on each mandible. Adults have the labrum fused to the head capsule, movable hind coxae with usually visible trochantins, and five (not six) ventral abdominal plates (ventrites), but share with Myxophaga and Adephaga certain wing-folding features, lack of any cervical sclerites, and an external prothoracic pleuron. In contrast to Myxophaga, the pretarsus and tarsus are unfused.

Adephaga is diverse, second in size only to Polyphaga, and includes ground beetles, tiger beetles, whirligigs, predaceous diving beetles and wrinkled bark beetles, amongst others. Larval mouthparts generally are adapted for liquid feeding, with a fused labrum and no mandibular mola. Adults have the notopleural sutures visible on the prothorax, and have six visible abdominal sterna, with the first three fused into a single ventrite that is divided by the hind coxae. Pygidial defence glands are widespread in adults. The most speciose adephagan family is Carabidae, or ground beetles, with a predominantly predaceous feeding habit. Rhysodidae, or wrinkled bark beetles, are thought to feed on slime moulds (Mycetozoa). Adephaga also includes the aquatic families Dytiscidae, Gyrinidae, Haliplidae and Noteridae (see Box 10.3). Morphology suggests that Adephaga is sister group to the combined Myxophaga and Polyphaga, although nucleotide sequences suggest various arrangements of the suborders, depending on the gene and the taxon sampling.

Myxophaga is a clade of small, primarily riparian aquatic beetles, comprising extant families Lepiceridae, Torridincolidae, Hydroscaphidae and Sphaeriusidae (= Microsporidae), united by the synapomorphic fusion of the pretarsus and tarsus, and pupation occurring in the last-larval exuviae. The three-segmented larval antenna, five-segmented larval legs with a single pretarsal claw, fusion of trochantin with the pleuron, and ventrite structure support a sister-group relationship of Myxophaga with the Polyphaga.

Polyphaga contains the majority (>90% of species) of beetle diversity, with almost 350,000 described species. The suborder includes the well-known rove beetles (Staphylinoidea), scarabs and stag beetles (Scarabaeoidea), metallic wood-boring beetles (Buprestoidea), click beetles and fireflies (Elateroidea), as well as the diverse Cucujiformia, including fungus beetles, grain beetles, ladybird beetles, darkling beetles, blister beetles, longhorn beetles, leaf beetles and weevils. The prothoracic pleuron is not visible externally, but is fused with the trochantin and remnant internally as a "cryptopleuron". Thus, one suture between the notum and the sternum is visible in the prothorax in polyphagans, whereas two sutures (the sternopleural and notopleural) often are visible externally in other suborders (unless secondary fusion between the sclerites obscures the sutures, as in *Micromalthus*). The transverse fold of the hind wing never crosses the media posterior (MP) vein, cervical sclerites are present, and hind coxae are mobile and do not divide the first ventrite. Female polyphagan beetles have telotrophic ovarioles, which is a derived condition within beetles.

The internal classification of Polyphaga involves several superfamilies, whose constituents are relatively stable, although some smaller families (whose rank even is disputed) are allocated to different clades by different authors. Large superfamilies include Staphylinoidea, Scarabaeoidea, Hydrophiloidea, Buprestoidea, Byrrhoidea, Elateroidea, Bostrichoidea, and the grouping Cucujiformia. This latter includes the vast majority of phytophagous (plant-eating) beetles, united by cryptonephric Malpighian tubules of the normal type, the eye with a cone ommatidium with open rhabdom, and lack of functional spiracles on the eighth abdominal segment. Constituent superfamilies of Cucujiformia are Lymexyloidea, Cleroidea, Cucujoidea, Tenebrionoidea, Chrysomeloidea and Curculionoidea. Evidently, adoption of a phytophagous lifestyle correlates with speciosity in beetles, with Cucujiformia, especially weevils (Curculionoidea), resulting from a major radiation (see section 8.6).

Order Strepsiptera (strepsipterans) (see also Taxobox 23)

Strepsipterans form an enigmatic order showing extreme sexual dimorphism. The male's head has bulging eyes, usually comprising few large facets, and lacks ocelli; the antennae are flabellate or branched, with four to seven segments. The fore wings are stubby and lack veins, whereas the hind wings are broadly fan-shaped, with few radiating veins; the legs lack the trochanter and often also claws. Females are either coccoid-like or larviform (paedomorphic), wingless and usually retained in a pharate (cloaked) state, protruding from the host. The first-instar larva is a triungulin, without antennae and mandibles, but with three pairs of thoracic legs; subsequent instars are maggot-like, lacking mouthparts or appendages. The male pupa develops within a puparium formed from the cuticle of the last larval instar.

The phylogenetic position of Strepsiptera has been subject to much speculation because modifications associated with their endoparasitoid lifestyle mean that few characteristics are shared with possible relatives. Adult male strepsipterans resemble Coleoptera in some wing features and in having posteromotor flight (using only metathoracic wings); plus their first-instar larvae (called triungula) and adult males look rather like those of certain Coleoptera, notably parasitic Ripiphoridae. Phylogenetic studies of nuclear protein-coding genes and whole genomes support the Strepsiptera either as the sister group of Coleoptera or as a group within Coleoptera. A relationship of Strepsiptera with Diptera, as suggested by some earlier molecular evidence and by haltere development, has been refuted. The fore-wing-derived halteres of strepsipterans are gyroscopic organs of equilibrium, with the same functional role as the halteres of Diptera (although the latter are derived from the hind wing). Strepsiptera has undergone much morphological and molecular evolution, and is highly derived relative to other taxa. Such long-isolated evolution of the genome can create a problem known as "long-branch attraction", in which nucleotide sequences may converge solely by chance mutations with those of an unrelated taxon (in this case Diptera) with a similarly long independent evolution.

The order is clearly old, since a well-preserved adult male has been described from 100-million-year-old Cretaceous amber. Accepted internal relationships of Strepsiptera have the Mengenillidae (with free-living females and apterygote hosts) sister to all other strepsipterans, which are placed in the clade Stylopidia united by many morphological features and endoparasitism of the adult female and use of pterygote hosts. The recently described Bahiaxenidae (known only from an adult male from Brazil) is considered sister to all other strepsitpterans, based on male morphology.

Order Diptera (true flies) (see also Box 10.1 and Taxobox 24)

Diptera are readily recognized by the development of hind (metathoracic) wings as balancers, or halteres (halters), and in the larval stages by a lack of true legs and an often maggot-like appearance. Venation of the fore (mesothoracic), flying wings ranges from complex to extremely simple. Mouthparts range from biting-and-sucking (e.g. biting midges and mosquitoes) to "lapping"-type, with paired pseudotracheate labella functioning as a sponge (e.g. house flies). Dipteran larvae lack true legs, although they have various kinds of locomotory apparatus, ranging from unsegmented pseudolegs to creeping welts on maggots. The larval head capsule may be complete, partially undeveloped, or completely absent, as in a maggot head that consists only of the internal sclerotized mandibles ("mouth hooks") and supporting structures.

Diptera are sister group to Mecoptera + Siphonaptera in the Antliophora. The fossil record shows abundance and diversity in the mid-Triassic, with some suspects from as early as the Permian perhaps better allocated to "Mecopteroids".

Traditionally, Diptera had two suborders, Nematocera (crane flies, midges, mosquitoes and gnats) with a slender, multisegmented antennal flagellum, and stouter Brachycera ("higher flies", including hover flies, blow flies and dung flies) with a shorter, stouter and fewer-segmented antenna. However, Brachycera is sister to only part of "Nematocera", and thus "Nematocera" is paraphyletic.

Internal relationships amongst Diptera are becoming better understood, although with some notable exceptions. Ideas concerning the oldest nodes in dipteran phylogeny remain inconsistent. The complex wing venation of Tipulidae (or Tipulomorpha) was taken to indicate an early branch, but the incomplete larval head capsule is in keeping with the more derived position from molecular studies. The latter evidence supports some enigmatic, small aquatic families (Deuterophlebiidae and Nymphomyiidae) as the possible sister group of all the rest of Diptera.

The well-supported Culicomorpha group comprises mosquitoes (Culicidae) and their relatives (Corethrellidae, Chaoboridae, Dixidae) and their sister group, the black flies, midges and their relatives (Simuliidae, Thaumaleidae, Ceratopogonidae, Chironomidae), as depicted in the vignette at the beginning of this chapter. Bibionomorpha, comprising the fungus gnats and relatives (Mycetophilidae, Bibionidae, Anisopodidae and others), is well supported on morphological and molecular data, but may include the Cecidomyiidae (gall midges).

Monophyly of Brachycera, comprising "higher flies", is established by features including the larva having the posterior part of an elongate head contained within the prothorax, a divided mandible and no premandible, and in the adult by eight or fewer antennal flagellomeres, two or fewer palp segments, and separation of the male genitalia into two parts (epandrium and hypandrium). Proposed relationships of Brachycera are always to a subgroup within "Nematocera", with growing support for a sister relationship to the Bibionomorpha. Brachycera contains four equivalent groups with poorly resolved relationships: Tabanomorpha (with a brush on the larval mandible and the larval head retractile); Stratiomyomorpha (with shared modified mandibular-maxillary apparatus and filtering with grinding apparatus, and two families with larval cuticle calcified and pupation in last-larval instar exuviae); Xylophagomorpha (with a distinctive elongate, conical, strongly sclerotized larval head capsule, and abdomen posteriorly ending in a sclerotized plate with terminal hooks); and Muscomorpha (adults with tibial spurs absent, flagellum with no more than four flagellomeres, and female cercus single-segmented). This latter speciose group contains Asiloidea (robber flies, bee flies and relatives) and Eremoneura (Empidoidea and Cyclorrhapha). Eremoneura is a strongly supported clade based on wing venation (loss or fusion of vein M_4 and closure of anal cell before margin), presence of ocellar setae, unitary palp and genitalic features, plus larval stage with only three instars and maxillary reduction. Cyclorrhaphans, united by metamorphosis in a puparium formed by the larval skin of the last instar, include the Syrphidae (hover flies) and the many families of Schizophora, defined by the presence of a balloon-like ptilinum that everts from the frons to allow the adult to escape the puparium. Within Schizophora are the ecologically very diverse acalypterates, and the Calyptrata, the blow flies and relatives, including ectoparasitic bat flies and bird/mammal parasites.

Order Mecoptera (scorpionflies, hangingflies and snowfleas) (see also Taxobox 25)

Adult scorpionflies (Panorpidae and several other families), hangingflies (Bittacidae) and snowfleas (Boreidae) have an elongate, ventrally projecting rostrum, containing elongate, slender mandibles and maxillae and an elongate labium; the eyes are large and separated; the antennae are filiform and multisegmented. Mecopteran fore and hind wings are narrow, similar in size, shape and venation, but often are reduced (e.g. Boreidae, Apteropanorpidae). Larval scorpionflies and hangingflies have a heavily sclerotized head capsule, are mandibulate, and usually have eyes composed of groups of stemmata; there are short thoracic legs and prolegs usually are present on abdominal segments 1–8, with a suction disk or paired hooks on the terminal segment (10).

Although some adult Mecoptera resemble neuropterans, strong evidence supports a relationship to Siphonaptera (fleas), with Mecoptera + Siphonaptera sister to Diptera. Analysis of morphological data has not resolved relationships among mecopteran families and the fleas. The phylogenetic position of Nannochoristidae, a southern hemisphere mecopteran taxon, has a significant bearing on internal relationships within Antliophora (Diptera + Mecoptera + Siphonaptera). It has been suggested to be sister to all other mecopterans, but competing studies based on nucleotide sequence data suggest that Nannochoristidae is either: (i) sister to Boreidae + Siphonaptera (Fig. 7.7a); or (ii) part of a monophyletic Mecoptera (Fig. 7.7b). Analyses mostly of single-copy nuclear protein-coding genes or of transcriptomes support a monophyletic Mecoptera (i.e. Fig. 7.7b). Thus, we retain Nannochoristidae as a family, and Mecoptera and Siphonaptera as separate orders, pending resolution of this conflict. Further taxon sampling and additional data analyses are required to validate relationships among families within Mecoptera.

Order Siphonaptera (fleas) (see also Taxobox 26)

Adult fleas are bilaterally compressed, apterous ectoparasites of mammals and birds, with mouthparts specialized for piercing the host and sucking up blood; an unpaired labral stylet and two elongate serrate lacinial stylets lie together within a labial sheath, and mandibles are lacking. Fleas lack compound eyes, and the antennae lie in deep lateral grooves; wings are always absent; the body is armed with many posteriorly directed setae and spines, some of which form combs; the metathorax and hind legs are well developed, associated with jumping. Larval fleas are slender, legless and maggot-like, but with a well-developed, mandibulate head capsule and no eyes.

Early morphological studies suggested that the fleas were sister either to the Mecoptera or Diptera. Phylogenetic studies using data from single-copy nuclear protein-coding genes and from transcriptomes support a sister relationship of Siphonaptera and Mecoptera, with each monophyletic (Fig. 7.7b). A competing view, based on some molecular and morphological data, suggests a sister-group relationship to only part of Mecoptera, specifically the Boreidae (Fig. 7.7a; also see entry above for Mecoptera). Some of the shared features of fleas and snowfleas are found in female reproductive structures, sperm ultrastructure and proventriculus structure, as well as the presence of multiple sex chromosomes, a similar process of resilin secretion, the jumping ability of adults, and the production of a silken pupal cocoon. Some of these features may be plesiomorphies, and thus here the fleas are maintained as a separate order.

A molecular phylogenetic study of internal relationships of the fleas, using only boreids as outgroups, suggests that the family Tungidae (also called Hectopsyllidae; mostly parasitic on mammals) is sister to all other extant fleas, and that at least 10 of the 16 currently recognized families of fleas may be monophyletic, and that three are grossly paraphyletic; the monophyly of three others could not be assessed.

Order Trichoptera (caddisflies) (see also Box 10.4 and Taxobox 27)

The moth-like adult trichopteran has reduced mouthparts lacking any proboscis, but with three- to five-segmented maxillary palps and three-segmented labial palps. The antennae are multisegmented and filiform and often as long as the wings. The compound eyes are large, and there are two to three ocelli. The wings are haired or, less often, partially scaled, and differentiated from all but a few Lepidoptera by the looped anal veins in the fore wing and absence of a discal cell. The larva is aquatic, has fully developed mouthparts, three pairs of thoracic legs (each with at least five segments) and, except for terminal hook-bearing prolegs, the abdomen lacks the ventral prolegs characteristic of most lepidopteran larvae. The tracheal system is closed and associated with tracheal gills on most abdominal segments. The pupa also is aquatic, enclosed in a retreat often made of silk, with functional mandibles that aid in emergence from the sealed case.

Amphiesmenoptera (Trichoptera + Lepidoptera) is now unchallenged, supported by the shared ability of larvae to spin silk from modified salivary glands and by a large number of adult anatomical features. Proposed internal relationships within the Trichoptera range from stable and well supported, to unstable and anecdotal. Monophyly of suborder Annulipalpia (comprising families Hydropsychidae, Polycentropodidae, Philopotamidae and some close relatives) – is well supported by larval and adult morphology – including presence of an annulate apical segment of both adult maxillary and larval palps, absence of male phallic parameres, presence of papillae lateral to the female cerci, and in the larva by the presence of elongate anal hooks and reduced abdominal tergite 10. Annulipalpia includes the retreat-making groups that spin silken nets for food capture.

The monophyly of the suborder Integripalpia (comprising families Phryganeidae, Limnephilidae, Leptoceridae, Sericostomatidae and relatives) is supported by the absence of the *m* cross-vein, hind wings broader than fore wings especially in the anal area, female lacking both segment 11 and cerci, and larval character states, including usually complete sclerotization of the mesonotum, hind legs with lateral projection, lateral and mid-dorsal humps on abdominal segment 1, and short and stout anal hooks. In Integripalpia, larvae construct a tubular case made of a variety of materials in different groups and feed mostly as detrivores, or sometimes as predators or algivores, but rarely as herbivores.

Monophyly of a third putative suborder, Spicipalpia, is more contentious. Defined for a grouping of families Glossosomatidae, Hydroptilidae, Hydrobiosidae, Ptilocolepidae and Rhyacophilidae, uniting features are the spiculate apex of the adult maxillary and labial palps, the ovoid second segment of the maxillary palp and an eversible oviscapt (egg-laying appendage). Morphological and molecular evidence fail to confirm Spicipalpia monophyly, unless at least Hydroptilidae (the "micro-caddisflies") is removed. Sometimes these families are treated as part of Integripalpia. Larvae of Spicipalpia are either free-living predators or cocoon- or case-makers that feed on detritus and algae.

All possible relationships between Annulipalpia, Integripalpia and Spicipalpia have been proposed, sometimes associated with scenarios concerning the evolution of case making. An early idea that Annulipalpia are sister to a paraphyletic Spicipalpia + monophyletic Integripalpia finds support from some morphological and molecular data.

Order Lepidoptera (butterflies and moths) (see also Taxobox 28)

Adult heads bear a long, coiled proboscis formed from greatly elongated maxillary galeae; large labial palps usually are present but other mouthparts are absent, except that mandibles are present primitively in some groups. The compound eyes are large and ocelli usually are present. The multisegmented antennae often are pectinate in moths and knobbed or clubbed in butterflies. The wings are covered completely with a double layer of scales (flattened modified macrotrichia), and the hind and fore wings are linked either by a frenulum, a jugum or simple overlap. Lepidopteran larvae have a sclerotized head capsule with mandibulate mouthparts, usually six lateral stemmata, and short three-segmented antennae. The thoracic legs are five-segmented with single claws, and the abdomen has 10 segments with short prolegs on some segments. Silk-gland products are extruded from a characteristic spinneret at the median apex of the labial prementum. The pupa usually is a chrysalis contained within a silken cocoon, but is naked in butterflies.

Lepidoptera is sister to Trichoptera (see entry for Trichoptera above). The early-branching events in the radiation of this large order are considered well enough resolved to serve as a test for the ability of particular nucleotide sequences to recover the expected phylogeny. Although more than 98% of the species of Lepidoptera belong in Ditrysia, the morphological diversity is concentrated in a small non-ditrysian grade. Three of the four suborders are species-poor lineages, each with just a single family (Micropterigidae, Agathiphagidae and Heterobathmiidae), and perhaps successively sister to the rest of Lepidoptera (but Micropterigidae and Agathiphagidae may be sister groups). These three groups lack the synapomorphy of the mega-diverse fourth suborder Glossata, namely the characteristically developed coiled proboscis formed from the fused galea (see Fig. 2.12). The highly speciose Glossata contains a comb-like branching pattern of many species-poor taxa, plus a species-rich grouping united by the larva (caterpillar) having abdominal prolegs with muscles and apical crochets (hooklets). This latter group contains the diverse Ditrysia, defined by the unique two genital openings in the female, one the ostium bursae on sternite 8 and the other the genitalia proper on sternites 9 and 10. Additionally, the wing coupling is always frenulate or amplexiform and not jugate, and the wing venation tends to be heteroneuran (with venation dissimilar between fore and hind wings). Trends in the evolution of Ditrysia include elaboration of the proboscis and the reduction to loss of maxillary palpi. One of the well-known groups of Ditrysia, supported as monophyletic by molecular data, is the Papilionoidea, the butterflies (comprising families Papilionidae, Hesperiidae, Hedylidae, Pieridae, Nymphalidae, Lycaenidae and Riodinidae) (Fig. 7.8). They are diurnal, except adults of the small Neotropical family Hedylidae, which are typically nocturnal.

Order Hymenoptera (ants, bees, wasps, sawflies and wood wasps) (see also Taxobox 29)

The mouthparts of adults range from being directed ventrally to forward projecting, and from generalized mandibulate to sucking and chewing, with mandibles often used for killing and handling prey, defence and

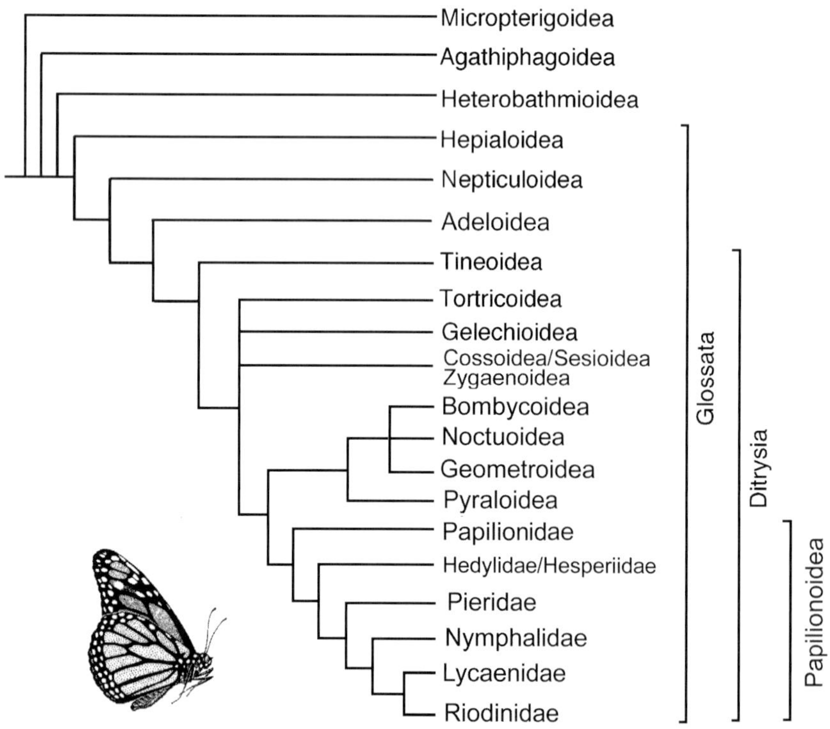

Fig. 7.8 Cladogram of postulated relationships of selected lepidopteran higher taxa, based on molecular data. (After Mutanen *et al.* 2010; Regier *et al.* 2013.)

nest building. The compound eyes often are large; the antennae are long, multisegmented and often prominently held forwardly or recurved dorsally. "Symphyta" (wood wasps and sawflies) has a conventional three-segmented thorax, but in Apocrita (ants, bees and wasps) the propodeum (abdominal segment 1) is included with the thorax to form a mesosoma. The wing venation is relatively complete in large sawflies, and reduced in Apocrita in correlation with body size, such that very small species of 1–2 mm have only one divided vein, or none. In Apocrita, the second abdominal segment (and sometimes also the third) forms a constriction, the petiole (Taxobox 29). Female genitalia include an ovipositor, comprising three valves and two major basal sclerites, which in aculeate Hymenoptera is modified as a sting associated with a venom apparatus.

Symphytan larvae are eruciform (caterpillar-like), with three pairs of thoracic legs bearing apical claws and with some abdominal legs. Apocritan larvae are apodous, with the head capsule frequently reduced, but with prominent strong mandibles.

Hymenoptera forms the sister group to all other holometabolous orders (Fig. 7.2), as shown by recent molecular analyses. Traditionally, Hymenoptera were treated as containing two suborders, Symphyta (wood wasps and sawflies) and Apocrita (wasps, bees and ants). However, Apocrita is sister to only one family of wood wasps, the Orussidae (parasitic wood wasps; Orussoidea), and thus "Symphyta" is a paraphyletic group (Fig. 7.9).

Within Apocrita, the monophyletic aculeate wasps (Aculeata) arose from within the parasitic wasps (the "Parasitica"). Major parasitic groups include Ichneumonoidea, Chalcidoidea, Cynipoidea and Proctotrupoidea, with relationships within each of the hyperdiverse Chalcidoidea and Ichneumonoidea being unresolved. Molecular studies are resolving relationships within aculeates (see Fig. 12.2), with Chrysidoidea (cuckoo wasps and relatives) sister to the rest, and Vespoidea (including paper wasps, potter wasps, yellowjackets and hornets) sister to a large clade containing the tiphioid-pompiloid wasps (including velvet ants, and spider and tiphiid wasps),

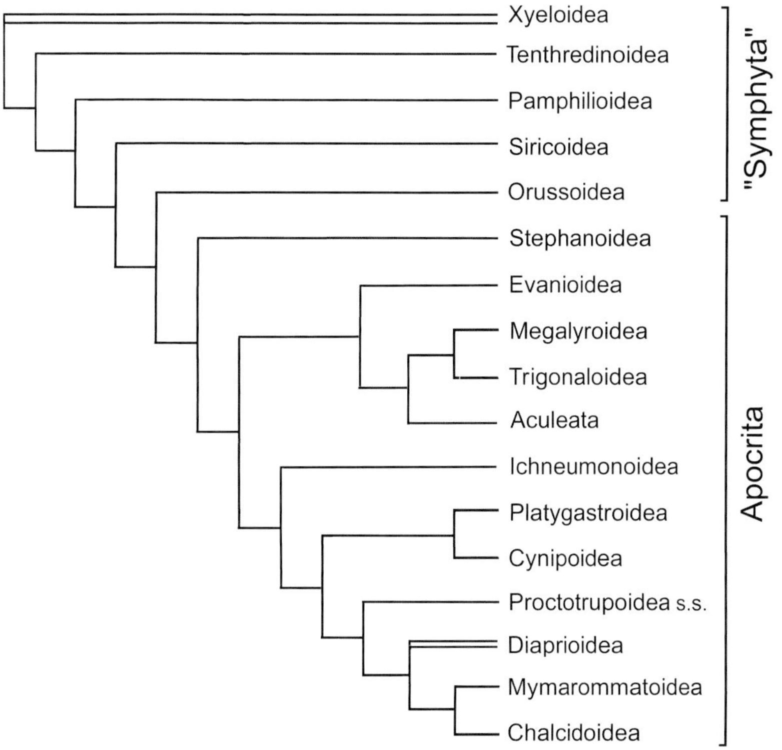

Fig. 7.9 Simplified tree of postulated relationships among higher taxa of Hymenoptera based on molecular and morphological data. Three species-poor taxa (Cephoidea, Xiphydrioidea and Ceraphronoidea) with unstable positions have been omitted; Xyeloidea and Diaprioidea may be paraphyletic, as indicated by double lines in the tree. *Sensu stricto* (abbreviated *s.s.*) means in the restricted sense. (After Klopfstein *et al.* 2013.)

the scolioid wasps (Scoliidae and Bradynobaenidae), the ants (Formicidae) and Apoidea (the bees plus apoid wasps). Apoidea comprises seven families of bees (the unranked taxon Anthophila) plus four families of apoid wasps, with the Anthophila sister to a group of crabronid wasps (making Crabonidae paraphyletic). The position of Formicidae is either as sister to Apoidea (see Fig. 12.2) or to Apoidea + Scolioidea; if the former is substantiated, then the shared behaviour of ants and apoids of collecting and transporting either arthropod prey (as in ants and apoid wasps) or pollen (as in bees) to a constructed nest can be interpreted as inherited directly from a common ancestor.

FURTHER READING

Andrew, D.R. (2011) A new view of insect-crustacean relationships II. Inferences from expressed sequence tags and comparisons with neural cladistics. *Arthropod Structure and Development* **40**, 289–302.

Beutel, R.G., Friedrich, F., Hörnschemeyer, T. *et al.* (2011) Morphological and molecular evidence converge upon a robust phylogeny of the megadiverse Holometabola. *Cladistics* **27**, 341–55.

Bitsch, C. & Bitsch, J. (2000) The phylogenetic interrelationships of the higher taxa of apterygote hexapods. *Zoologica Scripta* **29**, 131–56.

Blanke, A., Greve, C., Wipfler, B. *et al.* (2013) The identification of concerted convergence in insect heads corroborates Palaeoptera. *Systematic Biology* **62**, 250–63.

Buckman, R.S., Mound, L.A. & Whiting, M.F. (2013) Phylogeny of thrips (Insecta: Thysanoptera) based on five molecular loci. *Systematic Entomology* **38**, 123–33.

Bybee, S.M., Ogden, T.H., Branham, M.A. & Whiting, M.F. (2008) Molecules, morphology and fossils: a comprehensive approach to odonate phylogeny and the evolution of the odonate wing. *Cladistics* **23**, 1–38.

Cameron, S.L. (2014) Insect mitochondrial genomics: implications for evolution and phylogeny. *Annual Review of Entomology* **59**, 95–117.

Cameron, S.L., Sullivan, J., Song, H. *et al.* (2009) A mitochondrial genome phylogeny of the Neuropterida (lace-wings, alderflies and snakeflies) and their relationship to other holometabolous insect orders. *Zoologica Scripta* **38**, 575–90.

Cameron, S.L., Lo, N., Bourguignon, T. *et al.* (2012) A mitochondrial genome phylogeny of termites (Blattoidea: Termitoidae): robust support for interfamilial relationships and molecular synapomorphies define clades. *Molecular Phylogenetics and Evolution* **65**,163–73.

Cranston, P.S., Gullan, P.J. & Taylor, R.W. (1991) Principles and practice of systematics. In: *The Insects of Australia*, 2nd edn. (CSIRO), pp. 109–24. Melbourne University Press, Carlton.

Cryan, J.R. & Urban, J.M. (2012) Higher-level phylogeny of the insect order Hemiptera: is Auchenorrhyncha really paraphyletic? *Systematic Entomology* **37**, 7–21.

Damgaard, J., Klass, K.-D., Picker, M.D. & Buder, G. (2008) Phylogeny of the heelwalkers (Insecta: Mantophasmatodea) based on mtDNA sequences, with evidence for additional taxa in South Africa. *Molecular Phylogenetics and Evolution* **47**, 443–62.

Djernæs, M., Klass, K.-D., Picker, M.D. & Damgaard, J. (2012) Phylogeny of cockroaches (Insecta, Dictyoptera, Blattodea), with placement of aberrant taxa and exploration of out-group sampling. *Systematic Entomology* **37**, 65–83.

Dumont, H.J., Vierstraete, A. & Vanfleteren, J.R. (2010) A molecular phylogeny of Odonata. *Systematic Entomology* **35**, 6–18.

Everaerts, C., Maekawa, K., Farine, J.P. *et al.* (2008) The *Cryptocercus punctulatus* species complex (Dictyoptera: Cryptocercidae) in the eastern United States: comparison of cuticular hydrocarbons, chromosome number and DNA sequences. *Molecular Phylogenetics and Evolution* **47**, 950–9.

Friedemann, K., Spangenberg, R., Yoshizawa, K. & Beutel, R.G. (2013) Evolution of attachment structures in the highly diverse Acercaria (Hexapoda). *Cladistics* **30**, 170–201.

Giribet, G. & Edgecombe, G.D. (2012) Reevaluating the arthropod tree of life. *Annual Review of Entomology* **57**, 167–86.

Gregory, T.R. (2008) Understanding evolutionary trees. *Evolution: Education and Outreach* **1**, 121–37.

Grimaldi, D. & Engel, M.S. (2005) *Evolution of the Insects*. Cambridge University Press, Cambridge.

Hall, B.G. (2007) *Phylogenetic Trees Made Easy: A How-To Manual*, 3rd edn. Sinauer Associates, Sunderland, MA.

Holzenthal, R.W., Blahnik, R.J., Prather, A.L. & Kjer, K.M. (2007) Order Trichoptera Kirby, 1813 (Insecta), Caddisflies. In: *Linnaeus Tercentenary: Progress in Invertebrate Taxonomy* (eds Z.-Q. Zhang & W.A. Shear). *Zootaxa* **1668**, 639–98.

Inward, D.J.G., Vogler, A.P. & Eggleton, P. (2007) A comprehensive phylogenetic analysis of termites (Isoptera) illuminates key aspects of their evolutionary biology. *Molecular Phylogenetics and Evolution* **44**, 953–67.

Ishiwata, K., Sasaki, G., Ogawa, J. *et al.* (2011) Phylogenetic relationships among insect orders based on three nuclear protein-coding sequences. *Molecular Phylogenetics and Evolution* **58**, 169–80.

Janzen, D.H. & Hallwachs, W. (2011) Joining inventory by parataxonomists with DNA barcoding of a large complex tropical conserved wildland in northwestern Costa Rica. *PLoS ONE* **6**, e18123. doi: 10.1371/journal.pone.0018123.

Jarvis, K.J. & Whiting, M.F. (2006) Phylogeny and biogeography of ice crawlers (Insecta: Grylloblattodea) based on six molecular loci: designating conservation status for Grylloblattodea species. *Molecular Phylogenetics and Evolution* **41**, 222–37.

Johnson, B.R., Borowiec, M.L., Chiu, J.C. *et al.* (2013) Phylogenomics resolves evolutionary relationships among ants, bees, and wasps. *Current Biology* **23**, 2058–62.

Johnson, K.P., Yoshizawa, K. & Smith, V.S. (2004) Multiple origins of parasitism in lice. *Proceedings of the Royal Society of London B* **271**, 1771–6.

Klass, K.-D., Zompro, O., Kristensen, N.P. & Adis, J. (2002) Mantophasmatodea: a new insect order with extant members in the Afrotropics. *Science* **296**, 1456–9.

Klopfstein, S., Vilhelmsen, L., Heraty, J.M. *et al.* (2013) The hymenopteran tree of life: evidence from protein-coding genes and objectively aligned ribosomal data. *PLoS One* **8**(8), e69344. doi: 10.1371/journal.pone.0069344.

Kocarek, P., John, V. & Hulva, P. (2013) When the body hides the ancestry: phylogeny of morphologically modified epizoic earwigs based on molecular evidence. *PLoS ONE* **8**(6), e66900. doi: 10.1371/journal.pone.0066900.

Kristensen, N.P., Scoble, M.J. & Karsholt, O. (2007) Lepidoptera phylogeny and systematics: the state of inventorying moth and butterfly diversity. In: *Linnaeus Tercentenary: Progress in Invertebrate Taxonomy* (eds Z.-Q. Zhang & W.A. Shear). *Zootaxa* **1668**, 699–747.

Lambkin, C.L., Sinclair, B.J., Pape, T. *et al.* (2013) The phylogentic relationships among infraorders and superfamilies of Diptera based on morphological evidence. *Systematic Entomology* **38**, 164–79.

Lawrence, J.F., Ślipiński, A., Seago, A.E. *et al.* (2011) Phylogeny of Coleoptera based on morphological characters of adults and larvae. *Annales Zoologici* **61**, 1–217.

Lemey, P., Salemi, M. & Vandamme, A.-M. (eds) (2009) *The Phylogenetic Handbook: A Practical Approach to Phylogenetic Analysis and Hypothesis Testing*, 2nd edn. Cambridge University Press, New York.

Letsch, H. & Simon, S. (2013) Insect phylogenomics: new insights on the relationships of lower neopteran orders (Polyneoptera). *Systematic Entomology* **38**, 783–93.

Lo, N. & Eggleton, P. (2011) Termite phylogenetics and co-cladogenesis with symbionts. In: *Biology of Termites: A Modern Synthesis* (eds D.E. Bignell, Y. Roisin & N. Lo), pp. 27–50. Springer, Dordrecht.

Luan, Y.X., Mallatt, J.M., Xie, R.D. *et al.* (2005) The phylogenetic positions of three basal-hexapod groups (Protura, Diplura, and Collembola) based on ribosomal RNA gene sequences. *Molecular Biology and Evolution* **22**, 1579–92.

Mapondera, T.S., Burgess, T., Matsuki, M. & Oberprieler, R.G. (2012) Identification and molecular phylogenetics of the cryptic species of the *Gonipterus scutellatus* complex (Coleoptera: Curculionidae: Gonipterini). *Australian Journal of Entomology* **51**, 175–88.

McKenna, D.D. & Farrell, B.D. (2010) 9-genes reinforce the phylogeny of Holometabola and yield alternate views on the phylogenetic placement of Strepsiptera. *PLoS ONE* **5**(7), e11887. doi: 10.1371/journal.pone.0011887.

Miller, K.B., Hayashi, C., Whiting, M.F. *et al.* (2012) The phylogeny and classification of Embioptera (Insecta). *Systematic Entomology* **37**, 550–70.

Mutanen, M., Wahlberg, N. & Kaila, L. (2010) Comprehensive gene and taxon coverage elucidates radiation patterns in moths and butterflies. *Proceedings of the Royal Society B* **277**, 2839–48.

Niehuis, O., Hartig, G., Grath, S. *et al.* (2013) Genomic and morphological evidence converge to resolve the enigma of Strepsiptera. *Current Biology* **22**, 1309–13.

Ogden, T.H., Gattolliat, J.L., Sartori, M. *et al.* (2009) Towards a new paradigm in mayfly phylogeny (Ephemeroptera): combined analysis of morphological and molecular data. *Systematic Entomology* **34**, 616–34.

Omland, K.E., Cook, L.G. & Crisp, M.D. (2008) Tree thinking for all biology: the problem with reading phylogenies as ladders of progress. *BioEssays* **30**, 854–67.

Pohl, H. & Beutel, R.G. (2013) The Strepsiptera-Odyssey: the history of the systematic placement of an enigmatic parasitic insect order. *Entomologia* **1**:e4, 17–26. doi: 10.4081/entomologia.2013.e4.

Predel, R., Neupert, S., Huetteroth, W. *et al.* (2012) Peptidomics-based phylogeny and biogeography of Mantophasmatodea (Hexapoda). *Systematic Biology* **61**, 609–29.

Regier, J.C., Schultz, J.W., Zwick, A. *et al.* (2010) Arthropod relationships revealed by phylogenomic analysis of nuclear protein-coding sequences. *Nature* **463**, 1079–84.

Regier, J.C., Mitter, C., Zwick, A. *et al.* (2013) A large-scale, higher-level, molecular phylogenetic study of the insect order Lepidoptera (moths and butterflies). *PLoS One* **8**(3), e58568. doi: 10.1371/journal.pone.0058568.

Rehm, P., Borner, J., Meusemann, K. *et al.* (2011) Dating the arthropod tree based on large-scale transcriptome data. *Molecular Phylogenetics and Evolution* **61**, 880–7.

Rota-Stabelli, O., Daley, A.C. & Pisani, D. (2013) Molecular timetrees reveal a Cambrian colonization of land and a new scenario for ecdysozoan evolution. *Current Biology* **23**, 392–8.

Sasaki, G., Ishiwata, K., Machida, R. *et al.* (2013) Molecular phylogenetic analyses support the monophyly of Hexapoda and suggest the paraphyly of Entognatha. *BMC Evolutionary Biology* **13**, 236. doi: 10.1186/1471-2148-13-236.

Schlick-Steiner, B.C., Steiner, F.M. *et al.* (2010) Integrative taxonomy: a multisource approach to exploring biodiversity. *Annual Review of Entomology* **55**, 421–38.

Schuh, R.T. (2000) *Biological Systematics: Principles and Applications*. Cornell University Press, Ithaca, NY.

Sharkey, M.J. (2007) Phylogeny and classification of Hymenoptera. In: *Linnaeus Tercentenary: Progress in Invertebrate Taxonomy* (eds Z.-Q. Zhang & W.A. Shear). *Zootaxa* **1668**, 521–48.

Simon, S., Narechania, A., DeSalle, R. & Hadrys, H. (2012) Insect phylogenomics: exploring the source of incongruence using new transcriptomic data. *Genome Biology and Evolution*. **4**, 1295–309.

Svenson, G.J. & Whiting, M.F. (2009) Reconstructing the origins of praying mantises (Dictyoptera, Mantodea): the roles of Gondwanan vicariance and morphological convergence. *Cladistics* **25**, 468–514.

Thomas, J.A., Trueman, J.W.H., Rambaut, A. & Welch, J.J. (2013) Relaxed phylogenetics and the Palaeoptera problem: resolving deep ancestral splits in the insect phylogeny. *Systematic Biology* **62**, 285–97.

Trautwein, M.D., Weigmann, B.M., Beutel, R. *et al.* (2012) Advances in insect phylogeny at the dawn of the postgenomic era. *Annual Review of Entomology* **57**, 44–68.

Von Reumont, B.M., Jenner, R.A., Wills, M.A. *et al.* (2012) Pancrustacean phylogeny in the light of new phylogenomic data: support for Remipedia as the possible sister group of Hexapoda. *Molecular Biology and Evolution* **29**, 1031–45.

Whitfield, J.B. & Kjer, K.M. (2008) Ancient rapid radiations of insects: challenges for phylogenetic analysis. *Annual Review of Entomology* **53**, 449–72.

Whiting, M.F. (2002) Mecoptera is paraphyletic: multiple genes and phylogeny of Mecoptera and Siphonaptera. *Zoologica Scripta* **312**, 93–104.

Whiting, M.F., Whiting, A.S., Hastriter, M.W. & Dittmar, K. (2008) A molecular phylogeny of fleas (Insecta: Siphonaptera): origins and host associations. *Cladistics* **24**, 677–707.

Wiegmann, B.M., Trautwein, M.D., Kim, J.-W. *et al.* (2009) Single-copy nuclear genes resolve the phylogeny of the holometabolous insects. *BMC Biology* **7**, 34. doi: 10.1186/1741-7007-7-34.

Wiegmann, B.M., Trautwein, M.D., Winkler, I.S. *et al.* (2011) Episodic radiations in the fly tree of life. *Proceedings of the National Academy of Sciences* **108**, 5690–5.

Winterton, S.L., Hardy, N.B. & Wiegmann, B.M. (2010) On wings of lace: phylogeny and Bayesian divergence time estimates of Neuropterida (Insecta) based on morphological and molecular data. *Systematic Entomology* **35**, 349–78.

Yeates, D.K., Wiegmann, B.M., Courtney, G.W. *et al.* (2007) Phylogeny and systematics of Diptera: two decades of progress and prospects. In: *Linnaeus Tercentenary: Progress in Invertebrate Taxonomy* (eds Z.-Q. Zhang & W.A. Shear). *Zootaxa* **1668**, 565–90.

Yoshizawa, K. & Johnson, K.P. (2010) How stable is the "Polyphyly of Lice" hypothesis (Insecta: Psocodea)?: a comparison of the phylogenetic signal in multiple genes. *Molecular Phylogenetics and Evolution* **55**, 939–51.

Zhang, H.-L., Huang, Y., Lin, L.-L. *et al.* (2013) The phylogeny of the Orthoptera (Insecta) as deduced from mitogenomic gene sequences. *Zoological Studies* **52**, 37. http://www.zoologicalstudies.com/content/52/1/37

Zhang, Z.-Q. (ed.) (2011) Animal biodiversity: an outline of higher-level classification and survey of taxonomic richness. *Zootaxa* **3148**, 1–237.

Zhang, Z.-Q. (ed.) (2013) Animal biodiversity: an outline of higher-level classification and survey of taxonomic richness (Addenda 2013). *Zootaxa* **3703**(1), 1–82.

Zwick, P. (2000) Phylogenetic system and zoogeography of the Plecoptera. *Annual Review of Entomology* **45**, 709–46.

Chapter 8

INSECT EVOLUTION AND BIOGEOGRAPHY

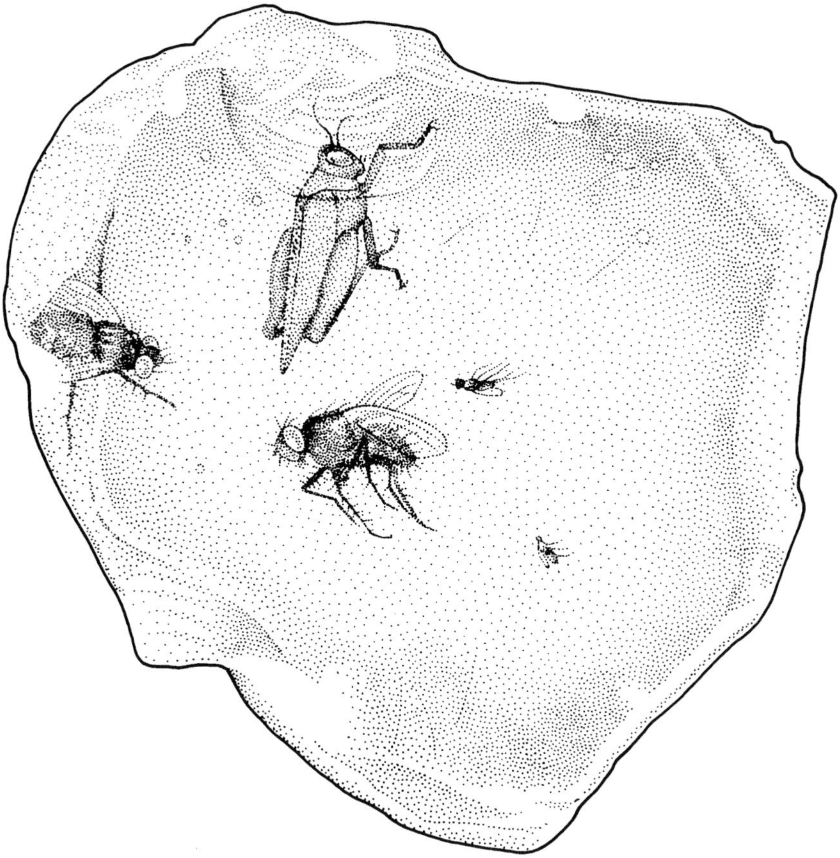

Fossils: insects in amber.

The insects have had a long history since the Hexapoda arose from within the Crustacea perhaps about five hundred million years ago (Ma). Since then, the Earth has undergone much evolution, from droughts to floods, from ice ages to aridity and heat. The gases of the Earth's atmosphere have varied in their proportions. Extra-terrestrial objects have collided with the Earth, and major extinction events have occurred periodically. Throughout this long time, insects have evolved to display the enormous modern diversity outlined in our opening chapter.

In this chapter, we introduce fossil and contemporary evidence for the ages associated with diversification of insects and their proposed relatives. We discuss the evidence for aquatic or terrestrial origins of the group, then address in detail some aspects of insect evolution that have been proposed to explain their success – their terrestriality, the origin of wings (and hence flight) and of metamorphosis. We summarize explanations for insect diversification, and we review patterns and causes for the distribution of insects on the planet – their biogeography. We conclude with a review of insects on Pacific islands, highlighting patterns and processes seen there as a more general explanation of insect radiations. The two boxes cover the estimation of dates using fossils and molecules (Box 8.1), and gigantism in extinct winged insects (Box 8.2).

8.1 RELATIONSHIPS OF THE HEXAPODA TO OTHER ARTHROPODA

The immense phylum Arthropoda, the joint-legged animals, includes several major lineages: the myriapods (centipedes, millipedes and their relatives), the chelicerates (horseshoe crabs and arachnids), the crustaceans (crabs, shrimps and their relatives) and the hexapods (the six-legged arthropods – the Insecta and their relatives). The onychophorans (velvet worms, lobopods) traditionally have been included in the Arthropoda, but now are considered to be the sister group. Recent morphological and molecular studies all support monophyly of arthropodization, and internal relationships of the major arthropod groups are becoming clearer and less controversial. Multiple, ever-larger molecular data sets support monophyly of each of Chelicerata, Mandibulata (= Myriapoda + Crustacea + Hexapoda), Pancrustacea (= Crustacea + Hexapoda), Myriapoda and Hexapoda. Crustacea is paraphyletic because Hexapoda was derived from within it. The relationship of Hexapoda to Crustacea is supported also by shared features of the ultrastructure of the nervous system (e.g. brain structure, neuroblast formation and axon development), the visual system (e.g. fine structure and development of the ommatidia and optic nerves) and of development, especially of segmentation. Major findings from molecular studies of embryology include similarities in developmental expression in hexapods and crustaceans, including of homeotic (developmental regulatory) genes such as *Dll* (*Distal-less*) in the mandible. The second antennae of Crustacea appear to be homologous with the labrum of hexapods, with both being derived from the third cephalic segment. Anatomical features of Hexapoda shared with Myriapoda must have been derived convergently during adoption of terrestriality.

The molecular data strongly supporting a paraphyletic Crustacea are less clear concerning the sister group to the Hexapoda. Depending on the study, Hexapoda are sister to one or more crustacean taxa, namely the Branchiopoda (fairy shrimps, tadpole shrimps and water fleas), Remipedia alone, or Remipedia + Cephalocarida (called Xenocarida). Recent analyses of large phylogenomic data sets favour Xenocarida as sister to Hexapoda (Fig. 8.1).

Palaeontologists identify the first crustacean fossils from the marine Upper Cambrian (505 Ma), whereas the first fossils that assuredly belong to the hexapods appear some 100 million years later, in terrestrial deposits (Rhynie Chert) from the Devonian (section 8.2.1). However, estimates based on analyses from multiple molecular data sets (see Box 8.1) suggest that the hexapod lineage diverged from crustacean ancestors in the Cambrian or Ordovician, but diversification of hexapods probably did not start until the Silurian or Devonian. The topology of the pancrustacean tree is important to understanding the early ecology and evolution of hexapods, especially whether postulated transitions from sea to land are direct or

The Insects: An Outline of Entomology, Fifth Edition. P.J. Gullan and P.S. Cranston.

Companion Website: www.wiley.com/go/gullan/insects

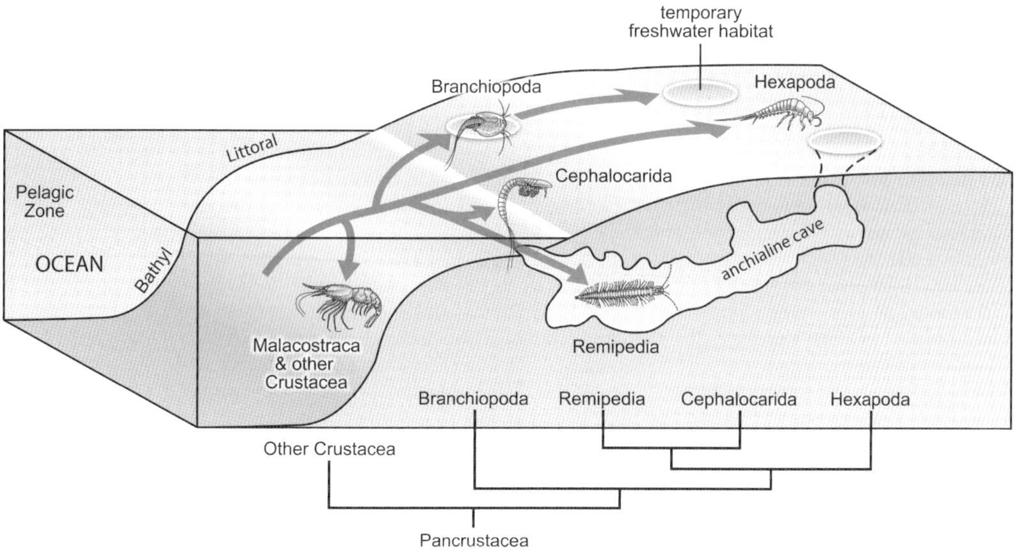

Fig. 8.1 One possible evolutionary scenario for Pancrustacea (Crustacea + Hexapoda), in which hexapods evolved from an ancestor shared with Remipedia (today in coastal aquifers) and Cephalocarida (today in benthic intertidal). Note that none of the depicted relationships are fully supported by all available molecular and morphological data. (After von Reumont *et al.* 2012).

via freshwater. Both Remipedia and Cephalocarida are marine crustaceans that inhabit coastal aquifers or the benthic intertidal zone, respectively, whereas extant Branchiopoda live almost entirely in freshwater (although the group probably had a marine origin). Thus, whichever crustacean taxon is the sister group of hexapods, a Cambrian origin implies that hexapods evolved in the sea, since all groups were marine at that time, and the terrestrialization of hexapods may have occurred directly from coastal habitats. However, if the postulated Cambrian origin of hexapods and lineages of their putative crustacean relatives is too early, then an alternative scenario can be envisaged in which hexapods originated during the Ordovician to Silurian from a common ancestor shared perhaps with branchiopod crustaceans that lived in freshwater. Whichever sister group and dating are correct, an early diversification of hexapods is congruent with the independent transition to land of chelicerates and myriapods. The earliest definite fossils of land plants (spores similar to those of liverworts) are from the Middle Ordovician (about 473–471 Ma), although their origin is likely to be earlier. This timing is consistent with arthropods colonizing land at the same time or slightly earlier than land plants.

8.2 THE ANTIQUITY OF INSECTS

8.2.1 The insect fossil record

The oldest fossils definitively recognized as hexapods are considered to be Collembola, including *Rhyniella praecursor*, known from about 400 Ma in the Lower Devonian of Rhynie, Scotland, and slightly younger archaeognathans from North America (Fig. 8.2). Two other Rhynie fossils may increase hexapod diversity of this period: *Rhyniognatha hirsti*, known from mouthparts only, may be the oldest "ectognathous" insect and possibly even an early pterygote; *Leverhulmia mariae* is argued to be an apterygote (of uncertain affinity), but equally could represent a fragmented crustacean. Tantalizing evidence from Lower Devonian fossil plants shows damage resembling that caused by the piercing-and-sucking mouthparts either of insects or mites. Earlier fossil evidence for Insecta or their relatives may be difficult to find because appropriate alluvial (river and lake) deposits containing fossils are scarce prior to the Devonian. Dated molecular phylogenies suggest an origin for hexapods much earlier than the Devonian, suggesting the existence of "ghost lineages", in which the group occurred but with fossils not yet found (Box 8.1).

Box 8.1 The difficulties with dating

A decade ago it was firmly believed that we were on the verge of understanding both the relationships and the timing of the evolution of most parts of the tree of life – and the origin and major diversifications of the hexapods (including the insects) were no exception. Molecular data held great promise (see section 7.1.1). Given this, it is appropriate to ask why we still lack definitive dates for major events in insect evolution.

The first difficulty in dating evolutionary events are the complexities involved in reconstructing the "correct" phylogenetic tree. When speciation events are closely spaced in time, the amount of phylogenetic signal, even from molecular data, may be small, leading to short internal tree branches that are difficult to resolve. If events are ancient, multiple substitutions can occur at the same position on the gene (homoplasy) – and these can be inferred only with probabilistic modelling. Proposed addition of massive new genomic data may not help if there is insufficient phylogenetic signal to allow the very oldest divergences to be resolved.

Even if we have confidence in our evolutionary trees, there are problems associated with applying dates to events. Methods for deriving ages of evolutionary events, for example the splitting of lineages at nodes from which descendants radiate, have progressed strongly in recent years. From the outset, given a phylogeny derived from morphological analysis (section 7.1.1), well-preserved fossils could be placed on the tree to give an approximation of the date of origination of a lineage of interest. The advent of molecular phylogenies allows estimation of the time of origin of lineages using models of the rates of variation in DNA (a "molecular clock"). From the earliest days, it has been clear that there is no exact clock-like rate of mutation allowing simple backtracking to date nodes in a tree. Increasingly complex models of how DNA evolves have been developed, using probability statistics to allow complex variation in rates of change in different lineages. Bayesian statistics (inference) can take into account uncertainty in the estimated phylogeny, and also can handle ranges of variation in uncertainty concerning how calibrations from fossils are implemented.

There are inherent problems with both approaches to dating (i.e. using fossils or models of DNA evolution), including difficulty in interpreting fossils. For example, if diagnostic morphology is obscured, placement of a fossil on a tree is more speculative than if the pertinent morphology is visible. Clearly, a fossil does not represent the oldest age of the taxon, but rather the period of its preservation. Thus, ages of fossils are best considered minimum age constraints for the groups that they represent. Accurately determining the ages of fossils themselves also is not without uncertainty, and there are examples of interpretations of the ages of critical fossil deposits being re-evaluated and adjusted by many millions of years. Additionally, geologists have modelled the influence of the biases in the fossil records, with some geographic regions and geological ages under- and over-represented.

Nonetheless, molecular models of insect evolution frequently infer older, sometimes very much older, dates of origin of taxa than the fossil record implies. Of course, in analyses, younger dates cannot be produced because the age of each fossil provides a constraint on the youngest record of its group's existence. Inevitably, given the imperfections of the geological record, many taxa will have existed for a period of time, perhaps quite lengthy, prior to the date of the oldest fossil – insects are no different. These differences between inferred age and fossil dates can be termed "ghosts" of taxa present but invisible as fossils. What we can hope for is targeted studies of geological periods that are currently under-represented (for example, the Lower Carboniferous), and poorly sampled fossil locations (much of the southern hemisphere, especially for Mesozoic amber). Combined with stronger supported phylogenetic hypotheses from additional molecular data, we should see date estimates from both data sources converge.

In the Carboniferous, an extensive radiation is shown by substantial insect fossil deposits in the Upper Carboniferous (= Mississippian in USA). Lower Carboniferous fossils are unknown, again because of lack of freshwater deposits. By some 300 Ma a probably monophyletic grouping of Palaeodictyopterida, comprising four now-extinct ordinal groups, the Palaeodictyoptera (Fig. 8.3), Megasecoptera, Dicliptera (= Archodonata) and Diaphanopterodea, was diverse. Palaeodictyopterids varied in size (with wingspans up to 56 cm), diversity (over 70 genera in 21 families are known), and in morphology, notably in mouthparts and wing articulation and venation. Most probably, they fed by piercing and sucking plants using their beak-like mouthparts, which in some species were about 3 cm long. Palaeodictyopterid nymphs are

Plate 1

(a) An atlas moth, *Attacus atlas* (Lepidoptera: Saturniidae), one of the largest of all lepidopterans, with a wingspan of about 24 cm and a larger wing area than any other moth; southern India and Southeast Asia (P.J. Gullan).

(b) The moon moth, *Argema maenas* (Lepidoptera: Saturniidae), is found in Southeast Asia and India; this female, from rainforest in Borneo, has a wingspan of about 15 cm (P.J. Gullan).

(c) Lord Howe Island stick-insect, *Dryococelus australis* (Phasmatodea: Phasmatidae), Lord Howe Island, Pacific Ocean, Australia (N. Carlile).

(d) A female of the Stephens Island giant weta, *Deinacrida rugosa* (Orthoptera: Anostostomatidae), Mana Island, New Zealand (G.H. Sherley; courtesy of New Zealand Department of Conservation).

(e) A Richmond birdwing, *Ornithoptera richmondia* (Lepidoptera: Papilionidae), and its cast pupal exuviae on a native pipevine (*Pararistolochia* sp.), eastern Australia (D.P.A. Sands).

(f) Owl butterfly, *Caligo memnon*, with two common morpho butterflies, *Morpho peleides* (both Nymphalidae), Cali Zoo, Colombia (P.J. Gullan).

(g) A cage of butterfly pupae awaiting eclosion, Penang Butterfly Farm, Malaysia (P.J. Gullan).

Plate 2

(a) Palm weevil grubs, *Rhynchophorus ferrugineus* (Coleoptera: Curculionidae), reared for human consumption from ground palm mash and pig pellets, Thailand (M.S. Hoddle).
(b) A "worm" or "phane"–the caterpillar of *Gonimbrasia belina* (Lepidoptera: Saturnidae)–feeding on the foliage of *Schotia brachypetala*, Limpopo Province, South Africa (R.G. Oberprieler).
(c) Witchety (witjuti) grub, a larva of *Endoxyla* (Lepidoptera: Cossidae) from a desert *Acacia* tree, Flinders Ranges, South Australia (P. Zborowski).
(d) Food insects at a market stall displaying silk-moth pupae (*Bombyx mori*), beetle pupae, and adult hydrophiloid beetles and water bugs (*Lethocerus indicus*), Lampang Province, northern Thailand (R.W. Sites).
(e) A dish of edible water bugs, *Lethocerus indicus* (Hemiptera: Belostomatidae), Lampang Province, northern Thailand (R.W. Sites).
(f) Edible stink bugs (Hemiptera: Tessaratomidae), at an insect market, Thailand (A.L. Yen).
(g) Repletes (see Fig. 2.4) of the honeypot ant, *Camponotus inflatus* (Hymenoptera: Formicidae), on an Aboriginal wooden dish, Northern Territory, Australia (A.L. Yen).
(h) Repletes of the honeypot ant, *Camponotus inflatus*, Northern Territory, Australia (A.L. Yen).

Plate 3

(a) A tropical butterfly, the five-bar swordtail, *Graphium antiphates* (Lepidoptera: Papilionidae), obtaining salts by imbibing sweat from a training shoe, Borneo (P.J. Gullan).

(b) A female katydid of an undescribed species of *Austrosalomona* (Orthoptera: Tettigoniidae), with a large spermatophore attached to her genital opening, northern Australia (D.C.F. Rentz).

(c) Pupa of a Christmas beetle, *Anoplognathus* sp. (Coleoptera: Scarabaeidae), removed from its pupation site in the soil, Canberra, Australia (P.J. Gullan).

(d) A teneral (newly moulted) giant burrowing cockroach, *Macropanesthia rhinoceris* (Blattodea: Blaberidae), Queensland, Australia (M.D. Crisp).

(e) Egg mass of *Tenodera australasiae* (Mantodea: Mantidae) with young mantid nymphs emerging, Queensland, Australia (D.C.F. Rentz).

(f) Eclosing (moulting) adult katydid of an *Elephantodeta* species (Orthoptera: Tettigoniidae), Northern Territory, Australia (D.C.F. Rentz).

(g) Overwintering monarch butterflies, *Danaus plexippus* (Lepidoptera: Nymphalidae), Mill Valley, California, USA (D.C.F. Rentz).

Plate 4

(a) A fossilized worker ant of *Pseudomyrmex oryctus* (Hymenoptera: Formicidae) in Dominican amber from the Miocene (P.S. Ward).

(b) Female (long snout) and male (short snout) of the cycad weevil, *Antliarhinus zamiae* (Coleoptera: Curculionidae), on seeds of *Encephalartos altensteinii* (Zamiaceae), South Africa (P.J. Gullan).

(c) The common milkweed grasshopper, *Phymateus morbillosus* (Orthoptera: Pyrgomorphidae), for which bright colours advertise toxicity acquired by feeding on milkweed foliage, Northern Cape, South Africa (P.J. Gullan).

(d) Mine of a scribbly gum moth, *Ogmograptis racemosa* (Lepidoptera: Bucculatricidae), on trunk of *Eucalyptus racemosa*, New South Wales, Australia (P.J. Gullan).

(e) Euglossine bees (Hymenoptera: Apidae: Euglossini) collecting fragrances from spadix of *Anthurium* sp. (Araceae), Ecuador (P.J. Gullan).

(f) A bush coconut or bloodwood apple gall of *Cystococcus pomiformis* (Hemiptera: Eriococcidae), cut open to show the cream-coloured adult female and her numerous, tiny nymphal male offspring covering the gall wall, northern Australia (P.J. Gullan).

(g) Close-up of the second-instar male nymphs of *C. pomiformis* feeding from the nutritive tissue lining the cavity of the maternal gall, northern Australia (P.J. Gullan).

Plate 5

(a) Coccoid-induced gall of *Apiomorpha pharetrata* (Hemiptera: Eriococcidae): dark compound gall of males attached to green gall of female, with ants collecting honeydew at orifice of female's gall, eastern Australia (P.J. Gullan).
(b) Aphid-induced galls of *Baizongia pistaciae* (Hemiptera: Aphididae: Fordinae) on turpentine tree, *Pistacia teredinthus*, Bulgaria (P.J. Gullan).
(c) Rose bedeguar gall of *Diplolepis rosae* (Hymenoptera: Cynipidae) on *Rosa* sp. (wild rose), Bulgaria (P.J. Gullan).
(d) A female thynnine wasp of *Zaspilothynnus trilobatus* (Hymenoptera: Tiphiidae) (on right) compared with flower of the sexually deceptive orchid *Drakaea glyptodon*, which attracts pollinating male wasps by mimicking the female wasp, Western Australia (R. Peakall).
(e) A male thynnine wasp of *Neozeleboria cryptoides* (Hymenoptera: Tiphiidae) attempting to copulate with the sexually-deceptive orchid *Chiloglottis trapeziformis*, Australian Capital Territory (R. Peakall).
(f) Myophily – pollination of mango flowers by a flesh fly, *Australopierretia australis* (Diptera: Sarcophagidae), northern Australia (D.L. Anderson).
(g) Hummingbird hawk moth, *Macroglossum stellatarum* (Lepidoptera: Sphingidae), on a thistle, Bulgaria (P.J. Gullan).
(h) Honey bee, *Apis mellifera* (Hymenoptera: Apidae), pollinating a passion flower, *Passiflora edulis*, Colombia (T. Kondo).

Plate 6

(a) Ovipositing parasitic wasps (Hymenoptera): a eurytomid (Eurytomidae, top) and cynipid (Cynipidae, right), on an oak apple gall on *Quercus*, Illinois, USA (A.L. Wild).
(b) Weaver ants, *Oecophylla smaragdina* (Hymenoptera: Formicidae), tending *Rastococcus* mealybugs (Hemiptera: Pseudococcidae), Thailand (T. Kondo).
(c) The huge queen termite (approx. 7.5 cm long) of *Odontotermes transvaalensis* (Blattodea: Termitoidae: Termitidae: Macrotermitinae) surrounded by her king (mid front), soldiers and workers, South Africa (the late J.A.L. Watson).
(d) A parasitic *Varroa* mite on a pupa of *Apis cerana* (Hymenoptera: Apidae) in a hive, Irian Jaya, New Guinea (D.L. Anderson).
(e) Ant (Hymenoptera: Formicidae) interactions: the smaller Argentine ant (*Linepithema humile*) attacks the much larger red imported fire ant (*Solenopsis invicta*), Austin, Texas, USA (A.L. Wild).
(f) An egg-parasitoid wasp, *Telenomus* sp. (Hymenoptera: Scelionidae), oviposits into an egg of an owl butterfly, *Caligo* sp. (Lepidoptera: Nymphalidae), Belize (A.L. Wild).

Plate 7

(a) A cryptic grasshopper, *Calliptamus* sp. (Orthoptera: Acrididae), Bulgaria (T. Kondo).
(b) A camouflaged late-instar caterpillar of *Plesanemma fucata* (Lepidoptera: Geometridae) resting on a eucalypt leaf so that its red dorsal line resembles the leaf midrib, eastern Australia (P.J. Gullan).
(c) A female webspinner of *Antipaluria urichi* (Embioptera: Clothodidae) defending the entrance of her gallery from an approaching male, Trinidad (J.S. Edgerly-Rooks).
(d) A snake-mimicking caterpillar of the spicebush swallowtail, *Papilio troilus* (Lepidoptera: Papilionidae), New Jersey, USA (D.C.F. Rentz).
(e) An adult moth of *Utetheisa ornatrix* (Lepidoptera: Arctiidae) emitting defensive froth containing pyrrolizidine alkaloids sequestered by larval feeding on *Crotalaria* (Fabaceae) (the late T. Eisner).
(f) A blister beetle, *Lytta polita* (Coleoptera: Meloidae), reflex-bleeding from the knee joints; the haemolymph contains the toxin cantharidin (the late T. Eisner).
(g) The cryptic adult moths of four species of *Acronicta* (Lepidoptera: Noctuidae): *A. alni*, the alder moth (top left); *A. leporina*, the miller (top right); *A. aceris*, the sycamore (bottom left); and *A. psi*, the grey dagger (bottom right) (D. Carter and R.I. Vane-Wright).
(h) Aposematic or mechanically protected caterpillars of the same four species of *Acronicta: A. alni* (top left); *A. leporina* (top right); *A. aceris* (bottom left); and *A. psi* (bottom right); showing the divergent appearance of the larvae compared with their drab adults (D. Carter and R.I. Vane-Wright).

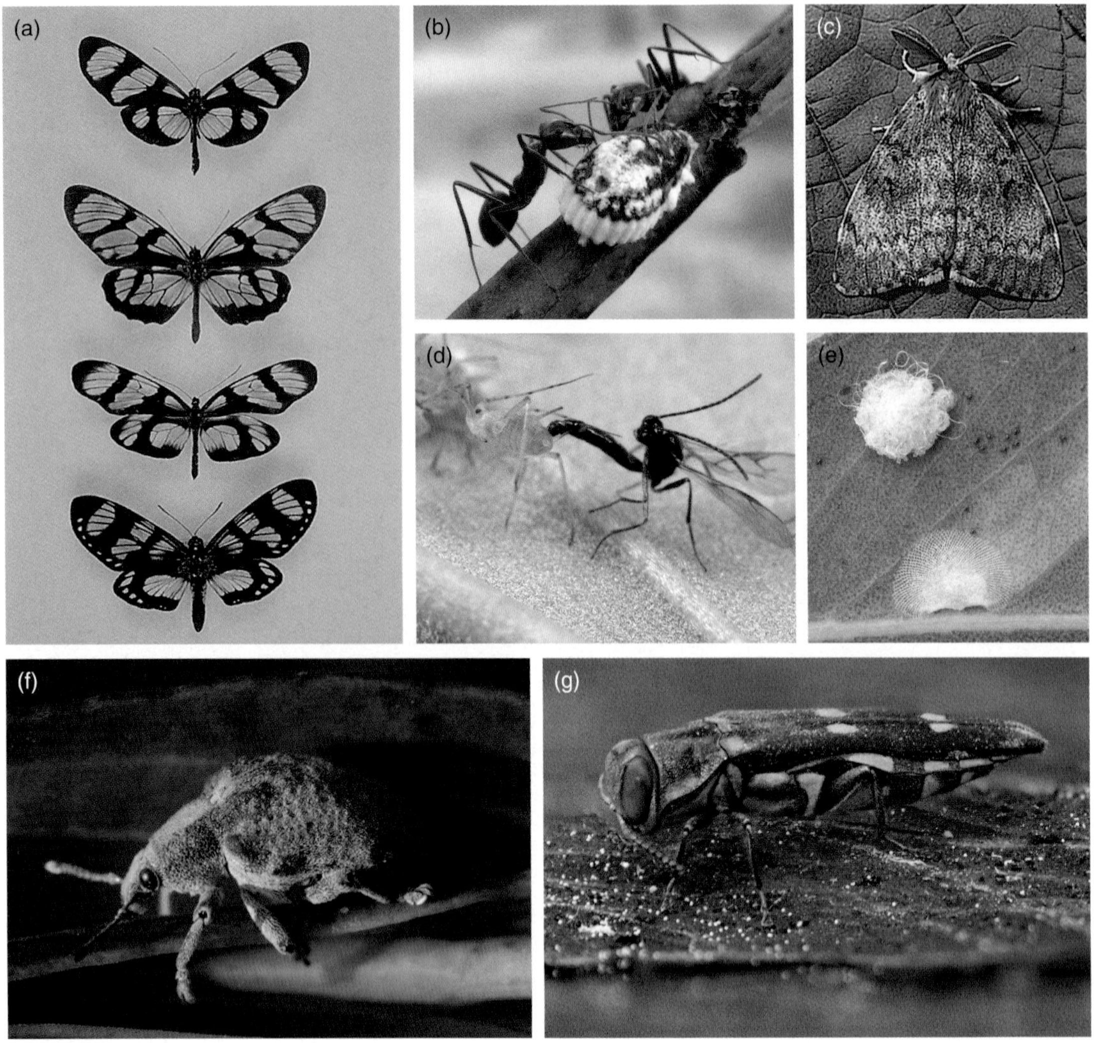

Plate 8

(a) One of Bates' mimicry complexes from the Amazon Basin involving species from three different lepidopteran families – the butterflies *Methona confusa confusa* (Nymphalidae: Ithomiinae) (top), *Lycorea ilione ilione* (Nymphalidae: Danainae) (second top) and *Patia orise orise* (Pieridae) (second from bottom), and a day-flying moth of *Gazera heliconioides* (Castniidae) (R.I. Vane-Wright).

(b) A mature cottony-cushion scale, *Icerya purchasi* (Hemiptera: Monophlebidae), with a partly formed ovisac, on the stem of an *Acacia* host, attended by meat ants of *Iridomyrmex* sp. (Formicidae), New South Wales, Australia (P.J. Gullan).

(c) Adult male gypsy moth, *Lymantria dispar* (Lepidoptera: Lymantriidae), New Jersey, USA (D.C.F. Rentz).

(d) A biological control wasp *Aphidius ervi* (Hymenoptera: Braconidae) attacking pea aphids, *Acyrthosiphon pisum* (Hemiptera: Aphididae), Arizona, USA (A.L. Wild).

(e) A circular lerp of the red gum lerp psyllid, *Glycaspis brimblecombei*, and a white lace lerp of *Cardiaspina albitextura* (Hemiptera: Psyllidae), on *Eucalyptus blakelyi*, Canberra, Australia; note the small brown eggs of *C. albitextura* attached to the leaf (M.J. Cosgrove).

(f) An adult of the eucalypt-damaging weevil, *Gonipterus platensis* (Coleoptera: Curculionidae), Western Australia (M. Matsuki).

(g) Adult beetle of the goldspotted oak borer, *Agrilus auroguttatus* (Coleoptera: Buprestidae), which threatens native oaks, southern California, USA (M. Lewis).

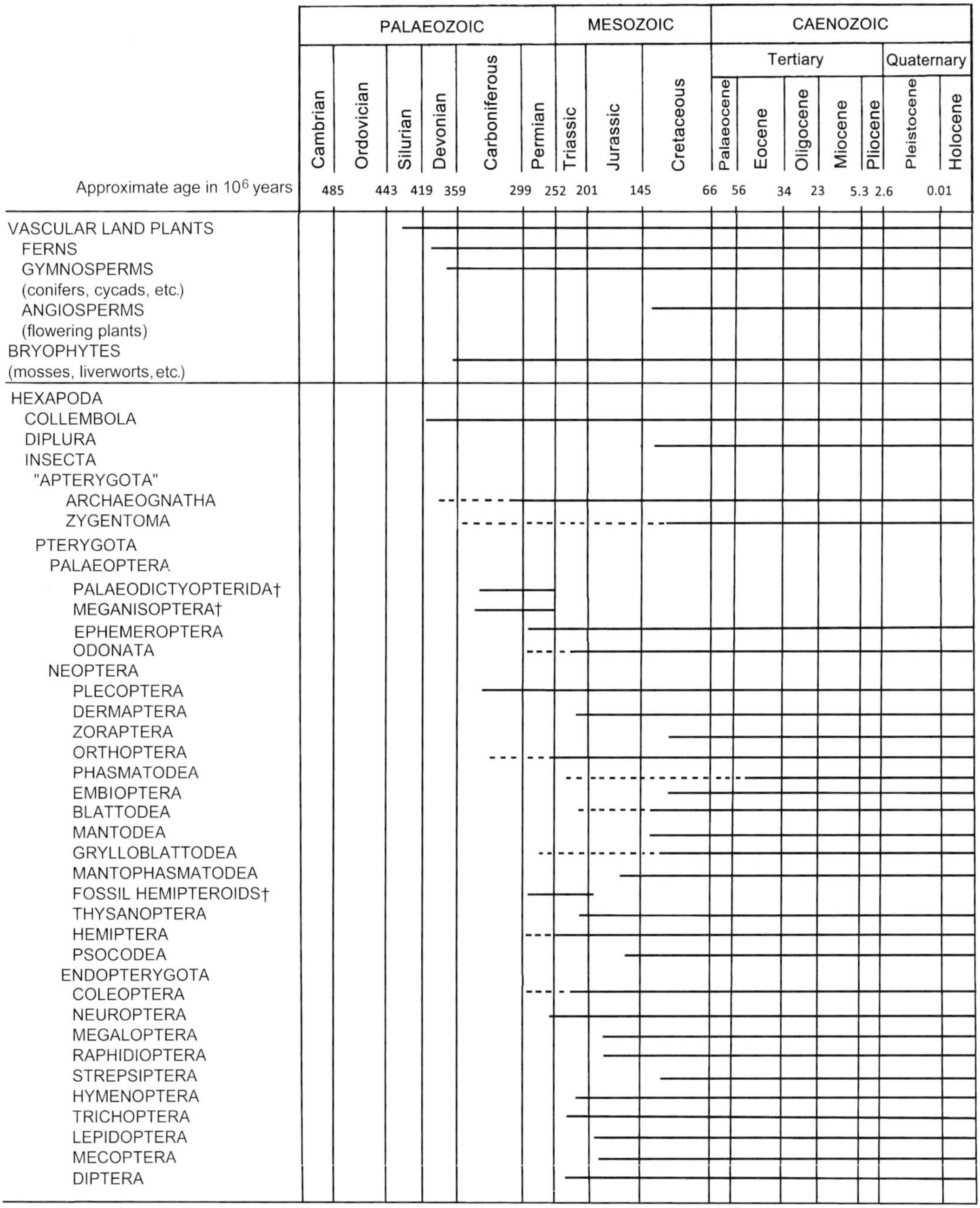

Fig. 8.2 The geological history of insects in relation to plant evolution. Taxa that contain only fossils are indicated by the symbol †. The record for extant orders is based on definite members of the crown group and does not include stem-group fossils; dashed lines indicate uncertainty in placement in the crown group. Thus, this chart does not include records of most early insect radiations; for example "roachoid" fossils occur in the Palaeozoic but are not part of the more narrowly defined Dictyoptera and Blattodea. Protura and Siphonaptera are not shown due to inadequacy of their fossil record; Isoptera is part of Blattodea. The placement of *Rhyniognatha* is unknown. (Insect fossil records have been interpreted from primary sources and after Grimaldi & Engel 2005; the date for the start of each geological period is from the International Commission on Stratigraphy; the Tertiary often is divided into the Palaeogene (66–23 Ma) and the Neogene (23–2.6 Ma.)

Box 8.2 There were giants – evolution of insect gigantism

We are used to insects being amongst the smaller animals we encounter, with many being inconspicuous in terms of size and lifestyles. However, back in the Carboniferous and Permian, some were very large, exemplified by the giant Bolsover dragonfly and a palaeodictyopteran on a *Psaronius* tree fern, depicted here (inspired by a drawing by Mary Parrish in Labandeira 1998). Late Palaeozoic Ephemeroptera and dragonfly-like Meganisoptera had wingspans of up to 45 cm and 71 cm, respectively. Gigantism at this time was evident, not only among insects but also in branchiopod crustaceans, certain other invertebrates, and some amphibians.

From the earliest recognition of gigantism in the insect fossil record, elevated atmospheric oxygen levels (above today's 20%) were suggested, although the geological processes responsible were unknown. Restriction of modern insects to a modest maximum size (see section 3.5.1) and shape (volume) seems to derive from the physical laws governing gas transport along the tracheae to the sites of uptake and release in cells (see Box 3.2). Flight is so highly demanding of oxygen that a large, winged insect in a modern oxygen atmosphere would require such dense tracheae that there would be inadequate space for the muscles and other internal organs.

New biogeochemical models infer elevated atmospheric oxygen during the Carboniferous and Permian as shown in the graph. Oxygen levels of at least 30%, half as much again above today's levels, imply a very different (and more flammable) world in the late Carboniferous to the mid-Permian. For insects, higher oxygen pressures promoted greater oxygen diffusion via tracheae, and reduced the demand for the thorax to be packed with tracheae to power flight. Furthermore, with elevated oxygen, air would be denser, further facilitating flight. This hypothesis for Palaeozoic gigantism has obvious appeal, although morphological and physiological consequences (including ability to actively ventilate tracheae) of alterations of gaseous composition need further study.

The drop in atmospheric oxygen from the mid-Permian onwards coincided with reduction in maximum size of winged insects, as expected. This trend continued through to the end-Permian mass extinction, when many insects became extinct, including those groups in which gigantism had been prevalent. However, the decrease in maximum size from the Permian to modern times seems too large to be explained by declining atmospheric oxygen alone. Furthermore, the postulated recurrence of high oxygen in the Cretaceous and early Tertiary was not associated with regain of large-sized insects. The explanation favoured for the subsequent disconnect between atmospheric oxygen levels and insect body size was the arrival of flying vertebrate predators, namely birds and bats.

These insectivores would have competed with larger insects, over-riding the effect of environmental change on body size. The air was no longer the exclusive preserve of winged insects. Thus, since the Tertiary, maximum sizes attained for insects such as dragonflies and mayflies are below the upper limits of their extinct distant ancestors, and less than their gas-exchange physiology would permit.

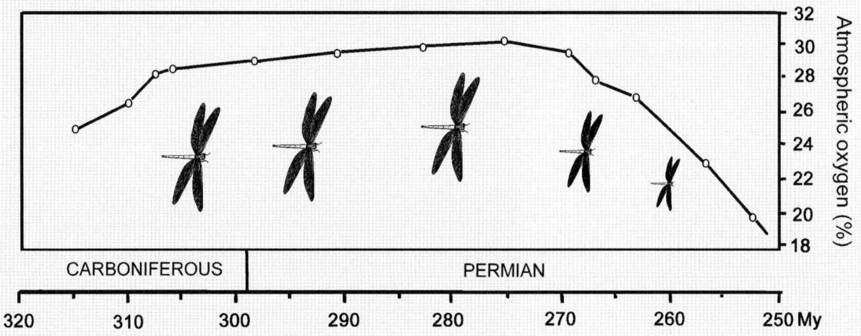

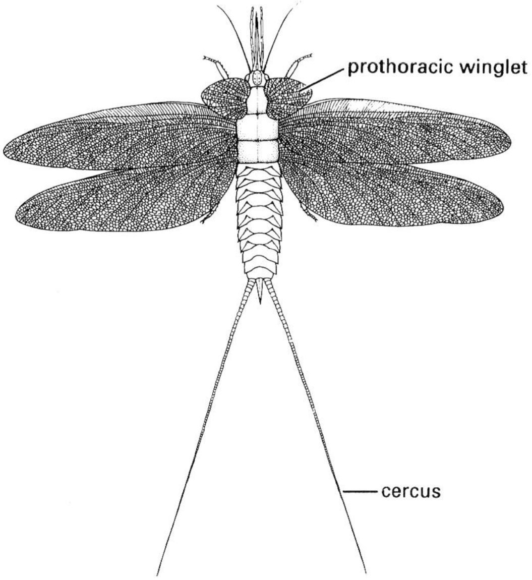

Fig. 8.3 Reconstruction of *Stenodictya lobata* (Palaeodictyoptera: Dictyoneuridae). (After Kukalová 1970.)

believed to have been terrestrial. Some Meganisoptera (griffenflies, perhaps a stem-group of Odonata and previously called "Protodonata") had not only meso- and metathoracic wings, but also winglets on the prothorax, as did Palaeodictyopterida, and in the Permian included insects with the largest wingspans ever recorded (Box 8.2). No extant orders of pterygotes are represented unambiguously by Carboniferous fossils; putative Ephemeroptera and Orthoptera, fossil hemipteroids and blattoids are treated best as stem-groups or paraphyletic groups lacking the defining features of any extant clade. Beak-like, piercing mouthparts and the expanded clypeus in some Carboniferous insects indicate early plant feeding.

The earliest definitive holometabolan fossils date from 328 to 318 Ma in Late Carboniferous coal deposits in the Illinois basin (USA) and Pas-de-Calais in northwestern France. Fossils representing taxonomically diverse groups are allocated to Antliophora, Mecopteroidea, early Coleoptera and early Hymenoptera or its stem group. Fossil larvae represent three major feeding strategies already evident at this time: an external-feeding active caterpillar, an internal-feeding legless larva, and an active, fully legged, predatory larva. Molecular-estimated time of origin of the Holometabola, and fossilized plant-feeding damage inferred as being caused by them, conjectures minimally a Lower Carboniferous origin of holometaboly. Diversification into the extant orders, however, did not commence until the Upper Carboniferous or the Permian.

In the Permian, gymnosperms (conifers and allies) became abundant in what had been a fern-dominated flora. Concurrently, a dramatic increase took place in hexapod ordinal diversity, with at least 30 insect orders known from the Permian. The evolution of plant-sucking hemipteroid insects may have been associated with the newly available plants with thin cortex and sub-cortical phloem. Other Permian insects included those that fed on pollen, another resource of previously restricted supply.

Many groups present in the Permian, including Ephemeroptera (both nymphs and adults), Plecoptera, early Dictyoptera, and probable stem-group Odonata, Orthoptera and Coleoptera, survived from this period. However, lineages such as Palaeodictyopterida and Meganisoptera disappeared at the end of the Palaeozoic. The Permian–Triassic boundary was a time of major extinction that particularly affected marine biota, and coincided with a dramatic reduction in diversity in taxa and feeding types within surviving insect orders.

The Triassic period (commencing about 252 Ma), although famed for the "dominance" by dinosaurs and pterosaurs, and the origin of the mammals, was a period of rapid diversification of the insects. The major orders of modern insects, except Lepidoptera (which has a poor fossil record), are well represented in the Triassic. Hymenoptera are seen first in this period, represented only by symphytans. The oldest extant families of many orders appeared, together with diversified taxa with aquatic immature (and some adult) stages, including modern Odonata, Heteroptera and many families of nematocerous Diptera. The Jurassic saw the first appearance of aculeate Hymenoptera, many nematoceran Diptera, and the first Brachycera. Triassic and Jurassic fossils include some excellent preserved material in fine-grained deposits, such as those of Solenhofen, the site of beautifully preserved insects and *Archaeopteryx*. The origin and subsequent diversification of birds (Aves) provided the first aerial competition for insects since the evolution of flight.

From the Cretaceous (145–66 Ma) and onwards throughout the subsequent Tertiary period (66–2.6 Ma) (now often subdivided into Palaeogene and Neogene), excellently preserved arthropod specimens are found in amber – a resinous plant secretion that trapped insects and hardened into a clear preservative (as illustrated in the vignette at the beginning of this chapter and **Plate 4a**). The high-quality preservation of whole insects in amber contrasts favourably with earlier imprints in rocks ("compression fossils") that may comprise little more than crumpled wings. Although many early fossil records of extant higher taxa (groups above species level) derive from these well-preserved amber specimens, inherent sampling biases occur. Easily trapped smaller and forest-dwelling taxa are over-represented. Amber of Cretaceous origin occurs in France, Spain, Lebanon, Jordan, Burma (Myanmar), Japan, Siberia, Canada and New Jersey, with a few less-diverse amber deposits in other parts of Eurasia. Insect compression fossils of Cretaceous age also are well represented in Eurasia. The biota of this period shows a numerical dominance of insects coincident with angiosperm (flowering plant) diversification. The first bee (in the broad sense) occurs in Burmese amber dated to about 100 Ma, and the first true bee, *Cretotrigona prisca*, is from the Late Cretaceous. Newly recognized amber from the southern continents, including India and northern Australia, should provide insights into insect evolution outside of the northern hemisphere.

Major mouthpart types of extant insects evolved prior to the angiosperm radiation, associated with insects feeding on earlier terrestrial plant radiations. The fossil record indicates the great antiquity of certain insect–plant associations. The lower Cretaceous of China (130 Ma) has revealed both early angiosperms and a distinctive fly belonging to Nemestrinidae, with a characteristic long proboscis associated with angiosperm pollination. Elsewhere, a fossil leaf of an ancient sycamore has the highly characteristic mine of the extant genus *Ectoedemia* (Lepidoptera: Nepticulidae), suggesting at least a 97-million-year association between the nepticulid moth and particular plants. The primarily phytophagous orders, Coleoptera and Lepidoptera, commenced their massive radiations in the Cretaceous. By 66 Ma, the insect fauna looks rather modern, with some fossils able to be allocated to extant genera. For many animals, notably the dinosaurs, the Cretaceous–Tertiary ("K–T") (now often called Cretaceous–Palaeogene, "K–Pg") boundary marked a major extinction event. Although it is generally believed that the insects entered the Tertiary with little extinction, recent studies show that although generalized insect–plant interactions survived, the prior high diversity of insect–plant associations was greatly attenuated. At least in the palaeobiota of southwestern North Dakota, a major ecological perturbation at 66 Ma set back specialist insect–plant associations.

Our understanding of Tertiary insects increasingly comes from amber from the Dominican Republic (see **Plate 4a**), dated to the Miocene (15–18 Ma), to complement the abundant and well-studied earlier amber that derives from Baltic Eocene deposits (37–52, mostly 43, Ma). Baltic ambers have been preserved and are now partially exposed beneath the northern European Baltic Sea and, to a lesser extent, the southern North Sea, brought to shore by periodic storms. Many attempts have been made to extract, amplify and sequence ancient DNA from fossil insects preserved in amber – an idea popularized by the movie *Jurassic Park*. Amber resin is argued to dehydrate specimens and thus protect their DNA from bacterial degradation. Although sequences from ancient DNA have been claimed for a variety of amber-preserved insects, including a termite (30 myo (million years old)), a stingless bee (25–40 myo) and a weevil (120–135 myo), no-one has been able to authenticate these ancient sequences. Degradation of DNA from amber fossils and contamination by fresh DNA may be insurmountable.

An unchallengeable outcome of ongoing studies of fossil insects is that many insect taxa, especially genera and families, are revealed as old – much more so than previously thought. At species level, all northern temperate, sub-Arctic and Arctic zone fossil insects dating from the last million years or so appear to be identical in morphology to existing species. Many of these are beetles (particularly their preserved elytra), but the situation seems no different amongst other insects. Pleistocene climatic fluctuations (glacial and interglacial cycles) evidently caused taxon range changes, via movements and extinctions of individuals, but resulted in the genesis of few, if any, new species, at least as defined on their morphology. The implication is that if species of insect are typically greater than a million years old, then insect higher taxa such as genera and families may be of immense age.

Modern microscopic palaeontological techniques can allow inference of age for insect taxa based on recognition of specific types of feeding damage caused to plants that have been fossilized. As seen above, leaf mining in ancient sycamore leaves is attributable to a genus of extant nepticulid moth, despite the absence of any preserved remains of the actual insect. In like manner, a cassidine beetle (Chrysomelidae) causing a unique type of grazing damage on young leaves of ginger plants (Zingiberaceae) has never been seen preserved as contemporary with the leaf fossils. The characteristic damage caused by their leaf chewing is recognizable, however, in the Late Cretaceous (c. 65 Ma) deposits from Wyoming, some 20 million years before any body fossil of a culprit beetle. To this day, these beetles specialize in feeding on young leaves of gingers and heliconias of the modern tropics.

Despite these valuable contributions made by fossils, it should be noted that:

1 character states observed in a fossil need not be ancestral;

2 fossils are not actual ancestral forms of later taxa;

3 the oldest known fossil of a group does not necessarily represent the phylogenetically earliest taxon;

4 the oldest known fossil of a group should not be assumed to represent the date of origin of that group (Box 8.1).

Nevertheless, fossil insects may show a stratigraphic (time) sequence of earliest-dated fossils reflecting early branches in the phylogeny. For example, fossils of Mastotermitidae are seen before those of the "higher" termites, fossils of midges, gnats and sand flies are seen before those of house and blow flies, and "primitive" moths before the butterflies.

Insect fossils may show that taxa currently restricted in distribution once were distributed more widely. Some such taxa include:

• Mastotermitidae (Blattodea: Termitoidae), now represented by one northern Australian species, is diverse and abundant in ambers from the Dominican Republic (Caribbean), Brazil, Mexico, USA, France, Germany and Poland dating from Cenomanian Cretaceous (c. 100 Ma) to lower Miocene (some 20 Ma);

• the biting-midge subfamily Austroconopinae (Diptera: Ceratopogonidae), now restricted to one extant species of *Austroconops* in Western Australia, was diverse in Lower Cretaceous Lebanese amber (Neocomian, 120 Ma) and Upper Cretaceous Siberian amber (90 Ma);

An emerging pattern stemming from ongoing study of amber insects, especially those dating from the Cretaceous, is the former presence in the north of groups now restricted to the south. We might infer that the modern distributions, often involving Australia, are relictual due to differential extinction in the north. Perhaps such patterns relate to northern extinction at the K–T (= K–Pg) boundary due to bolide ("meteorite") impact. Evidently, some insect taxa presently restricted to the southern hemisphere but known from Dominican and Baltic ambers (15–52 Ma) did survive the K–T event, and regional extinction has occurred more recently.

The relationship of fossil-insect data to phylogeny derivation is complex. Although early fossil taxa seem often to precede phylogenetically later-branching ("more derived") taxa, it is methodologically unsound to assume so. Although phylogenies can be reconstructed from the examination of characters observed in extant material alone, fossils provide important information, not least allowing dating of the minimum age of origin of diagnostic derived character states and of clades. Optimally, all data, fossil and extant, can and should be reconciled into a single estimate of evolutionary history.

8.2.2 Living insect distributions as evidence for antiquity

Evidence from the current distribution (biogeography) implies antiquity of many insect lineages. The disjunct distribution, specific ecological requirements, and restricted vagility of insects in a number of genera suggest that their constituent species were derived from ancestors that existed prior to the continental movements of the Jurassic and Cretaceous periods (commencing some 155 Ma). For example, the occurrence of several closely related species from different lineages of chironomid midges (Diptera) only in southern Africa and Australia suggests that the ancestral taxon ranges were fragmented by separation of the continental masses during the breakup of the supercontinent Gondwana, giving a minimum age of 120 Ma for the separation of these midge lineages. Related Cretaceous amber fossils date from only slightly later than commencement of the southern continental breakup. Such estimates are substantiated by molecular dating studies of chironomids.

An intimate association between figs and fig wasps (see Box 11.1) has been subjected to molecular phylogenetic analysis for host figs and wasp pollinators. The radiations of both show episodes of colonization and radiation that largely track each other (co-speciation). Although earlier studies have explained the distribution of figs and fig wasps by Gondwanan vicariance, new data and analyses suggest that the mutualism originated in Eurasia in the late Cretaceous (c. 75 Ma) with the major lineages of figs and wasp pollinators splitting during the Tertiary and dispersing southwards, possibly tracking warmer climates. Subsequent diversifications occurred on different continents during the Miocene.

The woodroaches (Blattodea: *Cryptocercus*) have a disjunct distribution in Eurasia (seven species), western USA (one species), and the Appalachians of southeastern USA where there is cryptic diversity (probably five species; see Box 7.2). The species of *Cryptocercus* are near-indistinguishable in their morphology, but are distinctive in chromosome number, mitochondrial and nuclear sequences, and in their endosymbionts. *Cryptocercus* harbour endosymbiont bacteria in bacteriocytes of their fat bodies (see section 3.6.5). Phylogenetic analysis of the bacterial RNA sequences shows that they follow faithfully the branching pattern of their host cockroaches. Using an existing estimate of the tempo of molecular evolution, dates have been reconstructed for the major disjunctions in woodroach evolution. The earliest branch, the North American/east Asian separation (on either side of the Bering Strait), was dated at 59–78 Ma. These patterns evidently are much older than the Pleistocene glaciations, but instead are consistent with the replacement of forest by grassland across the Bering Strait in the early Tertiary. The lack of obvious differentiation of *Cryptocercus* suggests morphological stasis over a long time.

Such morphological conservatism and yet great antiquity of many insect species needs to be reconciled with the obvious species and genetic diversity discussed in Chapter 1. The occurrence of species assemblages in Pleistocene deposits that resemble those seen today (although not necessarily at the same geographical location) suggests considerable physiological, ecological and morphological constancy of species. In comparative terms, insects display slower rates of morphological evolution than is apparent in many larger animals such as mammals. For example, *Homo sapiens* is a mere 200,000 years old; and any grouping of humans and the two chimpanzee species is some 8–14 my old (based on a 2012 reassessment of divergence dates). In contrast, the species-rich genus *Drosophila* originated about 100 mya. Perhaps, therefore, the difference is due to mammals having undergone a recent radiation and yet already suffered major extinctions including significant losses in the Pleistocene. In contrast, insects underwent early and many subsequent radiations, each followed by relative stasis and persistence of lineages (see section 8.6).

8.3 WERE THE FIRST INSECTS AQUATIC OR TERRESTRIAL?

Arthropods evolved in the sea. This is based on evidence from the diverse arthropod forms preserved in Cambrian-age marine-derived deposits, such as the Burgess Shale in Canada and the Qiongzhusi Formation at Chengjiang in southern China. Our understanding of the evolution of the arthropods suggests that several colonizations of land took place, with one or more occurring independently within each of the arachnid, crustacean + hexapod, and myriapod lineages. The evolutionary scenario presented in section 8.1 allows us to infer that hexapod terrestriality evolved either directly from a group of marine crustaceans or possibly via freshwater if branchiopods alone are the sister group. The main evidence in support of a terrestrial origin for the Insecta derives from the fact that all extant non-pterygote insects (the apterygotes) and the other hexapods (Diplura, Collembola and Protura) are terrestrial. That is, all the early nodes (branching points) in the hexapod phylogenetic tree (see Fig. 7.2) are best estimated as being terrestrial, and there is no evidence from fossils (either by possession of aquatic features or from details of preservation site) to suggest that the ancestors of these groups were not terrestrial (although they may have been associated with the margins of aquatic habitats). Although the juveniles of five pterygote orders (Ephemeroptera, Odonata, Plecoptera, Megaloptera and Trichoptera) live almost exclusively in freshwater, the position of these orders in Fig. 7.2 does not allow inference of freshwater ancestry in protopterygotes.

Another line of evidence against an aquatic origin for the earliest insects is the difficulty in envisaging how a tracheal system could have evolved in water. In an aerial environment, simple invagination of external respiratory surfaces and subsequent internal elaboration probably gave rise to a tracheal system (as shown in Fig. 8.4a) that later served as a preadaptation for tracheal gas exchange in the gills of aquatic insects (as shown in Fig. 8.4b). Thus, gill-like structures could assume an efficient oxygen uptake function (more than just diffusion across the cuticle) only ***after*** the evolution of tracheae in a terrestrial ancestor. Therefore, the presence of tracheae would be a preadaptation for efficient gills. An alternative view is that the tracheae evolved in early hexapods living in fresh or highly saline water as a way of reducing ion loss/gain to the surrounding water

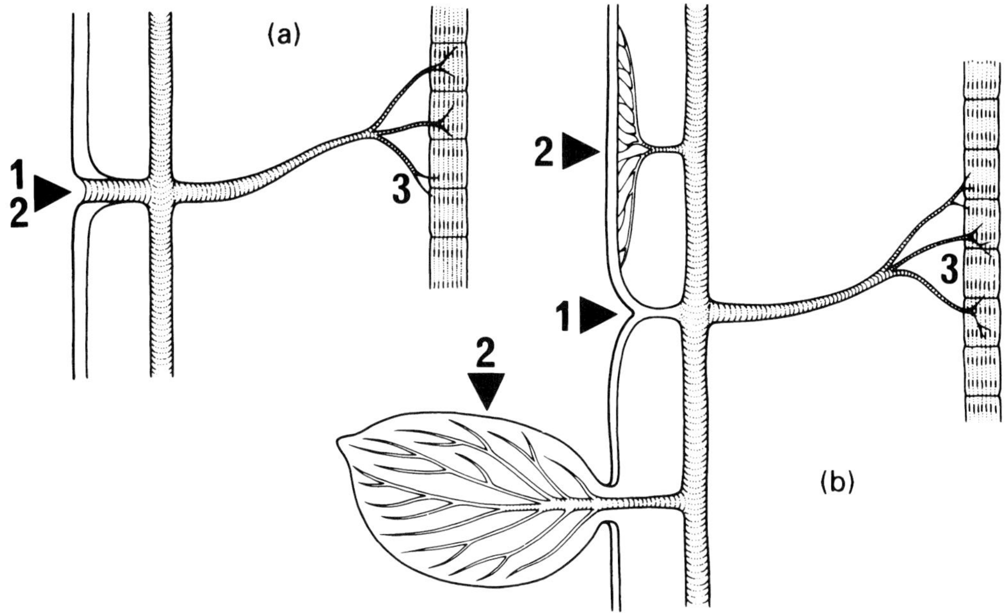

Fig. 8.4 Stylized tracheal system. (a) Oxygen uptake through invagination. (b) Invagination closed, with tracheal gas exchange through gill. 1, indicates point of invagination of the tracheal system; 2, indicates point for oxygen uptake; 3, indicates point for oxygen delivery, such as muscles. (After Pritchard *et al.* 1993.)

during oxygen uptake; by developing an air-filled space ("proto-tracheae") below the cuticle, ions in the haemolymph would be removed from direct contact with the cuticular epithelium of the gills. Whether hexapod tracheae evolved in freshwater or on land, it is most likely that the earliest hexapods living in moist environments obtained oxygen, at least in large part, by diffusion across the cuticle into the haemolymph, where gas transport was facilitated by the respiratory pigment haemocyanin, which reversibly binds with oxygen (see section 3.4.1). Haemocyanin, the major respiratory pigment in Crustacea including in putative sister groups to the hexapods (remipedes or malacostracans), occurs in modified form in Plecoptera, Zygentoma, and a few other insects (but not in Ephemeroptera, Odonata and the Holometabola). Tracheae (at least in *Drosophila*) arise from cells associated with the developing wing and leg, and homologues of the tracheal inducer gene are expressed in the gills of crustaceans, associating the evolution of tracheae with concurrent development of gills and/or wings in insects. Replacement of haemolymph-based oxygen transport with tracheal gas exchange may have provided aerodynamic advantages because the volume (and weight) of haemolymph is reduced in insects that use direct gas exchange via tracheae.

There is no single explanation as to why virtually all insects with aquatic immature development have retained an aerial adult stage. Certainly, retention has occurred independently in several lineages (such as a number of times within both the Coleoptera and Diptera). The suggestion that flight is a predator-avoidance mechanism seems unlikely, as predation could be avoided by a motile aquatic adult, as with so many crustaceans. It is conceivable that an aerial stage is retained to facilitate mating. Perhaps there are mechanical disadvantages to underwater copulation in insects, or perhaps mate-recognition systems may not function in water, especially if they are pheromonal or auditory.

8.4 EVOLUTION OF WINGS

The success of insects can be attributed largely to an evolutionary novelty: the wings. Flying insects are unusual in that no limbs lost their pre-existing function in the acquisition of flight (compared with bats and birds, in which fore-limbs were co-opted for flight). We cannot observe the origins of flight, and fossils (although relatively abundant) do not help much in interpretation, so hypotheses of the origins of flight have been speculative. Certainly, insect flight (wings) originated just once, at the node uniting the monophyletic ptergygota (see Fig. 7.2).

A long-standing hypothesis attributed the origin of the wings to **paranota** – lobes postulated to have been derived *de novo* from thoracic terga. Such lobes would not have been articulated, and thus tracheation, innervation, venation and musculature would have arisen secondarily, although no mechanism was known. This paranotal hypothesis was challenged by an "exite" hypothesis, inferring wing origination from pre-existing, serially repeated, mobile pleural structures, such as one (or two) appendages of the **epicoxa** (Fig. 8.5a), a basal leg segment. Each "proto-wing" or winglet was envisaged as formed from either an **exite** (outer) or a fused exite + **endite** (inner) lobe of the respective ancestral leg. In this hypothesis, proximal leg segments fused with the pleura, providing existing musculature, articulation and tracheation. Fossils show articulated, tracheated structures could occur on any body segment, with development strongest on the thorax (Fig. 8.5b).

Detailed embryological investigations are providing compelling evidence for the origin of wings. In a studied mayfly and certain other insects, wings derive partially from notal extensions located at the junction of the thoracic tergite and pleurite, but such paranotal swellings on the mesothorax and metathorax appear to incorporate elements of the coxopodite from the dorsal pleuron. A recent study of the expression of *vestigial*, the wing "master gene", in the prothorax of *Tribolium* (Coleoptera: Tenebrionidae), suggests a dual origin of wings from the merger of the carinated plate (a lateral ridge of the pronotum) and two pleural plates (derived from proximal leg segments). Wing development evidently involves both pronotum and leg-bases, implying that the two wing-origin hypotheses should be combined.

Molecular studies of developmental regulation (see Box 6.1) show that one or few changes in timing of downstream expression of homeotic patterning genes, such as sex-combs reduced (*Scr*), can vary repression of appendages (e.g. preventing expression on the prothorax). Unrepressed appendage development, notably on mesothoracic and metathoracic segments, allows development of wings. Activation of a range of development genes across different gradients (antero-posterior, dorso-ventral, proximal-distal) controls the expansion

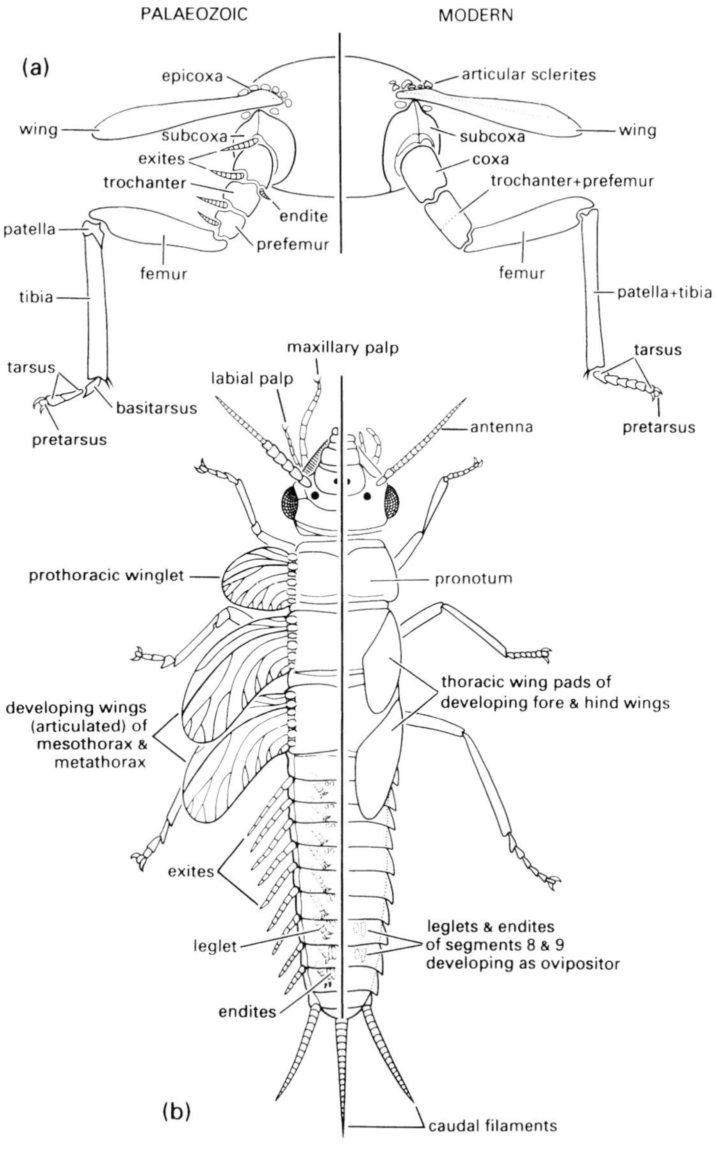

Fig. 8.5 Appendages of hypothetical primitive Palaeozoic (left of each diagram) and modern (right of each diagram) pterygotes (winged insects). (a) Thoracic segment of adult showing generalized condition of appendages. (b) Dorsal view of nymphal morphology. (Modified from Kukalová-Peck 1991; to incorporate ideas of J.W.H. Trueman (unpublished).)

of the leg-base (pleural) elements in the developing wing.

A dual origin "fusion" hypothesis of wing origin appears to contradict the proposal that wings derive from tracheated gills of an ancestral aquatic "proto-pterygote", in which a gas-exchange function was replaced by an aerodynamic one. The evident propensity for hexapods to develop (and lose) lateral appendages of the thorax and abdomen in response to similar developmental genetic cascades does not imply homology of structures (such as wings, gills and legs).

All hypotheses concerning early wings make a common assumption that proto-wings (or winglets) in adults originally had a non-flight function, as small winglets could have little or no use in flapping flight. Suggestions for preadaptive functions of proto-wings have included any (or all!) of the following: protection of the legs, covers for the spiracles, thermoregulation, sexual display, aid in concealment by breaking up the outline, or predator avoidance by extension of an escape jump by gliding. Aerodynamic function would come about only after enlargement of the wing.

The manner in which flight evolved remains speculative but, whatever the origin of winglets, some aerodynamic function evolved. Four routes to flight have been argued, via:

1 floating, in which small insects were assisted in passive dispersal by convection;

2 paragliding, in which winglets assisted in stable gliding or parachuting from trees and tall vegetation, perhaps after a powered leap;

3 running–jumping to flying;

4 surface sailing, in which the raised winglets allowed the adults of aquatic insects to skim across the water surface.

The first two hypotheses would apply equally to fixed, non-articulated winglets and to articulated but rigidly extended winglets. Articulated winglets and flapping flight can most easily be incorporated into the running–jumping scenarios of developing flight, although paragliding might have been a precursor of flapping flight. The "floating" route to flight suffers from the flaw that wings actually hinder passive dispersal, and selection would tend to favour diminution in body size and reduction in the wings, with commensurate increase in features such as long hairs. The third, running–jumping route, is unlikely, as no insect could attain the necessary velocity for flight originating from the ground, and only the scenario of a powered leap to allow limited gliding or flight is plausible.

The surface-sailing hypothesis requires articulated winglets, and suggests that surface sailing drove the evolution of wing length more than aerial gliding did. Thus, when proto-wings had reached certain dimensions then gliding or flapping flight may have been facilitated greatly. Some extant adult stoneflies (Plecoptera) can skim across water, holding their wings up as sails to increase speed. However, such skimming behaviour surely evolved subsequent to flight, since a terrestrial insect lifestyle appears to have evolved long prior to an aquatic one. Furthermore, fossil evidence for aquatic pterygotes appeared nearly 100 million years after the earliest known winged insects, despite the bias towards fossilization in freshwater sediments.

Aerodynamic theory has been applied to the problem of how large winglets had to be to give some aerodynamic advantage, and model insects have been constructed for testing in wind tunnels. Although a size-constrained and fixed-wing model lacks realism, even small structures give immediate advantage by allowing some retarding of velocity of the fall compared to an unwinged 1-cm-long insect model that lacks control in a glide. The possession of caudal filaments and/or paired cerci would give greater glide stability, particularly when associated with the reduction and eventual loss of posterior abdominal winglets. Additional control over gliding or flight would come with increase in body and wing size. Extant apterygote insects reveal morphologies and behaviours beneficial to the evolution of insect mastery of the air. Lepismatid silverfish (Zygentoma) can use their flattened bodies and outstretched antennae and caudal filaments to control attitude and to land the right way up after a fall. Machilids (Archaeognatha) perform controlled jumps and control landing using their appendages in a manner similar to silverfish, with the median terminal filament being especially important in steering the jump. There would have been high selective advantage for any insect that could attain forward movement by flapping of lateral body extensions. The ability to fold and flex the wings along the back (neoptery) may have given advantage to early pterygotes living in vegetation. Flight, including neoptery, had evolved by at least the Middle Carboniferous (315 Ma) and probably much earlier.

There is a basic structural division of the extant pterygotes into Neoptera, with complex wing articulation that allows folding of the wings backwards along the body, and Palaeoptera, with non-folding wings (section 7.4.2). Although the wing-bases differ such that some have argued that wings must have originated more than once, a basic pattern of venation common to all pterygotes exists irrespective of the articulation, implying a single origin of wings. The pterygote wing base involves many articulated sclerites: such a system is seen in some extinct palaeopterans and, in variably modified forms, in extant neopterans. Fossils of the extinct Diaphanopterodea show that these palaeopterans could flex their wings over the abdomen. Extant Ephemeroptera and Odonata have certain basal wing sclerites fused to stiffen the wing such that it cannot flex backwards. The nature of these fusions and different ones within the neopteran lineages, suggest many alterative pathways have developed from a common ancestor.

The origin of wing venation may have involved tracheated, supporting or strengthening ridges of the proto-wing. Alternatively, veins may have arisen along the course of haemolymph canals that supplied the winglets, in a manner seen in the gills of some aquatic insects. The basic venational pattern (section 2.4.2, Fig. 2.23) consists of eight veins, each arising from a basal blood sinus, named from anterior to posterior: precosta, costa, subcosta, radius, media, cubitus, anal and jugal. Each vein (perhaps excepting the media) branched basally into anterior concave and posterior convex components, with additional dichotomous branching away from the base, and a polygonal pattern of cells. Evolution of the insect wing has involved frequent reduction in the number of cells, development of bracing struts (cross-veins), selected increase in division of some veins, and reduction in complexity or complete loss of others. Although the thoracic muscles involved in flight can be homologized between palaeopterans and neopterans, some of the muscles used in direct and indirect powering of flight have diverged, as have the phases of wing beat (section 3.1.4). Much alteration in the function of wings has taken place, including the protection of the hind wings by modified fore wings (tegmina or elytra) in some groups. In some other groups, flight control has been increased by coupling the fore and hind wings as a single unit, and in Diptera by reduction of the metathoracic wings to halteres that function like gyroscopes.

8.5 EVOLUTION OF METAMORPHOSIS

Insects with a holometabolous life cycle, in which metamorphosis allows larval immature stages to be separated ecologically from the adult stage, are successful by any measure. The evolution of **holometaboly** (with larval juvenile instars highly differentiated from adults) from some form of **hemimetaboly** or from a winged **ametabolous** ancestor has been debated for a very long time. An early suggestion was that larvae of holometabous insects are the equivalent life stage of hemimetabolous nymphs, and that a distinctive pupa arose *de novo* as the immature and adult holometabolous insects diverged in their structure. A second early hypothesis was that larvae evolved from embryos ("precocious eclosion"). Thus, a pronymph (hatchling or pre-hatching stage, distinct from subsequent nymphal stages) would be the evolutionary precursor to the holometabolous larva, and the holometabolous pupa the sole remnant of all nymphal stages.

In ametabolous taxa (which form the earliest branches in the hexapod phylogeny), at each moult the subsequent instar is a larger version of the previous, and development is linear, progressive, and continuous. Even early flying insects, including Palaeodictyoptera (Fig. 8.3), had proportionally scaled winglets in all stages (sizes) of fossil nymphs, and thus were ametabolous. A distinctive earliest developmental stage, the **pronymph**, forms an exception to proportional nymphal development. The pronymph feeds only on yolk reserves, and can survive and move independently for some days after hatching. Hemimetabolous insects have a distinct pronymph, and differ from ametabolous taxa in that the one and only adult instar (with fully formed genitalia and wings) undergoes no further moulting.

The body proportions of the pronymph differ from those of subsequent nymphal stages, perhaps constrained by confinement inside the egg and by morphology to assist in hatching. The pterygote pronymph clearly is more than a highly miniaturized first-instar nymph. In certain orders (Blattodea, Hemiptera and Psocodea), the hatchling may be a pharate first-instar nymph, inside the pronymphal cuticle. At hatching, the nymph emerges from the egg, since the first moult occurs concurrently with eclosion. In Odonata and Orthoptera, the hatchling is the actual pronymph, which can undertake limited, often specialized, post-hatching movement to locate a potential nymphal development site before moulting to the first true nymphal stage.

The "precocious eclosion" hypothesis infers larval instars of Holometabola are homologous only to this pronymphal stage. The equivalents of hemimetabolan nymphal stages would be contracted into the holometabolous pupa, which thus would be the only nymphal stage in Holometabola. Some support for this hypothesis may come from differences between pronymphal, nymphal and larval cuticle, and timing of different cuticle formations relative to embryogenetic stages (**katatrepsis** – adoption of the final position in the egg – and dorsal closure; see section 6.2.1). However, strong contra evidence comes from similar numbers of moults (if embryonic ones are counted) in hemimetabolous and holometabolous insects, and especially from **evo-devo** studies showing similar modes of action of **juvenile hormone** (JH) throughout development of both groups.

Control of metamorphosis involves interactions between neuropeptides, ecdysteroids, and especially,

JH titres (section 6.3). The balance between controlling factors commences in the egg, and continues throughout development. Subtle differences in **heterochrony**, the timing of expression (activation) or constraint (suppression) of different genes involved in the processes of development (see Box 6.1), lead to very different outcomes. In post-embryonic hemimetabolous insects, continuous low JH exposure allows nymphal development to progress gradually towards the adult form. By contrast, elevated JH in the holometabolous embryo maintains the larval form because continued high JH suppresses maturation. In both forms of development, it is the cessation of JH prior to the final moult that allows production of the imaginal instar.

In some Holometabola, JH prevents any precocious production of adult features in the larva until the pupa. However, in others some adult features escape suppression by JH and commence development in larval instars. Such features include wings, legs, antennae, eyes and genitalia: their early expression is via groups of primordial cells that become **imaginal discs** – already differentiating for their final adult function (section 6.2, Fig. 6.4). With ability to vary the onset of differentiation of each adult organ in the larvae, there is scope for great variation and flexibility in life-cycle evolution, including capacity to greatly shorten any or all stages. For example, variation in JH expression can affect leg expression in any stage of the larva, from apodous (no expression of leg) to essentially fully developed (see Fig. 6.6).

Further insights into the evolution of metamorphosis have come from extension of studies on molecular development from model holometabolans (*Drosophila melanogaster*, *Bombyx mori*, *Manduca sexta* and *Tribolium castaneum*) to include hemimetabolans (*Blattella germanica*, *Oncopeltus fasciatus* and thrips) and the ametabolous firebrat (*Thermobia domestica*). Evidently, fundamental control of moulting by interplay between moult-inducing ecdysteroids and anti-metamorphosis action of JH commenced prior to the origin of hexapods. The evolution of a "stop" (cessation) of moulting at adult sexual maturity separates the ametabolous (and continuously moulting) hexapods from those that have a terminal moult to the adult (excluding the ephemeropteran winged **subimago**, which is a complex and unclear anomaly). Within the Pterygota, heterochronic difference in temporal expression and sensitivity to JH and downstream transcription factors (including Broad-complex (*Br-C*), Krüppel homologue 1 (*Kr*-h1) and especially the JH receptor Methoprene tolerant (*Met*)) differentiate all development types. In hemimetabolans, high expression of *BR-C* induced by elevated JH throughout all nymphal stages induces progressive differential growth of wings through each moult, with titres of JH and *BR-C* declining only at the final adult moult. In contrast, high titres of JH in holometabolans induce larval–larval moults, with *BR-C* expressed only when JH declines at the end of the last larval instar, correlated with onset of pupation and differential growth of adult structures. Shifts in the timing of JH and ecdysteroid peaks provide a mechanism for differential development, and a switch to inhibition of JH action upon *BR-C* expression in immature holometabolans perhaps was the key innovation in the transition from hemimetaboly to holometaboly.

Such studies help understand the convergent "holometaboly", sometimes termed neometaboly, of some Paraneoptera, specifically Thysanoptera (thrips) and certain Sternorrhyncha (Aleyrodoidea and males of Coccoidea). Thysanoptera show a near-hemimetabolan development, but two or three quiescent "pupal" stages are included, and wing development is discontinuous. In studied thrips, the profiles of JH, ecdysteroids and transcription factors *BR-C* and *Kr-h1* in the embryonic development resemble those of hemimetabolans, but in the post-embryo more resemble holometabolans, with elevated *BR-C* associated with onset of wing development in the larval–propupal moult. It seems that thrips development is convergently holometabolous, with similar regulatory genes involved.

These developmental studies undermine the precocious eclosion hypothesis as the explanation for differences between development types. Instead, it appears that holometabolan larvae are developmentally equivalent to hemimetabolan nymphs. The evolution of larvae undoubtedly led to the success of the Holometabola since their larvae have very different resource requirements from their adults and do not compete with them, whereas hemimetabolan nymphs and adults typically share the same lifestyles. The Holometabola thus decoupled larval and adult lives, and separated resources used in growth from those used for reproduction.

8.6 INSECT DIVERSIFICATION

An estimated half of all insect species chew, suck, gall or mine the living tissues of higher plants (**phytophagy**), yet only nine (of 28) extant insect orders are primarily phytophagous. This imbalance suggests that when a barrier to phytophagy (e.g. plant defences) is breached, an asymmetry in species number occurs, with the phytophagous lineage being much more speciose than the lineage of its closest relative (the sister group) of different feeding mode. For example, the tremendous diversification of the almost universally phytophagous Lepidoptera can be compared with that of its sister group, the relatively species-poor, non-phytophagous Trichoptera. Likewise, the enormous phytophagous beetle group Phytophaga (Chrysomeloidea plus Curculionoidea) is overwhelmingly more diverse than the entire Cucujoidea, the whole or part of which forms the sister group to the Phytophaga. Clearly, the diversifications of insects and flowering plants are related in some way, which we explore further in Chapter 11. By analogy, the diversification of phytophagous insects should be accompanied by the diversification of their insect parasites or parasitoids, as discussed in Chapter 13. Such parallel species diversifications clearly require that the phytophage or parasitod be able to seek out and recognize its host(s). Indeed, the high level of host-specificity observed for insects is possible only because of their highly developed sensory and neuromotor systems.

An asymmetry, similar to that of phytophagy compared with non-phytophagy, is seen if flightedness is contrasted to aptery. The monophyletic Pterygota (winged or secondarily apterous insects) are vastly more speciose than their immediate sister group, the Zygentoma (silverfish), or the totality of primitively wingless apterygotes. The conclusion is unavoidable: the gain of flight correlates with a radiation under any definition of the term. Flight allows insects the increased mobility necessary to use patchy food resources and habitats, and to evade non-winged predators. These abilities may enhance species survival by reducing the threats of extinction, but wings also allow insects to reach novel habitats by dispersal across a barrier and/or by expansion of their range. Thus, vagile pterygotes may be more prone to species formation by the two modes of geographical (**allopatric**) speciation: (i) small isolated populations formed by the vagaries of chance dispersal by winged adults may be the progenitors of new species; or (ii) the continuous range of widely distributed species may become fragmented into isolates by **vicariance** (range division) events such as vegetation fragmentation or geological changes.

New species arise as the genotypes of isolated populations diverge from those of parental populations. Such isolation may be phenological (temporal or behavioural), leading to **sympatric** speciation, as well as spatial or geographical, and host transfers or changes in breeding times are documented better for insects than for any other organisms. Host "races" of specialized phytophagous insects have been postulated to represent sympatric speciation in progress. The classic insect model of sympatric speciation has been the North American apple maggot fly, *Rhagoletis pomonella* (Tephritidae), which has a race that switched to apple (*Malus*) from the ancestral host, hawthorn (*Crataegus*), more than 150 years ago. Different populations of the apple maggot fly have been shown to respond preferentially to the fruit volatiles of their natal host plants (different *Crataegus* species). However, it now appears that some of the phenological variation that contributed to host-related ecological adaptation and reproductive isolation in sympatry of apple and hawthorn races actually arose in allopatry, specifically in Mexico, with subsequent gene flow into the United States leading to variation in diapause traits that facilitated shifts to new host plants, such as apple, that have differing fruiting times. Thus, some of the genetic changes that caused barriers to gene flow among races of the apple maggot fly evolved in geographic isolation, whereas other changes developed between populations in sympatry, leading to a mode of speciation that is neither strictly sympatric nor allopatric, but rather mixed or pluralistic. Although allopatric divergence appears to be the dominant mode of speciation in insects, factors contributing to taxon divergence also may evolve under other geographic conditions.

In addition to host specialization, highly competitive interactions – arms races – between males and females of one species may cause certain traits in one population to diverge rapidly from those of another population (see section 1.3.4). This kind of sexual selection, called sexual conflict, may contribute greatly to speciation rates in polyandrous insects (those in which females mate multiply with different males). Males benefit from adaptations that increase their paternity, such as achieving sperm precedence, increasing female short-term egg production and reducing female re-mating rate (section 5.7 and Box 5.6), whereas if female interests are compromised by males, females will evolve mechanisms to overcome male tactics, leading to post-mating sexual selection such as sperm competition and cryptic female choice. An extreme example of sexual conflict occurs in taxa with traumatic insemination (section 5.5). The differing post-mating interests of males and females can lead to rapid antagonistic coevolution of male and female reproductive morphology and physiology, which in turn may lead to rapid reproductive isolation of allopatric populations. Support for this hypothesis comes from comparisons of the species richness of pairs of related groups of insects that differ in post-mating sexual conflict, with each pair of contrasted taxa composed of a polyandrous and a monandrous group of species. Taxa in which females mate with many males have nearly four times as many species on average as related groups in which females mate with only one male.

The Holometabola (see section 7.4.2) contains the orders Diptera, Lepidoptera, Hymenoptera and Coleoptera (section 1.3), all of which have very high species richness (megadiversity). An explanation for their success lies in their metamorphosis (section 8.5), which allows the adult and larval stages to differ or overlap in phenology, depending upon timing of suitable conditions. Alternative food resources and/or habitats may be used by a sedentary larva and vagile adult, enhancing species survival by avoidance of intraspecific competition. Furthermore, deleterious conditions for some life-history stages, such as extreme temperatures, low water levels, or shortage of food, may be tolerated by a less susceptible life-history stage, for example a diapausing larva, non-feeding pupa, or migratory adult.

No single factor explains the astonishing diversification of the insects. An early origin and an elevated rate of species formation in association with the angiosperm radiation, combined with high species persistence through time, leave us with the great number of living species. We can obtain some ideas on the processes involved by study of selected cases of insect radiations in which the geological framework for their evolution is well known, as on some Pacific islands.

8.7 INSECT BIOGEOGRAPHY

Viewers of nature documentaries, biologically alert visitors to zoos or botanic gardens, and global travellers will be aware that different plants and animals live in different parts of the world. This is more than a matter of differing climate and ecology. Thus, Australia has suitable trees but no woodpeckers, tropical rainforests but no monkeys, and prairie grasslands without native ungulates. American deserts have cacti, but arid regions elsewhere that lack native cacti have a range of ecological analogues, including succulent euphorbs (Euphorbiaceae). The study of the distributions and the past historical and current ecological explanations for these distributions is the discipline of biogeography. Insects, no less than plants and vertebrates, show patterns of restriction to one geographic area (endemism) and entomologists have been, and remain, amongst the most prominent biogeographers. Our ideas on the biological relationships between the size of an area, the number of species that the area can support, and changes in species (turnover) in ecological time – called island biogeography – have come from the study of island insects (see section 8.8). Although islands typically are considered to be oceanic, the concept applies also to habitats isolated in metaphorical "oceans" of unsuitable habitat – such as montane peaks and plateaus arising from lowlands, or isolated forest remnants in agro-landscapes.

Entomologists have been prominent amongst those who have studied dispersal between areas, across land bridges, and along corridors, with ground beetle (carabid) specialists being especially prominent. An early paradigm of an Earth with static continents has shifted to one of dynamic movement powered by plate tectonics. Much of the evidence for faunas drifting along with their continents came from entomologists studying the distribution and evolutionary relationships of taxa shared exclusively among the modern disparate remnants of the once-united southern continental land mass (Gondwana). Within this cohort, those studying aquatic insects were especially prominent, perhaps because the adult stages are ephemeral and the immature stages so tied to freshwater habitats, that long distance trans-oceanic dispersal seemed an unlikely explanation for the many observed disjunct (discontinuous) distributions. Stoneflies, mayflies, dragonflies, and aquatic flies including midges (Diptera: Chironomidae) show southern hemisphere disjunct associations, even at low taxonomic levels (species groups, genera). These modern-day distributions imply that their ancestors must have been around, and subjected to Earth history events, in the Upper Jurassic and Cretaceous. Such findings imply that many groups must have been extant for at least 130 million years. Some ancient time-scales appear to be confirmed by fossil material, but molecular estimates of dates (based on rates of acquisition of mutations in molecules) vary in their support for great age.

On the finer scale, insect studies have played a major role in understanding the role of geography in processes of species formation and maintenance of local differentiation. Naturally, the genus *Drosophila* figures prominently, with its Hawai'ian radiation having provided valuable data. Studies of parapatric speciation – divergence of spatially separated populations that share a boundary – have involved detailed understanding of orthopteran, especially grasshopper, genetics and micro-distributions. Research on putative sympatric speciation has centred on the apple maggot fly, *Rhagoletis pomonella* (Tephritidae), for which barriers to gene flow appear to have evolved partly in allopatry and partly in sympatry (see section 8.6). The distribution modelling analyses outlined in section 17.1 exemplify some potential applications of ecological biogeographic rationales to relatively recent historical, environmental and climatic events that influence distributions. Entomologists using similar modelling tools to interpret recent fossil material from lake sediments have played a vital role in recognizing how insect distributions have tracked past environmental change, and allowed estimation of past climate fluctuations (section 17.1.2).

Strong biogeographic patterns in the modern fauna are becoming more difficult to recognize and interpret since humans have been responsible for the expansion of ranges of certain species and for the loss of much endemism, such that many of our most familiar insects are cosmopolitan (that is, virtually worldwide) in distribution. There are at least five explanations for this expansion of so many insects that once were of more-restricted distribution.

1 Human-loving (anthropophilic) insects, such as many cockroaches, silverfish and house flies, accompany humans virtually everywhere.

2 Humans create disturbed habitats wherever they live and some synanthropic (human-associated) insects, such as tramp ants (see Box 1.3), act rather like weedy plants and are able to take advantage of disturbed conditions better than native species can. (Synanthropy is a weaker association with humans than anthropophily.)

3 Insect (and other arthropod) external parasites (ectoparasites) and internal parasites (endoparasites) of humans and domesticated animals are often cosmopolitan.

4 Humans rely on agriculture and horticulture, with a few food crops cultivated very widely. Plant-feeding (phytophagous) insects associated with plant species that were once localized but now disseminated by humans can follow the introduced plants and may cause damage wherever the host plants grow. Many insects have been distributed in this way.

5 Insects have expanded their ranges by deliberate anthropogenic (aided by humans) introduction of selected species as biological control agents to control pest plants and animals, including other insects.

Attempts are made to restrict the shipment of agricultural, horticultural, forestry and veterinary pests through quarantine regulations (see Box 17.5), but much of the mixing of insect faunas took place before effective measures were implemented. Thus, pest insects tend to be identical throughout climatically similar parts of the world, meaning that applied entomologists must take a worldwide perspective in their studies.

8.8 INSECT EVOLUTION IN THE PACIFIC

Study of the evolution of insects (and other arthropods such as spiders) of oceanic islands, such as Hawai'i and the Galapagos, is comparable in importance to those of the perhaps more famous plants (e.g. Hawa'ian silverswords), birds (Hawai'ian honeycreepers and "Darwin's finches" of the Galapagos), land snails (Hawai'i) and lizards (Galapagos iguanas). The earliest and most famous island evolutionary studies of insects involved the Hawai'ian fruit flies (Diptera: Drosophilidae). This radiation has been revisited many times, but recent research has included evolutionary studies of certain crickets, microlepidopterans, carabid beetles, pipunculid flies, mirid bugs and damselflies.

Why this interest in the fauna of isolated island chains in the mid-Pacific? The Hawai'ian fauna is highly endemic, with an estimated 99% of its native arthropod species found nowhere else. The Pacific is an immense ocean, in which lies Hawai'i, an archipelago (island chain) some 3800 km distant from the nearest continental land mass (North America) or high islands (the Marquesas). The geological history, which is quite well known, involves continued production of new volcanic material at an oceanic hotspot located in the southeast of the youngest island, Hawai'i, whose maximum age is 0.43 myo. Islands lying to the northwest are increasingly older, having been transported to their current locations (Fig. 8.6) by the northwestern movement of the Pacific plate. The production of islands in this way is likened to a "conveyor belt" carrying islands away from the hotspot (which stays in the same relative position). Thus, the oldest

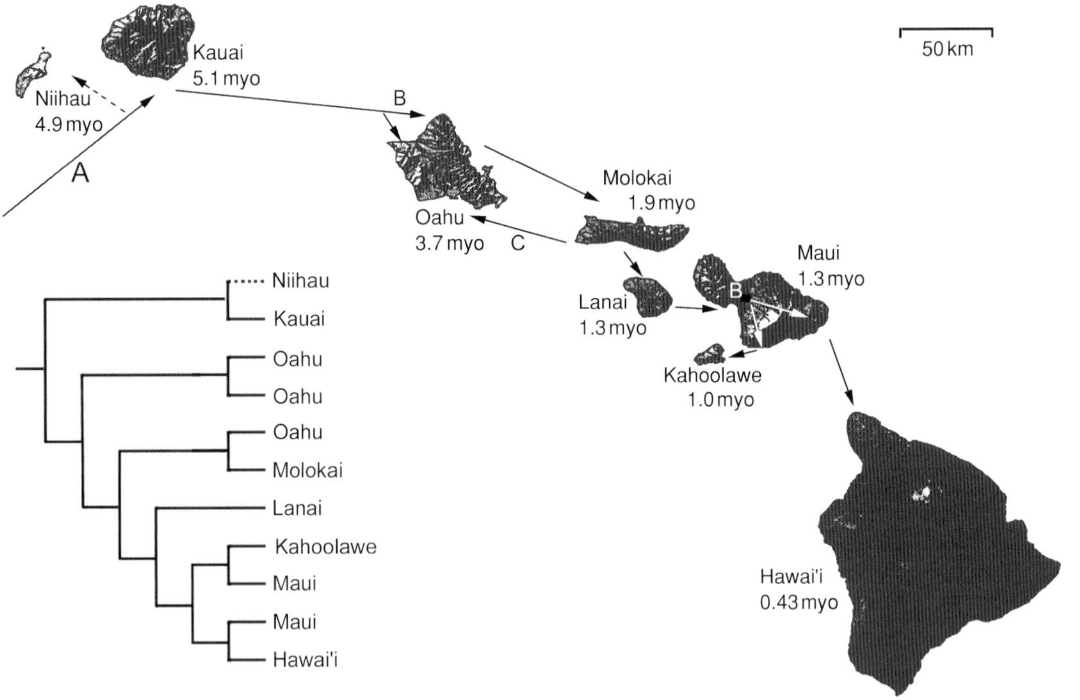

Fig. 8.6 Area cladogram showing phylogenetic relationships of hypothetical insect taxa with taxon names replaced by their areas of endemism in the Hawai'ian archipelago. The pattern of colonization and speciation of the insects on the islands is depicted by arrows showing the sequence and direction of events: A, founding; B, diversification within an island; C, back-colonization event. myo, million years old. Dashed line denotes extinct lineage.

existing above-water "high islands" (that is of greater elevation than a sand bar/atoll) are Niihau (aged 4.9 myo) and Kauai (c. 5.1 myo), positioned to the northwest. Between these two and Hawai'i lie Oahu (aged 3.7 myo), Molokai (1.9 myo), Maui and Lanai (1.3 myo), and Kahoolawe (1.0 myo). Undoubtedly, there have been older islands – some estimates are that the chain originated some 80 Ma – but only since about 23 Ma have there been continuous high islands for colonization.

These mid-oceanic and volcanic islands must have originated without a terrestrial biota, and thus present-day inhabitants must have descended from colonists. The great distance from source areas (other islands, the continents) implies that colonization is a rare event – and this is borne out in nearly all studies. The biota of islands is quite discordant (unbalanced) compared to that of continents. Major groups are missing, presumably by chance failure to arrive and flourish. Those that did arrive successfully and founded viable populations often speciated, and may exhibit quite strange biologies with respect to their ancestors. Thus, some Hawai'ian damselflies have terrestrial nymphs, in contrast to their aquatic habitats elsewhere; Hawai'ian geometrid moth caterpillars are predaceous, not phytophagous; otherwise marine midge larvae are found in freshwater torrents.

As a consequence of the rarity of founding events, most insect radiations have been identified as monophyletic, that is all members of the radiation belongs to one clade derived from a sole founder individual or population. For some clades, each species of the radiation is restricted to one island, whereas other ("widespread") species can be found on more than one island. Fundamental to understanding the history of the colonization and subsequent diversification is a phylogeny of relationships between the species in the clade. The Hawai'ian Drosophilidae lack widespread species (i.e. all are single-island endemics) and their relationships have been studied, first with morphology and more recently with molecular techniques. The history of this clade is straightforward. Species distributions generally are congruent with the geology, such that the colonists of older islands and older volcanoes (those of Oahu and Molokai) gave rise to descendants that have radiated more recently on the younger islands and younger volcanoes of Maui and Hawai'i. Similar scenarios of an older colonization with more recent radiation associated with island age are seen in the Hawai'ian Pipunculidae, damselflies, mirid bugs and some clades of cicadellid leafhoppers, and this probably is typical for the diversified biota. Where estimates have been made to date the colonization and radiation, it seems few, if any, originated prior to the currently oldest high island (c. 5 Ma), and sequential colonization seems to have been approximately contemporaneous with each newly formed island.

Insects such as black flies (Diptera: Simuliidae), which have larvae restricted to running water, cannot colonize islands until persistent streams and seepages form. As islands age in geological time, greatest environmental heterogeneity with maximum aquatic habitat diversity may occur in middle age, until senescence-induced erosion and loss of elevated areas cause extinction. In this island "middle age", speciation may occur on a single island as specializations in different habitats, as in Hawai'ian *Megalagrion* damselflies. In this clade, most speciation has been associated with existing ecological larval-habitat specialists (fast-running water, seepages, plant axils, or even terrestrial habitats) colonizing and subsequently differentiating on newly formed islands as they arose from the ocean and suitable habitats became available. However, on top of this pattern there can be radiations associated with different habitats on the same island, perhaps occurring very rapidly after initial colonization. Furthermore, examples exist showing recolonization from younger to older islands (back-founder events) that indicate substantial complexity in the evolution of some insect radiations on islands.

Sources for the original colonizers sometimes have been difficult to find because the offspring of Hawai'ian radiations often are very distinct from any prospective non-Hawai'ian relatives. However, the western or southwestern Pacific is a likely source for platynine carabids, *Megalagrion* damselflies and several other groups, and North America for some mirid bugs. In contrast, the evolution of the insect fauna of the Galapagos on the eastern side of the Pacific Ocean presents a rather different story to that of Hawai'i. Widespread insect species on the Galapagos predominantly are shared with Central or South America, and endemic species often have sister-group relationships with the nearest South American mainland, as is proposed for much of the fauna. The biting-midge (Ceratopogonidae) fauna derives apparently from many independent founding events, and similar findings come from other families of flies. Evidently, long-distance dispersal from the nearest continent outweighs *in-situ* speciation in generating the diversity of the Galapagos, compared to Hawai'i. Nonetheless, some estimates of arrival of founders are earlier than the currently oldest islands.

Orthopteroids of the Galapagos and Hawai'i show another evolutionary feature associated with island living – wing loss or reduction (aptery or brachyptery) in one or both sexes. Similar losses are seen in carabid beetles, with phylogenetic analyses postulating multiple losses. Furthermore, extensive radiations of certain insects in the Galapagos and Hawai'i are associated with underground habitats such as larva tubes and caves. Studies of the role of sexual selection – primarily female choice of mating partner (section 5.3) – suggest that this may have played an important role in species differentiation on islands, at least for crickets and fruit flies.

All islands of the Pacific are highly impacted by the arrival and establishment of non-native species, through introductions perhaps by continued over-water colonizations, but certainly associated with human commerce, including well-meaning biological control activities. Some accidental introductions, such as of tramp ant species (see Box 1.3) and a mosquito vector of avian malaria, have affected Hawai'ian native ecosystems detrimentally across many taxa. Even parasitoids introduced to control agricultural pests have spread to native moths in remote natural habitats (section 16.5). Our unique natural laboratories for the study of evolutionary processes are being destroyed apace.

FURTHER READING

Andrew, D.R. (2011) A new view of insect-crustacean relationships II. Inferences from expressed sequence tags and comparisons with neural cladistics. *Arthropod Structure and Development* **40**, 289–302.

Arnqvist, G., Edvardsson, M., Friberg, U. & Nilsson, T. (2000) Sexual conflict promotes speciation in insects. *Proceedings of the National Academy of Sciences* **97**, 10460–4.

Belles, X. (2011) Origin and evolution of insect metamorphosis. In: *Encyclopedia of Life Sciences*, pp. 1–11. John Wiley & Sons, Chichester. 10.1002/9780470015902.a0022854.

Clark-Hachtel C.M., Linz, D.M. & Tomoyasu Y. (2013) Insights into insect wing origin provided by functional analysis of *vestigial* in the red flour beetle, *Tribolium castaneum*. *Proceedings of the National Academy of Sciences* **110**, 16951–6.

Cranston, P.S. & Naumann, I. (1991) Biogeography. In: *The Insects of Australia*, 2nd edn (CSIRO), pp. 181–97. Melbourne University Press, Carlton.

Dudley, R. (1998) Atmospheric oxygen, giant Palaeozoic insects and the evolution of aerial locomotor performance. *Journal of Experimental Biology* **201**, 1043–50.

Engel, M.S., Davis, S.R. & Prokop, J. (2013) Insect wings: the evolutionary development of nature's first flyers. In: *Arthropod Biology and Evolution* (eds A. Minelli, G. Boxshall & G. Fusco), pp. 269–98. Springer, Berlin, New York.

Gillespie, R.G. & Roderick, G.K. (2002) Arthropods on islands: colonization, speciation, and conservation. *Annual Review of Entomology* **47**, 595–632.

Giribet, G. & Edgecombe, G.D. (2012) Reevaluating the arthropod tree of life. *Annual Review of Entomology* **57**, 167–86.

Grimaldi, D. & Engel, M.S. (2005) *Evolution of the Insects*. Cambridge University Press, Cambridge.

Harrison, J.F., Kaiser, A. & VandenBrooks, J.M. (2010) Atmospheric oxygen and the evolution of insect body size. *Proceedings of the Royal Society B* **277**, 1937–46.

Ho, S.Y.W. & Lo, N. (2013) The insect molecular clock. *Australian Journal of Entomology* **52**, 101–5.

Jordan, S., Simon, C. & Polhemus, D. (2003) Molecular systematics and adaptive radiation of Hawaii's endemic damselfly genus *Megalagrion* (Odonata: Coenagrionidae). *Systematic Biology* **52**, 89–109.

Kukalová-Peck, J. (1983) Origin of the insect wing and wing articulation from the arthropodan leg. *Canadian Journal of Zoology* **61**, 1618–69.

Kukalová-Peck, J. (1987) New Carboniferous Diplura, Monura, and Thysanura, the hexapod ground plan, and the role of thoracic side lobes in the origin of wings (Insecta). *Canadian Journal of Zoology* **65**, 2327–45.

Kukalová-Peck, J. (1991) Fossil history and the evolution of hexapod structures. In: *The Insects of Australia*, 2nd edn (CSIRO), pp. 141–79. Melbourne University Press, Carlton.

Labandeira, C.C. (2005) The fossil record of insect extinction: new approaches and future directions. *American Entomologist* **51**, 14–29.

Minakuchi, C., Tanaka, M., Miura, K. & Tanaka, T. (2010) Developmental profile and hormonal regulation of the transcription factors broad and Krüppel homolog 1 in hemimetabolous thrips. *Insect Biochemistry and Molecular Biology* **41**, 125–34.

Mitter, C., Farrell, B. & Wiegmann, B. (1988) The phylogenetic study of adaptive zones: has phytophagy promoted insect diversification? *American Naturalist* **132**, 107–28.

Pritchard, G., McKee, M.H., Pike, E.M. *et al.* (1993) Did the first insects live in water or in air? *Biological Journal of the Linnean Society* **49**, 31–44.

Regier, J.C., Schultz, J.W., Zwick, A. *et al.* (2010) Arthropod relationships revealed by phylogenomic analysis of nuclear protein-coding sequences. *Nature* **463**, 1079–84.

Rehm, P., Borner, J., Meusemann, K. *et al.* (2011) Dating the arthropod tree based on large-scale transcriptome data. *Molecular Phylogenetics and Evolution* **61**, 880–7.

Resh, V.H. & Cardé, R.T. (eds) (2009) *Encyclopedia of Insects*, 2nd edn. Elsevier, San Diego, CA. [Particularly see articles on biogeographical patterns; fossil record; island biogeography and evolution.]

Rota-Stabelli, O., Daley, A.C. & Pisani, D. (2013) Molecular timetrees reveal a Cambrian colonization of land and a new scenario for ecdysozoan evolution. *Current Biology* **23**, 392–8.

Truman, J.W. & Riddiford, L.M. (1999) The origins of insect metamorphosis. *Nature* **401**, 447–52.

Truman, J.W. & Riddiford, L.M. (2002) Endocrine insights into the evolution of metamorphosis in insects. *Annual Review of Entomology* **33**, 467–500.

von Reumont, B.M., Jenner, R.A., Wills, M.A. *et al.* (2012) Pancrustacean phylogeny in the light of new phylogenomic data: support for Remipedia as the possible sister group of Hexapoda. *Molecular Biology and Evolution* **29**, 1031–45.

Wheat, C.W. & Wahlberg, N. (2013) Phylogenomic insights into the Cambrian explosion, the colonization of land and the evolution of flight in arthropods. *Systematic Biology* **62**, 93–109.

Whitfield, J.B. & Kjer, K.M. (2008) Ancient rapid radiations of insects: challenges for phylogenetic analysis. *Annual Review of Entomology* **53**, 449–72.

Xie, X., Rull, J., Michel, A.P. *et al.* (2007) Hawthorn-infesting populations of *Rhagoletis pomonella* in Mexico and speciation mode plurality. *Evolution* **61**, 1091–105.

Chapter 9

GROUND-DWELLING INSECTS

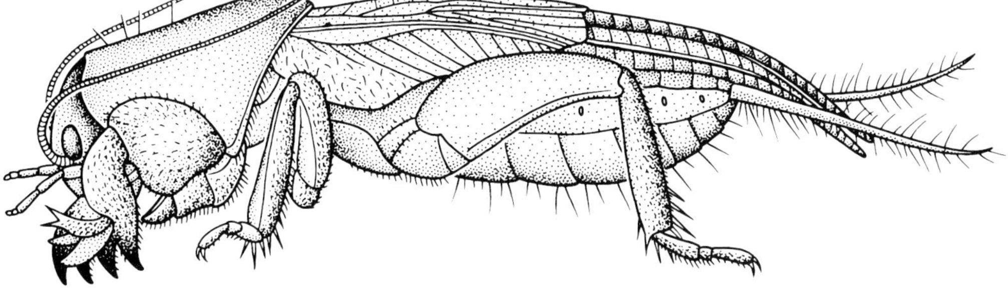

A mole cricket. (After Eisenbeis & Wichard 1987.)

A profile of a typical soil shows an upper layer of recently derived vegetative material, termed **litter**, overlying more decayed material that intergrades with **humus**-enriched organic soils. These organic materials lie above mineralized soil layers, which vary with local geology as well as with climatic factors such as present and past rainfall and temperature. Particle size and soil moisture are important influences on the micro-distributions of subterranean organisms. The decompositional habitat, comprising decaying wood, leaf litter, carrion and dung, is an integral part of the soil system. The processes of decay of vegetation and animal matter, and return of nutrients to the soil, involve many organisms. Notably, fungal hyphae and fruiting bodies provide a medium exploited by many insects, and all faunas associated with decompositional substrates include insects and other hexapods. Common ground-dwelling groups are the non-insect hexapods (Collembola, Protura and Diplura), primitively wingless bristletails and silverfish (Archaeognatha and Zygentoma), and several orders or other taxa of pterygote insects (including Blattodea – the cockroaches and termites, Dermaptera – the earwigs, Coleoptera – many beetles, and Hymenoptera – particularly ants and some wasps).

In this chapter, we consider the ecology and taxonomic range of soil and decompositional faunas in relation to the differing macrohabitats of soil and decaying vegetation and humus, dead and decaying wood, dung and carrion. Although root-feeding insects consume living plants, we discuss this poorly studied guild here rather than in Chapter 11. We survey the importance of insect–fungal interactions and examine two intimate associations, and provide a description of a specialized subterranean habitat (caves). The chapter ends with a discussion of some uses of terrestrial hexapods in environmental monitoring. Four boxes deal with selected topics, namely the use of soldier flies in composting (Box 9.1), the tactics used by ground-nesting wasps (specifically beewolves) to protect their brood from microorganisms (Box 9.2), the role of termites in engineering parts of the landscape in southern Africa (Box 9.3), and the biology of root-feeding "ground pearls" – the nymphs of certain margarodid scale insects (Box 9.4).

9.1 INSECTS OF LITTER AND SOIL

Litter is fallen vegetative debris, comprising decaying leaves, twigs, wood, fruit and flowers. The incorporation of recently fallen vegetation into the humus layer of the soil involves degradation by microorganisms, such as bacteria, protists and fungi. The actions of nematodes, earthworms and terrestrial arthropods, including crustaceans, mites, and a range of hexapods (Fig. 9.1), break down large particles and deposit finer particles as faeces. Mites (Acari), termites (Termitoidae), ants (Formicidae) and many beetles (Coleoptera) are important arthropods of litter and humus-rich soils. The immature stages of many insects, including beetles, flies (Diptera) and moths (Lepidoptera), may be abundant in litter and soils. For example, in Australian forests and woodlands, the eucalypt leaf litter is consumed by larvae of many oecophorid moths and certain chrysomelid leaf beetles. Decaying vegetation, and often plant roots and fungi, are consumed by a range of detritivorous fly larvae, such as those of soldier flies (Stratiomyidae; Box 9.1), March flies (Bibionidae) and fungus gnats (Sciaridae) (see also sections 9.1.1 and 9.5). The soil fauna also includes many species of non-insect hexapods – the springtails (Collembola), proturans (Protura) and diplurans (Diplura) – and many species of bristletails (Archaeognatha) and silverfish (Zygentoma). A number of Blattodea, Orthoptera and Dermaptera occur only in terrestrial litter – a habitat to which many species of several minor orders of insects, the Zoraptera, Embioptera and Grylloblattodea, are restricted. Permanently or regularly waterlogged soils, such as marshes and riparian (stream marginal) habitats, intergrade into fully aquatic habitats (Chapter 10) and show some faunal similarities.

In a soil profile, the transition from the upper, recently fallen litter to the lower well-decomposed litter to the humus-rich soil below may be gradual. Certain arthropods may be confined to a particular layer or depth, and show a distinct behaviour and morphology appropriate to that depth. For example, amongst the Collembola, species of Onychiuridae live in deep soil layers and have reduced appendages, are blind and white, and lack a furcula, the characteristic springing organ of collembolans. At intermediate soil depths,

The Insects: An Outline of Entomology, Fifth Edition. P.J. Gullan and P.S. Cranston.

Companion Website: www.wiley.com/go/gullan/insects

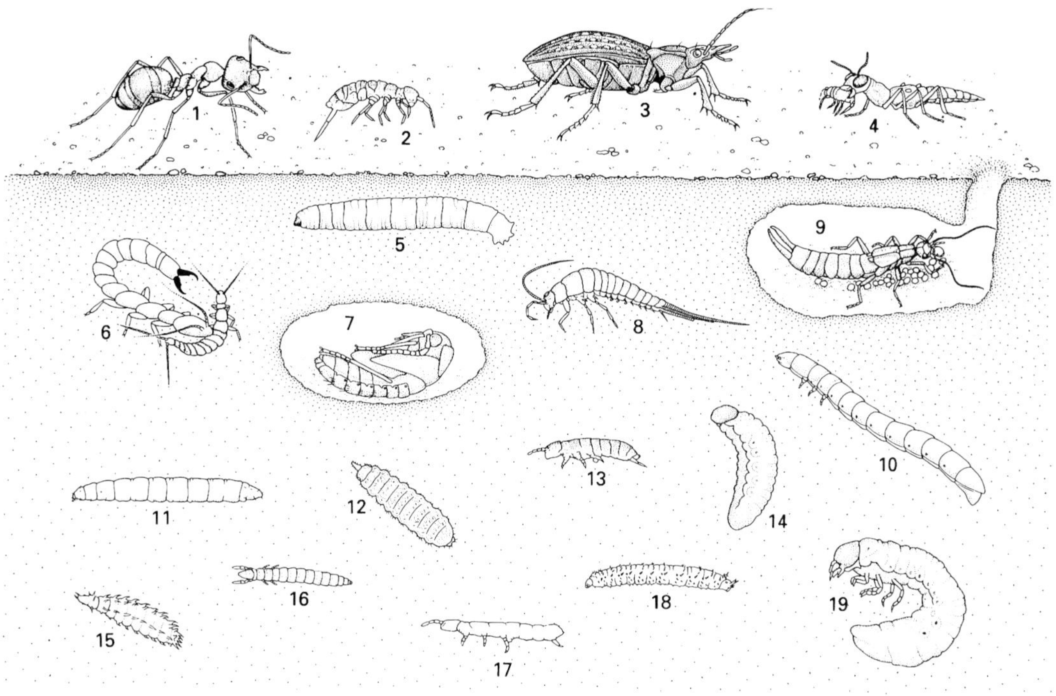

Fig. 9.1 Diagrammatic view of a soil profile showing some typical litter and soil insects and other hexapods. Note that organisms living on the soil surface and in litter have longer legs than those found deeper in the ground. Organisms occurring deep in the soil usually are legless or have reduced legs; they are unpigmented and often blind. The organisms depicted are: (1) worker of a wood ant (Hymenoptera: Formicidae); (2) springtail (Collembola: Isotomidae); (3) ground beetle (Coleoptera: Carabidae); (4) rove beetle (Coleoptera: Staphylinidae) eating a springtail; (5) larva of a crane fly (Diptera: Tipulidae); (6) japygid dipluran (Diplura: Japygidae) attacking a smaller campodeid dipluran; (7) pupa of a ground beetle (Coleoptera: Carabidae); (8) bristletail (Archaeognatha: Machilidae); (9) female earwig (Dermaptera: Labiduridae) tending her eggs; (10) wireworm, larva of a tenebrionid beetle (Coleoptera: Tenebrionidae); (11) larva of a robber fly (Diptera: Asilidae); (12) larva of a soldier fly (Diptera: Stratiomyidae); (13) springtail (Collembola: Isotomidae); (14) larva of a weevil (Coleoptera: Curculionidae); (15) larva of a muscid fly (Diptera: Muscidae); (16) proturan (Protura: Sinentomidae); (17) springtail (Collembola: Isotomidae); (18) larva of a March fly (Diptera: Bibionidae); (19) larva of a scarab beetle (Coleoptera: Scarabaeidae). (Individual organisms after various sources, especially Eisenbeis & Wichard 1987.)

Hypogastrura has simple eyes, and short appendages, with the furcula being shorter than half the body length. In contrast, Collembola such as *Orchesella* that live amongst the superficial leaf litter have larger eyes, longer appendages, and an elongate furcula, more than half as long as the body.

Soil insects show distinctive morphological variations. Larvae of some insects have well-developed legs to permit active movement through the soil, and pupae frequently have spinose transverse bands that assist movement to the soil surface for eclosion. Many adult soil-dwelling insects have reduced eyes, and their wings are protected by hardened fore wings, or are reduced (brachypterous), or lost altogether (apterous), or, as in the reproductives of ants and termites, are shed after the dispersal flight (deciduous, or caducous). Flightlessness (that is either through primary absence or secondary evolutionary loss of wings) in ground-dwelling organisms may be countered by jumping as a means of evading predation: the collembolan furcula is a spring mechanism, and the alticine Coleoptera ("flea-beetles") and terrestrial Orthoptera can leap to safety. Jumping is of little value in subterranean organisms, instead fore legs may be

Box 9.1 Soldier flies can recycle waste

Most of us live in cities and have little or no contact with waste management or agriculture, even on a backyard-scale. In some countries, a local government service collects our separated domestic "green" waste, such as kitchen scraps and garden material, but for many households it joins general waste destined for landfill sites. Increasingly, urban waste management needs to be re-assessed because landfill sites are finite, non-renewable and expensive. People growing home crops know that garden (yard) and kitchen waste can enhance their soil fertility via humus produced by worm activity in composters (vermiculture). Such systems can be scaled up for community use. However, even well-managed worm-based systems depend upon some pre-sorting (little or no citrus or animal products) and can suffer from by-products of nuisance flies such as *Musca domestica* (house fly) and Drosophilidae (fruit/vinegar flies), and failure at particular air, water or temperature conditions.

A potentially valuable aid in rapid breakdown of organic waste is based on a stratiomyid (Diptera), the black soldier fly *Hermetia illucens*. These attractive flies are common in temperate to tropical parts of the world, where they occur naturally in decomposing vegetation and sometimes in carrion. Open composting systems often contain the detritivorous larvae of these flies, and under certain conditions soldier fly larvae can dominate and exclude worms and all other invertebrates. Advantages include the voracious larvae with a dietary breadth that is wider than worms, and a tolerance of lower oxygen levels, higher pH, higher temperatures and the presence of animal waste products. Furthermore, the mature larvae and the prepupae of these reared stratiomyids provide valuable supplemental animal feeds, including for chickens and farmed fish, and a live pet-food supply for amphibians and reptiles. The prepupae are rich in proteins and lipids, and contain appreciable phosphorus and calcium from the cuticle (as is typical of all stratiomyids).

For domestic systems maintained to produce compost, using stratiomyids in the treatment may be disadvantageous, as most nutrients become recycled into the fly larvae and pupae, and the soil-enhancing residues are modest relative to worm-based production. However, for urban and farming waste, including putrid refuse high in nitrogen such as pig, poultry and coffee production wastes, rapid and hygienic treatment may prevail over compost production.

The natural life cycle involves a fertilized female fly laying between 500 and 1000 eggs close to moist decomposing material, with larval development in the medium through five instars, the last of which is a non-feeding but active prepupal stage. Optimally, this immature life cycle takes two weeks, but longer at lower temperatures. The prepupal larva, which can be 20 mm long, migrates away from the food source, seeking a dry sheltered site suitable for pupariation (pupation within the sclerotized cuticle of the last larval instar; section 6.2.3). The emerged adults live for a week or so, during which mating occurs (as shown here, with the female on the left; after a photograph by Muhammad Mahdi Karim, 2009, www.micro2macro.net), and oviposition commences the new cycle.

In commercial use of these flies for waste-disposal and production of larvae for feedstock, these characteristics are harnessed in fabricated waste disposal units (such as illustrated here for a bucket composter that produces fly larvae from food scraps; after Black Soldier Fly Blog 2012). Flies can colonize naturally or eggs can be purchased for "seeding" the culture. A great advantage of this system is the total exclusion of longer-lived disease-bearing flies that thrive in unmanaged wastes. In trials running three units for a year, an input of up to 100 kg daily of restaurant and supermarket waste was converted into larvae at a digestion rate of 15 kg m^{-2} of unit surface per day, at a conversion rate of 20% wet (24% dry) weight of original input. Bioconversion is fastest at high temperatures, aided by heat generated by larval respiration, until their maximum heat tolerance is reached and larvae disperse to cooler

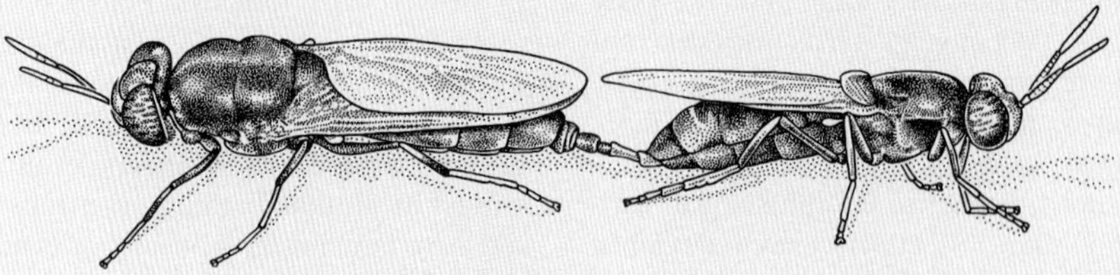

parts of the medium. In cool winters, activity slows, but can be enhanced in a greenhouse – however, sub-zero (°C) temperatures kill larvae. Production designs can incorporate larval ramps that permit the migrating larvae to self-harvest into a container.

Our own experience in California with a regular "compost" bin that naturally developed black soldier fly populations was that in summer, all household waste, including animal refuse such as smaller bones and otherwise intractable plant material (citrus and avocado skins, wine grape residues), was consumed (audibly sometimes) in quick time by a 5–10 cm thick layer of hungry larvae. Activity slowed in winter, but our only loss of larvae was due to our lengthy absence during a protracted drought. In 10 years, the contents of the bin never rose to half-full and we never emptied it. House flies were never present, and exiting prepupae and hidden puparia were sought-after delicacies by blue jays, skunks and occasional raccoons.

Australian research suggests that moderate-scale systems would be valuable for remote locations, including tropical/subtropical islands where resorts generate food waste that otherwise would need to be shipped off at high expense to mainland landfill. And native birds love the output!

modified for digging (Fig. 9.2) as **fossorial** limbs in groups that construct tunnels, such as mole crickets (as depicted in the vignette at the beginning of this chapter), immature cicadas, and many beetles.

The distribution of subterranean insects changes seasonally. The constant temperatures encountered at greater soil depths are attractive in winter to avoid low aboveground temperatures. The level of water in the soil is important in governing both vertical and horizontal distributions. Frequently, larvae of subterranean insects that live in moist soils seek drier sites for pupation, reducing the risks of fungal disease during the

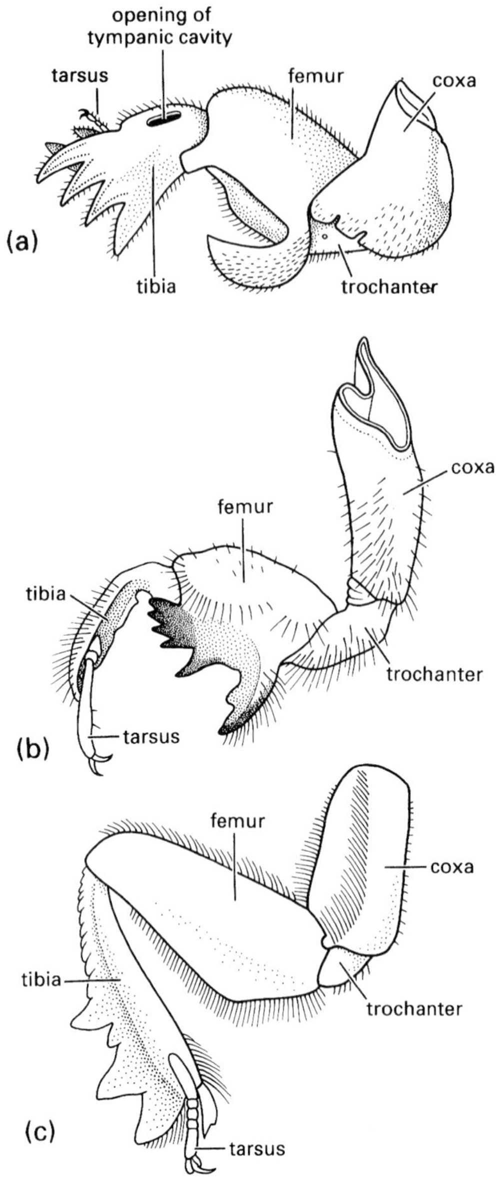

Fig. 9.2 Fossorial fore legs of: (a) a mole cricket of *Gryllotalpa* (Orthoptera: Gryllotalpidae); (b) a nymphal periodical cicada of *Magicicada* (Hemiptera: Cicadidae); and (c) a scarab beetle of *Canthon* (Coleoptera: Scarabaeidae). ((a) After Frost 1959; (b) after Snodgrass 1967; (c) after Richards & Davies 1977.)

immobile pupal stage. The subterranean nests of ants usually are located in drier areas, or the nest entrance is elevated above the soil surface to prevent flooding during rain, or the whole nest may be elevated to avoid excess ground moisture. Location and design of the nests of ants and termites is very important to the regulation of humidity and temperature because, unlike social wasps and bees, they cannot ventilate their nests by fanning, although they can migrate within nests or, in some species, between them. The passive regulation of the internal nest environment is exemplified by termites of *Amitermes* (see Fig. 12.10) and also *Macrotermes* (see Fig. 12.11), which maintain an internal environment suitable for the growth of particular fungi that serve as food (section 9.5.3 and section 12.2.4).

A constant threat to soil-dwelling insects is the risk of infection by microorganisms, especially pathogenic fungi. Thus, many ground-nesting ants protect themselves and their brood using antibiotic secretions produced from the metapleural glands on their thorax (Taxobox 29). Specialist ground-nesting ants of the tribe Attini control fungal disease in their nests with antibiotics produced by symbiotic bacteria cultivated in cavities of the ants' cuticle (section 9.5.2). The European beewolf, *Philanthus triangulum*, and other digger wasps of the genus *Philanthus* (Crabronidae), also use symbiotic bacteria to protect their offspring from infection during development within nest burrows in soil (Box 9.2). The mutualistic bacteria of both attine ants and beewolves belong to the Actinomycetales, a group characterized by the ability to synthesize a suite of antibacterial and antifungal chemicals. Most earwigs (Dermaptera) live in moist environments and must contend with high microbial exposure. Recent analysis of the abdominal defensive gland secretions of several earwigs (including the European earwig, *Forficula auricularia*) identified 1,4-benzoquinones that had antimicrobial activity against tested bacteria and entomopathogenic fungi. It is anticipated that further study of insect–microbe symbioses among ground-dwelling insects will reveal diverse mechanisms to fend off pathogens. One result may be the identification of novel antibiotics with application in human medicine.

Many soil-dwelling hexapods derive their nutrition from ingesting large volumes of soil containing dead and decaying vegetable and animal debris, and associated microorganisms. These bulk-feeders, known as

Box 9.2 Antimicrobial tactics to protect the brood of ground-nesting wasps

Ground-dwelling insects are particularly susceptible to infection by fungi and bacteria, due to the suitable humidity and temperature of soil burrows. Although the evolution of mechanisms for protecting against microbes should be favoured in such insects (section 9.1), very few examples are known. The discovery of symbiotic bacteria associated with a ground-nesting wasp suggests an important type of mutualism that may be widespread in other soil-nesting insects, particularly Hymenoptera. The European beewolf, *Philanthus triangulum*, is a large digger wasp (Crabronidae) that constructs nest burrows in sandy soil and provisions brood cells with paralyzed honey bees as food for her larvae (as shown on the left, with a female beewolf dragging a captured bee). The female beewolf smears the ceiling of each brood cell with a whitish substance that she secretes from glands in her antennae. These glands are present in antennomeres 4 to 8 (as shown in the middle drawing of a longitudinal section of the antenna, after Goettler *et al*., 2007). The secretion has dual functions: (i) it provides a cue for the orientation of the cocoon spun by the larval beewolf to later facilitate emergence of the adult beewolf from its brood cell; and (ii) it becomes incorporated into the beewolf cocoon and inhibits microbial infection during the overwintering diapause. The white substance consists mostly of symbiotic bacteria of the genus *Streptomyces* (Actinomycetales, the actinomycete bacteria), which are cultured in the female beewolf's antennal glands (shown enlarged in the drawing on the right, after Kaltenpoth *et al*., 2005). The bacteria are believed to produce antibiotics that control the growth of other microorganisms; larval beewolves deprived of the white secretion suffer high levels of mortality. Behavioural observations point to a vertical transmission of the bacteria from beewolf mother to offspring. Closely related *Streptomyces* bacteria occur in other species of *Philanthus*, suggesting an early origin of the beewolf–*Streptomyces* mutualism.

Another antimicrobial tactic practiced by the female of the European beewolf retards fungal degradation of the honey-bee prey that she stores for her larvae. By licking the surface of the paralyzed bees prior to laying an egg in each brood cell, the female beewolf applies large amounts of secretion from her post-pharyngeal glands (PPG) onto the prey. Unsaturated hydrocarbons in the PPG secretion help to preserve the bee prey, by preventing the condensation of water on the bee cuticle, thus creating conditions unsuitable for the germination of fungal spores. No direct, chemically mediated antifungal effect of the PPG secretion has been detected. In male beewolves, the PPG secretion functions as a scent mark for their territories and as an attractant to females, whereas the homologous gland in ants produces the colony odour. Research to investigate the PPG in other aculeate Hymenoptera is needed to understand the evolution and functions of this gland.

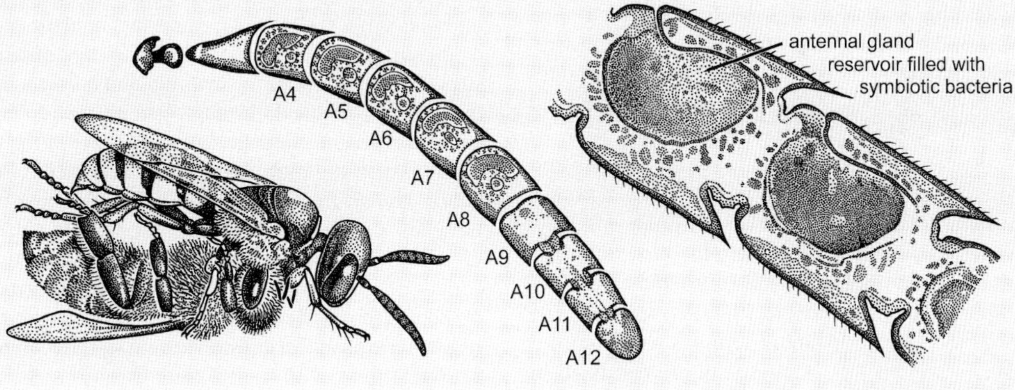

saprophages or **detritivores**, include hexapods such as some Collembola, beetle larvae, and certain termites (Termitoidae: Termitinae, including *Termes* and relatives). These termites apparently have endogenous cellulases and a range of gut microbes, and appear able to digest cellulose from the humus layers of the soil. Copious excreta (faeces) may be produced, and these organisms clearly play a significant role in structuring soils of the tropics and subtropics.

Arthropods that consume humic soils inevitably encounter plant roots. The fine parts of roots often associate with fungal mycorrhizae and rhizobacteria, forming a zone called the **rhizosphere**. Bacterial and fungal densities are an order of magnitude higher in soil close to the rhizosphere compared with soil distant from roots, and microarthropod densities are correspondingly higher close to the rhizosphere. The selective grazing of Collembola, for example, can curtail growth of fungi that are pathogenic to plants, and their movements aid in transport of beneficial fungi and bacteria to the rhizosphere. Furthermore, interactions between microarthropods and fungi in the rhizosphere and elsewhere may aid in mineralization of nitrogen and phosphates, making these elements available to plants.

Some groups of ground-dwelling insects, such as many species of ants and termites, derive their food mostly from the soil surface, rather than from within the soil. Among termites, for example, there are two main feeding groups: (i) the soil feeders (mentioned above); and (ii) the litter (including grass) feeders. Litter-feeding termites collect live or dead plant material, whereas ground-nesting ants collect live or dead arthropods, seeds, and/or sugar-rich liquids such as honeydew and nectar, depending on ant species and season. Major groups of ground-dwelling termites are the subterranean termites (Rhinotermitidae, especially *Coptotermes* and *Reticulitermes*) and the mound builders (mostly Termitidae such as *Macrotermes* species, but also a few Rhinotermitidae). The nesting and feeding habits of termites are discussed further in section 12.2.4, and ants are discussed in section 12.2.3. In many ecosystems, termites act as engineers, creating heterogeneous habitat and playing significant roles in physical mixing of the soil profile and in nutrient cycling (Box 9.3). In savannah landscapes, termite mounds (such as those of *Macrotermes* and *Odontotermes* species) often form islands of fertility associated with higher forage quality of the grasses and woody vegetation and thus creating hotspots for foraging by vertebrate herbivores.

9.1.1 Root-feeding insects

Although 50–90% of plant biomass may be below ground, herbivores feeding out-of-sight on plant roots have been neglected in studies of insect–plant interactions. Root-feeding activities have been difficult to quantify in space and time, even for charismatic taxa like the periodical cicadas (*Magicicada* spp.) of the deciduous forests of eastern North America. However, these root-xylem feeding cicada nymphs have the highest reported biomass per unit area of any native terrestrial animal and can reduce tree growth, as measured by wood growth rings, by up to 30% compared with non-infested trees. The damaging effects caused by root chewers and miners, such as larvae of hepialid and ghost moths and certain beetles including wireworms (Elateridae), false wireworms (Tenebrionidae), weevils (Curculionidae), scarabs or white grubs (Scarabaeidae), flea-beetles and galerucine Chrysomelidae, may become evident only if the aboveground plants collapse. However, death is one end of a spectrum of responses, with some plants responding with increased aboveground growth to root grazing, others being neutral (perhaps through resistance), and others sustaining non-lethal damage. Sap-sucking insects on the plant roots, such as some aphids (see Box 11.2) and scale insects (Box 9.4), cause loss of plant vigour, or death, especially if insect-damaged necrotized tissue is invaded secondarily by fungi and bacteria. If the nymphs of periodical cicadas occur in orchards they can cause serious damage, but the nature of the relationship with the roots upon which they feed remains poorly known (see also section 6.10.5). Increasingly, however, it is being shown that the feeding activities of soil-dwelling insect herbivores can influence the community dynamics of plants and the performance of aboveground herbivores. This is because plant responses typically are systemic, and signals can move between the roots and aerial plant parts.

Root-feeding insects exploit a range of plant chemical cues to locate their food plant in the soil. Emissions of carbon dioxide (CO_2) from respiring roots diffuse rapidly in the soil and may cause soil-living insects to search more intensely. Limited studies suggest that certain soil insects do orientate to a CO_2 source but that other root exudates may be more important

Box 9.3 Ecosystem engineering by southern African termites

Termites are the most important **ecosystem engineers** in many places, including in harsh environments, due to their wide foraging and their ability to exert control over the temperature and humidity of their own living space. They are allogenic engineers, in that they modify the environment by physically altering the form of the habitat.

Recently, termites have been confirmed as the causative agents of the enigmatic landscape formations called "fairy circles" – evenly spaced discs bare of vegetation in the centre but with a peripheral ring or belt of tall and dense perennial grasses (as illustrated here; after Juergens 2013). The origin of these circles has been long debated, although termites or ants have been the favoured culprits. Circles occur in arid grasslands on sandy soils in parts of the eastern edge of the Namib Desert in Namibia and adjoining parts of Angola and South Africa, and are found amidst a matrix of annual or short-lived grasses. Circle sizes vary, from a mean diameter of nearly 4 m in the south to 35 m in the north of the distribution, with a maximum diameter of almost 50 m. Individual circles persist for decades and perhaps for a few hundred years. Confirmation of the role of termites came from a study involving several years of sampling at sites throughout the known distribution of fairy circles. Only the sand termite, *Psammotermes allocerus* (Termitoidae: Rhinotermitidae), was associated consistently with most circles, including newly formed ones. Some 80–100% of circles contained sand-termite nests and their underground tunnel-like galleries as well as characteristic sheets of thin, cemented sand built over grass material to protect their aboveground foraging. Newly formed circles were observed to contain the burrows of sand termites, which kill the grass by damaging the roots. Soil beneath the centre of the circles has higher moisture content than soil outside the circle perimeter, and termite burrowing within circles of any age probably facilitates rainwater accumulation. Furthermore, the termites are thought to kill the roots of any germinating grass within the circle, as well as feeding on the perennial grasses on the inner edge of the peripheral belt, resulting in widening of the circle. The removal of water-transpiring plants inside the circles would promote water retention, providing suitable soil conditions for the termites, and this perennial water supply also would allow growth of long-lived grass at the periphery. Thus, the termites create and manage a perennial grass resource that would facilitate their survival in the desert, even in extreme drought years.

A different study, involving more-restricted sampling, found an association between the fairy circles and the black pugnacious ant *Anoplolepis steingroeveri* (Hymenoptera: Formicidae), which was more abundant in the circles than in the surrounding matrix. At the main study site in Namibia, these ants excavated and damaged grass roots to

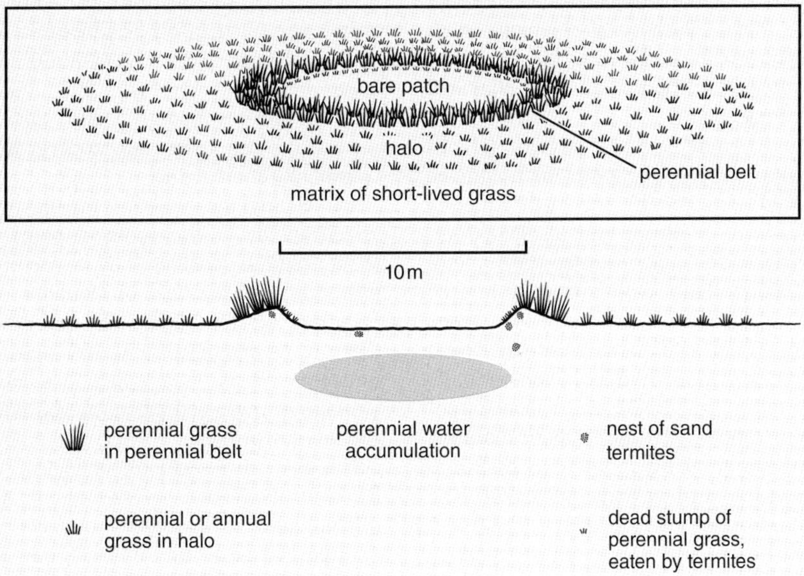

access honeydew-producing planthoppers (Hemiptera: Meenoplidae) that feed on the roots. Although in some places these ants may play a secondary role in keeping the circles free of vegetation, they are nomadic and not responsible for circle formation.

Another African example of ecosystem engineering by termites occurs farther south in the winter rainfall area of southwest South Africa. Here, the giant earth mounds of the southern harvester termite, *Microhodotermes viator* (Termitoidae: Hodotermitidae), form a prominent and widespread landscape feature. Mature or senescent mounds, called "heuweltjies", are usually capped with a sand layer, and can reach up to 32 m in diameter and 2.5 m in height. $\delta^{14}C$ dating and trace fossil data suggest that these large mounds can be at least 4000 years old. The termites are both herbivores and detritivores, as they feed by harvesting fresh vegetation and by foraging on the faeces of other animals. Soils on heuweltjies typically are more aerated, retain more water for longer following rain, and have higher nutrient levels and more biogeochemical activity than off-mound soils. Consequently, the on-mound vegetation can differ markedly from that which occurs away from the mounds. Another characteristic of these mounds is their generally evenly spaced distribution pattern, with mound density varying depending on soil fertility and vegetation type, under the influence of rainfall. The termites and their mounds play a keystone role in energy cycling, providing habitat and food resources for aardvarks and a range of other more conspicuous animals, and promoting species richness and turnover in the landscape.

signals and may alter the insect's CO_2 response. Nearly 100 different, mostly volatile, compounds that may mediate host-plant location have been identified from roots in the rhizosphere. Most of these chemicals are attractants, but a few are repellents, or may attract or repel depending on concentration. Once insects start feeding on roots, some compounds, such as sugars, act as phagostimulants, whereas plant secondary metabolites, such as phenolics, can deter feeding. Despite evidence that a diversity of chemicals elicit behavioural responses in root-feeding insects, there has been little attempt to manipulate root chemical ecology in pest management.

Root-feeding insects may be major pests of agriculture and horticulture. For example, the main damage to continuous field corn (maize) in North America is due to the western, northern and Mexican rootworms, the larvae of *Diabrotica* species (Coleoptera: Chrysomelidae). Corn plants with rootworm-injured roots are more susceptible to disease and water stress, and have decreased yield, leading to an estimated total loss plus control costs of more than $1 billion annually in the United States. In Europe, the cabbage root fly, *Delia radicum* (Diptera: Anthomyiidae), damages brassicas (cabbages and relatives) due to its root- and stem-feeding larvae (called cabbage maggots). The larvae of certain fungus gnats (Diptera: Sciaridae) are pests in mushroom farms and injure roots of house and greenhouse plants. The black vine weevil, *Otiorhynchus sulcatus* (Coleoptera: Curculionidae), is a serious pest of cultivated trees and shrubs in Europe and North America. Larvae injure both the crown and roots of a range of plants, and can maintain population reservoirs in weedy as well as cultivated areas. This weevil's pest status is aggravated by its high fecundity and the difficulty of detecting its early presence, including in nursery plants or soil being transported. Similarly, the larvae of *Sitona* weevils, which feed on the roots and nitrogen-fixing nodules of clover and other legumes in Europe and the United States, are particularly troublesome in both legume crops and forage pastures. Obtaining estimates of yield losses due to these larvae is difficult as their presence may be underestimated or even go undetected, and measuring the extent of underground damage is technically challenging.

Soil-feeding detritivorous insects probably do not selectively avoid the roots of plants. Thus, where there are high densities of fly larvae that ingest soil in pastures, such as those of Tipulidae (leatherjackets), Sciaridae (black fungus gnats) and Bibionidae (March flies), roots are injured by their activities. There are frequent reports of such activities causing economic damage in managed pastures, golf courses and turf-production farms.

The use of insects as biological control agents for control of alien/invasive plants has emphasized phytophages of aboveground parts, such as seeds and leaves (see section 11.2.7), but tends to neglect root-damaging taxa. Even with increased recognition of their importance, 10 times as many aboveground biological control agents are released compared to

Box 9.4 Ground pearls

In parts of Africa, the encysted nymphs ("ground pearls") of certain subterranean scale insects are sometimes made into necklaces by the local people. These nymphal insects have few cuticular features, except for their spiracles and sucking mouthparts. They secrete a transparent or opaque, glassy or pearly covering that encloses them, forming spherical to ovoid "cysts" of greatest dimension 1–8 mm, depending on species. Ground pearls belong to several genera of Margarodidae (Hemiptera), including *Eumargarodes*, *Margarodes*, *Neomargarodes*, *Porphyrophora* and *Promargarodes*. They occur worldwide in soils among the roots of grasses, especially sugarcane, and grape vines (*Vitis vinifera*). They may be abundant and their nymphal feeding can cause loss of plant vigour and death; in lawns, feeding results in brown patches of dead grass. In South Africa they are serious vineyard pests; in Australia different species reduce sugarcane yield; and in the southeast United States one species is a turf-grass pest.

Plant damage is caused mostly by the female insects because many species are parthenogenetic, or at least males have never been found, and when males are present they are smaller than the females. There are three distinct phases of female development (as illustrated here for *Margarodes* (= *Sphaeraspis*) *capensis*): after De Klerk *et al*. 1982): the first-instar nymph disperses in the soil, seeking a feeding site on roots, where it moults to the first of probably two cyst stages (only one is shown in the illustration); the adult female emerges from the nymphal cyst between spring and autumn (depending on species) and, in species with males, comes to the soil surface where mating occurs. The female then buries back into the soil, digging with its large fossorial fore legs. The fore-leg coxa is broad, the femur is massive, and the tarsus is fused with the strongly sclerotized claw. In parthenogenetic species, females may never leave the soil. Adult females have no mouthparts and do not feed; in the soil, they secrete a waxy mass of white filaments – an ovisac, which surrounds their several hundred eggs.

Although ground pearls can feed via their thread-like stylets, which protrude from the cyst, encysted nymphs of most species are capable of prolonged dormancy (up to 17 years has been reported for one species). Often, the encysted nymphs can be kept dry in the laboratory for one to several years and still are capable of "hatching" as adults. This long life and ability to rest dormant in the soil, together with resistance to desiccation, mean that they are difficult to eradicate from infested fields, and even crop rotations do not eliminate them effectively. Furthermore, the protection afforded by the cyst wall and subterranean existence makes insecticidal control largely inappropriate. Many of these curious pestiferous insects are probably African and South American in origin and, prior to quarantine restrictions, may have been transported within and between countries as cysts in soil or on rootstocks.

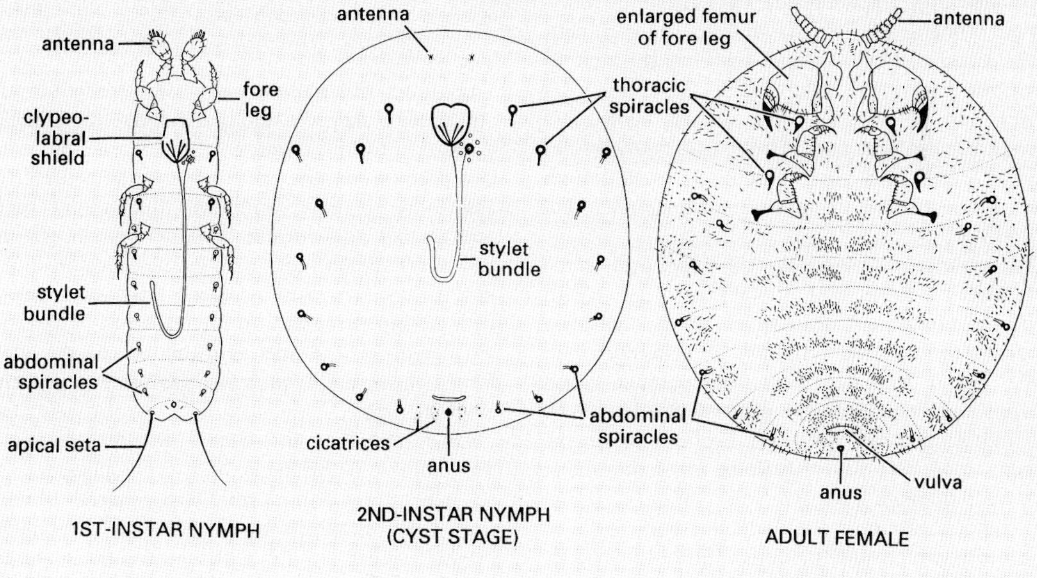

root feeders. By the year 2000, over 50% of released root-feeding biological control agents contributed to the suppression of target invasive plants; in comparison only about 33% of the aboveground biological control agents contributed some suppression of their host plant. Coleoptera, particularly Curculionidae and Chrysomelidae, appear to be the most successful root-feeding control agents, whereas Lepidoptera and Diptera are less so.

9.2 INSECTS AND DEAD TREES OR DECAYING WOOD

Insects may play a role in the transmission of pathogenic fungi causing decay of the wood of dead trees or the death of living host trees. For example, Dutch elm disease kills elm trees and is caused by a fungus transmitted by beetles (section 4.3.3; Box 17.4), and a fungal pathogen of pine trees is transmitted by *Sirex* wood wasps (see Box 11.3). Continued decay of such infected trees and those that die of natural causes often involves further interactions between insects and fungi.

Wood-boring weevils (Curculionidae) belonging to the subfamilies Scolytinae and Platypodinae are commonly called bark or ambrosia beetles, depending on their type and their degree of dependence on associated fungi. The association varies from the transmission of phytopathogens to host trees, to enrichment of the wood diet with fungal mycelia, through to **mycophagy** (fungus eating) in the case of ambrosia beetles. Bark beetles (most, but not all, Scolytinae) reproduce in the inner bark of trees, with many species entering dead, weakened or dying trees, although some attack and kill living trees (section 4.3.2 and section 4.3.3).

The ambrosia beetles (Platypodinae and some Scolytinae) are involved in a notable association with ambrosia fungus and dead wood, which has been popularly termed "the evolution of agriculture" in beetles. Adult beetles excavate tunnels (often called galleries), predominantly in dead wood (Fig. 9.3), although some attack live wood. Beetles mine in the phloem, wood, twigs or woody fruits, which they infect with wood-inhabiting ectosymbiotic "ambrosia" fungi that they transfer in special cuticular pockets called **mycangia**, which store the fungi during the insects' aestivation or dispersal. Ambrosia fungi appear dependent on their beetle hosts for transport and inoculation.

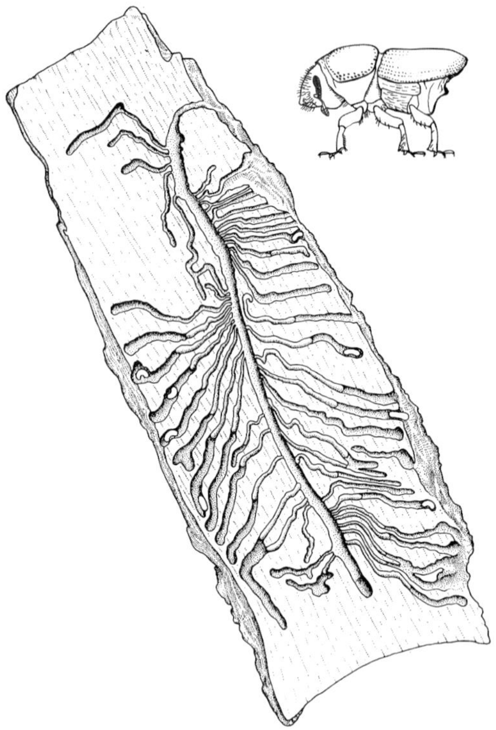

Fig. 9.3 A plume-shaped tunnel excavated by the bark beetle *Scolytus unispinosus* (Coleoptera: Curculionidae: Scolytinae), showing eggs at the ends of a number of galleries; enlargement shows an adult beetle. (After Deyrup 1981.)

These fungi, which come from a wide taxonomic range, curtail plant defences and break down wood, making it more nutritious. However, both larvae and adults feed primarily on the extremely nutritious fungi. The association between ambrosia fungus and beetles appears to be very ancient, perhaps originating as long ago as 60 million years with gymnosperm host trees, but with subsequent increased diversity associated with multiple transfers to angiosperms. Today there are more than 3000 species of ambrosia beetles, which form a feeding guild consisting of several independently evolved groups.

Some mycophagous insects, including beetles of the families Lathridiidae and Cryptophagidae, are strongly attracted to recently burned forest, to which they carry fungi in mycangia. The cryptophagid beetle *Henoticus serratus*, which is an early colonizer of burned forest in some areas of Europe, has deep depressions on the underside of its pterothorax (Fig. 9.4),

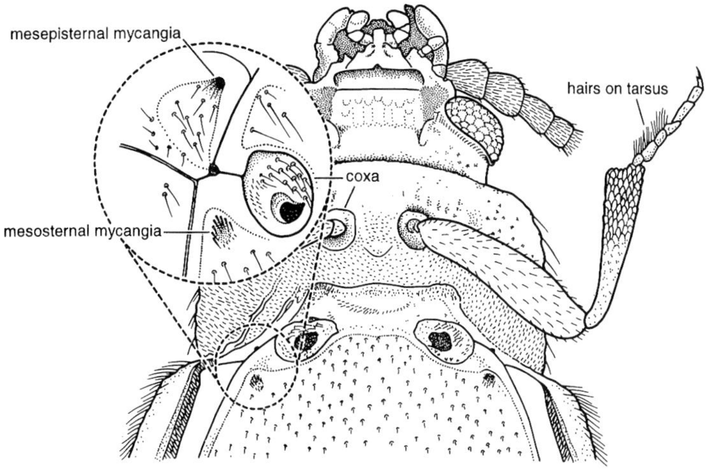

Fig. 9.4 Underside of the thorax of the beetle *Henoticus serratus* (Coleoptera: Cryptophagidae), showing the depressions, called mycangia, which the beetle uses to transport fungal material that inoculates new substrate on recently burnt wood. (After drawing by Göran Sahlén in Wikars 1997.)

from which glandular secretions and material of the ascomycete fungus *Trichoderma* have been isolated. The beetle probably uses its legs to fill its mycangia with fungal material, which it transports to newly burnt habitats as an inoculum. Ascomycete fungi are important food sources for many **pyrophilous** insects, i.e. species strongly attracted to burning or newly burned areas or which occur mainly in burned forest for a few years after the fire. Some predatory and wood-feeding insects are also pyrophilous. A number of pyrophilous heteropterans (Aradidae), flies (Empididae and Platypezidae) and beetles (Carabidae and Buprestidae) have been shown to be attracted to the heat or smoke of fires, and often from a great distance. Species of jewel beetle (Buprestidae: *Melanophila* and *Merimna*) locate burnt wood by sensing the infrared radiation typically produced by forest fires (section 4.2.1).

Fallen, rotten timber provides a valuable resource for a wide variety of detritivorous insects if they can overcome the problems of living on a substrate rich in cellulose and deficient in vitamins and sterols. Wood-feeding termites can live entirely on this diet, either through the use of cellulase enzymes in their digestive systems and gut symbionts (section 3.6.5), or with the assistance of fungi (section 9.5.3). Cockroaches and termites produce endogenous cellulase that allows digestion of cellulose from the diet of rotting wood. Other **xylophagous** (wood-eating) strategies of insects include very long life cycles with slow development, and probably the use of xylophagous microorganisms and fungi as food.

9.3 INSECTS AND DUNG

The excreta or dung produced by vertebrates can be a rich source of nutrients. In the grasslands and rangelands of North America and Africa, large ungulates produce substantial volumes of fibrous and nitrogen-rich dung that contains many bacteria and protists. Insect **coprophages** (dung-feeding organisms) and associated predators, parasitoids and fungivores depend upon this resource (Fig. 9.5). Certain higher flies – such as the Scathophagidae, Muscidae (notably the worldwide house fly, *Musca domestica*, the Australian *M. vetustissima* and the widespread tropical buffalo fly, *Haematobia irritans*), Faniidae and Calliphoridae – oviposit or larviposit into freshly laid dung. Development can be completed before the medium becomes too desiccated. Within the dung medium, predatory fly larvae (notably other species of Muscidae) can seriously reduce survival of coprophages. However, in the absence of predators or disturbance of the dung, larvae developing in dung in pastures can give rise to nuisance-level populations of flies.

The insects primarily responsible for disturbing dung, and thereby limiting fly breeding in the medium, are dung beetles, belonging to the family Scarabaeidae. Not all larvae of scarabs use dung: some ingest general soil organic matter, whereas some others are herbivorous on plant roots. However, many are coprophages. In Africa, where many large herbivores produce large volumes of dung, several thousand species of scarabs show a wide variety of coprophagous behaviours.

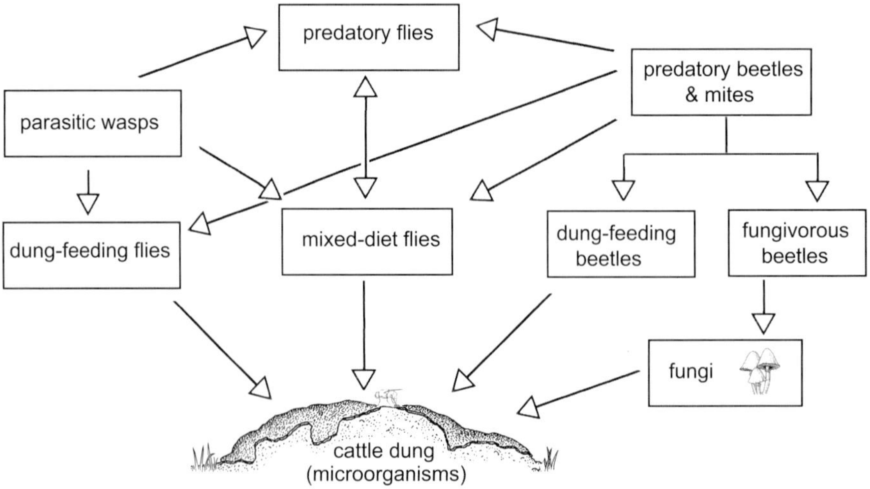

Fig. 9.5 Interactions among the arthropod groups that are common in cattle dung: the larvae of dung-feeding flies (from several families) feed on microorganisms; early-instar larvae of mixed-diet flies feed on microorganisms, but in later instars become predatory; larvae of predatory flies feed only on other insects; adult dung-feeding beetles (mainly Scarabaeidae) probably feed mostly on microorganisms in the fluids of fresh dung, whereas their larvae breakdown the plant fibres in ingested dung with the help of symbiotic gut bacteria; predatory beetles (such as Staphylinidae) feed on other insects, especially immature stages of flies; fungus-feeding beetles colonize pats in their later stages of decomposition after fungi have developed; larvae of wasps are mostly parasitoids on dung-breeding flies. (After Jochmann *et al.* 2011.)

Many can detect dung as it is deposited by a herbivore, and from the time that it falls to the ground, invasion is very rapid. Many individuals arrive, perhaps up to many thousands for a single fresh elephant dropping. Most dung beetles excavate networks of tunnels immediately beneath or beside the pad (also called a pat), and pull down pellets of dung (Fig. 9.6). Other beetles excise a chunk of dung and move it some distance to an excavated chamber, also often within a network of tunnels. This movement from pad to nest chamber may occur either by head-butting an unformed lump, or by rolling moulded spherical balls over the ground to the burial site. The female lays eggs into the buried pellets, and the larvae develop within the faecal food ball, eating fine and coarse particles. The adult scarabs also may feed on dung, but only on the fluids and finest particulate matter. Some scarabs are generalists and utilize virtually any dung encountered, whereas others specialize according to texture, wetness, pad size, fibre content, geographical area and climate. The activities of a range of scarab species ensure that all dung is buried rapidly.

The largest source of dung today is from our domestic animals. Agriculture, and particularly cattle farming, is a major contributor to anthropogenic greenhouse gases, mainly carbon dioxide (CO_2), methane (CH_4) and nitrous oxide (N_2O). Dung pads (pats) are one source of these gases, and an experimental study has shown that dung beetles (*Aphodius* spp.) mediate the gas fluxes from dung. Summed over the course of the study, total emissions of CH_4 were significantly lower and those of N_2O were significantly higher in pads with beetles than in those without beetles. There was no cumulative difference in emitted CO_2 levels from pads with and without beetles, but fresh dung pads (first week) emitted higher amounts of CO_2 in the presence of beetles. The mechanism for these gaseous modifications may be the beetle behaviour of digging holes, which can dry the dung and aerate the deeper parts of the pads, leading to increased aerobic (and reduced anaerobic) decomposition and reduced methane production; the microbes that produce methane require anaerobic conditions.

In Australia, a continent in which native ungulates are absent, native dung beetles cannot exploit the volume and texture of dung produced by introduced domestic cattle, horses and sheep. As a result, dung once lay around in pastures for prolonged periods, reducing the quality of pasture and allowing the development of prodigious numbers of nuisance flies.

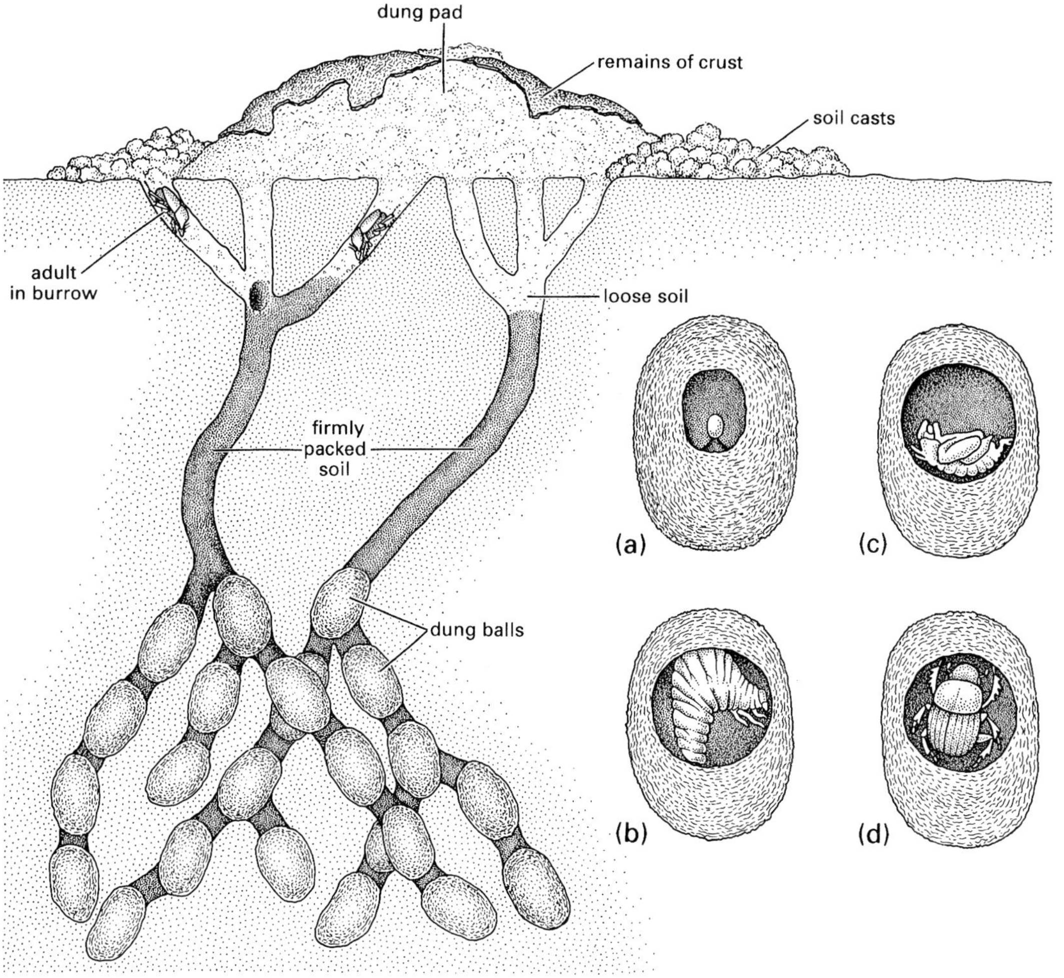

Fig. 9.6 A pair of dung beetles of *Onthophagus gazella* (Coleoptera: Scarabaeidae) filling in the tunnels that they have excavated below a dung pad. The inset shows an individual dung ball within which beetle development takes place: (a) egg; (b) larva, which feeds on the dung; (c) pupa; and (d) adult just prior to emergence. (After Waterhouse 1974.)

Introduction of alien dung beetles from Africa and Mediterranean Europe successfully accelerated dung burial in many regions.

Before the drastic loss of the American bison, *Bison bison* (Bovidae), from the plains of North America, their dung would have been habitat for a diversity of coprophilous beetles and flies. Each bison is estimated to have produced about 25 kg of dung per day. Did the bison's dung-feeding fauna go extinct, or move to the dung of the closely related but introduced cattle, *Bos taurus*? A recent Canadian experiment supports the latter hypothesis. Given the choice of dung from bison or cattle fed similar diets and cattle fed a different (barley silage) diet, both native and introduced (European) beetles were more responsive to the vertebrate animals' diets than to their species identity.

In tropical rainforests, an unusual guild of dung beetles has been recorded foraging in the tree canopy on every subcontinent. These specialist coprophages have been studied best in Sabah, Borneo, where a few species of *Onthophagus* collect the faeces of primates (such as gibbons, macaques and langur monkeys) from the foliage, form it into balls and push the balls over the edge of leaves. If the balls catch on the foliage below, then the dung-rolling activity continues until the ground is reached.

An increasing problem for the suite of dung-feeding insects associated with pastures is the widespread use of chemicals for the control of gastro-intestinal parasites, such as nematodes. These parasites often infest vertebrate herbivores (principally sheep and cattle) reared for human consumption. The most widely used anthelmint(h)ic ("anti-worm") chemicals are the macrocyclic lactones such as ivermectin. These anthelmintics are given to domesticated herbivores topically, orally or subcutaneously, and can pass through the animals into the dung, where they can exert toxic effects on non-target insects such as dung beetles and flies. Although these chemical residues do breakdown in the dung over time, their presence can retard colonization of cowpats, especially by Diptera. Residues have the potential to exert both lethal and sub-lethal effects on the dung fauna, but their impact depends on a range of factors, including timing and intensity of chemical treatments, the type of chemical used, and the abundance and behaviour of the dung-living insects. Ivermectin has been shown to reduce the developmental rate and survival of dung beetle larvae. One consequence of ivermectin use is a reduction in the decomposition rate of dung from treated vertebrates, sometimes comparable to rates when invertebrates are excluded from dung. Such ecological and non-target effects of chemicals such as ivermectin should be of concern to regulatory agencies.

9.4 INSECT–CARRION INTERACTIONS

Where ants are abundant, invertebrate corpses are discovered and removed rapidly by widely scavenging and efficient ant workers. Vertebrate corpses (carrion) support a wider diversity of organisms, of which many are insects. These form a **succession** – a non-seasonal, directional and continuous sequential pattern of populations of species colonizing and being eliminated as carrion decay progresses (see Fig. 15.2). The nature and timing of the succession depends upon the size of the corpse, seasonal and ambient climatic conditions, and the surrounding non-biological (edaphic) environment, such as soil type. The organisms involved in the succession vary according to whether they are upon or within the carrion, in the substrate immediately below the corpse, or in the soil at an intermediate distance below or away from the corpse. Furthermore, each succession will comprise different species in different geographical areas, even in places with similar climates. This is because few species are very widespread in distribution, and each biogeographic area has its own specialist carrion faunas. However, the broad taxonomic categories of cadaver specialists are similar worldwide.

The first stage in carrion decomposition, initial decay, involves only microorganisms already present in the body, but within a few days the second stage, called putrefaction, begins. About two weeks later, amidst strong odours of decay, the third, black putrefaction stage begins, followed by a fourth, butyric fermentation stage, in which the cheesy odour of butyric acid is present. This terminates in an almost dry carcass, and the fifth stage, slow dry decay, completes the process, leaving only bones.

The typical sequence of corpse **necrophages**, saprophages and their parasites is often referred to as following "waves" of colonization. The first wave involves certain blow flies (Diptera: Calliphoridae) and house flies (Muscidae) that arrive within hours or a few days at most. The second wave is of sarcophagids (Diptera) and additional muscids and calliphorids that follow shortly thereafter, as the corpse develops an odour. All these flies either lay eggs or larviposit on the corpse. The principal predators on the insects of the corpse fauna are staphylinid, silphid and histerid beetles, and hymenopteran parasitoids may be entomophagous on all the above hosts. At this stage, blow fly activity ceases as their larvae leave the corpse and pupate in the ground. When the fat of the corpse turns rancid, a third wave of species enters this modified substrate, notably more dipterans, such as certain Phoridae, Drosophilidae and *Eristalis* rat-tailed maggots (Syrphidae) in the liquid parts. As the corpse becomes butyric, a fourth wave of cheese-skippers (Diptera: Piophilidae) and related flies use the body. A fifth wave occurs as the ammonia-smelling carrion dries out, and adult and larval Dermestidae and Cleridae (Coleoptera) become abundant, feeding on keratin. In the final stages of dry decay, some tineid larvae ("clothes moths") feed on any remnant hair.

Immediately beneath the corpse, larvae and adults of the beetle families Staphylinidae, Histeridae and Dermestidae are abundant during the putrefaction stage. However, the normal, soil-inhabiting groups are absent during the carrion phase, and only slowly return as the corpse enters late decay. The predictable sequence of colonization and extinction of carrion insects allows forensic entomologists to estimate the age of a corpse, which can have medico-legal implications in homicide investigations (section 15.4).

9.5 INSECT–FUNGAL INTERACTIONS

9.5.1 Fungivorous insects

Fungi and, to a lesser extent, slime moulds are eaten by many insects, termed **fungivores** or **mycophages**, which belong to a range of orders. Amongst insects and other hexapods that use fungal resources, Collembola and larval and adult Coleoptera and Diptera are numerous. Two feeding strategies can be identified: (i) **microphages** gather small particles such as spores and hyphal fragments, or use more liquid media; whereas (ii) **macrophages** use the fungal material of fruiting bodies, which must be torn apart with strong mandibles. The relationship between fungivores and the specificity of their fungus feeding varies. Insects that develop as larvae in the fruiting bodies of large fungi are often obligate fungivores, and may even be restricted to a narrow range of fungi; whereas insects that enter such fungi late in development or during actual decomposition of the fungus are more likely to be saprophagous or generalists than specialist mycophages. Longer-lasting macrofungi such as the pored mushrooms, Polyporaceae, have a higher proportion of mono- or oligophagous associates than ephemeral and patchily distributed mushrooms such as the gilled mushrooms (Agaricales).

Smaller and more cryptic fungal food resources also are used by insects, but the associations tend to be less well studied. Yeasts are naturally abundant on live and fallen fruits and leaves, and **fructivores** (fruit-eaters), such as larvae of certain nitidulid beetles and drosophilid fruit flies, are known to seek and eat yeasts. Apparently, fungivorous drosophilids that live in decomposing fruiting bodies of fungi also use yeasts, and specialization on particular fungi may reflect variations in preferences for particular yeasts. The fungal component of lichens is probably used by grazing larval lepidopterans and adult plecopterans.

Amongst the Diptera that utilize fungal fruiting bodies, the Mycetophilidae (fungus gnats) are diverse and speciose, and many appear to have oligophagous relationships with fungi from amongst a wide range used by the family. The use by insects of subterranean fungal bodies in the form of mycorrhizae and hyphae within the soil is poorly known. The phylogenetic relationship of the Sciaridae (Diptera) to the mycetophilid "fungus gnats", and evidence from commercial mushroom farms, all suggest that sciarid larvae normally eat fungal mycelia. Other dipteran larvae, such as certain phorids and cecidomyiids, feed on commercial mushroom mycelia and associated microorganisms, and may also use this resource in nature.

A Southeast Asian rainforest ant, *Euprenolepis procera*, specializes in harvesting the fruiting bodies of a range of naturally growing fungi. These mushrooms represent a nutritionally suboptimal and spatiotemporally unpredictable food source, to which these ants appear to have adapted via their nomadic lifestyle and fungal processing within the nest. Unlike the attine ants (section 9.5.2), workers of *E. procera* do not cultivate any fungi but temporarily store and manipulate fungal material in piles within the nest, where it may ferment and alter in nutritive value. These ants have unknown effects on fungal diversity and distributions in the forest, but may be important dispersers of fungal spores, including of the mycorrhizal fungi that have mutualistic associations with many rainforest plants.

9.5.2 Fungus farming by leaf-cutter ants

The subterranean ant nests of the genus *Atta* (at least 15 species) and the rather smaller colonies of *Acromyrmex* (over 30 species) are amongst the major earthen constructions in Neotropical rainforest. The largest nests of *Atta* species involve excavation of some 36 metric tonnes (40 US tons) of soil. Both *Atta* and *Acromyrmex* are members of a tribe of myrmecine ants, the Attini, in which the larvae have an obligate dependence on symbiotic fungi for food. Other genera of Attini have monomorphic workers (of a single morphology) and cultivate fungi on dead vegetable matter, insect faeces (including their own and, for example, caterpillar "frass"), flowers and fruit. In contrast, *Atta* and *Acromyrmex*, the more derived genera of Attini, have polymorphic workers of several different kinds or castes (section 12.2.3) that exhibit an elaborate range of behaviours including cutting living plant tissues, hence the name "leaf-cutter ants". In *Atta*, the largest worker ants excise sections of live vegetation with their mandibles (Fig. 9.7a) and transport the pieces to the nest (Fig. 9.7b). During these processes, the working

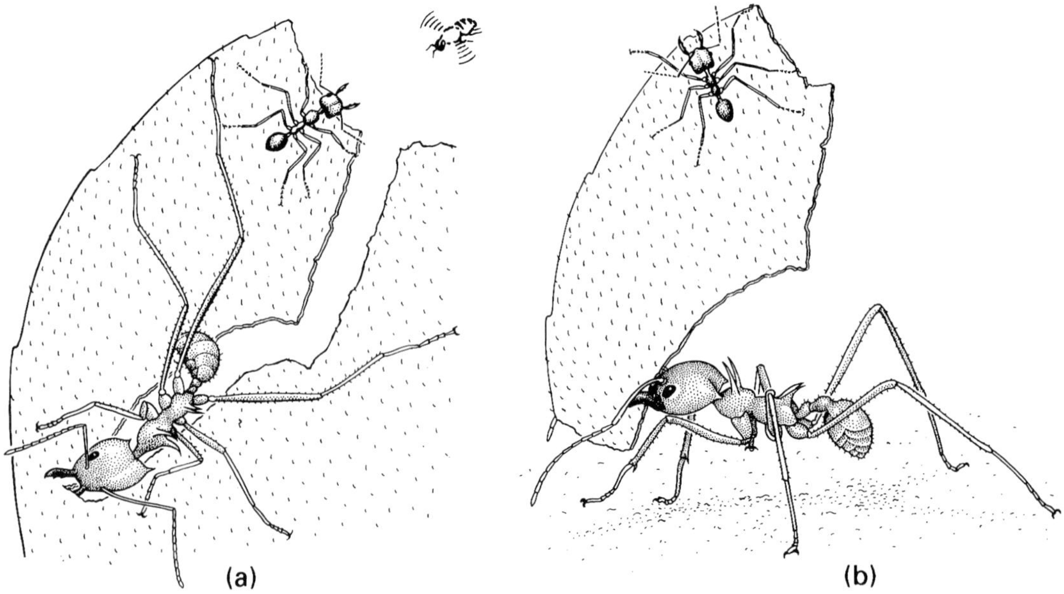

Fig. 9.7 The fungus gardens of the leaf-cutter ant, *Atta cephalotes* (Formicidae), require a constant supply of leaves. (a) A medium-sized worker, called a media, cuts a leaf with its serrated mandibles, while a minor worker guards the media from a parasitic phorid fly (*Apocephalus*) that lays its eggs on living ants. (b) A guarding minor hitchhikes on a leaf fragment carried by a media. (After Eibl-Eibesfeldt & Eibl-Eibesfeldt 1967.)

ant has its mandibles full, and may be the target of attack by a particular parasitic phorid fly (illustrated in the top right of Fig. 9.7a). The smallest worker is recruited as a defender, and is carried on the leaf fragment.

When the material reaches the leaf-cutter nest, other individuals lick any waxy cuticle from the leaves and macerate the plant tissue with their mandibles. The mash is then inoculated with a cocktail of faecal enzymes from the hindgut. This initiates digestion of the fresh plant material, which acts as an incubation medium for a fungus known only from these "fungus gardens" of leaf-cutter ants. Another specialized group of workers tends the gardens by inoculating new substrate with fungal hyphae and removing other species of undesirable fungi in order to maintain a monoculture. Control of alien fungi and bacteria is facilitated by pH regulation (4.5–5.0) and by antibiotics, including those produced by mutualistic actinomycete bacteria of *Pseudonocardia* (formerly thought to be *Streptomyces*) and other bacteria present in a biofilm associated with ant cuticle, and other apparently beneficial bacteria in the fungus gardens. The bacteria help to control a disease of the ants' fungal gardens caused by *Escovopsis* fungi, as well as protect the ants from fungal infection.

In darkness, and at optimal humidity and a temperature close to 25°C, the cultivated fungal mycelia produce nutritive hyphal bodies called gongylidia. These are not sporophores, and appear only to provide food for ants in a mutualistic relationship in which the fungus gains access to the controlled environment. Gongylidia are manipulated easily by the ants, providing food for adults, and are the exclusive food eaten by larval attine ants. Digestion of fungi requires specialized enzymes, which include chitinases produced by ants from their labial glands.

A single origin of fungus domestication might be expected, given the vertical transfer of fungi by transport in the mouth of the founding gyne (new queen) and regurgitation at the new site. However, molecular phylogenetic studies of the fungi across the diversity of attine ants show more than one domestication from free-living stocks, although the ancestral symbiosis, and thus ant agriculture, is at least 50 million years old. Almost all attine ant groups domesticate fungi belonging to the basidiomycete tribe Leucocoprineae, propagated as a mycelium, or occasionally as unicellular yeast. Although each attine nest has a single species of fungus, amongst different nests of a single species, a range of fungus species can be tended. Obviously, some ant species can change their fungus when a new nest is constructed, perhaps when a colony is founded by more than one queen (pleiometrosis). However, among the leaf-cutter ants (*Atta* and *Acromyrmex*), which originated 8–12 million years ago, all species appear to share a single derived species of cultivated *Leucoagaricus* fungus.

Leaf-cutter ants dominate the ecosystems in which they occur; some grassland *Atta* species consume as much vegetation per hectare as do domestic cattle, and certain rainforest species are estimated to cause up to 80% of all leaf damage and to consume up to 17% of all leaf production. The system effectively converts plant cellulose to usable carbohydrate, with at least 45% of the original cellulose of fresh leaves converted by the time the spent substrate is ejected into a dung store as refuse from the fungus garden. However, fungal gongylidia contribute only a modest fraction of the metabolic energy of the ants because most of the energy requirements of the colony are provided by adults feeding on plant sap from chewed leaf fragments.

Leaf-cutter ants may be termed highly polyphagous, as they use between 50 and 70% of all Neotropical rainforest plant species. However, as the adults feed on the sap of fewer species, and the larvae are monophagous on fungus, the term polyphagy strictly may be incorrect. The key to the relationship is the ability of the worker ants to harvest from a wide variety of sources, and the cultivated fungus to grow on a wide range of hosts. Coarse texture and latex production by leaves can discourage attines, and chemical defences may play a role in deterrence. However, leaf-cutter ants have adopted a strategy to evade plant defensive chemicals that act on the digestive system: they use the fungus to digest the plant tissue. The ants and fungus co-operate to break down plant defences, with the ants removing protective leaf waxes that deter fungi, and the fungi converting cellulose indigestible to the ants into carbohydrates.

9.5.3 Fungus cultivation by termites

The terrestrial microfauna of tropical savannahs (grasslands and open woodlands) and some forests of the Afrotropical and Oriental (Indo-Malayan) zoogeographic regions can be dominated by a single subfamily of Termitidae, the Macrotermitinae. These termites may form conspicuous aboveground mounds up to 9 m high, but more often their nests consist of huge underground structures. Abundance, density and production of macrotermitines may be very high and, with estimates of a live biomass of 10 g m^{-2}, termites consume over 25% of all terrestrial litter (wood, grass and leaf) produced annually in some West African savannahs.

The litter-derived food resources are ingested, but not digested, by the termites. The food is passed rapidly through the gut, and the undigested material comprising the faeces is added to comb-like structures within the nest. The combs may be located within many small subterranean chambers, or one large central hive or brood chamber. Upon these combs of faeces, a *Termitomyces* fungus develops. The fungi are restricted to Macrotermitinae nests, or occur within the bodies of termites. The combs are constantly replenished and older parts eaten, on a cycle of 5–8 weeks. Fungus action on the termite faecal substrate raises the nitrogen content of the substrate from about 0.3% until in the asexual stages of *Termitomyces* it may reach 8%. These asexual spores (mycotêtes) are eaten by the termites, as well as the nutrient-enriched older comb. Although some species of *Termitomyces* have no sexual stage, others develop aboveground basidiocarps (fruiting bodies, or "mushrooms") at a time that coincides with colony-founding forays of termites from the nest. A new termite colony is inoculated with the fungus by means of asexual or sexual spores transferred in the gut of the founder termite(s).

Termitomyces lives as a monoculture on termite-attended combs, but if the termites are removed experimentally or a termite colony dies out, or if the comb is extracted from the nest, many other fungi invade the comb and *Termitomyces* dies. Termite saliva has some antibiotic properties but there is little evidence for these termites being able to reduce local competition from other fungi. It seems that *Termitomyces* is favoured in the fungal comb by the remarkably constant microclimate at the comb, with a temperature of 30°C and scarcely varying humidity, together with an acid pH of 4.1–4.6. The heat generated by fungal metabolism is regulated appropriately via a complex circulation of air through the passageways of the nest, as illustrated for the aboveground nest of the African *Macrotermes natalensis* (see Fig. 12.11).

The origin of the mutualistic relationship between termite and fungus seems not to derive from joint attack on plant defences, in contrast to the ant–fungus interaction seen in section 9.5.2. Termites are associated closely with fungi, and fungus-infested rotting wood is likely to have been an early food preference. Termites can digest complex substances such as pectins and chitins, and there is good evidence that they have endogenous cellulases, which break down dietary cellulose. However, the Macrotermitinae have shifted some of their digestion to *Termitomyces* outside of the gut. The fungus facilitates conversion of plant compounds to more nutritious products, and probably allows a wider range of cellulose-containing foods to be consumed by the termites. Thus, the macrotermitines successfully use the abundant resource of dead vegetation.

9.6 CAVERNICOLOUS INSECTS

Caves often are perceived as extensions of the subterranean environment, resembling deep soil habitats in the lack of light and the uniform temperature, but differing in the scarcity of food. Food sources in shallow caves include roots of terrestrial plants, but in deeper caves all plant material originates from stream-derived debris. In many caves, nutrient supplies come from fungi and the faeces (guano) of bats and certain cave-dwelling birds, such as swiftlets in the Orient.

Cavernicolous (cave-dwelling) insects include those that seek refuge from adverse external environmental conditions – such as moths and adult flies, including mosquitoes, that hibernate to avoid winter cold, or aestivate to avoid summer heat and desiccation. **Troglobiont or troglobite** insects are restricted to caves, and often are phylogenetically related to soil-dwelling ones. The troglobite assemblage may be dominated by Collembola (especially the family Entomobryidae), and other important groups include the Diplura (especially the family Campodeidae), orthopterans (including cave crickets, Rhaphidophoridae) and beetles (chiefly carabids, but including fungivorous silphids).

In Hawai'i, past and present volcanic activity produces a spectacular range of "lava tubes" of different isolation in space and time from other volcanic caves. Here, studies of the wide range of troglobitic insects and spiders living in lava tubes have helped us to gain an understanding of the possible rapidity of morphological divergence rates under these unusual conditions. Even caves formed by very recent lava flows such as on Kilauea have endemic or incipient species of *Caconemobius* cave crickets.

Dermaptera and Blattodea may be abundant in tropical caves, where they are active in guano deposits. In Southeast Asian caves, a troglobite earwig is ectoparasitic on roosting bats. Associated with cavernicolous vertebrates are many more-conventional ectoparasites, such as hippoboscid, nycteribid and streblid flies, and fleas and lice.

9.7 ENVIRONMENTAL MONITORING USING GROUND-DWELLING HEXAPODS

Human activities such as agriculture, forestry and pastoralism have resulted in the simplification of many terrestrial ecosystems. Attempts to quantify the effects of such practices – for the purposes of conservation assessment, classification of land-types, and monitoring of impacts – have tended to be phytosociological, emphasizing the use of vegetation mapping data. More recently, data on vertebrate distributions and communities have been incorporated into surveys for conservation purposes.

Although arthropod diversity is estimated to be very great (section 1.3), it is rare for data derived from this group to be available routinely in conservation and monitoring. There are several reasons for this neglect. First, when "**flagship species**" species elicit public reaction to a conservation issue, such as loss of a particular habitat, these organisms are predominantly furry mammals, such as pandas and koalas, or birds; rarely are they insects. Excepting perhaps some butterflies, insects often lack the necessary charisma in the public perception.

Second, insects generally are difficult to sample in a comparable manner within and between sites. Abundance and diversity fluctuate on a relatively short time-scale, in response to factors that may be little understood. In contrast, vegetation often shows less temporal variation; and with knowledge of mammal seasonality and of the migration habits of birds, the temporal variations of vertebrate populations can be taken into account.

Third, arthropods often are more difficult to identify accurately because of the numbers of taxa and some deficiencies in taxonomic knowledge (discussed in section 18.3). Whereas competent mammalogists, ornithologists or field botanists might expect to identify to species level, respectively, all mammals, birds and plants of a geographically restricted area (outside the tropical rainforests), no entomologist could aspire to do so. Nonetheless, aquatic biologists routinely sample and identify all macroinvertebrates (mostly insects) in regularly surveyed aquatic ecosystems, for purposes including monitoring of deleterious change in environmental quality (section 10.5). Comparable studies of terrestrial systems, with objectives such as establishment of rationales for conservation and the detection of habitat- or pollution-induced changes, are undertaken in some countries. The problems outlined above have been addressed in the following ways.

Some charismatic insect species have been highlighted, often under "endangered-species" legislation designed with vertebrate conservation in mind. These species predominantly have been lepidopterans, and much has been learnt of the biology of selected species. However, from the perspective of site classification for conservation purposes, the structure of selected soil and litter communities has greater realized and potential value than any single-species study. Sampling problems are alleviated by using a single collection method, often that of pitfall trapping, but sometimes by the extraction of arthropods from litter samples by a variety of means (see section 18.1.2). Pitfall traps collect mobile terrestrial arthropods by capturing them in containers filled with preserving fluid and sunken level with the substrate. Traps can be aligned along a transect or dispersed according to a standard quadrat-based sampling regime. According to the sample size required, traps can be left *in situ* for several days or for up to a few weeks. Depending on the sites surveyed, arthropod collections may be dominated by Collembola (springtails), Formicidae (ants) and Coleoptera, particularly ground beetles (Carabidae), Tenebrionidae, Scarabaeidae and Staphylinidae, with some terrestrial representatives of many other orders.

Taxonomic difficulties often are alleviated by selecting (from among the organisms collected) one or more higher taxonomic groups for species-level identification. The carabids often are selected for study because of the diversity of species sampled, pre-existing ecological knowledge, and availability of taxonomic keys to species level, although these are largely restricted to temperate northern hemisphere taxa. Some carabid species are almost exclusively predatory and can be important for biological control of pasture or crop pests, but many species are omnivorous, and some can consume large quantities of seeds, including of weeds in agricultural systems. The presence and abundances of particular carabid species can change depending upon farm management practices, and thus the responses of ground beetles can be used to monitor the biological effects of different plantings or soil treatments. Similarly, carabids have been suggested as potential bioindicators of forest management programmes. A 15-year study of carabid species collected from pitfall transects in a diversity of habitats across the United Kingdom showed an overall decline in abundance, with about three-quarters of the species suffering declines, which were highest in montane sites. Populations in southern downland and in woodlands and hedgerows were stable or increased. This study emphasized the need to assess population trends for carabids, and other widespread taxa, across both regions and habitats in order to assess biodiversity losses adequately.

Ants have been the focus of a number of environmental monitoring studies, especially in Australia, South Africa and North America. These abundant and predominantly ground-living insects are relatively easy to identify to genus level, and species diversity can be estimated by sorting the specimens into morphologically distinct groups. The preferred sampling technique for ants is pitfall trapping because, compared with other methods, these traps usually capture the largest number of species and show the highest congruence with ant collections made by other methods. Although it is quite fast to efficiently sample ground-dwelling ants in the field, the subsequent sorting and identifying in the laboratory is time consuming. However, field studies have shown ants to be useful bioindicators of habitat disturbance, as evidenced by changes in their species richness and abundance.

Studies to date are ambivalent concerning correlates between species diversity (including taxon richness) established from vegetation survey and those from terrestrial insect trapping. Evidence from the well-documented British biota suggests that plant diversity does not necessarily predict insect diversity. However, a study in more natural, less human-affected environments in southern Norway showed congruence between carabid faunal indices and those obtained by vegetation and bird surveys. Further studies are required into the nature of any relationships between terrestrial insect species richness and diversity data obtained by conventional biological survey of selected plants and vertebrates.

FURTHER READING

Beynon, S.A. (2012) Potential environmental consequences of administration of anthelmintics to sheep. *Veterinary Parasitology* **189**, 113–24.

Blossey, B. & Hunt-Joshi, T.R. (2003) Belowground herbivory by insects: influence on plants and aboveground herbivores. *Annual Review of Entomology* **48**, 521–47.

Brooks, D.R., Bater, J.E., Clark, S.J. *et al.* (2012) Large carabid beetle declines in a United Kingdom monitoring network increases evidence for a widespread loss in insect biodiversity. *Journal of Applied Ecology* **49**, 1009–19.

Dindal, D.L. (ed.) (1990) *Soil Biology Guide*. John Wiley & Sons, Chichester.

Edgerly, J.S. (1997) Life beneath silk walls: a review of the primitively social Embiidina. In: *The Evolution of Social Behaviour in Insects and Arachnids* (eds J.C. Choe & B.J. Crespi), pp. 14–25. Cambridge University Press, Cambridge.

Eisenbeis, G. & Wichard, W. (1987) *Atlas on the Biology of Soil Arthropods*. Springer-Verlag, Berlin.

Gasch, T., Schott, M., Wehrenfennig, C., Düring, R.-A. & Vilcinskas, A. (2013) Multifunctional weaponry: the chemical defenses of earwigs. *Journal of Insect Physiology* **59**, 1186–93.

Hoffman, B.D. (2010) Using ants for rangeland monitoring: global patterns in the responses of ant communities to grazing. *Ecological Indicators* **10**, 105–11.

Hölldobler, B. & Wilson, E.O. (2010) *The Leafcutter Ants: Civilization by Instinct*. W.W. Norton & Company, New York, London.

Hopkin, S.P. (1997) *Biology of Springtails*. Oxford University Press, Oxford.

Hunter, M.D. (2001) Out of sight, out of mind: the impacts of root-feeding insects in natural and managed systems. *Agricultural and Forest Entomology* **3**, 3–9.

Jochmann, R., Blanckenhorn, W.U., Bussière, L. *et al.* (2011) How to test nontarget effects of veterinary pharmaceutical residues in livestock dung in the field. *Integrated Environmental Assessment and Management* **7**, 287–96.

Johnson, S.N. & Murray, P.J. (eds) (2008) *Root feeders – An Ecosystem Perspective*. CAB International, Wallingford.

Johnson, S.N. & Nielsen, U.N. (2012) Foraging in the dark – chemically mediated host plant location by belowground insect herbivores. *Journal of Chemical Ecology* **38**, 604–14.

Johnson, S.N., Hiltpold, I. & Turlings, T.C.J. (eds) (2013) *Advances in Insect Physiology: Behavior and Physiology of Root Herbivores*, Vol. **45**, 264 pp. Academic Press.

Jouquet, P., Traoré, S., Choosai, C. *et al.* (2011) Influence of termites on ecosystem functioning. *Ecosystem services provided by termites. European Journal of Soil Biology* **47**, 215–22.

Juergens, N. (2013) The biological underpinnings of Namib Desert fairy circles. *Science* **339**, 1618–21.

Kaltenpoth, M., Göttler, W., Herzner, G. & Strohm, E. (2005) Symbiotic bacteria protect wasp larvae from fungal infection. *Current Biology* **15**, 475–9.

Larochelle, A. & Larivière, M.-C. (2003) *A Natural History of the Ground-Beetles (Coleoptera: Carabidae) of America North of Mexico*. Faunistica 27. Pensoft Publishers, Sofia.

Lövei, G.L. & Sunderland, K.D. (1996) Ecology and behaviour of ground beetles (Coleoptera: Carabidae). *Annual Review of Entomology* **41**, 231–56.

McGeoch, M.A. (1998) The selection, testing and application of terrestrial insects as bioindicators. *Biological Reviews* **73**, 181–201.

Nardi, J.B. (2007) *Life in the Soil: A Guide for Naturalists and Gardeners*. The University of Chicago Press, Chicago, IL.

New, T.R. (1998) *Invertebrate Surveys for Conservation*. Oxford University Press, Oxford.

Penttilä, A., Slade, E.M., Simojoki, A. *et al.* (2013) Quantifying beetle-mediated effects on gas fluxes from dung pats. *PLoS ONE* **8**(8), e71454. doi: 10.1371/journal.pone.0071454

Picker, M.D., Hoffman, M.T. & Leverton, B. (2006) Density of *Microhodotermes viator* (Hodotermitidae) mounds in southern Africa in relation to rainfall and vegetative productivity gradients. *Journal of Zoology* **271**, 37–44.

Resh, V.H. & Cardé, R.T. (eds) (2009) *Encyclopedia of Insects*, 2nd edn. Elsevier, San Diego, CA. [Particularly see articles on cave insects; Collembola; soil habitats.]

Samuels, R.I., Mattoso, T.C. & Moreira, D.D.O. (2013) Leaf-cutting ants defend themselves and their gardens against parasite attack by deploying antibiotic secreting bacteria. *Communicative and Integrative Biology* **6**, e23095. doi: 10.4161/cib.23095

Schultz, T.R. & Brady, S.G. (2008) Major evolutionary transitions in ant agriculture. *Proceedings of the National Academy of Sciences* **105**, 5435–40.

Sutton, G., Bennett, J. & Bateman, M. (2014) Effects of ivermectin residues on dung invertebrate communities in a UK farmland habitat. *Insect Conservation and Diversity* **7**, 64–72.

Chapter 10

AQUATIC INSECTS

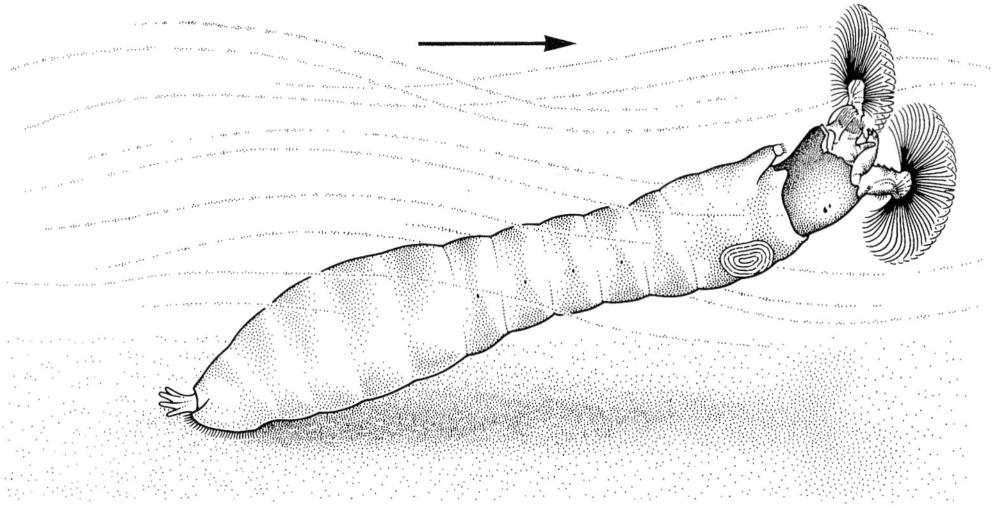

A black-fly larva in the typical filter-feeding posture. (After Currie 1986.)

Every inland waterbody, whether a river, stream, seepage or lake, supports a biological community. The most familiar components often are the vertebrates, such as fish and amphibians. However, at least at the macroscopic level, invertebrates provide the highest number of individuals and species, and the highest levels of biomass and production. In general, the insects dominate freshwater aquatic systems, where only nematodes can approach the insects in terms of species numbers, biomass and productivity. Crustaceans may be abundant, but are rarely diverse in species, in saline (especially temporary) inland waters. Some representatives of nearly all orders of insects live in water, and there have been many invasions of freshwater from the land. Recent studies reveal a diversity of aquatic diving beetles (Dytiscidae) in aquifers (underground waters). Insects have been almost completely unsuccessful in marine environments, with a few sporadic exceptions such as some water-striders (Hemiptera: Gerridae) and larval dipterans.

This chapter surveys the successful insects of aquatic environments, and considers the variety of mechanisms they use to obtain scarce oxygen from the water. Some of their morphological and behavioural modifications to life in water are described, including how they resist water movement, and a classification based on feeding groups is presented. The use of aquatic insects in biological monitoring of water quality is reviewed, and the few insects of the marine and intertidal zones are discussed. Boxes summarize information on the aquatic Diptera (Box 10.1), Hemiptera (Box 10.2), Coleoptera (Box 10.3) and Neuropterida (Box 10.4), and highlight the interchange of insects between aquatic and terrestrial systems, and thus the interactions of the **riparian** zone (the land that interfaces with the waterbody) with the aquatic environment (Box 10.5).

10.1 TAXONOMIC DISTRIBUTION AND TERMINOLOGY

The orders of insects that are almost exclusively aquatic in their immature stages are the Ephemeroptera (mayflies; Taxobox 4), Odonata (damselflies and dragonflies; Taxobox 5), Plecoptera (stoneflies; Taxobox 6) and Trichoptera (caddisflies; Taxobox 27). Amongst the major insect orders, Diptera (Box 10.1 and Taxobox 24) have many aquatic representatives in the immature stages, and a substantial number of Hemiptera (Box 10.2 and Taxobox 20) and Coleoptera (Box 10.3 and Taxobox 22) have at least some aquatic stages, and in the less speciose minor orders, all Megaloptera and some Neuroptera develop in freshwater (Box 10.4 and Taxobox 21). Some Hymenoptera parasitize aquatic prey, but these, together with certain collembolans, orthopteroids and other predominantly terrestrial frequenters of damp places, are considered no further in this chapter.

Aquatic entomologists often (correctly) restrict use of the term **larva** to the immature (i.e. post-embryonic and prepupal) stages of holometabolous insects; **nymph** (or **naiad**) is used for the pre-adult hemimetabolous insects, in which the wings develop externally. However, for the odonates, the terms larva, nymph and naiad have been used interchangeably, perhaps because the sluggish, non-feeding, internally reorganizing, final-instar odonate has been likened to the pupal stage of a holometabolous insect. Although the term "larva" is being used increasingly for the immature stages of all aquatic insects, we accept new ideas on the evolution of metamorphosis (section 8.5) and therefore use the terms larva and nymphs in their strict sense, including for immature odonates.

Some aquatic adult insects, including notonectid bugs and dytiscid beetles, can use atmospheric oxygen when submerged. Other adult insects are fully aquatic, such as several naucorid bugs and hydrophilid and elmid beetles, and can remain submerged for extended periods and obtain respiratory oxygen from the water. However, by far the greatest proportion of the adults of aquatic insects are aerial, and it is only their nymphal or larval (and often pupal) stages that live permanently below the water surface, where oxygen must be obtained whilst out of direct contact with the atmosphere. This ecological division of life history allows the exploitation of two different habitats, although there are a few insects that remain aquatic throughout their lives. Exceptionally, *Helichus*, a genus of dryopid beetles, has terrestrial larvae and aquatic adults.

The Insects: An Outline of Entomology, Fifth Edition. P.J. Gullan and P.S. Cranston.

Companion Website: www.wiley.com/go/gullan/insects

Box 10.1 Aquatic immature Diptera (true flies)

Aquatic larvae are typical of many Diptera, especially in the "Nematocera" group (Taxobox 24). There may be over 10,000 aquatic species in several families, including the speciose Chironomidae (non-biting midges), Ceratopogonidae (biting midges), Culicidae (mosquitoes; Fig. 10.2) and Simuliidae (black flies) (see the vignette at the beginning of this chapter). Dipterans are holometabolous and their larvae are commonly worm-like, as illustrated here for the third-instar larvae of (from top to bottom) *Chironomus* (Chironomidae), *Chaoborus* (Chaoboridae), a biting midge or no-see-um (Ceratopogonidae) and *Dixa* (Dixidae) (after Lane & Crosskey 1993). Diagnostically, they have unsegmented prolegs, variably distributed on the body. Primitively, the larval head is complete and sclerotized and the mandibles operate in a horizontal plane. In more derived groups, the head is progressively reduced, ultimately (in the maggot) with the head and mouthparts atrophied to an internalized cephalopharyngeal skeleton. The larval tracheal system may be closed, with cuticular gaseous exchange, including via gills, or be open, with a variety of spiracular locations, including sometimes a spiracular connection to the atmosphere through a terminal, elongate respiratory siphon. Spiracles, if present, function as a plastron, holding a gas layer in the mesh-like atrium structure. There are usually three or four (in black flies up to 10) larval instars (see Fig. 6.1). Pupation predominantly occurs underwater: the pupa is non-mandibulate, with appendages fused to the body; a puparium is formed in derived groups (few of which are aquatic) from the tanned retained third-instar larval cuticle. Emergence at the water surface may involve use of the cast exuviae as a platform (Chironomidae and Culicidae), or through the adult rising to the surface in a bubble of air secreted within the pupa (Simuliidae).

Development time varies from 10 days to over one year, with many multivoltine species; adults may be ephemeral to long-lived. At least some dipteran species occur in virtually every aquatic habitat, from the marine coast, salt lagoons and sulphurous springs to fresh and stagnant waterbodies, and from temporary containers to rivers and lakes. Temperatures tolerated range from 0°C for some species up to 55°C for a few species that inhabit thermal pools (section 6.6.2). The environmental tolerance to pollution shown by certain taxa is of value in biological indication of water quality.

The larvae show diverse feeding habits, ranging from filter feeding (as shown in Fig. 2.18), through algal grazing and saprophagy to micropredation.

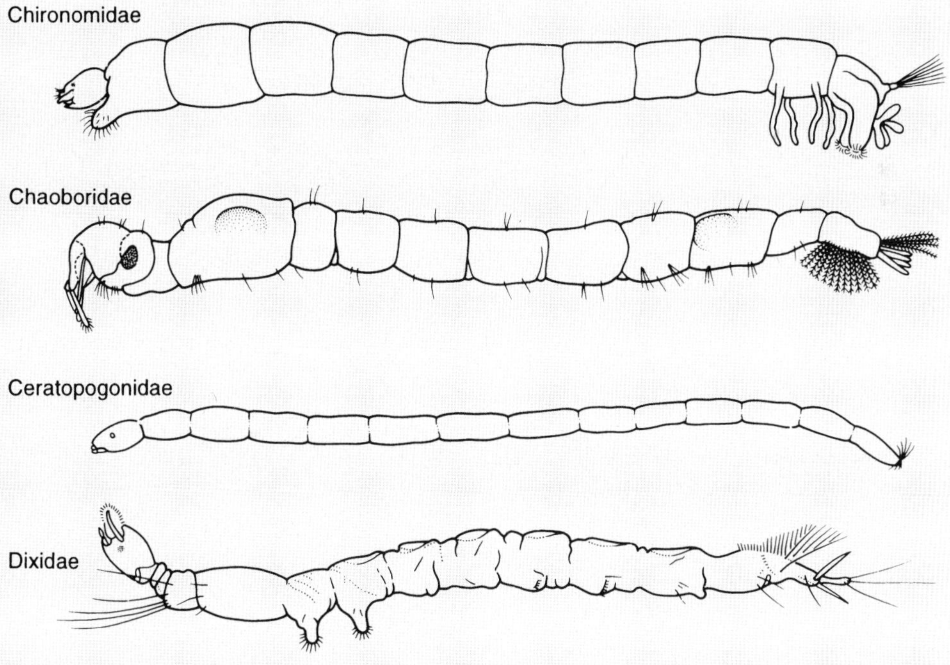

Box 10.2 Aquatic Hemiptera (true bugs)

Amongst the hemimetabolous insects, the order Hemiptera (Taxobox 20) has the most diversity in aquatic habitats. There are about 4000 aquatic and semi-aquatic (including marine) species in about 20 families worldwide, belonging to three heteropteran infraorders (Gerromorpha, Leptopodomorpha and Nepomorpha). These possess the subordinal characteristics of mouthparts modified as a rostrum (beak) and fore wings as hemelytra. All aquatic nymphs are spiraculate, with a variety of gaseous-exchange mechanisms. Nymphs have one, and adults have two or more, tarsal segments. The antennae are three- to five-segmented and are obvious in semi-aquatic groups but inconspicuous in aquatic ones. There is often reduction, loss and/or polymorphism of wings. However, many flighted aquatic hemipterans, especially corixids and gerrids, are highly dispersive, undertaking migrations to avoid unfavourable conditions and to seek newly created ponds. There are five (rarely four) nymphal instars, and species are often univoltine. Gerromorphs (water-striders, represented here by *Gerris*) scavenge or are predatory on the water surface. Diving taxa are either predatory – for example the backswimmers (Notonectidae) such as *Notonecta*, the water-scorpions (Nepidae) such as *Nepa*, and giant water bugs (Belostomatidae) (see Box 5.8) – or phytophagous detritivores – for example as in some water-boatmen (Corixidae) such as *Corixa*.

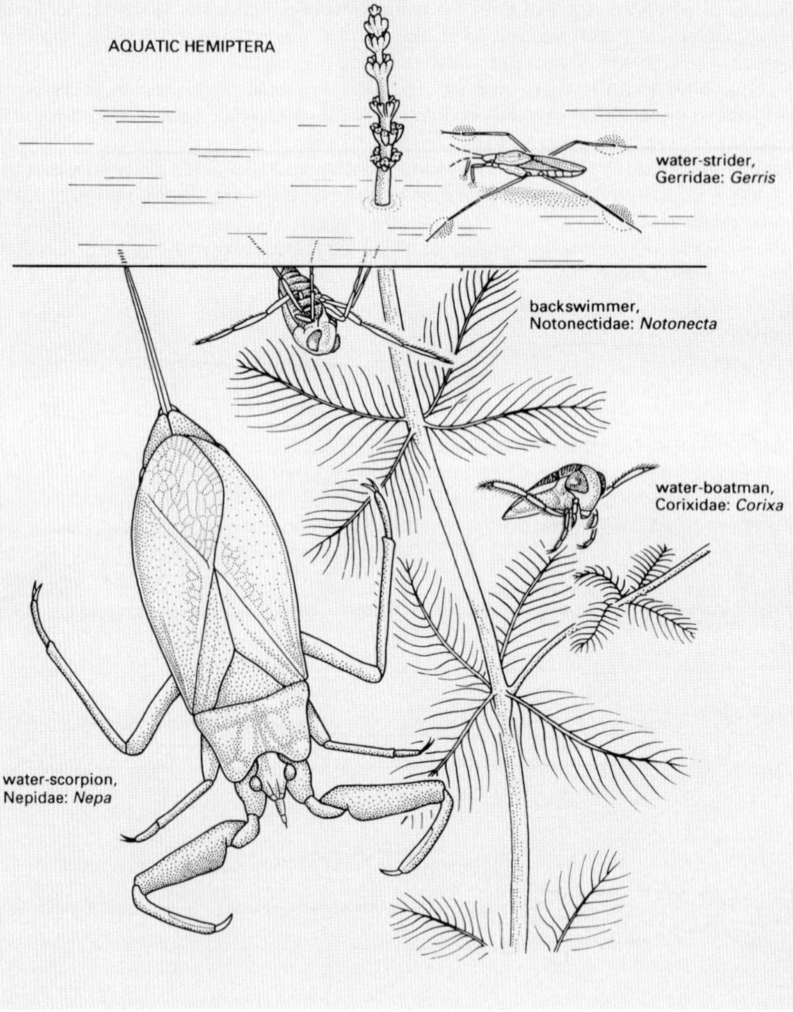

Box 10.3 Aquatic Coleoptera (beetles)

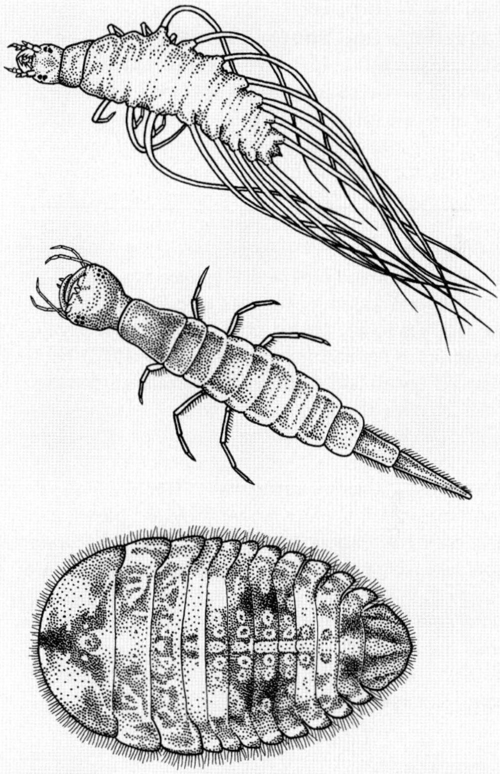

The diverse holometabolous order Coleoptera (Taxobox 22) contains over 5000 aquatic species (although these form less than 2% of the world's described beetle species). About 10 families are exclusively aquatic as both larvae and adults, an additional few are predominantly aquatic as larvae and terrestrial as adults or very rarely with terrestrial larvae and aquatic adults (notably in Dryopidae), and several more have only sporadic aquatic representation. Major families of Coleoptera that are predominantly aquatic in larval or both larval and adult stages are the Gyrinidae (whirligig beetles), Dytiscidae (predaceous diving beetles, with larva illustrated here in the top figure), Haliplidae (crawling water beetles), Hydrophilidae (water scavenger beetles; with larva illustrated in the middle figure), Scirtidae (marsh beetles), Psephenidae (water pennies; with the characteristically flattened larva illustrated in the bottom figure) and Elmidae (riffle beetles). Adult beetles have the mesothoracic wings modified diagnostically as rigid elytra (see Fig. 2.24d; Taxobox 22). Gaseous exchange in adults usually involves temporary or permanent air stores. The larvae are very variable, but all have a distinct sclerotized head with strongly developed mandibles and two- or three-segmented antennae. They have three pairs of jointed thoracic legs, and lack abdominal prolegs. The tracheal system is open but there is a variably reduced spiracle number in most aquatic larvae; some have lateral and/or ventral abdominal gills, sometimes hidden beneath the terminal sternite. Pupation is terrestrial (except in some Psephenidae in which it occurs within the waterbody), and the pupa lacks functional mandibles. Although aquatic Coleoptera exhibit diverse feeding habits, both larvae and adults of most species are predatory or scavengers.

10.2 THE EVOLUTION OF AQUATIC LIFESTYLES

Hypotheses concerning the origin of wings in insects (section 8.4) have different implications regarding the evolution of aquatic lifestyles. The paranotal hypothesis suggests that the "wings" originated in adults of a terrestrial insect for which immature stages may have been aquatic or terrestrial. Some proponents of the preferred exite–endite theory speculate that the progenitor of the pterygotes had aquatic immature stages. Support for the latter hypothesis appears to come from the fact that two of the oldest orders of Pterygota (mayflies and odonates) are aquatic, in contrast to the terrestrial apterygotes; but the aquatic habits of Ephemeroptera and Odonata cannot have been primary, as the tracheal system indicates a preceding terrestrial stage (section 8.3).

Whatever the origins of the aquatic mode of life, all proposed phylogenies of the insects demonstrate that it must have been adopted, adopted and lost, and readopted in several lineages, through geological time. The multiple independent adoptions of aquatic lifestyles are particularly evident in the Coleoptera and

Box 10.4 Aquatic Neuropterida

Aquatic Neuroptera (lacewings and spongillaflies)

The lacewings (order Neuroptera) are holometabolous, and all stages of most species are terrestrial predators (Taxobox 21). However, all spongillaflies (Sisyridae), representing approximately 60 species, have aquatic larvae. In the small family Nevrorthidae, larvae live in running waters, and some larval Osmylidae live in damp marginal riparian habitats. Nevrorthids are found only in Japan, Taiwan, the European Mediterranean and Australia. Freshwater-associated osmylids are found in Australia and East Asia, and the family is totally absent from the Nearctic. Sisyrid larvae (as illustrated here; after CSIRO 1970) have elongate stylet-like mandibles, filamentous antennae, paired ventral abdominal gills, and lack terminal prolegs. The pupa has functional mandibles. The eggs are laid on branches and the undersides of leaves in trees overhanging running water. The hatching larvae drop into the water where they are planktonic and seek out a sponge host. The larvae feed upon sponges using their stylet-mouthparts to suck out fluids from the living cells. There are three larval instars, with rapid development, and they may be multivoltine. The late final-instar larva leaves the water and pupation takes place in a silken cocoon on vegetation some distance from the water.

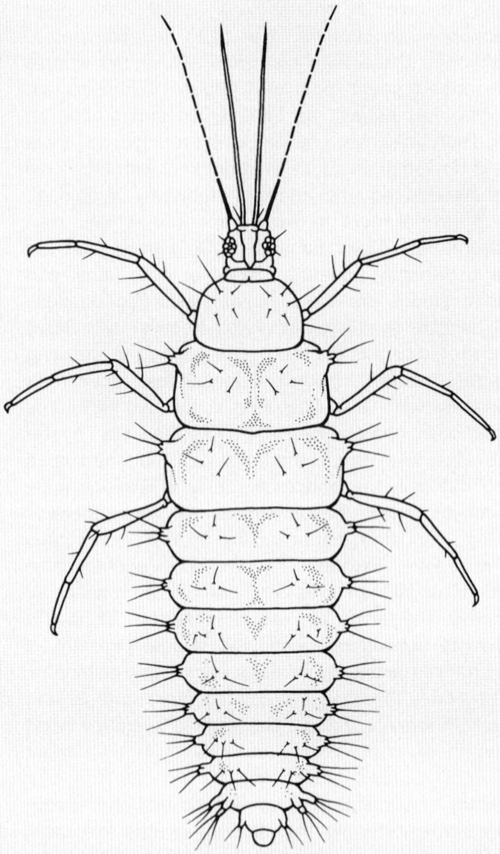

sisyrid larva

Aquatic Megaloptera (alderflies, dobsonflies, fishflies)

Megalopterans (Taxobox 21) are holometabolous, with about 350 described species in two families worldwide – Sialidae (alderflies, with larvae up to 3 cm long) and the larger Corydalidae (dobsonflies and fishflies, with larvae up to 10 cm long). Two

Megaloptera: Corydalidae: *Archichauliodes*

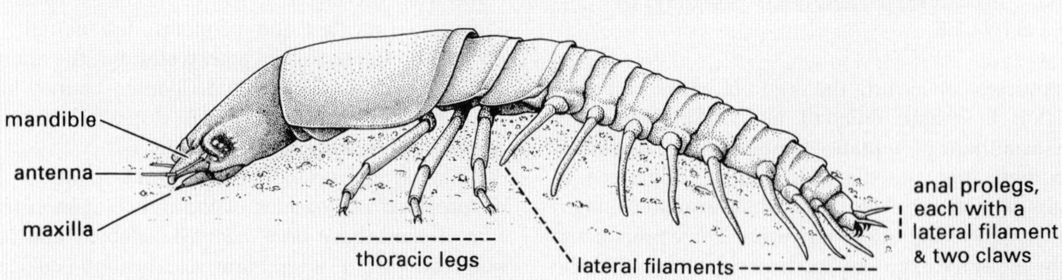

subfamilies of Corydalidae are recognized – Corydalinae (the dobsonflies) and Chauliodinae (the fishflies). Diversity of megalopterans is especially high in China and Southeast Asia, and in Amazonia and the Andes. Larval Megaloptera are prognathous, with well-developed mouthparts, including three-segmented labial palps (similar-looking gyrinid beetle larvae have one- or two-segmented palps). They are spiraculate, with gills consisting of four- to five-segmented (Sialidae) or two-segmented lateral filaments on the abdominal segments. The larval abdomen terminates in an unsegmented median caudal filament (Sialidae) or a pair of anal prolegs (as shown here for a species of *Archichauliodes* (Corydalidae)). The larvae (sometimes called hellgrammites) have 10–12 instars and take at least one year, usually two or more, to develop. Some Pacific coastal (notably Californian) megalopteran larvae can survive stream drying by burrowing beneath large boulders, whereas others appear to survive without such a behavioural strategy. Pupation occurs away from water, usually in damp substrates. The larvae are sit-and-wait predators or scavengers, in lotic and lentic waters, and are intolerant of pollution.

Diptera, with aquatic taxa distributed amongst many families across each of these orders. In contrast, the immature stages of all species of Ephemeroptera and Plecoptera are aquatic, and in the Odonata, the only exceptions to an almost universal aquatic lifestyle are the terrestrial nymphs of a few species.

Movement from land to water causes physiological problems, the most important of which is the requirement for oxygen. The following section considers the physical properties of oxygen in air and water, and the mechanisms by which aquatic insects obtain an adequate supply.

10.3 AQUATIC INSECTS AND THEIR OXYGEN SUPPLIES

10.3.1 The physical properties of oxygen

Oxygen comprises 200,000 ppm (parts per million) of air, but in aqueous solution its concentration is only about 15 ppm in saturated cool water. Energy at the cellular level can be provided by anaerobic respiration, but it is inefficient, providing 19 times less energy per unit of substrate respired than aerobic respiration. Although insects such as bloodworms (certain chironomid midge larvae) survive extended periods of almost anoxic conditions, most aquatic insects must obtain oxygen from their surroundings in order to function effectively.

The proportions of gases dissolved in water vary according to their solubilities: the amount is inversely proportional to temperature and salinity, and proportional to pressure, decreasing with elevation. In **lentic** (standing) waters, diffusion through water is very slow; it would take years for oxygen to diffuse several metres from the surface in still water. This slow rate, combined with the oxygen demand from microbial breakdown of submerged organic matter, can totally deplete the oxygen on the bottom (benthic anoxia). However, the oxygenation of surface waters by diffusion is enhanced by turbulence, which increases the surface area, forces aeration, and mixes the water. If this turbulent mixing is prevented, such as in a deep lake with a small surface area or one with extensive sheltering vegetation or under extended ice cover, anoxia can be prolonged or permanent. Living under these circumstances, benthic insects must tolerate wide annual and seasonal fluctuations in oxygen availability.

Oxygen levels in **lotic** (flowing) conditions can reach 15 ppm, especially in cold water. Equilibrium concentrations may be exceeded if photosynthesis generates locally abundant oxygen, such as in macrophyte- and algal-rich pools in sunlight. However, when this vegetation respires at night, oxygen is consumed, leading to a decline in dissolved oxygen. Aquatic insects must cope with a diurnal range of oxygen tensions.

10.3.2 Gaseous exchange in aquatic insects

The gaseous exchange systems of insects depend upon oxygen diffusion, which is rapid through the air, slow through water, and even slower across the cuticle. Eggs of aquatic insects absorb oxygen from water with the assistance of a chorion (section 5.8). Large eggs may have the respiratory surface expanded by elaborated horns or crowns, as in water-scorpions

(Hemiptera: Nepidae). The large eggs of giant water bugs (Hemiptera: Belostomatidae) uptake oxygen assisted by unusual male parental tending of the eggs (see Box 5.8).

Although insect cuticle is very impermeable, gas diffusion across the body surface may suffice for the smallest aquatic insects, such as some early-instar larvae or all instars of some dipteran larvae. Larger aquatic insects, with respiratory demands equivalent to spiraculate air-breathers, require either augmentation of gas-exchange areas or some other means of obtaining increased oxygen, because the reduced surface area to volume ratio precludes dependence upon cutaneous gas exchange.

Aquatic insects show several mechanisms to cope with the much lower oxygen levels in aqueous solutions. Aquatic insects may have open tracheal systems with spiracles, as do their air-breathing relatives. These may be either polypneustic (8–10 spiracles opening on the body surface) or oligopneustic (one or two pairs of open, often terminal spiracles), or closed and lacking direct external connection (section 3.5, Fig. 3.11).

10.3.3 Oxygen uptake with a closed tracheal system

Simple cutaneous gaseous exchange in a closed tracheal system suffices for only the smallest aquatic insects, such as early-instar caddisflies (Trichoptera). For larger insects, although cutaneous exchange can account for a substantial part of oxygen uptake, other mechanisms are needed.

A prevalent means of increasing surface area for gaseous exchange is by **gills** – tracheated cuticular lamellar extensions from the body. These are usually abdominal (ventral, lateral or dorsal) or caudal, but may be located on the mentum, maxillae, neck, at the base of the legs, around the anus, as in some Plecoptera (Fig. 10.1), or even within the rectum, as in dragonfly nymphs. Tracheal gills are found in the immature stages of Odonata, Plecoptera, Trichoptera, aquatic Megaloptera and Neuroptera, some aquatic Coleoptera, a few Diptera and pyralid lepidopterans, and probably reach their greatest morphological diversity in the Ephemeroptera.

In interpreting these structures as gills, it is important to demonstrate that they do function in oxygen uptake. In experiments with nymphs of *Lestes* (Odonata: Lestidae), the huge caudal gill-like lamellae of some individuals were removed by being broken at the site of natural **autotomy**. Both gilled and ungilled individuals were subjected to low-oxygen environments in closed-bottle respirometry, and survivorship was assessed. The three caudal lamellae of this odonate met all criteria for gills, namely:

- a large surface area;
- being moist and vascular;

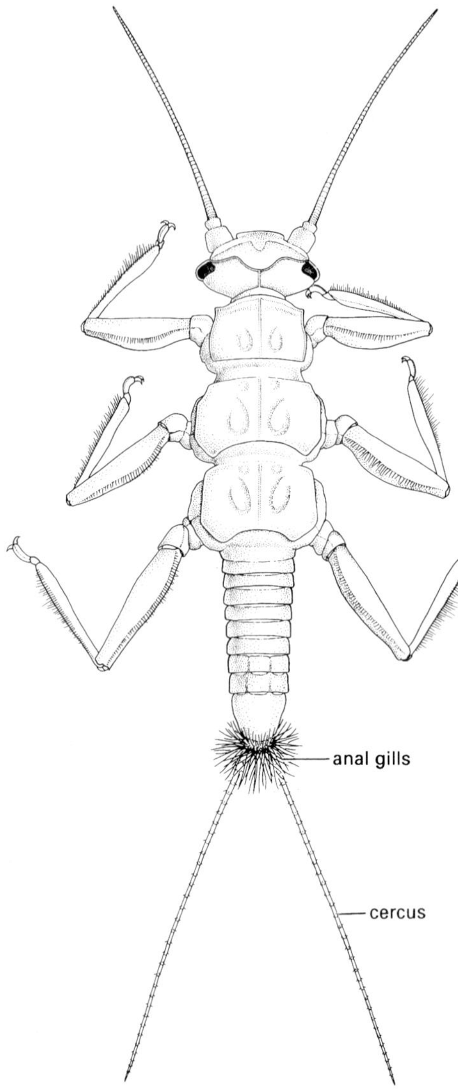

Fig. 10.1 A stonefly nymph (Plecoptera: Gripopterygidae) showing filamentous anal gills.

- able to be ventilated;
- responsible normally for 20–30% of oxygen uptake.

However, with experimentally increased temperature and reduction in dissolved oxygen, the gills accounted for increased oxygen uptake, until a maximum uptake of 70%. At this high level, the proportion of oxygen uptake via the lamellae equalled the proportion of gill surface to total body surface area. At low temperatures (< 12°C) and with dissolved oxygen at the environmental maximum of 9 ppm, the gills of the lestid accounted for very little oxygen uptake; cuticular uptake was presumed to be dominant. When *Siphlonurus* mayfly nymphs were tested similarly, at 12–13°C the gills accounted for 67% of oxygen uptake, which was proportional to their fraction of the total surface area of the body.

Dissolved oxygen can be extracted using respiratory pigments. These pigments are almost universal in vertebrates, but also are found in some invertebrates and even in plants and protists. Amongst the aquatic insects, some larval chironomids (bloodworms) and a few notonectid bugs possess haemoglobins. These molecules are homologous (same derivation) to the haemoglobin of vertebrates such as ourselves. The haemoglobins of vertebrates have a low affinity for oxygen, i.e. oxygen is obtained from a high-oxygen aerial environment and unloaded in muscles in an acid (carbonic acid from dissolved carbon dioxide) environment – the Bohr effect. Where environmental oxygen concentrations are consistently low, as in the virtually anoxic and often acidic sediments of lakes, the Bohr effect would be counterproductive. In contrast to vertebrates, chironomid haemoglobins have a high affinity for oxygen. Chironomid midge larvae can saturate their haemoglobins through undulating their bodies within their silken tubes or substrate burrows to permit the minimally oxygenated water to flow over the cuticle. Oxygen is unloaded when the undulations stop, or when recovery from anaerobic respiration is needed. The respiratory pigments allow a much more rapid oxygen release than is available by diffusion alone.

10.3.4 Oxygen uptake with an open spiracular system

For aquatic insects with open spiracular systems, there is a range of possibilities for obtaining oxygen. Many immature stages of Diptera can obtain atmospheric oxygen by suspending themselves from the water meniscus, in the manner of a mosquito larva and pupa (Fig. 10.2). There are direct connections between the atmosphere and the spiracles in the terminal respiratory siphon of the larva, and in the thoracic respiratory organ of the pupa. Any insect that uses atmospheric oxygen is independent of low levels of dissolved oxygen, such as occur in rank or stagnant waters. This independence from dissolved oxygen is particularly prevalent amongst larvae of flies, such as ephydrids, one species of which can live in oil-tar ponds, and certain pollution-tolerant hover flies (Syrphidae), the "rat-tailed maggots".

Several other larval Diptera and psephenid beetles have cuticular modifications surrounding the spiracular openings, which function as gills, to allow an increase in the extraction rate of dissolved oxygen without spiracular contact with the atmosphere. An unusual source of oxygen is the air stored in roots and stems of aquatic macrophytes. Aquatic insects, including the immature stages of some mosquitoes, hover flies and *Donacia* (a genus of chrysomelid beetles), can use this source. In *Mansonia* mosquitoes, the spiracle-bearing larval respiratory siphon and pupal thoracic respiratory organ both are modified for piercing plants.

Temporary air stores (compressible gills) are common means of storing and extracting oxygen. Many adult dytiscid, gyrinid, helodid, hydraenid and hydrophilid beetles, and both nymphs and adults of many belostomatid, corixid, naucorid and pleid hemipterans use this method of enhancing gaseous exchange. The gill is a bubble of stored air, in contact with the spiracles by various means, including subelytral retention in adephagan water beetles (Fig. 10.3), and fringes of specialized hydrofuge hairs on the body and legs, as in some polyphagan water beetles. When the insect dives from the surface, air is trapped in a bubble in which all gases start at atmospheric equilibrium. As the submerged insect respires, oxygen is used up and the carbon dioxide produced is lost due to its high solubility in water. Within the bubble, as the partial pressure of oxygen drops, more diffuses in from solution in water, but not rapidly enough to prevent continued depletion in the bubble. Meanwhile, as the proportion of nitrogen in the bubble increases, it diffuses outwards, causing diminution in the size of the bubble. This contraction in size gives rise to the term "compressible gill". When the bubble has become too small, the insect replenishes the air by returning to the surface.

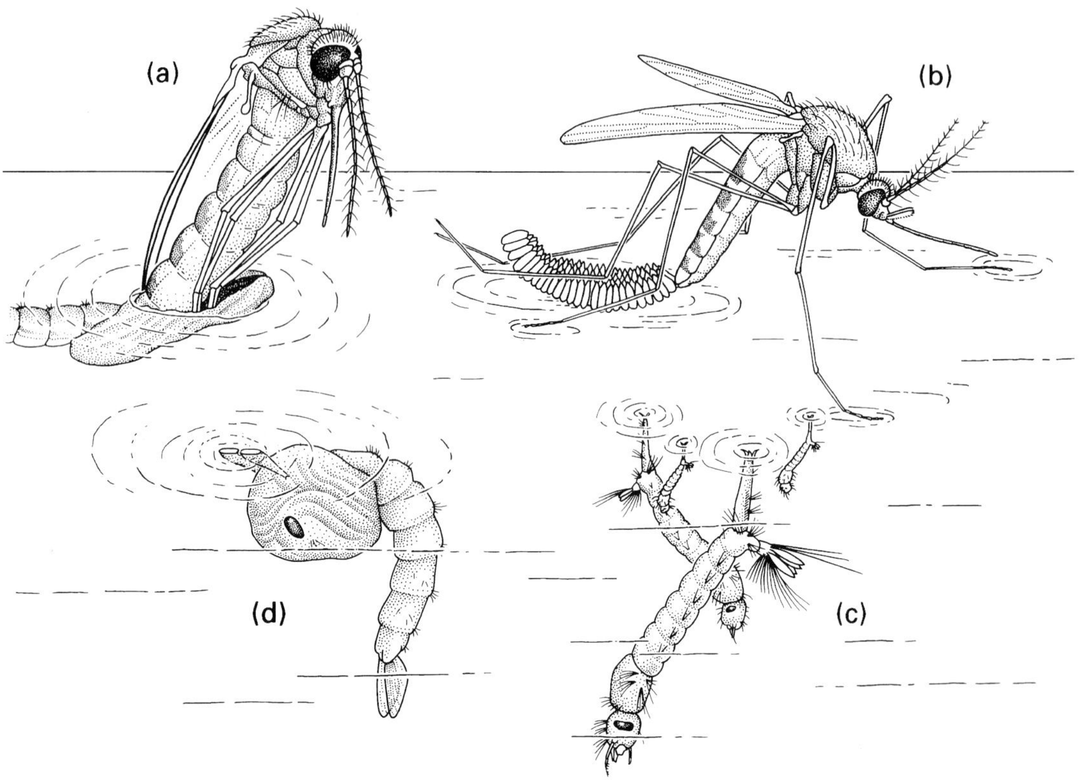

Fig. 10.2 The life cycle of the mosquito *Culex pipiens* (Diptera: Culicidae): (a) adult emerging from its pupal exuviae at the water surface; (b) adult female ovipositing, with her eggs adhering together as a floating raft; (c) larvae obtaining oxygen at the water surface via their siphons; (d) pupa suspended from the water meniscus, with its respiratory horn in contact with the atmosphere. (After Clements 1992.)

The longevity of the bubble depends upon the relative rates of consumption of oxygen and of gaseous diffusion between the bubble and the surrounding water. A maximum of eight times more oxygen can be supplied from the compressible gill than was in the original bubble. However, the available oxygen varies according to the amount of exposed surface area of the bubble and the prevailing water temperature. At low temperatures, the metabolic rate is lower, more gases remain dissolved in water, and the gill is long lasting. Conversely, at higher temperatures, metabolism is higher, less gas is dissolved, and the gill is less effective.

Another air-bubble gill, the **plastron**, allows some insects to use a permanent air store, termed an "incompressible gill". Water is held away from the body surface by hydrofuge hairs or a cuticular mesh, which leaves a gas layer permanently in contact with the spiracles. Most of the gas is relatively insoluble nitrogen but, a gradient is set up in response to metabolic use of oxygen, and more oxygen diffuses from water into the plastron. Most insects with such a gill are relatively sedentary, as the gill is not very effective in responding to sustained high oxygen demand. Adults of some curculionid, dryopid, elmid, hydraenid and hydrophilid beetles, nymphs and adults of naucorid bugs, and pyralid moth larvae use this mode of oxygen extraction. The spiracles of aquatic Diptera larvae, such as craneflies (Tipulidae) and stratiomyids, operate as plastrons, with gases maintained in the supportive mesh within the spiracular atrium, and with external hydrophobic hairs repelling water.

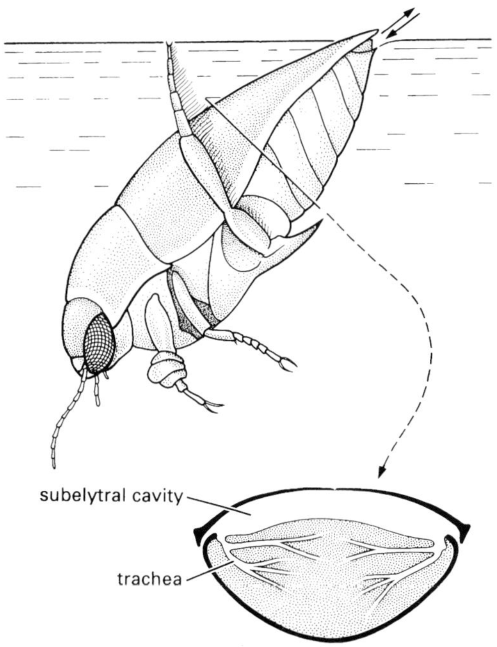

Fig. 10.3 A male water beetle of *Dytiscus* (Coleoptera: Dytiscidae) replenishing its store of air at the water surface. Below is a transverse section of the beetle's abdomen showing the large air store below the elytra and the tracheae opening into this air space. Note that the tarsi of the fore legs are dilated to form adhesive pads that are used to hold the female during copulation. (After Wigglesworth 1964.)

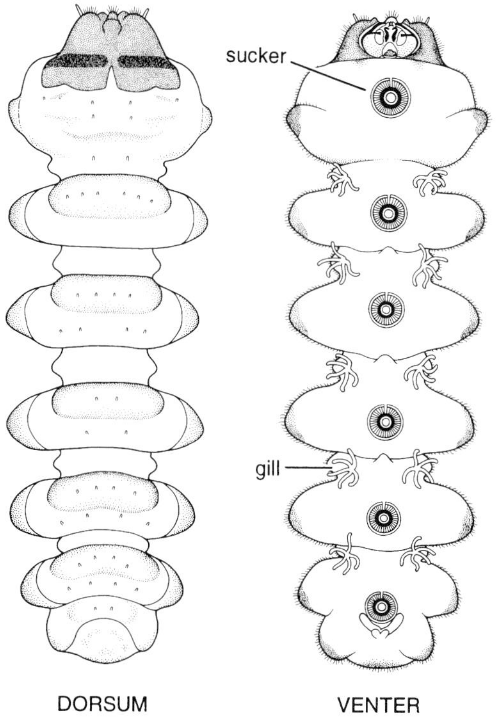

Fig. 10.4 Dorsal (left) and ventral (right) views of the larva of *Edwardsina polymorpha* (Diptera: Blephariceridae); the venter has suckers, which the larva uses to adhere to rock surfaces in fast-flowing water.

10.3.5 Behavioural ventilation

A consequence of the slow diffusion rate of oxygen through water is the development of an oxygen-depleted layer of water that surrounds the gaseous uptake surface, whether it be the cuticle, gill or spiracle. Aquatic insects exhibit a variety of ventilation behaviours that disrupt this oxygen-depleted layer. Cuticular gaseous diffusers undulate their bodies in tubes (Chironomidae), cases (young caddisfly larvae) or under shelters (young lepidopteran larvae) to produce fresh currents across the body. This behaviour continues even in later-instar caddisflies and lepidopterans in which gills are developed. Many ungilled aquatic insects select their positions in the water to allow maximum aeration by current flow. Some dipterans, such as blephariceric (Fig. 10.4) and deuterophlebiid larvae, are found only in torrents; ungilled simuliids, plecopterans and case-less caddisfly larvae are found commonly in high-flow areas. The very few sedentary aquatic insects with gills, notably black-fly (simuliid) pupae, some adult dryopid beetles and the immature stages of a few lepidopterans, maintain local high oxygenation by positioning themselves in areas of well-oxygenated flow. For mobile insects, swimming actions, such as leg movements, prevent the formation of a low-oxygen boundary layer.

Although most gilled insects use natural water flow to bring oxygenated water to them, they may also undulate their bodies, beat their gills, or pump water in and out of the rectum, as in anisopteran nymphs. In lestid zygopteran nymphs (for which gill function is discussed in section 10.3.3), ventilation is assisted by "pull-downs" (or "push-ups") that effectively move oxygen-reduced water away from the gills. When dissolved oxygen is reduced through a rise in temperature,

Siphlonurus nymphs elevate the frequency and increase the percentage of time spent beating gills.

10.4 THE AQUATIC ENVIRONMENT

The two different aquatic physical environments, the lotic (flowing) and lentic (standing), place different constraints on the organisms living therein. The following sections highlight these conditions, and discuss some of the morphological and behavioural modifications of aquatic insects. In addition, aquatic insects may respond to conditions external to the waterbody, such as plant and animal matter falling into the water from the riparian zone (Box 10.5).

10.4.1 Lotic adaptations

In lotic systems, the velocity of flowing water influences:

• substrate type, with boulders deposited in fast-flow and fine sediments in slow-flow areas;
• transport of particles, either as a food source for filter-feeders or, during peak flows, as scouring agents;
• maintenance of high levels of dissolved oxygen.

A stream or river contains heterogeneous microhabitats, with riffles (shallower, stony, fast-flowing sections) interspersed with deeper natural pools. Areas of erosion of the banks alternate with areas where sediments are deposited, and there may be areas of unstable, shifting sandy substrates. The banks may have trees (a vegetated riparian zone) or be unstable, with mobile deposits that change with every flood. Typically, where there is riparian vegetation, there will be local accumulations of drifted **allochthonous** (external to the stream) material, such as leaf packs and wood. In parts of the world where extensive pristine, forested catchments remain, the courses of streams often are periodically blocked by naturally fallen trees. Where the stream is open to light, and nutrient levels allow, **autochthonous** (produced within the stream) growth of plants and macroalgae (collectively called macrophytes) will occur. Aquatic flowering plants may be abundant, especially in chalk streams.

Characteristic insect faunas inhabit these various substrates, many with evident morphological modifications. Thus, those that live in strong currents (**rheophilic** species) tend to be dorsoventrally flattened, as in larvae of Psephenidae (Box 10.3), sometimes with laterally projecting legs. This is not strictly an adaptation to strong currents, as such modification is found in many aquatic insects. Nevertheless, the shape and behaviour minimizes or avoids exposure by allowing the insect to remain within a boundary layer of still water close to the surface of the substrate. However, fine-scale hydraulic flow of natural waters is complex, and the relationship between body shape, streamlining and current velocity is not straightforward.

Box 10.5 Aquatic–terrestrial insect fluxes

Ecologists have tended to treat the freshwater and terrestrial environments as separate systems, with different organisms, processes and research methods appropriate to each. However, it is increasingly clear that interchanges ("fluxes") between the two are important in transferring energy (food, nutrients) between them. The most visible exchange is due to adult insects emerging from lakes and rivers and moving to land, where they boost the diet of riparian predators, especially fly-catching birds and bats, spiders that spin webs to capture prey, and ground-dwelling arthropods such as carabid beetles. In turn, such a localized increase in predators can affect all food webs in the riparian zone. The reverse cycle involves riparian terrestrial insects falling into streams and rivers, especially during sporadic flood events, together with spent adults after mating and oviposition. This allochthonous resource may provide seasonally a substantial input into the diet of fish, such as salmonids, that catch their food from drift. In addition, many beetles and even ants may forage on the margins of waterbodies for localized post-flood strandings of surface-drifting terrestrial insects. The very variable extent and seasonality of these resource subsidies from streams to the riparian zone and vice-versa is becoming better understood through some clever manipulations of the subsidies. Furthermore, the ability to determine the diets of predators and scavengers using the different isotopic signatures of terrestrial arthropods and aquatic insects has helped recognize these subsidies. What remains unclear is the importance of quality of the riparian zone (intact, disturbed or invaded) to the strength of these predominantly insect-based fluxes.

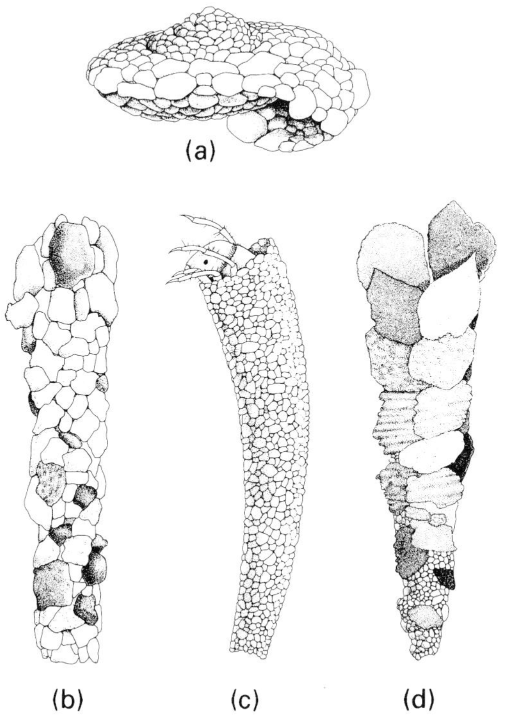

Fig. 10.5 Portable larval cases of representative families of caddisflies (Trichoptera): (a) Helicopsychidae; (b) Philorheithridae; (c) and (d) Leptoceridae.

The cases constructed by many rheophilic caddisflies assist in streamlining or otherwise modifying the effects of flow. The variety of shapes of the cases (Fig. 10.5) must act as ballast against displacement. Several aquatic larvae have suckers (Fig. 10.4) that allow the insect to stick to quite smooth exposed surfaces, such as rock-faces on waterfalls and cascades. Silk is widely produced, allowing maintenance of position in fast flow. Black-fly larvae (Simuliidae) (see the vignette at the beginning of this chapter) attach their posterior claws to a silken pad that they spin on a rock surface. Others, including hydropsychid caddisflies (Fig. 10.6) and many chironomid midges, use silk in constructing retreats. Some spin silken mesh nets to trap food brought into proximity by the stream flow.

Many lotic insects are smaller than their counterparts in standing waters. Their size, together with flexible body design, allows them to live amongst the cracks and crevices of boulders, stones and pebbles in the bed (**benthos**) of the stream, or even in unstable, sandy substrates. Another means of avoiding the current is to live in accumulations of leaves (leaf packs) or to mine in immersed wood. Many beetles and specialist dipterans, such as crane-fly larvae (Diptera: Tipulidae), use these substrates.

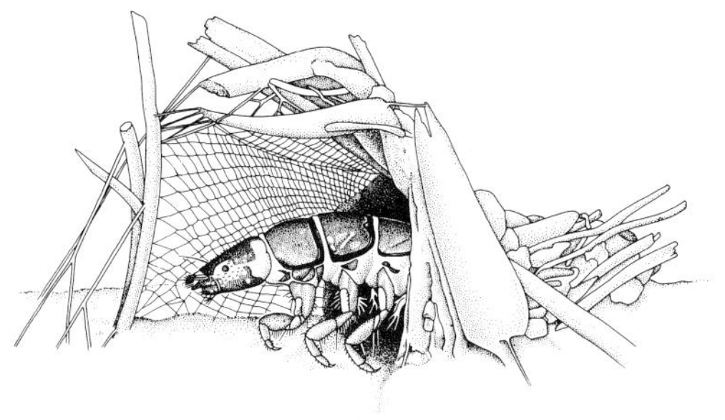

Fig. 10.6 A caddisfly larva (Trichoptera: Hydropsychidae) in its retreat; the silk net is used to catch food. (After Wiggins 1978.)

Two behavioural strategies are more evident in running waters than elsewhere. The first is the strategic use of the current to allow **drift** from an unsuitable location, with the possibility of finding a more suitable patch. Predatory aquatic insects frequently drift to locate aggregations of prey. Many other insects, such as stoneflies and mayflies, notably *Baetis* (Ephemeroptera: Baetidae), may show a diurnal periodic pattern of drift. "Catastrophic" drift is a behavioural response to physical disturbance, such as pollution or severe flow episodes. An alternative response, of burrowing deep into the substrate (the **hyporheic** zone), is a second particularly lotic behaviour. In the hyporheic zone, variations in flow regime, temperature, and perhaps in predation pressure, can be avoided, although food and oxygen availability may be diminished.

10.4.2 Lentic adaptations

With the exception of wave action at the shore of larger bodies of water, the effects of water movement cause little or no difficulty for aquatic insects that live in lentic environments. However, oxygen availability is more of a problem and lentic taxa show a greater variety of mechanisms for enhanced oxygen uptake compared with lotic insects.

The lentic water surface is used by many more species (the **neustic** community of **semi-aquatic** insects) than the lotic surface, because the physical properties of surface tension in standing water that can support an insect are disrupted in turbulent flowing water. Water-striders (Hemiptera: Gerroidea: Gerridae and Veliidae) are amongst the most familiar neustic insects that exploit the surface film (Box 10.2). They use hydrofuge (water-repellent) hair piles on the legs and venter to avoid breaking the film. Water-striders move with a rowing motion, and they locate prey items (and in some species, mates) by detecting vibratory ripples on the water surface. Certain staphylinid beetles use chemical means to move around the meniscus, by discharging from the anus a detergent-like substance that releases local surface tension and propels the beetle forwards. Some elements of this neustic community can be found in still-water areas of streams and rivers, and related species of Gerromorpha can live in estuarine and even oceanic water surfaces (section 10.8).

Underneath the meniscus of standing water, the larvae of many mosquitoes feed (see Fig. 2.18), and hang suspended by their respiratory siphons (Fig. 10.2), as do certain crane flies and stratiomyids (Diptera). Whirligig beetles (Gyrinidae) (Fig. 10.7) can straddle the interface between water and air, with an upper unwettable surface and a lower wettable one. Uniquely, each of their eyes is divided such that the upper part can observe the aerial environment, and the lower half can see underwater.

Between the water surface and the benthos, planktonic organisms live in a zone divisible into an upper limnetic zone (i.e. penetrated by light) and a deeper profundal zone (below effective light penetration). The most abundant planktonic insects belong to *Chaoborus* (Diptera: Chaoboridae); these "phantom midges" undergo diurnal vertical migration, and their predation on *Daphnia* is discussed in section 13.4. Other insects, such as diving beetles (Dytiscidae) and many hemipterans such as Corixidae, dive and swim actively through this zone in search of prey. The profundal zone generally lacks planktonic insects, but may support an abundant benthic community, predominantly of chironomid midge larvae, most of which possess haemoglobin. Even the profundal benthic zone of some deep lakes, such as Lake Baikal in Siberia, supports some midges, although at eclosion the pupa may have to rise more than 1 km to the water surface.

In the littoral zone, in which light reaches the benthos and macrophytes can grow, insect diversity is at its maximum. Many differentiated microhabitats are available, and physicochemical factors are less restricting than in the dark, cold, and perhaps anoxic, conditions of the deeper waters.

10.5 ENVIRONMENTAL MONITORING USING AQUATIC INSECTS

Aquatic insects form assemblages that vary with their geographical location, according to historical biogeographic and ecological processes. Within a more restricted area, such as a single lake or river drainage, the community structure derived from within this pool of locally available organisms is constrained largely by physicochemical factors of the environment. Amongst the important factors that govern which species live in a particular waterbody, variations in oxygen availability obviously lead to different insect communities. For example, in low-oxygen conditions, perhaps caused by oxygen-demanding sewage pollution, the community is typically species-poor, and differs in composition

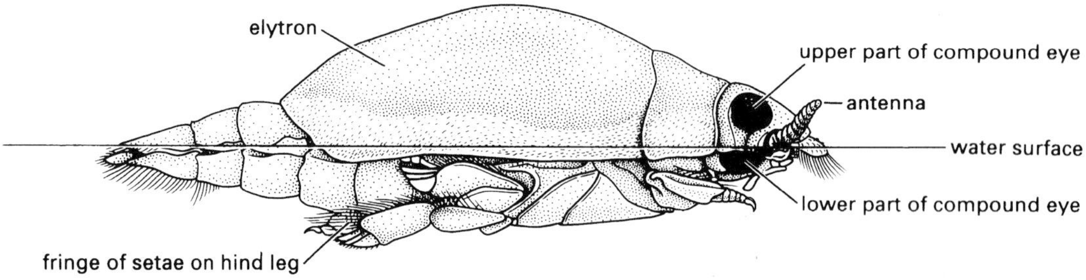

Fig. 10.7 The whirligig beetle, *Gyretes* (Coleoptera: Gyrinidae), swimming on the water surface. Note that the divided compound eye allows the beetle to see both above and below water simultaneously; hydrofuge hairs on the margin of the elytra repel water. (After White *et al.* 1984.)

from a comparable well-oxygenated system, as might be found upstream of a pollution site. Similar changes in community structure can be seen in relation to other physicochemical factors such as temperature, sediment and substrate type, and, of increasing concern, pollutants such as pesticides, acidic materials and heavy metals.

All of these factors, subsumed under the term "water quality", can be measured physicochemically, but such monitoring requires:

- knowledge of which of the hundreds of substances to monitor;
- understanding interactions among pollutants (which often exacerbate or multiply the effects of any compound alone);
- continuous monitoring to detect pollutants that may be intermittent, such as nocturnal release of industrial waste products.

We often know little about which substances are significant biologically; even with such knowledge, continuous monitoring of more than a few is difficult and expensive. If these impediments could be overcome, the important question remains: what are the biological effects of pollutants? Organisms and communities that are exposed to aquatic pollutants integrate multiple present and immediate-past environmental effects. Increasingly, insects are used in the description and classification of aquatic ecosystems, and in the detection of deleterious effects of human activities. For the latter purpose, aquatic insect communities (or a subset of the animals that comprise an aquatic community) are used as surrogates: their observed responses give early warning of damaging changes.

In this **biological monitoring** of aquatic environments, the advantages of using insects include:

- ability to select amongst the many insect taxa in any aquatic system, according to the resolution required;
- availability of many ubiquitous or widely distributed taxa, allowing elimination of non-ecological reasons as to why a taxon might be missing from an area;
- functional importance of insects in aquatic ecosystems, ranging from secondary producers to top predators;
- ease and lack of ethical constraints in sampling aquatic insects, giving sufficient numbers of individuals and taxa to be informative, and yet still be able to be handled;
- ability to identify most aquatic insects to a meaningful level;
- predictability and ease of detection of responses of many aquatic insects to disturbances, such as particular types of pollution.

Typical responses observed when aquatic insect communities are disturbed include:

- increased abundance of certain mayflies, such as Caenidae with protected abdominal gills, and caddisflies, including filter-feeders such as Hydropsychidae, as particulate material (including sediment) increases;
- increase in numbers of haemoglobin-possessing bloodworms (Chironomidae) as dissolved oxygen is reduced;
- loss of stonefly nymphs (Plecoptera) as water temperature increases;
- substantial reduction in diversity with pesticide run-off;
- increased abundance of a few species but a general loss of diversity with elevated nutrient levels (organic enrichment, or **eutrophication**).

More subtle community changes can be observed in response to less overt pollution sources, but it can be difficult to separate environmentally induced changes from natural variations in community structure. Another area for caution is that some larger insects with aquatic juvenile stages are highly vagile as adults and can fly great distances to habitats unsuitable for immature development. Thus, advocacy of recording adult odonates as evidence of suitable quality of adjacent aquatic habitats may be misplaced. Even confirmed oviposition and some development of nymphs in a waterbody does not verify suitability – the only certain confirmation is evidence of a completed life-cycle by discovery of nymphal exuviae remaining after adult emergence.

10.6 FUNCTIONAL FEEDING GROUPS

Although aquatic insects are used widely in the context of applied ecology (section 10.5), it may not be possible, necessary, or even instructive, to make detailed species-level identifications. Sometimes, the taxonomic framework is inadequate to allow identification to this level, or time and effort do not permit resolution. In most aquatic entomological studies there is a necessary trade-off between maximizing ecological information and reducing identification time. Two solutions to this dilemma involve summary by subsuming taxa into: (i) more readily identified higher taxa (e.g. families, genera), or (ii) functional groupings based on feeding mechanisms ("functional feeding groups").

The first strategy assumes that a higher taxonomic category summarizes a consistent ecology or behaviour amongst all member species, and indeed this is evident from some of the broad summary responses noted above. However, many closely related taxa diverge in their ecologies, and higher-level aggregates thus contain a diversity of responses. In contrast, functional groupings need make no taxonomic assumptions, but use mouthpart morphology as a guide to categorizing feeding modes. The following categories are generally recognized, with some further subdivisions used by some workers:

- shredders feed on living or decomposing plant tissues, including wood, which they chew, mine or gouge;
- collectors feed on fine particulate organic matter by filtering particles from suspension (see the vignette at the beginning of the chapter of a filter-feeding black-fly larva of the *Simulium vittatum* complex, with body twisted and cephalic feeding fans open) or fine detritus from sediment;
- scrapers feed on attached algae and diatoms by grazing solid surfaces;
- piercers feed on cell and tissue fluids from vascular plants or larger algae, by piercing and sucking the contents;
- **predators** feed on living animal tissues by engulfing and eating whole or parts of animals, or piercing prey and sucking body fluids;
- **parasites** feed on living animal tissue of any stage of another organism as external or internal parasites.

Functional feeding groups traverse taxonomic groups. For example, the grouping "scrapers" includes some convergent larval mayflies, caddisflies, lepidopterans and dipterans, and within Diptera there are examples of each functional feeding group.

Changes in functional groups associated with human activities that affect waterways include:

- reduction in shredders with loss of riparian habitat and reduced allochthonous inputs;
- increase in scrapers with increased periphyton development resulting from enhanced light and nutrient entry;
- increase in filtering collectors below impoundments, such as dams and reservoirs, associated with increased fine particles from upstream standing waters.

Some summary data suggest a sequential downstream change in proportions of functional feeding groups, related to differing energy inputs into the flowing aquatic system (called the **river continuum concept**). In forested headwaters with a tree-shaded riparian zone, photosynthesis is restricted and energy derives more from high inputs of allochthonous materials (leaves, wood, etc.). Here, shredders such as some stoneflies and caddisflies tend to predominate, breaking up large matter into finer particles. Further downstream, collectors such as larval black flies (Simuliidae) and hydropsychid caddisflies filter particles generated upstream and add finer particles in the form of their faeces. Where the waterway becomes broader with enhanced photosynthesis in the mid-reaches, algae and diatoms (periphyton) increase and serve as food on hard substrates for scrapers, and macrophytes provide a resource for piercers. Predators tend only to track the localized abundance of food resources. Morphological attributes broadly associate with each of these groups, as grazers in fast-flowing areas tend to be active, flattened and current resisting, compared with the sessile, clinging filterers; scrapers have characteristic robust, wedge-shaped mandibles. Although this hypothesis of a downstream spiral of nutrients is appealing, carbon and nitrogen stable isotope signatures in aquatic insects ("you are what you eat") suggest a dominant role for more locally-derived sources in most river systems.

10.7 INSECTS OF TEMPORARY WATERBODIES

At a geological time-scale, all waterbodies are temporary. Lakes fill with sediment, become marshes, and eventually dry out completely. Erosion reduces the catchments of rivers, and their courses change. These historical changes are slow compared with the lifespan of insects and have little impact on the aquatic fauna, apart from a gradual alteration in environmental conditions. However, in certain parts of the world, waterbodies may fill and dry out on a much shorter time-scale. This is particularly evident where rainfall is very seasonal or intermittent, or where high temperatures cause elevated evaporation rates. Rivers may run during periods of predictable seasonal rainfall, such as the "winterbournes" on chalk downland in southern England that flow only during, and immediately following, winter rainfall. Others may flow only intermittently, after unpredictable heavy rains, such as streams of the arid zone of central Australia and deserts of the western United States. Temporary bodies of standing waters may last for as little as a few days, as in water-filled footprints of animals, rocky depressions, pools beside a falling river, or in impermeable clay-lined pools filled by flood or snow-melt.

Even though temporary, these habitats are very productive and teem with life. Aquatic organisms appear almost immediately after the formation of such habitats. Amongst the macroinvertebrates, crustaceans are numerous, and many insects thrive in ephemeral waterbodies. Some insects lay eggs into a newly formed aquatic habitat within hours of its filling, and it seems that gravid females of these species are transported to such sites over long distances, associated with the frontal meteorological conditions that bring the rainfall. An alternative to colonization by the adult is the deposition by the female of desiccation-resistant eggs into the dry site of a future pool. This behaviour is seen in some odonates and many mosquitoes, especially of the genus *Aedes*. Development of the diapausing eggs is induced by environmental factors that include wetting, perhaps requiring several consecutive immersions (section 6.5).

A range of adaptations is shown amongst insects living in ephemeral habitats compared with their relatives in permanent waters. First, development to the adult often is more rapid, perhaps because of increased food quality and lowered interspecific competition. Second, development may be staggered or asynchronous, with some individuals reaching maturity very rapidly, thereby increasing the possibility of at least some adult emergence from a short-lived habitat. Associated with this is a greater variation in size of adult insects from ephemeral habitats – with metamorphosis hastened as a habitat diminishes. Certain larval midges (Diptera: Chironomidae and Ceratopogonidae) can survive drying of an ephemeral habitat by resting in silk- or mucus-lined cocoons amongst the debris at the bottom of a pool, or by complete dehydration (section 6.6.2). In a cocoon, desiccation of the body can be tolerated, and development continues when the next rains fill the pool. In the dehydrated condition, temperature extremes can be withstood.

Persistent temporary pools develop a fauna of predators, including immature beetles, bugs and odonates, which are the offspring of aerial colonists. These colonization events are important in the genesis of faunas of newly flowing intermittent rivers and streams. In addition, immature stages present in remnant water beneath the streambed may move into the main channel, or colonists may be derived from permanent waters with which the temporary water connects. It is a frequent observation that novel flowing waters are colonized initially by a single species, often otherwise rare, that rapidly attains high population densities and then declines rapidly, with the development of a more complex community, including predators.

Temporary waters are often saline because evaporation concentrates salts, and this type of pool develops communities of specialist saline-tolerant organisms. However, few, if any, species of insect living in saline inland waters also occur in the marine zone – nearly all of the former have freshwater relatives.

10.8 INSECTS OF THE MARINE, INTERTIDAL AND LITTORAL ZONES

The estuarine and subtropical and tropical mangrove zones are transitions between fresh and marine waters. Here, the extremes of the truly marine environment, such as wave and tidal actions, and some osmotic effects, are ameliorated. Mangroves and "saltmarsh" communities (such as *Spartina*, *Sarcocornia*, *Halosarcia* and *Sporobolus*) support a complex phytophagous insect fauna on the emergent vegetation. In intertidal substrates and tidal pools, biting flies (mosquitoes and biting midges) are abundant and may be diverse. At the littoral margin, species of any of four families of hemipterans stride on the surface, some venturing onto the open water. A few other insects, including some *Bledius* staphylinid beetles, cixid fulgoroid bugs and root-feeding *Pemphigus* aphids, occupy the zone of prolonged inundation by saltwater. This fauna is restricted compared with freshwater and terrestrial ecosystems.

Splash-zone pools on rocky shores have salinities that vary because of rainwater dilution and solar concentration. They can be occupied by many species of corixid bugs and several larval mosquitoes and crane flies. Flies and beetles are diverse on sandy and muddy marine shores, with some larvae and adults feeding along the strandline, often aggregated on and under stranded seaweeds.

Within the intertidal zone, which lies between high and low neap-tide marks, the period of tidal inundation varies with the location within the zone. The insect fauna of the upper level is indistinguishable from the strandline fauna. At the lower end of the zone, in conditions that are essentially marine, crane flies, chironomid midges, and species of several families of beetles occur. The female of a remarkable Australasian marine trichopteran (Chathamiidae: *Philanisus plebeius*) lays its eggs in a starfish coelom. The early-instar caddisflies feed on starfish tissues, but later free-living instars construct cases of algal fragments.

Three lineages of chironomid midges are amongst the few insects that have diversified in the marine zone. *Telmatogeton* (see Fig. 6.12) is common in mats of green algae, such as *Ulva*, and occurs worldwide, including many isolated oceanic islands. In Hawai'i, the genus has re-invaded freshwater. The ecologically convergent *Clunio* also is found worldwide. In some species, adult emergence from marine rock pools is synchronized by the lunar cycle to coincide with the lowest tides. A third lineage, *Pontomyia*, ranges from intertidal to oceanic, with larvae found at depths of up to 30 m on coral reefs.

The only insects on open oceans are pelagic water-striders (*Halobates*), which have been sighted hundreds of kilometres from shore in the Pacific Ocean. The distribution of these insects coincides with mid-oceanic accumulations of flotsam, where food of terrestrial origin supplements a diet of marine chironomid midges.

Physiology is unlikely to be a factor restraining diversification in the marine environment because so many different taxa are able to live in inland saline waters and in various marine zones. When living in highly saline waters, submerged insects can alter their osmoregulation to reduce chloride uptake, and increase the concentration of their excretion through Malpighian tubules and rectal glands. In the pelagic water-striders, which live on the surface film, contact with saline waters must be limited.

As physiological adaptation appears to be a surmountable problem, explanations for the failure of insects to diversify in the sea must be sought elsewhere. The most likely explanation is that the insects originated well after other invertebrates, such as the Crustacea and Mollusca, had already dominated the sea. The advantages to terrestrial (including freshwater) insects of internal fertilization and flight are superfluous in the marine environment, where gametes can be shed directly into the sea, and the tide and oceanic currents aid dispersal. Notably, of the few successful marine insects, many have modified wings or have lost them altogether.

FURTHER READING

Andersen, N.M. (1995) Cladistic inference and evolutionary scenarios: locomotory structure, function, and performance in water striders. *Cladistics* **11**, 279–95.

Cover, M. & Resh, V. (2008) Global diversity of dobsonflies, fishflies, and alderflies (Megaloptera; Insecta) and spongillaflies, nevrorthids, and osmylids (Neuroptera; Insecta) in freshwater. *Hydrobiologia* **595**, 409–17.

Dudgeon, D. (1999) *Tropical Asian Streams: Zoobenthos, Ecology and Conservation*. Hong Kong University Press, Hong Kong.

Eriksen, C.H. (1986) Respiratory roles of caudal lamellae (gills) in a lestid damselfly (Odonata: Zygoptera). *Journal of the North American Benthological Society* **5**, 16–27.

Eriksen, C.H. & Moeur, J.E. (1990) Respiratory functions of motile tracheal gills in Ephemeroptera nymphs, as exemplified by *Siphlonurus occidentalis* Eaton. In: *Mayflies and Stoneflies: Life Histories and Biology* (ed. I.C. Campbell), pp. 109–18. Kluwer Academic Publishers, Dordrecht.

Lancaster, J. & Briers, R. (eds.) (2008) *Aquatic Insects: Challenges to Populations*. CAB International, Wallingford.

Lancaster, J. & Downes, B.J. (2013) *Aquatic Entomology*. Oxford University Press, Oxford.

Merritt, R.W., Cummins, K.W. & Berg, M.B. (eds.) (2008) *An Introduction to the Aquatic Insects of North America*, 4th edn. Kendall/Hunt Publishing Co., Dubuque, IO.

Rosenberg, D.M. & Resh, V.H. (eds.) (1993) *Freshwater Biomonitoring and Benthic Macroinvertebrates*. Chapman & Hall, London.

Thorp, J.H. & Rogers, D.C. (2010) *Field Guide to Freshwater Invertebrates of North America*. Academic Press, Boston, MA.

Tundisi, J.G. & Tundisi, T.M. (2011) *Limnology*. CRC Press, Boca Raton, FL.

Wichard, W., Arens, W. & Eisenbeis, G. (2002) *Biological Atlas of Aquatic Insects*. Apollo Books, Stenstrup.

Chapter 11

INSECTS AND PLANTS

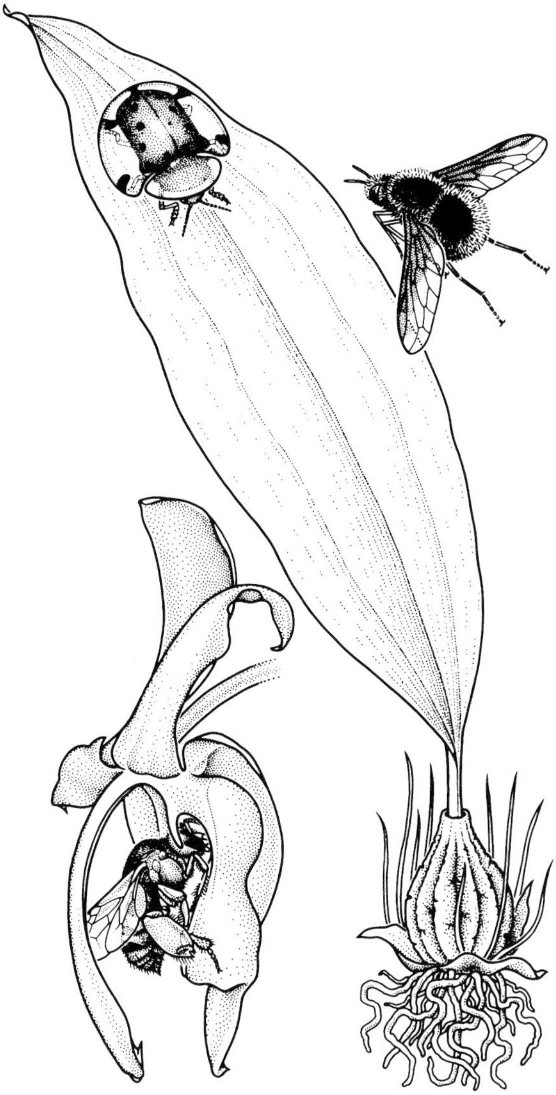

Specialized, plant-associated Neotropical insects. (After various sources.)

Insects and plants share ancient associations that date from the Carboniferous, some 300 million years ago (see Fig. 8.2). Evidence of insect damage preserved in fossilized plant parts indicates a diversity of types of **phytophagy** (plant-feeding) by insects, associated with tree and seed ferns from Late Carboniferous coal deposits. Prior to the origin of the now dominant angiosperms (flowering plants), the diversification of other seed plants, namely conifers, seed ferns, cycads and (extinct) bennettiales, provided the template for radiation of insects with specific plant-feeding associations. Both weevils (see **Plate 4b**) and thrips associate with cycads today. However, the major diversification of insects became manifest later, in the Cretaceous period. At this time, angiosperms dramatically increased in diversity in a radiation that displaced the previously dominant plant groups of the Jurassic period. Interpreting the early evolution of the angiosperms is contentious, partly because of the paucity of fossilized flowers prior to the period of radiation, and also because of the apparent rapidity of the origin and diversification within the major angiosperm families. However, according to estimates of their phylogeny, the earliest angiosperms might have been insect-pollinated, perhaps by beetles. Beetles (Coleoptera) pre-dated angiosperms and many beetles feed today on fungi, fern spores, or pollen of other non-angiosperm taxa such as cycads. As this feeding type preceded the angiosperm radiation, it can be seen as a preadaptation for angiosperm pollination. The ability of flying insects to transport pollen from flower to flower on different plants is fundamental to cross-pollination. Other than the beetles, the most significant and diverse present-day pollinator taxa belong to three orders – the Diptera (flies), Hymenoptera (wasps and bees) and Lepidoptera (moths and butterflies). Pollinator taxa within these orders are poorly represented in the fossil record until late in the Cretaceous. Although insects almost certainly pollinated cycads and other primitive plants, insect pollinators may have promoted speciation in angiosperms, through pollinator-mediated isolating mechanisms.

As seen in Chapter 9, many modern-day non-insect hexapods and apterygote insects scavenge in soil and litter, predominantly feeding on decaying plant material. The earliest true insects probably fed similarly. This manner of feeding certainly brings soil-dwelling insects into contact with plant roots and subterranean storage organs, but specialized use of plant aerial parts by sap sucking, leaf chewing, and other forms of phytophagy arose later in the phylogeny of the insects. Feeding on living tissues of higher plants presents problems that are experienced neither by the scavengers living in the soil or litter, nor by predators. First, to feed on leaves, stems or flowers, a phytophagous insect must be able to gain and retain a hold on the vegetation. Second, the exposed phytophage may be subject to greater desiccation than an aquatic or litter-dwelling insect. Third, a diet of plant tissues (excluding seeds) is nutritionally inferior in protein, sterol and vitamin content compared with food of animal or microbial origin. Last, but not least, plants are not passive victims of phytophages, but have evolved a variety of means to deter herbivores. These include physical defences, such as spines, spicules or sclerophyllous (hardened) tissue, and/or chemical defences that may repel, poison, reduce food digestibility, or otherwise adversely affect insect behaviour and/or physiology. Despite these barriers, about half of all living insect species are phytophagous, and the exclusively plant-feeding Lepidoptera, Curculionidae (weevils), Chrysomelidae (leaf beetles), Agromyzidae (leaf-mining flies) and Cynipidae (gall wasps) are very speciose. Plants represent an abundant resource and insect taxa that can exploit this have flourished in association with plant diversification (section 1.3.4).

This chapter begins with a consideration of the evolutionary interactions among insects and their plant hosts. We then go on to describe the vast array of interactions of insects and living plants, which can be grouped into three categories, defined by the effects of the insects on the plants. Phytophagy (herbivory) includes leaf chewing, sap sucking, seed predation, gall induction, and mining the living tissues of plants (section 11.2). Included here is a section concerning xylophagous (wood-eating) insects that feed on live trees (Box 11.3), but root-feeding insects are covered in section 9.1.1. The second category of interactions is important to plant reproduction and involves mobile insects that transport pollen between conspecific plants (pollination) or seeds to suitable germination sites (myrmecochory). These interactions

The Insects: An Outline of Entomology, Fifth Edition. P.J. Gullan and P.S. Cranston.

Companion Website: www.wiley.com/go/gullan/insects

are mutualistic because the insects obtain food or some other resource from the plants that they service (section 11.3). The first two of these categories of interactions are exemplified in the vignette at the beginning of this chapter, which shows a euglossine bee pollinator at work on the flower of a *Stanhopea* orchid, a chrysomelid beetle feeding on the orchid leaf, and a pollinating bee fly hovering nearby. The third category of insect–plant interaction involves insects that live in specialized plant structures and provide their host with either nutrition or defence against herbivores, or both (section 11.4). Such mutualisms, like the nutrient-producing fly larvae that live unharmed within the pitchers of carnivorous plants, are unusual but provide fascinating opportunities for evolutionary and ecological studies. There is a vast literature dealing with insect–plant interactions and the interested reader should consult the Further reading list at the end of this chapter.

The four boxes in this chapter cover the figs and their fig wasps (Box 11.1), grape phylloxera (Box 11.2), insects and the wood of live trees (Box 11.3), and *Salvinia* water weeds and their phytophagous weevils (Box 11.4).

11.1 COEVOLUTIONARY INTERACTIONS BETWEEN INSECTS AND PLANTS

Reciprocal interactions over evolutionary time between phytophagous insects and their food plants, or between pollinating insects and the plants they pollinate, have been described as **coevolution**. This term, coined by P.R. Ehrlich and P.H. Raven in 1964 from a study of butterflies and their host plants, originally was defined broadly, but now several modes of coevolution are recognized. These differ in the emphasis placed on the specificity and reciprocity of the interactions.

Specific or pair-wise coevolution refers to the evolution of a trait of one species (such as an insect's ability to detoxify a poison) in response to a trait of another species (such as the elaboration of the poison by the plant), which in turn evolved originally in response to the trait of the first species (i.e. the insect's food preference for that plant). This is a strict mode of coevolution, as reciprocal interactions between specific pairs of species are postulated. The outcomes of such coevolution may be evolutionary "arms races" between eater and eaten, or convergence of traits in mutualisms so that both members of an interacting pair appear perfectly adapted to each other. Reciprocal evolution between the interacting species may contribute to at least one of the species becoming subdivided into two or more reproductively isolated populations (as exemplified by figs and fig wasps; Box 11.1), thereby generating species diversity.

Another mode, **guild** or **diffuse coevolution**, describes reciprocal evolutionary change among groups, rather than pairs, of species. Here, the criterion of specificity is relaxed so that a particular trait in one or more species (e.g. of flowering plants) may evolve

Box 11.1 Figs and fig wasps

Figs belong to the large, mostly tropical genus *Ficus* (Moraceae) of at least 800 species. Each species of fig (except for the self-fertilizing, cultivated edible fig) has a complex obligatory mutualism with one or more species of pollinator. These pollinators all are fig wasps belonging to the hymenopteran family Agaonidae, which comprises numerous species in 20 genera. Each fig tree produces a large crop of 500–1,000,000 fruit (from inflorescences called syconia) as often as twice a year, but each synconium requires the action of at least one wasp in order to set seeds. Fig species are either dioecious (with male syconia on separate plants to those bearing the female syconia) or monoecious (with both male and female flowers in the same syconium), with monoecy being the ancestral condition. The following description of the life cycle of a fig wasp in relation to fig flowering and fruiting applies to monoecious figs, such as *F. macrophylla* (illustrated here, after Froggatt 1907; Galil & Eisikowitch 1968).

The female wasp enters the fig syconium via the ostiole (small hole), pollinates the female flowers lining the spheroidal cavity inside, oviposits in some of them (always short-styled ones) and dies. Each wasp larva develops within the ovary of a flower, which becomes a gall flower. Female flowers (usually long-styled ones) that escape wasp oviposition form seeds. About a month after oviposition, wingless male wasps emerge from their seeds and mate with female wasps still within the fig ovaries. Shortly after, the female wasps emerge from their seeds, gather pollen from another lot of flowers within the syconium (which is now in the male phase), and depart the mature fig to locate another conspecific fig tree in the phase of fig development suitable for oviposition. Different fig trees in a population are in different sexual stages, but all figs on one tree are synchronized. Species-specific volatile attractants produced by the trees allow very accurate, error-free location of another fig tree by the wasps.

Phylogenetic studies suggest that the mutualism of figs and fig-wasp pollinators arose only once: both interacting groups are monophyletic and have had a long-term co-diversification. The association is ancient, originating perhaps 75 million years ago in Eurasia. Early studies suggested parallel cladogenesis and coadaptation of genera of pollinating wasps and their respective sections of *Ficus*. Although it is difficult to resolve the deeper nodes of phylogenetic trees of *Ficus* and agaonids, recent work confirms the simultaneous radiation of the two lineages and their coordinated dispersal, but processes such as host switching and lineage extinction mean that strict **co-speciation** is not the sole pattern. For any given fig and wasp pair, reciprocal selection pressures may result in matching of fig and fig-wasp traits. For example, the sensory receptors of the wasp respond only to the volatile chemicals of its host fig, and the size and morphology of the guarding scales of the fig ostiole allow entry only to a fig wasp of the "correct" size and shape. It is likely that divergence in a local population of either fig or fig wasp, whether by genetic drift or selection, will induce **coevolutionary** change in the other. Host-specificity provides reproductive isolation among both wasps and figs, so coevolutionary divergence among populations is likely to lead to speciation. In many cases, the pollination system may be more complicated because there are at least 50 fig species with more than one pollinating wasp species. The amazing diversity of *Ficus* and Agaonidae may be a consequence of these coevolutionary interactions.

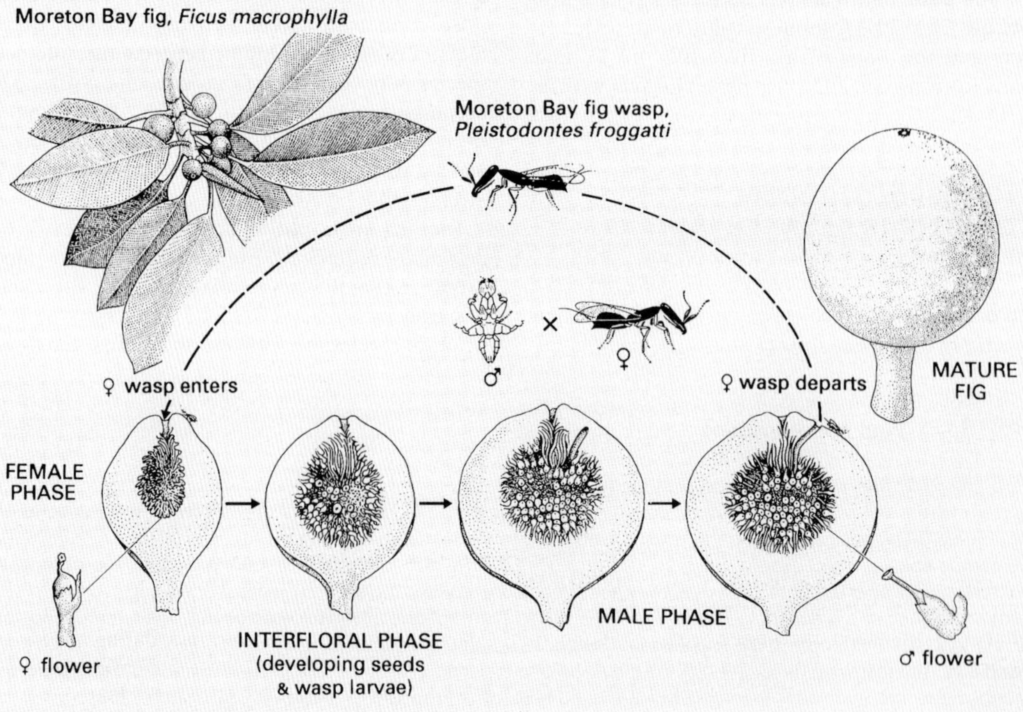

in response to a trait or suite of traits in several other species (e.g. as in several different, perhaps distantly related, pollinating insects).

These are the main modes of coevolution that relate to insect–plant interactions, but clearly they are not mutually exclusive. The study of such interactions is beset by the difficulty in demonstrating unequivocally that any kind of coevolution has occurred. Evolution takes place over geological time, and hence the selection pressures responsible for changes in "coevolving" taxa can be inferred only retrospectively, principally from correlated traits of interacting organisms. Specificity of interactions among living taxa can be demonstrated or refuted far more convincingly than can historical reciprocity in the evolution of the traits of these same taxa. For example, by careful observation, a flower bearing its nectar at the bottom of a very deep tube may be shown to be pollinated exclusively by a particular fly or moth species with a proboscis of appropriate length (e.g. Fig. 11.7), or a hummingbird with a particular length and curvature of its beak. Specificity of such an association between any individual pollinator species and plant is an observable fact, but the matching flower tube depth and mouthpart morphology are mere correlation and only suggest coevolution (section 11.3.1).

11.2 PHYTOPHAGY (OR HERBIVORY)

As outlined in the introduction to this chapter, insects face several challenges in order to feed on plants. The most important issues may be the suboptimal diet provided by plant material, and the defences that have evolved in plants to deter or reduce herbivory. Often, the tissues that have the highest nutritional value are the most defended by the plant, such as young shoots or leaves. There is good evidence that the performance (growth or fecundity) and abundance of many herbivorous insects is correlated to the nutritional status (especially available nitrogen content) of the plant tissue on which the insects feed. Variation in plant nutrients also can affect the amount of damage to a host plant, as insects usually eat more to compensate for a lower quality diet. A number of recent studies have shown the important role that bacterial endosymbionts play in the nutrition of phytophagous insects (section 3.6.5), especially sap-sucking ones that have a suboptimal diet, such as aphids (Hemiptera: Aphidoidea) feeding from phloem, or spittlebugs (Hemiptera: Cercopoidea) feeding from xylem. These bacteria can convert non-essential amino acids, which can be abundant is host-plant tissues, to essential amino acids that are crucial to insect wellbeing. Some sap-feeding insects can process enormous quantities of plant fluids in order to meet their nutritional needs. Thus, insects overcome the unfavourable carbon to nitrogen ratio of plant materials in several ways.

The majority of plant species support complex faunas of herbivores, each of which may be defined in relation to the range of plant taxa used. Thus, **monophages** are specialists that feed on one plant taxon, **oligophages** feed on a few plant taxa, and **polyphages** are generalists that feed on many plant groups. The adjectives for these feeding categories are **monophagous**, **oligophagous** and **polyphagous**. Gall-inducing cynipid wasps (Hymenoptera) exemplify monophagous insects, as nearly all species are host-plant specific; furthermore, all cynipid wasps of the tribe Rhoditini induce their galls only on roses (*Rosa*) (see Fig. 11.5d and **Plate 5c**) and almost all species of Cynipini form their galls only on oaks (*Quercus*) (see Fig. 11.5c and **Plate 6a**). The monarch or wanderer butterfly, *Danaus plexippus* (Lepidoptera: Nymphalidae), is an example of an oligophagous insect, with larvae that feed on various milkweeds, predominantly species of *Asclepias*. The polyphagous gypsy moth, *Lymantria dispar* (Lepidoptera: Lymantriidae) (see **Plate 8c**), feeds on a wide range of tree genera and species, and the Chinese wax scale, *Ceroplastes sinensis* (Hemiptera: Coccidae), is truly polyphagous, with its recorded host plants belonging to about 200 species in at least 50 families. In general, most phytophagous insect groups, except Orthoptera, tend to be specialized in their feeding.

Although it is useful to categorize insects based on their dietary specialization, in practice there are difficulties with the application of the terms monophage and oligophage. There is a graded spectrum of feeding habits, from very specialist to very generalist, and strict categories may not be meaningful. Furthermore, a particular insect species may have different preferences in different parts of its range or even during its development, and individuals of a single population may vary in their food preferences. Nevertheless, understanding the dietary specialization or breadth of herbivorous insects, especially of pest species, is important to many insect–plant studies (e.g. section 11.2.7). Whether an insect taxon evolves to be a specialist or a generalist must depend on the interplay of many factors, including sensory inputs used to locate hosts, the requirement for particular nutrients, and the ability to overcome or tolerate plant defences.

Many plants appear to have broad-spectrum defences against a very large suite of enemies, including insect and vertebrate herbivores and pathogens. These primarily physical or chemical defences are discussed in section 16.6 in relation to host-plant resistance to insect pests. Spines or pubescence on stems and leaves, silica or sclerenchyma in leaf tissue, or leaf shapes that aid camouflage are amongst the physical attributes of plants that may deter some herbivores. Furthermore, in addition to the chemicals considered essential to plant function, most plants contain compounds whose role generally is assumed to be defensive, although these chemicals may have, or once may have had, other metabolic functions or simply be metabolic waste products. Such chemicals are often called **secondary plant compounds**, noxious phytochemicals or **allelochemicals**. A huge array exists, including phenolics (such as tannins), terpene or terpinoid compounds (essential oils), alkaloids, cyanogenic glycosides, and sulphur-containing glucosinolates. The levels and kinds of these chemicals may vary within a plant species, due either to genetic or environmentally induced differences among individual plants. For example, even adjacent trees of the same *Eucalyptus* species can incur very different levels of foliage damage during an insect outbreak if the trees differ in their leaf terpenoid characteristics. The anti-herbivore action of many of these compounds has been demonstrated or inferred. For example, in *Acacia*, the loss of the otherwise widely distributed cyanogenic glycosides in those species that harbour mutualistic stinging ants implies that the secondary plant chemicals do have an anti-herbivore function in those many species that lack ant defences.

In terms of plant defence, secondary plant compounds may act in one of two ways. At a behavioural level, these chemicals may repel an insect or inhibit feeding and/or oviposition. At a physiological level, they may poison an insect or reduce the nutritional content of its food. However, the same chemicals that repel some insect species may attract others, either for oviposition or feeding (thus acting as kairomones; section 4.3.3). Such insects, thus attracted, are said to be adapted to the chemicals of their host plants, either by tolerating, detoxifying, or even sequestering them. An example is the monarch butterfly, *D. plexippus*, which usually oviposits on milkweed plants, many of which contain toxic cardiac glycosides (cardenolides), which the feeding larva can sequester for use as an anti-predator device (section 14.4.3 and section 14.5.2). Some other herbivorous insects (chrysomelid beetles, lygaeid bugs and an agromyzid fly) that also feed on cardenolide-containing plants display that same molecular adaptation that allows monarch butterfly larvae to feed on milkweeds, a clear case of convergent evolution.

Secondary plant compounds have been classified into two broad groups based on their inferred biochemical actions: (i) qualitative or toxic, and (ii) quantitative. The former are effective poisons in small quantities (e.g. alkaloids, cyanogenic glycosides), whereas the latter are believed to act in proportion to their concentration, being more effective in greater amounts (e.g. tannins, resins, silica). In practice, there probably is a continuum of biochemical actions, and tannins are not simply digestion-reducing chemicals but have more complex anti-digestive and other physiological effects. However, for insects that are specialized to feed on particular plants containing any secondary plant compound(s), these chemicals actually can act as phagostimulants. Furthermore, the narrower the host-plant range of an insect, the more likely that it will be repelled or deterred by non-host-plant chemicals, even if these substances are not noxious if ingested.

The observation that some kinds of plants are more susceptible to insect attack than others also has been explained by the relative apparency (obviousness) of the plants. Thus, large, long-lived, clumped trees are very much more apparent to an insect than are small, annual, scattered herbs. Apparent plants tend to have quantitative secondary compounds, with high metabolic costs in their production. Unapparent plants often have qualitative or toxic secondary compounds, produced at little metabolic cost. Human agriculture often turns unapparent plants into apparent ones when monocultures of annual plants are cultivated, with corresponding increases in insect damage.

Another consideration is the predictability of resources sought by insects, such as the suggested predictability of the presence of new leaves on a eucalypt tree or creosote bush in contrast to the erratic spring flush of new leaves on a deciduous tree. However, the question of what is the predictability (or apparency) of plants to insects is essentially untestable. Furthermore, insects can optimize the use of intermittently abundant resources by synchronizing their life cycles to environmental cues identical to those used by the plant.

Variation in herbivory rates within a plant species may correlate with the qualitative or quantitative chemical defences of individual plants (see above), including those that are induced (section 11.2.1). Another important correlate of variation in herbivory concerns the nature and quantities of resources (i.e. light, water, nutrients) available to individual plants. One hypothesis is that insect herbivores feed preferentially on stressed plants (e.g. affected by water-logging, drought, or nutrient deficiency), because stress can alter plant physiology in ways beneficial to insects. Alternatively, insect herbivores may prefer to feed on vigorously growing plants, or plant parts, in resource-rich habitats. Evidence for and against both is available. Thus, gall-inducing phylloxera (Box 11.2) prefers fast-growing meristematic tissue found in rapidly extending shoots of its healthy native vine host. In apparent contrast, the larva of *Dioryctria albovitella* (the pinyon pine cone and shoot boring moth; Pyralidae) attacks the growing shoots of nutrient-deprived and/or water-stressed pinyon pine (*Pinus edulis*) in preference to adjacent, less-stressed trees. Experimental alleviation of water stress has been shown to reduce rates of infestation, and enhance pine growth. Examination of a wide range of resource studies leads to the following partial explanation: boring and sucking insects seem to perform better on stressed plants, whereas gall inducers and chewing insects are adversely affected by plant stress. Additionally, performance of chewers may be reduced more on stressed, slow-growing plants than on stressed, fast growers.

The occurrence of butterflies with larvae adapted to toxic milkweeds, and the diverse suite of Australian insects that feed on terpene-rich eucalypt leaves, suggests that even well-defended food resources can become available to the specialist herbivore. Evidently, no single hypothesis (model) of herbivory is consistent with all observed patterns of temporal and spatial variation within plant individuals, populations and communities. However, all models of current herbivory theory make two assumptions, both of which are difficult to substantiate. These are:

1 damage by herbivores is a dominant selective force on plant evolution;

2 food quality has a dominant influence on the abundance of insects and the damage they cause.

Hybrid zones between two plant species of the same genus provide a useful system for addressing questions relating to herbivory. The food quality of hybrid plants can be higher than that of the parental plants, as a result of less efficient or more variable chemical defences and/or higher nutritive value of the genetically "rearranged" hybrids. There is ample evidence that herbivore species diversity and abundances are often higher on hybrid than on parental plants, although lower abundances have been recorded for some seed herbivores due to reduced seed set of hybrid hosts. However, evidence for herbivore pressure being higher on hybrids is often equivocal. In addition to plant quality, the fact that natural enemies can have an important role in regulating herbivore populations often is overlooked in studies of insect–plant interactions. The predators and parasitoids of phytophagous insects can be affected indirectly by the chemistry of the plants consumed by the phytophages. Thus, levels of herbivory are better considered as the outcome of tri-trophic interactions.

Many studies have demonstrated that phytophagous insects can impair plant growth, both in the short term and the long term. These observations have led to the suggestion that host-specific herbivores may affect the relative abundances of plant species by reducing the competitive abilities of host plants. The occurrence of induced defences (section 11.2.1) supports the idea that it is advantageous for plants to deter herbivores. In contrast with this view is the controversial hypothesis that "normal" levels of herbivory may be advantageous or selectively neutral to plants. Some degree of pruning, pollarding or mowing may increase (or at least not reduce) overall plant reproductive success by altering growth form or longevity and thus lifetime seed set. The important evolutionary factor is lifetime reproductive success, although most assessments of herbivore effects on plants involve only measurements of plant production (biomass, leaf number, etc.).

A major problem with all herbivory hypotheses is that they have been founded largely on studies of leaf-chewing insects, as the damage caused by these insects is easier to measure, and factors involved in defoliation are more amenable to experimentation than for other types of herbivory. The effects of sap-sucking, leaf-mining, gall-inducing and root-feeding insects may be as important, although, except for some agricultural and horticultural pests, they are generally poorly understood. The activities and effects on plants of leaf-chewing, sap-sucking, leaf-mining and boring, and gall-inducing insects are covered below, but root-feeding insects are dealt with in section 9.1.1.

Box 11.2 The grape phylloxera

An example of the complexity of a galling life cycle, host-plant resistance and even naming of an insect is provided by the grape phylloxera, sometimes called the grape louse. This aphid's native range is temperate–subtropical, from eastern North America and the southwest including Mexico, on a range of species of wild grapes (Vitaceae: *Vitis* spp.). Its complete life cycle is **holocyclic** (producing both sexual and asexual morphs). In its native range, its life cycle commences with the hatching of an overwintering egg, which develops into a **fundatrix** that crawls from the vine bark to a developing leaf where a pouch gall is formed in the rapidly growing meristematic tissue (as shown here, after several sources). Several generations of further apterous female offspring (sometimes called gallicolae) are produced and either continue to use the maternal gall or induce their own. As the nutrient status of the vine changes towards the end of the season, some of the wingless females (sometimes called radicicolae) migrate downwards to the roots. It is root-feeding wingless females that survive the winter, when vine leaves are shed along with their leaf galls. In the soil, these females induce nodose and tuberose galls (swellings) on the subapices of young roots (as illustrated here for the asexual life cycle) that act as nutrient sinks. In autumn/fall, in those biotypes with sexual stages, alates (**sexuparae**) are produced that fly from the soil to the stems of the vine, where they give rise to apterous, non-feeding **sexuales**. These mate, and each female lays a single overwintering egg. Within the natural range of aphid and host, the plants appear to show little damage from phylloxera, except perhaps in the late season, in which limited growth provides only a little new meristematic tissue for the explosive increase in leaf galls.

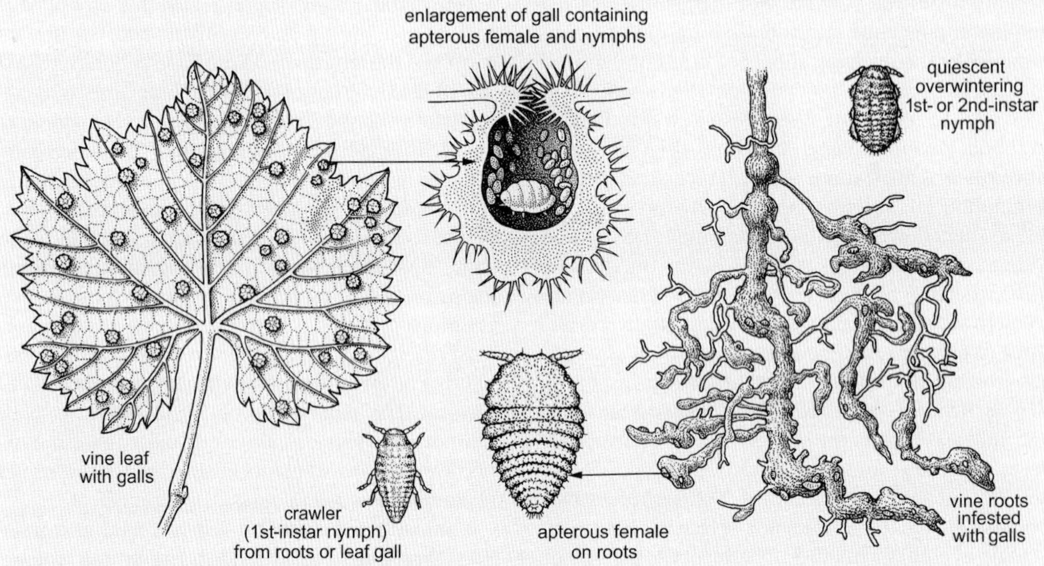

This life cycle shows modifications in its introduced range (e.g. Australia, Europe and South Africa), involving loss of the sexual and aerial stages, with persistence owing mostly or entirely to parthenogenetic populations (**anholocyclic** life cycle). Also involved are dramatic deleterious effects on the host vines by phylloxera feeding. This is of major economic importance when the host is *Vitis vinifera*, the native grape vine of the Mediterranean and Middle East. In the mid-19th century, American vines carrying phylloxera were imported into Europe; these devastated European grapes, which had no resistance to the aphid. Damage is principally through roots rotting under heavy loads of galls rather than sucking *per se*, and generally there is no aerial gall-inducing stage. The shipment from the eastern United States to France, by Charles Valentine Riley, of a natural enemy, the mite *Tyroglyphus*

phylloxerae, in 1873 was the first intercontinental attempt to control a pest insect. However, eventual control was achieved by grafting the already very diverse range of European grape cultivars (cépages such as Cabernet, Pinot Noir or Merlot) onto phylloxera-resistant rootstocks of North American *Vitis* species. Some *Vitis* species are not attacked by phylloxera, and in others the infestation starts and is either tolerated at a low level or rejected. Resistance (section 16.6) is mainly a matter of the speed at which the plant can produce inhibitory complex compounds from naturally produced phenolics, which can isolate each developing tuberose gall. Recently, it seems that some genotypes of phylloxera have circumvented certain resistant rootstocks, and resurgence may be expected.

The history of the scientific name of grape phylloxera is nearly as complicated as the life cycle. The name phylloxera now strictly refers only to species of the genus *Phylloxera* (family Phylloxeridae), which are mainly on *Juglans* (walnuts), *Carya* (pecans) and relatives. In the past, the grape phylloxera was known as *Phylloxera vitifoliae* and also as *Viteus vitifoliae* (under which name it is still known in parts of Europe and sometimes in China), but it is increasingly accepted that the genus name should be *Daktulosphaira* if a separate genus is warranted. This single species (*D. vitifoliae*) has a very wide range of behaviours, associated with different host species and cultivars and different geographic regions, but no morphometric, molecular or behavioural traits correlate well with any of the reported "biotypes" of *D. vitifoliae*. This is to be expected from a species that forms essentially clonal lineages on its non-native hosts.

11.2.1 Induced defences

Secondary plant compounds (noxious phytochemicals, or allelochemicals) deter herbivores, or at least reduce the suitability of many plants to some herbivores. Depending on plant species, such chemicals may be present in the foliage at all times, only in some plant parts, or only in some parts during development, such as during the growth period of new leaves. Such **constitutive defences** provide the plant with continuous protection, at least against non-adapted phytophagous insects. If defence is costly (in energetic terms) and if insect damage is intermittent, plants would benefit from being able to turn on their defences only when insect feeding occurs. There is good experimental evidence that in some plants damage to the foliage or roots induces chemical changes in the existing or future leaves or roots, which adversely affect insects. This phenomenon is called **induced defence** if the induced chemical response benefits the plant. However, sometimes the induced chemical changes may lead to greater foliage consumption by lowering food quality for herbivores, which thus eat more to obtain the necessary nutrients.

Both short-term (or rapidly induced) and long-term (or delayed) chemical changes have been observed in plants as responses to herbivory. For example, proteinase-inhibitor proteins are produced rapidly by some plants in response to wounds caused by chewing insects. These proteins can reduce the palatability of the plant to some insects. In other plants, the production of phenolic compounds may be increased, either for short or prolonged periods, within the wounded plant part or sometimes the whole plant. Alternatively, the longer-term carbon–nutrient balance may be altered, to the detriment of herbivores.

Such induced chemical changes have been demonstrated for studied plants, but their function(s) has been difficult to demonstrate, especially as herbivore feeding is not always deterred. Sometimes, induced chemicals, especially plant volatiles, may benefit the plant indirectly, not by reducing herbivory but by attracting natural enemies of the insect herbivores (section 4.3.3). Thus, herbivore-induced plant volatiles can act as cues both to other herbivores and to potential natural enemies. Recent advances in molecular techniques have allowed better understanding of the function of induced responses. These are elicited by "oxylipin signalling", involving chemicals related to lipoxygenases – common plant enzymes produced during wounding (amongst other effects). For example, jasmonic acid, which is one of the best-studied plant oxylipins, activates induced defences against chewing insects. Intriguing studies that "knocked-out" (inhibited expression of) selected genes of the signalling system in native tobacco demonstrated increased vulnerability of the genetically "silenced" plants to specialist herbivores and attraction of novel non-adapted phytophages.

When herbivore feeding induces the production of plant volatiles, there can be reduced rates of attraction of other herbivores to the damaged plants. Responses, however, are diverse, because even closely related herbivore and natural enemy species may respond in different ways to the same herbivore-induced plant volatiles, and the amount and composition of the blend of volatiles can vary depending on the identity and number of herbivores doing the damage. Also, the nature of the volatiles can be influenced by the plant species, its genotype, its age and conditions of growth. Nevertheless, it has been demonstrated experimentally that parasitoids can respond to blends of these induced volatiles and discriminate among plants damaged by different herbivores. In addition, herbivores can alter their oviposition behaviour in response to herbivore-induced volatiles, for example by avoiding laying eggs on the damaged part of the plant.

Although insect herbivore populations in the field are regulated by an array of factors, community-wide responses to variation in plant chemistry do occur. One response is likely to be the phenomenon referred to in popular literature as "talking trees", but better termed "eavesdropping", to describe an undamaged plant's reception of volatile chemicals emitted by herbivore-damaged neighbouring plants, inducing an increased resistance to herbivory by the receiver. Recent work even shows that an aphid-induced signal in one individual plant can be transmitted to uninfested plants via arbuscular mycorrhizal fungi that form a mycelial network connecting roots of neighbouring plants. In response to this signal, neighbouring plants are able to rapidly activate chemical defences that deter aphids.

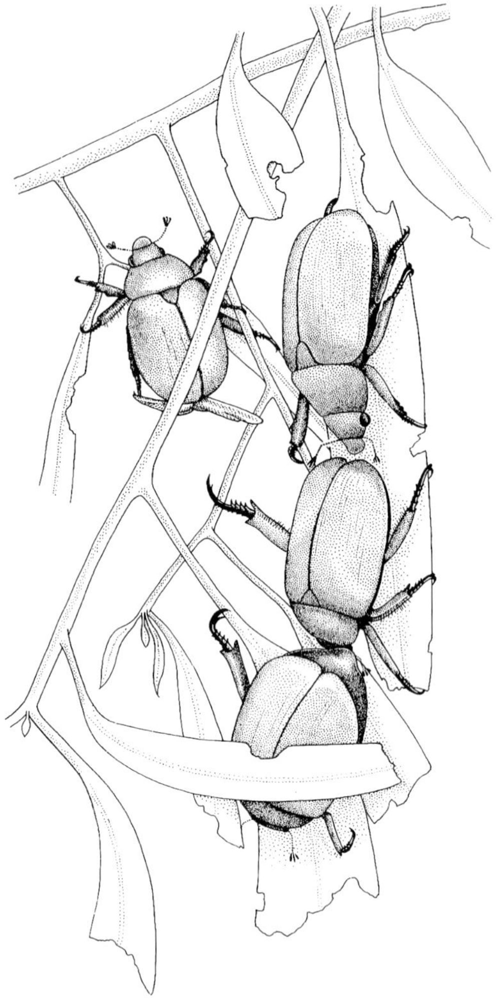

Fig. 11.1 Christmas beetles of *Anoplognathus* (Coleoptera: Scarabaeidae) on the chewed foliage of a eucalypt tree (Myrtaceae).

11.2.2 Leaf chewing

The damage caused by leaf-chewing insects is readily visible compared, for example, with that of many sap-sucking insects. Furthermore, the insects responsible for leaf-tissue loss are usually easier to identify than the small larvae of species that mine or gall plant parts. By far the most diverse groups of leaf-chewing insects are the Lepidoptera and Coleoptera. Most moth and butterfly caterpillars and many beetle larvae and adults feed on leaves, although plant roots, shoots, stems, flowers or fruits often are eaten as well. Certain Australian adult scarabs, especially species of *Anoplognathus* (Coleoptera: Scarabaeidae; commonly called Christmas beetles) (Fig. 11.1), can cause severe defoliation of eucalypt trees. The most important foliage-eating pests in north temperate forests are lepidopteran larvae, such as those of the gypsy moth, *Lymantria dispar* (Lymantriidae). Other important groups of leaf-chewing insects worldwide are the Orthoptera (most species) (see **Plate 4c**) and Hymenoptera (most Symphyta). The stick-insects (Phasmatodea) generally have only minor impact as leaf chewers, although outbreaks of the spur-legged stick-insect, *Didymuria violescens* (Taxobox 11), can defoliate eucalypts in Australia. A number of Australian phytophagous insects now damage eucalypts that are grown as forestry trees outside of their native Australia (see Box 17.3), but the main ones destructive of foliage are eucalypt snout weevils (see Box 7.1) and chrysomelid beetles.

High levels of herbivory result in economic losses to forest trees and other plants, so reliable and repeatable methods of estimating damage are desirable. Most methods rely on estimating leaf area lost due to leaf-chewing insects. This can be measured directly from foliar damage, either by once-off sampling, or monitoring marked branches, or by destructively collecting separate samples over time ("spot sampling"), or indirectly by measuring the production of insect **frass** (faeces). These sorts of measurements have been undertaken in several forest types, from rainforests to xeric (dry) forests, in many countries worldwide. Herbivory levels tend to be surprisingly uniform. For temperate forests, most values of proportional leaf area missing range from 3 to 17%, with a mean value of 8.8 ± 5.0% ($n = 38$) (values from Landsberg & Ohmart 1989). Data collected from rainforests and mangrove forests reveal similar levels of leaf area loss (range 3–15%, with mean 8.8 ± 3.5%). However, during outbreaks, especially of introduced pest species, defoliation levels may be very high and even lead to plant death. For some plant taxa, herbivory levels may be high (20–45%) even under natural, non-outbreak conditions.

Levels of herbivory, measured as leaf area loss, differ among plant populations or communities for a number of reasons. The leaves of different plant species vary in their suitability as insect food because of variations in nutrient content, water content, type and concentrations of secondary plant compounds, and degree of sclerophylly (toughness or hardness). Such differences may occur because of inherent differences among plant taxa, and/or may relate to the maturity and growing conditions of the individual leaves and/or the plants sampled. Assemblages in which the majority of the constituent tree species belong to different families (such as in many north temperate forests) may suffer less damage from phytophages than those that are dominated by one or a few genera (such as Australian eucalypt/acacia forests). In the latter systems, specialist insect species may be able to transfer relatively easily to new, closely related plant hosts. Favourable conditions thus may result in considerable insect damage to all or most tree species in a given area. In diverse (multigenera) forests, oligophagous insects are unlikely to switch to species unrelated to their normal hosts. Furthermore, there may be differences in herbivory levels within any given plant population over time as a result of seasonal and stochastic factors, including variability in weather conditions (which affects both insect and plant growth) or plant defences induced by previous insect damage. Such temporal variation in plant growth and response to insects can bias herbivory estimates made over a restricted time period.

11.2.3 Plant mining and boring

A range of insect larvae reside within and feed on the internal tissues of living plants. Some are **miners**, feeding just below the plant epidermis (the outermost protective tissue layer). Leaf-mining species live between the two epidermal layers of a leaf, and their presence can be detected externally after the area that they have fed upon dies, often leaving a thin layer of dry epidermis. This leaf damage appears as tunnels, blotches or blisters (Fig. 11.2). Tunnels may be straight (linear) to convoluted, and often widen throughout their course (Fig. 11.2a) as a result of larval growth during development. Generally, larvae that live in the confined space between the upper and lower leaf epidermis are flattened. Their excretory material, frass, is left in the mine as black or brown pellets (Fig. 11.2a,b,c,e) or lines (Fig. 11.2f).

The leaf-mining habit has evolved independently in only four holometabolous orders of insects: the Diptera, Lepidoptera, Coleoptera and Hymenoptera. The most common types of leaf miners are larval flies and moths. Some of the most prominent leaf mines result from the larval feeding of agromyzid flies (Fig. 11.2a–d). Agromyzids are virtually ubiquitous; there are about 2500 species, all of which are exclusively phytophagous. Most are leaf miners, although some mine stems and a few occur in roots or flower heads. Some anthomyiids and a few other fly species also mine leaves. Lepidopteran leaf miners (Fig. 11.2e–g) mostly belong to the families Gracillariidae, Gelechiidae, Incurvariidae, Lyonetiidae, Nepticulidae and Tisheriidae. The habits of leaf-mining moth larvae are diverse, with many variations in types of mines, methods of feeding, frass disposal, and larval morphology. Generally, the larvae are more specialized than those of other leaf-mining orders and are very dissimilar to their non-mining relatives. A number of moth species have habits that intergrade with gall inducing and leaf rolling. Leaf-mining Hymenoptera principally belong to the sawfly superfamily Tenthredinoidea, with most leaf-mining species forming blotch mines. Leaf-mining Coleoptera are represented by certain species of jewel beetles (Buprestidae), leaf beetles (Chrysomelidae) and weevils (Curculionoidea).

Leaf miners can cause economic damage by attacking the foliage of fruit trees, vegetables, ornamental

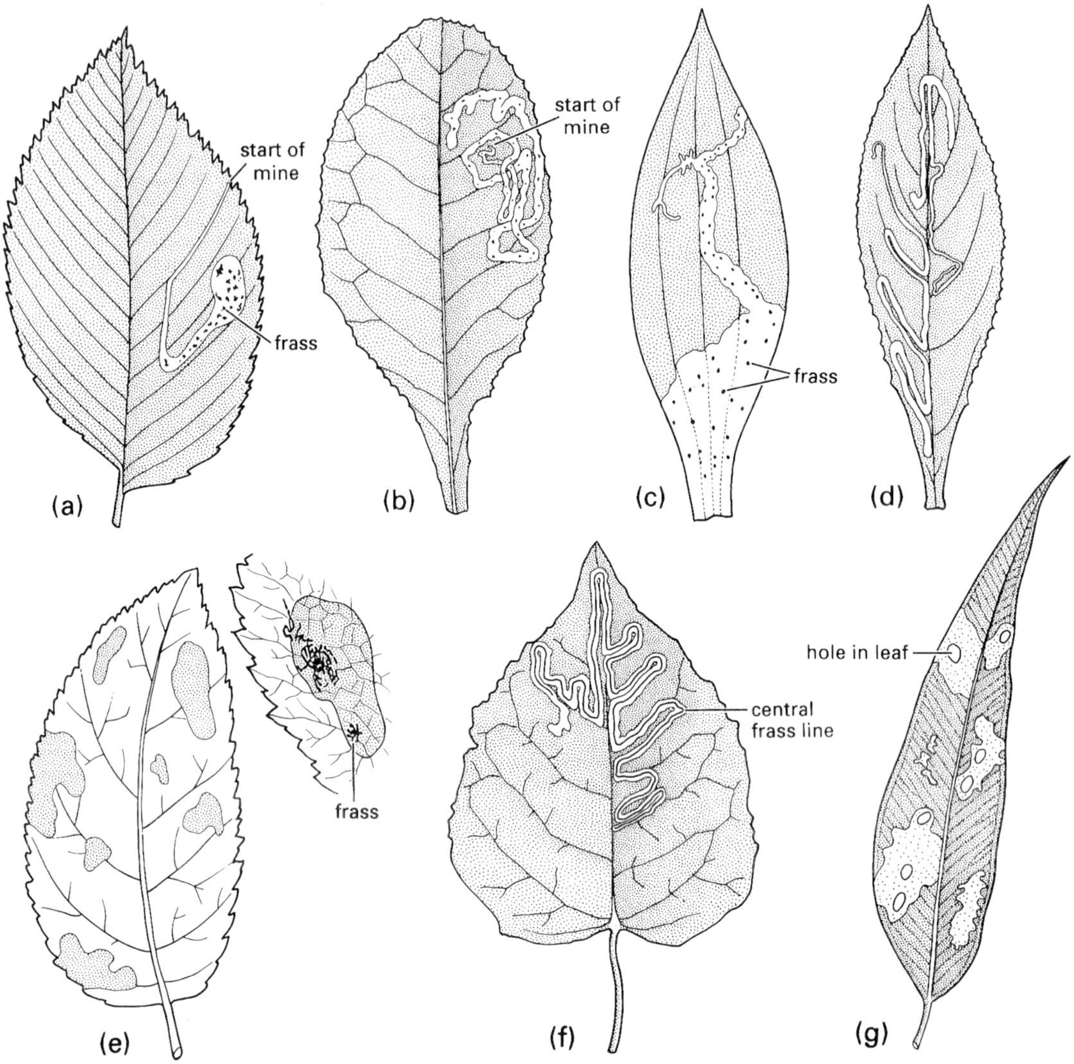

Fig. 11.2 Leaf mines: (a) linear-blotch mine of *Agromyza aristata* (Diptera: Agromyzidae) in leaf of an elm, *Ulmus americana* (Ulmaceae); (b) linear mine of *Chromatomyia primulae* (Agromyzidae) in leaf of a primula, *Primula vulgaris* (Primulaceae); (c) linear-blotch mine of *Chromatomyia gentianella* (Agromyzidae) in leaf of a gentian, *Gentiana acaulis* (Gentianaceae); (d) linear mine of *Phytomyza senecionis* (Agromyzidae) in leaf of a ragwort, *Senecio nemorensis* (Asteraceae); (e) blotch mines of the apple leaf miner, *Lyonetia prunifoliella* (Lepidoptera: Lyonetiidae), in leaf of apple, *Malus* sp. (Rosaceae); (f) linear mine of *Phyllocnistis populiella* (Lepidoptera: Gracillariidae) in leaf of poplar, *Populus* (Salicaceae); (g) blotch mines of jarrah leaf miner, *Perthida glyphopa* (Lepidoptera: Incurvariidae), in leaf of jarrah, *Eucalyptus marginata* (Myrtaceae). ((a,e–f) After Frost 1959; (b–d) after Spencer 1990.)

plants and forest trees. The citrus leaf miner, *Phyllocnistis citrella* (Gracillariidae), has spread around the world as a serious pest because its larvae mine the leaves of *Citrus* and related plants. The spinach leaf miner (or mangold fly), *Pegomya hyoscyami* (Diptera: Anthomyiidae), causes commercial damage to the leaves of spinach and beet. The larvae of the birch leaf miner, *Fenusa pusilla* (Hymenoptera: Tenthredinidae), produce blotch mines in birch foliage in northern North America, where this sawfly is considered a serious pest. In Australia, certain eucalypts are prone to the attacks of leaf miners, which can cause unsightly damage. The leaf blister sawflies (Hymenoptera: Pergidae: *Phylacteophaga*) tunnel in and blister the foliage of some species of *Eucalyptus* and related genera of Myrtaceae. The larvae of the jarrah leaf miner, *Perthida glyphopa* (Lepidoptera: Incurvariidae), feed in the leaves of jarrah, *Eucalyptus marginata*, causing blotch mines and then holes after the larvae have cut leaf discs for their pupal cases (Fig. 11.2g). Jarrah is an important timber tree in Western Australia, and the feeding of these leaf miners can cause serious leaf damage in vast areas of eucalypt forest.

Mining sites are not restricted to leaves, and some insect taxa display a diversity of habits. For example, different species of *Marmara* (Lepidoptera: Gracillariidae) not only mine leaves but some burrow below the surface of stems, or in the joints of cacti, and a few even mine beneath the skin of fruit. One species that typically mines the cambium of twigs even extends its tunnels into leaves if conditions are crowded. An iconic Australian phenomenon, the "scribbles" on the trunks of several smooth-barked eucalypt trees (see **Plate 4d**), derives from the activities of insects beneath the bark, feeding on growing tissue. Only recently has the biology become understood: a moth larva (one or more of up to 14 species of genus *Ogmograptis*, family Bucculatricidae) bores a zig-zag tunnel through and immediately beneath the developing cork cambium. The early-instar larva makes long irregular loops, followed by a more regular zigzag, which doubles back after a narrow turning loop. At this stage, the living cambium of the tree starts to produce cork and also produces localized scar tissue in response to the feeding of the caterpillar. This doubled-up tunnel becomes filled with highly nutritious, thin-walled cells, which are fed upon by the final-stage larva that re-mines the track, and when fully fed it drops to the ground to pupate. All this developmental activity has taken place out of sight beneath the old bark. When the bark is shed, the "ghost mine" of its past occupant is revealed, but the maker of the "scribbles" already has departed.

Stem mining, or feeding in the superficial layer of twigs, branches or tree trunks (as in the examples above), can be distinguished from stem boring, in which the insect feeds deep in the plant tissues. Stem boring is just one form of plant **boring**, which includes a broad range of habits that can be subdivided according to the part of the plant eaten and whether the insects are feeding on living or dead and/or decaying plant tissues. The latter group of saprophytic insects is discussed in section 9.2 and is not dealt with further here. The former group includes larvae that feed in buds, fruits, nuts, seeds, roots, stalks and wood. Stalk borers, such as the wheat stem sawflies (Hymenoptera: Cephidae: *Cephus* species) and the European corn borer (Lepidoptera: Pyralidae: *Ostrinia nubilalis*) (Fig. 11.3a), attack grasses and more succulent plants.

Wood borers feed in the roots, twigs, stems and/or trunks of living woody plants, where they may eat the bark, phloem, sapwood or heartwood (Box 11.3). The wood-boring habit is typical of many Coleoptera, especially the larvae of longicorn (or longhorn) beetles (Cerambycidae) (see Box 17.3), jewel beetles (Buprestidae (see Box 17.4) and weevils (Curculionoidea), and some Lepidoptera (e.g. Hepialidae and Cossidae; see Fig. 1.3) and Hymenoptera. The root-boring habit is well developed in the Lepidoptera, but moth larvae rarely differentiate between the wood of trunks, branches or roots. Many insects damage plant storage organs by boring into tubers, corms and bulbs. Some of the most effective biological control agents for weedy plants (section 11.2.7) are specialist root-boring insects, such as larvae of the flea beetle, *Longitarsus jacobaeae* (Chrysomelidae), on invasive ragwort, *Senecio jacobaea*.

The reproductive output of many plants is reduced or destroyed by the feeding activities of larvae that bore into and eat the tissues of fruits, nuts or seeds. Fruit borers include:

- Diptera (especially Tephritidae, such as the apple maggot, *Rhagoletis pomonella*, and the Mediterranean fruit fly, *Ceratitis capitata*);
- Lepidoptera (e.g. some tortricids, such as the oriental fruit moth, *Grapholita molesta*, and the codling moth, *Cydia pomonella*; Fig. 11.3b);
- Coleoptera (particularly certain weevils, such as the plum curculio, *Conotrachelus nenuphar*).

Weevil larvae also are common occupants of seeds and nuts, and many species are pests of stored grain (section 11.2.6).

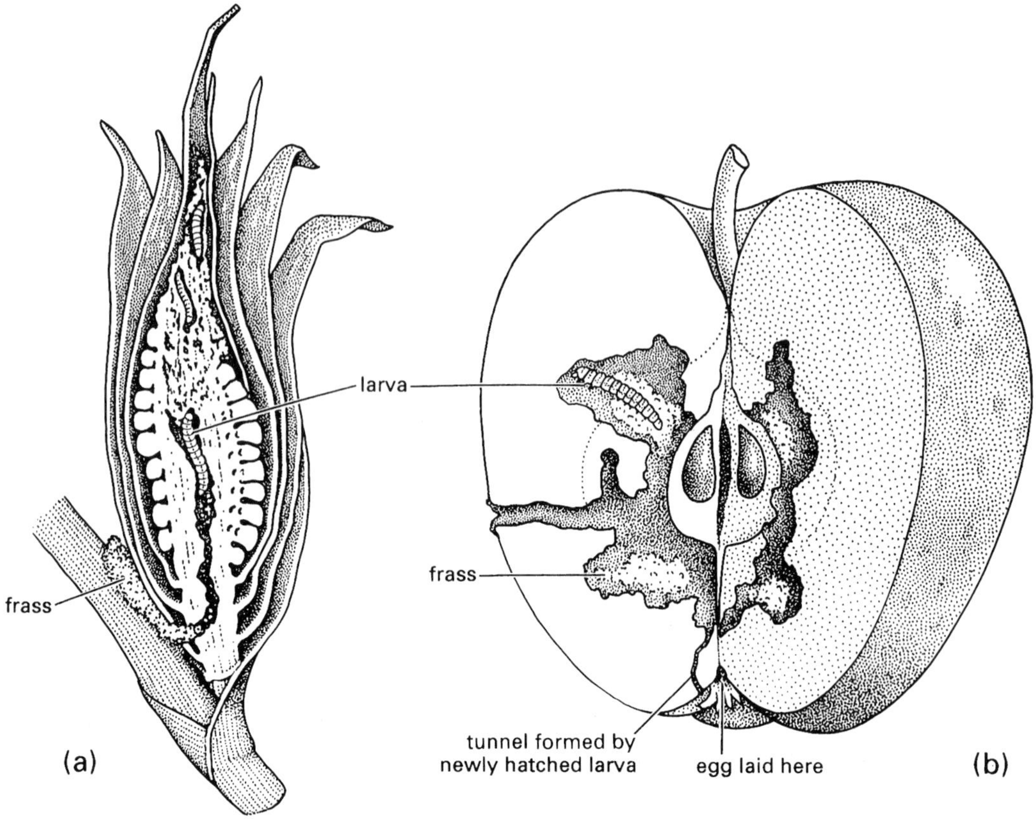

Fig. 11.3 Plant borers: (a) larvae of the European corn borer, *Ostrinia nubilalis* (Lepidoptera: Pyralidae), tunnelling in a corn stalk; (b) a larva of the codling moth, *Cydia pomonella* (Lepidoptera: Tortricidae), inside an apple. (After Frost 1959.)

11.2.4 Sap sucking

The feeding activities of insects that chew or mine leaves and shoots cause obvious damage. In contrast, structural damage caused by sap-sucking insects often is inconspicuous, as the withdrawal of cell contents from plant tissues usually leaves the cell walls intact. Damage to the plant may be difficult to quantify, even though the sap sucker drains plant resources (by removing phloem or xylem contents), causing loss of condition such as retarded root growth, fewer leaves, or less overall biomass accumulation compared with unaffected plants. These effects may be detectable with confidence only by controlled experiments in which the growth of infested and uninfested plants is compared. Certain sap-sucking insects do cause conspicuous tissue necrosis either by transmitting diseases, especially viral ones, or by injecting toxic saliva, whereas others induce obvious tissue distortion or growth abnormalities called galls (section 11.2.5).

Most sap-sucking insects belong to the Hemiptera. All hemipterans have long, thread-like mouthparts consisting of appressed mandibular and maxillary stylets forming a bundle lying in a groove in the labium (Taxobox 20). The maxillary stylets contain a salivary canal that directs saliva into the plant, and a food canal through which plant juice or sap is sucked up into the insect's gut. Only the stylets enter the tissues of the host plant (Fig. 11.4a). They may penetrate superficially into a leaf or deeply into a plant stem or leaf midrib, following either an intracellular or intercellular path, depending on species. The feeding site reached by the stylet tips may be in the parenchyma (e.g. some immature scale insects and many Heteroptera), the phloem (e.g. most aphids, mealybugs, soft scales and leafhoppers) or the xylem (e.g. spittle bugs and cicadas).

Box 11.3 Insects and the wood of live trees

Living trees provide ample foliage for phytophagous insects, and differ from annual plants and crops perhaps only in the density and heterogeneous structure of leaves available for herbivorous insects, greater height above the ground of the canopy, and for non-deciduous trees, the longer availability of leaves. If we look at the guilds of insects associated with plants, there are few that specialize on trees. Trees have multiple evolutionary origins: most families of plants contain some trees, and few contain only trees. For insects, what makes trees special is the presence of a substantial volume of wood – the inert, hard, fibrous material lying beneath the vascular cambium, located in the trunk (stem), roots and branches. Wood is an organic composite of cellulose fibres, which provide tension, embedded in a matrix of lignin, which resists compression. Heartwood that develops at the core of most trees is dead, and the paler, sometimes green, sapwood that surrounds the heartwood and comprises functional vessels, is considered "live". Physically, both heartwood and sapwood tend to be hard and dense, providing only refractory materials for insects. Nonetheless, a substantial suite of insects depend upon wood – these are termed **xylophages** even if wood is not fed upon directly.

One means of accessing nutritionally poor lignin and cellulose is for the xylophage to infect the wood with fungi, which promote its breakdown to digestible fungal matter. Many of these fungi are pathogenic, leading to death of trees and making the substrate more tractable for insect colonization when defensive tannin production and resin flow cease. For example, wood wasps of the genera *Sirex* and *Urocercus* (Hymenoptera: Siricidae) introduce *Amylostereum* fungal spores that are stored in invaginated intersegmental sacs connected to the ovipositor. During oviposition, spores and mucous are injected into the sapwood of trees, notably *Pinus* species, causing mycelial infection. The infestation causes locally drier conditions around the xylem, which is optimal for development of *Sirex* larvae. The fungal disease in Australia and New Zealand can cause death of trees already stressed by climate or damaged by fire. The role of bark beetles (Coleoptera: Curculionidae: Scolytinae: *Scolytus* spp.) in the spread of the fungus causing Dutch elm disease is discussed in section 4.3.3. Other fungal diseases that are transmitted by insects to live or dead trees encourage decay of the wood, and the proliferating fungi are food for the insects (section 9.2). Healthy living trees, however, mount a defence (such as resin release) against an attack from bark beetles. However, bark beetles that kill trees, such as *Dendroctonus frontalis* (southern pine beetle), can do so without the help of pathogenic fungi.

Although some termites (subfamily Macrotermitinae) use fungi to pre-digest wood (section 9.5.3), most use either symbiotic flagellates (protists) in their hindguts or produce their own cellulases to breakdown already dead wood (section 3.6.5). However, certain termites, termed "live wood" termites, attack living wood and can be pests especially of tea grown in the tropics.

Although xylophagous insects may be highly cryptic, most of them, especially the larger beetle and moth larvae, are evident by their frass, which reveals by sight and odour that the wood contains a potential host or prey. Visual predators such as woodpeckers locate hidden insects also by sound (tapping on the wood to hear movements of insects within), and a range of birds prize off bark to reveal xylophagous insects beneath, and some even have beaks modified to enter the insect's retreat to reach the contents. Predatory, and especially parasitoid, insects use chemoreception of semiochemicals (section 4.3.3), including terpenes emitted by damaged pines, to locate their hidden xylophagous prey. Access to the hidden target is possible for specialist hymenopterans with long ovipositors, such as portrayed in Fig. 5.11. The existence of a suite of parasitoids that control xylophages in their native ranges provides a source of such agents for biological control of alien xylophagous pests of forestry. The cryptic nature of immature xylophages means that many evade inspection in wood products, and thus continue to spread globally, particularly via packing materials used to transport goods in poorly regulated trade (see section 17.4).

The feeding site, whether parenchyma cells or vascular tissue, can change during the development of a single hemipteran species. In addition to a hydrolyzing type of saliva, many species produce a solidifying saliva that forms a sheath around the stylets as they enter and penetrate the plant tissue. This sheath can be stained in tissue sections and allows the feeding tracks to be traced to the feeding site (Fig. 11.4b,c). There are three feeding strategies in hemipterans: (i) stylet-sheath; (ii) cell-rupture feeding; and (iii) osmotic-pump. These three feeding methods are described in section 3.6.2, and the gut specializations of hemipterans for dealing with a watery diet are discussed in Box 3.3. Many species of plant-feeding Hemiptera are considered serious agricultural and horticultural pests. Loss of sap leads to wilting, distortion, or stunting of shoots. Movement of the insect between host plants can lead to the efficient transmission of plant viruses and other diseases, especially by aphids and whiteflies. The sugary waste (**honeydew**) eliminated from the anus of phloem-feeding Hemiptera, particularly coccoids, is used by black sooty moulds, which contaminate leaves and fruits and can impair photosynthesis. Honeydew also serves as an important carbohydrate source for hymenopterans, especially ants (section 11.4.1; see **Plate 6b**), which often defend the hemipterans from their natural enemies.

Psylloids (Hemiptera: Psylloidea) feed by sucking sap, mostly of woody dicotyledonous plants, from a variety of tissue sources, including phloem, xylem and mesophyll parenchyma, depending on species. The nymphs either are free-living, gall-inducing (Fig. 11.5f) or lerp-forming. A **lerp** is a protective cover under which a nymph lives and feeds. The lerp-forming habit is typical of many Australian species that feed on eucalypt foliage. The appearance of the lerp is genus (and sometimes species) specific; for example, nymphs of *Glycaspis* species make conical lerps, whereas those of *Cardiaspina* construct lace-like lerps. Lerp-forming nymphs discharge a solidifying starchy or sugary material from their anus and construct their lerp by building a framework above their body as they rotate at

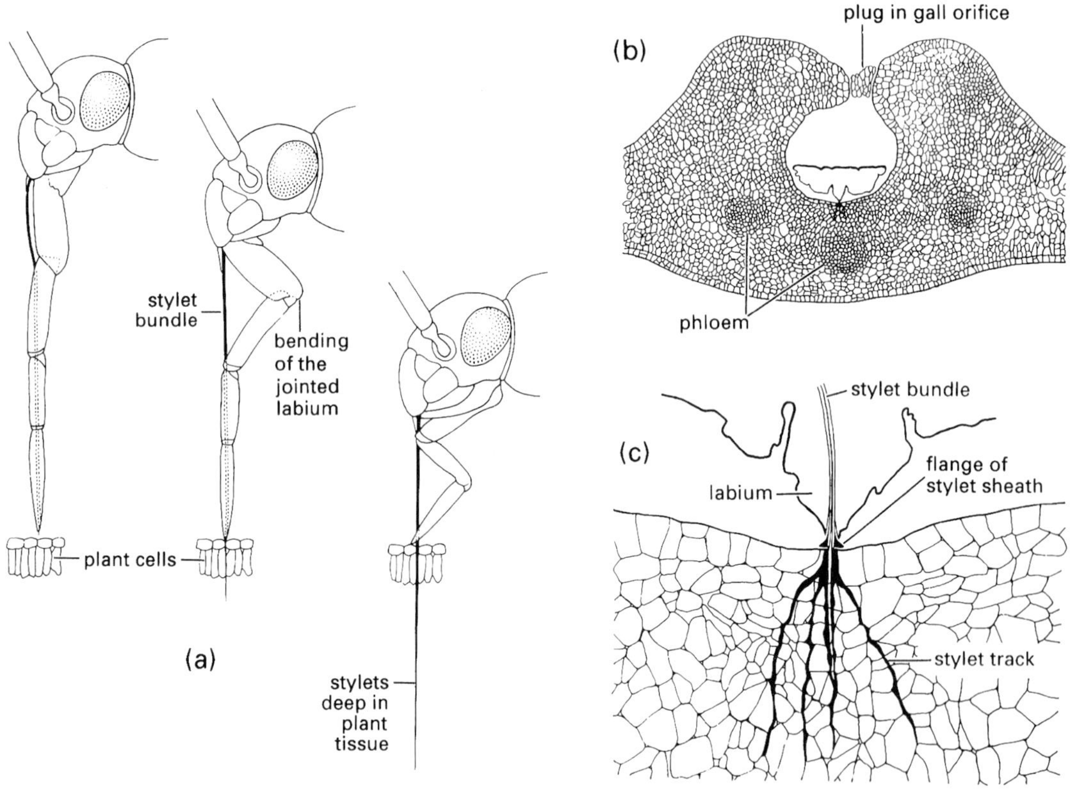

Fig. 11.4 Feeding in phytophagous Hemiptera: (a) penetration of plant tissue by a mirid bug, showing bending of the labium as the stylets enter the plant; (b) transverse section through a eucalypt leaf gall containing a feeding nymph of a scale insect, *Apiomorpha* (Eriococcidae); (c) enlargement of the feeding site of (b), showing multiple stylet tracks (formed of solidifying saliva), resulting from probing of the parenchyma. ((a) After Poisson 1951.)

a spot on the leaf, but most also produce typical sugary honeydew as well. Several Australian lerp-forming species are now pests of plantation and ornamental eucalypts in North and South America and Europe (see Box 17.3 and **Plate 8e**).

Thrips (Thysanoptera) that feed by sucking plant juices penetrate the tissues using their stylets (see Fig. 2.15) to pierce the epidermis and then rupture individual cells below. Damaged areas discolour and the leaf, bud, flower or shoot may wither and die. Plant damage typically is concentrated on rapidly growing tissues, so that flowering and leaf flushing may be seriously disrupted. Some thrips inject toxic saliva during feeding or transmit viruses, such as the *Tospovirus* (Bunyaviridae) carried by the pestiferous western flower thrips, *Frankliniella occidentalis*. A few hundred thrips species have been recorded attacking cultivated plants, but only 14 species (all Thripidae) are known to transmit tospoviruses.

Outside the Hemiptera and Thysanoptera, the sap-sucking habit is rare in extant insects. Many fossil species, however, had a rostrum with piercing-and-sucking mouthparts. Palaeodictyopteroids (see Fig. 8.3), for example, probably fed by imbibing juices from plant organs.

11.2.5 Gall induction

Insect-induced plant **galls** result from a very specialized type of insect–plant interaction in which the morphology of plant parts is altered, often substantially and characteristically, by the influence of the insect. Generally, galls are defined as pathologically developed cells, tissues or organs of plants that have arisen by hypertrophy (increase in cell size) and/or hyperplasia (increase in cell number) as a result of stimulation from foreign organisms. Some galls are induced by viruses, bacteria, fungi, nematodes and mites, but insects cause many more. The study of plant galls is called **cecidology**, gall-causing animals (insects, mites and nematodes) are **cecidozoa**, and galls induced by cecidozoa are referred to as **zoocecidia**. Cecidogenic insects account for about 2% of all described insect species, with perhaps 13,000 species known. Although galling is a worldwide phenomenon across most plant groups, global survey shows an eco-geographical pattern, with gall incidence being more frequent in vegetation with a sclerophyllous habit, or at least on plants in wet–dry seasonal environments.

On a world basis, the principal cecidozoa in terms of number of species are representatives of just three orders of insects – the Hemiptera, Diptera and Hymenoptera. In addition, about 300 species of mostly tropical Thysanoptera (thrips) are associated with galls, although not necessarily as inducers, and some species of Coleoptera (mostly weevils) and microlepidoptera (small moths) induce galls. Most hemipteran galls are elicited by Sternorrhyncha, in particular aphids, coccoids and psylloids. There are thousands of galling sternorrhynchans and their galls are structurally diverse. Those of gall-inducing eriococcids (Coccoidea: Eriococcidae) can be very large (e.g. the "bloodwood apples" of *Cystococcus* species; section 1.8.1; see **Plate 4f,g**) and often exhibit spectacular sexual dimorphism, with galls of female insects much larger and more complex than those of their conspecific males (e.g. galls of *Apiomorpha*; Fig. 11.5a,b; see **Plate 5a**). Worldwide, there are several hundred gall-inducing coccoid species in about 10 families, about 350 gall-forming Psylloidea, mostly in two families, and perhaps 700 gall-inducing aphid species distributed among the three families, Phylloxeridae (Box 11.2), Adelgidae and Aphididae. The huge galls of *Baizongia pistaciae* (Aphididae: Fordinae) on *Pistacia teredinthus* each contain hundreds of individual aphids (see **Plate 5b**).

The Diptera contains the highest number of gall-inducing species, perhaps thousands, but the probable number is uncertain because many dipteran gall inducers are poorly known taxonomically. Most cecidogenic flies belong to one family of more than 6000 species, the Cecidomyiidae (gall midges), and induce simple or complex galls on leaves, stems, flowers, buds and even roots. The other fly family that includes some important cecidogenic species is the Tephritidae, in which gall inducers mostly affect plant buds, often of the Asteraceae. Galling species of both cecidomyiids and tephritids are of actual or potential use for biological control of some weeds. Three superfamilies of wasps contain large numbers of gall-inducing species. Cynipoidea contains the gall wasps (Cynipidae, at least 1400 species), which are among the best-known gall insects in Europe and North America, where hundreds of species form often extremely complex galls, especially on oaks (Fig. 11.5c; see **Plate 6a**) and roses (Fig. 11.5d; see **Plate 5c**). Tenthredinoidea has a number of gall-forming sawflies, such as *Pontania* species (Tethredinidae) (Fig. 11.5g). Chalcidoidea includes several families of gall inducers, especially species in the

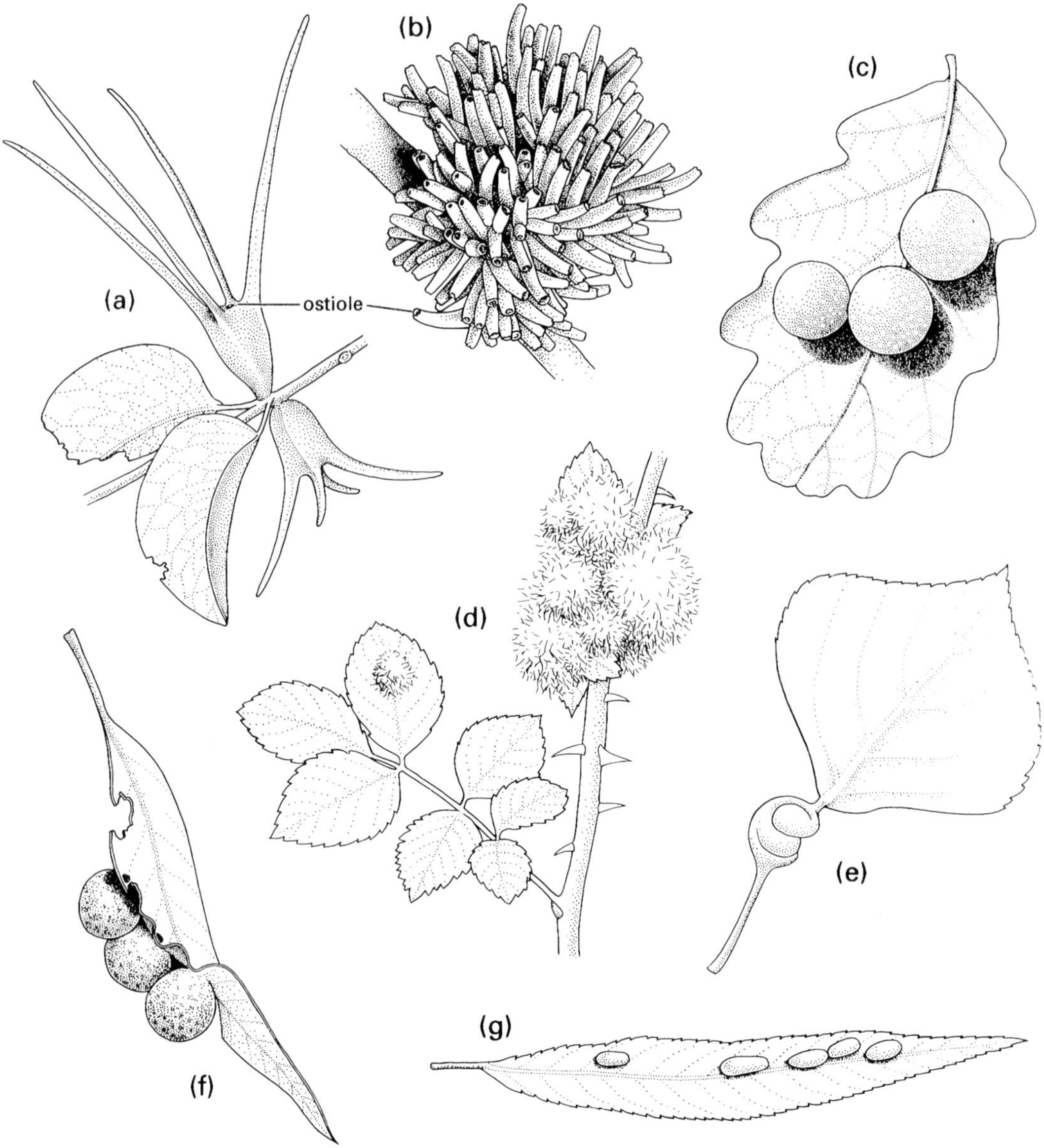

Fig. 11.5 A variety of insect-induced galls: (a) two coccoid galls, each formed by a female of *Apiomorpha munita* (Hemiptera: Eriococcidae) on the stem of *Eucalyptus melliodora*; (b) a cluster of galls, each containing a male of *A. munita* on *E. melliodora*; (c) three oak cynipid galls formed by *Cynips quercusfolii* (Hymenoptera: Cynipidae) on a leaf of *Quercus* sp.; (d) rose bedeguar galls formed by *Diplolepis rosae* (Hymenoptera: Cynipidae) on *Rosa* sp.; (e) a leaf petiole of lombardy poplar, *Populus nigra*, galled by the aphid *Pemphigus spirothecae* (Hemiptera: Aphididae); (f) three psyllid galls, each induced by a nymph of *Glycaspis* sp. (Hemiptera: Psyllidae) on a eucalypt leaf; (g) willow bean galls of the sawfly *Pontania proxima* (Hymenoptera: Tenthredinidae) on a leaf of *Salix* sp. ((d–g) After Darlington 1975.)

Agaonidae (fig wasps; Box 11.1), Eurytomidae and Pteromalidae.

There is enormous diversity in the patterns of development, shape and cellular complexity of insect galls (Fig. 11.5). They range from relatively undifferentiated masses of cells ("indeterminate" galls) to highly organized structures with distinct tissue layers ("determinate" galls). Determinate galls usually have a shape that is specific to each insect species. Cynipids, cecidomyiids and eriococcids form some of the most histologically complex and specialized galls; these galls have distinct tissue layers or types that may bear little resemblance to the plant part from which they are derived. Among the determinate galls, different shapes correlate with mode of gall formation, which is related to the initial position and feeding method of the insect (as discussed below). Some common types of galls are:

- covering galls, in which the insect becomes enclosed within the gall, either with an opening (ostiole) to the exterior, as in coccoid galls (Fig. 11.5a,b; see **Plate 5a**), or without any ostiole, as in cynipid galls (Fig. 11.5c);
- filz galls, which are characterized by their hairy epidermal outgrowths (Fig. 11.5d; see **Plate 5c**);
- roll and fold galls, in which differential growth provoked by insect feeding results in rolled or twisted leaves, shoots or stems, which are often swollen, as in many aphid galls (Fig. 11.5e);
- pouch galls, which develop as a bulge of the leaf blade, forming an invaginated pouch on one side and a prominent bulge on the other, as in many psylloid galls (Fig. 11.5f);
- mark galls, in which the insect egg is deposited inside stems or leaves so that the larva is completely enclosed throughout its development, as in sawfly galls (Fig. 11.5g);
- pit galls, in which a slight depression, sometimes surrounded by a swelling, is formed where the insect feeds;
- bud and rosette galls, which vary in complexity and cause enlargement of the bud or sometimes multiplication and miniaturization of new leaves, forming a pine-cone-like gall.

Gall induction may involve two separate processes: (i) initiation; and (ii) subsequent growth and maintenance of structure. Usually, galls can be stimulated to develop only from actively growing plant tissue. Therefore, galls are initiated on young leaves, flower buds, stems and roots, and rarely on mature plant parts. Some complex galls develop only from undifferentiated meristematic tissue, which becomes moulded into a distinctive gall by the activities of the insect. Development and growth of insect-induced galls (including, if present, the nutritive cells upon which some insects feed) depend upon continued stimulation of the plant cells by the insect. Gall growth ceases if the insect dies or reaches maturity. It appears that gall insects, rather than the plants, control most aspects of gall formation, largely via their feeding activities. Thus, the induced gall phenotype is an extension of the galling insect's phenotype.

The mode of feeding differs in different taxa as a consequence of fundamental differences in mouthpart structure. The larvae of gall-inducing beetles, moths and wasps have biting and chewing mouthparts, whereas larval gall midges and nymphal aphids, coccoids, psylloids and thrips have piercing and sucking mouthparts. Larval gall midges have vestigial mouthparts and largely absorb nourishment by suction. Thus, these different insects mechanically damage and deliver chemicals (or perhaps genetic material, see below) to the plant cells in a variety of ways.

Little is known about what stimulates gall induction and growth. Wounding and plant hormones (such as cytokinins) appear important in indeterminate galls, but the stimuli are probably more complex for determinate galls. Oral secretions, anal excreta, and accessory gland secretions have been implicated in different insect–plant interactions that result in determinate galls. The best-studied compounds are the salivary secretions of Hemiptera. Salivary substances, including amino acids, auxins (and other plant growth regulators), phenolic compounds and phenol oxidases, in various concentrations, may have a role either in gall initiation and growth or in overcoming the defensive necrotic reactions of the plant. Plant hormones, such as auxins and cytokinins, must be involved in cecidogenesis but it is equivocal whether these hormones are produced by the insect, by the plant as a directed response to the insect, or are incidental to gall induction. In certain complex galls, such as those of eriococcoids and cynipids, it is conceivable that the development of the plant cells is redirected by semiautonomous genetic entities (viruses, plasmids or transposons) transferred from the insect to the plant. Thus, the initiation of such galls may involve the insect acting as a DNA or RNA donor, as in some wasps that parasitize insect hosts (see Box 13.1). Unfortunately, in comparison with anatomical and physiological studies of galls, genetic investigations of gall induction are almost non-existent.

The gall-inducing habit may have evolved either from plant mining and boring (especially likely for Lepidoptera, Hymenoptera and certain Diptera) or from sedentary surface feeding (as is likely for Hemiptera, Thysanoptera and cecidomyiid Diptera). It is believed to be beneficial to the insects, rather than a defensive response of the plant to insect attack. All gall insects derive their food from the tissues of the gall and also some shelter or protection from natural enemies and adverse conditions of temperature or moisture. The relative importance of these environmental factors to the origin of the galling habit is difficult to ascertain because current advantages of gall living may differ from those gained in the early stages of gall evolution. Clearly, most galls are "sinks" for plant assimilates – the nutritive cells that line the cavity of wasp and fly galls contain higher concentrations of sugars, protein and lipids than ungalled plant cells. Thus, one advantage of feeding on gall rather than normal plant tissue is the availability of high-quality food. Moreover, for sedentary surface feeders, such as aphids, psylloids and coccoids, galls furnish a more protected microenvironment than the normal plant surface. Some cecidozoa may "escape" from certain parasitoids and predators that are unable to penetrate galls, particularly galls with thick woody walls.

Other natural enemies, however, specialize in feeding on gall-living insects or their galls and sometimes it is difficult to determine which insects were the original inhabitants. Some galls are remarkable for the association of an extremely complex community of species, other than the gall causer, belonging to diverse insect groups. These other species may be either parasitoids of the gall inducer (i.e. parasites that cause the eventual death of their host; Chapter 13) or inquilines ("guests" of the gall inducer) that obtain their nourishment from tissues of the gall. In some cases, gall inquilines cause the original inhabitant to die through abnormal growth of the gall; this may obliterate the cavity in which the gall inducer lives or prevent emergence from the gall. If two species are obtained from a single gall or a single type of gall, one of these insects must be a parasitoid or an inquiline. There are even cases of hyperparasitism, in which the parasitoids themselves are subject to parasitization (section 13.3.1).

11.2.6 Seed predation

Plant seeds usually contain higher levels of nutrients than other tissues, providing for the growth of the seedling. Specialist seed-eating insects use this resource. Notable seed-eating insects are many beetles (described below), harvester ants (especially species of *Messor*, *Monomorium* and *Pheidole*) that store seeds in underground granaries, bugs (many Coreidae, Lygaeidae, Pentatomidae, Pyrrhocoridae and Scutelleridae) that suck out the contents of developing or mature seeds, and a few moths (such as some Gelechiidae and Oecophoridae).

Harvester ants are ecologically significant seed predators. These are the dominant ants in terms of biomass and/or colony numbers in deserts and dry grasslands in many parts of the world. Usually, the species are highly polymorphic, with the larger individuals possessing powerful mandibles capable of cracking open seeds. Seed fragments are fed to larvae, but probably many harvested seeds escape destruction, either by being abandoned in stores or by germinating quickly within the ant nests. Thus, seed harvesting by ants, which could be viewed as exclusively detrimental, actually may carry some benefits to the plant through dispersal and provision of local nutrients to the seedling.

An array of beetles (especially Curculionidae and bruchine Chrysomelidae) develop entirely within individual seeds or consume several seeds within one fruit. Some bruchine seed beetles, particularly those attacking leguminous food plants such as peas and beans, are serious pests. Species that eat dried seeds are pre-adapted to be pests of stored products such as pulses and grains. Adult beetles typically oviposit onto the developing ovary or the seeds or fruits, and some larvae then mine through the fruit and/or seed wall or coat. The larvae develop and pupate inside seeds, thus destroying them. Successful development usually occurs only in the final stages of maturity of seeds. Thus, there appears to be a "window of opportunity" for the larvae; a mature seed may have an impenetrable seed coat but if young seeds are attacked, the plant can abort the infected seed or even the whole fruit or pod if little investment has been made in it. Aborted seeds and those shed to the ground (whether mature or not) generally are less attractive to seed beetles than those retained on the plant, but evidently stored-product pests have no difficulty in developing within cast (i.e. harvested and stored) seeds. The larvae of the granary weevil, *Sitophilus granarius* (Taxobox 22), and rice weevil, *S. oryzae*, develop inside dry grains of corn, wheat, rice and other plants. Weevils of *Antliarhinus zamiae* attack the gametophytes of ovules on cones of their cycad hosts (genus *Encephalartos*,

Zamiaceae) (see **Plate 4b**). This is a very rare feeding specialization, probably because cycad gametophyte tissues are defended both chemically and by several layers of protective tissues. The female weevil uses her exceedingly long snout (rostrum) to drill a deep hole to access the gametophyte tissue and then turns around and inserts her ovipositor to lay eggs. The rostrum is believed to be the longest relative to body size of any beetle and can be up to 20 mm long.

Plant defence against seed predation includes the provision of protective seed coatings or toxic chemicals (allelochemicals), or both (as in the cycads mentioned above). Another strategy is the synchronous production by a single plant species of an abundance of seeds, often separated by long intervals of time. Seed predators either cannot synchronize their life cycle to the cycle of glut and scarcity, or are overwhelmed and unable to find and consume the total seed production.

11.2.7 Insects as biological control agents for weeds

Weeds are simply plants that are growing where they are not wanted. Some weed species are of little economic or ecological consequence, whereas the presence of others results in significant losses to agriculture or causes detrimental effects in natural ecosystems. Most plants are weeds only in areas outside their native distribution, where suitable climatic and edaphic conditions, usually in the absence of natural enemies, favour their growth and survival. Sometimes, exotic plants that have become weeds can be controlled by introducing host-specific phytophagous insects from the area of origin of the weed. This is called **classical biological control** of weeds and it is analogous to the classical biological control of insect pests (as described in detail in section 16.5). Another form of **biological control**, called **augmentation** (section 16.5), involves increasing the natural level of insect enemies of a weed and thus requires mass rearing of insects for inundative release. This method of controlling weeds is unlikely to be cost-effective for most insect–plant systems. The tissue damage caused by introduced or augmented insect enemies of weeds may limit or reduce vegetative growth (as shown for the weed discussed in Box 11.4), prevent or reduce reproduction, or make the weed less competitive than other plants in the environment.

A classical biological control programme involves a sequence of steps that include biological as well as socio-political considerations. Each programme is initiated with a review of available data (including taxonomic and distributional information) on the weed, its plant relatives, and any known natural enemies. This forms the basis for assessment of the nuisance status of the target weed and a strategy for collecting, rearing, and testing the utility of potential insect enemies. Regulatory authorities must then approve the proposal to attempt control of the weed. Next, foreign exploration and local surveys must determine the potential control agents attacking the weed, both in its native and introduced ranges. The weed's ecology, especially in relation to its natural enemies, must be studied in its native range. The host-specificity of potential control agents must be tested, either inside or outside the country of introduction and, in the former case, always in quarantine. The results of these tests will determine whether the regulatory authorities approve the importation of the agents for subsequent release or only for further testing, or refuse approval. If approved and the agent is imported, there is a period of rearing in quarantine to eliminate any imported diseases or parasitoids, prior to mass rearing in preparation for field release. Release is dependent on the quarantine procedures being approved by the regulatory authorities. After release, the establishment, spread and effect of the insects on the weed must be monitored. If weed control is attained at the initial release site(s), the spread of the insects is assisted by manual distribution to other sites.

In most countries, modern host-specificity testing of insect herbivores that are potential agents for controlling a weed is rigorous, and involves determining whether the target insect(s) will feed and reproduce on beneficial or native plants related to the weed, for example in the same genus or family. However, the reliability of host-range tests done in confinement (quarantine facilities) is an issue. Sometimes the target insect will feed on native plants under artificial conditions but not have any impact in the field. For example, recent retrospective testing of the feeding and oviposition activities of the two beetle species released in 1943 and 1965 in New Zealand for control of St John's wort, *Hypericum perforatum* (Clusiaceae) (see below), indicated that native *Hypericum* species were suitable hosts for the beetles. Nevertheless, there is no evidence for impact of these control agents on native plants in the wild. Thus, if the beetles were being considered for introduction today, current risk assessment methods would have concluded that these agents were risky introductions and, had release not

Box 11.4 Salvinia and phytophagous weevils

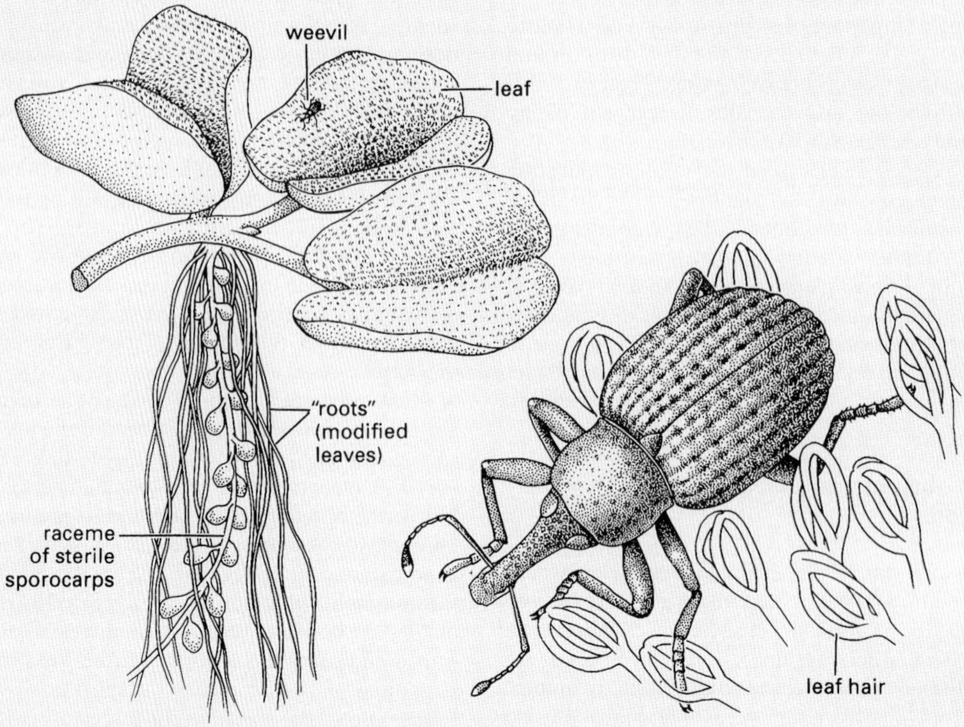

The floating aquatic fern salvinia (Salviniaceae: *Salvinia molesta*) (illustrated here, after Sainty & Jacobs 1981) has spread by human agency since 1939 to many tropical and subtropical lakes, rivers and canals throughout the world. Salvinia colonies consist of ramets (units of a clone) connected by horizontal branching rhizomes. Growth is favoured by warm, nitrogen-rich water. Conditions suitable for vegetative propagation and the absence of natural enemies in its non-native range have allowed very rapid colonization of large expanses of freshwater. Salvinia becomes a serious weed because its thick mats completely block waterways, choking the flow, and disrupting the livelihood of people who depend on them for transport, irrigation and food (especially fish, rice, sago palms, etc.). This problem was especially acute in parts of Africa, India, Southeast Asia and Australasia, including the Sepik River in Papua New Guinea. Expensive manual and mechanical removal and herbicides could achieve limited control but, by the early 1980s, some 2000 km^2 of water surface was covered by this invasive plant. The potential of biological control was recognized in the 1960s, although it was slow to be used (for reasons outlined below) until the 1980s, when outstanding successes were achieved in most areas where biological control was attempted. Choked lakes and rivers became open water again.

The phytophagous insect responsible for this spectacular control of *S. molesta* is a tiny (2 mm long) weevil (Coleoptera: Curculionidae) called *Cyrtobagous salviniae* (shown enlarged in the drawing on the right, after Calder & Sands 1985). Adult weevils feed on salvinia buds, whereas larvae tunnel through buds and rhizomes as well as feeding externally on roots. The weevils are host-specific, have a high searching efficiency for salvinia, and can live at high population densities without intraspecific interference stimulating emigration. These characteristics allow the weevils to control salvinia effectively.

Initially, biological control of salvinia failed because of unforeseen taxonomic problems with the weed and the weevil. Prior to 1972, the weed was thought to be *Salvinia auriculata*, which is a South American species fed upon by the weevil *Cyrtobagous singularis*. Even when the weed's correct identity was established as *Salvinia molesta*, it was not until 1978 that its native range was discovered to be southeast Brazil. Weevils feeding there on *S. molesta* were believed to be conspecific with *C. singularis* feeding on *S. auriculata*. However, after preliminary testing and subsequent success in controlling *S. molesta*, the weevil was recognized as specific to *S. molesta*, new to science and named as *C. salviniae*.

The benefits of control to people living in Africa, Asia, the Pacific and other warm regions are substantial, whether measured in economic terms or as savings in human health and social systems. For example, villages in Papua New Guinea that were abandoned because of salvinia have been reoccupied. Similarly, the environmental benefits of eliminating salvinia infestations are great, as this weed is capable of reducing a complex aquatic ecosystem to a virtual monoculture. Now, control by this weevil is benefiting aquatic systems in the United States, especially the southeast states where *S. molesta* was introduced in the 1990s through the aquarium and landscape trades.

The economics of salvinia control have been studied only in Sri Lanka, where a cost–benefit analysis showed returns on investment of 53 : 1 in terms of cash and 1678 : 1 in terms of hours of labour. Appropriately, the team responsible for the ecological research that led to biological control of salvinia was recognized by the award of the UNESCO Science Prize in 1985. Taxonomists made essential contributions by establishing the true identities of the salvinias and the weevils.

gone ahead, New Zealand would have foregone a significant biological control success.

There have been some outstandingly successful cases of deliberately introduced insects controlling invasive weeds. The control of St John's wort (also see above) is an excellent example. A century ago, this weed was reported first in northern California near the Klamath River. In its native range in Europe and adjacent regions, this non-invasive plant has provided herbal remedies for centuries, but is harmful to sheep, cattle and horses. In contrast, in North America, what became known as the "Klamath weed" spread rapidly, until by 1944 it was estimated to render 2 million acres of rangelands virtually valueless through unsuitability for cattle. The same weed had invaded Australia, where scientists had shown that phytophagous beetles imported from England and Europe potentially controlled the plant. Although a similar effort was proposed in the United States as early as 1929, it was not until the mid-1940s that permission was granted to release these species (from Australia, since war prevented import direct from England). Two leaf beetles, *Chrysolina quadrigemina* and *C. hyperici* (Coleoptera: Chrysomelidae), made it through quarantine, and from release sites in northern California spread rapidly, and with human intervention, onward throughout the rangelands of the western United States. Within a decade, these agents had reduced the massive infestation to a sporadic roadside sighting. Savings in the early years was estimated in the millions of dollars, and continue. Control with these two species persists in Australia (at sites suitable to the beetles) and in New Zealand, and has been attained in South Africa with just one of the *Chrysolina* species.

Similar spectacular success has been attained for control of the waterweed salvinia by the *Cyrtobagous* weevil in many countries (Box 11.4), and of prickly pear cacti, *Opuntia* species (Cactaceae), in Australia and, to a lesser extent, South Africa by the larvae of the cactus moth, *Cactoblastis cactorum* (Lepidoptera: Pyralidae). An Australian native paperbark tree, *Melaleuca quinquenervia* (Myrtaceae), is an aggressive weed that forms monospecific stands in the Everglades wetlands in Florida, USA, and in the Caribbean. The deliberate introduction and successful establishment of three Australian host-specific insects, the weevil *Oxyops vitiosa* (Coleoptera: Curculionidae), the psyllid *Boreioglycaspis melaleucae* (Hemiptera: Psyllidae) and the stem-galling midge *Lophodiplosis trifida* (Diptera: Cecidomyiidae), has led to greatly reduced flowering, seed production, and seedling growth and survival of *M. quinquenervia*. Reduced reproduction caused by insect herbivory, combined with herbicide treatments and mechanical harvesting of large paperbark stands, has led to at least a halving of the area of infestation in the Everglades.

On the whole, however, the chances of successful biological control of weeds by released phytophagous organisms are not high, and vary in different circumstances, often unpredictably. Furthermore, biological control systems that are highly successful and appropriate for weed control in one geographical region may be potentially disastrous in another region. For example, in Australia, which has no native cacti, *Cactoblastis* was used safely and effectively to almost completely destroy vast infestations of *Opuntia* cactus. However, in the 1950s this moth also was introduced into the Caribbean, and throughout the region there is now increased likelihood of extinction of native cactus species. Since 1989 it has been spreading in the southeast United States from an introduction or natural dispersal into Florida, reaching South Carolina in 2002 and moving west to Alabama (2004), Mississippi (2008) and Louisiana (2009). Despite attempts to limit the spread of this pest, it arrived in Isla Mujeres, Mexico, in 2006, but was eradicated. The moth potentially threatens many native species of *Opuntia* in the unique cacti-dominated ecosystems of southwest North America and especially Mexico. Although there is no current effective control method for *C. cactorum*, the use of the sterile insect technique (section 16.10) is a possibility.

In general, perennial weeds of uncultivated areas are well suited to classical biological control, because long-lived plants, which are predictable resources, are generally associated with host-specific insect enemies. Cultivation, however, can disrupt these insect populations. In contrast, augmentation of insect enemies of a weed may be best suited to annual weeds of cultivated land, where mass-reared insects could be released to control the plant early in its growing season. Some recent analyses have assessed the effectiveness of a number of classical biological control programmes against weedy plants and, although few provide substantial weed control, the successful programmes are highly beneficial. There is evidence that biological control agents do cause some reduction in plant size and mass, as well as flower and seed production, and thus target weed density, with weevils and chrysomelid beetles often more effective than other agents. The reasons for variation or failure in control of weeds are diverse. Modelling of weedy plant traits in relation to the quantitative impacts of the biological control programmes suggested that aquatic or wetland, asexual plants that are not weeds in their native range are most effectively controlled, whereas sexually reproducing terrestrial plants that are weedy in the native areas were poorly controlled. Thus, future prediction of the success or failure of control should be facilitated by knowledge of weed traits, as well as phytophage ecology and/or behaviour. However, some research suggests that plants that go weedy outside of their natural range can evolve over time to show reduced resistance (decreased chemical defences) and increased tolerance (e.g. by rapid growth and greater reproduction) to herbivory. As a consequence, even if deliberately introduced specialist insect herbivores reach high numbers, they may not exert control of the plants. There is a need for more long-term evaluations after release of biological control agents in order to rigorously assess the effectiveness of introduction programmes.

In addition to the uncertainty of success of classical biological control programmes, the control of certain weeds can cause potential conflicts of interest. Sometimes, not everyone may consider the target a weed. For example, in Australia, a dominant pasture weed, the introduced *Echium plantagineum* (Boraginaceae), is called "Paterson's curse" by those who consider it an agricultural weed and "Salvation Jane" by some pastoralists and beekeepers who regard it as a source of fodder for livestock and nectar for bees. A second type of conflict may arise if the natural phytophages of the weed are oligophagous rather than monophagous, and thus may feed on a few species other than the target weed. In this case, the introduction of insects that are not strictly host-specific may pose a risk for beneficial and/or native plants in the proposed area of introduction of the control agent(s). This is called a direct negative non-target impact of a biological control agent. For example, some of the five species of beetles and a moth introduced into Australia as potential control agents for *E. plantagineum* may feed on other native boraginaceous plants. The risks of damage to non-target species need to be assessed carefully prior to releasing foreign insects for the biological control of a weed. Some introduced phytophagous insects may become pests in their new habitat.

There is another environmental concern that applies even to very host-specific natural enemies of weeds. There may be non-target effects on native insects due to the phenomenon of apparent competition, which is competition due to shared natural enemies. Therefore, if an introduced biological control herbivore and a native insect herbivore in the same habitat are both fed on by a native predator, then an increase in abundance of the control agent is likely to cause an increase in abundance of the predator, which in turn may consume more of the native herbivore. A documented example of this is provided by the attempted biological control of the woody weed bitou, *Chrysanthemoides monilifera* ssp. *rotundata* (Asteraceae), in Australia using a highly host-specific seed herbivore of bitou, *Mesoclanis polana* (Diptera: Tephritidae). This fly is not very effective in reducing seed set of bitou, but it is host or prey to a number of natural enemies (parasitoids and a predator) that it shares with native seed herbivore species of native Australian plants. A food-web study showed that the abundance of the introduced *M. polana* was significantly negatively correlated with the abundance and species richness of a local native insect assemblage. There were local losses of up to 11 species of dipteran seed herbivores and their parasitoids associated with high abundance of *M. polana*. The lesson here is that biological control programmes need to give more attention to the network of species interactions that link biological agents to native species, because an agent that is attacked by local natural enemies could cause a cascade of effects through the food web and impact native species at several trophic levels. If an introduced biological control agent is ineffective in controlling its target weed (as in the case of *M. polana* and bitou), then the presence of that agent may exacerbate the already negative effects on native organisms due to the weed.

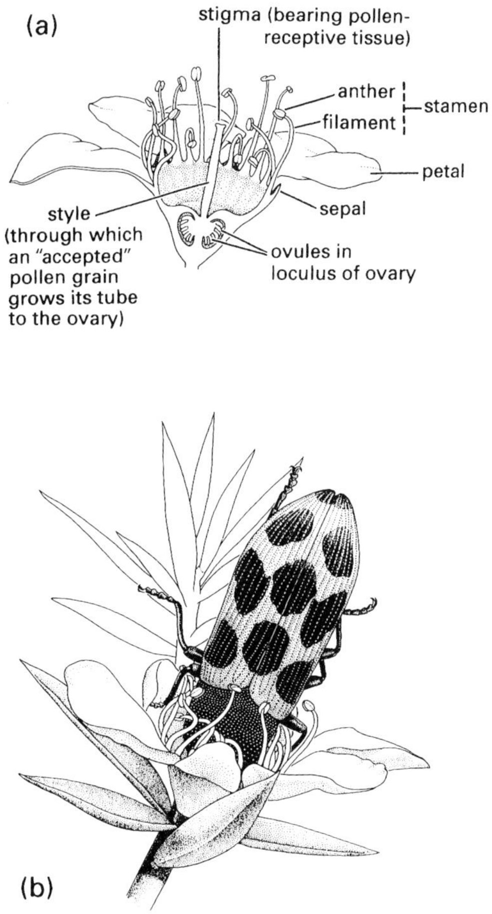

Fig. 11.6 Anatomy and pollination of a tea-tree flower, *Leptospermum* (Myrtaceae): (a) diagram of a flower showing the parts; (b) a jewel beetle, *Stigmodera* sp. (Coleoptera: Buprestidae), feeding from a flower.

11.3 INSECTS AND PLANT REPRODUCTIVE BIOLOGY

Insects are intimately associated with plants. Agriculturalists, horticulturalists and gardeners are aware of their role in damage and disease dispersal. However, certain insects are vitally important to many plants, assisting in their reproduction, through pollination, or their dispersal, through spreading their seeds.

11.3.1 Pollination

Sexual reproduction in plants involves **pollination** – the transfer of pollen (male germ cells in a protective covering) from the anthers of a flower to the stigma (Fig. 11.6a). A pollen tube grows from the stigma down the style to an ovule in the ovary where it fertilizes the egg. Pollen generally is transferred either by an animal pollinator or by the wind. Transfer may be from anthers to stigma of the same plant (either of the same flower or a different flower) (self-pollination), or between flowers on different plants (with different genotypes) of the same species (cross-pollination). Animals, especially insects, pollinate most flowering plants. It is argued that the success of the angiosperms relates to the development of these interactions. The benefits of insect pollination (**entomophily**) over wind pollination (**anemophily**) include:

- increase in pollination efficiency, including reduction of pollen wastage;
- successful pollination under conditions unsuitable for wind pollination;

• maximization of the number of plant species in a given area (as even rare plants can receive conspecific pollen carried into the area by insects).

Within-flower self-pollination also brings some of these advantages, but continued selfing induces deleterious homozygosity, and rarely is a dominant fertilization mechanism.

Generally, it is advantageous to a plant for its pollinators to be specialist visitors that faithfully pollinate only flowers of one or a few plant species. Pollinator constancy, which may initiate the isolation of small plant populations, is especially prevalent in the Orchidaceae – the most speciose family of vascular plants. For example, some Neotropical orchids are pollinated exclusively by male euglossine bees, with the flowers of a given orchid attracting male bees of only one or a few species out of many present in the habitat. The orchids do not produce nectar, but their sticky pollen unit (**pollinarium**) becomes attached to male bees that visit the orchid flower to collect fragrances that are used in the bee's own reproductive behaviour. Euglossine males also visit other plants, such as *Anthurium* (Araceae), to collect floral fragrances (see **Plate 4e**).

The major **anthophilous** (flower-frequenting) taxa among insects are the beetles (Coleoptera), flies (Diptera), wasps, bees and ants (Hymenoptera), thrips (Thysanoptera), and butterflies and moths (Lepidoptera). These insects visit flowers primarily to obtain nectar and/or pollen, but even some predatory insects may pollinate the flowers that they visit. Nectar primarily consists of a solution of sugars, especially glucose, fructose and sucrose. Pollen often has a high protein content plus sugar, starch, fat, and traces of vitamins and inorganic salts. In the case of a few bizarre interactions, male hymenopterans are attracted neither by pollen nor by nectar but by the resemblance of certain orchid flowers in shape, colour and odour to their conspecific females (see **Plate 5d,e**). In attempting to mate (**pseudocopulate**) with the insect-mimicking flower, the male inadvertently pollinates the orchid with pollen that adhered to his body during previous pseudocopulations. Pseudocopulatory pollination is common among Australian thynnine wasps (Tiphiidae), but also occurs in a few other wasp groups, some bees, and rarely in ants.

Cantharophily (beetle pollination) may be the oldest form of insect pollination. Beetle-pollinated flowers often are white or dull coloured, strong smelling, and regularly bowl- or dish-shaped (Fig. 11.6). Beetles mostly visit flowers for pollen, although nutritive tissue or easily accessible nectar may be utilized, and the plant's ovaries usually are well protected from the biting mouthparts of their pollinators. The major beetle families that commonly or exclusively contain anthophilous species are the Buprestidae (jewel beetles; Fig. 11.6b), Cantharidae (soldier beetles), Cerambycidae (longicorn or longhorn beetles), Cleridae (checkered beetles), Dermestidae, Lycidae (net-winged beetles), Melyridae (soft-winged flower beetles), Mordellidae (tumbling flower beetles), Nitidulidae (sap beetles) and Scarabaeidae (scarabs).

Myophily (fly pollination) occurs when flies visit flowers to obtain nectar (see **Plate 5f**), although hover flies (Syrphidae) feed chiefly on pollen rather than nectar. Fly-pollinated flowers tend to be less showy than other insect-pollinated flowers but may have a strong smell, often malodorous. Flies generally utilize many different sources of food and thus their pollinating activity is irregular and unreliable. However, their sheer abundance and the presence of some flies throughout the year mean that they are important pollinators for many plants. Both dipteran groups (Nematocera and Brachycera) contain anthophilous species. Among the Nematocera, mosquitoes and bibionids are frequent blossom visitors, and predatory midges, principally of *Forcipomyia* species (Ceratopogonidae), are essential pollinators of cocoa flowers. Pollinators are more numerous in the Brachycera, in which at least 30 families are known to contain anthophilous species. Major pollinator taxa are the Bombyliidae (bee flies), Syrphidae and muscoid families.

Most members of the Lepidoptera feed from flowers using a long, thin proboscis. In the speciose Ditrysia (the "higher" Lepidoptera), the proboscis is retractile (see Fig. 2.12 and **Plate 5g**), allowing feeding and drinking from sources distant from the head. Such a structural innovation may have contributed to the radiation of this successful group, which contains 98% of all lepidopteran species. Flowers pollinated by butterflies and moths often are regular, tubular and sweet smelling. **Phalaenophily** (moth pollination) typically is associated with light-coloured, pendant flowers that have nocturnal or crepuscular anthesis (opening of flowers); whereas **psychophily** (butterfly pollination) is typified by red, yellow or blue, upright flowers that have diurnal anthesis.

Many members of the large order Hymenoptera visit flowers for nectar and/or pollen. The Apocrita, which contains most of the wasps (as well as bees and ants), is more important than the Symphyta (sawflies) in terms of **sphecophily** (wasp pollination). Many pollinators are found in the superfamilies Ichneumonoidea and Vespoidea. Fig wasps (Chalcidoidea: Agaonidae) are highly specialized pollinators of the hundreds of species of figs (discussed in Box 11.1). Ants (Formicidae) are rather poor pollinators, although **myrmecophily** (ant pollination) is known for a few plant species. Ants are commonly anthophilous (flower loving), but rarely pollinate the plants that they visit. Two hypotheses, perhaps acting together, have been postulated to explain the paucity of ant pollination. First, ants are flightless, often small, and their bodies frequently are smooth, and thus they are unlikely to facilitate cross-pollination because the foraging of each worker is confined to one plant, they often avoid contact with the anthers and stigmas, and pollen does not adhere easily to them. Second, the metapleural glands of ants produce secretions that spread over the integument and inhibit fungi and bacteria, but also can affect pollen viability and germination. Some plants actually have evolved mechanisms to deter ants; however, a few, especially in hot dry habitats, appear to have evolved adaptations to ant pollination.

Generally, bees are regarded as the most important group of insect pollinators. They collect nectar and pollen for their brood as well as for their own consumption. There are almost 20,000 named species of bees worldwide and all are anthophilous. Plants that depend on **melittophily** (bee pollination) often have bright (yellow or blue), sweet-smelling flowers with nectar guides – lines (often visible only as ultraviolet light) on the petals that direct pollinators to the nectar. There is wide variation in the range of host plants visited, with most eusocial bees (such as honey bees and bumble bees) exhibiting **polylecty** (collecting pollen from the flowers of a variety of unrelated plants) and most other bees being **oligolectic** (with specialized pollen preferences, often the pollen of one plant genus). Very few bee species are **monolectic**, specializing in collecting pollen from just one plant species. However, even polylectic bees tend to visit just one kind of flower during a single foraging trip. Such flower constancy (or floral consistency) promotes plant pollination since an individual bee will fly between flowers of one plant species.

The main bee pollinator worldwide is the honey bee, *Apis mellifera* (Apidae) (see **Plate 5h**). The pollination services provided by this bee are extremely important for many crop plants (section 1.2), but serious problems can be caused in natural ecosystems that have been invaded by introduced honey bees. The bees compete with native insect and bird pollinators by depleting nectar and pollen supplies, and may disrupt pollination by displacing the specialist pollinators of native plant species. For example, in Australia, managed and feral honey bees have been shown to compete with nectar-feeding birds and to reduce seed set of some native plants, although honey-bee pollinating activities can enhance seed set of other native plants for which the natural pollinators have declined substantially due to a variety of causes. Thus, the situation is complex, and removing exotic honey bees from natural habitats can have both positive and negative effects on native species.

In the warmer areas of some countries, such as Australia and Brazil, hives of native stingless bees (Apidae: Meliponini) often are kept for honey or just as a hobby. Meliponines are harmless to people and domestic animals and resistant to the parasites and diseases of honey bees. Crops such as macadamia, mangoes, lychee, avocados, blueberry and strawberries all benefit from their pollination and their short flight ranges keep them within the crop.

In agroecosystems, the presence of both native and introduced honey bees is beneficial to crop production, and bee pollination can improve more than fruit yield in some crops. For example, recent research on strawberries in Europe has demonstrated that bee pollination produces fruit of superior quality and shelf life compared with fruits pollinated by wind or selfing. Wild native bees and honey bees are complementary in pollinating strawberries. Together, their activities result in berries that are redder, firmer and heavier, with reduced sugar–acid ratios and fewer deformities. These characteristics mean a higher market value and a longer shelf life of bee-pollinated strawberries compared with other pollination methods. The researchers calculated that bee pollination was worth almost half the value of the European Union's strawberry crop, valued at US$2.90 billion in 2009. However, maintenance of bee diversity in and near agricultural crops is challenging because the crop environment can be highly unfavourable for pollinators due to use of insecticides and often the planting of a single crop over large areas, with weedy annuals killed by herbicides, allowing no diversity of flower types or flowering times.

A global decline in populations of many important pollinators is causing concern in both natural and agricultural systems. In particular, losses of honey bees from managed hives has lead to recognition of a syndrome called colony collapse disorder (see Box 12.3), and increasing public awareness of the benefits of pollinator services, including from non-honey-bee pollinators. Long-term monitoring of native pollinator populations is required to establish baseline data, and to distinguish natural fluctuations (perhaps due to climate) from losses due to environmental changes such as habitat destruction, introductions of diseases and use of pesticides.

Insect–plant interactions associated with pollination are clearly mutualistic. The plant is fertilized by appropriate pollen, and the insect obtains food (or sometimes fragrances) produced by the plant, often specifically to attract the pollinator. It is clear that plants may experience strong selection as a result of insects. In contrast, in most pollination systems, evolution of the pollinators may have been little affected by the plants that they visited. For most insects, any particular plant is just another source of nectar or pollen, and even insects that appear to be faithful pollinators over a short observation period may utilize a range of plants in their lifetime. Nevertheless, symmetrical influences do occur in some insect–plant pollination systems, as evidenced by the specializations of each fig-wasp species to the fig species that it pollinates (Box 11.1), and by correlations between moth proboscis (tongue) lengths and flower depths for a range of orchids and some other plants. For example, the Malagasy star orchid, *Angraecum sesquipedale*, has floral spurs that may exceed 30 cm in length, and has a pollinator with a tongue length of more than 20 cm, a giant hawk moth, *Xanthopan morganii praedicta* (Sphingidae) (Fig. 11.7). Only this moth can reach the nectar at the apex of the floral spurs and, during the process of pushing its head into the flower, it pollinates the orchid. This is cited often as a spectacular example of a coevolved "long-tongued" pollinator, whose existence had been predicted by Charles Darwin and Alfred Russel Wallace, who knew of the long-spurred flower but not the hawk moth. However, the interpretation of this relationship as coevolution has been challenged with the suggestion that the long tongue evolved in the nectar-feeding moth to evade (by distance-keeping and feeding in hovering flight) ambushing predators (e.g. spiders) lurking in other less-specialized flowers

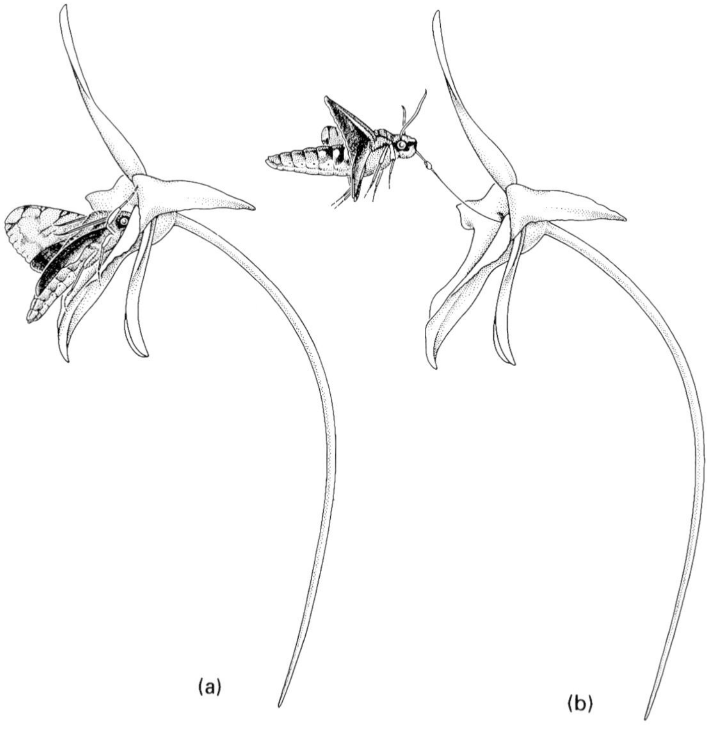

Fig. 11.7 A male hawk moth of *Xanthopan morganii praedicta* (Lepidoptera: Sphingidae) feeding from the long floral spur of a Malagasy star orchid, *Angraecum sesquipedale*: (a) full insertion of the moth's proboscis; (b) upward flight during withdrawal of the proboscis with the orchid pollinium attached. (After Wasserthal 1997.)

frequented by *X. morgani*. In this interpretation, pollination of *A. sesquipedale* follows a host-shift of the pre-adapted pollinator, with only the orchid showing adaptive evolution. The specificity of location of pollinia (pollen masses) on the tongue of *X. morgani* seems to argue against the pollinator-shift hypothesis, but detailed field study is required to resolve the controversy. Unfortunately, this rare Malagasy insect–plant system is threatened because its natural rainforest habitat is being destroyed.

11.3.2 Myrmecochory: seed dispersal by ants

Many ants are seed predators that harvest and eat seeds (section 11.2.6). Seed dispersal may occur when seeds are accidentally lost in transport or seed stores are abandoned. Some plants, however, have very hard seeds that are inedible to ants and yet many ant species actively collect and disperse them, a phenomenon called **myrmecochory**. These seeds have food bodies, called **elaiosomes**, with special chemical attractants that stimulate ants to collect them. Elaiosomes are seed appendages that vary in size, shape and colour, and contain nutritive lipids, proteins and carbohydrates in varying proportions. These structures have diverse derivations from various ovarian structures in different plant groups. The ants, each gripping the elaiosome with their mandibles (Fig. 11.8), carry the entire seed back to their nest, where the elaiosomes are removed and typically fed to the ant larvae. The hard seeds are then discarded, intact and viable, either in an abandoned gallery of the nest, or close to the nest entrance in a refuse pile.

Myrmecochory is a worldwide phenomenon, but is disproportionately prevalent in three plant

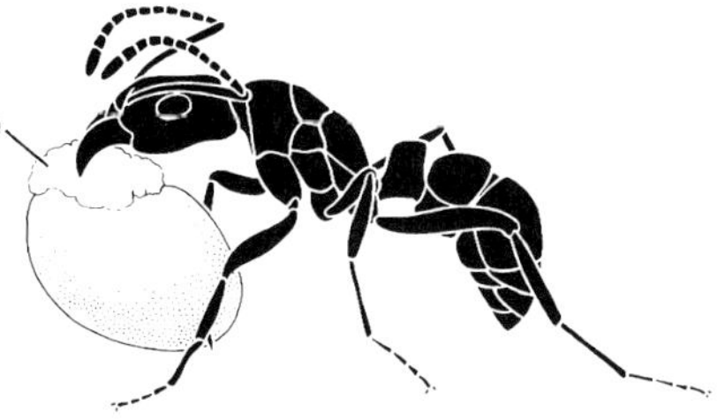

Fig. 11.8 An ant of *Rhytidoponera tasmaniensis* (Hymenoptera: Formicidae) carrying a seed of *Dillwynia juniperina* (Fabaceae) by its elaiosome (seed appendage).

assemblages: (i) early flowering herbs in the understorey of north temperate mesic forests; (ii) perennials in Australian and southern African sclerophyll vegetation; and (iii) an eclectic assemblage of tropical plants. Myrmecochorous plants number more than 11,000 species (at least 4.5% of all angiosperms) distributed amongst more than 80 plant families. They thus represent an ecological, rather than a phylogenetic, group, although they are predominantly legumes. Myrmecochory has evolved independently at least 100 times in angiosperms, and ant-dispersed lineages are mostly more species-rich than their non-myrmecochorous sister groups.

This association is of obvious benefit to the ants, for which the elaiosomes represent food; and the mere existence of the elaiosomes is evidence that the plants have become adapted for interactions with ants. Myrmecochory may reduce intraspecific and/or interspecific competition amongst plants by removing seeds to new sites. Seed removal to underground ant nests provides protection from fire or seed predators, such as some birds and small mammals, and other seed-eating insects. Following fires, South African fynbos (plant) community structure varies according to the presence of different seed dispersing ants (see Box 1.3). Furthermore, ant nests are rich in plant nutrients, making them better microsites for seed germination and seedling establishment. Myrmecochory may be beneficial to plants growing in low-nutrient soils. However, no universal explanation for myrmecochory should be expected, as the relative importance of factors responsible for myrmecochory must vary according to plant species and geographical location.

Myrmecochory can be called a mutualism, but specificity and reciprocity do not characterize the association. There is no evidence that any myrmecochorous plant relies on a single ant species to collect its seeds. Similarly, there is no evidence that any ant species has adapted to collect the seeds of one particular myrmecochorous species. Of course, ants that harvest elaiosome-bearing seeds could be called a guild, and the myrmecochorous plants of similar form and habitat also could represent a guild. However, it is highly unlikely that myrmecochory represents an outcome of diffuse or guild coevolution, as no reciprocity can be inferred. Elaiosomes are just food items to ants, which display no obvious adaptations to myrmecochory. Thus, this fascinating form of seed dispersal appears to be the result of plant evolution, as a result of selection from ants in general, and not of coevolution of plants and specific ants.

11.4 INSECTS THAT LIVE MUTUALISTICALLY IN SPECIALIZED PLANT STRUCTURES

A great many insects live within plant structures, in bored-out stems, leaf mines or galls, but these insects create their own living spaces by destruction or physiological manipulation. In contrast, some plants have specialized structures or chambers that house mutualistic insects and form in the absence of these guests. Two types of these specialized insect–plant interactions are discussed below.

11.4.1 Ant–plant interactions involving domatia

Domatia (little houses) may be hollow stems, tubers, swollen petioles or thorns, which are used by ants either for feeding or as nest sites, or both. True domatia are cavities that form independently of ants, such as in plants grown in glasshouses from which ants are excluded. It may be difficult to recognize true domatia in the field because ants often take advantage of natural hollows and crevices such as tunnels bored by beetle or moth larvae. Plants with true domatia, called ant plants or **myrmecophytes**, often are trees, shrubs or vines of the secondary regrowth or understorey of tropical lowland rainforest.

Ants benefit from association with myrmecophytes through provision of shelter for their nests and readily available food resources. Food comes either directly from the plant through food bodies or extrafloral nectaries (Fig. 11.9a), or indirectly via honeydew-producing hemipterans living within the domatia (Fig. 11.9b). Food bodies are small nutritive nodules on the foliage or stems of ant plants. Extrafloral nectaries (EFNs) are glands that produce sugary secretions (usually also containing amino acids) attractive to ants and other insects. Plants with EFNs often occur in temperate areas and lack domatia, for example many Australian *Acacia* species, whereas plants with food bodies nearly always have domatia, and some plants have both EFNs and food bodies. Many myrmecophytes, however, lack both of the latter structures and instead the ants "farm" soft scales or mealybugs (Coccoidea: Coccidae or Pseudococcidae) for their honeydew (sugary excreta derived from phloem on which they feed) and possibly cull them to obtain protein. Like EFNs and food bodies, coccoids can draw the ants into a closer relationship with the plant by providing a resource on that plant.

Obviously, myrmecophytes receive some benefits from ant occupancy of their domatia. The ants may provide protection from herbivores and plant competitors or supply nutrients to their host plant. Some ants aggressively defend their plant against grazing mammals, remove herbivorous insects, and prune or detach other plants, such as epiphytes and vines that grow on their host. This extremely aggressive behaviour is demonstrated by ants of *Pseudomyrmex* that protect *Acacia* in tropical America. Rather than protection, some myrmecophytes derive mineral nutrients and nitrogen from ant-colony waste via absorption through the inner surfaces of the domatia. Such plant "feeding" by ants, called **myrmecotrophy**, can be documented by following the fate of a radioactive label placed in ant prey. Prey is taken into the domatia, eaten, and the remains are discarded in refuse tunnels; the label ends up in the leaves of the plant. Myrmecotrophy occurs in the epiphytic *Myrmecodia* (Rubiaceae) (Fig. 11.10), species of which occur from Southeast Asia to Papua New Guinea and northern Australia.

The majority of ant–plant associations may be opportunistic and unspecialized, although some tropical and subtropical ants (e.g. some *Pseudomyrmex* and *Azteca* species) are totally dependent on their particular host plants (e.g. *Acacia* or *Triplaris* and *Cecropia* species, respectively) for food and shelter. Likewise, if deprived of their attendant ants, these myrmecophytes decline. These relationships clearly are obligatorily mutualistic and, no doubt, others remain to be documented.

11.4.2 Phytotelmata: plant-held water containers

Many plants support insect communities in structures that retain water. The containers formed by water retained in leaf axils of many bromeliads ("tank-plants"), gingers and teasels, for example, or in rot-holes of trees, appear incidental to the plants. Others, namely the pitcher plants, have a complex architecture, designed to lure and trap insects, which are digested in the container liquid (Fig. 11.11).

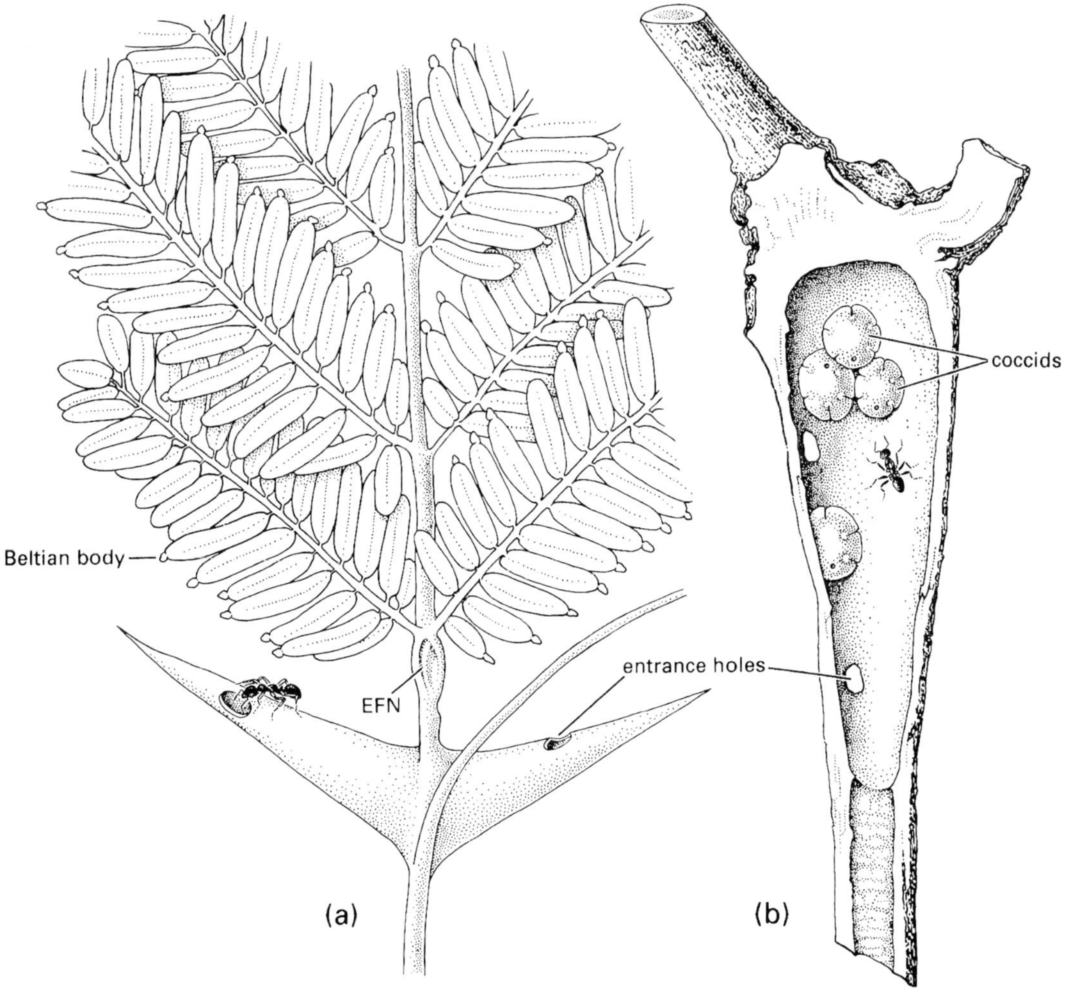

Fig. 11.9 Two myrmecophytes showing the domatia (hollow chambers) that house ants, and the food resources available to the ants. (a) A Neotropical bull's-horn acacia, *Acacia sphaerocephala* (Fabaceae), with hollow thorns, food bodies and extrafloral nectaries (EFNs), which are used by the resident *Pseudomyrmex* ants. (b) A hollow swollen internode of *Kibara* (Monimiaceae) with scale insects of *Myzolecanium kibarae* (Hemiptera: Coccidae), which excrete honeydew that is eaten by the resident ants, *Anonychomyrma scrutator*. ((a) After Wheeler 1910; (b) after Beccari 1877.)

The pitcher plants are a convergent grouping of the American Sarraceniaceae, Old World Nepenthaceae, and Australian endemic Cephalotaceae. They generally live in nutrient-poor soils. Odour, colour and nectar entice insects, predominantly ants, into modified leaves – the "pitchers". Guard hairs and slippery walls prevent exit, and thus the prey cannot escape and it drowns in the pitcher liquid, which contains digestive enzymes secreted by the plant.

This apparently inhospitable environment provides the home for a few specialist insects that live above the fluid, and many more living as larvae within. The adults of these insects can move in and out of the pitchers with impunity. Mosquito and midge larvae are the most common inhabitants, but other fly larvae of more than 12 families have been reported worldwide, and odonates, spiders, and even a stem-mining ant occur in Southeast Asian pitchers. Many of these insect

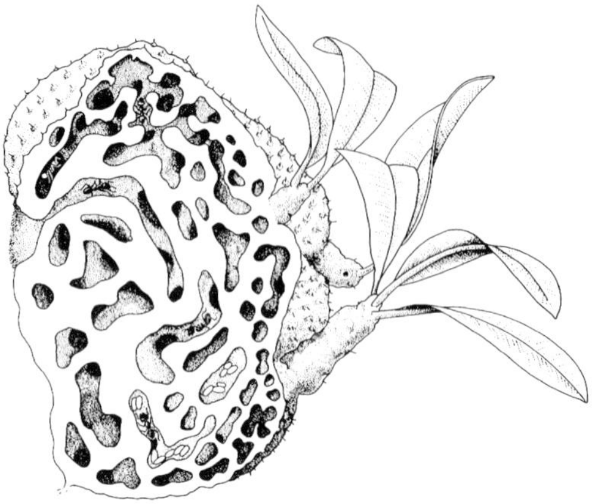

Fig. 11.10 A tuber of the epiphytic myrmecophyte *Myrmecodia beccarii* (Rubiaceae), cut open to show the chambers inhabited by ants. Ants live in smooth-walled chambers and deposit their refuse in warted tunnels, from which nutrients are absorbed by the plant. (After Monteith 1990.)

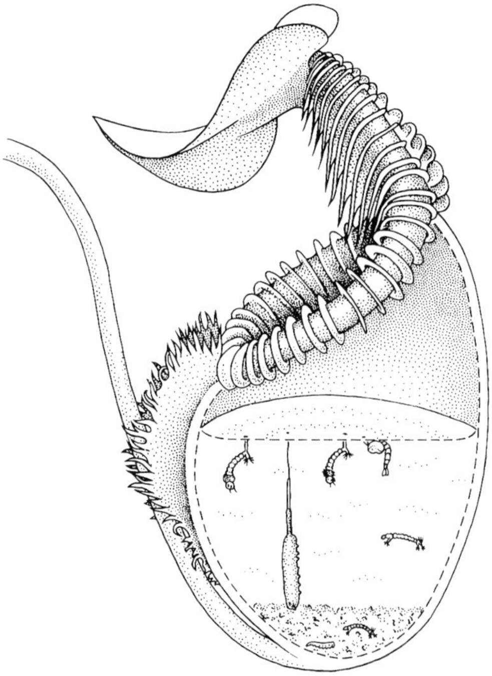

Fig. 11.11 A pitcher of *Nepenthes* (Nepenthaceae) cut open to show fly (Diptera) inquilines in the fluid: (clockwise from the top left) two mosquito larvae, a mosquito pupa, two chironomid midge larvae, a small maggot and a large rat-tailed maggot.

inquilines live in a mutualistic relationship with the plant, digesting trapped prey and microorganisms, and defaecating nutrients in a form readily available to the plant. Another unusual pitcher plant associate is a *Camponotus* ant that nests in the hollow tendrils of the pitcher plant *Nepenthes bicalcarata* in Borneo. The ants feed on large, trapped prey or mosquito larvae, which they haul from the pitchers, and thus benefit the plant by preventing the accumulation of excess prey, which can lead to putrefaction of pitcher contents.

FURTHER READING

Carvalheiro, L.G., Buckley, Y.M., Ventim, R. *et al.* (2008) Apparent competition can compromise the safety of highly specific biocontrol agents. *Ecology Letters* **11**, 690–700.

Center, T.D., Purcell, M.F., Pratt, P.D., *et al.* (2012) Biological control of *Melaleuca quinquenervia*: an Everglades invader. *BioControl* **57**, 151–65.

Clewley, G.D., Eschen, R., Shaw, R.H. & Wright, D.J. (2012) The effectiveness of classical biological control of invasive plants. *Journal of Applied Ecology* **49**, 1287–95.

Cruaud, A., Rønsted, N., Chantarasuwan, B. *et al.* (2012) An extreme case of plant-insect codiversification: figs and fig-pollinating wasps. *Systematic Biology* **61**, 1029–47.

Forneck, A. & Huber, L. (2009) (A)sexual reproduction – a review of life cycles of grape phylloxera, *Daktulosphaira vitifoliae*. *Entomologia Experimentalis et Applicata* **131**, 1–10.

Hare, J.D. (2011) Ecological role of volatiles produced by plants in response to damage by herbivorous insects. *Annual Review of Entomology* **56**, 161–80.

Horak, M., Day, M.F., Barlow, C., Edwards, E.D. *et al.* (2012) Systematics and biology of the iconic Australian scribbly gum moths *Ogmograptis* Meyrick (Lepidoptera: Bucculatricidae) and their unique insect-plant interaction. *Invertebrate Systematics* **26**, 357–98.

Huxley, C.R. & Cutler, D.F. (eds.) (1991) *Ant–Plant Interactions*. Oxford University Press, Oxford.

Julien, M., McFadyen, R. & Cullen, J. (eds.) (2012) *Biological Control of Weeds in Australia*. CSIRO Publishing, Collingwood.

Kaplan, I., Halitschke, R., Kessler, A. *et al.* (2008) Constitutive and induced defenses to herbivory in above- and belowground plant tissues. *Ecology* **89**, 392–406.

Klatt, B.K., Holzschuh, A., Westphal, C. *et al.* (2013) Bee pollination improves crop quality, shelf life and commercial value. *Proceedings of the Royal Society of London B* **281**, 20132440. doi: org/10.1098/rspb.2013.2440.

Landsberg, J. & Ohmart, C. (1989) Levels of insect defoliation in forests: patterns and concepts. *Trends in Ecology and Evolution* **4**, 96–100.

Lengyel, S., Gove, A.D., Latimer, A.M. *et al.* (2010) Convergent evolution of seed dispersal by ants, and phylogeny and biogeography in flowering plants: a global survey. *Perspectives in Plant Ecology, Evolution and Systematics* **12**, 43–55.

Nilsson, L.A. (1998) Deep flowers for long tongues. *Trends in Ecology and Evolution* **13**, 259–60.

Ode, P.J. (2006) Plant chemistry and natural enemy fitness: effects on herbivore and natural enemy interactions. *Annual Review of Entomology* **51**, 163–85.

Patiny, S. (ed.) (2012) *Evolution of Plant–Pollinator Relationships*. Cambridge University Press, Cambridge.

Price, P.W. (2003) *Macroevolutionary Theory on Macroecological Patterns*. Cambridge University Press, Cambridge.

Price, P.W., Denno, R.F., Eubanks, M.D. *et al.* (2011) *Insect Ecology: Behavior*, Populations and Communities. Cambridge University Press, Cambridge.

Raman, A., Schaefer, C.W. & Withers, T.M. (eds.) (2004) *Biology, Ecology, and Evolution of Gall-Inducing Arthropods*. Science Publishers, Enfield, NH.

Resh, V.H. & Cardé, R.T. (eds.) (2009) *Encyclopedia of Insects*, 2nd edn. Elsevier, San Diego, CA. [Particularly see articles on phytophagous insects; plant–insect interactions; pollination and pollinators; Sternorrhyncha.]

Room, P.M. (1990) Ecology of a simple plant–herbivore system: biological control of *Salvinia*. *Trends in Ecology and Evolution* **5**, 74–9.

Schaller, A. (ed.) (2008) *Induced Plant Resistance to Herbivory*. Springer Science+Business Media B.V.

Schoonhoven, L.M., Van Loon, J.J.A. & Dicke, M. (eds.) (2005) *Insect–Plant Biology*. Oxford University Press, Oxford.

Sharma, A., Khan, A.N., Subrahmanyam, S. *et al.* (2014) Salivary proteins of plant-feeding hemipteroids – implication in phytophagy. *Bulletin of Entomological Research* **104**, 117–136.

Six, D.L. & Wingfield, M.J. (2011) The role of phytopathogenicity in bark beetle–fungus symbioses: a challenge to the classical paradigm. *Annual Review of Entomology* **56**, 255–72.

Speight, M.R., Hunter, M.D. & Watt, A.D. (2008) *Ecology of Insects. Concepts and Applications*, 2nd edn. Wiley-Blackwell, Oxford.

Thompson, J.N. (1994) *The Coevolutionary Process*. University of Chicago Press, Chiacgo, IL.

Van Driesche, R.G., Carruthers, R.I., Center, T. *et al.* (2010) Classical biological control for the protection of natural ecosystems. *Biological Control Supplement* **1**, S2–S33. doi: 10.1016/j.biocontrol.2010.03.003.

Walters, D.R. (2011) *Plant Defense: Warding off Attack by Pathogens, Herbivores, and Parasitic Plants*. Wiley-Blackwell, Chichester.

Wasserthal, L.T. (1997) The pollinators of the Malagasy star orchids *Angraecum sesquipedale, A. sororium* and *A. compactum* and the evolution of extremely long spurs by pollinator shift. *Botanica Acta* **110**, 343–59.

White, T.C.R. (1993) *The Inadequate Environment. Nitrogen and the Abundance of Animals*. Springer-Verlag, Berlin.

Willmer, P. (2011) *Pollination and Floral Ecology*. Princeton University Press, Princeton, NJ.

Chapter 12

INSECT SOCIETIES

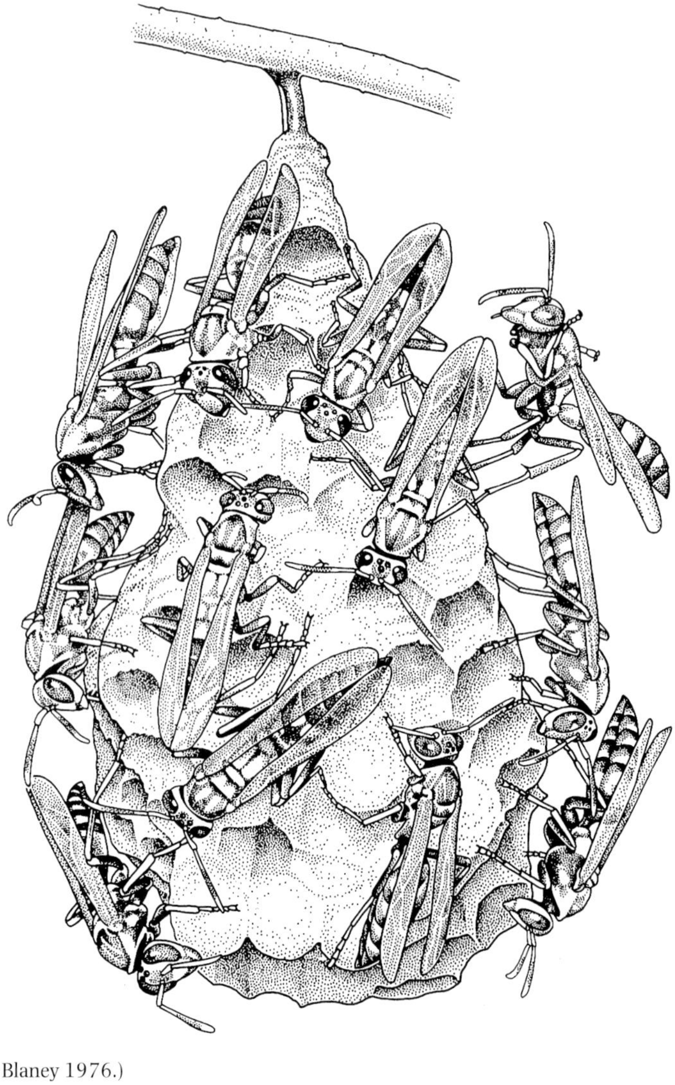

Vespid wasp nest. (After Blaney 1976.)

The study of insect social behaviours is a popular entomological topic and there is a voluminous literature, ranging from the popular to the highly theoretical. The proliferation of some insects, notably the ants and termites, is attributed to the major change from a solitary lifestyle to a social one.

Social insects are ecologically successful and are important to us. Ecologically dominant "tramp" ants threaten our agriculture and outdoor behaviour, and affect biodiversity (see Box 1.3). Leaf-cutter ants (*Atta* spp.) are the major herbivores in the Neotropics, and in southwest United States deserts, harvester ants take as many seeds as do mammals. Termites turn over at least as much soil as do earthworms in many tropical regions. The numbers of social insects can be astonishing: a Japanese supercolony of the ant *Formica yessensis* was estimated at 306 million workers and over 1 million queens, dispersed over 2.7 km^2 amongst 45,000 interconnected nests. In West African savannah, densities of up to 20 million resident ants per hectare have been estimated, and single nomadic colonies of driver ants (*Dorylus* sp.) may attain 20 million workers. The value of honey bees in pollination of crops and honey production is estimated in the tens of billions of US dollars per annum globally. Social insects clearly affect our lives.

A broad definition of social behaviour could include all insects that interact in any way with other members of their species. However, entomologists limit **sociality** to a more restricted range of co-operative behaviours. Amongst the social insects, we can recognize **eusocial** ("true social") insects, which co-operate in reproduction and have division of reproductive effort, and **subsocial** ("below social") insects, which have less strongly developed social habits, falling short of extensive co-operation and reproductive partitioning. **Solitary** insects exhibit no social behaviours.

Eusociality is defined by three traits:

1 Division of labour, with a **caste** system involving sterile or non-reproductive individuals assisting those that reproduce.

2 Co-operation among colony members in tending the young.

3 Overlap of generations capable of contributing to colony functioning.

Eusociality occurs in all ants and termites and some bees and wasps, such as the vespine paper wasps depicted in the vignette at the beginning of this chapter. Subsociality is a more widespread phenomenon, known to have arisen independently in 13 orders of insects, including some cockroaches, embiopterans, thysanopterans, hemipterans, beetles and hymenopterans. As insect lifestyles become better known, forms of subsociality may be found in yet more orders. The term "presociality" often is used for social behaviours that do not fulfil the strict definition of eusociality. However, the implication that presociality is an evolutionary precursor to eusociality is not always correct, and the term is best avoided.

In this chapter we discuss subsociality and then detail eusociality in bees, wasps, ants, termites and ambrosia beetles. We conclude with some ideas concerning the origins and success of eusociality. In boxes, we introduce the dance language of bees (Box 12.1), Africanized honey bees (Box 12.2), "colony collapse disorder" (Box 12.3), and discuss social insects in urban situations (Box 12.4).

12.1 SUBSOCIALITY IN INSECTS

12.1.1 Aggregation

Non-reproductive aggregations of insects, such as the gregarious overwintering of monarch butterflies at specific sites in Mexico and California, are social interactions. Many tropical butterflies form roosting aggregations, particularly in **aposematic** species (distasteful and with warning signals, including colour and/or odour). Aposematic phytophagous insects often form conspicuous feeding aggregations, sometimes using pheromones to lure conspecific individuals to a favourable site (section 4.3.2). A solitary aposematic insect runs a greater risk of being encountered by a naïve predator (and being eaten by it) than if it is a member of a conspicuous group. Belonging to a conspicuous social grouping, either of the same or several species, provides benefits by the sharing of protective warning colouration and the education of local predators.

The Insects: An Outline of Entomology, Fifth Edition. P.J. Gullan and P.S. Cranston.

Companion Website: www.wiley.com/go/gullan/insects

12.1.2 Parental care as a social behaviour

Parental care may be considered to be a social behaviour; although few insects, if any, show a complete lack of parental care – eggs are not deposited randomly. Females select an appropriate oviposition site, affording protection to the eggs and ensuring an appropriate food resource for the hatching offspring. The ovipositing female may protect the eggs in an ootheca, or deposit them directly into suitable substrate with her ovipositor, or modify the environment, as in nest construction. Parental care conventionally is seen as post-oviposition and/or post-hatching attention, including the provision and protection of food resources for the young. A convenient basis for discussing parental care is to distinguish between care with and without nest construction.

Parental care without nesting

For most insects, the highest mortality occurs in the egg and first instar, and many insects tend these stages until the more-mature larvae or nymphs can better fend for themselves. The orders of insects in which tending of eggs and young is most frequent are the Blattodea, Orthoptera and Dermaptera (orthopteroid orders), Embioptera, some Psocodea (only some bark lice), Thysanoptera, Hemiptera, Coleoptera and Hymenoptera. There has been a tendency to assume that subsociality is a precursor of termite eusociality, as the eusocial termites are related to cockroaches. The phylogenetic position (see Fig. 7.4) and social behaviour, including parental care, of the subsocial cockroach family Cryptocercidae gives insights into the origin of sociality, discussed in more detail in section 12.4.2.

Egg and early-instar attendance is predominantly a female role; yet paternal guarding is known in some Hemiptera, notably amongst some tropical assassin bugs (Reduviidae) and giant water bugs (Belostomatidae). After each copulation, the female belostomatid oviposits small batches of eggs onto the dorsum of the male. The eggs are tended in various ways by the male (see Box 5.8) and die if neglected. There is no tending of belostomatid nymphs, but in some other hemipterans the female (or in some reduviids, the male) may guard at least the early-instar nymphs. In these species, experimental removal of the tending adult increases losses of eggs and nymphs as a result of parasitization and/or predation. Other functions of parental care include keeping the eggs free from fungi, maintaining appropriate conditions for egg development, herding the young, and sometimes actually feeding them.

Unusually, certain treehoppers (Hemiptera: Membracidae) have "delegated" parental care of their young to ants. Ants obtain honeydew from treehoppers, which are protected from their natural enemies by the presence of the ants. In the presence of protective ants, brooding females prematurely may cease to tend a first brood and raise a second one. Another species of membracid will abandon its eggs in the absence of ants and seek a larger treehopper aggregation, where ants are in attendance, before laying another batch of eggs.

Many wood-mining beetles show advanced subsocial care that verges on the nesting described in the following section and on eusociality. For instance, all Passalidae (Coleoptera) live in communities of larvae and adults, with the adults chewing dead wood to form a substrate for the larvae to feed upon. Some ambrosia beetles (Curculionidae: Platypodinae) prepare galleries for their offspring (section 9.2), where the larvae feed on cultivated fungus and are defended by a male that guards the tunnel entrance (see also section 12.2.5).

Parental care with solitary nesting

Nesting is a social behaviour in which eggs are laid in a pre-existing or newly constructed structure to which the parent(s) bring food supplies for the young. Nesting, as thus defined, is seen in only five insect orders. Nest builders amongst the subsocial Orthoptera, Dermaptera, Coleoptera and Hymenoptera are discussed below; the nests of eusocial Hymenoptera and the prodigious mounds of the eusocial termites are discussed later in this chapter.

Earwigs of both sexes overwinter in a nest. In spring, the male is ejected when the mother starts to tend the eggs (see Fig. 9.1). In some species, mother earwigs forage and provide food for the young nymphs. Mole crickets and other ground-nesting crickets exhibit somewhat similar behaviour. A greater range of nesting behaviours is seen in the beetles, particularly in the dung beetles (Scarabaeidae) and carrion beetles (Silphidae). For these insects, the attractiveness of the short-lived, scattered, but nutrient-rich dung (and carrion) food resource induces competition. Upon location of a fresh source, dung beetles bury it to prevent drying out or being ousted by a competitor (section 9.3; Fig. 9.6). Some scarabs roll the dung away from its source; others coat the dung with clay. Both sexes co-operate, but the female is mostly responsible for burrowing and preparation of the larval food source. Eggs are laid on the buried dung and in some species no further interest is taken. However, parental care is well developed in others, commonly with maternal attention to fungus reduction, and removal or exclusion of conspecifics and ants by paternal defence.

Amongst the Hymenoptera, subsocial nesting is restricted to some **aculeate** Apocrita, especially within Vespoidea and Apoidea (Fig. 12.2); these wasps and bees are the most prolific and diverse nest builders amongst the insects. Excepting bees, nearly all these insects are parasitoids, in which adults attack and immobilize arthropod prey upon which the young feed. Wasps demonstrate a series of increasingly complex prey handling and nesting strategies, from using the prey's own burrow (e.g. many Pompilidae), to building a simple burrow following prey capture (a few Sphecidae), to construction of a nest burrow before prey-capture (most Sphecidae). In bees and masarine wasps, pollen replaces arthropod prey as the food source that is collected and stored for the larvae. Nest complexity in the aculeates ranges from a single burrow provisioned with one food item for one developing egg, to linearly or radially arranged multicellular nests. The primitive nest site was probably a pre-existing burrow, with the construction medium later being soil or sand. Further specializations involved the use of plant material – stems, rotten wood, and even solid wood by carpenter bees (Xylocopini) – and free-standing constructions of chewed vegetation (Megachilidae) and mud (Eumenidae). A range of natural materials are used in making and sealing cells, including mud and plants, and saps, resins and oils secreted by plants as rewards for pollination, and even the wax adorning soft scale insects. An extraordinary cellophane-like substance produced by Dufour's gland is used by Colletidae to provide a tough, durable, waterproof inner cell lining. In some subsocial nesters such as mason wasps (Eumeninae), many individuals of one species may aggregate, building their nests close together.

Parental care with communal nesting

Communal nesting may occur when conditions for nest construction are scarce and scattered throughout the environment. Even under apparently favourable conditions, many subsocial and all eusocial hymenopterans share nests. Communal nesting may arise if daughters nest in their natal nest, enhancing utilization of nesting resources and encouraging mutual defence against parasites. However, communal nesting in subsocial species allows "antisocial" or selfish behaviour, with frequent theft or takeover of nest and prey, so that extended time defending the nest against others of the same species may be required. Furthermore, the same cues that lead the wasps and bees to communal nesting sites easily can direct specialized nest parasites to the location. Examples of communal nesting in subsocial species are known or presumed to be the case in the Sphecidae, and in bees among Halictidae, Megachilidae and Andrenidae.

After oviposition, female bees and wasps remain in their nests, often until the next generation emerges as adults. Usually they guard, but also may remove faeces and generally maintain nest hygiene. The supply of provisions to the nest may be through mass provisioning, as in many communal sphecids and subsocial bees, or replenishment, as seen in the various vespid wasps that return with new prey as their larvae develop.

Subsocial aphids and thrips

Certain aphids belonging to the subfamilies Pemphiginae and Hormaphidinae (Hemiptera: Aphididae) have a sacrificial sterile soldier caste, consisting of some first- or second-instar nymphs that exhibit aggressive behaviour and never develop into adults. Soldiers are pseudoscorpion-like, as a result of body sclerotization and enlarged anterior legs, and will attack intruders using either their frontal horns (anterior cuticular projections) (Fig. 12.1) or feeding stylets (mouthparts) as piercing weapons. These modified individuals may defend good feeding sites against competitors or defend their colony against predators. As the offspring are produced by parthenogenesis, soldiers and normal nymphs from the same mother aphid should be genetically identical, favouring the evolution of these non-reproductive and apparently altruistic soldiers (as a result of increased inclusive fitness via kin selection; section 12.4). A similar phenomenon occurs in other related aphid species, but in this case all nymphs become temporary soldiers, which later moult into normal, non-aggressive individuals that reproduce. These unusual aphid polymorphisms have led some researchers to claim that the Hemiptera is a third insect order displaying eusociality. Although these few aphid species clearly have a reproductive division of labour, they do not appear to fulfil the other attributes of eusocial insects, as overlap of generations capable of

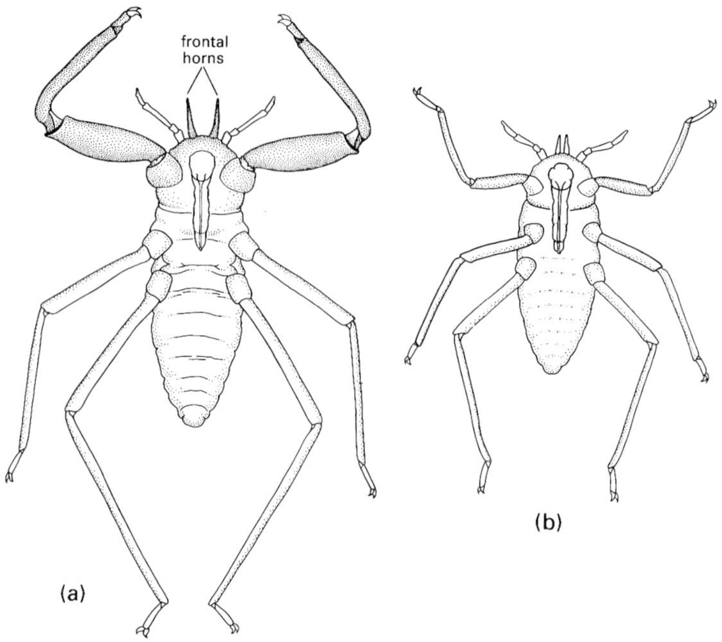

Fig. 12.1 First-instar nymphs of the subsocial aphid *Pseudoregma alexanderi* (Hemiptera: Hormaphidinae): (a) pseudoscorpion-like soldier; (b) normal nymph. (After Miyazaki 1987b.)

contributing to colony labour is equivocal and tending of offspring does not occur. Here we consider these aphids to exhibit subsocial behaviour.

A range of subsocial behaviours is seen in a few species of several genera of thrips (Thysanoptera: Phlaeothripidae). At least in the gall thrips, the level of sociality appears to be similar to that of the aphids discussed above. Thrips sociality is well developed in a bark-dwelling species of *Anactinothrips* from Panama, in which thrips live communally, co-operate in brood care, and forage with their young in a highly co-ordinated fashion. However, this species has no obvious non-reproductive females, and all adults may disappear before the young are full-grown. Evolution of subsocial behaviours in *Anactinothrips* may bring advantages to the young in group foraging, as feeding sites, although stable over time, are patchy and difficult to locate. In several species of Australian gall thrips, females show polymorphic wing reduction associated in some species with very enlarged fore legs. This "soldier" morph is more frequent amongst the first young to develop, which are differentially involved in defending the gall against intrusion by other species of thrips, and appear to be incapable of dispersing or of inducing galls. As in most thrips, sex determination is via haplodiploidy, with galls founded by a single female producing polymorphic offspring, and with establishment of multiple generations. Self-sacrificing defence by some individuals is favoured by demonstrated high relatedness of the offspring (altruism; section 12.4). Generational overlap is modest, and soldiers defend their siblings and their offspring rather than their mother (who has died). Soldiers reproduce, but at a much lower rate than the foundress. Such examples are valuable in showing the circumstances under which co-operation might have evolved.

Quasisociality and semisociality

Division of reproductive labour is restricted to the subsocial aphids amongst the insect groups discussed above: all females of all the other subsocial insects can reproduce. However, within the social Hymenoptera, females show variation in fecundity, or reproductive division of labour. This variation ranges from fully reproductive (the subsocial species described above), through reduced fecundity (many halictid bees), the laying of only male eggs (workers of bumble bees, *Bombus*), sterility (workers of *Aphaenogaster* ants), to super-reproductives (queens of honey bees, *Apis*).

This range of female behaviours is reflected in the classification of social behaviours in the Hymenoptera. Thus, in **quasisocial** behaviour, a communal nest consists of members of the same generation all of which assist in brood rearing, and all females are able to lay eggs, even if not necessarily at the same time. In **semisocial** behaviour, the communal nest similarly contains members of the same generation co-operating in brood care, but there is division of reproductive labour, with some females (queens) laying eggs, whereas their sisters act as workers and rarely lay eggs. This differs from eusociality only in that the workers are sisters to the egg-laying queens, rather than daughters, as is the case in eusociality. As in the primitive condition in eusocial hymenopterans, there is no morphological (size or shape) difference between queens and workers.

Any or all the subsocial behaviours discussed above may be evolutionary precursors of eusociality. It is clear that solitary nesting is the ancestral behaviour, with communal nesting (and additional subsocial behaviours) having arisen independently in many lineages of aculeate hymenopterans.

12.2 EUSOCIALITY IN INSECTS

Eusocial insects have a division of labour in their colonies, involving a caste system comprising a restricted reproductive group of one or several **queens**, aided by **workers** – non-reproductive individuals that assist the reproducers – and in termites and many ants, an additional defensive **soldier** group. There may be further division into subcastes that perform specific tasks. The most specialized members of some castes, such as queens and soldiers, may lack the ability to feed themselves. The tasks of workers therefore include bringing food to these individuals as well as to the **brood** – the developing offspring.

The primary differentiation is female from male. In eusocial Hymenoptera, which have a **haplodiploid** genetic system, queens control the sex of their offspring. Releasing stored sperm fertilizes haploid eggs, which develop into diploid female offspring, whereas unfertilized eggs produce male offspring. At most times of the year, reproductive females (queens, or **gynes**) are rare compared with sterile female workers. Males do not form castes and may be infrequent and short-lived, dying soon after mating. In termites, males and females may be equally represented, with both sexes contributing to the worker caste. A single male termite, the **king**, permanently attends the gyne.

Members of different castes, if derived from a single pair of parents, are close genetically and may be morphologically similar, or, as a result of environmental influence, may be morphologically very different, in an environmental polymorphism termed **polyphenism**. Individuals within a caste (or subcaste) often differ behaviourally, in what is termed **polyethism**, either by an individual performing different tasks at different times in its life (age polyethism), or by individuals within a caste specializing on certain tasks during their lives. The intricacies of social insect caste systems can be considered in terms of the increasing complexity demonstrated in the Hymenoptera, and concluding with the remarkable systems of the termites (Termitoidae). The characteristics of these two groups, which contain the majority of the eusocial species, are given in Taxobox 16 and Taxobox 29.

12.2.1 Hymenopterans showing primitive eusociality

Hymenopterans exhibiting primitive eusociality include polistine vespids (paper wasps of the genus *Polistes*), stenogastrine wasps, and even one sphecid (Fig. 12.2). In these wasps, all individuals are morphologically similar and live in colonies that seldom last more than one year. The colony is often founded by more than one gyne, but rapidly becomes **monogynous**, i.e. dominated by one queen, with other foundresses either departing the nest or remaining but reverting to a worker-like state. The queen establishes a dominance hierarchy physically by biting, chasing and begging for food, with the winning queen gaining monopoly rights to egg-laying and initiation of cell construction. Dominance may be incomplete, with non-queens laying some eggs: the dominant queen may eat these eggs or allow them to develop as workers to assist the colony. The first brood of females produced by the colony is of small workers, but subsequent workers increase in size as nutrition improves and as worker assistance in rearing increases. Sexual retardation in subordinates is reversible: if the queen dies (or is removed experimentally), either a subordinate foundress takes over, or if none is present, a high-ranking worker can mate (if males are present) and lay fertile eggs. Some other species of primitively eusocial wasps are **polygynous**, retaining several

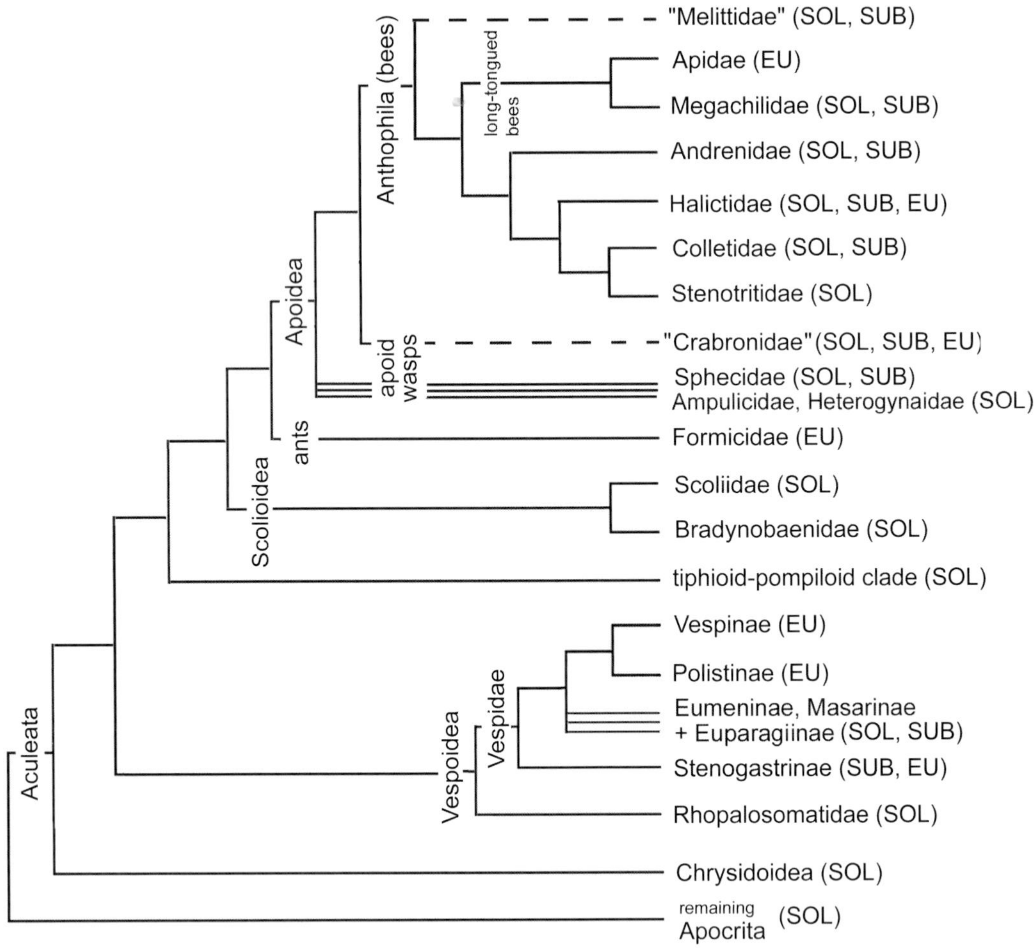

Fig. 12.2 Cladogram showing proposed relationships among selected aculeate Hymenoptera to depict the multiple origins of sociality (SOL, solitary; SUB, subsocial; EU, eusocial). Apoidea includes all bees and some wasp families. Relationships within non-social aculeates are not depicted. Possible non-monophyletic families are shown in quotes on a dashed branch. (Adapted from various sources, including Hines *et al.* 2007; Debevec *et al.* 2012; Danforth *et al.* 2013; Johnson *et al.* 2013.)

functional queens throughout the duration of the colony; whereas others are serially polygynous, with a succession of functional queens.

Bees exhibiting primitively eusociality, such as certain species of Halictidae (Fig. 12.2), have a similar breadth of behaviours. In female castes, differences in size between queens and workers range from little or none to no overlap in their sizes. Bumble bees (Apidae: *Bombus* spp.) found colonies through a single gyne, often after a fight to the death between gynes vying for a nest site. The first brood consists only of workers, which are dominated by the queen physically, by aggression and by eating of any worker eggs, and by means of pheromones that modify the behaviour of the workers. In the absence of the queen, or late in the season as the queen's physical and chemical influence wanes, workers can undergo ovarian development. The queen eventually fails to maintain dominance over those workers that have commenced ovarian development, and the queen either is killed or driven from the nest. When this happens, workers are unmated but they can produce male offspring from their haploid eggs. Gynes are thus derived solely from the fertilized eggs of the queen.

12.2.2 Hymenopterans showing specialized eusociality: wasps and bees

The highly eusocial hymenopterans comprise the ants (family Formicidae) and some wasps, notably Vespinae, and many bees, including all Apidae (Fig. 12.2 and Fig. 12.3). Bees are derived from apoid wasps, and differ from wasps in anatomy, physiology and behaviour in association with their dietary specialization. Most bees provision their larvae with nectar and pollen rather than animal material. Morphological adaptations of bees associated with pollen collection include plumose (branched) hairs, and usually a widened hind tibia adorned with hairs in the form of a brush (**scopa**) or, in bumble bees and honey bees, a fringe surrounding a concavity (the **corbicula**, or pollen basket) (Fig. 12.4). Pollen collected on the body hairs is groomed by the legs and transferred to the mouthparts, scopae or corbiculae. The diagnostic features and the biology of all hymenopterans are covered in Taxobox 29, which includes an illustration of the morphology of a worker vespine wasp and a worker ant.

Colony and castes in eusocial wasps and bees

The female castes are dimorphic, differing markedly in their appearance. Generally, the queen is larger than any worker, as in vespines such as the Eurasian wasps (*Vespula vulgaris* and *V. germanica*) and honey bees (*Apis* spp.). The typical eusocial wasp queen has a differentially (allometrically) enlarged gaster (abdomen). In worker wasps, the bursa copulatrix is small, preventing mating, even though in the absence of a queen their ovaries will develop.

In the vespine wasps, the colony-founding queen, or gyne, produces only workers in the first brood. Immediately after these are hatched, the queen wasp ceases to forage and devotes herself exclusively to reproduction. As the colony matures, subsequent broods include increasing proportions of males, and finally gynes are produced late in the season from larger cells than those from which workers are produced.

The tasks of vespine workers include:

- distribution of protein-rich food to larvae and carbohydrate-rich food to adult wasps;
- cleaning cells and disposal of dead larvae;
- ventilation and air-conditioning of the nest by wing-fanning;
- nest defence by guarding entrances;
- foraging outside for water, sugary liquids and insect prey;
- construction, extension and repair of the cells and inner and outer nest walls with wood pulp, which is masticated to produce paper.

Each worker can carry out any of these tasks, but often there is an age polyethism: newly emerged workers tend to remain in the nest, engaged in construction and food distribution. A middle-aged foraging period follows, which may be partitioned into

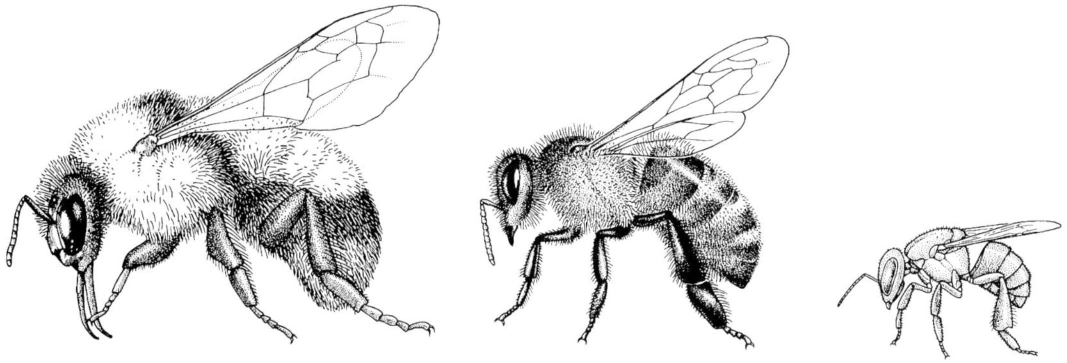

Fig. 12.3 Worker bees from three eusocial genera, from left, *Bombus* (bumble bees), *Apis* (honey bees) and *Trigona* (stingless bees) (all in Apidae), superficially resemble each other in morphology, but they differ in size and ecology, including their pollination preferences and level of eusociality. (After various sources, especially Michener 1974.)

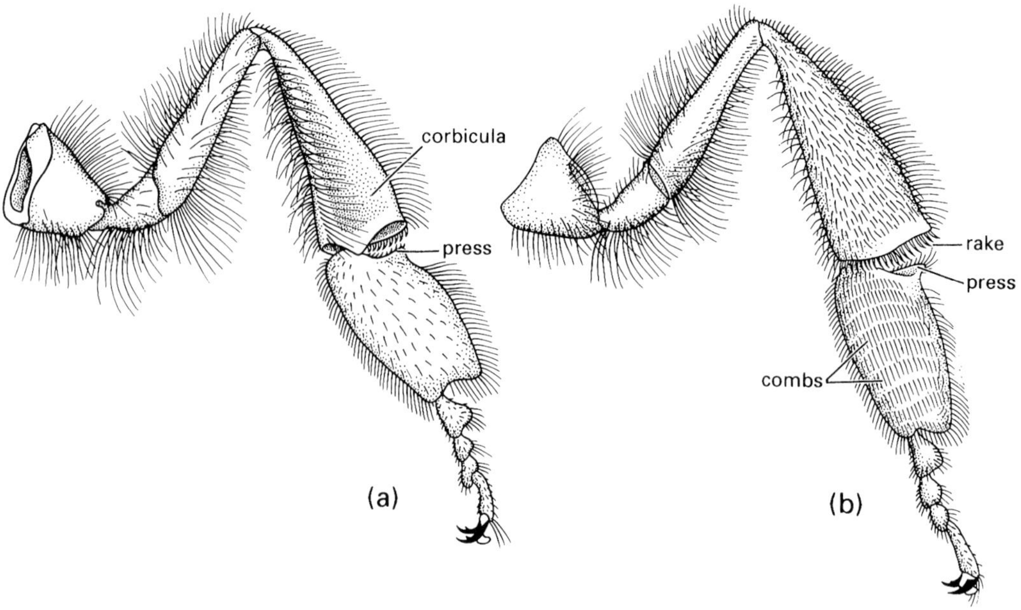

Fig. 12.4 The hind leg of a worker honey bee, *Apis mellifera* (Hymenoptera: Apidae): (a) outer surface showing corbicula, or pollen basket (consisting of a depression fringed by stiff setae), on the tibia, and the press on the basitarsus that pushes the pollen into the basket; (b) the inner surface with the combs and rakes that manipulate pollen into the press prior to packing. (After Snodgrass 1956; Winston 1987.)

wood-pulp collection, predation and fluid-gathering phases. In old age, guarding duties dominate. As newly recruited workers are produced continuously, the age structure allows flexibility in performing the range of tasks required by an active colony. There are seasonal variations, with foraging occupying much of the time of the colony in the founding period, with fewer resources – or a lesser proportion of workers' time – devoted to these activities in the mature colony. Male eggs are laid in increasing numbers as the season progresses, perhaps by queens, or by workers on whom the influence of the queen has waned.

The biology of the honey bee, *Apis mellifera*, is extremely well studied because of the economic significance of honey and the relative ease of observing honey-bee behaviour (Box 12.1). Workers differ from queens in being smaller, possessing wax glands, having a pollen-collecting apparatus comprising pollen combs and a corbicula on each hind leg, in having a barbed sting that cannot be retracted after use, and in some other features associated with the tasks that workers perform. The queen's sting is scarcely barbed and is retractable and reusable, allowing repeated assaults on pretenders to her position as queen. Queens have a shorter proboscis than workers and lack several glands.

Honey-bee workers are more or less monomorphic, but exhibit polyethism. Thus, young workers tend to be "hive bees", engaged in within-hive activities, such as nursing larvae and cleaning cells, and older workers are foraging "field bees". Seasonal changes are evident, such as the 8–9-month longevity of winter bees, compared with the 4–6-week longevity of summer workers. Juvenile hormone (JH) is involved in these behavioural changes, with levels of JH rising from winter to spring, and also in the change from hive bees to field bees. Honey-bee worker activities correlate with seasons, notably in the energy expenditure involved in thermoregulation of the hive.

Caste differentiation in honey bees, as in eusocial hymenopterans generally, is largely **trophogenic**, i.e. determined by the quantity and quality of the larval diet. In species that provision each cell with enough food to allow the egg to develop to the pupa and adult without further replenishment, differences in the food quantity and quality provided to each

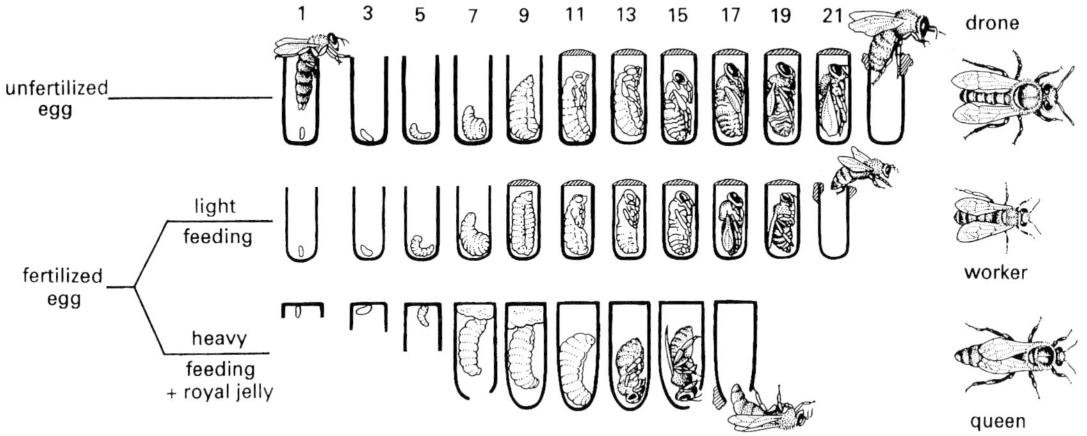

Fig. 12.5 Development of the honey bee, *Apis mellifera* (Hymenoptera: Apidae), showing the factors that determine differentiation of the queen-laid eggs into drones, workers, and queens (on the left), and the approximate developmental times (in days) and stages for drones, workers and queens (on the right). (After Winston 1987.)

cell determine how the larva will develop. In honey bees, although cells are constructed according to the type of caste that is to develop within them, the caste is determined neither by the egg laid by the queen, nor by the cell itself, but by food supplied by workers to the developing larva (Fig. 12.5). The type of cell guides the queen as to whether to lay fertilized or unfertilized eggs, and identifies to the worker which type of rearing (principally food) is to be supplied to the occupant. Food given to future queens, known as "royal jelly", differs from worker food in having high sugar content and being composed predominantly of mandibular gland products, namely pantothenic acid and biopterin. Royal jelly also contains a protein, called **royalactin**, that induces the differentiation of honey-bee larvae into queens. Eggs and larvae up to three days old can differentiate into queens or workers according to upbringing. However, by the third day, a potential queen has been fed royal jelly at up to 10 times the rate of less-rich food supplied to a future worker. At this stage, if a future queen is transferred to a worker cell for further development, she will become an intercaste, a worker-like queen. Transfer of a three-day-old larva reared as a worker into a queen cell gives rise to a queen-like worker, still retaining the pollen baskets, barbed stings, and mandibles of a worker. After four days of appropriate feeding, the castes are fully differentiated and transfers between cell types result in either retention of the early-determined outcome or failure to develop.

Trophogenic effects link with endocrine effects, as nutritional status is associated with corpora allata activity. Clearly, JH levels correlate with polymorphic caste differentiation in eusocial insects, but there is much specific and temporal variation in JH titres, and no common pattern of control is yet evident. Epigenetic effects, such as are induced by differential feeding (royal jelly) on caste development, probably operate via activation or silencing of gene expression associated with the larval development trajectory.

The queen maintains control over the workers' reproduction principally through **pheromones**. The mandibular glands of queens produce a compound identified as (*E*)-9-oxodec-2-enoic acid (9-ODA), but the intact queen inhibits worker ovarian development more effectively than this active compound. A second pheromone has been found in the gaster of the queen, and this, together with a second component of the mandibular gland, effectively inhibits ovarian development. Queen recognition by the rest of the colony involves a pheromone disseminated by attendant workers that contact the queen and then move about the colony as messenger bees. Also, as the queen moves around on the comb whilst ovipositing into the cells, she leaves a trail of footprint pheromone. Production of queens takes place in cells that are distant from the effects of the queen's pheromone control, as occurs when nests become very large. Should the queen die, the volatile pheromone signal dissipates rapidly, and the workers become aware of the absence. Honey bees

Box 12.1 The dance language of bees

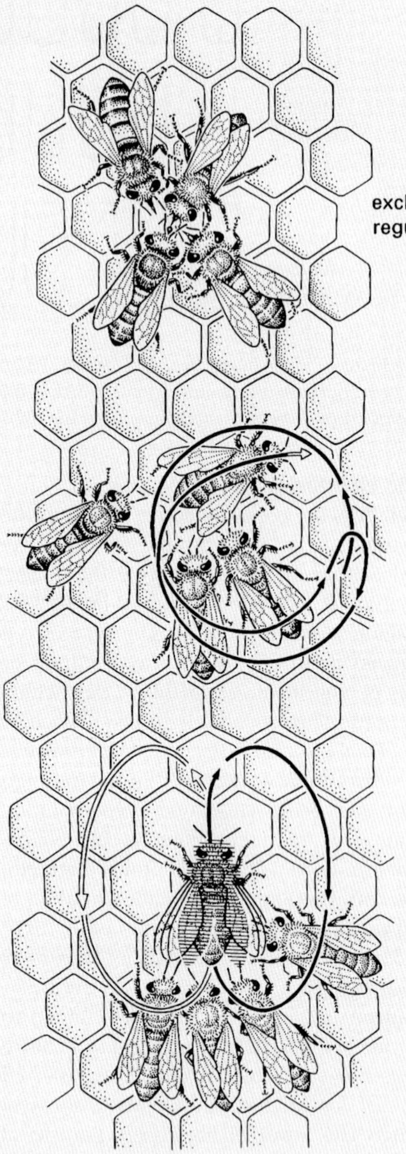

Honey bees have impressive communication abilities. Their ability to communicate forage sites to their nest-mates first was recognized when a marked worker provided with an artificial food source was allowed to return to its hive, and then prevented from returning to the food. The rapid appearance at the food source of other workers indicated that information concerning the resource had been transferred within the hive. Subsequent observations using a glass-fronted hive showed that foragers often performed a dance on return to the nest. Other workers followed the dancer, made antennal contact and tasted regurgitated food, as depicted here in the upper illustration (after Frisch 1967). Olfactory communication alone could be discounted by experimental manipulation of food sources, and the importance of dancing became recognized. Variations within different dances allow communication and recruitment of workers to close or distant food sources, and to food versus prospective nest sites. The purpose and messages associated with three dances – the round, waggle and dorsoventral abdominal vibrating (DVAV) – have become well understood.

Nearby food is communicated by a simple round dance, involving the incoming worker exchanging nectar and making tight circles, with frequent reversals, for a few seconds to a few minutes, as shown in the central illustration (after Frisch 1967). The quality of nectar or pollen from the source is communicated by the vigour of the dance. Although no directionality is conveyed, 89% of 174 workers contacted by the dancer during a round dance were able to find the novel food source within five minutes, probably by flying in ever-increasing circles until the local source is found.

More-distant food sources are identified by a waggle dance, which involves abdomen shaking during a figure-of-eight circuit, shown in the lower illustration (after Frisch 1967), as well as food sharing. Informative characteristics of the dance include the length of the straight part (measured by the number of comb cells traversed), the dance tempo (number of dances per unit time), the duration of waggling and noise production (buzzing) during the straight-line section, and the orientation of the straight run relative to gravity. Messages conveyed are the energy required to get to the source (rather than absolute distance), quality of the forage, and direction relative to the sun's position (see Box 4.4). This interpretation of the significance and information content of the waggle dance was challenged by some experimentalists, who ascribed food-site location entirely to odours particular to the site and borne by the dancer. The claim mainly centred on the protracted time taken for the worker observers of the dance to locate a specific site. The duration matches more the time expected for a bee to take to locate an odour plume and subsequently zig-zag up the plume (see Fig. 4.7), compared with direct flight from bearings provided in the dance. Following some well-designed studies, it is now evident that food finding is as effective and efficient when the experimental source is placed upwind as when it is placed downwind. Furthermore, although experienced workers can locate food by odour, the waggle dance serves to communicate information to naïve workers that allows them to head in the correct general direction. Close to the food source, specific odour does appear to be significant, and the final stages of orientation may be the slow part of location (particularly in experimental set-ups, with non-authentic food sources stationed beside human observers).

The function of the vibration dance (DVAV) differs from the round and waggle dances in regulating the daily and seasonal foraging patterns in relation to fluctuating food supply. Workers vibrate their bodies, particularly their abdomens, in a dorsoventral plane, usually whilst in contact with another bee. Vibration dances peak at times of the day and season when the colony needs to be primed for increased foraging, and these dances act to recruit workers into the waggle-dance area. Vibration dancing with queen contact appears to lessen the inhibitory capacity of the queen, and is used during the period when queen rearing is taking place. Cessation of this kind of vibration dancing may result in the queen departing with a swarm, or in the mating flight of new queens.

Communication of a suitable site for a new nest differs somewhat from communication of a food source. The returning scout dances without any nectar or pollen, and the dance lasts for 15–30 minutes rather than the 1–2 minutes of the forager's dance. At first, several scouts returning from various prospective new sites will all dance, with differences in tempo, angle and duration that indicate the different directions and quality of the sites, as in a waggle dance. More scouts then fly out to prospect, and some sites are rejected. Gradually, a consensus is attained, as shown by one dance that indicates the agreed site.

have very strongly developed chemical communication, with specific pheromones associated with mating, alarm and orientation, as well as with colony recognition and regulation. Physical threats are rare, and are used only by young gynes towards workers.

Males, termed **drones**, are produced throughout the life of the honey-bee colony, either by the queen or perhaps by workers with developed ovaries. Males contribute little to the colony, living only to mate: their genitalia are ripped out after copulation and they die.

It must be emphasized that *A. mellifera* consists of a number of different genetic groups adapted to conditions in different parts of the world and named as subspecies. These subspecies display differing characteristics of morphology, physiology and behaviour, including worker reproductive potential and aggression. For example, in southern Africa two native taxa are recognized – the coastal Cape honey bee (*A. m. capensis*) and the more widespread African honey bee (*A. m. scutellata*) – and human movement of hives has lead to the Cape honey bee causing problems for colonies of the African honey bee (see section 12.3). The African honey bee was introduced deliberately to Brazil, escaped into the wild, and has spread north, causing alarm in North America (Box 12.2). Other species of *Apis*, such as *A. cerana* (the Asiatic or Eastern honey bee) and *A. dorsata* (the giant honey bee), have not been as well studied as *A. mellifera* but continue to provide sources of honey and beeswax to local people in regions where they occur.

Nest construction in eusocial wasps

The founding of a new colony of eusocial vespid wasps takes place in spring, following the emergence of an overwintering queen. After her departure from the natal colony the previous autumn/fall, the new queen mates, but her ovarioles remain undeveloped during the temperature-induced winter quiescence or facultative diapause. As spring temperatures rise, queens leave hibernation and feed on nectar or sap, and the ovarioles grow. The resting site, which may be shared by several overwintering queens, is not a prospective site for foundation of the new colony. Each queen scouts individually for a suitable cavity and fighting may occur if sites are scarce.

Nest construction begins with the use of the mandibles to scrape wood fibres from sound or, more rarely, rotten wood. The wasp returns to the nest site using visual cues, carrying the wood pulp masticated with water and saliva in the mandibles. This pulpy paper is applied to the underside of a selected support at

Box 12.2 Africanized honey bees

The native range of the honey bee *Apis mellifera* extends from southern-most Africa to Britain, northern Europe, and eastward to Iran and the Urals. Before humans started to mix populations, four genetic lineages probably existed in geographic allopatry (non-overlapping), although many more subspecies and/or races have been named. Of the four lineages, three (west European, east European and north African) had reached the New World for apicultural purposes in the middle of the 20th century. During the 1950s, to combat low yields of honey by the existing European honey bees, Brazilian apiarists, with assistance from government agencies, sought to introduce the southern African subspecies (now *A. m. scutellata*) to the country. In 1956, queens of this more tropical-adapted subspecies collected from South Africa and Tanzania were introduced successfully to apicultural research hives. Despite a "queen guard" on each hive, some queens accidentally escaped and these bees spread rapidly around Brazil. Bees of this subspecies are behaviourally different from existing honey bees in the New World, with an increased tendency to swarm, migrate, abscond, and to be more aggressive, including more belligerent nest/hive defence within a wider radius, and they allocate a higher proportion of individuals to guarding. These propensities led the media (including movie-makers) and some apiculturists to coin the term "killer" bees. Contributing to this fear of "Africanized" bees among North Americans was the pace of spread (up to 500 km per year) from the introduction site in Brazil south into Argentina and northwards through meso-America, reaching the US-Mexico border in about 1990.

However, all is not as predicted. What molecular genetics shows is that the rapid advance was of honey bees that were genetically mostly African. Although *A. m. scutellata* does hybridize with European honey bees, it also displaces them by a greater rate of increase, aggressiveness and nest-usurpation. Africanized honey bees now occur across the southern United States but their northward spread is limited by cold winter conditions. Also, commercial honey-bee hives in the United States are maintained by "re-queening" with European queens, since those with African or Africanized queens are much more difficult to handle. The spread in the southern United States is slower than in tropical and central America, perhaps because of resistance by beekeepers, a less tropical climate, and maybe the effects of colony collapse disorder (Box 12.3). Far fewer human deaths from bee attacks have been recorded compared with predictions, although a few deaths have occurred. At the source of the invasion, Brazilian apiarists have selected for less-aggressive lines of African bees. The fears invoked by warnings of the "killer" bees have not matched the early hype.

Interestingly, in contrast to the success of *A. m. scutellata* in the Americas, this bee has become the victim of social parasitism in northern South Africa, where a strain of the Cape honey bee, *A. m. capensis*, has been devastating colonies of the African honey bee in its native range (see section 12.3).

the top of the cavity. From this initial buttress, the pulp is formed into a descending pillar, upon which is suspended ultimately the embryonic colony of 20–40 cells (Fig. 12.6). The first two cells, rounded in cross-section, are attached, and then an umbrella-like envelope is formed over the cells. The envelope is elevated by about the width of the queen's body above the cells, allowing the queen to rest there, curled around the pillar. The developing colony grows by the addition of further cells, now hexagonal in cross-section and wider at the open end, and by either extension of the envelope or construction of a new one. The queen forages only for building material at the start of nest construction. As the larvae develop from the first cells, both liquid and insect prey are sought to nourish the developing larvae, although wood pulp continues to be collected for further cell construction. This first embryonic phase of the life of the colony ceases as the first workers emerge.

As the colony grows, further pillars are added, providing support to more lateral areas where brood-filled cells are aligned in **combs** (series of adjoining cells aligned in parallel rows). The early cells and envelopes become overgrown, and their materials may be reused in later construction. In a subterranean nest, the occupants may have to excavate soil and even small stones to allow colony expansion, resulting in a mature nest (as in Fig. 12.6), which may contain as many as 12,000 cells. The colony has some independence from external temperature, as thoracic heating through wing beating and larval feeding can raise temperature, and high temperature can be lowered by directional

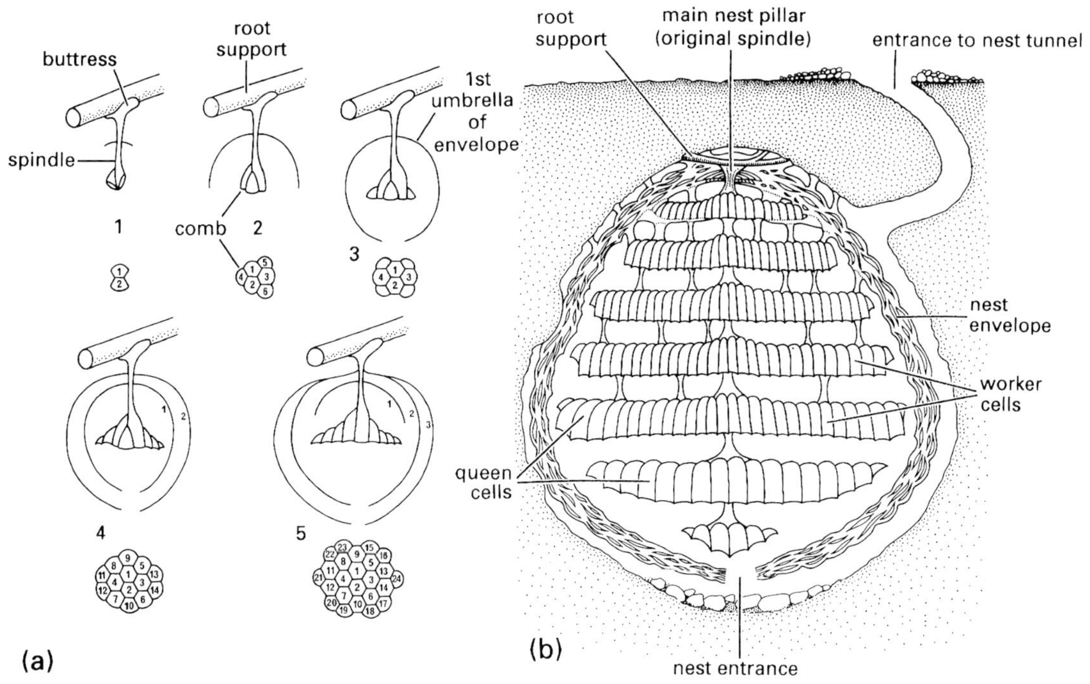

Fig. 12.6 The nest of the common European wasp, *Vespula vulgaris* (Hymenoptera: Vespidae): (a) initial stages (1–5) of nest construction by the queen (the embryonic phase of the colony's life); (b) a mature nest. (After Spradbery 1973.)

fanning or by evaporation of liquid applied to the pupal cells.

At the end of the season, males and gynes (potential queens) are produced and are fed with larval saliva and prey brought into the nest by workers. As the old queen fails and dies, and gynes emerge from the nest, the colony declines rapidly and the nest is destroyed as workers fight and larvae are neglected. Potential queens and males mate away from the nest, and the mated female seeks a suitable overwintering site.

Nesting in honey bees

In honey bees, new colonies are initiated if the old one becomes too crowded. When a bee colony becomes too large and the population density too high, a founder queen, accompanied by a swarm of workers, seeks a new nest site. Because workers cannot survive long on the honey reserves carried in their stomachs, a suitable site must be found quickly. Scouts may have started the search several days before formation of the swarm. If a suitable cavity is found, the scout returns to the cluster and communicates the direction and quality of the site by a dance (Box 12.1). Optimally, a new site should be beyond the foraging territory of the old nest, but not so distant that energy is expended in long-distance flight. Bees from temperate areas select enclosed nest sites in cavities of about 40 litres in volume, whereas more-tropical bees choose smaller cavities or nest outside.

Following consensus over the nest site, workers start building a nest using wax. **Wax** is unique to social bees, and is produced by workers that metabolize honey in fat cells located close to the wax glands. These modified epidermal cells lie beneath **wax mirrors** (overlapping plates) ventrally on the fourth to seventh abdominal segments. Flakes of wax are extruded beneath each wax mirror and protrude slightly from each segment of a worker that is actively producing wax. Wax is quite malleable at the ambient nest temperature of 35°C, and when mixed with saliva can be manipulated for cell construction. At nest foundation, workers already may have wax protruding from the abdominal wax glands. They start to construct combs of back-to-back hexagonal cells in a parallel series, or comb. Wax pillars and bridges separate combs from one another. A thick cell base of wax is extended into a thin-walled cell of remarkably constant dimensions, despite a series of workers being involved in construction. In contrast to other social insects such as the vespids described above, cells do not hang downwards but are angled at about 13° above the horizontal, thereby preventing loss of honey. The precise orientation of the cells and comb derives from the bees' ability to detect gravity through the proprioceptor hair plates at the base of their necks. Although removal of the hair plates prevents cell construction, worker bees could construct serviceable cells under conditions of weightlessness in space.

Unlike most other bees, honey bees do not chew up and reuse wax: once a cell is constructed it is permanently part of the nest, and cells are reused after the brood has emerged or the food contents have been used. Cell sizes vary, with small cells used to rear workers, and larger ones for drones (Fig. 12.5). The larvae produce silk, from their modified salivary glands (labial glands), to reinforce the wax cells in which they pupate. Later in the life of the nest, elongate conical cells in which queens are reared are constructed at the bottom and sides of the nest. The brood develops, and pollen is stored in lower and more central cells, whereas honey is stored in upper and peripheral cells. Workers form honey primarily from nectar taken from flowers, but also from secretions from extrafloral nectaries, or insect-produced honeydew. Workers carry nectar to the hive in honey stomachs, from which it may be fed directly to the brood and to other adults. However, most often it is converted to honey by enzymatic digestion of the sugars to simpler forms and reduction of the water content by evaporation, before storage in wax-sealed cells until required to feed adults or larvae. It has been calculated that in 66,000 bee-hours of labour, 1 kg of beeswax can be formed into 77,000 cells, which can support the weight of 22 kg of honey. An average colony requires about 60–80 kg of honey per annum.

Unlike wasps, honey bees do not hibernate with the arrival of the lower temperatures of temperate winter. Colonies remain active through the winter, but foraging is curtailed and no brood is reared. Stored honey provides an energy source for activity and heat generation within the nest. As outside temperatures drop, the workers cluster together, heads inwards, forming an inactive layer of bees on the outside, and warmer, more active, feeding bees on the inside. Despite the prodigious stores of honey and pollen, a long or extremely cold winter may kill many bees.

Beehives are artificial constructions that resemble feral honey-bee nests in some dimensions, notably the distance between the combs. When given wooden frames separated by an invariable natural spacing interval of 9.6 mm, honey bees construct their combs within the frame without formation of the internal waxen bridges needed to separate the combs of a feral nest. This width between combs is approximately the space required for bees to move unimpeded on both combs. The ability to remove frames allows the apiculturalist (beekeeper) to examine and remove the honey, and replace the frames in the hive. The ease of construction allows the building of several ranks of boxes. The hives can be transported to suitable locations without damaging the combs.

Although the apiculture industry has developed through commercial production of honey, lack of native pollinators in monocultural agricultural systems has led to increasing reliance on transported hive bees to ensure the pollination of crops as diverse as nuts (especially almonds), soybeans, fruits, clover, canola, alfalfa and other fodder crops. In the United States alone, in the past 20 years some 2.5 million colonies of *A. mellifera* have been rented annually for pollination purposes, and the value to US agriculture attributable to honey-bee pollination is estimated to be upwards of US$15 billion per year. Yield losses of over 90% of many fruit, seed and nut crops would occur without honey-bee pollination. The current threat to pollination services by colony collapse disorder (Box 12.3) is imposing a substantial risk to a range of bee-pollinated crops. Furthermore, the role in pollination of the many species of eusocial and solitary native bees remains inadequately recognized, but can be important, especially close to areas of natural vegetation (for example, see section 11.3.1 for the role of bees in pollinating strawberries).

12.2.3 Specialized hymenopterans: ants

Ants (Formicidae) form a well-defined, highly specialized group, probably sister to Apoidea (Fig. 12.2). The morphology of a worker ant of *Formica* is illustrated in Taxobox 29.

Colony and castes in ants

All ants are social and their species are polymorphic. There are two major female castes, the reproductive queen and the workers, usually with complete dimorphism between them. Many ants have monomorphic workers, but others have distinct subcastes called, according to their size, **minor, media or major workers**. Although workers may form clearly different morphs, more often there is a gradient in size. Workers are never winged, but queens have wings that are shed after mating, as do males, which die after mating. Winged individuals are called **alates**. Polymorphism in ants is accompanied by polyethism, with the queen's role restricted to oviposition, and the workers performing all other tasks. If workers are monomorphic, there may be temporal or age polyethism, with young workers undertaking internal nurse duties and older ones foraging outside the nest. If workers are polymorphic, the subcaste with the largest individuals, the major workers, usually has a defensive or soldier role.

The workers of certain ants, such as the fire ants (*Solenopsis*), have reduced ovaries and are irreversibly sterile. In others, workers have functional ovaries and may produce some or all of the male offspring by laying haploid (unfertilized) eggs. In some species, when the queen is removed, the colony continues to produce gynes from fertilized eggs previously laid by the queen, and males from eggs laid by workers. In the black garden ant, *Lasius niger*, a queen-produced pheromone (a cuticular hydrocarbon found on the queen and her eggs) has been shown to act both as a primer to reduce ovarian activation of the workers and as a releaser to reduce worker aggression towards the queen and her eggs. The inhibition by the queen of her daughter workers is quite striking in the African weaver ant, *Oecophylla longinoda*. A mature colony of up to half a million workers, distributed amongst as many as 17 nests, is prevented from reproduction completely by a single queen. Workers, however, do produce male offspring in nests that lie outside the influence (or territory) of the queen. Queens prevent the production of reproductive eggs by workers, but may allow the laying of specialized **trophic** eggs that are fed to the queen and/or larvae. By this means, the queen not only prevents any reproductive competition, but directs much of the protein in the colony towards her own offspring.

Caste differentiation is largely trophogenic (diet-determined), involving biased allocation of volume and quality of food given to the larvae. A high-protein diet promotes differentiation of gyne/queen, and a less rich, more dilute diet leads to differentiation of workers. The queen generally inhibits the development of gynes indirectly, by modifying the feeding behaviour of workers towards female larvae, which have the potential to differentiate as either gynes or workers. In *Myrmica*, large, slowly developing larvae will become gynes, so stimulation of rapid development and early metamorphosis of small larvae, or food deprivation and irritating of large larvae by biting to accelerate development, both induce differentiation as workers. When queen influence wanes, either through the increased size of the colony, or because the inhibitory pheromone is impeded in its circulation throughout the colony, gynes are produced at some distance from the queen. There is also a role for JH in caste differentiation. JH tends to induce queen development during egg and larval stages, and induces production of major workers from already differentiated workers.

According to a seasonal cycle, ant gynes mature to winged reproductives, or alates, and remain in the nest in a sexually inactive state until external conditions are suitable for departing the nest. At the appropriate time, they make their nuptial flight, mate and attempt to found a new colony.

Nesting in ants

The subterranean soil nests of *Myrmica* and the mounds of plant debris of *Formica* are typical temperate ant nests. Colonies are founded when a mated queen sheds her wings and overwinters, sealed into a newly dug nest that she will never leave. In spring, the queen lays some eggs and feeds the hatched larvae by stomodeal or oral **trophallaxis**, i.e. regurgitation of liquid food from her internal food reserves. Colonies develop slowly whilst worker numbers build up, and a nest may be many years old before alates are produced.

Colony foundation by more than one queen, known as **pleometrosis**, appears to be fairly widespread, and the digging of the initial nest may be shared, as in the honeypot ant *Myrmecocystus mimicus*. In this species and others, multi-queen nests may persist as polygynous colonies, but monogyny commonly arises through dominance of a single queen, usually following rearing of the first brood of workers. Polygynous nests often are associated with opportunistic use of ephemeral resources, or persistent but patchy resources.

Box 12.3 Colony collapse disorder

Beekeeping is ancient: humans have long maintained honey bees, many of which escape and form "feral" colonies (for effects on native ecosystems, see section 11.3.1). In the United States, a decline in feral bees has been recorded since the 1970s, seemingly associated with the spread of an alien Asian *Varroa* mite (section 12.3). Commercial apiaries (beehives) also declined, despite availability of an acaricide (= miticide) to control *Varroa*. From 2005–6, honey-bee decline accelerated dramatically, with around a third of all hives estimated to be lost each year – and nearly half of all US beekeepers left the field. The name "colony collapse disorder" (CCD) was coined, although in both the United States and western Europe somewhat similar historical declines, from unknown causes, had been documented previously. Symptoms of CCD are abandoned hives empty of live adults and without dead bees, but with "capped brood" present and the honey and pollen stores intact. Curiously, other honey bees are slow to usurp these food reserves, and hive pests (moths and beetles) also delay their attack. Attention also was drawn to ongoing loss of native-bee diversity in parts of North America and Europe, and declines in native bumble bees in western Europe.

Honeybee services require seasonal transport of hives around the United States, for example more than half of all in the country (over 1.5 million hives from out of state) are required to pollinate the spring Californian almond crop. Management practices are required to match crop and season, and include supplemental winter feeding and long-distance transport of pollinators (as shown in the map here, after Mairson 1993). Actual and potential losses to pollination services worth multibillion US dollars (section 1.2) led to many research projects, amidst an atmosphere of media speculation about the damaging effects on food production of "a world without insect pollinators".

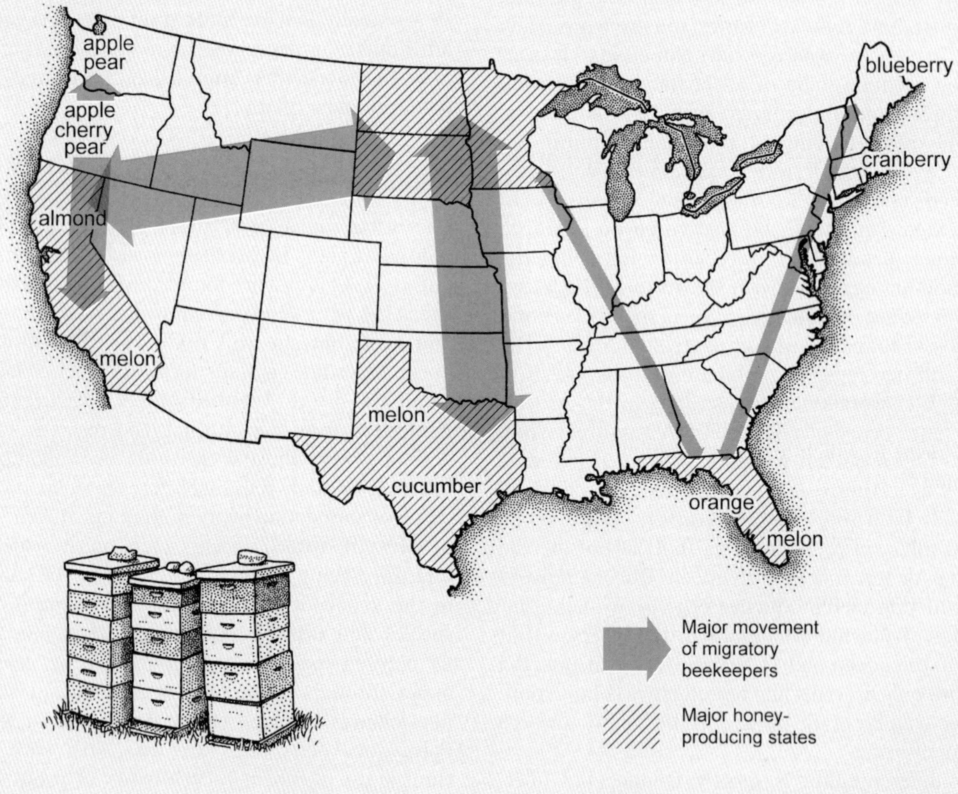

Many potential causes for CCD have been suggested. Several have been rejected, and others suspected as correlated, but few have been verified as causative. Amongst single-cause candidates are:

- *Varroa* mites, especially the virulent *V. destructor* from the Asian *Apis cerana*, which has switched to *A. mellifera*;
- viruses, such as Israel acute paralysing virus (IAPV), transmissible by *Varroa*, and shown to cause shivering, paralysis and death of worker bees away from the hive;
- the microsporidian fungus *Nosema*, with a western species *N. apis* becoming replaced by an eastern-derived species *N. ceranae*.

From the outset, a single biological cause seemed unlikely, and indeed a series of studies now report synergistic negative effects linked to many factors. For example, suspicion that supplemental feeding of high-fructose corn syrup to wintering hives was involved in CCD has been confirmed, due not to the honey substitute directly, but rather the lack of beneficial chemicals found in dietary honey, such as *p*-coumaric acid derived from pollens that specifically induce (upregulate) genes involved in detoxification of the widely used in-hive acaricide, coumaphos. Other studies have shown agonistic, additive toxicity effects between acaricides, detected widely in bees and bee products, and antimicrobial and antifungal compounds applied to manage the "health" of the colony. In an experiment documenting the source and composition of pollens returned by workers to colonies located to pollinate the crops named on the accompanying map, bees often appeared to find most pollen from weeds and wildflowers, rather than from the target cucumbers, blueberries, etc. Pollen from corbiculae (bees' pollen baskets) contained high levels of fungicides and miticides, as well as many different pesticides, including some worrying examples of above median lethal doses of certain insecticides. Consuming pollen with high fungicide and miticide loads increased bees' susceptibility to *Nosema* infection, as did exposure to some (but not all) pesticides.

Popular suspicion for induction of CCD has fallen on neonicotinoid insecticides, including imidacloprid (section 16.4.1 and Box 16.4), which are used increasingly as broad-spectrum pesticides and formulated as systematic insecticides (used, for example, against emerald ash borer; Box 17.4). Although cleared from lethality to honey bees at recommended doses, this highly neuroactive chemical has sublethal effects on bees at doses (even at a few parts per billion dilution) that result from routine applications and reflect real-world field conditions. At these concentrations, synergistic changes in the immune and neural systems may be expected, especially in hives with a wide range of other harmful chemicals present. However, arguing responsibility to this alone, amongst many chemical insults experienced by pollinator insects, seems unjustified given the evidence for a role of many beekeepers' own management practices.

The woven nests of *Oecophylla* species are well-known, complex structures (Fig. 12.7). These African and Asian/Australian weaver ants have extended territories that workers continually explore for any leaf that can be bent. A remarkable collaborative construction effort follows, in which leaves are manipulated into a tent-shape by linear ranks of workers, often involving "living chains" of ants that bridge wide gaps between the leaf edges. Another group of workers take larvae from existing nests and carry them, held delicately between their mandibles to the construction site. There, larvae are induced to produce silk threads from their well-developed silk glands and a nest is woven linking the framework of leaves.

Living plant tissues provide a location for nests of ants such as *Pseudomyrmex ferrugineus*, which nests in the expanded thorns of the Central American bull's-horn acacia trees (see Fig. 11.9a). In such mutualisms involving plant defence, plants benefit by deterrence of phytophagous animals by the ants, as discussed in section 11.4.1.

Foraging efficiency of ants can be very high. A typical mature colony of European red ants (*Formica polyctena*) is estimated to harvest about 1 kg of arthropod food per day. The legionary, or army, and driver ants are popularly known for their voracious predatory activities. These ants, which predominantly belong to the subfamilies Ecitoninae and Dorylinae, alternate cyclically between sedentary (**statary**) and migratory or nomadic phases. In the latter phase, a nightly **bivouac** is formed, which often is no more than an exposed cluster of the entire colony. Each morning, the millions-strong colony moves *in toto*, bearing the larvae. The advancing edge of this massive group raids and forages on a wide range of terrestrial arthropods, and group predation allows even large prey items to

Fig. 12.7 Weaver ants of *Oecophylla* making a nest by pulling together leaves and binding them with silk produced by larvae that are held in the mandibles of worker ants. (After CSIRO 1970; Hölldobler 1984.)

be overcome. After some two weeks of nomadism, a statary period commences, during which the queen lays 100,000–300,000 eggs in a statary bivouac. This is more sheltered than a typical overnight bivouac, perhaps within an old ants' nest, or beneath a log. In the three weeks before the eggs hatch, larvae of the previous oviposition complete their development to emerge as new workers, thus stimulating the next migratory period.

Not all ants are predatory. Some ants harvest grain and seeds (myrmecochory; section 11.3.2) and others, including the extraordinary honeypot ants, feed almost exclusively on insect-produced honeydew, including that of scale insects tended inside nests (section 11.4.1). Workers of honeypot ants return to the nest with crops filled with honeydew, which is fed by oral trophallaxis to selected workers called **repletes**. The abdomen of repletes are so distensible that they become virtually immobile "honey pots" (see Fig. 2.4), which act as food reserves for all in the nest.

12.2.4 Termitoidae (former order Isoptera, termites)

All termites (Termitoidae) are eusocial. Their diagnostic features and biology are summarized in Taxobox 16.

Colony and castes in termites

In contrast to the adult and female-only castes of holometabolous eusocial Hymenoptera, the castes of the hemimetabolous termites involve immature stages and equal representation of the sexes. However, before castes are discussed further, terms for termite immature stages must be clarified. Termitologists refer to the developmental instars of reproductives as nymphs, more properly called brachypterous nymphs; and the instars of sterile lineages as larvae, although strictly the latter are apterous nymphs.

The termites often are divided into "higher" and "lower" termites, but the implication of these terms is inappropriate. "Higher" refers to the monophyletic, species-rich family Termitidae; "lower" encompasses a paraphyletic, evolutionary grade (see section 7.1.1) comprising all other families.

The Termitidae differ from other termites in the following manners:

• Members of the Termitidae typically lack the symbiotic flagellates found in the hindgut of "lower termites"; these protists (protozoa) secrete enzymes (including cellulases) that may contribute to the breakdown of gut contents. One subfamily of Termitidae uses a cultivated fungus to pre-digest food.

• Termitidae have an elaborate and rigid caste system. For example, termitid queens undergo extraordinary **physogastry**, in which the abdomen is distended to 500–1000% of its original size (Fig. 12.8; see **Plate 6c**); in most other termites there is little or no distension of the queen's abdomen.

All termite colonies contain a pair of **primary reproductives** – the queen and king, which are former alate (winged) adults from an established colony. The king and queen are long-lived and mate repeatedly through their lives. Upon loss of the primary reproductives, potential replacement reproductives occur (in some species a small number may be ever-present). These individuals, called **supplementary reproductives**, or **neotenics**, are arrested in their development, either with wings present as buds (brachypterous neotenics) or without wings (apterous neotenics, or **ergatoids**), and can take on the reproductive role if the primary reproductives die.

In contrast to these reproductives, or potentially reproductive castes, the colony is dominated numerically by sterile termites of both sexes that function as workers and soldiers. Soldiers have distinctive, heavily sclerotized heads, with large mandibles or with a strongly produced snout (or **nasus**) through which sticky defensive secretions are ejected. Two classes, major and minor soldiers, may occur in some species. Workers are unspecialized, weakly pigmented and poorly sclerotized, giving rise to the popular name of "white ants".

Caste differentiation pathways are portrayed best in the more rigid system of the Termitidae, which can be contrasted with the greater plasticity shown by other termites. In *Nasutitermes exitiosus* (Termitidae: subfamily Nasutitermitinae) (Fig. 12.8), two different developmental pathways exist; one leads to reproductives; and the other (which is further subdivided) gives rise to sterile castes. This differentiation may occur as early as the first larval stage, although some castes may not be recognizable morphologically until later moults. The reproductive pathway (on the left in Fig. 12.8) is relatively constant between termite taxa, and typically gives rise to alates – the winged reproductives that leave the colony, mate, disperse and found new colonies. In *N. exitiosus*, no neotenics are formed; replacement for lost primary reproductives comes from amongst alates retained in the colony. Other *Nasutitermes* show even greater developmental plasticity.

The sterile (neuter) lineages are complex and variable between different termite species. In *N. exitiosus*, two categories of second-instar larvae can be recognized according to size differences, probably relating to sexual dimorphism, although which sex belongs to which size category is unclear. In both lineages, a subsequent moult produces a third-instar nymph of the worker caste, either small or large according to the pathway. These third-instar workers have the potential (**competency**) to develop into a soldier (via an intervening **presoldier** instar) or remain as workers through several more moults. The sterile pathway of *N. exitiosus* involves larger workers continuing to grow at successive moults, whereas the small worker ceases to moult beyond the fourth instar. Those that moult to become presoldiers and then soldiers develop no further.

Non-Termitidae ("lower") termites are more flexible and exhibit more routes to differentiation. Many lack a true worker caste, although often possess a functionally equivalent "child-labour" or "false worker" caste, the **pseudergates**, composed of either nymphs whose wing buds have been eliminated (regressed) by moulting or, less frequently, brachypterous nymphs, or even undifferentiated larvae. Unlike the "true" workers of the Termitidae, pseudergates are developmentally plastic and retain the capacity to differentiate into other castes by moulting. Differentiation of nymphs from larvae, and reproductives from pseudergates, may not be possible until a relatively late instar is reached. If there is sexual dimorphism in the sterile line, the larger workers often are male, but workers may be monomorphic. This may be through the absence of sexual dimorphism, or more rarely, because only one sex is represented. Moults can give rise to:

• morphological change within a caste;
• no morphological advance (stationary moult);
• change to a new caste (such as from a pseudergate to a reproductive);
• saltation to a new morphology, missing a normal intermediate instar;

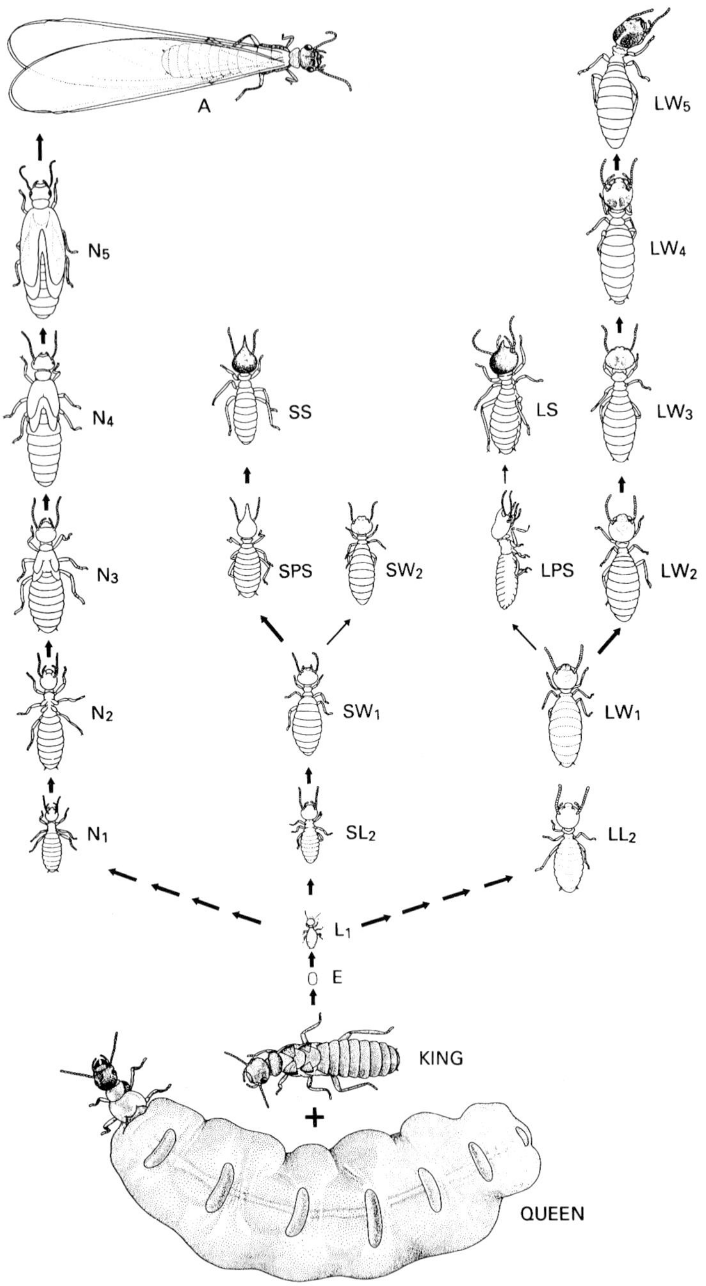

Fig. 12.8 Developmental pathways of the termite *Nasutitermes exitiosus* (Termitidae). Heavy arrows indicate the main lines of development, light arrows indicate the minor lines. A, alate; E, egg; L, larva; LL, large larva; LPS, large presoldier; LS, large soldier; LW, large worker; N, nymph; SL, small larva; SPS, small presoldier; SS, small soldier; SW, small worker. The numbers indicate the stages. (Pathways based on Watson & Abbey 1985.)

• supplementation, adding an instar to the normal route;
• reversion to an earlier morphology (such as a pseudergate from a reproductive), or a presoldier from any nymph, late-instar larva, or pseudergate.

Instar determination is impossibly difficult in the light of these moulting potentialities. The only inevitability is that a presoldier must moult to a soldier.

Pseudergates can co-exist with true workers in some termites; clearly, pseudergates are not evolutionary precursors to true workers, and have evolved through different developmental pathways. From phylogenetic studies, the true worker caste seems to have had multiple evolutionary origins, although it remains unclear if pseudergates arose prior to workers.

Certain unusual termites lack soldiers. Even the universal presence of only one pair of reproductives has exceptions; multiple primary queens cohabit in some colonies of some Termitidae.

Individuals in a termite colony are derived from one pair of parents. Therefore, genetic differences existing between castes either must be sex-related or due to differential expression of the genes. Gene expression is under complex multiple and synergistic influences entailing hormones (including neurohormones), interactions between colony members and external environmental factors (epigenetics). Termite colonies are very structured and have high homeostasy – caste proportions are restored rapidly after experimental or natural disturbance, by recruitment of individuals of appropriate castes and elimination of individuals excess to colony needs. Homeostasis is controlled by several pheromones that act specifically upon the corpora allata and more generally on the rest of the endocrine system. In the well-studied *Kalotermes*, primary reproductives inhibit differentiation of supplementary reproductives and alate nymphs. Presoldier formation is inhibited by soldiers, but stimulated through pheromones produced by reproductives.

Pheromones that inhibit reproduction are produced inside the body by reproductives and disseminated to pseudergates by proctodeal **trophallaxis**, i.e. by feeding on anal excretions. Transfer of pheromones to the rest of the colony is by oral trophallaxis. This was demonstrated experimentally in a *Kalotermes* colony by removing reproductives and dividing the colony into two halves with a membrane. Reproductives were reintroduced, orientated within the membrane such that their abdomens were directed into one half of the colony and their heads into the other. Only in the "head-end" part of the colony did pseudergates differentiate as reproductives; inhibition continued at the "abdomen-end". Painting the protruding abdomen with varnish eliminated any cuticular chemical messengers but failed to remove the inhibition on pseudergate development. In constrast, when the anus was blocked, pseudergates became reproductive, thereby verifying anal transfer. The inhibitory pheromones produced by both queen and king have complementary or synergistic effects: a female pheromone stimulates the male to release inhibitory pheromone, whereas the male pheromone has a lesser stimulatory effect on the female. Production of primary and supplementary reproductives involves removal of these pheromonal inhibitors produced by functioning reproductives.

Increasing recognition of the role of JH in caste differentiation comes from observations such as the differentiation of pseudergates into soldiers after injection or topical application of JH or implantation of the corpora allata of reproductives. Some of the effects of pheromones on colony composition may be due to JH production by the primary reproductives. Caste determination in Termitidae originates as early as the egg, during maturation in the ovary of the queen. As the queen grows, the corpora allata undergoes hypertrophy and may attain a size 150 times greater than the gland of the alate. The JH content of eggs also varies, and it is possible that a high JH level in the egg causes differentiation to follow the sterile lineage. This route is enforced if the larvae are fed proctodeal foods (or trophic eggs) that are high in JH, whereas a low level of JH in the egg allows differentiation along the reproductive pathway. In all termites, worker and soldier differentiation from the third-instar larva is under further hormonal control, as demonstrated by the induction of individuals of these castes by JH application.

Nesting in termites

In the warmer parts of the temperate northern hemisphere, drywood termites (Kalotermitidae, especially *Cryptotermes*) are most familiar because of the structural damage that they cause to timber in buildings. Termites are pests of drywood and dampwood in the subtropics and tropics, but in these regions termites may be more familiar through their spectacular mound nests. In the timber pests, colony size may be no greater than a few hundred termites, whereas in the mound formers (principally species of Termitidae and some Rhinotermitidae), several million individuals

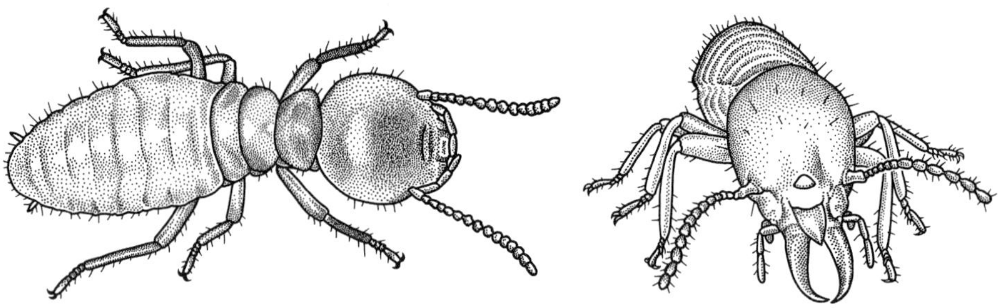

Fig. 12.9 The Formosan subterranean termite, *Coptotermes formosanus* (Blattodea: Termitoidae: Rhinotermitidae): worker (left) and soldier (right).

may be involved. The Formosan subterranean termite, *Coptotermes formosanus* (Rhinotermitidae) (Fig. 12.9), lives mostly in underground nests and can form huge colonies of up to 8 million individuals. It is a serious invasive pest in the southeast United States, where its activities probably contributed to the failure of the levees (flood walls) in the New Orleans floods caused by Hurricane Katrina. This termite has undermined many dam walls in its native China.

Termites can be grouped into four broad categories of nesting behaviour: (i) nesting and eating in a single piece of wood, or several closely associated pieces; (ii) starting in a single piece of wood but searching for and eating other pieces and moving the nest to another piece once the first is eaten; (iii) nesting separately from their food (not necessarily wood) and foraging to locate food; and (iv) continuously mobile with no permanent nest, as found in soil feeders that consume decomposed plant material. The invasive pest species of termites all belong to the first two nesting categories.

In all termites, a new nest is founded by a male and female, following the nuptial flight of alates. A small cavity is excavated into which the pair seal themselves. Copulation takes place in this royal cell, and then egg laying commences. The first offspring are workers, which are fed on regurgitated wood or other plant matter, primed with gut symbionts, until they are old enough to feed themselves and enlarge the nest. Early in the life of the colony, production is directed towards workers, with later production of soldiers to defend the colony. As the colony matures, perhaps not until it is 5–10 years old, production of reproductives commences. This involves differentiation of alate sexual forms at the appropriate season for swarming and foundation of new colonies.

Tropical termites can use virtually all cellulose-rich food sources, above and below the ground, from grass tussocks and fungi to living and dead trees. Workers radiate from the mound, often in subterranean tunnels, less often in above-ground, pheromone-marked trails, in search of materials. In the subfamily Macrotermitinae (Termitidae), fungi are raised in combs of termite faeces within the mound, and the complete culture of fungus and excreta is eaten by the colony (section 9.5.3). These fungus-tending termites form the largest termite colonies known, with estimated millions of inhabitants in some East African species.

The giant mounds of tropical termites mostly belong to species in the Termitidae. As the colony grows through production of workers, the mound is enlarged by addition of layers of soil and termite faeces until century-old mounds attain massive dimensions. Diverse mound architectures characterize different termite species; for example, the "magnetic mounds" of *Amitermes meridionalis* in northern Australia have a narrow north–south and broad east–west orientation, and can be used like a compass (Fig. 12.10). Orientation relates to thermoregulation, as the broad face of the mound receives maximum exposure to the warming of the early and late sun, and the narrowest face is presented to the high and hot midday sun. Aspect is not the only means of temperature regulation: intricate internal design, especially in fungus-farming *Macrotermes* species, allows circulation of air to give microclimatic control of temperature and carbon dioxide (Fig. 12.11).

12.2.5 Eusocial ambrosia beetles (Coleoptera: Curculionidae)

The association of ambrosia beetles (Platypodinae and some Scolytinae) with fungi that break down wood is discussed in section 9.2. These beetles live within a

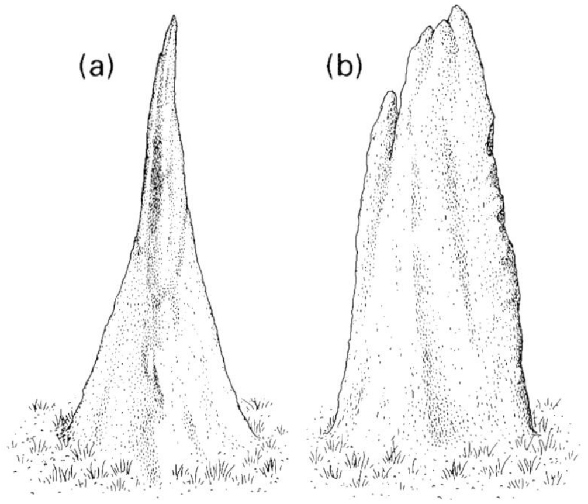

Fig. 12.10 A "magnetic" mound of the debris-feeding termite *Amitermes meridionalis* (Termitidae) showing: (a) the north–south view, and (b) the east–west view. (After Hadlington 1987).

tree in a colony founded by a sole female that bores the initial tunnel, inoculates it with fungi carried in her mycangia (section 9.2) and lays her eggs. All offspring of a colony derive from this sole individual, giving high relatedness amongst offspring. This relatedness, combined with the early observations of colony demographics (particularly of the platypodine *Austroplatypus incompertus*) showing retention of next generation's females (overlap of generations), demonstrates a form of sociality.

Recent studies of *Xyleborinus saxesenii* (Scolytinae: Xyloborini) show a unique division of labour also is involved. Being holometabolous, ambrosia beetle larvae differ greatly from their adults morphologically, yet there is little or no ecological differentiation as all stages rely on xylofungivory. Instead, it is the roles of larvae and adults in the colony that differ. Larvae dig and enlarge galleries and collect refuse ("balling" of frass from within the tunnels), without adult supervision. Adult beetles protect the brood, shuffle balled frass to the entrance, maintain the fungal crop and block the gallery entrance (usually by the foundress only). All stages and sexes groom to remove external fungus from the body, without which overgrowth is lethal.

Female offspring depart to found their own colony only once the colony reaches a high enough ratio of females to dependent brood. Taking into account all these social behaviours, the ambrosia beetle system evidently is eusocial, even though daughters do not sacrifice their breeding for relatives.

12.3 INQUILINES AND PARASITES OF SOCIAL INSECTS

The abodes of social insects provide many other insects with a hospitable place for their development. The term **inquiline** refers to an organism that shares the home of another. This covers a vast range of organisms that have some kind of obligate relationship with another organism, in this case a social insect. Complex classification schemes involve categorization of the insect host and the known or presumed ecological relationship between inquiline and host (e.g. myrmecophile, termitoxene). However, two alternative divisions appropriate to this discussion involve the degree of integration of the inquiline lifestyle with that of the host. Thus, **integrated inquilines** are incorporated into their hosts' social lives by behavioural modification of both parties, whereas **non-integrated inquilines** are adapted ecologically to the nest, but do not interact socially with the host. Predatory inquilines may negatively affect the host, whereas other inquilines may merely shelter within the nest, or give benefit, such as by feeding on nest debris.

Integration may be achieved by mimicking the chemical cues used by the host in social communication (such as pheromones), or by tactile signalling that releases social behavioural responses, or both. The term **Wasmannian mimicry** covers some or all chemical or tactile mimetic features that allow the mimic to be accepted by a social insect, but the distinction from other forms of mimicry (notably Batesian; section 14.5.1) is unclear. Wasmannian mimicry may, but need not, include imitation of the body form. Conversely, mimicry of a social insect may not imply inquilinism – the ant mimics shown in Fig. 14.12 may gain some protection from their natural enemies as a result of their ant-like appearance, but are not symbionts or nest associates.

The breaking of the social insect chemical code occurs through the ability of an inquiline to produce appeasement and/or adoption chemicals – the messengers that social insects use to recognize one another and to distinguish themselves from intruders. Caterpillars of *Phengaris* (formerly *Maculinea*) *arion* (the large blue butterfly; see Box 1.4) and congeners that develop in the nests of red ants (*Myrmica* spp.) as inquilines or parasites surmount the nest defences of the host ants in various ways. The butterfly larvae secrete semiochemicals that closely mimic the cuticular hydrocarbons of the ant larvae and, when they beg for food, are fed by worker ants. Additionally, the

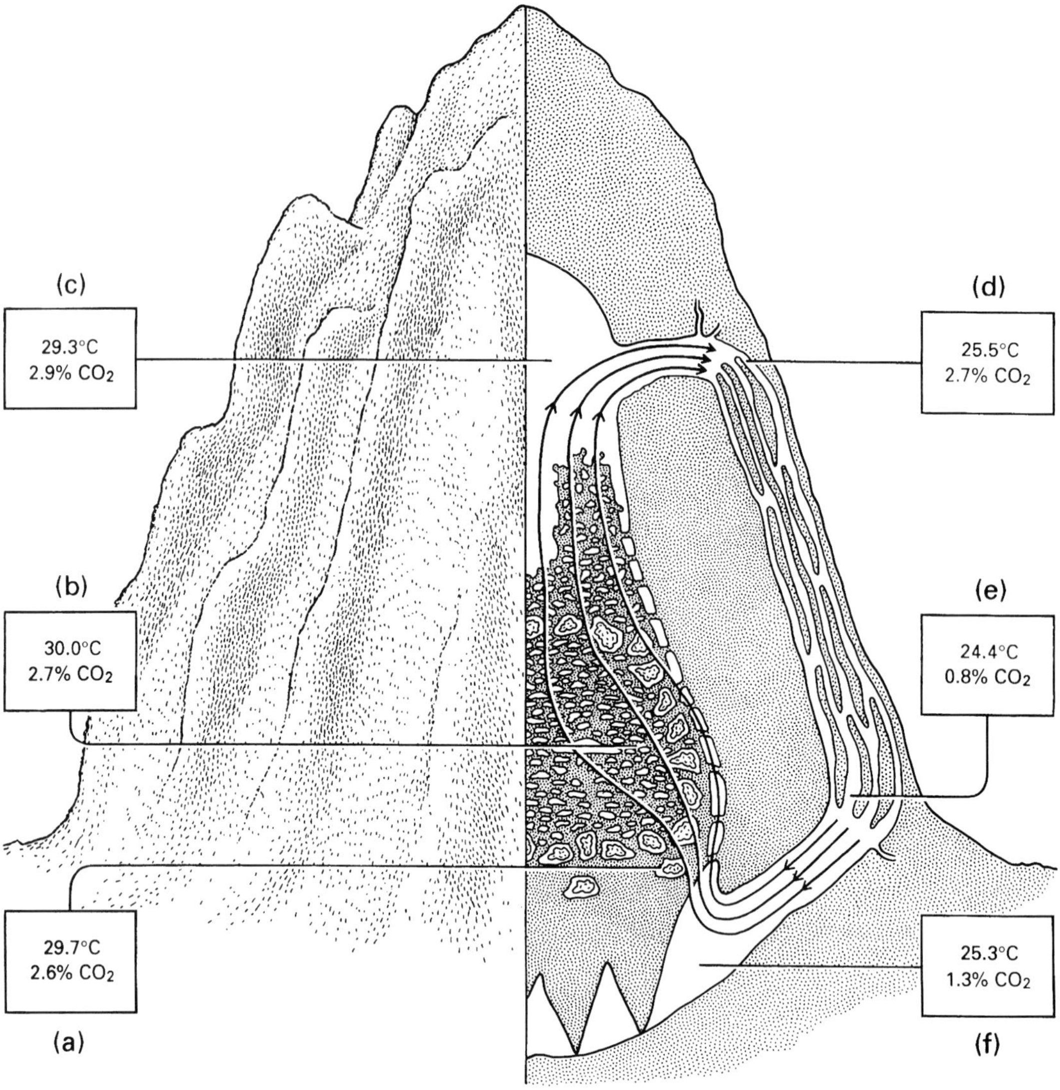

Fig. 12.11 Section through the mound nest of the African fungus-farming termite *Macrotermes natalensis* (Termitidae) showing how air circulating in a series of passageways maintains favourable culture conditions for the fungus at the bottom of the nest (a) and for the termite brood (b). Measurements of temperature and carbon dioxide are shown in the boxes for the following locations: (a) the fungus combs; (b) the brood chambers; (c) the attic; (d) the upper part of a ridge channel; (e) the lower part of a ridge channel; and (f) the cellar. (After Lüscher 1961.)

butterfly larvae and pupae make pulsed sounds that at least in *Phengaris* (formerly *Maculinea*) *rebeli* mimic those of the queen ant, and thus the intruders are treated like royalty.

Certain staphylinid beetles also use chemical mimicry, for example *Atemeles pubicollis*, which lives as a larva in the nest of the European ant, *Formica rufa*. The staphylinid larva produces a glandular secretion that induces brood-tending ants to groom the alien. Food is obtained by adoption of the begging posture of an ant larva, in which the larva rears up and contacts the adult ant's mouthparts, provoking a release of regurgitated food. The diet of the staphylinid is supplemented by predation on larvae of ants and of their own species. Pupation and adult eclosion take place in the *Formica rufa* nest. However, this ant ceases activity in winter and during this period the staphylinid seeks alternative shelter. Adult beetles leave the wooded *Formica* habitat and migrate to the more open grassland habitat of *Myrmica* ants. When a *Myrmica* ant is encountered, secretions from the "appeasement glands" are offered that suppress the aggression of the ant, and then the products of glands on the lateral abdomen attract the ant. Feeding on these secretions appears to facilitate "adoption", as the ant subsequently carries the beetle back to its nest, where the immature adult overwinters as a tolerated food-thief. In spring, the reproductively mature adult beetle departs for the woods to seek out another *Formica* nest for oviposition.

Amongst the inquilines of termites, many show convergence in shape in terms of physogastry (dilation of the abdomen), which is also seen in queen termites. In the curious case of flies of *Termitoxenia* and relatives (Diptera: Phoridae), the physogastric females from termite nests were the only stage known for so long that published speculation was rife that neither larvae nor males existed. It was suggested that the females hatched directly from huge eggs, were brachypterous throughout their lives (hitching a ride on termites for dispersal), and, uniquely amongst the endopterygotes, the flies were believed to be protandrous hermaphrodites, functioning first as males, then as females. The truth is more prosaic: sexual dimorphism in the group is so great that wild-caught, flying males had been placed in a different taxonomic group. The females are winged, but shed all but the stumps of the anterior veins after mating, before entering the termitarium. Although the eggs are large, short-lived larval stages exist. As the post-mated female is **stenogastrous** (with a small abdomen), physogastry must develop whilst in the termitarium. Thus, *Termitoxenia* is only a rather unconventional fly, well adapted to the rigours of life in a termite nest, in which its eggs are treated by the termites as their own, and with attenuation of the vulnerable larval stage, rather than the possessor of a unique suite of life-history features.

Inquilinism is not restricted to non-social insects that breach the colony defences and abuse the hospitality of social insects. **Social parasitism** occurs among species of Hymenoptera, with a few hundred species of ants, some vespine wasps and honey bees known to parasitize their relatives. For example, some ants may live as temporary or even permanent social parasites in the nests of other ant species. A reproductive female inquiline gains access to a host nest and either kills (temporary parasites) or tolerates (permanent inquilines) the resident queen ant. In the case of temporary social parasites, the intruder queen produces workers, which eventually take over the nest. In contrast, for permanent inquilines, the usurper produces only reproductives – the inquiline's worker caste is eliminated – and the nest survives only until the workers of the host ant species die off.

In a further twist of the complex social lives of ants, some species are slave-makers; they capture larvae and pupae from the nests of other species and take them to their own nest where they are reared to be slave workers. This phenomenon, known as **dulosis**, occurs in several inquiline ant species, all of which found their colonies by parasitism.

The phylogenetic relationships between ant hosts and ant inquilines reveal an unexpectedly high proportion of instances in which host and inquiline belong to sister species (i.e. are each other's closest relatives), and many more are congeneric close relatives. One possible explanation envisages the situation in which daughter species formed in isolation come into secondary contact after mating barriers have developed. If no differentiation of colony-identifying chemicals has taken place, it is possible for one species to invade the colony of the other undetected, and parasitization is facilitated.

In southern Africa, social parasitism is affecting hives of African honey bees (*Apis mellifera scutellata*) that are being invaded by a parasitic form of a different subspecies, the Cape honey bee (*A. m. capensis*). The invader workers, which do little work, produce diploid female eggs by thelytoky. These evade the regular policing of the colony, presumably by chemical mimicry of the queen pheromone. The colony is destroyed rapidly by these social parasites, which can then move on to invade another hive. This "capensis problem" is a consequence of commercial beekeepers moving hives of Cape honey bees from their natural areas on the coast to inland regions where only African honey bees occurred naturally.

Non-integrated inquilines are exemplified by hover flies of the genus *Volucella* (Diptera: Syrphidae), the adults of which are Batesian mimics of either *Polistes* wasps or of *Bombus* bees. Female flies appear free to fly in and out of hymenopteran nests, and lay eggs whilst walking over the comb. Hatching larvae drop to the bottom of the nest where they scavenge on fallen detritus and fallen prey. Another syrphid, *Microdon*, has a myrmecophilous larva so curious that it was described first as a mollusc, then as a coccoid. It lives unscathed amongst nest debris (and perhaps sometimes as a predator on young ant larvae), but the emerged adult is recognized as an intruder. Non-integrated inquilines include many predators and parasitoids whose means of circumventing the defences of social insects are largely unknown.

Social insects also support a few parasitic arthropods. For example, varroa and tracheal mites (Acari) and the bee louse, *Braula coeca* (Diptera: Braulidae; section 13.3.3), all live on honey bees (Apidae: *Apis* spp.). The extent of colony damage caused by the tracheal mite *Acarapis woodi* is controversial, but infestations of *Varroa destructor* are resulting in serious declines in honey-bee populations in most parts of the world (see Box 12.3). Varroa mites feed externally on the bee brood (see **Plate 6d**) leading to deformation and death of the bees. Low levels of mite infestation are difficult to detect and it can take several years for a mite population to build to a level that causes extensive damage to the hive. Bee health is further threatened by viral diseases that varroa can transmit. Some *Apis* species, such as *A. cerana* (the Asiatic or Eastern honey bee), are more resistant to varroa than *A. mellifera*. *Apis cerana* is a natural host of *V. destructor* (and of *Nosema ceranae*, the microsporidian parasite that has become a pest of *A. mellifera*) and effectively removes mites by careful grooming behaviour so that hives are not devastated.

12.4 EVOLUTION AND MAINTENANCE OF EUSOCIALITY

At first impression, the complex social systems of hymenopterans and termites bear a close resemblance and it is tempting to suggest a common origin. However, examination of the phylogeny presented in Chapter 7 (see Fig. 7.2) shows that these two groups, and the social aphids and thrips, are distantly related and a single evolutionary origin is inconceivable. Thus, the possible routes for the origin of eusociality in Hymenoptera and Termitoidae are examined separately, followed by a discussion on the maintenance of social colonies.

12.4.1 The origins of eusociality in Hymenoptera

According to estimates derived from the proposed phylogeny of the Hymenoptera, eusociality has arisen independently in wasps, bees and ants (Fig. 12.2), with multiple origins within wasps and bees, and some losses by reversion to solitary behaviour. Comparisons of life histories between living species with different degrees of social behaviour allow extrapolation to possible historical pathways from solitariness to sociality. Three possible routes have been suggested and, in each case, communal living is seen to provide benefits through sharing the costs of nest construction and defence of offspring. Thus, nest construction and provisioning (either with prey or pollen) are key prerequisites for the evolution of eusociality.

The first suggestion envisages a monogynous (single queen) subsocial system, with eusociality developing through the queen remaining associated with her offspring through increased maternal longevity.

In the second scenario, involving semisociality and perhaps applicable only to certain bees, several unrelated females of the same generation associate and establish a colonial nest in which there is some reproductive division of labour, with an association that lasts only for one generation.

The third scenario involves elements of the previous two, with a communal group comprising related females (rather than unrelated) and multiple queens (in a polygynous system), within which there is increasing reproductive division. The association of queens and daughters arises through increased longevity.

These life-history-based scenarios must be considered in relation to genetic theories concerning eusociality, notably concerning the origins and maintenance by selection of **altruism** (or self-sacrifice in reproduction). Ever since Darwin, there has been debate about altruism – why should some individuals (non-reproductive workers) sacrifice their reproductive potential for the benefit of others?

Four proposals for the origins of the extreme reproductive sacrifice seen in eusociality are discussed below. Three proposals are partially or completely compatible with one another, but group selection, the first considered, seems incompatible. In this case, selection is argued to operate at the level of the group: an efficient colony with an altruistic division of reproductive labour will survive and produce more offspring than one in which rampant individual self-interest leads to anarchy. Although this scenario aids in understanding the maintenance of eusociality once it is established, it contributes little, if anything, to explaining the origin(s) of reproductive sacrifice in non-eusocial or subsocial insects. The concept of group selection operating on pre-eusocial colonies runs counter to the view that selection operates on the genome, and hence the origin of altruistic individual sterility is difficult to accept under group selection. It is amongst the remaining three proposals, namely kin selection, maternal manipulation, and mutualism, that the origins of eusociality are more usually sought.

The first, kin selection, stems from recognition that Darwinian or classical fitness – the direct genetic contribution to the gene pool by an individual through its offspring – is only part of the contribution to an individual's total, or inclusive or extended, fitness. An additional indirect contribution, termed the **kinship component**, must be included. This is the contribution to the gene pool made by an individual that assists and enhances the reproductive success of its kin. Kin are individuals with similar or identical genotypes derived from the relatedness due to having the same parents. In the Hymenoptera, kin relatedness is enhanced by the haplodiploid genetic system. In this system, males are haploid, so that each sperm (produced by mitosis) contains 100% of the paternal genes. In contrast, the egg (produced by meiosis) is diploid, containing only half the maternal genes. Thus, daughter offspring, produced from fertilized eggs, share all their father's genes, but only half of their mother's genes. Because of this, full sisters (i.e. those with the same father) share on average three-quarters of their genes. Therefore, sisters share more genes with each other than they would with their own female offspring (50%). Under these conditions, the inclusive fitness of a sterile female (worker) is greater than its classical fitness. As selection operating on an individual should maximize its inclusive fitness, a worker should invest in the survival of her sisters, the queen's offspring, rather than in the production of her own female young.

However, haplodiploidy alone is an inadequate explanation for the origin of eusociality, because altruism does not arise solely from relatedness. Haplodiploidy is universal in hymenopterans and kinship has encouraged repeated eusociality, but eusociality is not universal in the Hymenoptera. Furthermore, other haplodiploid insects, such as thrips, are not eusocial, although there may be social behaviour. Other factors promoting eusociality are recognized in Hamilton's rule, which emphasizes the ratio of costs and benefits of altruistic behaviour as well as relatedness. The conditions under which selection will favour altruism can be expressed as follows:

$$rB - C > 0$$

for which r is the coefficient of relatedness, B is the benefit gained by the recipient of altruism, and C is the cost suffered by the donor of altruism. Thus, variations in benefits and costs modify the consequences of the particular degrees of relatedness expressed in Fig. 12.12, although these factors are difficult to quantify.

Kinship calculations assume that all offspring of a single mother in the colony share an identical father, and this assumption is implicit in the kinship scenario for the origin of eusociality. At least in higher eusocial insects, queens may mate multiply with different males, and thus r-values are less than predicted by the monogamous model. This effect impinges on maintenance of an already existing eusocial system, discussed in section 12.4.3. Whatever, the opportunity to help relatives, in combination with high relatedness through haplodiploidy, predisposes insects to eusociality.

At least two further ideas concern the origins of eusociality. The first involves maternal manipulation of offspring (both behaviourally and genetically), such that by reducing the reproductive potential of certain offspring, parental fitness may be maximized by assuring reproductive success of a few select offspring. Most female Aculeata can control the sex of offspring through fertilizing the egg or not, and are able to vary offspring size through the amount of food supplied, making maternal manipulation a plausible option for the origin of eusociality.

A further, well-supported scenario emphasizes the roles of competition and mutualism. This envisages individuals acting to enhance their own classical fitness, with contributions to the fitness of neighbours

Fig. 12.12 Relatedness of a given worker to other possible occupants of a hive. (After Whitfield 2002.)

	sister	half-sister	own son	son of full sister	queen's son (brother)	son of half-sister
worker	0.75	0.375	0.5	0.375	0.25	0.125

arising only incidentally. Each individual benefits from colonial life through communal defence by shared vigilance against predators and parasites. Thus, mutualism (including the benefits of shared defence and nest construction) and kinship encourage the establishment of group living. Differential reproduction within a familial-related colony confers significant fitness advantages on all members through their kinship.

In conclusion, the three scenarios are not mutually exclusive, but are compatible in combination, with kin selection, female manipulation, and mutualism acting in concert to encourage evolution of eusociality.

The Vespinae illustrate a trend to eusociality commencing from a solitary existence, with nest-sharing and facultative labour division being a derived condition. Further evolution of eusocial behaviour is envisaged as developing through a dominance hierarchy that arose from female manipulation and reproductive competition among the nest-sharers: the "winners" are queens and the "losers" are workers. From this point onwards, individuals act to maximize their fitness and the caste system becomes more rigid. As the queen and colony acquire greater longevity and the number of generations retained increases, short-term monogynous societies (those with a succession of queens) become long-term, monogynous, **matrifilial** (mother–daughter) colonies. Exceptionally, a derived polygynous condition may arise in large colonies, and/or in colonies where queen dominance is relaxed.

The evolution of sociality from solitary behaviour should not be seen as unidirectional, with the eusocial bees and wasps at a "pinnacle". Recent phylogenetic studies show many reversions from eusocial to semisocial and even to solitary lifestyles. Such reversions have occurred in halictid and allodapine bees. These losses demonstrate that even with haplodiploidy predisposing towards group living, unsuitable environmental conditions can counter this trend, with selection able to act against eusociality.

12.4.2 The origins of eusociality in termites

In contrast to the haplodiploidy of Hymenoptera, termite sex is determined universally by an XX–XY chromosome system, and thus there is no genetic predisposition toward kinship-based eusociality. Furthermore, and in contrast to the widespread subsociality of hymenopterans, the lack of any intermediate stages on the route to termite eusociality has obscured its origin. Subsocial behaviours in some mantids and cockroaches (the nearest relatives of the termites) have been proposed to be an evolutionary precursor to the eusociality in termites. Notably, behaviour in the family Cryptocercidae, which is sister branch to the termite lineage (see Fig. 7.4), demonstrates how reliance on a nutrient-poor food source and adult longevity might predispose to social living. The internal symbiotic organisms needed to aid in the digestion of a cellulose-rich, but nutrient-poor, diet of wood is central to this argument. The need to transfer symbionts to replenish supplies lost at each moult encourages unusual levels of intracolony interaction through trophallaxis. Furthermore, transfer of symbionts between members of successive generations requires overlapping generations. Trophallaxis, slow growth induced by the poor diet, and parental longevity, act together to encourage group cohesion. These factors, together with patchiness of adequate food resources such as rotting logs, can lead to colonial life, but do not readily explain altruistic caste origins. When an individual gains substantial benefits from successful foundation of a colony, and where there is a high degree of intracolony relatedness (as is found in some termites), eusociality may arise. However, the origin of eusociality in termites remains much less clear-cut than in eusocial hymenopterans.

12.4.3 Maintenance of eusociality – the police state

As we have seen, workers in social hymenopteran colonies forgo their reproduction and raise the brood of their queen, in a system that depends upon kinship – proximity of relatedness – to "justify" their sacrifice. Once non-reproductive castes have evolved (theoretically under conditions of single paternity), the requirement for high relatedness may be relaxed if workers lack any opportunity to reproduce, through mechanisms such as chemical control by the queen. Nonetheless, sporadically, and especially when the influence of the queen wanes, some workers may lay their own eggs. These "non-queen" eggs are not allowed to survive: the eggs are detected and eaten by a "police force" of other workers. This is known from honey bees, certain wasps, and some ants, and may be quite widespread, although uncommon. For example, in a typical honey-bee hive of 30,000 workers, on average only three have functioning ovaries. Although other workers threaten these individuals, they can be responsible for up to 7% of the male eggs in any colony. The eggs of these workers lack chemical odours produced by the queen, and thus they can be detected and are eaten by the policing workers with such efficiency that only 0.1% of a honey-bee colony's males derive from a worker as a mother.

Hamilton's rule (section 12.4.1) provides an explanation for the policing behaviour. The relatedness of a sister to her sister (worker to worker) is $r = 0.75$, which is reduced to $r = 0.375$ if the queen has multiply mated (as does occur). An unfertilized egg of a worker, if allowed to develop, becomes a son to which his mother's relatedness is $r = 0.5$. This kinship value is greater than to her half-sisters ($0.5 > 0.375$), thus providing an incentive to escape queen control. However, from the perspective of the other workers, their kinship to the son of another worker is only $r = 0.125$, "justifying" the killing of a half-nephew (another worker's son), and tending the development of her sisters ($r = 0.75$) or half-sisters ($r = 0.375$) (relationships portrayed in Fig. 12.12). The evolutionary benefits to any worker derive from raising the queen's eggs and destroying her sisters' eggs. However, when the queen's strength wanes or she dies, the pheromonal repression of the colony ceases, anarchy breaks out, and the workers all start to lay eggs.

Outside the extreme rigidity of the honey-bee colony, a range of policing activities can be seen. In colonies of ants that lack clear division into queens and workers, a hierarchy exists, with only certain individuals' reproduction tolerated by nest-mates. Although enforcement involves violence towards an offender, such regimes have some flexibility, since there is regular ousting of the reproductives. Even for honey bees, as the queen's performance diminishes and her pheromonal control wanes, workers' ovaries develop and rampant egg laying takes place. Workers of some vespids discriminate between offspring of a singly-mated or a promiscuous queen, and behave according to kinship. Presumably, polygynous colonies at some stage have allowed additional queens to develop, or to return and be tolerated, providing possibilities for invasiveness by relaxed inter-nest interactions (as found in many tramp ants; see Box 1.3). The inquilines discussed in section 12.3 and Box 1.4 evidently evade policing efforts, mostly by chemical mimicry of the host species.

12.5 SUCCESS OF SOCIAL INSECTS

Social insects attain numerical and ecological dominance in many regions. In Box 1.3 and Box 12.4 we give some examples of social insects becoming pests by their dominance and/or invasiveness. Social insects tend to occur in high abundance at low latitudes and low elevations, and their activities are conspicuous in summer in temperate areas, or year-round in subtropical to tropical climates. As a generalization, the most abundant and dominant social insects are the most derived phylogenetically, and have the most complex social organization.

Three qualities of social insects contribute to their competitive advantage, all of which derive from the caste system that allows multiple tasks to be performed. Firstly, the tasks of foraging, feeding the queen, caring for offspring, and maintenance of the nest can be performed simultaneously by different groups, rather than sequentially as in solitary insects. Performing tasks in parallel means that one activity does not jeopardize another, thus the nest is not vulnerable to predators or parasites whilst foraging is taking place. Furthermore, individual errors have little or no consequence in parallel operations compared with those performed serially. Secondly, the ability of the colony to marshal all workers can overcome serious difficulties that a solitary insect cannot deal with, such as defence against a much larger or more numerous predator, or construction of a nest under unfavourable conditions. Thirdly, the specialization of function associated with castes allows some homeostatic regulation, including holding of food reserves in some castes (such as honeypot ants) or in developing larvae, and behavioural control of temperature and other microclimatic conditions within the nest. The ability to vary the proportion of individuals allocated to a particular caste allows distribution of resources according to differing demands of season and colony age. The widespread use of a variety of pheromones allows control to be exerted, even over millions of individuals. However, within the apparently rigid caste system, there is scope for many different life histories to have evolved, from nomadic army ants to parasitic inquilines.

Box 12.4 Social insects as urban pests

Humans increasingly live in cities – urban areas of high population densities compared with rural or agricultural settings – and divorced from more natural areas. Urban living means less exposure to, and hence tolerance of, insects than in most rural areas, where a greater diversity of insects occur. Additionally, certain insects, especially those with social lifestyles, have adapted well to cities, where they are often referred to as pests because they compete for resources. These pests vary with location – structural damaging termites and invasive ants are most common in warm temperate to tropical regions, whereas yellow jackets (several species of vespine wasps) can plague temperate areas seasonally. Our concerns over these insects (and others such as bed bugs and cockroaches) that share our urban environments mean that urban pest control and eradication is a burgeoning industry.

About 80 species of termites are serious pests of our wooden structures, with an additional 100 or so causing nuisance somewhere in the world. The cost of termite damage and control worldwide is estimated at upwards of US$40 billion annually. Most damaging are subterranean termites, especially *Coptotermes*, *Reticulitermes* and *Odontotermes*, which make galleries in soil to forage for food. Thus, the best termite control is attained by isolating timber of buildings from soil contact by a concrete slab with a barrier of long-lasting insecticide or repellent beneath and around the slab. If breaches occur, treatment involves removal of infected wood, and either soil treatment with insecticide or a baiting and monitoring programme. A range of commercial baits are available, and termites are conditioned to using baiting stations prior to their replacement with baits containing an active insecticide (especially chitin synthesis inhibitors; section 16.4.2). Recently, a weather-durable bait has been developed that does not require more than annual monitoring, reducing costs of repeat visits to an infested site.

More visible to urban dwellers, and interacting with us differently, are many ants that share our surroundings. Globally, more than 40 species are pests associated with our housing, other buildings and gardens or yards. Pest ants include those discussed in Box 1.3 as "tramp ants", especially the near-global Argentine ant, *Linepithema humile* (see **Plate 6e**), and the pharaoh ant, *Monomorium pharaonis*. Included in the list of pests are other disturbance-loving and ecologically dominant ants, some of which are native where they are pests. The diverse fire ants (*Solenopsis* species) have large colonies, high densities and aggressive stinging behaviour, with *S. invicta* (the red imported fire ant; **Plate 6e**) causing justifiable complaint. Another feature of many nuisance ants is their tendency to "farm" (that is, tend, manage and protect) for their honeydew (section 11.2.4) any sap-sucking hemipterans feeding on ornamental and garden plants, including indoor plants. Ant control by synthetic pheromones and disruption of trails has had limited success, and the best control, although expensive, has come from baits laced with slow-acting insecticides that are returned to the nest by foraging workers.

Just a few social wasps cause nuisance, mainly species of *Vespula* (*V. germanica*, *V. pensylvanica* and *V. vulgaris*) that are invasive outside their natural ranges. *Vespula pensylvanica* from the western United States is established now in Hawai'i, and the Eurasian *V. vulgaris* and *V. germanica* have spread to North America and many places in the southern hemisphere. Away from natural control, including by severe winters, numbers can build dramatically in their new ranges, especially where winters are mild. A nuisance duration of a few weeks in the home range can extend to half a year elsewhere, with mid-summer numbers so high that outdoor activities are curtailed and tourists head home. Deaths resulting from stings number in the high hundreds globally and annually, and as with reactions to fire ants, an anaphylactic reaction in sensitized people may be responsible. Particular care must be taken in drinking sweet beverages outdoors as *Vespula* wasps are attracted to these liquids. Destruction of nests (with care) reduces nuisance, but trapping of adults is ineffectual unless done intensively. The use of baits laced with fipronil insecticide (section 16.4.1) can eliminate or greatly reduce colony activity, but no registered wasp baits formulated with this chemical are available to the public.

Control of pest ants, termites and wasps tends to be on a property-by-property basis (rather than a wider area), by often poorly skilled operators or naïve house owners. Eradication of pests from one property is ineffective if a neighbouring source remains. The huge diversity of insecticides available to the public surely leads to inappropriate or excessive use, resulting in death of non-target beneficial insects (including honey bees) and toxic runoff into waterways.

Notwithstanding popular alarm at bees and wasps, encouraged by several movies, there can be positive interactions with some social insects in our cities. A growing movement to tend hive bees has developed in London (UK), both Melbourne and Sydney (Australia), and several cities of North America. Numbers have grown such that there are fears that nectar sources may be inadequate to support so many urban apiarists. Just how urban bees and the honey they produce interact with urban pesticide use needs more study.

FURTHER READING

Abe, T., Bignell, D.E. & Higashi, M. (eds) (2000) *Termites: Evolution, Sociality, Symbioses and Ecology*. Springer, Berlin.

Allsopp, M. (2004) Cape honeybee (*Apis mellifera capensis* Eshscholtz) and varroa mite (*Varroa destructor* Anderson & Trueman) threats to honeybees and beekeeping in Africa. *International Journal of Tropical Insect Science* **24**, 87–94.

Barbero, F., Thomas, J.A., Bonelli, S. *et al.* (2009) Queen ants make distinctive sounds that are mimicked by a butterfly social parasite. *Science* **323**, 782–5.

Bignell, D.E., Roisin Y. & Lo, N. (eds) (2011) *Biology of Termites: A Modern Synthesis*. Springer, Dordrecht, New York.

Choe, J.C. & Crespi, B.J. (eds.) (1997) *Social Behavior in Insects and Arachnids*. Cambridge University Press, Cambridge.

Costa, J.T. (2006) *The Other Insect Societies*. Belknap Press of Harvard University Press, Cambridge, MA.

Crozier, R.H. & Pamilo, P. (1996) *Evolution of Social Insect Colonies: Sex Allocation and Kin Selection*. Oxford University Press, Oxford.

Danforth, B.N., Cardinal, S., Praz, C. *et al.* (2013) The impact of molecular data on our understanding of bee phylogeny and evolution. *Annual Review of Entomology* **58**, 57–78.

Dyer, F.C. (2002) The biology of the dance language. *Annual Review of Entomology* **47**, 917–49.

Eggleton, P. & Tayasu, I. (2001) Feeding groups, lifetypes and the global ecology of termites. *Ecological Research* **16**, 941–60.

Evans, T.A., Forschler, B.T. & Grace, J.K. (2013) Biology of invasive termites: a worldwide review. *Annual Review of Entomology* **58**, 455–74.

Farooqui, T. (2013) A potential link among biogenic amines-based pesticides, learning and memory, and colony collapse disorder: a unique hypothesis. *Neurochemistry International* **62**, 122–36.

Goulson, D. (2013) Neonicotinoids and bees. What's all the buzz? *Significance* **10**(3), 6–11.

Henderson, G. (2008) The termite menace in New Orleans: did they cause the floodwalls to tumble? *American Entomologist* **54**, 156–62.

Hölldobler, B. & Wilson, E.O. (1990) *The Ants*. Springer-Verlag, Berlin.

Hölldobler, B. & Wilson, E.O. (2008) *The Superorganism: The Beauty, Elegance, and Strangeness of Insect Societies*. W.W. Norton & Co., New York.

Holman, L., Jørgensen, C.G., Nielsen, J. & d'Ettorre, P. (2010) Identification of an ant queen pheromone regulating worker sterility. *Proceedings of the Royal Society B* **277**, 3793–800.

Itô, Y. (1989) The evolutionary biology of sterile soldiers in aphids. *Trends in Ecology and Evolution* **4**, 69–73.

Kamakura, M. (2011) Royalactin induces queen differentiation in honeybees. *Nature* **473**, 478–83.

Korb, J. & Hartfelder, K. (2008) Life history and development – a framework for understanding developmental plasticity in lower termites. *Biological Reviews* **83**, 295–313.

Kranz, B.D., Schwarz, M.P., Morris, D.C. & Crespi, B.J. (2002) Life history of *Kladothrips ellobus* and *Oncothrips rodwayi*: insight into the origin and loss of soldiers in gall-inducing thrips. *Ecological Entomology* **27**, 49–57.

Legendre, F., Whiting, M.F., Bordereau, C. *et al.* (2008) The phylogeny of termites (Dictyoptera: Isoptera) based on mitochondrial and nuclear markers: implications for the evolution of the worker and pseudergate castes, and foraging behaviors. *Molecular Phylogenetics and Evolution* **48**, 615–27.

Lenior, A., D'Ettorre, P., Errard, C. & Hefetz, A. (2001) Chemical ecology and social parasitism in ants. *Annual Review of Entomology* **46**, 573–99.

Quicke, D.L.J. (1997) *Parasitic Wasps*. Chapman & Hall, London.

Rabeling, C. & Kronauer, D.J.C. (2013) Thelytokous parthenogenesis in eusocial Hymenoptera. *Annual Review of Entomology* **58**, 273–92.

Resh, V.H. & Cardé, R.T. (eds.) (2009) *Encyclopedia of Insects*, 2nd edn. Elsevier, San Diego, CA. [Particularly see articles on *Apis* species; beekeeping; caste; dance language; division of labor in insect societies; Hymenoptera; Isoptera; parental care; sociality.]

Rust, M.K. & Su, N.-Y. (2012) Managing social insects of urban importance. *Annual Review of Entomology* **57**, 355–75.

Sammataro, D., Gerson, U. & Needham, G. (2000) Parasitic mites of honey bees: life history, implications and impact. *Annual Review of Entomology* **45**, 519–48.

Schneider, S.S., DeGrandi-Hoffman, G. & Smith, D.R. (2004) The African honey bee: factors contributing to a successful biological invasion. *Annual Review of Entomology* **49**, 351–76.

Schwarz, M.P., Richards, M.H, & Danforth, B.N. (2007) Changing paradigms in insect social evolution: insights from halictine and allodapine Bees. *Annual Review of Entomology* **52**, 127–50.

Van Engelsdorp, D., Evans, J.D., Saegerman, C. *et al.* (2009) Colony Collapse Disorder: A descriptive study. *PLoS ONE* **4**, e6481. doi: 10.1371/journal.pone.0006481.

Wong, J.W.Y., Meunier, J. & Kölliker, M. (2013) The evolution of parental care in insects: the roles of ecology, life history and the social environment. *Ecological Entomology* **38**, 123–37.

Chapter 13

INSECT PREDATION AND PARASITISM

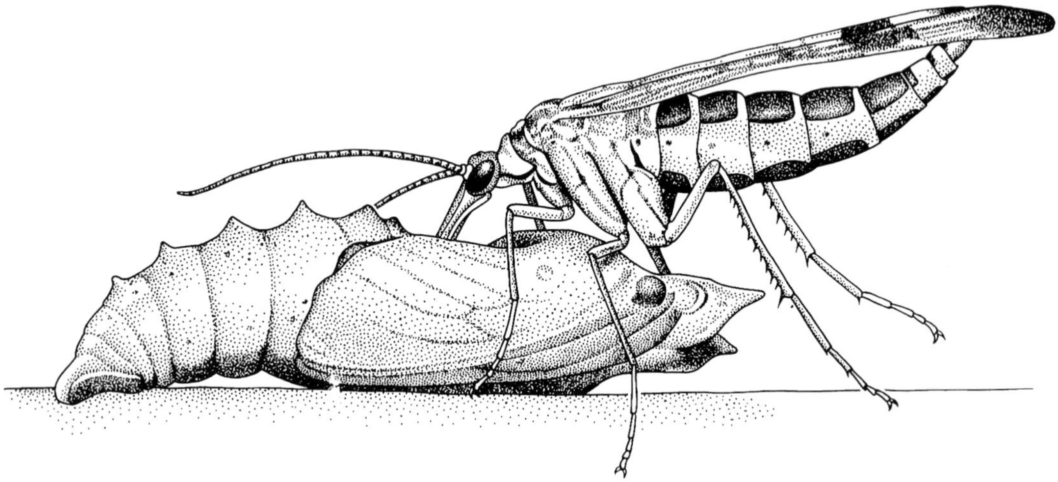

Scorpionfly feeding on a butterfly pupa. (After a photograph by P.H. Ward & S.L. Ward.)

Many insects are phytophagous, feeding directly on primary producers, the algae and higher plants. These abundant insects are fed upon by a range of carnivorous organisms. Individuals within this broad carnivorous group may be categorized as follows. A **predator** kills and consumes a number of **prey** animals during its life. **Predation** involves interactions in space and time between predator foraging and prey availability, although often it is treated in a one-sided manner as if predation is what the predator does. Animals that live at the expense of another single animal (a **host**) that dies as a result are called **parasitoids**; they may live externally (**ectoparasitoids**) or internally (**endoparasitoids**). Those that live at the expense of another animal (also a host) that they do not kill (or rarely significantly harm) are **parasites**, which likewise can be internal (**endoparasites**) or external (**ectoparasites**). A host attacked by a parasitoid or parasite is **parasitized**, and **parasitization** is the condition of being parasitized. **Parasitism** describes the relationship between parasitoid or parasite and the host. Predators, parasitoids and parasites, although defined above as if distinct, may not be so clear-cut, as parasitoids may be viewed as specialized predators.

By some estimates, about 25% of insect species are predatory or parasitic in feeding habit in some life-history stage. Representatives from amongst nearly every order of insects are predatory, with adults and immature stages of the Odonata, Mantophasmatodea, Mantodea and the Neuropterida orders (Neuroptera, Megaloptera and Raphidioptera), and adults of the Mecoptera, being almost exclusively predatory. These orders are considered in Taxobox 5, Taxobox 13, Taxobox 14, Taxobox 21 and Taxobox 25. The vignette at the beginning of this chapter depicts a female mecopteran, *Panorpa communis* (Panorpidae), feeding on a dead pupa of a small tortoiseshell butterfly, *Aglais urticae*. A number of species in the primarily phytophagous orders Hemiptera (Taxobox 20) and Coleoptera (Taxobox 22) are predatory, and a few bugs are blood feeders. The Hymenoptera (Taxobox 29) are speciose, with a preponderance of parasitoid taxa that almost exclusively use invertebrate hosts. The uncommon Strepsiptera (Taxobox 23) are unusual in being endoparasitoids in other insects. Parasites that are of medical or veterinary importance, such as lice, many Diptera and adult fleas, are considered in Chapter 15 and in Taxobox 18, Taxobox 24 and Taxobox 26, respectively.

Insects are amenable to field and laboratory studies of predator–prey interactions as they are easy to manipulate, may have several generations a year, and show a range of predatory and defensive strategies and life histories. Furthermore, studies of predator–prey and parasitoid–host interactions are fundamental to biological control strategies for pest insects. Attempts to model predator–prey interactions mathematically often emphasize parasitoids, as some simplifications can be made. These include the ability to simplify search strategies, as only the adult female parasitoid seeks hosts, and the number of offspring per unit host remains relatively constant from generation to generation.

In this chapter, we show how predators, parasitoids and parasites **forage**, i.e. locate and select their prey or hosts. We look at morphological modifications of predators for handling prey, and how some of the prey defences covered in Chapter 14 are overcome. The means by which parasitoids overcome host defences and develop within their hosts is examined, and different strategies of host use by parasitoids are explained. The use by certain parasitoid Hymenoptera of viruses or virus-like particles to overcome host immunity is considered in Box 13.1. The host use and specificity of ectoparasites is discussed from a phylogenetic perspective. We conclude by considering the relationships between predator/parasitoid/parasite and prey/host abundances and evolutionary histories.

13.1 PREY/HOST LOCATION

Insects forage in a stereotyped sequence that leads a predator or host-seeker toward a resource and, on contact, enables the insect to recognize and use it. Various stimuli along the route elicit an appropriate ensuing response, involving either action or inhibition. The foraging strategies of predators, parasitoids and parasites involve trade-offs between profits or benefits (the quality and quantity of resource obtained) and cost (in the form of time expenditure, exposure to suboptimal or adverse environments and the risks of

The Insects: An Outline of Entomology, Fifth Edition. P.J. Gullan and P.S. Cranston.

Companion Website: www.wiley.com/go/gullan/insects

being eaten). Recognition of the time component is important, as all time spent in activities other than reproduction can be viewed, in an evolutionary sense, as time wasted.

In an optimal foraging strategy, the difference between benefits and costs is maximized, either through increasing nutrient gain from prey capture or reducing effort expended to catch prey, or both. Choices available are:

- where and how to search;
- how much time to expend in fruitless search in one area before moving;
- how much (if any) energy to expend in capture of sub-optimal food, once located.

A primary requirement is that the insect be in the appropriate habitat for the resource sought. For many insects this may seem trivial, especially if development takes place in the area that contained the resources used by the parental generation. However, circumstances such as seasonality, climatic vagaries, ephemerality or major resource depletion, may necessitate local **dispersal** or perhaps major movement (**migration**) in order to reach an appropriate location.

Even in a suitable habitat, resources usually are not distributed evenly but occur in more or less discrete microhabitat clumps, termed **patches**. Insects show a gradient of responses to these patches. At one extreme, the insect waits in a suitable patch for prey or host organisms to appear. A predator may be camouflaged or apparent, and a trap may be constructed. At the other extreme, the prey or host is actively sought within a patch. As seen in Fig. 13.1, the waiting strategy is economically effective, but time-consuming; the active strategy is energy intensive, but time-efficient; and trapping lies intermediate between these two. Patch selection is vital to successful foraging.

13.1.1 Sitting and waiting

Sit-and-wait predators find a suitable patch and wait for mobile prey to come within striking range. As the vision of many insects limits them to recognition of movement rather than precise shape, a motionless sit-and-wait predator may be unobserved by its prey. Nonetheless, amongst those that wait, many have some form of camouflage (**crypsis**). This may be defensive, being directed against highly visual predators such as birds, rather than evolved to mislead invertebrate prey. Cryptic predators modelled on a feature that is of no interest to the prey (such as tree bark, lichen, a twig or even a rock) can be distinguished from those that model on a feature of some significance to prey, such as a flower that acts as an insect attractant.

In an example of the latter case, the Malaysian mantid *Hymenopus bicornis* closely resembles the red flowers of the orchid *Melastoma polyanthum* amongst which it rests. Flies are encouraged to land, assisted by the presence of marks resembling flies on the body of the mantid: larger flies that land are eaten by the mantid. In another related example of aggressive foraging mimicry, the African mantid *Idolum* resembles a flower due to petal-shaped, coloured outgrowths of the prothorax and the coxae of the anterior legs. Butterflies and flies attracted to this hanging "flower" are snatched and eaten.

Ambushers include cryptic, sedentary insects such as mantids, which prey fail to distinguish from the plant background. Although these predators take passing invertebrates, often they locate close to flowers, to take advantage of the increased visiting rate of flower feeders and pollinators.

Odonate nymphs, which are major aquatic predators, are classic ambushers and rest concealed in

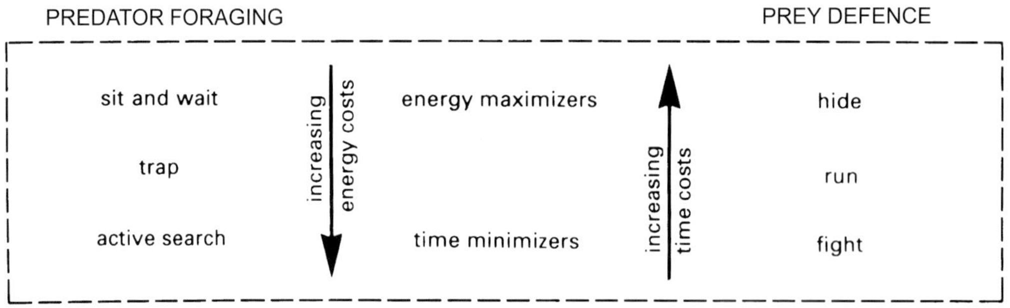

Fig. 13.1 The basic spectrum of predator foraging and prey defence strategies, varying according to costs and benefits in both time and energy. (After Malcolm 1990.)

submerged vegetation or in the substrate, awaiting prey. If waiting fails to provide food, the hungry insect may change to a more active searching mode. This energy expenditure may bring the predator into an area of higher prey density. In running waters, many predators drift passively with the current to relocate, perhaps induced by local prey shortage.

Sitting-and-waiting strategies are not restricted to cryptic and slow-moving predators. Fast-flying, diurnal, visual, rapacious predators such as many robber flies (Diptera: Asilidae) and adult odonates perch prominently on vegetation. From these conspicuous locations they detect passing flying insects using their excellent sight. With rapid and accurately controlled flight, the predator makes only a short foray to capture appropriately sized prey. This strategy combines energy saving, through not needing to fly incessantly in search of prey, with time efficiency, as prey is taken only from outside the immediate area of reach of the predator.

A sit-and-wait technique that involves greater energy expenditure is the use of traps to ambush prey. Although spiders are the prime exponents of this method, in the warmer parts of the world the pits of certain larval antlions (Neuroptera: Myrmeleontidae) (Fig. 13.2a,b) are familiar traps. The larvae either dig pits directly or form them by spiralling backwards into soft soil or sand. Trapping effectiveness depends upon the steepness of the sides, the diameter and the depth of the pit, which vary with species and instar. The larva waits, buried at the base of the conical pit, for passing prey to fall in. Escape is prevented physically by the slipperiness of the slope, and the larva may also flick sand at prey to restrict its defensive movements before dragging it underground. The location, construction and maintenance of the pit are vitally important to capture efficiency, but construction and repair are energetically very expensive. Experimentally, it has been shown that even starved Japanese antlions (*Myrmeleon bore*) would not relocate their pits to an area where prey was provided artificially. Instead, larvae of this species of antlion reduce their metabolic rate to tolerate famine, even if death by starvation is the result.

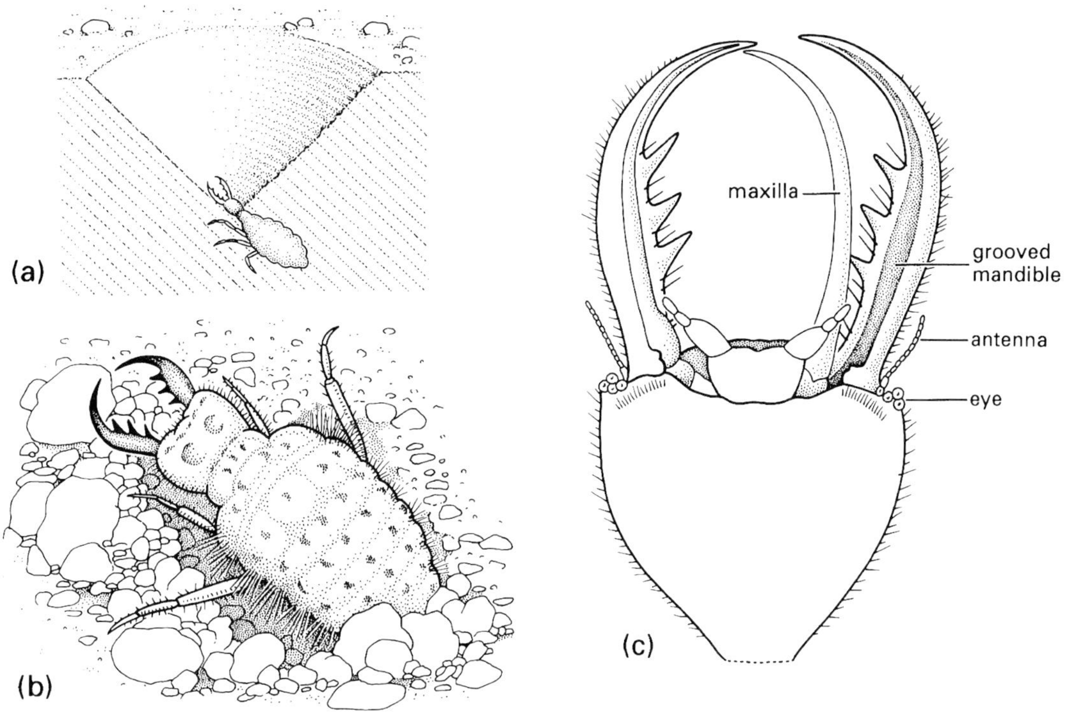

Fig. 13.2 An antlion of *Myrmeleon* (Neuroptera: Myrmeleontidae): (a) larva in its pit in sand; (b) detail of dorsum of larva; (c) detail of ventral view of larval head showing how the maxilla fits against the grooved mandible to form a sucking tube. (After Wigglesworth 1964.)

In holometabolous ectoparasites, such as fleas and parasitic flies, immature development takes place away from their vertebrate hosts. Following pupation, the adult must locate an appropriate host. Since the eyes are reduced or absent in many of these ectoparasites, vision cannot be used. Furthermore, as many are flightless, mobility is restricted. In fleas and some Diptera, in which larval development often takes place in the nest of a host vertebrate, the adult insect waits quiescent in the pupal cocoon until the presence of a host is detected. Quiescence may extend to a year or longer, as in the cat flea (*Ctenocephalides felis*) – a familiar phenomenon to humans who enter an empty dwelling that previously housed flea-infested cats. The stimuli to cease dormancy include some or all of: vibration; rise in temperature; increased carbon dioxide; or another stimulus generated by the host.

In contrast, the hemimetabolous lice spend their lives entirely on a host, with all developmental stages being ectoparasitic. Any transfer between hosts is either through phoresy (section 13.1.3) or when host individuals make direct contact, as from mother to young within a nest.

13.1.2 Active foraging

Active searching in suitable patches for prey or hosts requires energetic foraging. Movements associated with foraging and with other locomotory activities, such as seeking a mate, are so similar that the "motivation" may be recognized only in retrospect, by resultant prey capture or host finding. The locomotory search patterns used to locate prey or hosts are those described for general orientation in section 4.5, and comprise non-directional (random) and directional (non-random) locomotion.

Random, or non-directional foraging

The foraging of larval coccinellid beetles and syrphid flies amongst their clumped aphid prey illustrates several features of random food searching. The larvae advance, stop periodically, and "cast" about by swinging their raised anterior bodies from side to side. Subsequent behaviour depends upon whether or not an aphid is encountered. If no prey is encountered, motion continues, interspersed with casting and turning at a fundamental frequency. However, if contact is made and feeding has taken place or if the prey is encountered and lost, searching intensifies with an enhanced frequency of casting and, if the larva is in motion, increased turning or direction changing. Actual feeding is unnecessary to stimulate this more concentrated search: an unsuccessful encounter is adequate. For early-instar larvae that are very active but have limited ability to handle prey, this stimulus to search intensively near a lost feeding opportunity is important to survival.

Most laboratory-based experiments and models of foraging are derived from single species of walking predators, frequently assumed to encounter a single species of prey randomly distributed within selected patches. Only in an agricultural monoculture with a single pest controlled by one predator are these assumptions realistic. However, despite the limitations of most models, certain findings appear to have general biological relevance.

An important consideration is that the time allocated to different patches by a foraging predator depends upon the criteria for leaving a patch. Four mechanisms have been recognized to trigger departure from a patch:

1 a certain number of food items have been encountered (fixed number);
2 a certain time has elapsed (fixed time);
3 a certain searching time has elapsed (fixed searching time);
4 the prey capture rate falls below a certain threshold (fixed rate).

The fixed-rate mechanism has been favoured by modellers of optimal foraging, but even this is likely to be a simplification if the forager's responsiveness to prey is non-linear (e.g. declines with exposure time) and/or derives from more than simple prey encounter rate, or prey density. Differences between predator–prey interactions in simplified laboratory conditions and the actuality of the field cause many problems, including failure to recognize variation in prey behaviour that results from exposure to predation (perhaps multiple predators). Furthermore, there are difficulties in interpreting the actions of polyphagous predators, including the causes of predator/parasitoid/parasite behavioural switching between different prey animals or hosts.

Non-random, or directional foraging

Several more specific, directional means of host finding can be recognized, including the use of chemicals, sound and light. Experimentally these are rather difficult to establish and to separate, and it may be that the use of these cues is very widespread, if little understood. Of the variety of cues available, many insects probably use more than one, depending upon distance or proximity to the resource sought. Thus, the European crabronid wasp *Philanthus* (see also Box 9.2), which catches only bees, relies initially on vision to locate moving insects of appropriate size. Only bees, or other insects to which bee odours have been applied experimentally, are captured, indicating a role for odour when near the prey. However, the sting is applied only to actual bees, and not to bee-smelling alternatives, demonstrating a final tactile recognition.

Not only may a stepwise sequence of stimuli be necessary, as seen above, but also appropriate stimuli may have to be present simultaneously in order to elicit appropriate behaviour. Thus, *Telenomus heliothidis* (Hymenoptera: Scelionidae), an egg parasitoid of *Heliothis virescens* (Lepidoptera: Noctuidae), will investigate and probe at appropriate-sized round glass beads that simulate *Heliothis* eggs, if they are coated with female moth proteins. However, the scelionid makes no response to glass beads alone, or to female moth proteins applied to improperly shaped beads.

Predatory assassin bugs of *Salyavata* species (Hemiptera: Reduviidae), which hunt on *Nasutitermes* nests in Costa Rica, are lured to termites mending holes in their carton nests. After seizing and sucking dry its first termite victim, the bug nymph uses a novel method to "fish" for further unwary termite workers. It jiggles the empty carcass of the first termite victim near another termite, which grabs the proffered bait and is pulled out of the safety of the nest and consumed. The termite-baiting process continues until the bug is satiated or the termites complete repairs and seal their nest. The bodies and legs of the *Salyavata* nymphs are camouflaged by bits of termite nest carton (see Box 14.2 on African assassin bugs, which use similar disguise). This physical and probably chemical concealment may trick the defending termite soldiers, which never respond to the bugs but defend vigorously if an experimenter offers bait with forceps.

Chemicals

Insect communication is dominated by chemicals, especially by pheromones (section 4.3). Ability to detect the chemical odours and messages produced by prey or hosts (**kairomones**) allows specialist predators and parasitoids to locate these resources. Certain parasitic tachinid flies and braconid wasps can locate their respective stink bug or coccoid host by tuning to their hosts' long-distance sex attractant pheromones. Several unrelated parasitoid hymenopterans use the aggregation pheromones of their bark and timber beetle hosts. Chemicals emitted by stressed plants, such as terpenes produced by pines when attacked by insects, act as **synomones** (communication chemicals that benefit both producer and receiver); for example, certain pteromalid (Hymenoptera) parasitoids locate their hosts, the damage-causing scolytine timber beetles, in this way. Some species of tiny wasps (Trichogrammatidae) that are egg endoparasitoids (see Fig. 16.2) can locate the eggs laid by their preferred host moth by the sex attractant pheromones released by the moth. Furthermore, there are several examples of parasitoids that locate their specific insect larval hosts by **frass** odours – the smells of their faeces. Chemical location is particularly valuable when hosts are concealed from visual inspection, for example when encased in plant or other tissues.

Chemical detection need not be restricted to tracking volatile compounds produced by the prospective host. Thus, many parasitoids searching for phytophagous insect hosts initially are attracted, at a distance, to host-plant chemicals, in the same manner that the phytophage located the resource. At close range, chemicals produced by the feeding damage and/or frass of phytophages may allow precise targeting of the host. Once located, the acceptance of a host as suitable is likely to involve similar or other chemicals, judging by the increased use of rapidly vibrating antennae in sensing the prospective host.

Blood-feeding adult insects locate their hosts using cues that include chemicals emitted by the host. Females of many biting flies can detect increased carbon dioxide levels associated with animal exhalation and fly upwind towards the source. Highly host-specific biters can detect subtle odours: thus, human-biting black flies (Diptera: Simuliidae) respond to components of human exocrine sweat glands. Both sexes of tsetse flies (Diptera: Glossinidae) track the odour of exhaled breath, notably carbon dioxide, octanols, acetone and ketones emitted by their preferred cattle hosts.

Sound

The sound signals produced by animals, including those made by insects to attract mates, have been utilized by some parasites to locate their preferred hosts acoustically. Thus, the blood-sucking females of *Corethrella* (Diptera: Corethrellidae) locate their host, hylid tree frogs, by following the frogs' calls. The details of the host-finding behaviour of ormiine tachinid flies are considered in detail in Box 4.1. Flies of two other dipteran species are attracted by host songs: females of the larviparous tachinid *Euphasiopteryx ochracea* locate the male crickets of *Gryllus integer*, and the sarcophagid *Colcondamyia auditrix* finds its male cicada host, *Okanagana rimosa*, in this manner. This allows precise deposition of the parasitic immature stages in, or close to, the hosts in which they are to develop.

Predatory biting midges (Ceratopogonidae) that prey upon swarm-forming flies, such as midges (Chironomidae), appear to use cues similar to those used by their prey to locate the swarm; cues may include the sounds produced by wing-beat frequency of the members of the swarm. Vibrations produced by their hosts can be detected by ectoparasites, notably amongst the fleas. Certain parasitoids can detect at close range the substrate vibration produced by the feeding activity of their hosts. Thus, *Biosteres longicaudatus*, a braconid hymenopteran endoparasitoid of a larval tephritid fruit fly (Diptera: *Anastrepha suspensa*), detects vibrations made by the larval movement and feeding within the fruit. These sounds act as a behavioural releaser, stimulating host-finding behaviour and acting as a directional cue for their concealed hosts.

Light

Bioluminescence produced by larvae of the Australian cave-dwelling mycetophilid flies of *Arachnocampa* species and their New Zealand counterpart, *Arachnocampa luminosa*, lures small flies to sticky threads suspended from the cave ceiling. Luminescence (section 4.4.5), as with all communication systems, provides scope for abuse; in the following example, luminescent courtship signalling between beetles is misappropriated. Carnivorous female lampyrids of some *Photurus* species, in an example of aggressive foraging mimicry, can imitate the flashing signals of females of up to five other firefly species. The males of these different species flash their responses and are deluded into landing close by the mimetic female, whereupon she devours them. The mimicking *Photurus* female will eat the males of her own species, but cannibalism is avoided or reduced as the *Photurus* female is most piratical only after mating, at which time she becomes relatively unresponsive to the signals of males of her own species.

13.1.3 Phoresy

Phoresy is the transport of one or more individuals by a larger individual of another species. This relationship benefits the carried and does not affect the carrier directly, although in some cases its progeny may be disadvantaged (as we shall see below). Phoresy provides a means of finding a new host or food source. An often observed example involves ischnoceran lice (Psocodea) transported by the winged adults of *Ornithomyia* (Diptera: Hippoboscidae). Hippoboscidae are blood-sucking ectoparasitic flies and *Ornithomyia* occurs on many avian hosts. When a host bird dies, lice can reach a new host by attaching themselves by their mandibles to a hippoboscid, which flies away, likely to a new host. However, lice are highly host-specific but hippoboscids are much less so, and the chances of any hitchhiking louse arriving at an appropriate host may be low. In some other associations, such as a biting midge *Forcipomyia* (Diptera: Ceratopogonidae) found on the thorax of various adult dragonflies in Borneo, it is difficult to determine whether the hitchhiker actually is parasitic or merely phoretic.

Amongst the egg-parasitizing hymenopterans (notably the Scelionidae, Trichogrammatidae and Torymidae), some attach themselves to adult females of the host species, thereby gaining immediate access to the eggs at oviposition. *Mantibaria manticida* (Scelionidae), an egg parasitoid of the European praying mantid (*Mantis religiosa*), is phoretic, predominantly on female hosts. The adult wasp sheds its wings and may feed on the mantid, and therefore can be an ectoparasite. It moves to the wing bases and amputates the female mantid's wings and then oviposits into the mantid's egg mass whilst it is frothy, before the ootheca hardens. Individuals of *M. manticida* that are phoretic on male mantids may transfer to the female during mating. Certain chalcid hymenopterans (including species of Eucharitidae) have mobile planidium larvae that actively seek worker ants, on which they attach, thereby gaining transport to the ant nest. Here the remainder of the immature life cycle comprises typical sedentary grubs that develop within ant larvae or pupae.

The human bot fly, *Dermatobia hominis* (Diptera: Oestridae) of the Neotropical region (Central and South America), which causes myiasis (section 15.1) of humans and cattle, shows an extreme example of phoresy. The female fly does not find the vertebrate host herself, but uses the services of blood-sucking flies, particularly mosquitoes and muscoid flies. The female bot fly, which produces up to 1000 eggs in her lifetime, captures a phoretic intermediary and glues around 30 eggs to its body in such a way that flight is not impaired. When the intermediary finds a vertebrate host on which it feeds, an elevation of temperature induces the bot fly's eggs to hatch rapidly and the larvae transfer to the host where they penetrate the skin via hair follicles and develop within the resultant pus-filled boil.

13.2 PREY/HOST ACCEPTANCE AND MANIPULATION

During foraging, there are some similarities in location of prey by a predator and of a host by a parasitoid or parasite. When contact is made with the potential prey or host, its acceptability must be established by checking the identity, size and age of the prey/host. For example, many parasitoids reject old larvae, which are close to pupation. Chemical and tactile stimuli are involved in specific identification and in subsequent behaviours including biting, ingestion and continuance of feeding. Chemoreceptors on the antennae and ovipositor of parasitoids are vital in chemically detecting host suitability and exact location.

Different manipulations follow acceptance: the predator attempts to eat suitable prey, whereas parasitoids and parasites exhibit a range of behaviours regarding their hosts. A parasitoid either oviposits (or larviposits) directly or subdues and may carry the host elsewhere, for instance to a nest, prior to the offspring developing within or on it. An ectoparasite needs to gain a hold and obtain a meal. The different behavioural and morphological modifications associated with prey and host manipulation are covered in separate sections below, from the perspectives of predator, parasitoid and parasite.

13.2.1 Prey manipulation by predators

When a predator detects and locates suitable prey, it must capture and restrain it before feeding. As predation has arisen many times, and in nearly every order, the morphological modifications associated with this lifestyle are highly convergent. Nevertheless, in most predatory insects the principal organs used in capture and manipulation of prey are the legs and mouthparts. Typical **raptorial** legs of adult insects are elongate and bear spines on the inner surface of at least one segment (Fig. 13.3). Prey is captured by closing the spinose segment against another segment, which may itself be spinose, i.e. the femur against the tibia, or the tibia against the tarsus. As well as spines, there

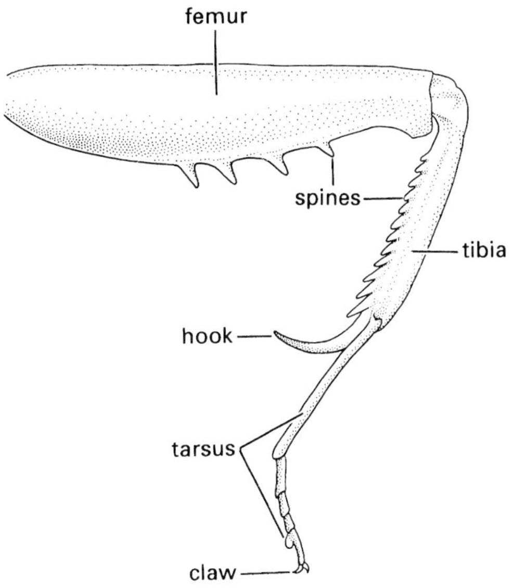

Fig. 13.3 Distal part of the leg of a mantid showing the opposing rows of spines that interlock when the tibia is drawn upwards against the femur. (After Preston-Mafham 1990.)

may be elongate spurs on the apex of the tibia, and the apical claws may be strongly developed on the raptorial legs. In predators with leg modifications, usually it is the anterior legs that are raptorial, but some hemipterans also employ the mid legs, and scorpionflies (see Box 5.1) grasp prey with their hind legs.

Mouthpart modifications associated with predation are of two principal kinds: (i) incorporation of a variable number of elements into a tubular rostrum to allow piercing and sucking of fluids; or (ii) development of strengthened and elongate mandibles. Mouthparts modified as a rostrum (Taxobox 20) are seen in bugs (Hemiptera), and function in sucking fluids from plants or from dead arthropods (as in many gerrid bugs) or in predation on living prey, as in many other aquatic insects, including species of Nepidae, Belostomatidae and Notonectidae. Amongst the terrestrial bugs, assassin bugs (Reduviidae), which use raptorial fore legs to capture other terrestrial arthropods, are major predators. They inject toxins and proteolytic saliva into captured prey, and suck the body fluids through the rostrum. Similar hemipteran mouthparts are used in blood sucking, as demonstrated by *Rhodnius prolixus*, a reduviid famous for its role in experimental insect physiology, and the family Cimicidae, including the bed bug, *Cimex lectularius*.

In the Diptera, mandibles are vital for wound production in the blood-sucking Nematocera (mosquitoes, midges and black flies) but have been lost in other flies, including those that have regained the blood-sucking habit. Thus, in the stable flies (*Stomoxys*) and tsetse flies (*Glossina*), for example, alternative mouthpart structures have evolved; some specialized mouthparts of blood-sucking Diptera are described in section 2.3.1 and illustrated in Fig. 2.13 and Fig. 2.14.

Many predatory larvae and some adults have hardened, elongate and apically pointed mandibles capable of piercing durable cuticle. Larval neuropterans (lacewings and antlions) have a slender maxilla and a sharply pointed and grooved mandible, which are pressed together to form a composite sucking tube (Fig. 13.2c). The composite structure may be straight, as in active pursuers of prey, or curved, as in the sit-and-wait ambushers such as antlions. Liquid may be sucked (or pumped) from the prey, using a range of mandibular modifications after enzymatic predigestion has liquefied the contents (**extra-oral digestion**).

An unusual morphological modification for predation is seen in the larvae of Chaoboridae (Diptera) that use modified antennae to grasp their planktonic cladoceran prey. Odonate nymphs capture passing prey by striking with a highly modified labium (Fig. 13.4), which is projected rapidly outwards by release of hydrostatic pressure, rather than by muscular means.

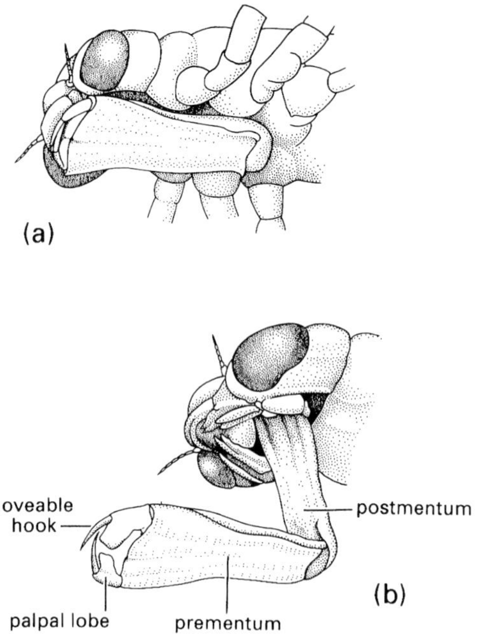

Fig. 13.4 Ventrolateral view of the head of a dragonfly nymph (Odonata: Aeshnidae: *Aeshna*) showing the labial "mask": (a) in folded position; and (b) extended during prey capture, with opposing hooks of the palpal lobes forming claw-like pincers. (After Wigglesworth 1964.)

13.2.2 Host acceptance and manipulation by parasitoids

The two orders with greatest numbers and diversity of larval parasitoids are the Diptera and Hymenoptera. Two basic approaches are displayed once a potential host is located, though there are exceptions. In the first, as seen in many hymenopterans, the adult seeks the larval development site (see **Plate 6f**). In contrast, in many dipterans, a first-instar planidium larva often makes the close-up host contact. Parasitic hymenopterans use sensory information from the elongate and constantly mobile antennae to precisely locate even a hidden host. The antennae and specialized ovipositor (see Fig. 5.11) bear sensilla that allow host acceptance and accurate oviposition, respectively. Modification of the ovipositor as a sting in the aculeate Hymenoptera permits behavioural modifications (section 14.6), including larval provisioning with food captured by the adult and maintained alive in a paralyzed state.

Endoparasitoid dipterans, including the Tachinidae, may oviposit (or in larviparous taxa, deposit a larva) onto the cuticle or directly into the host. In several distantly related families, a convergently evolved "substitutional" ovipositor (section 2.5.1 and section 5.8) is used. Frequently, however, the parasitoid's egg or larva is deposited onto a suitable substrate and the mobile planidium larva finds its host. Thus, *Euphasiopteryx ochracea*, a tachinid that responds phonotactically to the call of a male cricket, actually deposits larvae around the calling site, and these larvae locate and parasitize, not only the vocalist, but also other crickets attracted by the call. **Heteromorphosis**, in which the first-instar larva morphologically and behaviourally differs from subsequent larval instars (which are sedentary parasitic maggots), is common amongst parasitoids.

Certain parasitic and parasitoid dipterans and some hymenopterans use their aerial flying skills to gain access to a potential host. Some can intercept their hosts in flight, whereas others lunge at an alert and defended target. Some of the inquilines of social insects (section 12.3) can enter the nest via an egg laid upon a worker whilst it is active outside the nest. For example, certain phorid flies, lured by ant odours, may be seen darting at ants, seeking to oviposit on them. A West Indian leaf-cutter ant (*Atta* sp.) cannot defend itself from such attacks whilst bearing leaf fragments in its mandibles. Some defence is provided by stationing a guard on the leaf during transport; the guard is a small (minima) worker (see Fig. 9.7) that uses its jaws to threaten any approaching phorid fly.

The success of attacks of such insects against active and well-defended hosts demonstrates great rapidity in host acceptance, probing and oviposition. This may contrast with the sometimes leisurely manner of many parasitoids of sessile hosts, such as scale insects, pupae or immature stages that are restrained within confined spaces, such as plant tissue, and unguarded eggs.

13.2.3 Overcoming host immune responses

Insects that develop within the body of other insects must cope with the active immune responses of the host. An adapted or compatible parasitoid is not eliminated by the cellular immune defences of the host. These defences protect the host by acting against incompatible parasitoids, pathogens and biotic matter that may invade the host's body cavity. Host immune responses entail mechanisms for: (i) recognizing introduced material as non-self; and (ii) inactivating,

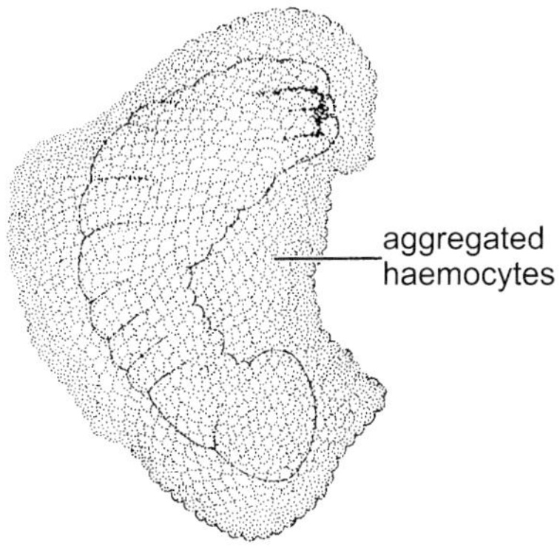

Fig. 13.5 Encapsulation of a living larva of *Apanteles* (Hymenoptera: Braconidae) by the haemocytes of a caterpillar of *Ephestia* (Lepidoptera: Pyralidae). (After Salt 1968.)

suppressing or removing the foreign material (see section 3.4.3). The usual host reaction to an incompatible parasitoid is **encapsulation**, i.e. surrounding the invading egg or larva by an aggregation of haemocytes (Fig. 13.5). The haemocytes become flattened onto the surface of the parasitoid, and phagocytosis commences as the haemocytes build up, eventually forming a capsule that surrounds and kills the intruder. This type of reaction rarely occurs when parasitoids infect their normal hosts, presumably because the parasitoid or some factor(s) associated with it alters the host's ability to recognize the parasitoid as foreign and/or to respond to it. Even the eggs of some insects, such as the moth *Manduca sexta*, can mount an immune response against a generalist egg parasitoid such as *Trichogramma*, resulting in poor survival of the parasitoid. Parasitoids that cope successfully with the host immune system do so in one or more of the following ways:

- Avoidance – for example, ectoparasitoids feed externally on the host (in the manner of predators), egg parasitoids lay into host eggs that may have no or a weak immune response, and many other parasitoids at least temporarily occupy host organs (such as the brain, a ganglion, a salivary gland or the gut) and thus escape the immune reaction of the host haemolymph.
- Evasion – this includes molecular mimicry (the parasitoid is coated with a substance similar to host proteins and is not recognized as non-self by the host), cloaking (e.g. the parasitoid may insulate itself in a membrane or capsule, derived from either embryonic membranes or host tissues; see also "subversion" below), and/or rapid development in the host.
- Destruction – the host immune system may be blocked by attrition of the host such as by gross feeding that weakens host defence reactions, and/or by destruction of responding cells (the host haemocytes).
- Suppression – host cellular immune responses may be suppressed by viruses associated with the parasitoids (Box 13.1); often suppression is accompanied by reduction in host haemocyte counts and other changes in host physiology.
- Subversion – in many cases parasitoid development occurs despite host response; for example, physical resistance to encapsulation is known for wasp parasitoids, and in dipteran parasitoids the host's haemocytic capsule is subverted for use as a sheath that the fly larva keeps open at one end by vigorous feeding. In many parasitic Hymenoptera, the serosa or trophamnion associated with the parasitoid egg

fragments into individual cells that float free in the host haemolymph, where they enlarge to form teratocytes (giant cells) that secrete hormones that assist in overwhelming host defences.

Obviously, the various ways of coping with host immune reactions are not discrete and most adapted parasitoids probably use a combination of methods to allow development within their respective hosts. Parasitoid–host interactions at the level of cellular and humoral immunity are complex and vary greatly among different taxa. Understanding of these systems is developing rapidly, with new insights into parasitoid genomes and coevolved associations between insects and viruses. A further level of complexity is added by evidence that some host plants of lepidopteran larvae can contain secondary plant compounds (such as pyrrolizidine alkaloids) that may reduce the viability of parasitoids. Although a diet rich in these compounds also impacts the growth of the caterpillars, there may be great benefit of such parasitoid-induced mortality when the risk of parasitization is high.

13.3 PREY/HOST SELECTION AND SPECIFICITY

As we have seen in Chapters 9–11, insects vary in the breadth of food sources that they use. Likewise, some predatory insects are **monophagous**, utilizing a single species of prey, others are **oligophagous**, using few species, and many are **polyphagous**, feeding on a variety of prey species. As a broad generalization, predators are mostly polyphagous, as a single prey species rarely will provide adequate resources. However, sit-and-wait (ambush) predators, by virtue of their chosen location, may have a restricted diet – for example, antlions may predominantly trap small ants in their pits. Furthermore, some predators select gregarious prey, such as certain eusocial insects, because the predictable behaviour and abundance of this prey allows monophagy. Although these prey insects may be aggregated, often they are aposematic and chemically defended. Nonetheless, if the defences can be countered, these predictable and often abundant food sources permit predator specialization.

Predator–prey interactions are not discussed further here; the remainder of this section concerns the more complicated host relations of parasitoids and parasites. In referring to parasitoids and their range of hosts, the terminology of monophagous, oligophagous and polyphagous is applied, as for phytophages and predators. However, a parallel terminology exists for parasites: **monoxenous** parasites are restricted to a single host, **oligoxenous** to a few, and **polyxenous** parasites avail themselves of many hosts. In the following sections, we discuss first the variety of strategies for host selection by parasitoids, followed by the ways in which a parasitized host may be manipulated by the developing parasitoid. In the final section, patterns of host use by parasites are discussed, with particular reference to co-speciation.

13.3.1 Host use by parasitoids

Parasitoids require only a single individual in which to complete development, they always kill their immature host, and rarely are parasitic in the adult stage. Insect-eating (**entomophagous**) parasitoids show a range of strategies for development on their selected insect hosts. The larva may be ectoparasitic, developing externally, or endoparasitic, developing within the host. Eggs (or larvae) of ectoparasitoids are laid close to or upon the body of the host, as are sometimes those of endoparasitoids. However, in the latter group, more often the eggs are laid within the body of the host, using a piercing ovipositor (in hymenopterans; see **Plates 6f** and **Plate 8d**) or a substitutional ovipositor (in parasitoid dipterans). Certain parasitoids that feed within host pupal cases or under the covers and protective cases of scale insects and the like actually are ectophages (external feeding), living internal to the protection but external to the insect host body. These different feeding modes give different exposures to the host immune system, with endoparasitoids encountering and ectoparasitoids avoiding host defences (section 13.2.3). Ectoparasitoids often are less host specific than endoparasitoids, as they have a less intimate association with the host than do endoparasitoids, which must counter species-specific variations of the host immune system.

Parasitoids may be solitary on or in their host, or gregarious. The number of parasitoids that can develop on a host relates to the size of the host, its post-infected longevity, and the size (and biomass) of the parasitoid. Development of several parasitoids in one individual host arises commonly through the female ovipositing several eggs on a single host or, less often, by **polyembryony**, in which a single egg laid by the mother divides and can give rise to numerous offspring (section 5.10.3). Gregarious parasitoids appear able to regulate the clutch size in relation to the quality and size of the host.

Most parasitoids **host discriminate**; i.e. they can recognize, and generally reject, hosts that are parasitized already, either by themselves, their conspecifics or another species. Distinguishing unparasitized from parasitized hosts generally involves a marking pheromone placed internally or externally on the host at the time of oviposition.

However, some parasitoids do not reject already parasitized hosts. In **superparasitism**, a host receives multiple eggs either from a single individual or from several individuals of the same parasitoid species, although the host cannot sustain the total parasitoid burden to maturity. The outcome of multiple oviposition is discussed in section 13.3.2. Theoretical models, some of which have been substantiated experimentally, imply that superparasitism will increase:

- as unparasitized hosts are depleted;
- as parasitoid numbers searching any patch increase;
- in species with high fecundity and small eggs.

Some adaptive benefits may derive from the strategy. For individual parasitoids, superparasitism may be adaptive when hosts are scarce, with avoidance being adaptive if hosts are abundant. Very direct benefits accrue to a solitary parasitoid that uses a host that can encapsulate a parasitoid egg (section 13.2.3). Here, a first-laid egg may use all the host haemocytes, and subsequent egg(s) thereby may escape encapsulation. However, the idea that superparasitism is adaptive is contradicted by finding that a virus causes behavioural changes in an infected parasitoid wasp (Figitidae: *Leptopilina boulardi*) so that it oviposits into already parasitized hosts (*Drosophila* larvae), allowing the virus to transfer to uninfected parasitoid larvae inside the host fly. Other cases of superparasitism require more careful scrutiny.

In **multiparasitism**, a host receives eggs of more than one species of parasitoid. Multiparasitism occurs more often than superparasitism, perhaps because parasitoid species are less able to recognize the marking pheromones placed by species other than their own. Closely related parasitoids may recognize each other's marks, whereas more distantly related species may be unable to do so. However, secondary parasitoids, called **hyperparasitoids**, can detect odours left by a primary parasitoid, allowing accurate location of the site for the development of the hyperparasite.

Hyperparasitic development involves a secondary parasitoid developing at the expense of a primary parasitoid. Some insects are **obligate** hyperparasitoids, developing only within primary parasitoids, whereas others are **facultative** and may develop also as primary parasitoids. Development may be external or internal to the primary parasitoid host, with oviposition into the primary host in the former, or into the primary parasitoid in the latter (Fig. 13.6). External feeding is frequent, and hyperparasitoids are predominantly restricted to the host larval stage, sometimes the pupa; hyperparasitoids of eggs and adults of primary parasitoid hosts are very rare.

Hyperparasitoids belong to two families of Diptera (certain Bombyliidae and Conopidae), two families of Coleoptera (a few Ripiphoridae and Cleridae), and notably the Hymenoptera, principally amongst 11 families of the superfamily Chalcidoidea, in four subfamilies of Ichneumonidae and in Figitidae (Cynipoidea). Hyperparasitoids are absent among the Tachinidae and surprisingly do not seem to have evolved in certain parasitic wasp families such as Braconidae, Trichogrammatidae and Mymaridae. Within the Hymenoptera, hyperparasitism has evolved several times, each originating in some manner from primary parasitism. Facultative hyperparasitism demonstrates the ease of the transition. Hymenopteran hyperparasitoids attack a wide range of hymenopteran-parasitized insects, predominantly amongst the hemipterans (especially Sternorrhyncha), Lepidoptera and symphytans. Hyperparasitoids often have a broader host range than

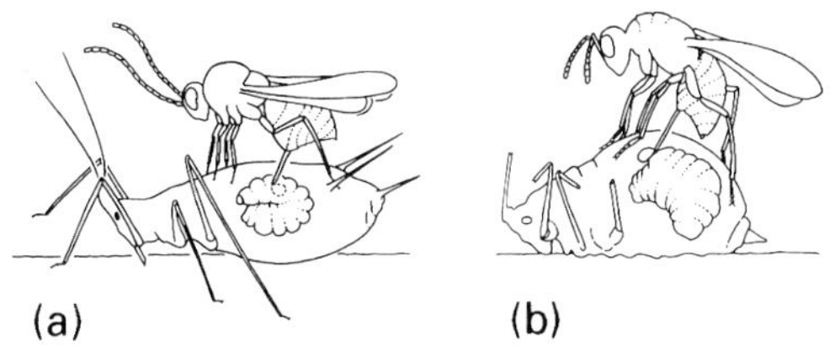

Fig. 13.6 Two examples of the ovipositional behaviour of hymenopteran hyperparasitoids of aphids: (a) endophagous *Alloxysta victrix* (Hymenoptera: Figitidae) ovipositing into a primary parasitoid inside a live aphid; (b) ectophagous *Asaphes suspensus* (Hymenoptera: Pteromalidae) ovipositing onto a primary parasitoid in a mummified aphid. (After Sullivan 1988.)

the frequently oligophagous or monophagous primary parasitoids. However, as with primary parasitoids, endophagous hyperparasitoids seem to be more host specific than those that feed externally, relating to the greater physiological problems experienced when developing within another living organism. Additionally, foraging and assessment of host suitability of a complexity comparable with that of primary parasitoids is known, at least for cynipoid hyperparasitoids of aphidophagous parasitoids (Fig. 13.7). As explained in section 16.5.1, hyperparasitism and the degree of host-specificity is fundamental information in biological control programmes.

13.3.2 Host manipulation and development of parasitoids

Parasitization may kill or paralyze the host, and the developing parasitoid, called an **idiobiont**, develops rapidly, in a situation that differs only slightly from predation. Of greater interest and much more complexity is the **koinobiont** parasitoid, which lays its egg(s) in a young host, which continues to grow, thereby providing an increasing food resource. Parasitoid development may be delayed until the host has attained a sufficient size. **Host regulation** is a feature of koinobionts, with certain parasitoids able to manipulate host physiology, including suppression of its pupation, allowing production of a "super host".

Many koinobionts respond to hormones of the host, as demonstrated by: (i) the frequent moulting or emergence of parasitoids in synchrony with the host's moulting or metamorphosis; and/or (ii) synchronization of diapause of host and parasitoid. It is uncertain whether, for example, host ecdysteroids act directly on the parasitoid's epidermis to cause moulting, or act indirectly on the parasitoid's own endocrine system to elicit synchronous moulting. Although the specific mechanisms remain unclear, some parasitoids undoubtedly disrupt the host endocrine system, causing developmental arrest, accelerated or retarded metamorphosis, or inhibition of reproduction in an adult host. This may arise through production of hormones (including mimetic ones) by the parasitoid, or through regulation of the host's endocrine system, or both. In cases of **delayed parasitism**, such as is seen in certain platygastrine and braconid hymenopterans, development of an egg laid in the host egg is delayed for up to a year, until the host is a late-stage larva. Host hormonal changes approaching metamorphosis are implicated in the stimulation of parasitoid development. Specific interactions between the endocrine systems of endoparasitoids and their hosts can limit the range of hosts utilized. Parasitoid-introduced viruses or virus-like particles (Box 13.1) may also modify host physiology and determine host range.

The host is not a passive vessel for parasitoids – as we have seen, the immune system can attack all but the adapted parasitoids. Furthermore, host quality (size and age) can induce variation in size, fecundity and even the sex ratio of emergent solitary parasitoids. Generally, more females are produced from high-quality (larger) hosts, whereas males are produced from poorer quality ones, including smaller and superparasitized hosts. Host aphids reared experimentally on deficient diets (lacking sucrose or iron) produced *Aphelinus* (Hymenoptera: Aphelinidae) parasitoids that developed more slowly, produced more males, and showed lowered fecundity and longevity. The young stages of an endophagous koinobiont parasitoid compete with the host tissues for nutrients from the haemolymph. Under laboratory conditions, if a parasitoid can be induced to oviposit into an "incorrect" host (by the use of appropriate kairomones), complete larval development often occurs, showing that haemolymph is adequate nutritionally for development of more than just the adapted parasitoid. Accessory gland secretions (which may include paralyzing venoms) are injected by the ovipositing female parasitoid with the eggs, and appear to play a role in regulation of the host's haemolymph nutrient supply to the larva. The specificity of these substances may relate to the creation of a suitable host.

In superparasitism and multiparasitism, if the host cannot support all parasitoid larvae to maturity, larval competition often takes place. Depending on the nature of the multiple ovipositions, competition may involve aggression between siblings, other conspecifics, or interspecific individuals. Fighting between larvae, especially in mandibulate larval hymenopterans, can result in deaths and **encapsulation**. Suppression by means of venoms, anoxia or food deprivation also may occur. Unresolved larval overcrowding in the host can result in a few weak and small individuals emerging, or no parasitoids at all if the host dies prematurely or resources are depleted before pupation. Gregariousness may have evolved from solitary parasitism in circumstances in which multiple larval development is permitted by greater host size. Evolution of

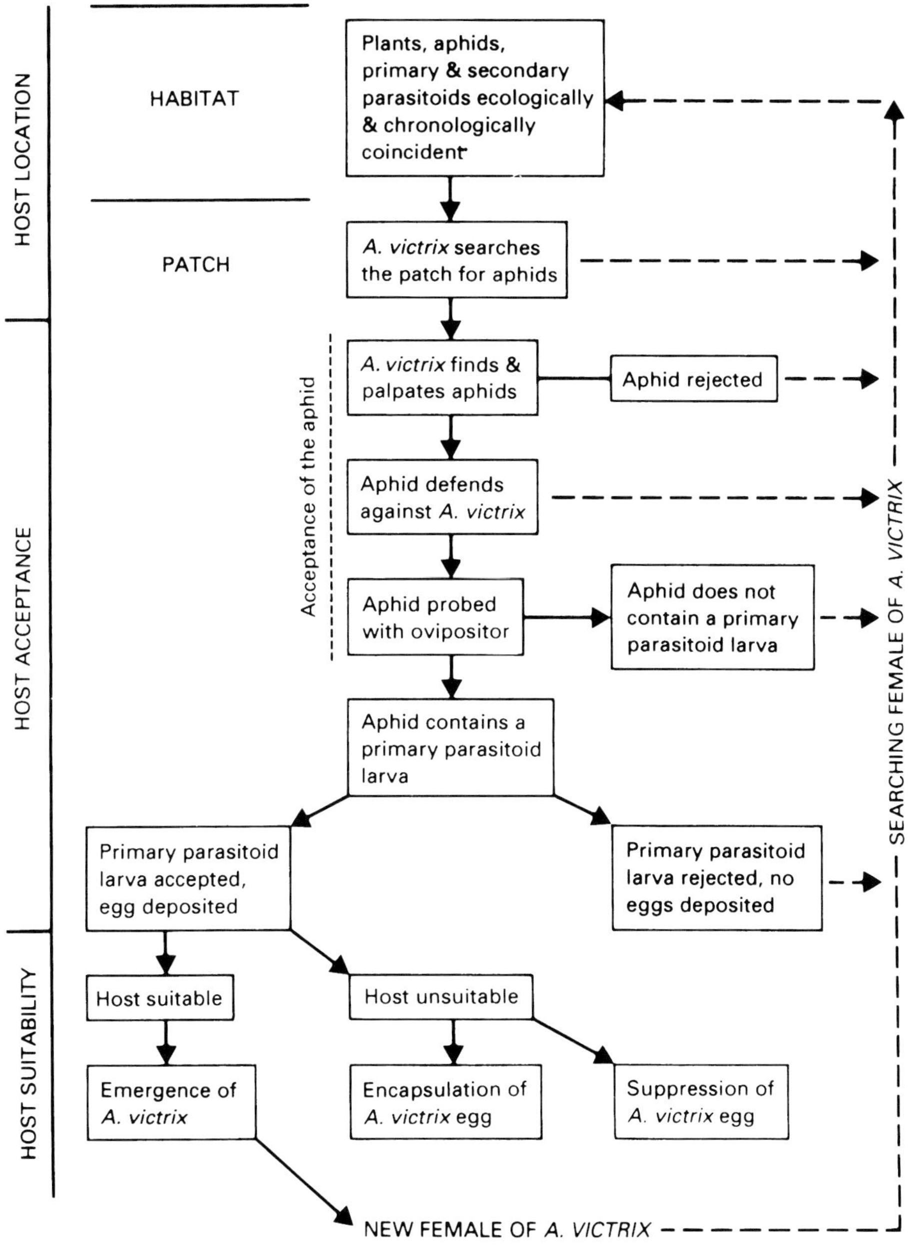

Fig. 13.7 Steps in host selection by the hyperparasitoid *Alloxysta victrix* (Hymenoptera: Figitidae). (After Gutierrez 1970.)

Box 13.1 Viruses, wasp parasitoids and host immunity

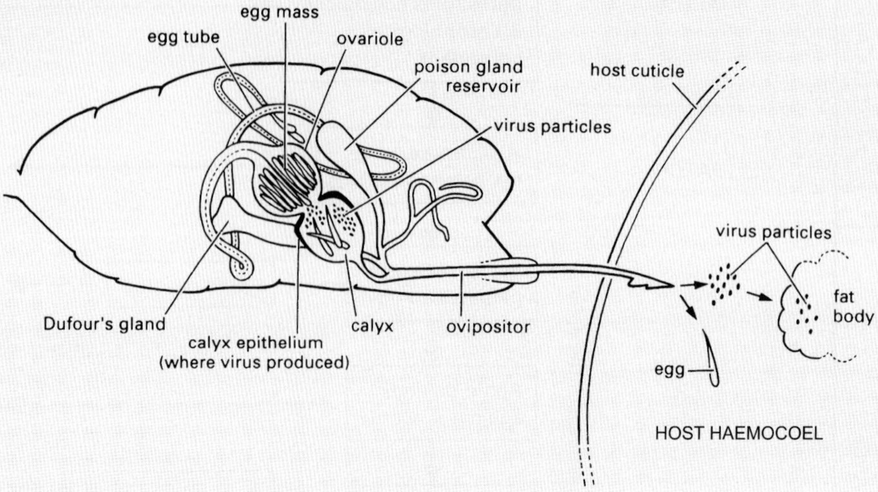

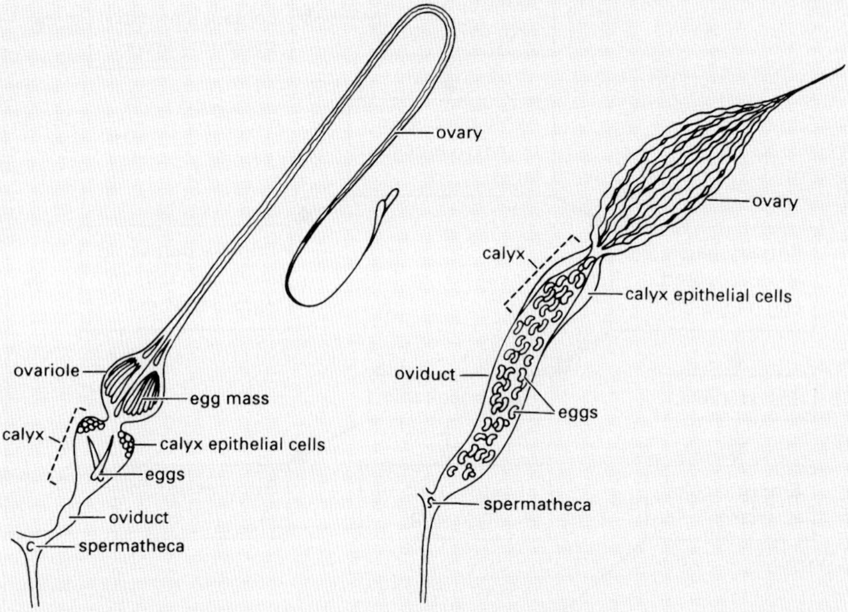

In certain endoparasitoid wasps in the families Ichneumonidae and Braconidae, the ovipositing female wasp injects the larval host not only with her egg(s), but also with accessory gland secretions and viruses or endogenous viral elements (as depicted in the upper drawing for the braconid *Toxoneuron* (formerly *Cardiochiles*) *nigriceps*, after Greany *et al*. 1984). The viruses belong to the Polydnaviridae (**polydnaviruses** or PDVs), characterized by their multipartite double-stranded circular DNA. In the wasp genomes, PDVs exist as integrated proviruses that are transmitted between wasp generations through the germline. Bracoviruses (the PDVs of braconids) differ from ichnoviruses (the PDVs of ichneumonids) in relation to their interaction with other wasp-derived factors in the parasitized host. The PDVs of different wasp species generally are considered to be distinct species of virus, although regulation of their replication is totally by the wasps, and they cannot multiply in the lepidopteran hosts. Within the braconids, PDVs occur only in one clade, the microgastroid group of subfamilies, and have coevolved with their wasp hosts. Virus-like particles (VLPs) (that resemble viruses but lack nucleic acids) are produced by several figitids and braconids and the ichneumonid *Venturia canescens*.

PDVs derive from viruses that have lost their independence, and represent an advanced stage of symbiotic integration into the wasp parasitoid's extended phenotype. The evolutionary association of ichnoviruses with ichneumonids is unrelated to the evolution of the braconid–bracovirus association. While bracoviruses are likely to have evolved from an ancient nudivirus, the different (but certainly independent) origin of ichnoviruses is as yet unknown.

All PDVs are involved in overcoming the host's immune reaction and altering host physiology to benefit the developing wasp at the expense of the host. For example, PDVs probably induce most changes in growth, development, behaviour and haemocytic activity observed in infected host larvae. The PDVs of many parasitoids (particularly braconids) require the presence of accessory factors, particularly venoms, to completely prevent encapsulation of the wasp egg by the host or to induce full symptoms in the host. Thus, PDVs play roles as both mutualists of the wasps, and as pathogens of the wasps' parasitized hosts.

Replication of PDVs takes place only in the calyx epithelium of the ovary of the female wasp's reproductive tract (as depicted for the braconid *Toxoneuron nigriceps* in the lower left drawing, and for the ichneumonid *Campoletis sonorensis* in the lower right drawing, after Stoltz & Vinson 1979) and is the only site of VLP-protein assembly (as in the ichneumonid *Venturia canescens*). The lumen of the wasp oviduct becomes filled with PDVs (or VLPs), which thus surround the wasp eggs. If VLPs or PDVs are removed artificially from wasp eggs, encapsulation occurs if the unprotected eggs are then injected into the host. If appropriate PDVs (or VLPs) are injected into the host with the washed eggs, encapsulation is prevented. The physiological mechanism for this protection is not clearly understood, although in the wasp *Venturia*, which coats its eggs in VLPs, it appears that molecular mimicry of a host protein by a VLP protein interferes with the lepidopteran host's immune recognition. The VLP protein is similar to a host haemocyte protein involved in recognition of foreign particles. In the case of PDVs, the process is more active and involves the expression of PDV-encoded gene products that interfere directly with the mode of action of lepidopteran haemocytes. PDVs seem also to contribute to various endocrine changes that occur in parasitized hosts.

gregariousness may be facilitated when the potential competitors for resources within a single host are relatives. This is particularly so in polyembryony, which produces clonal, genetically identical larvae (section 5.10.3).

The Strepsiptera (Taxobox 23) contains more than 600 species, all of which parasitize other insects. Strepsipterans were argued to be endoparasites, but their host interactions show that they actually are endoparasitoids. The characteristically aberrant bodies of their predominantly hemipteran and hymenopteran hosts are termed "stylopized", so-called for a common strepsipteran genus, *Stylops*. Within the host's body cavity, growth of the strepsipteran(s) causes malformations, including displacement of the internal organs, and the host's life span usually is lengthened. The host continues to grow and moult after parasitization, and holometabolous hosts can metamorphose. However, the host is castrated – the sexual organs degenerate, or fail to develop appropriately – and the host dies directly or indirectly due to parasitization, but only after emergence of the strepsipteran adult male or first-instar larvae. Host death may be due to atrophy of its internal organs or to infection by pathogens that enter through the strepsipteran exit hole. Parasitism by Strepsiptera is unlike that of any other parasitoid group. Strepsiptera have some of the characterisitics of koinobiont parasitoids in that the host continues to be mobile and to develop after parasitization, but stylopized hosts develop for much longer than the hosts of typical koinobionts. Also, unlike typical koinobionts, strepsipterans have a wide host range, and the strepsipteran family Myrmecolacidae exhibits a bizarre polymorphism of host use in that the males parasitize ants whereas the females parasitize mantids and orthopterans.

13.3.3 Patterns of host use and specificity in parasites

Insects ectoparasitic on vertebrate hosts are so significant to the health of humans and their domestic animals that we deal with them separately in Chapter 15, and medical issues will not be considered further here. In contrast to the radiation of ectoparasitic insects using vertebrate hosts and the immense numbers of species of insect parasitoids seen above, there are remarkably few insect parasites of other insects, or indeed, of other arthropods.

Although larval Dryinidae (Hymenoptera) develop parasitically part-externally and part-internally in hemipterans, the very few other insect–insect parasitic interactions involve only ectoparasitism. The Braulidae is a family of Diptera comprising some aberrant, mite-like flies belonging to two genera, *Braula* and *Megabraula*, intimately associated with *Apis* (honey bees). Larval braulids scavenge on pollen and wax in the hive, and the adults usurp nectar and saliva from the proboscis of the bee. This association certainly involves phoresy, with adult braulids always found on their hosts' bodies, but whether the relationship is ectoparasitic is open to debate. Likewise, the relationship of several genera of aquatic chironomid larvae with nymphal hosts, such as mayflies, stoneflies and dragonflies, ranges from phoresy to suggested ectoparasitism. These ectoparasites using insects show little host specificity at the species level, but this may not be so for insect parasites with vertebrate hosts.

Patterns of host-specificity and preferences of parasites raise some of the most fascinating questions in parasitology. For example, most orders of mammals bear lice (four suborders of the Psocodea), many of which are monoxenic or found amongst a limited range of hosts. Even some marine mammals, namely certain seals, have lice, although whales do not. No Chiroptera (bats) harbour lice, although they host many other ectoparasitic insects, including the Streblidae and Nycteribiidae – two families of ectoparasitic Diptera that are restricted to bats.

Some terrestrial hosts are free of all ectoparasites, others have very specific associations with one or a few guests, and in Panama the opossum *Didelphis marsupialis* has been found to harbour 41 species of ectoparasitic insects and mites. Although four or five of these are commonly present, none are restricted to the opossum, and the remainder are found on a variety of hosts, ranging from distantly related mammals to reptiles, birds and bats.

We can examine some principles concerning the different patterns of distribution of parasites and their hosts by examining close associations of parasites and hosts and relating these to ectoparasite–host relations in general.

The parasitic lice are obligate permanent ectoparasites, spending all their lives on their hosts, and lacking any free-living stage (Taxobox 18). Extensive surveys, such as one that showed that Neotropical birds averaged 1.1 lice species per host across 127 species and 26 families of birds, indicate that lice are highly monoxenous (restricted to one host species). A high level of co-speciation between louse and host might be expected and, in general, related animals have related lice. The widely quoted Fahrenholz's rule formally states that the phylogenies of hosts and parasites are identical, with every speciation event affecting hosts being matched by a synchronous speciation of the parasites, as shown in Fig. 13.8(a). It follows that:

- phylogenetic trees of hosts can be derived from the trees of their ectoparasites;
- ectoparasite phylogenetic trees are derivable from the trees of their hosts (the potential for circularity of reasoning is evident);
- the number of parasite species in the group under consideration is identical to the number of host species considered;
- no species of host has more than one species of parasite in the taxon under consideration;
- no species of parasite parasitizes more than one species of host.

Fahrenholz's rule has been tested for mammal lice selected from amongst the family Trichodectidae, for which robust phylogenetic trees, derived independently of any host mammal phylogeny, are available. Amongst a sample of these trichodectids, 337 lice species parasitize 244 host species, with 34% of host species parasitized by more than one trichodectid. Several possible explanations exist for these mismatches. Firstly, speciation may have occurred independently amongst certain lice on a single host (Fig. 13.8b): at least 7% of all events show independent speciation in sampled Trichodectidae. A second explanation involves secondary transfer of lice species to phylogenetically unrelated host taxa. Amongst extant species,

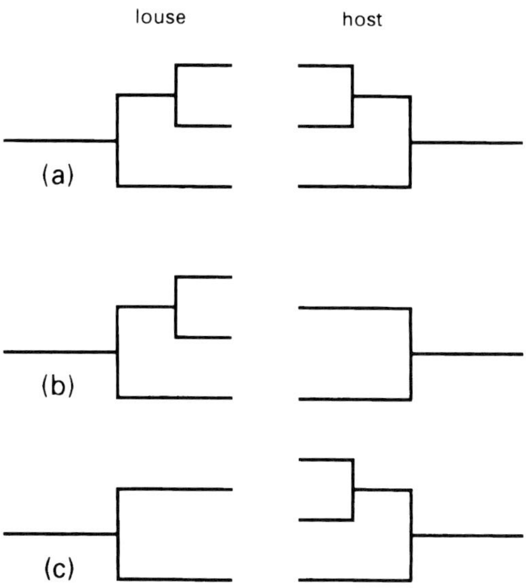

Fig. 13.8 Comparisons of louse and host phylogenetic trees: (a) adherence to Fahrenholz's rule; (b) independent speciation of the lice; (c) independent speciation of the hosts. (After Lyal 1986.)

when cases arising from human-induced unnatural host proximity are excluded (accounting for 6% of cases), unmistakable and presumed natural transfers (i.e. between marsupial and eutherian mammal, or bird and mammal) occur in about 2% of speciation events. However, hidden within the phylogenies of host and parasite are speciation events that involve lateral transfer between rather more closely-related host taxa, but these transfers fail to match precisely the phylogeny. Examination of the detailed phylogeny of the sampled Trichodectidae shows that a minimum of 20% of all speciation events are associated with distant and lateral secondary transfer, including historical transfers that lie deeper in the phylogeny.

In detailed examinations of relationships between a smaller subset of trichodectids and eight of their pocket gopher (Rodentia: Geomyidae) hosts, substantial concordance was claimed between trees derived from biochemical data for hosts and parasites, and some evidence was found for **co-speciation** (cospeciation) (identical patterns of speciation, measured by identical tree topology, in two unrelated but ecologically associated lineages). However, many hosts had two lice species, and most species of gopher actually have a substantial suite of associated lice. Furthermore, a minimum of three instances of lateral transfer (host switching) appeared to have occurred, in all cases between hosts with geographically contiguous ranges. Although many speciation events in these lice "track" speciation in the host, and some estimates even indicate similar ages of host and parasite species, it is evident from the Trichodectidae that strict co-speciation of host and parasite is not the sole explanation of the associations observed. Observations such as these have led to parasite–host dynamics being interpreted as a spectrum running from densely packed communities structured by high interspecies interactions to more isolated situations in which many niches are "vacant" and stochastic assembly rules apply. The appropriateness of island biogeography theory (section 8.7), with each host viewed as an "island", is unclear.

The reasons why apparently monoxenic lice sometimes do deviate from strict co-speciation apply equally to other ectoparasites, many of which show similar variation in complexity of host relationships. Deviations from strict co-speciation arise if host speciation occurs without commensurate parasite speciation (Fig. 13.8c). This resulting pattern of relationships is identical to that seen if one of two parasite sister taxa generated by co-speciation in concert with the host subsequently became extinct. Frequently, a parasite is not present throughout the complete range of its host, resulting perhaps from the parasite being restricted in range by environmental factors independent of those controlling the range of the host. Hemimetabolous ectoparasites, such as lice, which spend their entire lives on the host, might be expected to closely follow the ranges of their hosts, but there are exceptions in which the ectoparasite distribution is restricted by external environmental factors. For holometabolous ectoparasites, which spend some of their lives away from their hosts, such external factors will be even more influential in governing parasite range. For example, a homeothermic vertebrate may tolerate environmental conditions that cannot be sustained by the free-living stage of a poikilothermic ectoparasite, such as a larval flea. As speciation may occur in any part of the distribution of a host, host speciation may occur without necessarily involving the parasite. Furthermore, a parasite may show geographical variation within all or part of the host range that is incongruent with the variation of the host. If either or both variations lead to eventual species formation, there will be incongruence between parasite and host phylogeny.

Furthermore, poor knowledge of host and parasite interactions may result in misleading conclusions. A true host may be defined as one that provides the conditions for parasite reproduction to continue indefinitely. When there is more than one true host, there may be a principal (preferred) and an exceptional host, depending on the proportional frequencies of ectoparasite occurrence. An intermediate category may be recognized – the sporadic or secondary host – on which parasite development cannot normally take place, but an association arises frequently, perhaps through predator–prey interactions or environmental encounters (such as a shared nest). Small sample sizes and limited biological information can allow an accidental or secondary host to be mistaken for a true host, giving rise to a possible erroneous "refutation" of co-speciation. Extinctions of certain parasites and true hosts (leaving the parasite extant on a secondary host) will refute Fahrenholz's rule.

Even assuming perfect recognition of true host-specificity and knowledge of the historical existence of all parasites and hosts, evidently successful parasite transfers between hosts have taken place throughout the history of host–parasite interactions. For example, mapping of host associations onto a phylogenetic tree for fleas supports an ancestral association of fleas with mammals and four independent host transfers to birds. Co-speciation is fundamental to host–parasite relations, but the factors encouraging deviations must be considered. Predominantly, these concern: (i) geographical and social proximity of different hosts, allowing opportunities for parasite colonization of the new host; together with (ii) ecological similarity of different hosts, allowing establishment, survival and reproduction of the ectoparasite on the novel host. This **resource tracking** is in contrast with the **phyletic tracking** implied by Fahrenholz's rule. As with all matters biological, most situations lie somewhere along a continuum between these two extremes, and rather than forcing patterns into one category or the other, interesting questions arise from recognizing and interpreting the different patterns observed.

If all host–parasite relationships are examined, some of the factors that govern host-specificity can be identified:

- the stronger the life-history integration with that of the host, the greater the likelihood of monoxeny;
- the greater the vagility (mobility) of the parasite, the more likely it is to be polyxenous;
- the number of accidental and secondary parasite species increases with decreasing ecological specialization and with increase in geographical range of the host, as we saw earlier in this section for the opossum, which is widespread and unspecialized.

If a single host has several ectoparasites, there may be some ecological or temporal segregation on the host. For example, in haematophagous (blood-sucking) black flies (Simuliidae) that attack cattle, the belly is more attractive to certain species, whereas others feed only on the ears. *Pediculus humanus humanus* (or *P. humanus*) and *P. humanus capitis* (or *P. capitis*) (Psocodea: Anoplura), human body and head lice respectively, are ecologically separated examples of sibling taxa in which strong reproductive isolation is reflected by only slight morphological differences.

13.4 POPULATION BIOLOGY – PREDATOR/PARASITOID AND PREY/HOST ABUNDANCE

Ecological interactions between an individual, its conspecifics, its predators and parasitoids (and other causes of mortality), and its abiotic habitat are fundamentally important aspects of population dynamics. Accurate estimation of population density and its regulation is at the heart of population ecology, biodiversity studies, conservation biology, and monitoring and management of pests. A range of tools are available to entomologists to understand what factors influence population growth and survivorship, including sampling methods, experimental designs, and manipulations and modelling programmes.

Insects usually are distributed on a wider scale than investigators can survey in detail, and thus sampling must be used to allow extrapolation to the wider population. Sampling may be absolute, in which case all organisms in a given area or volume might be assessed, such as mosquito larvae per litre of water, or ants per cubic metre of leaf litter. Alternatively, relative measures, such as number of Collembola in pitfall trap samples or micro-wasps per yellow pan trap, may be obtained from an array of such trapping devices (section 18.1). Relative measures may or may not reflect actual abundances, with variables such as trap size, habitat structure, and insect behaviour and activity levels affecting "trappability" – the likelihood of capture. Measures may be integrated over time, for example a series of sticky, pheromone or continuous running light traps, or instantaneous snap-shots such as the inhabitants of a submerged freshwater rock, the contents of a timed sweep netting, or the knock-down from an insecticidal fogging of a tree's canopy. Instantaneous samples may be unrepresentative, whereas longer duration sampling can overcome some environmental variability.

Sampling design is the most important component in any population study, with stratified random designs providing power to interpret data statistically. Such a design involves dividing the study site into regular blocks (subunits), and within each of these blocks sampling sites are allocated randomly. Pilot studies can allow understanding of the variation expected, and the appropriate matching of environmental variables between treatments and controls for an experimental study. Although more widely used for vertebrate studies, mark-and-recapture methods have been effective for adult odonates, larger beetles and moths and, using fluorescent chemical dyes, for smaller pest insects.

A universal outcome of population studies is that the expectation that the number and density of individuals grows at an ever-increasing rate is met very rarely, perhaps only during short-lived pest outbreaks. Exponential growth is predicted because the rate of reproduction of insects potentially is high (hundreds of eggs per mother) and generation times are short – even with mortality as high as 90%, numbers increase dramatically. The equation for such geometric or exponential growth is:

$$\mathrm{d}N/\mathrm{d}t = rN$$

where N is population size or density, $\mathrm{d}N/\mathrm{d}t$ is the growth rate, and r is the instantaneous per capita rate of increase. At $r = 0$, rates of birth and death are equal and the population is static; if $r < 0$, the population declines; when $r > 0$, the population increases.

Growth continues only until a point at which some resource(s) become limiting, called the carrying capacity. As the population nears the carrying capacity, the rate of growth slows in a process represented by:

$$\mathrm{d}N/\mathrm{d}t = rN - rN^2/K$$

in which K, representing the carrying capacity, contributes to the second term, called environmental resistance. Although this basic equation of population dynamics underpins a substantial body of theoretical work, evidently natural populations persist in more narrowly fluctuating densities, well below the carrying capacity. Observed persistence over evolutionary time (section 8.2) allows the inference that, averaged over time, birth rate equals death rate.

Parasitism and predation are major influences on population dynamics as they affect death rate in a manner that varies with host density. Thus, an increase in mortality with density (positive density dependence) contrasts with a decrease in death rate with density (negative density dependence). A substantial body of experimental and theoretical evidence demonstrates that predators and parasitoids impose density dependent effects on components of their food webs, in a trophic cascade (see below). Experimental removal of the most important ("top") predator can induce a major shift in community structure, demonstrating that predators control the abundance of subdominant predators and certain prey species. Models of complex relationships between predators and prey frequently are motivated by a desire to understand interactions of native predators or biological control agents and target pest species.

Mathematical models may commence from simple interactions between a single monophagous predator and its prey. Experiments and simulations concerning the long-term trend in densities of each show regular cycles of predators and prey: when prey are abundant, predator survival is high; as more predators become available, prey abundance is reduced; predator numbers decrease as do those of prey; reduction in predation allows the prey to escape and rebuild numbers. The sinusoidal, time-lagged cycles of predator and prey abundances may exist in some simple natural systems, such as the aquatic planktonic predator *Chaoborus* (Diptera: Chaoboridae) and its cladoceran prey *Daphnia* (Fig. 13.9).

Examination of shorter-term feeding responses using laboratory studies of simple systems shows that predators vary in their responses to prey density. Early ecologists' assumptions of a linear relationship (increased prey density leading to increased predator feeding) have been superseded. A common functional response of a predator to prey density involves a gradual slowing of the rate of predation relative to increased prey density, until an asymptote is reached. This upper limit beyond which no increased rate of prey capture occurs is due to the time constraints of foraging and handling prey, in which there is a finite limit to the time spent in feeding

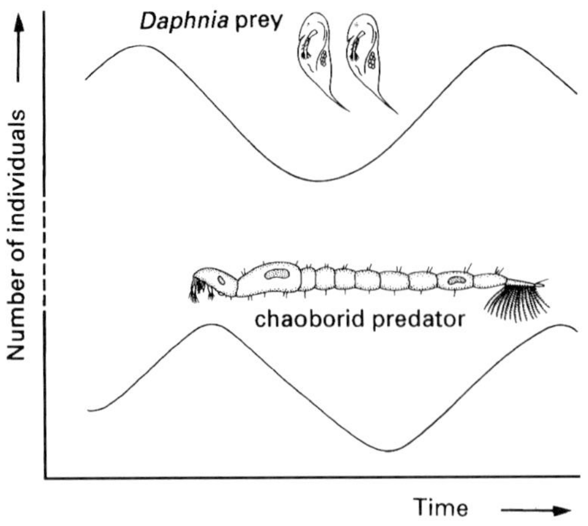

Fig. 13.9 An example of the regular cycling of numbers of predators and their prey: the aquatic planktonic predator *Chaoborus* (Diptera: Chaoboridae) and its cladoceran prey *Daphnia* (Crustacea).

activities, including a recovery period. The rate of prey capture does not depend upon prey density alone: individuals of different instars have different feeding rate profiles, and in poikilothermic insects there is an important effect of ambient temperature on activity rates.

Assumptions of predator monophagy can be biologically unrealistic and so more complex models include multiple prey items. Predator behaviour is based upon optimal foraging strategies involving simulated prey selection varying with changes in proportional availability of different prey items. However, predators may not switch between prey items based upon simple relative numerical abundance; other factors include differences in prey profitability (nutritional content, ease of handling, etc.), the hunger-level of the predator, and perhaps predator learning and development of a search-image for particular prey, irrespective of abundance.

Models of prey foraging and handling by predators, including more realistic choice between profitable and less profitable prey items, indicate that:

- prey specialization ought to occur when the most profitable prey is abundant;
- predators should switch rapidly from complete dependence on one prey to the other, with partial preference (mixed feeding) being rare;
- the actual abundance of a less-abundant prey should be irrelevant to the decision of a predator to specialize on the most abundant prey.

Improvements can be made concerning parasitoid searching behaviour that simplistically is taken to resemble a random-searching predator, independent of host abundance, the proportion of hosts already parasitized, or the distribution of the hosts. Parasitoids can identify and respond behaviourally to already-parasitized hosts. Furthermore, prey organisms (and hosts) are not distributed at random, but occur in patches, and within patches the density is likely to vary. As predators and parasitoids aggregate in areas of high resource density, interactions between predators/parasitoids (**interference**) become significant, perhaps rendering a profitable area less rewarding. Furthermore, **refuges** from predators and parasitoids may exist within a patch. Thus, amongst California red scale insects (Hemiptera: Diaspididae: *Aonidiella aurantii*) on citrus trees, those on the periphery of the tree may be up to 27 times more vulnerable to two species of parasitoids compared with individual scales in the centre of the tree, which thus may be termed a refuge. Similarly, the introduced cycad scale, *Aulacaspis yasumatsui* (Diaspididae), on cycad trees in Guam is controlled poorly by the introduced predator *Rhyzobius lophanthae* (Coleoptera: Coccinellidae) on leaves at ground level compared to higher on the plant. The ground-level leaves appear to be a refuge from the predator. Furthermore, the effectiveness of a refuge varies between taxonomic or ecological groups: external leaf-feeding insects support more parasitoid species than leaf-mining insects, which in turn support more than highly concealed insects such as root feeders or those living in structural refuges. These observations have important implications for the success of biological control programmes.

The direct effects of a predator (or parasitoid) on its prey (or host) translate into changes in the prey's or the host's energy supply (i.e. plants if the prey or host is a herbivore) in an interaction chain. The effects of resource consumption are predicted to cascade from the top consumers (predators or parasitoids) to the base of the energy pyramid via feeding links between inversely related trophic levels. The results of field experiments on such **trophic cascades** involving predator manipulation (removal or addition) in terrestrial arthropod-dominated food webs have been synthesized using meta-analysis. This involves the statistical analysis of a large collection of analysis results from individual studies for the purpose of integrating the findings. Meta-analysis found extensive support for the existence of trophic cascades, with predator removal mostly leading to increased densities of herbivorous insects and higher levels of plant damage. Furthermore, the amount of herbivory following relaxation of predation pressure was significantly higher in crop than in non-crop systems such as grasslands and woodlands. It is likely that "top-down" control (from predators) is more frequently observed in managed than in natural systems due to simplification of habitat and food-web structure in managed environments. These results suggest that natural enemies can be very effective in controlling plant pests in agroecosystems, and thus conservation of natural enemies (section 16.5.1) should be an important aspect of pest control.

13.5 THE EVOLUTIONARY SUCCESS OF INSECT PREDATION AND PARASITISM

In Chapter 11 we saw how the development of angiosperms and their colonization by specific plant-eating insects explained a substantial diversification of phytophagous insects relative to their non-phytophagous sister taxa. Analogous diversification of Hymenoptera in relation to adoption of a parasitic lifestyle exists, because numerous small groups form a "chain" on the phylogenetic tree outside the primarily parasitic sister group, the suborder Apocrita. The Orussoidea (with only one family, Orussidae) is the sister group to Apocrita, and probably all are parasitic on wood-boring insect larvae. However, the next prospective sister group lying in the (paraphyletic) "Symphyta" is a small group of wood wasps. This sister group is non-parasitic (as are the remaining symphytans) and species-poor with respect to the speciose combined Apocrita plus Orussoidea. This phylogeny implies that, in this case, adoption of a parasitic lifestyle was associated with a major evolutionary radiation. An explanation may lie in the degree of host restriction: if each species of phytophagous insect were host to a more or less monophagous parasitoid, then we would expect to see a diversification (radiation) of insect parasitoids that corresponded to that of phytophagous insects. Two assumptions need examination in this context – the degree of host-specificity and the number of parasitoids harboured by each host.

The question of the degree of monophagy amongst parasites and parasitoids is not answered conclusively. For example, many parasitic hymenopterans are extremely small, and the basic taxonomy and host associations are yet to be fully worked out. However, there is no doubt that the parasitic hymenopterans are extremely speciose (and molecular studies are revealing further undescribed cryptic or very similar species), and show a varying pattern of host-specificity, from strict monophagy to oligophagy. Amongst parasitoids within the Diptera, the species-rich Tachinidae are relatively general feeders, specializing only in hosts belonging to families or even ordinal groups. Amongst the ectoparasites, lice are predominantly monoxenic, as are many fleas and flies. However, even if several species of ectoparasitic insects were borne by each host species, because the vertebrates are not numerous, ectoparasites contribute relatively little to biological diversification in comparison with the parasitoids of insect (and other diverse arthropod) hosts.

Many hosts support multiple parasitoids, a phenomenon well known to lepidopterists who endeavour to rear adult butterflies or moths from wild-caught larvae – the frequency and diversity of parasitization is very high. Suites of parasitoid and hyperparasitoid species may attack the same species of host at different seasons, in different locations, and in different life-history stages. There are many records of more than 10 parasitoid species throughout the range of some widespread lepidopterans, and although this is true also for certain well-studied coleopterans, the situation is less clear for other orders of insects.

Finally, some evolutionary interactions between parasites and parasitoids and their hosts may be considered. Patchiness of potential host abundance throughout the host range seems to provide opportunity for increased specialization, perhaps leading to species formation within the guild of parasites/parasitoids. This can be seen as a form of niche differentiation, where the total range of a host provides a niche that is ecologically partitioned. Hosts may escape from parasitization within refuges within the range, or by modification of the life cycle, with the introduction of a phase that the parasitoid cannot track. Host diapause may be a mechanism for evading a parasitoid (or predator) that is restricted to continuous generations, with an extreme example of escape perhaps seen in the periodical cicada. These species of *Magicicada* grow concealed for many years as nymphs beneath the ground, with the very visible adults appearing only every 13 or 17 years. This cycle of a prime number of years may allow avoidance of parasitoids or predators that cannot adapt to hosts or prey with an unpredictable cyclical life history. Life-cycle shifts as attempts to evade predators may be important in species formation.

Strategies of prey/hosts and predators/parasitoids have been envisaged as evolutionary arms races, with a stepwise sequence of prey/host escape by evolution of successful defences, followed by radiation before the predator/parasitoid "catches-up", in a form of prey/host tracking. An alternative evolutionary model envisages both prey/host and predator/parasitoid evolving defences and circumventing them in near synchrony, in an evolutionarily stable strategy termed the "Red Queen" hypothesis (after the description in *Alice in Wonderland* of Alice and the Red Queen having to run faster and faster to just stay in the same place). Tests of each can be devised, and models for either can be justified, and it is unlikely that conclusive evidence will be found in the short term. What is clear is that parasitoids and predators do exert great selective pressure on their hosts or prey, and remarkable defences have arisen, as we shall see in the next chapter.

FURTHER READING

Asgari, S. & Rivers, D.B. (2011) Venom proteins from endoparasitoid wasps and their role in host-parasite interactions. *Annual Review of Entomology* **56**, 313–35.

Burke, G.R. & Strand, M.R. (2012) Polydnaviruses of parasitic wasps: domestication of viruses to act as gene delivery vectors. *Insects* **3**, 91–119.

Byers, G.W. & Thornhill, R. (1983) Biology of Mecoptera. *Annual Review of Entomology* **28**, 303–28.

Eggleton, P. & Belshaw, R. (1993) Comparisons of dipteran, hymenopteran and coleopteran parasitoids: provisional phylogenetic explanations. *Biological Journal of the Linnean Society* **48**, 213–26.

Feener, D.H. Jr, & Brown, B.V. (1997) Diptera as parasitoids. *Annual Review of Entomology* **42**, 73–97.

Godfray, H.C.J. (1994) *Parasitoids: Behavioural and Evolutionary Ecology*. Princeton University Press, Princeton, NJ.

Gundersen-Rindal, D., Dupuy, C., Huguet, E. & Drezen, J.-M. (2013) Parasitoid polydnaviruses: evolution, pathology and applications. *Biocontrol Science and Technology* **23**, 1–61.

Halaj, J. & Wise, D.H. (2001) Terrestrial trophic cascades: how much do they trickle? *The American Naturalist* **157**, 262–81.

Hassell, M.P. & Southwood, T.R.E. (1978) Foraging strategies of insects. *Annual Review of Ecology and Systematics* **9**, 75–98.

Herniou, E.A., Huguet, E., Thézé, J. *et al.* (2013) When parasitic wasps hijacked viruses: genomic and functional evolution of polydnaviruses. *Philosophical Transactions of the Royal Society B* **368**, 20130051. doi: 10.1098/rstb.2013.0051

Kathirithamby, J. (2009) Host-parasitoid associations in Strepsiptera. *Annual Review of Entomology* **54**, 227–49.

Lyal, C.H.C. (1986) Coevolutionary relationships of lice and their hosts: a test of Fahrenholz's Rule. In: *Coevolution and Systematics* (eds A.R. Stone & D.L. Hawksworth), pp. 77–91. Systematics Association, Oxford.

Pell, J.K., Baverstock, J., Roy, H.E. *et al.* (2008) Intraguild predation involving *Harmonia axyridis*: a review of current knowledge and future persepectives. *BioControl* **53**, 147–68.

Pennacchio, F. & Strand, M.R. (2006) Evolution of developmental strategies in parasitic Hymenoptera. *Annual Review of Entomology* **51**, 233–58.

Poirié, M., Carton, Y. & Dubuffet, A. (2009) Virulence strategies in parasitoid Hymenoptera as an example of adaptive diversity *Comptes Rendus Biologies* **332**, 311–20.

Quicke, D.L.J. (1997) *Parasitic Wasps*. Chapman & Hall, London.

Resh, V.H. & Cardé, R.T. (eds.) (2009) *Encyclopedia of Insects*, 2nd edn. Elsevier, San Diego, CA. [Particularly see articles on host seeking by parasitoids; Hymenoptera; hyperparasitism; parasitoids; predation and predatory insects.]

Schoenly, K., Cohen, J.E., Heong, K.L. *et al.* (1996) Food web dynamics of irrigated rice fields at five elevations in Luzon, Philippines. *Bulletin of Entomological Research* **86**, 451–66.

Strand, M.R. & Burke, G.R. (2012) Polydnaviruses as symbionts and gene delivery systems. *PLoS Pathogens* **8**(7), e1002757. doi: 10.1371/journal.ppat.1002757

Sullivan, D.J. (1987) Insect hyperparasitism. *Annual Review of Entomology* **32**, 49–70.

Symondson, W.O.C., Sunderland, K.D. & Greenstone, M.H. (2002) Can generalist predators be effective biocontrol agents? *Annual Review of Entomology* **47**, 561–94.

Vinson, S.B. (1984) How parasitoids locate their hosts: a case of insect espionage. In: Insect Communication (ed. T. Williams), pp. 325–48. Academic Press, London.

Chapter 14

INSECT DEFENCE

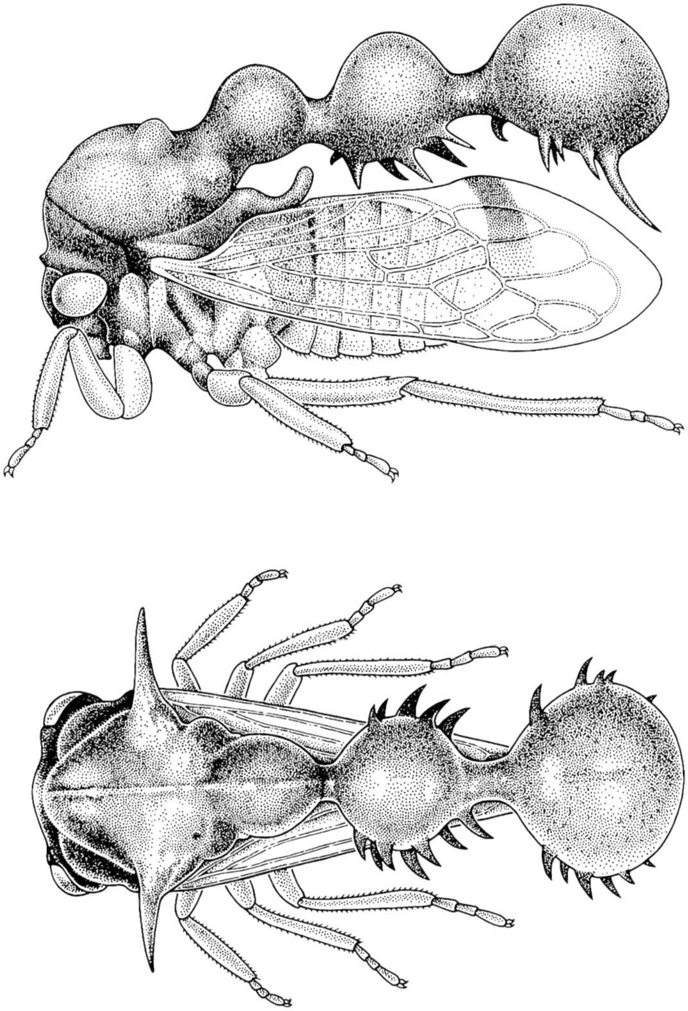

An African ant-mimicking membracid bug, shown in side and dorsal view. (After Boulard 1968.)

Although some humans eat insects (section 1.8.1), many "western" cultures are reluctant to use them as food; this aversion extends no further than humans. For very many organisms, insects provide a substantial food source because they are nutritious, abundant, diverse and found everywhere. Some animals, termed **insectivores**, rely almost exclusively on a diet of insects; omnivores may eat them opportunistically; and many herbivores unavoidably consume insects. Insectivores may be vertebrates or invertebrates, including arthropods – insects certainly eat other insects. Even plants lure, trap and digest insects; for example, pitcher plants (both New World Sarraceniaceae and Old World Nepenthaceae) digest arthropods, predominantly ants, in their fluid-filled pitchers (section 11.4.2), and the flypaper and Venus flytraps (Droseraceae) capture many flies. Insects, however, actively or passively resist being eaten, by means of a variety of protective devices – the insect defences – which are the subject of this chapter.

A review of the terms discussed in Chapter 13 is appropriate here. A **predator** is an animal that kills and consumes a number of prey animals during its life. Animals that live at the expense of another animal but do not kill it are **parasites**, which may live internally (**endoparasites**) or externally (**ectoparasites**). **Parasitoids** are those that live at the expense of one animal that dies prematurely as a result. The animal attacked by parasites or parasitoids is a **host**. All insects are potential **prey** or hosts to many kinds of predators (either vertebrate or invertebrate), parasitoids or, less often, parasites.

Many defensive strategies exist, including use of specialized morphology (as shown for the extraordinary, ant-mimicking membracid bug *Hamma rectum* from tropical Africa in the vignette at the beginning of this chapter), behaviour, noxious chemicals, and responses of the immune system. This chapter deals with aspects of defence that include death feigning, autotomy, crypsis (camouflage), chemical defences, aposematism (warning signals), mimicry, and collective defensive strategies. These are directed against a wide range of vertebrates and invertebrates but, because much study has involved insects defending themselves against insectivorous birds, the role of these particular predators is emphasized (Box 14.1). Three other boxes deal with the topics of predatory bugs that disguise themselves with a "backpack" of trash (Box 14.2), the chemical protection of insect eggs (Box 14.3), and the defence mechanism of bombardier beetles (Carabidae) (Box 14.4). Immunological defence against microorganisms is discussed in Chapter 3, and defences used against parasitoids are considered in Chapter 13.

A useful framework for discussion of defence and predation can be based upon the time and energy inputs to the respective behaviours. Thus, hiding, escape by running or flight, and defence by staying and fighting, involve increasing energy expenditure but diminishing costs in time expended (Fig. 14.1). Many

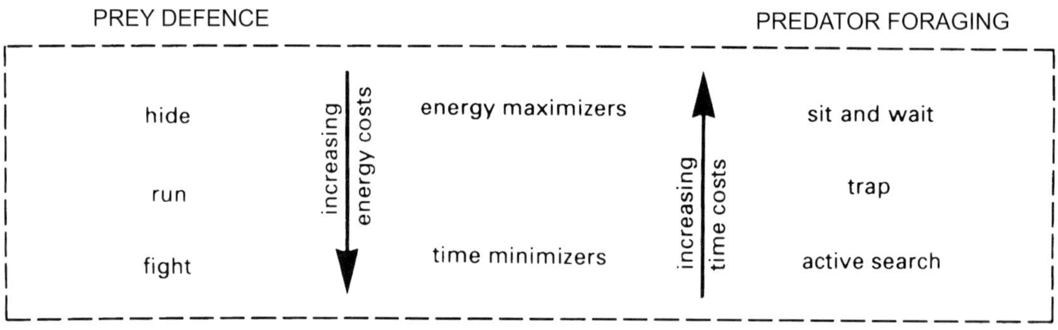

Fig. 14.1 The basic spectrum of prey defence strategies and predator foraging, varying according to costs and benefits in both time and energy. (After Malcolm 1990.)

The Insects: An Outline of Entomology, Fifth Edition. P.J. Gullan and P.S. Cranston.

Companion Website: www.wiley.com/go/gullan/insects

insects will change to another strategy if the previous defence fails: the scheme is not clear-cut and it has elements of a continuum.

14.1 DEFENCE BY HIDING

Visual deception may reduce the probability of being found by a natural enemy. A well-concealed **cryptic** insect that either resembles its general background or an inedible (neutral) object may be said to "mimic" its surroundings. In this book, **mimicry** (in which an animal resembles another animal that is recognizable by natural enemies) is treated separately (section 14.5). However, crypsis and mimicry can be seen as similar in that both arise when an organism gains in fitness through developing a resemblance (to a neutral or animate object) evolved under selection. In all cases, it is assumed that such defensive adaptive resemblance is under selection by predators or parasitoids, but, although maintenance of selection for accuracy of resemblance has been demonstrated for some insects, the origin can only be surmised.

Insect crypsis can take many forms. The insect may adopt **camouflage**, making it difficult to distinguish from the general background in which it lives (Box 14.1), by:

- resembling a uniform coloured background, such as a green geometrid moth on a leaf;
- resembling a patterned background, such as a grasshopper in vegetation or a mottled moth on tree bark (Fig. 14.2; see also **Plate 7a,g**);
- being countershaded – light below and dark above – as in some caterpillars and aquatic insects;
- having a pattern to disrupt the outline, as is seen in many moths that settle on leaf litter;
- having a bizarre shape to disrupt the silhouette, as demonstrated by some membracid leafhoppers.

In another form of crypsis, termed **masquerade** or **mimesis** to contrast with the camouflage described above, the organism deludes a predator by resembling an object that is a particular specific feature of its environment, but is of no inherent interest to a predator. This feature may be an inanimate object, such as the bird dropping resembled by young larvae of some butterflies, such as *Papilio aegeus* (Papilionidae), or an animate but neutral object – for example "looper" caterpillars (the larvae of geometrid moths) resemble twigs, others look like part of the leaf (see **Plate 7b**), some membracid bugs imitate thorns arising from a stem, and many stick-insects look very much like sticks and may even move like a twig in the wind. Many insects, notably amongst the lepidopterans and orthopterans, resemble leaves, even to the similarity in venation (Fig. 14.3) and appearing to be either dead or alive, mottled with fungus, or even partially eaten as if by a herbivore. However, interpretation of apparent resemblance to inanimate objects as simple crypsis may be revealed as more complex when subject to experimental manipulation (Box 14.2).

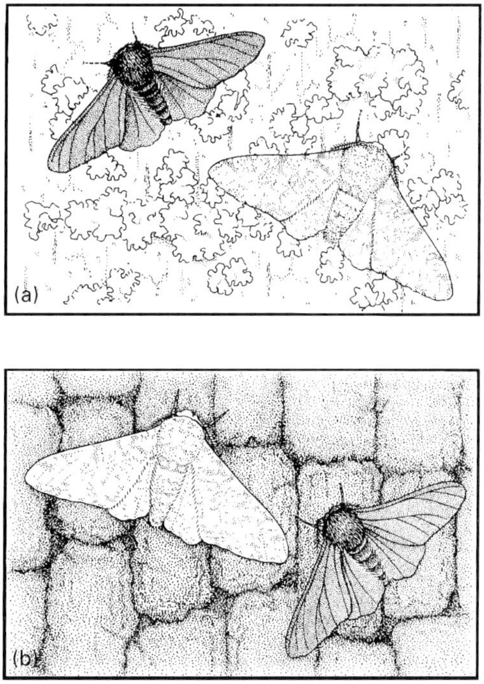

Fig. 14.2 Pale and melanic (*carbonaria*) morphs of the peppered moth *Biston betularia* (Lepidoptera: Geometridae) resting on: (a) pale, lichen-covered trunks; and (b) dark trunks.

Crypsis is a very common form of insect concealment, particularly in the tropics and amongst nocturnally active insects. It has low energetic costs, but relies on the insect being able to select the appropriate background. Experiments with two differently coloured morphs of *Mantis religiosa* (Mantidae), the European praying mantid, have shown that brown and green morphs placed against appropriate and inappropriate coloured backgrounds were fed upon in a highly selective manner by birds: they removed all "mismatched"

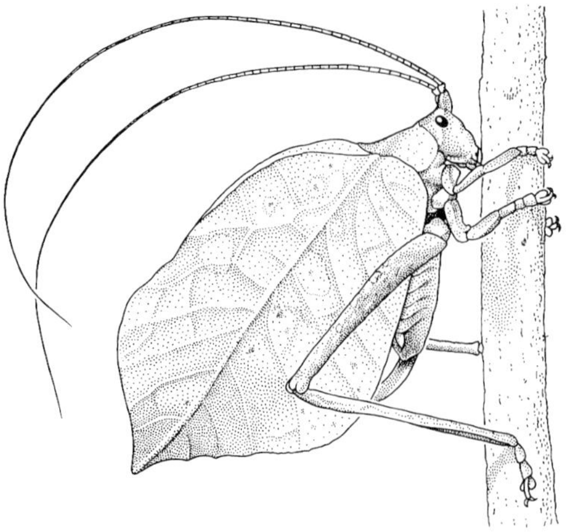

Fig. 14.3 A leaf-mimicking katydid, *Mimetica mortuifolia* (Orthoptera: Tettigoniidae), in which the fore wing resembles a leaf even to the extent of leaf-like venation and spots resembling fungal mottling. (After Belwood 1990.)

morphs and found no camouflaged ones. Even if the correct background is chosen, it may be necessary to orientate correctly: moths with disruptive outlines or with striped patterns resembling the bark of a tree may be concealed only if orientated in a particular direction on the trunk.

The Indomalayan orchid mantid, *Hymenopus coronatus* (Hymenopodidae), blends beautifully with the pink flower spike of an orchid, where it sits awaiting prey. The crypsis is enhanced by the close resemblance of the femora of the mantid's legs to the flower's petals. Crypsis enables the mantid to avoid detection by its potential prey (flower visitors) (section 13.1.1) as well as conceal itself from predators.

14.2 SECONDARY LINES OF DEFENCE

Little is known of the learning processes of inexperienced vertebrate predators, such as insectivorous birds. However, studies of the gut contents of birds show that cryptic insects are not immune from predation (Box 14.1). Once found for the first time (perhaps accidentally), birds subsequently seem able to detect cryptic prey via a "search image" for some element(s) of the pattern. Thus, having discovered that some twigs were caterpillars, American blue jays were observed to continue to peck at sticks in a search for food. Primates can identify stick-insects by one pair of unfolded legs alone, and will attack actual sticks to which phasmatid legs have been affixed experimentally. Clearly, subtle cues allow specialized predators to detect and eat cryptic insects.

Once the deception is discovered, the insect prey may have further defences available in reserve. In the energetically least demanding response, the initial crypsis may be exaggerated, as when a threatened masquerader falls to the ground and lies motionless. This behaviour is not restricted to cryptic insects: even visually obvious prey insects may feign death (**thanatosis**). This behaviour, used by many beetles (particularly weevils), can be successful because predators lose interest in apparently dead prey or may be unable to locate a motionless insect on the ground. Another secondary line of defence is to take flight and suddenly reveal a flash of conspicuous colour from the hind wings. Immediately on landing the wings are folded, the colour vanishes, and the insect is cryptic once more. This behaviour is common amongst certain orthopterans and underwing moths; the colour of the flash may be yellow, red, purple or, rarely, blue.

A third type of behaviour of cryptic insects upon discovery by a predator is the production of a **startle display**. One of the most common is to open the fore wings and reveal brightly coloured "eyes" that are usually concealed on the hind wings (Fig. 14.4). Experiments using birds as predators have shown that the more perfect the eye (with increased contrasting rings to resemble true eyes), the better the deterrence. Not all eyes serve to startle: perhaps a rather poor imitation of an eye on a wing may direct pecks from a predatory bird to a non-vital part of the insect's anatomy.

An extraordinary type of insect defence is the convergent appearance of part of the body to a feature of a vertebrate, albeit on a much smaller scale. Thus, the head of a species of fulgorid bug, commonly called the alligator bug (*Fulgora laternaria*), bears an uncanny resemblance to that of a caiman. The pupa of a particular lycaenid butterfly looks like a monkey head. Some tropical sphingid larvae assume a threat posture that, together with false eyespots that actually lie on the posterior abdomen, gives a snake-like impression. Similarly, the caterpillars of certain swallowtail butterflies bear a likeness to a snake's head (see **Plate 7d**). These resemblances may deter predators (such as birds that search by "peering about") by their startle effect, with the incorrect scale of the mimic being overlooked by the predator.

Box 14.1 Avian predators as selective agents for insects

Henry Bates, who was first to propose a theory for mimicry, suggested that natural enemies, such as birds, selected among different prey, such as butterflies, based upon an association between mimetic patterns and unpalatability. A century later, Henry Kettlewell argued that selective predation by birds on the peppered moth (Geometridae: *Biston betularia*) altered the proportions of dark- and light-coloured morphs (Fig. 14.2) according to their concealment (crypsis) on natural and industrially darkened trees upon which the moths rested by day. Amateur lepidopterists recorded that the proportion of the almost uniformly black ("melanic") *carbonaria* form dramatically increased as industrial pollution increased in northern England from the mid-19th century. Elimination of pale lichen on tree trunk resting areas was suggested to have made normal pale (*typica*) morphs more visible against the sooty, lichen-denuded trunks (as shown in Fig. 14.2b), and hence they were more susceptible to visual recognition by bird predators. This phenomenon, termed **industrial melanism**, often has been cited as a classic example of evolution through natural selection.

The peppered moth/avian predation story has been challenged for its experimental design and procedures, and possibly biased interpretation. Another complication is that, in addition to the extreme dark and pale forms of the peppered moth, in some regions of Britain there are intermediate-coloured *insularia* forms. However, detailed recent research (including observational and experimental data), especially by the late Michael Majerus, has confirmed that selective predation by birds is the major factor driving changes in frequency of typical and melanic moths. It seems that:

- birds are the major predators, with night-flying, pattern-insensitive bats unimportant predators;
- although some moths rest "exposed" to visual predators on trunks during the day, most settle higher in the canopy under branches or at branch junctions;
- there are no significant differences in the resting sites used by *typica* (non-melanic), *carbonaria* (full melanic) and *insularia* (intermediate melanic) forms;
- differential bird predation on melanic moths is sufficient to explain the rapid decline in moth melanism in post-industrial Britain, as improved air quality (particularly following the 1956 and 1968 Clean Air Acts) reduced soot on tree surfaces and improved camouflage of non-melanic moths;
- the melanic form in Britain originated from a single ancestral haplotype that is orthologous to a major wing patterning loci in certain other Lepidoptera, including that controlling mimicry forms in *Heliconius* butterflies.

Clearly, colour variation in *B. betularia* is under strong natural selection. However, experiments and modelling suggest that factors other than predation may have contributed to changes in melanic frequency. Either migration is higher than estimated or some non-visual advantage must accrue to melanics. Although there is no direct evidence in *B. betularia*, some melanics may have stronger immune defences. Also, in times of industrial pollution, melanism might confer protection from the toxic effects of pollutants, since melanin has a strong chelating action on metals.

In other systems, convincing evidence of bird predation comes from direct observation, inference from beak pecks on the wings of butterflies, and experiments with colour-manipulated daytime-flying moths. Thus, winter-roosting monarch butterflies (*Danaus plexippus*) are fed upon by black-backed orioles (Icteridae), which browse selectively on poorly defended individuals, and by black-headed grosbeaks (Fringillidae), which appear to be completely insensitive to the toxins. Specialized predators such as Old World bee-eaters (Meropidae) and Neotropical jacamars (Galbulidae) can deal with the stings of hymenopterans and the toxins of butterflies, respectively (the red-throated bee-eater, *Merops bullocki*, is shown here de-stinging a bee on a branch; after Fry *et al*. 1992). A similar suite of birds selectively feeds on noxious ants. The ability of these specialist predators to distinguish between varying pattern and edibility may make them selective agents in the evolution and maintenance of defensive mimicry.

Birds are observable insectivores for laboratory studies: their readily recognizable behavioural responses to unpalatable foods include head-shaking, disgorging of food, tongue-extending, bill-wiping, gagging, squawking and ultimately vomiting. For many birds, a single learning trial with noxious (Class I) chemicals appears to lead to long-term aversion to the particular insect, even with a substantial delay between feeding and illness. However, manipulative studies of bird diets are complicated by their fear of novelty (neophobia), which, for example, can lead

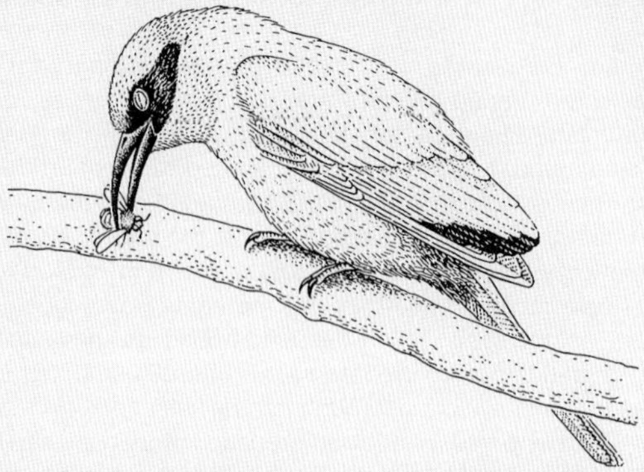

to rejection of prey with startling and frightening displays (section 14.2). Conversely, birds rapidly learn preferred items, as in Kettlewell's experiments in which birds quickly recognized both *Biston betularia* morphs on tree trunks in his artificial set-up.

Perhaps no insect has completely escaped the attentions of predators, and some birds can overcome even severe insect defences. For example, the lubber grasshopper (Acrididae: *Romalea guttata*) is large, gregarious and aposematic, and if attacked it squirts volatile, pungent chemicals accompanied by a hissing noise. The lubber is extremely toxic and is avoided by all lizards and birds except one, the loggerhead shrike (Laniidae: *Lanius ludovicanus*), which snatches its prey, including lubbers, and impales them "decoratively" upon spikes, with minimal handling time. These impaled items serve both as food stores and in sexual or territorial displays. *Romalea*, which are emetic to shrikes when fresh, become edible after two days of lardering, presumably by denaturation of the toxins. The impaling behaviour shown by most species of shrikes thus is preadaptive in permitting the loggerhead to feed upon an extremely well-defended insect. No matter how good the protection, there is no such thing as total defence in the arms race between prey and predator.

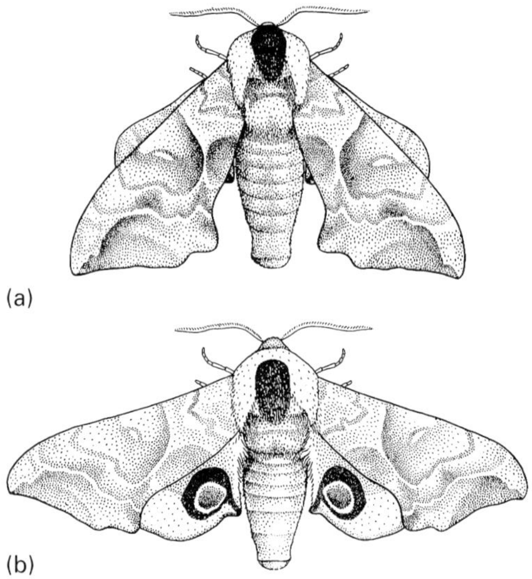

Fig. 14.4 The eyed hawk moth, *Smerinthus ocellatus* (Lepidoptera: Sphingidae). (a) The brownish fore wings cover the hind wings of a resting moth. (b) When the moth is disturbed, the black and blue eyespots on the hind wings are revealed. (After Stanek 1977.)

14.3 MECHANICAL DEFENCES

Morphological structures of predatory function, such as the modified mouthparts and spiny legs described in Chapter 13, also may be defensive, especially if a fight ensues. Cuticular horns and spines may be used in deterrence of a predator or in combating rivals for mating, territory or resources, as in *Onthophagus* dung beetles (section 5.3). For ectoparasitic insects, which are vulnerable to the actions of the host, body shape and sclerotization provide one line of defence. Fleas are laterally compressed, making these insects difficult to dislodge from host hairs. Biting lice are flattened dorsoventrally and are narrow and elongate, allowing them to fit between the veins of feathers, secure from preening by the host bird. Furthermore, many ectoparasites have resistant bodies, and the heavily sclerotized cuticle of certain beetles must act as a mechanical anti-predator device.

Many insects construct retreats or shelters that can deter a predator that fails to recognize the structure as containing anything edible or that is unwilling to eat construction materials. For example, webspinners (Embioptera) construct silken galleries (see **Plate 7c**) in which they live gregariously, protected from most predators, and which can be defended even from other embiopterans. The cases of caddisfly larvae (Trichoptera), constructed of sand grains, stones or organic fragments (see Fig. 10.5), may have originated in response to the physical environment of flowing water, but certainly have a defensive role. Similarly, a portable case of plant material bound with silk is constructed by the terrestrial larvae of bagworms (Lepidoptera: Psychidae). In both caddisflies and psychids, the case serves to protect during pupation. Some insects construct artificial shields; for example, the larvae of certain chrysomelid beetles decorate themselves with their faeces. The larvae of a few lacewings and reduviid bugs cover themselves with lichens and detritus and/or the sucked-out carcasses of their insect prey, which can act as barriers to a predator, and also may disguise them from prey (Box 14.2).

Box 14.2 Backpack bugs – dressed to kill?

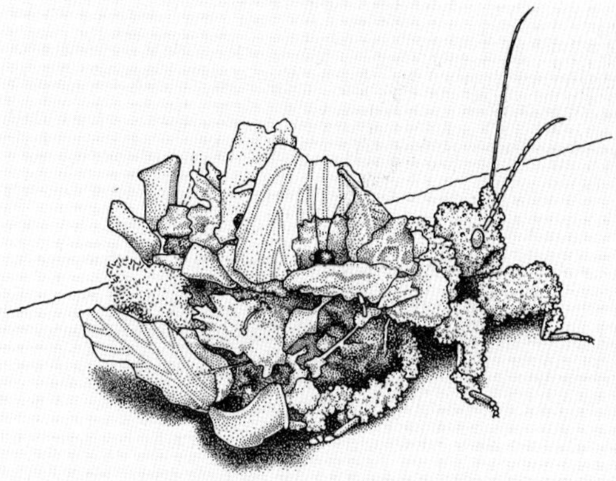

Certain West African predatory assassin bugs (Hemiptera: Reduviidae) decorate themselves with a coat of dust that they adhere to their bodies with sticky secretions from abdominal setae. To this undercoat, the nymphal instars (of several species) add vegetation and cast skins of prey items, mainly ants and termites. The resultant "backpack" of trash can be much larger than the animal itself (as in this illustration derived from a photograph by M. Brandt). It had been assumed that the bugs are mistaken by their predators or prey for an innocuous pile of debris; but rather few examples of such deceptive camouflage have been tested critically.

In the first behavioural experiment, investigators Brandt and Mahsberg (2002) exposed bugs to predators typical of their surroundings, namely spiders, geckos and centipedes. Three groups of bugs were tested experimentally: (i) naturally occurring ones with dustcoat and backpack; (ii) individuals only with a dustcoat; and (iii) naked ones lacking both dustcoat and backpack. Bug behaviour was unaffected, but the predators' reactions varied: spiders were slower to capture the individuals with backpacks than individuals of the other two groups; geckos also were slower to attack backpack wearers; and centipedes never attacked backpackers, although they ate most of the nymphs without backpacks. The implied anti-predatory protection certainly includes some visual disguise, but only the gecko is a visual predator: spiders are tactile predators, and centipedes hunt using chemical and tactile cues. Backpacks are conspicuous more than cryptic, but they confuse visual, tactile and chemical-orientating predators by looking, feeling and smelling wrong for a prey item.

Next, differently dressed bugs and their main prey, ants, were manipulated. Studied ants responded to individual naked bugs much more aggressively than they did to dustcoated or backpack-bearing nymphs. The backpack did not diminish the risk of hostile response (taken as equating to "detection") beyond that to the dustcoat alone, rejecting any idea that ants may be lured by the odour of dead conspecifics included in the backpack. One trialled prey item, an army ant, is highly aggressive but blind, and although unable to detect the predator visually, it responded as did other prey ants – with aggression directed preferentially towards naked bugs. Evidently, any covering confers "concealment", but not by the visual protective mechanism assumed previously.

Thus, what appeared to be simple visual camouflage proved more a case of disguise to fool chemical- and touch-sensitive predators and prey. Additional protection is provided by the bugs' abilities to shed their backpacks. While collecting research specimens, Brandt and Mahsberg observed that bugs readily vacated their backpacks in an inexpensive autotomy strategy, resembling the metabolically expensive lizard tail-shedding. Such experimental research undoubtedly will shed more light on other cases of visual camouflage/predator deception.

The waxes and powders secreted by many hemipterans (such as scale insects, woolly aphids, whiteflies and fulgorids) may function to entangle the mouthparts of a potential arthropod predator, but also have a waterproofing role. The larvae of many ladybird beetles (Coccinellidae) are coated with white wax, thus resembling their mealybug prey. This may be a disguise to protect them from ants that tend the mealybugs.

Body structures themselves, such as the scales of moths and caddisflies, can protect as they detach readily to allow the escape of a slightly denuded insect from the jaws of a predator, or from the sticky threads of spiders' webs or the glandular leaves of insectivorous plants such as the sundews. A mechanical defence that seems at first to be maladaptive is **autotomy**, the shedding of limbs, as demonstrated by stick-insects (Phasmatodea) and perhaps crane flies (Diptera: Tipulidae). The upper part of the phasmatid leg has the trochanter and femur fused, with no muscles running across the joint. A special muscle breaks the leg at a weakened zone in response to a predator grasping the leg. Immature stick-insects and mantids can regenerate lost limbs at moulting, and even certain autotomized adults can induce an adult moult at which the limb can regenerate.

Secretions of insects can have a mechanical defensive role, acting as a glue or slime that ensnares predators or parasitoids. Certain cockroaches have a permanent slimy coat on the abdomen that confers protection. Lipid secretions from the **cornicles** (also called **siphunculi**) of aphids may gum-up predator mouthparts or small parasitic wasps. Termite soldiers have a variety of secretions available to them in the form of cephalic glandular products, including terpenes that dry on exposure to air to form a resin. In *Nasutitermes* (Termitidae), the secretion is ejected via the nozzle-like nasus (a pointed snout or rostrum) as a quick-drying fine thread that impairs the movements of a predator such as an ant. This defence counters arthropod predators but is unlikely to deter vertebrates. Mechanical-acting chemicals are only a small selection of the total insect armoury that can be mobilized for chemical warfare.

14.4 CHEMICAL DEFENCES

Chemicals play vital roles in many aspects of insect behaviour. In Chapter 4 we considered the use of **pheromones** in many forms of communication, including alarm pheromones elicited by the presence of a predator. Similar chemicals, called **allomones**, that benefit the producer and harm the receiver, play important roles in the defences of many insects, notably amongst many Heteroptera and Coleoptera. The relationship between defensive chemicals and those used in communication may be very close, sometimes with the same chemical fulfilling both roles. Thus, a noxious chemical that repels a predator can alert conspecific insects to the predator's presence and may act as a stimulus to action. In the energy–time dimensions shown in Fig. 14.1, chemical defence lies towards the energetically expensive but time-efficient end of the spectrum. Chemically defended insects tend to have high apparency to predators, i.e. they are usually non-cryptic, active, often relatively large, long-lived and frequently aggregated or social in behaviour. Often they signal their distastefulness by **aposematism** – warning signalling that often involves bold colouring (see **Plate 7e,h**), but may include odour, or even sound or light production.

14.4.1 Classification by function of defensive chemicals

Amongst the diverse range of defensive chemicals produced by insects, two classes of compounds can be distinguished by their effects on a predator. Class I defensive chemicals are noxious because they irritate, hurt, poison or drug individual predators. Class II chemicals are innocuous, being essentially anti-feedant chemicals that merely stimulate the olfactory and gustatory receptors, or are aposematic indicator odours. Many insects use mixtures of the two classes of chemicals and, furthermore, Class I chemicals in low concentrations may give Class II effects. Contact by a predator with Class I compounds results in repulsion through, for example, emetic (sickening) properties or induction of pain and, if this unpleasant experience is accompanied by odorous Class II compounds, predators learn to associate the odour with the encounter. This conditioning results in the predator learning to avoid the defended insect at a distance, without the dangers (to both predator and prey) of having to feel or taste it.

Class I chemicals include both immediate-acting substances, which the predator experiences through handling the prey insect (which may survive the attack), and chemicals with delayed, often systemic, effects including vomiting or blistering. In contrast to immediate-effect chemicals sited in particular organs and applied topically (externally), delayed-effect chemicals are distributed more generally within the insect's tissues and haemolymph, and are tolerated systemically. Whereas a predator evidently learns rapidly to associate immediate distastefulness with particular prey (especially if it is aposematic), it is unclear how a predator identifies the cause of nausea some time after the predator has killed and eaten the toxic culprit, and what benefits this action brings to the victim. Experimental evidence from birds shows that at least these predators are able to associate a particular food item with a delayed effect, perhaps through taste when regurgitating the item. Too little is known of feeding in insects to understand if this applies similarly to predatory insects. Perhaps a delayed poison that fails to protect an individual from being eaten evolved through the education of a predator by a sacrifice, thereby allowing differential survival of relatives (section 14.6).

14.4.2 The chemical nature of defensive compounds

Class I compounds are much more specific and effective against vertebrate than arthropod predators. For example, birds are more sensitive than arthropods to toxins such as cyanides, cardenolides and alkaloids. Cyanogenic glycosides are produced or acquired by zygaenid moths (Zygaenidae), *Leptocoris* bugs (Rhopalidae) and *Acraea* and *Heliconius* butterflies (Nymphalidae). Cardenolides are very prevalent, occurring notably in monarch or wanderer butterflies (Nymphalidae), certain cerambycid and chrysomelid beetles, lygaeid bugs, pyrgomorphid grasshoppers (see **Plate 4c**) and even an aphid (*Aphis nerii*). A variety of alkaloids similarly are acquired convergently in many coleopterans and lepidopterans.

Possession of Class I emetic or toxic chemicals is very often accompanied by aposematism, particularly colouration directed against visual-hunting diurnal predators. However, visible aposematism is of limited use at night, and the sounds emitted by nocturnal moths, such as certain Arctiidae when challenged by bats, are aposematic, warning the predator of a distasteful meal (section 4.1.4). Furthermore, the bioluminescence emitted by certain larval beetles (Phengodidae, Lampyridae and their relatives; section 4.4.5) is an aposematic warning of distastefulness.

Class II chemicals tend to be volatile and reactive organic compounds with low molecular weight, such as aromatic ketones, aldehydes, acids and terpenes. Examples include the stink-gland products of Heteroptera and the many low molecular weight substances, such as formic acid emitted by ants. Bitter-tasting but non-toxic compounds such as quinones are common Class II chemicals. Many defensive secretions are complex mixtures that can involve synergistic effects. Thus, the carabid beetle *Helluomorphoides* emits a Class II compound, formic acid, that is mixed with *n*-nonyl acetate, which enhances skin penetration of the acid, giving a Class I painful effect.

The role of these Class II chemicals in aposematism, warning of the presence of Class I compounds, was considered above. In another role, these Class II chemicals may be used to deter predators such as ants that rely on chemical communication. For example, prey such as certain termites, when threatened by predatory ants, release mimetic ant alarm pheromones, thereby inducing inappropriate ant behaviours of panic and nest defence. In another case, ant-nest inquilines (section 12.3), which might provide prey to their host ants, are unrecognized as potential food because they produce chemicals that appease ants.

Class II compounds alone appear unable to deter many insectivorous birds. For example, blackbirds (Turdidae) will eat notodontid (Lepidoptera) caterpillars that secrete a 30% formic acid solution; many birds actually encourage ants to secrete formic acid into their plumage in an apparent attempt to remove ectoparasites (so-called "anting").

14.4.3 Sources of defensive chemicals

Many defensive chemicals, notably those of phytophagous insects, are derived from the host plant upon which the larvae (Fig. 14.5; Box 14.3) and, less commonly, the adults feed. Frequently, a close association is observed between restricted host-plant use (**monophagy** or **oligophagy**) and the possession of a chemical defence. An explanation may lie in a coevolutionary "arms race" in which a plant develops toxins to deter phytophagous insects. A few phytophages overcome the defences and thereby become specialists able to detoxify or sequester the plant toxins. These specialist herbivores can recognize their preferred host plants, develop on them, and use the plant toxins (or metabolize them to closely related compounds) for their own defence.

Box 14.3 Chemically protected eggs

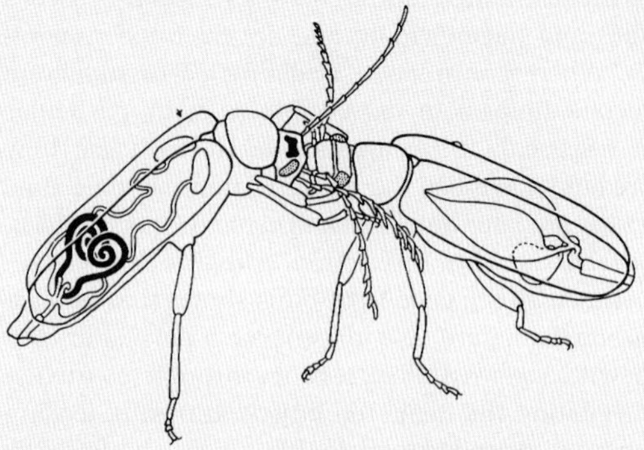

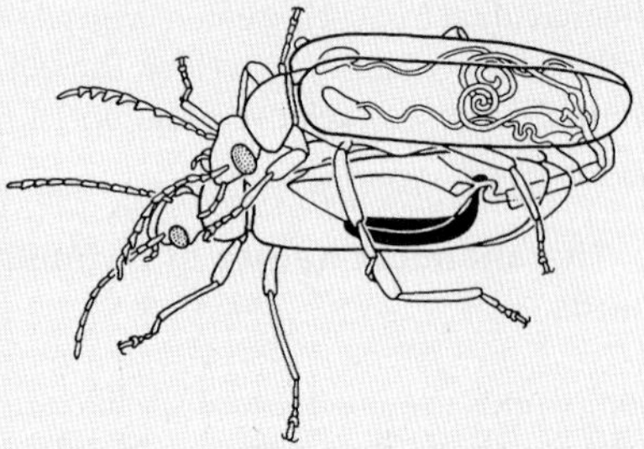

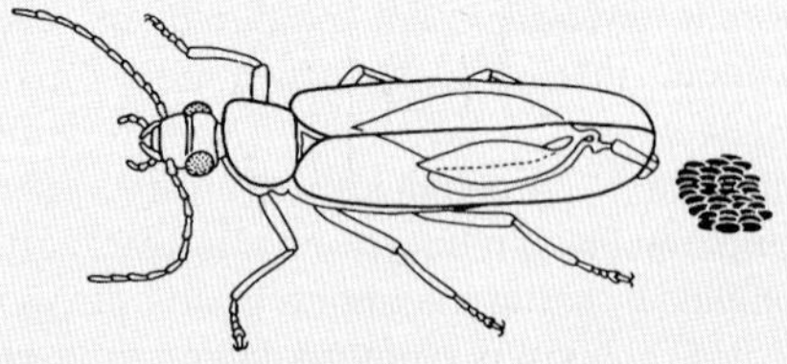

Some insect eggs can be protected by parental provisioning of defensive chemicals, as seen in certain arctiid moths and some butterflies. Pyrrolizidine alkaloids from the larval food plants are passed by the adult males to the females via seminal secretions, and the females transmit them to the eggs, which become distasteful to predators. Males advertise their possession of the defensive chemicals via a courtship pheromone derived from, but different to, the acquired alkaloids. In at least two of these lepidopteran species, it has been shown that males are less successful in courtship if deprived of their alkaloid.

Amongst the Coleoptera, certain species of Meloidae and Oedemeridae can synthesize cantharidin, and others, particularly species of Anthicidae and Pyrochroidae, can sequester it from their food. Cantharidin ("Spanish fly") is a sesquiterpene with very high toxicity due to its inhibition of protein phosphatase, an important enzyme in glycogen metabolism. The chemical is used for egg-protective purposes, and certain males transmit this chemical to the female during copulation. In *Neopyrochroa flabellata* (Pyrochroidae) males ingest exogenous cantharidin and use it both as a pre-copulatory "enticing" agent and as a nuptial gift. During courtship, the female samples cantharidin-laden secretions from the male's cephalic gland (as shown in the top illustration, after Eisner *et al.* 1996a,b) and will mate with cantharidin-fed males but reject males devoid of cantharidin. The male's glandular offering represents only a fraction of his systemic cantharidin; much of the remainder is stored in his large accessory gland and passed, presumably with the spermatophore, to the female during copulation (as shown in the middle illustration). Eggs are impregnated with cantharidin (probably in the ovary) and, after oviposition, egg batches (bottom illustration) are protected from coccinellid beetles and probably also other predators, such as ants and carabid beetles.

An unsolved question is where do the males of *N. flabellata* acquire their cantharidin from under natural conditions? They may feed on adults or eggs of the few insects that can manufacture cantharidin and, if so, might *N. flabellata* and other cantharidiphilic insects (including certain bugs, flies and hymenopterans, as well as beetles) be selective predators on each other?

Although some aposematic insects are closely associated with toxic food plants, certain insects can produce their own toxins. For example, amongst the Coleoptera, blister beetles (Meloidae) synthesize cantharidin, jewel beetles (Buprestidae) make buprestin, and some leaf beetles (Chrysomelidae) can produce cardiac glycosides. The very toxic staphylinid *Paederus* synthesizes its own blistering agent, paederin. Many of these chemically defended beetles are aposematic (e.g. Coccinellidae and Meloidae) and will reflex-bleed their haemolymph from the femoro-tibial leg joints if handled (see **Plate 7f**). Experimentally, it has been

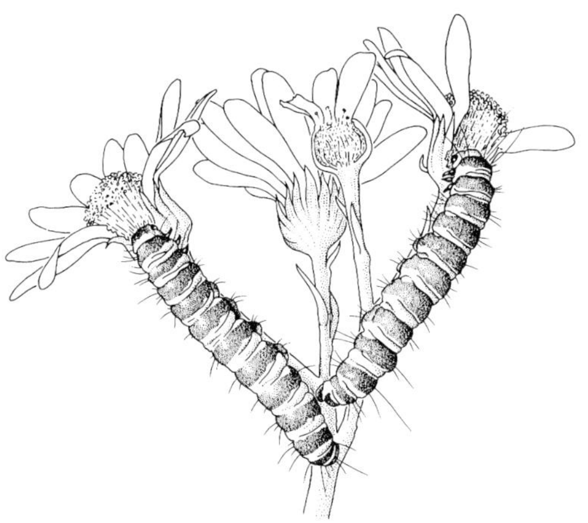

Fig. 14.5 The distasteful and warningly coloured caterpillars of the cinnabar moth, *Tyria jacobaeae* (Lepidoptera: Arctiidae), on ragwort, *Senecio jacobaeae*. (After Blaney 1976.)

shown that certain insects that sequester cyanogenic compounds from plants can still synthesize similar compounds if transferred to toxin-free host plants. If this ability preceded the evolutionary transfer to the toxic host plant, the possession of appropriate biochemical pathways may have preadapted the insect to using them subsequently in defence.

A bizarre means of obtaining a defensive chemical is used by *Photurus* fireflies (Lampyridae). Many fireflies synthesize deterrent lucibufagins, but *Photurus* females cannot do so. Instead they mimic the flashing sexual signal of *Photinus* females, thus luring male *Photinus* fireflies, which they eat to acquire their defensive chemicals.

Defensive chemicals, either manufactured by the insect or obtained by ingestion, may be transmitted between conspecific individuals of the same or a different life stage. Eggs may be especially vulnerable to natural enemies because of their immobility and it is not surprising that some insects endow their eggs with chemical deterrents (Box 14.3). This phenomenon may be more widespread among insects than is recognized currently.

14.4.4 Organs of chemical defence

Endogenous defensive chemicals (those synthesized within the insect) generally are produced in specific glands and stored in a reservoir (Box 14.4). Release is through muscular pressure or by evaginating the organ, rather like turning the fingers of a glove inside-out. The Coleoptera have developed a wide range of glands, many eversible, that produce and deliver defensive chemicals. Many Lepidoptera use urticating (itching) hairs and spines to inject venomous chemicals into predators. Venom injection by social insects is covered in section 14.6.

In contrast to these endogenous chemicals, exogenous toxins, derived from external sources such as foods, are usually incorporated in the tissues or the haemolymph. This makes the complete prey unpalatable, but requires the predator to test at close range in order to learn, in contrast to the distant effects of many endogenous compounds. However, the larvae of some swallowtail butterflies (Papilionidae) that feed

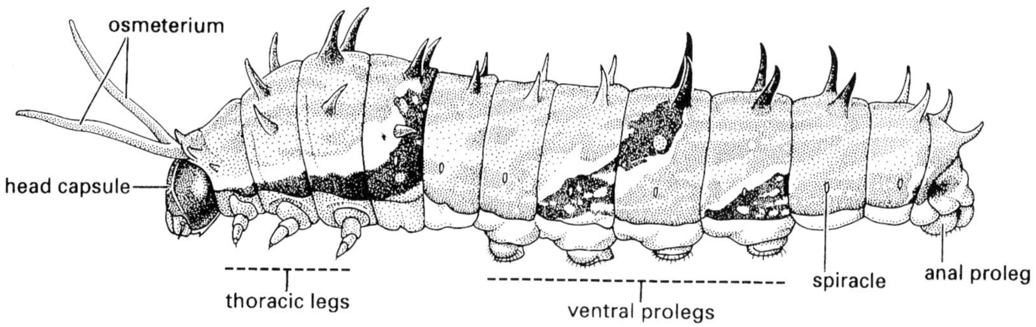

Fig. 14.6 A caterpillar of the orchard swallowtail butterfly, *Papilio aegeus* (Lepidoptera: Papilionidae), with the osmeterium everted behind its head. Eversion of this glistening, bifid organ occurs when the larva is disturbed, and is accompanied by a pungent smell.

Box 14.4 Insect binary chemical weapons

The common name of bombardier beetles (Carabidae: including genus *Brachinus*) derives from observations of early naturalists that the beetles released volatile defensive chemicals that appeared like a puff of smoke, accompanied by a "popping" noise resembling gunfire. The spray, released from the anus and able to be directed by the mobile tip of the abdomen, contains *p*-benzoquinone, a deterrent of vertebrate and invertebrate predators. This chemical is not stored, but when required is produced explosively from components held in paired glands. Each gland is double, comprising a muscular-walled compressible inner chamber containing a reservoir of hydroquinones and hydrogen peroxide, and a thick-walled outer chamber containing oxidative enzymes. When threatened, the beetle contracts the reservoir, and releases the contents through the newly opened inlet valve into the reaction chamber. Here an exothermic reaction takes place, resulting in the liberation of *p*-benzoquinone at a temperature of 100°C.

Studies on a Kenyan bombardier beetle, *Stenaptinus insignis* (illustrated here, after Dean *et al.* 1990), showed that the discharge is pulsed: the explosive chemical oxidation produces a build-up of pressure in the reaction chamber, which closes the one-way valve from the reservoir, thereby forcing discharge of the contents through the anus (as shown by the beetle directing its spray at an antagonist in front of it). This relieves the pressure, allowing the valve to open, permitting refilling of the reaction chamber from the reservoir (which remains under muscle pressure). Thus, the explosive cycle continues. By this mechanism, a high-intensity pulsed jet is produced by the chemical reaction, rather than requiring extreme muscle pressure. Humans discovered the principles independently and applied them to engineering (as pulse jet propulsion) some millions of years after the bombardier beetles developed the technique!

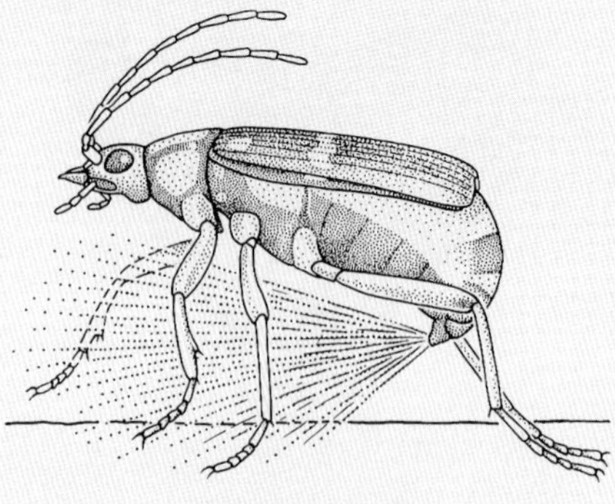

bombardier beetle

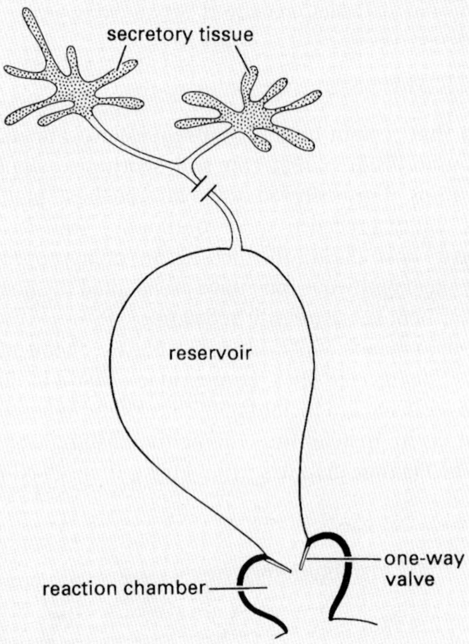

upon distasteful food plants concentrate the toxins and secrete them into a thoracic pouch called an **osmeterium**, which is everted if the larvae are touched. The colour of the osmeterium often is aposematic and reinforces the deterrent effect on a predator (Fig. 14.6). Larval sawflies (Hymenoptera: Pergidae), colloquially called "spitfires", store eucalypt oils, derived from the leaves that they eat, within a diverticulum of their fore gut and ooze this strong-smelling, distasteful fluid from their mouths when disturbed (Fig. 14.7).

14.5 DEFENCE BY MIMICRY

The theory of **mimicry**, an interpretation of the close resemblances of unrelated species, was an early

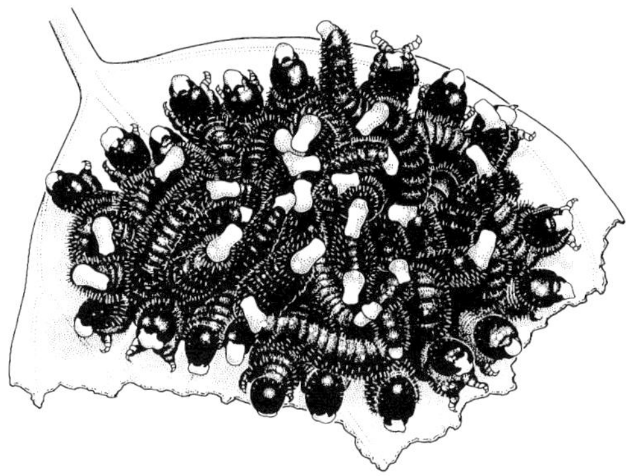

Fig. 14.7 An aggregation of sawfly larvae (Hymenoptera: Pergidae: *Perga*) on a eucalypt leaf. When disturbed, the larvae bend their abdomens in the air and exude droplets of sequestered eucalypt oil from their mouths.

application of the theory of Darwinian evolution. Henry Bates, a naturalist studying in the Amazon in the mid-19th century, observed that many similar butterflies, all slow-flying and brightly marked, seemed to be immune from predators. Although many species were common and related to each other, some were rare, and belonged to fairly distantly related families (see **Plate 8a**). Bates believed that the common species were chemically protected from attack, and this was advertised by their aposematism – high apparency (behavioural conspicuousness) through bright colour and slow flight. The rarer species, he thought, probably were not distasteful, but gained protection by their superficial resemblance to the protected ones. On reading the views that Charles Darwin proposed, Bates realized that his own theory of mimicry involved evolution through natural selection. Poorly protected species gain increased protection from predation by differential survival of subtle variants that more resembled protected species in appearance, smell, taste, feel or sound. The selective agent is the predator, which preferentially eats the inexact mimic. Since that time, mimicry has been interpreted in the light of evolutionary theory and studies of insects, particularly butterflies, have remained central to mimicry theory and manipulation.

An understanding of the defensive systems of mimicry (and crypsis; section 14.1) can be gained by recognizing three basic components: the **model**, the **mimic**, and an **observer** that acts as a selective agent. These components are related to each other through signal generating and receiving systems, of which the basic association is the warning signal given by the model (e.g. aposematic colour that warns of a sting or bad taste) and perceived by the observer (e.g. a hungry predator). The naïve predator must associate aposematism and consequent pain or distaste. When learnt, the predator subsequently will avoid the model. The model clearly benefits from this coevolved system, in which the predator can be seen to gain by not wasting time and energy chasing inedible prey.

Once such a mutually beneficial system has evolved, it is open to manipulation by others. The third component is the mimic: an organism that parasitizes the signalling system through deluding the observer, for example by false warning colouration. If this provokes a reaction from the observer similar to true aposematic colouration, the mimic is dismissed as unacceptable food. It is important to realize that the mimic need not be perfect, but only must elicit the appropriate avoidance response from the observer. Thus, only a limited subset of the signals given by the model may be required. For example, the black and yellow banding of venomous wasps is an aposematic colour pattern that is displayed by countless species from amongst many orders of insects. The exactness of the match, at least to our eyes, varies considerably. This may be due to subtle differences between several different venomous models, or it may reflect the inability of the observer to discriminate: if only yellow and black banding is required to deter a predator there may be little or no selection to refine the mimicry more fully.

14.5.1 Batesian mimicry

In these mimicry triangles, each component has a positive or negative effect on each of the others. In **Batesian mimicry**, an aposematic inedible model has an edible mimic. The model suffers by the mimic's presence because the aposematic signal aimed at the observer is diluted as the chances increase that the observer will taste an edible individual and fail to learn the association between aposematism and distastefulness. The mimic gains both from the presence of the protected model and the deception of the observer. As the mimic's presence disadvantages the model, interaction with the model is negative. The observer benefits by avoiding the noxious model, but misses a meal through failing to recognize the mimic as edible.

A Batesian mimicry system benefits the mimic only if it remains relatively rare. However, if the model declines in numbers relative to the mimic, or the mimic becomes more abundant, then the protection given to the mimic by the model will wane because the naïve observer increasingly encounters and tastes edible mimics. Evidently, some palatable butterfly mimics adopt different models throughout their range. For example, the African or mocker swallowtail, *Papilio dardanus*, is highly polymorphic, with up to five mimetic morphs in Uganda (central Africa) and several more throughout its wide range. This polymorphism allows a larger total population of *P. dardanus* without prejudicing (by dilution) the successful mimetic system, as each morph can remain rare relative to its Batesian model. In this case, and for many other mimetic polymorphisms, males retain the basic colour pattern of the species and only amongst females in some populations does mimicry of such a variety of models occur. The conservative male pattern may result from sexual selection to ensure recognition of the male by conspecific females of all morphs for mating, or by other conspecific males in territorial contests. An additional consideration concerns the effects of differential predation pressure on females (by virtue of their slower flight and conspicuousness at host plants) – meaning females may gain more by mimicry relative to the differently behaving males.

Larvae of the Old World tropical butterfly *Danaus chrysippus* (Nymphalidae: Danainae) feed predominantly on milkweeds (Apocynaceae) from which they can sequester cardenolides, which are retained to the aposematic, chemically protected adult stage. A variable but often high proportion of *D. chrysippus* develop on milkweeds lacking bitter and emetic chemicals, and the resulting adult is unprotected. These are intraspecific Batesian **automimics** of their protected relatives. Where there is an unexpectedly high proportion of unprotected individuals, this situation may be maintained by parasitoids that preferentially parasitize noxious individuals, perhaps using their cardenolides as kairomones in host finding. The situation is complicated further, because unprotected adults, as in many *Danaus* species, actively seek out sources of pyrrolizidine alkaloids from plants to use in production of sex pheromones; these alkaloids also may render the adult less palatable.

14.5.2 Müllerian mimicry

In a contrasting set of relationships, called **Müllerian mimicry**, the model(s) and mimic(s) are all distasteful and warningly coloured, and all benefit from coexistence, as observers learn from tasting any individual. Unlike Batesian mimicry, in which the system is predicted to fail as the mimic becomes relatively more abundant, Müllerian mimicry systems gain through enhanced predator learning when the density of component distasteful species increases. "Mimicry rings" of species may develop, in which organisms from distant families, and even orders, acquire similar aposematic patterns, although the source of protection varies greatly. In the species involved, the warning signal of the co-models differs markedly from that of their close relatives, which are non-mimetic.

Interpretation of mimicry may be difficult, particularly in distinguishing protected from unprotected mimetic components. For example, a century after discovery of one of the seemingly strongest examples of Batesian mimicry, the classical interpretation seems flawed. The system involves two North American danaine butterflies, *Danaus plexippus*, the monarch or wanderer, and *D. gilippus*, the queen, which are chemically defended models, each of which is mimicked by a morph of the nymphaline viceroy butterfly (*Limenitis archippus*) (Fig. 14.8). Historically, larval food plants and taxonomic affiliation suggested that viceroys were palatable, and therefore Batesian mimics. This interpretation was overturned by experiments in which isolated butterfly abdomens were fed to natural predators (wild-caught redwing blackbirds). The possibility that feeding by birds might be affected by previous exposure to aposematism was excluded by removal of the aposematically patterned butterfly wings. Viceroys were found to be at least as unpalatable as monarchs, with queens being least unpalatable. At least in the Florida populations and with this particular predator, the system is interpreted now as Müllerian, either with the viceroy as model, or with the viceroy and monarch acting as co-models, and the queen being a less well chemically protected member that benefits through the asymmetry of its palatability relative to the others. Few such appropriate experiments to assess palatability, using natural predators and avoiding

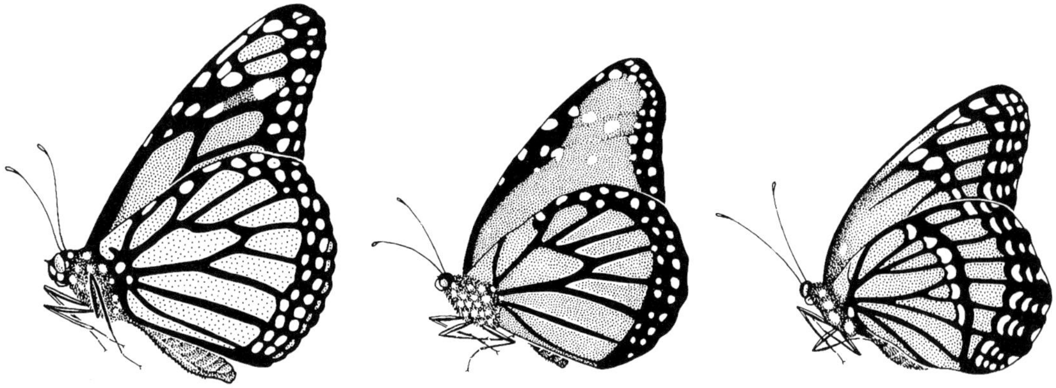

Fig. 14.8 Three nymphalid butterflies that are Müllerian co-mimics in Florida: (a) the monarch or wanderer (*Danaus plexippus*); (b) the queen (*D. gilippus*); (c) the viceroy (*Limenitis archippus*). (After Brower 1958.)

problems of previous learning by the predator, have been reported and clearly more are required.

If all members of a Müllerian mimicry complex are aposematic and distasteful, then it can be argued that an observer (predator) is not deceived by any member – and this can be seen more as shared aposematism than mimicry. More likely, as seen above, distastefulness is unevenly distributed, in which case some specialist observers may find the least well-defended part of the complex to be edible. Such ideas suggest that true Müllerian mimicry may be rare and/or dynamic and represents one end of a spectrum of interactions.

14.5.3 Mimicry as a continuum

The strict differentiation of defensive mimicry into two forms – Müllerian and Batesian – can be questioned, although each gives a different interpretation of the ecology and evolution of the components, and makes dissimilar predictions concerning life histories of the participants. For example, mimicry theory predicts that in aposematic species there should be:

- limited numbers of co-modelled aposematic patterns, reducing the number that a predator has to learn;
- behavioural modifications to "expose" the pattern to potential predators, such as conspicuous display rather than crypsis, and diurnal rather than nocturnal activity;
- long post-reproductive life, with prominent exposure to encourage the naïve predator to learn of the distastefulness on a post-reproductive individual.

All of these predictions appear to be true in some or most systems studied. Furthermore, theoretically there should be variation in polymorphism, with selection enforcing aposematic uniformity (monomorphism) in Müllerian cases, but encouraging divergence (mimetic polymorphism) in Batesian cases (section 14.5.1). Sex-limited (female-only) mimicry and divergence of the model's pattern away from that of the mimic (evolutionary escape) are also predicted in Batesian mimicry. Although these predictions are met in some mimetic species, there are exceptions to all of them. Polymorphism certainly occurs in Batesian mimetic swallowtails (Papilionidae), but is much rarer elsewhere, even within other butterflies; furthermore, there are polymorphic Müllerian mimics such as the viceroy. It is suggested now that some relatively undefended mimics may be fairly abundant relative to the distasteful model and need not have attained abundance via polymorphism. It is argued that this can arise and be maintained if the major predator is a generalist that requires only to be deterred relative to other more palatable species.

A complex range of mimetic relationships are based on mimicry of lycid beetles, which are often aposematically odouriferous and warningly coloured. The black and orange Australian lycid *Metriorrhynchus rhipidius* is protected chemically by odorous methoxyalkylpyrazine, and by bitter-tasting compounds and acetylenic antifeedants. Species of *Metriorrhynchus* provide models for mimetic beetles from at least six distantly related families (Buprestidae, Pythidae, Meloidae, Oedemeridae, Cerambycidae and Belidae) and at least one moth. All these mimics are convergent in colour; some have nearly identical alkylpyrazines and distasteful chemicals; others share the alkylpyrazines but have different distasteful chemicals; and some have the odourous chemical but appear to lack any distasteful chemicals. These aposematically coloured insects form a mimetic series. The oedemerids clearly are Müllerian mimics, modelled precisely on the local *Metriorrhynchus* species and differing only in using cantharidin as an antifeedant. The cerambycid mimics use different repellent odours, whereas the buprestids lack warning odour but are chemically protected by buprestins. Finally, pythids and belids are Batesian mimics, apparently lacking any chemical defences. After careful chemical examination, what appears to be a model with many Batesian mimics, or perhaps a Müllerian ring, is revealed to demonstrate a complete range between the extremes of Müllerian and Batesian mimicry.

Although the extremes of the two prominent mimicry systems are well studied, and in some texts appear to be the only systems described, they are but two of the possible permutations involving the interactions of model, mimic and observer. Further complexity ensues if model and mimic are the same species, as in **automimicry**, or in cases where sexual dimorphism and polymorphism exist. All mimicry systems are complex, interactive and never static, because population sizes change and relative abundances of mimetic species fluctuate so that density-dependent factors play an important role. Furthermore, the defence offered by shared aposematic colouring, and even shared distastefulness, can be circumvented by specialized predators able to learn and locate the warning, overcome the defences and eat selected species in the mimicry complex. Evidently, consideration of mimicry theory demands recognition of the role of predators as flexible, learning, discriminatory, coevolving and coexisting agents in the system (Box 14.1).

14.6 COLLECTIVE DEFENCES IN GREGARIOUS AND SOCIAL INSECTS

Chemically defended, aposematic insects are often clustered rather than uniformly distributed through a suitable habitat. Thus, unpalatable butterflies may live in conspicuous aggregations as larvae and as adults; the winter congregation of migratory adult monarch butterflies in California (see **Plate 3g**) and Mexico is an example. Many chemically defended hemipterans aggregate on individual host plants, and some vespid wasps congregate conspicuously on the outside of their nests (as shown in the vignette at the beginning of Chapter 12). Orderly clusters occur in the phytophagous larvae of sawflies (Hymenoptera: Pergidae; Fig. 14.7) and some chrysomelid beetles that form defended circles (**cycloalexy**). Some larvae lie within the circle and others form an outer ring, with either their heads or abdomens directed outwards, depending upon which end secretes the noxious compounds. These groups often make synchronized displays of head and/or abdomen bobbing, which increase the apparency of the group.

Formation of such clusters is sometimes encouraged by the production of aggregation pheromones by early arriving individuals (section 4.3.2), or may result from the young failing to disperse after hatching from one or several egg batches. Benefits to the individual from the clustering of chemically defended insects may relate to the dynamics of predator training. However, these also may involve kin selection in subsocial insects, in which aggregations comprise relatives that benefit at the expense of an individual "sacrificed" to educate a predator.

This latter scenario for the origin and maintenance of group defence certainly seems to apply to the eusocial Hymenoptera (ants, bees and wasps), as seen in Chapter 12. In these insects, and in the termites (Blattodea: Termitoidae), defensive tasks are undertaken usually by morphologically modified individuals called soldiers. In all social insects, the focus for defensive action is the nest, and the major role of the soldier caste is to protect the nest and its inhabitants. Nest architecture and location is often a first line of defence, with many nests buried underground or hidden within trees, with a few, easily defendable entrances. Exposed nests, such as those of savannah-zone termites, often have hard, impregnable walls.

Termite soldiers can be male or female, have weak sight or be blind, and have enlarged heads (sometimes exceeding the rest of the body length). Soldiers may have well-developed jaws, or be **nasute**, with small jaws but an elongate "nasus" or rostrum. They may protect the colony by biting, by chemical means or, as in *Cryptotermes*, by **phragmosis** – the blocking of access to the nest with their modified heads. Amongst the most serious adversaries of termites are ants, and complex warfare takes place between the two. Termite soldiers have developed an enormous battery of chemicals, many produced in highly elaborated frontal and salivary glands. For example, in *Pseudacanthotermes spiniger* the salivary glands fill nine-tenths of the soldier's abdomen, and soldiers of *Globitermes sulphureus* are filled to bursting with sticky yellow fluid used to entangle the predator – and the termite – usually fatally. This suicidal phenomenon is seen also in some *Camponotus* ants, which use hydrostatic pressure in the gaster to burst the abdomen and release sticky fluid from the huge salivary glands.

Some of the specialized defensive activities used by termites have developed convergently amongst ants. Thus, the soldiers of some formicines, notably the subgenus *Colobopsis*, and several myrmecines show phragmosis, with modifications of the head to allow the blocking of nest entrances (Fig. 14.9). Nest entrances are made by minor workers and are of such a size that the head of a single major worker (soldier) can seal it; in others such as the myrmecine *Cephalotes*, the entrances can be larger and a formation of guarding blockers may be required to act as "gatekeepers". A further defensive strategy of these myrmecines is for the head to be covered with a crust of secreted filamentous material, such that the head is camouflaged when it blocks a nest entrance on a lichen-covered twig.

Most soldiers use their strongly developed mandibles in colony defence as a means of injuring an attacker. A novel defence in termites involves elongate mandibles that snap against one another, as we might snap our fingers. A violent movement is produced as the pent-up elastic energy is released from the tightly appressed mandibles (Fig. 14.10a). In *Capritermes*

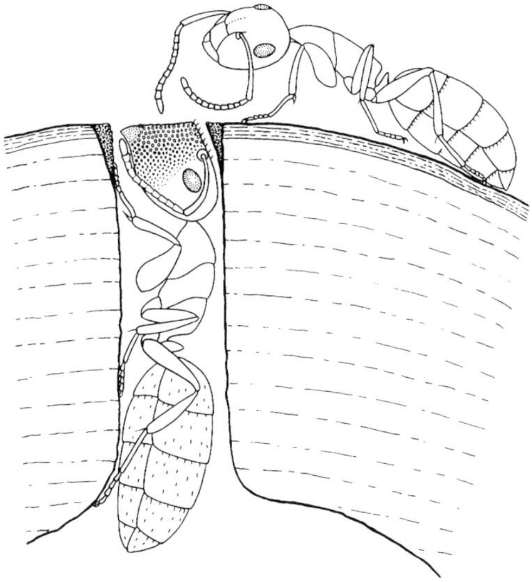

Fig. 14.9 Nest guarding by the European ant *Camponotus* (*Colobopsis*) *truncatus* (Hymenoptera: Formicidae): a minor worker approaching a soldier that is blocking a nest entrance with her plug-shaped head. (After Hölldobler & Wilson 1990; from Szabó-Patay 1928.)

and *Homallotermes*, the mandibles are asymmetric (Fig. 14.10b) and the released pressure results in the violent movement of only the right mandible; the bent left one, which provides the elastic tension, remains immobile. These soldiers can only strike to their left! The advantage of this defence is that a powerful blow can be delivered in a confined tunnel, in which there is inadequate space to open the mandibles wide enough to obtain conventional leverage on an opponent.

Major differences between termite defences and those of social hymenopterans are the restriction of the defensive caste to females in Hymenoptera, and the frequent use of venom injected through an ovipositor modified as a sting (Fig. 14.11). Whereas parasitic hymenopterans use this weapon to immobilize prey, in social aculeate hymenopterans it is a vital weapon in defence against predators. Many subsocial and all social hymenopterans co-operate to sting an intruder *en masse*, thereby escalating the effects of an individual attack and deterring even large vertebrates. The

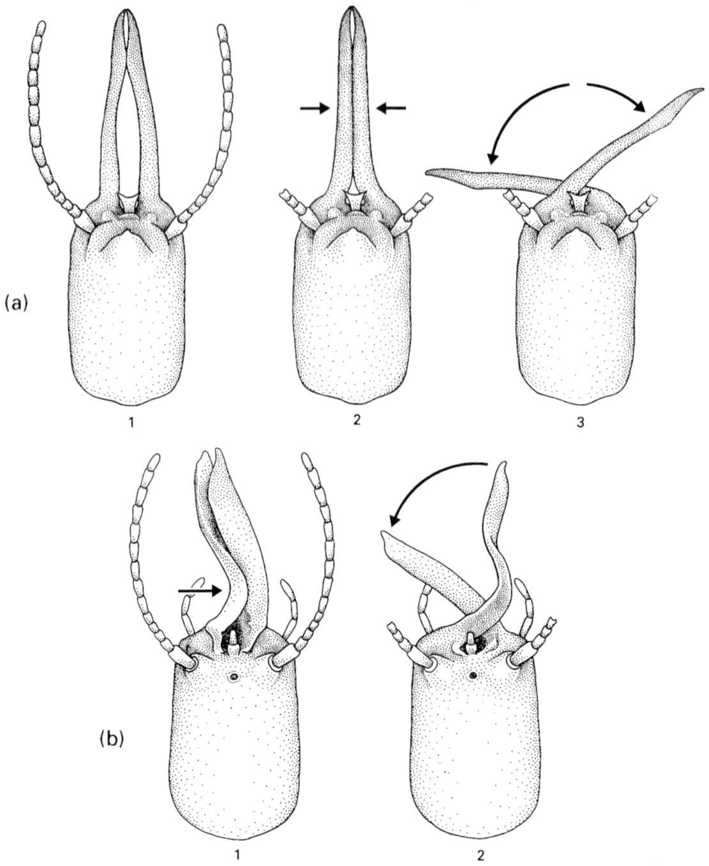

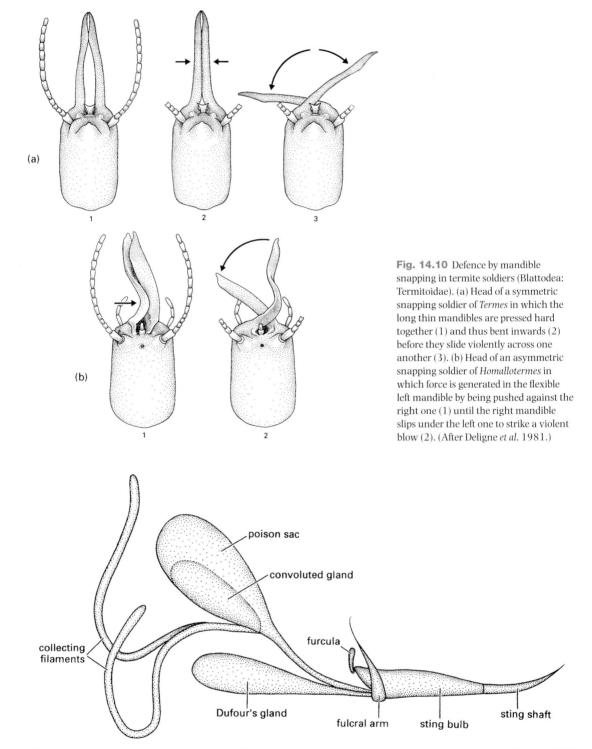

Fig. 14.10 Defence by mandible snapping in termite soldiers (Blattodea: Termitoidae). (a) Head of a symmetric snapping soldier of *Termes* in which the long thin mandibles are pressed hard together (1) and thus bent inwards (2) before they slide violently across one another (3). (b) Head of an asymmetric snapping soldier of *Homallotermes* in which force is generated in the flexible left mandible by being pushed against the right one (1) until the right mandible slips under the left one to strike a violent blow (2). (After Deligne *et al.* 1981.)

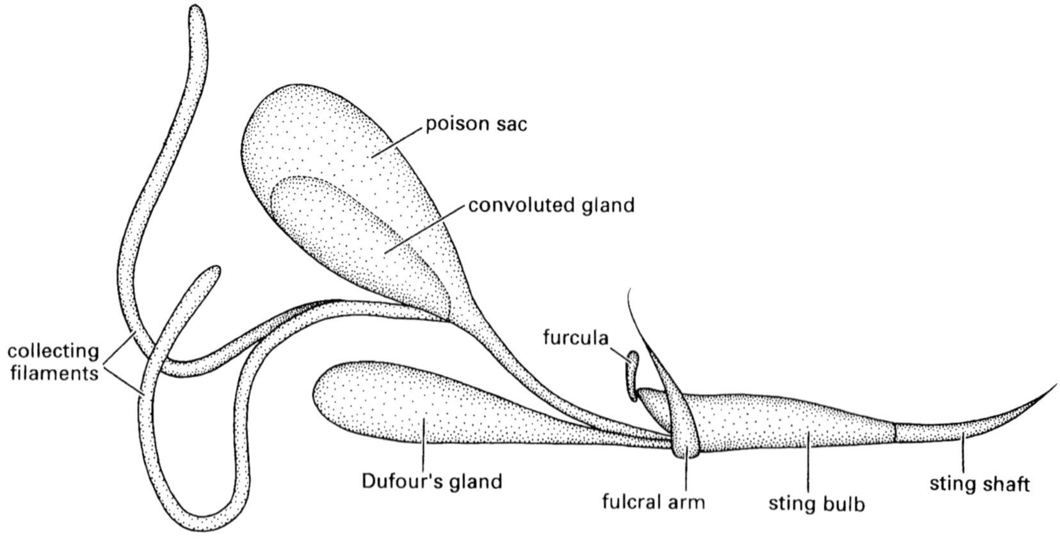

Fig. 14.11 Diagram of the major components of the venom apparatus of a social aculeate wasp. (After Hermann & Blum 1981.)

sting is injected into a predator through a lever (the **furcula**) acting on a fulcral arm, although fusion of the furcula to the sting base in some ants leads to a less manoeuvrable sting.

Venoms include a wide variety of products, many of which are polypeptides. Biogenic amines such as any or all of histamine, dopamine, adrenaline (epinephrine) and noradrenaline (norepinephrine) (and serotonin in wasps) may be accompanied by acetylcholine, and several important enzymes including phospholipases and hyaluronidases (which are highly allergenic). Wasp venoms have a number of vasopeptides – pharmacologically active kinins that induce vasodilation and relax smooth muscle in vertebrates. Non-formicine ant venoms comprise either similar materials of proteinaceous origin or a pharmacopoeia of alkaloids, or complex mixtures of both types of component. In contrast, formicine venoms are dominated by formic acid.

Venoms are produced in special glands sited on the bases of the inner valves of the ninth abdominal segment, comprising free filaments and a reservoir store, which may be simple or contain a convoluted gland (Fig. 14.11). The outlet of Dufour's gland enters the sting base ventral to the venom duct. The products of this gland in eusocial bees and wasps are poorly known, but in ants Dufour's gland is the site of synthesis of an astonishing array of hydrocarbons (over 40 in one species of *Camponotus*). These exocrine products include esters, ketones and alcohols, and many other compounds used in communication and defence.

The sting is reduced and lost in some social hymenopterans, notably the stingless bees and formicine ants. Alternative defensive strategies have arisen in these groups; thus many stingless bees mimic stinging bees and wasps, and use their mandibles and defensive chemicals if attacked. Formicine ants retain their venom glands, which disperse formic acid through an **acidophore**, often directed as a spray into a wound created by the mandibles.

Other glands in social hymenopterans produce additional defensive compounds, often with communication roles, and including many volatile compounds that serve as alarm pheromones. These stimulate one or more defensive actions: they may summon more individuals to a threat; mark a predator so that the attack is targeted; or, as a last resort, they may encourage the colony to flee from the danger. Mandibular glands produce alarm pheromones in many insects, and also substances that cause pain when they enter wounds caused by the mandibles. The metapleural glands in some species of ants produce compounds that defend against microorganisms in the nest through antibiotic action. Both sets of glands may produce sticky defensive substances, and a wide range of pharmacological compounds has been under study to determine possible human benefit.

Even well-defended insects can be parasitized by mimics, and the best of chemical defences can be breached by a predator (Box 14.1). Although the social insects have some of the most elaborate defences seen in the Insecta, they remain vulnerable. For example, many insects model themselves on social insects, with representatives of many orders converging morphologically on ants (Fig. 14.12), particularly with regard to the waist constriction and wing loss, and even kinked antennae. Some of the most extraordinary ant-mimicking insects are tropical African bugs of the genus *Hamma* (Membracidae), as exemplified by *H. rectum* depicted in the vignette at the beginning of this chapter.

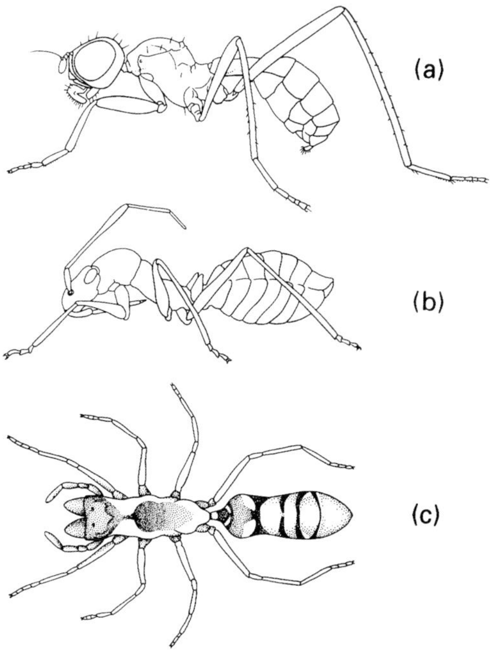

Fig. 14.12 Three ant mimics: (a) a fly (Diptera: Micropezidae: *Badisis*); (b) a bug (Hemiptera: Miridae: Phylinae); (c) a spider (Araneae: Clubionidae: *Sphecotypus*). ((a) After McAlpine 1990; (b) after Atkins 1980; (c) after Oliveira 1988.)

The aposematic yellow-and-black patterns of vespid wasps and apid bees provide models for hundreds of mimics throughout the world. Not only are these communication systems of social insects parasitized, but so are their nests, which provide many parasites and inquilines with a hospitable place for their development (section 12.3).

Defence must be seen as a continuing coevolutionary process, analogous to an "arms race", in which new defences originate or are modified and then are selectively breached, stimulating improved defences.

FURTHER READING

Brandt, M. & Mahsberg, D. (2002) Bugs with a backpack: the function of nymphal camouflage in the West African assassin bugs: *Paredocla* and *Acanthiaspis* spp. *Animal Behaviour* **63**, 277–84.

Cook, L.M. & Saccheri, I.J. (2013) The peppered moth and industrial melanism: evolution of a natural selection case study. *Heredity* **110**, 207–12.

Cook, L.M., Grant, B.S., Saccheri, I.J. & Mallet, J. (2012) Selective bird predation on the peppered moth: the last experiment of Michael Majerus. *Biology Letters* **8**, 609–12.

Dossey, A.T. (2010) Insects and their chemical weaponry: new potential for drug discovery. *Natural Product Reports*. **27**, 1737–57.

Eisner, T. (2003) *For the Love of Insects*. The Belknap Press, Harvard University Press, Cambridge, MA.

Eisner, T. & Aneshansley, D.J. (1999) Spray aiming in the bombardier beetle: photographic evidence. *Proceedings of the National Academy of Sciences of the USA* **96**, 9705–9.

Eisner, T., Eisner, M. & Siegler, M. (2007) *Secret Weapons: Defenses of Insects, Spiders, Scorpions, and Other Many-Legged Creatures*. The Belknap Press, Harvard University Press, Cambridge, MA.

Gross, P. (1993) Insect behavioural and morphological defenses against parasitoids. *Annual Review of Entomology* **38**, 251–73.

McIver, J.D. & Stonedahl, G. (1993) Myrmecomorphy: morphological and behavioural mimicry of ants. *Annual Review of Entomology* **38**, 351–79.

Moore, B.P. & Brown, W.V. (1989) Graded levels of chemical defense in mimics of lycid beetles of the genus *Metriorrhynchus* (Coleoptera). *Journal of the Australian Entomological Society* **28**, 229–33.

Pasteels, J.M., Grégoire, J.-C. & Rowell-Rahier, M. (1983) The chemical ecology of defense in arthropods. *Annual Review of Entomology* **28**, 263–89.

Resh, V.H. & Cardé, R.T. (eds.) (2009) *Encyclopedia of Insects*, 2nd edn. Elsevier, San Diego, CA. [Particularly see articles on aposematic coloration; chemical defense; defensive behavior; industrial melanism; mimicry; venom.]

Riley, P.A. (2013) A proposed selective mechanism based on metal chelation in industrial melanic moths. *Biological Journal of the Linnean Society* **109**, 298–301.

Ritland, D.B. (1991) Unpalatability of viceroy butterflies (*Limenitis archippus*) and their purported mimicry models, Florida queens (*Danaus gilippus*). *Oecologia* **88**, 102–8.

Starrett, A. (1993) Adaptive resemblance: a unifying concept for mimicry and crypsis. *Biological Journal of the Linnean Society* **48**, 299–317.

Turner, J.R.G. (1987) The evolutionary dynamics of Batesian and Muellerian mimicry: similarities and differences. *Ecological Entomology* **12**, 81–95.

Vane-Wright, R.I. (1976) A unified classification of mimetic resemblances. *Biological Journal of the Linnean Society* **8**, 25–56.

Waldbauer, G. (2012) *How Not to Be Eaten: The Insects Fight Back*. University of California Press, Berkeley, CA.

Wickler, W. (1968) *Mimicry in Plants and Animals*. Weidenfeld & Nicolson, London.

Papers in *Biological Journal of the Linnean Society* (1981) **16**, 1–54 [Includes a shortened version of H.W. Bates's classic 1862 paper.]

Chapter 15

MEDICAL AND VETERINARY ENTOMOLOGY

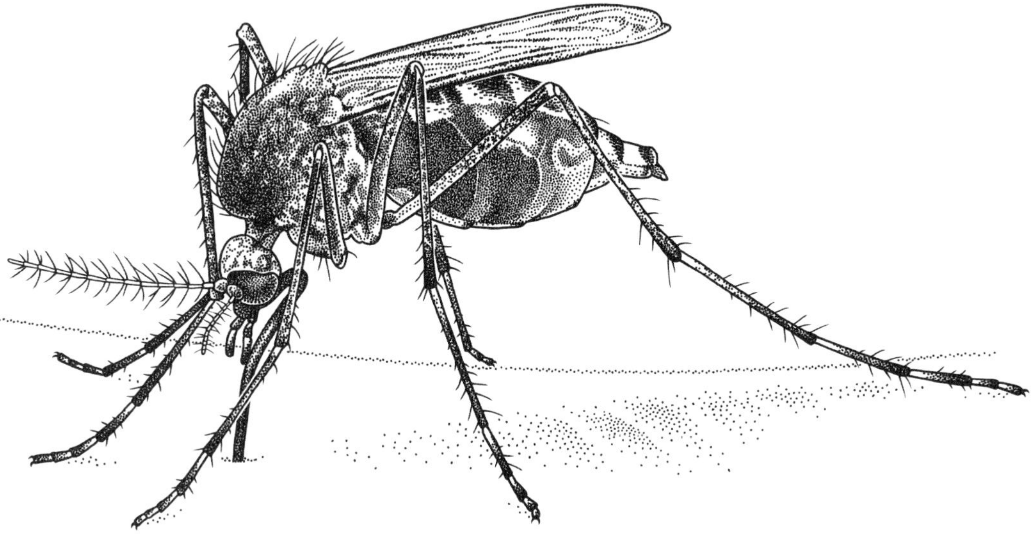

Feeding adult female of *Aedes aegypti*.

Insects affect agricultural and horticultural crops, but to many people their important impact is in transmission of disease to us and our domestic animals. The number of insect species involved is not large, but those that transmit disease (**vectors**), cause wounds, inject venom, or create nuisance have serious social and economic consequences. Thus, the study of the veterinary and medical impact of insects is a major scientific discipline.

Medical and veterinary entomology differs from other areas of entomological pursuit. First, the frequent motivation (and funding) for study is usually the insect-borne human or animal disease, rather than the insect itself. Second, a scientist specializing in this field must understand not only the insect vector of disease, but also the biology of host and parasite. Third, most practitioners do not restrict themselves to insects, but also study other arthropods, notably ticks, mites and perhaps spiders and scorpions.

For brevity in this chapter, we refer to medical entomologists as those who study all arthropod-borne diseases, including diseases of livestock. The insect, though a vital cog in the chain of disease, need not be the central focus of medical research. Medical entomologists usually work in multidisciplinary teams that may include medical practitioners and researchers, epidemiologists, virologists and immunologists, and ought to include those with skills in insect control.

In this chapter, we discuss insects and disease, and provide details of transmission of malaria, as an exemplar of a major insect-borne disease. We review diseases in which insects play an important role: some are well controlled, but others currently threaten human health – we discuss dengue and West Nile virus as case studies. A section concerning forensic entomology is included, and we finish with discussions of entomophobia, insect venoms, and allergic reactions and urtication caused by insects. The chapter includes boxes dealing with the life cycle of *Plasmodium* (Box 15.1), the *Anopheles gambiae* complex (Box 15.2), the use of insecticide-treated bed nets (Box 15.3), and three insect threats to our health and welfare: the emerging global threat of dengue fever throughout the tropics and subtropics (Box 15.4), West Nile fever in the United States (Box 15.5) and the resurgence of bed bugs (Box 15.6).

15.1 INSECTS AS CAUSES AND VECTORS OF DISEASE

In the tropics and subtropics, the role of insects in transmitting protists, viruses, bacteria and nematodes is clear. These pathogens cause many important and widespread human diseases, including malaria, dengue, yellow fever, onchocerciasis (river blindness), leishmaniasis (oriental sore, kala-azar), filariasis (elephantiasis) and trypanosomiasis (sleeping sickness).

The cause may be an arthropod itself, such as the human body or head louse (*Pediculus humanus humanus* and *P. humanus capitis*, respectively), which cause pediculosis, or the mite *Sarcoptes scabiei*, whose skin-burrowing activities cause the skin disease scabies. In **myiasis** (from *myia*, Greek for fly), the maggots or larvae of blow flies, house flies and their relatives (Diptera: Calliphoridae, Sarcophagidae and Muscidae) can develop in living flesh, either as primary agents or subsequently following wounding or damage by other arthropods, such as ticks and biting flies. If untreated, the animal victim may die. As death approaches and the flesh putrefies through bacterial activity, there may be a third wave of specialist fly larvae, and these colonizers are present at death. One particular form of myiasis affecting livestock is known as "strike", and is caused in the Old World by *Chrysomya bezziana* (Calliphoridae) and in the Americas by the New World screw-worm fly, *Cochliomyia hominivorax* (Calliphoridae) (see Fig. 6.6h and section 16.10). The name "screw-worm" derives from the distinct rings of setae on the maggot that resemble a screw. Many myiases, including screw-worm, can affect humans, particularly under conditions of poor hygiene. Other "higher" dipterans develop in mammals as endoparasitic larvae in the dermis, intestine or, as in the sheep nostril fly, *Oestrus ovis* (Oestridae), in the nasal and head sinuses. In many cattle-ranching regions, losses caused by fly-induced damage to hides and meat, and death as a result of myiases, may amount to many millions of dollars.

Even more frequent than direct injury by insects is their action as vectors, transmitting disease-inducing pathogens from one animal or human **host** to another. This transfer may be by mechanical or biological means. **Mechanical transfer** occurs, for example, when a mosquito transfers myxomatosis from rabbit to

The Insects: An Outline of Entomology, Fifth Edition. P.J. Gullan and P.S. Cranston.

Companion Website: www.wiley.com/go/gullan/insects

rabbit in the blood on its proboscis. Likewise, when a cockroach or house fly acquires bacteria when feeding on faeces, it may physically transfer some bacteria from its mouthparts, legs or body to human food, thereby transferring enteric diseases. The causative agent of the disease is transported passively from host to host, and does not increase in the vector. Usually in mechanical transfer, the arthropod is only one of several means of pathogen transfer, with poor public and personal hygiene often providing additional pathways.

In contrast, **biological transfer** is a tight association between insect vector, pathogen and host, and transfer never occurs naturally without all three components. The disease agent **replicates** (increases) within the vector insect, and there is often close specificity between vector and disease agent. The insect is thus a vital link in biological transfer, and efforts to curb disease nearly always involve attempts to reduce vector numbers. In addition, biologically transferred disease may be controlled by seeking to interrupt contact between vector and host, and by direct attack on the pathogen, usually whilst in the host. Disease control comprises a combination of these approaches, each of which requires detailed knowledge of the biology of all three components – vector, pathogen and host.

15.2 GENERALIZED DISEASE CYCLES

In all biologically transferred diseases, a biting (blood-feeding or sucking) adult arthropod, often an insect, particularly a true fly (Diptera), transmits a parasite from animal to animal, human to human, or from animal to human, or, very rarely, from human to animal. Some human pathogens (causative agents of human disease such as malaria parasites) can complete their parasitic life cycles solely within the insect vector and the human host. Human malaria exemplifies a **single-cycle disease,** involving *Anopheles* mosquitoes (Culicidae), malaria parasites and humans. Although related malaria parasites occur in animals, notably other primates and birds, these hosts and parasites are not involved in the human malarial cycle. Only a few human insect-borne diseases have single cycles, as in malaria, because these diseases require coevolution of pathogen and vector and *Homo sapiens*. As *H. sapiens* is of relatively recent evolutionary origin, there has not been time for the development of unique insect-borne diseases that require specifically a human rather than any other vertebrate for completion of the life cycle of the disease-causing organism.

In contrast to single-cycle diseases, many other insect-borne diseases that affect humans include a (non-human) vertebrate host, as for instance in yellow fever in monkeys, plague in rats, and leishmaniasis in desert rodents such as pack rats. Clearly, the non-human cycle is primary in these cases and the sporadic inclusion of humans in a **secondary cycle** is not essential to maintain the disease. However, when outbreaks do occur, these diseases can spread in human populations and may involve many cases.

Outbreaks in humans often stem from human actions, such as the spread of people into the natural ranges of the vector and animal hosts, which act as disease **reservoirs**. For example, yellow fever in native-forested Uganda (central Africa) has a "sylvan" (woodland) cycle, remaining within canopy-dwelling primates with the exclusively primate-feeding mosquito *Aedes africanus* as the vector. It is only when monkeys and humans coincide at banana plantations close to or within the forest that *Aedes bromeliae* (formerly *Ae. simpsoni*), a second mosquito vector that feeds on both humans and monkeys, can transfer jungle yellow fever to humans. In a second example, *Phlebotomus* sand flies (Psychodidae) depend upon arid-zone burrowing rodents and, in feeding, transmit *Leishmania* parasites between rodent hosts. This disfiguring ailment has dramatically increased in humans in war-torn Iraq and Afghanistan where people locate within a rodent reservoir. Unlike with yellow fever, there appears to be no change in vector when humans enter the cycle.

In epidemiological terms, the natural cycle is maintained in animal reservoirs: sylvan primates for yellow fever and desert rodents for leishmaniasis. Disease control clearly is complicated by the presence of these reservoirs in addition to a human cycle.

15.3 PATHOGENS

The disease-causing organisms transferred by the insect may be viruses (termed "arboviruses", an abbreviation of **arthropod-borne viruses**), bacteria (including rickettsias), protists or filarial nematode worms. Replication of these parasites in both vectors and hosts is required, and some complex life cycles have developed, notably amongst the protists and filarial nematodes. The presence of a parasite in the vector insect (which can be determined by dissection and microscopy and/or biochemical means) generally appears not to harm the insect. When the parasite is at an appropriate developmental stage, and following multiplication or replication (amplification and/or concentration in the vector), transmission can occur. Transfer of parasites from vector to host or vice versa takes place when the blood-feeding insect takes a meal from a vertebrate host. The transfer from host to previously uninfected vector is through parasite-infected blood. Transmission to a host by an infected insect usually is by injection along with anticoagulant salivary gland products that keep the wound open during feeding. However, transmission may also be through deposition of infected faeces close to the wound site. In the course of a disease, the time between infection with the pathogen and first symptoms is called the incubation period.

In the following survey of major arthropod-borne disease, malaria will be dealt with in some detail. Malaria is the most devastating and debilitating disease worldwide, and it illustrates a number of general points concerning medical entomology. This is followed by briefer sections reviewing the range of pathogenic diseases involving insects, arranged by parasite, from virus to filarial worm.

15.3.1 Malaria

The disease

Malaria affects more people, more persistently, throughout more of the world than any other insect-borne disease. According to the World Health Organization (WHO), 154 to 289 million new cases of malaria occurred in 2010, with annual deaths esimated to be some 660,000, mostly in sub-Saharan Africa. Although ongoing malaria control has reduced the incidence, about 100 countries and territories had ongoing exposure in 2011. In some countries, malaria has increased as a result of growing socio-economic inequality and civil unrest (wars), warranted concern over undesirable side-effects of dichlorodiphenyl-trichloroethane (DDT), increasing development of resistance of insects to modern pesticides and of malaria parasites to antimalarial drugs, and even from sales of counterfeit, often ineffective antimalarial drugs. Even in countries in which there is little or no transmission of malaria, the disease affects travellers, military personnel, migrants and refugees, with some 600 cases annually notified in Australia, and 1500 in the United States.

The parasitic protists that cause malaria are sporozoans, belonging to the genus *Plasmodium*. Five (or probably six; see below) species are responsible for human malarias; all are shared with at least one other primate species. Vectors of mammalian malaria always are *Anopheles* mosquitoes, belonging to many different species according to region.

The disease follows a course of a prepatent period between infective bite and patency, the first appearance of parasites (sporozoites; Box 15.1) in the erythrocytes

Box 15.1 Life cycle of *Plasmodium*

The malarial cycle, shown here modified after Kettle (1984) and Katz *et al.* (1989), commences with an infected female *Anopheles* mosquito taking a blood meal from a human host (H). Saliva contaminated with the sporozoite stage of the *Plasmodium* is injected (a). The sporozoite circulates in the blood until reaching the liver, where a pre- (or exo-) erythrocytic schizogonous cycle (b,c) takes place in the parenchyma cells of the liver. This leads to the formation of a large schizont, containing from 2000 to 40,000 merozoites, according to *Plasmodium* species. The pre-patent period of infection, which started with an infective bite, ends when the merozoites are released (c) to either infect more liver cells or enter the bloodstream and invade the erythrocytes. Invasion occurs by the erythrocyte invaginating to engulf the merozoite, which subsequently feeds as a trophozoite (e) within a vacuole. The first and several subsequent erythrocyte schizogonous (d–f) cycles produce a trophozoite that becomes a schizont, which releases from 6 to 16 merozoites (f), which commence the repetition of the erythrocytic cycle. Synchronous release of merozoites from the erythrocytes liberates parasite products that stimulate the host's cells to release cytokines (a class of immunological mediators) and these provoke the fever and illness of a malaria attack. Thus, the duration of the erythrocyte schizogonous cycle is the duration of the interval between attacks (i.e. 48 h for tertian, 72 h for quartan).

After several erythrocyte cycles, some trophozoites mature to two types of gametocytes (g,h), a process that takes eight days for *P. falciparum*, but only four days for *P. vivax*. If a female *Anopheles* (M) feeds on an infected human host at this stage in the cycle, she ingests blood that contains erythrocytes, some of which contain both types of gametocytes. Within a susceptible mosquito, the erythrocyte is disposed of and both types of gametocytes (i) develop further: half are female gametocytes, which remain large and are termed macrogametes; the other half are males, which divide into eight-flagellate microgametes (j), which rapidly deflagellate (k), and seek and fuse with a macrogamete to form a **zygote** (l). All this sexual activity has taken place in a matter of 15 min or so. While within the female mosquito, the blood meal passes towards the midgut. Here, the initially inactive zygote becomes an active ookinete (m) that burrows into the epithelial lining of the midgut to form a mature oocyst (n–p).

Asexual reproduction (sporogony) now takes place within the expanding oocyst. In a temperature-dependent process, numerous nuclear divisions give rise to sporozoites. Sporogony does not occur below 16°C or above 33°C, thus explaining the temperature limitations for *Plasmodium* development noted in section 15.3.1. The mature oocyst may contain 10,000 sporozoites, which are shed into the haemocoel (q), from whence they migrate into the mosquito's salivary glands (r). This sporogonic cycle takes a minimum of 8–9 days and produces sporozoites that are active for up to 12 weeks, which is several times the complete life expectancy of the mosquito. At each subsequent feeding, the infective female *Anopheles* injects sporozoites into the next host along with the saliva containing an anticoagulant, and the cycle recommences.

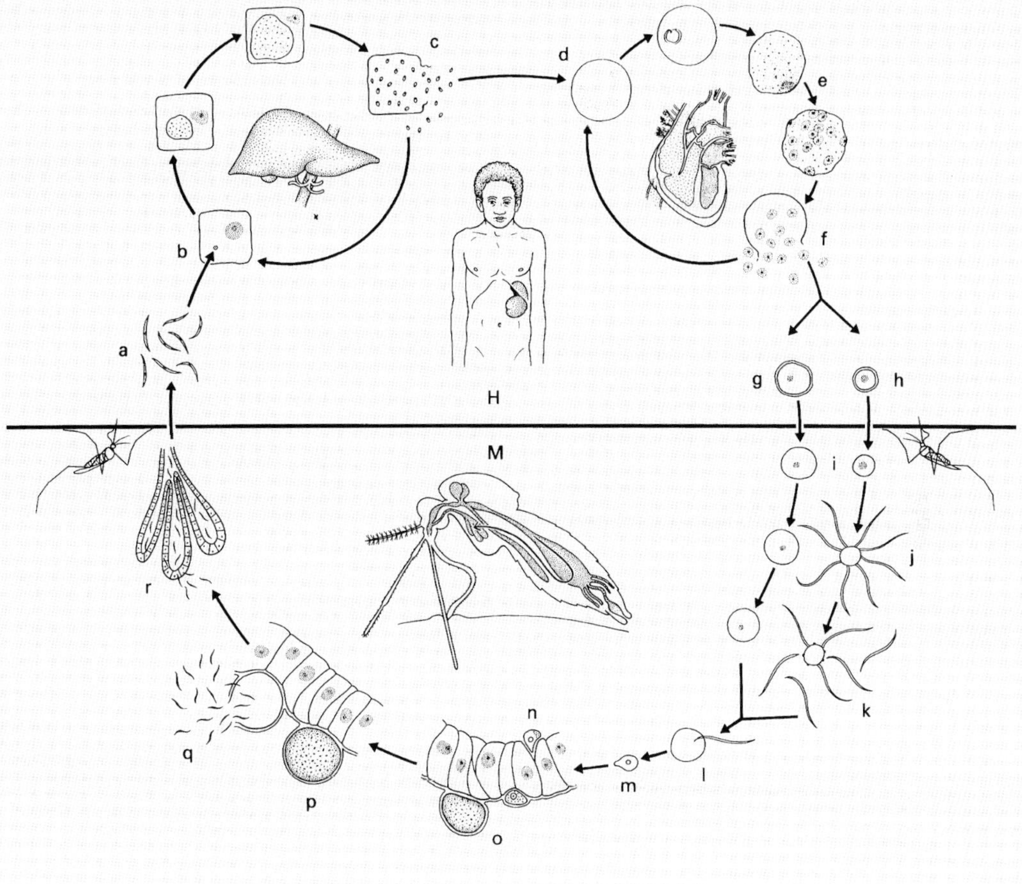

(red blood cells). The first clinical symptoms define the end of an incubation period, some 9–14 (*P. falciparum*) to 18–40 (*P. malariae*) days after infection. Periods of fever followed by severe sweating recur cyclically, and follow several hours after synchronous rupture of infected erythrocytes (see below). The spleen is characteristically enlarged. The five main malaria parasites each induce rather different symptoms:

1 *Plasmodium falciparum*, or malignant tertian malaria, kills many untreated sufferers through, for example, cerebral malaria or renal failure. Fever recurrence is at 48 h intervals (tertian is Latin for third day, the name for the disease being derived from the sufferer having a fever on day one, being normal on day two, with fever recurrent on the third day). *Plasmodium falciparum* is limited by a minimum 20°C isotherm and is thus most common in the warmest areas of the world.

2 *Plasmodium vivax*, or benign tertian malaria, is less serious and rarely kills. However, it is more widespread than *P. falciparum*, and has a wider temperature tolerance, extending as far as the 16°C summer isotherm. Recurrence of fever is every 48 h, and the disease may persist for up to eight years, with relapses some months apart.

3 *Plasmodium malariae* is known as quartan malaria, and is a more widespread but rarer parasite than *P. falciparum* or *P. vivax*. If allowed to persist for an extended period, death occurs through chronic renal failure. Recurrence of fever is at 72 h, hence the name quartan (fever on day one, recurrence on the fourth day). It is persistent, with relapses occurring up to half a century after the initial attack.

4 *Plasmodium ovale* is a tertian malaria very similar morphologically to *P. vivax*, with relapses at intervals from 17 days to more than two years. The contribution of *P. ovale* to human malaria morbidity probably has been underestimated. Recent genetic study has identified two non-recombining sympatric forms (or species) of *P. ovale*. Studies are required to determine if the two forms differ in their relapse periodicity and frequency.

5 *Plasmodium knowlesi* is a parasite of monkeys in Southeast Asia, but in humans is often misdiagnosed as *P. malariae*. It is an emerging infection, being first reported in humans in 1965, and now accounting for the majority of human malaria cases in some parts of Southeast Asia. For example, it is major cause of severe and fatal malaria in Malaysian Borneo.

Malaria epidemiology

Malaria exists in many parts of the world but the incidence varies from place to place. As with other diseases, malaria is said to be **endemic** in an area when it occurs at a relatively constant incidence by natural transmission over successive years. Categories of endemicity have been recognized based on the incidence and severity of symptoms (spleen enlargement) in both adults and children. An **epidemic** occurs when the incidence in an endemic area rises or a number of cases of the disease occur in a new area. Malaria is stable when there is little seasonal or annual variation in the disease incidence, and transmission is by a strongly **anthropophilic** (human-loving) *Anopheles* vector species. Stable malaria usually is associated with the pathogen *P. falciparum* in the warmer areas of the world, where sporogeny is rapid. In contrast, unstable malaria is associated with sporadic epidemics, often with a short-lived and more **zoophilic** (preferring other animals to humans) mosquito vector that may occur in massive numbers. More often, *P. vivax* is the pathogen, ambient temperatures are lower, and sporogeny is slower.

Disease transmission depends upon the potential of each vector to transmit the particular disease. This involves complex relationships between:

- vector distribution;
- vector abundance;
- life expectancy (survivorship) of the vector;
- predilection of the vector to feed on humans (anthropophily);
- feeding rate of the vector;
- vector competence.

With reference to *Anopheles* and malaria, these factors can be detailed as follows.

Vector distribution

Anopheles mosquitoes occur almost worldwide, with the exception of cold-temperate areas, and there are over 450 known species. However, the species of human-pathogenic *Plasmodium* are transmitted significantly in nature only by about 30 species of *Anopheles*. Some species have very local significance, others can be infected experimentally but have no natural role, and perhaps 75% of *Anopheles* species are rather refractory (intolerant) to malaria. A few vectors are important in stable malaria, whereas others become involved only in epidemic spread of unstable malaria. Vectorial status varies due to hidden sibling species and forms that, although lacking morphological differences, differ in biology, including in their epidemiological significance, as in the *An. gambiae* complex (Box 15.2).

Vector abundance

Anopheles development is temperature dependent, as it is in *Aedes aegypti* (see Box 6.2). Where winter temperatures force hibernation of adult females, there may be only one or two generations per year, with generation times of perhaps six weeks at 16°C, but life cycles can be as short as 10 days in tropical conditions. Under optimal conditions, with batches of over 100 eggs laid every two to three days, and a development time of 10 days, 100-fold increases in adult *Anopheles* can take place within 14 days.

Since *Anopheles* larvae develop in water, rainfall significantly governs numbers. The dominant African malaria vector, *An. gambiae* (in the restricted sense; Box 15.2), breeds in short-lived pools that require replenishment; increased rainfall obviously increases the number of *Anopheles* breeding sites. On the other hand, rivers where other *Anopheles* species develop in lateral pools or streambed pools during a low- or no-flow period will be scoured out by excessive wet-season rainfall. Adult survivorship clearly is related to elevated humidity and, for the female, availability of blood meals and a source of carbohydrate.

Vector survival rate

The duration of the adult life of the infective female *Anopheles* mosquito is of great significance in its effectiveness as a disease transmitter. If a mosquito dies within eight or nine days of an initial infected blood meal, no sporozoites will have become available and no malaria is transmitted onwards. The age of a mosquito can be estimated from the physiological age based on the ovarian "relicts" left by each ovarian cycle (section 6.9.2). Physiological age and the duration of the sporogonic cycle (Box 15.1) allows calculation of the proportion of each *Anopheles* vector population of sufficient age to be infective. In African *An. gambiae* (in the restricted sense; Box 15.2), three ovarian cycles are completed before infectivity is detected. Maximum transmission of *P. falciparum* to humans occurs in *An. gambiae* that has completed four to six ovarian cycles. Old individuals form only 16% of the population, but constitute 73% of infective individuals. Clearly, adult life expectancy (demography) is important in epidemiological calculations. Raised humidity prolongs adult life, and the most important cause of mortality is desiccation.

Anthropophily of the vector

A female *Anopheles* mosquito must feed at least twice to act as a vector; once to gain the pathogenic *Plasmodium* and a second time to transmit the disease onward. Host preference is the term for the propensity of a vector mosquito to feed on a particular host species. In malaria, the host preference for humans (anthropophily) rather than alternative hosts (zoophily) is crucial to human malaria epidemiology. Stable malaria is associated with strongly anthropophilic vectors that may never feed on other hosts. In these circumstances, the probability of two consecutive meals being taken from a human is very high, and disease transmission can take place even when mosquito densities are low. In contrast, if the vector has a low rate of anthropophily (a low probability of human feeding), the probability of consecutive blood meals being taken from humans is slight, and human malarial transmission by this particular vector is correspondingly low. Transmission will take place only when the vector is very numerous, as in epidemics of unstable malaria.

Feeding interval

The frequency of feeding by the female *Anopheles* vector is important in disease transmission. This frequency can be estimated from mark–release–recapture data or from survey of the ovarian-age classes of indoor resting mosquitoes. Although it is assumed that one blood meal is needed to mature each batch of eggs, some mosquitoes may mature a first egg batch without a meal, and some anophelines require two meals. Already-infected vectors may experience difficulty in feeding to satiation at one meal because of blockage of the feeding apparatus by parasites, and may probe many times. This, as well as disturbance during feeding by an irritated host, may lead to feeding on more than one host.

Vector competence

Even if an uninfected *Anopheles* feeds on an infectious host, either the mosquito may not acquire a viable infection, or the *Plasmodium* parasite may fail to replicate within the vector. Furthermore, the mosquito may not transmit the infection onwards at a subsequent

Box 15.2 *Anopheles gambiae* complex

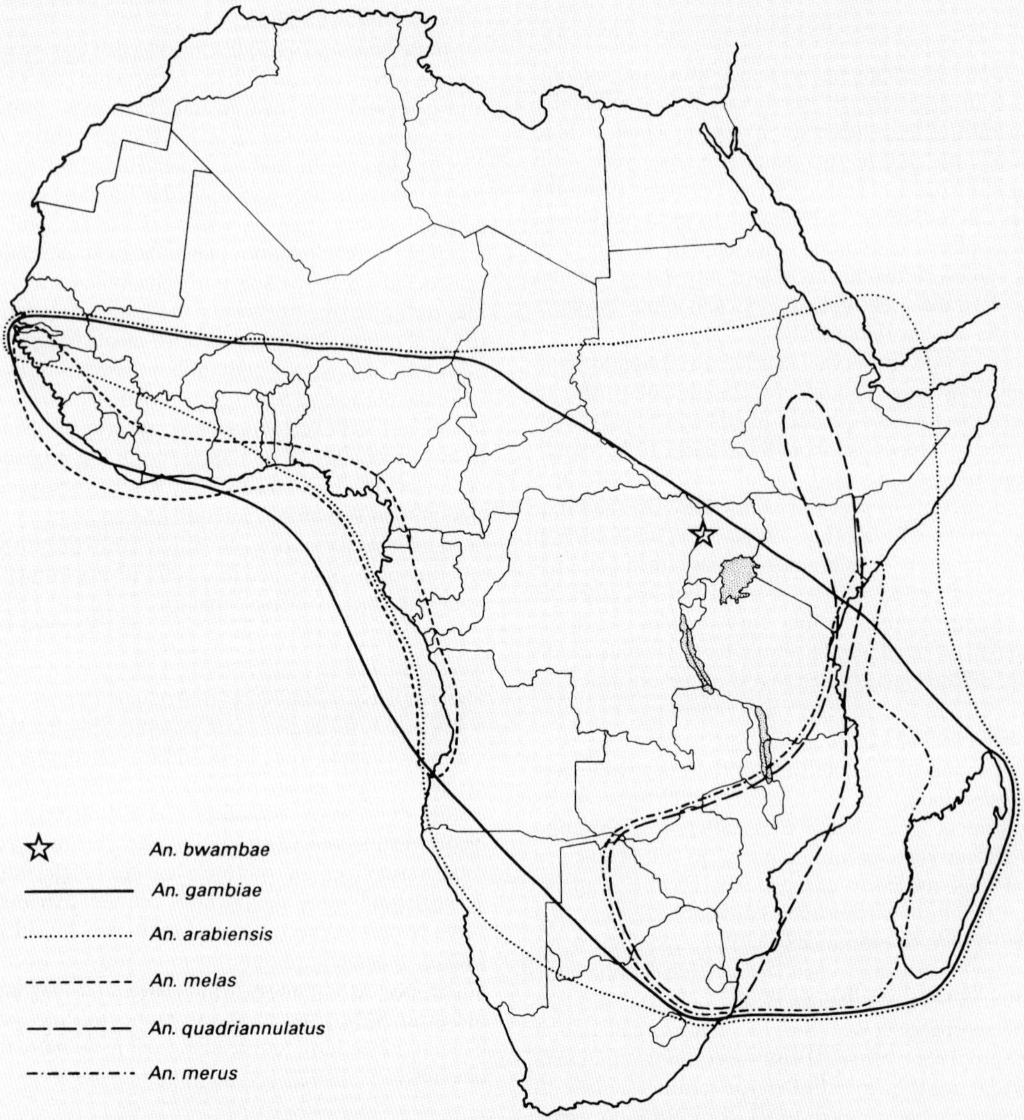

In the early days of African malariology, the common, predominantly pool-breeding *Anopheles gambiae* (Diptera: Culicidae) was found to be a highly anthropophilic, very efficient vector of malaria, virtually throughout the continent. Subtle variation in morphology and biology suggested, however, that more than one species might be involved. Initial investigations allowed morphological segregation of West African *An. melas* and East African *An. merus*; both breed in saline waters, unlike the freshwater-breeding *An. gambiae*. Reservations remained as to whether the latter comprised a single species, and further studies involving meticulous rearing from single egg masses, cross-fertilization and examination of fertility of thousands of hybrid offspring indeed revealed discontinuities in the *An. gambiae* gene pool. These were interpreted as supporting four species, a view that was substantiated by banding patterns of the larval salivary gland and ovarian nurse-cell giant chromosomes, and by protein electrophoresis. Even with reliable cytologically determined specimens, morphological features do not allow segregation of the component species of the freshwater members of the *An. gambiae* complex of sibling (or cryptic) species.

Anopheles gambiae is restricted now to one widespread African taxon, with recognized forms: *An. arabiensis* was recognized for a sibling taxon that in many areas is sympatric with *An. gambiae sensu stricto*; *An. quadriannulatus* is an East and southern African sibling; and *An. bwambae* is a rare and localized taxon from hot mineral pools in Uganda. The maximum distributional limit of each sibling species is shown here on the map of Africa (data from White 1985). The siblings differ markedly in their vectorial status: *An. gambiae sensu stricto* is largely **endophilic** (feeding indoors), but heavy selection in West Africa exerted by the use of bed nets is favouring an **exophilic** form (one that feeds outdoors); *An. arabiensis* also is endophilic and both species are highly anthropophilic vectors of malaria and bancroftian filariasis. However, when cattle are present, *An. arabiensis* shows increased zoophily, much reduced anthropophily and an increased tendency to exophily. In contrast to these two sibling species, *An. quadriannulatus* is entirely zoophilic and does not transmit disease of medical significance to humans. *Anopheles bwambae* is a very localized vector of malaria and is endophilic if native huts are available.

Today, species within the *Anopheles gambiae* complex can be identified using DNA-based methods. Furthermore, some chromosomal forms within *An. gambiae s.s.* exhibit molecular (DNA) differences. Understanding such variation is important because of epidemiologically significant differences between forms.

meal. Thus, there is scope for substantial variation, both within and between species, in the competence to act as a disease vector. Allowance must also be made for the density, infective condition, and demographics of the human host, as immunity to malaria increases with age.

Vectorial capacity

The **vectorial capacity** of a given *Anopheles* vector to transmit malaria in a circumscribed human population can be modelled. This involves a relationship between the:

- number of female mosquitoes per person;
- daily biting rate on humans;
- daily mosquito survival rate;
- time between mosquito infection and sporozoite production in the salivary glands;
- vectoral competence;
- the human recovery rate from infection.

This vectorial capacity must be related to some estimate concerning the biology and prevalence of the parasite when modelling disease transmission, and in monitoring disease control programmes. In malarial studies, the infantile conversion rate (ICR), the rate at which young children develop antibodies to malaria, may be used. In Nigeria (West Africa), the Garki Malaria Project found that over 60% of the variation in the ICR derived from the human-biting rate of the two dominant *Anopheles* species. Only 2.2% of the remaining variation is explained by all other components of vectorial capacity.

Control of malaria

Clearly, from the information outlined above, two broad strategies could control malaria: (i) reduction in the numbers of mosquitoes; or (ii) fighting the disease itself. Programmes dedicated to the eradication of malaria have eliminated the disease from the United States, Europe and the former USSR. However in the tropics, with year-round mosquitoes, resistance to insecticides by mosquitoes, and public resistance to the chemicals involved, control declined. Widespread "blanket" spraying programmes used to control malarial mosquitoes are deemed costly and ineffective. More effective (and less environmentally harmful) insecticidal use involves spraying houses with a contact insecticide that kills adult mosquitoes when they rest on the walls. Of particular interest is the juvenoid pyriproxyfen (section 16.4.2) that even in aquatic dilutions of a few parts per million prevents larval and pupal metamorphosis. Nonetheless, outside urban areas, such control measures cannot eradicate at an acceptable environmental or economic cost. A compelling means to interrupt mosquito transmission of malaria is argued to be the use of long-lasting insecticide-impregnated bed nets to protect sleepers and kill or sterilize incoming female mosquitoes on contact (Box 15.3). Worryingly though, this technique selects for vector strains that are exophilic, with transmission increasingly outside of the houses.

An alternate or concurrent strategy to control malaria is to restrict the disease – the fewer people with malaria, the smaller the risk of a feeding mosquito transmitting the disease onward. An infected person will not transmit the disease if treated within the first 10 days of sickness. In most of the developed world where malaria is recognized and treated clinically within this period, malaria transmission remains rare or non-existent. Malaria in China fell from around 8,000,000 cases per year historically to 2743 autochthonous cases in 2012, thanks to an aggressive detection and treatment programme. In contrast, in much of the developing world where malaria is endemic, medical facilities and budgets are inadequate to treat the many thousands of cases that develop each year.

With a vaccine not yet available, and despite the increased resistance of *Plasmodium* to drugs, appropriate prophylactics do give high personal protection and should be used in accordance with up-to-date information such as provided by the (US) Centers for Disease Control and Prevention.

Box 15.3 Bed nets

Where malaria is a major cause of death, mostly in Sub-Saharan Africa, the disease and its associated mortality seems almost intractable. We know from elsewhere that mosquito vectors can persist but that the disease can be reduced or eliminated. Eradication of the vectors is not necessary if other factors such as biting rate can be reduced. Bed nets have a long history of use, but an infected mosquito can feed where the occupant contacts the net, or can enter a damaged net, and thus they have been incompletely effective. Application of a residual synthetic pyrethroid insecticide to nets to kill, repel or decrease mosquito longevity (and additionally to deter other pest insects) has been effective. With high adoption, both the numbers and the average lifespan of mosquitoes are reduced, and the community can benefit even with incomplete use of bed nets. Results may be similar to those obtained by indoor residual spraying, but some problems are common to both strategies – namely the need for follow-up (retreatment) and development of resistance. Pyrethroid insecticides licensed for use with nets are degraded by sunlight and washing, and although retreatment is simple, this has proved to be major impediment to effective and complete implementation of insecticide-treated bed nets.

Long-lasting insecticide-treated nets (LLINs), which retain lethal pyrethroid insecticide concentrations for at least three years, have been developed in response to retreatment issues. Under World Heath Organization (WHO) standards, pyrethroid either is incorporated into polyethylene from which the netting is made and the insecticide migrates to the fibre surface as the insecticide is lost, or is bound to the surface in a resin-based polymer used to coat the polyester netting coating. Chemical release is unaffected by multiple washings, and up to eight years of effective action is claimed. Trials confirm strongly reduced incidence of malaria, for example in Kenyan and Ugandan villages, and in malarial Papua New Guinea, with effective protection coming from adoption by a high proportion of the population at risk. However, with high adoption come problems with mosquitoes developing resistance to pyrethroid insecticides. Now a twin chemical approach is proposed, with addition of the insect growth regulator pyriproxyfen (section 16.4.2) to the mix. This juvenoid has several modes of action – it can sterilize female mosquitoes by contact, shorten their lifespan, and be transported to oviposition sites to retard the immature stage development there.

There is ongoing debate as to how the necessary high coverage with nets is to be achieved. Some agencies, including the US Centers for Disease Control and Prevention, advocate free provision to all exposed people, supporting a public-health intervention to reduce death and disease across the complete community. However, as is understood by the many agencies associated with malaria prevention, if LLINs become the major public-health intervention in Africa, there will need to be sustained political and financial commitments from politicians of the countries involved, international donors and the makers of low-cost nets. Furthermore, the evening behaviours of both humans and mosquitoes, and peoples' time spent in bed and use (or not) of bed nets in different countries in relation to socio-economic status, must be understood. Already there is some evidence in high-use bed-net areas that nocturnal indoor-biting mosquitoes are being replaced by an **exophilic** form that bites humans earlier in the evening and after rising in the morning, outdoors, or inside, when bed nets are not in use.

15.3.2 Arboviruses

Viruses that multiply in an invertebrate vector and a vertebrate host are termed arboviruses (from *ar*thropod-*bo*rne *viruses*). This definition excludes the mechanically transmitted viruses, such as the myxoma virus that causes myxomatosis in rabbits. There is no viral amplification in myxomatosis vectors such as the rabbit flea, *Spilopsyllus cuniculi* (Siphonaptera: Pulicidae), and, in Australia, in *Anopheles* and *Aedes* mosquitoes. Arboviruses are united by their ecologies, notably their ability to replicate in an arthropod. It is an unnatural grouping rather than one based upon virus phylogeny, as arboviruses belong to several virus families. These include some Bunyaviridae, Reoviridae and Rhabdoviridae, and notably, many Flaviviridae and Togaviridae. *Alphavirus* (Togaviridae) includes exclusively mosquito-transmitted viruses, notably the agents of equine encephalitides, Ross River virus in Australia and Chikungunya disease. This latter has emerged from Africa and Asia into Réunion and neighbouring Indian Ocean islands, Yemen and across the Mediterranean into Italy and France (section 17.2.3). Members of *Flavivirus* (Flaviviridae), which includes yellow fever, dengue, Japanese encephalitis, West Nile and other encephalitis viruses, are borne by mosquitoes or ticks.

Yellow fever exemplifies a flavivirus life cycle. A similar cycle to the African sylvan (forest) one seen in section 15.2 involves a primate host in Central and South America, although with different mosquito vectors from those in Africa. Sylvan transmission to humans does happen, as occurred in Ugandan banana plantations, but the disease makes its greatest fatal impact in urban epidemics. The urban and peri-domestic insect vector in Africa and the Americas is the female of the yellow-fever mosquito, *Aedes (Stegomyia) aegypti*. This mosquito acquires the virus by feeding on a human yellow-fever sufferer in the early stages of disease, from 6 h pre-clinical to four days later. The viral cycle in the mosquito is 12 days long, after which the yellow-fever virus reaches the mosquito saliva and remains there for the rest of the mosquito's life. With every subsequent blood meal, the female mosquito transmits virus-contaminated saliva. Infection results, and yellow-fever symptoms develop in the host within a week. An urban disease cycle originates from individuals infected with yellow fever from the sylvan (rural) cycle moving to an urban environment. Here, disease outbreaks may persist, such as those in which hundreds or thousands of people have died, including in New Orleans as recently as 1905. In South America, monkeys may die of yellow fever, but African ones are asymptomatic: perhaps Neotropical monkeys have yet to develop tolerance to the disease. The common urban mosquito vector of yellow fever and dengue is *Ae. aegypti*. Molecular phylogenetic studies show that both the disease yellow fever and the vector mosquito evolved in West Africa and were transported to Brazil at about the same time, probably aboard the earliest Portuguese slaving ships. However, the range of *Ae. aegypti* is greater than that of the disease, as this mosquito also is present in the southern United States, where it is spreading, and in Australia and much of Asia. Only India has susceptible but, as yet, uninfected monkey hosts of the disease.

Flaviviruses also cause dengue – an insect-borne disease that is intensifying and spreading across much of the tropics (Box 15.4). Other Flaviviridae affecting humans and transmitted by mosquitoes cause several diseases called encephalitis (or encephalitides), because in clinical cases inflammation of the brain occurs. Each encephalitis has a preferred mosquito host, often a *Culex* species. The reservoir hosts for these diseases vary, and for encephalitis include wild birds, with amplification cycles in domestic mammals, for example pigs for Japanese encephalitis. Horses can carry togaviruses, giving rise to the name for a subgroup of diseases termed "equine encephalitides". West Nile virus, belonging to the Japanese encephalitis virus complex, invaded North America from Africa and Mediterranean Europe in the recent past (Box 15.5).

Several flaviviruses are transmitted by ixodid ticks, including more viruses that cause encephalitis and haemorrhagic fevers in humans, but more significantly in domestic animals. Bunyaviruses may be tick-borne, notably haemorrhagic diseases of cattle and sheep, particularly when conditions encourage an explosion of tick numbers and disease alters from normal hosts (**enzootic**) to epidemic (**epizootic**) conditions. Mosquito-borne bunyaviruses include African Rift Valley fever, which can produce high mortality amongst African sheep and cattle during mass outbreaks.

Amongst the Reoviridae, bluetongue virus (an *Orbivirus*) is the best known, most debilitating and most significant economically. The disease, which has become almost worldwide, has 26 different serotypes and causes tongue ulceration (hence "bluetongue") and an often-terminal fever in sheep. Bluetongue

Box 15.4 Dengue – an emerging insect-borne disease

Dengue and the related dengue haemorrhagic and dengue shock fever, also known as break-bone fever, are tropical diseases caused by infection with a mosquito-borne virus of the genus *Flavivirus* (Flaviviridae). The four serotypes (DENV1–4) of *Flavivirus* are different enough from each other to offer no cross-protection, thus allowing later infection by a different strain. The symptoms of classic dengue include rash, headache, and muscle and joint aches (hence break-bone), with some gastrointestinal symptoms. Subsequent infection by a second serotype can lead to serious haemorrhagic (bleeding) or shock symptoms. The febrile period (duration of fever) is about a week, in which the infection can be passed to a feeding mosquito. The principal vector is *Aedes aegypti* (Diptera: Culicidae), shown in the vignette at the beginning of this chapter. This day-biting, peri-domestic, largely tropical mosquito breeds in small water containers, especially in urban environments. Another mosquito species, *Ae. albopictus*, also an excellent vector for the disease, has been spreading from Asia around the global tropics/subtropics, including the Americas, for the past two decades. By the end of the 20th century, dengue was affecting at least 50 million people per year, with hundreds of thousands exhibiting the haemorrhagic and shock forms. In the present century, major dengue outbreaks have been reported already throughout Southeast Asia, the western Pacific, the Caribbean, much of Mesoamerica and South America including Brazil, plus several tropical African countries. Newer areas, such as Hawai'i and Florida (USA) and tropical northern Australia, have experienced outbreaks. The worst news is that in some locations all four serotypes may be present, as in the Indian epidemic of 2006, producing an increase in morbidity and death.

As yet there is no vaccine against dengue, and control of the mosquitoes is the major means to fight the spread of the disease. Standard measures include public education to remove small water containers, plus civic aerial spraying of insecticides to reduce adult longevity. Three novel approaches may prove effective. (i) An ovipositing female mosquito, especially of species such as *Ae. aegypti* that lay one or a few eggs successively, may transfer a lethal dose of the juvenoid pyriproxyfen between consecutive containers, as shown by trials in a graveyard in which flower vases support high populations of vector mosquitoes. (ii) Release of laboratory-reared adults of *Ae. aegypti* carrying specific strains of *Wolbachia* bacteria (section 5.10.4) into natural populations could spread the *Wolbachia* via eggs, and thus halve the lifespan of the adult mosquito, thereby reducing transmission rates. (iii) Another innovative method is to spread the container-dwelling copepod crustacean, *Mesocyclops*, which has a mosquito-eating larva – a technique that has provided good control in trials in Queensland, Australia, and shows promise for community-based, cost-effective control in Vietnam.

The emergence, or re-emergence, of dengue is due to: (i) increased urbanization, with inadequate piped water encouraging the use of inappropriate storage containers; (ii) discarded containers, including used vehicle tyres, providing mosquito breeding sites; (iii) human mobility, including international travel; (iv) breakdown of civic schemes to control mosquitoes; and (v) the increasing resistance of mosquitoes to insecticides. Some predictions of global climate change imply that drought, which encourages increased water storage around residences, will contribute to continued expansion of the disease, as poverty precludes mosquito-proofing these containers.

is one of the few diseases in which biting midges of *Culicoides* (Ceratopogonidae) are the sole vectors of an arbovirus of major significance. Bluetongue has spread north and west of its normal circum-Mediterranean and African distribution, reaching the Netherlands in 2007 and the United Kingdom (in cattle imported from France) in late 2008, and it continues northwards (see section 17.2.2).

Studies of the epidemiology of arboviruses have been complicated by the discovery that some viruses may persist between generations of the vector. Thus, La Crosse virus, a bunyavirus that causes encephalitis in the United States, can pass from the adult mosquito through the egg (**vertical** or **transovarial transmission**) to the larva, which overwinters in a near-frozen tree-hole. The first emerging female mosquito of the spring generation can transmit La Crosse virus to chipmunk, squirrel or human, with its first meal of the year. Vertical transmission is substantiated in other cases, including Japanese encephalitis

Box 15.5 West Nile virus – an arbovirus disease emergent in North America

West Nile virus (WNV) belongs to the Japanese encephalitis antigenic complex of the genus *Flavivirus*, family Flaviviridae. There are another 10 members of this complex, including Murray Valley encephalitis (Australia), St Louis encephalitis (USA), Japanese encephalitis (Asian and West Pacific, but spreading), and several poorly known viruses, any of which could become emergent. All are transmissible by mosquitoes (Diptera: Culicidae), and many cause fever and sometimes-fatal illnesses in humans.

WNV was isolated first in Uganda in the West Nile district (hence the name). By the 1950s, patients, birds and mosquitoes in Egypt were infected. The virus or antibodies in vertebrates had been identified from much of Africa and Eurasia by the mid-1990s, with outbreaks sporadically reported even in temperate Europe (e.g. Romania 1996–7). Symptoms in humans include flu-like fever, with a small percentage (>15%) developing meningeal or encephalitis symptoms. Recovery is usually complete, but aches may persist. Death is rare, and limited largely to the elderly.

The principal vectors of WNV are mosquitoes, mainly bird-feeding species of the genus *Culex*. Many species potentially are involved, and different species predominate in diverse geographic and ecological settings. Birds, especially those that associate with wetland harbouring populations of bird-feeding mosquitoes, are natural reservoirs and the principal hosts of the disease. Infected migratory birds may survive to spread the disease seasonally. In Eurasia, two basic types of cycles exist – a rural or "sylvatic" cycle involving wild, often wetland, birds and ornithophilic (bird-feeding) mosquitoes, and an urban cycle of domestic or human-associated birds and mosquitoes, particularly members of the *Culex pipiens* "complex", that feed both on birds and humans.

In Africa and Eurasia, exposed birds largely tolerate the disease asymptomatically, and outbreaks in humans have been limited to a few hundred cases in any one instance. Horses are very susceptible to WNV, and some European outbreaks have affected these animals differentially, notably in France in the 1960s and 2000. Approximately one-third of all affected animals die, and near half the survivors show residual symptoms six months later. Vaccines for horses have been developed and their use is strongly recommended in endemic areas. Both horses and humans are dead-ends for the disease; only in birds is the virus amplified for onward transmission by a feeding mosquito.

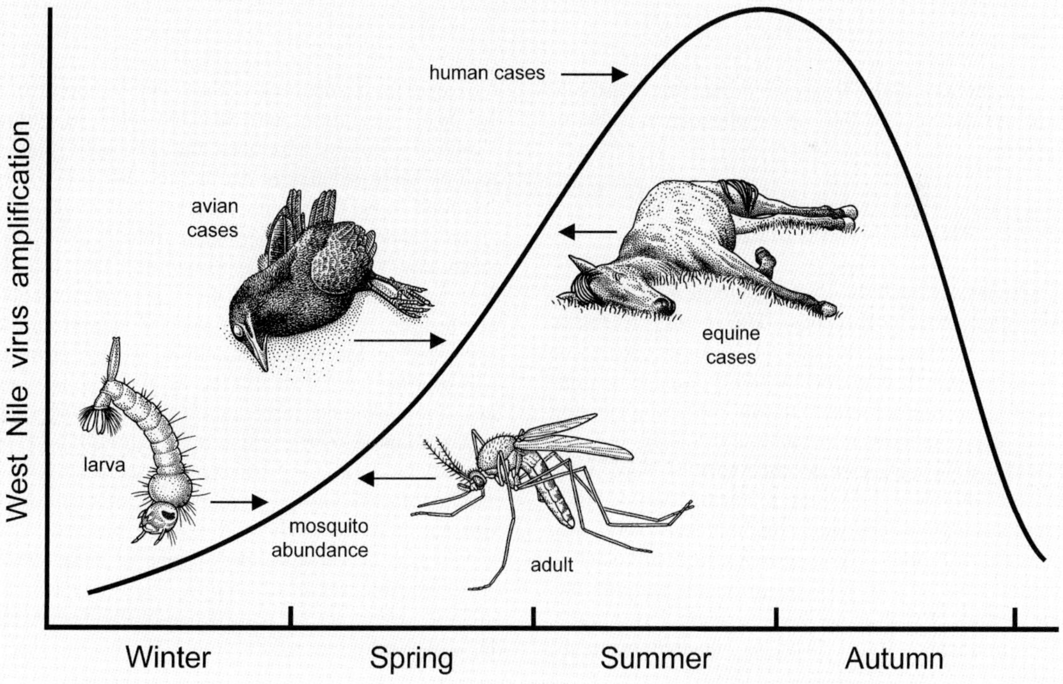

In the New World (western hemisphere), WMV was unknown until 1999 when it appeared first in the New York urban area, with encephalitis reported in humans, horses and domestic pets. The virus may have entered either in an infected bird or mosquito; the virus strain resembled an isolate found in Israel in the previous year. The spread was very rapid, reaching all contiguous states of the United States, most of Canada, and with sporadic cases in Mexico, the Caribbean and Central America, in less than a decade. The strain (as in Israel) is more virulent than has been usual in Eurasia, and caused especially high mortality rates in infected birds, especially American robins (Turdidae) and crows (Corvidae). Their mass deaths acted as an early indicator of the arrival of the virus in a new area. Unusually, the disease has survived the harsh winters experienced in northern states, and the disease is now endemic across the continent. Human mortality was initially around 4% of reported cases (over 1000 deaths from 27,000 infected by the end of August 2008). However, this mortality rate is overestimated since many mild cases were unreported. Deaths are notably amongst older, immuno-compromised or diabetic patients. Despite the availability of horse-vaccine, many horses were infected and died.

The seasonal cycle of the disease in North America, shown in the graph (after an illustration by W.K. Reisen), involves increasing numbers of mosquito larvae in spring as temperatures rise, with emergent adult mosquitoes becoming infected and amplifying the virus as the season progresses. First signs of the disease are seen in deaths of birds, especially corvids, starting in late spring, followed by sickness and mortality in horses. The first human cases tend to appear when virus amplification in mosquitoes has risen towards its mid-summer peak. By late summer, mosquito numbers diminish, and the disease cycle attenuates in autumn/fall.

As in the eastern hemisphere, many different American mosquitoes can transmit the WNV, especially species of *Culex* that feed on birds, humans and horses. The "rural" cycle often involves *Culex tarsalis* in wetlands, including agricultural areas of rice production such as in the central valley of California, and an urban/suburban cycle involving peri-domestic mosquitoes of the *C. pipiens* group that feed on birds and humans. Research generated in response to this newly emergent disease shows complex interactions between the North American landscape (including irrigated agriculture and urbanization), population dynamics of both hosts and vectors, and a likely strong response to climate, such as to ambient temperatures and rainfall patterns.

Numbers peaked near 10,000 cases (264 deaths) nationwide in 2003, but rose again in 2012 to over 5000 cases and 243 deaths. Evidently, North Americans will have to live with a rather virulent WNV and associated health-care costs, and with the still poorly understood environmental consequences of high mortality amongst local birds.

in mosquitoes of *Culex tritaenorhynchus* and West Nile virus in *Culex* species, but the significance in epidemiology needs verification.

15.3.3 Rickettsias and plague

Rickettsias are bacteria (Proteobacteria: Rickettsiales) associated with arthropods. The genus *Rickettsia* includes virulent pathogens of humans. *Rickettsia prowazekii*, which causes epidemic typhus, has influenced world affairs as much as any politician, causing the deaths of millions of refugees and soldiers in times of social upheaval, such as the years of Napoleonic invasion of Russia, and those following World War I. Typhus symptoms are headache, high fever, spreading rash, delirium and aching muscles, and in epidemic typhus from 10% to 60% of untreated patients die. The vectors of typhus are lice (Taxobox 18), notably the human body louse, *Pediculus humanus humanus*, if treated as a subspecies (Psocodea: Anoplura). In the past, infestation with lice indicated unsanitary conditions, but in developed nations, after years of decline, the incidence of lice is increasing. Although the head louse (either an ecotype of *P. humanus*, or the subspecies *P. humanus capitis*), pubic louse (*Pthirus pubis*) and some fleas experimentally can transmit *R. prowazekii*, they are of little or no epidemiological significance. After the rickettsias of *R. prowazekii* have multiplied in the louse epithelium, they rupture the cells and are voided in the faeces. Because the louse dies, the rickettsias are demonstrated to be rather poorly adapted to the louse host. Human hosts are infected by scratching infected louse faeces (which remain infective for up to two months after deposition) into the itchy site where the louse has fed. There is evidence of low-level persistence of rickettsias in those who recover from typhus. These people act as endemic reservoirs for resurgence of the disease, and domestic and a few wild animals also may be disease reservoirs. Lice are also vectors of relapsing fever, a spirochaete disease that historically occurred together with epidemic typhus.

Other rickettsial diseases include murine typhus, transmitted by flea vectors, scrub typhus through trombiculid mite vectors, and a series of spotted fevers, termed tick-borne typhus. Many of these diseases have a wide range of natural hosts, with antibodies to the widespread American Rocky Mountain spotted fever (*Rickettsia rickettsii*) reported from numerous bird and mammal species. Throughout the range of the disease from Canada to Brazil, several species of ticks with broad host ranges are involved, with transmission to vertebrates solely via tick saliva during feeding.

Bartonellosis (Oroya fever or Carrion's disease) is a bacterial infection caused by *Bartonella bacilliformis* (Proteobacteria: Rhizobiales) and transmitted by South American phlebotomine sand flies (Diptera: Psychodidae), with symptoms of exhaustion, anaemia and high fever, followed by wart-like eruptions on the skin.

Plague is a rodent–flea–rodent disease caused by the bacterium *Yersinia pestis* (Proteobacteria: Enterobacteriales). Plague-bearing fleas are principally *Xenopsylla cheopis* (Siphonaptera: Pulicidae), which is ubiquitous between 35°N and 35°S, but also include *X. brasiliensis* in India, Africa and South America, and *X. astia* in Southeast Asia. Although other species, including *Ctenocephalides felis* and *C. canis* (cat and dog fleas, also in the family Pulicidae), can transmit plague, they play a minor role. The major vector fleas occur especially on peri-domestic (house-dwelling) species of *Rattus*, such as the black rat (*R. rattus*) and the brown rat (*R. norvegicus*). Reservoirs for plague in specific localities include the lesser bandicoot rat (*Bandicota bengalensis*) in India, rock squirrels (*Spermophilus* spp.) in the western United States and the related suslik (ground squirrel, *Citellus pygmaeus*) in Eurasia, gerbils (*Meriones* spp.) in the Middle East, and different gerbils (*Tatera* spp.) in India and South Africa. Between plague outbreaks, the bacterium circulates within some or all of these rodents without evident mortality, thus providing long-term, silent, reservoirs of infection.

When humans become involved in plague outbreaks (such as the pandemic known as the "Black Death" that ravaged Europe during the 14th century), mortality may approach 90% in undernourished people and around 25% in previously healthy people. The plague epidemiological cycle commences amongst rats, with fleas naturally transmitting *Y. pestis* between peri-domestic rats. In an outbreak of plague, when the preferred-host brown rats die, some infected fleas move on to and eventually kill the secondary preference, black rats. As *X. cheopis* readily bites humans, infected fleas switch host again in the absence of the rats. Plague is a particular problem where rat (and flea) populations are high, as occurs in overcrowded, unsanitary urban conditions. Outbreak conditions require appropriate preceding conditions of mild temperatures and high humidity, which encourage build-up of flea populations by increased larval survival and adult longevity. Thus, natural variations in the intensity of plague epidemics relate to the previous year's climate. Even during prolonged plague outbreaks, periods of fewer cases used to occur when hot, dry conditions prevented recruitment, because flea larvae are very susceptible to desiccation, and low humidity reduced adult survival in the subsequent year.

During its infective lifetime, a flea varies in its ability to transmit plague, according to internal physiological changes induced by *Y. pestis*. If the flea takes an infected blood meal, *Y. pestis* increases in the proventriculus and midgut and may form an impassable plug. Further feeding involves a fruitless attempt by the pharyngeal pump to force more blood into the gut, with the result that a contaminated mixture of blood and bacteria is regurgitated. However, the survival time of *Y. pestis* outside the flea (of no more than a few hours) suggests that mechanical transmission is unlikely. More likely, even if the proventricular blockage is alleviated, is that it fails to function properly as a one-way valve, and at every subsequent attempt at feeding, the flea regurgitates a contaminated mixture of blood and pathogen into the feeding wound of each successive host.

15.3.4 Protists other than malaria

Some of the most important insect-borne pathogens are protists (protozoans), which affect a substantial proportion of the world's population, particularly in subtropical and tropical areas. Malaria has been covered in detail in section 15.3.1, and two important flagellate protists of medical significance are described here.

Trypanosoma

Trypanosoma is a large genus of parasites of vertebrate blood that are transmitted usually by blood-feeding "higher" flies. However, throughout South America, blood-feeding triatomine bugs ("kissing bugs"), notably

Rhodnius prolixus and *Triatoma infestans* (Hemiptera: Reduviidae), transmit trypanosomes of *Trypanosoma cruzi* that cause Chagas' disease. Symptoms of the disease, also called American trypanosomiasis, are predominantly fatigue, with cardiac and intestinal problems if untreated. The disease, which affects at least 7–8 million people in the Neotropics, especially in rural areas, caused over 10,000 deaths in 2010. From a public-health perspective, in the United States, an estimated 300,000 of the millions of Latino migrants have the disease, mostly from infections acquired elsewhere. However, the near absence of vector-borne cases of the disease in the southern United States is surprising given the high percentages of infected triatomines as assessed by DNA data in Tucson in southern Arizona.

Other such diseases, termed **trypanosomiasis**, include sleeping sicknesses transmitted to African humans and their cattle by tsetse flies (species of *Glossina*) (Fig. 15.1). In this and other diseases, the development cycle of the *Trypanosoma* species is complex. Morphological change occurs in the protist as it migrates from the tsetse-fly gut, around the posterior free end of the peritrophic membrane, then anteriorly to the salivary gland. Transmission to human or cattle host is through injection of saliva. Within the vertebrate, symptoms depend upon the species of trypanosome: in humans, a vascular and lymphatic infection is followed by an invasion of the central nervous system that gives rise to "sleeping" symptoms, followed by death.

Leishmania

A second group of flagellates belong to the genus *Leishmania*, which includes parasites that cause internal visceral or disfiguring external ulcerating diseases of humans and dogs. The vectors are exclusively phlebotomines (Diptera: Psychodidae) – small to minute sand flies that can pass through mosquito netting and, in view of their usual very low biting rates, have impressive abilities to transmit disease. Most infections are in animals such as desert and forest rodents, canids (including semi-feral dogs) and hyraxes, with humans becoming involved as they move into areas that are naturally home to these animal reservoirs. The disease is found in nearly 90 countries, and some two million new cases are diagnosed each year, with approximately 12 million people infected at any given time. Visceral leishmaniasis (also known as kala-azar) inevitably kills if untreated; cutaneous leishmaniasis disfigures and leaves scars; mucocutaneous leishmaniasis destroys the mucous membranes of the mouth, nose and throat. Infections of leishmaniasis have been acquired by troops serving in Iraq and Afghanistan.

15.3.5 Filariases

Two of the five main debilitating diseases transmitted by insects are caused by nematodes, namely filarial worms of the family Onchocercidae. The diseases are bancroftian and brugian filariases, commonly termed elephantiasis and onchocerciasis (or river blindness). Other filariases cause minor ailments in humans. *Dirofilaria immitis* (canine heartworm) is one of the few significant veterinary diseases caused by this type of parasite. These filarial nematodes depend on *Wolbachia* bacteria for embryo development, and thus infection can be reduced or eliminated with antibiotics (see also section 5.10.4).

Bancroftian and brugian filariasis

Two worms, *Wuchereria bancrofti* and *Brugia malayi*, are responsible for over a hundred million active cases

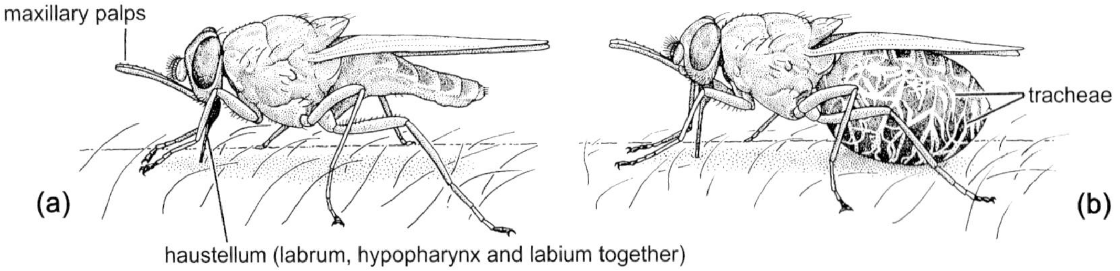

Fig. 15.1 A tsetse fly, *Glossina morsitans* (Diptera: Glossinidae): (a) at the commencement of feeding; and (b) fully engorged with blood. Note that the tracheae are visible through the abdominal cuticle in (b). (After Burton & Burton 1975.)

of filariasis worldwide, with *B. timori* causing human filariasis only in parts of Indonesia. The worms live in the lymphatic system, causing debilitation and oedema, culminating in extreme swellings of the lower limbs or genitals, called elephantiasis. Although the disease is less often seen in the extreme form, the number of sufferers is increasing as one major vector, the worldwide peri-domestic mosquito, *Culex pipiens quinquefasciatus*, increases. The World Health Organization (WHO) is co-ordinating efforts to eradicate filariasis, including using anti-filarial drugs.

The cycle starts with uptake of small microfilariae with blood taken up by the vector mosquito. The microfilariae move from the mosquito gut through the haemocoel into the flight muscles, where they mature into infective larvae. The 1.5 mm long larvae migrate through the haemocoel into the mosquito head where, when the mosquito next feeds, they rupture the labella and invade the host through the puncture of the mosquito bite. In the human host, the larvae mature slowly over many months. The sexes are separate, and pairing of mature worms must take place before further microfilariae are produced. These microfilariae cannot mature without the mosquito phase. Cyclical (nocturnal periodic) movement of microfilariae into the peripheral circulatory system may make them more available to feeding mosquitoes. The vectors are mostly species of *Culex*, *Aedes*, *Anopheles* and *Mansonia*.

Onchocerciasis

Onchocerciasis kills few, but debilitates millions of people (99% of whom live in sub-Saharan Africa) by scarring their eyes, which leads to blindness. The common name of "river blindness" refers to the impact of the disease on people living alongside rivers where the insect vectors, *Simulium* black flies (Diptera: Simuliidae), live in flowing waters. The pathogen is a filarial worm, *Onchocerca volvulus*, in which the female is up to 50 mm long and the male is smaller at 20–30 mm. The adult filariae live in subcutaneous nodules and are relatively harmless, although they cause skin irritation. It is the microfilariae that cause the damage to the eye when they invade the tissues and die there. The major black-fly vector is one of the most extensive complexes of sibling species: "*Simulium damnosum*" has at least 30 cytologically determined species known from West and East Africa; in South America comparable sibling species diversity in *Simulium* vectors is apparent. The larvae, which are common filter feeders in flowing waters, can be controlled, but the strongly migratory adults re-invade readily. The African Programme for Onchocerciasis Control (APOC) uses a network of community paramedics to distribute the anti-filarial drug ivermecten to reduce the disease. Although loss of vision is irreversible, the prevalence of the disease is declining. It remains to be seen if the reduction can be sustained after proposed cessation of the programme in 2015.

15.4 FORENSIC ENTOMOLOGY

As seen in section 15.1, some flies develop in living flesh, with two waves discernible: primary colonizers that cause initial myiases, with secondary myiases developing in pre-existing wounds. A third wave may follow before death. This ecological **succession** results from changes in the attractiveness of the substrate to different insects. An analogous succession of insects occurs in a corpse following death (section 9.4), with a somewhat similar course taken whether the corpse is a guinea pig (Fig. 15.2), pig, rabbit or human. Although the process of corpse decomposition is continuous, for utility it is divided into a series of stages (up to nine, depending on author and region) characterized by corpse physical features and the associated assemblage of insects. Four named stages are shown in Fig. 15.2, but one nomenclature used for human corpses has five stages: fresh, bloated, decay, post-decay and skeletal. The rather predictable succession of insects in corpses has been used for medico-legal purposes by **forensic entomologists** as a faunistic method to assess the elapsed time (and even prevailing environmental conditions) since death for human corpses. It is important to note that the succession of arthropods inhabiting a corpse (whether human or animal) is not a progression from one discrete assemblage of organisms to another, but a gradual gain and loss of arthropod taxa continuously with time.

The generalized sequence of colonization is as follows. A fresh corpse is rapidly visited by a first wave of *Calliphora* (blow flies) and *Musca* (house flies), which oviposit or drop live larvae onto the cadaver. Their subsequent development to mature larvae (which depart the corpse to pupariate away from the larval development site) is temperature dependent. Given knowledge of the particular species, the larval development times at different temperatures, and the ambient temperature at the corpse, an estimate of the age of a corpse may be made, perhaps accurate to within

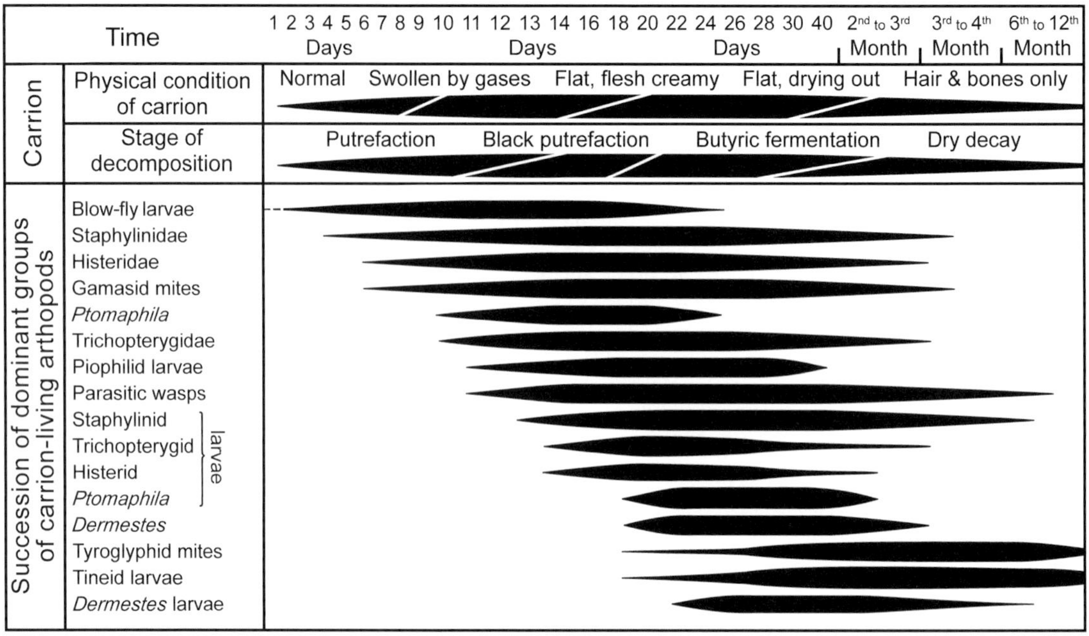

Fig. 15.2 The stages of carcass (carrion) decomposition associated with a succession of arthropod groups in guinea-pig carcasses during spring in a woodland habitat in Perth, Australia. Variation in the thickness of each band indicates the approximate relative abundance within the groups at different times. (After Bornemisszza 1957.)

half a day if fresh, but with diminishing accuracy with increasing exposure.

As the corpse ages, cheese-skipper larvae (Diptera: Piophilidae) appear, along with or followed by larvae and adults of *Dermestes* (Coleoptera: Dermestidae). As the body becomes drier, it often is colonized by a sequence of different dipteran larvae, including those of Drosophilidae (fruit flies) and *Eristalis* (Diptera: Syrphidae: the rat-tailed maggot, a hover fly). After some months, when the corpse is completely dry, more species of Dermestidae appear and several species of clothes moth (Lepidoptera: Tineidae) scavenge the desiccated remnants.

This simple outline is confounded by a number of factors including:

1 geography, with different insect species (though perhaps relatives) present in different regions, especially if considered on a continental scale;

2 difficulty in identifying the early stages, especially of blow-fly larvae, to species;

3 variation in ambient temperatures, with direct sunlight and high temperatures speeding the succession (even leading to rapid mummification), and shelter and cold conditions retarding the process;

4 variation in exposure of the corpse, with burial, even partial, slowing the process considerably, and with a very different entomological succession;

5 variation in cause and site of death, with death by drowning and subsequent degree of exposure on the shore giving rise to a different **necrophagous** fauna from those infesting a terrestrial corpse, with differences between freshwater and marine stranding.

Problems with identification of larvae using morphology are being alleviated using DNA-based approaches. Entomological forensic evidence has proved crucial to post-mortem investigations. Forensic entomological evidence has been particularly successful in establishing disparities between the location of a crime scene and the site of discovery of the corpse, and between the time of death (perhaps homicide) and subsequent availability of the corpse for insect colonization.

15.5 INSECT NUISANCE AND PHOBIA

Our perceptions of nuisance may be little related to the role of insects in disease transmission. Insect nuisance is often perceived as a product of high densities of a particular species, such as bush flies (*Musca vetustissima*) in rural Australia, or ants and silverfish around the house. Most people have a more justifiable avoidance of filth-frequenting insects such as blow flies and cockroaches, biters such as some ants, and venomous stingers such as some ants, bees and wasps. Many serious disease vectors are rather uncommon and have inconspicuous behaviours, aside from their biting habits, such that the lay public may not perceive them as particular nuisances.

Harmless insects and arachnids sometimes arouse reactions such as unwarranted phobic responses (**arachnophobia** or **entomophobia** or **delusory parasitosis**). These cases may cause time-consuming and fruitless inquiry by medical entomologists, when the more appropriate investigations ought to be psychological. Nonetheless, some sufferers of persistent "insect bites" and skin rashes, for which no physical cause can be established, actually suffer from undiagnosed local or widespread infestation with microscopic mites. In these circumstances, diagnosis of delusory parasitosis, through medical failure to identify the true cause, and referral to psychological counselling is unhelpful, to say the least.

There are, however, some insects that transmit no disease, but feed on blood and whose attentions almost universally cause distress – bed bugs and head lice. *Cimex lectularius* (Hemiptera: Cimicidae), the cosmopolitan common bed bug, and *C. hemipterus*, the tropical bed bug, are resurging pests (Box 15.6).

Box 15.6 Bed bugs resurge

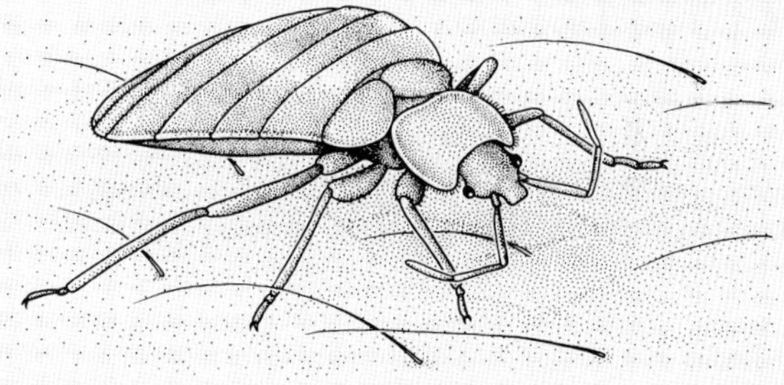

Depending on the age and location of readers of this book, bed bugs could be either a historical problem, long ago eliminated, a "horror story" from a tropical beachside backpacker hostel, or, increasingly, an unwanted souvenir from a conference hotel or an upmarket cruise, or even an infestation in one's own home. Until the late 1990s, reports of bed bugs (one is shown here on the skin of its host; after Anon 1991) were scarce, as they had been since the 1940s when modern insecticides reduced their incidence to rarity. Evidently though, these blood-sucking hemipterans are now returning with a vengeance. Reports in North America, Australia, and parts of Europe imply a massive surge in complaints, including litigation by affected hotel guests. A 2011 survey in the United States reported that one in five Americans had had an infestation in their house or knew someone who had encountered bed bugs, and that public awareness of the problem had resurged along with the bugs.

Explanations for this recurrence of bed bugs include the phasing out of general insect control and its replacement with targeted control of specific pests, resistance to some insecticides, and the reluctance to treat bedrooms. For example, the change in cockroach control from use of broad-spectrum surface insecticides to use of selective baits reduces collateral exposure of bed bugs. At least in the tropical bed bug *Cimex hemipterus* (Hemiptera: Cimicidae) in Africa, there is evidence of resistance to synthetic pyrethroids, although resistance to older insecticides has been long known in the more temperate common bed bug *Cimex lectularius*. When bed bugs began to resurge at the beginning of the 21st century, pesticide operatives had little experience with these "historic" pests, and were unfamiliar with the symptoms (mysterious bites, specks of defaecated blood on sheets, and a distinctive sweet and sickly odour) or with the cryptic daytime aggregation sites of the bugs. Another problem is that infestations may be concealed, especially by those engaged in the hospitality industry fearing loss of trade (but thereby exposing themselves to litigation!). A novel method of discovering bed bugs is the use of trained dogs, which can detect even the presence of eggs.

The resurgence in bed bugs is associated no doubt with increased modern air travel, and with the daytime refuge of bugs in dark places, including in travellers' luggage, especially when left lying, accessible to insects, on a floor by day. One can assume that the prevalence of infestations in some backpacker accommodation is associated with long-distance travel, which facilitates transport between bed-bug-infested hostels on different continents. Quarantine intercepts of bed bugs at airports often are associated with fabrics, and with household possessions such as furniture.

The tropical bed bug occurs mostly within the 30° latitudes and the common bed bug in temperate areas, although both species can occur outside their normal range and sometimes together. The United States mainland appears to have only *C. lectularius*, but *C. hemipterus* could invade as it has in northeast Australia. Head lice (either an ecotype of *Pediculus humanus* or the subspecies *P. humanus capitis*) also cause parents much concern, probably mainly due to the social stigma, although their presence does cause itching. Despite numerous products to kill head lice, they persist because of insecticide resistance and/or ineffective or inconsistent treatment (e.g. many treatments kill adult lice but not their eggs) and are acquired easily by direct head-to-head contact.

15.6 VENOMS AND ALLERGENS

15.6.1 Insect venoms

Some people's earliest experiences with insects are memorable for their pain. Although the sting of the females of many social hymenopterans (bees, wasps and ants) can seem unprovoked, it is an aggressive defence of the nest. The delivery of venom is through the sting, a modified female ovipositor (Fig. 14.11). The honey-bee sting has backwardly directed barbs that allow only one use, as the bee is fatally damaged when it leaves the sting and accompanying venom sac in the wound as it struggles to retract the sting. In contrast, wasp and ant stings are smooth, can be retracted, and are capable of repeated use. In some ants, the ovipositor sting is greatly reduced, and venom is either sprayed around liberally or it can be directed with great accuracy into a wound made by the jaws. The venoms of social insects are discussed in more detail in section 14.6.

15.6.2 Blister and urtica (itch)-inducing insects

Some toxins produced by insects can cause injury to humans, even though they are not inoculated through a sting. Blister beetles (Meloidae) contain toxic chemicals, cantharidins, which are released if the beetle is crushed or handled. Cantharidins cause blistering of the skin and, if taken orally, inflammation of the urinary and genital tracts, which gave rise to its notoriety (as "Spanish fly") as a supposed aphrodisiac. Rove beetles of the genus *Paederus* (Staphylinidae) produce potent contact poisons, including paederin (or pederin), that cause delayed onset of severe blistering and long-lasting ulceration.

Lepidopteran caterpillars, notably moths, are a frequent cause of skin irritation, or urtication (a description derived from a similarity to the reaction to nettles, genus *Urtica*). Some species have hollow spines containing the products of a subcutaneous venom gland, which are released when the spine is broken. Other species have setae (bristles and hairs) containing toxins, which cause intense irritation when the setae contact human skin. Urticating caterpillars include the processionary caterpillars (Thaumetopoeidae) and some cup moth larvae (Limacodidae). The pine and oak processionary caterpillars (*Thaumetopoea pityocampa* and *T. processionea*, respectively) of the Palaearctic live in groups and forage from tent-like silk nests that they spin in pine trees. Australian processionary caterpillars (*Ochrogaster*) combine frass (dry insect faeces), cast larval skins and shed hairs into bags, either suspended in or at the base of trees and bushes. If the bag is damaged by contact or by high wind, urticating hairs are widely dispersed. In parts of South America, the setae of caterpillars of the taturana, *Lonomia obliqua* (Saturniidae), are hollow and contain anti-coagulant venom with an enzyme that destroys blood cells, proteins and connective tissue. Reactions in humans range from itching, intestinal problems, kidney failure, brain haemorrhage and even death, although an anti-venom has reduced the death rate. This severe haemorrhagic syndrome, sometimes called lonomiasis, is most frequent in southern Brazil, where deforestation is believed to have reduced the natural enemies of the taturana.

The pain caused by hymenopteran stings may last a few hours, urtication may last a few days, and the most ulcerated beetle-induced blisters may last some weeks. However, increased medical significance of these injurious insects comes when repeated exposure leads to allergic disease in some humans.

15.6.3 Insect allergenicity

Insects and other arthropods are often implicated in allergic disease, which occurs when exposure to some arthropod allergen (a moderate-sized molecular weight chemical component, usually a protein) triggers excessive immunological reaction in some exposed people or animals. Those who handle insects in their occupations, such as in entomological rearing facilities, tropical fish-food production or research laboratories, frequently develop allergic reactions to one or more of a range of insects. Mealworms (beetle larvae of *Tenebrio* spp.), bloodworms (larvae of *Chironomus* spp.), locusts and blow flies have all been implicated. Stored products infested with astigmatic mites give rise to allergic diseases such as baker's and grocer's itch. The most significant arthropod-mediated allergy arises through the faecal material of house-dust mites (*Dermatophagoides pteronyssinus* and *D. farinae*), which are ubiquitous and abundant in houses throughout many regions of the world. Exposure to naturally occurring allergenic arthropods and their products may be underestimated, although the role of house-dust mites in allergy is now well recognized.

The venomous and urticating insects discussed above can cause greater danger when some sensitized (previously exposed and allergy-susceptible) individuals are exposed again, as anaphylactic shock is possible, with death occurring if untreated. Individuals showing indications of allergic reaction to hymenopteran stings must take appropriate precautions, including allergen avoidance and carrying adrenaline (epinephrine).

FURTHER READING

Amendt, J., Campobasso, C.P., Goff, M.L. & Grassberger, M. (eds) (2010) *Current Concepts in Forensic Entomology*. Springer, Dordrecht, Heidelberg, London, New York.

Battisti, A., Holm, G., Fagrell, B. & Larsson, S. (2011) Urticating hairs in arthropods: their nature and medical significance. *Annual Review of Entomology* **56**, 203–20.

Bonilla, D.L., Durden, L.A., Eremeeva, M.E. & Dasch, G.A. (2013) The biology and taxonomy of head and body lice – implications for louse-borne disease prevention. *PLoS Pathogens* **9**(11), e1003724. doi: 10.1371/journal.ppat.1003724

Byrd, J.H. & Castner, J.L. (eds) (2009) *Forensic Entomology: The Utility of Arthropods in Legal Investigations*, 2nd edn. CRC Press, Boca Raton, FL.

Doggett, S.L., Dwyer, D.E., Peñas, P.F. & Russell, R.C. (2012) Bed bugs: clinical relevance and control options. *Clinical Microbiology Reviews* **25**, 164–92.

Eldridge, B.F. & Edman, J.D. (eds) (2003) *Medical Entomology: A Textbook on Public Health and Veterinary Problems Caused by Arthopods*, 2nd edn. Springer, Berlin.

Gennard, D. (2012) *Forensic Entomology: An Introduction*, 2nd edn. Wiley-Blackwell, Chichester.

Goff, M.L. (2009) Early post-mortem changes and stages in decomposition in exposed cadavers. *Experimental and Applied Acarology* **49**, 21–36.

Guzman, M.G., Halstead, S.B., Artsob, H. *et al.* (2010) Dengue: a continuing global threat. *Nature Reviews Microbiology* **8**, S7–S16.

Hinkle, N.C. (2000) Delusory parasitosis. *American Entomologist* **46**, 17–25.

Kramer, L.D., Styer, L.M. & Ebel, G.D. (2008) A global perspective on the epidemiology of West Nile virus. *Annual Review of Entomology* **53**, 61–81.

Lehane, M.J. (2005) *Biology of Blood-sucking Insects*, 2nd edn. Cambridge University Press, Cambridge.

Lockwood, J.A. (2008) *Six-legged Soldiers: Using Insects as Weapons of War*. Oxford University Press, New York.

Mullen, G.R. & Durden, L.A. (eds) (2009) *Medical and Veterinary Entomology*, 2nd edn. Academic Press, San Diego, CA.

Reinhardt, K. & Siva-Jothy, M.T. (2007) Biology of the bed bugs (Cimicidae). *Annual Review of Entomology* **52**, 351–74.

Resh, V.H. & Cardé, R.T. (eds) (2009) *Encyclopedia of Insects*, 2nd edn. Elsevier, San Diego, CA. [Particularly see articles on bed bugs; blood sucking; bubonic plague; delusory parasitosis; dengue; forensic entomology; lice, human; malaria; medical entomology; veterinary entomology; yellow fever; zoonoses, arthropod borne.]

Russell, R.C., Otranto, D.P. & Wall, R.L. (2013) *Encyclopedia of Medical and Veterinary Entomology*. CAB International, Wallingford.

Villet, M.H. (2011) African carrion ecosystems and their insect communities in relation to forensic entomology. *Pest Technology* **5**(1), 1–15.

Chapter 16

PEST MANAGEMENT

Biological control of aphids by coccinellid beetles. (After Burton & Burton 1975.)

Insects become pests when they conflict with our welfare, aesthetics or profits. For example, otherwise innocuous insects can provoke severe allergic reactions in sensitized people, and reduction or loss of food-plant yield is a universal result of insect-feeding activities and pathogen transmission. Pests thus have no particular ecological significance but are defined from a purely anthropocentric point of view. Insects may be pests of people either directly through disease transmission (Chapter 15), or indirectly by affecting our domestic animals, cultivated plants, or timber reserves. From a conservation perspective, introduced insects become pests when they displace native species, often with ensuing effects on other non-insect species in the community. Some introduced and behaviourally dominant ants, such as the big-headed ant, *Pheidole megacephala*, and the Argentine ant, *Linepithema humile*, impact negatively on native biodiversity on many islands, including those of the tropical Pacific (see Box 1.3). Honey bees (*Apis mellifera*) outside their native range form feral nests and, although they are generalists, may out-compete local insects. Native insects usually are efficient pollinators of a smaller range of native plants than are honey bees, and their loss may lead to reduced seed set. Research on insect pests relevant to conservation biology is increasing, but remains modest compared to a vast literature on pests of our crops, garden plants, and forest trees.

In this chapter, we deal predominantly with the occurrence and control of insect pests of agriculture, including horticulture and silviculture, and with the management of insects of medical and veterinary importance. Many of the topics discussed are relevant also to urban entomology – the study of insects and arachnids that affect people, their pets and their property in urban settings (e.g. nuisance flies, termites, wood-boring beetles, fleas and garden pests). We commence with a discussion of what constitutes a pest, how damage levels are assessed, and why insects become pests. Next, the effects of insecticides and problems of insecticide resistance are considered, prior to an overview of integrated pest management (IPM). The remainder of the chapter discusses the principles and methods of management applied in IPM, namely: (i) chemical control, including insect growth regulators, neuropeptides and the rapidly expanding use of neonicotinoid insecticides; (ii) biological control using natural enemies (such as the coccinellid beetles shown eating aphids in the vignette at the beginning of this chapter) and microorganisms; (iii) host-plant resistance; (iv) mechanical, physical, and cultural control; (v) the use of attractants such as pheromones; and finally (vi) genetic control of insect pests. Seven boxes cover topics of special interest, namely exotic crop pests in the United States (Box 16.1), *Bemisia* whiteflies (Box 16.2), the cottony-cushion scale *Icerya purchasi* (Box 16.3), neonicotinoid insecticides (Box 16.4), the cassava mealybug *Phenacoccus manihoti* (Box 16.5), the glassy-winged sharpshooter *Homalodisca vitripennis* (Box 16.6), and the Colorado potato beetle *Leptinotarsa decemlineata* (Box 16.7). A more comprehensive list than for most other chapters is provided as further reading because of the importance and breadth of topics covered in this chapter.

16.1 INSECTS AS PESTS

16.1.1 Assessment of pest status

The pest status of an insect population depends on the abundance of individuals, as well as the type of nuisance or injury that the insects inflict. Injury is the usually deleterious effect of insect activities (mostly feeding) on host physiology, whereas damage is the measurable loss of host usefulness, such as yield quality or quantity, or aesthetics. Host injury (or insect number used as an injury estimate) does not necessarily inflict detectable damage, and even if damage occurs it may not result in appreciable economic loss. Sometimes, however, the damage caused by even a few individual insects is unacceptable, as in fruit infested by codling moth or fruit fly. Other insects must reach high or plague densities before becoming pests, as in locusts feeding on pastures. Most plants tolerate considerable leaf or root injury without significant loss of vigour. Unless these plant parts are harvested (e.g. leaf or root vegetables) or are the reason for sale (e.g. indoor plants), certain levels of insect feeding on these parts should be more tolerable than for fruit, which most consumers wish to be blemish-free. Often the effects of insect feeding may be merely cosmetic (such as small

The Insects: An Outline of Entomology, Fifth Edition. P.J. Gullan and P.S. Cranston.

Companion Website: www.wiley.com/go/gullan/insects

marks on the fruit surface), and consumer education is more desirable than expensive controls. As market competition demands high standards of appearance for food and other commodities, assessments of pest status often require socio-economic as much as biological judgements.

Pre-emptive measures to counter the threat of arrival of particular novel insect pests are sometimes taken. Generally, however, control becomes economic only when insect density or abundance cause (or are expected to cause if uncontrolled) financial loss of productivity or marketability greater than the costs of control. Quantitative measures of insect density (section 13.4) allow assessment of the pest status of different insect species associated with particular agricultural crops. In each case, an **economic injury level (EIL)** is determined as being the pest density at which the loss caused by the pest equals in value the cost of available control measures or, in other words, the lowest population density that will cause economic damage. The formula for calculating the EIL includes four factors:

1 costs of control;
2 market value of the crop;
3 yield loss attributable to a unit number of insects;
4 effectiveness of the control;
and is as follows:

$$\mathrm{EIL} = C/VDK$$

in which EIL is pest number per production unit (e.g. insects ha^{-1}), C is cost of control measure(s) per production unit (e.g. \$ ha^{-1}), V is market value per unit of product (e.g. \$ kg^{-1}), D is yield loss per unit number of insects (e.g. kg reduction of crop per n insects), and K is the proportionate reduction of insect population caused by control measures.

The calculated EIL will not be the same for different pest species on the same crop, or for a particular insect pest on different crops. The EIL also may vary depending on environmental conditions, such as soil type or rainfall, as these can affect plant vigour and compensatory growth. Control measures normally are instigated before the pest density reaches the EIL, as there may be a time lag before the measures become effective. The density at which control measures should be applied to prevent an increasing pest population from attaining the EIL is referred to as the **economic threshold (ET)** (or an "action threshold"). Although the ET is defined in terms of population density, it actually represents the time for instigation of control measures. It is set explicitly at a different level from the EIL and thus is predictive, with pest numbers being used as an index of the time when economic damage will occur.

Insect pests may be described as being one of the following:

- Non-economic, if their populations are never above the EIL (Fig. 16.1a).
- Occasional pests, if their population densities exceed the EIL only under special circumstances (Fig. 16.1b), such as atypical weather or inappropriate use of **insecticides**.
- Perennial pests, if the general equilibrium population density of the pest is close to the ET so that pest population density reaches the EIL frequently (Fig. 16.1c).
- Severe or key pests, if their numbers (in the absence of controls) always are higher than the EIL (Fig. 16.1d). Severe pests must be controlled if the crop is to be grown profitably.

The EIL fails to consider the influence of variable external factors, including the role of natural enemies, resistance to insecticides, and the effects of control measures in adjoining fields or plots. Nevertheless, the virtue of the EIL is its simplicity, with management depending on the availability of decision rules that can be comprehended and implemented with relative ease. The concept of the EIL was developed primarily as a means for more sensible use of insecticides, and its application is confined largely to situations in which control measures are discrete and curative, i.e. chemical or microbial insecticides. Often EILs and ETs are difficult or impossible to apply, due to the complexity of many agroecosystems and the geographic variability of pest problems. More-complex models and dynamic thresholds are needed, but these require years of field research.

The discussion above applies principally to insects that directly damage an agricultural crop. For forest pests, estimation of almost all of the components of the EIL is difficult or impossible, and EILs are relevant only to short-term forest products such as Christmas trees. Furthermore, if insects are pests because they can transmit (vector) diseases of animals or plants (e.g. the Asian citrus psyllid that transmits the disease "citrus greening"; see Box 16.1), then the ET may be their first appearance. The threat of a virus affecting crops or livestock and spreading via an insect vector requires constant vigilance for the appearance of the vector and the presence of the virus. With the first occurrence

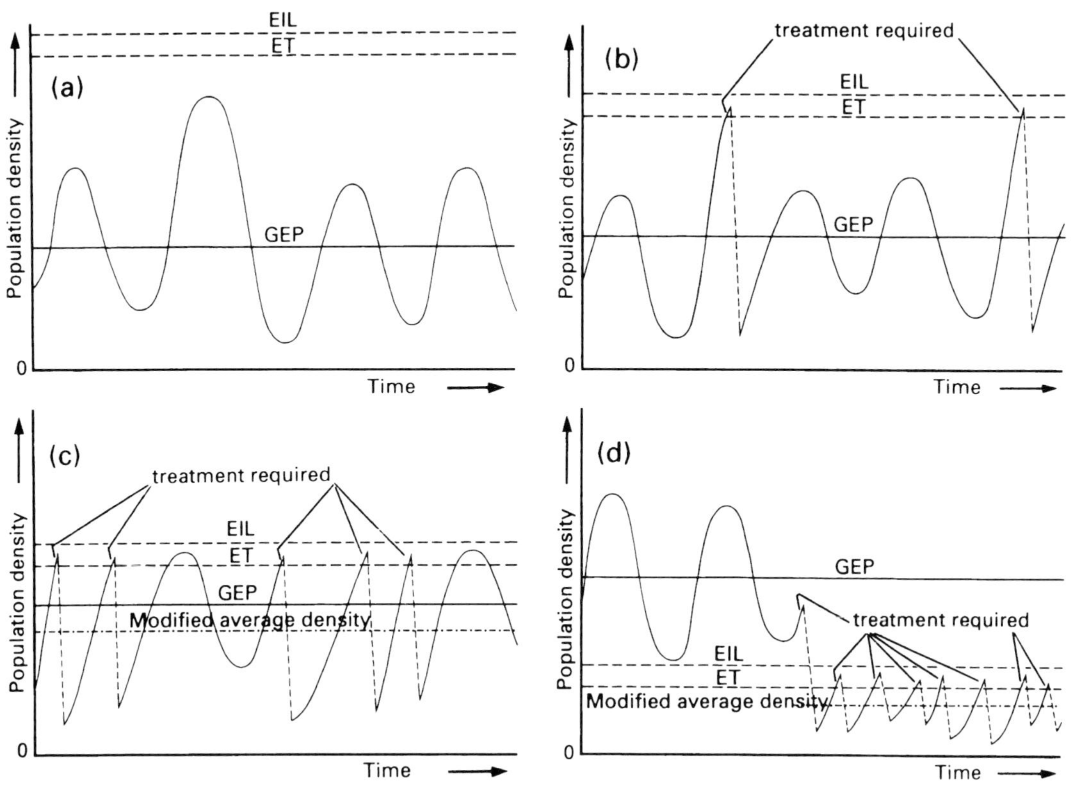

Fig. 16.1 Schematic graphs of the fluctuations of theoretical insect populations in relation to their general equilibrium position (GEP), economic threshold (ET) and economic injury level (EIL). From comparison of the GEP with the ET and EIL, insect populations can be classified as: (a) non-economic pests if population densities never exceed the ET or EIL; (b) occasional pests if population densities exceed the ET and EIL only under special circumstances; (c) perennial pests if the GEP is close to the ET so that the ET and EIL are exceeded frequently; or (d) severe or key pests if population densities always are higher than the ET and EIL. In practice, as indicated here, control measures are instigated before the EIL is reached. (After Stern *et al.* 1959.)

of either vector or disease symptoms, precautions may need to be taken. For economically very serious disease, and often in human health, precautions are taken before any ET is reached, and insect vector and virus population monitoring and modelling is used to estimate when pre-emptive control is required. Calculations such as the **vectorial capacity**, referred to in Chapter 15, are important in allowing decisions concerning the need and appropriate timing for pre-emptive control measures. However, in human insect-borne disease, such rationales often are replaced by socio-economic ones, in which levels of vector insects that are tolerated in less-developed countries or rural areas are perceived as requiring action in developed countries or in urban communities.

A limitation of the EIL is its unsuitability for multiple pests, as calculations become complicated. However, if injuries from different pests produce the same type of damage, or if effects of different injuries are additive rather than interactive, then the EIL and ET may still apply. The ability to make management decisions for a pest complex (many pests in one crop) is an important part of integrated pest management (IPM) (section 16.3).

16.1.2 Why insects become pests

Insects may become pests for one or more reasons. First, some previously harmless insects become pests

Box 16.1 Exotic insect pests of crops in the United States

Global commerce ("free trade") has brought with it accidental passengers, including both potential and actual pestilential insects of our crops and ornamental plants (see section 17.4). The inevitable newly arrived and established pests must be surveyed, and control measures planned and executed. In this box, we discuss one long-standing and two emergent insect threats to agriculture in the United States as examples of this issue.

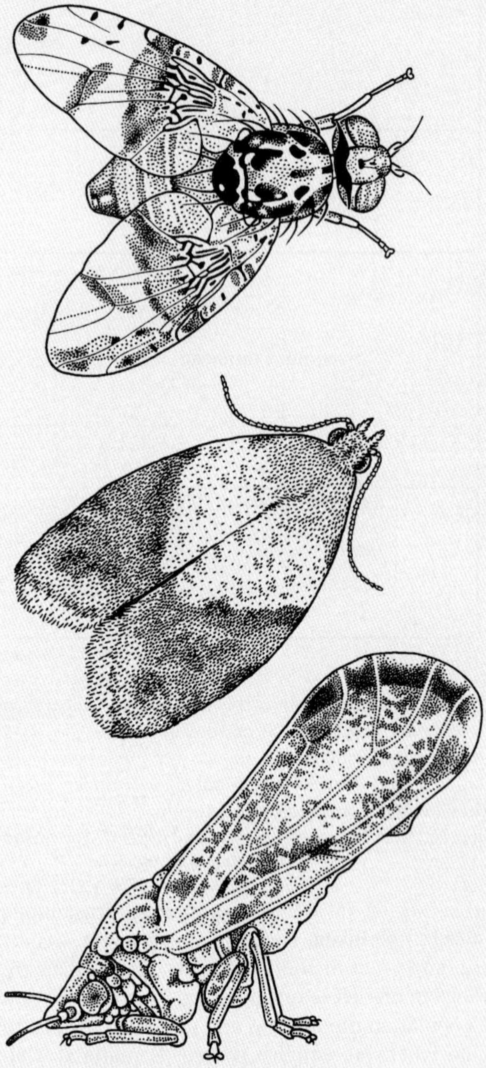

Tephritid fruit flies

Fruit flies (family Tephritidae) include some of the most troublesome agricultural pests, causing actual or potential damage to many commercial horticultural products by their larval development in produce (section 11.2.3). Economic damage to growers comes not only from crop losses, but also loss of export income from quarantine restrictions imposed by importer countries lacking the pests. Although there are native fruit-damaging tephritid flies in the United States, most problematic species for agriculture come from elsewhere. The situation is particularly bad in Hawai'i. Here, an invasion sequence of melon fly (*Bactrocera cucurbitae*) in 1895, Mediterranean fruit fly or "medfly" (*Ceratitis capitata*) in 1907, oriental fruit fly (*Bactrocera dorsalis*) in 1945 after World War II, and Malaysian fruit fly (*Bactrocera latifrons*) in 1983 has devastated a diverse and valuable tropical agricultural sector. As a trade centre, Hawai'i may act in turn as a potential source for such pests onwards elsewhere around the Pacific, including to mainland United States. In California, from 1954 to 2012, 17 exotic fruit-fly species have been detected, and over 240 eradication projects have been undertaken. Most detections have involved medfly, oriental fruit fly and Mexican fruit fly (*Anastrepha ludens*). The medfly (shown in the top figure) potentially is one of the most destructive pests known to agriculture, since it can attack over 250 species of fruits and vegetables. Although showing a preference for soft, fleshy fruits like peaches, apricots and cherries, almost any fruit and most vegetables raised in the temperate to subtropical United States can serve as host to the developing larvae.

Adult or larval medflies are found in most years by the California state-wide monitoring programmes, and eradication of outbreaks involves SIT (sterile insect technique), bait sprays, fruit stripping and increased trapping. However, there is controversy over whether the species actually is eliminated, or is maintained at populations below detection levels. The distinction is important: estimated annual losses of US$1.3 to $1.8 billion could be incurred in agricultural trade were this pest to become (or be declared to have become) a permanent presence in

California. Recent analyses of historical capture patterns and modelling the probability of recapture of fruit flies in California suggest that some species, including medfly, Mexican fruit fly and oriental fruit fly, are established and widespread. In spite of these findings, the high phytosanitary standards practised in California should ensure that its trading partners will continue to classify its exports as risk-free in terms of fruit flies. However, should the state stop the funding for monitoring for tephritids and elimination of detected infestations, the trade consequences will be dire.

Light brown apple moth (LBAM)

The light brown apple moth, *Epiphyas postvittana* (Lepidoptera: Tortricidae) (shown in the middle figure), is a leaf-roller native to Australia, where its larvae are generalist herbivores that feed on a diversity of dicotyledonous plants, including both natives and commercial crops. In New Zealand, where it is well established, LBAM feeds on most fruit crops, vegetables and ornamentals, both outdoors and in greenhouses. Larvae damage fruit and foliage, with later instars attaching leaves to fruit by silk webbing, beneath which they graze the fruit surface, causing aesthetic damage. In contrast, in Hawai'i where LBAM has been present for more than a century, horticultural damage is modest, and in the United Kingdom the species has been present without major economic consequences for 70 years. Assessments of pest risk for LBAM in the United States predicted high likelihood of establishment, with commensurate severe consequences of establishment for US agricultural and natural ecosystems, including financial losses due to quarantine bans that would be imposed by produce-importing countries such as Japan. Seemingly inevitably, in 2007 the species was recognized in California (by one of the very few lepidopterists skilled enough to do so). Given its known broad phytophagy and the value of horticulture to the state, control efforts were ramped up quickly. Quarantine, nursery inspections and control measures derived from the New Zealand experience, based on synthesized LBAM pheromones, were put in place to attempt to eradicate the pest. Unfortunately, the species clearly had been present already for some time, as it was found in numerous counties around the Bay Area of northern and central California. Extensive over-flying of population centres for aerial spraying of pheromone engendered public alarm and concerns for the "clean, green, organic" image of the state, amidst disputes about just how damaging LBAM would be. By 2012, the moth had spread to nearly 20, mostly coastal, counties in California but, despite predictions, it has not reached outbreak levels, probably due to resident natural enemies, especially generalist wasp parasitoids that move to LBAM from other moth hosts. Furthermore, its further spread may be restricted by climatic factors, particularly temperature and aridity.

Asian citrus psyllid and huanglongbing (HLB)(citrus greening)

The plant disease known as huanglongbing (HLB or yellow dragon disease) or citrus greening in the United States, and under names such as leaf mottle yellows, citrus chlorosis or citrus dieback in its original Asian range, is impacting citrus crops in the United States and elsewhere. The pathogen, a motile phloem-limited bacterium *Candidatus* Liberibacter asiaticus (Alphaproteobacteria), is vectored by the Asian citrus psyllid, *Diaphorina citri* (Hemiptera: Psyllidae) (shown in the bottom figure). This psyllid also can vector closely related bacteria and, in Africa, the African or 2-spotted citrus psyllid, *Trioza erytreae*, vectors *Candidatus* Liberibacter africanus. The disease, as the common names imply, affects citrus fruit and leaf colour, reduces quality, flavour and production (as does the black sooty mould growing on the sugars of the honeydew-producing sapsucker). Huanglongbing causes citrus and closely related rutaceous trees, such as orange jasmine, to decline and die. The disease was described first in 1929, reported in China in 1943, and the African species observed in 1947. The Asian citrus psyllid was found first in Brazil in the 1940s, and spread to Guam, Puerto Rico and Hawai'i, and these areas were quarantined for citrus. Nevertheless, this vector of the disease reached Florida in 1998 and now infests most citrus-growing areas of the United States, as well as much of South America and the Caribbean. The disease itself started to affect fruit quality in Florida citrus groves in 2005, leading to reduced fruit yield and size, with thick peel that retains some of the unripe green colour (hence "citrus greening"). Management of the psyllid in the United States includes heavy reliance on insecticides (leading to some field-evolved resistance) as well as use of natural enemies, particularly the ectoparasitoid wasp *Tamarixia radiata* (Eulophidae) that was introduced to Florida. Another primary parasitoid of *D. citri*, the endoparasitoid wasp *Diaphorencyrtus aligarhensis* (Encyrtidae), has failed to establish in Florida, and does not compete well where *T. radiata* is present. Infestation rates in Florida are lower than expected, but the eulophid spread quickly, including making its way accidentally to Texas.

Box 16.2 *Bemisia tabaci* – a pest species complex

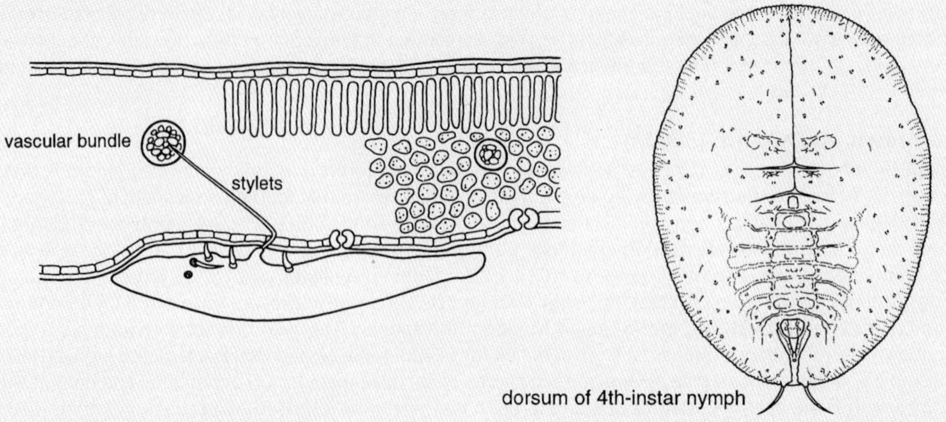

Bemisia tabaci, often called the tobacco or sweetpotato whitefly, is a polyphagous and predominantly tropical–subtropical whitefly (Hemiptera: Aleyrodidae) that feeds on numerous fibre (particularly cotton), food and ornamental plants. Nymphs suck phloem sap from minor veins (as illustrated diagrammatically on the left of the figure, after Cohen *et al.* 1998). Their thread-like mouthparts (section 11.2.4; Fig. 11.4) must contact a suitable vascular bundle in order for the insects to feed successfully. The whiteflies cause plant damage by inducing physiological changes in some hosts, such as irregular ripening in tomato and silverleafing in squash and zucchini (courgettes), by fouling with honeydew and subsequent sooty mould growth, and by the transmission of numerous begomoviruses (Geminiviridae) that cause plant disease. It appears that the primary endosymbionts of *B. tabaci* (Gammaproteobacteria: *Candidatus* Portiera aleyrodidarum) mediate the transmission of begomoviruses by these whiteflies, similar to the enhanced luteovirus transmission due to the *Buchnera* endosymbionts of aphids (section 3.6.5).

Infestations of *B. tabaci* have increased in severity since the early 1980s, owing to intensive continuous cropping with heavy reliance on insecticides, and the related spread of what was considered once to be a virulent form of the insect but now has been shown to be a morphologically indistinguishable sibling species. The area of origin of this pest, often called *B. tabaci* biotype B, is the Middle East–Asia Minor region. Since the 1990s, certain entomologists (especially in the United States) have recognized the severe pest as a separate species, *B. argentifolii*, the silverleaf whitefly (the fourth-instar nymph or "puparium" is depicted on the right, after Bellows *et al.* 1994), so-named because of the leaf symptoms it causes in squash and zucchini. *Bemisia argentifolii* exhibits minor and labile cuticular differences from other forms or biotypes of *B. tabaci*, but no reliable morphological features have been found to separate them. However, clear allozyme, mitochondrial and other genetic information allow recognition of many segregates of *B. tabaci*. Some segregates show variable reproductive incompatibility, as shown by crossing experiments.

The sudden appearance and spread of this apparently new pest, *B. tabaci* biotype B or *B. argentifolii* or the "Middle East–Asia Minor 1" genetic group, highlights the importance of recognizing fine taxonomic and biological differences among economically significant insect taxa. Recent analyses, mostly of *cytochrome c oxidase* subunit I sequences, suggest that *B. tabaci* is a sibling species complex of more than 35 genetically distinct entities, some of which include more than one of the almost 40 recognized biotypes. A convincing argument has been made to abandon the misleading classification into biotypes, and instead recognize the genetically and probably reproductively distinct entities as morphologically indistinguishable species.

It is possible that strong selection, resulting from heavy insecticide use, may select for particular species of whitefly (or their bacterial symbionts) that are more resistant to the chemicals. Effective biological control of *Bemisia* whiteflies is possible using host-specific parasitoid wasps, such as *Encarsia* and *Eretmocerus* species (Aphelinidae). However, the intensive and frequent application of broad-spectrum insecticides adversely affects biological control. Even the so-called B biotype can be controlled if insecticide use is reduced.

after their accidental (or intentional) introduction to areas outside their native range (e.g. the palm weevils discussed in Box 1.5), where they escape from the controlling influence of their natural enemies. Such range extensions have allowed many previously innocuous phytophagous insects to flourish as pests, usually following the deliberate spread of their host plants through human cultivation. Second, an insect may be harmless until it becomes a vector of a plant or animal (including human) pathogen. For example, mosquito vectors of malaria and filariasis occur in the United States, England and Australia, but the diseases are currently absent. Third, native insects may become pests if they move from native plants onto introduced ones; such host switching is common for polyphagous and oligophagous insects. For example, the oligophagous Colorado potato beetle switched from other solanaceous host plants to potato, *Solanum tuberosum*, during the 19th century (Box 16.7), and some polyphagous larvae of *Helicoverpa* and *Heliothis* (Lepidoptera: Noctuidae) have become serious pests of cultivated cotton and other crops within the native range of the moths. Other polyphagous plant-feeding insects, such as the light brown apple moth and various tephretid fruit flies (Box 16.1), which made host shifts to cultivated plants in their native range, have become serious pests in other countries following accidental introductions. New pests are appearing continually, and some well-known ones are spreading concomitant with expanding global trade (section 17. 4).

A fourth, related, problem is that the simplified, virtually monocultural, ecosystems in which our food crops and forest trees are grown and our livestock are raised create dense aggregations of predictably available resources that encourage the proliferation of specialist and some generalist insects. Certainly, the pest status of many native noctuid caterpillars is elevated by the provision of abundant food resources. Moreover, natural enemies of pest insects generally require more diverse habitat or food resources and are discouraged from agro-monocultures. Fifth, in addition to large-scale monocultures, other farming or cultivating methods can lead to previously benign species or minor pests becoming major pests. Cultural practices such as continuous cultivation without a fallow period allow build-up of insect pest numbers. The inappropriate or prolonged use of insecticides can eliminate natural enemies of phytophagous insects, while inadvertently selecting for insecticide resistance in the latter. Released from natural enemies, other previously non-pest species sometimes increase in numbers until they reach ETs. These problems of insecticide use are discussed in more detail in section 16.2.

Sometimes, the primary reason why a minor nuisance insect becomes a serious pest is unclear, at least until extensive research has been conducted. Such a change in status may occur suddenly, and none of the conventional explanations given above may appear totally satisfactory either alone or in combination. An example is the rise to notoriety of the silverleaf whitefly, which is variously known as *Bemisia tabaci* biotype B, *B. argentifolii*, or the Middle East–Asia Minor 1 group (see Box 16.2). Although this whitefly now appears to be a genetically distinct species that has been transported accidentally around the world, its pest status may be linked more to its ability to induce deleterious physiological changes in host plants than to more efficient transmission of viruses or lack of natural enemies in areas to which it has been introduced.

16.2 THE EFFECTS OF INSECTICIDES

The chemical insecticides developed during and after World War II initially were effective and cheap. Farmers came to rely on the new chemical methods of pest control, which rapidly replaced traditional forms of chemical, cultural and biological control. The 1950s and 1960s were times of an insecticide boom, but use continued to rise and insecticide application is still the single main pest-control tactic employed today. Although pest populations are suppressed by insecticide use, undesirable effects include the following:

1 Selection for insects that are genetically resistant to the chemicals (section 16.2.1).

2 Destruction of non-target organisms, including pollinators, the natural enemies of the pests, and soil arthropods.

3 **Pest resurgence** – as a consequence of effects **1** and **2**, a dramatic increase in numbers of the targeted pest(s) can occur (e.g. severe outbreaks of cottony-cushion scale as a result of dichloro-diphenyl-trichloroethane (DDT) use in California in the 1940s (Box 16.3)). If the natural enemies recover much more slowly than the pest population, the latter can exceed levels found prior to insecticide treatment.

4 **Secondary pest outbreak** – a combination of suppression of the original target pest and effects **1** and **2** can lead to insect species previously not considered

pests being released from control and becoming major pests.
5 Adverse environmental effects, resulting in contamination of soils, water systems, and the produce itself with chemicals that accumulate biologically (especially in vertebrates) as the result of biomagnification through food chains.
6 Dangers to human health either directly from the handling and consumption of insecticides or indirectly via exposure to environmental sources.

Box 16.3 The cottony-cushion scale

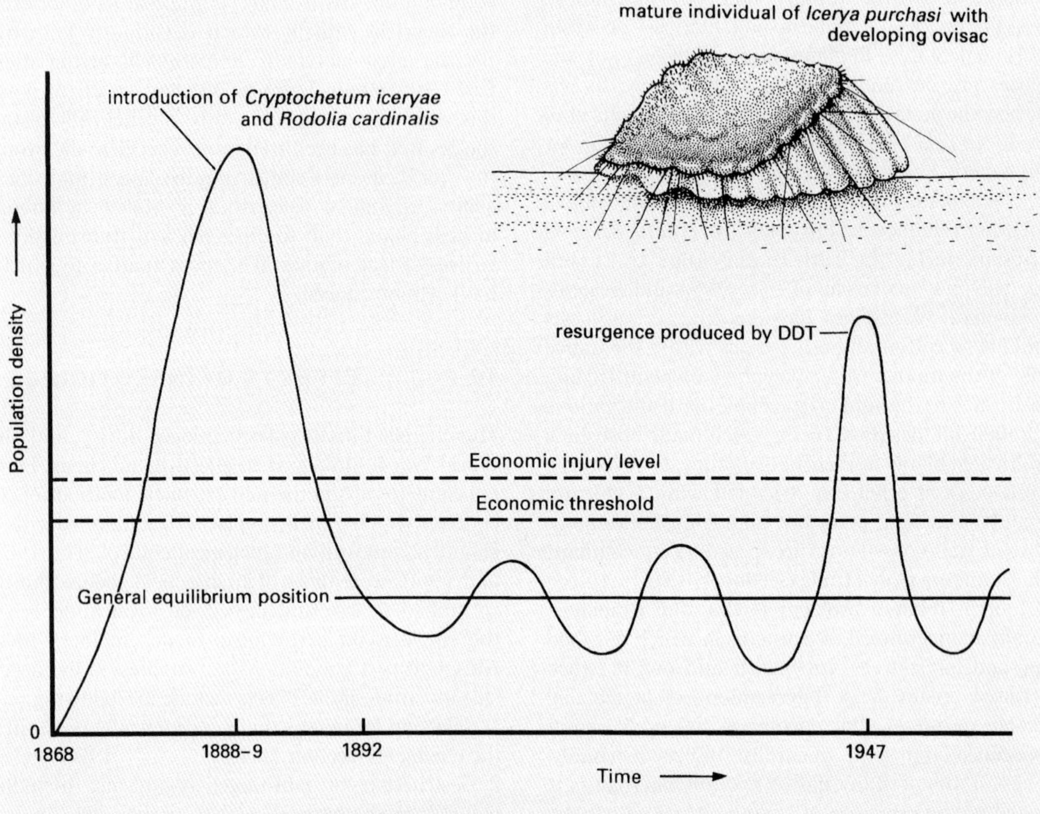

An example of a spectacularly successful classical biological control system is the control of infestations of the cottony-cushion scale, *Icerya purchasi* (Hemiptera: Monophlebidae; see Plate 8b), in Californian citrus orchards from 1889 onwards, as illustrated in the accompanying graph (after Stern *et al*. 1959). Control has been interrupted only by DDT use, which killed natural enemies and allowed resurgence of cottony-cushion scale.

The hermaphroditic, self-fertilizing adult of this scale insect produces a very characteristic fluted white ovisac (see inset on graph and Plate 8b), under which several hundred eggs are laid. This mode of reproduction, in which a single immature individual can establish a new infestation, combined with polyphagy and capacity for multivoltinism in warm climates, makes the cottony-cushion scale a potentially serious pest. In Australia, the country of origin of the cottony-cushion scale, populations are kept in check by natural enemies, especially ladybird beetles (Coleoptera: Coccinellidae) and parasitic flies (Diptera: Cryptochetidae).

Cottony-cushion scale was first noticed in the United States in about 1868 on a wattle (*Acacia*) growing in a park in northern California. By 1886, it was devastating the new and expanding citrus industry in southern California.

Initially, the native home of this pest was unknown, but correspondence between entomologists in the United States, Australia and New Zealand identified Australia as the source. The impetus for the introduction of exotic natural enemies came from C.V. Riley, Chief of the Division of Entomology of the US Department of Agriculture. He arranged for A. Koebele to collect natural enemies of the pest in Australia and New Zealand, from 1888 to 1889, and ship them to D.W. Coquillett for rearing and release in Californian orchards. Koebele obtained many cottony-cushion scales infected with flies of *Cryptochetum iceryae* and also coccinellids of *Rodolia cardinalis*, the vedalia ladybird (or ladybug). Mortality during several shipments was high, and only about 500 vedalia beetles arrived alive in the United States; these were bred and distributed to all Californian citrus growers, with outstanding results. The vedalia beetles ate their way through infestations of cottony-cushion scale, the citrus industry was saved, and biological control became popular. The parasitic fly was largely forgotten in these early days of enthusiasm for coccinellid predators. Thousands of flies were imported as a result of Koebele's collections, but establishment from this source is doubtful. Perhaps the major or only source of the present populations of *C. iceryae* in California was a batch sent in late 1887 by F. Crawford of Adelaide, Australia, to W.G. Klee, the California State Inspector of Fruit Pests, who made releases near San Francisco in early 1888, before Koebele ever visited Australia.

Today, both *R. cardinalis* and *C. iceryae* control populations of *I. purchasi* in California, with the beetle dominant in the hot, dry inland citrus areas and the fly most important in the cooler coastal regions; interspecific competition can occur if conditions are suitable for both species. Furthermore, the vedalia beetle, and to a lesser extent the fly, have been introduced successfully into many countries worldwide wherever *I. purchasi* has become a pest, including the Galapagos Islands where native plants were threatened. Both predator and parasitoid have proved to be effective regulators of cottony-cushion scale numbers, presumably owing to their specificity and efficient searching ability, aided by the limited dispersal and aggregative behaviour of their target scale insect. Unfortunately, few subsequent biological control systems involving coccinellids have enjoyed the same success.

Despite increased insecticide use, damage by insect pests has increased; for example, insecticide use in the United States increased 10-fold from about 1950 to 1985, whilst the proportion of crops lost to insects roughly doubled (from 7% to 13%) during the same period. Such figures do not mean that insecticides have not controlled insects, because non-resistant insects clearly are killed by chemical poisons. Rather, an array of factors accounts for this imbalance between pest problems and control measures. Human trade has accelerated the spread of pests to areas outside the ranges of their natural enemies. Selection for high-yield crops often inadvertently has resulted in susceptibility to insect pests. Extensive monocultures are commonplace, with reduction in sanitation and other cultural practices such as crop rotation. Finally, aggressive commercial marketing of chemical insecticides has led to their inappropriate use, perhaps especially in developing countries.

16.2.1 Insecticide resistance

Insecticide **resistance** is the result of selection of individuals that are predisposed genetically to survive exposure to an insecticide. Tolerance, the ability of an individual to survive an insecticide, implies nothing about the basis of survival. Field-evolved resistance is the genetically based decrease in susceptibility of a population to an insecticide (a toxin) caused by exposure to that insecticide in the field. Over the past few decades, more than 700 species of arthropod pests have developed resistance to one or more insecticides, as well as resistance to toxins that have been genetically engineered into major crop plants (see section 16.6.1).

The tobacco or silverleaf whitefly (Box 16.2), the Colorado potato beetle (Box 16.7), and the diamondback moth (see discussion of *Bacillus thuringiensis* (Bt) in section 16.5.2) are resistant to virtually all chemicals available for control. Chemically based pest control of these and many other pests may soon become virtually ineffectual because many show cross- or multiple resistance. **Cross-resistance** is the phenomenon of a resistance mechanism for one insecticide giving tolerance to another. **Multiple resistance** is the occurrence in a single insect population of more than one defence mechanism against a given compound, or the resistance to several compounds due to expressing multiple resistance mechanisms. The difficulty of distinguishing cross-resistance from multiple resistance presents a major challenge to research on insecticide resistance. The gene loci involved in resistance may interact in a synergistic, antagonistic or additive manner. Mechanisms of insecticide resistance include:

- increased behavioural avoidance, as some insecticides, such as neem and pyrethroids, can repel insects;
- physiological changes, such as sequestration (deposition of toxic chemicals in specialized tissues), reduced cuticular permeability (penetration), or accelerated excretion;
- biochemical detoxification (called **metabolic resistance**) mediated by specialized enzymes, such as esterases, mono-oxygenases and glutathione S-transferases;
- increased tolerance as a result of decreased sensitivity to the presence of the insecticide at its target site (called **target-site resistance**).

The tobacco budworm, *Heliothis virescens* (Lepidoptera: Noctuidae), a major pest of cotton in the United States, exhibits behavioural, penetration, metabolic and target-site resistance.

Phytophagous insects, especially polyphagous ones, frequently develop resistance more rapidly than their natural enemies. Polyphagous herbivores may be preadapted to evolve insecticide resistance because they have general detoxifying mechanisms for secondary compounds encountered among their host plants. Certainly, detoxification of insecticidal chemicals is the most common form of insecticide resistance. Furthermore, insects that chew plants or consume non-vascular cell contents appear to have a greater ability to evolve pesticide resistance compared with phloem- and xylem-feeding species. Resistance has developed also under field conditions in some arthropod natural enemies (e.g. some lacewings, parasitic wasps and predatory mites), although few have been tested. Intraspecific variability in insecticide tolerances has been found among certain populations subjected to differing insecticide doses.

Insecticide resistance in the field is based on relatively few or single genes (monogenic resistance), i.e. owing to allelic variants at just one or two loci. Field applications of chemicals designed to kill all individuals lead to rapid evolution of resistance because strong selection favours novel variants, such as a very rare allele for resistance present at a single locus. In contrast, laboratory selection often is weaker, producing polygenic resistance. Single-gene insecticide resistance could be due also to the very specific modes of action of certain insecticides, which allow small changes at the target site to confer resistance.

Management of insecticide resistance requires a programme of controlled use of chemicals with the primary goals of: (i) avoiding or (ii) slowing the development of resistance in pest populations; (iii) causing resistant populations to revert to more susceptible levels; and/or (iv) fostering resistance in selected natural enemies. The tactics for resistance management can involve maintaining reservoirs of susceptible pest insects (either in refuges or by immigration from untreated areas) to promote dilution of any resistant genes, varying the dose or frequency of insecticide applications, using less-persistent chemicals, and/or applying insecticides as a rotation or sequence of different chemicals or as a mixture. The optimal strategy for retarding the evolution of resistance is to use insecticides only when control by natural enemies fails to curtail economic damage. Furthermore, resistance monitoring should be an integral component of management, as it allows the anticipation of problems and assessment of the effectiveness of operational management tactics.

Recognition of the problems discussed above, cost of insecticides, and also a strong consumer reaction to environmentally damaging agronomic practices and chemical contamination of produce have led to the current development of alternative pest-control methods. In some countries and for certain crops, chemical controls increasingly are being integrated with, and sometimes replaced by, other methods.

16.3 INTEGRATED PEST MANAGEMENT

Historically, **integrated pest management (IPM)** was promoted first during the 1960s as a result of the failure of chemical insecticides, notably in cotton production, which in some regions required at least 12 sprayings per crop. IPM philosophy is to limit economic damage to the crop and simultaneously to minimize adverse effects on non-target organisms in the crop and surrounding environment and on consumers of the produce. Successful IPM requires a thorough knowledge of the biology of the pest insects, their natural enemies, and the crop to allow rational use of a variety of cultivation and control techniques under differing circumstances. If pesticides are applied as part of IPM, then economic or treatment thresholds should be used, and their effects on natural enemies monitored. The key concept in IPM is integration of (or compatibility among) pest-management tactics. The factors that regulate populations of insects (and other organisms) are varied and interrelated in complex ways. Thus, successful IPM requires an understanding of both population processes (e.g. growth and reproductive capabilities, competition, and effects of predation and parasitism) and the effects of environmental factors (e.g. weather, soil conditions, disturbances such as fire, and availability of water, nutrients and shelter), some of which are largely stochastic in nature and may have predictable or unpredictable effects on insect populations. The most advanced form of IPM also takes into consideration societal and environmental costs and benefits within an ecosystem context when making management decisions. Efforts are made to conserve the long-term health and productivity of the ecosystem, with a philosophy approaching that of organic farming. One of the rather few examples of this advanced IPM is insect pest management in tropical irrigated rice, in which there is co-ordinated training of farmers by other farmers, and field research involving local communities in implementing successful IPM. Worldwide, other functional IPM systems include the field crops of cotton, alfalfa (lucerne) and citrus in certain regions, and many greenhouse crops.

Despite the economic and environmental advantages of IPM, implementation of advanced IPM systems has been slow, even in many developed countries. Often what is called IPM is simply "integrated pesticide management" (sometimes called first-level IPM), with pest consultants monitoring crops to determine when to apply insecticides. Universal reasons for lack of adoption of advanced IPM include:

- lack of sufficient data on the ecology of many insect pests and their natural enemies;
- requirement for knowledge of EILs for each pest of each crop;
- requirement for interdisciplinary research in order to obtain the above information;
- risks of pest damage to crops associated with IPM strategies;
- apparent simplicity of total insecticidal control combined with the marketing pressures of pesticide companies;
- necessity of training farmers, agricultural extension officers, foresters and others in new principles and methods; and, most importantly,
- the dilemma of choosing between IPM activities for individual fields that provide the best short-term economic return and IPM activities that are better applied regionally and have longer-term benefits, including environmental ones that apply to more than the individual farm.

Successful IPM often requires extensive biological research. Such applied research is unlikely to be financed by many industrial companies because IPM may reduce their insecticide market. However, IPM does incorporate the use of chemical insecticides, albeit at a reduced level, although its main focus is the establishment of a variety of other methods of controlling insect pests. These usually involve modifying the insect's physical or biological environment or, more rarely, entail changing the genetic properties of the insect. Thus, the control measures that can be used in IPM include: insecticides, biological control, cultural control, plant resistance improvement, and techniques that interfere with the pest's physiology or reproduction, namely genetic (e.g. sterile insect technique; section 16.10), semiochemical (e.g. pheromone), and insect growth-regulator control methods. The remainder of this chapter discusses the various principles and methods of insect pest control that could be employed in IPM systems.

16.4 CHEMICAL CONTROL

Despite the hazards of conventional insecticides, some use is unavoidable. However, careful chemical choice and application can reduce ecological damage. Carefully timed suppressant doses can be delivered at vulnerable stages of the pest's life cycle or when a pest population is about to explode in numbers. Appropriate and efficient use requires a thorough knowledge of the pest's field biology and an appreciation of the differences among available insecticides.

An array of chemicals has been developed for the purposes of killing insects. These enter the insect body either by penetrating the cuticle, called contact action or dermal entry, by inhalation into the tracheal system, or by oral ingestion into the digestive system. Most contact poisons also act as stomach poisons if ingested by the insect, and toxic chemicals that are ingested by the insect after translocation through a host are referred to as **systemic insecticides**. Fumigants used for controlling insects are inhalation poisons. Some chemicals may act simultaneously as inhalation, contact and stomach poisons. Chemical insecticides generally have an acute effect, and their mode of action (i.e. method of causing death) is via the nervous system, either by inhibiting acetylcholinesterase (an essential enzyme for transmission of nerve impulses at synapses) or by acting directly on the nerve cells. Most synthetic insecticides (including pyrethroids) are nerve poisons. Other insecticidal chemicals affect the developmental or metabolic processes of insects, either by mimicking or interfering with the action of hormones, or by affecting the biochemistry of cuticle production. Included after the sections on insecticides and insect growth regulators are two subsections covering the potential use of neuropeptides and RNA interference in insect control. Both of these methods involve genetic engineering of plants or viruses as delivery systems.

16.4.1 Insecticides (chemical poisons)

Chemical insecticides may be synthetic or natural products. Natural plant-derived products, usually called botanical insecticides, include:

- alkaloids, including nicotine from tobacco;
- rotenone and other rotenoids from the tissues of legumes (Fabaceae);
- pyrethrins, derived from flowers of *Tanacetum cinerariifolium* (formerly in *Pyrethrum* and then *Chrysanthemum*);
- essential oils derived from aromatic plants;
- neem, i.e. extracts of the tree *Azadirachta indica* (Meliaceae). These have a long history of use as insecticides. The neem tree is renowned, especially in India, Pakistan and some areas of Africa, for its anti-insect properties. The abundance of neem trees in many developing countries means that resource-poor farmers can have access to non-toxic insecticides for controlling crop and stored-product pests.

Insecticidal alkaloids have been used since the 1600s, and pyrethrum since at least the early 1800s. Although nicotine-based insecticides have been phased out for reasons including high mammalian toxicity and limited insecticidal activity, the new-generation nicotinoids or neonicotinoids, which are synthetic pesticides modelled on natural nicotine, have a large market, in particular the systemic insecticide imidacloprid, which is used against a range of insect pests (see below and Box 16.4). Neonicotinoids selectively target the nicotinic acetylcholine receptors (nAChRs) in the insect central nervous system and cause paralysis and death, often within a few hours. Rotenoids are mitochondrial poisons that kill insects by respiratory failure, but they also poison fish, and must be kept out of waterways. Pyrethrins (and the structurally related synthetic pyrethroids) are especially effective against lepidopteran larvae, kill on contact even at low doses, and have low environmental persistence. An advantage of most pyrethrins and pyrethroids, and also neem derivatives, is their much lower mammalian and avian toxicity compared with synthetic insecticides, although pyrethroids are highly toxic to fish. A number of insect pests have developed resistance to pyrethroids. Extracts of neem seed kernels and leaves act as repellents, antifeedants and/or growth disruptants. The main active compound in kernels is azadirachtin (AZ), a limonoid. Neem derivatives can repel, prevent settling and/or inhibit oviposition, inhibit or reduce food intake, interfere with the regulation of growth (as discussed in section 16.4.2), as well as reduce the fecundity, longevity and vigour of adults. In recent years, there has been increasing interest in using essential oils as low-risk insecticides. These oils are produced usually by steam distillation of plant material, and contain mixtures of volatile terpenes and, to a lesser extent, phenolics that can have insecticidal, repellent and/or growth-reducing effects on insects. The essential oils act upon contact or inhalation and have neurotoxic effects on insects. Registration requirements have slowed the insecticidal use of essential oils, but orange oil is allowed in France for control of selected whitefly problems, and some other products are marketed in the United States.

Spinosyns form a unique type of insecticide based on metabolites of the naturally occurring soil bacteria, *Saccharopolyspora spinosa* and *S. pogona*. Currently, there are two kinds of commercial product: spinosad that contains a mixture of spinosyn A and spinosyn D, and spinetoram that is a mixture of two semi-synthetic spinosyn derviatives. The spinosyns affect the nervous system of insects, in a distinct but related way to that of neonicotinoids. Spinosad is relatively fast acting (death in 1–2 days) and kills insects on contact or after ingestion. Spinosad degrades rapidly on exposure to light, has low toxicity to mammals, slight to moderate toxicity to birds, fish and aquatic invertebrates, and may have reduced field toxicity to beneficial insects. Some field and laboratory studies have demonstrated insect resistance to spinosyns.

Box 16.4 Neonicotinoid insecticides

The tobacco plant *Nicotiana tabacum* was introduced from the Americas to Europe in the mid-16th century for smoking, but within a century its utility as an insecticide and insect repellent was well recognized. The plant family to which *Nicotiana* belongs, the nightshades (Solanaceae), contain many alkaloids, some of which are psychotropic, whereas others are variably toxic. Nicotine, a potent stimulant comprising up to 6% of the dry weight of tobacco, is inducible in plants as an anti-herbivore chemical defence. Until the 1950s, nicotine insecticide derived from tobacco industry waste was used widely as a crop dust or aqueous sulphate, with maximum annual production of some 2500 tons. Curiously, although derived directly from plants, nicotine has never been allowed in organic production.

Due to toxicity to animals (including humans), costs of production and availability of alternative synthetic insecticides, use of nicotine declined. However, its mode of action in binding to cholinergic receptors in the insect nervous system has been emulated in the biochemical synthesis of new nicotine-like chemicals, the neonicotinoids. These are argued to pose a lower threat to mammals and the environment, and are flexible in application, including water-soluble applications or seed coatings, with uptake by the plant as a systematic pesticide. That is, the insecticide is absorbed by the roots from soil or from foliar spray and becomes distributed throughout all plant tissues. Neonicotinoids often are applied as a coating to seeds prior to planting: in major production areas of the United States and Canada nearly all corn (maize) and soya bean seeds are so treated, irrespective of actual insect threat status. Benefits claimed are reduced damage from soil insects that feed on roots, above- and below-ground phloem-feeding bugs, and leaf and stem-eating beetles.

Since neonicotenoids became available about two decades ago, a range of such pesticides, especially imidacloprid, have come to dominate insect control in arable agro-ecosystems, horticulture and domestic garden applications. This period coincides with growing recognition of losses of beneficial insects, especially honey bees (see colony collapse disorder, CCD, Box 12.3), but including also bumble bees and other non-target insects that do not feed directly on treated plants. For many environmentalists, neonicotinoids pose threats comparable to those associated with DDT half-a-century ago, with widespread negative repercussions in ecosystems. For scientists (and most politicians) concerned with enhancing cheap food production for an ever-increasing global population, the risks are unproven, small and/or manageable.

Early approvals for these chemicals, for example by the US Environmental Protection Agency (EPA), seem to have been gained without adequate tests for effects of ultra-low doses on non-target organisms and various other sublethal effects – relying on industry studies alone. Recent experimental exposure of honey bees to neonicotinoid levels comparable to those found in nectar and pollen of plants visited in agro-landscapes induced disorientation, altered olfactory memory and caused other neurological symptoms seen in CCD. The decline in insects, for example, in the United Kingdom and much of Western Europe, is unarguable, as is the decline in many insectivorous birds – yet it is near impossible to tease apart the causes, since ongoing agricultural intensification and loss of habitat diversity is visible to all. It is in continental Europe that restrictions on neonicotinoid insecticides are most prevalent. After unilateral action by several countries, based largely on circumstantial evidence, the European Food Safety Authority produced a peer-reviewed report concerning safe use of three neonicotinoids. As a result, the European Commission recommended, as a precaution, to cease seed treatment, soil application (granules) and foliar treatment in crops attractive to bees, and a majority of member countries agreed. In the United States, the EPA is in the process of re-registering neonicotinoids, under pressure from a coalition of beekeepers, conservationists, sustainable agronomists, bird watchers and environmentally aware senators. The information gaps remain large, and the necessary experiments on honey bees with realistic (nanogram) quantities of these chemicals remain to be made.

Disconcertingly, studies of long-term fields planted with a succession of treated crops show that a high percentage of the chemicals applied are not taken up by target crop plants, but can accumulate in the soil and reach groundwater and running waters, where non-target insects are impacted. Plants (e.g. dandelions) close to treated crop fields acquire systematic doses and can pass these onwards to pollinators that avoid the treated crop. Organisms directly feeding on exposed treated seeds die, as do bees exposed to wind drift of neonicotinoid dust. In accordance with the precautionary principle, use of neonicotinoids ought to be restricted, as the European Union suggests, pending thorough, peer-reviewed, study by independent researchers.

The other major classes of insecticides have no natural analogues. These are the synthetic carbamates (e.g. aldicarb, carbaryl, carbofuran, methiocarb, methomyl, propoxur), organophosphates (or phosphate esters, e.g. chlorpyrifos, dichlorvos, dimethoate, malathion, parathion, phorate) and organochlorines (also called chlorinated hydrocarbons, e.g. aldrin, chlordane, DDT (**d**ichloro-**d**iphenyl-**t**richloroethane), dieldrin, endosulfan, gamma-hexachlorocyclohexane (lindane), heptachlor). Certain organochlorines (e.g. aldrin, chlordane, dieldrin, endosulfan and heptachlor) are known as cyclodienes because of their chemical structure. A relatively new class of insecticides is the phenylpyrazoles (or fiproles, with some similarities to DDT), of which the widely used fipronil, sold under brand names such as Frontline®, MaxForce®, Regent® and Termidor®, is registered for a variety of insecticidal uses.

Most synthetic insecticides are broad spectrum in action, i.e. they have non-specific killing action, and most act on the insect (and incidentally on the mammalian) nervous system. Organochlorines are stable chemicals and persistent in the environment, have a low solubility in water but a moderate solubility in organic solvents, and accumulate in mammalian body fat. Their use is banned in many countries and they are unsuitable for use in IPM. Organophosphates may be highly toxic to mammals but are not stored in fat and, being less environmentally damaging and non-persistent, are suitable for use in IPM. They usually kill insects by contact or upon ingestion, although some are systemic in action, being absorbed into the vascular system of plants so that they kill most phloem-feeding insects. Because they are non-persistent, their application must be timed carefully to ensure efficient kill of pests. Carbamates usually act by contact or stomach action, more rarely by systemic action, and they have short to medium persistence. Neonicotinoids, such as imidacloprid, thiamethoxam and clothianidin, are highly toxic to insects due to their blockage of nicotinic acetylcholine receptors (nAChRs), less toxic to mammals, and relatively non-persistent (probably less than a year, and less if exposed to light). Imidacloprid (marketed under trade names such as Admire®, Advantage®, Confidor®, Gaucho® and Merit®) kills insects on contact or by ingestion, has high acute and residual activity against sucking and some chewing insects, is translocated within plants, and can be used as a foliar spray, trunk injection, seed treatment or soil application. In addition to its agricultural and horticultural uses, imidacloprid is used widely in urban environments to control cockroaches, termites and fleas. The phenylpyrazole insecticide fipronil is a contact and stomach poison that acts as a potent inhibitor of gamma-aminobutyric acid (GABA) regulated chloride channels in neurons of insects, but is less potent in vertebrates. However, this poison and its degradates are moderately persistent and one photo-degradate (a breakdown product produced by exposure to light) appears to have an acute toxicity to mammals that is about 10 times that of fipronil itself. Although human and environmental health concerns are associated with its use, it is very effective in controlling many soil and foliar insects, for treating seed, and as a bait formulation to kill ants, vespid wasps, termites and cockroaches. Fipronil in baits is taken back to the nests of social insects by foraging workers and can kill the entire colony.

In addition to the chemical and physical properties of insecticides, their toxicity, persistence in the field, and method of application are influenced by how they are formulated. Formulation refers to what and how other substances are mixed with the active ingredient, and largely constrains the mode of application. Insecticides may be formulated in various ways, including as solutions or emulsions, as unwettable powders that can be dispersed in water, as dusts or granules (i.e. mixed with an inert carrier) or as gaseous fumigants. Formulation may include abrasives that damage the insect cuticle and/or baits that attract the insects (e.g. fipronil often is mixed with fishmeal bait to attract and poison pest ants and wasps). The same insecticide can be formulated in different ways according to the application requirements, such as aerial spraying of a crop versus domestic use.

There is a growing concern about the sublethal effects of pesticides on non-target organisms, especially plant pollinators (e.g. effects on honey-bee behaviour; see Box 12.3) and the natural enemies of pests (e.g. effects on the development of predatory insects fed aphids treated with a neem derivative). Such effects include physiological or behavioural changes in individuals that survive exposure to a pesticide. Each country has regulations for testing for the side effects of pesticides on beneficial species prior to chemical registration and use. The traditional method of testing involves determining a median lethal dose (LD_{50} or the dose required to kill half of the individuals of the test sample) or lethal concentration (LC_{50}) estimate for the test species. Efforts also have been directed to selecting chemicals with the lowest non-target effects based on LD_{50} values, but sublethal effects generally are not considered by regulatory authorities. However, there is an expanding literature on non-target effects of pesticides, and methods are being developed to detect various chemically mediated changes to the physiology (including biochemistry, neurophysiology, development, longevity, fecundity and sex ratio) and behaviour (including learning, feeding, oviposition, mobility and navigation/orientation) of non-target insects exposed to insecticides. Pesticide regulation procedures need to be modified to test for sublethal effects in addition to mortality of non-target organisms.

16.4.2 Insect growth regulators

Insect growth regulators (IGRs) are compounds that affect insect growth via interference with metabolism or development. They offer a high level of efficiency against specific stages of many insect pests, with a low level of mammalian toxicity. The two most commonly used groups of IGRs are distinguished by their mode of action. (i) Chemicals that interfere with the normal maturation of insects by disturbing the hormonal control of metamorphosis are the juvenile hormone mimics, also called juvenoids, juvenile hormone analogues or JHAs (e.g. fenoxycarb, hydroprene, methoprene, pyriproxyfen). These synthetic analogues of **juvenile hormone** (JH) halt development so that the insect either fails to reach the adult stage or the resulting adult is sterile and malformed. As juvenoids deleteriously affect adults rather than immature insects, their use is most appropriate to species in which the adult rather than the larva is the pest, such as fleas, mosquitoes and ants. (ii) The chitin synthesis inhibitors (e.g. buprofezin, cyromazine, diflubenzuron, hexaflumuron, lufenuron, triflumuron) prevent the formation of **chitin**, which is a vital component of insect cuticle. Many conventional insecticides cause a weak inhibition of chitin synthesis, but the benzoylureas (also known as benzoylphenyl ureas or acylureas, of which diflubenzuron and triflumuron are examples) strongly inhibit formation of cuticle. Insects exposed to chitin synthesis inhibitors usually die at or immediately after ecdysis. Typically, the affected insects shed the old cuticle partially or not all, and, if they do succeed in escaping from their exuviae, their body is limp and easily damaged as a result of the weakness of the new cuticle.

IGRs, which are fairly persistent indoors, usefully control insect pests in storage silos and domestic premises. Typically, juvenoids are used in urban pest control, and inhibitors of chitin synthesis have greatest application in controlling beetle pests of stored grain. However, some IGRs (e.g. pyriproxyfen) have been used also in field crops, for example in citrus in southern Africa. This use has led to severe secondary pest outbreaks because of their adverse effects on natural enemies, especially coccinellids but also wasp parasitoids. In the United States, in the citrus-growing areas of California, the IGRs pyriproxyfen and buprofezin are part of the strategy to control California red scale (Diaspididae: *Aonidiella aurantii*), but their use is timed to reduce non-target effects on the predatory coccinellids that control several scale pests. Buprofezin, diflubenzuron and pyriproxyfen have been shown to be effective against the Asian citrus psyllid, *Diaphorina citri* (see Box 16.1), which is developing resistance to insecticides used in Florida, but possible effects on parasitoids of *D. citri* must be evaluated. The application of methoprene (often used as a mosquito larvicide) to wetlands can have effects on non-target insects and other arthropods.

Neem derivatives are another group of growth regulatory compounds with significance in insect control. Their ingestion, injection or topical application disrupts moulting and metamorphosis, with the effect depending on the insect and the concentration of chemical applied. Treated larvae or nymphs fail to moult, or the moult results in abnormal individuals in the subsequent instar; treated late-instar larvae or nymphs generally produce deformed and non-viable pupae or adults. These physiological effects of neem derivatives are not fully understood, but are believed to result from more than one mode of action. The main active principle of neem, azadirachtin (AZ), appears to act by blocking transport and release from the brain of the peptide prothoracicotropic hormone (PTTH) that controls ecdysteroid release, thus preventing the usual moult-initiating rise in ecdysteroid titre. Importantly, cell division also is blocked at the prometaphase stage, leading to many abnormalities in rapidly dividing tissues, such as wing buds, testes and ovaries.

The newest group of IGRs developed for commercial use comprises the bisacylhydrazines or moulting hormone mimics, often called moult-accelerating compounds (e.g. tebufenozide, methoxyfenozide and halofenazide), which are ecdysone agonists that disrupt moulting by binding to the ecdysone receptor protein. They are toxic to susceptible insects mainly when ingested (rather than by contact), and the resulting "hyperecdysonism" triggers a **moult** attempt in a susceptible larva. However, because of the amount and stability of the bisacylhydrazine in the insect's haemolymph, various genes involved in the moulting process cannot be expressed. For example, eclosion hormone (which triggers ecdysis to complete a normal moult, see section 6.3) is repressed and the larva cannot complete its moult; it is trapped in its old cuticle. Also, new cuticle formed in the presence of the bisacylhydrazine is usually malformed. Thus, bisacylhydrazines derail the moulting process at the molecular, ultrastructural and physiological levels, leading to premature death. These compounds have been used successfully against immature insect pests, especially lepidopterans, and have no adverse effects on natural enemies due to their taxon-selective action. Once more ecdysone receptors (EcRs) are identified from additional pest species, it will become increasingly feasible to develop further moult-accelerating insecticides that safely and selectively control particular pests.

There are a few other types of IGRs, such as the anti-juvenile hormone analogues (e.g. precocenes), but these currently have little potential in pest control. Anti-juvenile hormones disrupt development by accelerating termination of the immature stages.

16.4.3 Neuropeptides and insect control

Neuropeptides regulate most aspects of insect development, metabolism, homeostasis and reproduction (summarized in Table 3.1). Although neuropeptides are unlikely to be used as insecticides *per se*, novel approaches to insect control may come from understanding their chemistry and biological actions. Neuroendocrine manipulation involves disrupting the general hormone process of synthesis–secretion–transport–action–degradation. For example, a synthetic peptide mimic that blocks or overstimulates, either at the release or receptor (target) site, could alter the secretion of a neuropeptide and affect the fitness of the insect. The protein nature of neuropeptides makes them amenable to manipulation using recombinant DNA technology and genetic engineering. Such synthesized neuropeptides produced by transgenic crop plants or bacteria that express neuropeptide genes would need to penetrate either the insect gut or cuticle. Transgenic plants containing double-stranded (ds) RNA against insect genes have been trialled with some success, including in cotton against bollworms and in maize (corn) against western corn rootworm.

Manipulation of insect viruses appears promising for control, since neuropeptide or "antineuropeptide" genes can be incorporated into the genome of insect-specific viruses. These then act as expression vectors of the genes to produce and release the insect hormone(s) within infected insect cells. Baculoviruses may be used in this way, especially in Lepidoptera. Normally, such viruses cause slow or limited mortality in their host insect (section 16.5.2), but their efficacy might be improved by creating an endocrine imbalance that reduces fitness or kills infected insects more quickly or increases viral-mediated mortality among infected insects. An advantage of neuroendocrine manipulation is that some neuropeptides seem to be specific to insects or arthropods, or even species-specific: a property that would reduce deleterious effects on many non-target organisms.

16.4.4 RNA interference and insect control

An exciting tool that promises highly specific insect control involves use of externally generated (exogenous) RNA sequences to interfere with ("silence" or "knockdown") expression of a target gene via the complementary internal (endogenous) messenger RNA. This technique depended initially on discovery of the mechanisms in living cells to regulate expression of messenger (m) RNA. Building on sequencing of the complete genome of *Drosophila melanogaster* and annotation of the function of its genes, then sequencing of genomes of *Apis mellifera* (the honey bee) and *Tribolium castaneum* (the red flower beetle), many genes were found to be identical, with assumed homologous functions across all insects. For targeted genes, double-stranded (ds) RNA can be generated, and transferred to the insect where it interacts with the target gene to prevent its expression. Delivery of the dsRNA such that expression of the target gene is blocked throughout the insect remains problematic. This is especially so outside of the laboratory, where soaking the target insect in dsRNA or microinjection into the haemocoel are not possible. The way forward in control of pest insects in broad acre crops (i.e. those grown on a large scale) is likely to be via continuous delivery, through ingestion of host plants made transgenic to deliver specific dsDNA.

16.5 BIOLOGICAL CONTROL

Regulation of the abundance and distributions of species is influenced strongly by the activities of naturally occurring enemies, namely **predators**, **parasites/parasitoids**, pathogens and/or competitors. In most managed ecosystems, these biological interactions are severely restricted or disrupted in comparison with natural ecosystems, and certain species escape from their natural regulation and become pests. In **biological control** (or biocontrol), deliberate human intervention attempts to restore some balance, by introducing or enhancing the natural enemies of target organisms such as insect pests or weedy plants. One advantage of natural enemies is their host-specificity, but a drawback (shared with other control methods) is that they do not eradicate pests. Thus, biological control may not necessarily alleviate all economic consequences of pests, but control systems are expected to reduce the abundance of a target pest to below economic threshold levels. In the case of weeds, natural enemies include phytophagous insects (biological control of weeds is discussed in section 11.2.7). Several approaches to biological control are recognized, but these categories are not discrete and published definitions vary widely, leading to some confusion. Such overlap is recognized in the following summary of the basic strategies of biological control.

Classical biological control involves the importation and establishment of natural enemies of exotic pests, and is intended to achieve control of the target pest with little further assistance. This form of biological control is appropriate when insects that spread or are introduced (usually accidentally) to areas outside their natural range become pests mainly because of the absence of adapted or coevolved natural enemies. Three examples of successful classical biological control are outlined in Box 16.3, Box 16.5 and Box 16.6. Despite the many beneficial aspects of this control strategy, negative environmental impacts can arise through ill-considered introductions of exotic natural enemies. Many introduced agents have failed to control pests; for example, over 60 predators and parasitoids have been introduced into northeastern North America with little effect thus far on the target gypsy moth, *Lymantria dispar* (Lymantriidae) (see **Plate 8c**). Some introductions have exacerbated pest problems, whereas others have become pests themselves. Exotic introductions generally are irreversible, and non-target species can suffer worse consequences from efficient non-native predators and parasitoids than from chemical insecticides, which are unlikely to cause total extinctions of native insect species.

Box 16.5 Taxonomy and biological control of the cassava mealybug

Cassava (manioc or tapioca – *Manihot esculenta*) is a staple food crop for 200 million Africans. In 1973, a new mealybug (Hemiptera: Pseudococcidae) was found attacking cassava in central Africa. Named in 1977 as *Phenacoccus manihoti*, this pest spread rapidly, until by the early 1980s it was causing production losses of over 80% throughout tropical Africa. The origin of the mealybug was considered to be the same as the original source of cassava – the Americas. In 1977, the apparent same insect was located in Central America and northern South America, and parasitic wasps attacking it were found. However, as biological control agents they failed to reproduce on the African mealybugs.

Working from existing collections and fresh samples, taxonomists quickly recognized that two closely related mealybug species were involved. The one infesting African cassava proved to be from central South America, and not from further north. When the search for natural enemies was switched to central South America, the true *P. manihoti* was eventually found in the Paraguay basin, together with an encyrtid wasp, *Apoanagyrus* (formerly known as *Epidinocarsis*) *lopezi*. This wasp gave spectacular biological control when released in Nigeria, and by 1990 had been established successfully in 26 African countries and had spread to more than 2.7 million km^2. The mealybug is now considered to be under almost complete control throughout its range in Africa.

When the mealybug outbreak first occurred in 1973, although it was clear that this was an introduction of Neotropical origin, the detailed species-level taxonomy was insufficiently refined, and the search for the mealybug and its natural enemies was misdirected for three years. The search was redirected thanks to taxonomic research. The savings were enormous: by 1988, the total expenditure on attempts to control the pest was estimated at US$14.6 million. In contrast, accurate species identification has led to an estimated annual benefit of at least US$200 million, and this financial saving may continue indefinitely.

Box 16.6 Glassy-winged sharpshooter biological control – a Pacific success

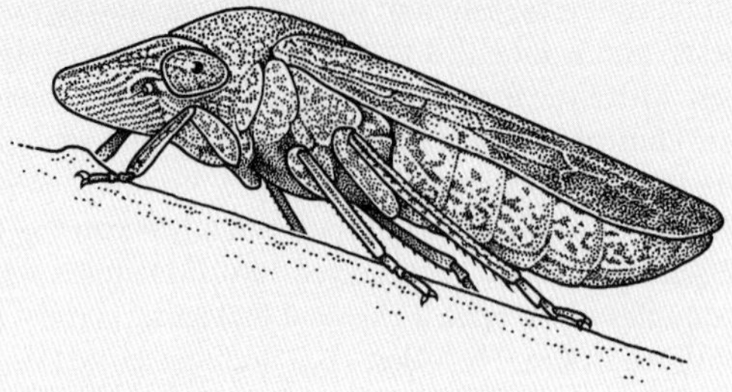

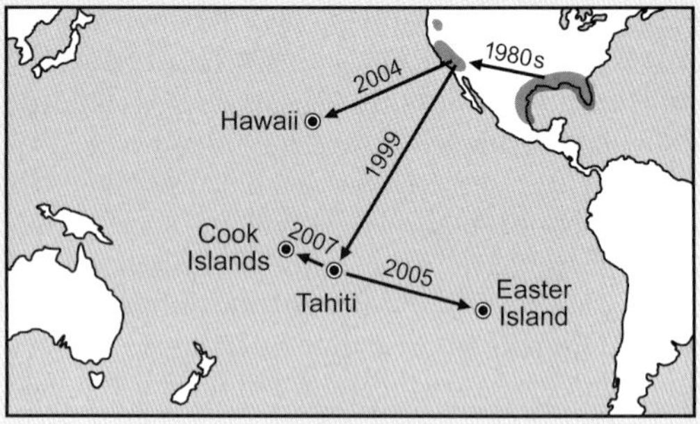

Homalodisca vitripennis, the glassy-winged sharpshooter (GWSS) (Hemiptera: Cicadellidae), which is native to the southeast of the United States and northeast Mexico, is causing trepidation outside its native range. This large (up to 13 mm long) cicadellid sucks xylem sap, and may vector a bacterium, *Xylella fastidiosa*, that can be lethal to certain plants including grape vines, in which it is called Pierce's disease. Smaller, native sharpshooters can transmit the disease into young shoots of vines, but the diseased parts are pruned annually in routine viticultural maintenance. The larger GWSS can attack mature woody stems, injecting the bacterium deeper into tissues that are not pruned after harvest, causing an incurable disease that is lethal due to water stress.

GWSS entered southern California in the late 1980s, probably as egg masses on horticultural trade plants. It has spread northward, hosted by citrus in which damage is modest, but can transfer onto a wide range of host plants, especially in the urban environment. A novel strain of *X. fastidiosa* transmitted by GWSS is increasingly damaging oleander, causing leaf wilt and death of this important Californian landscape plant, especially along the thousands of miles of median divide on highways. GWSS also transferred into the Pacific – first to French Polynesia (Tahiti in 1999, via plants that evaded quarantine), then to Hawai'i in 2004, the very remote Easter Island in 2005, and the Cook Islands in 2007 (as shown in the top map, with dates from Petit *et al*. 2008). In the tropical conditions of French Polynesia, sharpshooters bred year-round and infestations quickly became immense – causing the phenomenon of the "mouche pisseuse" – with liquid excreta raining from urban trees. Exports and native vegetation were threatened, and the risk of onward movement was very high. The insects spread to many of the French Polynesian islands, perhaps with accidental transmission by local people carrying plant material bearing the eggs. Aircraft, boats and associated freight containers were potential sources for long-distance dispersal.

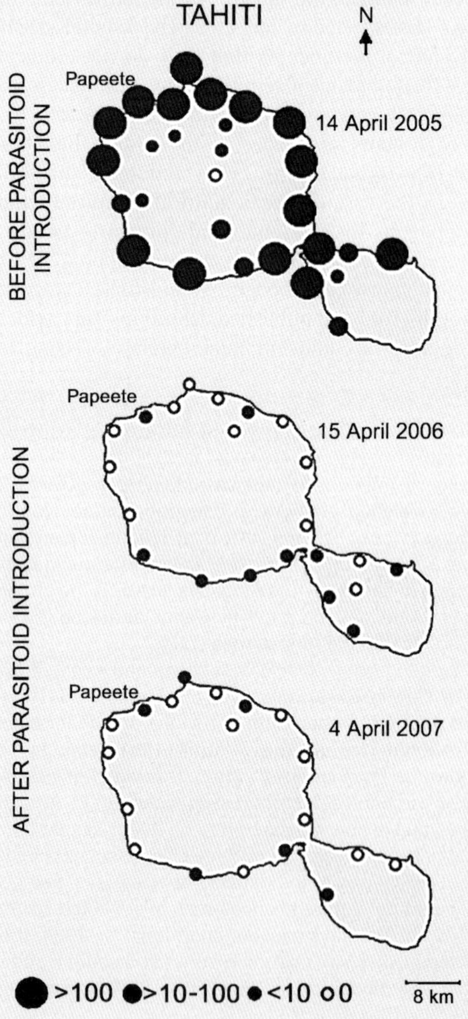

In response to this outbreak, a host-specific egg parasitoid, a tiny wasp – *Gonatocerus ashmeadi* (Hymenoptera: Mymaridae) – was considered for release. The absence of any close relative to the GWSS in the Pacific region meant that safe release, after rearing and evaluating in quarantine for approximately 12 months, appeared assured and, in May 2005, a first release of the natural enemies was made at a single site on Tahiti. Within four months, the biological control agents had spread 5 km from the release site, both with and against the prevailing winds. After only seven months there had been a 98% decrease in GWSS, as measured by nymphal abundance in sweep-net catches, and this was sustained for more than two years post-establishment (as shown here in the maps of Tahiti, after Grandgirard *et al.* 2009). Back calculation showed the rate of spread of *G. ashmeadi* averaged 4 m per day, and it took less than a year for this egg parasitoid to spread throughout all of French Polynesia (it had taken about six years for GWSS to attain the same distribution) (data from Petit *et al.* 2008, 2009).

Of course, following this astonishing success, it may be asked why such control cannot be used in California to prevent the threat to the multimillion dollar wine industry should GWSS reach the premium viticulture areas. The answer to the success of both GWSS and its parasitoid in the Pacific is their continuous development in the more uniform and favourable tropical conditions. In the Mediterranean climate of the southwestern United States, the parasitoids have to survive a "down-time" in winter, when no GWSS eggs are available for several months, and then in spring have to "catch-up" with the synchronized emergence of the pest. The lag time may prohibit successful control via egg parasitoids, unless alternative control agents with life cycles better synchronized to that of GWSS are found.

Introduced biological control agents can have both direct (discussed here) and indirect (e.g. see apparent competition in section 11.2.7), unintended effects on native fauna (and flora). There are documented cases of introduced biological control agents annihilating native invertebrates. A number of endemic Hawai'ian insects (target and non-target) have become extinct, apparently largely as a result of biological control introductions. The endemic snail fauna of Polynesia has been almost completely replaced by accidentally as well as deliberately introduced alien species. The introduction of the fly *Bessa remota* (Tachinidae) from Malaysia to Fiji, which led to extinction of the target coconut moth, *Levuana iridescens* (Zygaenidae), has been argued to be a case of biological control induced extinction of a native species. However, this seems to be an oversimplified interpretation, and it remains unclear as to whether the pest moth was indeed native to Fiji or an adventitious insect of no economic significance elsewhere in its native range. Moth species most closely related to *L. iridescens* predominantly occur from Malaysia to New Guinea, but their systematics is poorly understood. Even if *L. iridescens* had been native to Fiji, habitat destruction, especially replacement of native palms with coconut palms, also may have affected moth populations that probably underwent natural fluctuations in abundance.

At least 84 parasitoids of lepidopteran pests have been released in Hawai'i, with 32 becoming established, mostly on pests at low elevation in agricultural areas. Suspicions that native moths were being impacted in natural habitats at higher elevation have been confirmed in part. In a massive rearing exercise, over 2000 lepidopteran larvae were reared from the remote, high-elevation Alaka'i Swamp on Kauai, producing either adult moths or emerged parasitoids, each of which was identified and categorized as native or introduced. Levels of parasitization, based on the emergence of adult parasitoids, was approximately 10% each year, higher based on dissections of larvae, and rose to 28% for biological control agents in certain native moth species. Some 83% of parasitoids belonged to one of three biological control species (two braconids and an ichneumonid), and there was some evidence that these competed with native parasitoids. These substantial non-target effects appear to have developed over many decades, but the progression of the incursion into native habitat and hosts was not documented.

The emerald ash borer, *Agrilus planipennis* (Buprestidae), is an alien pest of ash trees in North America (discussed in Box 17.4). Three parasitoid wasp species have been released widely there in an attempt to control this jewel beetle. However, there is a diverse native fauna of *Agrilus* species in the United States and Mexico, and two of the introduced wasps are known to parasitize native *Agrilus* species to some extent. The long-term consequences for populations of the native beetles due to these introduced natural enemies is unknown.

The ultimate measure of the impact of a classical biological control agent is its effects on the population sizes of both the target pest and non-target species. Unfortunately, it is rare for population-level impacts to be estimated, even for the target species. Usually, there is insufficient population data from before and after release of a biological control agent to determine even the likely impacts on the pest (or weed), and generally little or nothing is known of population sizes of non-target species. Sometimes, non-target attacks are not anticipated, and so pre-release data are unavailable.

A controversial form of biological control, sometimes referred to as **neoclassical biological control**, involves the importation of non-native species to control native ones. Such new associations have been suggested to be very effective at controlling pests because the pest has not coevolved with the introduced enemies. Unfortunately, the species that are most likely to be effective neoclassical biological control agents because of their ability to utilize new hosts are also those most likely to be a threat to non-target species. An example of the possible dangers of neoclassical biological control is provided by the work of Jeffrey Lockwood, who campaigned against the introduction of a parasitic wasp and an entomophagous fungus from Australia as control agents of native rangeland grasshoppers in the western United States. Potential adverse environmental effects of such introductions include the suppression or extinction of many non-target grasshopper species, with probable concomitant losses of biological diversity and existing weed control, and disruptions to food chains and plant community structure. The inability to predict the ecological outcomes of neoclassical introductions means that they are high risk, especially in systems where the exotic agent is free to expand its range over large geographical areas.

Polyphagous agents have the greatest potential to harm non-target organisms, and native species in tropical and subtropical environments may be especially vulnerable to exotic introductions because, in comparison with temperate areas, biotic interactions can be more important than abiotic factors in regulating their populations. Sadly, the countries and states that may have most to lose from inappropriate introductions are those with the most lax quarantine restrictions and few or no protocols for the release of alien organisms.

Biological control agents that are present already or are non-persistent may be preferred for release. **Augmentation** is sometimes used as a general term for the supplementation of existing natural enemies, including **periodic release** of those that do not establish permanently but nevertheless are effective for a while after release. Periodic releases may be made regularly during a season so that the natural enemy population is gradually increased (augmented) to a level at which pest control is very effective. Augmentation or periodic release may be achieved in one of two ways, inoculation or inundation, although in some systems a distinction between the two methods may be inapplicable, especially if the nature of the control is difficult to determine. **Inoculation** (also called inoculative release or inoculation biological control) is the periodic release of a natural enemy unable either to survive indefinitely or to track an expanding pest range. Control depends on the progeny of the natural enemies, rather than the original release. Examples of this strategy include *Trichogramma* and *Encarsia* wasps that can be mass reared and released into glasshouses where their progeny provide season-long control, or certain insect pathogens that multiply but do not persist permanently. **Inundation** (also called inundative release or inundation biological control) resembles insecticide use as control is achieved by the individuals released or applied, rather than by their progeny; control is relatively rapid but short-term. Clear examples of inundation are entomopathogens, such as certain bacteria and fungi, used as microbial insecticides (section 16.5.2). For cases in which short-term control is mediated by the original release and pest suppression is maintained for a period by the activities of the progeny of the original natural enemies (as for *Chrysoperla carnea*), then the control process is neither strictly inoculative nor inundative. Augmentative releases are particularly appropriate for pests that combine good dispersal abilities with high reproductive rates – features that make them unsuitable candidates for classical biological control. The success of augmentative biological control is demonstrated by the wide-scale use of a range of natural enemies for controlling arthropods pests in greenhouses in Europe, concomitant with substantial reductions in pesticide use.

Conservation biological control is another broad strategy of biological control that aims to protect and/or enhance the activities of natural enemies and thus reduce the effects of pests. In some ecosystems this may involve preservation of existing natural enemies through practices that minimize disruption to natural ecological processes. For example, the IPM systems for rice in Southeast Asia encourage management practices, such as reduction or cessation of insecticide use, that interfere minimally with the predators and parasitoids that control rice pests such as brown planthopper (*Nilaparvata lugens*). The potential of biological control is much higher in tropical than in temperate countries because of high arthropod diversity and year-round activity of natural enemies. Complex arthropod food webs and high levels of natural biological control have been demonstrated in tropical irrigated rice fields. Furthermore, for many crop systems, environmental manipulation (also called ecological engineering) can greatly enhance the impact of natural enemies in reducing pest populations. Typically, this involves altering the habitat available to insect predators and parasitoids to improve conditions for their growth and reproduction by the provision or maintenance of shelter (including overwintering sites), alternative foods and/or oviposition sites. For example, flowering plants, such as sesame, can be planted close to rice crops in Asia to provide a nectar source for parasitoid wasps that attack brown planthoppers. Similarly, the effectiveness of entomopathogens of insect pests sometimes can be improved by altering environmental conditions at the time of application, such as by spraying a crop with water to elevate the humidity during release of fungal pathogens.

All biological control systems should be underpinned by sound taxonomic research on both pest and natural enemy species. Failure to invest adequate resources in systematic studies can result in incorrect identifications of the species involved, and ultimately may cost more in time and resources than any other step in the biological control system. The value of taxonomy in biological control is exemplified by the cassava mealybug in Africa (Box 16.5) and the management of *Salvinia* weed (see Box 11.4).

The next two subsections cover more-specific aspects of biological control by natural enemies. Natural enemies are divided somewhat arbitrarily into arthropods (section 16.5.1) and smaller, non-arthropod organisms (section 16.5.2) that are used to control various insect pests. In addition, many vertebrates, especially birds, mammals and fish, are insect predators and their significance as regulators of insect populations should not be underestimated. However, as biological control agents, the use of vertebrates is limited because most are dietary generalists and their times and places of activity are difficult to manipulate. An exception may be the mosquito fish, *Gambusia*, which has been released in many subtropical and tropical waterways worldwide in an effort to control the immature stages of biting flies, particularly mosquitoes. Although some control has been claimed, competitive interactions have been severely detrimental to small native fishes. Birds, as visually hunting predators that influence insect defences, are discussed in Box 14.1.

16.5.1 Arthropod natural enemies

Entomophagous arthropods may be predatory or parasitic. Most **predators** are either other insects or arachnids, particularly spiders (order Araneae) and mites (Acarina, also called Acari). Predatory mites are important in regulating populations of phytophagous mites, including the pestiferous spider mites (Tetranychidae). Some mites that parasitize immature and adult insects or feed on insect eggs are potentially useful control agents for certain scale insects, grasshoppers, and stored-product pests. Spiders are diverse and efficient predators, with a much greater impact on insect populations than mites, particularly in tropical ecosystems. The role of spiders may be enhanced in IPM by preservation of existing populations or habitat manipulation for their benefit, but their lack of feeding specificity is restrictive. Predatory beetles (Coleoptera: notably Coccinellidae and Carabidae) and lacewings (Neuroptera: Chrysopidae and Hemerobiidae) have been used successfully in biological control of agricultural pests, but many predatory species are polyphagous and inappropriate for targeting particular pest insects. Entomophagous insect predators may feed on several or all stages (from egg to adult) of their prey, and each predator usually consumes several individual prey organisms during its life, with the predaceous habit often characterizing both immature and adult instars. The biology of predatory insects is discussed in Chapter 13 from the perspective of the predator.

The other major type of entomophagous insect is parasitic as a larva and free-living as an adult. The larva develops either as an endoparasite within its insect host or externally as an ectoparasite. In both cases, the host is consumed and killed by the time that the fully fed larva pupates in or near the remains of the host. Such insects, called **parasitoids**, all are holometabolous insects and most are wasps (Hymenoptera: especially superfamilies Chalcidoidea, Ichneumonoidea and Platygasteroidea) or flies (Diptera: especially

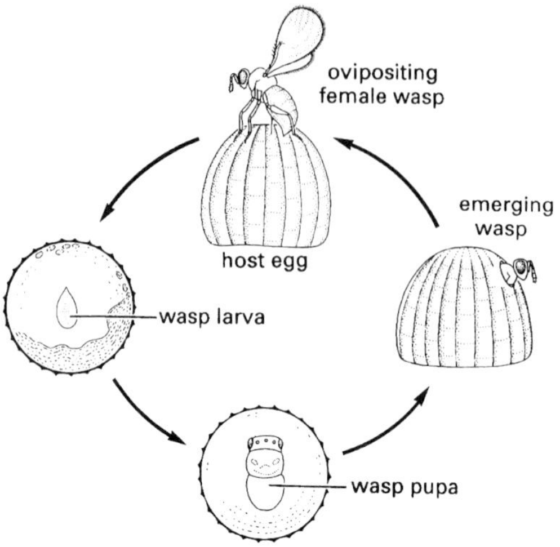

Fig. 16.2 Generalized life cycle of an egg parasitoid. A tiny female wasp of a *Trichogramma* species (Hymenoptera: Trichogrammatidae) oviposits into a lepidopteran egg; the wasp larva develops within the host egg, pupates, and emerges as an adult, often with the full life cycle taking only one week. (After van den Bosch & Hagen 1966.)

the Tachinidae). The Chalcidoidea contains about 20 families and perhaps 100,000–500,000 species (mostly undescribed), of which many are parasitoids, including egg parasitoids such as the Mymaridae and Trichogrammatidae (Fig. 16.2), and the speciose ecto- and endoparasitoid Aphelinidae and Encyrtidae, which are biological control agents, especially of mealybugs (Box 16.5), other scale insects, aphids and whiteflies. The Ichneumonoidea includes two speciose families, the Braconidae (see **Plate 8d**) and Ichneumonidae, which contain numerous parasitoids mostly feeding on insects and often exhibiting quite narrow host-specificity. The Platygasteroidea contains the Platygasteridae, which are parasitic on insect eggs and larvae, and the Scelionidae (see **Plate 6f**), which parasitize the eggs of insects and spiders. Parasitoids from many of these wasp groups have been utilized for biological control, whereas within the Diptera only the tachinids have been used as biological control agents.

Parasitoids often are parasitized themselves by secondary parasitoids, called **hyperparasitoids** (section 13.3.1), which may reduce the effectiveness of the primary parasitoid in controlling the primary host – the pest insect. In classical biological control, usually great care is taken to exclude the natural hyperparasitoids of primary parasitoids, and also the parasitoids and specialized predators of other introduced exotic natural enemies. However, some highly efficient natural enemies, especially certain predatory coccinellids, sometimes eliminate their food organisms so effectively that their own populations die out, with subsequent uncontrolled resurgence of the pest. In such cases, limited biological control of the pest's natural enemies may be warranted. More commonly, exotic parasitoids that are imported free of their natural hyperparasitoids are utilized by indigenous hyperparasitoids in the new habitat, with varying detrimental effects on the biological control system. Little can be done to overcome this latter problem, except to test the host-switching abilities of some indigenous hyperparasitoids prior to introductions of the natural enemies. Of course, the same problem applies to introduced predators, which may become subject to parasitization and predation by indigenous insects in the new area. Such hazards of classical biological control systems result from the complexities of food webs, which can be unpredictable and difficult to test in advance of introductions.

Some positive management steps can facilitate long-term biological control. For example, there is clear evidence that providing a stable, structurally and floristically diverse habitat near or within a crop can foster the numbers and effectiveness of predators and parasitoids. Habitat stability is naturally higher in perennial systems (e.g. forests, orchards and ornamental gardens) than in annual or seasonal crops (especially monocultures), because of differences in the duration of the crop. In unstable systems, the permanent provision or maintenance of ground cover, hedgerows, or strips or patches of cultivated or remnant native vegetation enable natural enemies to survive unfavourable periods, such as winter or harvest time, and then reinvade the next crop. Shelter from climatic extremes, particularly during winter in temperate areas, and provision of alternative food resources (when the pest insects are unavailable) are essential to the continuity of predator and parasitoid populations. In particular, the free-living adults of parasitoids generally require different food sources from their larvae, such as nectar and/or pollen from flowering plants. Thus, appropriate cultivation practices can contribute significant benefits to biological control. Diversification of agroecosystems also can provide refuges for pests, but densities are likely to be low, with damage only significant for crops with low EILs. For these crops, biological control must be integrated with other methods of IPM.

Pest insects must contend not only with predators and parasitoids but also with competitors. Competitive interactions appear to have little regulative influence on most phytophagous insects, but may be important for species that utilize spatially or temporally restricted resources, such as rare or dispersed prey/host organisms, dung, or animal carcasses. Interspecific competition can occur within a guild of parasitoids or predators, particularly for generalist feeders and facultative hyperparasitoids, and may inhibit biological control agents.

Biological control using natural enemies is particularly successful within the confines of greenhouses (glasshouses) or within certain crops. The commercial use of inundative and seasonal inoculative releases of natural enemies is common in many greenhouses, orchards and fields in Europe and the United States. In Europe, more than 80 species of natural enemies are available commercially, with the most commonly sold arthropods being various species of parasitoid wasps (including *Aphidius*, *Encarsia*, *Leptomastix* and *Trichogramma* spp.), predatory insects (especially coccinellid beetles such as *Cryptolaemus montrouzieri* and *Hippodamia convergens*, and mirid (*Macrolophus*) and anthocorid (*Orius* spp.) bugs) and predatory mites (*Amblyseius* and *Hypoaspis* spp.).

16.5.2 Microbial control

Microorganisms include bacteria, viruses and small eukaryotes (e.g. protists, fungi and nematodes). Some are insect pathogens, usually killing their host, and of these many are host-specific to a particular insect genus or family. Infection is from spores, viral particles, or organisms that persist in the insect's environment, often in the soil. These pathogens enter insects by several routes. Entry via the mouth (*per os*) is common for viruses, bacteria, nematodes and protists. Cuticular and/or wound entry occurs in fungi and nematodes; the spiracles and anus are other sites of entry for fungi and nematodes. Viruses and protists also can infect insects via the female ovipositor or during the egg stage. The microorganisms then multiply within the living insect but have to kill it to release more infectious spores, particles or, in the case of nematodes, juveniles. Disease is common in dense insect populations (pest or non-pest) and under environmental conditions suitable to the microorganisms. At low host density, however, disease incidence is often low as a result of lack of contact between the pathogens and their insect hosts.

Microorganisms that cause diseases in natural or cultured insect populations can be used as biological control agents in the same way as other natural enemies (section 16.5.1). The usual strategies of control are appropriate, namely:

- classical biological control (i.e. an introduction of an exotic pathogen such as the bacterium *Paenibacillus* (formerly *Bacillus*) *popilliae* established in the United States for the control of the Japanese beetle *Popillia japonica* (Scarabaeidae));
- augmentation via either:
  - (i) inoculation (e.g. a single treatment that provides season-long control, as in the fungus *Lecanicillium longisporum* used against *Myzus persicae* aphids in glasshouses), or
  - (ii) inundation (i.e. entomopathogens such as *Bacillus thuringiensis* used as microbial insecticides; see subsection on Bacteria, below);
- conservation of entomopathogens through manipulation of the environment (e.g. raising the humidity to enhance the germination and spore viability of fungi).

Some disease organisms are fairly host-specific (e.g. viruses), whereas others, such as fungal and nematode species, often have wide host ranges but possess different strains that vary in their host adaptation. Thus, when formulated as a stable microbial insecticide, different species or strains can be used to kill pest species with little or no harm to non-target insects. In addition to virulence for the target species, other advantages of microbial insecticides include their compatibility with other control methods and the safety of their use (they are non-toxic and non-polluting). For some **entomopathogens** (insect pathogens), further advantages include the rapid onset of feeding inhibition in the host insect, stability and thus long shelf-life, and often the ability to self-replicate and thus persist in target populations. Obviously, not all of these advantages apply to every pathogen; many have a slow action on host insects, with efficacy being dependent on suitable environmental conditions (e.g. high humidity or protection from sunlight) and appropriate host age and/or density. The very selectivity of microbial agents also can have practical drawbacks, as when a single crop has two or more unrelated pest species, each requiring separate microbial control. All entomopathogens are more expensive to produce than chemicals, and the cost is even higher if several agents must be used. However, bacteria, fungi and nematodes that can be mass-produced in liquid fermenters (*in vitro* culture) are much cheaper to produce than those microorganisms (most viruses and protists) requiring living hosts (*in vivo* techniques). Some of the problems with the use of microbial agents are being overcome by research on formulations and mass-production methods.

Insects can become resistant to microbial pathogens, as evidenced by the early success in selecting honey bees and silkworms resistant to viral, bacterial and protist pathogens. Furthermore, many pest species exhibit significant intraspecific genetic variability in their responses to all major groups of pathogens. The current rarity of significant field resistance to microbial agents probably results from the limited exposure of insects to pathogens, rather than any inability of most pest insects to evolve resistance. Of course, unlike chemicals, pathogens do have the capacity to coevolve with their hosts, and over time there is likely to be a constant trade-off between host resistance, pathogen virulence, and other factors such as persistence.

Each of the five major groups of microorganisms (viruses, bacteria, protists, fungi and nematodes) has different applications in insect pest control. Insecticides based on the bacterium *Bacillus thuringiensis* have been used most widely, but entomopathogenic fungi, nematodes and viruses have specific and often highly successful applications. Although protists, especially microsporidia such as *Nosema*, are responsible for natural disease outbreaks in many insect populations and can be appropriate for classical biological control, they have less potential commercially than other microorganisms because of their typical low pathogenicity (infections are chronic rather than acute) and the present difficulty of large-scale production for most species.

Nematodes

Nematodes from four families, the Mermithidae, Heterorhabditidae, Steinernematidae and Neotylenchidae, include useful or potentially useful control agents for insects. The infective stages of entomopathogenic nematodes (often called EPNs) are usually applied inundatively, although establishment and continuing control is feasible under particular conditions. Genetic engineering of nematodes is expected to improve their biological control efficacy (e.g. increased virulence), production efficiency, and storage capacity. However, entomopathogenic nematodes are susceptible to desiccation, which restricts their use to moist environments.

Mermithid nematodes are large and infect their host singly, eventually killing it as they break through the cuticle. They kill a wide range of insects, but aquatic larvae of black flies and mosquitoes are prime targets for biological control by mermithids. A major obstacle to their use is the requirement for *in vivo* production, and their environmental sensitivity (e.g. to temperature, pollution and salinity).

Heterorhabditids and steinernematids are small, soil-dwelling nematodes, associated with symbiotic gut bacteria (of the genera *Photorhabdus* and *Xenorhabdus*) that are pathogenic to host insects, killing them by septicaemia. In conjunction with their respective bacteria, nematodes of *Heterorhabditis* and *Steinernema* can kill their hosts within two days of infection. They can be mass-produced easily and cheaply and applied with conventional equipment, and have the advantage of being able to search for their hosts. The infective stage is the third-stage juvenile (or dauer stage) – the only stage found outside the host. Host location is an active response to chemical and physical stimuli. Although these nematodes are best at controlling soil pests, some plant-boring beetle and moth pests can be controlled as well. Recently, commercially available *Steinernema* species have been shown to have potential as biological agents for small hive beetle (*Aethina tumida*: Nitidulidae), which has become a bee-hive pest outside of its native Africa. Mole crickets (Gryllotalpidae: *Scapteriscus* spp.) are soil pests that can be infected with nematodes by being attracted to acoustic traps containing infective-phase *Steinernema scapterisci*, and then being released to inoculate the rest of the cricket population.

The Neotylenchidae contains the parasitic *Beddingia siricidicola* (formerly *Deladenus siricidicola*), which is one of the biological control agents of the sirex wood wasp, *Sirex noctilio* (Hymenoptera: Siricidae) – a serious pest of forestry plantations of *Pinus* spp. in Australia, South Africa and South America, and which recently invaded North America. The juvenile nematodes infect larvae of *S. noctilio*, leading to sterilization of the resulting adult female wasp. This nematode has two completely different forms – one with a parasitic life cycle completely within the sirex wood wasp, and the other with a number of cycles feeding within the pine tree on the fungus introduced by the ovipositing wasp. The fungal feeding cycle of *B. siricidicola* is used to mass-culture the nematode, and thus obtain infective juvenile nematodes for classical biological control purposes.

Fungi

Fungi are the most common disease organisms in insects, with approximately 750 species known to infect arthropods, although only a few dozen naturally infect agriculturally and medically important insects. At least 12 species of entomopathogenic fungi (those pathogenic to insects) have been used to develop about 170 pest-control products. Fungal spores that contact and adhere to an insect germinate and send out hyphae. These penetrate the cuticle, invade the haemocoel and cause death, either rapidly owing to release of toxins, or more slowly owing to massive hyphal proliferation that disrupts insect body functions. The fungus then sporulates, releasing spores that can establish infections in other insects; and thus the fungal disease may spread through the insect population.

Sporulation and subsequent spore germination and infection of entomopathogenic fungi often require moist conditions. Although formulation of fungi in oil improves their infectivity at low humidity, water requirements may restrict the use of some species to particular environments, such as soil, glasshouses or tropical crops. Despite this limitation, the main advantage of fungi as control agents is their ability to infect insects by penetrating the cuticle at any developmental stage. This property means that insects of all ages and feeding habits, even sapsuckers, are susceptible to fungal disease. However, fungi can be difficult to mass-produce, and the storage life of some fungal products can be limited unless kept at low temperature. A novel application method uses felt bands containing living fungal cultures applied to the tree trunks or branches, as is done in Japan using a strain of *Beauveria brongniartii* against longhorn beetle borers in citrus and mulberry. Useful species of entomopathogenic fungi belong to genera such as *Beauveria*, *Entomophthora*, *Hirsutella*, *Isaria*, *Metarhizium*, *Nomuraea* and *Verticillium*. Many of these fungi overcome their hosts after very little growth in the insect haemocoel, in which case peptide toxins are believed to cause death.

Entomopathogenic fungi primarily have been used inundatively as **mycoinsecticides**. *Lecanicillium lecanii* and *L. longisporum* (both formerly *Verticillium lecanii*) are used commercially to control aphids and scale insects in European glasshouses. *Entomophthora* species also are useful for aphid control in glasshouses. Species of *Beauveria* and *Metarhizium*, known as white and green muscardines, respectively (depending on the colour of the spores), are pathogens of soil pests, such as termites and beetle larvae, and can affect other insects, such as spittle bugs of sugarcane and certain moths that live in moist microhabitats. One *Metarhizium* species, *M. acridum* (formerly *M. anisopliae* var. *acridum*) has been developed as a successful mycoinsecticide for locusts and other grasshoppers in Africa. In freshwater environments, some aquatic fungi such as *Coelomomyces* species and *Lagenidium giganteum* can cause high levels of mortality in mosquitoes, and *L. giganteum* has been developed as a commercially available biological control agent due to its ease of culture.

Bacteria

Bacteria rarely cause disease in insects, although saprophytic bacteria, which mask the real cause of death, frequently invade dead insects. Relatively few bacteria are used for pest control, but several have proved to be useful entomopathogens against particular pests. *Paenibacillus popilliae* is an obligate pathogen of scarab beetles (Scarabaeidae) and causes milky disease (named for the white appearance of the body of infected larvae). Ingested spores germinate in the larval gut and lead to septicaemia. Infected larvae and adults are slow to die, which means that *P. popilliae* is unsuitable as a microbial insecticide, but the disease can be transmitted to other beetles by spores that persist in the soil. Thus, *P. popilliae* is useful in biological control by introduction or inoculation, although it is expensive to produce. Two species of *Serratia* are responsible for amber disease in the scarab *Costelytra zealandica*, a pest of pastures in New Zealand, and have been developed for scarab control. *Bacillus sphaericus* has a toxin that kills mosquito larvae. The strains of *Bacillus thuringiensis* have a broad spectrum of activity against larvae of many species of Lepidoptera, Coleoptera and aquatic Diptera, but can be used only as inundative insecticides because of lack of persistence in the field.

Bacillus thuringiensis, usually called Bt, was isolated first from diseased silkworms (*Bombyx mori*) by a Japanese bacteriologist, Shigetane Ishiwata, about a century ago. He deduced that a toxin was involved in the pathogenicity of Bt and, shortly afterwards, other researchers demonstrated that the toxin was a protein present only in sporulated cultures, was absent from culture filtrates, and thus was not an exotoxin. Of the many isolates of Bt, a number have been commercialized for insect control. Bt is produced in large, liquid fermenters and formulated in various ways, including as dusts and granules that can be applied to plants as aqueous sprays. Currently, the most widely used isolate of Bt is available in numerous commercial products used to control lepidopteran pests in forests, and in vegetable and field crops.

Bt forms spores, each containing a proteinaceous inclusion called a crystal, which is the source of the toxins that cause most larval deaths. The mode of action of Bt varies among different susceptible insects. In some species, insecticidal action is associated with the toxic effects of the crystal proteins alone (as for some moths and black flies). However, in many others (including a number of lepidopterans), the presence of the spore enhances toxicity substantially, and in a few insects, death results from septicaemia following spore germination in the insect midgut, rather than from the toxins. For insects affected by the toxins, paralysis occurs in mouthparts, the gut, and often the body, so that feeding is inhibited. Upon ingestion by a larval insect, the crystal is dissolved in the midgut, releasing proteins called delta-endotoxins. These proteins are protoxins that must be activated by proteases in the insect's alkaline midgut before they can interact with receptors in gut epithelium and disrupt its integrity, after which the insect ultimately dies. Early-instar larvae generally are more susceptible to Bt than are older larvae or adult insects. Bt is harmless to mammals, including humans, which have an acidic gut and lack the gut receptors for toxin binding.

Effective control of insect pests by Bt depends on the following factors:

- the insect population being uniformly young, so as to be susceptible;
- active feeding of insects so that they consume a lethal dose;
- evenness of spraying of Bt;
- persistence of Bt, especially lack of denaturation by ultraviolet light;
- suitability of the strain and formulation of Bt for the insect target.

Different Bt isolates vary greatly in their insecticidal activity against a given insect species, and a single Bt isolate usually displays very different activity in different insects. There are about 100 recognized Bt subspecies (or serovars), based on serotype and certain biochemical and host-range data. This Bt classification scheme has been argued to be less functional than one based on the crystal toxins (or genes), which directly determine the level and range of Bt activity. The current nomenclature for Bt delta-endotoxins (or crystal proteins) is based solely on hierarchical clustering using amino acid sequence identity, and each Cry or Cyt protein has a unique name with four ranks, for example Cry4Ba1 or Cyt2Ba6. Cry toxins are activated by host-insect proteases following binding with specific receptors on the midgut epithelium (as described above), whereas Cyt toxins insert into the epithelium by interacting directly with membrane lipids. In addition, some cultures of Bt produce vegetative insecticidal proteins (Vips) or beta-exotoxins; the latter are effective against various insects, including larvae of the Colorado potato beetle, but have mammalian toxicity. Thus, the nature and insecticidal effects of the various isolates of Bt are far from simple, and further research on the modes of action of the toxins is desirable, especially for understanding the basis of potential and actual resistance to Bt.

Bt products have been used increasingly for control of various Lepidoptera (such as caterpillars in forests and on vegetable crops) since 1970. For the first two decades of use, resistance was rare or unknown, except in a stored-grain moth (Pyralidae: *Plodia interpunctella*). The first insect to show resistance in the field was a major plant pest, the diamondback moth (Plutellidae: *Plutella xylostella*), which feeds on plants in the Brassicaceae (cabbages and relatives). Watercress growers in Japan and Hawai'i complained that Bt had reduced ability to kill this pest, and by 1989 further reports of resistant moths in Hawai'i were confirmed in areas where frequent high doses of Bt had been used. Similarly in Japan, by 1988 an extremely high level of Bt resistance was found in moths in greenhouses where watercress had been grown year-round, with a total of 40–50 applications of Bt over three to four years. Moths resistant to Bt also have been reported in mainland United States, Central America, Thailand, the Philippines and other parts of Asia. Furthermore, laboratory studies and field reports have indicated that other insect species have naturally evolved, or could be bred, to show differing levels of resistance. Bt resistance mechanisms of the diamondback moth have been shown to derive from a single gene that confers resistance to several different Bt toxins.

Resistance involves mutations in Bt toxin receptors or proteases. A different issue is the physiological induction of Bt tolerance (or induced resistance) in larvae that have been previously exposed to low doses of Bt toxins. Furthermore, immune induction and Bt tolerance can be transmitted to offspring maternally (via the eggs). This tolerance has been demonstrated in a few lepidopteran species, and appears to be an immune response involving increased enzyme activity in the larval midgut and haemolymph. Recent work has identified genes in *Heliothis armigera* that are up-regulated in Bt-tolerant individuals. Such findings also may help explain the reduction in susceptibility to Bt toxins that occurs as larvae age.

Problems with chemical insecticides stimulated interest in the use of Bt products as an alternative method of pest control. In addition to conventional applications of Bt, genetic engineering with Bt genes has produced transgenic plants ("Bt plants") that manufacture their own protective toxins (section 16.6.1), such as Bollgard II® cotton that expresses two Bt toxins, and transgenic varieties of corn and soybean that are grown widely in the United States and elsewhere. Optimism has led to the belief that insects are unlikely to develop extremely high levels of Bt resistance in the field, as a result of both instability of resistance and dilution by immigrants from susceptible populations (also see section 16.6.1). Strategies to prevent or slow down the evolution of resistance to Bt are the same as those used to retard resistance to synthetic insecticides. Obviously, the continued success of Bt products and the benefits of technological advances will depend on appropriate use, as well as understanding and limiting resistance to the Bt crystal proteins.

Viruses

Many viruses infect and kill insects, but those with potential for insect control are from just three viral groups, all with proteinaceous inclusion bodies, which enclose the virions (virus particles). These "occluded" viral species are considered safe because they have been found only in arthropods and appear unable to replicate in vertebrates or vertebrate cell cultures, although distant relatives of two of these groups have wider host ranges. Many "non-occluded" viruses that infect insects are considered unsafe for pest control because of their lack of specificity and possible adverse side effects (such as infection of vertebrates and/or beneficial insects). A relatively new group of insect viruses, the ascoviruses (Ascoviridae), are known to infect only certain lepidopteran caterpillars, causing a chronic, fatal disease. However, they are transmitted from host to host only by female parasitoid wasps at oviposition, and so would be difficult to use for pest control.

The useful entomopathogenic groups are the nuclear polyhedrosis viruses (NPVs), granulosis viruses (GVs) (both belonging to Baculoviridae – the baculoviruses or BVs), the cytoplasmic polyhedrosis viruses (cypoviruses or CPVs) (Reoviridae: *Cypovirus*) and the entomopoxviruses (EPVs) (Poxviridae: Entomopoxvirinae). Baculoviruses replicate within the nuclei of the host cells, whereas the CPVs and EPVs replicate in the host cell cytoplasm. Baculoviruses have DNA genomes and are found mostly in endopterygotes, such as moth and beetle larvae, which become infected when they ingest the inclusion bodies with their food. Inclusion bodies dissolve in the high pH of the insect midgut and release the virion(s) (Fig. 16.3). These infect the gut epithelial cells and usually spread to other tissues, particularly the fat body. The inclusion bodies of NPVs are usually very stable and may persist in the environment for years (if protected from ultraviolet light, as in the soil), increasing their utility as biological control agents or microbial insecticides. The host-specificity of different viruses also influences their potential usefulness as pest-control agents; some baculoviruses (such as the *Helicoverpa* NPV) are specific to an insect genus. CPVs have RNA genomes and have been found in a few hundred insect species, mainly of Lepidoptera and Diptera. Their inclusion bodies are less stable than those of NPVs. EPVs have large DNA genomes and infect a wide range of hosts in the Orthoptera, Lepidoptera, Coleoptera and Diptera, but individual viral isolates generally have a narrow host range. Infection of insect cells follows a similar path to that of baculoviruses. At present, baculoviruses of lepidopteran pests comprise the vast majority of viruses registered for use as biopesticides.

For certain pests, viral insecticides provide feasible alternatives to chemical controls, but several factors may restrict the usefulness of different viruses. Ideally, viral insecticides should be host-specific, virulent, kill quickly, persist for a reasonable time in the environment after application, and be easy to provide in large amounts. CPVs fulfil these requirements poorly, whereas the other viruses score better on these criteria, although they are inactivated by ultraviolet light within hours or days, often they kill larvae slowly and/or have a low virulence, and production costs

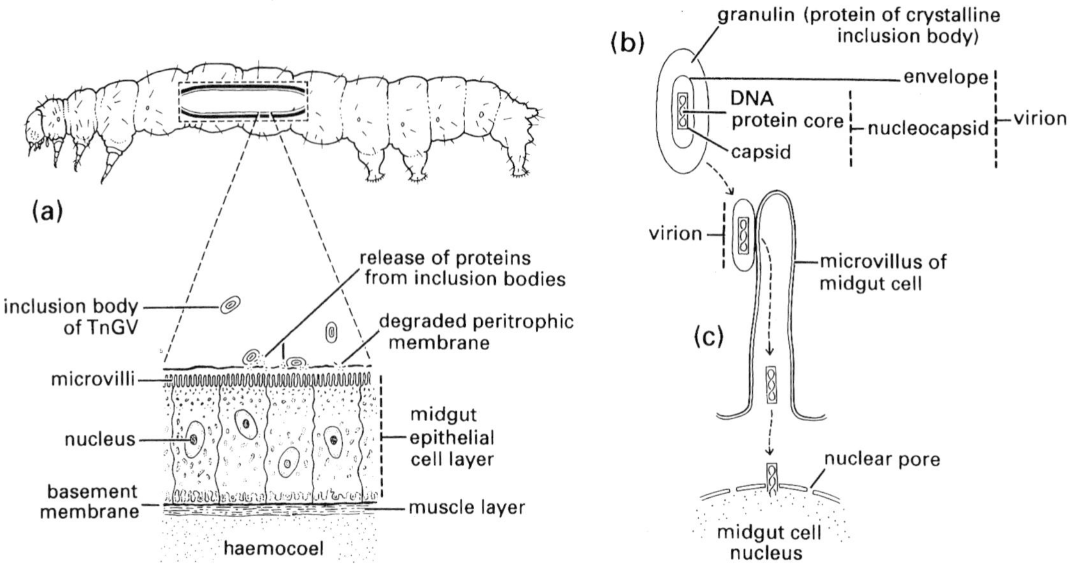

Fig. 16.3 The mode of infection of insect larvae by baculoviruses. (a) A caterpillar of the cabbage looper, *Trichoplusia ni* (Lepidoptera: Noctuidae), ingests the viral inclusion bodies of a granulosis virus (called TnGV) with its food and the inclusion bodies dissolve in the alkaline midgut, releasing proteins that destroy the insect's peritrophic membrane, allowing the virions access to the midgut epithelial cells. (b) A granulosis virus inclusion body with virion in longitudinal section. (c) A virion attaches to a microvillus of a midgut cell, where the nucleocapsid discards its envelope, enters the cell and moves to the nucleus in which the viral DNA replicates. The newly synthesized virions then invade the haemocoel of the caterpillar where viral inclusion bodies are formed in other tissues (not shown). (After Entwistle & Evans 1985; Beard 1989.)

can be high. Viral pesticides are produced mostly by *in vivo* or small-scale *in vitro* methods, which have been expensive because of the costs of rearing the host larvae, although recent low-cost *in vivo* production methods greatly improve the cost/benefit ratio for producing baculovirus pesticides. Also, the use of new tissue-culture technology has significantly reduced the very high cost of *in vitro* production methods. Potency problems may be overcome by genetic engineering to increase either the speed of action or the virulence of naturally occurring viruses, such as the baculoviruses that infect the heliothine pests (Lepidoptera: Noctuidae: *Helicoverpa* and *Heliothis* spp.) of cotton. The presence of particular proteins appears to enhance the action of baculoviruses; viruses can be modified to produce much more protein or the gene controlling protein production can be added to viruses that lack it. There has been considerable commercial interest in the manufacture of toxin-producing viral insecticides by inserting genes encoding insecticidal products, such as insect-specific neurotoxins or proteases, into baculoviruses, and field trials in recent years have shown that recombinant baculovirus is non-persistent in the environment and appears to have minimal, if any, non-target effects. However, commercialization of genetically modified (GM) baculovirus is limited, apparently due to perceived market concerns for environmental safety. Clearly, any GM viruses must be evaluated carefully prior to their wide-scale application, but such biological pesticides may be safer and more effective than many chemical pesticides to which insects are developing ever-increasing resistance.

Insect pests that damage valuable crops, such as bollworms of cotton and sawflies of coniferous forest trees, are suitable for viral control because substantial economic returns offset the large costs of development (including genetic engineering) and production. The other way in which insect viruses could be manipulated for use against pests is to transform the host plants so that they produce the viral proteins that damage the gut lining of phytophagous insects. This is analogous to the engineering of host-plant resistance by incorporating foreign genes into plant genomes using the crown-gall bacterium as a vector (section 16.6.1).

16.6 HOST-PLANT RESISTANCE TO INSECTS

Plant resistance to insects consists of inherited genetic qualities that result in one plant being less damaged than another (susceptible one) that is subject to the same conditions but lacks these qualities. Plant resistance is a relative concept, as spatial and temporal variations in the environment influence its expression and/or effectiveness. Generally, the production of plants resistant to particular insect pests is accomplished by selective breeding for resistance traits. The three functional categories of plant resistance to insects are:

1 antibiosis, in which the plant is consumed and adversely affects the biology of the phytophagous insect;
2 antixenosis, in which the plant is a poor host, deterring any insect feeding;
3 tolerance, in which the plant is able to withstand or recover from insect damage.

Antibiotic effects on insects range from mild to lethal, and antibiotic factors include toxins, growth inhibitors, reduced levels of nutrients, sticky exudates from glandular trichomes (hairs) and high concentrations of indigestible plant components such as silica and lignin. Antixenosis factors include plant chemical repellents and deterrents, pubescence (a covering of simple or glandular trichomes), surface waxes and foliage thickness or toughness – all of which may deter insect colonization. Tolerance involves only plant features and not insect–plant interactions, as it depends only on a plant's ability to outgrow or recover from defoliation or other damage caused by insect feeding. These categories of resistance are not necessarily discrete – any combination may occur in one plant. Furthermore, selection for resistance to one type of insect may render a plant susceptible to another or to a disease.

Selecting and breeding for host-plant resistance can be an extremely effective means of controlling pest insects. The grafting of susceptible *Vitis vinifera* cultivars onto naturally resistant American vine rootstocks confers substantial resistance to grape phylloxera (see Box 11.2). At the International Rice Research Institute (IRRI), numerous rice cultivars have been developed with resistance to all of the major insect pests of rice in southern and Southeast Asia. Some cotton cultivars are tolerant of the feeding damage of certain insects, whereas other cultivars have been developed for their chemicals (such as gossypol) that inhibit insect growth. In general, there are more cultivars of insect-resistant cereal and grain crops than insect-resistant vegetable or fruit crops. The former often have a higher value per hectare and the latter have a low consumer tolerance of any damage but, more importantly, resistance factors can be deleterious to food quality.

Conventional methods of obtaining host-plant resistance to pests are not always successful. Despite more than 50 years of intermittent effort, no commercially suitable potato varieties resistant to the Colorado potato beetle (Chrysomelidae: *Leptinotarsa decemlineata*) have been developed. Attempts to produce potatoes with high levels of toxic glycoalkaloids mostly have stopped, partly because potato plants with high foliage levels of glycoalkaloids often have tubers rich in these toxins, resulting in risks to human health. Breeding potato plants with glandular trichomes also may have limited utility, because of the ability of the beetle to adapt to different hosts. The most promising resistance mechanism for control of the Colorado potato beetle on potato is the production of genetically modified potato plants that express a foreign gene for a bacterial toxin that kills many insect larvae (Box 16.7).

16.6.1 Genetic engineering of host-plant resistance and the potential problems

Molecular biologists have used genetic engineering techniques to produce insect-resistant varieties of a number of crop plants, including corn (maize), cotton, tobacco, tomato, potato and rice, that can manufacture foreign antifeedant or insecticidal proteins under field conditions. The genes encoding these proteins are obtained from bacteria or other plants and are inserted into the recipient plant mostly via one of two common methods: (i) the biolistic (or particle gun) method, using a DNA-coated metal fibre or particle to pierce the cell wall and transport the gene into the nucleus; or (ii) via a plasmid of the crown-gall bacterium, *Agrobacterium tumefaciens*. This bacterium can move part of its own DNA into a plant cell during infection because it possesses a tumour-inducing (Ti) plasmid containing a piece of DNA that can integrate into the chromosomes of the infected plant. Ti plasmids can be modified by removal of their tumour-forming capacity, and useful foreign genes, such as insecticidal toxins, can be inserted. These plasmid vectors are introduced

Box 16.7 The Colorado potato beetle

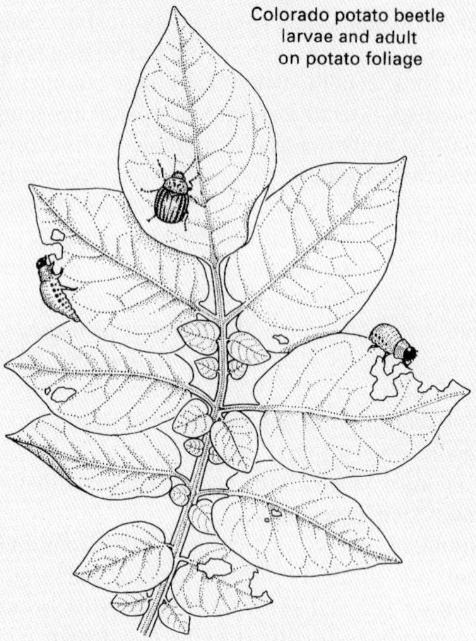

Leptinotarsa decemlineata (Coleoptera: Chrysomelidae), commonly known as the Colorado potato beetle, is a striking beetle (illustrated here, after Stanek 1969) that has become a major pest of cultivated potatoes in the northern hemisphere. Originally probably native to Mexico, it expanded its host range about 170 years ago, and then spread into Europe from North America in the 1920s, and is still expanding its range. For example, it entered China from Kazakhstan in the 1990s and has spread eastward by more than 40 km per year, threatening Chinese potato production. Its present hosts are about 20 species in the family Solanaceae, especially *Solanum* spp. and in particular *S. tuberosum*, the cultivated potato. Other occasional hosts include *Solanum lycopersicum*, the cultivated tomato, and *Solanum melongena*, eggplant. Adult beetles are attracted by volatile chemicals released by the leaves of *Solanum* species, on which they feed and lay eggs. Female beetles live for about two months, in which time they can lay a few thousand eggs each. Larvae defoliate potato plants (as illustrated here), resulting in yield losses of up to 100% if damage occurs prior to tuber formation. The Colorado potato beetle is the most important defoliator of potatoes and, where it is present, control measures are necessary if crops are to be grown successfully.

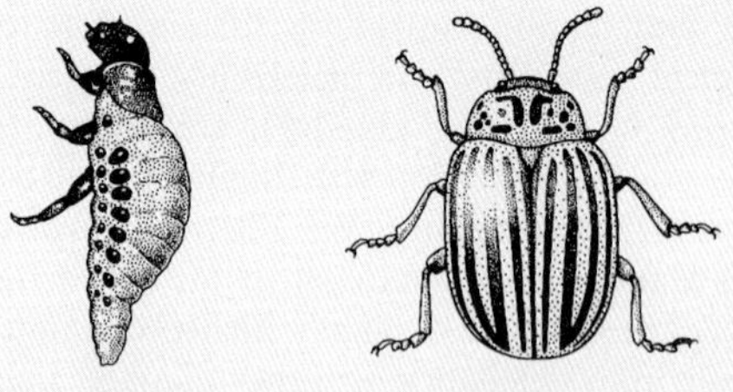

Insecticides effectively controlled the Colorado potato beetle until it developed resistance to DDT in the 1950s. Since then the beetle has developed resistance to each new insecticide (including synthetic pyrethroids and, most recently, imidacloprid, a neonicotinoid) at progressively faster rates. Currently, most beetle populations are resistant to all traditional insecticides, although some of the neonicotinid insecticides still control resistant populations. A semicarbazone insecticide, metaflumizone, has potential use if there is no cross-resistance from neonicotinoids. Feeding can be inhibited by application to leaf surfaces of antifeedants, including neem products and certain fungicides; however, deleterious effects on the plants and/or slow suppression of beetle populations has made antifeedants unpopular. Cultural control, via rotation of crops, delays infestation of potatoes and can reduce the build-up of early-season beetle populations. Diapausing adults mostly overwinter in the soil of fields where potatoes were grown the previous year and are slow to colonize new fields because much post-diapause dispersal is by walking. However, populations of second-generation beetles may or may not be reduced in size compared with those in non-rotated crops. Attempts to produce potato varieties resistant to the Colorado potato beetle have failed to combine useful levels of resistance (either from chemicals or glandular hairs) with a commercially suitable product. Even biological control has been unsuccessful because known natural enemies generally do not reproduce rapidly enough nor individually consume sufficient prey to regulate populations of the Colorado potato beetle effectively, and most natural enemies cannot survive the cold winters of temperate potato-growing areas. However, mass rearing and augmentative releases of certain predators (e.g. two species of pentatomid bugs) and an egg parasitoid (a eulophid wasp) may provide substantial control. Sprays of bacterial insecticides can produce effective microbial control if applications are timed to target the vulnerable early-instar larvae. Two strains of the bacterium *Bacillus thuringiensis* produce toxins that kill the larvae of Colorado potato beetle. An effective integrated control programme involves early-season biopesticide applications of fast-acting Bt to control early-instar larvae, followed by applications of the slow-acting fungus *Beauveria bassiana* (see section 16.5.2) against any late-instar larvae that survive Bt treatment. The bacterial genes, especially *cry3a*, responsible for producing the Cry3A toxin of *B. thuringiensis* ssp. *tenebrionis* (= *B.t.* var. *san diego*), have been genetically engineered into potato plants by inserting the genes into another bacterium, *Agrobacterium tumefaciens*, which is capable of inserting its DNA into that of the host plant. Remarkably, these transgenic potato plants are resistant to both adult and larval stages of the Colorado potato beetle, and also produce high-quality potatoes. However, their use has been restricted by consumer pressure to reject transgenic potatoes [see box 2 in Sanahuja *et al.* (2011) in Further reading] and because Bt plants do not deter certain other pests that still must be controlled with insecticides. Of course, even if Bt potatoes did become popular, the Colorado potato beetle may rapidly develop resistance to the Cry3A toxin.

into plant cell cultures, from which the transformed cells are selected and regenerated as whole plants.

Insect control via resistant genetically modified (transgenic) plants has several advantages over insecticide-based control methods, including continuous protection (even of plant parts inaccessible to insecticide sprays), elimination of the financial and environmental costs of unwise insecticide use, and cheaper costs associated with modification of a new crop variety compared to development of a new chemical insecticide. Whether such genetically modified (GM) plants (also called biotech crops) lead to greater or reduced environmental and human safety is currently a highly controversial issue. Problems with GM plants that produce foreign toxins include complications concerning registration and patent applications for these new biological entities, and the potential for the development of resistance in the target insect populations. For example, insect resistance to the toxins of *Bacillus thuringiensis* (Bt) (section 16.5.2) is to be expected after continuous exposure to these proteins in transgenic plant tissue. Since their commercialization in 1996, transgenic Bt crops have been planted on a cumulative total of more than 400 million hectares (about 1 billion acres) worldwide. To date, five of 13 major insect pest species targeted by Bt crops have evolved resistance (i.e. reduced crop efficacy reported) to field-planted Bt corn and Bt cotton (Fig. 16.4). Moreover, several other insect species show evidence of field-evolved resistance to Bt crops, and thus it is essential to proactively improve resistance management to prolong the efficacy of GM crops.

Refuges of plants that do not express the toxins appear to slow resistance by allowing the few resistant pests to mate with abundant susceptible individuals produced on refuge plants lacking the Bt toxins. Hybrid progeny of such matings will die on Bt crops if inheritance of resistance is a recessive trait, and especially if a high dose of toxins is ingested by hybrid larvae on Bt plants, and thus the evolution of resistance in the pest population is slowed. Bt-resistant insects also may experience lower fitness (survival, development time and body mass) relative to susceptible insects in refuges where they are not exposed to Bt toxins. The use of refuges in association with transgenic crops is mandated as a resistance strategy in the United States, Australia and many other countries. Resistance to Bt toxins also might be slowed by using transgenic plants that produce two or more distinct toxins, and by restricting expression of the toxins to certain plant parts (e.g. the bolls of cotton rather than the whole cotton plant) or to tissues damaged by insects. Many Bt crops, called **pyramids**, now produce two or more different Bt toxins that are active against the same pest. It has been shown that resistance to pyramids evolves more rapidly if two-toxin plants are grown concurrently with one-toxin plants. The explanation is that one-toxin plants select for insect resistance to each toxin separately, so that two-toxin plants become ineffective. A specific limitation of plants modified to produce Bt toxins is that the spore, and not just the toxin, must be present for maximum Bt activity with some pest insects.

The recent incursion of the polyphagous Old World cotton bollworm, *Helicoverpa armigera* (Lepidoptera: Noctuidae), into the Americas (specifically the State of Matto Grosso in Brazil) has serious consequences for crops, including for Bt agriculture. Currently, enormous areas of Bt soybeans and Bt corn, and a smaller area of Bt cotton (mostly expressing just

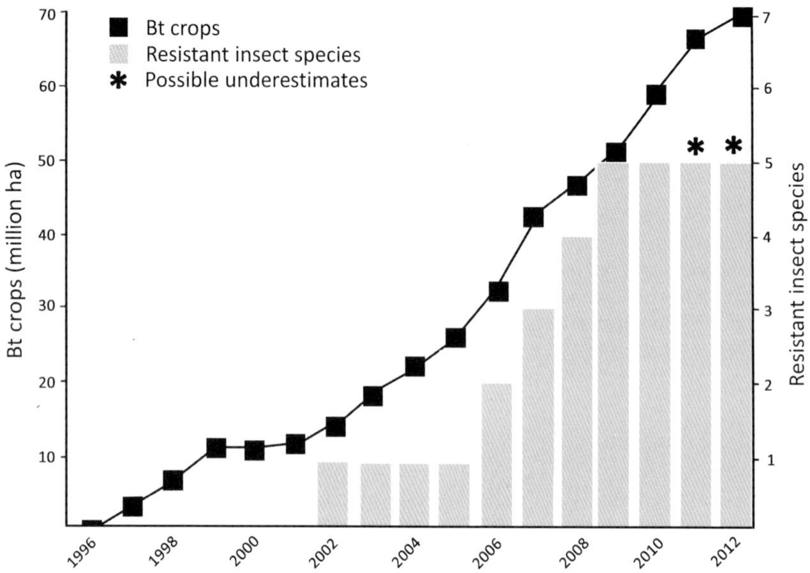

Fig. 16.4 Area of global planting of Bt crops annually and cumulative number of insect species showing field-evolved resistance associated with reduced efficacy of Bt crops. The area of Bt crops increased from 1.1 million hectares (ha) in 1996 to 66 million hectares in 2011. * Indicates possible underestimate pending publications reporting resistance. (After Tabashnik *et al.* 2013.)

the Cry1 toxin), are grown in Brazil. The resistance management practices, such as maintaining Bt-free refuges and tilling soil to kill any resistant pupae, which often are mandated elsewhere, are not in place in Brazil. Given the capacity for *H. armigera* to evolve resistance to both insecticides and Bt crops, and the year-round presence of suitable host plants, this invasive pest is likely to spread, including probably northwards into Central America and the United States. The economic and environmental impacts will be substantial.

It is possible that plant resistance based on toxins (**allelochemicals**) from genes transferred to plants might result in exacerbation rather than alleviation of pest problems. At low concentrations, many toxins are more active against natural enemies of phytophagous insects than against their pest hosts, adversely affecting biological control. Alkaloids and other allelochemicals ingested by phytophagous insects affect development of, or are toxic to, parasitoids that develop within hosts containing them, and can kill or sterilize predators. In some insects, allelochemicals sequestered whilst feeding pass into the eggs with deleterious consequences for egg parasitoids. Furthermore, allelochemicals can increase the tolerance of pests to insecticides by selecting for detoxifying enzymes that lead to cross-reactions to other chemicals. Most other plant resistance mechanisms decrease pest tolerance to insecticides, and thus improve the possibilities of using pesticides selectively to facilitate biological control.

In addition to the hazards of inadvertent selection of insecticide resistance, there are several other environmental risks resulting from the use of transgenic plants. First, there is the concern that genes from the modified plants may transfer to other plant varieties or species, leading to increased weediness in the recipient of the transgene, or the extinction of native species by hybridization with transgenic plants. Second, the transgenic plant itself may become weedy if genetic modification improves its fitness in certain environments. Third, non-target organisms, such as beneficial insects (pollinators and natural enemies) and other non-pest insects, may be affected by human use of transgenic plants. This risk may occur in two main ways: (i) via accidental ingestion of genetically modified plants, including their pollen; and (ii) by indirect effects due to habitat changes in genetically modified croplands. A potential hazard to populations of the monarch butterfly, *Danaus plexippus*, from their larvae eating milkweed foliage dusted with pollen from Bt corn has attained some notoriety. Milkweeds, the host plants of the monarch larvae, and commercial cornfields commonly grow in close proximity in the United States. Following detailed assessment of the distance and Bt content of pollen drift, the exposure of caterpillars to corn pollen was quantified. A comprehensive risk assessment concluded that the threat to the butterfly populations was low. In contrast, a highly significant threat to monarchs is the enormous increase in area of US farmland planted to corn and soybean that has been genetically modified to tolerate herbicides. As a consequence, the weeds that used to grow in and around corn and soybean in the corn belt of the Midwestern United States have been almost eliminated. The consequent loss of milkweeds used by the larvae, and of flowering annual weeds used as nectar sources by the adult butterflies, has lead to a reduction of critical breeding sites for the monarch. Estimates suggest that perhaps 50–60 million hectares (up to 150 million acres) of breeding habitat have been lost. Such loss is further compounded by rapid expansion of farmland onto former grassland and other habitats that were a source of milkweeds. These habitat changes in the United States are contributing to a continuing and dramatic decline in numbers of monarch butterflies. The monarch has received much attention because it is a charismatic, **flagship species** (section 1.7), and similar effects on populations of numerous other insects are unlikely to be noticed so readily.

16.7 PHYSICAL CONTROL

Physical control refers to non-chemical, non-biological methods that destroy pests or make the environment unsuitable for the entry or survival of pests. Most of these control methods may be classified as passive (e.g. fences, trenches, traps, oils and inert dusts) or active (e.g. mechanical, impact and thermal treatments). Physical control measures generally are limited to confined environments such as glasshouses, food storage structures (e.g. silos) and domestic premises, although certain methods, such as exclusion barriers or trenches, can be employed in fields of crops. The best-known mechanical method of pest control is the "fly swatter", but the sifting and separating procedure used in flour mills to remove insects is another example. An obvious method is physical exclusion, such as packaging of food products, semi-hermetic sealing of grain silos, or placing mesh screens on glasshouses. In addition, products may be treated or stored under controlled conditions of temperature (low or high), atmospheric gas composition (e.g. low oxygen (O_2) or high carbon dioxide (CO_2) or both), or low relative humidity, which can kill or reduce reproduction of insect pests. For stored-product pests, a controlled atmosphere of low O_2 and high CO_2 leads to higher insect mortality than either treatment alone, and CO_2 levels higher than 40% are required for efficient killing with CO_2 alone. Ionizing radiation can be used as a quarantine treatment for insects inside exported fruit, and hot-water immersion of mangoes has been used to kill immature tephritid fruit flies. The use of certain physical control methods are increasing and often replacing methyl bromide (also called bromomethane), which was used as a soil sterilant and still is used as a fumigant for many stored and exported/imported products, but is meant to be phased out (as required by the Montreal Protocol) because it depletes ozone in the atmosphere.

Traps that use long-wave ultraviolet light (e.g. "insect-o-cutors" or "zappers", which lure flying insects towards an electrified metal grid) or adhesive surfaces can be effective in domestic or food-retail buildings or in glasshouses, but should not be used outdoors because of the likelihood of catching native or introduced beneficial insects. One study of the insect catches from electric traps in suburban yards in the United States showed that insects from more than a hundred non-target families were killed; about half of the insects caught were non-biting aquatic insects, over 13% were predators and parasitoids, and only about 0.2% was nuisance biting flies.

16.8 CULTURAL CONTROL

Subsistence farmers have utilized cultural methods of pest control for centuries, and many of their techniques are applicable to large-scale as well as small-scale, intensive agriculture. Typically, cultural practices involve reducing insect populations in crops by one or a combination of the following techniques: crop rotation, tillage or burning of crop stubble to disrupt pest life cycles, careful timing or placement of plantings to avoid synchrony with pests, destruction of wild plants that harbour pests, and/or cultivation of non-crop plants to conserve natural enemies, and use of pest-free root-stocks and seeds. Intermixed plantings of several crops (called intercropping or polyculture) may reduce crop apparency (plant apparency hypothesis) or resource concentration for the pests (resource concentration hypothesis), increase protection for susceptible plants growing near resistant plants (associational resistance hypothesis) and/or promote natural enemies (the natural enemies hypothesis). Recent agroecology research has compared densities of insect pests and their natural enemies in monocultures and polycultures (including di- and tricultures) to test whether the success of intercropping can be explained better by a particular hypothesis; however, the hypotheses are not mutually exclusive and there is some support for each one.

In medical entomology, cultural control methods consist of habitat manipulations, such as draining marshes and removal or covering of water-holding containers to limit larval breeding sites of disease-transmitting mosquitoes, and covering rubbish dumps to prevent access and breeding by disease-disseminating flies, as well as using bed nets to prevent disease-carrying mosquitoes biting sleeping people (see Box 15.3). Examples of cultural control of livestock pests include removal of dung that harbours pestiferous flies, and simple walk-through traps that remove and kill flies resting on cattle. These exclusion and trapping methods also could be classified as physical methods of control.

16.9 PHEROMONES AND OTHER INSECT ATTRACTANTS

Insects use a variety of chemical odours, called **semiochemicals**, to communicate within and between species (section 4.3.2). **Pheromones** are particularly important chemicals used for signalling between members of the same species – these are often mixtures of two, three or more components, which, when released by one individual, elicit a specific response in another individual. Other members of the species, for example prospective mates, arrive at the source. Naturally derived or synthetic pheromones, especially sex pheromones, can be used in pest management to misdirect the behaviour and prevent reproduction of pest insects. The pheromone is released from point-source dispensers, often in association with traps that are placed in the crop. The strength of the insect response depends upon dispenser design, placement and density. The rate and duration of pheromone emission from each dispenser depends upon the method of release (e.g. from impregnated rubber, microcapsules, capillaries or wicks), strength of formulation, original volume, surface area from which it is volatilized, and longevity and/or stability of the formulation. Male lures, such as cuelure, trimedlure and methyl eugenol (sometimes called **parapheromones**), which are strongly attractive to many male tephritid fruit flies, can be dispensed in a manner similar to pheromones. Methyl eugenol is thought to attract males of the oriental fruit fly *Bactrocera dorsalis* because of the benefit its consumption confers on their mating success (see under "Sex pheromones" in section 4.3.2). Sometimes, other attractants, such as food baits or oviposition site lures, can be incorporated into a pest management scheme to function in a manner analogous to pheromones (and parapheromones), as discussed below.

There are three main uses for insect pheromones (and sometimes other attractants) in horticultural, agricultural and forest management: monitoring, attraction–annihilation, and mating disruption. Insect pheromones are used in monitoring, initially to detect the presence of a particular pest and then to give some measure of its abundance. A trap containing the appropriate pheromone (or other lure) is placed in the susceptible crop and checked at regular intervals for the presence of any individuals of the pest lured to the trap. In most pest species, females emit sex pheromone to which males respond, and thus the presence of males of the pest (and by inference, females) can be detected even at very low population densities, allowing early recognition of an impending outbreak. Knowledge of the relationship between trap-catch size and actual pest density allows a decision about when the economic threshold (section 16.1.1) for the crop will be reached, and thus facilitates the efficient use of control measures, such as insecticide application. Some kind of monitoring is an essential part of IPM.

Attraction–annihilation (also called mass annihilation), in which individuals of the targeted pest species are lured and removed from the population, is another method of using pheromones in pest management. Two approaches are recognized. **Mass trapping** uses semiochemical lures, usually pheromones, to attract the pest insects to the source, where they are entrapped in various ways (e.g. in large-capacity traps, on sticky paper or onto an electrocutor grid). This method has been used against a range of forest, agricultural and orchard pests. A closely related strategy called **lure-and-kill** (or attract-and-kill) differs from mass trapping in the method of killing the insects. Lure-and-kill systems use insecticides, or sometimes sterilants or pathogens, in combination with the lure (or a bait, in the case of crawling insect pests such as ants and cockroaches). This strategy has been used mostly in cotton fields and orchards. Lures may be light (e.g. ultraviolet), colour (e.g. yellow is a common attractant) or semiochemicals, such as pheromones or odours produced by the mating or oviposition site (e.g. dung), host plant or host animal, or empirical attractants (e.g. fruit fly chemical lures). Sometimes, the lure is more attractive than any other substance used by the insect, as with methyl eugenol for tephritid fruit flies. The effectiveness of the attraction–annihilation technique appears to be inversely related to the population density of the pest and the size of the infested area. Thus, this method is likely to be most effective for control of non-resident insect pests that become abundant through annual or seasonal immigration, or pests that are geographically restricted or always present at low density. Pheromone mass-trapping systems have been undertaken mostly for certain moths, such as the gypsy moth (Lymantriidae: *Lymantria dispar*) (see **Plate 8c**), using their female sex pheromones, and for bark and ambrosia beetles (Curculionidae: Scolytinae) using their aggregation pheromones (section 4.3.2). An advantage of this technique for scolytines is that both sexes are caught. Success has been difficult to demonstrate because of the difficulties of designing controlled, large-scale experiments. Nevertheless, mass trapping appears effective in isolated gypsy moth populations and at low scolytine beetle densities. If beetle populations are high, even removal of part of the pest population may be beneficial, because in tree-killing beetles there is a positive feedback between population density and damage.

The third method of practical pheromone use involves sex pheromones and is called **mating disruption** (previously sometimes called male confusion, which as we shall see is an inappropriate term). It has been applied very successfully in the field to a number of moth species, such as the pink bollworm (Gelechiidae: *Pectinophora gossypiella*) in cotton, the spruce budworm (Tortricidae: *Choristoneura fumiferana*) in Canadian boreal forests, the Oriental fruit moth or peach moth (Tortricidae: *Grapholita molesta*) in stone-fruit orchards and the tomato pinworm (Gelechiidae: *Keiferia lycopersicella*) in tomato fields, and is one of the strategies used to control light brown apple moth (Tortricidae: *Epiphyas postvittana*) (Box 16.1). Typically, numerous synthetic pheromone dispensers are placed within the crop so that the level of female sex pheromone in the orchard or field becomes higher than the background level. A reduction in the number of males locating female moths means fewer matings and a lowered population in subsequent generations. The exact behavioural or physiological mechanisms responsible for mating disruption are far from resolved for most pest species but relate to altered behaviour in males and/or females. Disruption of male behaviour may be through habituation – temporary modifications within the central nervous system – rather than adaptation of the receptors on the antennae or confusion resulting in the following of false plumes. The high background levels promoted by use of synthetic pheromones also may mask (camouflage) the natural pheromone plumes of the females so that males can no longer differentiate them, and/or reduce mating by competing with the natural sources. The continuous presence of synthetic pheromone sometimes may advance the daily rhythm of male response so that males fly before females are receptive. All four possible mechanisms of mating disruption (listed above) may occur in the pink bollworm. Understanding the mechanism(s) of disruption is important for production of the appropriate type of formulation and quantities of synthetic pheromone needed to cause disruption, and thus control.

All of thc above three uses of insect pheromones have been used most successfully for certain moth, beetle and fruit fly pests. Pest control using pheromones appears most effective for species that: (i) are highly dependent on chemical (rather than visual) cues for locating dispersed mates or food sources; (ii) have a limited host range; and (iii) are resident and relatively sedentary, so that locally controlled populations are not constantly supplemented by immigration. Advantages of using pheromone mass trapping or mating disruption include:

- non-toxicity, leaving fruit and other products free of toxic chemicals (insecticides);
- application may be required only once or a few times per season;
- confinement of suppression to the target pest, unless predators or parasitoids use the pest's own pheromone for host location;
- enhancement of biological control (except for the circumstance mentioned in the previous point).

The limitations of pheromone use include the following:

- high selectivity and therefore no effect on other primary or secondary pests;
- cost-effective only if the target pest is the main pest for which insecticide schedules are designed;
- requirement that the treated area be isolated or large enough to avoid mated females flying in from untreated crops;

• requirement for detailed knowledge of pest biology in the field (especially of flight and mating activity), as timing of application is critical to successful control if continuous costly use is to be avoided;
• the possibility that artificial use will select for a shift in natural pheromone preference and production, as has been demonstrated for some moth species.

The latter three limitations apply also to pest management using chemical or microbial insecticides; for example, appropriate timing of insecticide applications is particularly important to target vulnerable stages of the pest, to reduce unnecessary and costly spraying, and to minimize detrimental environmental effects.

16.10 GENETIC MANIPULATION OF INSECT PESTS

Genetic control involves spreading, by inheritance or mating, factors that reduce pest damage. A heritable element is introduced into the target pest population via genetically modified members of the same pest species. Thus, genetic control strategies are highly species-specific, but rely on the mate-seeking behaviour of the modified insects, which must disperse and actively seek wild members of their species. Genetic control strategies either are self-limiting, in that the genetic modification is maintained in the target population only by further releases of modified insects, or are self-sustaining, in that the modification is intended to persist indefinitely in the target population. Most self-limiting strategies involve the use of sterile male insects (as discussed below), are safe and have been successful in suppression or elimination of target populations. In contrast, self-sustaining strategies rely on a selfish DNA element (such as *Wolbachia* bacteria) as a gene driver to spread a novel trait through an insect population. Such methods are under development in mosquitoes (Culicidae), with the aim of introducing novel traits, such as reduced ability of pest mosquitoes to transmit disease pathogens. Such strategies, however, are controversial due to the risk of releasing a genetic modification that may have unintended and permanent consequences in wild populations.

Cochliomyia hominivorax (Calliphoridae), the New World screw-worm fly, is a devastating pest of livestock in tropical America, laying eggs into wounds, where the larvae cause myiasis (section 15.1) by feeding in the growing, suppurating wounds of living animals, including humans. The fly perhaps was present historically in the United States, but seasonally spread into the southern and southwestern states, where substantial economic losses of stock hides and carcasses required a continuing control campaign. As the female of *C. hominivorax* mates only once, control can be achieved by swamping the population with infertile males, so that the first male to arrive and mate with each female is likely to be sterile, and the resultant eggs inviable. This **sterile insect technique (SIT)** (once called the sterile male technique or sometimes the sterile insect release method, SIRM) in the Americas depends upon mass-rearing facilities, sited in Mexico, where billions of screw-worm flies are reared in artificial media of blood and casein. The larvae (see Fig. 6.6h) drop to the floor of the rearing chambers, where they form a puparium. At a crucial time, after gametogenesis, sterility of the developing adult is induced by gamma-irradiation of the five-day-old puparia. This treatment sterilizes the males, and although the females cannot be separated in the pupal stage and are also released, irradiation prevents their ovipositing. The released sterile males mix with the wild population, and with each mating the fertile proportion diminishes, with eradication a theoretical possibility.

The technique has eradicated the screw-worm fly, first from Florida, then Texas and the western United States, and more recently from Mexico, from whence reinvasions of the United States once originated. The goal to create a fly-free buffer zone from Panama northwards has been attained, with progressive elimination from Central American countries and releases continuing in a permanent "sterile fly barrier" in eastern Panama. In 1990, when *C. hominivorax* was introduced accidentally to Libya (North Africa), the Mexican facility was able to produce enough sterile flies to prevent the establishment of this potentially devastating pest.

The impressive cost/benefit ratio of screw-worm control and eradication using the sterile insect technique has induced the expenditure of substantial sums in attempts to control similar economic pests. Other examples of successful pest insect eradications involving sterile insect releases are the Mediterranean fruit fly or "medfly", *Ceratitis capitata* (Tephritidae), from Mexico and northern Guatemala, the melon fly, *Bactrocera cucurbitae* (Tephritidae), from the Ryukyu Archipelago of Japan, and the Queensland fruit fly, *Bactrocera tryoni*, from Western Australia. The frequent lack of success of other ventures can be attributed to difficulties with one or more of the following:

- inability to mass culture the pest;
- lack of competitiveness of sterile males, including discrimination against captive-reared sterile males by wild females;
- genetic and phenotypic divergence of the captive population so that the sterile insects mate preferentially with each other (assortative mating);
- release of an inadequate number of males to swamp the females;
- failure of irradiated insects to mix with the wild population;
- poor dispersal of the sterile males from the release site, and rapid reinvasion of wild types.

Attempts have been made to introduce deleterious genes into pest species that then can be mass cultured and released, with the intention that the detrimental genes spread through the wild population. The reasons for the failure of these attempts are likely to include those cited above for many sterile insect releases, particularly their lack of competitiveness, together with genetic drift and recombination that reduces the genetic effects. However, new transgenic approaches mean that the use of genetically modified insects in pest control is becoming feasible. It is possible to create transgenic insects that can be sexed genetically using a heritable marker or that carry a fluorescent protein marker that allows them to be distinguished readily from non-modified insects. An advanced derivative of the sterile insect technique is the **r**elease of **i**nsects carrying a **d**ominant **l**ethal marker (**RIDL**). RIDL insects have a genetic modification that causes their progeny to die, but the RIDL adults live and reproduce normally if fed a diet containing a supplement (an antibiotic). This approach has been applied to Mediterranean fruit fly, *Ceratitis capitata*, and certain lepidopterans and mosquitoes, and to female as well as male insects, but has not been used yet on wild pest populations.

FURTHER READING

Alphey, L. (2014) Genetic control of mosquitoes. *Annual Review of Entomology* **59**, 205–24.

Altieri, M.A. (1991) Classical biological control and social equity. *Bulletin of Entomological Research* **81**, 365–9.

Barratt, B.I.P., Howarth, F.G., Withers, T.M. *et al.* (2010) Progress in risk assessment for classical biological control. *Biological Control* **52**, 245–54.

Beech, C.J., Koukidou, M., Morrison, N.I. & Alphey, L. (2012) Genetically modified insects: science, use, status and regulation. *Collection of Biosafety Reviews* **6**, 66–124.

Boyer, S., Zhang, H. & Lempérière, G. (2012) A review of control methods and resistance mechanisms in stored-product pests. *Bulletin of Entomological Research* **102**, 213–29.

Brewer, M.J. & Goodell, P.B. (2012) Approaches and incentives to implement integrated pest management that addresses regional and environmental issues. *Annual Review of Entomology* **57**, 41–59.

Caltagirone, L.E. (1981) Landmark examples in classical biological control. *Annual Review of Entomology* **26**, 213–32.

Caltagirone, L.E. & Doutt, R.L. (1989) The history of the vedalia beetle importation to California and its impact on the development of biological control. *Annual Review of Entomology* **34**, 1–16.

De Barro, P.J., Liu, S.-S., Boykin, L.M. & Dinsdale, A.B. (2011) *Bemisia tabaci*: a statement of species status. *Annual Review of Entomology* **56**, 1–19.

Desneux, N., Decourtye, A. & Delpuech, J.-M. (2007) The sublethal effects of pesticides on beneficial arthropods. *Annual Review of Entomology* **52**, 81–106.

Dhadialla, T.S. (ed.) (2012) *Advances in Insect Physiology: Insect Growth Disruptors*. Vol. **43**. Academic Press, London.

Dillman, A.R., Chaston, J.M., Adams, B.J. *et al.* (2012) An entomopathogenic nematode by any other name. *PLoS Pathogens* **8**(3), e1002527. doi: 10.1371/journal.ppat.1002527

Ehler, L.E. (2006) Integrated pest management (IPM): definition, historical development and implementation, and the other IPM. *Pest Management Science* **62**, 787–9.

Flint, M.L. & Dreistadt, S.H. (1998) *Natural Enemies Handbook. The Illustrated Guide to Biological Pest Control*. University of California Press, Berkeley, CA.

Gibert, L.I. & Gill, S.S. (eds) (2010) *Insect Control: Biological and Synthetic Agents*. Academic Press, London.

Gilbert, L.I., Iatrou, K. & Gill, S.S. (eds) (2005) *Comprehensive Molecular Insect Science* Vol. 6, *Control*. Elsevier Pergamon, Oxford.

Goulson, D. (2013) An overview of the environmental risks posed by neonicotinoid insecticides. *Journal of Applied Ecology* **59**, 977–87.

Grafton-Cardwell, E.E., Stelinski, L.L. & Stansly, P.A. (2013) Biology and management of Asian citrus psyllid, vector of the huanglongbing pathogens. *Annual Review of Entomology* **58**, 413–32.

Grandgirard, J., Hoddle, M.S., Petit, J.N. *et al.* (2008) Engineering an invasion: classical biological control of the glassy-winged sharpshooter, *Homalodisca vitripennis*, by the egg parasitoid *Gonatocerus ashmeadi* in Tahiti and Moorea, French Polynesia. *Biological Invasions* **10**, 135–48.

Hajek, A. (2004) *Natural Enemies: An Introduction to Biological Control*. Cambridge University Press, Cambridge.

Hogg, B.N., Wang, X.G., Levy, K. *et al.* (2013) Complementary effects of resident natural enemies on the suppression of the introduced moth *Epiphyas postvittana*. *Biological Control* **64**, 125–31.

Hoy, M.A. *Insect Molecular Genetics: An Introduction to Principles and Applications*, 3rd edn. Academic Press, London & San Diego, CA.

Isman, M.B. (2006) Botanical insecticides, deterrents, and repellents in modern agriculture and an increasingly regulated world. *Annual Review of Entomology* **51**, 45–66.

Jervis, M. (ed.) (2007) *Insects as Natural Enemies: A Practical Perspective*. Springer, Dordrecht.

Jonsson, M., Wratten, S.D., Landis, D.A. & Gurr, G.M. (2008) Advances in conservation biological control of arthropods. *Biological Control* **45**, 172–5. [Part of a special issue of this journal dealing with conservation biological control.]

Kirst, H.A. (2010) The spinosyn family of insecticides: realizing the potential of natural products research. *The Journal of Antibiotics* **63**, 101–11.

Lockwood, J.A. (1993) Environmental issues involved in biological control of rangeland grasshoppers (Orthoptera: Acrididae) with exotic agents. *Environmental Entomology* **22**, 503–18.

Louda, S.M., Pemberton, R.W., Johnson, M.T. & Follett, P.A. (2003) Nontarget effects – the Achille's heel of biological control? Retrospective analyses to reduce risk associated with biocontrol introductions. *Annual Review of Entomology* **48**, 365–96.

Morales-Ramos, J., Rojas, G. & Shapiro-Ilan, D.I. (eds.) (2014) *Mass Production of Beneficial Organisms: Invertebrates and Entomopathogens*. Academic Press, London.

Neuenschwander, P. (2001) Biological control of the cassava mealybug in Africa: a review. *Biological Control* **21**, 214–29.

Papadopoulos, N.T., Plant, R.E. & Carey, J.R. (2013) From trickle to flood: the large-scale, cryptic invasion of California by tropical fruit flies. *Proceedings of the Royal Society B* **280**, 0131466. doi: 10.1098/rspb.2013.1466.

Pedigo, L.P. & Rice, M.E. (2006) *Entomology and Pest Management*, 5th edn. Pearson Prentice-Hall, Upper Saddle, NJ.

Radcliffe, E.B., Hutchison, W.D. & Cancelado, R.E. (eds) (2009) *Integrated Pest Management: Concepts, Tactics, Strategies and Case Studies*. Cambridge University Press, Cambridge.

Rahman, A.M., Roberts, H.L.S., Sarjan, M. *et al.* (2004) Induction and transmission of *Bacillus thuringiensis* tolerance in the flour moth *Ephestia kuehniella*. *Proceedings of the National Academy of Sciences* **101**, 2696–9.

Regnault-Roger, C., Vincent, C. & Arnason, J.T. (2012) Essential oils in insect control: low-risk products in a high-stakes world. *Annual Review of Entomology* **57**, 405–24.

Resh, V.H. & Cardé, R.T. (eds) (2009) *Encyclopedia of Insects*, 2nd edn. Elsevier, San Diego, CA. [See articles on biological control of insect pests; genetically modified plants; insecticide and acaricide resistance; integrated pest management; pathogens of insects; pheromones; physical control of insect pests; sterile insect technique.]

Robinson, W. (2005) *Urban Insects and Arachnids: A Handbook of Urban Entomology*. Cambridge University Press, Cambridge.

Roy, H.E., Vega, F.E., Chandler, D. *et al.* (eds) (2010) *The Ecology of Fungal Entomopathogens*. Springer, Dordrecht. [Previously published in *BioControl* **55**(1).]

Sanahuja, G., Banakar, R., Twyman, R.M. *et al.* (2011) *Bacillus thuringiensis*: a century of research, development and commercial applications. *Plant Biotechnology Journal* **9**, 283–300.

Shelton, A. (ed.) (2013) Biological Control: A Guide to Natural Enemies in North America. http://www.nysaes.cornell.edu/

Simberloff, D. (2012) Risks of biological control for conservation purposes. *BioControl* **57**, 263–76.

Suckling, D.M. & Brockerhoff, E.G. (2010) Invasion biology, ecology, and management of the light brown apple moth (Tortricidae). *Annual Review of Entomology* **55**, 285–306.

Tabashnik, B.E., Brévault, T. & Carrière, Y. (2013) Insect resistance to Bt crops: lessons from the first billion acres. *Nature Biotechnology* **31**, 510–21.

Tay, W.T., Soria, M.F., Walsh, T. *et al.* (2013) A brave New World for an Old World pest: *Helicoverpa armigera* (Lepidoptera: Noctuidae) in Brazil. *PLoS ONE* **8**(11), e80134. doi: 10.1371/journal.pone.0080134

Thacker, J.R.M. (2002) *An Introduction to Arthropod Pest Control*. Cambridge University Press, Cambridge.

Van Driesche, R., Hoddle, M. & Center, T. (2008) *Control of Pests and Weeds by Natural Enemies: An Introduction to Biological Control*. Blackwell Publishing, Malden, MA.

Van Emden, H. (2013) *Handbook of Agricultural Entomology*. Wiley-Blackwell, Chichester.

Vincent, C., Hallman, G., Panneton, B. & Fleurat-Lessard, F. (2003) Management of agricultural insects with physical control methods. *Annual Review of Entomology* **48**, 261–81.

Wajnberg, E. & Colazza, S. (eds) (2013) *Chemical Ecology of Insect Parasitoids*. Wiley-Blackwell, Chichester.

Williams, D.F. (ed.) (1994) *Exotic Ants: Biology, Impact, and Control of Introduced Species*. Westview Press, Boulder, CO.

Witzgall, P., Kirsch, P. & Cork, A. (2010) Sex pheromones and their impact on pest management. *Journal of Chemical Ecology* **36**, 80–100.

Zhang, H., Li, H.C. & Miao, X-X. (2013) Feasibility, limitation and possible solutions of RNAi-based technology for insect pest control. *Insect Science* **20**, 15–30.

Chapter 17

INSECTS IN A CHANGING WORLD

Insects can take advantage of our transport system.

We live with change in our surroundings and climate, and in human demographics (especially population growth). The consequences of the dominance of our species, with its ever-increasing and disproportionate demand for Earth's resources, continue to increase. Equally, since their origin some 400 million years ago, the insects have been exposed to remarkable changes due to geological and climatic events and diverse biotic interactions. Major extinctions have been caused by extraterrestrial (bolide) impacts, massive volcanic activity, and both global heating and cooling. Over tens of millions of years, continents have united and divided, oceans have formed and disappeared, and the proportions of atmospheric gases have fluctuated. Undoubtedly, amounts of "greenhouse" gases (carbon dioxide (CO_2), methane and the nitrous oxides), together with their impacts on climate, have varied in the geological past, in concert with storage or release of carbon from oceans and vegetation. Such changes have had consequences for life on Earth, with cascading effects across trophic levels.

The complexity of gradual and long-term change is that any one person's observations are inadequate to draw firm conclusions regarding cause and effect. So it is with climate data. The near 1°C rise in global mean surface temperatures since the start of the 20th century is undoubted. However, despite long-term data on land and ocean temperatures, rainfall, sea levels and extent of ice at the poles, global temperature-induced changes are subtle and inconsistent in the short term. Within our lifetimes we are aware of hotter years, drier and wetter seasons, and especially in recent years, more extreme climate events. Multi-year droughts, and "hundred-year" floods are part of the "new normal". Many long-established glaciers, including those curiosities on equatorial (tropical) mountains in Africa, Indonesia and New Guinea, soon will be lost completely. A very visible result of climate change is seen on a hike to the face of glaciers in New Zealand's mountains, which, due to glacial retreat, are several kilometres more distant than only two decades ago.

Although there is uncertainty concerning precise details of locations and degree of regional responses to global warming, data predicting long-term patterns of climate change clearly show enhanced human-induced greenhouse gases, plus land-use changes. In this chapter, we look at how insects already are responding to climatic change and environmental/habitat modifications, and our movement of insects and hosts around the globe. Naturally, most research has been undertaken on insect responses to climate change with regard to our welfare, and that of our animals and crops. The importance of modelling in prediction of how change will affect insects is addressed. We discuss increased biotic globalization, including of nuisance, agricultural and medically significant insects, and explore what changes already have been detected, and speculate as to the future for insects and ultimately ourselves, our crops and our diseases. We conclude by observing the need for an increasing role of entomologists in biosecurity. Boxes cover the modelling of fruit-fly distributions after new establishment (Box 17.1), the expansion of a coffee pest under climate change (Box 17.2), the ever-expanding suite of insects associated with eucalypts (Box 17.3), how insects change our urban landscapes (Box 17.4), and how biosecurity is changing with global trade (Box 17.5). The vignette at the beginning of this chapter, in appreciation of the cartoon style of Canadian insect illustrator Barry Flahey (http://www.magma.ca/~bflahey/lead.htm), highlights some pest insects known to hitch-hike via our transportation.

17.1 MODELS OF CHANGE

In Chapter 6 we saw some ways in which environmental factors affect insect development. Here we examine some predictive models of how insect abundance and distribution can change with abiotic climatic factors. Such models are used in: (i) predicting the potential distribution if a species "escapes" its native range (how far will it expand?); (ii) predicting what may happen under global-warming scenarios; and (iii) reconstructing past climates (or past species' distributions during a previous climate) from current insect distribution data. Increasingly, models are tested for veracity against the observed range changes to predict future or past changes.

Several aspects are less often included in climate models yet govern climate responses, especially of insects. The first is the incidence and duration of

The Insects: An Outline of Entomology, Fifth Edition. P.J. Gullan and P.S. Cranston.

Companion Website: www.wiley.com/go/gullan/insects

diapause (section 6.5). Diapause onset and cessation may be triggered by photoperiod – day length – and hence fixed by calendar rather than climatic variables. However, for many insects, entry to or exit from diapause, or both, is influenced by climate, notably temperature, which evidently does vary between years and will do so in the longer term under climate change. Diapause, which can be obligatory or facultative, is much more prevalent where resources fluctuate seasonally, and is rarer in less seasonal and in tropical climates. Low winter temperatures interact with diapause and survivorship. Milder winters will encourage higher survivorship (reduced death) during diapause, but less snow cover may counter survivorship by failing to provide a thermal protection against lethal fluctuating cold–thaw cycles. Milder winter and spring temperatures may set up a timing conflict with a fixed photoperiod trigger for cessation of diapause. Resumption of development may be decoupled from the availability of host plants and other resources. Predicted higher maximum temperatures approach or even exceed lethal temperatures for many insects.

Another aspect concerns generation time and number (**voltinism**; section 6.4). Climatic changes have direct effects on insect development, which depends largely on temperature (within upper and lower limits), since insects are **poikilotherms**. Shorter generation times at elevated temperatures, especially if development commences earlier in the year, will increase the number of generations for many insects, and thus the propensity for increased population size in the absence of commensurate increase in predation, parasitization or disease.

17.1.1 Modelling climate and insect distributions

The abundance of any **ectothermic** species is determined largely by proximate ecological factors, including the population densities of natural enemies and competitors (section 13.4) and interactions with habitat, food availability and climate. Although the distributions of insect species result from these ecological factors, there is also a historical component. Ecology determines whether a species can live in an area; history determines whether it does, or ever had the chance to live there. This difference relates to timing; given enough time, an ecological factor becomes a historical factor. In the context of present-day studies of where invasive insects occur and what the limits of their spread might be, history accounts for the original or native distribution. Knowledge of ecology may allow prediction of potential or future distributions under changed environmental conditions (e.g. as a result of "greenhouse effects") or as a result of accidental (or intentional) dispersal by humans. Thus, ecological knowledge of insect pests and their natural enemies, especially information on how climate influences their development, is vital for the prediction of pest outbreaks and for successful pest management.

Many models pertain to the population biology of economically important insects, especially those affecting major crop systems in western countries. One example of a climatic model of arthropod distribution and abundance is the computer-based system called CLIMEX (developed by R.W. Sutherst and G.F. Maywald), which predicts an insect's potential relative abundance and distribution around the world using ecophysiological data and the known current geographical distribution. An annual "ecoclimatic index" (EI), describing the climatic favourability of a given location for permanent colonization of an insect species, is derived from a climatic database combined with estimates of the insect's response to temperature, moisture and day length. The EI is calculated as follows (Fig. 17.1). A population growth index (GI) is determined first from weekly values averaged over a year to obtain a measure of the potential for population increase. This GI is estimated from data on the seasonal incidence and relative abundance across the species' range. Second, the GI is reduced by incorporation of four stress indices, which are measures of cold, heat, dry and wet.

Commonly, the existing geographical distribution and seasonal incidence of a pest species are known but biological data pertaining to climatic effects on development are scanty. Fortunately, the limiting effects of climate on a species can be estimated or derived from the current geographical distribution. Climatic tolerances are inferred from the climate of the sites where the species is known to occur and are described by the stress indices of the CLIMEX model. The values of the stress indices are adjusted progressively until the CLIMEX predictions match the observed distribution of the species. Naturally, other information on the climatic tolerances of the species should be incorporated where possible because the above procedure assumes that the present distribution is limited by climate alone, which is an oversimplification.

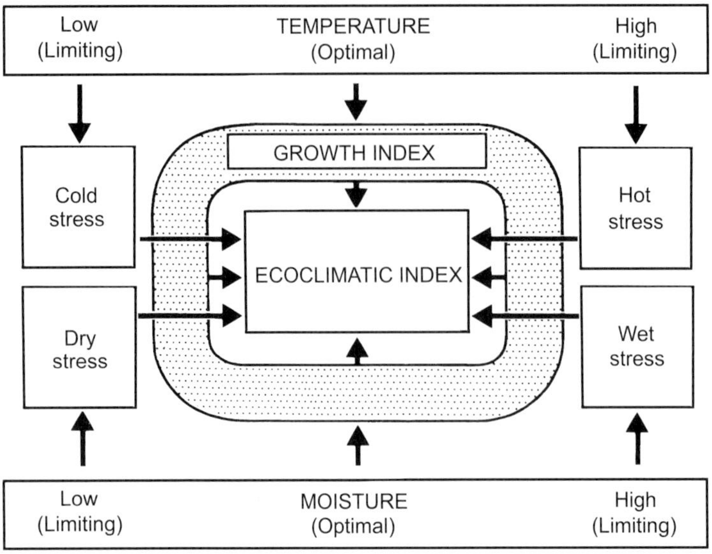

Fig. 17.1 Flow diagram depicting the derivation of the "ecoclimatic index" (EI) as the product of the population growth index and four stress indices. The EI value describes the climatic favourability of a given location for a given species. Comparison of EI values allows different locations to be assessed for their relative suitability to a particular species. (After Sutherst & Maywald 1985.)

Such climatic modelling has been carried out for many pest insects, including fruit flies (Box 17.1), Russian wheat aphid, the Colorado potato beetle, screw-worm flies, biting flies of *Haematobia* species, dung beetles, invasive fire ants, bark beetles, and coffee weevils (Box 17.2). The output is valuable in applied entomology, notably in epidemiology, quarantine, control of pest insects, and entomological management of weeds and animal pests (including other insects).

In reality, detailed information on ecological performances may never be attained for many taxa, although such data are essential for the ecophysiological distribution models described above. Nonetheless, demand exists for models of distribution in the absence of ecological performance data. Given these practical constraints, a class of modelling has been developed that accepts distribution point data as surrogates for "performance (process) characteristics" of organisms. These points are defined bioclimatically, and potential distributions can be modelled using some flexible procedures. Analyses assume that current species distributions are restricted (constrained) by bioclimatic factors. Models allow estimation of potential constraints on species distribution in a stepwise process. First, the sites at which a species occurs are recorded and the climate estimated for each data point, using a set of bioclimatic measures based on appropriate existing weather-station records. Annual precipitation, seasonality of precipitation, driest quarter precipitation, minimum temperature of the coldest period, maximum temperature of the warmest period, and elevation appear to be particularly influential, with significance in determining the distribution of ectothermic organisms. From this information, a bioclimatic profile is developed from the pooled climate per site estimates, providing a profile of the range of climatic conditions at all sites for the species. Next, the bioclimatic profiles so produced are matched with climate estimates at other mapped sites across a regional grid to identify all other locations with similar climates. Specialized software then can be used to measure similarity of sites, with comparison being made via a digital elevation model with fine resolution. All locations within the grid with similar climates to the species-profile form a predicted bioclimatic domain. This is represented spatially (mapped) as a "predicted potential distribution" for the taxon under consideration, in which isobars (or colours) represent different degrees of confidence in the prediction of presence.

17.1.2 Climate and historic insect range changes

Modelling techniques lend themselves to back-tracking, allowing reconstruction of past species' distributions based on previous climate and/or reconstruction of

Box 17.1 Modelling distributions of fruit flies

Bioclimatic models have been used to simulate the responses of pest insects should they arrive in a new area. The Queensland fruit fly (Australian Q-fly), *Bactrocera tryoni* (Diptera: Tephritidae), is a highly polyphagous and potentially invasive pest that can affect many commercial fruits. The female oviposits into maturing fruit, laying a clutch of eggs. Larval feeding combined with bacterial growth causes rotting and destruction of the fruit. Even minor damage in an orchard restricts interstate and international marketing of the fruit.

The known response of *B. tryoni* to climatic parameters in Australia has been used to extrapolate to North America, should the species gain entry there. First, using CLIMEX (section 17.1.1, Fig. 17.1), the response of *B. tryoni* to Australia's climate was modelled (after Sutherst & Maywald 1991). Growth and stress indices of CLIMEX are inferred from maps of the geographical distribution and from estimates of the relative abundance of this species in different parts of its native range. The map of Australia depicts the ecoclimatic indices (EI) describing the favourableness of each site for permanent colonization by *B. tryoni*. The area of each circle is proportional to its EI. Crosses indicate sites that the fly could not permanently colonize.

The potential survival of *B. tryoni* as an immigrant pest in North America has been predicted using CLIMEX by climate matching with the fly's native Australian range. Accidental transport of this fly could lead to its establishment at the point of entry or it might be taken to other areas with climates more favourable to its persistence. Should *B. tryoni* become established in North America, the eastern seaboard from New York to Florida and west to Kansas, Oklahoma, and Texas in the United States and much of Mexico are predicted to be most at risk. Canada and most of the central and western United States are unlikely to support permanent colonization. High risk of infestation by *B. tryoni* is only in certain regions of the continent, and quarantine authorities have been maintaining appropriate vigilance.

Sophisticated models incorporating additional parameters are being developed. An initial application allows assessment of how the "Q-fly" and biosecurity personnel might respond if the fly were introduced to Western Australia (where it has invaded and been eradicated previously). The model incorporates detailed life-cycle modelling of cohorts of insects, accounting for ambient temperature, landscape heterogeneity and stochastic dispersal of the flies, and incorporating host variation according to season. Such a dynamic model, with some fixed entomological and geographic data and other stochastic elements (allowing variation in dispersal and life-cycle timing), allows assessment in real time of likely spread, and the effectiveness of control strategies.

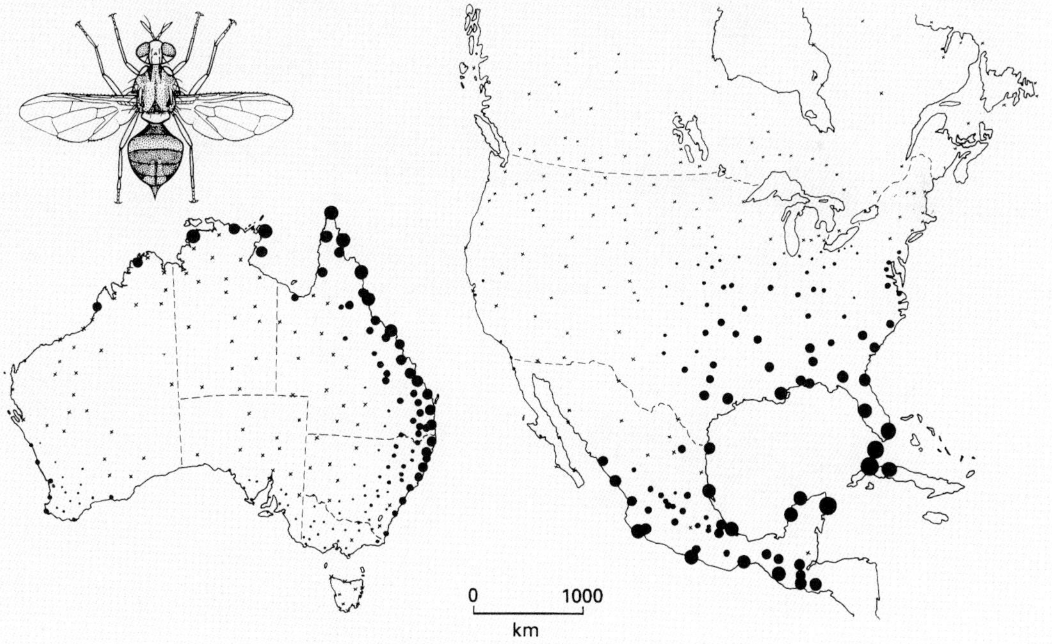

Box 17.2 Trouble brewing? A beetle threat to coffee

Could a humble insect bring one of the world's most lucrative drug trades to a halt? Not a leaf beetle chewing coca leaves, or a bug sucking sap of marijuana, or a tobacco moth – but a humble weevil with an addiction to coffee beans. Billions of cups of coffee are drunk daily, providing the major source of caffeine and, for many of us, the preferred recreational drug. But now it seems that the source of the world's favourite drink is threatened by an unexpected side effect of climate change. Worst-case scenarios predict that all places where the plants grow optimally (for premium coffee) will no longer be suitable by 2080 under climate change, compounded by the growing menace of an insect seed predator.

Our coffee comes from the berries (seeds) of a bush belonging to the genus *Coffea*, which originated in highland Ethiopia in northeast Africa and subsequently was relocated to Yemen on the opposite side of the Gulf of Aden (see map). The stimulatory effects from the high caffeine content reputedly were discovered by goat-herders observing the hyperactivity of their flocks after browsing the bushes.

When introduced to Europe in the 16th and 17th centuries, coffee consumption surged. Demand grew such that to supplement the original, premium coffee species, *C. arabica*, a second species of coffee (*C. canephora* = *C. robusta*) was brought into production. This less-desirable species, a mainstay of instant coffee, is native to lowland African humid forests, and although hardier than *C. arabica,* cannot tolerate either temperatures above

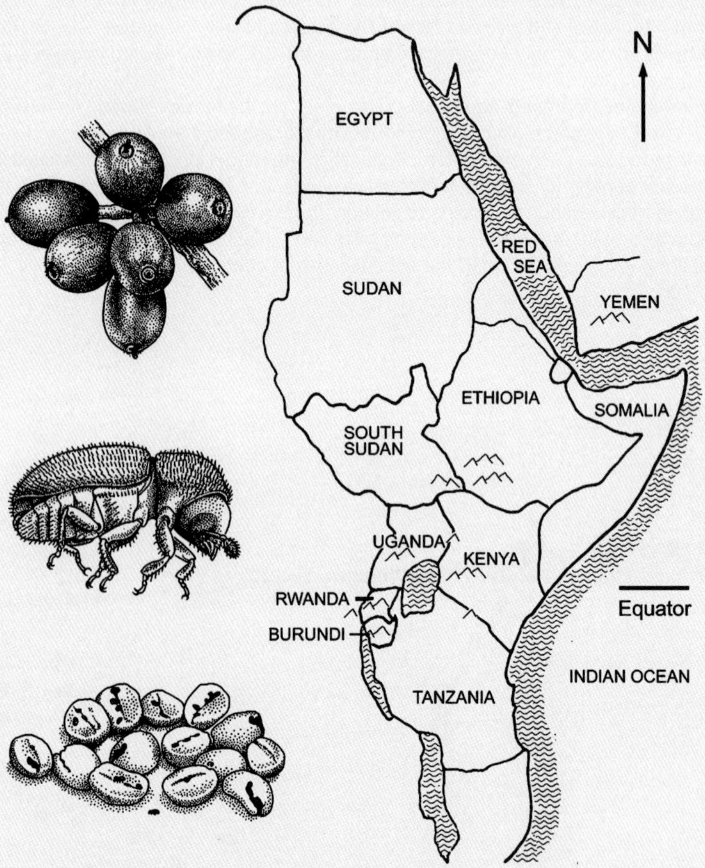

26°C or prolonged dry conditions. Between these two species, coffee is now produced in climatically suitable areas throughout the tropical and subtropical world. Currently, some 26 million farmers worldwide depend on coffee to provide an income from the global US$15 billion export industry.

All this is threatened by elevated temperatures and altered rainfall patterns already evident in coffee-growing areas, perhaps especially in East Africa. Temperature and humidity changes especially challenge *C. arabica*, which is grown between 1000 and 2000 m elevation under shade trees, where ambient temperatures range narrowly from 18 to 21°C. Stressed trees yield fewer and lower quality beans.

However, climate change alone is not the whole story. The coffee berry borer, *Hypothenemus hampei* (Coleoptera: Curculionidae, Scolytinae), is an ambrosia beetle (adult is illustrated on the middle left) with females that bore into coffee berries (illustrated top left) and excavate galleries to lay eggs. Larval feeding increases the damage to the beans (shown bottom left with several adults). The beetles resist (actually break down) caffeine, which is toxic to other insects. Damage to the valuable berry is exacerbated by the entry, via the bore holes, of bacterial and fungal pathogens, one or more of which are likely to be a specific associate of the beetles, vectored in fungus-carrying structures on or in the beetles' bodies. Arabica coffee is the beetle's favoured host, but other coffee species sometimes are affected. The beetle originates in tropical Africa, but one consequence of the world trade in commodities (as we have seen elsewhere in this book) is the redistribution of pests, and the coffee berry borer is now in all coffee production areas worldwide, excepting China and Nepal. Furthermore, with elevated temperatures in the growing areas of *C. arabica* in East Africa, the beetle increasingly overlaps with the distribution range of arabica coffee and has increased its number of generations per year. Studies show that in a decade the weevil has extended its elevational range upwards by 300 m on Tanzania's Mount Kilimanjaro, where temperatures previously had restrained its distribution to the lowest slopes. The same has been observed in Indonesia as temperatures rise.

Predictions based on a well-accepted climate-change scenario for east Africa, using the climatic model CLIMEX, imply that the pest will spread rapidly to all premium coffee-growing areas in eastern Africa, from Ethiopia to Kenya, and furthermore, the number of weevil generations per year may increase to ten. This will reduce the income of subsistence farmers throughout the continent (and presumably elsewhere as the borer takes hold), increase demand for pesticides and reduce availability of premium organic coffee, result in more use of lower quality robusta coffee, and inevitably increase the cost of our daily coffee fix.

past climates based on postglacial fossil remains representing past distributional information. Such studies originated with pollen (palynology) from lakes and mires, in which rather broad assemblages of pollen, which often include a handful of "indicator species" that facilitate tracking changes in vegetation through time and across landscapes, are associated with reconstructed past climates. More-refined data come from preserved insects, including beetles (especially their elytra) and the head capsules of larval chironomid flies. Short-lived organisms, such as insects, respond more rapidly to climatic events than do trees. Extrapolation from inferred bioclimatic controls governing the present-day distributional ranges of insect species to those same taxa preserved hundreds or thousands of years ago allows reconstructions of previous climates. For example, major features from the late Quaternary period include a rapid recovery from extreme conditions at the peak of the last glaciation (14,500 years ago), with intermittent reversals to colder periods in a general warming trend. Such data derived from fossil insect models refines cruder signals provided from pollen. Non-biological data providing verification for such insect-based reconstructions has come from independent chemical signals and congruence with a Younger Dryas cold period (11,400–10,500 years ago), and documented records in human history such as a medieval 12th century warm event and the 17th century Little Ice Age when "Ice Fairs" were held on the frozen River Thames. Inferred temperature changes range from 1°C to 6°C, sometimes over just a few decades.

17.2 ECONOMICALLY SIGNIFICANT INSECTS UNDER CLIMATE CHANGE

Confirmation of past temperature-associated biotic changes lends support for modelling future range changes. For example, estimates have been made for

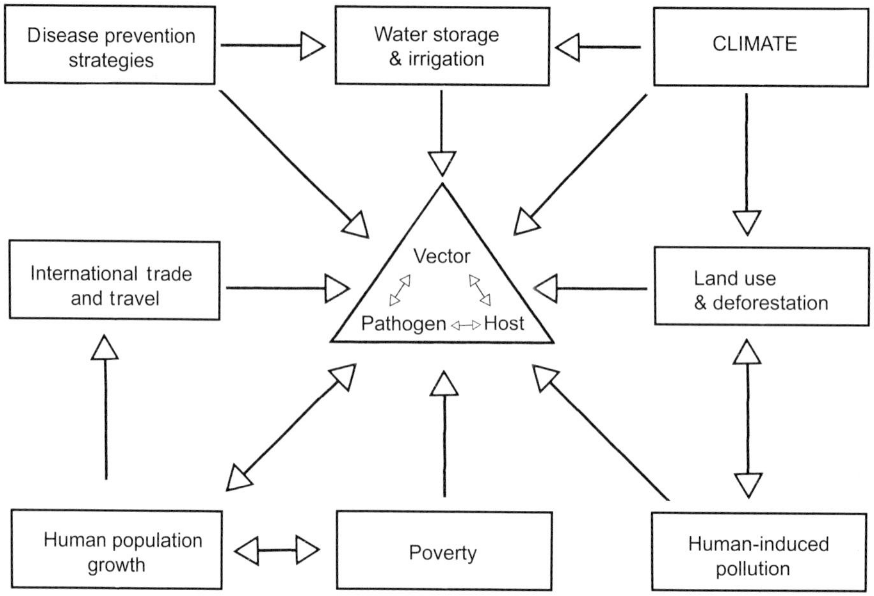

Fig. 17.2 The complexity of interactions between some of the many environmental factors that affect the vector–pathogen–host epidemiological cycle. (Adapted from Sutherst 2004; Tabachnick 2010.)

disease-transmitting vector mosquitoes and biting midges under different climate-change scenarios. These have ranged from naïve estimates of increased range of disease vectors into populated areas currently disease-free (where vectors exist in the absence of the virus) to increasingly sophisticated models accounting for altered development rates for vector and arbovirus, and altered environments for larval development. Linkage with climate is extremely complex, and studies remain speculative if based on simple models lacking detail concerning major factors. Even projections of major changes in insect-borne disease transmission can lack crucial detail. Because of some uncertainty in models, and lack of clarity concerning predicted future climate, certain policy makers continue to deny the existence or biotic significance of climate change. However, Europe certainly warmed 0.8°C in the 20th century, and realistic expectations are for a further global increase of between 2.1°C and 4.6°C in the 21st century, along with commensurate variation in other climatic factors such as seasonality and rainfall. Predicted altered distributions of insects are evident, especially from studies of individual species of butterflies and odonates, and of insects that affect our health and that of our crops and animals.

Climate (whether changing or not) undoubtedly has a profound and overriding control over the potential of each and every vector to transmit disease. Amongst the factors discussed in section 15.3.1, environment (via climate) directly affects vector distribution, vector abundance, life expectancy (survivorship) of the vector, and vector competence. Amongst the factors that affect these complex vector–pathogen–host cycles, elevated temperatures will decrease vector generation time, elevate vector population growth rate, diminish the incubation period for pathogens, and increase the period during which transmission can occur – all increasing the potential for transmission. However, this is countered by a decrease in adult longevity and shortened life history of the vector. In the well-studied West Nile virus disease cycle in the United States (see Box 15.5), responses of the mosquito *Culex pipiens quinquefasciatus* to temperature were found to be non-linear – as such, vector competence under changed environmental conditions could not be easily predicted. For insect vectors with aquatic immature stages, predicted changes in severity and frequency of droughts and floods will affect populations in differing and largely unpredictable ways. Add genetic variability of the vector, plus adaptation to environmental change, and host demographic changes with climate, and we can see that prediction is fraught with difficulty. Figure 17.2 summarizes the complex interactions between some (of many) environmental factors that affect vector–pathogen–host epidemiological cycles.

Turning to the relationships between plant–insect interactions and environmental factors, again we find great complexity, with many variables affecting the host (plant) under climate-change scenarios. Thus, with increased CO_2, elevated temperatures and more frequent water stress, plants alter resources for herbivores, including insects feeding below and above ground. Estimates that plants will produce more biomass, but be of lower nutritional quality (nutrient-depleted leaves, for example) suggest herbivorous insects will be smaller or require longer development time (or both) and will be less fit. A changed climate will alter phenology (timing) of resource availability, especially in temperate host plants, the timing of insect herbivore activity with respect to food availability (both leaves and flowers), and affect the suite of predators and parasitoids that track their prey. Each will respond differently, according to their physiological preferences and tolerances. Although this potential de-synchronization of parts of complex interactions again is difficult to model and predict, some progress has been made in manipulated experimental systems and using "natural experiments" in which "greenhouse effects" are seen in agroecosystems.

17.2.1 Future agricultural health

Agriculture will be affected by changes in temperature, rainfall (amount and seasonality), water availability, absolute amount of CO_2 available for photosynthesis, and the amount of solar radiation. All factors are expected to change over time, sometimes benefitting, sometimes retarding crop production. For such an important economic activity, models proliferate, although few consider the impacts of change on the insects – both pest and beneficial – associated with agriculture. Evidently, increased pest-insect activity will lead to increased demand for insecticides, adding costs to production, as would research to establish more-appropriate biological control agents. Loss of effective control of pests by their natural enemies could arise if some crop pests find an "escape" from their more climatically constrained natural enemies, comprising specialist predators and parasitoids.

Effective biological control depends upon the activity of predators and adult parasitoids, the susceptibility of the particular pest species, and the health status of the host plant (e.g. flushing of new growth, production of seeds and fruit, etc.). The first two factors are vulnerable to alteration in diapause in pests or parasitoids. If each uses a different cue to terminate diapause, the tightly synchronized life histories required for sustainable control will be decoupled. If each interacting species responds differently to climate change, certain host plants may escape pests, and pests independently may evade parasitoids.

Theoretically, earlier emergence of adult pest hosts under warmer conditions would mean that eggs are laid earlier, thereby evading the timely attention of egg parasitoids such as trichogrammatid wasps. Under heightened temperature and lower humidity, parasitic wasps may fail to exert any control, and generally parasitoids seem to be more sensitive to climatic variability than their hosts. Studies of tritrophic interactions between plants, lepidopteran larvae and their parasitoids suggest that phenological asynchrony occurs between parasitoids and their hosts when extreme weather events such as floods and droughts cause temporal shifts in caterpillar populations, making them less available to specialized parasitoids that are in their host-seeking searching phase.

Examples of escape from parasitoids include the aphid *Aphis gossypii*, which is predicted to become a more serious pest under heightened CO_2 levels, with greater survival in part due to extended development times for its major coccinellid beetle predator (*Propylaea japonica*) under these altered conditions. The cassava mealybug, which is generally well controlled by one or more encyrtid hymenopterans (see Box 16.5), responds to drought-stress of its host cassava plants by enhancing its rate of encapsulation of parasitoids (section 13.2.3). The combination of drought stress and enhanced resistance of the mealybug to parasitoids can lead to greater losses of this staple crop for many rural Africans.

Elevated temperatures can mean earlier breaking of dormancy (diapause) and seasonal advancement of egg-laying (at least for species with temperature rather than day-length control of these behaviours), and also more rapid development and potentially more generations. For **multivoltine** pests such as aphids, cornborers and the grape berry moth, *Paralobesia viteana* (Lepidoptera: Tortricidae), additional generations are predicted if degree-days accumulate more rapidly under climate change. For the global pest of pome fruit, the codling moth, *Cydia pomonella*, in Switzerland the chance of a second generation rises from less than 20% to between 70 and 100%, with heightened risk even of a third generation within the next half century at current temperature predictions. Although diapause in codling moth is induced by decreasing photoperiod (day length), selection for onset of diapause at shorter photoperiod may occur, as in some other lepidopterans. The longer season and increased numbers of codling moth under climate change inevitably will need altered management of this pest.

17.2.2 Future animal health

A range of diseases of our livestock (mammals and birds reared for human consumption) have been expanding their geographic range in recent times. In some cases this expansion has been related to climate change, but in other cases to greater global movements of humans and vectors. Elevated temperatures can allow more rapid development of pathogenic organisms, especially in the ectothermic insect vectors of arboviruses (see section 15.3.2). A well-studied example concerns bluetongue virus (BTV), which causes a disease in ruminants (sheep and occasionally cattle). The virus is transmitted by biting midges (Ceratopogonidae: *Culicoides* spp.), especially *Culicoides imicola*. The disease was present in Africa, Asia, Australia, South America and North America, but only in recent times has it expanded northwards into Europe, causing death in sheep and cattle. The introduction of new virus serotypes, a change in vector species, and elevated temperatures all may have contributed to expansion of the range in which this disease now occurs. Although the major sweep into southern and central Europe started in the late 1990s in concert with a warming climate, the jump to northwest Europe appears to have taken place in 2006, a year in which climate was calculated (retrospectively) to have been optimal for midge expansion. Applying such a model to include predicted future climate change implies an inevitable increased risk of BTV continuing its expansion northwards in Europe, with commensurate deaths and costs of management. However, other researchers consider that other non-climate factors are at play, including a role for the risky movement of African game to European wildlife parks – bringing BTV into the range of *Culicoides* species that can transmit it from asymptomatic large African mammals to European deer and susceptible sheep.

However, in a remarkable parallel expansion, the equine disease African horse sickness, vectored by the same blood-feeding midge, has expanded from its endemic area in sub-Saharan Africa to the northeast (India and Pakistan) and Iberia (Spain and Portugal) in the west. This suggests that spread of this arbovirus relates primarily to range expansion of the vector(s), due to alteration of prevalence or seasonality, whether directly or indirectly due to climate change. The same applies to other vectors of diseases that affect our health.

17.2.3 Future human health

Altered climate surely has a role in increased insect-borne pathogens affecting the spread of human diseases. Our rapid-changing and increasingly connected world encourages human-induced spread of diseases and their vectors. Increased temperatures and altered rainfall patterns lead to regions becoming newly suitable for more tropical and subtropical vectors. The arrival of exotic diseases is enhanced by many of our activities, including migration of humans infected with pathogens. Thus, malaria entered the New World via the 15th–16th-century slave trade from its origin in West Africa, and was transmitted by a range of existing *Anopheles* mosquito vectors on its arrival in the Americas. In the early 1930s, a member of the highly vectorial *An. gambiae* complex, almost certainly *An. arabiensis* (see Box 15.2), arrived in Brazil via human-mediated transport from Africa. Only by the chance coincidence with a yellow-fever eradication programme in Brazil was this anopheline vector eliminated. Similar and novel scenarios for insect-borne disease spread continue to be created.

Chikungunya fever, which broke out recently in Europe (Italy and France), exempifies the coincidence of contributing factors. First, a most effective vector mosquito, *Aedes albopictus*, has expanded from its native range in Southeast Asia (hence its name of Asian tiger mosquito) to much of the subtropical and tropical world. Range expansion has been enhanced by its biology – the eggs are laid in small water containers that also are capable of accidental or intentional movement around the globe. The arrival of this most invasive species in North America was via unregulated trade in used vehicle tyres, stored outdoors and then shipped for recycling. Water retained in these tyres supported high densities of eggs and larvae of *Ae. albopictus*, which persisted across the Pacific as tyres were shipped to the United States. Entry was essentially unregulated, and the species quickly found a ready home in similar "habitats" throughout the southern United States. Similar trade appears to be have been involved in transfer to central and South America, Europe and beyond. An alternative dispersal route in water associated with "lucky bamboo" plants imported from Asia to the United States and Europe has allowed a form of *Ae. albopictus* to establish in European horticultural hothouses. In Italy, numbers of potential vector mosquitoes grew rapidly in the first decade of the 21st century to attain nuisance proportions. The problem of human biting is exacerbated by the propensity of *Ae. albopictus* to transfer arboviruses when the female feeds (section 15.5). When chikungunya virus, known previously as a rare disease in Africa, came to Italy with African migrants, *Ae. albopictus* – an effective vector – already was established. A "new" disease had arrived, and became established due to changed human migration patterns, altered distribution of a potent vector, and a warming European climate that allowed both vector and arbovirus to persist and spread.

Undoubtedly, climate change has also played a role in the current rapid rate of spread of dengue (see Box 15.4), due to increased human water storage against increased drought, but without the containers being mosquito-proofed. As shown in Fig. 17.2, climate change interacts with human ecology to change the potential course of insect-mediated diseases, including dengue.

17.3 IMPLICATIONS OF CLIMATE CHANGE FOR INSECT BIODIVERSITY AND CONSERVATION

17.3.1 Range change

Researchers studying insect responses to ongoing change in climate suggest that insects are expected to "track" their preferred climate by altering their range to higher latitudes and/or elevations in response to increased temperature. A study of species of western European butterflies (limited to non-migrants, and excluding monophagous and/or geographically restricted taxa) is quite conclusive. Northward extension of ranges is evident for many taxa (63% of 57 species), whereas only two species shifted south. For the many butterfly species for which boundaries moved, an observed range shift of from 35 to 240 km in the past 30–100 years coincided quite closely with the (north) polewards movement of the isotherms over the period. That such changes have been induced by a modest temperature increase of <1°C surely provides evidence for the dramatic effects of the ongoing "global warming" over the remainder of this century.

An entire order of insects may be at particular risk from climate change – the Grylloblattodea. All North American species (genus *Grylloblatta*) depend on snowfields, ice caves and peri-glacial habitats, all of which are under threat from global warming. Equally, the East Asian diversity is restricted to cool temperate and montane areas. In western North America, the rate of glacier retreat is predicted to increase two- to four-fold compared with the current, already high rate of loss. Recent searches have failed to find several Californian *Grylloblatta* species, apparently due to climate-induced loss of their habitats. However, for East Asian grylloblattids, molecular evolutionary studies suggest that these insects have responded more to the historic creation and loss of land-bridges by sea-level changes between Japan and the continent due to cyclical climatic change during the Pliocene and Pleistocene periods, than to the direct effects of altered climate.

In terms of insect conservation, the ability of insects to respond to climate change by migrating to track suitable climate is constrained by prevailing geography. Thus, insects that increase their elevational range will be limited if the mountains are too low to provide appropriate cooler conditions. Although most research concerns temperate insects, even tropical geometrid moths and ants studied along an altitudinal transect in Costa Rica may face range-shift gaps and potential extinction with a predicted 1000 m elevational range shift of isotherms. Equally, continents are bounded by coastlines and inhospitable oceans that prevent expanded ranges. We might expect narrowly range-constrained insects to contract their populations, even to become extinct, when faced with elevated temperatures. Conversely, those that can overcome barriers to their movement may become more abundant (perhaps as "invasives") and interact negatively with residents. Both outcomes occur. Additionally, even within larger areas of habitat, human interference with such habitats (fragmentation) means that possibilities for accommodating to elevated temperature are constrained by landscape barriers to range change. Constraints include continued land conversion to "hostile" agricultural monocultures, and expansion of urban landscapes that prevent appropriate range-shift responses of insects unable to tolerate such changed land use. Solutions can lie in increasing connectivity of small fragmented areas of suitable habitat ("patches") to allow reconnection of isolated and endangered populations into more sustainable **metapopulations** via corridors of suitable habitat in a matrix of unsuitable environments. Such conservation schemes include re-establishment of native riparian vegetation along rivers and streams, and of heathland corridors and coppiced woodlands, especially for some charismatic butterfly species. These will allow many insects to track their optimal climate "envelope", with fewer barriers to movement.

17.3.2 Temporal changes and asynchrony of mutual interactions

Climate warming can lead to early emergences of insect species or populations in the spring and summer. For example, long-term records from the Central Valley of California show that the first flight date of 70% of 23 butterfly species advanced by an average of 24 days over a period of 31 years, with climate factors explaining most of the variation in this trend. An associated phenomenon is the increasing phenological asynchrony seen in some insect–plant systems, with insect activity becoming mismatched with availability of their host plants. Similar changes in the activity periods of host insects and their parasitoids may lead to higher or lower levels of parasitization, depending on individual temperature responses of the interacting species. Such trophic asynchronies may have serious ecological consequences and lead to species declines or extinctions.

From a conservation perspective, flowering plant and specialist pollinator biodiversity may be threatened if the flight period of adult pollinators (especially lepidopterans, hymenopterans and dipterans) becomes asynchronous with the flowering of plants with which they have close, even obligate, relationships. We have evidence of major advancement of flowering of many spring plants in the northern hemisphere in response to less than 1°C of warming. This will not be an issue for generalist pollinators like honey bees, but may be so for the large suite of specialists, such as solitary bees. Of course, this scenario is predicated on the pollinators responding to different environmental signals to that of the flowering of the plant – if both become earlier in tandem, all is well and good. However, it is still unclear whether phenological mismatch or identical response will be the norm.

17.4 GLOBAL TRADE AND INSECTS

Only a few generations ago we relied on produce from a limited local area. Vegetables and fruit were grown regionally, eaten seasonally, and preserved for the lean times. Our timber for construction and firewood came from forests surrounding our domiciles. Nonetheless, some exotic commodities have been traded for hundreds of years, including spices from the Far East. Exotic plants for our horticulture and agriculture were gathered from remote areas of the globe. With human population growth outstripping local resources, or with people living where few such supplies were available to begin with, commodity trade increased. We have became ever more interdependent, especially in the so-called developed world. We demanded, and have come to expect, year-round availability of seasonal products from our stores, wherever we live. We can assume that as soon as we stockpiled and traded grains, stored-product pests took a ride with the produce. Initially, many imported crops from the New World, such as potatoes, tomatoes, corn and peppers, came to Europe without their pests, as transport was of seeds, presumably pest-free by luck rather than by design. Only centuries later have some pests of these plants arrived (e.g. the Colorado potato beetle, Box 16.7).

Many insect pests have also spread around the world on ornamental plants and cut flowers. The floriculture trade easily can distribute small flower- and foliage-feeding insects such as thrips (Thysanoptera), mealybugs and other scale insects (Hemiptera: Coccoidea), whiteflies (Hemiptera: Aleyrodoidea) and psylloids (Hemiptera: Psylloidea). For example, poinsettias (*Euphorbia pulcherrima*; Euphorbiaceae), which are widely distributed and sold in North America at Christmas, are excellent hosts of introduced *Bemisia* whiteflies (see Box 16.2), and the poinsettia thrips, *Echinothrips americanus* (Thripidae), which is native to eastern North America, has travelled with the horticultural trade to much of Europe and is gaining a foothold in Asia. Asian tiger mosquitoes, *Aedes albopictus* (Diptera: Culicidae), have been transported around the world associated with water of "lucky bamboo" plants, *Dracaena braunii* (Asparagaceae). This plant is popularly believed to bring happiness and prosperity to the owners, but often it brings dengue-transmitting mosquitoes instead.

It is not only edible produce and ornamental plants that are traded widely – for example, our houses are constructed with timber even when we live in treeless, arid places. Much of the world's construction timber comes from only a few species in the diverse pine genus. Two species, *Pinus radiata*, which once grew only in coastal central California, and *Pinus patula*, from the southeast highlands of Mexico, dominate plantation forests, including in Australia and southern Africa, respectively. Both species are highly affected by *Sirex* wood wasps presumed to have been introduced from Europe with timber and firewood. Globally, cellulose fibre (historically utilized for paper and packaging but with increasingly broad uses as a food additive, an inactive ingredient in prescription drugs, and as a component of rayon, etc.) depends largely upon a few species of *Eucalyptus*, originally from southeast Australian forests. These tree species, which have been pest-free for over 100 years, are increasingly harmed by trade-introduced pest insects (Box 17.3). Timber, forest products and agriculture seem particularly susceptible to globally transported pests – the schematic in Fig. 17.3, based on the world's major timber pests, but applicable more widely, shows the dominant features leading to their globalization. Massively increased trade, homogenization of the world's important crops and forestry species, and failure to intercept insects (including those associated not only with living materials but also with packing materials such as wooden pallets and similar cheap raw wood) is leading to a global pool of insect pests waiting to evade biosecurity and expand their pest ranges. Unrestricted trade undoubtedly has allowed a diversity of livewood- and deadwood-boring insects entry to far shores, including North America. The emerald ash borer (Box 17.4) exemplifies an introduction route and subsequent spread, belated discovery and difficulties in managing this tree destroyer. More extreme climate events, broadly predicted by current models, stress forest trees and heighten susceptibility to pest insects such as bark and ambrosia beetles (section 9.2).

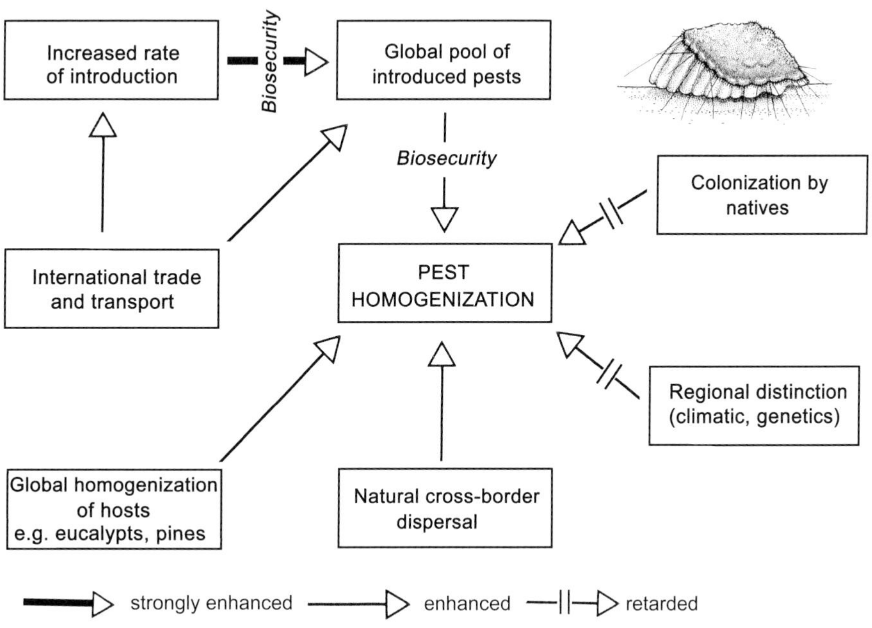

Fig. 17.3 Interactions between factors associated with insect-pest homogenization. Inset: *Icerya purchasi* (Hemiptera: Monophlebidae), a global pest scale insect (see Box 16.3). (Adapted from Garnas *et al.* 2012.)

Box 17.3 Global eucalypts and their pests

The gum trees and their relatives, the "eucalypts" (Myrtaceae), form a diverse group of Australian trees, the largest genus of which, *Eucalyptus*, has species of great timber value especially as pulpwood for cellulose production (for paper and other diverse, value-added products; see text). Commercially important species have been cultivated not only in Australia outside their native ranges, but over much of the globe. The attractive mature shape of many species has encouraged their use as urban landscape trees, notably in California. The young foliage of certain ornamental species constitutes an important source for the floral trade; other species are cultivated for their essential oils, used in herbal medicines, fragrances, insect repellents, or antimicrobials. One appealing factor for worldwide cultivation of eucalypts was their escape from the attentions of a suite of native Australian phytophagous insects (e.g. sections 11.2.2–5), whose damage reduces growth rates, detracts from timber qualities, the mature tree shape and the aesthetics of ornamental foliage, and in some cases causes tree mortality.

For about a century, eucalypts in the United States (including Hawai'i) remained pest-free. However, mostly since the 1980s, a suite of phytophagous insects from Australia have become established there. These include leaf-eating chrysomelid beetles, gall wasps, many psyllids and two longhorn borers belonging to the genus *Phoracantha* (Cerambycidae). The first recognized arrival, the eucalypt longhorn borer, *P. semipunctata* (illustrated in Taxobox 22), has been introduced accidentally into virtually all regions of the world in which *Eucalyptus* is grown. A subsequently introduced borer, *P. recurva*, has a similar biology but, whereas *P. semipunctata* was controlled rapidly by an egg parasitoid wasp (Encyrtidae) from Australia, *P. recurva* is less controlled and has come to dominate in California. A eucalypt snout beetle (*Gonipterus* sp., Curculionidae) (see Plate 8f) has established widely, but is partially controlled by an introduced Australian egg parasitoid wasp (Mymaridae), whose efficacy may be compromised by the recent discovery of cryptic species of the weevil (see Box 7.1). Perhaps the most rapidly spreading, and perhaps most devastating pest, the blue gum chalcid (*Leptocybe invasa*, Eulophidae), has reached Florida after worldwide spread to eucalypt-growing countries in less than a decade. This **thelytokous** wasp induces galls on foliage and young stems of a number of eucalypt species, including the commercially important *E. camaldulensis* and *E. globulus*, and currently there is no suitable control measure. Many of these pests were established already outside of Australia, and evidently entered the United States by indirect routes, whereas others appear to have arrived directly from Australia, perhaps with the help of human agency. Ongoing arrivals of pests, especially lerp-building psyllids (see Plate 8e), means that the search for biological control agents in the native range must continue.

Unexpectedly, some native phytophagous insects, perhaps especially in South America, can and do move over to plantation and ornamental eucalypts. These insects either are generalist herbivores, or feed especially on plants in the family Myrtaceae that are related to the eucalypts. Ironically, this means that the home of *Eucalyptus* diversity is threatened now by introduction of alien pests acquired in the expanded range of the commercial plantations. Globalization can be a two-way process – with export of eucalypt insect pests and import of others. Vigilance over trade, by enhanced quarantine procedures, will be necessary to limit the impact of inevitable continued transfer of insects.

As with transport of used car tyres, dirty industrial machinery has allowed exotic insects, especially tramp ants (see Box 1.3), to hitch rides to other parts of the world. Unless biosecurity/quarantine vigilance is enhanced, we can expect ever-more such pests, which can create habitat change of similar magnitude to that expected under an altered climate alone.

Naturally, all movement of produce runs the risk of transporting insects (see examples in Box 16.1 and Box 17.3). Released from the natural controls of their home range, many insects are predisposed to become "invasive" in a new location. This is especially so if the insects encounter monocultures of an appropriate host in a "pest-free" state. In the past, and still in countries that value their pest-free status highly, governments sought/seek to protect from invasive pests by border biosecurity (Box 17.5) via risk assessment and selective inspection of "at-risk" transport. Such "biosecurity" is easier to operate in single political entities (countries) with secure borders formed by coastlines, such as Australia, New Zealand and Japan, but much more difficult for countries with extensive land borders with neighbours. As well as savings in control costs, including pesticides, the benefits of certifiably

Box 17.4 Alien insects change landscapes

Native forests and rural landscapes are part of our culture and heritage – from the pine-dominated forests of the north, the redwoods of California, the cedars of the Middle East, to the wattles and eucalypts of Australia. Urban streetscapes of native and exotic trees can mean much for urbanites in providing shade and aesthetics. Old, large exotic trees such as redwoods and oaks can gain heritage status even in the new lands. While loss by clear felling of complete forests, such as in the rainforests of Asia and South America, rightly claim attention of conservationists, the insidious losses of individual tree species can be as troubling. It is difficult now to recall how important elms (*Ulmus* species) once were in Western Europe, as portrayed in the early 19th century by the English landscape artist, John Constable. In the 1960s, a fungal infection of elm trees (Dutch elm disease) transmitted by two species of *Scolytus* beetles (Curculionidae) reached England. Within a decade, 75% (over 20 million) of trees were dead and a dominant tree was lost from the countryside. Dutch elm disease caused similar devastation to elms in much of North America, where elms also are damaged seriously by elm leaf beetles, *Xanthogaleruca luteola* (Chrysomelidae), and most recently by the Japanese beetle *Popillia japonica* (Scarabaeidae).

Landscape trees in North America are seriously threatened by the emerald ash borer, *Agrilus planipennis* (Buprestidae), a jewel beetle native to Asia. The beetle feeds on ash trees (*Fraxinus* spp.): on the foliage as adults and under the bark as larvae (as shown here; partly after Cappaert *et al*. 2005). The beautiful iridescent emerald green adult beetles use a wider range of host tree species in Japan. In the United States, ashes are popular woodlot trees and provide bower-like avenues in suburbs. In 2002, responding to local deaths of ash trees, entomologists found the ash borer near Detroit in southeast Michigan. Subsequently, in less than a decade, many million of ash trees died in the "core area". Despite rapid installation of a quarantine zone, with fines for transporting wood, tens of millions more ash trees are being lost in an ever-expanding region including southern Ontario, Michigan, Ohio, Illinois, Indiana, Pennsylvania, Missouri, Virginia and West Virginia. The costs to municipalities, property owners and nurseries has run into the millions of dollars, and the aesthetics of tree-lined suburban streets, as well as biodiversity values of woodlots, has been lost. The emerald ash borer attacks all species of North American ashes, albeit with different intensity, and the threat of a broadened host range, as is found in Japan, remains possible. European ash trees are now threatened from a population of ash borer established in the Moscow area and spreading west and southwards.

Based on genetic studies, emerald ash borer is native to China, apparently exported in wood used in pallets and packing. From cross-dating techniques using dendrochronology ("tree rings") in ash trees killed by buprestids, the emerald ash borer appears to have been in the United States for a decade prior to first detection. Such a lag time between arrival and detection seems common to many such invasions, and obviously hinders control. For valuable ornamental trees, drenching with a systematic insecticide (imidacloprid; section 16.4.1) will save a tree, although

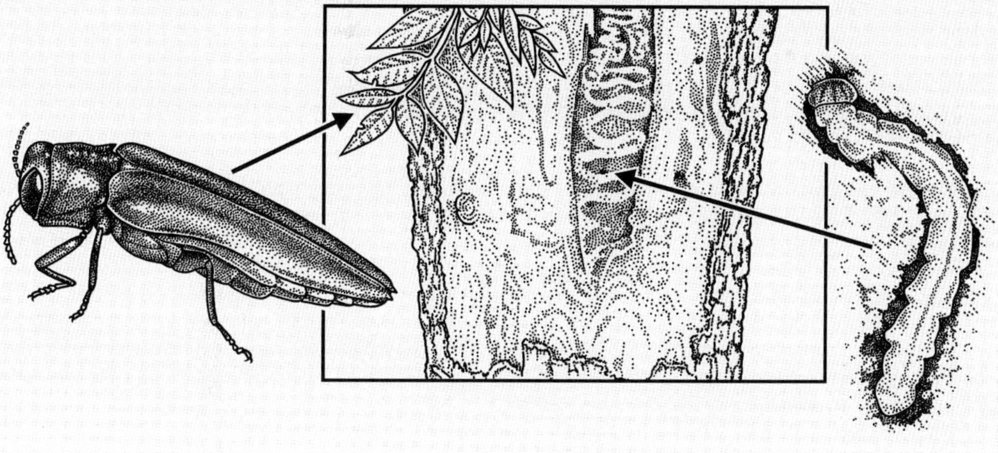

with poorly understood negative environmental effects. Such strategies are unsuitable for extensive woodlots and efforts have been made to introduce classical biological control agents (section 16.5) from the native range in China. The most promising is the gregarious larval endoparasitoid, *Tetrastichus planipennisi* (Eulophidae). In the United States, a native parasitoid wasp, *Atanycolus hicoriae* (Braconidae), will develop within the ash borer, but its life cycle is too poorly synchronized to effect control. Interestingly, *Cerceris fumipennis* (Crabronidae), a native solitary wasp that provisions its nests with buprestid beetles, may assist in locating populations that otherwise are very cryptic almost until death of the tree.

Another species of *Agrilus*, the goldspotted oak borer (*Agrilus auroguttatus*) (see **Plate 8g**), may be poised to destroy many of California's oaks. Early evidence of oak deaths in San Diego County attributed to effects of drought has been found subsequently to be due to this borer. This beautiful buprestid is native to arid southwest Arizona, where it appears to be under control by natural enemies. Inter-state movement was probably associated with timber cut for firewood.

New invasive species, numbers of which appear each year, exemplify the risks of accidental movement of damaging species through poorly regulated commerce. Even the most diligent inspections can miss beetles living in timber used for packing materials, or in firewood, but the eradication costs ("shutting the stable door") are immense compared to initial prevention.

Box 17.5 Insects and biosecurity – an Australian perspective

Biosecurity is a term coined recently for all aspects of the protection of the economy, environment and human health from negative impacts associated with entry, establishment or spread of exotic pests (including weeds) and diseases. As defined, it is broader than quarantine, as it includes pests and weeds that are not vectors of disease yet can cause great economic or environmental damage.

Australia, as an island continent, has a relatively pest- and disease-free status that provides an advantage in global markets. The national economy has a strong primary produce (agriculture, fisheries and forestry) emphasis and relies on exports for its standard of living. Protection of its status through biosecurity policies can serve as a model for balancing need for trade and protection of natural and agricultural resources.

With costs estimated in the billions of dollars for the incursion of a single species of tramp ant (see Box 1.3), precautions need to be taken to prevent incursions. Being relatively pest-free, Australia justifiably could restrict entry of many agricultural products from less-favoured parts of the world. Here, the World Trade Organization has a role to play in providing rules governing biosecurity and restrictions to free trade. Thus, although nations are allowed to adopt measures necessary for the protection of human, animal and plant life or health, such measures must be "science-based, not more trade restrictive than required and not arbitrarily or unjustifiably discriminatory against trading partners". The concept of an "Appropriate Level of Protection" applies. For Australia (and New Zealand), the appropriate level is set high to defend the substantially pest-free status, with risk of incursion reduced to near zero. However, this is a source of dispute, as trading partners claim that it is designed to protect primary producers against "fair competition".

Against this background, many countries employ entomologists (and phytologists and related scientists) to provide the science behind the risk assessment exercises on a "per pest" and "per crop" basis. Biological factors assessed for prospective pest insects include native range and behaviour, propensity to spread naturally or by human actions, likelihood of interception and detection, likelihood of establishment and spread from a source, and possibilities of biological control and/or other forms of eradication. Predictive models such as CLIMEX (Box 17.1) are integral to assessment of the impact of a novel insect arrival. Entomological expertise is meshed with socio-political and economic assessments to produce a risk analysis for any given potential invasive insect. Such a biosecurity procedure differs from quarantine in shifting from the difficult to enforce "zero risk" to a calculated "manageable risk" of incursion.

Risk analysis allows identification and listing of "most unwanted" insects and production of educational fact sheets for their recognition and actions to be taken. Australia's current list includes Asian gypsy moth (Lymantriidae: *Lymantria dispar*), Asian longhorn beetle (Cerambycidae: *Anoplophora glabripennis*), Asian tiger mosquito (Culicidae: *Aedes albopictus*), Formosan termite (Rhinotermitidae: *Coptotermes formosanus*) and khapra beetle (Dermestidae: *Trogoderma granarium*). These insects span a wide range of biologies, but are united by being serious pests elsewhere, and in the devastation they would cause by breaching biosecurity.

Biosecurity according to this scheme requires well-trained entomologists with substantial knowledge of: (i) the diagnostics of a wide variety of pest and potential pest insects; (ii) modern taxonomy, including molecular techniques, to distinguish natives from new arrivals; and (iii) sampling designs to survey imported materials of all kinds optimally and efficiently. In this book, we have shown examples of the failure to distinguish pests from relatives (of significance in effective biological control), and also of the continuing breaches of biosecurity and the emergent pests such as those mentioned in Box 16.1. Evidently, there will be an ongoing requirement for biosecurity personnel skilled in entomology, for detection and control, including (hopefully) biological control programmes.

pest-free commodities lies especially with high-quality, high-value exports. The verified or potential presence of an unwanted pest in a crop will render it only of local value as any recipient country would want to exclude pests and would refuse entry of contaminated produce. The costs of eradication of an introduced pest insect may run into millions of dollars (see Box 1.3), if indeed elimination can be attained. Clearly, prevention is best, and border vigilance is essential, yet the World Trade Organization rules governing "free trade" favour unfettered trade over biosecurity concerns. Can we put a price on native plants and insects in their native landscapes, or must we accept global homogenization, with weedy biota everywhere?

FURTHER READING

Auger-Rozenberg, M.-A. & Roques, A. (2012) Seed wasp invasions promoted by unregulated seed trade affect vegetal and animal biodiversity. *Integrative Zoology* **7**, 228–46.

Bale, J.S. & Hayward, S.A.L. (2010) Insects overwintering in a changing climate. *The Journal of Experimental Biology* **213**, 980–94.

Bentz, B.J., Régnière, J., Fettig, C.J. *et al.* (2010) Climate change and bark beetles of the Western United States and Canada: direct and indirect effects. *BioScience* **60**, 602–13.

Cornelissen, T. (2011) Climate change and its effects on terrestrial insects and herbivory patterns. *Neotropical Entomology* **40**, 155–63.

Garcia-Adeva, J.J., Botha, J.H. & Reynolds, M. (2012). A simulation modelling approach to forecast establishment and spread of *Bactrocera* fruit flies. *Ecological Modelling* **227**, 93–108.

Garnas, J.R., Hurley, B.P., Slippers, B. & Wingfield, M.J. (2012) Biological control of forest plantation pests in an interconnected world requires greater international focus. *International Journal of Pest Management* **58**, 211–23.

Hoffmann, A.A. (2010) A genetic perspective on insect climate specialists. *Australian Journal of Entomology* **49**, 93–103.

Jaramillo, J., Muchugu, E., Vega, F.E. *et al.* (2011) Some like it hot: the influence and implications of climate change on coffee berry borer (*Hypothenemus hampei*) and coffee production in East Africa. *PLoS One* **6**(9), e24528. doi: 10.1371/journal.pone.0024528

Paine T.D., Steinbauer, M.J. & Lawson, S.A. (2011) Native and exotic pests of *Eucalyptus*: a worldwide perspective. *Annual Review of Entomology* **56**, 181–201.

Simberloff, D. (2012) Risks of biological control for conservation purposes. *BioControl* **57**, 263–76.

Sutherst, R.W. (2004) Global change and human vulnerability to vector-borne diseases. *Clinical Microbiology Reviews* **17**, 136–73.

Tabachnick, W.J. (2010) Challenges in predicting climate and environmental effects on vector-borne disease episystems in a changing world. *The Journal of Experimental Biology* **213**, 946–54.

Thomson, L.J., Macfadyen, S. & Hoffmann, A.A. (2010) Predicting the effects of climate change on natural enemies of agricultural pests. *Biological Control* **52**, 296–306.

Zavala, J.A., Nabity, P.D. & DeLucia, E.H. (2013) An emerging understanding of mechanisms governing insect herbivory under elevated CO_2. *Annual Review of Entomology* **58**, 79–97.

Chapter 18

METHODS IN ENTOMOLOGY: COLLECTING, PRESERVATION, CURATION AND IDENTIFICATION

Alfred Russel Wallace collecting butterflies. (After various sources, especially van Oosterzee 1997; Gardiner 1998.)

For many entomologists, questions of how and what to collect and preserve are determined by the research project (see also section 13.4). Choice of methods may depend upon the target taxa, life-history stage, geographical scope, kind of host plant or host animal, disease vector status, and most importantly, sampling design and cost-effectiveness. One factor common to all such studies is the need to communicate the information unambiguously to others, not least concerning the identity of the study organism(s). Undoubtedly, this will involve identification of specimens to provide names (section 1.4), which are necessary not only to tell others about the work, but also to provide access to previously published studies on the same, or related, insects. Identification requires material to be appropriately preserved so as to allow recognition of morphological features that vary among taxa and life-history stages, or to allow later extraction and amplification of DNA if a barcoding approach is used. After identifications have been made, the specimens remain important, and even have added value, and it is important to preserve some material (**vouchers**) for future reference. As information grows, it may be necessary to revisit the specimens to confirm identity, or to compare with later-collected material.

In this chapter, we review a range of collecting methods, mounting and preservation techniques, and specimen curation, and discuss methods and principles of identification, including DNA-based methods.

18.1 COLLECTION

Those who study many aspects of vertebrate and plant biology can observe and manipulate their study organisms in the field, identify them and, for larger animals, also capture, mark and release them with few or no harmful effects. Amongst the insects, these techniques are available perhaps only for butterflies and dragonflies, and the larger beetles and bugs. Most insects can be identified reliably only after collection and preservation. Naturally, this raises ethical considerations, and it is important to:

- collect responsibly;
- obtain the appropriate permit(s);
- ensure that **voucher specimens** are deposited in a well-maintained museum collection.

Responsible collecting means collecting only what is needed, avoidance or minimization of habitat destruction, and making the specimens as useful as possible to all researchers by providing labels with detailed collection data. In many countries or in designated reserve areas, permission is needed to collect insects. It is the collector's responsibility to apply for permits and fulfil the demands of any permit-issuing agency. Furthermore, if specimens are worth collecting in the first place, they should be preserved as a record of what has been studied. Collectors should ensure that all specimens (in the case of taxonomic work) or at least representative voucher specimens (in the case of ecological, genetic or behavioural research) are deposited in a recognized museum. Voucher specimens from surveys or experimental studies may be vital to later research.

Depending upon the project, collection methods may be active or passive. Active collecting involves searching the environment for insects, and may be preceded by periods of observation before obtaining specimens for identification purposes. Active collecting tends to be quite specific, allowing targeting of the insects to be collected. Passive collecting involves erection or installation of traps, lures or extraction devices, and entrapment depends upon the activity of the insects themselves. This is a much more general type of collecting, being relatively unselective in terms of what is captured.

18.1.1 Active collecting

Active collecting may involve physically picking individuals from the habitat, using fingers, a fine-hair brush, forceps or an aspirator (also known in Britain as a pooter). Such techniques are useful for relatively slow-moving insects, such as immature stages and sedentary adults that may be incapable of flying or reluctant to fly. Insects revealed by searching particular habitats, as in turning over stones, removing tree bark, or observed at rest by night, are all amenable to direct picking in this manner. Night-flying insects can be selectively picked from a light sheet – a piece of white cloth with an ultraviolet light suspended above it

The Insects: An Outline of Entomology, Fifth Edition. P.J. Gullan and P.S. Cranston.

Companion Website: www.wiley.com/go/gullan/insects

(but be careful to protect eyes and skin from exposure to ultraviolet light).

Netting has long been a popular technique for capturing active insects. The vignette at the beginning of this chapter depicts the naturalist and biogeographer Alfred Russel Wallace attempting to net the rare butterfly *Graphium androcles*, in Ternate in 1858. Most insect nets have a handle about 50 cm long and a bag held open by a hoop of 35 cm diameter. For fast-flying, mobile insects such as butterflies and flies, a net with a longer handle and a wider mouth is appropriate, whereas a net with a narrower mouth and a shorter handle is sufficient for small and/or less agile insects. The net bag should always be deeper than the diameter so that the insects caught may be trapped in the bag when the net is twisted over. Nets can be used to capture insects whilst on the wing, or by using sweeping movements over the substrate to capture insects as they take wing on being disturbed, as for example from flower heads or other vegetation. Techniques of beating (sweeping) the vegetation require a stouter net than those used to intercept flight. Some insects when disturbed drop to the ground: this is especially true of beetles. The technique of beating vegetation whilst a net or tray is held beneath allows the capture of insects that exhibit this defensive behaviour. Indeed, it is recommended that even when seeking to pick individuals from exposed positions a net or tray be placed beneath for the inevitable specimen that will evade capture by dropping to the ground (where it may be impossible to locate). Nets should be emptied frequently to prevent damage to the more fragile contents by more massive objects. Emptying depends upon the methods to be used for preservation. Selected individuals can be removed by picking or aspiration, or the complete contents can be emptied into a container, or onto a white tray from which targeted taxa can be removed (but beware of fast fliers departing).

The above netting techniques can be used in aquatic habitats, though specialist aquatic nets tend to be of different materials from those used for terrestrial insects, and of smaller size (resistance to dragging a net through water is much greater than through air). Choice of mesh size is an important consideration – the finer mesh net required for capture of a small aquatic larva (compared with an adult beetle) provides more resistance to being dragged through the water than a wider mesh. Aquatic nets are usually shallower and triangular in shape, rather than the circular shape used for trapping active aerial insects. This allows for more effective use in aquatic environments.

18.1.2 Passive collecting

Many insects live in microhabitats from which they are difficult to extract – notably in leaf litter and similar soil debris or in deep tussocks of vegetation. Physical inspection of the habitat may be difficult, and in such cases the behaviour of the insects can be used to separate them from the vegetation, detritus or soil. Particularly useful are negative phototaxic and thermotaxic and positive hygrotaxic responses, in which the target insects move away from a source of strong heat and/or light along a gradient of increasing moisture, at the end of which they are concentrated and trapped. The Tullgren funnel (sometimes called a Berlese funnel) comprises a large (e.g. 60 cm diameter) metal funnel tapering to a replaceable collecting jar. Inside the funnel a metal mesh supports the sample of leaf litter or vegetation. A well-fitting lid containing illuminating lights is placed just above the sample and sets up a heat and humidity gradient that drives the live animals downwards in the funnel until they drop into the collecting jar, which contains ethanol or other preservative.

The Winkler bag operates on similar principles, with drying of organic matter (litter, soil, leaves) forcing mobile animals downwards into a collecting chamber. The device consists of a wire frame enclosed with cloth that is tied at the top to ensure that specimens do not escape and to prevent invasion by scavengers, such as ants. Pre-sieved organic matter is placed into one or more mesh sleeves, which are hung from the metal frame within the bag. The bottom of the bag tapers into a screw-on plastic collecting jar containing either preserving fluid, or moist tissue paper for live material. Winkler bags are hung from a branch or from rope tied between two objects, and operate via the drying effects of the sun and wind. However, even mild windy conditions cause much detritus to fall into the jar, thus defeating the major purpose of the trap. They are extremely light, require no electric power and are very useful for collecting in remote areas, although when housed inside buildings or in areas subject to rain or high humidity they can take many days to dry completely, and thus extraction of the fauna may be slow.

Separating bags rely on the positive phototaxic (light) response of many flying insects. The bags are made from thick calico, with the upper end fastened to a supporting internal ring on top of which is a clear Perspex lid; they are either suspended on strings or supported on a tripod. Collections made by general sweeping or from a specific habitat are introduced by quickly tipping the net contents into the separator and closing the lid. Those mobile (flying) insects that are attracted to light will fly to the upper, clear surface, from which they can be collected with a long-tubed aspirator inserted through a slit in the side of the bag.

Insect flight activity is seldom random, and it is possible for an observer to recognize more frequently used routes and to place barrier traps to intercept the flight path. Margins of habitats (ecotones), stream edges, and gaps in vegetation are evidently more utilized routes. Traps that rely on the interception of flight activity and the subsequent predictable response of certain insects include Malaise traps and window traps. The Malaise trap is a kind of modified tent in which insects are intercepted by a transverse barrier of net material. Those that seek to fly or climb over the vertical face of the trap are directed by this innate response into an uppermost corner and from there into a collection jar, usually containing liquid preservative. A modified Malaise trap, with a fluid-filled gutter added below, can be used to trap and preserve those insects whose natural reaction is to drop when contact is made with a barrier. Based on similar principles, the window trap consists of a window-like vertical surface of glass, Perspex or black fabric mesh, with a gutter of preserving fluid lying beneath. Only insects that drop on contact with the window are collected when they fall into the preserving fluid. Both traps are conventionally placed with the base to the ground, but either trap can be raised above the ground, for example into a forest canopy, and still function appropriately.

On the ground, interception of crawling insects can be achieved by sinking containers into the ground to rim-level such that active insects fall in and cannot climb out. These pitfall traps vary in size, and may feature a roof to restrict dilution with rain and preclude access by inquisitive vertebrates (Fig. 18.1). Trapping can be enhanced by construction of a fence-line to guide insects to the pitfall, and by baiting the trap. Specimens can be collected dry if the container contains insecticide and crumpled paper, but more usually they are collected into a low-volatile liquid, such as propylene glycol or ethylene glycol, and water, of varying composition, depending on the frequency of visitation to empty the contents. Adding a few drops of detergent to the pitfall trap fluid reduces the surface tension and prevents the insects from floating on the surface of the liquid. Pitfall traps are used routinely to estimate species richness and relative abundances of ground-active insects. However, strong biases in trapping success must be considered when comparing sites of differing habitat structure (density of

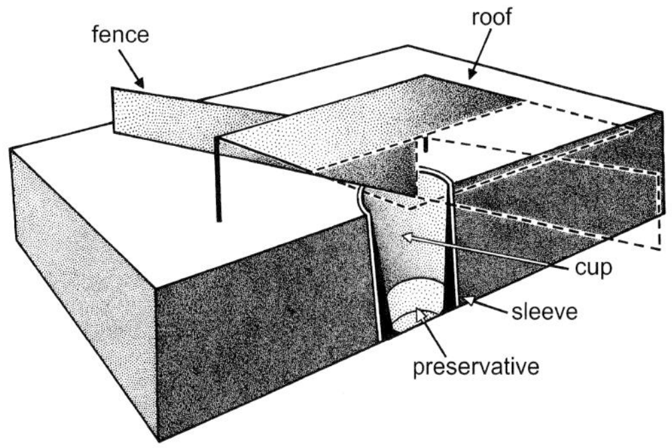

Fig. 18.1 A diagrammatic pitfall trap cut away to show the in-ground cup filled with preserving fluid. (After an unpublished drawing by A. Hastings.)

vegetation). This is because the probability of capture of an individual insect (trappability) is affected by the complexity of the vegetation and/or substrate that surrounds each trap. Habitat structure should be measured and controlled for in such comparative studies. Trappability is affected also by the activity levels of insects (due to their physiological state, weather, etc.), their behaviour (e.g. some species avoid traps or escape from them), and by trap size (e.g. small traps may exclude larger species). Thus, the capture rate (C) for pitfall traps varies with the population density (N) and trappability (T) of the insect according to the equation $C = TN$. Usually, researchers are interested in estimating the population density of captured insects or in determining the presence or absence of species, but such studies will be biased if trappability varies between study sites or over the time interval of the study. Similarly, comparisons of the abundances of different species will be biased if one species is more trappable than another.

Many insects are attracted by baits or lures, placed in or around traps; these can be designed as "generic" to lure many insects, or "specific", designed for a single target. Pitfall traps, which trap a broad spectrum of mobile ground insects, can have their effectiveness increased by baiting with meat (for carrion attraction), dung (for **coprophagous** insects such as dung beetles), fresh or rotting fruit (for certain Lepidoptera, Coleoptera and Diptera), or **pheromones** (for specific target insects such as fruit flies). A sweet, fermenting mixture of alcohol plus brown sugar or molasses can be daubed on surfaces to lure night-flying insects, a method termed "sugaring". Carbon dioxide and volatiles such as butanol can be used to lure vertebrate-host-seeking insects such as mosquitoes and horse flies.

Colours differentially attract insects: yellow is a strong lure for many hymenopterans and dipterans. This behaviour is exploited in yellow pan traps, which are simple yellow dishes filled with water and a surface-tension reducing detergent, and placed on the ground to lure flying insects to death by drowning. Outdoor swimming pools act as giant pan traps.

Light trapping (see section 18.1.1 for light sheets) exploits the attraction to light of many nocturnal flying insects, particularly to the ultraviolet light emitted by fluorescent and mercury vapour lamps. After attraction to the light, insects may be picked or aspirated individually from a white sheet hung behind the light, or they may be funnelled into a container such as a tank filled with egg carton packaging. There is rarely a need to kill all insects arriving at a light trap, and live insects may be sorted and inspected for retention or release.

In flowing water, strategic placement of a stationary net to intercept the flow will trap many organisms, including live immature stages of insects that may otherwise be difficult to obtain. Generally, a fine mesh net is used, secured to a stable structure such as bank, tree or bridge, to intercept the flow in such a way that drifting insects (either deliberately or by dislodgement) enter the net. Another passive-trapping technique used in water involves emergence traps, which are generally large inverted cones, into which adult insects fly on emergence. Such traps also can be used in terrestrial situations, such as over detritus or dung, etc.

18.2 PRESERVATION AND CURATION

Most adult insects are pinned or mounted and stored dry, although the adults of some orders and all soft-bodied immature insects (eggs, larvae, nymphs, pupae or puparia) are preserved in vials of 70–80% ethanol (ethyl alcohol) or mounted onto microscope slides. Pupal cases, cocoons, waxy coverings and exuviae may be kept dry and either pinned, mounted on cards or points, or, if delicate, stored in gelatin capsules or in preserving fluid.

18.2.1 Dry preservation

Killing and handling prior to dry mounting

Insects that are intended to be pinned and stored dry are best killed either in a killing bottle or tube containing a volatile poison, or in a freezer. Freezing avoids the use of chemical killing agents but it is important to place the insects into a small, airtight container to prevent drying out and to freeze them for at least 12–24 h. Frozen insects must be handled carefully and properly thawed before being pinned, otherwise the brittle appendages may break off. The safest and most readily available liquid killing agent is ethyl acetate, which although flammable, is not especially dangerous unless directly inhaled. It should not be used in an enclosed room. More-poisonous substances, such as cyanide and chloroform, should be avoided by all except the most experienced entomologists. Ethyl acetate killing containers are made by pouring a thick mixture of plaster of Paris and water into the bottom of a tube or wide-mouthed bottle or jar to a depth of 15–20 mm; the plaster must be completely dried before use. To "charge" a killing bottle, a small amount of ethyl acetate is poured onto and absorbed by the plaster, which can then be covered with tissue or cellulose wadding. With frequent use, particularly in hot weather, the container will need to be recharged regularly by adding more ethyl acetate. Crumpled tissue placed in the container will prevent insects from contacting and damaging each other. Killing bottles should be kept clean and dry, and insects should be removed as soon as they die to avoid colour loss. Moths and butterflies should be killed separately to avoid them contaminating other insects with their scales. For details of the use of other killing agents, refer to either Martin (1977) or Upton & Mantle (2010) under Further reading.

Dead insects exhibit *rigor mortis* (stiffening of the muscles), which makes their appendages difficult to handle, and it is usually better to keep them in the killing bottle or in a hydrated atmosphere for 8–24 h (depending on size and species) until they have relaxed (see below), rather than pin them immediately after death. It should be noted that some large insects, especially weevils, may take many hours to die in ethyl acetate vapours, and a few insects do not freeze easily and thus may not be killed quickly in a normal household freezer.

It is important to eviscerate (remove the gut and other internal organs of) large insects or gravid females (especially cockroaches, grasshoppers, katydids, mantids, stick-insects and very large moths), otherwise the abdomens may rot and the surface of the specimens go greasy. Evisceration, also called gutting, is best carried out by making a slit along the side of the abdomen (in the membrane between the terga and sterna) using fine, sharp scissors and removing the body contents with a pair of fine forceps. A mixture of 3 parts talcum powder and 1 part boracic acid can be dusted into the body cavity, which in larger insects may be stuffed carefully with cotton wool.

The best preparations are made by mounting insects while they are fresh, and insects that have dried out must be relaxed before they can be mounted. Relaxing involves placing the dry specimens in a water-saturated atmosphere, preferably with a mould deterrent, for one to several days depending on the size of the insects. A suitable relaxing box can be made by placing a wet sponge or damp sand in the bottom of a plastic container or a wide jar and closing the lid firmly. Most smaller insects will be relaxed within 24 h, but larger specimens will take longer, during which time they should be checked regularly to ensure they do not become too wet.

Pinning, staging, pointing, carding, spreading and setting

Specimens should be mounted only when they are fully relaxed, i.e. when their legs and wings are freely movable, rather than stiff or dry and brittle. All dry-mounting methods use entomological macropins – these are stainless steel pins, mostly 32–40 mm long, and come in a range of thicknesses and with either a solid or a nylon head. *Never use dressmakers' pins for mounting insects*; they are too short and too thick. There are three widely used methods for mounting insects and the choice of the appropriate method depends on the kind of insect and its size, as well as the purpose of mounting. For scientific and professional collections, insects are either pinned directly with a macropin, micropinned or pointed, as follows.

Direct pinning

This involves inserting a macropin, of appropriate thickness for the insect's size, directly through the insect's body; the correct position for the pin varies among insect orders (Fig. 18.2; section 18.2.4) and it is important to place the pin in the suggested place to avoid damaging structures that may be useful in identification. Specimens should be positioned about three-quarters of the way up the pin, with at least 7 mm protruding above the insect to allow the mount to be gripped below the pin head using entomological forceps (which have a broad, truncate end) (Fig. 18.3). Specimens then are held in the desired positions on a piece of polyethylene foam or a cork board until they dry, which may take up to three weeks for large specimens. A desiccator or other artificial drying methods are recommended in humid climates, but oven temperature should not rise above 35°C.

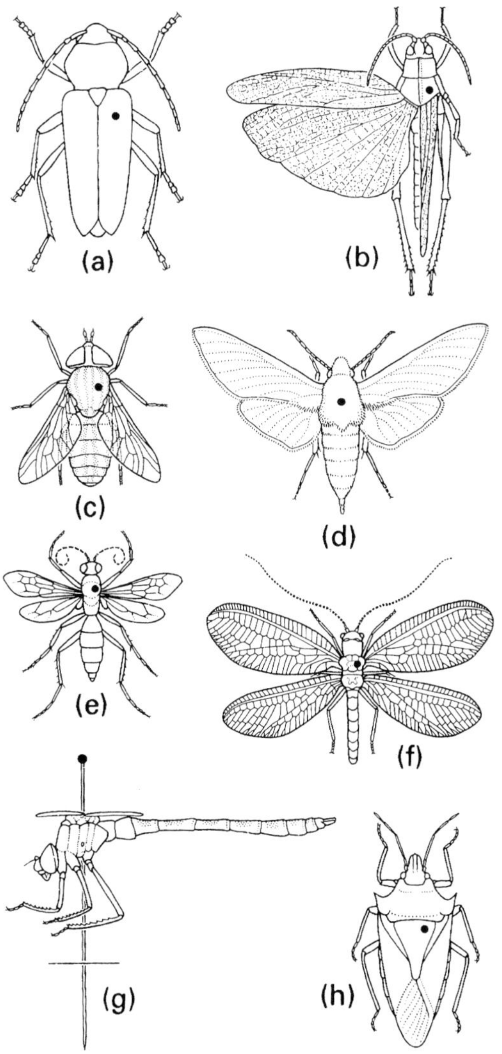

Fig. 18.2 Pin positions for representative insects: (a) larger beetles (Coleoptera); (b) grasshoppers, katydids and crickets (Orthoptera); (c) larger flies (Diptera); (d) moths and butterflies (Lepidoptera); (e) wasps and sawflies (Hymenoptera); (f) lacewings (Neuroptera); (g) dragonflies and damselflies (Odonata), lateral view; (h) bugs, cicadas, leafhoppers and planthoppers (Hemiptera: Heteroptera, Cicadomorpha and Fulgoromorpha).

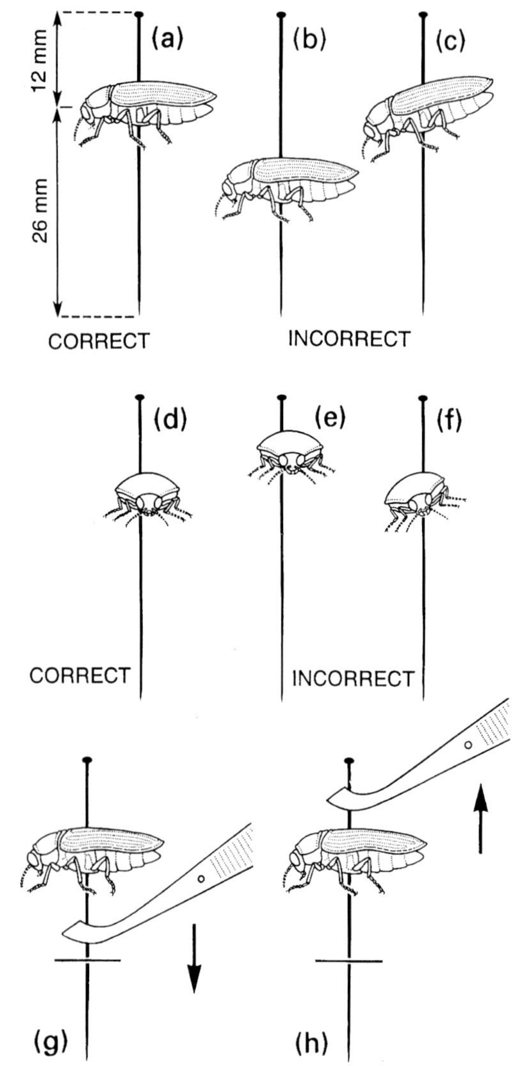

Fig. 18.3 Correct and incorrect pinning: (a) insect in lateral view, correctly positioned; (b) too low on pin; (c) tilted on long axis, instead of horizontal; (d) insect in front view, correctly positioned; (e) too high on pin; (f) body tilted laterally and pin position incorrect. Handling insect specimens with entomological forceps: (g) placing specimen mount into foam or cork; (h) removing mount from foam or cork. ((g,h) After Upton 1991.)

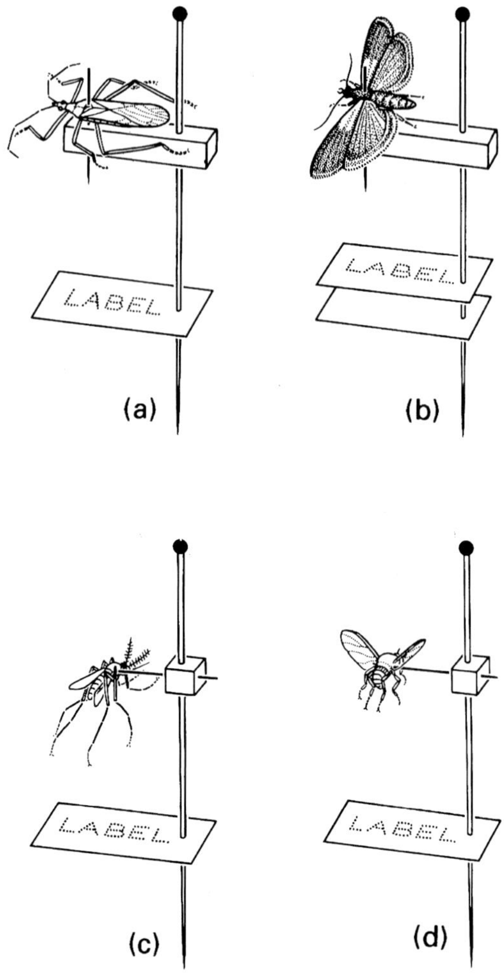

Fig. 18.4 Micropinning with stage and cube mounts: (a) a small bug (Hemiptera) on a stage mount, with position of pin in thorax as shown in Fig. 18.2h; (b) moth (Lepidoptera) on a stage mount, with position of pin in thorax as shown in Fig. 18.2d; (c) mosquito (Diptera: Culicidae) on a cube mount, with thorax impaled laterally; (d) black fly (Diptera: Simuliidae) on a cube mount, with thorax impaled laterally. (After Upton 1991.)

Micropinning (staging or double mounting)
This is used for many small insects and involves pinning the insect with a micropin to a stage that is mounted on a macropin (Fig. 18.4a,b); micropins are very fine, headless, stainless steel pins, from 10 to 15 mm long, and stages are small square or rectangular strips of white polyporus pith or synthetic equivalent. The micropins are inserted through the insect's body in the same positions as used in macropinning. Small wasps and moths are mounted with their bodies parallel to the stage with the head facing away from the macropin, whereas small beetles, bugs and flies are pinned with their bodies at right angles to the stage and to the left of the macropin. Some very small and delicate insects that are difficult to pin, such as mosquitoes and other small flies, are pinned to cube mounts; a cube of pith is mounted on a macropin and a micropin is inserted horizontally through the pith so that most of its length protrudes, and the insect then is impaled ventrally or laterally (Fig. 18.4c,d).

Pointing
This is used for small insects that would be damaged by pinning (Fig. 18.5a) (but *not* for small moths because the glue does not adhere well to scales, nor flies because important structures are obscured), for very sclerotized, small to medium-sized insects (especially weevils and ants) (Fig. 18.5b,c) whose cuticle is too hard to pierce with a micropin, or for mounting small specimens that are already dried. Points are made from small triangular pieces of white cardboard that either can be cut out with scissors or punched out using a special point punch. Each point is mounted on a stout macropin that is inserted centrally near the base of the triangle and the insect is then glued to the tip of the point using a minute quantity of water-soluble glue, for example based on gum arabic. The head of the insect should be to the right when the apex of the point is directed away from the person mounting. For most very small insects, the tip of the point should contact the insect on the vertical side of the thorax below the wings. Ants are glued to the upper apex of the point, and two or three points, each with an ant from the same nest, can be placed on one macropin. For small insects with a sloping lateral thorax, such as beetles and bugs, the tip of the point can be bent downwards slightly before applying the glue to the upper apex of the point.

Carding
For hobby collections or display purposes, insects (especially beetles) are sometimes carded, which involves gluing each specimen, usually by its venter, to

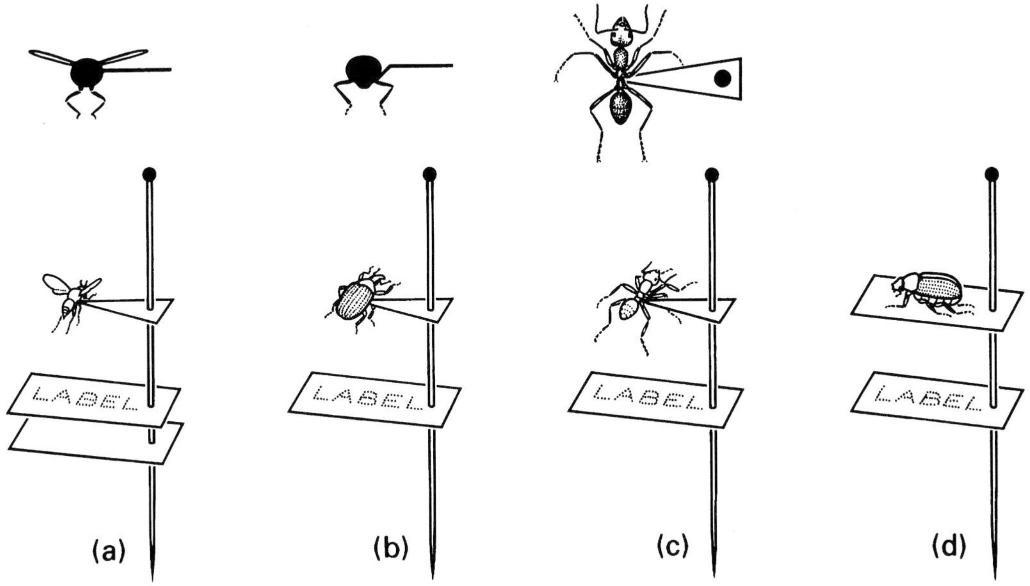

Fig. 18.5 Point mounts: (a) a small wasp; (b) a weevil; (c) an ant. Carding: (d) a beetle glued to a card mount. (After Upton 1991.)

a rectangular piece of card through which a macropin passes (Fig. 18.5d). Carding is not recommended for adult insects because structures on the underside are obscured by being glued to the card; however, carding may be suitable for mounting exuviae, pupal cases, puparia or scale covers.

Spreading and setting

It is important to display the wings, legs and antennae of many insects during mounting because features used for identification are often on the appendages. Specimens with open wings and neatly arranged legs and antennae also are more attractive in a collection. Spreading involves holding the appendages away from the body while the specimens are drying. Legs and antennae can be held in semi-natural positions using pins as supports (Fig. 18.6a) and the wings can be opened and held out horizontally on a setting board using pieces of tracing paper, cellophane, greaseproof paper, etc. (Fig. 18.6b). Setting boards can be constructed from pieces of polyethylene foam or soft cork glued to sheets of plywood or masonite; several boards with a range of groove and board widths are needed to hold insects of different body sizes and wingspans. Insects must be left to dry thoroughly before removing the pins and/or setting paper, but it is essential to keep the collection data associated correctly with each specimen during drying. A permanent data label must be placed on each macropin below the mounted insect (or its point or stage) after the specimen is removed from the drying or setting board. Sometimes two labels are used – an upper one for the collection data and a second, lower label for the taxonomic identification. See section 18.2.5 for information on the data that should be recorded.

18.2.2 Fixing and wet preservation

Most eggs, nymphs, larvae, pupae, puparia and soft-bodied adults are preserved in liquid because drying usually causes them to shrivel and rot. The most commonly used preservative for the long-term storage of insects is ethanol (ethyl alcohol) mixed in various concentrations (but usually 75–80%) with water. Isopropanol (propan-2-ol or isopropyl alcohol)

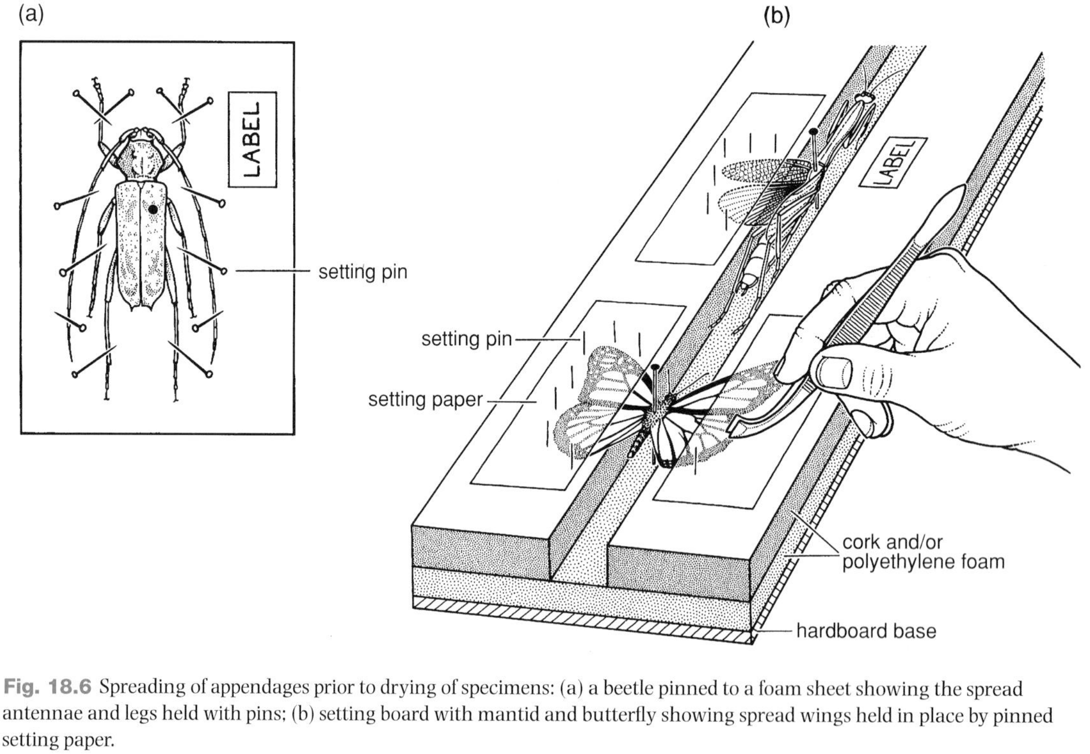

Fig. 18.6 Spreading of appendages prior to drying of specimens: (a) a beetle pinned to a foam sheet showing the spread antennae and legs held with pins; (b) setting board with mantid and butterfly showing spread wings held in place by pinned setting paper.

can be substituted for ethanol, including for specimens intended for DNA analysis (section 18.3.3). However, aphids and scale insects are often preserved in lactic-alcohol, which is a mixture of 2 parts 95% ethanol and 1 part 75% lactic acid, because this liquid prevents them from becoming brittle and facilitates subsequent maceration of body tissue prior to slide mounting. Most immature insects will shrink, and pigmented ones will discolour if placed directly into ethanol. Immature and soft-bodied insects, as well as specimens intended for study of internal structures, must first be dropped alive into a fixative solution prior to liquid preservation. All fixatives contain ethanol and glacial acetic acid, in various concentrations, combined with other liquids. Fixatives containing formalin (40% formaldehyde in water) should never be used for specimens intended for slide mounting (as internal tissues harden and will not macerate), but are ideal for specimens intended for histological study. Specimens intended for DNA extraction should not be preserved in fixatives containing acids, kerosene or formalin. Recipes for some fixatives are:

KAA – 2 parts glacial acetic acid, 10 parts 95% ethanol and 1 part kerosene (dye free).

Carnoy's fluid – 1 part glacial acetic acid, 6 parts 95% ethanol and 3 parts chloroform.

AGA – 1 part glacial acetic acid, 6 parts 95% ethanol, 4 parts water and 1 part glycerol.

FAA – 1 part glacial acetic acid, 25 parts 95% ethanol, 20 parts water and 5 parts formalin.

Pampel's fluid – 2–4 parts glacial acetic acid, 15 parts 95% ethanol, 30 parts water and 6 parts formalin.

Each specimen or collection should be stored in a separate glass vial or bottle that is sealed to prevent evaporation. The data label (section 18.2.5) should be inside the vial to prevent its separation from the specimen. Vials can be stored in racks or, to provide greater protection against evaporation, they can be placed inside a larger jar containing ethanol.

18.2.3 Microscope slide mounting

The features that need to be seen for the identification of many of the smaller insects (and their immature stages) often can be viewed satisfactorily only under the higher magnification of a compound microscope. Specimens must therefore be mounted either whole on glass microscope slides or dissected before mounting. Furthermore, the discrimination of minute structures may require the staining of the cuticle to differentiate the various parts, or the use of special microscope optics such as phase- or interference-contrast microscopy. There is a wide choice of stains and mounting media, and the preparation methods largely depend on which type of mountant is employed. Mountants are either aqueous gum-chloral-based (e.g. Hoyer's mountant, Berlese fluid) or resin-based (e.g. Canada balsam, Euparal). The former are more convenient for preparing temporary mounts for some identification purposes but deteriorate (often irretrievably) over time, whereas the latter are more time consuming to prepare but are permanent and thus are recommended for taxonomic specimens intended for long-term storage.

Prior to slide mounting, the specimens generally are "cleared" by soaking in either alkaline solutions (e.g. 10% potassium hydroxide (KOH) or 10% sodium hydroxide (NaOH)) or acidic solutions (e.g. lactic acid or lactophenol) to macerate and remove the body contents. Hydroxide solutions are used where complete maceration of soft tissues is required, and are most appropriate for specimens that are to be mounted in resin-based mountants. In contrast, most gum-chloral mountants continue to clear specimens after mounting and thus gentler macerating agents can be used or, in some cases, very small insects can be mounted directly into the mountant without any prior clearing. After hydroxide treatment, specimens must be washed in a weak acidic solution to halt the maceration. Cleared specimens are mounted directly into gum-chloral mountants, but must be stained (if required) and dehydrated thoroughly prior to placing in resin-based mountants. If a specimen is to be stained (e.g. in acid fuchsin or chlorazol black E), then it is placed, prior to dehydration, in a small dish of stain for the length of time required to produce the desired depth of colour. Dehydration involves successive washes in a graded alcohol (usually ethanol) series, with several changes in absolute alcohol. A final wash in isopropanol (isopropyl alcohol) is recommended because this alcohol is hydrophilic and will remove all trace of water from the specimen.

The last stage of mounting is to put a drop of the mountant centrally on a glass slide, place the specimen in the liquid, and carefully lower a cover slip (cover glass) onto the preparation. A small amount of mountant on the underside of the cover slip will help to reduce the likelihood of bubbles in the preparation. The slides should be maintained in the flat (horizontal) position during drying, which can be hastened in an oven at 40–45°C; slides prepared using aqueous mountants should be oven dried for only a few days but resin-based mountants may be left for several weeks (Canada balsam mounts may take many months to harden unless oven dried). If longer-term storage of gum-chloral slides is required, then they must be "ringed" with an insulating varnish to give an airtight seal around the edge of the cover slip. Finally, it is essential to label each dried slide mount with the collection data and, if available, the identification (section 18.2.5). For more-detailed explanations of slide-mounting methods, refer to Upton (1993), Upton & Mantle (2010) or Brown (1997) under Further reading.

18.2.4 Habitats, mounting and preservation of individual orders

The following list is alphabetical by order, and gives a summary of the usual habitats or collection methods, and recommendations for mounting and preserving each kind of insect or other hexapod. Insects that are to be pinned and stored dry can be killed either in a freezer or in a killing bottle (section 18.2.1); the list also specifies those insects that should be preserved in ethanol or fixed in another fluid prior to preservation (section 18.2.2). Generally, 75–80% ethanol is suggested for liquid storage, but the preferred strength often differs between collectors and depends on the kind of insect. For detailed instructions on how to collect and preserve different insects, refer to the publications in the Further reading list at the end of this chapter.

Archaeognatha (Microcoryphia; archaeognathans or bristletails)

These occur in leaf litter, under bark, or similar situations. Collect into and preserve in 80% ethanol.

Blattodea (cockroaches or roaches)

These are ubiquitous, found in sites ranging from peri-domestic to native vegetation, including caves and burrows; they are predominantly nocturnal. Eviscerate large specimens, and pin through the centre of the metanotum – wings of the left side may be spread. They may also be preserved in 80% ethanol.

Blattodea: Termitoidae (former order Isoptera; termites, "white ants")

Collect termites from colonies in mounds, on live or dead trees, or below ground. Preserve all castes available in 80% ethanol.

Coleoptera (beetles)

Beetles are found in all habitats. Pin adults and store dry; pin through the right elytron near its front so that the pin emerges between the mid and hind legs (Fig. 18.2a, Fig. 18.3 and Fig. 18.6a). Mount smaller specimens on card points with the apex of the point bent down slightly (Fig. 18.5b) and contacting the posterior lateral thorax between the mid and hind pair of legs. Immature stages are preserved in fluid (stored in 85–90% ethanol, preferably after fixation in KAA or Carnoy's fluid).

Collembola (springtails)

These are found in soil, litter and at water surfaces (fresh and intertidal). Collect into 95–100% ethanol and preserve on microscope slides.

Dermaptera (earwigs)

Favoured locations include litter, under bark or logs, dead vegetation (including along the shoreline) and in caves; exceptionally they are ectoparasitic on bats. Pin through the right elytron and with the left wings spread. Collect a representative sample of immature stages into Pampel's fluid and then 75–80% ethanol.

Diplura (diplurans)

These occur in damp soil under rocks or logs. Collect into 75% ethanol; preserve in 75% ethanol or slide mount.

Diptera (true flies)

Flies are found in all habitats. Pin adult specimens and store dry, or preserve in 75% ethanol; pin most adults to right of centre of the mesothorax (Fig. 18.2c); stage or cube mount smaller specimens (Fig. 18.4c,d) (card pointing not recommended). Collect immature stages into Pampel's fluid or fix in hot water (larger specimens) or 75–80% ethanol (smaller specimens). Slide mount smaller adults and the larvae and pupae of some families.

Embioptera (Embiidina, Embiodea; embiopterans or webspinners)

Typical locations for the silken galleries of webspinners are in or on bark, lichens, rocks or wood. Preserve and store in 75–80% ethanol or slide mount; winged adults can be pinned through the centre of the thorax with wings spread.

Ephemeroptera (mayflies)

Adults occur beside water. Preserve in 75% ethanol (preferably after fixing in Carnoy's fluid or FAA) or pin through the centre of the thorax with the wings spread. Immature stages are aquatic. Collect these into and preserve in 75–80% ethanol or first fix in Carnoy's fluid or FAA, or store dissected on slides or in microvials.

Grylloblattodea (Grylloblattaria or Notoptera; grylloblattids, ice crawlers or rock crawlers)

These can be collected on or under rocks, or on snow or ice. Preserve specimens in 75% ethanol (preferably after fixing in Pampel's fluid), or slide mount.

Hemiptera (bugs)

The Cicadomorpha (cicadas, leafhoppers, spittle bugs), Fulgoromorpha (Fulgoroidea or planthoppers) and Heteroptera (true bugs) are associated with their host plants or are predaceous and free-living; aquatic forms also have these habits. Preserve the adults dry; pin through the scutellum or thorax to the right of centre (Fig. 18.2h); spread the wings of cicadas and fulgoroids, point or stage smaller specimens (Fig. 18.4a). Preserve nymphs in 80% ethanol.

The Aphidoidea (aphids) and Coccoidea (scale insects, mealybugs) are found associated with their host plants, including leaves, stems, roots and galls. Store nymphs and adults in lactic-alcohol or 80% ethanol, or dry on a plant part; slide mount to identify.

Aleyrodoidea (whiteflies) are associated with their host plants. The sessile final-instar nymph ("puparium") or its exuviae ("pupal case") are of taxonomic importance. Preserve all stages in 80–95% ethanol; slide mount puparia or pupal cases.

The Psylloidea (psyllids, lerps) are associated with host plants; rear nymphs to obtain adults. Preserve nymphs in 80% ethanol, dry mount galls or lerps (if present). Preserve adults in 80% ethanol or dry mount on points; slide mount dissected parts.

Hymenoptera (ants, bees, wasps, sawflies and wood wasps)

Hymenoptera are ubiquitous, and many are parasitic, in which case the host association should be retained. Collect bees, sawflies and wasps into 80% ethanol or pin and store dry – pin larger adults to the right of centre of the mesothorax (Fig. 18.2e) (sometimes with the pin angled to miss the base of the fore legs); point mount smaller adults (Fig. 18.5a); slide mount if very small. Immature stages should be preserved in 80% ethanol, often after fixing in Carnoy's fluid or KAA. Ants require stronger ethanol (90–95%); point a series of ants from each nest, with each ant glued on to the upper apex of the point between the mid and hind pairs of legs (Fig. 18.5c); two or three ants from a single nest can be mounted on separate points on a single macropin.

Lepidoptera (butterflies and moths)

Lepidoptera are ubiquitous. Collect by netting and (especially moths) at a light. Pin vertically through the thorax and spread the wings so that the hind margins of the fore wings are at right angles to the body (Fig. 18.2d and Fig. 18.6b). Microlepidopterans are best micropinned (Fig. 18.4b) immediately after death. Immature stages are killed in KAA or boiling water, and transferred to 80–90% ethanol. Butterflies can be stored dry in envelopes with wings closed and later relaxed and the wings set.

Mantodea (mantids, mantises or praying mantids)

These are generally found on vegetation, sometimes attracted to light at night. Rear the nymphs to obtain adults. Eviscerate larger specimens. Pin between the wing bases and set the wings on the left side (Fig. 18.6b).

Mantophasmatodea (heelwalkers)

These are found on mountains in Namibia by day, and also at lower elevations in South Africa at night. Pin mid-thorax, or preserve in 80–90% ethanol.

Mecoptera (hangingflies, scorpionflies and snowfleas)

Mecoptera often occur in damp habitats, near streams or wet meadows. Pin adults to the right of centre of the thorax with the wings spread. Alternatively, all stages may be fixed in KAA, FAA or 80% ethanol, and preserved in 80% ethanol.

Megaloptera (alderflies, dobsonflies and fishflies)

These are usually found in damp habitats, often near streams and lakes. Pin adults to the right of centre of the thorax with the wings spread. Alternatively, all stages can be fixed in FAA or 80% ethanol, and preserved in ethanol.

Neuroptera (lacewings, owlflies and antlions)

Neuroptera are ubiquitous, associated with vegetation, sometimes in damp places. Pin adults to the right of centre of the thorax with the wings spread (Fig. 18.2f) and the body supported. Alternatively, preserve in 80% ethanol. Immature stages are fixed in KAA, Carnoy's fluid or 80% ethanol, and preserved in ethanol.

Odonata (damselflies and dragonflies)

Although generally found near water, adult odonates may disperse and migrate; the nymphs are aquatic. If possible, keep the adult alive and starve for 1–2 days before killing (this helps to preserve body colours after death). Pin through the mid-line of the thorax between the wings, with the pin emerging between the first and second pair of legs (Fig. 18.2g); set the wings with the front margins of the hind wings at right angles to the body (a good setting method is to place the newly pinned odonate upside down with the head of the pin pushed into a foam drying board). Preserve immature stages in 80% ethanol; the exuviae should be placed on a card associated with adult.

Orthoptera (grasshoppers, locusts, katydids and crickets)

Orthoptera are found in most terrestrial habitats. Remove the gut from all but the smallest specimens, and pin vertically through the right posterior quarter of the prothorax, spreading the left wings (Fig. 18.2b). Nymphs and soft-bodied adults should be fixed in Pampel's fluid then preserved in 75–80% ethanol.

Phasmatodea (phasmids, stick-insects or walking sticks)

These are found on vegetation, usually nocturnally (sometimes attracted to light). Rear the nymphs to obtain adults, and remove the gut from all but the smallest specimens. Pin through the base of the mesothorax with the pin emerging between bases of the mesothoracic legs, spread the left wings, and fold the antennae back along the body. Preserve eggs from females by fixing in hot water or at 50°C in an oven; store eggs dry in association with female.

Plecoptera (stoneflies)

Adult plecopterans are restricted to the proximity of aquatic habitats. Net or pick from the substrate, infrequently attracted to light. Nymphs are aquatic, being found especially under stones. Pin adults through the centre of the thorax with the wings spread, or preserve in 80% ethanol. Immature stages are preserved in 80% ethanol, or dissected on slides or in microvials.

Protura (proturans)

Proturans are most easily collected by extracting from litter using a Tullgren funnel. Collect into and preserve in 80% ethanol, or slide mount.

Psocodea: "Phthiraptera" (chewing lice and sucking lice)

Lice can be seen on their live hosts by inspecting the plumage or pelt, and can be removed using an ethanol-soaked paintbrush. Lice depart recently dead hosts as the temperature drops – and can be picked from a dark cloth background. Ectoparasites also can be removed from a live host by keeping the host's head free from a bag enclosing the rest of the body and containing chloroform to kill the parasites, which can be shaken free, leaving the host unharmed. Legislation concerning the handling of vertebrate hosts and of chloroform make this is specialized technique. Lice are preserved in 80% ethanol and slide mounted.

Psocodea: "Psocoptera" (bark lice and book lice)

Psocids occur on foliage, bark and damp wooden surfaces, sometimes in stored products. Collect with an aspirator or ethanol-laden paintbrush into 80% ethanol; slide mount small specimens.

Raphidioptera (snakeflies)

These are typically found in damp habitats, often near streams and lakes. Pin adults or fix in FAA or 80% ethanol; immature stages are preserved in 80% ethanol.

Siphonaptera (fleas)

Fleas can be removed from a host bird or mammal by methods similar to those outlined above for parasitic lice. If free-living in a nest, use fine forceps or an alcohol-laden brush. Collect adults and larvae into 75–80% ethanol; preserve in ethanol or by slide mounting; on slide, mount on left side with legs extended.

Strepsiptera (strepsipterans)

Adult males are winged, whereas females and immature stages are endoparasitic, especially in leafhoppers and planthoppers (Hemiptera) and Hymenoptera. Preserve in 80% ethanol or by slide mounting.

Thysanoptera (thrips)

Thrips are common in flowers, fungi, leaf litter and some galls. Collect adults and nymphs into AGA or 60–90% ethanol, and preserve by slide mounting.

Trichoptera (caddisflies)

Adult caddisflies are found beside water and attracted to light, and immature stages are aquatic in all waters. Pin adults through the right of centre of the mesonotum with the wings spread, or preserve in 80% ethanol. Immature stages are fixed in FAA or 75% ethanol, and preserved into 80% ethanol. Micro-caddisflies and dissected nymphs are preserved by slide mounting.

Zoraptera (zorapterans or angel insects)

These occur in rotten wood and under bark, with some found in termite nests. Preserve in 75% ethanol or slide mount.

Zygentoma (silverfish)

Silverfish are peri-domestic, and also occur in leaf litter, under bark, in caves and with termites and ants. They are often nocturnal, and elusive to normal handling. Collect by stunning with ethanol, or using Tullgren funnels; preserve with 80% ethanol.

18.2.5 Curation

Labelling

Even the best-preserved and displayed specimens are of little or no scientific value without associated data such as location, date of capture, and habitat. Such information should be uniquely associated with the specimen. Although this can be achieved by a unique numbering or lettering system associated with a notebook or computer file, it is essential that it appears also on a permanently printed label associated with the specimen. The following is the minimal information that should be recorded, preferably into a field notebook at the time of capture rather than from memory later.

- Location – usually in descending order from country and state (your material may be of more than local interest), town, or distance from map-named location. Include map-derived names for habitats such as lakes, ponds, marshes, streams, rivers, forests, mountains, etc.
- Co-ordinates – preferably using a global positioning system (GPS) and citing latitude and longitude rather than non-universal metrics. Increasingly, these locations are used in geographic information systems (GIS) and climate-derived models that depend upon accurate ground positioning.
- Elevation – derived from map or GPS as elevational accuracy has increased.

• Date – usually in sequence of day in Arabic numerals, month preferably in abbreviated letters or in Roman numerals (to avoid the ambiguity of, say, 9.11.2001 – which is 9th November in many countries but 11th September in others), and year, from which the century is best not omitted. Thus, 2.iv.1999 and 2 Apr. 1999 are acceptable alternatives.
• Collector's identity, brief project identification, and any codes that refer to notebook.
• Collection method, any host association or rearing record, and any microhabitat information.

On another label, record details of the identity of the specimen, including the name of the person who made the identification and the date on which it was made. It is important that subsequent examiners of the specimen know the history and timing of previous study, notably in relation to changes in taxonomic concepts in the intervening period. If the specimen is used in taxonomic description, such information should also be appended to pre-existing labels or additional label(s). It is important never to discard previous labels – transcription may lose useful evidence from handwriting and, at most, vital information on status, location, etc. Assume that all specimens valuable enough to conserve and label have potential scientific significance into the future, and thus print labels on high-quality acid-free paper using permanent ink – which can be provided now by high-quality laser printers.

Care of collections

Collections start to deteriorate rapidly unless precautions are taken against pests, mould, and vagaries of temperature and humidity. Rapid alteration in temperature and humidity should be avoided, and collections should be kept in as dark a place as possible because light causes fading. Application of some insecticides may be necessary to kill pests such as *Anthrenus*, "museum beetles" (Coleoptera: Dermestidae), but use of all dangerous chemicals should conform to local regulations. Deep freezing (below −20°C for 48 h) also can be used to kill any pest infestation. Vials of ethanol should be securely capped, with a triple-ring nylon stopper if available, and preferably stored in larger containers of ethanol. Larger ethanol collections must be maintained in separate, ventilated, fireproof areas. Collections of glass slides preferably are stored horizontally, but with major taxonomic collections of groups preserved on slides, some vertical storage of *well-dried* slides may be required on grounds of costs and space saving.

Other than small personal ("hobby") collections of insects, it is good scientific practice to arrange for the eventual deposition of collections into major local or national institutions such as museums. This guarantees the security of valuable specimens, and enters them into the broader scientific arena by facilitating the sharing of data, and the provision of loans to colleagues and fellow scientists.

18.3 IDENTIFICATION

Identification of insects is at the heart of almost every entomological study, but this is not always recognized. Rather too often a survey is made for one of a variety of reasons (e.g. ranking diversity of particular sites or detecting pest insects), but with scant regard to the eventual need, or even core requirement, to identify the organisms accurately. There are several possible routes to attaining accurate identification, of which the most satisfying may be to find an interested taxonomic expert in the insect group(s) under study. Often this person is assumed to have time available and willingness to undertake the exercise solely out of interest in the project and the insects collected. If this possibility was ever commonplace, it is no longer so because the pool of expertise has diminished and pressures upon remaining taxonomic experts have increased. A more satisfactory solution is to incorporate the identification requirements into each research proposal at the outset of the investigation, including producing a realistic budget for the identification component. Even with such planning, there may be some further problems. There may be:
• logistical constraints that prevent timely identification of mass (speciose) samples (e.g. canopy fogging samples from rainforest, vacuum sampling from grassland), even if the taxonomic skills are available;
• no entomologists who are both available and have the skills required to identify all, or even selected groups, of the insects that are encountered;
• no specialist with knowledge of the insects from the area in which your study takes place – as seen in Chapter 1, entomologists are distributed in an inverse manner to the diversity of insects;
• no specialists able or prepared to study the insects collected because the condition or life-history stage of the specimens prevents ready identification.

There is no single answer to such problems, but certain difficulties can be minimized by early consultation with local experts or with relevant published information, by collecting the appropriate life-history stages, by preserving material correctly, and by making use of vouchered material. It should be possible to advance the identification of specimens using taxonomic publications, such as field guides and keys, which are designed for this purpose.

18.3.1 Identification keys

The output of taxonomic studies usually includes keys for determining the names (i.e. for identification) of organisms. Traditionally, keys involve a series of questions, concerning the presence, shape or colour of a structure, which are presented in the form of choices. For example, one might have to determine whether the specimen has wings or not – in the case that the specimen of interest has wings then all possibilities without wings are eliminated. The next question might concern whether there is one or two pairs of wings, and if there are two pairs, whether one pair of wings is modified in some way relative to the other pair. This means of proceeding by a choice of one out of two (couplets), thereby eliminating one option at each step, is termed a "dichotomous key" because at each consecutive step there is a dichotomy, or branch. One works down the key until eventually the choice is between two alternatives that lead no further: these are the terminals in the key, which may be of any rank (section 1.4) – families, genera or species. This final choice gives a name and although it is satisfying to believe that this is the "answer", it is necessary to check the identification against some form of description. An error in interpretation early on in a key (by either the user or the compiler) can lead to correct answers to all subsequent questions but a wrong final determination. However, an erroneous conclusion can be recognized only following comparison of the specimen with some "diagnostic" statements for the taxon name that was obtained from the key.

Sometimes a key may provide several choices at one point, and as long as each possibility is mutually exclusive (i.e. all taxa fall clearly into one of the multiple choices), this can provide a shorter route through the available choices. Other factors that can assist in helping the user through such keys is to provide clear illustrations of what is expected to be observed at each point. Of necessity, as we discuss in the introduction to the Glossary, there is a language associated with the morphological structures that are used in keys. This nomenclature can be rather off-putting, especially if different names are used for structures that appear to be the same, or very similar, between different taxonomic groups.

A good illustration can be worth a thousand words – but nonetheless there are also lurking problems with illustrated keys. It can be difficult to relate a drawing of a structure to what is seen in the hand or under the microscope. Photography, which seems to be an obvious aid, actually can hinder because it is always tempting to look at the complete organism or structure (and in doing so to recognize or deny overall similarity to the study organism) and fail to see that the key requires only a particular detail. Another major difficulty with any branching key, even if well illustrated, is that the compiler enforces the route through the key – and even if the feature required to be observed is elusive, the structure must be recognized and a choice made between alternatives in order to proceed. There is little or no room for error by compiler or user. Even the best constructed keys may require information on a structure that the best intentioned user cannot see – for example, a choice in a key may require assessment of a feature of one sex, and the user has only the alternative sex, or an immature specimen. The answer undoubtedly requires a different approach, using the power of computers to allow multiple points of access to data used for identification. Instead of a dichotomous structure, the compiler builds a matrix of all features that in any way can help in identification, and allows the user to select (with some guidance available for those that want it) which features to examine. Thus, it may not matter if a specimen lacks a head (through damage), whereas a conventional key may require assessment of the antennal features at an early stage. Using a computer-based, so-called interactive key, it may be possible to proceed using options that do not involve "missing" anatomy, and still make an identification. Possibilities of linking illustrations and photographs, with choices of looking "like this, or this, or this", rather than dichotomous choice, can allow efficient movement through less-constrained options than paper keys. Computer keys proceed by elimination of possible answers until one (or a few) possibilities remain – at which stage detailed descriptions may be called up to allow optimal comparisons. The ability to attach compendious information concerning the included taxa allows confirmation of identifications against illustrations and summarized diagnostic features. Furthermore, the compiler can attach all manner of biological data about the organisms, plus references.

Advances such as these, as implemented in proprietary software such as Lucid (www.lucidcentral.com), suggest that interactive keys inevitably will be the preferred method by which taxonomists present their work to those who need to identify insects.

18.3.2 Unofficial taxonomies and voucher specimens

As explained elsewhere in this book, the sheer diversity of the insects means that even some fairly commonly encountered species are not described formally yet. Probably only in Britain and Ireland can it really be said that the total fauna is described and recognizable using identification keys. Elsewhere, the undescribed and unidentifiable proportion of the fauna can be substantial. This is an impediment to understanding how to separate species and communicate information about them. In response to the lack of formal names and keys, some "informal" taxonomies have arisen, which bypass the time-consuming formal distinguishing and naming of species. Although these taxonomies are not intended to be permanent, they do fulfil a need and can be effective. One practical system is the use of voucher numbers or codes as unique identifiers of species or **morphospecies** (estimates of species based on morphological criteria), following comparative morphological analysis across the complete geographical range of the taxa but prior to the formal act of publishing names as Latin binomens (section 1.4). If the informal name is in the form of a species name, these are referred to as manuscript names, and sometimes they never do become published. The use of unpublished names causes future nomenclatural confusion and so it is preferable to use voucher numbers or codes. In this system, taxa can be compared across their distributional and ecological range in an identical manner to taxa provided with formal names.

In narrower treatments, informal codes refer only to the biota of a limited region, typically in association with an inventory (survey) of a restricted area. The codes allocated in these studies typically represent morphospecies, which may not have been compared with specimens from other areas. Furthermore, the informal coded units may include taxa that may have been described formally from elsewhere. This system suffers lack of comparability of units with those from other areas – it is impossible to assess beta diversity (species turnover with distance). Furthermore, vouchers (or morphospecies) may or may not correspond to real biological units – although strictly this criticism applies to a greater or lesser extent to all forms of taxonomic arrangements. For simple number-counting exercises at sites, with no further questions being asked of the data, a morphospecies voucher system can approximate reality, unless confused by, for example, polymorphism, cryptic species, or unassociated life-history stages.

Essential to all informal taxonomies is the need to retain voucher specimens for each segregate. This allows contemporary and future researchers to integrate informal taxa into the standardized system, and retain the association of biological information with the names, be they formal or informal. In many cases where informality is advocated, ignorance of the taxonomic process is at the heart – but in others, the sheer number of readily segregated morphospecies that lack formal identification requires such an approach.

18.3.3 DNA-based identifications and voucher specimens

Insect DNA is acquired for population studies, to assist with species delimitation (Box 7.3) or for phylogenetic purposes and, as recently publicized, may be used for DNA-based identification in which the sequence of base pairs of parts of one or more genes is used as the main criterion for identifying species, a technique called **DNA barcoding**. Identification is based on the observation that some genes show variations (mutations) at a rate that allows clustering of individuals from the "same species" and provides increasing distinction from more distantly related species. The successful use of DNA barcoding for insect identification requires a comprehensive reference database (library) of sequences from authoritatively identified specimens representing the diversity of the taxon under study and against which a sequence from the organism to be identified can be compared. Compilation of such DNA libraries is expensive and time consuming, in large part due to the enormous effort needed to acquire and identify the reference specimens. The process has been facilitated by extracting DNA from accurately identified and appropriately preserved museum specimens. There are a number of major projects to "barcode" groups of insects, and such projects purport to maintain voucher collections of their barcoded specimens, which is essential for quality control (see below).

The most frequently used barcode is the 5' end of the *cytochrome c oxidase* subunit 1 mitochondrial region (*COI* or *cox*1), which can be obtained reasonably easily and cheaply for most animals, including insects. The *COI* sequences of even closely related insects usually differ by several percent, allowing identification of the species. However, for some insect groups, *COI* does not resolve species-level differences and other gene regions need to be used. Furthermore, the interpretation of mitochondrial sequences may be complicated by the inadvertent amplification of paralogous nuclear copies and heritable endosymbionts such as *Wolbachia*.

Another application of DNA barcodes is to associate the immature and adult stages of an insect species (section 7.1.2) when only an immature life stage is collected and rearing to the identifiable adult is impossible or difficult. This might be especially useful for the identification of a pest collected only as a larva.

A related and very recent application of DNA identification is in the new field of **ecogenomics**, which has been developed and used mostly for understanding the microbial biosphere. This technology mainly uses sequence-based methods that rely on cheap high-throughput sequencing of DNA and RNA extracted directly from environmental samples, such as soil cores or water samples. In entomology, ecogenomics could define insect biodiversity at the DNA level and use this knowledge to quantify the functions and interactions of organisms at an ecosystem level and relate these to ecological and evolutionary processes. For example, thousands of individual insects could be collected from localities or habitats of interest, and some of the DNA of each specimen could be sequenced to allow segregation of the specimens into genetically distinct entities or putative species. Although scientific names would not necessarily be matched with specimens, it would be possible to screen and compare a large number of survey or sample sites to determine whether, for example, any genetic entities were potentially unique to one or a limited number of sites, or if any entities were always found together. Currently, an ambitious project utilizing ecogenomics is analysing insect samples collected in isolated rainforest pockets of Australia's northwest Kimberley region in order to assess the invertebrate and metacommunity structure of this habitat type, and ultimately to improve conservation and management of the region.

The optimal preservation of insects for subsequent DNA extraction, amplification and sequencing usually requires fresh specimens preserved and stored in a freezer, ideally at −80°C, or in absolute ethanol and refrigerated. Museum specimens, if relatively fresh and appropriately preserved, can retain DNA, but many do not, or the DNA is too degraded for adequate use. Although DNA can be amplified from small parts of a preserved specimen, many museums are reluctant to allow valuable old and rare specimens, especially types, to be sampled destructively.

Last but not least, it is essential that the nucleotide sequences obtained from specimens are archived in appropriate long-term repositories (e.g. GenBank), and that appropriate **voucher specimens** are retained and, if possible, most or part of the actual specimens from which the DNA is extracted. For example, DNA can be extracted from a single leg of larger insects or, for smaller insects such as lice, thrips, aphids and scale insects, there are methods for obtaining DNA from the whole specimen while retaining the relatively intact cuticle as the voucher. Preservation of voucher specimens and their secure long-term storage (in museum collections) is essential because these specimens fulfil an archival role, allowing subsequent verification of identification and reassessment of studies that have used the specimens.

FURTHER READING

Regional texts for identifying insects

Africa

Picker, M., Griffiths, C. & Weaving, A. (2005) *Field Guide to Insects of South Africa*. Updated edition. Struik Publishers, Cape Town.

Scholtz, C.H. & Holm, E. (eds)(1985) *Insects of Southern Africa*. University of Pretoria, Pretoria.

Australia

CSIRO (1991) *The Insects of Australia*, 2nd edn. Vols I and II. Melbourne University Press, Carlton.

Zborowski, P. & Storey, R. (2010) *A Field Guide to Insects in Australia*, 3rd edn. New Holland Publishers Australia, Chatswood.

Europe

Chinery, M. (2012) *Insects of Britain and Western Europe*, 3rd edn. Bloomsbury Publishing, London.

Gibbons, B. (1996) *Field Guide to Insects of Great Britain and Northern Europe*. Crowood Press, Wiltshire.

Richards, O.W. & Davies, R.G. (1977) *Imms' General Textbook of Entomology*, 10th edn. Vol. 1: *Structure, Physiology and Development*; Vol. 2: *Classification and Biology*. Chapman & Hall, London.

The Americas

Arnett, R.H. Jr. (2000) *American Insects – A Handbook of the Insects of America North of Mexico*, 2nd edn. CRC Press, Boca Raton, FL.
Arnett, R.H. & Thomas, M.C. (2001) *American Beetles*, Vol. I: *Archostemata, Myxophaga, Adephaga, Polyphaga: Staphyliniformia*. CRC Press, Boca Raton, FL.
Arnett, R.H., Thomas, M.C., Skelley, P.E. & Frank, J.J. (2002) *American Beetles*, Vol. II: *Polyphaga: Scarabaeoidea through Curculionoidea*. CRC Press, Boca Raton, FL.
Hogue, C.L. (1993) *Latin American Insects and Entomology*. University of California Press, Berkeley, CA.
Johnson, N.F. & Triplehorn, C.A. (2005) *Borror and DeLong's Introduction to the Study of Insects*, 7th edn. Brooks/Cole, Belmont, CA [now Cengage Learning, Independence, KY].
Merritt, R.W., Cummins, K.W. & Berg, M.B. (eds) (2008) *An Introduction to the Aquatic Insects of North America*, 4th edn. Kendall/Hunt Publishing Co., Dubuque, IA.

Identification of immature insects

Chu, H.F. & Cutkomp, L.K. (1992) *How to Know the Immature Insects*. William C. Brown Communications, Dubuque, IA.
Stehr, F.W. (ed.) (1987) *Immature Insects*, Vol. 1. Kendall/Hunt Publishing, Dubuque, IA. [Deals with non-insect hexapods, apterygotes, Trichoptera, Lepidoptera, Hymenoptera, plus many small orders.]
Stehr, F.W. (ed.) (1991) *Immature Insects*, Vol. 2. Kendall/Hunt Publishing, Dubuque, IA. [Deals with Thysanoptera, Hemiptera, Megaloptera, Raphidioptera, Neuroptera, Coleoptera, Strepsiptera, Siphonaptera and Diptera.]

Collecting and preserving methods

Brown, P.A. (1997) A review of techniques used in the preparation, curation and conservation of microscope slides at the Natural History Museum, London. The Biology Curator, Issue 10, special supplement, 34 pp.
Covell, C.V., Jr. (2009) Collection and preservation. In: *Encyclopedia of Insects*, 2nd edn (eds V.H. Resh & R.T. Cardé), pp. 201–06. Elsevier, San Diego, CA.
Gibb, T. & Oseto, C. (2005) *Arthropod Collection and Identification: Laboratory and Field Techniques*. Academic Press, Burlington, MA.
Martin, J.E.H. (1977) Collecting, preparing, and preserving insects, mites, and spiders. In: *The Insects and Arachnids of Canada*, Part 1. Canada Department of Agriculture, Biosystematics Research Institute, Ottawa.
McGavin, G.C. (1997) *Expedition Field Techniques. Insects and Other Terrestrial Arthropods*. Expedition Advisory Centre, Royal Geographical Society, London.
Melbourne, B.A. (1999) Bias in the effect of habitat structure on pitfall traps: an experimental evaluation. *Australian Journal of Ecology* **24**, 228–39.
New, T.R. (1998) *Invertebrate Surveys for Conservation*. Oxford University Press, Oxford.
Oman, P.W. & Cushman, A.D. (2005) *Collection and Preservation of Insects*. Fredonia Books, Amsterdam.
Upton, M.S. (1993) Aqueous gum-chloral slide mounting media: an historical review. *Bulletin of Entomological Research* **83**, 267–74.
Upton, M.S. & Mantle, B.L. (2010) *Methods for Collecting, Preserving and Studying Insects and Other Terrestrial Arthropods*, 5th edn. The Australian Entomological Society Miscellaneous Publication No. 3, Canberra. 81 pp.

Museum collections

Arnett, R.H. Jr., Samuelson, G.A. & Nishida, G.M. (1993) *The Insect and Spider Collections of the World*, 2nd edn. Flora & Fauna Handbook No. 11. Sandhill Crane Press, Gainesville, FL. [hbs.bishopmuseum.org/codens/codens-r-us.html]
Nishida, G.M. (2009) Museums and Display Collections. In: *Encyclopedia of Insects*, 2nd edn (eds V.H. Resh & R.T. Cardé), pp. 680–84. Elsevier, San Diego, CA.

TAXOBOXES

Estimates of species numbers are mostly from Zhang (2013) and are for extant described species unless otherwise stated.

Taxobox 1 Entognatha: non-insect hexapods (Collembola, Diplura and Protura)

The Collembola, Protura and Diplura have been united as the "Entognatha", based on similar mouthpart morphology in which mandibles and maxillae are enclosed in folds of the head (except when everted for feeding). Although the monophyly of the Entognatha has been disputed, some molecular data provide support for this group, which can be treated as a class within the Hexapoda and of equal rank to Insecta. We treat these orders together here, although the entognathy of these taxa may not to be homologous and these non-insect hexapods may not form a monophyletic group (section 7.2). All have indirect fertilization – males deposit sperm bundles or stalked spermatophores, which are picked up from the substrate by unattended females. For phylogenetic considerations concerning these three taxa, see section 7.2 and section 7.3.

Collembola (springtails)

The springtails are non-insect hexapods, and include about 8000 described species in 30 extant families, but the true species diversity may be much higher. Small (usually 2–3 mm, but up to 17 mm) and soft-bodied, their body varies in shape from globular to elongate (as illustrated here for *Isotoma* and *Sminthurinus*; after Fjellberg 1980), and is pale or often characteristically pigmented grey, blue or black. The eyes and/or ocelli are often poorly developed or absent; the antennae have four to six segments. Behind the antennae usually there is a pair of post-antennal organs, which are specialized sensory structures (believed by some to be the remnant apex of the second antenna of crustaceans). The entognathous mouthparts comprise elongate maxillae and mandibles enclosed by pleural folds of the head; maxillary and labial palps are present as remnants. The legs each comprise one or two apparent subcoxal segments, coxa, trochanter, femur and a tibiotarsus to which is attached a claw and an empodial appendage. The six-segmented abdomen has a sucker-like ventral tube (the collophore), a retaining hook (the retinaculum) and a furca (sometimes called furcula; forked jumping organ, usually three-segmented) on segments 1, 3 and 4, respectively, with the gonopore on segment 5 and the anus on segment 6; cerci are absent. The ventral tube is the main site of water and salt exchange and thus is important to fluid balance, but also can be used as an adhesive organ. The springing organ (furca), formed by fusion of a pair of appendages, is longer in surface-dwelling species than those living within the soil (as shown in Fig. 9.1). In general, jump length is correlated with furca length, and some species can spring up to 10 cm. Among hexapods, collembolan eggs uniquely are microlecithal (lacking large yolk reserves) and holoblastic (with complete cleavage). The immature instars are similar to the adults, developing epimorphically (with a constant segment number); maturity in males is attained after five moults but usually after more in females, and moulting continues for life. Springtails are most abundant

The Insects: An Outline of Entomology, Fifth Edition. P.J. Gullan and P.S. Cranston.

Companion Website: www.wiley.com/go/gullan/insects

in moist soil and litter, where they are major consumers of decaying vegetation, but also they occur in caves, in fungi, as commensals with ants and termites, on still water surfaces and in the intertidal zone. Most species feed on fungal hyphae and/or spores or dead plant material, whereas some species eat other small invertebrates. Many collembolan species can digest plant and fungal tissues, but it is unclear if the enzymes involved (cellulase, chitinase and trehalase) are produced by the springtails themselves or by microorganisms in their gut. Only a very few species are injurious to living plants, for example, the "lucerne flea" *Sminthurus viridis* (Sminthuridae) damages the tissues of crops such as lucerne and clover and can cause economic injury. Springtails can reach extremely high densities (e.g. 10,000–100,000 individuals m^{-2}) and are ecologically important in adding nutrients to the soil via their faeces, and in facilitating decomposition processes, for example by stimulating and inhibiting the activities of different microorganisms. Specifically, their selective grazing can affect both the vertical distribution of fungal species and the rate of fungal decomposition of litter material.

Diplura (diplurans)

The diplurans are non-insect hexapods, with nearly 1000 species in up to 10 families. They are small to medium sized (2–5 mm, exceptionally up to 50 mm), mostly unpigmented and weakly sclerotized. They lack eyes and their antennae are long, moniliform and multi-segmented. The mouthparts are entognathous and the mandibles and maxillae are well developed, with their tips visible protruding from the pleural fold cavity; the maxillary and labial palps are reduced. The thorax is little differentiated from the abdomen and bears legs each comprising five segments. The abdomen is 10-segmented, with some segments having small styles and protrusible vesicles; the gonopore is between segments 8 and 9 and the anus is terminal; the cerci are filiform (as illustrated here for

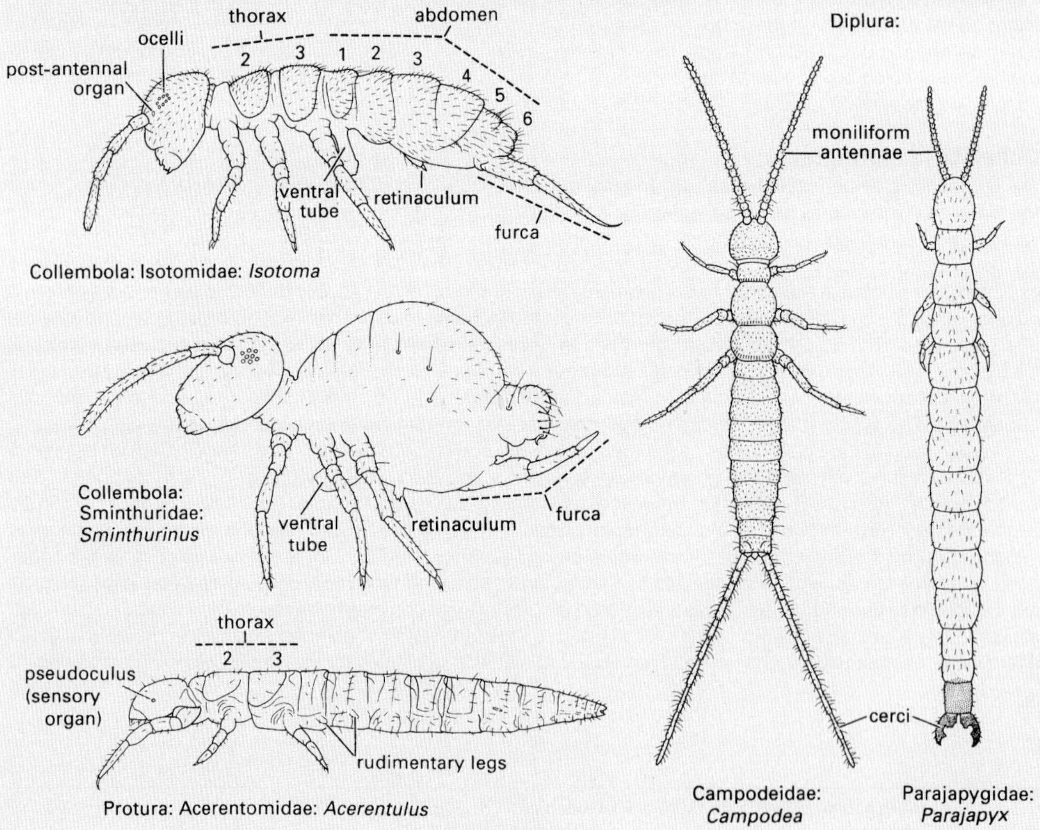

Campodea; after Lubbock 1873) to forceps-like (as in *Parajapyx* shown here; after Womersley 1939). Development of the immature forms is epimorphic, with moulting continuing through life. Some species are gregarious, and females of certain species tend the eggs and young. Diplurans are generally omnivorous, but some feed on live and decayed vegetation and japygid diplurans are predators.

Protura (proturans)

The proturans are non-insect hexapods, with over 800 described species in seven families. Nearly half the species are placed in the genus *Eosentomon* (Eosentomidae). Proturans are very small (<2 mm long), delicate, elongate, pale to white or almost unpigmented, with a fusiform body and conically shaped head. The thorax is poorly differentiated from the abdomen. Eyes and antennae are lacking, and the mouthparts are entognathous, consisting of slender mandibles and maxillae, slightly protruding from a pleural fold cavity; maxillary and labial palps are present. The thorax is weakly developed and bears legs each comprising five segments; the anterior legs are held forward (as shown here for *Acerentulus*; after Nosek 1973), fulfilling an antennal sensory function. The adult abdomen is 12-segmented, with the gonopore between segments 11 and 12 and a terminal anus; cerci are absent. Immature development is anamorphic (with segments added posteriorly during development). Proturans are cryptic, found exclusively in soil, moss and leaf litter. Their biology is little known, but they probably feed on decaying plant material and fungi (especially hyphal contents); some species are known to feed on mycorrhizal fungi.

Taxobox 2 Archaeognatha (Microcoryphia; archaeognathans or bristletails)

The bristletails are primitively wingless insects, with some 500 species in two extant families. Development is ametabolous and moulting continues for life. They are moderate sized, 6–25 mm long, elongate and cylindrical. The head is hypognathous and bears large compound eyes that are in contact dorsally; three ocelli are present; the antennae are multisegmented. The mouthparts are partially retracted into the head and include elongate, monocondylar (single-articulated) mandibles and elongate, seven-segmented maxillary palps. The thorax is humped with segments subequal and unfused and with poorly developed pleura; the legs have large coxae each bearing a style and the tarsi are two- or three-segmented. The abdomen is 11-segmented and continues the thoracic contour; segments 2–9 bear ventral muscle-containing styles (representing limbs), whereas segments 1–7 have one or two pairs of protrusible vesicles medial to the styles (fully developed only in mature individuals). In females, the gonapophyses of segments 8 and 9 form an ovipositor. A long, multisegmented caudal appendix dorsalis, located mediodorsally on the tergum of segment 11, forms an epiproct extension lying between the paired cerci and dorsal to the genitalia. The paired multisegmented cerci are shorter than the median caudal appendage (as shown here for *Petrobius maritima*; after Lubbock 1873).

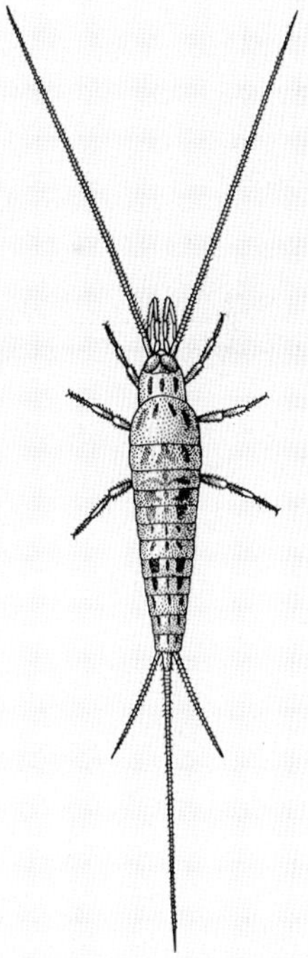

bristletail, Archaeognatha: Machilidae: *Petrobius maritima*

Fertilization is indirect, with sperm droplets attached to silken lines produced from the male gonapophyses, or stalked spermatophores are deposited on the ground, or, more rarely, sperm are deposited on the female's ovipositor. Bristletails often are active nocturnally, feeding on litter, detritus, algae, lichens and mosses, and sheltering beneath bark or in litter during the day. They can run fast and jump, using the arched thorax and flexed abdomen to spring considerable distances.

Bristletails are similar superficially to silverfish (Zygentoma), but differ in pleural structures and quite fundamentally in their mouthpart morphology. These two orders represent the surviving remnants of a wider radiation of primitively flightless insects. Phylogenetic relations are discussed in section 7.4.1 and depicted in Fig. 7.2.

Taxobox 3 Zygentoma (silverfish)

Silverfish are primitively wingless insects, with nearly 600 species in five extant families. Development is ametabolous and moulting continues for life. Their bodies are moderately sized (5–30 mm long) and dorsoventrally flattened, often with silvery scales. The head is hypognathous to slightly prognathous; compound eyes are absent or reduced to isolated ommatidia, and there may be one to three ocelli present; the antennae are multisegmented. The mouthparts are mandibulate and include dicondylous (= dicondylar or double-articulated) mandibles and five-segmented maxillary palps. The thoracic segments are subequal and unfused with poorly developed pleura; the legs have large coxae and two- to five-segmented tarsi. The abdomen is 11-segmented and continues the taper of the thorax with segments 7–9 at least, but sometimes 2–9, bearing ventral muscle-containing styles; mature individuals may have a pair of protrusible vesicles medial to the styles on segments 2–7, although these are often reduced or absent. In females, the gonapophyses of segments 8 and 9 form an ovipositor. A long, multisegmented caudal appendix dorsalis, located mediodorsally on the tergum of segment 11, forms an epiproct between the cerci and dorsal to the genitalia. The paired elongate multisegmented cerci are nearly as long as the median caudal appendage (as shown here for *Lepisma saccharina*; after Lubbock 1873).

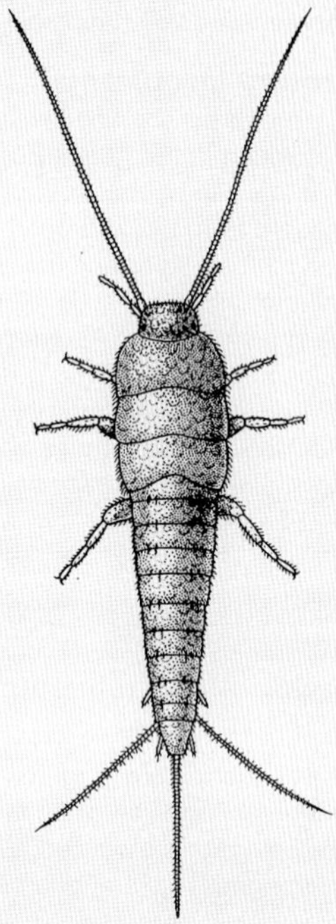

silverfish, Zygentoma: Lepismatidae: *Lepisma saccharina*

Fertilization is indirect, via flask-shaped spermatophores that females pick up from the substrate. Many silverfish live in litter or under bark; some are subterranean or are cavernicolous, but some species can tolerate low humidity and high temperatures of arid areas; for example, there are desert-living lepismatid silverfish in the sand dunes of the Namib Desert in southwestern Africa, where they are important detritivores. Some other zygentoman species live in mammal burrows, a few are commensals in nests of ants and termites, and several species are familiar synanthropic insects, living in human dwellings. These include *L. saccharina*, *Ctenolepisma longicaudata* (silverfishes) and *Thermobia domestica* (= *Lepismodes inquilinus*; the firebrat), which eat materials such as paper, cotton and plant debris, using their own cellulase to digest the cellulose.

Orders Zygentoma and Archaeognatha represent the surviving remnants of a wider radiation of primitively flightless insects. Insects of both groups are similar superficially but differ in pleural structures, and silverfish have dicondylous mandibles (like winged insects, the Pterygota) in contrast to the monocondylous mandible of bristletails. Phylogenetic relations are discussed in section 7.4.1 and depicted in Fig. 7.2.

Taxobox 4 Ephemeroptera (mayflies)

The mayflies constitute a small order of over 3000 described species in about 40 families, with highest genus diversity in the relatively poorly studied Neotropical region. Adults have reduced mouthparts and large compound eyes, especially in males, and three ocelli. Their antennae are filiform, sometimes multisegmented. The thorax, particularly the mesothorax, is enlarged for flight, with large triangular fore wings and smaller hind wings (as illustrated here for an adult male of the ephemerid *Ephemera danica*; after Stanek 1969, Elliott & Humpesch 1983), which are sometimes much reduced or absent. Males have elongate fore legs used to seize the female during the mating flight. The abdomen is 10-segmented, typically with three long, multisegmented, caudal filaments consisting of a pair of lateral cerci and usually a median terminal filament.

Development is hemimetabolous. Nymphs have 12–45 aquatic instars, with fully developed mandibulate mouthparts. Developing wings are visible in older nymphs (as shown here for a leptophlebiid nymph). Gas exchange is aided by a closed tracheal system lacking spiracles, with abdominal lamellar gills on some segments, sometimes elsewhere, including on the maxillae and labium. Nymphs have three usually filiform caudal filaments consisting of paired cerci and a variably reduced (rarely absent) median terminal filament. The penultimate instar or subimago (subadult) is fully winged and flies or crawls.

The subimago and adult are non-feeding and short-lived. Exceptionally, the subimagos mate and the adult stage is omitted. Imagos typically form mating swarms, sometimes of thousands of males, over water or nearby landmarks.

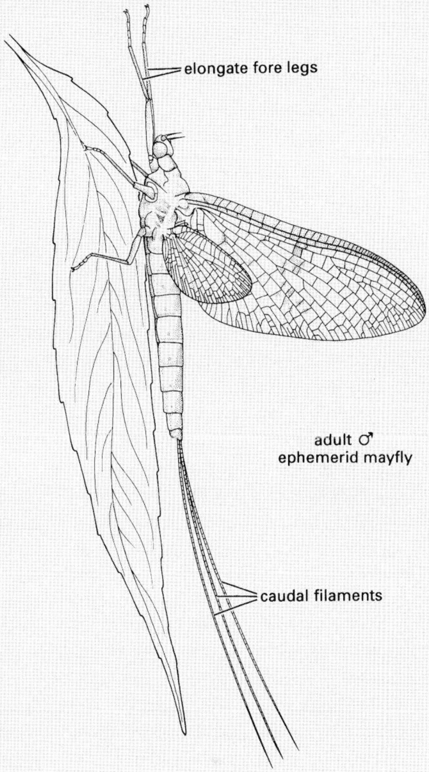

adult ♂ ephemerid mayfly

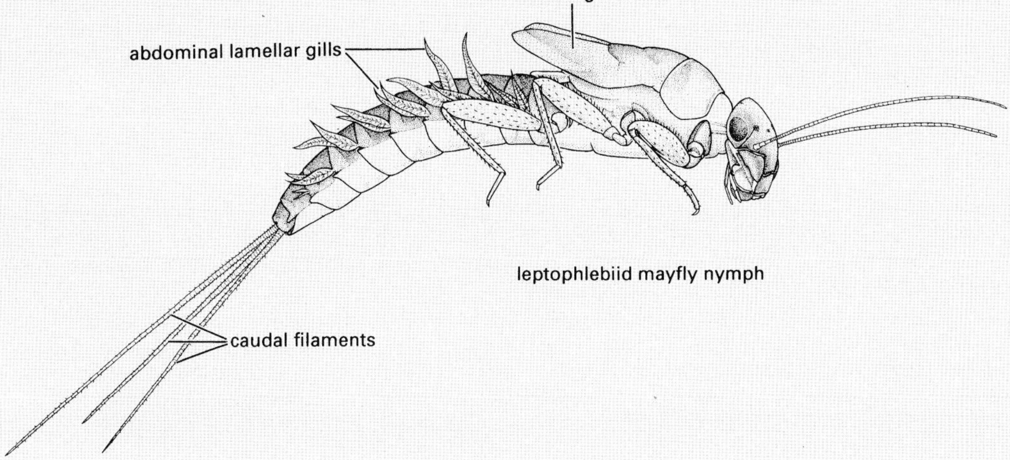

leptophlebiid mayfly nymph

Copulation usually takes place in flight, and eggs are laid in water by the female either dipping her abdominal apex below the surface or crawling under the water.

Nymphs graze on periphyton (algae, diatoms, aquatic fungi) or collect fine detritus; some are predatory on other aquatic organisms. Development takes from 16 days, to over one year in cold and high-latitude waters; some species are multivoltine. Nymphs occur predominantly in well-oxygenated cool fast-flowing streams, with fewer species in slower rivers and cool lakes; some tolerate elevated temperatures, organic enrichment, or increased sediment loads.

Phylogenetic relationships are discussed in section 7.4.2 and depicted in Fig. 7.2 and Fig. 7.3.

Taxobox 5 Odonata (damselflies and dragonflies)

These conspicuous insects comprise a small, largely tropical order containing about 6000 described species, with about one-half belonging to the suborder Zygoptera (damselflies) and the remaining species to the suborder Epiprocta (dragonflies), comprising the two extant species of Epiophlebiidae (formerly treated as suborder Anisozygoptera) sister to the speciose Anisoptera (section 7.4.2). The adults are medium to large (from <2 cm to >15 cm long, with a maximum wingspan of 17 cm in the South American giant damselfly (Pseudostigmatidae: *Mecistogaster*)). They have a mobile head with large, multifaceted compound eyes, three ocelli, short bristle-like antennae and mandibulate mouthparts. The thorax is enlarged to accommodate the flight muscles of two pairs of elongate membranous wings that are richly veined. The slender 10-segmented abdomen terminates in clasping organs in both sexes; males possess secondary genitalia on the venter of the second to third abdominal segments; females often have an ovipositor at the ventral apex of the abdomen. In adult zygopterans, the eyes are widely separated and the fore and hind wings are equal in shape with narrow bases (as illustrated in the top right figure for a lestid, *Austrolestes*; after Bandsma & Brandt 1963). Anisopteran adults have eyes either contiguous or slightly separated, and their wings have characteristic closed cells called the triangle (T) and hypertriangle (ht) (Fig. 2.24b); the hind wings are considerably wider at the base than are the fore wings (as illustrated in the top left figure for a libellulid dragonfly, *Sympetrum*; after Gibbons 1986). Odonate nymphs have a variable number of up to 20 aquatic instars, each with fully developed mandibulate mouthparts, including an extensible grasping labium or "mask" (Fig. 13.4). The developing wings are visible in older nymphs. The tracheal system is closed and lacks spiracles, but specialized gas-exchange surfaces are present on the abdomen as external gills (Zygoptera) or internal folds in the rectum (Anisoptera; Fig. 3.11f). Zygopteran nymphs (such as the lestid illustrated on the lower right; after CSIRO 1970) are slender, with the head wider than the thorax and the apex of the abdomen with three (rarely two) elongate tracheal gills (caudal lamellae). Anisopteran nymphs (such as the libellulid illustrated on the lower left; after CSIRO 1970) are more stoutly built, with the head rarely much broader than the thorax and the abdominal apex characterized by an anal pyramid consisting of three short projections and a pair of cerci in older nymphs. Many anisopteran nymphs rapidly eject water from their anus – "jet propulsion" – as an escape mechanism.

Prior to mating, the male fills his secondary genitalia with sperm from the primary genital opening on the ninth abdominal segment. At mating, the male grasps the female by her neck or prothorax and the pair fly in tandem, usually to a perch. The female then bends her abdomen forwards to connect to the male's secondary genitalia, thus forming the "wheel" position (as illustrated in Box 5.5). The male may displace sperm of a previous male before transferring his own (Box 5.5), and mating may last from seconds to several hours, depending on species. Egg-laying may take place with the pair still in tandem. The eggs (Fig. 5.10) are laid onto a water surface, into water, mud, sand or plant tissue, depending on species. After eclosion, the hatchling ("pronymph") immediately moults to the first true nymph, which is the first feeding stage.

The nymphs are predatory on other aquatic organisms, whereas the adults catch terrestrial aerial prey. At metamorphosis (Fig. 6.8), the pharate adult moves to the water/land surface where atmospheric gaseous exchange commences; then it crawls from the water, anchors terrestrially and the imago emerges from the cuticle of the final-instar nymph. The imago (adult) is long-lived, active and aerial. Nymphs occur in all waterbodies, particularly in well-oxygenated, standing waters, but elevated temperatures, organic enrichment or increased sediment loads are tolerated by many species.

Phylogenetic relations are discussed in section 7.4.2 and depicted in Fig. 7.2 and Fig. 7.3.

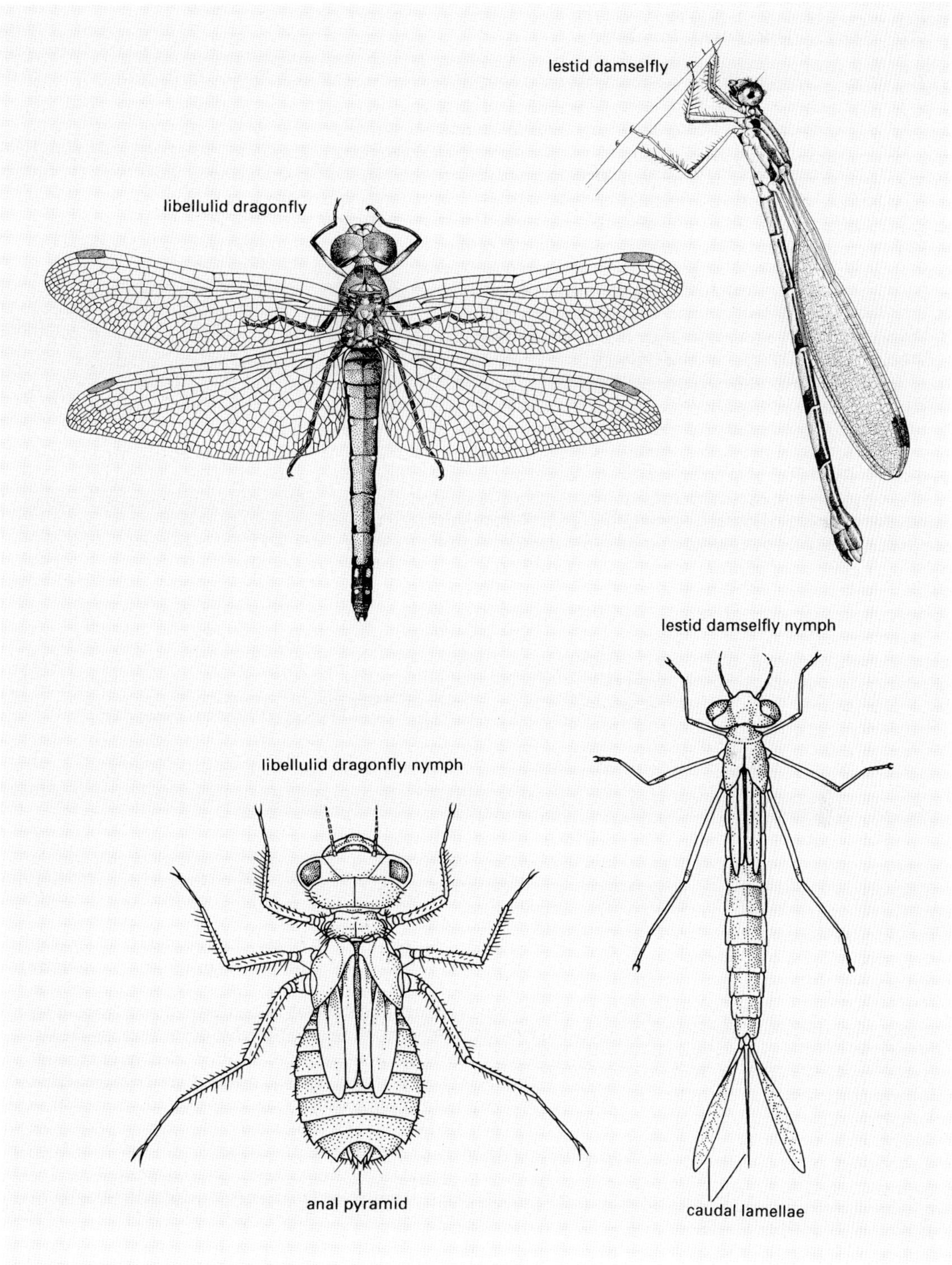
lestid damselfly
libellulid dragonfly
lestid damselfly nymph
libellulid dragonfly nymph
anal pyramid
caudal lamellae

Taxobox 6 Plecoptera (stoneflies)

The stoneflies constitute a minor and often cryptic order of 16 families, with nearly 4000 species worldwide, predominantly in temperate and cool areas. They are hemimetabolous, with adults resembling winged nymphs. The adult is mandibulate with filiform antennae, bulging compound eyes, and two or three ocelli. The thoracic segments are subequal, and the fore and hind wings are membranous and similar (except the hind wings are broader), with the folded wings partly wrapping the abdomen and extending beyond the abdominal apex (as illustrated for an adult of the Australian gripopterygid, *Illiesoperla*); however, aptery and brachyptery are frequent. The legs are unspecialized and the tarsi comprise three segments. The abdomen is soft and 10-segmented, with vestiges of segments 11 and 12 serving as paraprocts, cerci and epiproct, a combination of which serve as male accessory copulatory structures, sometimes in conjunction with the abdominal sclerites of segments 9 and 10. The nymphs have 10–24, rarely as many as 33, aquatic instars, with fully developed mandibulate mouthparts; the wing pads are first visible in half-grown nymphs. The tracheal system is closed, with simple or plumose gills on the basal abdominal segments or near the anus (Fig. 10.1) – sometimes extrusible from the anus – or on the mouthparts, neck or thorax, or lacking altogether. The cerci are usually multisegmented and there is no median terminal filament.

Stoneflies usually mate during daylight; some species drum the substrate with their abdomen prior to mating. Eggs are dropped into water, laid in a jelly on water, or laid underneath stones in water or into damp crevices near water. Eggs may diapause. Nymphal development may take several years in some species.

Nymphs may be omnivores, detritivores, herbivores or predators. Adults feed on algae, lichen, higher plants and/or rotten wood; some may not eat. Mature nymphs crawl to the water's edge where adult emergence takes place. Nymphs occur predominantly on stony or gravelly substrates in cool water, mostly in well-aerated streams, with fewer species in lakes. Generally, they are very intolerant of organic and thermal pollution.

Phylogenetic relationships are discussed in section 7.4.2 and depicted in Fig. 7.2.

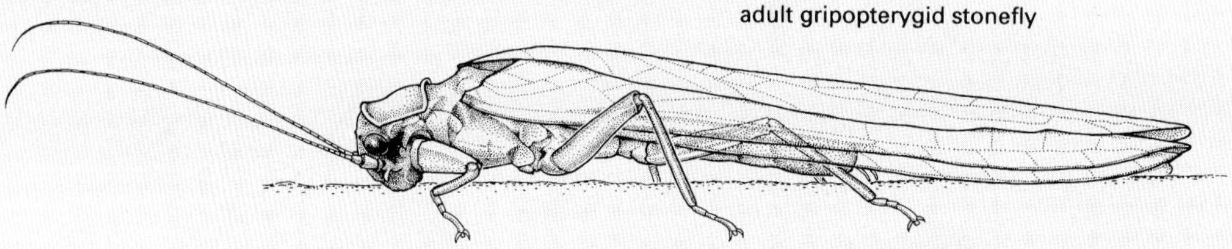

Taxobox 7 Dermaptera (earwigs)

The earwigs comprise a worldwide order containing almost 2000 described species, usually placed in 11 families, but with unstable classification. They are hemimetabolous, with small to moderately sized (4–25 mm long) elongate bodies that are dorsoventrally flattened. The head is prognathous; the compound eyes may be large, small or absent, and ocelli are absent. The antennae are short to moderate in length and filiform with segments elongate; there are fewer antennal segments in immature individuals than in the adult. The mouthparts are mandibulate (section 2.3.1; Fig. 2.10). The legs are relatively short, and the tarsi are three-segmented with the second tarsomeres short. The prothorax has a shield-like pronotum, and the meso- and metathoracic sclerites are of variable size. Earwigs are apterous or, if winged, their fore wings are small and leathery with smooth, unveined tegmina; the hind wings are large, membranous and semi-circular (as illustrated here for an adult male of the common European earwig, *Forficula auricularia*) and when at rest are folded fan-like and then longitudinally, protruding slightly from beneath the tegmina; hind-wing venation is dominated by the anal fan of branches of A_1 and cross-veins. The abdominal segments are telescoped (terga overlapping), with 10 visible segments in the male and eight in the female, terminating in prominent cerci modified into forceps; the latter are often heavier, larger and more curved in males than in females.

Copulation is end-to-end and male spermatophores may be retained in the female for some months prior to fertilization. Oviparous species lay eggs often in a burrow in debris (Fig. 9.1), guard the eggs and lick them to remove fungus. The female may assist the nymphs to hatch from the eggs and may care for them until the second or third instar. Maturity is attained after four or five moults. The two semi-parasitic or epizoic groups, Arixeniidae and Hemimeridae, exhibit pseudoplacental viviparity (section 5.9).

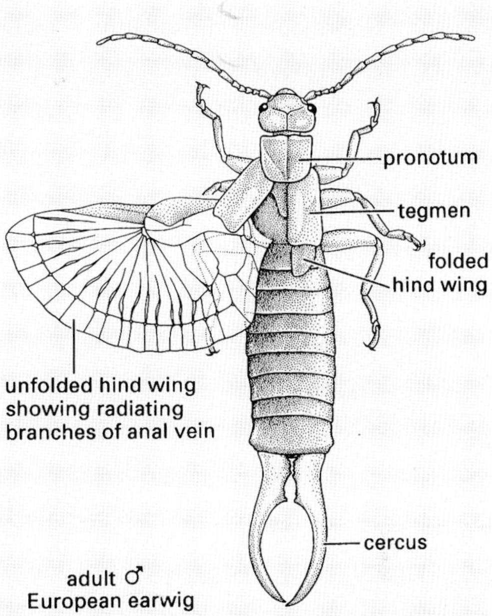

Earwigs are mostly cursorial and nocturnal, with most species rarely flying. Feeding is predominantly on dead and decaying vegetable and animal matter, with some predation and some damage to living vegetation, especially in gardens. A small number are commensals or ectoparasites of bats in Southeast Asia (family Arixeniidae) or semi-parasites of African rodents (family Hemimeridae): earwigs in both groups are blind, apterous and with curved or rod-like forceps. The forceps of free-living earwigs are used for manipulating prey, for defence and offense, and in some species for grasping the partner during copulation. The common name "earwig" may derive from a supposed predilection for entering ears, or from a corruption of "ear wing" referring to the shape of the wing, but these are unsupported.

Phylogenetic relationships are discussed in section 7.4.2 and depicted in Fig. 7.2.

Taxobox 8 Zoraptera (zorapterans or angel insects)

These insects comprise the single genus *Zorotypus*, sometimes subdivided into seven genera, containing nearly 40 described species found worldwide in tropical and warm temperate regions except Australia. They are small (<4 mm long) and rather termite-like. The head is hypognathous and compound eyes and ocelli are present in winged species but absent in apterous species. The antennae are moniliform and nine-segmented, and the mouthparts are mandibulate with five-segmented maxillary palps and three-segmented labial palps. The subquadrate prothorax is larger than the similar-shaped meso- and metathorax. The wings are polymorphic; some forms are apterous in both sexes, whereas other forms are alate, with two pairs of paddle-shaped wings with reduced venation and smaller hind wings (as illustrated here for *Zorotypus hubbardi*; after Caudell 1920). The wings are shed as in ants and termites. The legs have well-developed coxae, expanded hind femora bearing stout ventral spines and two-segmented tarsi, each with two claws. The 11-segmented abdomen is short and rather swollen, with cerci comprising just a single segment. The male genitalia are asymmetric.

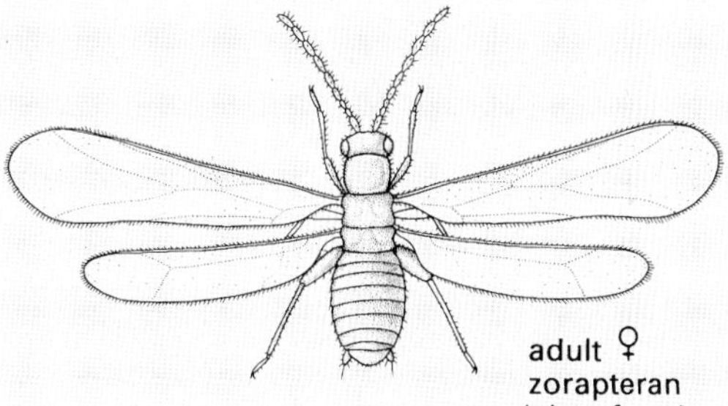

The immature stages are polymorphic according to wing development. Zorapterans are gregarious, occurring in leaf litter, rotting wood or near termite colonies, eating fungi and perhaps small arthropods. Phylogenetically they are enigmatic, with several suggested relationships within the Polyneoptera (see section 7.4.2 and Fig. 7.2).

Taxobox 9 Orthoptera (grasshoppers, locusts, katydids and crickets)

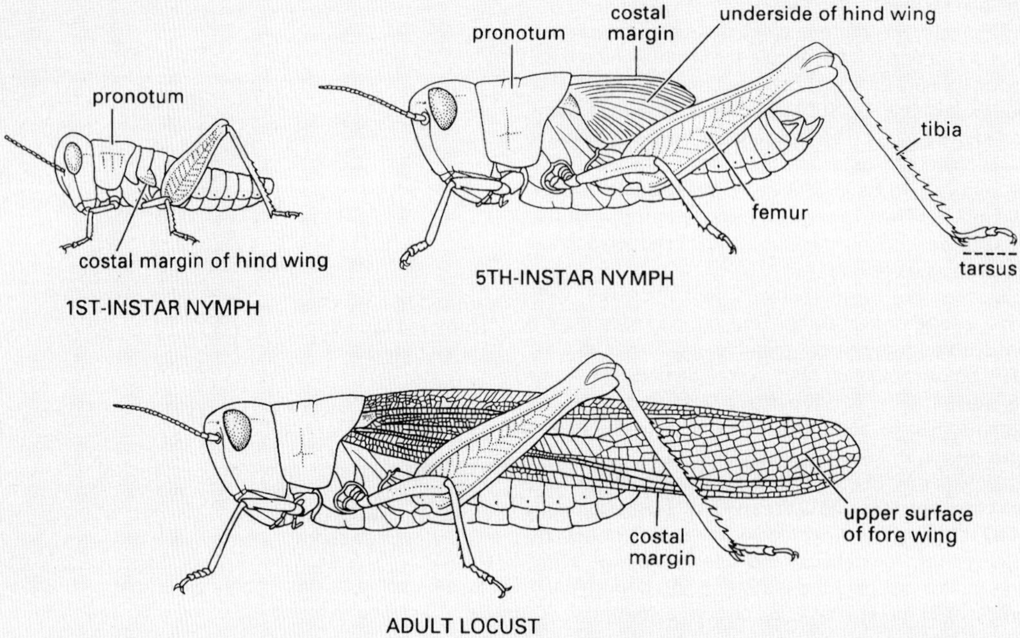

The Orthoptera is a worldwide order of almost 24,000 described species in up to 40 families (the classification is unstable) and comprising two suborders: Caelifera (grasshoppers and locusts) and Ensifera (katydids and crickets). Orthopterans have hemimetabolous development, and are typically elongate cylindrical, medium-sized to large (up to 12 cm long), with enlarged hind legs for jumping. They are hypognathous and mandibulate, and have well-developed compound eyes; ocelli may be present or absent. The antennae are multisegmented. The prothorax is large, with a shield-like pronotum curving over the pleura; the mesothorax is small, and the metathorax large. The fore wings form narrow, leathery tegmina; the hind wings are broad, with numerous longitudinal and cross-veins, folded beneath the tegmina by pleating. Aptery and brachyptery are frequent. The legs are often elongate and slender, and the hind legs large, usually saltatorial; the tarsi have 1–4 segments. The abdomen has 8–9 annular visible segments, with two or three terminal segments reduced. Females have a well-developed appendicular ovipositor (Fig. 2.25b,c; Box 5.2). The cerci each consist of a single segment.

Courtship may be elaborate and often involves communication by sound production and reception (section 4.1.3 and section 4.1.4). In copulation, the male is astride the female, with mating sometimes prolonged for many hours. Ensiferan eggs are laid singly into plants or soil, whereas Caelifera use their ovipositor to bury batches of eggs in soil chambers. Egg diapause is frequent. Nymphs resemble small adults except in the lack of development of wings and genitalia, but apterous adults may be difficult to distinguish from nymphs. In all winged species, nymphal wing pad orientation changes between moults (as illustrated here for a locust); in early instars the wing pad rudiments are laterally positioned with the costal margin ventral, until prior to the penultimate nymphal instar they rotate about their base so that the costal margin is dorsal and the morphological ventral surface is external; the hind wing then overlaps the fore wing (as in the fifth-instar nymph illustrated here). During the moult to the adult, the wings resume their normal position with the costal margin ventral. This wing pad "rotation", otherwise known only in Odonata, is unique to the Orthoptera amongst the Polyneoptera orders.

Caelifera are predominantly day-active, fast-moving, visually acute, terrestrial herbivores, and include some destructive insects such as migratory locusts (section 6.10.5; Fig. 6.14; Plate 4c and Plate 7a). Ensifera are more often night-active, camouflaged or mimetic, and are predators, omnivores or phytophages, and have long antennae (usually more than 30 segments) (Plate 1d and Plate 3b,f).

Phylogenetic relationships are considered in section 7.4.2 and depicted in Fig. 7.2.

Taxobox 10 Embioptera (Embiidina, Emboidea; embiopterans or webspinners)

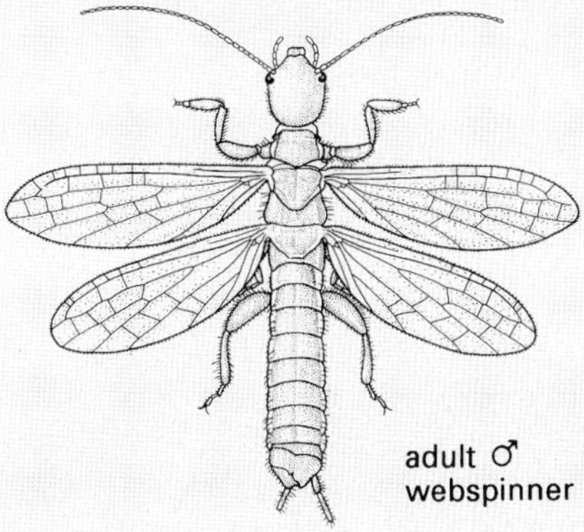

There are over 450 described species of embiopterans (and perhaps up to an order of magnitude more remain undescribed), in perhaps 13 families, worldwide. Small to moderately sized, they have an elongate, cylindrical body, somewhat flattened in males. The head is prognathous, and the compound eyes are reniform (kidney-shaped), larger in males than females; ocelli are absent. The antennae are multisegmented, and the mouthparts are mandibulate. The quadrate prothorax is larger than the meso- or metathorax. All females and some males are apterous, and, if present, the wings (illustrated here for *Embia major*; after Imms 1913) are characteristically soft and flexible, with blood sinus veins stiffened for flight by haemolymph pressure. The legs are short, with three-segmented tarsi; the basal segment of each fore tarsus is swollen and contains silk glands, whereas the hind femora are swollen with strong tibial muscles. The abdomen is 10-segmented, with only the rudiments of segment 11; the cerci comprise two segments and are responsive to tactile stimuli. The female external genitalia are simple, whereas the male genitalia are complex and asymmetrical. Immature stages resemble the adults except for their wings and genitalia.

During copulation, the male holds the female with his prognathous mandibles and/or his asymmetrical cerci. Webspinners live gregariously in silken galleries, spun with the tarsal silk glands (present in all instars); their galleries occur in leaf litter, beneath stones, on rocks, on tree trunks, or in cracks in bark and soil, often around a central retreat. Their food comprises litter, moss, bark and dead leaves. The galleries are extended to new food sources, and the safety of the gallery is left only when mature males disperse to new sites, where they mate, do not feed, and sometimes are cannibalized by females (Plate 7c). The eggs and young nymphs are tended by the female parent within the gallery, with the female of some species feeding the nymphs with chewed-up plant material. Webspinners readily reverse within their galleries, for example when threatened by a predator.

Phylogenetic relationships are discussed in section 7.4.2 and depicted in Fig. 7.2.

Taxobox 11 Phasmatodea (phasmids, stick-insects or walking sticks)

The Phasmatodea is a worldwide, predominantly tropical order of more than 3000 described species, and lacks a phylogenetically based classification. They have hemimetabolous development and are elongate cylindrical (Plate 1c) and stick-like or flattened and often leaf-like in form, up to >30 cm in body length (*Phobaeticus chani*, the longest species, has a body length of up to 36 cm and a total length, including outstretched legs, of up 57 cm, and is from Borneo). Phasmids have mandibulate mouthparts. The compound eyes are anterolaterally placed and relatively small, and ocelli occur only in winged species, often only in males. The antennae range from short to long, with 8–100 segments. The prothorax is small, and the mesothorax and metathorax are elongate if winged, shorter if apterous. The wings, when present, are functional in males but are often reduced in females; many species are apterous in both sexes. The fore wings form leathery tegmina, whereas the hind wings are broad, with a network of numerous cross-veins and the anterior margin toughened as a remigium that protects the folded wing. The legs are elongate, slender, gressorial, with five-segmented tarsi; they can be shed in defence (section 14.3) and may be regenerated at a nymphal moult. The abdomen is 11-segmented, with segment 11 often forming a concealed supra-anal plate in males or a more obvious segment in females; the male genitalia are concealed and asymmetrical. The cerci are variably lengthened and consist of a single segment.

In copulation (which is often prolonged), the smaller male is astride the female, as illustrated here for the spur-legged stick-insect, *Didymuria violescens* (Phasmatidae). The eggs often resemble seeds (as shown here in the

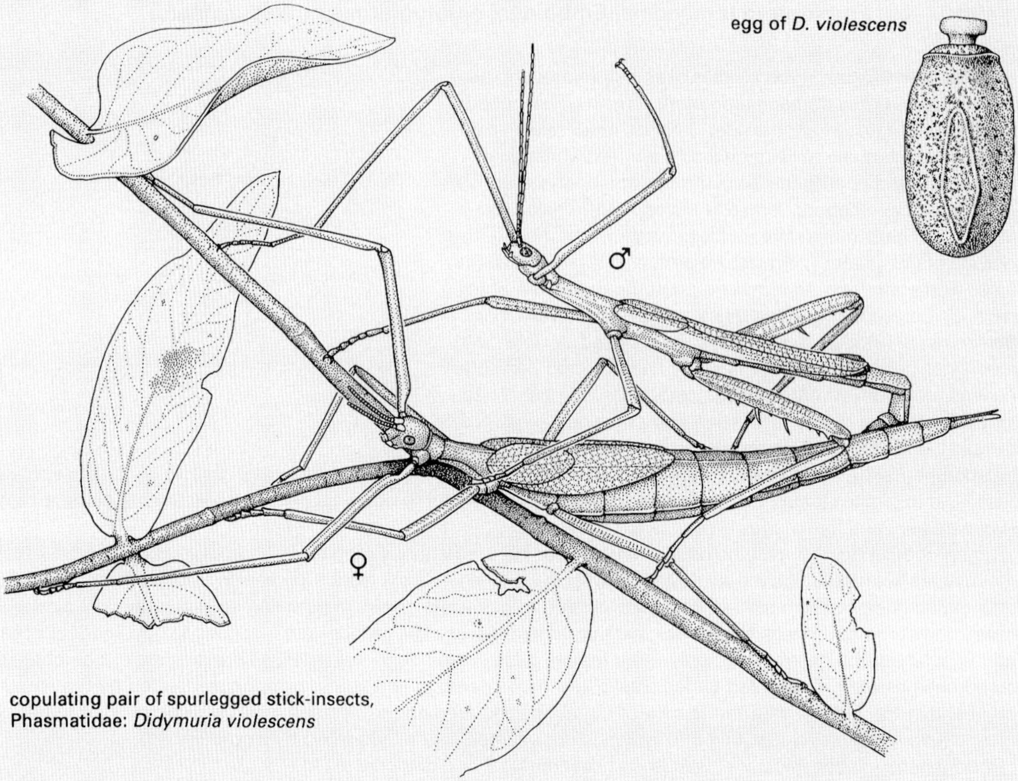

copulating pair of spurlegged stick-insects, Phasmatidae: *Didymuria violescens*

enlargement of the egg of *D. violescens*; after CSIRO 1970) and are deposited singly, glued on vegetation, or dropped to the ground; there may be a lengthy egg diapause. Nymphal phasmids mostly resemble adults except in their lack of wing and genitalia development, the absence of ocelli, and the fewer antennal segments.

Phasmatodea are phytophagous, and predominantly resemble (mimic) various plant features such as stems, sticks and leaves. In conjunction with crypsis, phasmids demonstrate an array of anti-predator defences, ranging from general slow movement, grotesque and often asymmetrical postures, to death feigning (section 14.1 and section 14.2) and, in a number of species, ejection of noxious chemicals from prothoracic glands.

Phylogenetic relationships are considered in section 7.4.2 and depicted in Fig. 7.2.

Taxobox 12 Grylloblattodea (Grylloblattaria or Notoptera; grylloblattids, ice crawlers or rock crawlers)

Grylloblattodea comprise a single extant family, Grylloblattidae, with 32 described species, restricted to western North America and central to eastern Asia including Japan, but have high fossil diversity, with over 500 extinct species in many families. North American species are particularly tolerant of cold and may live at high elevations on glaciers and snow banks; Asian species inhabit montane areas and cool temperate forests. Grylloblattodea are moderately sized insects (14–35 mm long) with an elongate, pale, cylindrical body that is soft and pubescent.

The head is prognathous and the compound eyes are reduced or absent; ocelli are absent. The antennae are multisegmented, and the mouthparts mandibulate. The quadrate prothorax is larger than the meso- or metathorax; wings are absent. The legs are cursorial, with large coxae and five-segmented tarsi. The abdomen has 10 visible segments and the rudiments of segment 11, with five- to 10-segmented cerci. The female has a short ovipositor and the male genitalia are asymmetrical.

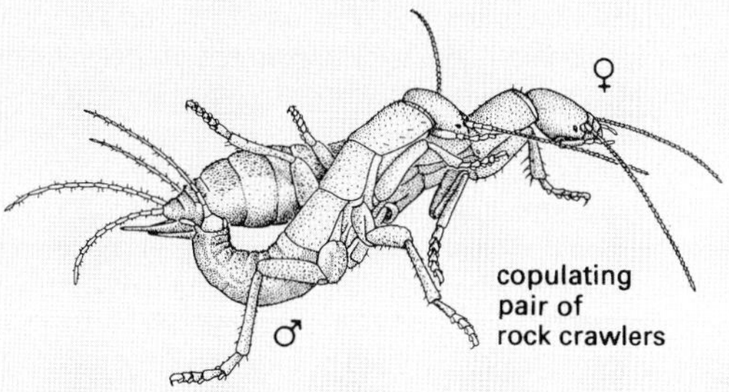

Copulation takes place side-by-side, with the male on the right, as illustrated here for a common Japanese species, *Galloisiana nipponensis* (after Ando 1982). Eggs may diapause up to a year in damp wood or soil under stones. Nymphs, which resemble adults, develop slowly through eight instars. The typical lifespan is estimated to be five years, but may be much longer in some species. North American ice crawlers, genus *Grylloblatta*, are active by day and night at low temperatures, feeding on dead arthropods and organic material, notably from the surface of ice and snow in spring snow melt, within caves (including ice caves), in alpine soil and damp places such as beneath rocks. Rapid loss of such habitats due to climate change means that the ranges of some species are contracting substantially, leading to conservation concern for these montane species (section 17.3).

Phylogenetic relationships are discussed in section 7.4.2 and depicted in Fig. 7.2.

Taxobox 13 Mantophasmatodea (heelwalkers)

The discovery of a previously unrecognized order of insects is an unusual event. In the 20th century, only two orders were newly described: Zoraptera in 1913 and Grylloblattodea in 1932. The opening of the 21st century saw a flurry of scientific and popular media interest concerning the unusual discovery and subsequent recognition of a new order, the Mantophasmatodea.

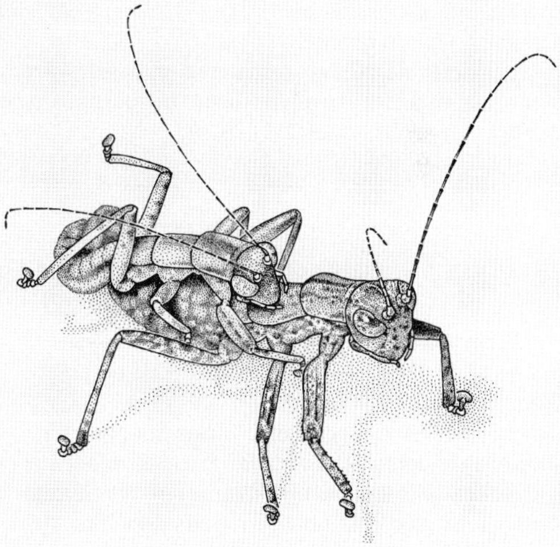

The first formal recognition of this new taxon was from a specimen preserved in 45-million-year-old Baltic amber, which bore a superficial resemblance to a stick-insect or a mantid, but evidently belonged to neither. Shortly thereafter, a museum specimen from Tanzania and another from Namibia were discovered, and comparison with more fossil specimens, including adults, showed that the fossil and recent insects were related. Further museum searches and appeals to curators uncovered specimens from rocky outcrops in Namibia. An expedition found the living insects in several Namibian localities, and subsequently many specimens were identified in historic and recent collections from succulent karoo, nama karoo and fynbos vegetation of South Africa.

Currently, there are two or three extant families of Mantophasmatodea, with at least three extinct and about 12 extant genera (there is disagreement on delimitation of genera and families), and about 20 described extant species (plus several undescribed), now restricted to southwestern Africa (Namibia and South Africa) and Tanzania in eastern Africa. The higher classification is changing as new species are discovered and relationships are investigated (see entry for this order in section 7.4.2).

Mantophasmatodeans are moderate-sized (up to 2.5 cm long in extant species, 1.5 cm long in fossil species) hemimetabolous insects, with a hypognathous head with generalized mouthparts (mandibles with three small teeth) and long slender antennae with 26–32 segments and a bend distally. The prothoracic pleuron is large and exposed, not covered by pronotal lobes. Each tergum of the thorax narrowly overlaps and is smaller than the previous. All species are apterous, without any rudiments of wings. The coxae are elongate, the fore and mid femora are somewhat broadened and with bristles or short spines ventrally; the fore and mid tibiae each bear two rows of spines. The tarsi are five-segmented with euplantulae on the basal four, the ariolum is very large and, characteristically, the distal tarsomere is held off the substrate (hence the common name "heelwalkers"). The hind legs are elongate and can be used in making small jumps. Male cerci are prominent (as on the male shown in the Appendix; after a photograph by M.D. Picker), clasping, and do not form a differentiated articulation with the 10th tergite. Female cerci are one-segmented and short. The ovipositor projects beyond the short subgenital lobe and there is no protective operculum (plate below ovipositor) as occurs in phasmids.

Heelwalkers communicate with each other for species recognition and mate location via substrate vibrations that they produce by drumming their abdomen repeatedly. Copulation may be prolonged (up to three days uninterrupted) and, at least in captivity, the male often is eaten after mating. The male mounts the female with his genitalia engaged from her right-hand side, as shown here for a copulating pair of South African heelwalkers (after a photograph by S.I. Morita). Eggs are laid in pods made of sand grains cemented by a water-resistant secretion. The life cycle is not well known, although the resistant egg stage survives the dry season, and nymphal development coincides with the wetter months of the year. The moulted cuticle is eaten after ecdysis. At least some Namibian species are diurnal, whereas South African species are nocturnal. Heelwalkers are either ground dwelling or live on shrubs or in grass clumps. They usually occur singly or as mating pairs. All heelwalkers are predatory, feeding for example on small flies, bugs and moths, and hence the alternative common name of "gladiators". Raptorial femora are grooved to receive the spined tibiae during prey capture; at rest the raptorial limbs are not folded. Most species exhibit considerable colour variation from light green to dark brown. Males generally are smaller and of a different colour to females.

Most molecular evidence suggests that heelwalkers comprise the sister group to Grylloblattodea, which is one of the suggested relationships based on morphology (see section 7.4.2 and Fig. 7.2).

Taxobox 14 Mantodea (mantids, mantises or praying mantids)

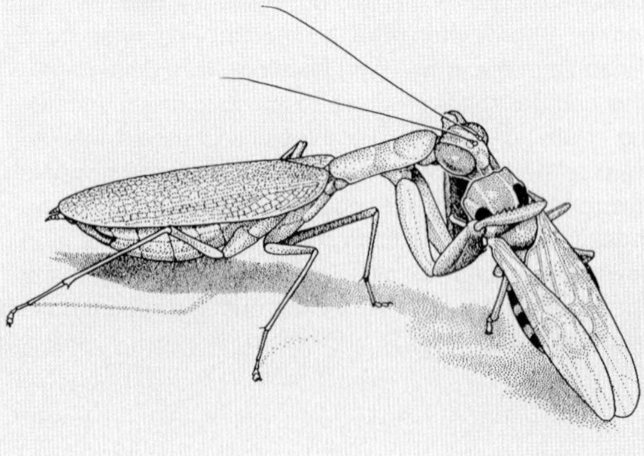

The Mantodea is an order of about 2400 species of moderate to large (1–15 cm long) hemimetabolous predators classified in eight to 15 families, with unstable family-level classification. Males are generally smaller than females. The head is small, triangular and mobile, with slender antennae, large, widely separated eyes, and mandibulate mouthparts. The thorax comprises an elongate, narrow prothorax and shorter (almost subquadrate) meso- and metathorax. The fore wings form leathery tegmina, with the anal area reduced; the hind wings are broad and membranous, with long veins unbranched and many cross-veins. Aptery and subaptery are frequent. The fore legs are raptorial (Fig. 13.3 and as illustrated here for a mantid of a *Tithrone* species holding and eating a fly; after Preston-Mafham 1990), whereas the mid and hind legs are elongate for walking. On the abdomen, the 10th visible segment bears variably segmented cerci. The ovipositor is predominantly internal; the external male genitalia are asymmetrical.

Eggs are laid in an ootheca (**Plate 3e**) produced from accessory gland frothy secretions that harden on contact with the air. Some females guard their ootheca. First-instar nymphs do not feed, but moult immediately. As few as three or as many as 12 instars follow; the nymphs resemble adults except for lack of wings and genitalia. Adult mantids are sit-and-wait predators (see section 13.1.1) that use their fully mobile head and excellent sight to detect prey. Female mantids sometimes consume the male during or after copulation (Box 5.3); males often display elaborate courtship.

Mantodea are undoubtedly the sister group to the Blattodea (cockroaches and termites), forming the Dictyoptera grouping (Fig. 7.2 and Fig. 7.4).

Taxobox 15 Blattodea: roach families (cockroaches or roaches)

The concept of Blattodea (also called Blattaria) has been broadened (see below and also Fig. 7.4) to include both cockroaches and termites (see Taxobox 16). This box deals only with the characteristics of cockroaches, which comprise about 4500 described species in five or more families worldwide. They are hemimetabolous, with small to large (<3 mm to >100 mm), dorsoventrally flattened bodies. The head is hypognathous, and the compound eyes may be moderately large to small, or absent in cavernicolous species; ocelli are represented by two pale spots. The antennae are filiform and multisegmented, and the mouthparts are mandibulate. The prothorax has an enlarged, shield-like pronotum, often covering the head; the meso- and metathorax are rectangular and subequal. The fore wings (Fig. 2.24c) are sclerotized as tegmina, protecting the membranous hind wings; each tegmen lacks an anal lobe, and is dominated by branches of veins R and CuA. In contrast, the hind wings have a large anal lobe, with many branches in the R, CuA and anal sectors; at rest they lie folded fan-like beneath the tegmina. Wing reduction is frequent. The legs are often spinose (Fig. 2.21) and have five-segmented tarsi. The large coxae abut each other and dominate the ventral thorax. The abdomen can have 10 visible segments, with the subgenital plate of the male (sternum 9) often bearing one or a pair of styles, and concealing segment 11, which is represented only by paired paraprocts. The cerci comprise from one to usually many segments. The male genitalia are asymmetrical, and the female's ovipositor valves are concealed inside a genital atrium.

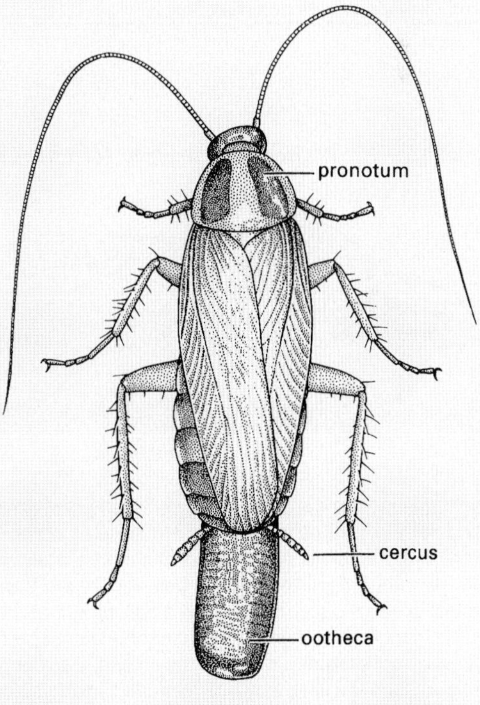

adult ♀ cockroach

Mating in cockroaches may involve stridulatory courtship, both sexes may produce sex pheromones, and the female may mount the male prior to end-to-end copulation. Eggs generally are laid in a purse-shaped ootheca, comprising two parallel rows of eggs with a leathery enclosure (section 5.8), which may be carried externally by the female (as illustrated here for a female of *Blatella germanica*; after Cornwell 1968). Certain species demonstrate a range of forms of ovoviviparity in which a variably reduced ootheca is retained within the reproductive tract in a "uterus" (or brood sac) during embryogenesis, often until nymphal hatching; true viviparity is rare. Parthenogenesis occurs in a few species. Nymphs develop slowly, resembling small apterous adults.

Cockroaches are amongst the most familiar insects, owing to the widespread human-associated habits of some 30 species, including *Periplaneta americana* (the American cockroach), *B. germanica* (the German cockroach) and *B. orientalis* (the Oriental cockroach). These nocturnal, malodorous, disease-carrying, refuge-seeking, peridomestic roaches are unrepresentative of the wider diversity. Typically, cockroaches are tropical, either nocturnal or diurnal, and sometimes arboreal, with some cavernicolous species. Cockroaches include solitary and gregarious species; woodroaches (*Cryptocercus*) live in family groups. Cockroaches mostly are saprophagous scavengers, but some eat wood and use enteric protists to break it down. Almost all cockroaches, and the termite *Mastotermes* (see below), house endosymbiont bacteria (*Blattabacterium*) in specialized calls in their fat body; the bacteria, which probably recycle nitrogenous wastes, are vertically transmitted (i.e. via the eggs).

Phylogenetic relations are discussed in section 7.4.2 and depicted in Fig. 7.2 and Fig. 7.4. *Cryptocercus* has long been known to have termite-like features, such as sociality and digestion of cellulose via protists, and this similarity reflects actual relationships, with termites having arisen from within Blattodea (Fig. 7.4). We here treat termites as derived cockroaches, because otherwise the Blattodea would be paraphyletic, however we discuss termites in a separate box (Taxobox 16).

Taxobox 16 Blattodea: epifamily Termitoidae (former order Isoptera; termites, "white ants")

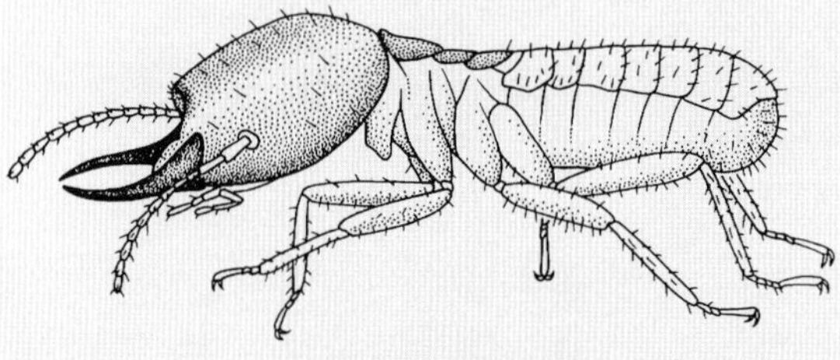

soldier of *Coptotermes*

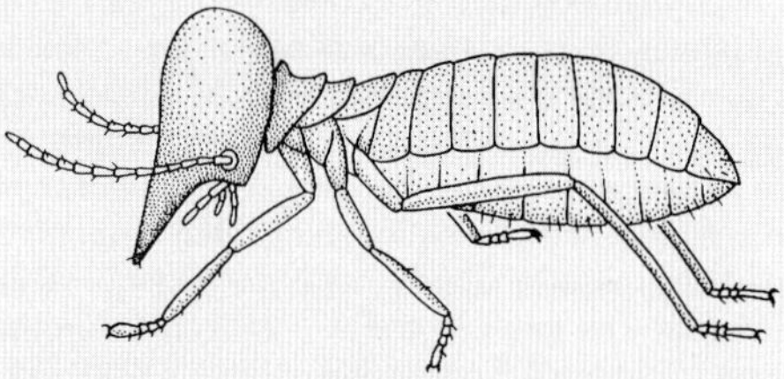

soldier of *Nasutitermes*

The termites form a small autapomorphic clade of about 3000 described species of hemimetabolous neopterans, living socially with polymorphic caste systems of reproductives, workers and soldiers (section 12.2.4; Fig. 12.8; Plate 6c). All stages are small to moderately sized (even winged reproductives are usually <20 mm long). The head is hypognathous or prognathous, and the mouthparts typically blattoid and mandibulate, but varying between castes: soldiers often have bizarre development of their mandibles or possess a nasus (as illustrated on the left for the mandibulate *Coptotermes* and on the right for the nasute *Nasutitermes*; after Harris 1971). The compound eyes are frequently reduced, and the antennae are long and multisegmented, with a variable number of segments. The wings are membranous with restricted venation; fore and hind wings are similar, except in *Mastotermes*, which has complex venation and an expanded hind-wing anal lobe. All castes have a pair of one- to five-segmented cerci terminally. External genitalia are highly reduced, except in *Mastotermes* in which the female has a reduced blattoid ovipositor and the male a membranous copulatory organ.

Internal anatomy is dominated by a convoluted alimentary tract, including an elaborated hindgut containing symbiotic bacteria and, in all species except most Termitidae, also flagellates (section 9.5.3 discusses fungal culture by Macrotermitinae). Food exchange between individuals (trophallaxis) is the sole means of replenishment of symbionts to young and newly moulted individuals, and is one explanation of the universal eusociality of termites.

Nests may be galleries or more complex structures within wood, such as rotting timber or even a sound tree, or above-ground nests (termitaria) such as prominent earth mounds (Fig. 12.10 and Fig. 12.11). Termites feed predominantly on cellulose-rich material; many harvest grasses and return the food to their subterranean nests or above-ground nest mounds. Termites are important decomposer organisms, especially in tropical lowland ecosystems, where they can constitute 95% of the soil insect biomass. In the tropics, termite biomass commonly ranges from 70 to 110 kg ha^{-1}, based on abundances of 2000–7000 individuals m^{-2}, but sometimes much higher. Such values are comparable to the biomass of vertebrate herbivores in the same ecosystems. Although termites play essential roles in maintaining the function and diversity of many habitats, they are best known as pests of our crops, trees and wood (Box 12.4). However, fewer than 200 species are known pests.

Termites belong to the order Blattodea, within a clade called Dictyoptera. Generally, seven to nine families and 16 subfamilies of termites are recognized, with about 70% of species belonging to the Termitidae. For phylogenetic relationships of termites and the Dictyoptera, see section 7.4.2 and Fig. 7.2 and Fig. 7.4.

Taxobox 17 Psocodea: "Psocoptera" (bark lice and book lice)

The order Psocodea (previously treated as a superorder) contains the former orders "Psocoptera" (psocids, or bark lice and book lice; nearly 6000 species) and "Phthiraptera" (chewing lice and sucking lice; see Taxobox 18), which together comprise nearly 11,000 described species. There are seven suborders of Psocodea, including two of bark lice (Psocomorpha and Trogiomorpha) and one containing both bark lice and book lice (Troctomorpha). The latter three suborders are the non-parasitic lice. They are common but cryptic, minute to small insects (1–10 mm long). Development is hemimetabolous with five or six nymphal instars. They have a large and mobile head, and large compound eyes; three ocelli are present in winged species, but absent in apterous ones. The antennae are usually 13-segmented and filiform. The mouthparts have asymmetrical chewing mandibles, rod-shaped maxillary laciniae and reduced labial palps. The thorax varies according to the presence of wings. The pronotum is small, whereas the meso- and metanotum are larger. The legs are gressorial and slender. The wings are often reduced or absent (as shown here for the book louse *Liposcelis entomophila* (Liposcelididae); after Smithers 1982). When present, the wings are membranous, with reduced venation, with the hind wing coupled to the larger fore wing in flight and at rest, when the wings are held roof-like over the abdomen (as shown here for the bark louse *Amphigerontia contaminata* (Psocidae); after Badonnel 1951). The abdomen has 10 visible segments, with the 11th represented by a dorsal epiproct and paired lateral paraprocts. Cerci are always absent.

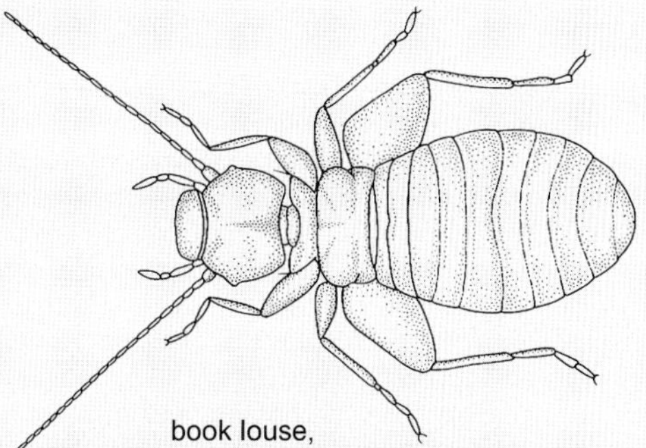

book louse,
Liposcelidae: *Liposcelis entomophilus*

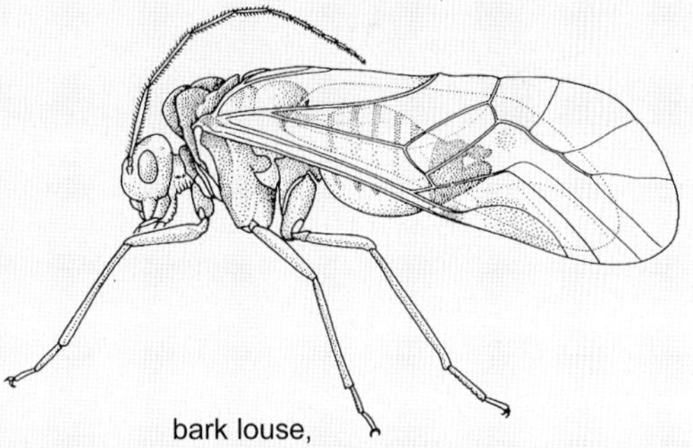

bark louse,
Psocidae: *Amphigerontia contaminata*

Courtship often involves a nuptial dance, followed by spermatozoa transfer via a spermatophore. Eggs are laid in groups or singly onto vegetation or under bark, in sites where nymphs subsequently develop. Parthenogenesis is common, and may be obligatory or facultative. Viviparity is known in at least one genus.

Adults and nymphs feed on fungi (hyphae and spores), lichens, algae or insect eggs, or are scavengers on dead organic matter. Some species are solitary; others may be communal, forming small groups of adults and nymphs beneath webs.

Phylogenetic relationships of Psocodea are considered in section 7.4.2 and depicted in Fig. 7.2 and Fig. 7.5.

Taxobox 18 Psocodea: "Phthiraptera" (chewing lice and sucking lice)

Four of the seven suborders of the order Psocodea are parasitic on homeothermic vertebrates and formerly comprised the order "Phthiraptera" – the parasitic lice. There are about 5000 species of these lice, which are highly modified, apterous, dorsoventrally flattened ectoparasites, as typified by *Werneckiella equi*, the horse louse (Ischnocera: Trichodectidae) illustrated here. The four suborders of parasitic lice are: Rhynchophthirina (a small group found only on elephants, wart hogs and bush pigs), Amblycera and Ischnocera (these three taxa form the chewing or biting lice, formerly called Mallophaga), and Anoplura (sucking lice). Development is hemimetabolous. Mouthparts are mandibulate in Amblycera, Ischnocera and Rhynchophthirina, and beak-like for piercing and sucking in Anoplura (Fig. 2.16), which also lack maxillary palps. The eyes are either absent or reduced; the antennae are either held in grooves (Amblycera) or extended, filiform (and sometimes modified as claspers) in Ischnocera and Anoplura. The thoracic segments are variably fused, and completely fused in Anoplura. The legs are well developed and stout, with one or two-segmented tarsi and strong claws used in grasping host hair or fur. Eggs are laid on the hair or feathers of the host. The nymphs resemble smaller, less pigmented adults, and all stages live on the host.

Lice are obligate ectoparasites, lacking any free-living stage and occurring on all orders of birds and most orders of mammals (with the notable exception of bats). Ischnocera and Amblycera feed on bird feathers and mammal skin, with a few amblycerans feeding on blood. Anoplura feed solely on mammal blood.

The degree of host-specificity amongst lice is high, and many monophyletic groups of lice occur on monophyletic groups of hosts. However, host speciation and parasite speciation do not match precisely, and historically many transfers have taken place between ecologically proximate but unrelated taxa (section 13.3.3). Furthermore, even when louse and host phylogenies match, a lag in timing between host speciation and lice differentiation may be evident, although gene transfer must have been interrupted simultaneously.

As with most parasitic insects, some lice are involved in disease transmission. *Pediculus humanus humanus* (suborder Anoplura), the human body louse, is one vector of typhus (section 15.3.3). It is notable that *P. humanus capitis*, the human head louse and *Pthirus pubis*, the pubic louse (illustrated on the right in the louse diagnosis in the Appendix), are insignificant typhus vectors, although often co-occurring with the body louse.

The parasitic lice evolved from free-living lice, probably twice. Phylogenetic relationships of Psocodea are considered in section 7.4.2 and depicted in Fig. 7.2 and Fig. 7.5.

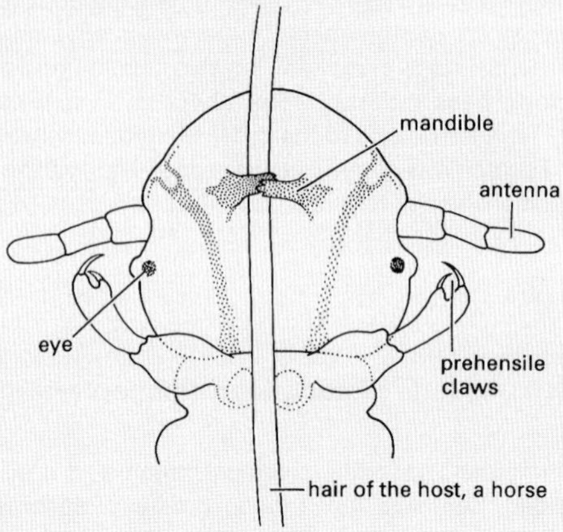

VENTRAL VIEW OF HEAD

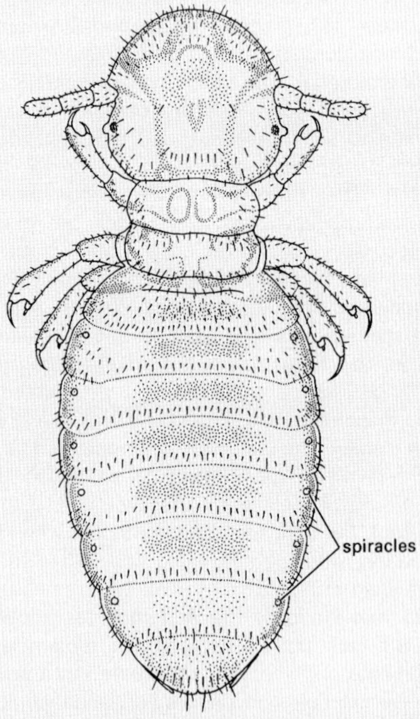

DORSAL VIEW OF LOUSE

Taxobox 19 Thysanoptera (thrips)

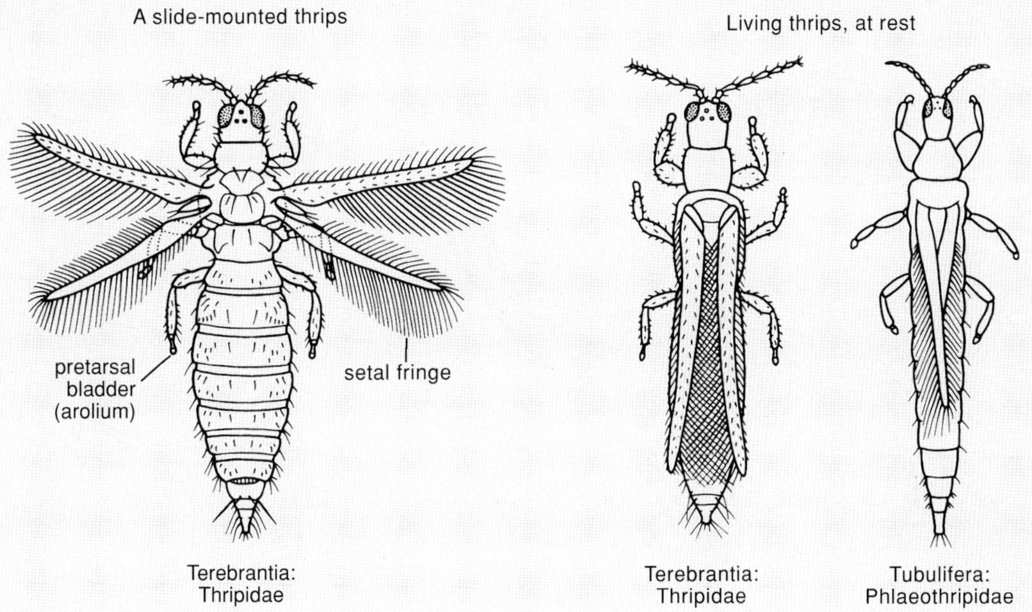

The Thysanoptera is a worldwide order of minute to small insects (from 0.5 mm to a maximum length of 15 mm, but usually 1–3 mm), comprising nearly 6000 species in two suborders: Terebrantia with eight families (including the speciose Thripidae) and Tubulifera with one family (the speciose Phlaeothripidae). Their development is holometabolous, but convergent with that seen in the Holometabola (see section 8.5). The body is slender and elongate, and the head is elongate and usually hypognathous. The mouthparts (Fig. 2.15a) comprise the maxillary laciniae formed as grooved stylets, with the right mandible atrophied and the left mandible acting as a further stylet; the maxillary stylets form a feeding tube. The compound eyes range from small to large, and there are three ocelli in fully winged forms. The antennae are four- to nine-segmented and anteriorly directed. Thoracic development varies according to the presence of wings; fore and hind wings are similar and narrow with a long setal fringe (as illustrated on the left for a terebrantian thrips; after Lewis 1973). At rest the wings usually are parallel in Terebrantia (middle figure) but overlap in Tubulifera (right figure); microptery and aptery occur, and intraspecific wing and body polymorphisms are frequent. The legs are short and adapted for walking, sometimes with the fore legs raptorial and the hind legs saltatory; the tarsi are each one- or two-segmented, and the pretarsus has an apical protrusible adhesive arolium (bladder or vesicle). The abdomen is 11-segmented (though with only 10 segments visible). In males, the genitalia are concealed and symmetrical. In females, the cerci are absent; the ovipositor is serrate in Terebrantia, but chute-like and retracted internally in Tubulifera.

Eggs are laid into plant tissue (Terebrantia) or into crevices or exposed vegetation (Tubulifera). The first- and second-instar nymphs resemble small adults except for lack of wings and genitalia, and are usually called “larvae”; instars 3–4 (Terebrantia) or 3–5 (Tubulifera) are resting or pupal stages, during which significant tissue reconstruction takes place. Female thrips are diploid, whereas males (if present) are haploid, produced from unfertilized eggs. Arrhenotokous parthenogenesis is common; thelytoky is rare (section 5.10.1).

The ancestral feeding mode of thrips probably was fungal feeding, and about half of the species feed only on fungi, mostly hyphae. Most other thrips primarily are phytophages, feeding on flowers or leaves and including some gall inducers, and there are a few predators. Plant-feeding thrips use their single mandibular stylet to pierce a hole through which the maxillary stylets are inserted. The contents of single cells are sucked out one at a time; pollen-feeding thrips similarly remove the contents of individual pollen grains, but spore-feeding thrips imbibe whole spores. Several cosmopolitan thrips species (e.g. western flower thrips, *Frankliniella occidentalis*) act as vectors of viruses that damage plants. Thrips may aggregate in flowers, where they may act as pollinators. Subsocial behaviour, including parental care, is exhibited by a few thrips (section 12.1.2).

Phylogenetic relationships are considered in section 7.4.2 and depicted in Fig. 7.2.

Taxobox 20 Hemiptera (bugs, moss bugs, cicadas, leafhoppers, planthoppers, spittle bugs, treehoppers, aphids, jumping plant lice, scale insects and whiteflies)

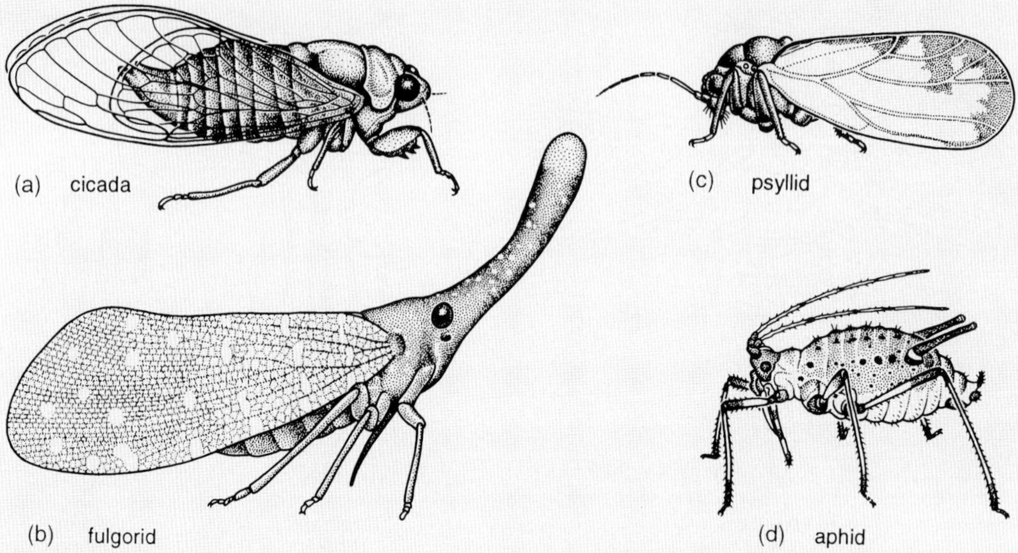

The Hemiptera is distributed worldwide, and is the most diverse of the non-endopterygote orders, with just over 100,000 described species in about 145 families, although many more species await discovery and description, especially among tropical leafhoppers. Historically, it was divided into two suborders (sometimes treated as orders): Heteroptera (bugs) and "Homoptera" (cicadas, leafhoppers, planthoppers, spittle bugs, treehoppers, aphids, jumping plant lice (= psylloids), scale insects (= coccoids) and whiteflies). However, homopterans represent a grade of organization (a paraphyletic rather than a monophyletic group). Currently, five (or sometimes fewer) suborders can be recognized (Fig. 7.6): (i) Heteroptera, the "true" bugs, with more than 40,000 described species; (ii) Coleorrhyncha, the moss bugs (family Peloridiidae), with fewer than 30 species; (iii) Cicadomorpha (= Clypeorrhyncha; cicadas, leafhoppers, treehoppers and spittle bugs) with about 30,000 described species; (iv) Fulgoromorpha (= Archaeorrhyncha; planthoppers) with about 12,000 described species; and (v) Sternorrhyncha (aphids, jumping plant lice, scale insects and whiteflies) with more than 16,000 species. The Cicadomorpha and Fulgoromorpha are collectively called the Auchenorrhyncha (and often treated as a suborder), a group supported by morphological and molecular data, although some authors have questioned its monophyly. Four hemipterans are illustrated here in lateral view: (a) the mountain or New Forest cicada, *Cicadetta montana* (Cicadidae) (after drawing by Jon Martin in Dolling 1991); (b) a green lantern bug, *Pyrops sultan* (Fulgoridae), from Borneo (after Edwards 1994); (c) the psyllid *Psyllopsis fraxini* (Psyllidae), which deforms leaflets of western European ash trees (after drawing by Jon Martin in Dolling 1991); and (d) an apterous viviparous female of the aphid *Macromyzus woodwardiae* (Aphididae) (after Miyazaki 1987a).

Hemipteran compound eyes are often large, and ocelli may be present or absent. Antennae vary from short with few segments to filiform and multisegmented. The mouthparts comprise mandibles and maxillae modified as needle-like stylets, lying in a beak-like grooved labium (as shown for a pentatomid heteropteran in (e) and (f)), collectively forming a rostrum or proboscis. The stylet bundle contains two canals, one delivering saliva, the other uptaking fluid (as shown in (f)); there are no palps. The thorax often consists of large pro- and mesothorax, but a small metathorax. Both pairs of wings often have reduced venation; some hemipterans are apterous, and rarely there may be just one pair of wings (in male scale insects, which have hamulohalteres on the metathorax). The legs are frequently gressorial, sometimes raptorial, often with complex pretarsal adhesive structures. The abdomen is variable, and cerci are absent.

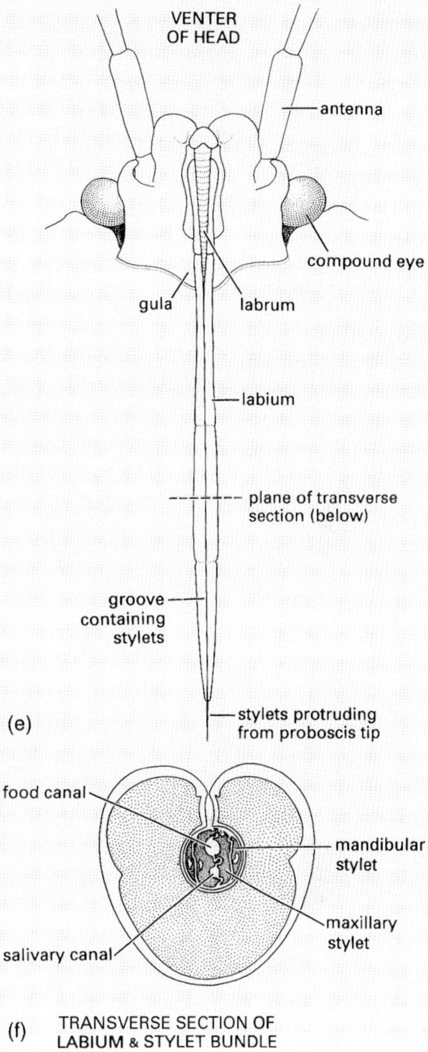

Most Heteroptera hold their head horizontally, with the rostrum anteriorly distinct from the prosternum (although the rostrum may be in body contact at coxal bases and on anterior abdomen). When at rest, the wings are usually folded flat over the abdomen (Fig. 5.8). The fore wings usually are thickened basally and membranous apically to form hemelytra (Fig. 2.24e). Heteroptera mostly have abdominal scent glands. Apterous heteropterans can be identified by the rostrum arising from the anteroventral region of the head and the presence of a large gula. Non-heteropterans hold the head deflexed, with the complete length of the rostrum appressed to the prosternum, directed posteriorly, often between the coxal bases. They have membranous wings that usually rest roof-like over the abdomen; apterous species are identified by the absence of a gula and the rostrum arising from the posteroventral head (Auchenorrhyncha) or near the prosternum (Sternorrhyncha). Mouthparts are absent in some aphids, and in some female and all male scale insects.

Nymphal Heteroptera (Fig. 6.2) and auchenorrhynchans resemble adults except in the lack of development of the wings and genitalia. However, immature Sternorrhyncha show much variation in a range of complex life cycles. Many aphids exhibit parthenogenesis (section 5.10.1), usually alternating with seasonal sexual reproduction. The immature stages of Aleyrodoidea (whiteflies) and Coccoidea (scale insects) may differ greatly from adults, with larviform stages followed by a quiescent, non-feeding "pupal" stage, in convergently acquired holometaboly.

The ancestral feeding mode is piercing and sucking plant tissue (Fig. 11.4) and many species induce galls on their host plants (section 11.2.5; Box 11.2; **Plate 4f** and **Plate 5.a–c**). The nymphs of certain jumping plant lice (Psylloidea) live under lerps (**Plate 8e**) that they construct from their anal excreta. All hemipterans have large salivary glands and an alimentary canal modified for absorption of liquids, with a filter chamber to remove water (Box 3.3). Many hemipterans rely exclusively on living plant sap (from phloem or xylem, and sometimes parenchyma). Elimination of large quantities of honeydew by phloem-feeding Sternorrhyncha provides the basis for mutualistic relationships with ants. Many hemipterans exude waxes (Fig. 2.5), which form powdery, enclosing or plate-like protective covers. The non-phytophagous Heteroptera comprise many predators, some scavengers, a few haematophages (blood-feeders) and some necrophages (consumers of dead prey), with the last trophic group including successful colonizers of aquatic environments (Box 5.8 and Box 10.2) and some of the few insects to live in the oceans (section 10.8).

Phylogenetic relationships are considered in section 7.4.2 and depicted in Fig. 7.2 and Fig. 7.6.

Taxobox 21 Neuropterida: Neuroptera (lacewings, owlflies and antlions), Megaloptera (alderflies, dobsonflies and fishflies) and Raphidioptera (snakeflies)

Members of these three small neuropteroid orders have holometabolous development, and are mostly predators. Approximate numbers of described extant species are: over 6500 for Neuroptera (lacewings, owlflies and antlions) in almost 20 families; about 350 in Megaloptera in two families worldwide – Sialidae (alderflies, with adults 10–15 mm long) and the larger Corydalidae (dobsonflies and fishflies, with adults up to 75 mm long); and about 200 in Raphidioptera (snakeflies) in two families – Inocelliidae and Raphidiidae.

Adults have multisegmented antennae, large separated eyes and mandibulate mouthparts. The prothorax may be larger than the meso- and metathorax, which are about equal in size. The legs may be modified for predation. Fore and hind wings are similar in shape and venation, with folded wings often extending beyond the abdomen. The abdomen lacks cerci.

Neuroptera

Adult Neuroptera (illustrated in Fig. 6.13 and the Appendix, and exemplified here by an owlfly, *Ascalaphus* sp. (Ascalaphidae); after a photograph by C.A.M. Reid) possess wings typically with numerous cross-veins and "twigging" at ends of veins; many are predators, but nectar, honeydew and pollen are consumed by some species. Neuropteran larvae (Fig. 6.6d) are usually specialized, active predators, with prognathous heads and slender, elongate mandibles and maxillae combined to form piercing and sucking mouthparts (Fig. 13.2c); all have a blind-ending hindgut. Larval dietary specializations include spider egg-masses (for Mantispidae), freshwater sponges (for Sisyridae, the spongillaflies, with larva illustrated in Box 10.4) or soft-bodied hemipterans such as aphids and scale

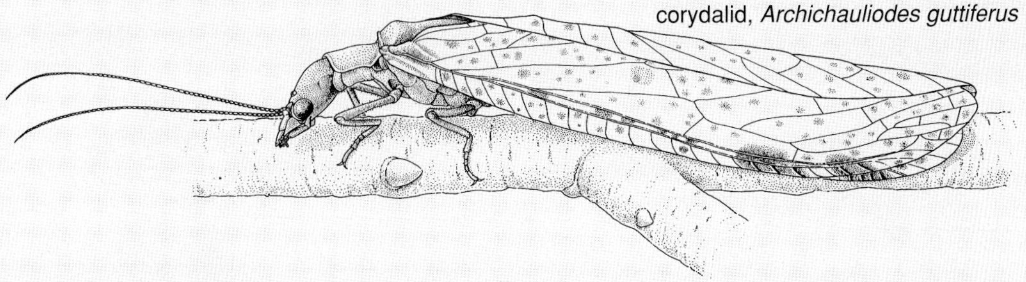

corydalid, *Archichauliodes guttiferus*

insects (for Chrysopidae, Hemerobiidae and Coniopterygidae). Pupation is terrestrial, within shelters spun with silk from Malpighian tubules. The pupal mandibles are used to open a toughened cocoon.

Megaloptera

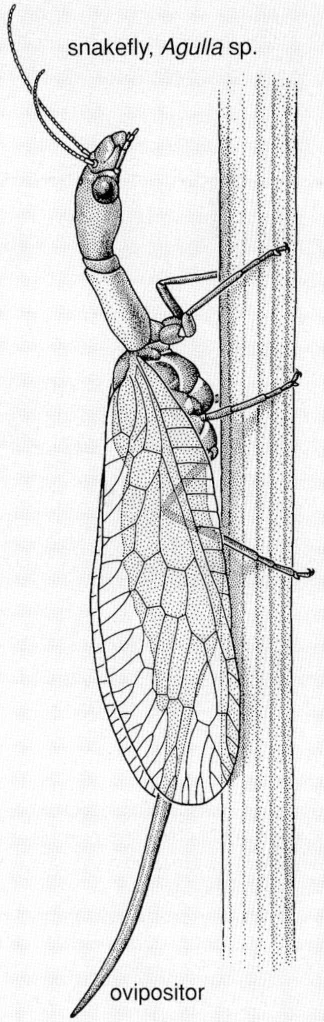

snakefly, *Agulla* sp.

Megaloptera are predatory and scavenging only in the aquatic larval stage (see Box 10.4 and Appendix for illustrations) – although the adults have strong mandibles, these are not used in feeding. Adults (such as the corydalid, *Archichauliodes guttiferus*, illustrated here) closely resemble neuropterans, except for the presence of a large pleated anal field on the hind wing that infolds when the wings are at rest over the back. The abdomen is soft. The larvae (sometimes called hellgrammites) have 10–12 instars and take at least one year, usually two or more, to develop, and are intolerant of pollution. Larvae are prognathous, with well-developed mouthparts, including three-segmented labial palps; they are spiraculate, with gills consisting of four- to five-segmented (Sialidae) or two-segmented lateral filaments on the abdominal segments; the larval abdomen terminates in an unsegmented median caudal filament (Sialidae) or a pair of anal prolegs (as shown in Box 10.4 for Corydalidae). The pupa is beetle-like (Fig. 6.7a), except that it has mobility due to its free legs, and has a head similar to that of the larva, including functional mandibles. Pupation is away from water, often in chambers in damp soil under stones, or in damp timber.

Raphidioptera

Raphidioptera are terrestrial predators both as adults and larvae, and occur only in the northern hemisphere and almost exclusively in the Holarctic region. The adult is mantid-like, with an elongate prothorax – as shown here for the female snakefly of an *Agulla* sp. (Raphidiidae) (after a photograph by D.C.F. Rentz) – and mobile head used to strike, snake-like, at prey. The adult female has an elongate ovipositor. The larva (illustrated in the Appendix) has a large prognathous head and a sclerotized prothorax that is slightly longer than the membranous meso- and metathorax. The number of larval instars is variable, typically 10–11 but sometimes more, and larval development takes from one to several years. A period of low temperature appears to be necessary to induce either pupation or adult eclosion. The pupa is mobile.

The Megaloptera, Raphidioptera and Neuroptera are treated here as separate orders, however, some authorities include the Raphidioptera in the Megaloptera, and all three form the Neuropterida. Phylogenetic relationships are considered in section 7.4.2 and depicted in Fig. 7.2.

Taxobox 22 Coleoptera (beetles)

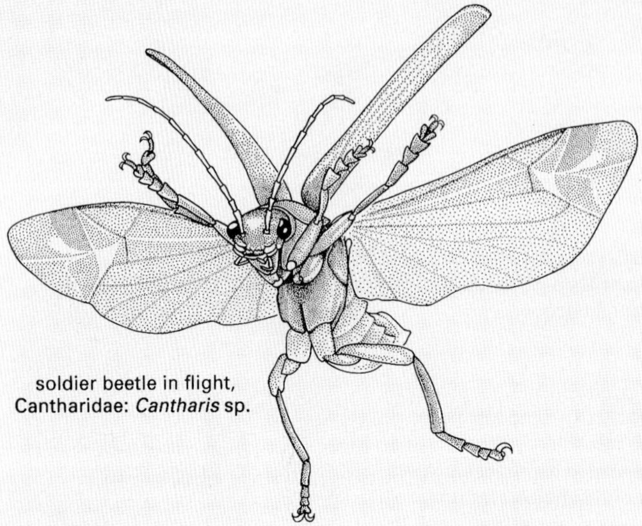

soldier beetle in flight, Cantharidae: *Cantharis* sp.

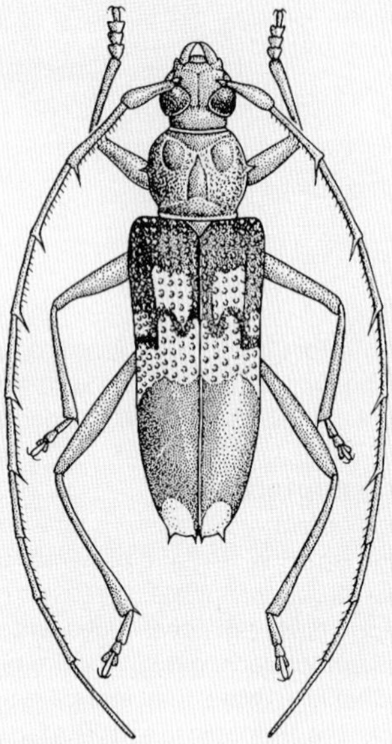

longhorn beetle, Cerambycidae: *Phoracantha semipunctata*

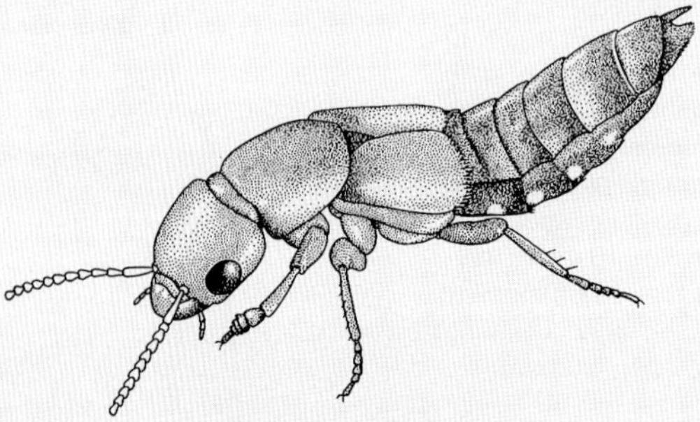

rove beetle, Staphylinidae: *Staphylinus caesareus*

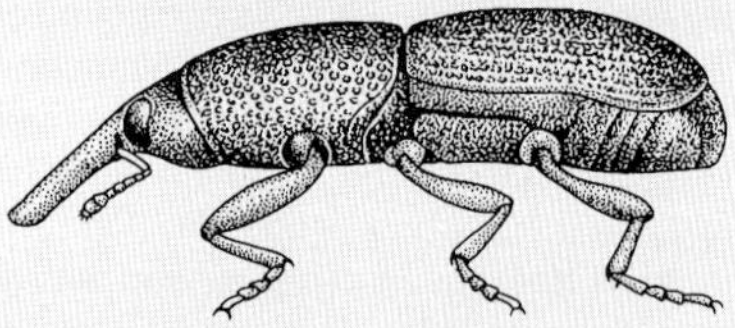

granary weevil, Curculionidae: *Sitophilus granarius*

The Coleoptera is probably the largest order of insects, with some 390,000 described species in four suborders (Archostemata, Myxophaga, Adephaga and the speciose Polyphaga). Although the family-level classification is unstable, some 500 families and subfamilies are recognized. Adult beetles range from small to very large, but are usually heavily sclerotized, sometimes even armoured, and often compact. Development is holometabolous. The mouthparts are mandibulate, and compound eyes range from well developed (sometimes even meeting medially) to absent; ocelli are usually absent. The antennae comprise 11 or frequently fewer segments (exceptionally with 20 segments in male Rhipiceridae). The prothorax is distinct, large and extends laterally beyond the coxae; the mesothorax is small (at least dorsally), and fused to the metathorax to form the wing-bearing pterothorax. The fore wings are modified as sclerotized, rigid elytra (Fig. 2.24d), whose movement may assist in lift or may be restricted to opening and closing before and after flight; the elytra cover the hind wings and the abdominal spiracles, thus facilitating control of water loss. The hind wings are longer than the elytra when extended for flight (as illustrated on the upper left for a soldier beetle, *Cantharis* sp. (Cantharidae); after Brackenbury 1990), and have variably reduced venation, much of which is associated with complex pleating to allow the wings to be folded longitudinally and transversely beneath the elytra even if the latter are reduced in size, as in rove beetles (Staphylinidae) such as *Staphylinus caesareus* (illustrated on the lower left; after Stanek 1969). The legs are very variably developed, with coxae that are sometimes large and mobile; the tarsi are primitively five-segmented, although often with a reduced number of segments, and bear variously shaped claws and adhesive structures (Fig. 10.3). Sometimes the legs are fossorial (Fig. 9.2c) for digging in soil or wood, or modified for swimming (Fig. 10.3 and Fig. 10.7) or jumping. The abdomen is primitively nine-segmented in females, and 10-segmented in males, with at least one terminal segment retracted; the sterna are usually strongly sclerotized, often more so than the terga. Females have a substitutional ovipositor, whereas the male external genitalia are primitively trilobed (Fig. 2.26b). Cerci are absent.

Larvae exhibit a wide range of morphologies, but most can be recognized by the sclerotized head capsule with opposable mandibles and their usually five-segmented thoracic legs. They can be distinguished from similar-looking lepidopteran larvae by the lack of ventral abdominal crochet-bearing prolegs and lack of a median labial silk gland. Similar-looking symphytan wasp larvae can be distinguished as they have prolegs on abdominal segments 2–7. Beetle larvae vary in body shape and leg structure; some are apodous (lacking any thoracic legs; Fig. 6.6g; **Plate 2a**), whereas legged larvae may be campodeiform (prognathous with long thoracic legs; Fig. 6.6e), eruciform (grub-like with short legs) or scarabaeiform (grub-like but long-legged; Fig. 6.6f). The tracheal system of beetle larvae is open and typically has nine pairs of spiracles, but with variable reduction in larvae of most aquatic species, which often have gills (Box 10.3). Pupation is often in a specially constructed cell or chamber (Fig. 9.1 and Fig. 9.6), rarely in a cocoon spun from silk from Malpighian tubules or exposed, as in coccinellids (Fig. 6.7j).

Beetles occupy virtually every conceivable habitat, including freshwater, a few marine and intertidal habitats and, above all, every plant microhabitat from external foliage (Fig. 11.1), flowers, buds, stems, bark and roots, to internal sites such as in galls in any living plant tissue, or in any kind of dead material in all its various states of decomposition. Saprophagy and fungivory are fairly common, and dung and carrion are exploited (section 9.3 and section 9.4, respectively). Few beetles are parasitic, but carnivory is frequent, occurring in nearly all Adephaga and many Polyphaga, including Lampyridae (fireflies) and many Coccinellidae (ladybird beetles; vignette at beginning of Chapter 16, Fig. 5.9). Some herbivorous Chrysomelidae (leaf beetles) and Curculionidae (weevils; **Plate 4b** and **Plate 8f**) are widely introduced as biological control agents of weedy plants, but other species of these two families are pests of plants (such as the Colorado potato beetle, Box 16.7). Coccinellidae have been used as biological control agents for aphid and coccoid pests of plants (Box 16.3). Some beetles are significant pests of roots (section 9.1.1) in pastures and crops (especially larval Scarabaeidae), of timber (including Buprestidae (Box 17.4; **Plate 8g**) and especially Cerambycidae such as *Phoracantha semipunctata*, illustrated on the upper right; after Duffy 1963) and of stored products (such as the granary weevil, *Sitophilus granarius* (Curculionidae), illustrated on the lower right). These last beetles tend to be adapted to dry conditions and thrive on stored grains, cereals, pulses and dried animal material such as skins and leather. Aquatic Coleoptera (Box 10.3; Fig. 10.3 and Fig. 10.7) exhibit diverse feeding habits, but both larvae and adults of most species are predatory.

Phylogenetic relationships are considered in section 7.4.2 and depicted in Fig. 7.2.

Taxobox 23 Strepsiptera (strepsipterans)

The Strepsiptera is an order of just over 600 species of highly modified endoparasitoids with extreme sexual dimorphism. The male (top right figure; after CSIRO 1970) has a large head and bulging compound eyes often with only a few facets, and no ocelli. Antennae of the male are flabellate or branched, with four to seven segments. The pro- and mesothorax are small; the fore wings stubby and without veins, and the hind wings broad, fan-shaped and with few radiating veins. The unusual twisted hind wings of the flying adult male gave rise to one common name, "twisted-wing parasites". The legs lack a trochanter and often also claws. An elongate metanotum overlies the anterior part of the tapering abdomen. The female is either coccoid-like or larviform, wingless and usually retained in a pharate (cloaked) state as an endoparasitoid protruding from the host (as illustrated in ventral view and longitudinal section in the left figures; after Askew 1971). The triungulin (first-instar larva; bottom right figure) has three pairs

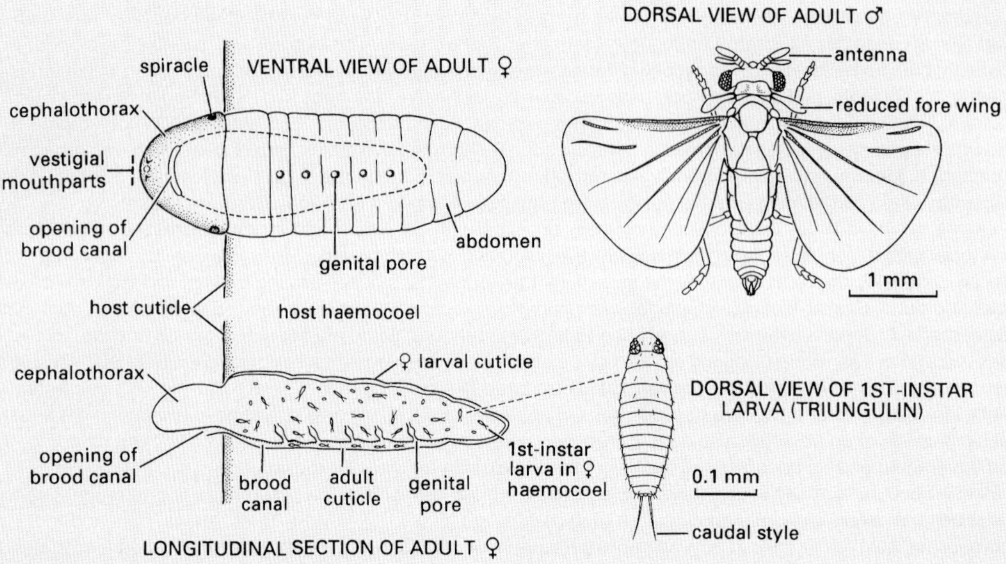

of thoracic legs, but lacks antennae and mandibles; subsequent instars are maggot-like, lacking mouthparts or appendages. The male pupa is exarate (appendages free from its body) and adecticous (has immovable mandibles) within a puparium formed from the cuticle of the final larval instar. Most females (except in Mengenillidae) become adult without an obvious pupal instar.

Strepsipterans are parasitic on other insects belonging to seven orders, most commonly of Hemiptera and Hymenoptera. Strepsipterans of the family Mengenillidae are unusual in that they parasitize Zygentoma (Lepismatidae), both males and females depart the host to pupate and adult females are free-living (although still wingless). Host insects infected by strepsipterans are said to be "stylopized" and suffer morphological, physiological and developmental abnormalities and, although not killed prematurely, they do not reproduce, and die after strepsipteran departure. Strepsipteran eggs hatch within the mother (via haemocoelous ovoviviparity) and active triungula (the mobile first-instar larvae) emerge via a brood canal (as shown here on the bottom left) and seek out a host, usually in its immature stage. In Stylopidae that parasitize hymenopterans, triungula leave the maternal host whilst on flowers, and from here seek a suitable adult bee or wasp to gain a ride to the nest, where they enter a host egg or larva.

Larvae enter the host by enzymatically softening the cuticle, followed by an immediate moult to a maggot-like instar larva that develops as an endoparasitoid (see section 13.3.2 for discussion of why Strepsiptera are considered to be endoparasitoids). For all families except Mengenillidae and Bahiaxenidae, the pupa protrudes from the host's body; the male emerges by pushing off a cephalothoracic cap, but in most strepsipteran groups the female remains within the cuticle. The virgin female releases pheromones to lure free-flying males, one of which copulates, inseminating through the brood canal on the female cephalothorax.

Nine extant families of Strepsiptera and another three families known only from fossils are recognized. Phylogenetic relationships of this order are controversial and are considered in section 7.4.2 and depicted in Fig. 7.2.

Taxobox 24 Diptera (true flies)

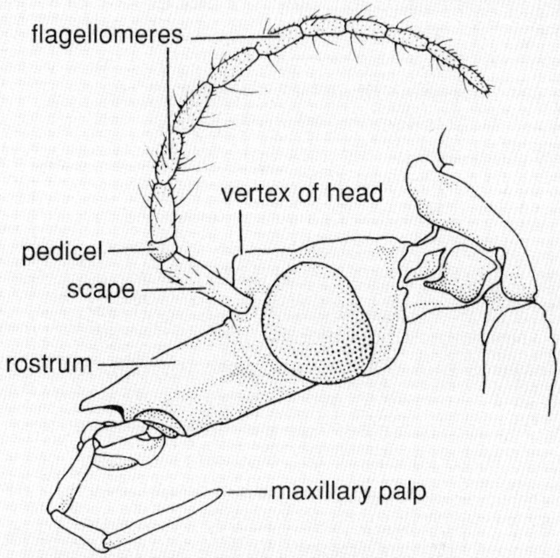

(a) crane fly, *Tipula*

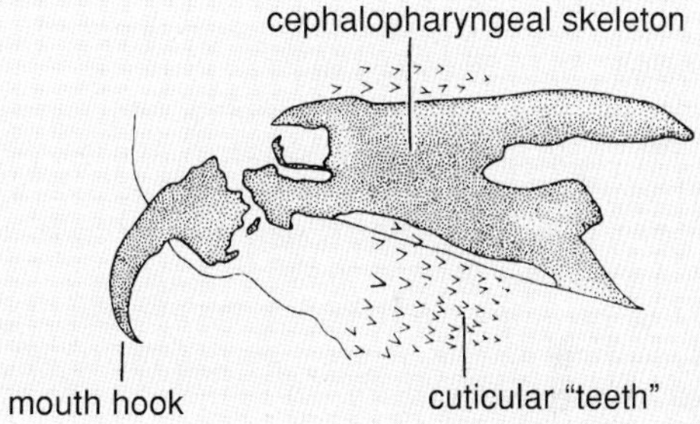

(b) Old World screw-worm fly maggot, *Chrysomya bezziana*

The Diptera is an order containing perhaps almost 160,000 described species, in about 150 families, with several thousands of species of medical or veterinary importance. Development is holometabolous. Adult flies typically have a mobile head, large compound eyes and ventrally directed mouthparts that often are formed as a proboscis – a tubular sucking organ comprising elongate mouthparts with the labium enclosing or supporting the other mouthparts (usually including the labrum); modifications of the mouthparts for a biting function, usually correlated with blood feeding, are found in many groups. Adult mouthparts are described in section 2.3.1 and illustrated in Fig. 2.13 and Fig. 2.14. Adult flies are characterized by well-developed mesothoracic wings and metathoracic wings as halteres (balancers) (Fig. 2.24f). In conjunction with flight powered from mesothoracic wings alone, the prothorax and metathorax are reduced in size – only in the few flightless (apterous) flies is the mesothorax reduced. The legs can be highly modified, but all have five tarsomeres. The abdomen has 11 segments, although much reduction occurs, especially in many brachyceran families in which the abdomen is stout in contrast to an elongate and slender form. Female flies lack a true ovipositor comprising valves, but a substitutional ovipositor, comprising telescoping terminal segments, is recognized in some cyclorrhaphans.

The larvae lack true legs (Fig. 6.6h) and their head structure ranges from a complete sclerotized capsule to acephaly, with no external capsule and only an internal skeleton. The pupae are adecticous and obtect, or exarate in a puparium (Fig. 6.7e,f).

The paraphyletic "Nematocera" comprises crane flies, mosquitoes, midges, gnats and relatives; these have slender antennae with upwards of six flagellomeres and a three- to five-segmented maxillary palp (illustrated for a crane fly (Tipulidae: *Tipula*) in (a); after McAlpine 1981). Brachycera contains heavier-built flies, including hover flies, bee flies, dung flies and blow flies, that have more solid, often shorter antennae with fewer (less than seven) flagellomeres, often with terminal arista (Fig. 2.19i), and the maxillary palps have only one or two segments. Within the Brachycera, schizophoran Cyclorrhapha use a ptilinum to aid emergence from the puparium.

Fly larvae have a wide variety of habits. Many "nematoceran" larvae are aquatic (Fig. 2.18; Box 10.1), and brachyceran larvae show a phylogenetic radiation into drier and more specialized larval habits, including phytophagy, predation and parasitization of other arthropods, and in living flesh of vertebrates (myiasis, section 15.1). Myiasis-inducing maggots have a much-reduced head, but with sclerotized mouthparts known as mouth hooks (illustrated for a third-instar larva of the Old World screw-worm fly, *Chrysomya bezziana* (Calliphoridae), in (b); after Ferrar 1987), which scrape the living flesh of the host. Adult flies mostly suck liquids, including nectar from flowers (**Plate 5f**), sometimes honeydew from sap-sucking insects, or blood and other animal body fluids, and some do not feed.

Phylogenetic relationships are considered in section 7.4.2 and depicted in Fig. 7.7.

Taxobox 25 Mecoptera (hangingflies, scorpionflies and snowfleas)

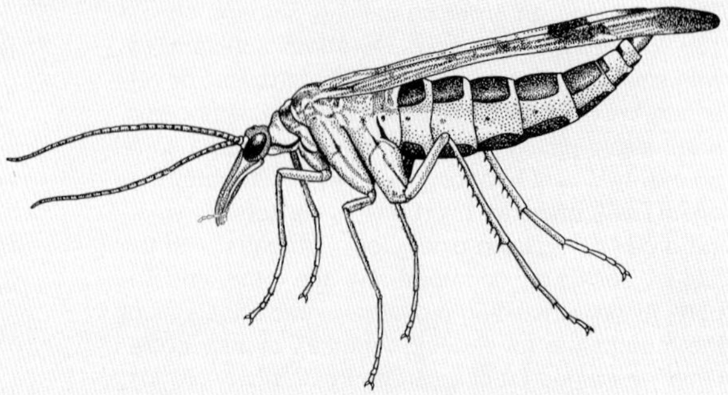

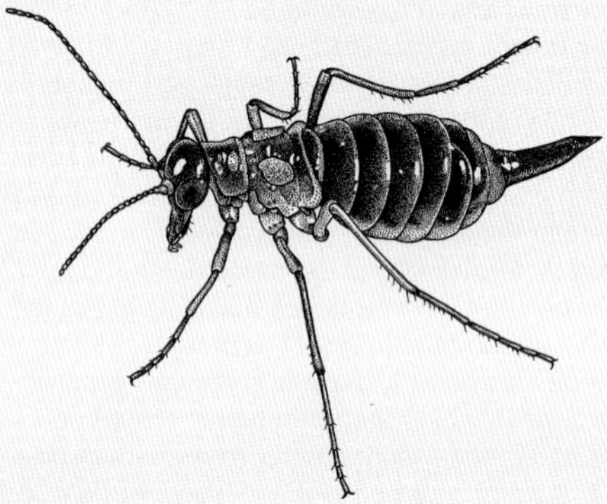

The Mecoptera is an order of about 400 described species in nine families, with common names associated with the two largest families – Bittacidae (hangingflies, see Box 5.1) and Panorpidae (scorpionflies, as illustrated here in the top figure and in the Appendix) – and with the Boreidae (snowfleas, also called snow scorpionflies, exemplified in the bottom figure by the female of *Boreus brumalis*; after a photograph by Tom Murray). Development is holometabolous. Adults have an elongate hypognathous rostrum; their mandibles and maxillae are elongate, slender and serrate; the labium is elongate. They have large, separated eyes and filiform, multisegmented antennae. The prothorax may be smaller than the equally developed meso- and metathorax, each with a scutum, scutellum and postscutellum visible. The fore and hind wings are narrow and of similar size, shape and venation; they are often reduced or absent, as in the vestigial wings of Boreidae, which are modified in males to grasp the female during copulation. The legs may be modified for predation. The abdomen is 11-segmented, with the first tergite fused to the metathorax. The cerci have one or two segments. Larvae possess a heavily sclerotized head capsule, are mandibulate, and have eyes composed of groups of stemmata (up to 30 in Panorpidae; indistinct in Nannochoristidae). Their thoracic segments are about equal in size, and the short thoracic legs have a fused tibia and tarsus and a single claw. Prolegs usually occur on abdominal segments 1–8, and the (10th) terminal segment bears either paired hooks or a suction disc. The pupa (Fig. 6.7b) is immobile, exarate and mandibulate.

The dietary habits of mecopterans vary among families and often between adults and larvae within a family. The Bittacidae are predatory as adults but saprophagous as larvae; Panorpidae are scavengers, probably feeding mostly on dead arthropods, as both larvae and adults. Less is known of the diets of the other families, but saprophagy and phytophagy, including moss feeding (for boreids), have been reported.

Copulation in certain mecopterans is preceded by elaborate courtship procedures that may involve nuptial feeding (Box 5.1). Oviposition sites vary, but known larval development is predominantly in moist litter, or aquatic in Gondwanan Nannochoristidae.

Recent phylogenetic studies provide competing hypotheses on relationships of mecopterans. Either Mecoptera is monophyletic and sister to Siphonaptera (fleas), or Mecoptera is paraphyletic unless the fleas are included. In this book, the ordinal status of each of Mecoptera and Siphonaptera is maintained. These relationships are discussed in section 7.4.2 and depicted in Fig. 7.2 and Fig. 7.7.

Taxobox 26 Siphonaptera (fleas)

The Siphonaptera is a monophyletic group of some 2000 species, all adults of which are highly modified, apterous and laterally compressed ectoparasites of birds and mammals. Development is holometabolous. The mouthparts (Fig. 2.17) are modified for piercing and sucking, without mandibles but with a stylet derived from the epipharynx and two elongate, serrate lacinial blades within a sheath formed from the labial palps. The gut has a salivary pump to inject saliva into the wound and cibarial and pharyngeal pumps to suck up blood. Compound eyes are absent, and ocelli range from absent to well developed. Each antenna lies in a deep lateral groove. The body has many backwardly directed setae and spines; some may be grouped into combs (ctenidia) on the gena (part of the head) and thorax (especially the prothorax). The large metathorax houses the hind-leg muscles, which power the prodigious leaps made by these insects. The legs are long and strong, terminating in strong claws for grasping host hairs.

The large eggs are laid predominantly into the host's nest, where free-living worm-like larvae (illustrated in the Appendix) develop on material such as skin debris shed from the host. High temperatures and humidity are required for development by many fleas, including those on domestic cats (*Ctenocephalides felis*) (illustrated here), dogs (*C. canis*) and humans (*Pulex irritans*). The pupa is exarate and adecticous in a loose cocoon. Adults of both sexes take blood from a host, some species being monoxenous (restricted to one host), but many others being polyxenous (occurring on several to many hosts). The plague flea *Xenopsylla cheopis* belongs to the latter group, with polyxeny facilitating transfer of plague from rat to human host (section 15.3.3). Fleas transmit some other diseases of minor significance from other mammals to humans, including murine typhus and tularemia, but apart from plague, the most common human health threat is from allergic reaction to frequent bites from the fleas of our pets, *C. felis* and *C. canis*.

Fleas predominantly use mammalian hosts, with relatively few birds having fleas, and these being derived from several different lineages of mammal fleas. Some hosts (e.g. *Rattus fuscipes*) have been reported to harbour more than 20 different species of flea, and conversely, some fleas have over 30 recorded hosts, so host-specificity is clearly much less than for lice.

Traditionally, fleas have been treated as the order Siphonaptera, but one molecular phylogenetic study, including many taxa of fleas and mecopterans, found Siphonaptera sister to the Boreidae (snowfleas) and thus part of the order Mecoptera. However, more recent studies using different molecular data found no evidence for fleas to be part of Mecoptera. Phylogenetic relationships are discussed further in section 7.4.2 and depicted in Fig. 7.7.

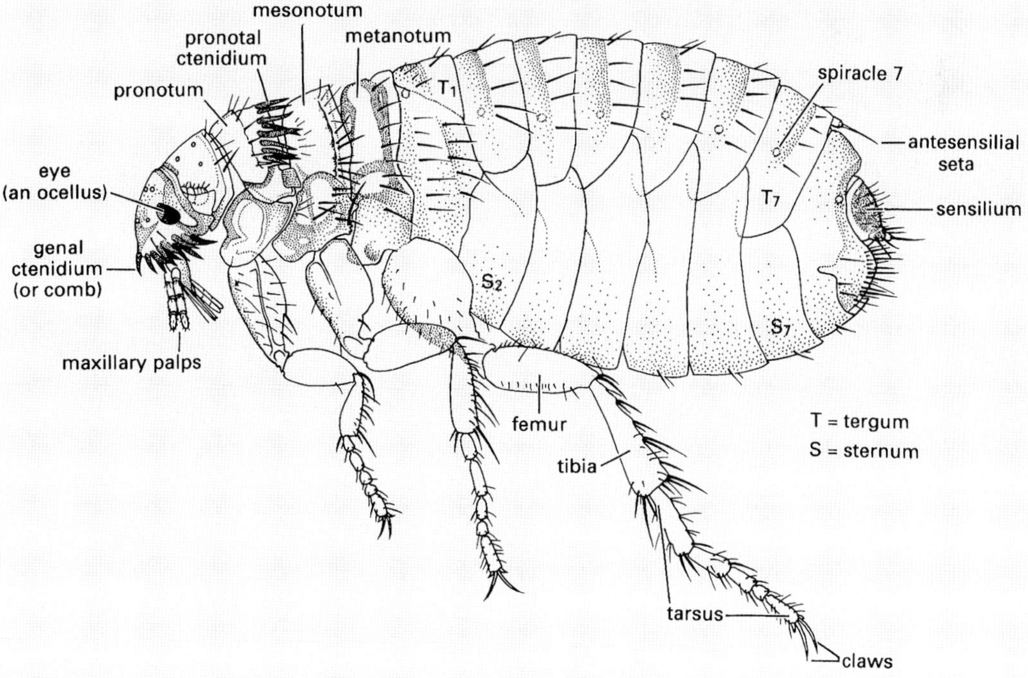

Taxobox 27 Trichoptera (caddisflies)

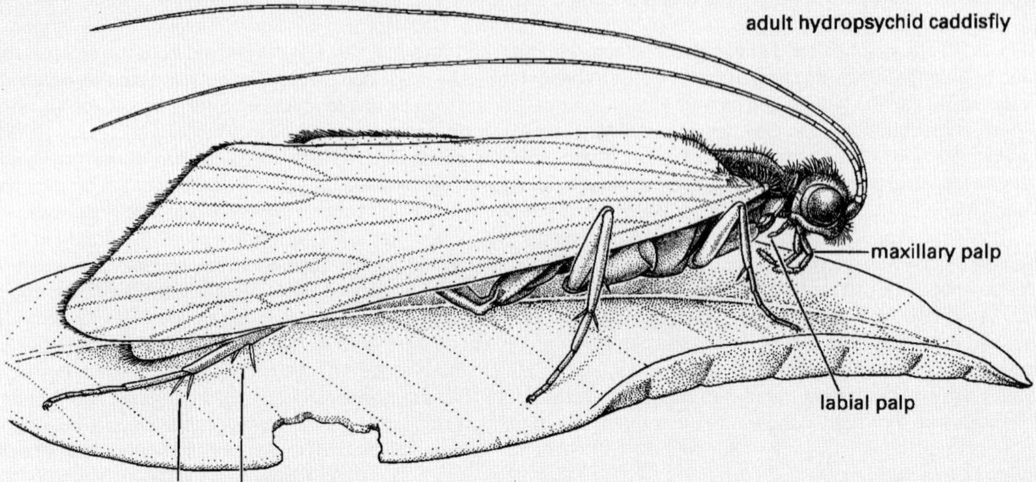

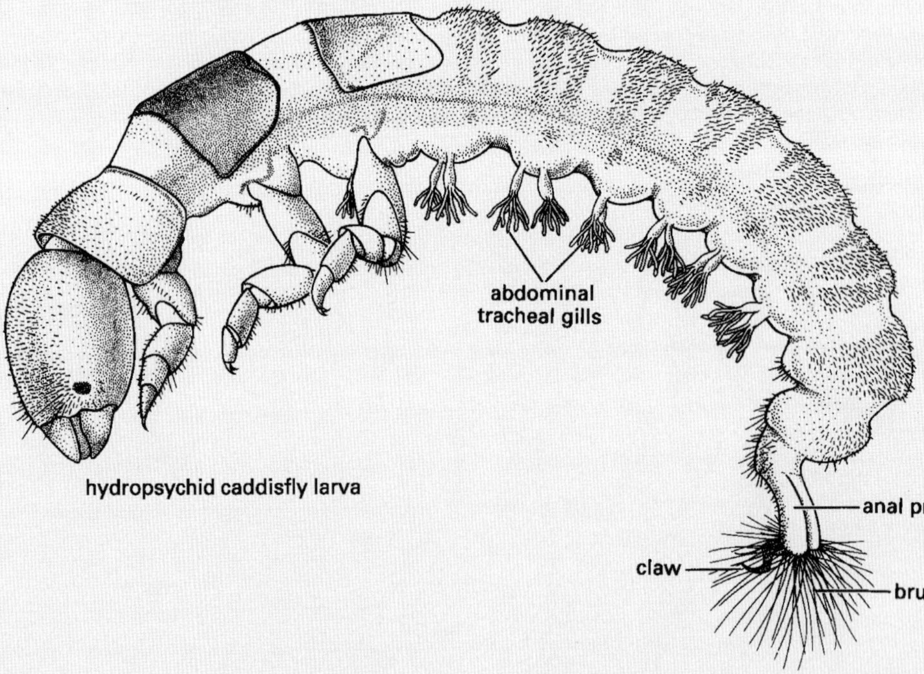

Caddisflies comprise an order of about 14,500 described species in some 45 families and occur worldwide. They are holometabolous, with a moth-like adult (as illustrated here for a hydropsychid), and usually are covered in hairs; setal warts (setose protuberances) often occur on the dorsum of the head and thorax. The head has reduced mouthparts, but with three- to five-segmented maxillary palps and three-segmented labial palps (cf. the proboscis of most Lepidoptera). The antennae are multisegmented and filiform, often as long as or longer than the wings. There are large compound eyes and two or three ocelli. The prothorax is smaller than the meso- or metathorax, and the wings are haired or sometimes partially scaled; they can be distinguished from lepidopteran wings by their different venation, including anal veins looped in the fore wings and no discal cell. The abdomen typically is 10-segmented, with the male terminalia more complex (often with claspers) than in the female.

The larva has five to seven aquatic instars, with fully developed mouthparts and three pairs of thoracic legs, each with at least five segments. Most abdominal segments lack the ventral prolegs seen in most lepidopteran larvae; only the terminal segment bears a pair of hook-bearing prolegs. The tracheal system is closed, with tracheal gills often on most or all nine abdominal segments (as illustrated here for a hydropsychid, *Cheumatopsyche* sp.), and sometimes associated with the thorax or anus. Gas exchange is cuticular, enhanced by ventilatory undulation of the larva in species that occupy a tubular case or retreat. The pupa is aquatic, enclosed in a silken retreat or case, with large functional mandibles to chew free from the pupal case or cocoon; it also has free legs with setose mid-tarsi to swim to the water surface; its gills coincide with the larval gills. Eclosion involves the pharate adult swimming to the water surface, where the pupal cuticle splits; the exuviae are used as a floating platform.

Caddisflies are predominantly univoltine, with development exceeding one year at high latitudes and elevations. The larvae either make saddle-, purse- or tube-cases (Fig. 10.5) or are free-living, including net-spinning (Fig. 10.6); they exhibit diverse feeding habits and include predators, filterers and/or shredders of organic matter, and some grazers on macrophytes. Net-spinners are restricted to flowing waters, with case-makers frequent in standing waters. Adults may ingest nectar or water, but often do not feed.

Phylogenetic relationships are discussed in section 7.4.2 and depicted in Fig. 7.2.

Taxobox 28 Lepidoptera (butterflies and moths)

The Lepidoptera is one of the major insect orders, both in terms of size, with almost 160,000 described species in more than 120 families, and in terms of popularity, with many amateur and professional entomologists studying the order, particularly the butterflies. Three of the four suborders contain few species and lack the characteristic proboscis of the largest suborder, Glossata, which contains the speciose series Ditrysia, defined by unique abdominal features, especially in the genitalia. Adult lepidopterans range in size from very small (some microlepidopterans) to large (see **Plate 1a,b,e,f**, **Plate 3a**, **Plate 5g**, **Plate 7g** and **Plate 8c**), with wingspans up to 30 cm. Development is holometabolous (see the vignette at the beginning of Chapter 6). The head is hypognathous, bearing a long coiled proboscis (Fig. 2.12) formed from greatly elongated maxillary galeae; large labial palps are usually present, whereas other mouthparts are absent, although mandibles are primitively present. The compound eyes are large, and ocelli and/or chaetosemata (paired sensory organs lying dorsolateral on the head) are frequent. The antennae are multisegmented, often pectinate in moths (Fig. 4.6) and knobbed or clubbed in butterflies. The prothorax is small, with paired dorsolaterally placed plates (patagia), whereas the mesothorax is large and bears a scutum and scutellum, and a lateral tegula protects the base of each fore wing. The metathorax is small. The wings are completely covered with a double layer of scales (flattened modified macrotrichia), and hind and fore wings are linked by a frenulum, jugum or simple overlap. Wing venation consists predominantly of longitudinal veins with few cross-veins and some large cells, notably the discal (Fig. 2.24a). The legs are long and usually gressorial, with five tarsomeres. The abdomen is 10-segmented, with segment 1 variably reduced, and segments 9 and 10 modified as external genitalia (Fig. 2.25a). Internal female genitalia are very complex.

Pre-mating behaviour, including courtship, often involves pheromones (Fig. 4.7 and Fig. 4.8). Encounter between the sexes is often aerial, but copulation is on the ground or a perch. Eggs are laid on, close to, or, more rarely, within a larval host plant. Egg numbers and degree of aggregation are very variable. Diapause is common.

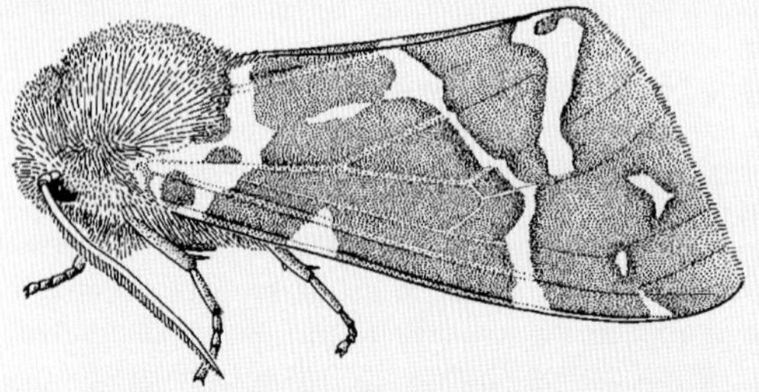

garden tiger moth, Arctiidae: *Arctia caja*

cabbage white butterfly, Pieridae: *Pieris rapae*

Lepidopteran larvae can be recognized by their sclerotized, hypognathous or prognathous head capsule, mandibulate mouthparts, usually six lateral stemmata (Fig. 4.9a), short three-segmented antennae, five-segmented thoracic legs with single claws, and 10-segmented abdomen with short prolegs on some segments (usually on 3–6 and 10, but may be reduced) (Fig. 6.6a,b and Fig. 14.6; Plate 2b). Silk-gland products are extruded from a characteristic spinneret at the median apex of the labial prementum. The pupa is usually contained within a silken cocoon, but naked in butterflies, typically adecticous and obtect (a chrysalis) (Fig. 6.7g–i), with only some abdominal segments unfused; the pupa is exarate in "primitive" groups.

Adult lepidopterans that feed utilize nutritious liquids, such as nectar, honeydew and other seepages from live and decaying plants, and a few species pierce fruit or suck blood. However, none suck sap from the vessels of live plants. Many species supplement their diet by feeding on nitrogenous animal wastes (Box 5.4). Most larvae feed exposed on higher plants and form the major insect phytophages; a few "primitive" species feed on non-angiosperm plants, and some feed on fungi. Several are predators and others are scavengers, notably amongst the Tineidae (wool moths).

The larvae are often cryptic (see Plate 7b), particularly when feeding in exposed positions, or warningly coloured (aposematic; Plate 7h) to alert predators to their toxicity (Chapter 14). Toxins derived from larval food plants often are retained by adults, which show anti-predator devices, including advertisement of non-palatability and defensive mimicry (section 14.5; Plate 8a).

Although the butterflies popularly are considered to be distinct from the moths, they form a clade that lies deep within the phylogeny of the Lepidoptera: butterflies are *not* the sister group to all moths. Butterflies are almost exclusively day flying, whereas most moths are active at night or dusk. In life, butterflies hold their wings together vertically above the body (as shown here on the right for a small white or cabbage white butterfly) in contrast to moths, which hold their wings flat or wrapped around the body (as shown on the left for the garden tiger moth); a few lepidopteran species have brachypterous adults and sometimes completely wingless adult females.

Phylogenetic relationships are considered in section 7.4.2 and depicted in Fig. 7.2 and Fig. 7.8.

Taxobox 29 **Hymenoptera (ants, bees, wasps, sawflies and wood wasps)**

The Hymenoptera is an order of more than 150,000 described extant species of holometabolous neopterans, classified traditionally in two suborders, the "Symphyta" (wood wasps and sawflies) (which is a paraphyletic group) and Apocrita (wasps, bees and ants). Within the Apocrita, the aculeate taxa (comprising ants, bees, and apoid, chrysidoid, scolioid, tiphioid-pomplioid and vespoid wasps) form a monophyletic group (Fig. 12.2), characterized by the use of the ovipositor for stinging prey (Plate 8d) or enemies rather than for egg laying. Adult hymenopterans range in size from minute (e.g. Trichogrammatidae, Fig. 16.2) to large (i.e. 0.15–120 mm long), and from slender (e.g. many Ichneumonidae) to robust (e.g. the bumble bee, Fig. 12.3). The head is hypognathous or prognathous, and the mouthparts range from generalized mandibulate to sucking and chewing, with mandibles in Apocrita often used for killing and handling prey, defence and nest building. The compound eyes are often large; ocelli may be present, reduced or absent. The antennae are long, multisegmented, and often prominently held forwardly or recurved dorsally. Symphytans have three conventional segments in the thorax, whereas the thoracic tagma in Apocrita includes the first abdominal segment (propodeum), combined in a mesosoma (or in ants, alitrunk) (as illustrated for workers of the wasp *Vespula germanica* and a *Formica* ant). The wings have reduced venation, and the hind wings have rows of hooks (hamuli) along the leading edge that couple with the hind margin of the fore wing in flight. Abdominal segment 2 (and sometimes also 3) of Apocrita forms a constriction, or petiole, followed by the remainder of the abdomen, or gaster. The female genitalia include an ovipositor, comprising three valves and two major basal sclerites, which may be long and highly mobile, allowing the valves to be directed vertically between the legs (Fig. 5.11). The ovipositor of aculeate Hymenoptera is modified as a sting associated with the venom apparatus (Fig. 14.11).

The eggs of endoparasitic species are often deficient in yolk, and sometimes each may give rise to more than one individual (polyembryony; section 5.10.3). Symphytan larvae are eruciform (caterpillar-like) (Fig. 6.6c), with three pairs of thoracic legs with apical claws and some abdominal legs; most are phytophagous, and there have been multiple evolutionary transitions from wood- to foliage-feeding among wood wasps. Apocritan larvae are apodous (Fig. 6.6i), with the head capsule frequently reduced but with prominent strong mandibles; the larvae may vary greatly in morphology during development (heteromorphosis). Apocritan larvae have diverse feeding habits and may be parasitic (section 13.3), gall inducing, or fed with prey or nectar and pollen by their parent (or, if a social species, by other colony members). There have been many evolutionary transitions to specialized phytophagy within Apocrita. Adult hymenopterans mostly feed on nectar (**Plate 5h**) or honeydew (**Plate 8b**), and sometimes drink haemolymph of other insects; only a few consume other insects.

Haplodiploidy allows a reproductive female to control the sex of offspring according to whether the egg is fertilized or not. Possible high relatedness amongst aggregated individuals facilitates well-developed social behaviours in many aculeate Hymenoptera. Female parthenogenesis, thelytoky, also occurs in some eusocial species, especially ants, and commonly is induced in solitary wasps infected by certain bacteria.

For phylogenetic relationships of the Hymenoptera, see section 7.4.2 and Fig. 7.2, Fig.7.9 and Fig. 12.2.

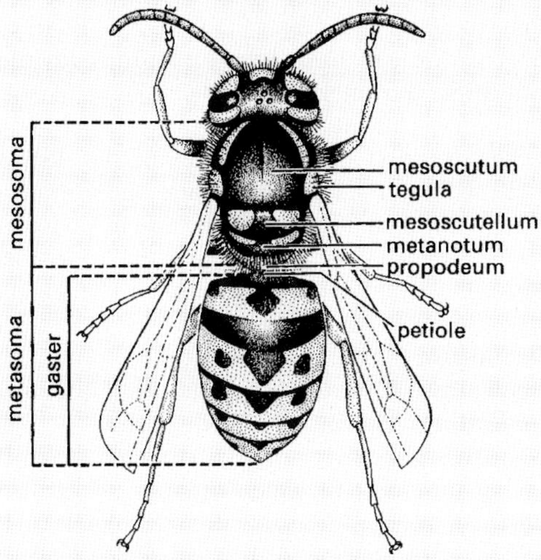

worker of the European wasp, *Vespula germanica*

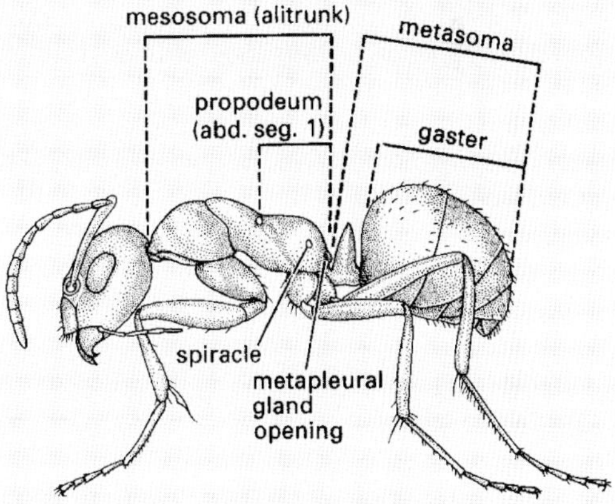

worker ant of *Formica subsericea*

GLOSSARY

Each scientific and technical field has a particular vocabulary: entomology is no exception. This is not an attempt by entomologists to restrict access to their science; it results from the need for precision in communication, whilst avoiding, for example, misplaced anthropocentric terms derived from human anatomy. Many terms are derived from Latin or Greek; when competence in these languages was a prerequisite for scholarship (including in the natural sciences), these terms were comprehensible to the educated, whatever their native language. The utility of these terms remains, although fluency in the languages from which they are derived does not. In this glossary, we have tried to define the terms we use in a straightforward manner to complement the definitions used on first mention in the main body of the book. Terms highlighted in **bold** in the main text are in this glossary. The glossary does not contain definitions of all words (e.g. insecticide names) or phrases that entomologists might use; please refer to the index if a word is not in the glossary. Comprehensive glossaries of entomological terms are provided by Nichols (1989) and Gordh & Headrick (2011); see Further reading at end of the glossary.

Bold within the text of an entry indicates a relevant cross-reference to another term. The following abbreviations are used:

adj. adjective
dim. diminutive
n. noun
pl. plural
sing. singular
Am. American spelling.

abdomen The third (posterior) major division (tagma) of an insect body.

acanthae Fine, unicellular, cuticular extensions (Fig. 2.6c).

accessory gland(s) A gland subsidiary to a major one; more specifically, a gland opening into the **genital chamber** (Fig. 3.1 & Fig. 3.20a,b).

accessory pulsatile organs Valved pumps aiding the circulation of **haemolymph** in appendages (antennae, legs, wings and mouthparts); *see also* **dorsal vessel**.

acclimation Physiological changes to a changed environment (especially temperature) that allow tolerance of more extreme conditions than are tolerated prior to acclimation.

acidophore A small circular opening (orifice) at the tip of the **gaster** of certain ants (the subfamily Formicinae); used to expel formic acid for defence or communication.

acrotergite The anterior part of a secondary segment, sometimes large (then called **postnotum**), often reduced (Fig. 2.7).

acrotrophic ovariole *See* **telotrophic ovariole**.

activation (in embryology) The initiation of embryonic development within the egg, which occurs when the first meiotic division in the oocyte is released from its blocked state; the stimuli for activation include entry of the sperm and oocyte compression during passage down the ovipositor.

aculeate Belonging to the aculeate Hymenoptera (Fig. 12.2) – ants, bees and wasps, in which the **ovipositor** is modified as a sting.

adecticous Describing a **pupa** with immovable **mandibles** (Fig. 6.7); *see also* **decticous**.

adenotrophic viviparity **Viviparity** (producing living offspring) in which there is no free-living larval stage; eggs develop within the female uterus,

The Insects: An Outline of Entomology, Fifth Edition. P.J. Gullan and P.S. Cranston.

Companion Website: www.wiley.com/go/gullan/insects

nourished by special **milk glands** until the larvae mature, at which stage they are laid and immediately pupate; occurring only in some Diptera (e.g. Hippoboscidae and *Glossina*).

adipocyte *See* **trophocyte**.

aedeagus (*Am.* edeagus) The male copulatory organ, variably constructed (sometimes refers to the **penis** alone) (Fig. 2.26b & Fig. 5.4).

aeriferous trachea **Trachea** with surface bearing a system of evaginated spiral tubules with permeable cuticle, allowing aeration of surrounding tissues, especially in the ovary.

aestivate To undergo **quiescence** or **diapause** during seasonal hot or dry conditions.

age-grading Determination of the physiological age of an insect.

air sac Any of the thin-walled, dilated sections of the **tracheae** (Fig. 3.11b).

alary muscles Paired muscles that support the heart region of the **dorsal vessel**.

alate Possessing wings.

alinotum The wing-bearing plate on the upper surface (**dorsum**) of the **mesothorax** or **metathorax** (Fig. 2.20).

alitrunk The fused thorax and first abdominal segment (**propodeum**) of adult ants (*see* **mesosoma**) (see Taxobox 29).

allelochemical A kind of **secondary plant compound** used by the plant to defend itself against herbivores or plant competitors; often functioning in interspecific communication; *see also* **allomone**, **kairomone**, **pheromone**, **synomone**.

allochthonous Originating from elsewhere, as of nutrients entering an aquatic ecosystem; *see also* **autochthonous**.

allometry (*adj.* allometric) The proportional change in the dimensions of one trait (e.g. head width) relative to another trait or to overall body size.

allomone A communication chemical that benefits the producer by the effect it invokes in the receiver.

allopatric Non-overlapping geographical distributions of organisms or taxa; *see also* **sympatric**.

altruism Behaviour often costly to the individual but beneficial to others.

ametaboly (*adj.* ametabolous) Lacking **metamorphosis**, i.e. with no change in body form during development to the adult, with the immature stages lacking only reproductive structures.

amnion (in embryology) The layer covering the **germ band** (Fig. 6.5).

amphimixis True sexual reproduction, in which each female inherits a **haploid** genome from both her mother and her father.

amphitoky (amphitokous parthenogenesis; deuterotoky) A form of **parthenogenesis** in which the female produces offspring of both sexes.

amplexiform coupling A form of wing-coupling in which there is extensive overlap between the fore wing and the hind wing, but without any specific coupling mechanism.

anal In the direction or position of the **anus**; near the anus or on the last abdominal segment.

anal area The posterior part of the wing, supported by the **anal vein**(s) (Fig. 2.22).

anal fold (vannal fold) A distinctive fold in the **anal area** of the wing (Fig. 2.22).

anal vein In wing venation, the seventh longitudinal **vein**, posterior to the **cubitus** (Fig. 2.23), and with anterior (AA) and posterior (PA) branches.

anamorphic Describing development in which the immature stages have fewer abdominal segments than the adult; *see also* **epimorphic**.

anautogenous Requiring a protein meal to develop eggs.

androdioecy The coexistence of **hermaphrodites** and males in one species..

anemophily Wind pollination.

anholocyclic (*adj.*) Of aphids, describing a life cycle in which the sexual phase is lost and reproduction is solely by **parthenogenesis**.

anhydrobiosis A form of **cryptobiosis** in which a living organism survives in an ametabolic state of extreme desiccation.

anlage A cell cluster in an immature individual that will give rise to a specific organ in the adult; *see also* **imaginal disc**.

annulate/annular A structure that is ringed or ring-like.

antecostal suture (intersegmental groove) A groove marking the position of the intersegmental fold between the primary segments (Fig. 2.7 & Fig. 2.20).

antenna (*pl.* antennae) Paired, segmented, sensory appendages, lying usually anterodorsally, on the head (Fig. 2.9, Fig. 2.10 & Fig. 2.19); derived from the second head segment.

antennomere A subdivision of the **antenna**.

anterior At or towards the front (Fig. 2.8).

anthophilous Flower-loving.

anthropogenic Caused by humans.

anthropophilic Associated with humans.
anus The posterior opening of the digestive tract (Fig. 2.25b & Fig. 3.13).
apical At or towards the apex (Fig. 2.8).
apneustic A gas-exchange system without functional **spiracles**; *see also* **oligopneustic**, **polypneustic**.
apocritan Belonging to the suborder of Hymenoptera (Apocrita) in which the first abdominal segment is fused to the thorax; *see also* **propodeum**.
apod A **larva** lacking true legs (Fig. 6.6); *see also* **oligopod**, **polypod**.
apode (*adj*. apodous) An organism without legs.
apodeme An ingrowth of the **exoskeleton**, tendon-like, to which muscles are attached (Fig. 2.25 & Fig. 3.2c); *see also* **apophysis**.
apolysial space The space between the old and the new **cuticle** that forms during **apolysis**, prior to **ecdysis**.
apolysis The separation of the old from the new **cuticle** during **moulting**.
apomixis **Parthenogenesis** in which eggs are produced mitotically (no meiosis); *see also* **automixis**.
apomorphy (synapomorphy) A derived feature (shared by two or more groups).
apophysis (*pl*. apophyses) An elongate **apodeme**, an internal projection of the **exoskeleton**.
aposematic Warning of unpalatability (especially toxicity), particularly using colour.
aposematism A communication system based on warning signals.
appendicular ovipositor The true **ovipositor** formed from appendages of segments 8 and 9; *see also* **substitutional ovipositor**.
appendix dorsalis The medial caudal appendage arising from the **epiproct**, lying above the **anus**; present in apterygotes, most mayflies and some fossil insects.
apposition eye A type of **compound eye** that gathers multiple images, one from each **ommatidium**, which is isolated from its neighbours by pigment.
apterous Wingless.
arachnophobia Fear of arachnids (spiders and relatives).
aroliar pad Pretarsal pad-like on ventral on ventral surface of **arolium** (Fig. 2.21).
arolium (*pl*. arolia) Pretarsal pad-like or sac-like structure(s) lying between the **claws** (Fig. 2.21).
arrhenotoky (arrhenotokous parthenogenesis) A form of **haplodiploidy** in which **haploid** male offspring are produced from unfertilized eggs.
arthrodial membrane Soft, stretchable **cuticle**, for example between sclerotized parts of segments (Fig. 2.4).
articular sclerites Separate, small, movable plates that lie between the body and a wing.
asynchronous muscle A muscle that contracts many times per nerve impulse, as in many flight muscles and those controlling the cicada tymbal.
atrium (*pl*. atria) A chamber, especially inside a tubular conducting system, such as the **tracheal system** (Fig. 3.10a).
attraction–annihilation A method of pest control in which individuals of the targeted species are lured and removed from the population; two approaches are **mass trapping** and **lure-and-kill**.
augmentation The supplementation of existing natural enemies by the release of additional organisms.
autapomorphy A feature unique to a taxonomic group; *see also* **apomorphy**, **plesiomorphy**.
autochthonous Originating from within, as of nutrients generated in an aquatic ecosystem, for example primary production; native; *see also* **allochthonous**.
automimicry A condition of **Batesian mimicry** in which palatable members (called automimics) of a species are defended by their resemblance to members of the same species that are chemically unpalatable.
automixis **Parthenogenesis** in which eggs are produced after meiosis, but ploidy is restored by various mechanisms; sometimes called meiotic parthenogenesis.
autotomy The shedding of appendage(s), notably for defence.
axillary area An area at the wing base bearing the wing articulation (Fig. 2.22).
axillary plates Two (anterior and posterior) articulating plates that are fused with the **veins** in a dragonfly or damselfly wing; the anterior supports the **costa**, the posterior supports the remaining veins; in Ephemeroptera there is only a posterior plate.
axillary sclerites Three or four **sclerites** that, together with the **humeral plate** and **tegula**, comprise the **articular sclerites** of the neopteran wing base (Fig. 2.23).

axon A nerve cell fibre that transmits a nerve impulse away from the cell body (Fig. 3.5); *see also* **dendrite**.

bacteriocyte A cell containing symbiotic microorganisms, scattered throughout the body, particularly within the **fat body**, or aggregated in organs called **bacteriomes** (also called mycetomes).

bacteriome An organ containing aggregations of **bacteriocytes** (also called mycetocytes), usually located in the **fat body** or gonads.

basal At or towards either the base or the main body, or closer to the point of attachment (Fig. 2.8).

basalare (*pl.* basalaria) A small **sclerite**, one of the **epipleurites** that lies anterior to the pleural wing process (Fig. 2.20); an attachment for the **direct flight** muscles.

basisternum The main **sclerite** of the **eusternum**, lying between the anterior **presternum** and posterior **sternellum** (Fig. 2.20).

Batesian mimicry A mimetic system in which a palatable species obtains protection from predation by resembling an unpalatable species; *see also* **Müllerian mimicry**.

benthos The bottom sediments of aquatic habitats and/or the organisms that live there.

biogeography The study of biotic distribution in space and time.

biological control (biocontrol) The human use of selected living organisms (including viruses), often called natural enemies, to suppress the population density or reduce the impact of plant or animal pest species. Biological control aims to make the pest less abundant or less damaging than it would be in the absence of the natural enemy (enemies).

biological monitoring Using plants or animals to detect changes in the environment.

biological transfer The movement of a disease organism from one **host** to another by one or more **vectors** in which there is a biological cycle of disease.

bioluminescence The production by an organism of cold light, commonly involving the action of an enzyme (luciferase) on a substrate (luciferin).

biosecurity Procedures intended to protect humans, their animals or the natural environment against disease agents or other harmful organisms (e.g. pests and weeds) that are considered "biosecurity risks".

biotype A biologically differentiated form of a purported single species.

bivoltine Having two generations in one year; *see also* **multivoltine**, **semivoltine**, **univoltine**.

bivouac An army ant camp during the mobile phase.

borer (*adj.* boring) A maker of burrows in dead or living tissue.

brachypterous Having shortened wings.

brain In insects, the supraoesophageal ganglion of the nervous system (Fig. 3.6), comprising **protocerebrum**, **deutocerebrum** and **tritocerebrum**.

brood A clutch of individuals that hatch at the same time from eggs produced by one set of parents.

bud (of wing) *See* **imaginal disc**.

bursa copulatrix The female **genital chamber** if functioning as a copulatory pouch; in Lepidoptera, the primary receptacle for sperm (Fig. 5.6).

bursicon A neuropeptide **hormone** that controls hardening and darkening of the **cuticle** after **ecdysis**.

caecum (*pl.* caeca) (*Am.* cecum) A blind-ending tube or sac (Fig. 3.1).

calliphorin A protein produced in the **fat body** and stored in the **haemolymph** of larval Calliphoridae (Diptera).

calyx (*pl.* calyces) A cup-like expansion, especially of the oviduct into which the **ovaries** open (Fig. 3.20a).

camouflage A form of **crypsis** in which an organism is indistinguishable from its background.

campaniform sensillum A mechanoreceptor that detects stress on the **cuticle**, comprising a dome of thin cuticle overlying one **neuron** per **sensillum**, located especially on joints (Fig. 4.2).

cantharophily Plant pollination by beetles.

cap cell The outermost cell of a sense organ such as a **chordotonal organ** (Fig. 4.3).

cardiopeptide A neuropeptide **hormone** that stimulates the **dorsal vessel** ("heart") causing **haemolymph** movement.

cardo The proximal part of the maxillary base (Fig. 2.10).

castes Morphologically distinctive groups of individuals within a single species of social insect, usually differing in behaviour.

caudal At or towards the **anal** (tail) end.

caudal filament One of two or three terminal filaments (Fig. 8.5; see also Taxobox 4).

caudal lamellae One of two or three terminal gills (see Taxobox 5).

cavernicolous (troglodytic; troglobiont) Living in caves.

cecidology The study of plant **galls**.

cecidozoa Gall-inducing animals.

cell (of wing) An area of the wing membrane partially or completely surrounded by **veins**; *see* **closed cell**, **open cell**.

cement layer The outermost layer of the **cuticle** (Fig. 2.1), often absent.

central nervous system In insects, the central series of **ganglia** extending the length of the body (Fig. 3.6); *see also* **brain**.

cephalic Pertaining to the head.

cercus (*pl.* cerci) One of a pair of appendages originating from abdominal segment 11 but usually visible as if on segment 10 (Fig. 2.25b).

cervical sclerite(s) Small **sclerite(s)** on the membrane between the head and thorax (actually the first thoracic segment) (Fig. 2.9).

chitin The major component of arthropod **cuticle**, a polysaccharide composed of linked units of *N*-acetyl-D-glucosamine (Fig. 2.2).

chloride cells Osmoregulatory cells found in the epithelium of the abdominal gills of aquatic insects.

chordotonal organs Sense organs (mechanoreceptors) that perceive vibrations, comprising one to several elongate cells called **scolopidia** (Fig. 4.3 & Fig. 4.4). Examples include the **tympanum**, **subgenual organ** and **Johnston's organ**.

chorion The outermost shell of an insect egg, which may be multilayered, including the exochorion, endochorion and wax layer (Fig. 5.10).

chronogram A phylogenetic tree for which branch lengths are proportional to evolutionary time.

cibarium The dorsal food pouch, lying between the **hypopharynx** and the inner wall of the **clypeus**, often with a muscular pump (Fig. 2.16a & Fig. 3.14).

circadian rhythms Repeated periodic behaviour with an interval of about 24 h.

clade A **monophyletic** group, comprising an ancestor and all of its descendants; a branch on a phylogenetic tree.

cladistics A classification system in which **clades** are the only permissible groupings.

cladogram A phylogenetic tree based on shared derived features (**synapomorphies**) of **taxa** and depicting ancestor–descendant relationships (gene alogy) with only the branching pattern of the phylo geny depicted (Fig. 7.1), i.e. not branch lengths.

classical biological control Long-term control of an exotic pest by one or more coevolved natural enemies introduced deliberately from the pest's area of origin.

classification The process of establishing, defining and ranking **taxa** (e.g. species) into a hierarchical series of groups; see also Table 1.1.

claval furrow A **flexion-line** on the wing that separates the **clavus** from the **remigium** (Fig. 2.22).

clavus An area of the wing delimited by the **claval furrow** and the posterior margin (Fig. 2.22 & Fig. 2.24e).

claw (pretarsal claw; unguis) A hooked structure on the distal end of the **pretarsus**, usually paired (Fig. 2.21); more generally, any hooked structure.

closed cell An area of the wing membrane bounded entirely by **veins**; *see also* **open cell**.

closed tracheal system A gas-exchange system comprising **tracheae** and **tracheoles** but lacking **spiracles** and therefore closed to direct contact with the atmosphere (Fig. 3.11d–f); *see also* **open tracheal system**.

clypeus The part of the insect head to which the **labrum** is attached anteriorly (Fig. 2.9 & Fig. 2.10); it lies below the **frons**, with which it may be fused in a **frontoclypeus** or separated by a suture.

coeloconic sensillum (sensillum coeloconica) An olfactory sensory structure that is sunken into a pit in the **cuticle** and detects odours via receptive **neurons** (Box 4.2).

coevolution Evolutionary interactions between two organisms, such as plants and pollinators, hosts and parasites; the degree of specificity and reciprocity varies; *see also* **guild coevolution**, **phyletic tracking**, **specific coevolution**.

colleterial (cement) glands Accessory glands of the female internal genitalia that produce secretions used to cement eggs to the substrate.

collophore The **ventral tube** of Collembola.

colon The **hindgut** between the **ileum** and **rectum** (Fig. 3.1 & Fig. 3.13).

comb In a social hymenopteran nest, a layer of regularly arranged cells (Fig. 12.6 & Box 12.1).

common (median) oviduct In female insects, the tubes leading from the fused **lateral oviducts** to the **vagina** (Fig. 3.20a).

competency (in termite development) The potential of a termite of one **caste** to become another, for example a worker to become a soldier.

compound eye An aggregation of **ommatidia**, each acting as a single facet of the eye (Fig. 2.9 & Fig. 4.10).

conjunctiva (conjunctival membrane) *See* **intersegmental membrane**.

connective Anything that connects; more specifically, the paired longitudinal nerve cords that connect the **ganglia** of the **central nervous system**.

conservation biological control Measures that protect and/or enhance the activities of natural enemies.

constitutive defence Part of the normal chemical composition; *see also* **induced defence**.

coprophage (*adj.* coprophagous) A feeder on dung or excrement (Fig. 9.5).

corazonin A neuropeptide, produced in the brain and nerve cord, that acts on **Inka cells** to promote release of **ecdysis triggering hormones**.

corbicula The pollen basket of bees (Fig. 12.4).

coremata (*sing.* corema) Eversible, thin-walled abdominal organs of male moths used for dissemination of sex **pheromone**.

corium A section of the heteropteran **hemelytron** (fore wing), differentiated from the **clavus** and membrane, usually leathery (Fig. 2.24e).

cornea The cuticle covering the eye or **ocellus** (Fig. 4.9 & Fig. 4.10).

corneagenous cell One of the translucent cells beneath the **cornea** that secretes and supports the corneal lens (Fig. 4.9).

cornicle (siphunculus) Paired tubular structures on the **abdomen** of aphids that discharge defensive lipids and alarm **pheromones**.

corpora allata (*sing.* corpus allatum) Paired endocrine glands associated with the stomodeal ganglia behind the **brain** (Fig. 3.8), the source of **juvenile hormone**.

corpora cardiaca (*sing.* corpus cardiacum) Paired glands lying close to the aorta and behind the **brain** (Fig. 3.8), acting as stores and producers of **neurohormones**.

cosmopolitan Distributed worldwide (or nearly so).

co-speciation (cospeciation) The speciation of one population in response to, and together with, the speciation of another with which it is associated, such as a specialist herbivore speciating with its host plant, or a parasitic insect with its vertebrate host.

costa (*adj.* costal) The most anterior longitudinal wing **vein**, running along the costal margin of the wing and ending near the apex (Fig. 2.23).

costal fracture A break or weakness in the costal margin in Heteroptera that divides the **corium**, separating the **cuneus** from the embolium (Fig. 2.24e).

coxa (*pl.* coxae) The proximal (basal) leg segment (Fig. 2.21).

crepuscular Active at low light intensities, dusk or dawn; *see also* **diurnal**, **nocturnal**.

crista acustica (crista acoustica) The main **chordotonal organ** of the tibial tympanal organ of katydids (Orthoptera: Tettigoniidae) (Fig. 4.4).

crochets Curved hooks, spines or spinules on a **proleg**.

crop The food storage area of the digestive system, posterior to the oesophagus (Fig. 3.1 & Fig. 3.13).

cross-resistance Resistance of an insect to one insecticide providing resistance to a different insecticide.

cross-veins Transverse wing veins that link the longitudinal **veins**.

crown-group The smallest **monophyletic** group that contains the last common ancestor of all extant members as well as all descendants of that ancestor.

crustacean cardioactive peptide (CCAP) A highly conserved peptide, found in crustaceans and insects, that appears to have multiple roles, including in moulting, digestion and cardiac control.

cryoprotection Mechanisms that allow organisms to survive periods of, often extreme, cold.

crypsis Camouflage by resemblance to environmental features.

cryptic Hidden, camouflaged, concealed.

cryptobiosis The state of a living organism during which there are no signs of life, and metabolism virtually ceases.

cryptonephric system A condition of the excretory system in which the **Malpighian tubules** form an intricate contact with the **rectum**, allowing production of dry excreta (see Box 3.4).

crystalline cone A hard crystalline body lying beneath the **cornea** in an **ommatidium** (Fig. 4.10).

crystalline lens A lens lying beneath the cuticle of the **stemma** of some insects (Fig. 4.9).

ctenidium (*pl.* ctenidia) A comb (see Taxobox 26).

cubitus (Cu) The sixth longitudinal **vein**, lying posterior to the **media**, often divided into an anterior two-branched CuA_1 and CuA_2 and a posterior unbranched CuP_1 (Fig. 2.23).

cuneus The distal section of the **corium** in the heteropteran wing (Fig. 2.24e).

cursorial Running or adapted for running.

cuticle The external skeletal structure, secreted by the **epidermis**, composed of **chitin** and protein, comprising several differentiated layers (Fig. 2.1).

cycloalexy Forming aggregations in defensive circles (Fig. 14.7).

cytoplasmic (reproductive) incompatibility Reproductive incompatibility arising from cytoplasmically inherited microorganisms that causes embryological failure; can be unidirectional or bidirectional.

Darwinian (classical) fitness The contribution of an individual to the gene pool through its offspring.

day-degrees (degree-days) A measure of physiological time, the product of time and temperature above a **threshold**.

decapitation Separation of the head from the body; beheading; used particularly in early hormone studies of insects.

deciduous Falling off, detaching (e.g. at maturity).

decticous Describing an **exarate** pupa in which the **mandibles** are articulated (Fig. 6.7); *see also* **adecticous**.

delayed parasitism Parasitism in which hatching of the **parasite** (or **parasitoid**) egg is delayed until the **host** is mature.

delusory parasitosis A psychotic illness in which parasitic infection is imagined.

dendrite A fine branch of a nerve cell (Fig. 3.5); *see also* **axon**.

dermal gland A unicellular epidermal gland that may secrete moulting fluid, cements, wax, etc., and sometimes **pheromones** (Fig. 2.1).

determinate Describing growth or development in which there is a distinctive final, adult, instar; *see also* **indeterminate**.

detritivore (*adj.* detritivorous) An eater of organic detritus of plant or animal origin.

deuterotoky *See* **amphitoky**.

deutocerebrum The middle part of the insect **brain**; the **ganglion** of the second segment, comprising antennal and olfactory lobes.

developmental threshold (growth threshold) The temperature below which no development takes place.

diapause Delayed development that is not the direct result of prevailing environmental conditions.

diapause hormone (DH) A **hormone** produced by neurosecretory cells in the **suboesophageal ganglion** that affects the timing of future development of eggs and, in the pupae of at least some insects, is involved in diapause termination.

dicondylous/dicondylar Describing an articulation (as of a mandible) with two points of articulation (condyles).

diplodiploidy (diploid males) The genetic system found in most insects in which each male receives a **haploid** genome from both his mother and his father, and these two genomes have equal probability of being transmitted through his sperm; *see also* **haplodiploidy**, **paternal genome elimination**.

diploid With two sets of chromosomes; *see also* **haploid**.

direct flight muscles Flight muscles that are attached directly to the wing (Fig. 3.4); *see also* **indirect flight muscles**.

discontinuous gas exchange The phenomenon of oxygen uptake and carbon dioxide release via the **spiracles** occurring in a three-phase cyclical pattern that includes periods of little or no gas exchange.

disjunct Widely separated ranges, as in populations or species geographically separated so as to prevent gene flow.

dispersal Movement of an individual or population away from its birth site.

distal At or near the farthest end from the attachment of an appendage (opposite to **proximal**) (Fig. 2.8).

diurnal Day active; *see also* **crepuscular**, **nocturnal**.

DNA barcoding A method that uses a short genetic marker (for insects usually part of the mitochondrial *COI* gene) from an organism's DNA to identify it as belonging to a particular species.

domatia Chambers produced by plants specifically to house certain arthropods, especially ants.

dorsal On the upper surface (Fig. 2.8).

dorsal closure The embryological process in which the dorsal wall of an embryo is formed by growth of the **germ band** to surround the yolk.

dorsal diaphragm The main fibromuscular septum that divides the **haemocoel** into the **pericardial** and **perivisceral sinuses** (compartments) (Fig. 3.9).

dorsal vessel The "aorta" and "heart", the main pump for **haemolymph**; a longitudinal tube lying in the dorsal pericardial sinus (Fig. 3.9).

dorsoventral The axis extending from the **dorsal** (upper) to **ventral** (lower) side.

dorsum The upper surface.

drift Passive movement caused by water or air currents.

drone The male bee, especially of honey bees and bumble bees, derived from an unfertilized egg.

Dufour's gland In **aculeate** hymenopterans, a sac opening into the poison duct near the sting (Fig. 14.11); the site of production of **pheromones** and/or poison components.

dulosis An extreme form of **social parasitism** in which there is a slave-like relationship between a parasitic ant species and the captured brood from another species.

Dyar's rule An observational "rule" governing the size increment found between successive **instars** of the same species (Fig. 6.12).

ecdysial lines Lines of weakness in the **cuticle** that allow splitting of the cuticle at the time of **moulting**.

ecdysis (*adj.* ecdysial) The final stage of **moulting**, the process of casting off the **cuticle** (Fig. 6.8).

ecdysis triggering hormones (PETH and ETH) Peptide hormones that initiate the behavioural sequence of **ecdysis** and that are secreted from the **Inka cells**.

ecdysone The steroid **hormone** secreted by the **prothoracic gland** that is converted to 20-hydroxyecdysone, which stimulates moulting fluid secretion.

ecdysteroid The general term for steroids that induce **moulting** (Fig. 5.13, Fig. 6.9 & Fig. 6.10).

ecdysterone An old term for 20-hydroxyecdysone, the major moult-inducing steroid.

eclosion The release of the adult insect from the cuticle of the previous **instar**; sometimes used of hatching from the egg.

eclosion hormones (EH) **Neurohormones** with several functions associated with adult **eclosion**, including increasing cuticle extensibility.

ecogenomics (environmental genomics) The application of molecular techniques to ecological and environmental science, including biodiversity studies.

economic injury level (EIL) The level at which economic pest damage equals the costs of pest control.

economic threshold (ET) The pest density at which control must be applied to prevent the **economic injury level** being reached.

ecosystem engineer An organism that creates or significantly modifies a habitat.

ectognathy (*adj.* ectognathy) Having exposed mouthparts.

ectoparasite A **parasite** that lives externally on and at the expense of another organism, which it does not kill; *see also* **ectoparasitoid**, **endoparasite**.

ectoparasitoid A **parasite** that lives externally on and at the expense of another organism, which it kills; *see also* **ectoparasite**, **endoparasitoid**.

ectoperitrophic space The space between the **peritrophic matrix/membrane** and the midgut wall (Fig. 3.16).

ectothermy (*adj.* ectothermic) The inability to regulate the body temperature relative to the surrounding environment.

edeagus (*Am.*) *See* **aedeagus**.

ejaculatory duct The duct that leads from the fused **vas deferens** to the **gonopore** (Fig. 3.20b), through which semen or sperm is transported.

elaiosome A food body forming an appendage on a plant seed (Fig. 11.8).

elytron (*pl.* elytra) The modified, hardened, fore wing of a beetle that protects the hind wing (Fig. 2.24d).

embolium The marginal area of the heteropteran wing, anterior to the **vein** R + M (Fig. 2.24e).

empodium (*pl.* empodia) A central spine or pad on the **pretarsus** of Diptera.

encapsulation A reaction of the **host** to an **endoparasitoid** in which the invader is surrounded by **haemocytes** that eventually form a capsule (Fig. 13.5).

endemic Describing a taxon or disease that is restricted to a particular geographical area.

endite An inwardly directed (**mesal**) appendage or lobe of a limb segment (Fig. 8.5).

endocrine gland A gland that secretes its product (usually a hormone) within the body, usually into the **haemolymph**.

endocuticle The flexible, unsclerotized inner layer of the **procuticle** (Fig. 2.1); *see also* **exocuticle**.

endoparasite A **parasite** that lives internally at the expense of another organism, which it does not kill; *see also* **ectoparasite**, **endoparasitoid**.

endoparasitoid A **parasite** that lives internally at the expense of another organism, which it kills; *see also* **ectoparasitoid**, **endoparasite**.

endoperitrophic space In the gut, the space enclosed within the **peritrophic matrix/membrane** (Fig. 3.16).

endophallus (vesica) The inner, eversible tube of the **penis** (Fig. 5.4); given different names in different insect groups.

endophilic Indoor loving, as of an insect that feeds inside a dwelling; *see also* **exophilic**.

endopterygote Describing development in which the wings form within pockets of the integument, with eversion taking place only at the larval–pupal moult, as in the **monophyletic** grouping Holometabola (=Endopterygota).

endosymbiont Intracellular symbionts, typically bacteria, that usually have a mutualistic association with their insect hosts.

endothermy (*adj.* endothermic) The ability to regulate the body temperature higher than the surrounding environment.

energids In an embryo, the daughter nuclei cleavage products and their surrounding cytoplasm.

entognathy (*adj.* entognathy) Having mouthparts hidden in a fold of the head.

entomopathogen A pathogen (disease-causing organism) that attacks insects particularly.

entomophage (*adj.* entomophagous) An eater of insects.

entomophily Pollination by insects.

entomophobia Fear of insects.

enzootic A disease present in a natural **host** within its natural range.

epicoxa A basal leg segment (Fig. 8.5), forming the **articular sclerites** in all extant insects, and believed to have borne the **exites** and **endites** that may have fused to form the evolutionary precursors of wings.

epicranial suture A Y-shaped line of weakness on the **vertex** of the head where the split at **moulting** occurs.

epicuticle The inextensible and unsupportive outermost layer of **cuticle**, lying outside the **procuticle** (Fig. 2.1).

epidemic An increase in number of cases of a disease above what is expected in an area or the spread of a disease from its **endemic** area and/or from its normal host(s).

epidermis The unicellular layer of ectodermally derived **integument** that secretes the **cuticle** (Fig. 2.1).

epigenetics The study of heritable changes in gene activity that are not caused by changes to DNA sequences; functionally relevant changes to the genome that do not involve changes to nucleotide sequences, but rather mechanisms such as DNA methylation or histone modification.

epimeron (*pl.* epimera) The posterior division of the **pleuron** of a thorax, separated from the **episternum** by the **pleural suture** (Fig. 2.20).

epimorphic Describing development in which the segment number is fixed in the embryo before hatching; *see also* **anamorphic**.

epipharynx The ventral surface of the **labrum**, a membranous roof to the mouth (Fig. 2.17).

epipleurite (1) The more dorsal of the **sclerites** formed when the **pleuron** is divided longitudinally. (2) One of two small sclerites of a wing-bearing segment: the anterior **basalare** and posterior **subalare** (Fig. 2.20).

epiproct The dorsal relic of segment 11 (Fig. 2.25b).

episternum (*pl.* episterna) The anterior division of the **pleuron**, separated from the **epimeron** by the **pleural suture** (Fig. 2.20).

epizootic Of a disease, when **epidemic** (there is an unusually high number of cases and/or deaths).

ergatoid (apterous neotenic) In termites, a supplementary reproductive derived from a worker, held in a state of arrested development and lacking wings, able to replace reproductives if they die; *see also* **neotenic**.

esophagus (*Am.*) *See* **oesophagus**.

euplantula (*pl.* euplantulae) A pad-like structure on the ventral surface of some **tarsomeres** of the leg.

eusocial Exhibiting co-operation in reproduction and division of labour, with overlap of generations.

eusternum (*pl.* eusterna) The dominant ventral plate of the thorax that frequently extends into the pleural region (Fig. 2.20).

eutrophication Nutrient enrichment, especially of water bodies.

evo-devo Informal name for the field of biology called evolutionary developmental biology, which compares the developmental processes of different organisms to estimate ancestral relationships among the organisms and how their developmental processes evolved.

evolutionary systematics A classification system in which both **clades** (**monophyletic** groups) and **grades** (**paraphyletic** groups) are recognized.

exaptation A morphological–physiological predisposition or preadaptation that allows a trait to evolve a new function.

exarate Describing a **pupa** in which the appendages are free from the body (Fig. 6.7), as opposed to being cemented; *see also* **obtect**.

exocrine gland A gland that secretes its product (e.g. a pheromone, venom or wax) to the outside of the body, usually via a duct.

excretion The elimination of metabolic wastes from the body, or their internal storage in an insoluble form.

exite An outer appendage or lobe of a limb segment (Fig. 8.5).

exocuticle The rigid, sclerotized outer layer of the **procuticle** (Fig. 2.1); *see also* **endocuticle**.

exophilic (*n.* exophily) Out-of-door loving, used of biting insects that do not enter buildings; *see also* **endophilic**.

exopterygote Describing development in which the wings form progressively in sheaths that lie externally on the dorsal or dorsolateral surface of the body.

exoskeleton The external, hardened, cuticular skeleton to which muscles are attached internally.

external genitalia Organs concerned specifically with mating and, in the female, also deposition of eggs, although they may be largely internal or retractable into the abdomen.

extra-oral digestion Digestion that takes place outside the organism, by secretion of salivary enzymes onto or into the food, with soluble digestive products then being sucked up.

facultative Not compulsory, optional behaviour.

fat body A loose or compact aggregation of cells, mostly **trophocytes** suspended in the **haemocoel**, responsible for metabolism, synthesis and storage of a range of compounds.

femur (*pl.* femora) The third segment of an insect leg, following the **coxa** and **trochanter**; often the stoutest leg segment (Fig. 2.21).

fermentation Breakdown of complex molecules by microbes, as of carbohydrates by yeast.

file A toothed or ridged structure used in sound production by **stridulation** through contact with a **scraper**.

filter chamber Part of the alimentary canal of many hemipterans, in which the anterior and posterior parts of the **midgut** are in intimate contact, forming a system in which most fluid bypasses the absorptive midgut (see Box 3.3).

fitness *See* **Darwinian (classical) fitness**, **inclusive (extended) fitness**.

flabellum In bees, the lobe at the tip of the **glossae** ("tongue") (Fig. 2.11).

flagellomere One of the subdivisions of a "multi-segmented" (actually multi-annulate) antennal **flagellum**.

flagellum The third part of an antenna, distal to the **scape** and **pedicel**; more generally, any whip or whip-like structure.

flagship species A species that represents an environmental cause, such as conservation of a particular habitat, or raises public awareness to promote conservation of the taxonomic group to which the flagship species belongs.

flexion-line A line along which a wing flexes (bends) when in flight (Fig. 2.22).

fluctuating asymmetry The level of deviation from absolute symmetry in a bilaterally symmetrical organism, argued to be due to variable stress during development.

fold-line A line along which a wing is folded when at rest (Fig. 2.22).

follicle The **oocyte** and follicular epithelium; more generally, any sac or tube.

follicular relic Relict morphological evidence left in the **ovary** showing that an egg has been laid (or resorbed), which may include dilation of the lumen and/or pigmentation.

food canal A canal anterior to the **cibarium** through which fluid food is ingested (Fig. 2.11 & Fig. 2.12).

forage To seek and gather food.

fore Anterior, towards the head.

fore wings The anterior pair of wings, usually on the **mesothorax**.

foregut (stomodeum) The part of the gut that lies between the mouth and the **midgut** (Fig. 3.13), derived from the ectoderm.

forensic entomologist A scientist who studies the role of insects in criminal matters.

fossorial Digging, or adapted for digging (Fig. 9.2).

frass Solid excreta of an insect, particularly a larva.

frenate coupling A form of wing-coupling in which one or more hind-wing structures (**frenulum**) attach to a retaining structure (**retinaculum**) on the fore wing.

frenulum Spine or group of bristles on the **costa** of the hind wing of Lepidoptera that locks with the fore-wing **retinaculum** in flight.

frons The single anteromedial **sclerite** of the insect head, usually lying between the epicranium and the **clypeus** (Fig. 2.9).

frontal sutures The lower arms of the **epicranial suture**, delimiting the **frons** (Fig. 2.10).

frontoclypeal suture (epistomal suture) A groove that runs across the insect's face, often separating the **frons** from the **clypeus** (Fig. 2.9).

frontoclypeus The combined **frons** and **clypeus**.

fructivore (*adj*. fructivorous) A fruit-eater.

fundatrix (*pl*. fundatrices) An apterous viviparous parthenogenetic female aphid that develops from the overwintering egg.

fungivore (*adj*. fungivorous) A fungus-eater.

furca (furcula) (1) The abdominal springing organ of Collembola (see Taxobox 1). (2) With the fulcral arm, the lever of the hymenopteran sting (Fig. 14.11).

galea The lateral lobe of the maxillary stipes (Fig. 2.10, Fig. 2.11 & Fig. 2.12).

gall An aberrant plant growth produced in response to the activities of another organism, often an insect (Fig. 11.5).

gametocyte A cell from which a gamete (egg or spermatozoon) is produced.

ganglion (*pl*. ganglia) A nerve centre; in insects, forming fused pairs of white, ovoid bodies lying in a row ventrally in the body cavity, linked by a double nerve cord (Fig. 3.1).

gas exchange The system of oxygen uptake and carbon dioxide elimination.

gas gills Specific gas-exchange surfaces on aquatic insects, often as abdominal lamellae, but may be present almost anywhere on the body.

gaster The swollen part of the **abdomen** of **aculeate** Hymenoptera, lying posterior to the **petiole** (waist) (see Taxobox 29).

gena (*pl*. genae) Literally, a cheek; on each side of the head, the part lying beneath the **compound eye**.

gene silencing The "switching off" of a gene by a mechanism other than genetic modification (e.g. by DNA methylation) – a process that can occur both in nature and in laboratory experiments.

genital chamber A cavity of the female body wall that contains the **gonopore** (Fig. 3.20a), also known as the **bursa copulatrix** if functioning as a copulatory pouch.

genitalia All ectodermally derived structures of both sexes associated with reproduction (copulation and, in females, fertilization and oviposition).

genus (*pl*. genera, *adj*. generic) The name of the taxonomic category ranked between species and family; a grouping of one or more species united by one or more derived features and therefore believed to be of a single evolutionary origin (i.e. **monophyletic**).

germ anlage (in embryology) The germ disc that denotes the first indication of a developing embryo (Fig. 6.5).

germ band (in embryology) The post-gastrulation band of thickened cells on the ventral gastroderm, destined to form the ventral part of the developing embryo (Fig. 6.5).

germarium The structure within an **ovariole** in which the oogonia give rise to **oocytes** (Fig. 3.20a).

giant axon A nerve fibre that conducts impulses rapidly from the sense organ(s) to the muscles.

gill A gas-exchange organ, found in various forms in aquatic insects.

glossa (*pl*. glossae) The "tongue", one of a pair of lobes on the inner apex of the **prementum** (Fig. 2.11).

gonapophysis (*pl*. gonapophyses) A valve (part of the shaft) of a female insect's **ovipositor** (Fig. 2.25); also in the genitalia of many male insects (Fig. 2.26a).

gonochorism Sexual reproduction in which males and females are separate individuals.

gonocoxite The base of an appendage, formed of **coxa + trochanter**, of a genital segment (8 or 9) (also called a **valvifer** in females) (Fig. 2.25b & Fig. 2.26a).

gonopore The opening of the genital duct; in the female the opening of the common oviduct (Fig. 3.20a), in the male the opening of the ejaculatory duct.

gonostyle The style (rudimentary appendage) of the ninth segment (Fig. 2.25), often functioning as a male clasper (Fig. 2.26a).

grade A **paraphyletic** group, one which does not include all descendants of a common ancestor, united by shared primitive features.

gregarious Forming aggregations.

gressorial Walking, or adapted for walking.

guild coevolution (diffuse coevolution) Concerted evolutionary change that takes place between interacting groups of organisms, as opposed to between two species; *see also* **specific coevolution**.

gula A ventromedial sclerotized plate on the head of **prognathous** insects (Fig. 2.10).

gyne A reproductive female hymenopteran, a **queen**.

haemo- (*Am.* hemo-) Referring to blood.

haemocoel (*Am.* hemocoel) The main body cavity of many invertebrates including insects, formed from an expanded "blood" system.

haemocoelous viviparity (*Am.* hemocoelous viviparity) Viviparity (producing live offspring) in which the immature stages develop within the **haemocoel** of the parent female, for example as in Strepsiptera.

haemocyte (*Am.* hemocyte) An insect blood cell.

haemolymph (*Am.* hemolymph) The fluid filling the **haemocoel**.

haematophage (*adj.* haematophagous) (*Am.* hematophage) An eater of blood (or similar fluid).

hair A cuticular extension, also called a **macrotrichium** or **seta**.

hair plate A group of sensory hairs that act as a **proprioceptor** for movement of articulating parts of the body (Fig. 4.2a).

haltere The modified hind wing in Diptera, acting as a balancer (Fig. 2.24f).

hamulate coupling A mechanism for coupling the fore and hind wings during flight that involves hooks (**hamuli**) on the anterior margin of the hind wing that couple with a fold on the fore wing.

hamuli (*adj.* hamulate) Hooks along the anterior (costal) margin of the hind wing of Hymenoptera that couple the wings in flight by catching on a fold of the fore wing.

hamulohaltere The highly reduced and modified hind wing in male scale insects (Hemiptera: Coccoidea).

haplodiploidy (*adj.* haplodiploid) A genetic system in which the male is either functionally or actually **haploid** and transmits only his mother's genome, and the female arises from fertilized eggs and is diploid; *see also* **arrhenotoky**, **diplodiploidy**, **paternal genome elimination**.

haploid With one set of chromosomes; *see also* **diploid**.

haustellate Sucking, as of mouthparts.

haustellum Tube-like sucking mouthparts, a **proboscis** or **rostrum**; used mostly in reference to mouthparts of some flies (Fig. 15.1).

head The anterior of the three major divisions (**tagmata**) of an insect body.

hemelytron (*pl.* hemelytra) The fore wing of Heteroptera, with a thickened basal section and membranous apical section (Fig. 2.24e).

hemimetaboly (*adj.* hemimetabolous) Development in which the body form gradually changes at each moult, with wing buds growing larger at each moult; incomplete **metamorphosis**; *see also* **holometaboly**.

hermaphroditism Having individuals (called hermaphrodites) that possess both testes and ovaries.

heterochrony Alteration in the relative timing of activation of different developmental pathways.

heteromorphosis (hypermetamorphosis) Undergoing a major change in morphology between larval instars, as from **triungulin** to grub.

hibernate To undergo **quiescence** or **diapause** during seasonal cold conditions.

hind At or towards the posterior.

hind wings The wings on the **metathorax**.

hindgut (proctodeum) The posterior section of the gut, extending from the end of the **midgut** to the **anus** (Fig. 3.13).

holocyclic In aphids, describing a (complete) life cycle in which reproduction typically consists of a generation of sexual morphs (sexuals or sexuales) and several generations of only parthenogenetic females.

holometaboly (*adj.* holometabolous) Development in which the body form abruptly changes at the pupal moult; complete **metamorphosis**, as in the group Holometabola (=Endopterygota); *see also* **hemimetaboly**.

homeosis (*adj.* homeotic) The genetic or developmental modification of a structure (e.g. an appendage) on one segment to resemble a morphologically similar or different structure on another segment (*see also* **serial homology**).

homeostasis Maintenance of a prevailing condition (physiological or social) by internal feedback.

homeothermy The maintenance of an even body temperature despite variation in the ambient temperature.

homology (*adj.* homologous) Morphological identity or similarity of a structure or other feature in two (or more) different groups (**taxa**) as a result of common evolutionary origin.

homoplasy Possession of similar or identical features due to convergent or parallel evolution in different groups (**taxa**), rather than by direct inheritance from a common ancestor.

honeydew A watery fluid mostly containing sugars derived from plant sap and eliminated from the anus of some Hemiptera.

hormone A chemical messenger that regulates some activity at a distance from the endocrine organ that produced it.

host An organism that harbours another, especially a **parasite** or **parasitoid**, either internally or externally.

host discriminate To choose between different **hosts**.

host regulation The ability of a **parasitoid** to manipulate the **host's** physiology.

humeral plate One of the articular **sclerites** of the neopteran wing base (Fig. 2.23); *see also* **axillary sclerites**, **tegula**.

humus Organic soil.

hydrostatic skeleton Turgid structural support provided by fluid pressure maintained by muscle contractions on a fixed volume of liquid, especially within larval insects.

hypermetamorphosis *See* **heteromorphosis**.

hyperparasite (*adj.* hyperparasitic) A **parasite** that lives upon another parasite.

hyperparasitoid A secondary **parasitoid** that develops upon another **parasite** or parasitoid.

hypognathous With the head directed vertically and mouthparts directed ventrally; *see also* **opisthognathous**, **prognathous**.

hypopharynx A median lobe of the preoral cavity of the mouthparts (Fig. 2.10).

hyporheic Living in the substrate beneath the bed of a water body.

idiobiont A **parasitoid** that prevents its **host** from developing any further, by paralysis or death; *see also* **koinobiont**.

ileum The second section of the **hindgut**, preceding the **colon** (Fig. 3.1 & Fig. 3.13).

imaginal disc (imaginal bud) Latent adult structure in an immature insect, visible as a group of undifferentiated cells (Fig. 6.4).

imago (*pl.* imagines *or* imagos) The adult insect.

inclusive (extended) fitness The contribution of an individual to the gene pool by enhanced success of its kin.

indeterminate Describing growth or development in which there is no distinctive final, adult instar, with no definitive terminal moult; *see also* **determinate**.

indirect flight muscles Muscles that power flight by deforming the thorax rather than directly moving the wings (Fig. 3.4); *see also* **direct flight muscles**.

indirect system (of flight) With power for flight coming from regular deformation of the thorax by **indirect flight muscles**, in contrast to predominantly from muscular connection to the wings.

induced defence A chemical change, deleterious to herbivores, induced in foliage or another plant part as a result of feeding damage.

industrial melanism The phenomenon of dark **morphs** occurring in a higher than usual frequency in areas in which industrial pollution darkens tree trunks and other surfaces upon which insects may rest.

Inka cells Endocrine cells associated with the tracheal system that produce and release **ecdysis triggering hormones** (PETH and ETH).

innate Describing behaviour requiring no choice or learning.

inner epicuticle The innermost layer of **epicuticle**, with the **procuticle** beneath it (Fig. 2.1); *see also* **outer epicuticle**.

inoculation (inoculative release; inoculation biological control) To infect with a disease by introducing it into the body fluids or tissues; in **biological control**, to periodically release a natural enemy that will reproduce and control the target pest for a certain period but not indefinitely.

inquiline An organism that lives in the home of another, sharing food; in entomology, used particularly of residents in the nests of social insects (*see also* **integrated inquiline**, **non-integrated inquiline**) or in plant galls induced by another organism.

insecticide A chemical used to kill, or attempt to kill, insects.

insectivore (*adj.* insectivorous) An insect eater; *see also* **entomophage**.

instar The growth stage between two successive moults.

integrated inquiline An **inquiline** that is incorporated into the social life of the host by behavioural modification of both inquiline and host; *see also* **non-integrated inquiline**.

integrated pest management (IPM) A pest management strategy that integrates the use of multiple suppressive tactics, often involving **biological control**, for optimizing control of pests in an economically and ecologically sound manner, by taking into account negative and positive impacts of pest control on producers, society and the environment.

integrative taxonomy A comprehensive approach to taxonomic research that utilizes information from multiple fields of study, such as behaviour, morphology, molecular biology and biogeography.

integument The epidermis plus **cuticle**; the outer covering of the living tissues of an insect.

interference (1) (in colours) Iridescent colours produced by variable reflection of light by narrowly separated surfaces (as in the scales of lepidopterans). (2) (in population dynamics) A reduction in the profitability of an otherwise high resource density caused by intra- and interspecific interactions between predators and parasitoids.

intermediate organ A **chordotonal organ** in the **subgenual organ** of the fore leg of some orthopterans (Fig. 4.4), associated with the **tympanum** and believed to respond to specific sound frequencies.

intermoult period *See* **stadium**.

interneuron (association neuron) A nerve cell (**neuron**) that forms a connection between other nerve cells, generally between sensory and motor neurons, and thus receiving and transmitting information.

intersegmental groove *See* **antecostal suture**.

intersegmental membrane (conjunctiva; conjunctival membrane) A membrane between segments, particularly of the **abdomen** (Fig. 2.7).

intersternite An intersegmental sternal plate posterior to the **eusternum**, known as the **spinasternum** except on the metasternum (Fig. 2.7).

inundation (inundative release; inundative biological control) Swamping a pest with large numbers of living control agents, with control deriving from the released organisms rather than from their progeny.

IPM system The operating system used by farmers in order to manage the control of crop pests in an environmentally and socially responsible manner; *see also* **integrated pest management**.

Johnston's organ A **chordotonal** (sensory) **organ** within the antennal **pedicel**.

jugal area (jugum; *pl.* juga) The posterobasal area of the wing, delimited by the **jugal fold** and the wing margin (Fig. 2.22).

jugal fold A **fold-line** of the wing, dividing the **jugal area** from the **clavus** (Fig. 2.22).

jugal vein In wing venation, the most posterior longitudinal **vein**, after the **anal vein** (Fig. 2.23), usually represented by one or two small veins in extant insects.

jugate coupling A mechanism for coupling the fore and hind wings in flight by a large **jugal area** of the fore wing overlapping the hind wing.

juvenile hormone (JH) A hormone, occurring in several forms based on a chain of 16–19 carbon atoms, that is released by the **corpora allata** into the **haemolymph** and is involved in many aspects of insect physiology, including modification of the expression of a moult.

kairomone A communication chemical that benefits the receiver and is disadvantageous to the producer; *see also* **allomone**, **synomone**.

katatrepsis Adoption by the embryo of the final position in the egg, involving moving from dorsal on ovum to ventral aspect.

keystone species A species that has a disproportionate effect on its environment relative to its abundance; keystone species are often **ecosystem engineers** (creating or modifying the habitat) or **predators**.

kinesis (*pl.* kineses) Movement of an organism in response to a stimulus, usually restricted to response to stimulus intensity only.

king The male primary reproductive in termites (Fig. 12.8).

kinship component An indirect contribution to an individual's **inclusive (extended) fitness** derived from increased reproductive success of the individual's kin (relatives) through the altruistic assistance of the individual.

koinobiont A **parasitoid** that allows its **host** to continue to develop while living within it; *see also* **idiobiont**.

labellum (*pl.* labella) In certain flies, paired lobes at the apex of the **proboscis**, derived from **labial palps** (see Fig. 2.13 & Fig. 2.14).

labial palp One- to five-segmented appendage of the **labium** (Fig. 2.9, Fig. 2.10 & Fig. 2.11).

labium (*adj.* labial) The "lower lip", forming the floor of the mouth, often with a pair of palps and two pairs of median lobes (Fig. 2.9 & Fig. 2.10); derived from the sixth head segment.

labrum (*adj.* labral) The "upper lip", forming the roof of the preoral cavity and mouth (Fig. 2.9, Fig. 2.10 & Fig. 2.11); probably derived from the third head segment.

lacinia The mesal lobe of the maxillary stipes (Fig. 2.10).

larva (*pl.* larvae) An immature insect after emerging from the egg, usually restricted to insects in which there is complete metamorphosis (**holometaboly**), but sometimes used for any immature insect that differs strongly from its adult; *see also* **nymph**.

larval paedogenesis Reproduction in a larval stage (known in some Diptera).

lateral At, or close to, the side (Fig. 2.8).

lateral oviducts In female insects, the paired tubes leading from the ovaries to the **common oviduct** (Fig. 3.20a).

laterosternite The result of a fusion of the **eusternum** and a pleural sclerite.

lerp The protective cover constructed from starchy or sugary anal discharges of nymphal psylloids (Hemiptera: Psylloidea); the nymph lives and feeds under its lerp.

lek A male mating aggregation associated with a defended territory that contains no resources other than available courting males.

lentic Of standing water.

ligation An experimental technique that isolates one part of the body of a living insect from another, usually by tightening a ligature; used particularly in early hormone studies of insects.

ligula The **glossae** plus **paraglossae** of the **prementum** of the **labium**, whether fused or separate.

litter A layer of dead vegetative matter overlying the soil.

longitudinal In the direction of the long axis of the body.

lotic Of flowing water.

lure-and-kill (attract-and-kill) The use of **pheromones** or other lures to attract pest insects to an insecticide or, more rarely, to sterilants or pathogens; *see also* **attraction–annihilation**.

macrophage An eater of large particles; *see also* **microphage**.

macrotrichium (*pl.* macrotrichia) A **trichoid sensillum**, also called a **seta** or **hair**.

maggot A legless larval insect, frequently a fly, and usually with a reduced head.

major worker (soldier) An individual of the largest-sized worker **caste** of termites and ants, specialized for defence; *see also* **media worker**, **minor worker**.

Malpighian tubules Thin, blind-ending tubules, originating near the junction of the **midgut** and **hindgut** (Fig. 3.1 & Fig. 3.13), predominantly involved in regulation of salt, water and nitrogenous waste excretion.

mandible (*adj.* mandibular) The jaws, either jaw-like in shape in biting and chewing (mandibulate) insects (Fig. 2.9 & Fig. 2.10), or modified as narrow **stylets** in piercing and sucking insects (Fig. 2.15); the first pair of jaws; derived from the fourth head segment.

mandibulate Possessing **mandibles**.

masquerade (mimesis) A form of **crypsis** in which an organism resembles a feature of its environment that is of no interest to a predator.

mass trapping The use of **pheromones** or other lures to attract pest insects to a trap or adhesive surface where they are confined and die; *see also* **attraction–annihilation**.

mating disruption A form of insect control in which synthetic sex **pheromones** (usually of the female) are maintained artificially at a higher level than the background so as to interfere with mate location.

matrifilial Describing **eusocial** hymenopterans whose colonies consist of mothers and their daughters.

maxilla (*pl.* maxillae) The second pair of jaws, jaw-like in chewing insects (Fig. 2.9 & Fig. 2.10), variously modified in others (Fig. 2.15); derived from the fifth head segment.

maxillary palp A one- to seven-segmented sensory appendage borne on the **stipes** of the **maxilla** (Fig. 2.9, Fig. 2.10 & Fig. 2.11).

mechanical transfer The movement of a disease organism from one **host** to another by passive transfer, with no biological cycle in the **vector**.

meconium The first excreta of a newly emerged adult following the pupal stage.

media In wing venation, the fifth longitudinal **vein**, lying between the **radius** and the **cubitus**, with a maximum of eight branches (Fig. 2.23).

media worker An individual of the medium-sized worker **caste** of termites and ants; *see also* **major worker**, **minor worker**.

medial Towards the middle.

median At or towards the middle (Fig. 2.8).

median flexion-line A **fold-line** that runs longitudinally through the approximate middle of the wing (Fig. 2.22).

melanism (*adj.* melanic) Darkening caused by increased pigmentation.

melittophily Pollination by bees.

mentum The ventral fused plate derived from the **labium** (Fig. 2.18).

mesal (medial) Nearer to the midline of the body.

mesenteron *See* **midgut**.

mesosoma The middle of the three major divisions (**tagmata**) of the insect body, equivalent to the **thorax**, but in **apocritan** Hymenoptera including the **propodeum**; called the **alitrunk** in adult ants (see Taxobox 29).

mesothorax The second (and middle) segment of the **thorax** (Fig. 2.20).

metabolic resistance The ability to avoid harm by biochemical detoxification of an insecticide.

metamorphosis The relatively abrupt change in body form between the end of immature development and the onset of the imaginal (adult) phase.

metapopulation A group of spatially separated populations of the same species that interact in some way.

metasoma In **apocritan** Hymenoptera, the **petiole** plus **gaster** (see Taxobox 29).

metathorax The third (and last) segment of the **thorax** (Fig. 2.20).

microlecithal Describing an egg lacking large yolk reserves.

microphage A feeder on small particles, such as spores; *see also* **macrophage**.

micropyle A minute opening in the **chorion** of an insect egg (Fig. 5.10), through which sperm enter.

microtrichium (*pl.* microtrichia) A subcellular cuticular extension, usually several to very many per cell (Fig. 2.6d).

microvillus (*pl.* microvilli) A small finger-like projection.

midgut (mesenteron) The middle section of the gut, extending from the end of the **proventriculus** to the start of the **ileum** (Fig. 3.1 & Fig. 3.13).

migration Directional movement to more appropriate conditions.

milk glands Specialized accessory glands in certain flies (e.g. Hippoboscidae and *Glossina*) that are adenotrophically viviparous (*see* **adenotrophic viviparity**), producing secretions fed upon by larvae.

mimesis Resemblance to an inedible object in the environment; *see also* **masquerade**.

mimic (*adj.* mimetic) In a mimicry system, the emitter of false signal(s) received by an **observer** such as a predator; an individual, population or species that resembles a **model**, usually another species or part thereof; *see also* **automimicry**, **Batesian mimicry**, **Müllerian mimicry**.

mimicry The resemblance of a **mimic** to a **model**, by which the mimic derives protection from predation provided to the model (e.g. by unpalatability).

miner (*adj.* mining) An insect that feeds below the epidermal surface of a plant leaf, stem, trunk or fruit, for example leaf miners feed in the mesophyll layer between the upper and lower epidermis of a leaf (Fig. 11.2).

minor worker An individual of the smaller-sized worker **caste** of termites and ants; *see also* **media worker**, **major worker**.

model In a **mimicry** system, the emitter of signal(s) received by the **observer** such as a predator; the organism resembled by a **mimic**, protected from predation, for instance, by distastefulness.

molar area The grinding surface of the **mandible** (Fig. 2.10).

monocondylar Describing an articulation (as of a **mandible**) with one point of articulation (condyle).

monogynous Describing a colony of **eusocial** insects dominated by one **queen**.

monolecty (*adj.* monolectic) The phenomenon of an insect (usually used in reference to bees) having very specialized pollen preferences and collecting from the flowers of just one plant species.

monophage (*adj.* monophagous) An eater of only one kind of food, used particularly of specialized **phytophages**; *see also* **oligophage**, **polyphage**.

monophyletic Describing a group (**taxon**) that includes all descendants of a single ancestor, recognized by the joint possession of shared derived feature(s).

monoxene (*adj.* monoxenous) A **parasite** restricted to one **host**.

morph A phenotypic form or variant, sometimes genetically determined.

morphospecies A species defined only on morphological criteria; or a putative species recognized by a non-specialist based on morphological criteria.

motor neuron A nerve cell with an **axon** that transmits stimuli from an **interneuron** to muscles (Fig. 3.5).

moult increment The increase in size between successive instars (Fig. 6.12).

moulting (moult) The formation of new **cuticle** followed by **ecdysis** (Fig. 6.9 & Fig. 6.10).

mouth hooks The head skeleton of the maggot larva of higher flies (see Taxobox 24).

Müllerian mimicry A **mimicry** system in which two or more unpalatable species obtain protection from predation by resembling each other; *see also* **Batesian mimicry**.

multiparasitism Parasitization of a **host** by two or more **parasites** or **parasitoids**.

multiple resistance The concurrent existence in a single insect population of two or more defence mechanisms against one insecticide.

multiporous Having several small openings.

multivoltine Having more than two generations in one year; *see also* **bivoltine**, **semivoltine**, **univoltine**.

mushroom body A cluster of **seminal vesicles** and **accessory gland** tubules forming a single mushroom-shaped structure, found in certain orthopteroid or blattoid insects.

mycangium (*pl.* mycangia) A special structure on the body, often a complex cuticular invagination in insects, that stores symbiotic fungi (usually in spore form) for transport and later use.

mycetocyte *See* **bacteriocyte**.

mycetome *See* **bacteriome**.

mycoinsecticide A fungus formulated (often in spore form) for use as an **insecticide**.

mycophage (*adj.* mycophagous) An eater of fungi; *see also* **fungivore**.

myiasis Disease or injury caused by feeding of larval flies on live flesh of humans or other animals.

myofibrils Contractile fibres that run the length of a muscle fibre, comprising actin sandwiched between myosin fibres.

myophily Plant pollination by flies.

myrmecochory The collection and dispersal of **elaiosome**-bearing seeds by ants.

myrmecophily Plant pollination by ants.

myrmecophyte ("ant plant") A plant that contains **domatia** to house ants (Fig. 11.9 & Fig. 11.10).

myrmecotrophy The feeding of plants by ants, notably through the waste products of an ant colony.

naiad An alternative name for the immature stages of aquatic hemimetabolous insects; *see also* **larva**, **nymph**.

nasus A nose, the snout of certain termite soldiers (**nasutes**).

nasute A **soldier** termite possessing a snout.

natatorial Swimming.

near-field The region of space very close to a sound source.

necrophage (*adj.* necrophagous) An eater of dead and/or decaying animals.

neoclassical biological control The use of exotic (introduced) **predators**, **parasitoids** or pathogens to control native pests.

neotenic In termites, a supplementary reproductive, arrested in its development, that has the potential to take on the reproductive role should the primary reproductives be lost; *see also* **ergatoid**.

neoteny (*adj.* neotenous) The retention of juvenile features in an adult by slowing of somatic (physical) development.

nephrocyte (pericardial cell) Cell that sieves the **haemolymph** for metabolic waste products.

neuroendocrine cells *See* **neurosecretory cells**.

neurohormone (neuropeptide) Any of the largest class of insect **hormones**, being small proteins secreted within different parts of the nervous system (Fig. 5.13).

neuron A nerve cell, comprising a cell body, **dendrite** and **axon** (Fig. 3.5 & Fig. 4.3).

neuropeptide *See* **neurohormone**.

neurosecretory cells (neuroendocrine cells) Modified **neurons** found throughout the nervous system (Fig. 3.8), producing insect **hormones** excepting **ecdysteroids** and **juvenile hormones**.

neuston (*adj.* neustic) The water surface.

nocturnal Night active; *see also* **crepuscular**, **diurnal**.

node A branch point on a phylogenetic tree; can be considered to represent the inferred most recent common ancestor of the descendent taxa.

nomenclature The science of naming (living organisms).

non-integrated inquiline An **inquiline** that is adapted ecologically to the nest of the host, but does not interact socially with the host; *see also* **integrated inquiline**.

notum (*pl.* nota) A thoracic **tergum**.

nulliparous Describing a female that has laid no eggs.

nymph An immature insect after emerging from the egg, usually restricted to insects in which there is incomplete metamorphosis (**hemimetaboly**); *see also* **larva**.

obligatory (obligate) Compulsory or exclusive; for example obligatory/obligate diapause is a resting stage that occurs in every individual of each generation of a univoltine insect.

observer In a **mimicry** system, the receiver (often a predator) of the signal(s) emitted by the **model** and **mimic**.

obtect Describing a **pupa** with body appendages fused (cemented) to the body; not free (Fig. 6.7); *see also* **exarate**.

occipital foramen The opening of the back of the head.

occiput The dorsal part of the posterior cranium (Fig. 2.9).

ocellus (*pl.* ocelli) The "simple" eye (Fig. 4.10b) of adult and nymphal insects, typically three in a triangle on the **vertex** of the head, with one median and two lateral ocelli (Fig. 2.9 & Fig. 2.11); the **stemma** of some holometabolous larvae.

oenocyte A cell associated with the **epidermis**, **haemocoel** or the **fat body**, probably with many functions, which are unclear in most insects, but they have important roles in lipid metabolism, as well as detoxification and developmental signalling in some insects.

oesophagus (*Am.* esophagus) The **foregut** that lies posterior to the **pharynx**, anterior to the **crop** (Fig. 2.16a, Fig. 3.1 & Fig. 3.13).

oligolecty (*adj.* oligolectic) The phenomenon of an insect (usually used in reference to bees) having specialized pollen preferences and collecting usually from the flowers of just one plant genus, although sometimes multiple genera in one family.

oligophage (*adj.* oligophagous) An eater of few kinds of food, for example several plant species within one genus or one family; used particularly of **phytophages**; *see also* **monophage**, **polyphage**.

oligopneustic Describing a gas-exchange system with one to two functional **spiracles** on each side of the body; *see also* **apneustic**, **polypneustic**.

oligopod A **larva** with legs on the **thorax** and not on the **abdomen** (Fig. 6.6); *see also* **apod**, **polypod**.

oligoxene (*adj.* oligoxenous) A **parasite** or **parasitoid** restricted to a few **hosts**.

ommatidium (*pl.* ommatidia) A single element of the **compound eye** (Fig. 4.10).

ontogeny The process of development from egg to adult.

oocyte An immature egg cell formed from the **oogonium** within the **ovariole**.

oogonium The first stage in the development in the **germarium** of an egg from a female germ cell.

ootheca (*pl.* oothecae) A protective surrounding for eggs (see Taxobox 15).

open cell An area of the wing membrane partially bounded by **veins** but including part of the wing margin; *see also* **closed cell**.

open tracheal system A gas-exchange system comprising **tracheae** and **tracheoles** and with **spiracular** contact with the atmosphere (Fig. 3.11a–c); *see also* **closed tracheal system**.

opisthognathous With the head deflexed such that the mouthparts are directed posteriorly, as in many Hemiptera; *see also* **hypognathous**, **prognathous**.

optical superposition eye A type of **compound eye** in which light sensitivity is enhanced because each **ommatidium** is not isolated optically from its neighbours by pigment; found in insects that are active at night.

osmeterium (*pl.* osmeteria) An eversible tubular pouch on the **prothorax** of some larval swallowtail butterflies (Lepidoptera: Papilionidae) (Fig. 14.6), used to disseminate volatile toxic, defensive compounds.

osmoregulation The regulation of water balance, maintaining the **homeostasis** of osmotic and ionic content of the body fluids.

ostium (*pl.* ostia) A slit-like opening in the **dorsal vessel** ("heart") present usually on each thoracic and the first nine abdominal segments; each ostium has a one-way valve that permits flow of **haemolymph** from the **pericardial sinus** into the **dorsal vessel** (Fig. 3.9).

outer epicuticle A layer of epicuticle, with the **inner epicuticle** beneath it (Fig. 2.1) and the wax layer and sometimes a cement layer above it.

ovarian cycle The length of time between successive ovipositions.

ovariole One of several ovarian tubes that form the **ovary** (Fig. 3.1a), each consisting of a **germarium**, a **vitellarium** and a stalk or pedicel (Fig. 3.20a).

ovary (*pl.* ovaries) One of the paired gonads of female insects, each comprising several **ovarioles**.

oviparity (*adj.* oviparous) Reproduction in which eggs are laid; *see also* **ovoviviparity**, **viviparity**.

ovipositor The organ used for laying eggs; *see also* **appendicular ovipositor**, **substitutional ovipositor**.

ovoviviparity Retention of the developing fertilized egg within the mother, usually considered to be a form of **viviparity** (producing live offspring) because the eggs often hatch as, or just before, the young emerge from the female's reproductive tract, but the only nutrition provided to each developing embryo is that inside the egg; *see also* **oviparity**.

paedogenesis (*adj.* paedogenetic) (*Am.* pedogenetic) Reproduction in an immature stage.

paedomorphosis (*adj.* paedomorphic) The phenomenon of a reproductive adult retaining juvenile features, such as an adult female that resembles a larva or nymph; *see also* **neoteny**.

pair-wise coevolution *See* **specific coevolution**.

palp (palpus; *pl.* palpi) A finger-like, usually segmented appendage of either the maxilla (**maxillary palp**) or the labium (**labial palp**) (Fig. 2.9 & Fig. 2.10).

panoistic ovariole An **ovariole** that lacks nurse cells; *see also* **polytrophic ovariole**, **telotrophic ovariole**.

paraglossa (*pl.* paraglossae) One of a pair of lobes on the **prementum** of the **labium**, lying outside the **glossae**, but **mesal** to the **labial palp** (Fig. 2.10).

paramere One of a pair of lobes lying lateral to the **penis**, forming part of the **aedeagus** (Fig. 2.26b).

paranota (*sing.* paranotum, *adj.* paranotal) Postulated lobes arising from the thoracic terga of an ancestral insect and from which, it has been argued, the wings (or part of them) derive.

parapheromone A chemical of anthropogenic origin, but structurally related to some natural pheromone component(s) that has a physiological or behavioural effect on insect pheromone communication; for example methyl eugenol functions as a strong lure to attract male tephritid fruit flies.

paraphyletic Describing a group (**grade**) that is evolutionarily derived from a single ancestor, but which does not contain all descendants, and recognized by the joint possession of shared primitive character state(s); rejected in **cladistics** but often accepted in **evolutionary systematics**; *see also* **monophyletic**, **polyphyletic**.

paraproct Ventral relic of segment 11 (Fig. 2.25b).

parasite An organism that lives at the expense of another (**host**), which it does not usually kill; *see also* **ectoparasite**, **endoparasite**, **parasitoid**.

parasitism The relationship between a **parasitoid** or **parasite** and its **host**.

parasitization The condition of being parasitized, by either a **parasitoid** or **parasite**.

parasitized Describing the state of a **host** that supports a **parasitoid** or **parasite**.

parasitoid A **parasite** that kills its **host**; *see also* **ectoparasitoid**, **endoparasitoid**.

parataxonomists (biodiversity technicians) Trained non-taxonomists who generally collect, rear, sort and mount collected specimens into recognizable taxonomic units (RTUs) based on morphological features. This process assists large-scale ecological and taxonomic projects and expedites rapid assessment of biodiversity.

parous Describing a female that has laid at least one egg.

parthenogenesis Development from an unfertilized egg; *see also* **amphitoky**, **arrhenotoky**, **paedogenesis**, **thelytoky**.

patch A discrete area of microhabitat.

paternal genome elimination Loss of the paternal genome during the development of an initially **diploid** male, so that his sperm carries only his mother's genes; a form of **haplodiploidy**.

pedicel (1) The stem or stalk of an organ. (2) The stalk of an **ovariole** (Fig. 3.20a). (3) The second antennal segment (Fig. 2.10). (4) The "waist" of an ant.

pedogenesis *See* **paedogenesis**.

penis (*pl.* penes) (phallus) The median intromittent organ (Fig. 2.26b & Fig. 3.20b), variously derived in different insect orders; *see also* **aedeagus**.

pericardial sinus The body compartment that contains the **dorsal vessel** ("heart") (Fig. 3.9).

perimicrovillar membrane An extracellular lipoprotein membrane that ensheaths the microvilli of the midgut cells of Hemiptera and Thysanoptera.

perineural sinus The ventral body compartment that contains the nerve cord, separated from the **perivisceral sinus** by the **ventral diaphragm** (Fig. 3.9).

periodic release The regular release of biological control agents that are effective in control but unable to establish permanently.

peripheral nervous system The network of nerve fibres and cells associated with the muscles.

peritreme A sclerotized plate surrounding an orifice, notably around a **spiracle**.

peritrophic matrix/membrane/envelope A thin sheath lining the midgut epithelium of many insects (Fig. 3.16).

perivisceral sinus The central body compartment, delimited by the **ventral** and **dorsal diaphragms**.

pest resurgence The rapid increase in numbers of a pest following cessation of control measures or resulting from development of **resistance** and/or elimination of natural enemies.

petiole A stalk; in **apocritan** Hymenoptera, the narrow second (and sometimes third) abdominal segment(s) that precedes the **gaster**, forming the "waist" (see Taxobox 29).

phalaenophily Plant pollination by moths.

phallobase In male genitalia, the support for the **aedeagus** (Fig. 2.26b & Fig. 5.4).

phallomere A lobe lateral to the **penis**.

pharate Within the cuticle of the previous **instar**; "cloaked".

pharynx The anterior part of the **foregut**, anterior to the **oesophagus** (Fig. 2.16a & Fig. 3.13).

phenetic Describing a classification system in which overall resemblance between organisms is the criterion for grouping; *see also* **cladistics**, **evolutionary systematics**.

pheromone A chemical used in communication between individuals of the same species, which releases a specific behaviour or development in the receiver. Pheromones have roles in aggregation, alarm, courtship, queen recognition, sex, sex attraction, spacing (epideictic or dispersion) and trail marking.

phoresy (*adj.* phoretic) The phenomenon of one individual being carried on the body of a larger individual of another species.

photoperiod The duration of the light (and therefore also dark) part of the 24 h daily cycle.

photoreceptor A sense organ that responds to light.

phragma (*pl.* phragmata) A plate-like **apodeme**, notably those of the **antecostal suture** of the thoracic segments that support the longitudinal flight muscles (Fig. 2.7d & Fig. 2.20).

phragmosis The closing of a nest opening with part of the body.

phyletic tracking Strict **coevolution** in which the phylogenies of each interacting taxon (e.g. host and parasite, plant and pollinator) match precisely, as the evolution of one partner tracks the other.

phylogenetic Relating to **phylogeny**.

phylogeny The evolutionary history (of a taxon).

phylogram A phylogenetic tree for which branch lengths are proportional to the number of character state changes along each branch.

physiological time A measure of development time based upon the amount of heat required rather than calendar time elapsed.

physogastry Having a swollen **abdomen**, as in mature **queen** termites (Fig. 12.8), ants and bees.

phytophage (*adj.* phytophagous) An eater of plants.

phytophagy The eating of plants.

plant resistance A range of mechanisms by which plants resist insect attack.

plasma The aqueous component of **haemolymph**.

plastron The air–water interface (or the air film itself) on an external surface of an aquatic insect, which is the site of gaseous exchange.

pleiotropic Describing a single gene that has multiple effects on morphology and physiology.

pleometrosis The foundation of a colony of social insects by more than one **queen**.

plesiomorphy (symplesiomorphy) An ancestral feature (shared by two or more groups).

pleural coxal process The anterior end of the **pleural ridge** providing reinforcement for the coxal articulation (Fig. 2.20).

pleural ridge The internal ridge dividing the **pterothorax** into the anterior **episternum** and posterior **epimeron**.

pleural suture The externally visible indication of the **pleural ridge**, running from the leg base to the **tergum** (Fig. 2.20).

pleural wing process The posterior end of the **pleural ridge** providing reinforcement for the wing articulation (Fig. 2.20).

pleurite The diminutive of **pleuron**; a subdivision of a pleuron.

pleuron (*pl.* pleura, *adj.* pleural) The lateral region of the body, bearing the limb bases.

poikilothermy (*adj.* poikilothermic) The inability to maintain an invariant body temperature independent of the ambient temperature.

poison glands A class of **accessory glands** that produce poison, as in the stings of Hymenoptera (Fig. 14.11).

pollinarium A unit comprising the pollen mass (or pollinium) plus other components (including a sticky part) that gets attached to an insect as it leaves the flower; found in orchids and a few other plants such as some milkweeds.

pollination The transfer of pollen from male to female flower parts.

polydnaviruses (PDVs) A group of viruses found in the ovaries of some parasitic wasps, involved in overcoming host immune responses when injected with the wasp eggs.

polyembryony The production of more than one (often many) embryos from a single egg, notably in parasitic insects.

polyethism Within a social insect **caste**, the division of labour either by specialization throughout the life of an individual or by different ages performing different tasks.

polygyny Social insects that have several **queens**, either at the same time or sequentially (serial polygyny).

polylecty (*adj.* polylectic) The phenomenon of an insect (usually used in reference to bees) collecting pollen from the flowers of many unrelated plant species.

polymorphic Describing a species with two or more variants (morphs).

polyphage (*adj.* polyphagous) An eater of many kinds of food, for example many plant species from a range of families; used particularly of **phytophages**.

polyphenism Environmentally induced differences between successive life stages or generations of a species, or different **castes** of social insects, lacking a genetic basis.

polyphyletic Describing a group that is evolutionarily derived from more than one ancestor, recognized by the possession of one or more features evolved convergently; rejected in **cladistics** and **evolutionary systematics**.

polypneustic Describing a gas-exchange system with at least eight functional **spiracles** on each side of the body; *see also* **apneustic**, **oligopneustic**.

polypod A type of **larva** with jointed legs on the **thorax** and **prolegs** on the **abdomen** (Fig. 6.6); *see also* **apod**, **oligopod**.

polytrophic ovariole An **ovariole** in which several nurse cells remain closely attached to each **oocyte** as it moves down the ovariole; *see also* **panoistic ovariole**, **telotrophic ovariole**.

polyxene (*adj.* polyxenous) A **parasite** or **parasitoid** with a wide range of **hosts**.

pore canals Fine tubules that run through the **cuticle** and carry epidermally derived compounds to the **wax canals** and thus to the epicuticular surface.

pore kettle The chamber within a chemoreceptor **sensillum** that has many pores (slits) in the wall (Box 4.2).

postcoxal bridge The pleural area behind the **coxa**, often fused with the **sternum** (Fig. 2.20).

posterior At or towards the rear (Fig. 2.8).

posterior cranium The posterior, often horseshoe-shaped, area of the head.

postgena The lateral part of the occipital arch posterior to the **postoccipital suture** (Fig. 2.9).

postmentum The proximal part of the **labium** (Fig. 2.10 & Fig. 13.4).

postnotum The posterior part of a **notum** on the **pterothorax**, bearing the **phragmata** that support longitudinal muscles (Fig. 2.7d & Fig. 2.20).

postocciput The posterior rim of the head behind the **postoccipital suture** (Fig. 2.9).

postoccipital suture A groove on the head that indicates the original head segmentation, separating the **postocciput** from the remainder of the head (Fig. 2.9).

post-tarsus *See* **pretarsus**.

precosta The most anterior wing **vein** (Fig. 2.23).

precoxal bridge The pleural area anterior to the **coxa**, often fused with the **sternum** (Fig. 2.20).

predation (1) Preying on other animals. (2) Interactions between predator foraging and prey availability.

predator An animal that eats more than one other animal during its life; *see also* **parasitoid**.

pre-ecdysis triggering hormone *See* **ecdysis triggering hormones**.

pregenital segments The first seven abdominal segments.

prementum The free distal end of the **labium**, usually bearing **labial palps**, **glossae** and **paraglossae** (Fig. 2.10 & Fig. 13.4).

prescutum The anterior third of the **alinotum** (either meso- or metanotum), in front of the **scutum** (Fig. 2.20).

presoldier In termites, an intervening stage between **worker** and **soldier**.

press The process on the proximal apex of the **tarsus** of a bee that pushes pollen into the **corbicula** (basket) (Fig. 12.4).

presternum A smaller **sclerite** of the **eusternum**, lying anterior to the **basisternum** (Fig. 2.20).

pretarsus (*pl.* pretarsi) **(post-tarsus)** The **distal** segment of the insect leg (Fig. 2.21).

prey A food item for a **predator**.

primary cycle In a disease, the cycle that involves the typical host(s); *see also* **secondary cycle**.

primary reproductives In termites, the king and queen founders of a colony (Fig. 12.8).

proboscis A general term for elongate mouthparts (Fig. 2.12); *see also* **haustellum**, **rostrum**.

procephalon (in embryology) The anterior head formed by fusion of the primitive anterior three segments (Fig. 6.5).

proctodeum (*adj.* proctodeal) *See* **hindgut**.

procuticle The thicker layer of **cuticle**, which in sclerotized cuticle comprises an outer **exocuticle** and inner **endocuticle**; lying beneath the thinner **epicuticle** (Fig. 2.1).

prognathous With the head horizontal and the mouthparts directed anteriorly; *see also* **hypognathous**, **opisthognathous**.

proleg An unsegmented leg of a **larva**.

pronotum The upper (dorsal) plate of the **prothorax**.

pronymph The post-embryonic form, either a hatchling or pre-hatchling, distinct from subsequent nymphal stages.

propodeum In **apocritan** Hymenoptera, the first abdominal segment if fused with the **thorax** to form a **mesosoma** (or **alitrunk** in ants) (see Taxobox 29).

proprioceptors Sense organs that respond to the position of body appendages or organs.

prothoracic glands The thoracic or cephalic glands (Fig. 3.8) that secrete **ecdysteroids** (Fig. 5.13).

prothoracicotropic hormone (PTTH) A neuropeptide **hormone** secreted by the brain that controls aspects of **moulting** and **metamorphosis** via action on the **corpora cardiaca**.

prothorax The first segment of the **thorax** (Fig. 2.20).

protocerebrum The anterior part of the insect **brain**, the **ganglia** of the first segment, comprising the ocular and associative centres.

protrusible vesicle (exsertile vesicle) A small sac or bladder, capable of being extended or protruded.

proventriculus (gizzard) The grinding organ of the **foregut** (Fig. 3.1 & Fig. 3.13).

proximal Describing the part of an appendage closer to or at the body (opposite to **distal**) (Fig. 2.8).

pseudergate In "lower" termites, the equivalent to the worker caste, comprising immature nymphs or undifferentiated larvae.

pseudocopulation The attempted copulation of an insect with a flower.

pseudoplacental viviparity Viviparity (producing live offspring) in which a **microlecithal** egg develops via nourishment from a presumed placenta.

pseudotrachea A ridged groove on the ventral surface of the **labellum** of some higher Diptera (Fig. 2.14a), used to uptake liquid food.

psychophily Plant pollination by butterflies.

pterostigma A pigmented (and denser) spot near the anterior margin of the fore and sometimes hind wings (Fig. 2.22, Fig. 2.23 & Fig. 2.24b).

pterothorax The enlarged second and third segments of the **thorax** bearing the wings in pterygotes.

ptilinum A sac everted from a fissure between the antennae of schizophoran flies (Diptera) that aids **puparium** fracture at emergence.

pubescent (*adj.*) Clothed in fine short **setae**.

puddling The action of drinking from pools, especially evident in butterflies, to obtain scarce salts.

pulvillus (*pl.* pulvilli) A bladder-like appendage of the **pretarsus** (Fig. 2.21).

pupa (*adj.* pupal) The term for an insect when it is in the inactive stage between the **larva** and adult in holometabolous insects; also termed a chrysalis in butterflies.

pupal paedogenesis Reproduction in the pupal stage (known in some Diptera).

puparium The hardened skin of the final-instar larva (in Strepsiptera and some brachyceran Diptera), in which the **pupa** forms, or the last nymphal instar in whiteflies (Aleyrodoidea).

pupation Becoming a **pupa**.

pylorus The anterior **hindgut** where the **Malpighian tubules** enter, sometimes indicated by a muscular valve.

pyramids (in genetically modified crops) A variety of transgenic plant (usually a Bt crop) engineered to produce two or more insecticidal toxins, as a strategy to delay the evolution of resistance in the pest insects.

pyrophily (*adj.* pyrophilous) Loving fire; thriving in a habitat that recently has been burnt.

quarantine A strict isolation imposed to prevent the spread or entry of disease or harmful organisms.

quasisocial Social behaviour in which individuals of the same generation co-operate and nest-share without division of labour.

queen A female belonging to the reproductive caste in **eusocial** or **semisocial** insects (Fig. 12.8), called a **gyne** in social Hymenoptera.

quiescence A slowing down of metabolism and development in response to adverse environmental conditions; *see also* **diapause**.

radius In wing venation, the fourth longitudinal **vein**, posterior to the **subcosta**; with a maximum of five branches R_{1-5} (Fig. 2.23).

rake (of bees) The process on the distal apex of the **tibia** of a **worker** bee that gathers pollen into the **press** (Fig. 12.4).

rank The classificatory level in a taxonomic hierarchy, for example species, genus, family, order.

raptorial Adapted for capturing prey by grasping.

rectal pad Thickened sections of the epithelium of the rectum involved in water uptake from the faeces (Fig. 3.17 & Fig. 3.18).

rectum (*adj.* rectal) The posterior part of the **hindgut** (Fig. 3.1 & Fig. 3.13).

reflex A simple response to a simple stimulus.

refractory period (1) The time interval during which a nerve will not initiate another impulse. (2) In reproduction, the period during which a mated female will not re-mate.

refuge A safe place, as in a refuge from parasitoids.

regulatory sequence A region of DNA or RNA that regulates the expression of genes.

releaser A particular stimulus whose signal stimulates a specific behaviour.

remigium The anterior part of the wing, usually more rigid than the posterior **clavus** and with more **veins** (Fig. 2.22).

reniform Kidney-shaped.

replete A specialized worker ant that is distended by liquid food and serves as a food store for the colony.

replicate (of disease organisms) To increase in numbers.

reservoir (of diseases) The natural host and geographical range.

resilin A rubber-like or elastic protein in some insect **cuticles**.

resistance The ability to withstand (e.g. temperature extremes, insecticides, insect attack).

resource tracking A relationship, for example between parasite and host or plant and pollinator, in which the evolution of the association is based on ecology rather than phylogeny; *see also* **coevolution**, **phyletic tracking**.

respiration (1) A metabolic process in which substrates (food) are converted to useful energy in cells, usually by oxidation (aerobic respiration). (2) Used inappropriately to mean breathing (**gas exchange**), as through spiracles or gas exchange across a thin cuticle.

retinaculum (1) The specialized hooks or scales on the base of the fore wing that lock with the **frenulum** of lepidopteran hind wings in flight. (2) The retaining hook of the springtail **furca** (spring) (see Taxobox 1).

retinula cell A nerve cell of the light receptor organs (**ommatidia**, **stemma** or **ocellus**) comprising a **rhabdom** of several **rhabdomeres** and connected by nerve axons to the optic lobe of the brain (Fig. 4.9).

rhabdom The central zone of the retinula consisting of microvilli filled with visual pigment; comprising **rhabdomeres** belonging to several different **retinula cells** (Fig. 4.10).

rhabdomere One of the seven or eight units comprising a **rhabdom**, the inner part of a **retinula cell** (Fig. 4.10).

rheophilic Liking running water.

rhizosphere A zone surrounding the roots of plants, usually richer in fungi and bacteria than elsewhere in the soil.

RIDL The **r**elease of **i**nsects carrying a **d**ominant **l**ethal marker that causes the death of their progeny; an advanced derivative of the **sterile insect technique**.

riparian Associated with or relating to the waterside.

river continuum concept A formulation of the idea that energy inputs into a river are **allochthonous** in the upper parts and increasingly **autochthonous** in the lower reaches.

RNA interference (RNAi) A process in which RNA molecules inhibit gene expression, typically by causing the destruction of specific messenger RNA (mRNA) molecules; sometimes called post-transcriptional gene silencing.

royalactin A protein found in royal jelly that induces **queen** development in honey-bee larvae by activating multiple changes, including an increased titre of **juvenile hormone**.

rostrum A snout-like projection of the head formed of the mouthparts (see Taxobox 20) or bearing the mouthparts at the end (as in weevils); *see also* **proboscis**.

salivarium (salivary reservoir) The cavity into which the **salivary gland** opens, between the **hypopharynx** and the **labium** (Fig. 3.1 & Fig. 3.14).

salivary gland The gland that produces saliva (Fig. 3.1).

saltatorial Adapted for jumping or springing.

saprophage (*adj.* saprophagous) An eater of decaying organisms.

sarcolemma The outer sheath of a striated muscle fibre.

scale A flattened **seta**; a unicellular outgrowth of the **cuticle**.

scape The first segment of the **antenna** (Fig. 2.10).

sclerite A plate on the body wall surrounded by membrane or sutures.

sclerophyllous (*n.* sclerophylly) (of plants) Bearing tough/hard leaves.

sclerotization Stiffening of the **cuticle** by cross-linkage of protein chains.

scolopale cell In a **chordotonal organ**, the **sheath cell** that envelops the **dendrite** (Fig. 4.3).

scolopidium (*pl.* scolopidia) In a **chordotonal organ**, the combination of three cells, the **cap cell**, **scolopale cell** and **dendrite** (Fig. 4.3).

scopa A brush of thick hair on the hind **tibia** of adult bees.

scraper The ridged surface drawn over a **file** to produce stridulatory sounds.

scutellum The posterior third of the **alinotum** (either meso- or metanotum), lying behind the **scutum** (Fig. 2.20).

scutum The middle third of the **alinotum** (either meso- or metanotum), in front of the **scutellum** (Fig. 2.20).

secondary cycle In a disease, the cycle that involves an atypical host; *see also* **primary cycle**.

secondary pest outbreak Previously harmless or very minor pest species becoming pests, usually following insecticide treatment for a primary pest.

secondary plant compound A plant chemical usually produced for defensive purposes since it is not used for normal plant growth and reproduction; *see also* **allelochemical**.

secondary segmentation Any segmentation that fails to match the embryonic segmentation; more specifically, the insect external skeleton in which each apparent segment includes the posterior (intersegmental) parts of the primary segment preceding it (Fig. 2.7).

sector A major wing **vein** branch and all of its subdivisions.

semi-aquatic Living in saturated soils, but not immersed in free water.

seminal vesicle Male sperm storage organ (Fig. 3.20b).

semiochemical Any chemical used in intra- and interspecific communication.

semisocial Describing social behaviour in which individuals of the same generation co-operate and nest-share with some division of reproductive labour.

semivoltine Having a life cycle (generation time) of greater than one year; *see also* **bivoltine**, **multivoltine**, **univoltine**.

sensillum (*pl.* sensilla) A sense organ, either simple and isolated, or part of a more complex organ.

sensory neuron A nerve cell that receives and transmits stimuli from the environment (Fig. 3.5).

serial homology The occurrence of identically derived features on different segments (e.g. the pair of legs on each thoracic segment).

serosa The membrane covering the embryo (Fig. 6.5).

seta (*pl.* setae) A cuticular extension, a **trichoid sensillum**; also called a **hair** or **macrotrichium**.

sexuals (sexuales) Sexually reproductive aphids (Aphidoidea) of either sex: oviparae (females) and males.

sexupara (*pl.* sexuparae) An **alate** parthenogenetic female aphid that produces both sexes of offspring.

sibling Full brother or sister.

single-cycle disease A disease involving one species of **host**, one **parasite** and one insect **vector**.

siphunculus (*pl.* siphunculi) *See* **cornicle**.

sister group The closest related group of the same taxonomic rank as the group under study.

smell The olfactory sense, the detection of airborne chemicals.

social parasitism The phenomenon of one social insect parasitizing the colony of another social insect.

sociality The condition of living in an organized community.

soldier In social insects, an individual worker belonging to a **caste** involved in colony defence; *see also* **major worker**.

solitary Non-colonial, occurring singly or in pairs.

species (*adj.* specific) In sexually reproducing organisms, a group of all individuals that can interbreed, mating within the group (sharing a gene pool) and producing fertile offspring, usually similar in appearance and behaviour (but see **polymorphic**) and sharing a common evolutionary history.

specific coevolution (pair-wise coevolution) Concerted evolutionary change that takes place between two interacting species, in which the evolution of a trait in one leads to reciprocal development of a trait in a second organism in a feature that evolved initially in response to a trait of the first species; *see also* **guild coevolution**.

sperm competition In multi-mated females, the syndrome by which sperm from one mating compete with sperm of other males to fertilize the eggs.

sperm precedence The preferential use by the female of the sperm of one mating (one male) over others.

spermatheca The female receptacle for sperm deposited during mating (Fig. 3.20a).

spermathecal gland A tubular gland off the **spermatheca**, producing nourishment for sperm stored in the spermatheca (Fig. 3.20a).

spermatophore An encapsulated package of sperm (spermatozoa) (Fig. 5.6).

spermatophylax In katydids, a proteinaceous part of the **spermatophore** eaten by the female after mating (Box 5.2).

sphecophily Plant pollination by wasps.

spina An internal **apodeme** of the intersegmental sternal plate called the **spinasternum**.

spinasternum An **intersternite** bearing a **spina** (Fig. 2.20), sometimes fused with the **eusterna** of the **prothorax** and **mesothorax**, but never the **metathorax**.

spine A multicellular and unjointed cuticular extension, often thorn-like (Fig. 2.6a).

spiracle (*adj.* spiracular) An external opening of the **tracheal system** (Fig. 3.10a).

spur An articulated **spine**.

stadium (*pl.* stadia) The period between moults, the **instar** duration or intermoult period.

startle display A display made by some **cryptic** insects upon discovery, involving exposure of a startling colour or pattern, such as eyespots.

statary The sedentary, stationary phase of the life of army ants.

stem-group The paraphyletic group of taxa close to but outside a particular **crown-group**; all members of a stem-group are extinct.

stemma (*pl.* stemmata) The "simple" eye of many larval insects, sometimes aggregated into a more complex visual organ.

stenogastrous Having a shortened or narrow **abdomen**; *see also* **physogastry**.

sterile insect technique (SIT) A means of controlling insects by swamping populations with large numbers of artificially sterilized males or sometimes females (once called the "sterile male technique" as it was applied originally just to males).

sternellum The small **sclerite** of the **eusternum**, lying posterior to the **basisternum** (Fig. 2.20).

sternite The diminutive of **sternum**; a subdivision of a sternum.

sternum (*pl.* sterna, *adj.* sternal, *dim.* sternite) The ventral surface of a segment (Fig. 2.7).

stipes The distal part of the **maxilla**, bearing a **galea**, a **lacinia** and a **maxillary palp** (Fig. 2.10).

stomodeum (*adj.* stomodeal) *See* **foregut**.

striated muscle Muscles in which myosin and actin filaments overlap to give a striated effect.

stridulation The production of sound by rubbing two rough or ridged surfaces together.

style In apterygote insects, small appendages on abdominal segments, homologous to abdominal legs.

stylet One of the elongate parts of piercing–sucking mouthparts (Fig. 2.15, Fig. 2.16 & Fig. 11.4), a needle-like structure.

subalare (*pl.* subalaria) A small **sclerite**, one of the **epipleurites** that lies posterior to the **pleural wing process**, forming an attachment for the **direct flight muscles** (Fig. 2.20).

subcosta In wing venation, the third longitudinal **vein**, posterior to the **costa** (Fig. 2.23).

subgenual organ A **chordotonal organ** on the proximal **tibia** that detects substrate vibration (Fig. 4.4).

subimaginal instar *See* **subimago**.

subimago (subimaginal instar) In Ephemeroptera, the winged penultimate instar; subadult.

suboesophageal ganglion The fused **ganglia** of the mandibular, maxillary and labial segments, forming a ganglionic centre beneath the **oesophagus** (Fig. 3.6 & Fig. 3.14).

subsocial Describing a social system in which adults look after immature stages for a certain period.

substitutional ovipositor An **ovipositor** formed from extensible posterior abdominal segments; *see also* **appendicular ovipositor**.

succession In ecology, the observed changes in species composition and abundance in an assemblage or community over time.

superlingua (*pl.* superlinguae) A lateral lobe of the **hypopharynx** (Fig. 2.10), the remnant of a leg appendage of the third **head** segment.

superparasitism The occurrence of more **parasitoids** within a **host** than can complete their development within the host.

superposition eye *See* **optical superposition eye**.

supplementary reproductive In termites, a potential replacement reproductive within its natal (birth) nest, which does not become **alate**; also called a **neotenic** or **ergatoid**.

supraoesophageal ganglion *See* **brain**.

suture An external groove that may show the fusion of two plates (**sclerites**) (Fig. 2.10).

swarm An aggregation of insects, often aerial, for the purposes of mating.

swarming The behaviour of forming aggregations for the purpose of mating.

symbiont An organism that lives in **symbiosis** with another.

symbiosis A long-lasting, close and dependent relationship between organisms of two different species, often of different kingdoms (such as insects and bacteria or protists).

sympatric Describing overlapping geographical distributions of organisms or taxa; *see also* **allopatric**.

synanthropic Associated with humans or their dwellings.

synapomorphy A shared derived state (condition) of a character.

synapse The site of approximation of two nerve cells at which they may communicate.

synchronous muscle A muscle that contracts once per nerve impulse.

syncytium (*adj.* syncytial) A multinucleate tissue without cell division.

synergism The enhancement of the effects of two substances that is greater than the sum of their individual effects.

synomone A communication chemical that benefits both the receiver and the producer; *see also* **allomone**, **kairomone**.

systematics The science of biological classification and diversity, i.e. taxonomy plus phylogenetics.

systemic insecticide An insecticide taken into the body of a host (plant or animal) that kills insects feeding on the host.

taenidium (*pl.* taenidia) The spiral thickening in the wall of a **trachea** that prevents collapse.

tagma (*pl.* tagmata) A group of segments that forms a major body unit (**head**, **thorax**, **abdomen**).

tagmosis The organization of the body into major units (**head**, **thorax**, **abdomen**).

tapetum A reflective layer at the back of the eye formed from small **tracheae**.

target-site resistance Increased tolerance by an insect to an insecticide through reduced sensitivity at the target site of the chemical.

tarsomere A subdivision of the **tarsus** (Fig. 2.21).

tarsus (*pl.* tarsi) The leg segment distal to the **tibia**, comprising one to five **tarsomeres** and apically bearing the **pretarsus** (Fig. 2.21).

taste Chemoreception of chemicals in a liquid dissolved form.

taxis (*pl.* taxes) An orientated movement of an organism.

taxon (*pl.* taxa) A taxonomic unit (species, genus, family, phylum, etc.).

taxonomy (*adj.* taxonomic) The theory and practice of describing, naming and classifying organisms.

tegmen (*pl.* tegmina) A leathery, hardened fore wing (Fig. 2.24c).

tegula One of the articular **sclerites** of the neopteran wing, lying at the base of the **costa** (Fig. 2.23); *see also* **axillary sclerites**, **humeral plate**.

telotrophic (acrotrophic) ovariole An **ovariole** in which the nurse cells are only within the germarium; the nurse cells remain connected to the **oocytes** by long filaments as the oocytes move down the ovariole; *see also* **panoistic ovariole**, **polytrophic ovariole**.

teneral The condition of a newly eclosed (moulted) adult insect, which is unsclerotized and unpigmented.

tentorial arms The **apophyses** or internal struts that form the **head** endoskeleton, often fused to form the **tentorium**.

tentorium The endoskeletal cuticular invaginations of the **head**, including anterior and posterior **tentorial arms**.

tergite The diminutive of **tergum**; a subdivision of the tergum.

tergum (*pl.* terga, *adj.* tergal, *dim.* tergite) The dorsal surface of a segment (Fig. 2.7).

terminalia The terminal abdominal segments involved in the formation of the genitalia.

testis (*pl.* testes) One of (usually) a pair of male gonads (Fig. 3.1b & Fig. 3.20b).

thanatosis Feigning death.

thelytoky (thelytokous parthenogenesis) A form of **parthenogenesis** in which females develop from unfertilized diploid eggs.

thorax The middle of the three major divisions (**tagma**) of the body, comprising **prothorax**, **mesothorax** and **metathorax** (Fig. 2.20).

threshold The minimum level of stimulus required to initiate (release) a response.

tibia (*pl.* tibiae) The fourth leg segment (from the body), following the **femur** (Fig. 2.21).

tonofibrillae Fibrils of **cuticle** that connect a muscle to the **epidermis** (Fig. 3.2).

tormogen cell The socket-forming epidermal cell associated with a **seta** (Fig. 2.6 & Fig. 4.1).

trachea (*pl.* tracheae) A tubular element of the insect gas-exchange system, within which air moves (Fig. 3.10 & Fig. 3.11).

tracheal system The insect gas-exchange system, comprising **tracheae**, **tracheoles** and **spiracles** (Fig. 3.10 & Fig. 3.11); *see also* **closed tracheal system**, **open tracheal system**.

tracheole The fine tubules of the insect gas-exchange system that contact respiring tissue (Fig. 3.10b).

transgenic plants Plants containing genes introduced from another organism by genetic engineering.

transovarial transmission *See* **vertical transmission**.

transverse At right angles to the longitudinal axis.

traumatic insemination In some Heteroptera (Hemiptera), including bed bugs, unorthodox mating behaviour in which the male punctures the female's cuticle with the **penis (phallus)** to deposit sperm into the **haemocoel** instead of utilizing the female reproductive tract.

triad (*adj.* triadic) A triplet of long wing **veins** (paired main veins and an intercalated longitudinal vein).

trichogen cell A hair-forming epidermal cell associated with a **seta** (Fig. 2.6 & Fig. 4.1).

trichoid sensillum A hair-like cuticular projection; a **seta**, **hair** or **macrotrichium** (Fig. 2.6b, Fig. 3.5 & Fig. 4.1).

tritocerebrum The posterior (or posteroventral) paired lobes of the insect **brain**, the **ganglia** of the third segment, functioning in handling the signals from the body.

triungulin An active, dispersive first-instar larva of some insects (e.g. Strepsiptera and some Coleoptera), including many that undergo **heteromorphosis**.

trochanter The second leg segment, following the **coxa** (Fig. 2.21).

trochantin A small **sclerite** anterior to the **coxa** (Fig. 2.20 & Fig. 2.21).

troglobite (troglobiont) An obligate cave-dweller.

trophallaxis (oral = stomodeal trophallaxis, anal = proctodeal trophallaxis) In social and subsocial insects, the transfer of alimentary (gut) fluid from one individual to another; may be mutual or unidirectional.

trophamnion In **parasitoids**, the enveloping membrane, derived from the host's **haemolymph**, that surrounds the multiple individuals that arise from a single egg derived by **polyembryony**.

trophic (1) Relating to food. (2) Describing an egg of a social insect that is degenerate and used in feeding other members of the colony.

trophic cascade The ecosystem-wide effects of the removal or introduction of predators on primary production via herbivores.

trophocyte (adipocyte) The dominant metabolic and storage cell of the **fat body**.

trophogenesis (*adj.* trophogenic) In social insects, the determination of **caste** type by differential feeding of the immature stages (in contrast to genetic determination of caste).

trypanosomiasis A disease caused by *Trypanosoma* protozoans (protists), transmitted to humans predominantly by reduviid bugs (Chagas' disease) or tsetse flies (sleeping sickness).

tymbal (tymbal organ) (*Am.* timbal) A stretched elastic membrane capable of sound production when flexed.

tympanum (*pl.* tympana) **(tympanal organ)** Any organ sensitive to vibration, comprising a tympanic membrane (thin cuticle), an air sac and a sensory **chordotonal organ** attached to the tympanic membrane (Fig. 4.4).

umbrella effect The protection provided to all species in a habitat due to the conservation of one important species, often either a **flagship** or a **keystone species**.

unguis (*pl.* ungues) A **claw** (Fig. 2.21).

unguitractor plate The ventral **sclerite** of the **pretarsus** that articulates with the **claws** (Fig. 2.21).

uniporous Having a single opening.

univoltine Having one generation in one year; *see also* **bivoltine**, **multivoltine**, **semivoltine**.

uric acid The main nitrogenous excretion product, $C_5H_4N_4O_3$ (Fig. 3.19).

uricotelism An excretory system based on **uric acid** excretion.
urocyte (urate cell) A cell that acts as a temporary store for urate excretion products.
vagility The propensity to move or disperse.
vagina A pouch-like or tubular genital chamber of the female genitalia.
valve (1) Generally, any unidirectional opening flap or lid. (2) In female genitalia, the blade-like structures comprising the ovipositor shaft (also called **gonapophysis**) (Fig. 2.25b).
valvifer In female insect genitalia, derivations of **gonocoxites** 8 and 9, supporting the valves of the ovipositor (Fig. 2.25b).
vannus The **anal area** of the wing anterior to the **jugal area** (Fig. 2.22).
vas deferens (*pl.* vasa deferentia) One of the ducts that carry sperm from the testes (Fig. 3.20b).
vector Literally "a bearer", specifically a **host** of a disease that transmits the pathogen to another species of organism.
vectorial capacity A mathematical expression of the probability of disease transmission by a particular **vector**.
veins The chitinous, hollow tube-like structures that support and strengthen insect wings; the main veins extend from the base of the wing to the outer margin.
venter The lower surface of the body.
ventilate To pass air or oxygenated water over a gas-exchange surface.
ventral Towards or at the lower surface (Fig. 2.8).
ventral diaphragm A membrane lying horizontally above the nerve cord in the body cavity, separating the **perineural sinus** from the **perivisceral sinus** (Fig. 3.9).
ventral nerve cord The chain of ventral **ganglia**.
ventral tube (collophore) In Collembola, a ventral sucker (see Taxobox 1).
ventriculus The tubular part of the **midgut**, the main digestive section of the gut (Fig. 3.13).
vertex The top of the **head**, posterior to the **frons** (Fig. 2.9).
vertical transmission The transmission of microorganisms between generations via the eggs.
vesica *See* **endophallus**.
vicariance Division of the range of a species by an Earth history event (e.g. ocean or mountain formation).
visceral (sympathetic) nervous system The nerve system that innervates the gut, reproductive organs, and tracheal system.
vitellarium The structure within the **ovariole** in which **oocytes** develop and yolk is provided to them (Fig. 3.20a).
vitelline membrane The outer layer of an **oocyte**, surrounding the yolk (Fig. 5.10).
vitellogenesis The process by which **oocytes** grow by yolk deposition.
viviparity The bearing of live young (i.e. post-egg hatching) by the female; *see also* **adenotrophic viviparity**, **haemocoelous viviparity**, **oviparity**, **ovoviviparity**, **pseudoplacental viviparity**.
voltinism The number of generations or broods per year.
voucher (voucher specimen) A specimen identified or used as part of ecological, genetic or behavioural research and retained for future reference, preferably deposited in a recognized museum.
vulva The external opening of the copulatory pouch (**bursa copulatrix**) or vagina of the female genitalia (Fig. 3.20a).
Wasmannian mimicry A form of **mimicry** that allows an insect of another species to be accepted into a social insect colony.
wax A complex lipid mixture giving waterproofing to the **cuticle** or providing covering or building material.
wax canals Fine tubes that transport lipids (wax filaments) from the pore canals to the surface of the **epicuticle** (Fig. 2.1).
wax layer The lipid or waxy layer outside the **epicuticle** (Fig. 2.1).
wax mirrors Overlapping plates on the venter of the fourth to seventh abdominal segments of social bees that serve to direct the wax flakes that are produced beneath each mirror.
weed Any organism "in the wrong place", particularly used of plants away from their natural range or invading human monocultural crops.
worker In social insects, a member of the sterile **caste** that assists the reproductives.
xylophagy (*adj.* xylophagous) The dietary habit of consuming primarily or only wood.
zoocecidia (*sing.* zoocecidium) Plant **galls** induced by animals such as insects, mites and nematodes, as opposed to those formed by the plant response to microorganisms.

zoophilic Preferring other animals to humans, especially used of feeding preference of blood-feeding insects.

zygote A fertilized egg; the union of two gametes; in malaria parasites (*Plasmodium* spp.) resulting from fusion of a microgamete and macrogamete (Box 15.1).

FURTHER READING

Gordh, G. & Headrick, D. (2011) *A Dictionary of Entomology*, 2nd edn. CSIRO Publishing, Collingwood.

Nichols, S.W. (1989) *The Torre-Bueno Glossary of Entomology*, 2nd edn. The New York Entomological Society in co-operation with the American Museum of Natural History, New York.

REFERENCES

The following articles and books are the sources for figures and data cited in the text and figure and table captions. For additional information on topics covered in any chapter, please refer to the Further reading section at the end of each chapter.

Alcock, J. (1979) Selective mate choice by females of *Harpobittacus australis* (Mecoptera: Bittacidae). *Psyche* **86**, 213–17.

Alstein, M. (2003) Neuropeptides. In: *Encyclopedia of Insects* (eds V.H. Resh & R.T. Cardé), pp. 782–5. Academic Press, Amsterdam.

Ando, H. (ed.) (1982) *Biology of the Notoptera*. Kashiyo-Insatsu Co. Ltd, Nagano, Japan.

Anon. (1991) *Ladybirds and Lobsters, Scorpions and Centipedes*. British Museum (Natural History), London.

Askew, R.R. (1971) *Parasitic Insects*. Heinemann, London.

Atkins, M.D. (1980) *Introduction to Insect Behaviour*. Macmillan, New York.

Austin, A.D. & Browning, T.O. (1981) A mechanism for movement of eggs along insect ovipositors. *International Journal of Insect Morphology and Embryology* **10**, 93–108.

Badonnel, A. (1951) Ordre des Psocoptères. In: *Traité de Zoologie: Anatomie, Systématique, Biologie*. Tome X. *Insectes Supérieurs et Hémiptéroïdes*, Fascicule II (ed. P.-P. Grassé), pp. 1301–40. Masson, Paris.

Bandsma, A.T. & Brandt, R.T. (1963) *The Amazing World of Insects*. George Allen & Unwin, London.

Bartell, R.J., Shorey, H.H. & Barton Browne, L. (1969) Pheromonal stimulation of the sexual activity of males of the sheep blowfly *Lucilia cuprina* (Calliphoridae) by the female. *Animal Behaviour* **17**, 576–85.

Barton Browne, L., Smith, P.H., van Gerwen, A.C.M. & Gillott, C. (1990) Quantitative aspects of the effect of mating on readiness to lay in the Australian sheep blowfly, *Lucilia cuprina*. *Journal of Insect Behavior* **3**, 637–46.

Bar-Zeev, M. (1958) The effect of temperature on the growth rate and survival of the immature stages of *Aedes aegypti* (L.). *Bulletin of Entomological Research* **49**, 157–63.

Beard, J. (1989) Viral protein knocks the guts out of caterpillars. *New Scientist* **124**(1696–7), 21.

Beccaloni, G. & Eggleton, P. (2013) Order Blattodea. In: *Animal Biodiversity: An Outline of Higher-level Classification and Survey of Taxonomic Richness* (Addenda 2013), (ed. Z.-Q. Zhang), pp. 46–8. *Zootaxa* **3703**(1), 1–82.

Beccari, O. (1877) Piante nuove o rare dell'Arcipelago Malese e della Nuova Guinea, raccolte, descritte ed illustrate da O. Beccari. *Malesia (Genova)* **1**, 167–92.

Bellows, T.S. Jr, Perring, T.M., Gill, R.J. & Headrick, D.H. (1994) Description of a species of *Bemisia* (Homoptera: Aleyrodidae). *Annals of the Entomological Society of America* **87**, 195–206.

Belwood, J.J. (1990) Anti-predator defences and ecology of neotropical forest katydids, especially the Pseudophyllinae. In: *The Tettigoniidae: Biology, Systematics and Evolution* (eds W.J. Bailey & D.C.F. Rentz), pp. 8–26. Crawford House Press, Bathurst.

Bennet-Clark, H.C. (1989) Songs and the physics of sound production. In: *Cricket Behavior and Neurobiology* (eds F. Huber, T.E. Moore & W. Loher), pp. 227–61. Comstock Publishing Associates (Cornell University Press), Ithaca, NY.

Binnington, K.C. (1993) Ultrastructure of the attachment of *Serratia entomophila* to scarab larval cuticle and a review of nomenclature for insect epicuticular layers. *International Journal of Insect Morphology and Embryology* **22**(2–4), 145–55.

The Insects: An Outline of Entomology, Fifth Edition. P.J. Gullan and P.S. Cranston.

Companion Website: www.wiley.com/go/gullan/insects

Birch, M.C. & Haynes, K.F. (1982) *Insect Pheromones*. Studies in Biology no. 147. Edward Arnold, London.

Black Soldier Fly Blog (2012) http://blacksoldierflyblog.com/ [Accessed 6 June 2013.]

Blaney, W.M. (1976) *How Insects Live*. Elsevier-Phaidon, Oxford.

Bonhag, P.F. & Wick, J.R. (1953) The functional anatomy of the male and female reproductive systems of the milkweed bug, *Oncopeltus fasciatus* (Dallas) (Heteroptera: Lygaeidae). *Journal of Morphology* **93**, 177–283.

Bornemissza, G.F. (1957) An analysis of arthropod succession in carrion and the effect of its decomposition on the soil fauna. *Australian Journal of Zoology* **5**, 1–12.

Borror, D.J., Triplehorn, C.A. & Johnson, N.F. (1989) *An Introduction to the Study of Insects*, 6th edn. Saunders College Publishing, Philadelphia, PA.

Boulard, M. (1968) Description de cinq Membracides nouveaux du genre *Hamma* accompagnée de précisions sur *H. rectum*. *Annales de la Societé Entomologique de France (N.S.)* **4**(4), 937–50.

Brackenbury, J. (1990) Origami in the insect world. *Australian Natural History* **23**(7), 562–9.

Bradley, T.J. (1985) The excretory system: structure and physiology. In: *Comprehensive Insect Physiology, Biochemistry, and Pharmacology*, Vol. 4: *Regulation. Digestion, Nutrition, Excretion* (eds G.A. Kerkut & L.I. Gilbert), pp. 421–65. Pergamon Press, Oxford.

Brandt, M. & Mahsberg, D. (2002) Bugs with a backpack: the function of nymphal camouflage in the West African assassin bugs: *Paredocla* and *Acanthiaspis* spp. *Animal Behaviour* **63**, 277–84.

Brower, J.V.Z. (1958) Experimental studies of mimicry in some North American butterflies. Part III. *Danaus gilippus berenice and Limenitis archippus floridensis*. *Evolution* **12**, 273–85.

Brower, L.P., Brower, J.V.Z. & Cranston, F.P. (1965) Courtship behavior of the queen butterfly, *Danaus gilippus berenice* (Cramer). *Zoologica* **50**, 1–39.

Burton, M. & Burton, R. (1975) *Encyclopedia of Insects and Arachnids*. Octopus Books, London.

Calder, A.A. & Sands, D.P.A. (1985) A new Brazilian *Cyrtobagous* Hustache (Coleoptera: Curculionidae) introduced into Australia to control salvinia. *Journal of the Entomological Society of Australia* **24**, 57–64.

Cappaert, D., McCullough, D.G., Poland, T.M. & Siegert, N.W. (2005) Emerald ash borer in North America: a research and regulatory challenge. *American Entomologist* **51**, 152–65.

Carroll, S.B. (1995) Homeotic genes and the evolution of arthropods and chordates. *Nature* **376**, 479–85.

Carroll, S.B. (2008) Evo-devo and an expanding evolutionary synthesis: a genetic theory of morphological evolution. *Cell* **134**, 25–36.

Caudell, A.N. (1920) Zoraptera not an apterous order. *Proceedings of the Entomological Society of Washington* **22**, 84–97.

Chapman, R.F. (1982) *The Insects. Structure and Function*, 3rd edn. Hodder and Stoughton, London.

Chapman, R.F. (1991) General anatomy and function. In: *The Insects of Australia*, 2nd edn (CSIRO), pp. 33–67. Melbourne University Press, Carlton.

Chapman, R.F. (2013) *The Insects. Structure and Function*, 5th edn (eds S.J. Simpson & A.E. Douglas). Cambridge University Press, Cambridge.

Cherikoff, V. & Isaacs, J. (1989) *The Bush Food Handbook*. Ti Tree Press, Balmain.

Chu, H.F. (1949) *How to Know the Immature Insects*. William C. Brown, Dubuque, IA.

Clements, A.N. (1992) *The Biology of Mosquitoes*, Vol. 1: *Development, Nutrition and Reproduction*. Chapman & Hall, London.

Cohen, A.C., Chu, C.-C., Henneberry, T.J. *et al.* (1998) Feeding biology of the silverleaf whitefly (Homoptera: Aleyrodidae). *Chinese Journal of Entomology* **18**, 65–82.

Cohen, E. (1991) Chitin biochemistry. In: *Physiology of the Insect Epidermis* (eds K. Binnington & A. Retnakaran), pp. 94–112. CSIRO Publications, Melbourne.

Common, I.F.B. (1990) *Moths of Australia*. Melbourne University Press, Carlton.

Common, I.F.B. & Waterhouse, D.F. (1972) *Butterflies of Australia*. Angus & Robertson, Sydney.

Cornwell, P.B. (1968) *The Cockroach*. Hutchinson, London.

Cox, J.M. (1987) Pseudococcidae (Insecta: Hemiptera). *Fauna of New Zealand* **11**, 1–228.

Coyne, J.A. (1983) Genetic differences in genital morphology among three sibling species of *Drosophila*. *Evolution* **37**, 1101–17.

Cryan, J.R. & Urban, J.M. (2012) Higher-level phylogeny of the insect order Hemiptera: is Auchenorrhyncha really paraphyletic? *Systematic Entomology* **37**, 7–21.

CSIRO (1970) *The Insects of Australia*, 1st edn. Melbourne University Press, Carlton.

CSIRO (1991) *The Insects of Australia*, 2nd edn. Melbourne University Press, Carlton.

Currie, D.C. (1986) An annotated list of and keys to the immature black flies of Alberta (Diptera: Simuliidae). *Memoirs of the Entomological Society of Canada* **134**, 1–90.

Daly, H.V., Doyen, J.T. & Ehrlich, P.R. (1978) *Introduction to Insect Biology and Diversity*. McGraw-Hill, New York.

Danforth, B.N., Cardinal, S., Praz, C., *et al.* (2013) The impact of molecular data on our understanding of bee phylogeny and evolution. *Annual Review of Entomology* **58**, 57–78.

Darlington, A. (1975) *The Pocket Encyclopaedia of Plant Galls in Colour*, 2nd edn. Blandford Press, Dorset.

Dean, J., Aneshansley, D.J., Edgerton, H.E. & Eisner, T. (1990) Defensive spray of the bombardier beetle: a biological pulse jet. *Science* **248**, 1219–21.

Debevec, A.H., Cardinal, S. & Danforth, B.N. (2012) Identifying the sister group to the bees: a molecular phylogeny of Aculeata with an emphasis on the superfamily Apoidea. *Zoological Scripta* **41**, 527–35.

De Klerk, C.A., Ben-Dov, Y. & Giliomee, J.H. (1982) Redescriptions of four vine infesting species of *Margarodes* Guilding (Homoptera: Coccoidea: Margarodidae) from South Africa. *Phytophylactica* **14**, 61–76.

Deligne, J., Quennedey, A. & Blum, M.S. (1981) The enemies and defence mechanisms of termites. In: *Social Insects*, Vol. II (ed. H.R. Hermann), pp. 1–76. Academic Press, New York.

Devitt, J. (1989) Honeyants: a desert delicacy. *Australian Natural History* **22**(12), 588–95.

Deyrup, M. (1981) Deadwood decomposers. *Natural History* **90**(3), 84–91.

Djernæs, M., Klass, K.-D., Picker, M.D. & Damgaard, J. (2012) Phylogeny of cockroaches (Insecta, Dictyoptera, Blattodea), with placement of aberrant taxa and exploration of out-group sampling. *Systematic Entomology* **37**, 65–83.

Dodson, G. (1989) The horny antics of antlered flies. *Australian Natural History* **22**(12), 604–11.

Dodson, G.N. (1997) Resource defence mating system in antlered flies, *Phytalmia* spp. (Diptera: Tephritidae). *Annals of the Entomological Society of America* **90**, 496–504.

Dolling, W.R. (1991) *The Hemiptera*. Natural History Museum Publications, Oxford University Press, Oxford.

Dow, J.A.T. (1986) Insect midgut function. *Advances in Insect Physiology* **19**, 187–328.

Downes, J.A. (1970) The feeding and mating behaviour of the specialized Empidinae (Diptera); observations on four species of *Rhamphomyia* in the high Arctic and a general discussion. *Canadian Entomologist* **102**, 769–91.

Duffy, E.A.J. (1963) *A Monograph of the Immature Stages of Australasian Timber Beetles (Cerambycidae)*. British Museum (Natural History), London.

Eastham, L.E.S. & Eassa, Y.E.E. (1955) The feeding mechanism of the butterfly *Pieris brassicae* L. *Philosophical Transactions of the Royal Society of London B* **239**, 1–43.

Eberhard, W.G. (1985) *Sexual Selection and Animal Genitalia*. Harvard University Press, Cambridge, MA.

Edwards, D.S. (1994) *Belalong: a Tropical Rainforest*. The Royal Geographical Society, London, and Sun Tree Publishing, Singapore.

Eibl-Eibesfeldt, I. & Eibl-Eibesfeldt, E. (1967) Das Parasitenabwehren der Minima-Arbeiterinnen der Blattschneider-Ameise (*Atta cephalotes*). *Zeitschrift für Tierpsychologie* **24**, 278–81.

Eidmann, H. (1929) Morphologische und physiologische Untersuchungen am weiblichen Genitalapparat der Lepidopteren. I. Morphologischer Teil. *Zeitschrift für Angewandte Entomologie* **15**, 1–66.

Eisenbeis, G. & Wichard, W. (1987) *Atlas on the Biology of Soil Arthropods*, 2nd edn. Springer-Verlag, Berlin.

Eisner, T., Smedley, S.R., Young, D.K., *et al.* (1996a) Chemical basis of courtship in a beetle (*Neopyrochroa flabellata*): cantharidin as precopulatory "enticing agent". *Proceedings of the National Academy of Sciences of the USA* **93**, 6494–8.

Eisner, T., Smedley, S.R., Young, D.K., *et al.* (1996b) Chemical basis of courtship in a beetle (*Neopyrochroa flabellata*): cantharidin as "nuptial gift". *Proceedings of the National Academy of Sciences of the USA* **93**, 6499–503.

Elliott, J.M. & Humpesch, U.H. (1983) A key to the adults of the British Ephemeroptera. *Freshwater Biological Association Scientific Publication* **47**, 1–101.

Encalada, A.C. & Peckarsky, B.L. (2007) A comparative study of the costs of alternative mayfly oviposition behaviors. *Behavioral Ecology and Sociobiology* **61**, 1437–48.

Entwistle, P.F. & Evans, H.F. (1985) Viral control. In: *Comprehensive Insect Physiology, Biochemistry and Pharmacology*, Vol. 12: *Insect Control* (eds G.A. Kerkut & L.I. Gilbert), pp. 347–412. Pergamon Press, Oxford.

Evans, E.D. (1978) Megaloptera and aquatic Neuroptera. In: *An Introduction to the Aquatic Insects of North America* (eds R.W. Merritt & K.W. Cummins), pp. 133–45. Kendall/Hunt, Dubuque, IA.

Everaerts, C., Maekawa, K., Farine, J.P., *et al.* (2008) The *Cryptocercus punctulatus* species complex (Dictyoptera: Cryptocercidae) in the eastern United States: comparison of cuticular hydrocarbons, chromosome number and DNA sequences. *Molecular Phylogenetics and Evolution* **47**, 950–9.

Ferrar, P. (1987) *A Guide to the Breeding Habits and Immature Stages of Diptera Cyclorrhapha*. Pt. 2, Entomonograph Vol. 8. E.J. Brill, Leiden , and Scandinavian Science Press, Copenhagen.

Filshie, B.K. (1982) Fine structure of the cuticle of insects and other arthropods. In: *Insect Ultrastructure*, Vol. 1 (eds R.C. King & H. Akai), pp. 281–312. Plenum, New York.

Fjellberg, A. (1980) *Identification Keys to Norwegian Collembola*. Utgitt av Norsk Entomologisk Forening, Norway.

Foldi, I. (1983) Structure et fonctions des glandes tégumentaires de Cochenilles Pseudococcines et de leurs sécrétions. *Annales de la Societé Entomologique de France (N.S.)* **19**, 155–66.

Freeman, W.H. & Bracegirdle, B. (1971) *An Atlas of Invertebrate Structure*. Heinemann Educational Books, London.

Frisch, K. von (1967) *The Dance Language and Orientation of Bees*. The Belknap Press of Harvard University Press, Cambridge, MA.

Froggatt, W.W. (1907) *Australian Insects*. William Brooks Ltd, Sydney.

Frost, S.W. (1959) *Insect Life and Insect Natural History*, 2nd edn. Dover Publications, New York.

Fry, C.H., Fry, K. & Harris, A. (1992) *Kingfishers, Bee-Eaters and Rollers*. Christopher Helm, London.

Futuyma, D.J. (1986) *Evolutionary Biology*, 2nd edn. Sinauer Associates, Sunderland, MA.

Gäde, G., Hoffman, K.-H. & Spring, J.H. (1997) Hormonal regulation in insects: facts, gaps, and future directions. *Physiological Reviews* **77**, 963–1032.

Galil, J. & Eisikowitch, D. (1968) Pollination ecology of *Ficus sycomorus* in East Africa. *Ecology* **49**, 259–69.

Gardiner, B.G. (1998) Editorial. *The Linnean* **14**(3), 1–3.
Garnas, J.R., Hurley, B.P., Slippers, B. & Wingfield, M.J. (2012) Biological control of forest plantation pests in an interconnected world requires greater international focus. *International Journal of Pest Management* **58**, 211–23.
Gibbons, B. (1986) *Dragonflies and Damselflies of Britain and Northern Europe*. Country Life Books, Twickenham.
Gilbert, L.I. (ed.) (2012) *Insect Molecular Biology and Biochemistry*. Academic Press, London.
Gilbert, L.I., Iatrou, K. & Gill, S.S. (eds) (2005) *Comprehensive Molecular Insect Science*. Vols 1–6, *Control*. Elsevier Pergamon, Oxford.
Goettler, W., Kaltenpoth, M. & Herzner, G. & Strohm, E. (2007) Morphology and ultrastructure of a bacteria cultivation organ: the antennal glands of female European beewolves, *Philanthus triangulum* (Hymenoptera, Crabronidae). *Arthropod Structure & Development* **36**, 1–9.
Grandgirard, J., Hoddle, M.S., Petit, J.N., *et al.* (2009) Classical biological control of the glassy-winged sharpshooter, *Homalodisca vitripennis*, by the egg parasitoid *Gonatocerus ashmeadi* in the Society, Marquesas, and Austral Archipelagos of French Polynesia. *Biological Control* **48**, 155–63.
Gray, E.G. (1960) The fine structure of the insect ear. *Philosophical Transactions of the Royal Society of London B* **243**, 75–94.
Greany, P.D., Vinson, S.B. & Lewis, W.J. (1984) Insect parasitoids: finding new opportunities for biological control. *BioScience* **34**, 690–6.
Gregory, T.R. (2008) Understanding evolutionary trees. *Evolution: Education and Outreach* **1**, 121–37.
Grimaldi, D. & Engel, M.S. (2005) *Evolution of the Insects*. Cambridge University Press, Cambridge.
Grimstone, A.V., Mullinger, A.M. & Ramsay, J.A. (1968) Further studies on the rectal complex of the mealworm *Tenebrio molitor* L. (Coleoptera: Tenebrionidae). *Philosophical Transactions of the Royal Society B* **253**, 343–82.
Gutierrez, A.P. (1970) Studies on host selection and host specificity of the aphid hyperparasite *Charips victrix* (Hymenoptera: Cynipidae). 6. Description of sensory structures and a synopsis of host selection and host specificity. *Annals of the Entomological Society of America* **63**, 1705–9.
Gwynne, D.T. (1981) Sexual difference theory: Mormon crickets show role reversal in mate choice. *Science* **213**, 779–80.
Gwynne, D.T. (1990) The katydid spermatophore: evolution of a parental investment. In: *The Tettigoniidae: Biology, Systematics and Evolution* (eds W.J. Bailey & D.C.F. Rentz), pp. 27–40. Crawford House Press, Bathurst.
Hadley, N.F. (1986) The arthropod cuticle. *Scientific American* **255**(1), 98–106.
Hadlington, P. (1987) *Australian Termites and Other Common Timber Pests*. New South Wales University Press, Kensington.
Harris, W.V. (1971) *Termites: Their Recognition and Control*, 2nd edn. Longman, London.
Haynes, K.F. & Birch, M.C. (1985) The role of other pheromones, allomones and kairomones in the behavioural responses of insects. In: *Comprehensive Insect Physiology, Biochemistry, and Pharmacology*, Vol. 9: *Behaviour* (eds G.A. Kerkut & L.I. Gilbert), pp. 225–55. Pergamon Press, Oxford.
Hely, P.C., Pasfield, G. & Gellatley, J.G. (1982) *Insect Pests of Fruit and Vegetables in NSW*. Inkata Press, Melbourne.
Hepburn, H.R. (1985) Structure of the integument. In: *Comprehensive Insect Physiology, Biochemistry and Pharmacology*, Vol. 3: *Integument, Respiration and Circulation* (eds G.A. Kerkut & L.I. Gilbert), pp. 1–58. Pergamon Press, Oxford.
Hermann, H.R. & Blum, M.S. (1981) Defensive mechanisms in the social Hymenoptera. In: *Social Insects*, Vol. II (ed. H.R. Hermann), pp. 77–197. Academic Press, New York.
Herms, W.B. & James, M.T. (1961) *Medical Entomology*, 5th edn. Macmillan, New York.
Hines, H.M., Hunt, J.H., O'Connor, T.K., *et al.* (2007) Multigene phylogeny reveals eusociality evolved twice in vespid wasps. *Proceedings of the National Academy of Sciences* **104**, 3295–99.
Hölldobler, B. (1984) The wonderfully diverse ways of the ant. *National Geographic* **165**, 778–813.
Hölldobler, B. & Wilson, E.O. (1990) *The Ants*. Springer-Verlag, Berlin.
Holman, G.M., Nachman, R.J. & Wright, M.S. (1990) Insect neuropeptides. *Annual Review of Entomology* **35**, 201–17.
Horridge, G.A. (1965) Arthropoda: general anatomy. In: *Structure and Function in the Nervous Systems of Invertebrates*, Vol. II (eds T.H. Bullock & G.A. Horridge), pp. 801–964. W.H. Freeman, San Francisco, CA.
Hungerford, H.B. (1954) The genus *Rheumatobates* Bergroth (Hemiptera–Gerridae). *University of Kansas Science Bulletin* **36**, 529–88.
Huxley, J. & Kettlewell, H.B.D. (1965) *Charles Darwin and His World*. Thames & Hudson, London.
Imms, A.D. (1913) Contributions to a knowledge of the structure and biology of some Indian insects. II. On *Embia major. sp. nov.*, from the Himalayas. *Transactions of the Linnean Society of London* **11**, 167–95.
Jobling, B. (1976) On the fascicle of blood-sucking Diptera. *Journal of Natural History* **10**, 457–61.
Jochmann, R., Blanckenhorn, W.U., Bussière, L. *et al.* (2011) How to test nontarget effects of veterinary pharmaceutical residues in livestock dung in the field. *Integrated Environmental Assessment and Management* **7**, 287–96.
Johnson, B.R., Borowiec, M.L., Chiu, J.C., *et al.* (2013) Phylogenomics resolves evolutionary relationships among ants, bees, and wasps. *Current Biology* **23**(20), 2058–62.
Johnson, K.P., Yoshizawa, K. & Smith, V.S. (2004) Multiple origins of parasitism in lice. *Proceedings of the Royal Society of London B* **271**, 1771–6.

Johnson, W.T. & Lyon, H.H. (1991) *Insects that Feed on Trees and Shrubs*, 2nd edn. Comstock Publishing Associates of Cornell University Press, Ithaca, NY.

Juergens, N. (2013) The biological underpinnings of Namib Desert fairy circles. *Science* **339**, 1618–21.

Kaltenpoth, M., Göttler, W., Herzner, G. & Strohm, E. (2005) Symbiotic bacteria protect wasp larvae from fungal infection. *Current Biology* **15**, 475–9.

Karim, M.M. (2009) Black soldier flies mating.jpg http://commons.wikimedia.org/wiki/File:Black_soldier_flies_mating.jpg [Accessed 6 June 2013]

Katz, M., Despommier, D.D. & Gwadz, R.W. (1989) *Parasitic Diseases*, 2nd edn. Springer-Verlag, New York.

Keeley, L.L. & Hayes, T.K. (1987) Speculations on biotechnology applications for insect neuroendocrine research. *Insect Biochemistry* **17**, 639–61.

Kettle, D.S. (1984) *Medical and Veterinary Entomology*. Croom Helm, London.

Klopfstein, S., Vilhelmsen, L., Heraty, J.M., *et al.* (2013) The hymenopteran tree of life: evidence from protein-coding genes and objectively aligned ribosomal data. *PLoS One* **8**(8), e69344. doi: 10.1371/journal.pone.0069344

Kukalová, J. (1970) Revisional study of the order Palaeodictyoptera in the Upper Carboniferous shales of Commentry, France. Part III. *Psyche* **77**, 1–44.

Kukalová-Peck, J. (1991) Fossil history and the evolution of hexapod structures. In: *The Insects of Australia*, 2nd edn (CSIRO), pp. 141–79. Melbourne University Press, Carlton.

Labandeira, C.C. (1998) Plant–insect associations from the fossil record. *Geotimes*, September 1998.

Landsberg, J. & Ohmart, C. (1989) Levels of insect defoliation in forests: patterns and concepts. *Trends in Ecology and Evolution* **4**, 96–100.

Lane, R.P. & Crosskey, R.W. (eds) (1993) *Medical Insects and Arachnids*. Chapman & Hall, London.

Lewis, T. (1973) *Thrips: Their Biology, Ecology and Economic Importance*. Academic Press, London.

Lindauer, M. (1960) Time-compensated sun orientation in bees. *Cold Spring Harbor Symposia on Quantitative Biology* **25**, 371–7.

Lloyd, J.E. (1966) Studies on the flash communication system in *Photinus* fireflies. *University of Michigan Museum of Zoology, Miscellaneous Publications* **130**, 1–95.

Loudon, C. (1989) Tracheal hypertrophy in mealworms: design and plasticity in oxygen supply systems. *Journal of Experimental Biology* **147**, 217–35.

Lubbock, J. (1873) *Monograph of the Collembola and Thysanura*. The Ray Society, London.

Lüscher, M. (1961) Air-conditioned termite nests. *Scientific American* **205**(1), 138–45.

Lyal, C.H.C. (1986) Coevolutionary relationships of lice and their hosts: a test of Fahrenholz's Rule. In: *Coevolution and Systematics* (eds A.R. Stone & D.L. Hawksworth), pp. 77–91. Systematics Association, Oxford.

Mairson, A. (1993) America's beekeepers: hives for hire. *National Geographic* **183** (May), 72–93.

Majer, J. (1985) Recolonisation by ants of rehabilitated mineral sand mines on North Stradbroke Island, Queensland, with particular reference to seed removal. *Australian Journal of Ecology* **10**, 31–4.

Malcolm, S.B. (1990) Mimicry: status of a classical evolutionary paradigm. *Trends in Ecology and Evolution* **5**, 57–62.

Matsuda, R. (1965) Morphology and evolution of the insect head. *Memoirs of the American Entomological Institute* **4**, 1–334.

McAlpine, D.K. (1990) A new apterous micropezid fly (Diptera: Schizophora) from Western Australia. *Systematic Entomology* **15**, 81–6.

McAlpine, J.F. (ed.) (1981) *Manual of Nearctic Diptera*, Vol. 1. Monograph No. 27. Research Branch, Agriculture Canada, Ottawa.

McAlpine, J.F. (ed.) (1987) *Manual of Nearctic Diptera*, Vol. 2. Monograph No. 28. Research Branch, Agriculture Canada, Ottawa.

McIver, S.B. (1985) Mechanoreception. In: *Comprehensive Insect Physiology, Biochemistry, and Pharmacology*, Vol. 6: *Nervous System: Sensory* (eds G.A. Kerkut & L.I. Gilbert), pp. 71–132. Pergamon Press, Oxford.

Mercer, W.F. (1900) The development of the wings in the Lepidoptera. *New York Entomological Society* **8**, 1–20.

Merritt, R.W., Craig, D.A., Walker, E.D., *et al.* (1992) Interfacial feeding behavior and particle flow patterns of *Anopheles quadrimaculatus* larvae (Diptera: Culicidae). *Journal of Insect Behavior* **5**, 741–61.

Michelsen, A. & Larsen, O.N. (1985) Hearing and sound. In: *Comprehensive Insect Physiology, Biochemistry, and Pharmacology*, Vol. 6: *Nervous System: Sensory* (eds G.A. Kerkut & L.I. Gilbert), pp. 495–556. Pergamon Press, Oxford.

Michener, C.D. (1974) *The Social Behavior of Bees*. The Belknap Press of Harvard University Press, Cambridge, MA.

Miyazaki, M. (1987a) Morphology of aphids. In: *Aphids: Their Biology, Natural Enemies and Control*, Vol. 2A (eds A.K. Minks & P. Harrewijn), pp. 1–25. Elsevier, Amsterdam.

Miyazaki, M. (1987b) Forms and morphs of aphids. In: *Aphids: Their Biology, Natural Enemies and Control*, Vol. 2A (eds A.K. Minks & P. Harrewijn), pp. 27–50. Elsevier, Amsterdam.

Moczek, A. & Emlen, D.J. (2000) Male horn dimorphism in the scarab beetle, *Onthophagus taurus*: do alternative reproductive tactics favour alternative phenotypes? *Animal Behaviour* **59**, 459–66.

Monteith, S. (1990) Life inside an ant-plant. *Wildlife Australia* **27**(4), 5.

Mutanen, M., Wahlberg, N. & Kaila, L. (2010) Comprehensive gene and taxon coverage elucidates radiation patterns in moths and butterflies. *Proceedings of the Royal Society B* **277**, 2839–48.

Nagy, L. (1998) Changing patterns of gene regulation in the evolution of arthropod morphology. *American Zoologist* **38**, 818–28.

Nijhout, H.F., Davidowitz, G. & Roff, D.A. (2006) A quantitative analysis of the mechanism that controls body size in *Manduca sexta*. *Journal of Biology* **5**, 1–16.

Nosek, J. (1973) *The European Protura*. Muséum D'Histoire Naturelle, Geneva.

Novak, V.J.A. (1975) *Insect Hormones*. Chapman & Hall, London.

Oliveira, P.S. (1988) Ant-mimicry in some Brazilian salticid and clubionid spiders (Araneae: Salticidae, Clubionidae). *Biological Journal of the Linnean Society* **33**, 1–15.

Omland, K.E., Cook, L.G. & Crisp, M.D. (2008) Tree thinking for all biology: the problem with reading phylogenies as ladders of progress. *BioEssays* **30**, 854–67.

Palmer, M.A. (1914) Some notes on life history of ladybeetles. *Annals of the Entomological Society of America* **7**, 213–38.

Peckarsky, B.L., Encalada, A.C. & Macintosh, A.R. (2012) Why do vulnerable mayflies thrive in trout streams? *American Entomologist* **57**, 152–64.

Petit, J.N., Hoddle, M.S., Grandgirard, J., *et al.* (2008) Short-distance dispersal behavior and establishment of the parasitoid *Gonatocerus ashmeadi* (Hymenoptera: Mymaridae) in Tahiti: Implications for its use as a biological control agent against *Homalodisca vitripennis* (Hemiptera: Cicadellidae). *Biological Control* **45**, 344–52.

Petit, J.N., Hoddle, M.S., Grandgirard, J., *et al.* (2009) Successful spread of a biocontrol agent reveals a biosecurity failure: elucidating long distance invasion pathways for *Gonatocerus ashmeadi* in French Polynesia. *BioControl* **54**, 485–95.

Pivnick, K.A. & McNeil, J.N. (1987) Puddling in butterflies: sodium affects reproductive success in *Thymelicus lineola*. *Physiological Entomology* **12**, 461–72.

Poisson, R. (1951) Ordre des Hétéroptères. In: *Traité de Zoologie: Anatomie, Systématique, Biologie*, Tome X: *Insectes Supérieurs et Hémiptéroïdes*, Fascicule II (ed. P.-P. Grassé), pp. 1657–803. Masson, Paris.

Preston-Mafham, K. (1990) *Grasshoppers and Mantids of the World*. Blandford, London.

Pritchard, G., McKee, M.H., Pike, E.M., *et al.* (1993) Did the first insects live in water or in air? *Biological Journal of the Linnean Society* **49**, 31–44.

Purugganan, M.D. (1998) The molecular evolution of development. *Bioessays* **20**, 700–11.

Raabe, M. (1986) Insect reproduction: regulation of successive steps. *Advances in Insect Physiology* **19**, 29–154.

Regier, J.C., Mitter, C., Zwick, A. *et al.* (2013) A large-scale, higher-level, molecular phylogenetic study of the insect order Lepidoptera (moths and butterflies). *PLoS One* **8**(3), e58568. doi: 10.1371/journal.pone.0058568

Resh, V.H. & Cardé, R.T. (eds) (2009) *Encyclopedia of Insects*, 2nd edn. Elsevier, San Diego, CA.

Richards, A.G. & Richards, P.A. (1979) The cuticular protuberances of insects. *International Journal of Insect Morphology and Embryology* **8**, 143–57.

Richards, G. (1981) Insect hormones in development. *Biological Reviews of the Cambridge Philosophical Society* **56**, 501–49.

Richards, O.W. & Davies, R.G. (1959) *Outlines of Entomology*. Methuen, London.

Richards, O.W. & Davies, R.G. (1977) *Imms' General Textbook of Entomology*, Vol. I: *Structure, Physiology and Development*, 10th edn. Chapman & Hall, London.

Riddiford, L.M. (1991) Hormonal control of sequential gene expression in insect epidermis. In: *Physiology of the Insect Epidermis* (eds K. Binnington & A. Retnakaran), pp. 46–54. CSIRO Publications, Melbourne.

Robert, D., Read, M.P. & Hoy, R.R. (1994) The tympanal hearing organ of the parasitoid fly *Ormia ochracea* (Diptera, Tachinidae, Ormiini). *Cell and Tissue Research* **275**, 63–78.

Rossel, S. (1989) Polarization sensitivity in compound eyes. In: *Facets of Vision* (eds D.G. Stavenga & R.C. Hardie), pp. 298–316. Springer-Verlag, Berlin.

Rumbo, E.R. (1989) What can electrophysiology do for you? In: *Application of Pheromones to Pest Control, Proceedings of a Joint CSIRO–DSIR Workshop, July 1988* (ed. T.E. Bellas), pp. 28–31. Division of Entomology, CSIRO, Canberra.

Sainty, G.R. & Jacobs, S.W.L. (1981) *Waterplants of New South Wales*. Water Resources Commission, New South Wales.

Salt, G. (1968) The resistance of insect parasitoids to the defence reactions of their hosts. *Biological Reviews* **43**, 200–32.

Samson, P.R. & Blood, P.R.B. (1979) Biology and temperature relationships of *Chrysopa* sp., *Micromus tasmaniae* and *Nabis capsiformis*. *Entomologia Experimentalis et Applicata* **25**, 253–9.

Schwabe, J. (1906) Beiträge zur Morphologie und Histologie der tympanalen Sinnesapparate der Orthopteren. *Zoologica, Stuttgart* **50**, 1–154.

Sivinski, J. (1978) Intrasexual aggression in the stick insects *Diapheromera veliei* and *D. covilleae* and sexual dimorphism in the Phasmatodea. *Psyche* **85**, 395–405.

Smedley, S.R. & Eisner, T. (1996) Sodium: a male moth's gift to its offspring. *Proceedings of the National Academy of Sciences of the USA* **93**, 809–13.

Smith, P.H., Gillott, C., Barton Browne, L. & van Gerwen, A.C.M. (1990) The mating-induced refractoriness of *Lucilia cuprina* females: manipulating the male contribution. *Physiological Entomology* **15**, 469–81.

Smith, R.L. (1997) Evolution of paternal care in the giant water bugs (Heteroptera: Belostomatidae). In: *The Evolution of Social Behavior in Insects and Arachnids* (eds J.C. Choe & B.J. Crespi), pp. 116–49. Cambridge University Press, Cambridge.

Smithers, C.N. (1982) Psocoptera. In: *Synopsis and Classification of Living Organisms*, Vol. 2 (ed. S.P. Parker), pp. 394–406. McGraw-Hill, New York.

Snodgrass, R.E. (1935) *Principles of Insect Morphology*. McGraw-Hill, New York.
Snodgrass, R.E. (1946) The skeletal anatomy of fleas (Siphonaptera). *Smithsonian Miscellaneous Collections* **104**(18), 1–89.
Snodgrass, R.E. (1956) *Anatomy of the Honey Bee*. Comstock Publishing Associates, Ithaca, NY.
Snodgrass, R.E. (1957) A revised interpretation of the external reproductive organs of male insects. *Smithsonian Miscellaneous Collections* **135**(6), 1–60.
Snodgrass, R.E. (1967) *Insects: Their Ways and Means of Living*. Dover Publications, New York.
Spencer, K.A. (1990) *Host Specialization in the World Agromyzidae (Diptera)*. Kluwer Academic Publishers, Dordrecht.
Spradbery, J.P. (1973) *Wasps: an Account of the Biology and Natural History of Solitary and Social Wasps*. Sidgwick & Jackson, London.
Stanek, V.J. (1969) *The Pictorial Encyclopedia of Insects*. Hamlyn, London.
Stanek, V.J. (1977) *The Illustrated Encyclopedia of Butterflies and Moths*. Octopus Books, London.
Stern, V.M., Smith, R.F., van den Bosch, R. & Hagen, K.S. (1959) The integrated control concept. *Hilgardia* **29**, 81–101.
Stoltz, D.B. & Vinson, S.B. (1979) Viruses and parasitism in insects. *Advances in Virus Research* **24**, 125–71.
Struble, D.L. & Arn, H. (1984) Combined gas chromatography and electroantennogram recording of insect olfactory responses. In: *Techniques in Pheromone Research* (eds H.E. Hummel & T.A. Miller), pp. 161–78. Springer-Verlag, New York.
Sullivan, D.J. (1988) Hyperparasites. In: *Aphids. Their Biology, Natural Enemies and Control*, Vol. B (eds A.K. Minks & P. Harrewijn), pp. 189–203. Elsevier, Amsterdam.
Sutherst, R.W. (2004) Global change and human vulnerability to vector-borne diseases. *Clinical Microbiology Reviews* **17**, 136–73.
Sutherst, R.W. & Maywald, G.F. (1985) A computerised system for matching climates in ecology. *Agriculture, Ecosystems and Environment* **13**, 281–99.
Sutherst, R.W. & Maywald, G.F. (1991) Climate-matching for quarantine, using CLIMEX. *Plant Protection Quarterly* **6**, 3–7.
Suzuki, N. (1985) Embryonic development of the scorpionfly, *Panorpodes paradoxa* (Mecoptera, Panorpodidae) with special reference to the larval eye development. In: *Recent Advances in Insect Embryology in Japan* (eds H. Ando & K. Miya), pp. 231–8. ISEBU, Tsukubo.
Szabó-Patay, J. (1928) A kapus-hangya. *Természettudományi Közlöny* **60**, 215–19.
Tabachnick, W.J. (2010) Challenges in predicting climate and environmental effects on vector-borne disease episystems in a changing world. *The Journal of Experimental Biology* **213**, 946–54.
Tabashnik, B.E., Brévault, T. & Carrière, Y. (2013) Insect resistance to Bt crops: lessons from the first billion acres. *Nature Biotechnology* **31**, 510–21.
Terra, W.R. & Ferreira, C. (1981) The physiological role of the peritrophic membrane and trehalase: digestive enzymes in the midgut and excreta of starved larvae of *Rhynchosciara*. *Journal of Insect Physiology* **27**, 325–31.
Thornhill, R. (1976) Sexual selection and nuptial feeding behavior in *Bittacus apicalis* (Insecta: Mecoptera). *American Naturalist* **110**, 529–48.
Trueman, J.W.H. (1991) Egg chorionic structures in Corduliidae and Libellulidae (Anisoptera). *Odonatologica* **20**, 441–52.
Upton, M.S. (1991) *Methods for Collecting, Preserving, and Studying Insects and Allied Forms*, 4th edn. Australian Entomological Society, Brisbane.
Uvarov, B. (1966) *Grasshoppers and Locusts*. Cambridge University Press, Cambridge.
van den Bosch, R. & Hagen, K.S. (1966) Predaceous and parasitic arthropods in Californian cotton fields. *Californian Agricultural Experimental Station Bulletin* **820**, 1–32.
van Oosterzee, P. (1997) *Where Worlds Collide. The Wallace Line*. Reed, Kew, Victoria.
von Reumont, B.M., Jenner, R.A., Wills, M.A. *et al.* (2012) Pancrustacean phylogeny in the light of new phylogenomic data: support for Remipedia as the possible sister group of Hexapoda. *Molecular Biology and Evolution* **29**, 1031–45.
van Voorthuizen, E.G. (1976) The mopane tree. *Botswana Notes and Records* **8**, 223–30.
Waage, J.K. (1986) Evidence for widespread sperm displacement ability among Zygoptera (Odonata) and the means for predicting its presence. *Biological Journal of the Linnean Society* **28**, 285–300.
Wasserthal, L.T. (1997) The pollinators of the Malagasy star orchids *Angraecum sesquipedale*, *A. sororium* and *A. compactum* and the evolution of extremely long spurs by pollinator shift. *Botanica Acta* **110**, 343–59.
Waterhouse, D.F. (1974) The biological control of dung. *Scientific American* **230**(4), 100–9.
Watson, J.A.L. & Abbey, H.M. (1985) Seasonal cycles in *Nasutitermes exitiosus* (Hill) (Isoptera: Termitidae). *Sociobiology* **10**, 73–92.
Wheeler, W.C. (1990) Insect diversity and cladistic constraints. *Annals of the Entomological Society of America* **83**, 91–7.
Wheeler, W.M. (1910) *Ants: Their Structure, Development and Behavior*. Columbia University Press, New York.
White, D.S., Brigham, W.U. & Doyen, J.T. (1984) Aquatic Coleoptera. In: *An Introduction to the Aquatic Insects of North America*, 2nd edn (eds R.W. Merritt & K.W. Cummins), pp. 361–437. Kendall/Hunt, Dubuque, IA.
White, G.B. (1985) *Anopheles bwambae* sp. n., a malaria vector in the Semliki Valley, Uganda, and its relationships with other sibling species of the *An. gambiae* complex (Diptera: Culicidae). *Systematic Entomology* **10**, 501–22.

Whitfield, J. (2002) Social insects: The police state. *Nature* **416**, 782–4.

Whiting, M.F. (2002) Phylogeny of the holometabolous insect orders: molecular evidence. *Zoologica Scripta* **31**, 3–15.

Wiegmann, B.M., Trautwein, M.D., Winkler, I.S. *et al.* (2011) Episodic radiations in the fly tree of life. *Proceedings of the National Academy of Sciences* **108**, 5690–5.

Wiggins, G.B. (1978) Trichoptera. In: *An Introduction to the Aquatic Insects of North America* (eds R.W. Merritt & K.W. Cummins), pp. 147–85. Kendall/Hunt, Dubuque, IA.

Wigglesworth, V.B. (1964) *The Life of Insects*. Weidenfeld & Nicolson, London.

Wigglesworth, V.B. (1972) *The Principles of Insect Physiology*, 7th edn. Chapman & Hall, London.

Wikars, L.-O. (1997) Effects of forest fire and the ecology of fire-adapted insects. PhD Thesis, Uppsala University, Sweden.

Williams, J.L. (1941) The relations of the spermatophore to the female reproductive ducts in Lepidoptera. *Entomological News* **52**, 61–5.

Wilson, M. (1978) The functional organisation of locust ocelli. *Journal of Comparative Physiology* **124**, 297–316.

Winston, M.L. (1987) *The Biology of the Honey Bee*. Harvard University Press, Cambridge, MA.

Womersley, H. (1939) *Primitive Insects of South Australia*. Government Printer, Adelaide.

Yoshizawa, K. & Johnson, K.P. (2010) How stable is the "Polyphyly of Lice" hypothesis (Insecta: Psocodea)?: a comparison of the phylogenetic signal in multiple genes. *Molecular Phylogenetics and Evolution* **55**, 939–51.

Youdeowei, A. (1977) *A Laboratory Manual of Entomology*. Oxford University Press, Ibadan.

Zacharuk, R.Y. (1985) Antennae and sensilla. In: *Comprehensive Insect Physiology, Biochemistry, and Pharmacology*, Vol. 6: *Nervous System: Sensory* (eds G.A. Kerkut & L.I. Gilbert), pp. 1–69. Pergamon Press, Oxford.

Zanetti, A. (1975) *The World of Insects*. Gallery Books, New York.

Zhang, Z.-Q. (2013) Phylum Arthropoda. In: *Animal Biodiversity: An Outline of Higher-level Classification and Survey of Taxonomic Richness* (Addenda 2013), (ed. Z.-Q. Zhang), pp. 17–26. *Zootaxa* **3703**(1), 1–82.

INDEX

Readers are referred also to the Glossary: glossary entries are not cross-referenced in this index. Parenthetical entries that follow non-italic entries almost always refer to either plural, adjectival and/or diminutive endings for morphological terms, but occasionally are abbreviations of terms or alternative names or terms. All *italicized* entries are the scientific names of insects, except where indicated by the names of other organisms, abbreviated where necessary as: Bact., bacterial name; Bot., botanical name; Mam., mammal name; Nem., nematode name; Prot., protist name; Vir., virus name. Some scientific names are followed by a parenthetical abbreviation(s) indicating the higher taxon (taxa) to which they belong [for example, *Chironomus* (-idae) refers to this genus and also the family Chironomidae] and page references may be to the higher taxon name only. Page numbers in *italics* refer to Figures; those in **bold** to Tables.

Abdomen
  appendages 56, 226
  distensible 33
  ganglia 64, 65, 213
  pumping movement 77, 107
  structure 52–55, 201
  tagmosis 33–5, 201
  tympanal organs 100, 123
Acacia (Bot.) 8, 22–3, 294, 299, 318–9, 339, 426
Acanthae 32
Acarapis woodi 348
Accessory gland
  JH effect on 68
  of parasitoid 366, *368*
  secretions of 133, 143, 129, 135, 138, 149–50, 153, 307, 386
  structure/function 57, *58*, 90–93, 507
Accessory pulsatile organs 71
Acclimation 123, 175–6, 186
Acerentulus 495
Acraea 385
Acrididae 30, 87, *89*, 103, 116, 179, 188, 382, 456, *Plate 7a*
Acromyrmex *265*, *267*
Acronicta Plates 7.7 & 7.8
Acrotergite 33
Acyrthosiphon pisum Plate 8d
Actinomycetales (Bact.) *254–5*
Aculeata 223–4, 255, 325, 327–8, 349, 362, 393–4, 524–5
Adelgidae 305
Adephaga 217–8, 492, 516–7
Adipocyte *see* Trophocyte
Adult 27–30, 33, 40–5, 168–72; *see also* Imago
  teneral 24, 168, *169*, *Plate 3d*
Aedeagus 54–5, **91**, 131–2, 136–8, 141, 199
Aedes 10, *127*, 174, 183, *185*, 287, 403, 407–8, 413
  Ae. aegypti 10, 154, 183, 185, *397*, 403, 408
  Ae. africanus 399
  Ae. albopictus 408, 466–7, 469, 473
  Ae. bromeliae 399
  eggs 146, 154, 157, 297
Aethina tumida *442*
Africa
  ants 15, 114, 269, 317, 323, 337, 339
  ant-mimicking bugs *377–8*, 395
  assassin bugs 383
  bed nets 406
  biocontrol of weeds 310–11
  bombardier beetle 388
  butterflies 12, 390
  cassava mealybug 435, 439
  coffee 462–3
  disease vectors 399, 400–13
  dung beetles 121, 261, *263*
  food insects 20–4
  ground pearls 259
  Kenya 12, 463
  Mantophasmatodea 211
  midges 177, 236
  myrmecochory 317
  Rift Valley fever 407
  termites 201, 250, 257–8, 267, 344, *346*
  weevils 198
  West Nile virus 409
African honey bee 333–4, 347
Agaonidae 291–2, 307, 315
Agaricales (Fungi) 265
Age determination 30, 157, 181–3
Aggregation 9, 19, 102–4, 114, 171, 300–1, 332, *366*, 369, 422
Agrilus spp. 437, 471–2, *Plate 8g*
Agrobacterium tumefaciens (Bact.) 445, 449
Agromyza aristata 300
Agromyzidae 290, 300
Agrotis infusa 22
Agulla 515
Air sac 75–77, 103–4
Alderfly *see* Neuroptera
Aleyrodidae *214*, 242, 424, 468, 485, 513, 547
Alimentary canal *see* Gut
Alinotum 46
Alkaloid 113, 294, 384–6, 390, 395, 430–1, 447, 450, *Plate 7e*
Allatostatin **70**
Allatotropin **70**
Allergy 5, 24, 395, 417, 419, 521

The Insects: An Outline of Entomology, Fifth Edition. P.J. Gullan and P.S. Cranston.

Companion Website: www.wiley.com/go/gullan/insects

Allogenic engineers 257
Allometry 157, 182, 329
Allomone 115–6, 384
Allomyrina dichotoma 13
Alloxysta victrix *365, 367*
Alphavirus (Vir.) 407
Alticinae 60, 251
Altruism 325–6, 348–50
Amber 13, 211, 214, 219, 227, 230–6, 443, 505, *see also* Fossil (insects)
Amblycera 213, 214, 510
Amblyseius (Acari) 441
Ambrosia beetles 86, 260, 324, 344–5, 453, 463, 469; *see also* Platypodinae; Scolytinae
America, North
  arboviruses 407–410
  biocontrol 311–2, 437
  biogeography 236, 245–6
  butterfly conservation 17
  cottony-cushion scale 426–7
  danaine butterflies 390
  food insects 22
  galling insects 305–6
  gypsy moth 293, 435, 453
  honeypot ants 23
  modelled fruit fly 422–3
  phytophagous insects 296, 307
  plague 411
  Rocky Mountain spotted fever 411
  screw-worm fly 454
  woodroach 199, 212, 236
  West Nile virus 407, 409–10, 464
  see also California; Canada; Hawai'i; Mexico
America, South/Central
  Acacia insects 304
  ants 265–7, *319*, 323
  arboviruses 407–410
  biocontrol 310–11
  bot flies 360–1
  butterflies 138, 222, 392
  food insects 23–4
  giant insects 9, 498
  screw-worm fly 454
  source of pests 17, 259, 352, 435
Ametaboly 27, 48, 158
Amitermes 254, 344–5
Ammonia *90*, 264
Amphiesmenoptera 216, 221
Amphigerontia contaminata 509
Amphimixis 150
Amylostereum (Fungi) 303
Anabrus simplex 133
Anactinothrips 326
Anal
  fold 217
  vein 221, 523
  wing sector 49, 51, 115, 207, 211, 221, 506
Anaphes spp. 198, 365
Anastrepha 360, 422
Andrenidae 325
Androdioecy 152
Angraecum sesquipedale (Bot.) 316
Anholocyclic 176
Anhydrobiosis 177
Anisoptera 51, 206, 498, 499
Anisozygoptera *see* Epiophlebiidae 206, 498
Anlage 160, 161, 164
Anomala 51
Anonychomyrma scrutator 319
Anopheles 10, 44, 86, 183, 200, 398–407, 413, 466
  An. arabiensis 404, 466
  An. bwambae 404–5
  An. gambiae 86, 183, 200, 398, 403–5, 466
  An. melas 404
  An. merus 404
  An. quadriannulatus 404–5
  An. quadrimaculatus 44
Anoplognathus 298, *Plate 3c*
Anoplolepis steingroeveri 257
Anoplophora glabripennis 473
Anoplura 41, 42, 213, 214, 372, 410, 510
Anostostomatidae 10, *Plate 1d*
Ant *see* Formicidae
Ant plants 318–*20*
Antenna (-ae)
  chemoreception 107–111
  electroantennogram 110–11
  flagellomere 43, 105, 219–20, 519
  function 98–9, 105, 107
  Johnston's organ 99
  morphology *35, 37–8*, 43, *45, 58*
Anthelmintic 264
Anthicidae 386
Anthomyiidae 258, 301
Anthrenus 488
Anthropophily 245, 402–3, 405
Anthurium (Bot.) 314, *Plate 4e*
Antibiosis 447
Antidiuretic peptide 70
Antifreeze 175
Antigonadotropin **70**, 154
Anting 385
Antipaluria urichi Plate 7c
Antixenosis 447
Antliarhinus zamiae 308, *Plate 4b*
Antlion 42, 217, *357*, 362, 364, 486, 514
Antliophora 216, 219, 220, 233
Anus 53, 57, 80, 85–*8, 132*
Aonidiella aurantii 187, 374, 433
Aphaenogaster 326
Aphelinus (-idae) 366, 424, 440
Aphid
  and ants 15
  biocontrol *418*, 419, 432, 440–1, 443
  pigment 33
  disease transmission 85
  endosymbionts 85–6, 192
  feeding/mouthparts 83, 296–7, 512–3
  galling 296, 305–8, *Plate 5b*
  hosts for hyperparasitoids *365–7*
  migration 179
  pheromones 114
  polyphenism 181
  preservation 463
  reproduction 150–1
  Russian wheat aphid 460
  soldiers 181, 325–6
  subsociality 325–6, 348
  temperature tolerance 176
  see also Phylloxera
Aphididae 114, 176, 305–6, 325, 512
Aphidius 441, *Plate 8d*
Aphidoidea 215, 293, 485
Aphis nerii 385
Apidae (-oidea) 11, 38, 195, 315, 328–31, 348
  see also Euglossine bees
Apiomorpha 80, 304, 305, 306, *Plate 5a*
Apis spp. 326, 329–34, 370, *Plate 6d*
  A. mellifera 38, 114, 140, 151, 315, 330–1, 333–4, 336, 339, 347–8, 419, 434, Plate 5h
  sex determination *140*
Apocephalus 266
Apocrita 223, 314, 325, 328, 375, 524–5
Apocynaceae (Bot.) 179, 390
Apodeme 47, 59
Apoidea 224, 325, 328, 336
Apolysis 169–72
Apophyses 36, 59
Aposematism *see* Colour, warning
Apparency 294, 384, 389, 392, 451
Apparent competition 312, 437
Aptery 176, 207, 209, 243, 247, 297, 500, 502, 506, 511
Apterygotes 27, 52
Aquatic insects
  conservation 14, 17
  evolution of lifestyles 237–8, 239
  excretion 90
  functional feeding groups 285–6
  Hawai'i 288
  immature terminology 165
  lentic adaptations 283–4

lotic adaptations 282–3
marine/intertidal/littoral 287–8
in monitoring 284–5
origins 236–8, 275–7
oxygen use 277–*91*
sound production 98–9
taxonomic groups 272
of temporary waterbodies 286–7
Arachnocampa 122, 360
Aradidae 261
Arbovirus 175, 384–5; *see also* under specific virus
Archaeognatha
ametaboly 159, 203–5, 495
fossil record 229, *231*
genitalia *54*
jumping 240
preservation 484
relationships 203–5
in soil 250–*1*
taxobox 495
see also Appendix
Archichauliodes guttiferus 551
Archostemata 217, 492, 516
Arctiidae 105, 385, 387, *Plate 7e*
Argema maenas Plate 1b
Argentine ant 15, 201, 352, 419
Arixeniidae 209, 501
Ariolum 47, 211, 214, 506
Aristolochia (Bot.) 17, 146
Arthrodial membrane 30
Asaphes lucens 365
Ascalaphus (-idae) 514
Asclepias (Bot.) 293
Ascovirus (Vir.) 445
Asilidae 251, 357
Asobara tabida 137
Atanycolus hicoriae 472
Atemeles pubicollis 346
Atlas moth *see Attacus atlas*
Atta 105, 116, 265–7, 323, 363
Attacus atlas Plate 1a
Attraction-annihilation 452
Auchenorrhyncha 214–5, 224, 512–3
Augmentation (biological control) 309, 312, 438
Aulacaspis yasumatsui 374
Australia
Acacia 318
ants 16, 339
biocontrol 311–3
bioluminescent flies 122, 360
butterfly conservation 12–13, 17
Christmas beetles 298
climatic modelling (CLIMEX) 459, 461, 463
cottony-cushion scale 426–7, Plate 8b
dengue 408
dung beetles 261–2
edible insects 11
phytophagous insects 295–6, 311
insect nuisance 261, 415–6, 423
magnetic termites 344
mimetic beetles 115, 391–2
myrmecochory 317–8
subsocial thrips 214, 326
sugarcane pests 259
Australopierretia australis Plate 5f
Austrolestes 498
Austroplatypus incompertus 345
Austrosalomona Plate 3b
Autotomy 278, 378, 383–4
Axil (-lary) 49, 52, 205–6
Axon 64, 96–8, 120, 228
Azadirachta indica (Bot.) 430
Azteca 318

Bacillus (Bact.)
B. sphaericus 443
B. thuringiensis 441–5, 449
see also Paenibacillus
Backswimmer *see* Notonectidae
Bactrocera spp. 16, 422, 452, 454, 461
Baculovirus (Vir.) 434, 445–6
Badisis 395
Baetis (-idae) 147, 283
Bagworm *see* Psychidae
Bahiaxenidae 219, 518
Baizongia pistaciae 305, *Plate 5b*
Bancroftian filariasis 405
Barathra brassicae 98
Bark beetle 114–6, *260*, 303, 460; *see also Ips grandicollis*; Scolytinae
Bartonellosis 411
Basalare (-ia) *46–7*
Basement membrane 28, 69, 80
Basisternum *46–7*
Bat (Chiroptera) 97, 101, 103, 105, 209, 232, 268, 282, 370, 385, 501
Bates, Henry 381, 389
Baumannia cicadellinicola (Bact.) 85
Beauveria (Fungi) 443, 449
Bed bug *see Cimex*
Beddingia siricidicola (Nem.) 442
Bed nets 405–7, 452
Bees *see Apis*
Africanized 334
fossil 234
see also Euglossine bees
Bee fly *see* Bombyliidae
Beetle *see* Coleoptera
Behaviour
aggregation 13, 112, 114–5, 126–7, 188, 323–4, 359, *389*, 392, 415, 453
classification 122–4
courtship 126–9, 132–5, 139, 148, 360, 386, 502, 507, 509, 520, 523
death feigning 378, 380, 504
eclosion 70, 165, 171, 522
foraging 107, 117–8, 177, 257–8, 263, 315, 329–30, 333–9, 344, 351, 355–8, 360–1, 373–4, *378*
and hearing 97, 102–3
innate 112, 123
kineses 122–3
mate attraction 116, 123, 126–8
migratory 174, 178–180, *188*, 252–3, 274, 284, 296, 339–40, 347, 356, 392
nesting *322*, 324–6, 329–331, *335–40*
oviposition 107, 114, 126–7, *131*, 133–4, 143–48, 160, 163, 174, *184*, 291–2, 298, 303, 324–5, 345–7
provisioning young 255, 324–5, 329–30, 348, 362, 471
puddling 135–6
releaser 112, 114, 116, 123, 337
rhythmic 118
searching 310, 356–9, 365, 374, 380–1
and sensory systems 9
taxes 123
and temperature 105–6, 174–7
ventilatory 281–2
see also Colour, warning; Sociality
Belidae 392
Belostoma (-atidae) 147–8, 324, 361, *Plate 2e*
Bembix 124
Bemisia 419, *424*–5, 468
Benthic insects 229, 277, 284
Bessa remota 437
Bibionidae (-omorpha) 51, 219, 250, 251, 258
Bioclimatic models 460–3
Biodiversity *see* Biological diversity
Biogeography 191, 212, 236, 244–7; *see also* Dispersal; Vicariance
Biological
clock 117–8
control 14, 198, 245, 247, 260, 301–5, 309–13, 366, 374, 418, 424–7, 435–47
diversity 3, 6–10, 12–16, 191, 323, 419, 467–8, 471, 491
Bioluminescence 122, 360
Biosecurity 472–3 *see also* Quarantine
Biosteres longicaudatus 360
Birds
as insect hosts 213–4, 220, 360, 370–2, 521

Birds (*continued*)
  as predators on insects 378–82, 439
  West Nile virus 409–10
Bisacylhydrazines 433–4
Biston betularia 180, 379, 381–2
Bittacus (-idae) *129*, *167*, 220, 520
Blaberidae 121, 212, *Plate 3d*
Black Death *see* Plague
Blattabacterium (Bact.) 90, 508
Blatella spp. 242, 508
Blatta (-idae) 47, *51*, *58*, 65, 87, 150
Blattaria 507
Blattodea
  cave dwelling 268
  fossil record 212, *231*
  in litter 250
  morphology 47, 50–52, *58*, 62, 65, 80, 507
  parental care 324
  pinning 484
  relationships 203, *205*, *208*, 211–2, 507
  species richness 6
  taxobox 507
  see also Cockroach, woodroach; Appendix
Bledius 287
Blephariceridae 281
Blood feeding
  by bugs 66, 77, 83, 85, 361, 411, 416–7, 513
  by fleas 220, 411, 521
  by flies 40–2, 77, 116–7, 359, 360, 362, 372, 399–411, *412*, 466, 519
  by lepidopterans 39, 524
  by lice 85, 510
  see also Haemolymph
Bloodwood *see Corymbia* (Bot.)
Bloodworm 14, 277, 279, 285, 417
Blow fly *see Calliphora*; *Lucilia*
Blue gum chalcid *see Leptocybe invasa*
Bluetongue 407–8, 466
Bogong moth 22
Bollworm *see Helicoverpa*; *Heliothis*
Bombardier beetle 378–8
Bombus 107, 326–9, 347
Bombyliidae 220, 291, 314, 519
Bombyx mori 5, 24, 86, 108, 174, 195, 242, 443, *Plate 2d*
Book lice *see* Psocodea
Boreioglycaspis melaleucae 311
Boreus (-eidae) 175, 220, 520, 521
Borneo 7, 263, 320, 360, 402, 503, 512
Brachinus 388
Brachycera 40, 219, 234, 314, 519
Braconidae 153, 363, 365, 368, 440, 472, *Plate 8d*
Bracovirus (Vir.) 369
Bradynobaenidae 224
Braula 348, 370
Brain 36, 57, 64, 66–7, 105, 117, 119, 153–4, 166, 169, 171, 174
Bristletail *see* Archaeognatha
Broad-complex (*Br-C.*) 242
Bromeliad (Bot.) 318
Brooding 147–8, 324
Bruchinae 138
Brugia (Nem.) 412
Brugian filariasis 412–3
Bt *see Bacillus thuringiensis*
Bucculatricidae 301
Buchnera (Bact.) 85–6, 424
Buds *see* Imaginal disc
Buffalo fly 261
Bug *see* Hemiptera
Bunyaviridae (Vir.) 305, 407
Buprestidae 59, 105, 261, 300–1, 313–4, 386, 392, 437, 471, 517, *Plate 8g*
Bursa copulatrix **91**–2, 132, 137
Bursicon 70, 168, 171
Bush fly *see Musca vetustissima*
Bushcricket *see* Katydid
Butterfly
  and climate change 467
  conservation 14–16, 450, 468
  defences 294, 345, 379–80, *387–392*
  definition 222, 523–4
  farming 12–13, *Plate 1g*
  houses 2, 12–14
  life cycle 156
  migration 16, 179
  mimicry 389–91, *Plate 8a*
  mouthparts *39*, 221
  pheromones 112–3
  as pollinators 314
  as prey *354–5*, 356, 375, 390
  puddling 135–6
  roosting *323*, 381
  wing 32, *51*, 62, *524*
  see also Lepidoptera

Cabbage white butterfly *see Pieris rapae*
Cacama valvata 106–7
Caconemobius 268
Cactoblastis 311–12
Caddisfly *see* Trichoptera
Caecum (-a) 82
Caelifera 209–10, 502
Caenidae 285
California
  butterflies 13, 16, 17, 323, 392, 468, *Plate 3g*
  crop pests 422–3, 425–7, 436–7
  ornamental trees 470–2, *Plate 8g*
  palmworms 21–2
  pollination services 5, 338
  ragweed 311
  red scale 187, 374, 433
Caligo spp. *Plate 1f, 6f*
Calliphora (-idae) 65, 84, 143, 150, 166–7, 261, 264, 398, 413, 454, 519; *see also* Maggot; Screw-worm fly
Calliphorin 84
Calliptamus Plate 7a
Callosobruchus maculatus 138
Calopteryx 141
Calyx *91–2*, 369
Cambrian 228–9, *231*, 237
Camouflage 29, 294, 356, 378–83, 453; *see also* Crypsis; Masquerade; Mimicry
Campodea (-idae) 268, 495
Campoletis sonorensis 369
Camponotus 23, 30, 320, 393, 395, *Plate 2g, h*
Canada 7, 16, 176, 179, 234, 237, 410–11, 431, 461
Canada balsam 483–4
Cantharidin 386, 416
Cantharis (-idae) 314, 516
Cantharophily 314
Canthon 254
Cap cell *99*
Capritermes 393
Carabus (–idae) 61, 137, 166, 176, 218, 251, 261, 269, 378, 388, 439
Carapacea 206
Carbamate 432
Carbon dioxide 77, 116–7, 148, 256, 262, 279, 346, 358–9, 451, 458, 477
Carboniferous 9–10, 48, 205, 230–3, 240, 290
Cardenolides (cardiac glycosides) 294, 385–6, 390
Cardiaspina 304, *Plate 8e*
Carding 479, 481–2
Cardinium (Bact.) 153
Cardiopeptide **70**, 168
Cardo 37–8
Carotenoids 33
Carpenter bee *see* Xylocopini
Carrion beetle *see* Silphidae
Cassava (*Manihot esculenta*) (Bot.) 419, 435, 439, 465
Cassava mealybug 435, 439, 465
Caste system
  in aphids 181, 325–6
  benefits 351
  definition 181, 323, 327
  in Hymenoptera 265, 327–8, 337, 347, 350

in termites 340–1, 343
see also Polyphenism
Cat flea *see Ctenocephalides canis*
Categories (taxonomic) 11
Caterpillar *see* Lepidoptera, larva
Caudal
filament 52, 205, 239, 240, 277, 497, 515
lamellae 278, 498, 499
Cave insects 117, 122, 268, 360
Cecidomyiidae 150, 151, 219, 305, 311
Cecidozoa 305, 308
Cecropia (Bot.) 318
Cecropins 73
Cement (colleterial)
gland **91**–2
layer 29
Central nervous system 64–5, 96, 98, 117, 171, 430, 453; *see also* Brain
Cephalotes (-aceae) (Bot.) 319, 393
Cephus 301
Cerambycidae 301, 314, 392, 470, 473, 517
Ceratitis capitata 301, 422, 454–5
Ceratocystis ulmi (Fungi) 115
Ceratopogonidae 40, 190, 219, 235, 247, 273, 287, 314, 360, 408, 466
Cerceris fumipennis 472
Cercopoidea *see* Spittle bug
Cercus (-i) 52–*3*, *58*, 98, 136, 240
"heart" 208
Ceroplastes sinensis 298
Chaetosemata 523
Chagas' disease 412
Chalcidoidea *223*, 305, 315, 360, 365, 439–40, 470
Chaoborus (-idae) *190*, 219, 273, 284, 362, 373–4
Chathamiidae 288
Checkered beetle *see* Cleridae
Cheese skipper *see* Piophilidae
Cheilomenes lunata 145
Chemical control of pests 419, 429–434
Chemical defences 384–388, 392, 395–6
Chemoreception 38, 45, 107–111, 303, 361
Cheumatopsyche 523
Chikungunya 398, 407, 466
Chiloglottis (Bot.) *Plate 5e*
China
fossils 211, 234, 237
insect feed 24
malaria 406
pest insects 423, 448, 471–2
termites 344
Chinese wax scale *see Ceroplastes sinensis*
Chionea 175
Chironomus (-idae) 99, 151, 158, 172, 175, 182, 190, 219, 244, 273, 281, 285–7, 360; *see also* Bloodworm
Chitin 5, 29, 30, 82, 169, 352, 433
Chitin synthesis inhibitors 352, 433
Chloride
cells 86, 203
channel 432
reduction 288
retention 86, 88
transport stimulating hormone **70**
Chlorinated hydrocarbons
see Organochlorines
Chorion 145, 148, 277
Choristoneura fumiferana 453
Chortoicetes terminifera 179
Christmas beetle 298, *Plate 3c*
Chromatomyia spp. *300*
Chromosome 140, 164, 194, 199, 224, 236, 350
Chrysalis *167*, 221, 524; *see also* Pupa
Chrysanthemoides (Bot.) 313
Chrysanthemum (Bot.) 430
Chrysidoidea 223, 328
Chrysobothrus femorata 59
Chrysolina spp. 311
Chrysomelidae (-oidea) 48, 60, 86, 138, 235, 256–8, 260, 290, 300–1, 308, 311, 386, 447–8, 471, 517; *see also Chrysolina*; *Leptinotarsa*
Chrysomya bezziana 398, 519
Chrysopa (-idae) 184, 438
Chrysoperla carnea *438*
Cibarium 38–40, 42, 80–*1*, 83, 213
Cicada 11, 12, 24, 48, 59, 77, 80–*1*, 85, 100, 104, 106, 126, 128, 172, 188, 214–5, *254*, 256, 360, 376, *512*; *see also* Cicadidae
Cicadetta montana 512
Cicadidae (-oidea, -omorpha) 48, 214–5, 254, 479, 485, 512
Cimex (-icidae) 85, 138, 153, 361, 415
Cinnabar moth *see Tyria jacobaeae*
Circadian rhythm 118–9, 179, 187
Circulatory system 57, 71, 83–4, 267, 337, 344
Citizen science 4, 20
Cladistics 193–4
Cladograms 193–4, *204–222*, *246*, *328*
Classification
of defensive chemicals 384–5
of insects, current status 10–11, 191
of pheromones 111–3
and taxonomy 191, 196–201
see also ordinal taxonomic boxes
Clava (-l) furrow/fold 439
Claw 47, 218, 259, 362, 493, 520
Cleridae 264, 314, 365
Click beetles *see* Elateridae
Climate
change 3, 14, 19, 408, 458–9, 462–7, 467–8, 505
effect on insect distribution 459–463
modelling 459, 461–6, 469–70, 472
CLIMEX (model) 459, 461–3, 472
Clothes moth *see* Tineidae
Club (of antenna) *45*, 221, 523
Clunio 172, 288
Clypeus 35–8, 40–1, 233
CO_2 *see* Carbon dioxide
Coccidae 293, 318–9
Coccinellidae 48, 145, 167, 374, 384, 386, 426, 439, 517
ladybird beetle (ladybug) 4, 10, *145*, 173, 179, 384, 426–7
Coccoidea (scale insects)
and ants 15, 18, 319, 340
chemical production 30, 359, 384
control 426–7, 439–40, 443, 517
development 151, 158, 168, 242
feeding 83, 256, 259, 302, *304*, 318, 468, 513
galling 305–8, *Plates 4f,g*, *5a*
morphology 43, 51, 53, 86
relationships 215, 512
reproduction 140, 150–1, 153
see also Ground pearls; *Icerya*; individual families
Cochliomyia hominivorax 398, 454
Cockroach
anthropophily 13–14, 245, 352, 508
behaviour 245, 323, 350
and disease 399, 415
excretion 85, 87, 90
feeding 78, 85–6, 261
giant burrowing cockroach *Plate 3d*
locomotion 60
morphology/anatomy 36, 46, *47*, *51*, 57–*8*, *65*, *75*, 80, 93, 212, 507
ootheca 149
relationships 212, 236, 250, 324, 508
reproduction 150
see also Dictyoptera; Blattodea; *Cryptocercus*
Codling moth *see Cydia pomonella*
Coelomomyces (Fungi) 443
Coevolution 9, 86, 244, 291–3, 316–8, 385–6
Coffea (Bot.) 462–3
Coffee berry borer 463
Colcondamyia auditrix 360
Coleoptera
aquatic 99, 176, 238, 275, 277–87, *281*, *284*
defences 115, 382–8, 416

Coleoptera (*continued*)
diversity 6–9, 247, 516
and flowers *313*–4
as food 20–4, *Plate 2a*
fossil record *231*, 233–5
fungivory 265, 268, 290
larvae 72, 151, *160*, 165–6, *251*, 275, 517
light production 121–*2*
locomotion 60–*1*, 251
morphology *26*, 48–51, *54*, *59*, *61*, 65, 88, 100, 166–*7*, 217–8, 516–7
ovipositor 54, *145*, 516
parental care *262*–3, 324–5
as pests 114–6, 258, 441, 444, 448–9, 460, 463
pinning/preservation *479*, 484
plant chewing/mining 235, 256, 290, *298*–301, 442
predatory 173, 358, *418*, 439, 441, 516
relationships 197, *204*, 216–8
seed predation (feeding) 85, 308–9
in soil monitoring 269
taxobox 516–7
wings 49–51, 516
wood boring 218, 260–*1*, 301, 419, 469
see also Appendix; individual families
Coleorrhyncha 214–5, 512
Collecting methods *474*–8
Collembola
cave dwelling 268
development 158
morphology 131, 202, 250–1
fossil record 229
fungivory 256, 265
preservation 484
relationships 202–3, *204*
in soil and litter 237, *251*, 259, 269, 372
taxobox 493–4
see also Appendix
Colleterial gland 149
Colletidae 325, 328
Colobopsis see Camponotus
Colon 82, *87*
Colony 15, 23, 255, 267, 308, 323–5, 327–52
Colony Collapse Disorder 316, 323, 334, 336, 338–9, 431
Colo-rectum 80
Colophospermum mopane (Bot.) 20
Colour
hormones in **70**
production 32–3, 90
and temperature 106–7
ultraviolet 29, 32, 101, 121, 315
vision 121
warning (aposematism) 115–6, 384–5, *387*–92, *Plate 4c, 7e,h*
Colorado potato beetle 419, 425, 427, 444, 447–9, 460, 468, 517
Comb
morphology 44–5, *330*–1, 521
nesting 267, 332, 334–6, 344, *346*, 348
Communication 9, 10
in honey bees 332–3
by luminescence 121, 126, 360
by odour/chemistry 21, 90, 96, 107, 109–12, 115, 127–8, 345, 359, 384–5, 395–6
by sound 96–7, 104–5, 502
Condylognatha 213–4
Coniopterygidae 515
Conjunctival membrane 33
Conopidae 365
Conotrachelus nenuphar 301
Conservation
and climate change 467–8
of natural enemies 439–41
effects of non-natives 15–16, 419
of insects 14–20, 268–270
Control (of pests)
biological 14, 198, 258–60, 301, 305, 309–313, 355, 366, 374, *418*, 426–7, 435–445, 465, 472–3
chemical 258, 338, 415, 424, 425–8, 428–433, 448–9
cultural 451
growth regulators 406, 429, 433–4
integrated 421, 428–9, 449
mechanical/physical 451
RNA interference 434–5
using lures/pheromones 113, 423, 429, 452–4
using sterile insect technique 454–5
Co-operation *see* Sociality
Copidosoma 152
Coprophagy 261–3, 477
Coptotermes 256, 344, 352, 473, 508
Copulation 131–4, *132*, *144*
cannibalistic 135
cantharidin transfer 396
"lock and key" hypothesis 136–7
mating plugs 141
nuptial feeding 136
physiology 153–4
pseudocopulation 116, 314, *Plate 5d, e*
traumatic insemination 138, 243
see also Courtship
Corazonin **70**
Corbicula 329–30
Coreidae 83, 106, 308
Coremata 113
Corethrella (-idae) *190*, 219, 359
Corium 51
Corixa (-idae) 24, 274, 284
Cornea 120
Cornicle *see* Siphunculi
Corpus (-ora)
allatum (-a) 66, 68–9, 153–4, 171–2, 174, 331, 343
cardiacum (-a) 66–8, 171
Corydalidae 276, 277, 514, 515
Corymbia (Bot.) 22
Co-speciation 192, 214, 236, 292, 370–2
Cossidae 22, 23, 167, 301, *Plate 2c*
Costa (-al) 49, *50*, 206, 241, 502
Costa Rica 7, 12, 13, 200, 359, 467
Costelytra zealandica 443
Cotton (Bot.) 184, 424–5, 428–9, 434, 445–9, 452–3
Cottony-cushion scale *see Icerya*
Courtship 97, 99, 104, 112–3, 121, 128–9, 132, 135–6, 148, 360, 386
feeding / gifts 129, 134–6, 386
internal courtship 144
see also Copulation; Taxoboxes
Coxa(-al) *46*–7, 52, 54–5, 60, 105, 201, 203
Crabro (-onidae) 65, 123, 254, 255, 328, 472
Cranefly *see* Tipulidae
Crataegus (Bot.) 243
Crawling 60
Creatonotus gangis 113
Crepuscular activity 118, 120, 314
Cretaceous 214, 219, *231*-6, 244, 290
Cretotrigona prisca 234
Crickets 57, 58, 102, 103, 104, 133, 134, 149, 249, 254, 362, 442
Crista acustica 100
Crop 23, 80–2
Crustacea 38, 228–9, 238, 288, 374
second antenna 228
Crustacean cardioactive peptide (CCAP) **70**, 168
Cryoprotection 175–6
Crypsis 356, 378–81, 389, 391, 504; *Plates 7a, b*, g; see *also* Camouflage
Cryptobiosis 177
Cryptocercus spp. 199, 208, 212, 224, 236, 508
Cryptochetum (-idae) spp. 426–7
Cryptolaemus montrouzieri 441
Cryptonephric system 88
Cryptophagidae 260–1
Cryptotermes 343, 393
Ctenocephalides spp. 358, 411, 521
Ctenidia 521

Ctenolepisma longicaudata 496
Cubital vein (cubitus) 49–*50*, 241
Culex 407
Cx. tarsalis 410
Cx. pipiens 152, 280, 409, 413, 464
Cx. p.quinquefasciatus 413, 464
Cx. tritaenorhynchus 410
Culicidae 40, 43–4, 99, 185, *190*, 219, 273, 280, 399, 404, 408–9, 454, 469, 473; *see also Anopheles; Culex*
Culicoides spp. 408, 466
Culicomorpha 219
Culture, and insects 11–13
Cuneus 51
Curation 191, 200, 474–5, 478, 487
Curculionidae (-oidea) 20, 48, 60, 114–5, 176, 218, 243, *251*, 258, 260, 290, 300, 301, 303, 308, 310–11, 324, 344, 453, 463, 470–1, 517, *Plates 2a, 4b, 8f*
Cuticle 5, 15, 24, 27–33, 39, 59–61, 68–9, 87, 96–100, 111, 117, 119, 136, 157, 165
elastic 60, 104
hydrocarbons 111, 188, 198–9, 201, 337, 345
Cyanogenic glycosides 294, 385
Cycad scale *see Aulacaspis yasumatsui*
Cycadothrips 106
Cycloalexy *389*, 392
Cyclorrhapha 220, 519
Cydia pomonella 301, 302, 419, 466
Cynipidae (-oidea) 223, 290, 305–6, 365, *Plate 5c, 6a*
Cynips quercusfolii 306
Cypovirus (Vir.) 445
Cyrtobagous spp. 310–11
Cystococcus 22, 305, *Plate 4f*
Cytoplasmic polyhedrosis virus (CPV) 445

Dactylopius coccus 5
Daktulosphaira see Phylloxera
Damselflies *see* Odonata: Zygoptera
Danaus spp. 3, 12, 16, 51, 113, 156, 179, 195, 293, 381, 390–1, 450, *Plate 3g*
Dance 118, 323, 332–5
Daphnia (Crustacea) 284, 37–4
Darwin, Charles 1, 2, 3, 5, 128, 130, 316, 348, 389
Day-degrees (degree-days) 57, 184–6
DDT 400, 425–6, 431–2, 448
Dead wood insects 18, 172, 260, 303, 324
Death feigning *see* Thanatosis
Decticus 101
DEET 108, 111
Defence
chemicals 341, 384–388, 392, 395–6
induced / constitutive 295, 297–8
in gregarious and social insects 324–7, 329, 334, 337, 341, 355, 363–4, 384–92
by haemolymph 71, 363–4
by hiding 379–80
mechanical 382–4
by mimicry 389–92
secondary 380
strategies figured *356, 378*
Deinacrida spp. 10, *Plate 1d*
Deladenus siricidicola (Nem.) 442
Delia radicum 258
Delusory parasitosis 415
Dendrite 64, 96, 99, 105, 108–9, 116
Dendroctonus spp. 114, 303
Dengue 398, 407–8, 467, 469
Dermal gland *28*, 30
Dermaptera
cave dwelling 268
diversity 6
fossil record *231*
in litter and soil 250–*1*, 254, 484
morphology 36, 49–50, 150
parental care 150, 324
pinning/preservation 484
relationships 207–9
taxobox 500
see also Appendix
Dermatobia hominis 360
Dermatophagoides (Acari) 417
Dermestes (-idae) 167, 264, 314, 414, 473, 488
Deuterophlebiidae 219
Development
arrest 366
of honey bee 331
biotic / crowding effects on 187–8
embryonic 160–3
evolution of 241–2
of gall 308
major patterns 158–60, *162*
molecular basis 163–5
mutagenic and toxic effects on 187
ovarian 143, 154, 328, 331
pathways in termites *342*–3
photoperiod effects on 186
temperature effects on 184–6
threshold 184–6
see also Metamorphosis
Devonian 203, 228, 229, *231*, 233
Diabrotica 258
Diamesa 175
Diamondback moth 427, 444
Diapause 70, 118, 172–9, 181, 186, 243, 255, 333, 366, 375, 523
and climate change 459, 465–6
hormone **70**
Diaphanopterodea 230, 240
Diapheromera velii 130
Diaphorina citri 423, 433
Dicliptera 230
Dictyoptera 207–9, 212, *231*, 234, 507, 509
Dictyopterus 65
Didelphis marsupialis (Mam.) 370
Didymuria violescens 298, 503
Digestive system 27, 57, 79–80, 83–4, 175, 261, 267
Dillwynia juniperina (Bot.) 317
Diopsidae 130
Dioryctria albovitella 295
Diplodiploidy 150
Diplolepis rosae 306, Plate 5c
Diplura 131, 202, 237, 250, 251, 268, 484
taxobox 494–5
see also Appendix
Diprionidae 166
Diptera
in biological control 313, 427, 437
chemoreception 107, 116–7
cyclorrhaphous puparium 70, 167, 220, 273, 454, 519
diversity 6–7, 244
ectoparasitic 370
and flowers 290, 313–4
fossil record *231*, 234, 290
fruit-boring 243, 245, 301
galling 305
haltere 51, 63, 97, 98, 164, 217, 219
as hyperparasitoids 365
immature stages 35, 60, 84, 165–6, 219, 258, 264, *271*, *273*, *280*, *281*, 398, 414, 519
leaf-mining 299–300
of medical importance 399, 404–5, 408–10, *412*
mouthparts 36, 39–*41*, 107
myiasis-causing 360, 398, 454, 519
as parasitoids 102–3, 362, 375
pinning/preservation *479–80*, 484
relationships *190*, *204*, 219–20
taxobox 519
see also Blood, feeding by flies; individual families
Dirofilaria immitis (Nem.) 412
Discal cell 50, 221, 523
Disease
endemicity 402, 406–8, 409–10
epidemic 402–3, 410–11
reservoirs 399, 407, 409–12

Disease (*continued*)
transmission 5, 41, 183, 398, 400, 402–12
vectors 40, 183, 247, 398–413, 420–1
Dispersal 52, 61, 114–5, 179, 181, 240–4, 247, 251, 312, 317–8, 347, 356, 436, 448, 455, 459, 461, 467
Ditrysia 222, 314, 523
Diurnal behaviour 106, 118–9, 222, 283–4, 357, 385, 391, 506, 508
Diversity 2–18, 136, 277
Division of labour 272, 323, 325–7, 345, 348
Dixidae *190*, 219, 273
DNA barcoding 7, 191, 193, 200, 490
Dobsonfly *see* Megaloptera
Domatia 318
Donacia 279
Dorcus curvidens 13
Dormancy **70**, 157, 186, 173, 180, 259, 358, 465, *see also* Diapause; Quiescence
Dorsal
closure 161, 241
diaphragm 71–*2*
vessel 69, 71–*2*, 76
Dorylus (-inae) 323, 339
Dracaena braunii (Bot.) 469
Dragonfly *see* Anisoptera; Odonata
Drakaea (Bot.) *Plate 5d*
Drone 140, *331*, 333, 336
Droseraceae (Bot.) 378
Drosophila (-idae)
D. bipectinata 138
D. mauritiana 136
D. melanogaster 5, 13, *136*, 242
D. simulans 136
courtship and mating 127–8, 138, 144
embryology 157, 160–1, 163–6, 238
genomics 5, 67, 69, 86, 142, 144, 157, 195, 434
radiation, Hawai'i 245–6
segment formation 163–5
sex determination 140
tracheae 75
Dryinidae 370
Dryococelus australis 14, *Plate 1c*
Drywood termites *see* Kalotermitidae
Dufour's gland 114, 116, 325, 395
Dulosis 347
Dung 4, 11, 121, 127, 130, 187, 250, 261–4, 324–5, 441, 452–3, 477–8
Dutch elm disease 115, 260, 303, 471
Dyar's rule 147, 181–2
Dynastes hercules 10
Dytiscus (-idae) 218, 272, 275, 281, 284

Earwig *see* Dermaptera
Ecdysis *see* Moulting
Ecdysis triggering hormones (ETH, pre-ETH) 68, 168, 171
Ecdysone 68, 84, 172, 181, 433–4
Ecdysteroid 67–9, **70**, 153–4, 168, *170–2*, 174, 188, 241–2, 366, 433
Echinothrips americanus 468
Echium plantagineum (Bot.) 312
Ecitoninae 339
Eclosion 118, 160, 165, 168, 183, 241–2, 284, 347, 498, 515, 523
hormones **70**, 168, *170–2*, 434
Ecoclimatic index (EI) 459–61
Economic injury level (EIL) 420–*1*
Economic threshold (ET) 420–*1*, 452
Ectoedemia 234
Ectognathy 202–3, 229
Ectoparasite 213, 220, 245, 268, 355, 358, 360–1, 363–4, 370–2, 375, 378, 382, 439
Ectothermy 106, 459–60, 466
Edwardsina polymorpha 281
Egg 145
attendance 148–9
canal 53, 149
inhibition of development 152–3
laying *145*, 146–7
movement down ovipositor 149
rafts, of mosquito 280
size and structure 145
see also Embryology; Ootheca; under each order
Ejaculatory duct 55, **91**, 93, 133, 215
Elaiosome 317
El Segundo blue 17
Elateridae 54, 121, 256
Electroantennogram 108–11
Elephantiasis 398, 412–3
Elephantodeta Plate 3f
Ellipura 202–3
Elytron (-a) *51*, 100, 193, 217, 225, 241, 275, *281*, *284*, 463, 516–7
hemelytra 51, 274, 513
subelytral cavity 178, 279
Embia major 503
Embiidina *see* Embioptera
Embioptera 207, 210, 250, 323–4, 382–3, 485, 503, *Plate 7c*
taxobox 503
see also Appendix
Embolium 51
Embryo (-ology) 5, 33, 36, 68, 86, 145, 148–53, 160–5, 228, 238, 241–2, 334–5, 363
Emerald ash borer *see Agrilus planipennis*
Empididae *127*, 261
Encapsulation 71–3, 363, 365–6, 369, 465
Encarsia 424, 438, 441
Encephalartos (Bot.) 308, Plate 4b
Encephalitis (-ides) 407–10
Encyrtidae 152, 423, 440, 470
Endite 52–5, 201, 238, 275
Endocrine system 64, 66–9, 84, 126, 133, 153, 170, 174, 186, 331, 343, 366
Endocuticle *28*, 169–172, 183
Endoparasite 245, 355, 369, 378, 439
Endoparasitoid 218, 355, 359–0, 362, 364–9, 423, 440, 472, 517–8
Endophallus 55, 131, 136–7
Endopterygota *see* Holometabola
Endothermy 106–7
Endoxyla Plate 2c
Ensifera 209–10, 502
Entognatha 202, 493
Entomobryidae 268
Entomology, defined 2
Entomopathogenic nematodes (EPNs) *see* Nematodes
Entomophagy 20, 24, 364, 438, 439, *Plate 2a-h*
Entomophily 313
Entomophobia 415
Entomophthora (Fungi) 443
Environment
aquatic 282–4
effects on development 183–8
extremes 174–8
induced variation 179–81, 187–8
monitoring 268–9, 284–5
Eocene 103, *231*, 235
Eosentomon 495
Ephemera danica 497
Ephemeroptera
classification 205–6
diagnostic features 206, 497
egg laying 146–7
fossil record *230*, 233
morphology 75, 240
nymphs 52, 272, 278
preservation 485
relationships 192, 205–6
subimago 158, 168, 206, 242, 497
taxobox 497–8
wing articulation 52, 238
see also Appendix
Ephestia 14, 363
Ephydra (-idae) 176

Epiblema scudderiana 175
Epicoxa *46–7*, 238
Epicuticle 28–30, 32, 111, 169, 171–2
Epidemic 402–3, 410–11
Epidinocarsis 435
Epimeron *46–7*
Epiophlebia 206
Epipharynx 36, 40–1, 43, 521
Epiphyas postvittana 423, 453
Epiproct 52, 206, 495, 496, 498, 500, 509
Epirrita autumnata 175
Episternum *46–7*
Eretmocerus 424
Erigonum parvifolium (Bot.) 17
Eriococcidae 22, 80, 304–6, *Plates 3f, g, 4a*
Eristalis 264, 414
Escovopsis (Fungi) 266
Essential oils 294, 430, 470
Eucalyptus (Bot.) 294, 300–1, 306, 470; *see also Corymbia*
Eucera 65
Eucharitidae 360
Euglossine bees 289, 291, 314, *Plate 4e*
Eumargarodes 259
Eumeninae 325, 328
Euphasiopteryx ochracea 360, 362
Euphilotes battoides 17
Euprenolepis procera 265
Eurema hecabe 152, 180
Euphorbia pulcherrima (Bot). 468
Euplantulae 48, 207, 506
Europe (-an)
  beewolf *see Philanthus*
  biocontrol agents 441–3
  corn borer *see Ostrinia nubilalis*
  pest insects 21, 176, 198, 258, 296, 301–2, 305, 448, 466–71, 512
  galling insects 305
  earwig *see Forficula auricularia*
  grapes and phylloxera 296–7
  insect conservation 3, 14, 18–20
  praying mantis *see Mantis religiosa*
  red ant *see Formica polyctena*
  wasp *see Vespula* spp.
Eurosta solidaginis 175
Eurytomidae 307, *Plate 6a*
Eusociality 327–45, 348–50; *see also* Sociality
Eusternum (-a) *46–7*
Evolution
  arthropods 183, 201–2, 228–9
  of aquatic lifestyles 275, 277
  developmental biology 157, 164–5, 217
  of genitalia 131, 138
  of metamorphosis 160, 241–2
  in the Pacific 245–7
  of plants and insects 236–8
  success of predation/parasitism 375–6
  of sociality 168, 214
  and systematics 191–7, 199–200
  of wings 206, 238–41
  see also Coevolution; Darwin, Charles; Phylogenetics
Excretory system **70**, 86–90, 178, 319
Exite 201, 238, 275
Exocrine (dermal) gland 30, 111, 114, 359, 395
Exocuticle 29–30, 32, 169, 171–2
Exophily 405–6
Exoskeleton 27, 57, 59–60, 73
External genitalia 52–5, 128, 131–3, 136, 138, 145, 210, 503, 508, 517, 523
Extrafloral nectaries 318
Exuviae 218, 220, 273, 280, 285, 433, 478, 482, 485–6, 523, *Plate 1e, 3f*
Eye 33, 36, 43, 64, 95, 117–121, 130; *see also* Ocellus; Stemma; Vision

Fahrenholz's rule 370–2
Fairy circles 257
Faniidae 261
Fat body 33, 57–*8*, 68–**70**, 76, 84–5, 89–90, 153, 445, 508
Faeces 80, 82, 86, 88, 90, 250, 256, 265, 267, 286, 344, 359, 383, 410, 416; *see also* Dung; Frass
Feeding
  of *Apis mellifera* 38–9
  aquatic functional groups 286
  on carrion 4, 250, 264–5, 324, 413–4, 477
  categories 38, 78
  on dung 261–4, 324
  filter chamber 79–80
  filter feeding 43–*4*, *271*, 273, 285–6
  fluid 39–42, 83
  on fungus 265–8
  macerate-and-flush 83
  methods 38–44, 78–84
  mouthparts 36–44
  nuptial 128–9, 134–6, 386, 520
  phytophagy (herbivory) 83, 242–3, 290, 293–309
  sap sucking 80, 85, 256, 302–5
  stereotyped behaviour 107
  stylet-sheath 83, 303–4
  see also Blood; Gut; ordinal taxoboxes
Female choice 128–131, 136–139, 144, 243, 247
Femur 47–8, 60, 203, 259, 361
Fenusa pusilla 301
Fertilization 92, 128–9, 134, 136–40, 144, 150, 153, 160, 268
  self-fertilisation 150–2
  indirect 131, 493–6
Ficus (Bot.) 236, 291–2
Fig *see Ficus*
Fig wasp 236, 291–2, 315, 316; *see also* Agaonidae
Figitidae 365, 367
Filariasis 398, 405, 412–3, 425
Filter
  chamber 79–80, 215, 513
  feeding 43–*4*, *271*, 273, 285–6
Fipronil 352, 432
Fire ants *Plate 6e, see Solenopsis*
Firebrat 242, 496
Firefly *see* Lampyridae
Fishfly *see* Megaloptera
Flabellum 39
Flagellum (flagellomere) 43, *45*, 99, 105, 213, 219–20, 519
Flavivirus (-idae) (Vir.) 407–9
Flavonoid 33
Flea beetle *see* Alticinae
Flight 48–52, 57, 61–3, 77, 96, 99, 101, 105–6, 121–4, 179, 218, 232, 238–41
  evolution 206, 208, 238–41
  flightlessness (aptery) 14, 52, 159, 179–80, 193, 220, 243, 251, 296, 340–1
  muscles 35, 46–8, 59, 61–3, 105–7, 176, 203
  path *112*, 114, 477
  tone 105
Fluctuating asymmetry 187
Flux (aquatic-terrestrial) 282–3
Fly *see* Diptera
Fog basking 178
Follicular relic 183
Food
  canal 38–41, 302
  ingestion 4, 38, 71, 77–83, 254, 361–2
  for domesticated animals 24, 252–3
  for humans 20–4, *Plate 2a-h*
Foraging 107, 117–8, 177, 257–8, 263, 315, 324, 326, 329–37, 339, 344, 351–2, 355–6, 358–61, 373–4
Forcipomyia 314, 360
Forensic entomology 265, 413–4
Forficula auricularia 36, 37, 254, 500
Formica spp. 323, 336, 337, 339, 346, 347, 525
Formicidae
  ant mimics 345, *377–8*, 395
  ant plants (myrmecophytes) 318–320

Formicidae (*continued*)
colony and castes 337
defence methods 392–6
eusociality 336–7
fossil *Plate 4a*
flower-visiting 315
fungus farming 265–7, 344, *346*
as human food 22–3, *Plate 2g, 2h*
leafcutter ants (*Atta*, Attini) 105, 265–7, 323, 363
morphology 223, 525
mutualism with *Acacia* 294
myrmecophily 18, 315
myrmecotrophy 318
nesting 18–19, 265–6, 308, 317, 337, 339–40, 360, 385
preserving 481
relationships 222–4, 328
slave-making (dulosis) 347
in soil monitoring 268–9
tending Hemiptera 214, 256–8, 304, 318–9, 324, 340, 352, 513, *Plate 6b, 8b*
trail-marking pheromones 114
tramp ants 13, 15–16, 245, 247, 323, 351, 352, 470, *Plate 6e*
Formosan subterranean termite *see Coptotermes formosanus*
Fossil (insects) 9, 45, 47–9, 103, 203–9, 211–4, 219, *227*–38, 240–1, 290, 463, *Plate 4a*; *see also* Amber
Frankliniella occidentalis 305, 511
Frass *see* Faeces
Fraxinus (Bot.) 471
Freeze tolerance / avoidance 175
Frons 36, 43, 220
Frontoclypeus (suture) 36
Fruit damaging 39, 116, 301, 304, 308, 360
Fruit fly *see Drosophila*; Drosophilidae; Tephritidae
Fulgora laternalia 380
Fulgoridae (-oidae, -omorpha) 121, 214–5, 287, 384, 479, 485, 512
Fungus
in biological control 443
farming 265–8, 344, *346*
fungivory 261, 265–8, 344–6, 517
gnat *see* Mycetophilidae; Sciaridae
Funicle 45
Furca (-ula) 202, 209, 250–1, 395, 493
Furcatergalia 206

Galea 37
Gall
induction 214, 293, *296*, 304–8
of scale insects 80, 305–6
types *306–7*
midges *see* Cecidomyiidae
wasps *see* Cynipidae
see also Phylloxera
images *Plates 4f, g, 5a-c, 6a*
Galloisiana nipponensis 505
Gambusia (Fish) 439
gamma-aminobutyric acid (GABA) 64, 432
Ganglion (-a) 36, 64–6, 105, 135, 174
Garden tiger moth *see Arctia caja*
Gas exchange 45, 57, 73–7, 148, 178, 203, 237–9, 278, 498, 523, *see also* Tracheal system
Gaster 329, 331, 393, 525
Gastrulation 160–1
Gazera heliconioides Plate 8a
Gelechiidae 299, 308, 453
Geminiviridae (Vir.) 424
Gena (-ae) 35–6, 521
Generation time 10, 13, 126, 128, 172–3, 180, 373, 403, 459, 464
Genetic / genomic
engineering 67, 430, 434, 442, 445–7
expression 36, 67, 86, 164, 183, 331, 343
sequencing 67, 69, 86, 90, 157, 192, 195–6, 200, 208, 214, 217, 219, 368, 434, 445–6
manipulation of pests 454–5
modification 450, 454–5
regulatory sequences 157, 165, 228, 242
Genitalia
diversity of form 136–9
external 52–5, 128, 131–3, 136, 138, 145, 210, 503, 508, 517, 523
internal 136–7, 200
species-specificity 53, 136–9
Gentiana acaulis (Bot.) 300
Geological time 230–2, 244–7, 275, 293, 458
Geometridae 166, 175, 180, 379, 381
Germ
anlage/disc 160–1
band 161–2
Germarium 92
Gerris (-idae, -omorpha) 99, 138, 179, 216, 272, 274, 284
Gigantism 10, 232–3
Gill
anal *278*, 498, 500
compressible 279–80
function 74, 237–8, 241, 279, 281–2
incompressible (plastron) 280
plecopteran 208, *278*
as proto-wings 208, 239–40
tracheal 74–5, 221, 275–9, 498
see also Caudal, lamellae; Plastron
Global
trade 425, 458, 468–473
warming 458, 467, *see also* Climate, change
Globitermes sulphureus 393
Glossa (-ae) 38–9
Glossata 222, 523
Glossina (-idae) 92, 362, 412
Glow worm 13, 17, 121; *see also Arachnocampa*; Lampyridae
Gluphisia septentrionis 135
Glycaspis 304, *306*, *Plate 8e*
Gonapophysis 53–4
Gondwana 236, 244
Gongylidia 266–7
Gonimbrasia belina 20, 23, *Plate 2b*
Gonipterus 197, 199, 470, *Plate 8f*
Gonopore 53, 55, 92–3, 133, 202–3, 493, 495
Gonostyle 54–5
Gracillariidae 299–301
Granary weevil *see Sitophilus granarius*
Graphium spp. 12, 476, *Plate 3a*
Grapholita molesta 301, 453
Grasshopper *see* Orthoptera
Green vegetable bug *see Nezara viridula*
Gregarious 323, 364–5, 366, 369, 382–3, 392–6, 472, 495
phase locust 112, 181
Gripopterygidae 208, *278*
Ground beetle *see* Carabidae
Ground pearls 259
Growth 68, **70**, 84–5, 157–8, 162, 166, 172–3, 181–4, 186–7, 373; *see also* Development
Gryllus (-idae) *58*, *120*, 133, *149–50*, 360
Grylloblatta (-odea) 207, 210–11, 250, 467, 485, 504–6
Gryllodes 134
Gryllotalpa (-idae) 48, *104*, 254, 442
Gula 37
Gut
foregut 27, 68, 80–2
hindgut 27, 79, 80–7, 266, 341, 508, 514
midgut 40, 78–87, 154, 213, 401, 411, 444–6
Gyne *see* Queen
Gypsy moth 106, 293, 298, 435, 453, 473, *Plate 8c*
Gyretes 284
Gyrinidae 99, 218, 275, 284

Haematobia irritans 261
Haemocoel 57, 60, 71, 84, 89, 98, 100, 138, 151, 368, 401, 413, 434, 443, 446
Haemocyanin 238
Haemocytes 69, 72, 363, 365, 369
Haemoglobin 33, 69, 71, 84, 279, 284–5
Haemolymph
bleeding 386, *Plate 7f*
circulation 57, 69–72, 107
constituents 57, 69–72
functions 39, 57, 66, 168
and immune system 69, 72–3, 363–6, *308*
osmoregulation 86–7, 288
role in moulting 168–9, 171–2
storage proteins 84–5
transport of hormones 66–8
Hair 29, 32, 48, 50, 60–2; 96–8, 107–8, 278, 280, 284; *see also* Seta
pencil 113
plate 97–*8*, 336
Halictidae 325–6, 328, 350
Halobates 288
Haltere 51, 63, 97, 98, 164, 217, 219
Hamma rectum 378
Hamuli 49, 525
Hamulohaltere 52, 214, 512
Hangingfly *see* Mecoptera
Haplodiploidy 140, 152–3, 326–7, 349–50, 525
Harmonia axyridis 4
Harpobittacus australis 129
Harvester ant 308
Hawai'i 15, 245–7, 268, 288, 352, 408, 422–3, 436–7, 444
Hawkmoth *see* Sphingidae
Head 33–45, 64, 71, 95
Hearing 96–103, 105
Heart 69–71, 107, 168; *see also* Dorsal, vessel
Heat-shock proteins 177
Heelwalker *see* Mantophasmatodea
Helichus 272
Heliconius 12, 138, 381, 385
Helicopsychidae *283*
Helicoverpa 174, 425, 445–6, 449
Heliothis 174, 359, 425, 428, 445–6; *see also Helicoverpa*
Hellgrammite 277, 515
Helodidae 279
Helluomorphoides 385
Hemelytron *51*
Hemimeridae 209, 501
Hemimetaboly 27, 48, 68, 146, 159–62, 165, 183, 207, 241–2, 272
Hemerobiidae *184*, 439, 515
Hemiptera
acoustic courtship 128
aquatic 99, 179, 272, 274, 278–9, 284, 287, *Plate 2d, e.*
digestive system 78–9, 82–6
diversity 6, 512
feeding 36, 41–3, 293, 302–4, 352
fossil record *231*, 233–4
galling 296–7, 304–8, 311, *Plate 4f, 5a, b*
life cycle *159*
mouthparts 36, 41, 43, 302, 361
parental care 324–6
peritrophic membrane 82–3
predatory 359, 361, 383, 512
pinning/preservation 485
relationships 212–5, 512
sap sucking 258–9, 302–5, 352, 361
seed feeding 105, 308
sharpshooter 85, 436–7
sound production 99, 104
taxobox 512–3
wing 51, 158, 512
see also Heteroptera; Appendix
Hemipteroid (assemblage) 55, 212, 233–4
Henoticus serratus 260–1
Hepialidae 22, 108, 301
Herbivory *see* Phytophagy
Hercules beetle 10
Hermaphroditism 150–2, 347, 426
Hermetia illucens 252
Hesperiidae (-ioidea) 135, 222
Heterochrony 242
Heteromorphosis 166, 525
Heteropeza pygmaea 151
Heteroptera 51, 83, 85, 116, 144, 147, 214–5, 302, 384–5, *479*, 512–3
Heterorhabditis (-idae) (Nem.) 442
Heuweltjies 258
Hexamerin 69
Hexapoda 27, 191, 197, 201–2, 228–9, 493
Hierodula membranacea 135
Hippoboscidae 150, 360
Hippodamia convergens 170, 179, 441
Hirsutella (Fungi) 443
Histeridae 264
Hodotermitidae 258
Holocyclic 176, 296
Holometabola (= Endopterygota) 160, 165–6, 207, 214, 216–7, 347
Holometaboly 27, 159–60, 207, 214, 233, 241–2, 513
Homallotermes 393–4
Homeostasis 68, **70**, 86, 89, 343, 434
Homeotic genes 164–5, 228, 238
Homology 27, 52, 90, 192, 208, 239
serial 27, 38, 192
Homoptera 214–5, 512
Honey 2, 5, 12, 23, 39, 315, 333–6, 338–9
Honey bee *see Apis mellifera*
Honeydew 15, 31, 80, 214, 256, 258, 304–5, 318–9, 324, 336, 340, 352, 423–4, 513, 514, 519, 524–5
Honeypot ant 11, 23, *30*, 337, 340, 351, *Plates 2g, h*; *see also Melophorus; Myrmecocystus*
Hormaphidinae 325, 326
Hormone
adipokinetic (AKH) 67, **70**
bursicon **70**, 168, 171
brain hormone *see* Hormone, prothoracicotropic
in caste differentiation 331, 337, 343
chloride-transport stimulating **70**
corazonin **70**, 168
crustacean cardioactive peptide **70**, 168
definition 66
diapause (DH) **70**, 174
ecdysis-triggering (ETH) **70**, 171
ecdysteroid 68, 153–4, 171–2
eclosion (EH) **70**, 168, *170*–1, 434
effects on parasitoids 366
juvenile (JH) 67–**70**, 130, 153, *170*–3, 241, 330, 433–4
JH mimics 433–1
melanization and reddish colouration (MRCH) **70**
moulting hormone *see* Ecdysone
oostatic (OH) **70**, 154
ovarian ecdysteroidogenic (OEH) **70**, 154
pigment-dispersing (PDH) **70**
pre-ecdysis triggering (PETH) **70**, 171
prothoracicotropic (PTTH) 67, 69–**70**, 170–3, 433
in reproduction **70**, 145, 153–*4*
see also Neuropeptide
Host
acceptance and manipulation 359, 361–3, *367*
definition 345, 355
disease transmission to 361–2
immune responses 73, 363–4, 445
location 14, 102, 166, 355–61, 442, 453
preference of parasite 403, 407, 411
selection and specificity 364–72
switching of parasites 292, 371, 425, 440
use by parasites/parasitoids 355, 364–6, 369, 370–2

Hour-degrees *see* Day-degrees
House dust mite *see Dermatophagoides pteronyssinus*
House fly *see Musca*; Muscidae
Hover flies *see* Syrphidae
Hox genes 164–5
Human
  bot fly 360–1
  body / head louse 398, 410, 416, 510
Humeral
  angle *49*
  plate 52, 206
  vein *50*
Humidity 96, 123, 148, 157, 177–8, 187, 254–5, 257, 266–7, 403, 411
Hyalophora cecropia 105
Hydraenidae 279–80
Hydrocarbons 29, 111–12, 188, 198–201, 224, 255, 345, 395
Hydrometra 65
Hydrophilidae (-oidea) 218, 275
Hydropsychidae 221, 283, 285
Hylobittacus apicalis 129
Hymenoptera
  in biological control 14, 116, 437, 439–40, 465, 525, *Plate 8d*
  defence methods 389, 392–6
  diversity 6, 524
  embryology 161
  flower-frequenting 314, *Plate 4e, 5d,e,h*
  fossil record *231*, 233–5, 290, *Plate 4a*
  galling 293, 308, *Plate 5c, 6a*
  haplodiploidy (arrhenotoky) 140, 152–3, 327, 349–50, 525
  as hyperparasitoids 308, 365–*7*, 375, 440–1
  immature stages 165–*7*, 524
  mining and boring 299–301
  morphology 36, 38, 49, 52–4, 65, 116, 222–3
  as parasitoids 264, 325, 347, 359–60 *Plate 6f*
  pinning/preservation 479, 485
  radiation due to parasitism 375–6
  relationships 217, 222–4, 524
  sociality 323–7, 348–51
  stings *149*, 331, 352, 381, 416–7, 525
  taxobox 524
  see also Appendix
Hymenopus spp. 13, 356, 380
Hypericum (Bot.) 309
Hypermetamorphosis 165–6
Hyperparasitism (-oid) 308, 365–*7*, 375, 440–1
Hypertriangle 498
Hypoaspis (Acari) 441
Hypogastrura 251
Hypognathy 35
Hypopharynx 36, 38, 40–1
Hypothenemus hampei 463

Icerya 151–2, 419, *426–7*, 469, *Plate 8b*
Ichneumonidae (-oidea) 149, 223, 315, 365, 368, 439, 440, 525
Ichnovirus (Vir.) 369
Identification 3–4, 6, 27, 38, 50, 174, 181, 475, 488–90, 492
  Interactive keys 198, 490
Idiobiont 366
Idolum 356
Ileum **70**, 80, 82, 86–7
Illiesoperla 500
Imago (imagines, imagos) 158–60, 168–9, 216
  imaginal disc (bud) 160–*1*, 164–6, 216, 242
  see also Adult
Imidacloprid 339, 430–2, 448, 471
Immune system 69, 72–3, 363–4, 366, 378
Incurvariidae 299–301
Indonesia 22, 413, 458, 463
Industrial melanism 381
Inka cells 68, 168
Inoculative release of natural enemies 438, 441
Inquiline 308, 345, 347
Insect conservation 14–20, 268–70, 439–41
Insect growth regulators (IGRs) 419, 430, 433–4
Insecta, defined 203
Insecticide
  broad spectrum 339, 415, 424, 432
  effects of 425–8, 430–4
  formulation 339, 352, 432, 441–2
  microbial 438, 441–3, 445, 448
  modes of action 428, 444
  resistance 405–6, 408, 415–6, 420, 423, 427–8, 430, 433, 444–51
  sublethal effects 339, 431–3
  tolerance 427–8, 445
  types 430–5
  use on brown planthopper 439
  viral 434, 441–2, 445–7
Insectivory 97, 101, 103, 233, 378, 380–1, 384–5, 431
Instar
  constrained number 147, 157
  definition 157
  determination 181, 343
  duration 158
  Dyar's rule 181–2
  first 161, 165–6, 217–8, 241, 259, 324, 326, 362, 369
  imaginal 158–60, 168–9, 216
  number 158
  Przibram's rule 182
  pupal 151, 160, 166, 519
  subimaginal instar 48, 158, 205–6, 242, 497
  see also Stadium
Integrated pest management (IPM) 419, 421, 428–9
Integument 28, 72, 160, 315
Interactions
  insect–carrion 264–5
  insect–dung 261–4
  insect–fungus 265–8
  insect–plant 291–3
International Code of Zoological Nomenclature 197
Intersegmental membrane / fold 30, 33, 46–7, 208, 303
Intertidal zone 118, 229, 272, 287–8, 484, 494, 517
Ips grandicollis 160
Iridomyrmex Plate 8b
Ischnocera 213, 510
Isoptera *see* Termitoidae
Isotoma (-idae) 251, 493
Israel 339, 410
Ivermectin 264

Japan
  antlions 357
  ants 323
  beetles 137, 441, 443, 471
  biocontrol 443–4
  diapause research 174
  encephalitis 407–9
  Grylloblattodea 467, 504–5
  insect pets 12–13
  trade 423, 470
Japygidae 251
Jasmonic acid 297
Jewel beetle *see* Buprestidae
Johnston's organ *see* Antenna
Jugal (wing) 49
Jumping 48, 61, 202, 209, 214–5, 220, 240, 251, 493, 502, 512–3, 517
Jurassic 211, *231*, 234–5, 236, 244, 290
Juvenile hormone (JH) *see* Hormone, juvenile; JH mimics

Kairomone 115, 294, 359, 366, 390
Kalotermes (-itidae) 343
Katydid
  leaf mimic *380*
  mating 133–4, 136, 144
  ovipositor 53, 149

preservation 478–9, 486
relationships 207, 209
sound production 102–4
sound reception 100–3
see also Orthoptera; Taxobox 9
Ked 150
Keiferia lycopersicella 453
Kerria lacca 5
Keystone species 5
khapra beetle *see Trogoderma granarium*
Kibara (Bot.) 319
Kinesis (kineses) 122–4
Koinobiont 366, 369
Krüppel homolog 1 (*Kr-h1*) 163, 172, 242

La Crosse virus 408
Labellum (-a) 40–1, 107
Labiduridae 251
Labium (-al)
of biting flies 40–1, 519
defined 38, 40
glands 82, 266, 336, 517
of Hemiptera 214–5, 302, 304, 512
illustrated *37–42*, 304, 362
of larval Odonata 42–3, 362, 498
palps 38–9, 42, 220–1, 493, 521, 523
Labrum (-al) 36–*7*
Lac insect 5
Lacewing *see* Neuroptera
Lacinia (-ae) *36–41*, 212, 214, 220, 509, 511, 521
Ladybird beetle/ladybug *see* Coccinellidae
Lampyridae 121–2, 385, 387, 517
Larva
definition 27, *160*, 165–7, 241–2
heteromorphosis 165–6, 362, 525
types 165–6
see also under each order
Larviposition 153
Lasius 112, 337
Leaf
chewing 235, 290, 295, 298–9
mining 235, 299–301, 374
Leaf beetle *see* Chrysomelidae
Leaf-cutter ant (*Atta*) 265–7, 323, 363
Lecanicillium (Fungi) 443
Leg
adult vs immature 45, 164
development 46–7, 164
of flea 521
fossorial 48, 253–4, 259, 517
functional types 48, *254*
modification in predators 361
morphology 47–8
movement 60–1
shedding (autotomy) 278, 378, 383–4
Leishmania (Prot.) 399, 412
Leishmaniasis 398–9, 412
Lek 127, 130
Lepidoptera
adult mouthparts 39, 43, 221
conservation 12, 18–19, 269
control of pests 425, 428, 430, 434, 437, 443–6, 449
corpora allata 68, 171
crypsis 379–*82*, *Plate 7a, g*
diversity 3, 6–7, 243, 290, *Plate 1a, b*
farming 12–13, *Plate 1g*
flower-frequenting 314, *316*, *Plate 5g*
as human food 20, *22–3*, *Plate 2c, d*
fossil record *231*, 234
fruit-boring *302*
larva (caterpillar) 165–6, 221, 383
leaf-chewing/mining 216, 293, 298–302, *Plate 1e, 4d*
parasitization 375
pheromones *112*–13, 116, 523
pinning/preservation *482*, 485
puddling 135–6, *Plate 3a*
pupae (chrysalis) 156, 167, 221, 354, 524
radiation 222
relationships 216, 221–2
scales 32, 50, 62, 90, 107, 221, 384, 478, 481
sex-determination 101
taxobox 523–4
wings 49–51
see also Butterfly; Moth; Appendix
Lepisma saccharina 496
Lepismodes inquilinus see Thermobius domestica
Leptinotarsa decemlineata 419, 447–8
Leptoceridae 221, 283
Leptocoris 385
Leptocybe invasa 470
Leptoglossus occidentalis 105
Leptomastix 441
Leptophlebiidae (-oidea) 206
Leptopilina boulardi 365
Leptopodomorpha 216, 274
Leptospermum (Bot.) 313
Lerp 304, 485, 513, *Plate 8e*
Lestes (-idae) 278
Lethocerus indicus 24, *147*, *Plate 2d, e*
Leverhulmia maraie 229
Levuana iridescens 437
Libellulidae *51*, *145*
Lice *see* Psocodea
Life cycle
of *Apis mellifera 331*
of *Chironomus 158*
of *Culex pipiens 280*
of *Danaus plexippus 156*
of *Ips grandicollis 160*
of *Leptinotarsa decemlineata 448*
of *Phengaris arion 19*
of *Nezara viridula 159*
of *Nasutitermes exitiosus 342*
of phylloxera *296*
of *Plasmodium 401*
of *Pleistodontes froggatti* (fig wasp) *292*
of *Trichogramma 440*
Life histories 157–69
Ligation 66
Light
production 121–2, 384
reception *see* Eye; Vision
ultraviolet 29, 32, 101, 315, 444–5, 451–2, 475–8
Limacodidae 416
Limenitis archippus 390–1
Linepithema humile 15, 352, 419, *Plate 6e*
Liposcelis (-idae) 213–4, 509
Lipophorin 69, 111
Lipoxygenases 297
Litter insects 250–6, 290, 372, 476, 484, 486–7
Littoral insects 284, 287–8
Locomotion 57–63, 358; *see also* Flight; Jumping; Swimming; Walking
Locust
allergy 417
development illustrated 502
excretion 87–9
as human food 22
internal structures 56
movement 62, 107
mycoinsecticide for 443
ootheca 149–50
osmoregulation 86–7
phase 112, 181, 187–8
sound production 103
see also Locusta migratoria; Orthoptera; *Schistocerca gregaria*
Locusta migratoria 30, 179, *188; see also* Locust
Longhorn/longicorn beetle *see* Cerambycidae
Longitarsus jacobaeae 301
Lonomia obliqua 416
Lophodiplosis trifida 311
Lord Howe Island 14
stick insect *Plate 1c, see Dryococelus australis*
Louse (lice) *see* Psocodea
Lubber grasshopper 382
Lucanidae 13, 38
Luciferase (-in) 121

Lucerne flea *see Sminthurus viridis*
Lucibufagin 387
Lucilia 65, 143
Lucky bamboo *see Dracaena braunii*
Lure-and-kill 86, 452–3
Luteovirus (Vir.) 86, 424
Lycaenidae 18, 222
Lycidae 65, 115, 314,
Lycorea ilione Plate 8a
Lygaeidae 83, *132*, 308
Lymantria (–idae) 293, 298, 435, 453, 473, *Plate 8c*
Lyonetia (-idae) 299–*300*
Lytta polita Plate 7f

Machilis (-idae) 54, 203, 251
Macroglossum stellatarum Plate 5g
Macrolophus 201, 441
Macromyzus woodwardiae 512
Macropanesthia rhinoceros Plate 3d
Macrotermes (-itinae) 201, 254, 256, 267, 344, 346
Macrotrichia 32, 50, 96, 221, 523, *see also* Hair; Seta
Macrozamia lucida (Bot.) 106
Maculinea see *Phengaris*
Maggot 35, 60, 84, 165, 219, 243, 258, 264, 273, 279, 398, 519; *see also* Rat-tailed maggot
Magicicada 188, 254, 256, 376
Malagasy star orchid *see Angraecum sesquipedale*
Malaria 400–6, 411
Mallophaga 213, 510; *see also* Ischnocera
Malpighian tubules 57, 80, 82, 86–9, 122, 175, 202, 213, 218, 288, 515, 517
Malus (Bot.) 243, 300
Mamestra configurata 176
Mandible
  of antlion 357, 362
  in defence 393–5
  dicondylar 203, 205–6, 496
  of Diptera 40, 43–4, 182, 219, 273
  glands 82, 115, 331, 395
  of Hemiptera 214
  homology 201
  of leaf-cutter ants 266
  monocondylar 495
  of predators 3, 42, 357, 362
  of pupa 167
  structure and function 36–40, 42, 201–2
  stylet 40–2, 214, 276, 302–5, 511–2
Mandibulata 228
Manduca sexta 13, 68, 168, 171, 173, 242, 363
Manihot esculenta (Bot.) 435
Mansonia 279, 413
Mantibaria manticida 360
Mantis (-idae) 360, 379
Mantispidae 514
Mantodea
  copulation 135
  flower-mimicking 13
  fossil record *231*
  nuptial feeding 129
  nymphs *Plate 3e*
  as pets 13
  pinning/preservation 485
  raptorial legs 48, 211, *361*, 506
  relationships 207, 211, 507
  sexual cannibalism 135
  taxobox 507
  see also Appendix
Mantophasmatodea
  discovery 505
  fossil record 211, *231*, 241, 505
  pinning/preservation 486
  relationships 211, 506
  sound 506
  taxobox 505–6
  see also Appendix
March fly *see* Bibionidae
Margarodes (-idae) 250, 259
Marine insects 118, 151, 172, 182, 246, 272–4, 287–8, 414, 517
Marmara 301
Mason wasp *see* Eumeninae
Masquerade 379–80
Mass trapping 452–3
Mastotermes (-itidae) 212, 508
Matibaria manticida 360
Mating *see* Copulation
Maxilla
  galea 38–9, 221, 523
  lacinia 39–41, 212, 214, 509, 511
  palp 42, 116, 203–5, 209, 221, 222, 495–6, 501, 519, 523
  stylets 40–2, 302, 511, 513
Mayfly *see* Ephemeroptera
Mealworm *see Tenebrio*
Mealybug 30–*1*, 192, 302, 318, 384, 435, 439, 465, 468, 485, *Plate 6b*; *see also* Pseudococcidae
Mechanoreception 96–105
Mecistogaster 498
Meconium 168
Mecoptera
  courtship and mating 128–9, 144, 520
  embryonic development 162
  fossil record *231*
  immature stages 165, 167, 520
  pinning/preservation 486
  as predators 354–5, 520
  relationships 216, 219–20, 520
  taxobox 520
  wing coupling 49, 520
  see also Appendix
Mediterranean fruit fly *see Ceratitis capitata*
Meenoplidae 258
Megabraula 370
Megachilidae 325, 328
Megaloptera
  aquatic larvae 167, 217, 237, 276–8, 515
  fossil record *231*
  pinning/preservation 486
  as predators 355
  pupa 167
  relationships 11, 216–7
  taxobox 514–5
  taxonomic rank 216–7, 515
  see also Appendix
Meganeuropsis americana 10
Meganisoptera *see* Protodonata
Megarhyssa nortoni 149
Megasecoptera 230
Meinertellidae 203
Melaleuca (Bot.) 311
Melanin (-ism) 30, 33, 70, 73, 181, 379, 381
Melanophila 105, 261
Melanoplus 22
Melastoma polyanthum (Bot.) 356
Meliponini 315
Melittophily 315
Meloidae 386, 392, 416, *Plate 7f*
Melophorus 23, 177
Melyridae 314
Membracidae (-oidea) 114, 324, 395
Mentum 38, 44, 278; *see also* Postmentum; Prementum
Merimna atrata 105
Mermithidae (Nem.) 442
Mesoclanis polana *313*
Mesocyclops (Crustacea) 408
Mesosoma 223, 525
Mesothorax 45–7, 50, 194, 214, 238, 497
Messor 308
Metamorphosis
  complete *see* Holometaboly
  definition 159–60
  description 166–8
  effects of parasitoids 366
  evolution 241–2
  hormonal control 66, 68, 170–3, 366
  hypermetamorphosis 165
  incomplete/partial *see* Hemimetaboly
  insecticide interference 405

and legs 48
onset 242
role in success 9–10, 241–3
see also Heteromorphosis; Hypermetamorphosis
Metapleural glands 115, 254, 315, 395
Metapterygota 205
Metarhizium (Fungi) 443
Metasoma 525
Metathorax 45, *47*, 50, 205, 238, 501
Methona confusa Plate 8a
Methoxyfenozide 433
Methyl eugenol 452
Metriorrhynchus 115, 391, 392
Mexico 12–13, 16, 22–4, 129, 235, 243, 258, 296, 312, 323, 334, 392, 410, 422–3, 436–7, 448, 454, 461, 469
Microcoryphia 203, 484, 495
Microdon 348
Microhodotermes viator 258
Micromalthus 151, 218
Micromus tasmaniae 184
Microorganisms
in biological control 438, 441–3, 445, 448, 454
defence against 72–3, 254–5, 266–7
insect transfer 86, 350
and nutrition (symbionts) 33, 85–6, 90, 215, 261, 293, 344, 508
and reproduction 152–3
see also Buchnera; *Wolbachia*
Micropezidae 395
Micropyle 140
Microsporidia 85, 153, 442
Microtrichia 32, 50, 96
Midges
biting *see* Ceratopogonidae
non-biting *see* Chironomidae
Migration 174, 178–180, *188*, 252–3, 274, 284, 296, 339–40, 347, 356, 392
Migratory locust *see Locusta migratoria*
Milk gland 92, 150
Milkweed *see* Apocynaceae; *Asclepias*
Milkweed bug *see Oncopeltus fasciatus*
Mimetica mortuifolia 380
Mimicry
aggressive 356, 360
of ants *377*–6, 295, 395
automimicry 390, 392
Batesian 345, 347, 389–392
butterflies 389–91, *Plate 8a*
chemical 112, 116, 180, 346–7, 351
components of system 180, 389
as continuum 391–2
defensive 388–92
definition 379
of lycid beetles 115, 391–2
molecular 363, 369
Müllerian 390–2
Wasmannian 345
Miocene *231*, 234–5
Miridae 51, 83, 201, *395*
Mite (Acari) 296, 305, 338–9, 398, 411, 415, 417, 428, 439, 441; *see also Dermatophagoides, Varroa*
Mitogenome 195, 214
Models (predictive)
in biology 5
climatic, for distribution 458–66, 469, 472, 487
foraging theory 295, 358
parasitoid / host 366–7
kinship 359
in molecular evolution 163, 195–6, 230, 242
origin of flight 232, 240
in population biology 181, 186, 243, 245, 420–1, 423
predator/prey 355, 372–4, 376, 381
Red Queen 376
superparasitism 365
temperature on development 459–60
vectorial capacity 405
Model (in mimicry) 180, 356, 389–96
Mole cricket 103–4, *249*, 253–*4*, 324, 442
Molecular
biology (genetics) 5, 57, 67, 69
clock / dating 230
data 191–2, 195–6, 200–24, 228, 230
phylogenetics 193–6
Monarch butterfly 3, 13, 16, *156*, 179, 293–4, 323, 381, 385, 390–*1*, 392, 450–1, Plate *3g, see also Danaus*
Monogyny 337
Monolecty 315
Monomorium 308, 352
Monophlebidae 151, 426, *Plate 8b*
Monophyly 160, 193
Monura 205
Moon moth *see Argema maenis*
Mopane
tree *see Colophospermum mopane*
"worm" 20, 22–*3*, *Plate 2b*
Mordellidae 314
Mormon cricket 133
Morpho peleides Plate 1f
Morphology 27–93
Mosquito *see* Culicidae
Mosquito fish 439
Moth
bat-avoidance 98, 103–5
in citizen entomology 3–4
crypsis 379–82, *Plate 7g*
difference from butterflies 524
as food 23
hawkmoth 39, 316, *Plate 5g*
hearing 100–1, 103
industrial melanism 381–2
pheromones 108–111, 453
pollination 314, 316
puddling 135
scribbly gum moths 301, *Plate 4d*
see also Bombyx mori; Lepidoptera; *Lymantria dispar*
Moulting 29, 36, 66, 68–9, 84–5, 157–8, 169–74, 242; *see also* Ecdysone; Ecdysteroid
Mouthparts
modification in predators 361–2
structure and function 35–43
see also under each order
Multiparasitism 365–6
Mumpa 20
Musca (-idae) 24, *41*, 48, 65, 252, 261, 413, 415
Muscardine (Fungi) 443
Muscle
alary 71
attachment 27, 59–60, 171, 183
control 57–9, 64, 73, 104
in feeding 39, 42, 80–*1*, 83
flight 35, 46–8, 61–3, 77, 105, 107, 176, 211
in locomotion 59–61, 208
in sound production 104
structure and function 22, 57–63
Mushroom body 93
Mutualism
ant–fungus interactions 265–7
ant–Hemiptera interactions 318
ant–plant interactions 293, 295, 305–8, 316–20
fig wasp–fig 236, 291–2, 315
insect–plant 291–3
myrmecochory 317–8, 340
myrmecophytes 318
in origin of sociality 348–50
in plant defence 339
Mycangia 260–*1*, 345
Mycetocytes 84–5
Mycetome 85
Mycetophagy *see* Fungus, fungivory
Mycetophilidae 122, 219, 265
Myiasis 398, 454, 519
Mymaridae 365, 437, 440, 470
Myofibrils 59
Myophily 314

Myriapoda 228
Myrmecochory 317, 340
Myrmecocystus 23, 337
Myrmecodia beccarii (Bot.) 320
Myrmecolacidae 369
Myrmecophily 18, 315
Myrmecophyte 318
Myrmecotrophy 318
Myrmeleon (–tidae) 42, 357
Myrmica spp. 18, 19, 337, 345, 347
Myrtaceae (Bot.) 298, 300–1, 311, 313, 470
Myxomatosis 398, 407
Myxophaga 217–8, 516
Myzolecanium kibarae 319
Myzus persicae 176, 441

Nabis kinbergii 184
Naiad *see* Nymph
Names, common and scientific 10, 457, 489–91
Namibia 177, 178, 257, 496
Nannochoristidae 216, 220, 520
Nasutitermes 341, 342, 359, 384, 508
Naucoridae 272, 279–80
Navigation 96, 121, 179, 433
Necrophagy 264, 414, 513
Nectar 314
Neem 430, 433
Nematocera 219, 273, 314, 362, 519
Nematodes
  in biological control 441–3
  causing feminization 152, 180
  insect vector-borne 398, 399, 412–3
Nemoria arizonaria 181
Neobellieria bullata 154
Neodermaptera 209
Neomargarodes 259
Neoptera 52, 203, 205, 207, 240
Neopyrochroa flabellata 386
Neoteny 150–1
Neotenic termite 341
Neotylenchidae (Nem.) 442
Neozeleboria cryptoides Plate 5e
Nepa (-idae, -omorpha) 147–8, 216, 274, 278, 361
Nepenthes (-aceae) (Bot.) 319–20, 378
Nephrocytes 71
Nepticulidae 234, 299
Nervous system 57–8, 63–67, 96–8, 117, 153, 170–1; *see also* Neurone
Nesting 250, 254–6, 324–7, 335–44; *see also* Hymenoptera; Isoptera
Net-winged beetles *see* Lycidae
Neuroendocrine
  cells 64, 66
  manipulation 434
  system 73, 126, 188
Neurohormone *see* Neuropeptide
Neurone 64–6, 96–8, 109, 119
Neuropeptides **70**, 168–71; *see also* Hormone
Neuroptera
  antlions 217, *357*, 362, 364, 486, 514
  in biological control 439
  fossil record *231*
  freshwater 272, 276–8
  larvae *166*, 276, *357*
  pinning/preservation 479, 486
  as predators 42, 48, 217, 355, 362
  relationships 217
  sound reception 100–1
  taxobox 514–5
  taxonomic rank 11, 514
  see also Appendix
Neuropterida 216–7, 276, 355, 514–5
Neurosecretory cells 64, 66, 153, 168, 171, 174
Neurotransmitters 64
New Orleans 344, 407
New Zealand
  amber disease of scarabs 443
  bioluminescent flies 13, 122, 360
  cicadas 104
  giant weta *Plate 1d*
  hot-spring insects 176
  insect conservation 13–14, 17, 309, 311, 470
  insect pests 16, 198, 303, 423, 427, 443
Nezara viridula 159
Nilaparvata lugens 104, 439
Nitidulidae 314, 442
Nitrogen excretion 89
Noctuidae 9, 22, 39, 137, 176, 359, 425, 428, 446, 449, *Plate 7g,h*
Nocturnal activity 96, 99, 101, 118, 120–1, 130, 188, 122, 285, 314, 379, 406, 478
Nomenclature *see* Names, common and scientific
Nomuraea (Fungi) 443
Nonoculata 202
Non-target effects
  biocontrol 312, 428, 435
  pesticides 264, 352, 425, 431–4
Nosema (Prot.) 339, 348, 442
Notodontidae 135
Notonecta (-idae) 24, 272, 274, 279, 361
Notoptera *see* Grylloblattodea
Notum (-a)
  alinotum 46
  mesonotum 221, 487
  metanotum 509, 517
  postnotum *33*, 46–*7*
  pronotum 45–6, 188, 209, 212, 238, 500, 502, 506–7, 509
Nuclear polyhedrosis virus (NPV) 445
Nulliparity 183
Nuptial
  dance 509
  feeding 134–5, 520
  flight 52, 105, 337, 344
  gift 128–9, 134–6, 386
Nurse cells 92, 153, 405
Nutrition 77–86, 134–5, 152–3, 160, 187–8, 192, 254, 265, 290–4, 303, 327, 331, 366
Nycteribiidae 150, 370
Nymph 27, 52, 61, 86, 146–7, *159*, 165–9, 181, 188; *see also* under each order
Nymphalidae 12, *51*, *113*, 222, 293, 385, 390, *Plates 1f, 3g, 8a*

Oak *see Quercus*
  apple gall *Plate 6a*
  borer 472, *Plate 8g*
Occiput (-ital) 35–6
Ocellus (-i) 119, 121
Ochrogaster 416
Ocymyrmex 177
Odonata
  classification 205–6, 498
  copulation 138, 141–*2*
  diagnostic features 206, 498
  egg structure 145
  fossil record *231*, 233–4
  larval prey capture 42, 356, *362*
  moult 165, *169*, 272
  pinning/preservation 479, 486
  predatory behaviour 356–7, 362
  relationships 206
  taxobox 498
  tracheal gills 278–9, 498
  wings *50*–2, 62, 205–6
  see also Appendix
Odontotermes transvaalensis 256, 352, *Plate 6c*
Odorant binding protein 109, 116
Oecophoridae 250, 308
Oecophylla spp. 22, 114, 337, 339–*40*, *Plate 6h*
Oedemeridae 386, 392
Oenocytes 71
Oenocytoids 71, 73
Oesophagus 80–*1*
Oestrus (-idae) 360, 398
Ogmograptis 301, *Plate 4d*
Ohomopterus 137
Okanagana rimosa 360

Olfaction 96, 110–1, 123, 127, 181, 332, 384, 431
Oligocene *231*
Oligolecty 315
Oligophagy 265, 293, 299, 312, 364, 366, 375, 385, 425
Oligoxeny 364
Ommatidae 217
Ommatidium (-a) 120–1, 218, 228, 496
Ommochrome 33
Onchocerca (Nem.) 41, 413
Onchocerciasis 413
Oncopeltus fasciatus 132, 179, 242
Onthophagus spp. 130, 182, 263, 382
Ontogeny *see* Development
Oocyte 70, 92, 143, 145, 153–4
Ootheca 135, 149, 360, 507–8, *Plate 3e*
Opisthognathy 35
Opuntia (Bot.) 5, 311–2
Orchesella 251
Orchid (-aceae) (Bot.) 116, *289*, 291, 314, *316*–7, 356, 380, *Plates 5d, e*
  orchid mantid 13, 380
Organochlorines 432
Organophosphates 432
Oriental
  cockroach 508
  fruit fly 422–3, 452–3
  fruit moth 301, 453
  termites 267
Orius 441
Ormia 102–3
Ornithomyia 360
Ornithoptera spp. 17, 146, *Plate 1e*
Orthoptera
  acoustic courtship 97, 128, 502
  conservation 14, 17, *Plate 1d*
  diversity 6
  fossil record 103, *231*, 233–4
  in Hawai'i 245
  as leaf chewers 298
  locomotion 60, 62
  mating 133–4, 522
  morphology 30, 58, 80, 149–*50*, 187–8, 380, 502
  eggs 149–50
  parental care 324
  pinning/preservation 479, 486
  relationships 209–10
  sound 97, 100–1, 103
  taxobox 502–3
  see also Locust; Appendix; *Plates 3b, f, 4b, 7a*
Otiorhynchus sulcatus 146, 258
Osmeterium 387–8
Osmoregulation 86–7, 288
Osmylidae 42, 166, 276
Ostrinia nubilalis 301–2
Otiorhynchus sulcatus 146, 258
Ovary
  maturing peptide **70**
  ovarian cycle 143, 154, 183, 403
  ovariole **91**–2, 153, 174, 218, 333
  oviduct **91**–2, 140–1, 144, 369
Oviparity 144–50, 501
Ovipore *53*
Oviposition
  cuticular effects 30
  definition 145
  deterrence 114
  illustrated *145*, *365*
  induction/control **70**, 96, 126, 133, 143–4
  of large eggs 147–8, 277–8, 521
  mistaken host 17
  multiple 365–6
  peptide *70*
  rate 138, 184
  sites 107, 126–7, *131*, 146–8, 151, 174, 324, 406, 439, 452–3, 520
Ovipositor 53–4, **91**, 107, *132*, 148–9, 207–12, 223, 303, 361, 362, 364, 393, 416, 441, 495–6, 498
  substitutional *53*, 149, 362, 364, 517, 519
  see also Sting
Ovisac 31, 259, 426, *Plate 8b*
Ovotestis 151
Ovoviviparity *see* Viviparity
Owl butterfly *see Caligo*
Oxygen
  availability/uptake in water 10, 147–8, 277–82
  availability in late Palaeozoic 232–*3*
  closed tracheal system 73, *75*, 278–9, 497
  open tracheal system 73, *75*, 77, *237*, 279–81
  passage through tracheae 59, 73, 75–7
  physical properties 148, 277
  reactive oxygen 73
  respiratory pigments *see* Haemoglobin
  transport 69, 73, 75
Oxyops vitiosa 311

Paederus 386, 416
Paedogenesis 150–1
Paenibacillus popilliae 441, 443
Palaeodictyoptera 230, *233*, 241
Palaeoptera 203, 205–6, 224, 226, *231*, 240
Palaeozoic 10, *231*–2, 234, 239, *241*
Palm weevil ("worm") 21–2, 425, *Plate 2a*
Palp *see* Labium, palps; Maxilla, palp
Pancrustacea 228–9
Panorpa (-idae) 129, 220, 355, 520
Panorpodes paradoxa 162
Papilio spp. 12, 32, 379, 387, 390
Papilionidae (-oidae) 12, 32, 167, 222, 379, 387, 391, *Plate 1e, 3a, 7d*
Papua New Guinea 12, 17, 310–11, 318, 406
Paraglossa 37
Parajapyx 495
Paralobesia viteana 466
Paramere 54–5, 221
Paraneoptera 207, 212–3, 214, 242
Pararistolochia (Bot.) 17, *Plate* 1e
Parasite
  definition 286, 355
  host specificity 213, 358–60, 464, 370–2
  parasite–host interactions 220, 243, 370–2
  of social insects 345–8, 363, 396
  see also Ectoparasite; Hyperparasitism; Parasitism
Parasitism 150, 152, 213–4, 220, 243, 286, 334, 345–8, 355, 365–76, 358–60, 363, 370–2, 396
Parasitization 324, 347, 355, 364, 366, 369, 375, 437, 440, 459, 468, 519
Parasitoid
  in biological control 14, 116, 247, 303, 309, 355, 359, 437, *440*, 448, 450, 465, 470
  definition 355
  of eggs 14, 359–60, 437, *440*, 448, 450, 465, 470, *Plate 6f*
  of gallers 308
  gregarious 364–5
  host acceptance and manipulation 361–3, 366–70
  host location 14, 102–3, 112, 116, 298, 303, 355, 360
  host use by 355, 364–6, 517–8
  overcoming host responses 348, 363–4
  phoretic 360
  primary 365–6, 423, 440
  secondary (hyperparasitoid) 365, 372, 440
  and synomones 115–6, 359
Parental care 146–7, 212, 324–5, 511
Parthenogenesis 15, 126, 150–3, 325, 508–9, 511, 513, 525
Passalidae 324
Passiflora (Bot.) *Plate 5h*
Patagium (-a) 523

Paternal genome elimination 140
Paterson's curse *see Echium plantagineum*
Pathogen *see* Microorganisms
Patia orise Plate 8a
Pectinophora gossypiella 453
Pedicel 37, 44, *91, 519*
Pediculus 42, 372, 398, 410, 416, 510
Pegomya hyoscyami 301
Peloridiidae 215, 512
Pemphigus (-inae) 287, 306, 325
Penis 54–5, 131, 138, 141, 142, 209
Pentatomidae 99, *144, 159*, 308, 448, 512
Peppered moth *see Biston betularia*
Perga (-idae) 99, 301, 388–9, 392
Periplaneta americana 47, 51, 58, 65, 87, 105, 508
Peritreme 73–*4*
Peritrophic matrix/membrane 82–4, 412, *446*
Permian 10, 206, 208, 211, 219, *231*–4
Perthida glyphopa 300–1
Pest insects 24, 113, 126, 184, 245, 297, 373, 406, 434, 439–42, 452–4, *457*–461
  and climate change 463–7
  integrated management (IPM) 355, 425, 428–9, 450
  resurgency 16, 297, 410, 415, 425–6, 440
  and trade 457–8, 465, 469, 472–3
Pesticides *see* Insecticide
Petrobius maritima 495
Phaedon cochleariae 86
Phagocytosis 71–2, 363
Phalaenophily 314
Phallobase (-mere) 55, *132*
Pharate condition 122, 130, 166–9, *171*–2, 216, 218, 241, 498, 517, 523
Pharynx 36, 39–41, 43, 80–*1*, 83
Phasmatodea
  autotomy 384
  as defoliators 298
  eggs 503–4
  fossil record *231*
  illustrated *125, 504*
  pinning/preservation 486
  as pets 13
  as prey 380
  relationships 209–10, 507
  taxobox 503
  see also Appendix
Phasmid/phasmatid *see* Phasmatodea
Pheidole 15, 278, 419
Phenacoccus manihoti 419, 435
Phengaris spp. 18, 345, 346
Phengodidae 121, 385
Phenoloxidase 72–3
Pheromone
  aggregation 112–5, 323, 392, 453
  alarm 114–5, 384–5, 395
  classification 111–6
  courtship 112–3, 123, 386, 523
  definition 111
  detection 109–10
  in electroantennogram 109
  epideictic (spacing) 114
  footprint 331
  lure-and-kill 452
  mass trapping of pests 452–3
  mating disruption of pests 419, 423, 453
  monitoring of pests 22, 452, 477
  parapheromone 113, 452
  perception by antenna *108*–9, 111
  in pest control 352, 423, 429, 452–3
  of queens 328, 331–3, 337, 347, 351
  sex 45, 123, 128–9, 135, 390, 452–3, 507
  sex attraction 104, 108, 112, 518
  synthesis 68, *70*
  in termites 343
  trail marking 114, 344
Philanisus plebeius 288
Philanthus 123, 254–5, 359
Philorheithridae *283*
Phlaeothripidae 214, 326, 511
Phlebotomus (-inae) 40, 399, 411–2
Phoracantha 470, 517
Phoresy 358, 360–1, 370
Phoridae 264, *266*, 347
Photinus 121–2, 387
Photoperiod 117, 119, 153, 172, 174, 179, 180–1, 186–7, 459, 466
Photorhabdus (Bact.) 442
Photuris 122, 360, 387
Phragma (-ata) *46–7*
Phragmosis 393
Phthiraptera (now in Psocodea)
  aptery 52, 158
  coevolution 370–1
  collecting/preserving 496–7
  and disease 410–11, 415–6, 510
  hosts (inc. humans) 358, 372, 410, 510
  mouthparts 41, *510*
  ovipositor/oviposition 53, 149
  phoresy 358, 360
  relationships 213–4, 372
  status (taxonomic) 11, 203–*4*, 212–3
  taxobox 510
  see also Appendix
Phylacteophaga 301
Phyllocnistis spp. 300–1
Phylloxera (-idae) 295–7, 447
Phylogeny
  and biogeography 244–7
  and bioluminescence 121
  and classification 191–201
  constraint 157
  and fossils 229–30, 233–6
  host and parasite 370–2
  insects and plants 290
  molecular phylogenetics 193–6, 211, 220, 229, 230, 236, 266, 407, 521
  and social insects 343, 347–8, 350–1
  see also Chapter 7; under each order
Phymateus morbillosus Plate 4c
Physiological time 184–7
Physogastry 341, 347, *Plate 6c*
Phytalmia mouldsi 131
Phytomyza senecionis 300
Phytophaga 243
Phytophagy
  aquatic 309
  coevolution of insects and plants 291–3
  and defence 339
  definition 242, 290
  fossil history 235, 290
  gall induction 305–8
  herbivory rates 295
  host seeking 115–6
  leaf chewing 286, 295–7, 298–9
  mining and boring 299–301
  performance 295
  sap sucking 302–5
  seed predation 106, 308–9
  species richness 9, 218, 242–3
Phytotelmata 318
Pieris (-idae) *39, 82*, 90, 152, *161*, 180, 222, *Plate 8a*
Pigment (colour) 22, 30, 32–3, 50, 57, 69–**70**, 90, 117, 157, 168, 171–2, 183, 279
  cells 119–20
  see also Haemocyanin; Haemoglobin; Retina
Pinning 479–81
Pinus (Bot.) 114, 295, 303, 442, 469
Piophilidae 264, 414
Pistacia (Bot.) 305, *Plate 5b*
Pitcher plants 318–9, 378
Plague 411
Planococcus citri 31
Planthopper 99, 104, 214–5, 258, 439, *479*, 485, 512
Plant
  apparency to insects 294, 384, 451
  coevolution with insects 291–3
  damage by insects 419–25
  defences vs insects 243, 260, 267–8, 294, 309, 339

galls 305–8
genetically modified 447, 449–51
herbivory levels 295
mining and boring by insects 299–301
myrmecochorous plants 290, 317–8
polycultures 451–2
resistance to insects 294, 296, 419, 429, 446–51
stress affecting herbivory 258, 295, 303, 359
Plasmodium (Prot.) 398, 400–3, 406
Plastron 273, 280–*1*
Platygasteridae (-oidea) 439–40
Platypezidae 261
Platypodinae 260, 324, 344
Plecoptera
adult feeding 265
fossil record *231*, 234
haemocyanin 69, 238
locomotion 240
nymphs 272, 277–*8*, 281, 285, 486, 500
oviposition 52, 149, 500
pinning/preservation 486
relationships *204*, 207–8
sound 104
taxobox 500
see also Appendix
Pleidae 279
Pleiometrosis 267
Pleistocene *231*, 235–6, 467
Plesanemma fucata Plate 7b
Pleuron (-a, -al) 34–5, *46–7*, 218, 238, 506
Pliocene *231*, 467
Plodia interpunctella 444
Plutella xylostella 444
Poecilometis 144
Poinsettia (Bot.) *see Euphorbia pulcherrima*
Poinsettia thrips *see Echinothrips americanus*
Pointing 481
Poison
gland 92, 114–5
sac *394*
see also Insecticide
Polistes (-inae) 327, 347
Pollen 313, 314, 329, 339
Pollination 106, 234, 290, 292, 313, 325, 329, 336
value of 5, 315–7, 323, 336, 338
Polydnavirus (PDV) *368–9*
Polyembryony 152, 365, 369, 525
Polyethism 327, 329–30, 337
Polygyny 327–8, 337, 348–51
Polylecty 315
Polymorphism
of aphids 181, 325
definition and diversity 179–81, 327, 369, 490
of locusts 188
in mimicry 390–2
of social insects 212, 265, 303, 331, 337, 508
of wings 179–80, 188, 274, 326, 501, 511
see also Polyethism; Polyphenism; Sex, sexual dimorphism
Polyneoptera 207–12
Polypedilum vanderplanki 177
Polyphaga 217–8, 516–7
Polyphenism 179, 180–1, 327
Polyporaceae (Fungi) 265
Pompilidae 325
Pontania proxima 306
Pontomyia 288
Popillia japonica 441, 471
Population biology 372–5, 459
Populus (Bot.) 300, 306
Pore
canal 30
kettle and tubule 108–9
Porphyrophora 259
Postantennal organs 493
Postcoxal bridge *46–7*
Postmentum 38, 42–3
Postnotum *46*
Potato *see Solanum tuberosum*
Precocious eclosion hypothesis 241–2
Precosta (vein) 49–*50*, 241
Precoxal bridge *46–7*
Predator
abundances 364, 372–5
in biological control 427, 433, 435, 439–41, 448
birds on insects 282, 378–82
concealment by/from 240, 356, 360–1, 379–83
defence against 350, 376, 378–96
definition 355
foraging 355–6
inquilines 345–8
learning 374, 380–1, 384, 390–2
in mimicry systems 388–93
predator–prey interactions 355, 358, 364, 372, 374
prey specificity 359–61, 364–6, 370, 372
refuge from 374–5
sit-and-wait 277, 356–7, 362, 364, 507
Prementum 38, 42–3, 221, 524
Prescutum 46–*7*
Preservation (of specimens) 484–7, 491
Presociality 323
Presoldier 341–3
Press 224, 491
Pretarsus 47–48, 203, 214, 218, 511
Prey
abundances 364, 372–3
acceptance and manipulation 363–4
definition 364, 378
location 116, 122, *127*, 283, 355–61
selection and specificity 370–4, 376
Prickly pear cactus *see Opuntia*
Primula vulgaris (Bot.) *300*
Proboscis 38–42, 107, 214, 221–2, 234, 293, 314, 316, 370, 399, 512, 519
Processionary caterpillars 416
Proctodeum 80
Proctotrupoidea *223*
Proctolin 64, **70**
Procuticle 28–30, 169
Prognathy 35–6, 38, 165
Promargarodes 259
Pronotum 43, 45–6, 209, 212, 238
Pronymph 162, 165, 241, 498
Propodeum 223, 525
Proprioceptor 100, 336
Prosorrhyncha 215
Prosternum 100, 102, 513
Prothorax (-acic) 45–6
gland 67
winglets 45
Protista (Protozoa)
in biological control 441–2
as gut symbionts 85, 303, 341, 508
in vertebrate disease 398–400, 411–2
Protodonata *see* Meganisoptera
Protozoa *see* Protista
Protrusible vesicle 202, 494–6
Protura
diagnostic features 202
fossil *231*
preservation 486
relationships 202–3
in soil fauna 237, 250–1
taxobox 495
see also Appendix
Proventriculus 80, 207, 212, 220, 411
Przibram's rule 183
Psammotermes allocerus 257
Psephenidae 275, 279, 282
Pseudacanthotermes spiniger 393
Pseudergate 341–3
Pseudococcidae *31*, 318, 435, *Plate 6b*
Pseudocopulation 116, 314, *Plate 5d, e*
Pseudomyrmex 318–9, 339, *Plate 4a*
Pseudonocardia (Bact.) 266
Pseudoregma alexanderi 326
Psocid *see* Psocodea
Psocodea
fossil record *231*
morphology 41–*2*, 50, 53, 509–10

Psocodea (*continued*)
parasites 85, 360, 370, 372, 410
parental care 324
preservation 486–7
relationships 41, 52, 203–5, 207, 212–4, 509–10
taxobox 509–10
viviparity 150
see also Appendix
Psocoptera *see* Psocodea
Psychidae 383
Psychodidae (-omorpha) 40, 399, 411–2
Psychophily 314
Psyllidae (-oidea) 215, 304–6, 311, 423, 468, 485, 512–3, *Plate 8e*
Psyllopsis fraxini 512
Pteromalidae 307, 365
Pterostigma 49
Pterygota 27, 52, 193, 202, 205, 242–3, 275, 496
Pthirus pubis 410, 510
Puddling 135
Pulex 43, 65, 521
Pulvillus (-i) 47–8, 60, 105
Pupa
diapause 173–4
paedogenesis 151
protein synthesis 84
pupal moult 160, 166, 171–2, 242
pupal instar 151, 160, 166, 518
pupariation 252
types *167*
see also *Plates 1g, 3c, 6d*
Puparium 151, 167, 218, 220, 273, 424, 454, 485, 518
tanning factor **70**
Pupation 12, 18, 69, 84, 146, 166, 177, 218, 242, 253, 273, 347, 358
Pygidicranidae 209
Pyralidae 295, 301–2, 311, *363*, 444
Pyramid 449
Pyrethrum (Bot.) 430
Pyrgomorphidae 385, *Plate 4c*
Pyriproxyfen 405–6, 408, 433
Pyrops sultan 512
Pyrrhocoridae 308
Pythidae 392

Quarantine 14, 16, 245, 259, 309, 311, 415, 422–3, 436–8, 451, 460–1, 470–2, *see also* Biosecurity
Queen
ant 15, 18–19, 112, 168, 266–7, 323, 337, 340, 346–7
bee/wasp 39, 140, 168, 325, 327–36, 348–51
definition 327
Red (model) 376
termite 30, 341–3
see also Monogyny; Polygyny
Queen Alexandra's birdwing
see Ornithoptera alexandrae
Queen butterfly *see Danaus gilippus*
Queensland fruit fly *see Bactrocera*
Quercus (oak) (Bot.) 181, 293, 305–6, 472, *Plate 6a*
Quiescence 118, 172–3, 176, 333, 358; *see also* Diapause
Quinone 30, 33, 73, 254, 385, 388

Rabbit flea *see Spilopsyllus cuniculi*
Radiation (diversification) 7, 9–10, 160, 165, 175, 197, 202, 208, 216–18, 222, 228, 230–*1*, 234–236, 243–7, 292, 314, 370, 375–6
Radius (-al) 49–*50*, 100, 241
Ragwort *see Senecio jacobeae*
Ramsdelepidion schusteri 10
Range change 235, 458, 460–3, 467–8
Raphidiidae 514–5
Raphidioptera 11, 216–7, 355, 487, 514–5
Rat-tailed maggot 264, 279, *320*, 414
Rectum (-al) 80, 87–9, 278, 281, 288, 498
pad *87*–9
uptake-oxygen 75, 278, 281
Reduviidae 30, 66, 324, 359, 361, 383, 412
Reflex 64, 122
bleeding 386, *Plate 7f*
Remigium 48–9, 503
Reoviridae (Vir.) 407, 445
Replete 23, *30*, 340, *Plate 2g, h*
Reproduction
asexual/atypical 150–2
control in *Lucilia cuprina* 143–4
effects of symbionts 152–3
fertilization 92, 128–9, 134, 136–40, 144, 148, 153, 160, 288
genitalic morphology 52–5, 92–3, *136*
gonochorism 150
hermaphroditism 150–2, 347, 426
hormones **70**, 145, 153–4
neotenic 341
non-receptivity 93, 129, 144
oviparity 144–50, 501
physiological control 126, 133, 141, 153–4
polyembryony 152, 365, 369, 525
vitellogenesis 92, 143, 153–4
viviparity 150–1, 209, 501, 508–9, 512, 518
see also Copulation; Courtship; Female choice; Haplodiploidy; Parthenogenesis
Reproductive system 90–3
female 92, *137*
male 92–3, *136*
Resilin 30, 60, 62, 220
Resin 39, 114, 234–5, 294, 303, 325, 384, 406, 483–4
Resistance
environmental 373
to gas movement 76
genetic engineering of 447, 449
to infection 73, 363, 465
to insecticides 400, 405–6, 408, 415–6, 420, 423, 425, 427–30, 433
of insects to Bt 442, 444–5, 449–50
of plants to insects 294, 296–8, 312, 447–51
Respiration 73, 77, 252, 277, 279
Resurgence
of disease 410
of pests 16, 297, 415, 425–6, 440
Reticulitermes 256, 352
Retina 117, 119–21
Retinaculum 49, 493
Retinula cells 117, 119–21
Rhabdom (-ere) 117, 119–21, 218
Rhagoletis pomonella 245, 301
Rhamphomyia nigrita 127
Rhaphidophoridae 268
Rheumatobates 139
Rhinotermitidae 212, 256–7, 343–4, 473
Rhipiceridae 516
Rhodnius 30, 66, 361, 412
Rhopalosiphum padi 176
Rhynchaenus fagi 176
Rhynchophorus 20–21, *Plate 2a*
Rhynchophthirina 510
Rhyniella praecursor 229
Rhyniognatha hirsti 229
Rhyparida 48
Rhysodinae 218
Rhyzobius lophanthae 374
Rhytidoponera tasmaniensis *317*
Rice (Bot.) 41, 308, 410, 429, 439, 447
Rice weevil *see Sitophilus oryzae*
Richmond birdwing butterfly
see Ornithoptera richmondia
Rickettsia (-ales) (Bact.) 410–11
RIDL 455
Ripiphoridae 217, 219, 365
River continuum concept 286

RNA interference (RNAi) 430, 434
Roach *see* Blattodea
Robber fly *see* Asilidae
Rock crawler *see* Grylloblattodea
Rocky Mountain spotted fever 411
Rodolia cardinalis 427
Romalea guttata 382
Rose (*Rosa*, -aceae) (Bot.) 293, *300*, *306*
  bedeguar gall *306*, *Plate 5c*
Ross River virus 407
Rostrum 39, 214–5, 220, 274, 305, 309, 361, 384, 393, 512–3, 520
Rotenone (-oid) 430
Rove beetles *see* Staphylinidae
Royalactin 331
Royal jelly 307, 331

Saccharopolyspora (Bact.) 430
Saliva (-ary)
  antibiotic 267
  canal 41, 302
  components 40, 82, 302, 361, 393
  in digestion 39–40, 80, 361
  in disease transmission 41, 305, 400–1, 405, 407, 411–2
  duct 38–41
  gland 57, 80–*1*, 336, 363, 393, 400–1, 405, 412, 513
  larval 335, 405
  in nest construction 333
  pump *81*, 521
  salivarium 38, 80
  solidifying 83, 303–*4*
  stylet 41, 83, 302, 512
  toxic 302, 361
Salix (Bot.) 306
Salmo (trout) 147, 282
Salvation Jane *see Echium plantagineum*
Salvinia (Bot.) 310–11, 439
Salyavata 359
Sampling 6–7, 178, 196–7, 207, 234, 269, 285, 299, 372–3, 473, 475, 488
Sand fly 40, 235, 399, 411–2; *see also* Ceratopogonidae; Phlebotominae
Sap beetle *see* Nitidulidae
Saprophagy 517
Sap sucking 302–5
Sarcolemma 59
Sarcophagidae 398
Sarcoptes scabiei 398
Sarraceniaceae (Bot.) 319, 378
Saturniidae 167, 416, *Plates 1a, b, 2b*
Sawfly 300–1, 306–7, *389*; *see also* Pergidae; Tenthredinidae
Scabies 398
Scale (seta) 32, 50, 62, 90, 107, 149–*50*, 221, 496, 523
Scale insect *see* Coccoidea
Scape 37, 43, 44
Scapteriscus acletus 104
Scarab *see* Scarabaeidae
Scarabaeidae
  control 441, 443, 471
  in dung 261–*3*
  flower-visitors 314
  fossorial fore leg 48
  horns 130
  human interest 11–14
  larvae *166*, *251*, 256, 261–3, 443, 517
  pupa, *Plate 3c*
  as pests 298, 441, 443, 471
  size 10
Scathophagidae 261
Scelionidae 359, 360, 440, *Plate 6f*
Schistocerca gregaria 87, 89
Sciaridae 250, 258, 265
Sclerite 30, 35, 47, 49, 157, 199
Sclerotization 29–30, 33, 35, 47, 50, 60, 69, 157, 167–71, 181
Scolopale/cell 99
Scolopidium (-a) 99–100, 102
Scolytus (-inae) 86, 114–5, 260, 303, 471
Scorpionfly *see* Mecoptera
Screw-worm fly 398, 454, 460, 519; *see also Chrysomya bezziana*
Scutelleridae 308, 520, 523
Scutum (-ellum) 46–*7*, 520, 523
Secondary
  disease cycle 399
  endosymbionts 192
  genitalia 138, 141, 498
  line of defence 380
  parasitoid 365, 372, 440
  pest outbreak 425, 433, 453
  plant compounds 294, 297, 299, 364
  segmentation 33
  sexual characteristics 128
Seed
  dispersal by ants *see* myrmecochory
  predation 8, 105–6, 256, 269, 301, 308–9, 317
Segmentation of body 33–5; *see also* Abdomen; Head; Tagma; Thorax
Semidalia unidecimnotata 173
Seminal
  fluid 93, 133, 145
  vesicle *91*–3, 133, 208
Semiochemicals 109, 111–6, 303, 345, 452–3; *see also* Allomone; Kairomone; Pheromone; Synomone
Senecio (Bot.) spp. *300*–1, *387*
Senescence 168, 247
Sensillum (-a)
  campaniform 97–*8*
  chordotonal 99–102
  definition 32, 97
  hair plate 97–*8*, 336
  multiporous 108–9
  olfactory 108–9
  of ovipositor 149–50, 362
  trichoid 32, 96–8, 111
  type and distribution 96
  uniporous 108
Sepsidae 34
Serratia (Bact.) 443
Seta 32, 43–5, 48, 52, 92, 96–8, 106, 108, 139, 214, 220, 416
Setisura 206, 330, 416
Sex
  (X-) chromosome 140
  conflict 134, 137–9, 243–4
  determination 139–40, 152, 326
  female choice 128–31, 136–9, 144, 243, 247
  male–male competition 38, *125*–*128*, 130, 137–9
  pheromone 45, 108, 111–4, 116, 123, 127–9, 135, 390, 452–3, 507
  ratio 140, 152–3, 366, 433
  retardation 327
  sexual cannibalism 134–5, 503
  sexual deception 113
  sexual dimorphism 128, 180, 218, 305, 341, 347, 392, 517
  sexual selection 5, 9, 126, 128–131, 136–9, 243, 247, 390
Sialidae *167*, 276–7, 514–5
Silk 5, 24, 43, 82, 168, 174, 210, 220–2, 276, 279, 283, 339–40, 382–3, 416, 423, 485, 495
  gland 82, 210, 221, 336, 503, 515–7, 523–4
  worm 24, 70, 108, 119, 186, 442–3; *see also Bombyx mori*
Silphidae 264, 268, 324
Silverfish *see* Zygentoma
Silverleaf whitefly *see Bemisia argentifolii*
Simulium (-idae) 40, 43, 190, 219, 247, 273, 283, 286, 359, 372, 413
Sinentomidae 251
Sinus (-es) 71, 109, 210, 241, 503
Siphlonurus 279, 282
Siphonaptera
  aptery 52, 158, 200
  defensive shape 382
  disease transmission 407, 410–11, 521
  hosts 102, 358, 372, 375, 407, 411

Siphonaptera (*continued*)
jumping 30, 60, 521
larval development 165, 358, 371, 521
mouthparts 41–*3*, 220
preservation 487
relationships *204*, 221, 216, 219–20, 520
taxobox 521
see also Appendix
Siphunculi 384
Sirex (-idae) 149, 260, 303, 442, 469
Sitona 258
Sister group 193–4, 375
Sisyridae 276, 514
Sitobion avenae 176
Sitophilus spp. 308, *516–7*
Sitotroga cerealella 14
Skeleton
cephalopharyngeal 207, 273, 519
endoskeleton 36, 59
external 57, 60, 73
hydrostatic 57, 60–1, 69
Skipper *see* Hesperiidae
Slave-making ant 347
Sleeping sickness *see* Trypanosomiasis
Slide (microscope) preparation 478, 483–4
Small hive beetle *see Aethina tumida*
Smell *see* Chemoreception; Olfaction
Smerinthus ocellatus 382
Sminthurus viridis 494
Snakefly *see* Raphidioptera
Snow flea *see* Boreidae
Social parasitism 334, 347
Sociality
care of eggs 145–8
definition 323
eusociality 323, 327–45
evolution of 214, 323, 325–7, 348–51
maintenance 350–1
quasi- and semisociality 326–7
subsociality 323–6
success 323, 351
Solanum (Bot.) spp. 425, 448
Soldier
in Aphididae 325–6
in Formicidae 327, 337
of termites 212, 341–4, 359, 384, 392–4, 508
in Thysanoptera 214, 326
Soldier beetle *see* Cantharidae
Soldier fly *see* Stratiomyidae
Solenopsis 15–6, 337, 352, *Plate 6e*
Solitary
habit 187–*8*, 323–4, 327–8, 336, 348–51, 508–9
locust phase 181, 187–8
nesting 324–5, 327
parasitism 365–6
Sound
in courtship 112, 128, 132, 208, 500
in location 99, 101–3, 123, 303
production 100, 103–5, 123, 345, 359–60
reception 99–100
ultrasound 101–5
warning 384–5
Southern pine beetle *see Dendroctonus frontalis*
Spanish fly 386, 416
Species recognition 10–11, 128, 137–8, 192, 197–201, 506
Sperm
competition 135, 140–4, 243
precedence 126, 141–2, 243
storage 90–**1**, 138–41, 143
transfer 129, 132–3, 138, 168
Spermatheca **91**–2, *132*, *137*–41
Spermatophore 54, 93, 131–8, 140, 386, *Plate 3b*; *see also* under each order
Spermatophylax 133–4
Sphaeriusidae 218
Sphecidae 124, 325, 327–*8*
Sphecotypus 395
Sphingidae 39, *166*, 168, *171*, *173*, *316*, 380, *382*, *Plate 5g*
Spicipalpia 170, 221
Spider (Araneae) 439
Spilopsyllus cuniculi 407
Spinosad 430
Spina/spinasternum 47
Spiracle 45, 52, 57, 66, 68, 73–7, 100–1, 178, 192, 203, 205
acoustic 100–*1*
plate *44*
Spitfire *see* Pergidae
Spittle bug 77, 80, 83, 214–5, 302, 443, 485, 512
Spongillafly *see* Sisyridae
Spongiphoridae 209
Springtail *see* Collembola
Spruce budworm *see Choristoneura fumiferana*
Stable fly *see Stomoxys*
Stadium (stadia) 158, 165–6, 168, 170, 181, 186–7
Staphylinidae (-oidea) 116, 218, *251*, *262*, 264, 269, 284, 287, 346–7, 386, 416, 517
Staphylinus caesareus 516
Startle display 380, *Plate 7d*
Steinernema (Nem.) 442
Stemma (-ata) 36, 43, 117, 119–21, 220, 520, 524
Stenaptinus insignis 388
Stenodictya lobata 233
Stenogastrinae 327
Sterile insect technique 312, 422, 429, 454–5
Sternorrhyncha *213*–5, 242, 305, 365, 512–3
Sternum (-a, -ellum, -ite) 35, 38, 47, 52–*3*, 62, 212, 218, 222
Stick-insect *see* Phasmatodea
Stigmodera 313
Stimulus (-i) 97–8, 107, 122–3, 358, 384
Sting 54, 114–6, *149*, 223, 294, 330–1, 352, 359, 362, 381, 389, 393–5, 415–7
Stingless bee 235, 315, *329*, 395; *see* Meliponini
Stink bug 144, *159*, 215, 359, *Plate 2f*
Stipes *37*–8, 213
Stomodeum *see* Gut, foregut
Stomoxys 41, 116, 362
Stonefly *see* Plecoptera
Stratiomyia (-idae, -omorpha) 76, 219, *251*–2, 280, 284
Streblidae 150, 268, 370
Strepsiptera
development 150–2, 165, 217, 369
as endoparasitoids 150, 152, 355, 369–70, 518
fossil record 219, *231*, 517
morphology 51, 86, 150, 218, 517
preservation 487
relationships 165, 192, 216–9, 518
taxobox 517–8
see also Appendix
Streptomyces see Pseudonocardia
Style
morphological 52, 54–5, 202, 203, 205, 212, 494–6, 507, 511–2, 521
of flower 291, 313
Stylet 40–1, 83, 214, 220, 259, 276, 302–5, 325
Stylops (-idae) 219, 369, 518
Subalare (-ia) *46*–7, 52
Subcostal vein (subcosta) 49–*50*, 241
Subgenual organ 99–*101*
Subimago 48, 158, 205–6, 242, 497
Submentum 38
Subsociality *see* Sociality
Success
of angiosperms 244, 313–4
of aquatic insects 187
of barcoding 200, 216
in biocontrol 310–12, 374
conservation 14, 16–19

of Endopterygota 160, 241–2, 244
of insects 9, 27, 57, 96, 123, 228, 238–43
of phytophagous insects 242–4
of predation/parasitism 116, 372, 375–6
in reproduction 113, 126, 128–30, 134, 138, 141
of social insects 323, 349, 351
Succession in corpse 264, 413–4
Sulcia muelleri (Bact.) 85, 215
Superficial layer 160, 301
Superlingua (-e) 36
Superparasitism 365–6
Sustainability 12–13, 17, 22, 24, 431, 465, 468
Swarm (-ing) 105, 126–8, 188, 333–5, 344, 360, 497
Sweetpotato whitefly *see Bemisia tabaci*
Swim (-ming) 48, 57, 60–1, 148, 208, 281, 284, 517, 523
Sylvan cycle 399, 407
Symbionts 33, 85–6, 152–3, 192, 236, 293, 344–5, 350, 424, 508
Sympetrum 119, 498–99
Symphyta 88, 165, 223, 234, 298, 315, 365, 375, 517, 524–5
Synanthropy 245, 496
Synapse 64, 117, 430
Synomone 115–6, 359
Syrphidae 220, 264, 279, 314, 347–8, 358, 414
Systematics 191–224

Tachinidae 102, 150, 362, 365, 375, 437, 440
Tactile mechanoreception 96–7
Taenidia 73, 75
Tagma (-osis) 33–5, 201, 525
Tanacetum cinerariifolium (Bot.) 430
Tannin 30, 70, 83, 294, 303
Tapetum 120
Tarsus (-omere) 47–8, 203, 208, 210–11, 213, 215, 218, 259, 330, 361; *see also* under each order
Taste 107, 115, 332, 384–92, *see also* Chemoreception
Taxis (-es) 123
Taxon (-onomy) 6–8, 10–**11**, 38, 52, 165, 191–201, 230, 233, 235–6, 243–4, 439
Tebufenozide 433
Tegmen (-ina) *51*, 103–4, 208–12, 214, 241, 500, 502–3, 506–7
Tegula 52, 523
Telenomus 359, *Plate 6f*
Teleogryllus commodus 58, 149
Telmatogeton 182, 288
Temperature
influence on activity 105, 123, 157, 168, 172–4, 333–6, 374
aquatic 273, 277–81, 285–7
and development 153, 158, 172, 358, 361
extreme tolerance 173–7
reception 105
regulation 105, 267, 334, 346, 351
Tenebrio (-nidae, -noidea) 14, 24, 76, 88, 106, 178, 218, 238, 251, 256, 417
Tenodera australasiae Plate 3e
Tenthredinidae (-oidea) 301, *306*
Tentorium 36, 207
Tephritidae 113, 127, 130–*1*, 138, 243, 245, 301, 305, 313, 422, 454, 461
Terebrantia 214, 511
Tergum (-a, -ite) 35, 47, 62–3, 111, 136, 238, 506
Termes (-itidae) 86, 212, 256, 267, 341–*2*, 343–*6*, 384, 508–9, *Plate 6c*
Termites *see* Termitoidae; Macrotermitinae.
Termitoidae
castes and instars 327, 340–4, *Plate 6c*
defence by soldiers 340–4, 359, 384, 392–4
ecological importance 250–1, 323, 508–9
eusociality 324, 327, 340 –4
fossil record *231*, 235
fungus cultivation 267–8, 303, 341, 344, 508
larvae and nymphs 340, *342*, *Plate 6c*
nesting 343–4, *346*
pest status 344, 473
preservation 484
relationships 207–*8*, 211–2, 509
taxobox 508–9
see also Appendix
Termitomyces (Fungi) 267–8
Termitoxenia 347
Terpene (-oid) 68, 114–6, 294–5, 303, 359, 384–6, 430
Tertiary (period) *231–6*
Testis (-es) 57, 91–2, 133, 151, 433
Tetrastichus planipennisi 472
Tettigoniidae (katydid) *53*, *101*, 133–4, 149, *380*, *Plate 3b, f*
Thanatosis 380
Thaumaleidiae *219*
Thaumetopoea (-idae) 416
Thelytoky 15, 151, 347, 511, 525
Thermobia domestica 242, 496
Thermoreception *see* Temperature
Thorax
appendages 47–52, *63*
ganglia 64–5, 105, 213
heating 106–7
muscle 42, 57–63
structure 45–52
tagmosis 33–5, 201, *494*
see also Leg; Pleuron; Sternum; Tergum; Wing
Thripidae *42*, 214, 305, 468, 511
Thrips *see* Thysanoptera
Thrips australis 42
Thymelicus lineola 135
Thymus praecox (Bot.) 18
Thysania agrippina 9
Thysanoptera
feeding 305, 511
flower-frequenting 106, 314, 511
fossil record *231*
galling 305, 308
"holometaboly" 242
mouthparts 41–*2*, 214
ovipositor 54, 149
parental care 324, 326
parthenogenesis 151
peritrophic membrane 82–3
introduced pests 305, 468–9, 511
preservation 487
relationships 41, 212–4
subsociality 324–6
taxobox 511
see also Appendix
Thysanura *see* Zygentoma
Tibia 47–8, 60, 100–*1*, 111, 203, 210, 329–*30*, 361, 503, 506, 520
Ticks (Acari) 398, 407, 411
Tiger moths *see* Arctiidae
Tineidae 264, 414, 524
Tiphiidae 223, 314, 524, *Plates 5d,e*
Tipula (-idae) 175, 219, 251, 258, 280, 283, 284, 519
Tisheriidae 299
Tithrone 507
Tobacco
budworm *see Heliothis virescens*
hornworm *see Manduca sexta*
whitefly *see Bemisia tabaci*
Togaviridae (Vir.) 407
Tolerance 123
of insects to insecticides 407, 427–8, 445, 450
of low oxygen 252
of plants to insects 312, 447, 465
to pollution 178, 273
of extreme temperatures 106, 174–7, 252, 402, 459
Tomato pinworm *see Keiferia lycopersicella*

Tongue *see* Glossa
Tonofibrillae 59
Tormogen cell 32, 96, 109
Tortricidae 175, 302, 423, 453, 466
Torymidae 360
Tospovirus (Vir.) 305
Toxoneuron nigriceps 369
Trachea (-eae, -eole) 73–6
acoustic 100–*1*
aeriferous 73
hypertrophy 76
proto-trachea 238
pseudotracheae 41, 219
Tracheal system 10, 45, 52, 57, 66, 68–69, 73–7, 103, 168, 202, 221, 237–8, 273, 275, 278–81, 497–8, 500, 517, 523
Transcriptional profiling 183
Transcriptome 195, 202, 207, 209–10, 213, 216–7, 220
Transgenic
insects 144, 455
plants 434–5, 449–50, 455
Transovarial transmission 86, 408
Trapping 21, 269, 352, 356–7, 372, 422, 452–3, 476–8
trappability 372, 477
Traumatic insemination 138, 243
Trehalose 69, 88, 175, 177
Triassic 219, *231*, 234
Tribolium 5, 13, 161, 163, 195, 238, 242, 434
Trichodectidae 370, 371, 510
Trichoderma (Fungi) 261
Trichogen cell 32, 96, 109
Trichogramma (-tidae) 14, 363, 438, 440–1
Trichoid sensillum *see* Sensillum
Trichoplusia 446
Trichoptera
fossil record *231*
larva 221, 278, *283*, 288, 383, 522–3
cf. Lepidoptera 221
marine 287–8
preservation 487
relationships 216, 221, 243, 523
taxobox 522–3
wings 49–50, 222
see also Appendix
Trictena atripalpis 108
Trigona 329
Triplaris (Bot.) 318
Tritrophic interactions 465
Triungulin 166, 218, 517
Trochanter 47, 52, 54–5, 60, 203, 208, 218, 394, 493, 517
Trochantin *46–7*, 217, 218
Troctomorpha 213, 214, 509
Trogiomorpha 213, 509
Trogoderma granarium 473
Trophallaxis 337, 340, 343, 350, 508
Trophamnion 152, 363
Trophic
cascade 373–5
eggs 337, 343
interactions 295, 313, 373–4, 458, 465, 468, 513
Trophocyte 84, 92, 153
Trypanosoma (Prot.) 398, 412
Trypanosomiasis 398, 412
Tsetse fly *see Glossina*; Glossinidae
Tubulifera 214, 511
Tymbal 104
Tympanum (-al organ) 100–1, 103
Typhlocybinae 83
Typhus 410
Tyria jacobaeae 387
Tyroglyphus phylloxerae (Acari) 296–7

Ulmus (Bot.) 115, 300, 471
Unguis *see* Claw
United States *see* America, North
Urea (urate, uric acid) 90, 178
Uricotelism 90
Urine 80, 82, 86–9, 135, 178
Urocercus 303
Urocytes 84–5
Urothemis 51
Urtica (Bot.) 416
Urtication 387, 398, 416–7
Utetheisa ornatrix Plate 7e

Vagina 91–2, 132, 137, 140
Valve 388, 395, 411
of dorsal vessel 71
of ovipositor 54, 132, 149, 212, 223, 507, 525
of spiracle 73–6
Valvifer 54
Vannus 49, 215
Varroa (Acari) 338–9, 348, *Plate 6d*
Vas deferens 92
Vector
abundance 402–3, 464
anthropophily 245, 402–5
capacity 405, 421
and climate change 464–7
competence 403, 464
of disease 40, 183, 247, 398–415, 420–1, 434, 436, 463–7
distribution 402–5, 408, 437, 458–67
feeding rate 402–3
of plant virus 420–1, 423, 425, 43
survival rate 403
Vedalia ladybird 427
Veliidae 284
Venation 27, 49–52, 206–23, 230, 238–41, 379–*80*; *see also* under each order
Venom 116, 223, 366, 369, 387, 389, 393–5, 398, 415–7, 525
Ventilation 69, 77, 97, 281–2, 329
Ventral diaphragm 71
Ventral tube 202, 493
Ventriculus 80–2
Venturia canescens 369
Vertex 35–6, 43
Vertical transmission 86, 408
Verticillium lecanii (Fungi) *see Lecanicillium*
Vesica *see* Endophallus
Vespula (-idae, -oidea) 223, 315, *322–5*, 327, 329, 333, *335–6*, 347, 350–2, 392, 396, 432, 524–5
Veterinary entomology 245, 355, 398–413
Vicariance 236, 243
Viceroy butterfly 390–1
Vision 117–122; *see also* Eye
Vitellarium 92
Vitelline membrane 145
Vitellogenesis 92, 143, 153–4
Vitellogenin 84, 153–4
Vitis vinifera (Bot.) 259, 296–7, 447
Viviparity
adenotrophic 92, 150
haemocoelous 150–1, 518
ovoviviparity 150, 508, 518
pseudoplacental 150, 501
Voltinism 157, 172–3, 179–80, 273, 426, 459, 465, 498, 523
Volucella 347–8
Voucher specimens 3–4, 475, 489–91
Vulva 92, 138, 145, 151

Waggle dance 332–3
Walking 27, 38, 45, 48, 60–1, 132, 211, 358, 448, 507, 511
random 123
sticks 130, 210, 486, 503; *see also* Phasmatodea
Wallace, Alfred Russel 5, 316, *474*, 476
Wasmannia auropunctata 15
Wanderer butterfly *see* Monarch butterfly
Wasp *see* Hymenoptera
Wasp (European) *see Vespula*
Waste disposal *see* Excretory system
Water bug, giant *see* Belostomatidae
Water-boatman *see* Corixidae

Water-scorpion *see* Nepidae
Water-strider *see* Gerromorpha
Wax
  bee's 39, 333, 335–6, 370
  cuticular 29–*31*, 170
  defensive 259, 384, 513
  egg 145, 148
  filament 30, 259, 325
  gland 30–*1*, 330, 335
  mirror 335
  scale *see Ceroplastes sinensis*
Weaver ant *see Oecophylla*
Webspinner *see* Embioptera
Weeds, biological control of 301, 305, 309–13, 274
Weevil *see* Curculionidae
Werneckiella equi 510
West Nile virus 398, 407, 409–10, 464
Western conifer seed bug *see Leptoglossus occidentalis*
Western flower thrips *see Frankliniella occidentalis*
Western pine beetle *see Dendroctonus brevicomis*
Weta 10, 14, 17, *Plate 1d*
Whirligig beetle *see* Gyrinidae
White ant *see* Termitoidae
White butterfly *see Pieris*
Wing
  buds 27, 45, 61, *161*, 209, 341, 433
  coupling 49, 215, 222, 241
  evolution 206, 238–41
  nomenclature 48–52
  winglets 45, 164, 233, 239–41
  see also Flight
Winkler bag 476
Wireworm 251, 256; *see also* Tenebrionidae
Witjuti (witchety/witchetty) grubs 11, 22–3, *Plate 2c*
Wolbachia (Bact.) 85–6, 151–3, 408, 412, 454, 491
Wood ant *251*, 346–7
Wood-boring 77, 85–6, 218, 260, 301, 375, 419, 469; *see also* Cerambycidae; Termitoidae
Woodroach *see Cryptocercus*
Wood wasp *see* Siricidae
Wool moth *see* Tineidae
Worker 15, 18–19, 23, 30, 38, 112, 114–5, 140, 181, 212, 265–7, 286, 315, 323, 326–52, 359–60, 363; *see also* Caste system
Wuchereria bancrofti (Nem.) 412

Xanthopan morgani 39, 316
Xanthogaleruca luteola 471
Xenonomia 207, 211
Xenopsylla cheopis 411, 521
Xenorhabdus (Nem.) 442
Xyleborinus saxesenii 345
Xylocopa (-ini) 107, 325
Xylophagomorpha 220
Xylophagy 261, 290, 303

Yellow fever 185, 398–9, 407, 466
Yellow fever mosquito *see Aedes, Ae. aegypti*
Yersinia pestis (Bact.) 411

Zaspilothynnus trilobatus Plate 5d
Zingiberaceae (Bot.) 235
Zoraptera 207–9, *231*, 250, 487, 501, *see also* Appendix
Zorotypus (-idae) 501
Zygaenidae 385, 437
Zygentoma 10, 52, 54, 68, 149, 205, *231*, 238, 240, 518
  development 159
  genitalia 54, 496
  mouthparts 205, 495–6
  preservation 487
  relationships 194, 202–3, 205, 496
  in soil 250
  taxobox 496
Zygoptera 206, 281, 498

APPENDIX: A REFERENCE GUIDE TO ORDERS

Summary of the diagnostic features of the adults and immature stages of the three non-insect hexapod orders and the 28 orders of insects. For more complete information on each order, refer to the taxoboxes that precede this appendix.

Collembola (springtails)	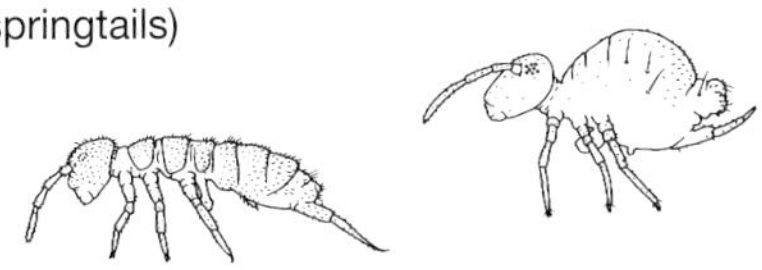	Small, wingless, entognathous (mouthparts within folds of head), antennae present, thoracic segments like those of abdomen, legs at least five-segmented, abdomen six-segmented with sucker-like ventral tube and forked jumping organ, without cerci; immature stages like small adult, with same number of segments.	Chapters 7, 9 Taxobox 1
Diplura (diplurans)	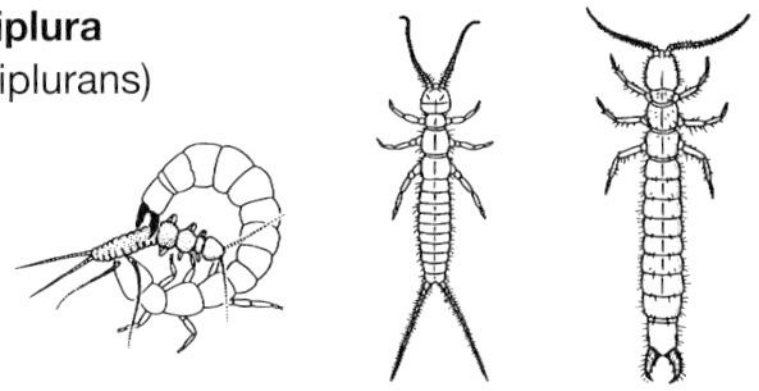	Small to medium, wingless, eyeless, entognathous, long antennae like string of beads, thoracic segments like those of abdomen, legs five-segmented, abdomen 10-segmented with some segments bearing small protrusions, terminal cerci filiform to forceps-like; immature stages like small adult.	Chapters 7, 9 Taxobox 1
Protura (proturans)	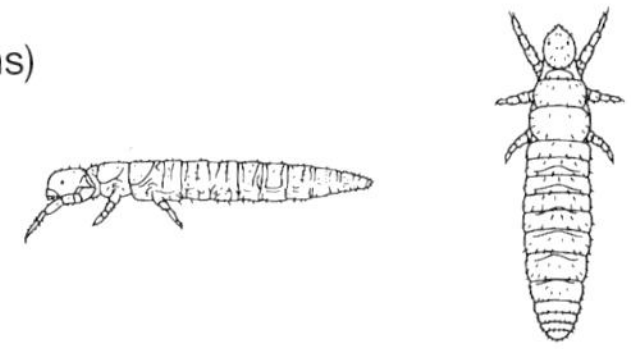	Very small, wingless, eyeless, without antennae, entognathous, fore legs held forward, thoracic segments like those of abdomen, legs five-segmented, adult abdomen 12-segmented without cerci; immature stages like small adult but with fewer abdominal segments.	Chapters 7, 9 Taxobox 1
Archaeognatha (or Microcoryphia) (archaeognathans or bristletails)	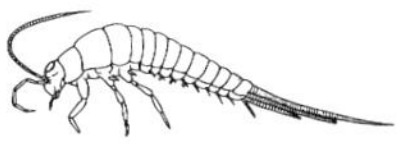 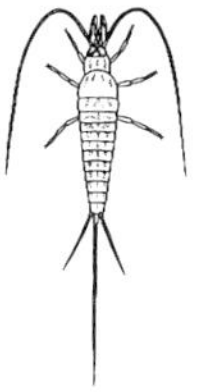	Medium, wingless, with humped thorax, hypognathous (mouthparts directed downwards), large compound eyes in near contact, some abdominal segments with paired styles and vesicles, with three "tails" – paired cerci shorter than single median caudal appendage; immature stages like small adult.	Chapters 7, 9 Taxobox 2

Zygentoma (silverfish) 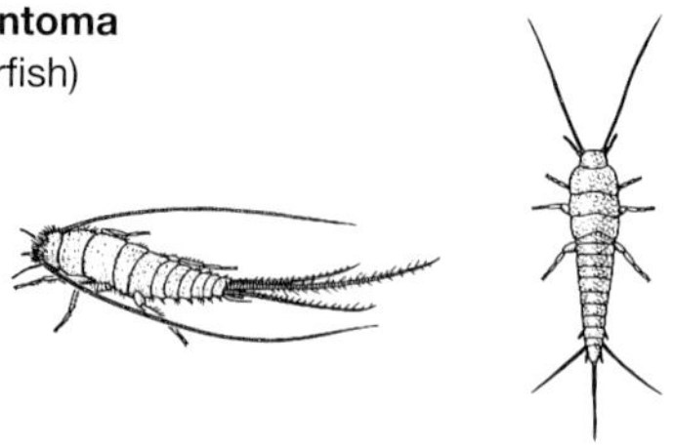	Medium, flattened, silvery-scaled, wingless, hypognathous to prognathous (mouthparts directed downwards to forwards), compound eyes small, widely separated or absent, some abdominal segments with ventral styles and sometimes vesicles, with three "tails" – paired cerci nearly as long as median caudal appendage; immature stages like small adult.	Chapters 7, 9 Taxobox 3
Ephemeroptera (mayflies) 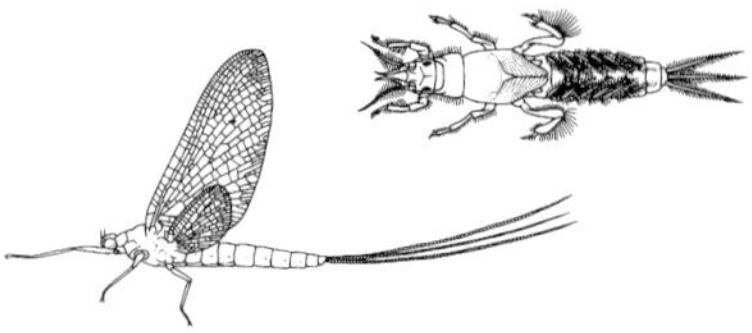	Small to large, winged with large triangular fore wings and smaller hind wings, mouthparts reduced, compound eyes large, short filiform antennae, abdomen slender compared to stout thorax, with three "tails" – paired cerci often as long as median caudal appendage; immature stages (nymphs) aquatic, with three "tails" and lateral abdominal gills; penultimate instar a winged subimago (subadult).	Chapters 7, 10 Box 5.7 Taxobox 4
Odonata (damselflies and dragonflies) 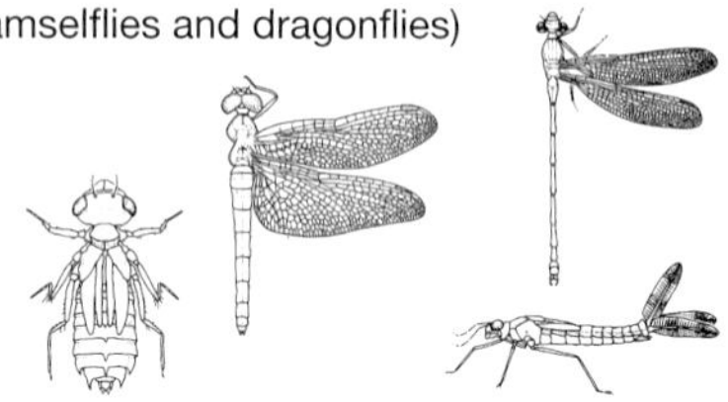	Medium to large, winged, with fore and hind wings equal (damselflies) or hind wings wider than fore wings (dragonflies), head mobile, with large compound eyes separated (damselflies) or nearly in contact (dragonflies), mouthparts mandibulate, antennae short, thorax stout, abdomen slender; immature stages (nymphs) aquatic, stout or narrow, with extensible labial "mask", terminal or rectal gills.	Chapters 7, 10 Box 5.5 Taxobox 5
Plecoptera (stoneflies) 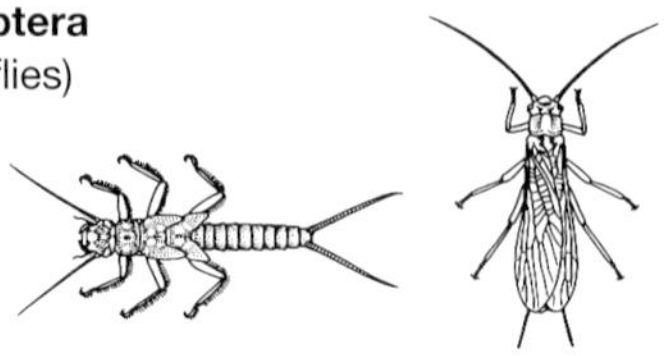	Medium, with fore and hind wings nearly equal (subequal) in size, at rest wings partly wrap abdomen and extend beyond abdominal apex but wing reduction frequent, legs weak, abdomen soft with filamentous cerci; immature stages (nymphs) aquatic, resembling wingless adults, often with gills on abdomen.	Chapters 7, 10 Taxobox 6
Dermaptera (earwigs) 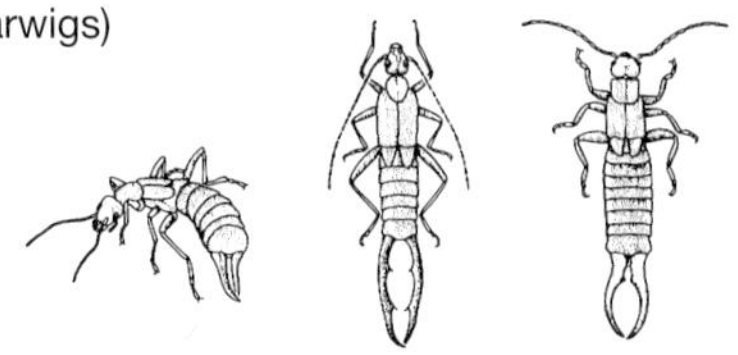	Small to medium, elongate and flattened, prognathous (mouthparts directed forwards), antennae short to moderate, legs short, if wings present the fore wings are small, leathery tegmina that conceal folded membranous and semi-circular hind wings, abdomen with overlapping terga, cerci typically modified as forceps; immature stages (nymphs) resemble small adults.	Chapters 7, 9 Taxobox 7

Zoraptera (zorapterans or angel insects) 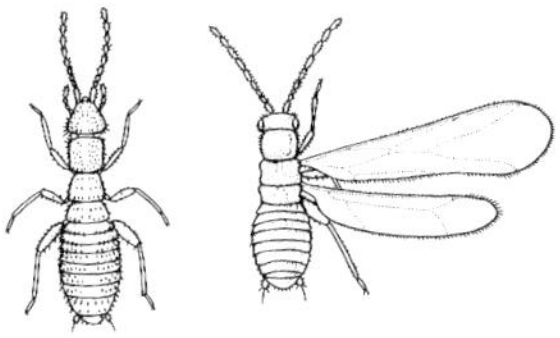	Small, termite-like, hypognathous, winged species with compound eyes and ocelli, wingless species lack eyes, if winged then wings with simple venation and readily shed, coxae well-developed, abdomen 11-segmented, short and swollen; immature stages (nymphs) resemble small adults.	Chapters 7, 9 Taxobox 8
Orthoptera (grasshoppers, locusts, katydids and crickets) 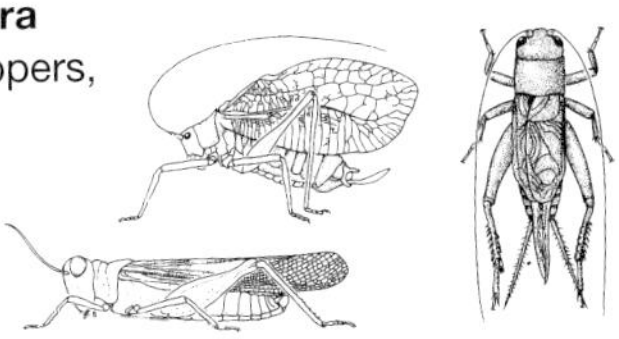	Medium to large, hypognathous, usually winged, fore wings forming leathery tegmina, hind wings broad, at rest pleated beneath tegmina, pronotum curved over pleura, hind legs often enlarged for jumping, cerci one-segmented; immature stages (nymphs) like small adults.	Chapters 5, 6, 7, 11 Box 5.2 Taxobox 9
Embioptera (or Embiidina or Embiodea) (embiopterans or webspinners) 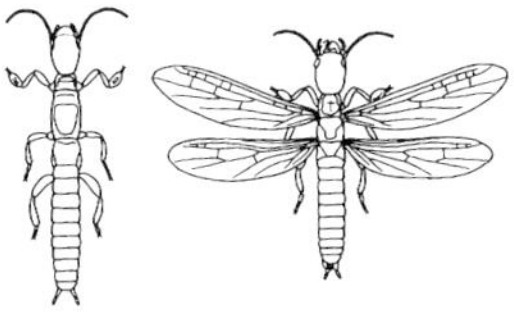	Small to medium, elongate, cylindrical, prognathous, compound eyes kidney-shaped, wingless in all females, some males with soft, flexible wings, legs short, basal fore tarsus swollen with silk gland, cerci two-segmented; immature stages (nymphs) like small adults.	Chapters 7, 9 Taxobox 10
Phasmatodea (phasmids, stick-insects or walking sticks) 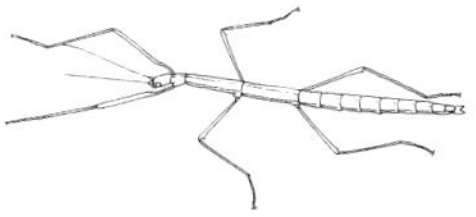 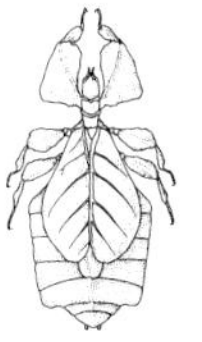	Medium to large, cylindrical stick-like or flattened leaf-like, prognathous, mandibulate, compound eyes small and laterally placed, fore wings form leathery tegmina, hind wings broad with toughened fore margin, legs elongate for walking, cerci one-segmented; immature stages (nymphs) like small adults.	Chapters 7, 11, 14 Taxobox 11
Grylloblattodea (or Grylloblattaria or Notoptera) (grylloblattids, ice crawlers or rock crawlers) 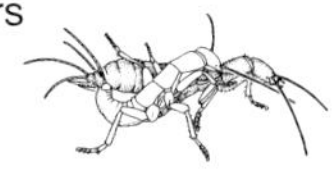 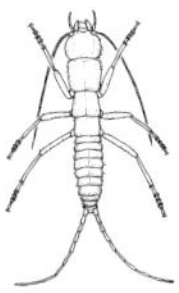	Medium, soft-bodied, elongate, pale, wingless and often eyeless, prognathous, with stout coxae on legs adapted for running, cerci five- to nine-segmented, female with short ovipositor; immature stages (nymphs) like small, pale adults.	Chapters 1, 7, 9 Taxobox 12
Mantophasmatodea (heelwalkers) 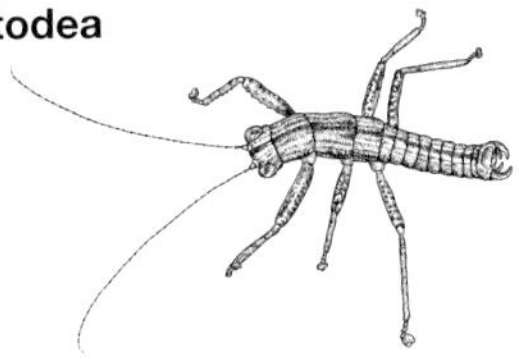	Small to medium, somewhat cylindrical, hypognathous, antennae long, multi-segmented, compound eyes large, fore and mid legs raptorial (adapted for grasping), wings absent, cerci small in female, prominent in male; immature stages (nymphs) resemble small adults.	Chapter 7 Taxobox 13

Mantodea (mantids, mantises or praying mantids) 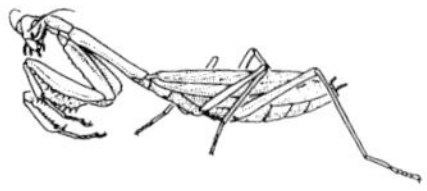 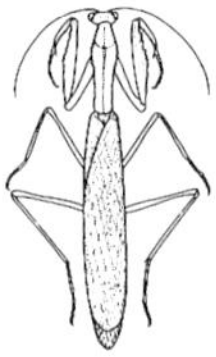	Moderate to large, head small, mobile and triangular, compound eyes large and separated, thorax narrow, fore wings form tegmina, hind wings broad, fore legs predatory (raptorial), mid and hind legs elongate; immature stages (nymphs) resemble small adults.	Chapters 7, 13 Box 5.3 Taxobox 14
Blattodea: roach families (cockroaches or roaches) 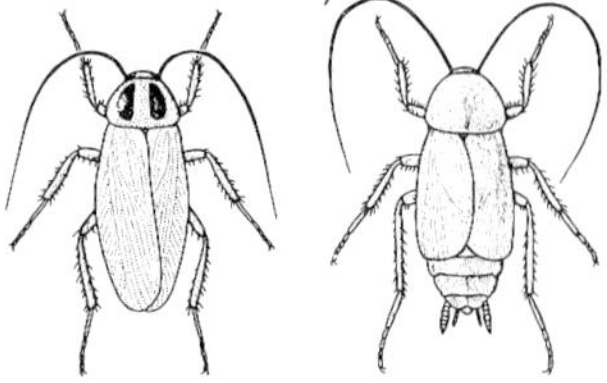	Small to large, dorsoventrally flattened, hypognathous, compound eyes well developed (except in cave dwellers), prothorax large and shield-like (may cover head), fore wings form leathery tegmina, protecting large hind wings, large anal lobe on hind wing, coxae large and abutting, cerci usually multisegmented; immature stages (nymphs) like small adults.	Chapters 7, 9 Taxobox 15
Blattodea: epifamily Termitoidae (former order Isoptera) (termites or "white ants") 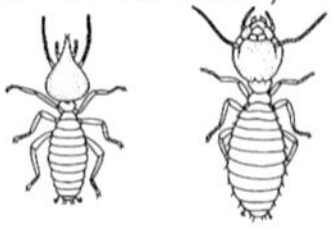 	Small to medium, mandibulate (with variable mouthpart development in different castes), antennae long, compound eyes often reduced, in winged forms fore and hind wings usually similar, often with reduced venation, body terminates in one- to five-segmented cerci; immature stages (nymphs) morphologically variable (polymorphic) according to caste.	Chapters 7, 12 Box 9.3, Box 12.4 Taxobox 16
Psocodea: "Psocoptera" (bark lice and book lice) 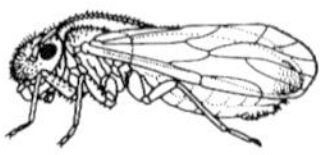 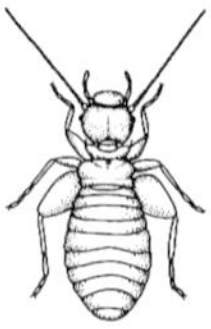	Small to medium, head large and mobile, chewing mouthparts asymmetrical, compound eyes large, antennae long and slender, wings often reduced or absent, if wings present venation simple, coupled in flight, held roof-like at rest, cerci absent; immature stages (nymphs) like small adults.	Chapter 7 Taxobox 17
Psocodea: "Phthiraptera" (chewing lice and sucking lice) 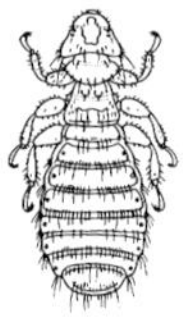 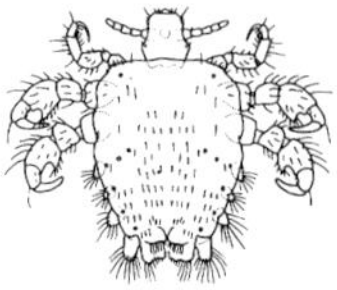	Small, dorsoventrally flattened, wingless, ectoparasites, mouthparts mandibulate or beak-like, compound eyes small or absent, antennae either in grooves or extended, legs stout with strong claw(s) for grasping host hair or feathers; immature stages (nymphs) like small, pale adults.	Chapters 7, 13, 15 Taxobox 18
Thysanoptera (thrips)  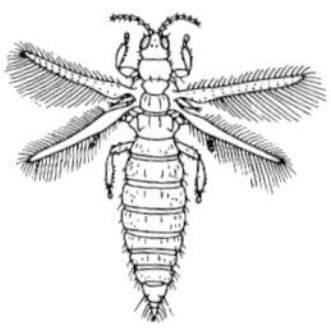	Small, slender, hypognathous with a feeding tube formed from three stylets – the maxillary laciniae plus the left mandible, with or without wings, if present wings subequal, strap-like, with long fringe; immature stages (nymphs) like small adults.	Chapters 7, 11 Taxobox 19

Hemiptera (bugs, moss bugs, cicadas, leafhoppers, planthoppers, spittle bugs, treehoppers, aphids, jumping plant lice, scale insects and whiteflies) 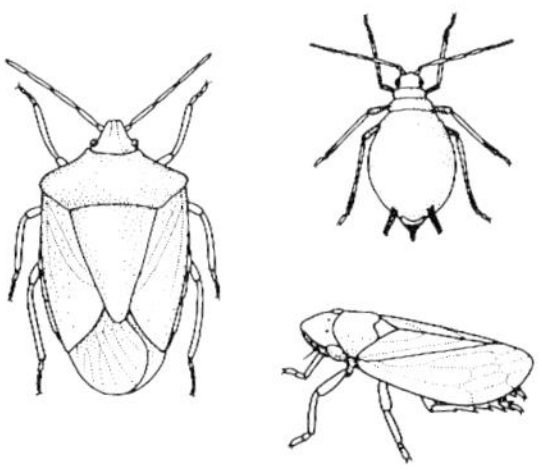	Small to large, mouthpart stylets lying in grooved labium to form proboscis (or rostrum) directed posteriorly at rest, lacking palps, fore wings may be membranous or thickened to form hemelytra (Heteroptera), wing reduction or absence is common; immature stages (nymphs) usually resemble small adults, except in whiteflies and male scale insects.	Chapters 7, 10, 11, 16 Box 3.3, Box 5.8, Box 9.4, Box 10.2, Box 11.2, Box 14.2, Box 15.6, Box 16.1, Box 16.2, Box 16.3, Box 16.5, Box 16.6 Taxobox 20
Neuroptera (lacewings, owlflies and antlions) 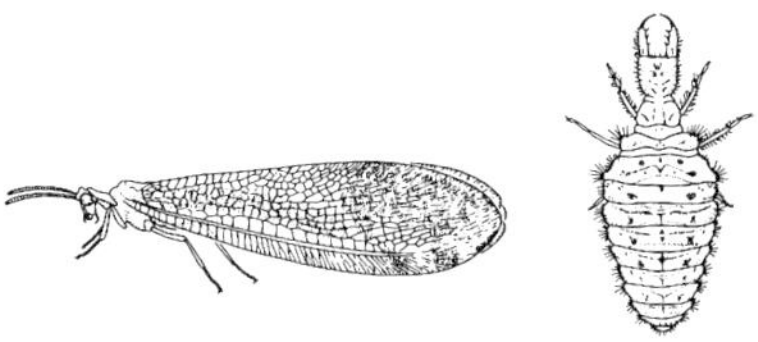	Medium, compound eyes large and separated, mandibulate, antennae multisegmented, prothorax often larger than meso- and metathorax, wings held roof-like over abdomen at rest, fore and hind wings subequal with numerous cross-veins and distal "twigging" of veins, without anal fold; immature stages (larvae) predominantly terrestrial, prognathous, with slender mandibles and maxillae usually forming piercing/sucking mouthparts, with jointed legs only on thorax, lacking abdominal gills.	Chapters 7, 13 Box 10.4 Taxobox 21
Megaloptera (alderflies, dobsonflies and fishflies) 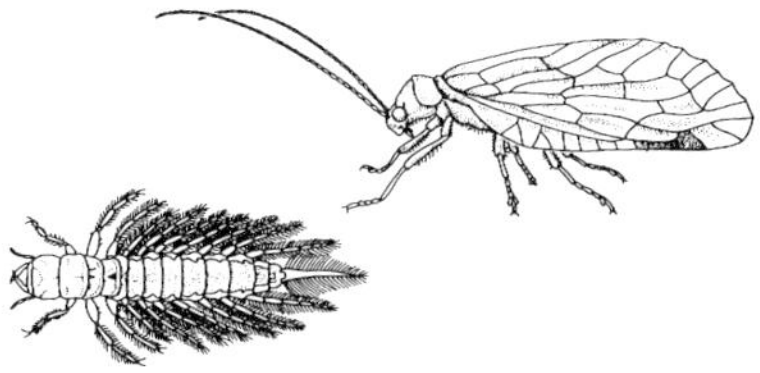	Medium to large, compound eyes large and separated, prognathous, mandibulate, antennae multisegmented, prothorax only slightly longer than meso- and metathorax, fore and hind wings subequal with anal fold in hind wing; immature stages (larvae) aquatic, prognathous, with stout mandibles, jointed legs only on thorax, with lateral abdominal gills.	Chapters 7, 13 Box 10.4 Taxobox 21
Raphidioptera (snakeflies) 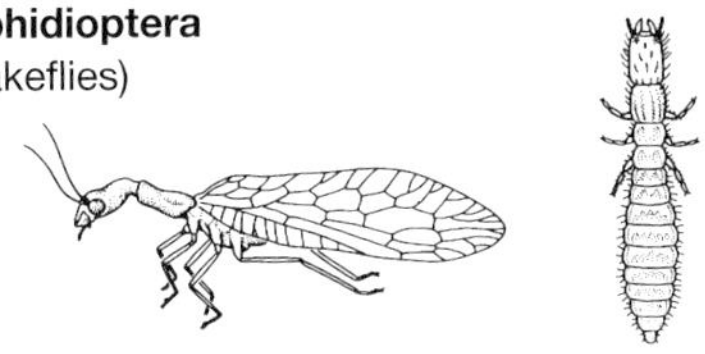	Medium, prognathous, mandibulate, antennae multisegmented, compound eyes large and separated, prothorax much longer than meso- and metathorax, fore wings rather longer than otherwise similar hind wings, without anal fold; immature stages (larvae) terrestrial, prognathous with jointed legs only on thorax, without abdominal gills.	Chapters 7, 13 Taxobox 21

Coleoptera (beetles) 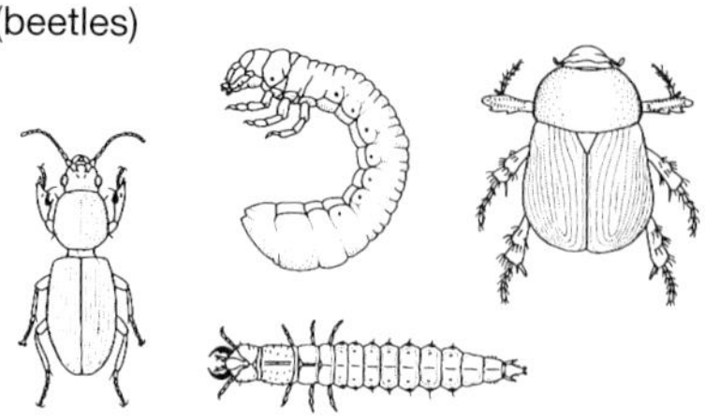	Small to large, often sturdy, compact and heavily sclerotized or armoured, mandibulate, with fore wings modified as rigid elytra covering folded membranous hind wings at rest, legs variously modified, with claws and adhesive structures; immature stages (larvae) terrestrial or aquatic, with sclerotized head capsule, opposable mandibles and usually five-segmented thoracic legs, without abdominal legs or labial silk glands.	Chapters 7, 10, 11, 14 Box 1.5, Box 3.2, Box 3.4, Box 10.3, Box 11.4, Box 14.3, Box 14.4, Box 16.7, Box 17.2, Box 17.3, Box 17.4 Taxobox 22
Strepsiptera (strepsipterans) 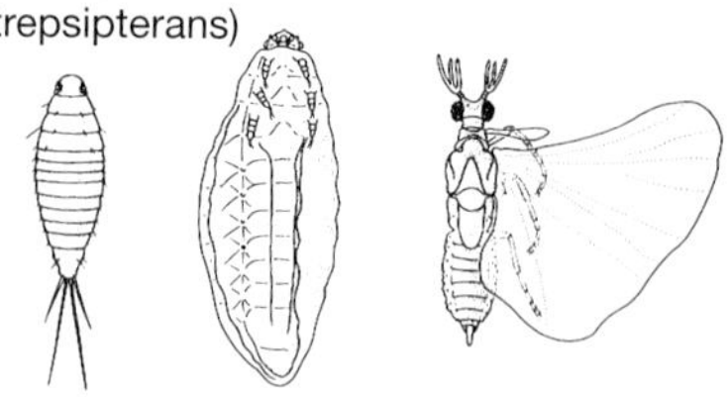	Small, aberrant endoparasitoids; male with large head, bulging eyes with few facets, antennae with branches, fore wings stubby, without veins, hind wings fan-shaped with few veins; female larviform, wingless, mostly retained in host; immature stages (larvae) initially a triungulin with three pairs of thoracic legs, later maggot-like without mouthparts.	Chapters 7, 13 Taxobox 23
Diptera (true flies) 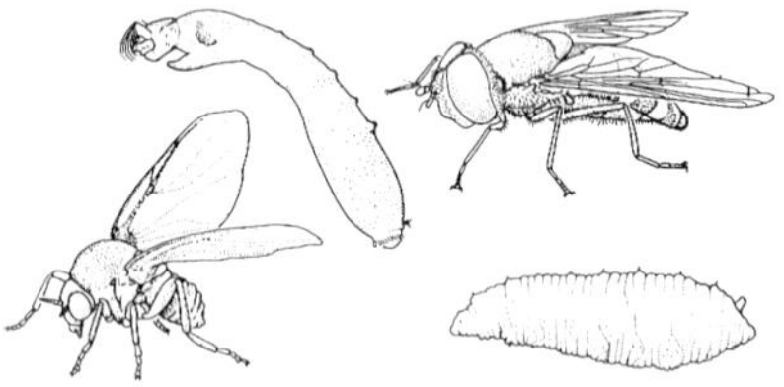	Small to medium, wings restricted to mesothorax, metathorax with balancing organs (halteres), mouthparts vary from non-functional, to biting and sucking; immature stages (larvae, maggots) variable, without jointed legs, with sclerotized head capsule or variably reduced ultimately to remnant mouth hooks.	Chapters 7, 10, 15 Box 4.1, Box 5.6, Box 6.2, Box 9.1, Box 10.1, Box 15.1, Box 15.2, Box 15.3, Box 15.4, Box 15.5, Box 16.1, Box 17.1 Taxobox 24
Mecoptera (hangingflies, scorpionflies and snowfleas) 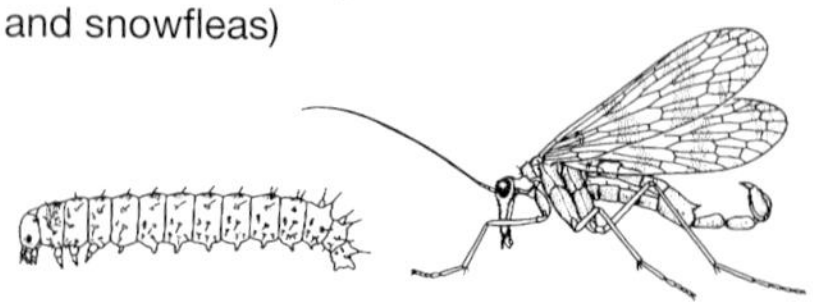	Medium, hypognathous with elongate rostrum formed from slender, serrate mandibles and maxillae and elongate labium, fore and hind wings narrow and subequal, legs raptorial; immature stages (larvae) mostly terrestrial, with heavily sclerotized head capsule, groups of simple eyes, short and jointed thoracic legs, abdomen usually with prolegs.	Chapters 5, 7, 13 Box 5.1 Taxobox 25

Siphonaptera (fleas) 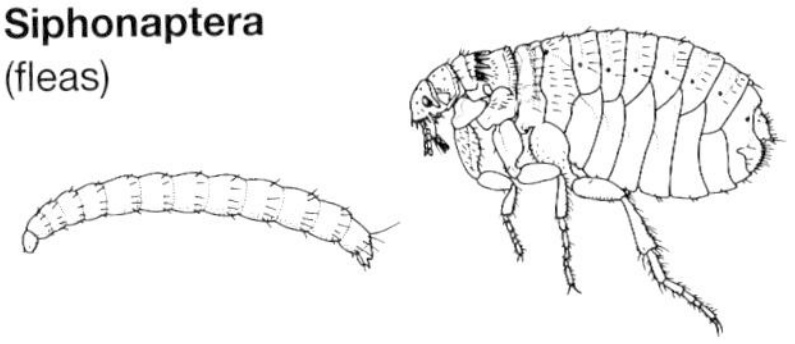	Small, highly modified, laterally compressed ectoparasites, mouthparts piercing and sucking, without mandibles, antennae lying in grooves, body with many backwardly directed setae and spines, some as combs, legs strong, terminating in strong claws for grasping host; immature stages (larvae) terrestrial, apodous (legless), with distinct head capsule.	Chapters 7, 15 Taxobox 26
Trichoptera (caddisflies) 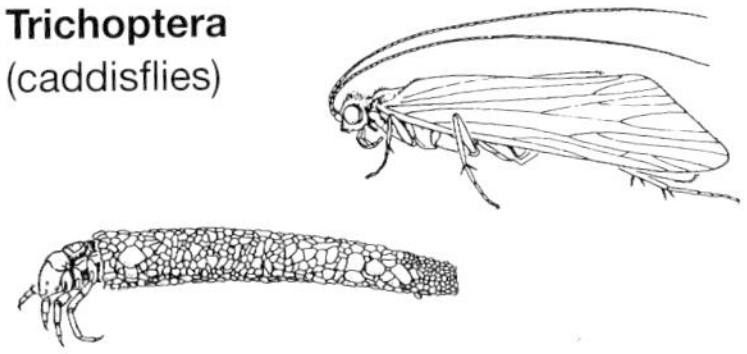	Small to large, with long, multisegmented antennae, reduced mouthparts (no proboscis) but well-developed maxillary and labial palps, hairy (or rarely scaly) wings, lacking discal cell and with fore wing anal veins looped (cf. Lepidoptera); immature stages (larvae) aquatic, often case-bearing, but many free-living, with three pairs of segmented thoracic legs and lacking abdominal prolegs.	Chapters 7, 10 Taxobox 27
Lepidoptera (butterflies and moths) 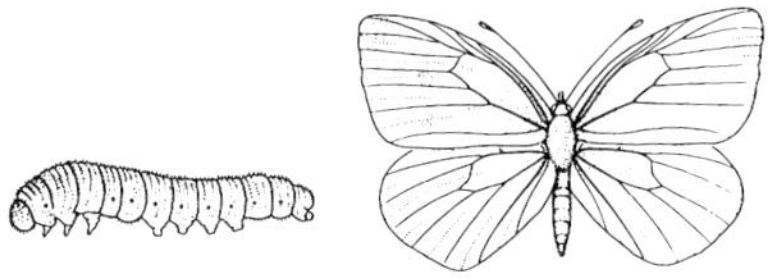	Small to large, hypognathous, nearly all with long coiled proboscis, antennae multisegmented and often comb-like (pectinate), clubbed in butterflies, wings with double layer of scales (flattened setae) and large cells including the discal; immature stages (larvae, caterpillars) with sclerotized mandibulate head, labial spinnerets producing silk, jointed legs on thorax, and some abdominal prolegs.	Chapters 7, 11, 14 Box 1.2, Box 1.4, Box 5.4, Box 14.1, Box 16.1 Taxobox 28
Hymenoptera (ants, bees, wasps, sawflies and wood wasps) 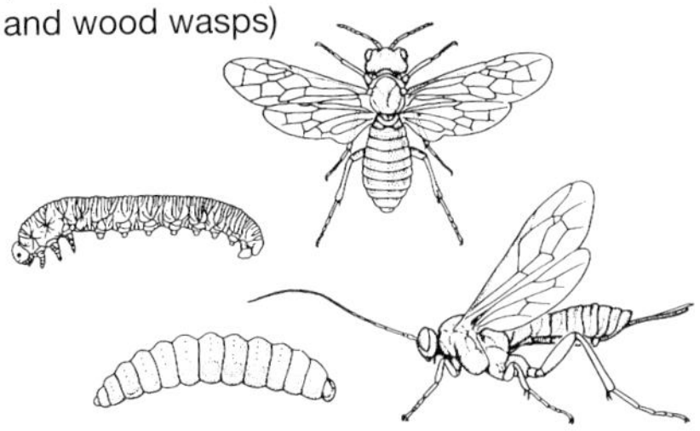	Minute to large, mouthparts mandibulate to sucking and chewing, antennae multisegmented often long and held forward, thorax either three-segmented or forms a mesosoma by incorporation of first abdominal segment in which case the abdomen is petiolate (waisted), wings with simple venation, fore and hind wings coupled together by hooks on hind wing; immature stages (larvae) very variable, many lack legs completely, all have distinct mandibles even if head is reduced.	Chapters 7, 12, 13, 14 Box 1.3, Box 9.2, Box 11.1, Box 12.1, Box 12.2, Box 12.3, Box 12.4, Box 13.1, Box 17.3 Taxobox 29